Prealgebra and Introductory Algebra

Instructor's Annotated Edition

Prealgebra and Introductory Algebra

Instructor's Annotated Edition

Richard N. Aufmann
Palomar College, California

Joanne S. Lockwood
New Hampshire Community Technical College

Houghton Mifflin Company
Boston   New York

Vice President, Publisher: Jack Shira
Senior Sponsoring Editor: Lynn Cox
Associate Editor: Melissa Parkin
Assistant Editor: Noel Kamm
Senior Project Editor: Nancy Blodget
Editorial Assistant: Sarah Driver
Art and Design Manager: Gary Crespo
Manufacturing Buyer: Florence Cadran
Senior Marketing Manager: Ben Rivera
Marketing Associate: Lisa Lawler

Cover photo: "Still life of Ced dice," © Nonstock Photography/Veer

Photo Credits: A complete list of photo credits appears on page 864, immediately following the appendix.

Printed in the U.S.A.

Library of Congress Control Number: 2005933972

Instructor's Annotated Edition:
ISBN 13: 978-0-618-60944-4
ISBN 10: 0-618-60944-X

For orders, use student text ISBNs:
ISBN 13: 978-0-618-60943-7
ISBN 10: 0-618-60943-1

2 3 4 5 6 7 8 9-WEB-10 09 08 07 06

Contents

v

11 **Rational Expressions** **569**

R Review Topics 827

Preface

Prealgebra and Introductory Algebra is designed to be a transition from the concrete aspects of arithmetic to the symbolic world of algebra and to provide mathematically sound and comprehensive coverage of the topics considered essential in an introductory algebra course. The text is designed not only to meet the needs of the traditional college student, but also to serve the needs of returning students whose mathematical proficiency may have declined during years away from formal education.

Two of the main challenges for students at this level are the manipulation of variables and the ability to translate verbal phrases into mathematical expressions. One reason for this difficulty is that students generally are not exposed to variables or to verbal phrases until they are enrolled in an algebra course. In *Prealgebra and Introductory Algebra,* we introduce variables and variable expressions in Chapter 1. For example, students are asked to "evaluate $a + b$ when $a = 67$ and $b = 29$." This strand continues throughout the early chapters of the text. In this way, students become comfortable with variables, variable expressions, and values of variable expressions before the presentation of strictly algebraic topics. Also, we introduce verbal phrases for operations as we introduce the operation. For instance, after addition concepts have been presented in Chapter 1, we provide exercises that say, for example, "Find the sum of 25 and 41." In this way, students are constantly confronted with verbal phrases and must make a connection between the phrase and a mathematical operation.

In *Prealgebra and Introductory Algebra,* we integrate approaches suggested by AMATYC. There is real sourced data in graphs and tables. Each chapter opens by illustrating and referencing a mathematical application within the chapter. The Index of Applications on the inside covers highlights, in an easily accessible location, the importance and scope of the applications of mathematics. At the end of each section, there are "Applying the Concepts" exercises that include writing, synthesis, critical thinking, and challenge problems. At the end of each chapter, there is a "Focus on Problem Solving" that introduces students to various problem-solving strategies. This is followed by "Projects and Group Activities" that can be used for cooperative learning activities. In keeping with the standards, a chapter on geometry and on one statistics and probability are included.

The hallmarks of the Aufmann developmental mathematics texts are incorporated in *Prealgebra and Introductory Algebra:* the interactive approach, an emphasis on problem-solving strategies, numerous applications, and the presentation of real data throughout. Accompanying this text are a computer tutorial, instructional DVDs, practice tests that students can use for self-assessment, and much more. The entire instructional package is a completely integrated learning system organized by objective and designed to promote student success.

Chapter Opening Features

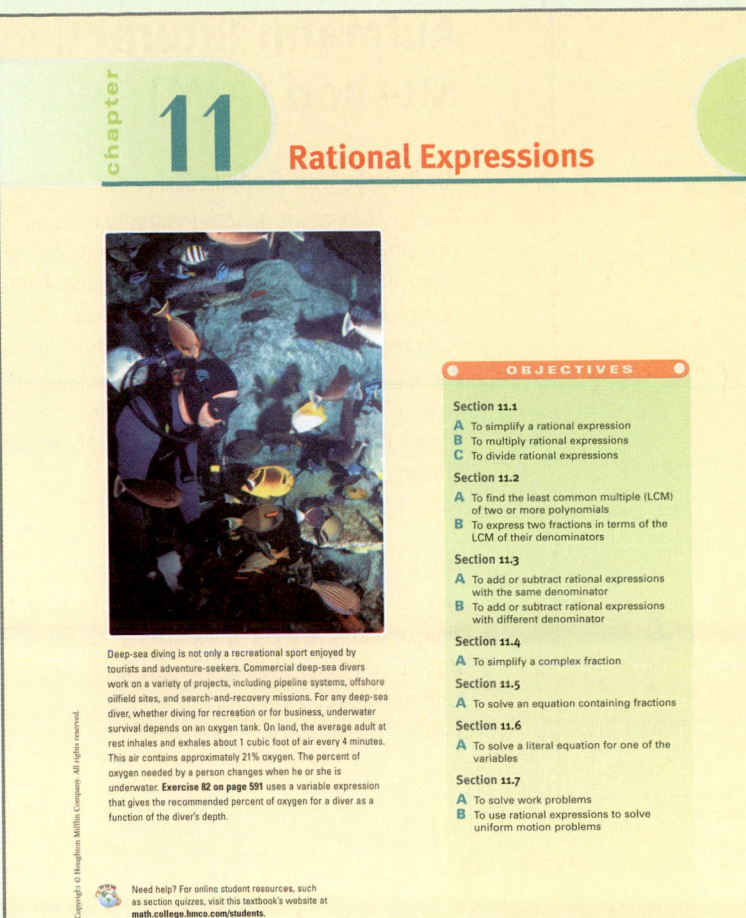

chapter

11 Rational Expressions

OBJECTIVES

Section 11.1
A To simplify a rational expression
B To multiply rational expressions
C To divide rational expressions

Section 11.2
A To find the least common multiple (LCM) of two or more polynomials
B To express two fractions in terms of the LCM of their denominators

Section 11.3
A To add or subtract rational expressions with the same denominator
B To add or subtract rational expressions with different denominator

Section 11.4
A To simplify a complex fraction

Section 11.5
A To solve an equation containing fractions

Section 11.6
A To solve a literal equation for one of the variables

Section 11.7
A To solve work problems
B To use rational expressions to solve uniform motion problems

Deep-sea diving is not only a recreational sport enjoyed by tourists and adventure-seekers. Commercial deep-sea divers work on a variety of projects, including pipeline systems, offshore oilfield sites, and search-and-recovery missions. For any deep-sea diver, whether diving for recreation or for business, underwater survival depends on an oxygen tank. On land, the average adult at rest inhales and exhales about 1 cubic foot of air every 4 minutes. This air contains approximately 21% oxygen. The percent of oxygen needed by a person changes when he or she is underwater. **Exercise 82 on page 591** uses a variable expression that gives the recommended percent of oxygen for a diver as a function of the diver's depth.

Need help? For online student resources, such as section quizzes, visit this textbook's website at **math.college.hmco.com/students**.

Page 569

Chapter Opener

Motivating chapter opener photos and captions have been included, illustrating and referencing a specific application from the chapter.

The at the bottom of the page lets students know of additional online resources at **math.college.hmco.com/students**.

Objective-Specific Approach

Each chapter begins with a list of learning objectives, which form the framework for a complete learning system. The objectives are woven throughout the text (i.e., Exercises, Prep Tests, Chapter Review Exercises, Chapter Tests, Cumulative Review Exercises) as well as throughout the print and multimedia ancillaries. This results in a seamless learning system delivered in one consistent voice.

Page 570

Prep Test and Go Figure

Prep Tests occur at the beginning of each chapter and test students on previously covered concepts that are required in the coming chapter. Answers are provided in the Answer Section. Objective references are also provided in case a student needs to review specific concepts.

The **Go Figure** problem that follows the *Prep Test* is a playful puzzle problem designed to engage students in problem solving.

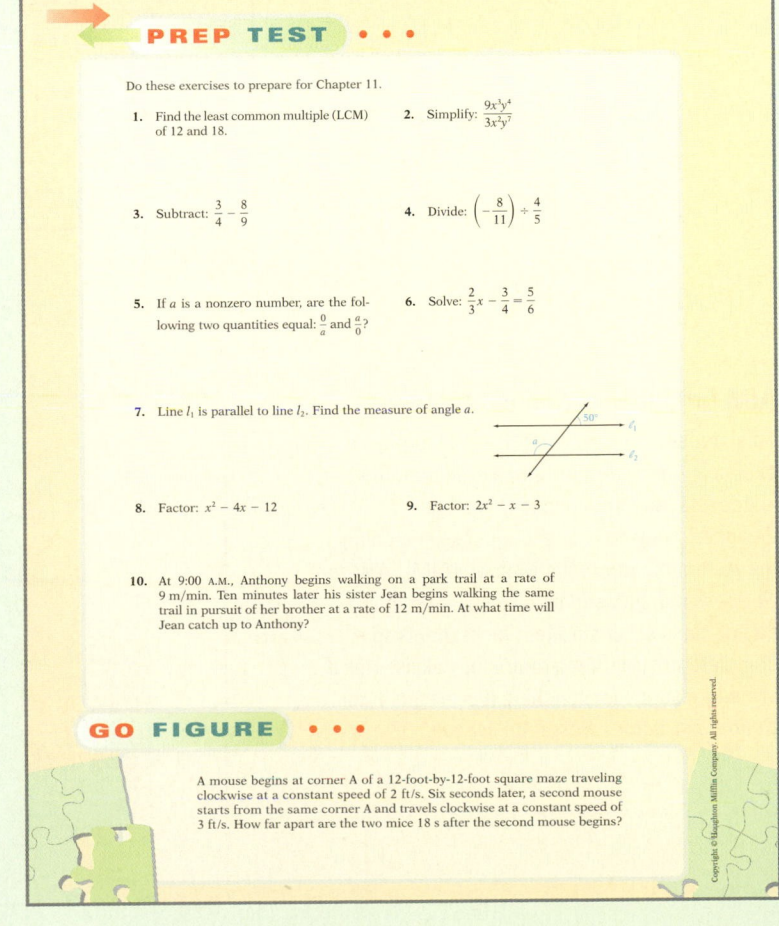

PREP TEST • • •

Do these exercises to prepare for Chapter 11.

1. Find the least common multiple (LCM) of 12 and 18.

2. Simplify: $\dfrac{9x^3y^4}{3x^2y^7}$

3. Subtract: $\dfrac{3}{4} - \dfrac{8}{9}$

4. Divide: $\left(-\dfrac{8}{11}\right) \div \dfrac{4}{5}$

5. If a is a nonzero number, are the following two quantities equal: $\dfrac{0}{a}$ and $\dfrac{a}{0}$?

6. Solve: $\dfrac{2}{3}x - \dfrac{3}{4} = \dfrac{5}{6}$

7. Line l_1 is parallel to line l_2. Find the measure of angle a.

8. Factor: $x^2 - 4x - 12$

9. Factor: $2x^2 - x - 3$

10. At 9:00 A.M., Anthony begins walking on a park trail at a rate of 9 m/min. Ten minutes later his sister Jean begins walking the same trail in pursuit of her brother at a rate of 12 m/min. At what time will Jean catch up to Anthony?

GO FIGURE • • •

A mouse begins at corner A of a 12-foot-by-12-foot square maze traveling clockwise at a constant speed of 2 ft/s. Six seconds later, a second mouse starts from the same corner A and travels clockwise at a constant speed of 3 ft/s. How far apart are the two mice 18 s after the second mouse begins?

Aufmann Interactive Method (AIM)

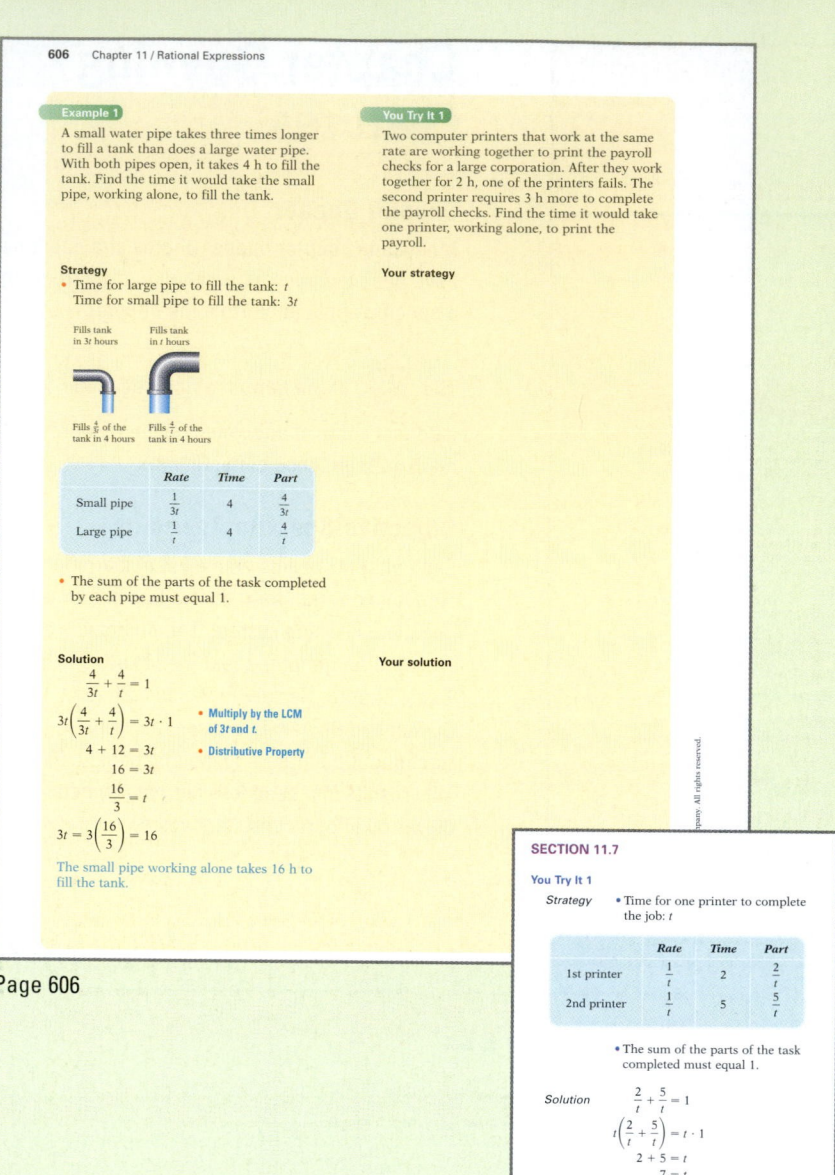

Page 606

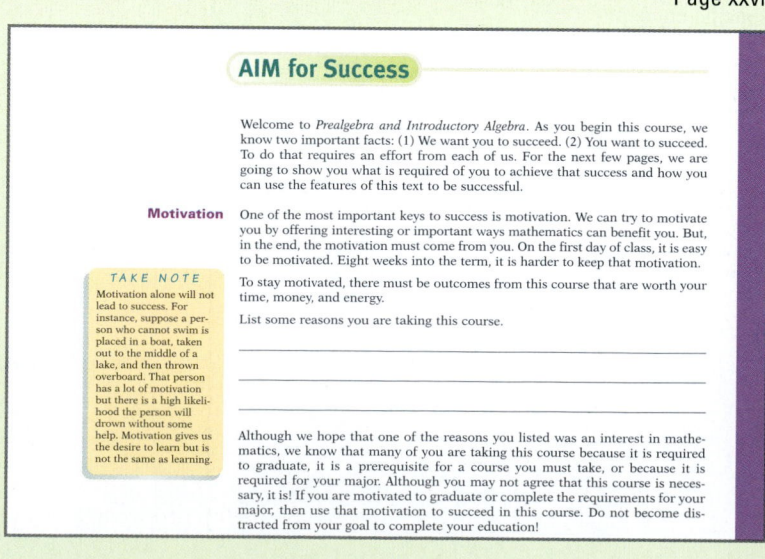

Page S29

An Interactive Approach

Prealgebra and Introductory Algebra uses an interactive style that provides a student with an opportunity to try a skill as it is presented. Each section is divided into objectives, and every objective contains one or more sets of matched-pair examples. The first example in each set is worked out; the second example, called "You Try It," is for the student to work. By solving this problem, the student actively practices concepts as they are presented in the text.

There are complete worked-out solutions to these examples in an appendix. By comparing their solutions to the solutions in the appendix, students obtain immediate feedback on, and reinforcement of, the concepts.

Page xxvii

AIM for Success Student Preface

This student "how to use this book" preface explains what is required of a student to be successful and how this text has been designed to foster student success, including the Aufmann Interactive Method (AIM). *AIM for Success* can be used as a lesson on the first day of class or as a project for students to complete to strengthen their study skills. There are suggestions for teaching this lesson in the *Online Instructor's Resource Manual*.

Problem Solving

Focus on Problem Solving

At the end of each chapter is a Focus on Problem Solving feature, which introduces the student to various successful problem-solving strategies. Strategies such as drawing a diagram, applying solutions to other problems, working backwards, inductive reasoning, and trial and error are some of the techniques that are demonstrated.

Page 775

Focus on Problem Solving

Deductive Reasoning

Deductive reasoning uses a rule or statement of fact to reach a conclusion. For instance, if two angles of one triangle are equal to two angles of another triangle, then the two triangles are similar. Thus any time we establish this fact about two triangles, we know that the triangles are similar. Below are two examples of deductive reasoning.

Given that $\triangle\triangle\triangle = \diamond\diamond\diamond\diamond$ and $\diamond\diamond\diamond\diamond = \acute{O}\acute{O}$, then $\triangle\triangle\triangle\triangle\triangle\triangle$ is equivalent to how many $\acute{O}$s?

Because three $\triangle$s = four $\diamond$s and four $\diamond$s = two $\acute{O}$s, three $\triangle$s = two $\acute{O}$s.

Six $\triangle$s is twice three $\triangle$s. We need to find twice two $\acute{O}$s, which is four $\acute{O}$s.

Therefore, $\triangle\triangle\triangle\triangle\triangle\triangle = \acute{O}\acute{O}\acute{O}\acute{O}$.

Lomax, Parish, Thorpe, and Wong are neighbors. Each drives a different type of vehicle: a compact car, a sedan, a sports car, or a station wagon. From the following statements, determine which type of vehicle each of the neighbors drives.

1. Although the vehicle owned by Lomax has more mileage on it than does either the sedan or the sports car, it does not have the highest mileage of all four cars. (Use X1 in the chart below to eliminate the possibilities that this statement rules out.)

2. Wong and the owner of the sports car live on one side of the street, and Thorpe and the owner of the compact car live on the other side of the street. (Use X2 to eliminate the possibilities that this statement rules out.)

3. Thorpe owns the vehicle with the most mileage on it. (Use X3 to eliminate the possibilities that this statement rules out.)

TAKE NOTE
To use the chart to solve this problem, write an X in a box to indicate that a possibility has been eliminated. Write a √ to show that a match has been found. When a row or column has 3 X's, a √ is written in the remaining open box in that row or column of the chart.

	Compact	Sedan	Sports Car	Wagon
Lomax	√	X1	X1	X2
Parish	X2	X2	√	X2
Thorpe	X2	X3	X2	√
Wong	X2	√	X2	

Lomax drives the compact car, Parish drives the sports car, Thorpe drives the station wagon, and Wong drives the sedan.

1. Given that $\ddagger\ddagger = \bullet\bullet\bullet\bullet\bullet$ and $\bullet\bullet\bullet\bullet\bullet = \Lambda\Lambda$, then $\ddagger\ddagger\ddagger\ddagger\ddagger$ = how many Λs?

2. Given that $\square\square\square\square\square = \acute{O}\acute{O}\acute{O}\acute{O}$ and $\acute{O}\acute{O}\acute{O}\acute{O} = \hat{I}\hat{I}$, then $\square\square\square$ = how many $\hat{I}$s?

3. Given that $\square\square\square\square = \Omega\Omega\Omega$ and $\Omega\Omega\Omega = \triangle\triangle$, then $\triangle\triangle\triangle\triangle$ = how many $\square$s?

4. Given that $\yen\yen\yen\yen\yen = \S\S$ and $\S\S = \hat{A}\hat{A}\hat{A}$, then $\hat{A}\hat{A}\hat{A}\hat{A}\hat{A}\hat{A}$ = how many $\yen$s?

Problem-Solving Strategies

The text features a carefully developed approach to problem solving that emphasizes the importance of *strategy* when solving problems. Students are encouraged to develop their own strategies—to draw diagrams, to write out the solution steps in words—as part of their solution to a problem. In each case, model strategies are presented as guides for students to follow as they attempt the "You Try It" problem. Having students provide strategies is a natural way to incorporate writing into the math curriculum.

Page 311

Example 4

A board 20 ft long is cut into two pieces. Five times the length of the shorter piece is 2 ft more than twice the length of the longer piece. Find the length of each piece.

Strategy
Let x represent the length of the shorter piece. Then $20 - x$ represents the length of the longer piece.

Make a drawing.

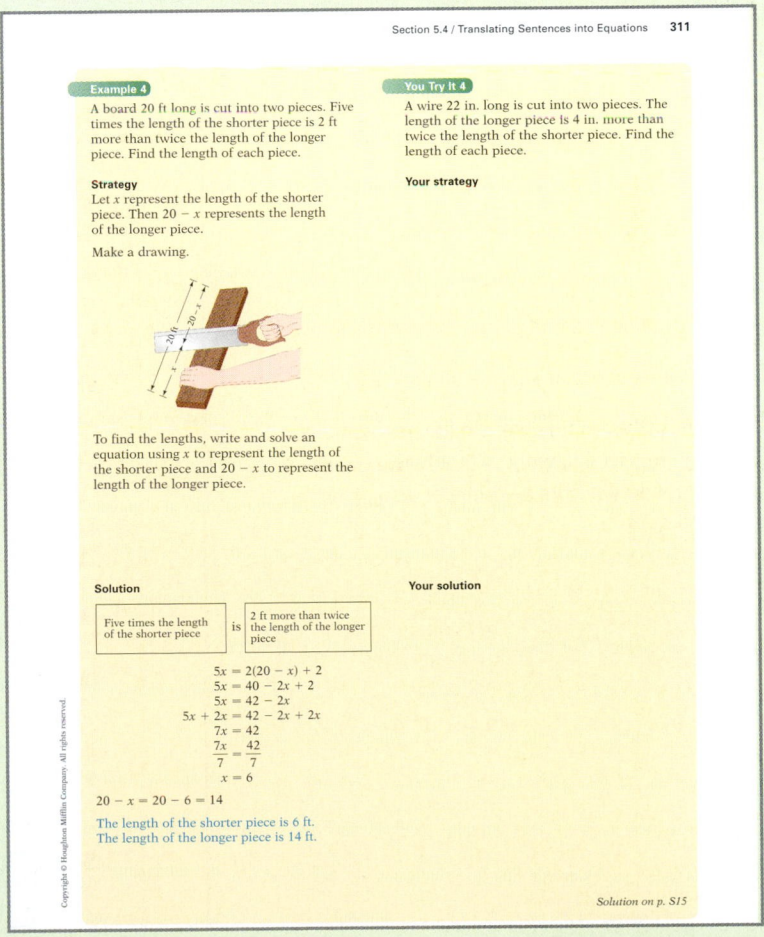

To find the lengths, write and solve an equation using x to represent the length of the shorter piece and $20 - x$ to represent the length of the longer piece.

Solution

Five times the length of the shorter piece	is	2 ft more than twice the length of the longer piece

$$5x = 2(20 - x) + 2$$
$$5x = 40 - 2x + 2$$
$$5x = 42 - 2x$$
$$5x + 2x = 42 - 2x + 2x$$
$$7x = 42$$
$$\frac{7x}{7} = \frac{42}{7}$$
$$x = 6$$

$20 - x = 20 - 6 = 14$

The length of the shorter piece is 6 ft.
The length of the longer piece is 14 ft.

You Try It 4

A wire 22 in. long is cut into two pieces. The length of the longer piece is 4 in. more than twice the length of the shorter piece. Find the length of each piece.

Your strategy

Your solution

Solution on p. S15

Real Data and Applications

Page 8

Objective D   To solve application problems and read statistical graphs

Graphs are displays that provide a pictorial representation of data. The advantage of graphs is that they present information in a way that is easily read.

A **pictograph** uses symbols to represent information. The symbol chosen usually has a connection to the data it represents.

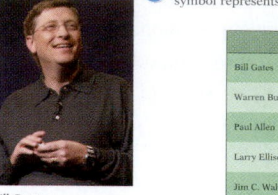

Bill Gates

Figure 1.1 represents the net worth of America's richest billionaires. Each symbol represents ten billion dollars.

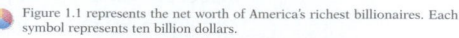

	Net Worth (in tens of billions of dollars)
Bill Gates	$ $ $ $ $
Warren Buffett	$ $ $ $
Paul Allen	$ $ $
Larry Ellison	$ $
Jim C. Walton	$ $

Figure 1.1   Net Worth of America's Richest Billionaires
Source: **www.Forbes.com**

From the pictograph, we can see that Bill Gates has the greatest net worth. Warren Buffett's net worth is $10 billion more than Paul Allen's net worth.

A typical household in the United States has an average after-tax income of $40,550. The **circle graph** in Figure 1.2 represents how this annual income is spent. The complete circle represents the total amount, $40,550. Each sector of the circle represents the amount spent on a particular expense.

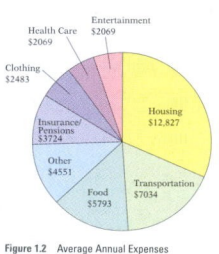

Health Care $2069
Entertainment $2069
Clothing $2483
Insurance/ Pensions $3724
Other $4551
Food $5793
Transportation $7034
Housing $12,827

From the circle graph, we can see that the largest amount is spent on housing. We can see that the amount spent on food ($5793) is less than the amount spent on transportation ($7034).

Figure 1.2   Average Annual Expenses in a U.S. Household
Source: American Demographics

Applications

One way to motivate an interest in mathematics is through applications. Wherever appropriate, the last objective of a section presents applications that require the student to use problem-solving strategies, along with the skills covered in that section, to solve practical problems. This carefully integrated applied approach generates student awareness of the value of algebra as a real-life tool.

Applications are taken from many disciplines, including agriculture, business, carpentry, chemistry, construction, education, finance, nutrition, real estate, sports, and weather.

Real Data

Real data examples and exercises, identified by , ask students to analyze and solve problems taken from actual situations. Students are often required to work with tables, graphs, and charts drawn from a variety of disciplines.

Page 16

Objective D   To solve application problems and read statistical graphs

88. **Baseball**   During his baseball career, Eddie Collins had a record of 743 stolen bases. Max Carey had a record of 738 stolen bases during his baseball career. Who had more stolen bases, Eddie Collins or Max Carey?

89. **Baseball**   During his baseball career, Ty Cobb had a record of 892 stolen bases. Billy Hamilton had a record of 937 stolen bases during his baseball career. Who had more stolen bases, Ty Cobb or Billy Hamilton?

90. **Nutrition**   The figure at the right shows the annual per capita turkey consumption in different countries.
a. What is the annual per capita turkey consumption in the United States?
b. In which country is the annual per capita turkey consumption the highest?

Britain	
Canada	
France	
Ireland	
Israel	
Italy	
U.S.	

Each represents 2 lb.

Per Capita Turkey Consumption
Source: National Turkey Federation

91. **The Arts**   The play *Hello Dolly* was performed 2844 times on Broadway. The play *Fiddler on the Roof* was performed 3242 times on Broadway. Which play had the greater number of performances, *Hello Dolly* or *Fiddler on the Roof*?

92. **The Arts**   The play *Annie* was performed 2377 times on Broadway. The play *My Fair Lady* was performed 2717 times on Broadway. Which play had the greater number of performances, *Annie* or *My Fair Lady*?

93. **Nutrition**   Two tablespoons of peanut butter contain 190 calories. Two tablespoons of grape jelly contain 114 calories. Which contains more calories, two tablespoons of peanut butter or two tablespoons of grape jelly?

94. **History**   In 1892, the diesel engine was patented. In 1844, Samuel F. B. Morse patented the telegraph. Which was patented first, the diesel engine or the telegraph?

Samuel F. B. Morse

95. **Geography**   The distance between St. Louis, Missouri, and Reno, Nevada, is 1892 mi. The distance between St. Louis, Missouri, and San Diego, California, is 1833 mi. Which is the shorter distance, St. Louis to Reno or St. Louis to San Diego?

96. **Movie Theaters**   The circle graph at the right shows the result of a survey of 150 people who were asked, "What bothers you most about movie theaters?"
a. Among the respondents, what was the most often mentioned complaint?
b. What was the least often mentioned complaint?

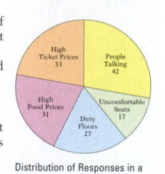

High Ticket Prices 23
People Talking 42
High Food Prices 31
Dirty Floors 27
Uncomfortable Seats 17

Distribution of Responses in a Survey

97. **Astronomy**   As measured at the equator, the diameter of the planet Uranus is 32,200 mi and the diameter of the planet Neptune is 30,800 mi. Which planet is smaller, Uranus or Neptune?

Page 8

Page 16

Student Pedagogy

Icons

The   at each objective head remind students that a video, a tutorial lesson, web resources, and solutions are available for that objective.

Point of Interest

These margin notes contain interesting sidelights about mathematics, its history, and its applications.

Study Tips

These margin notes remind students of study skills presented in the *AIM for Success;* some notes provide page references to the original descriptions. They also provide students with reminders of how to practice good study habits.

Key Terms and Concepts

Key terms, in bold, emphasize important terms. The key terms are also provided in a **Glossary** at the back of the text.

Key concepts are presented in orange boxes in order to highlight these important concepts and to provide for easy reference.

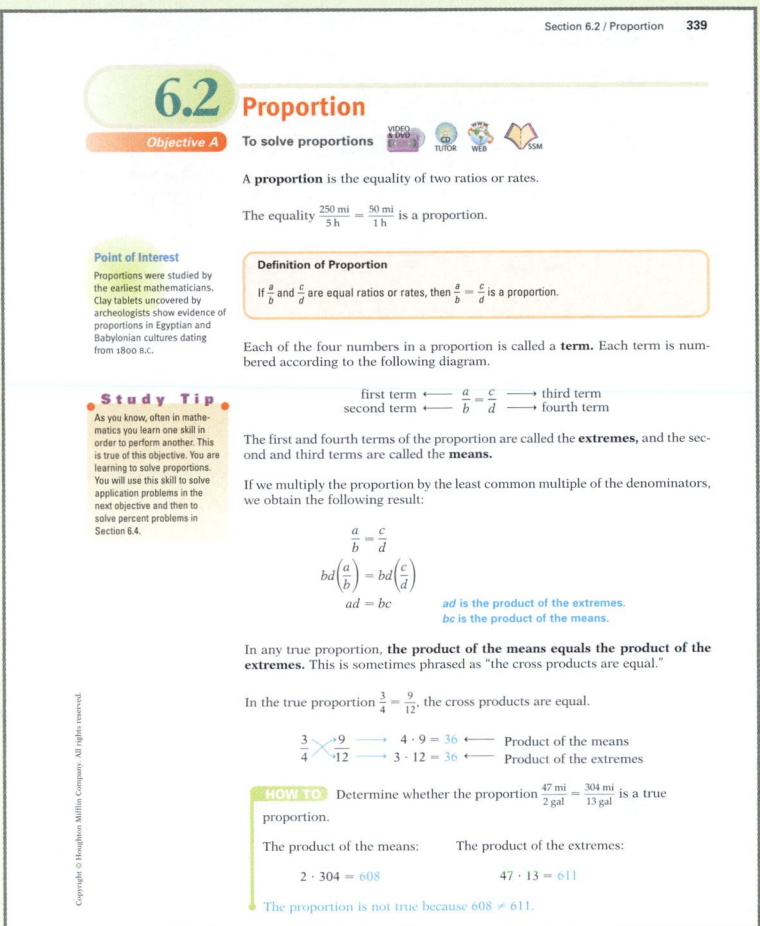

Page 339

Page 57

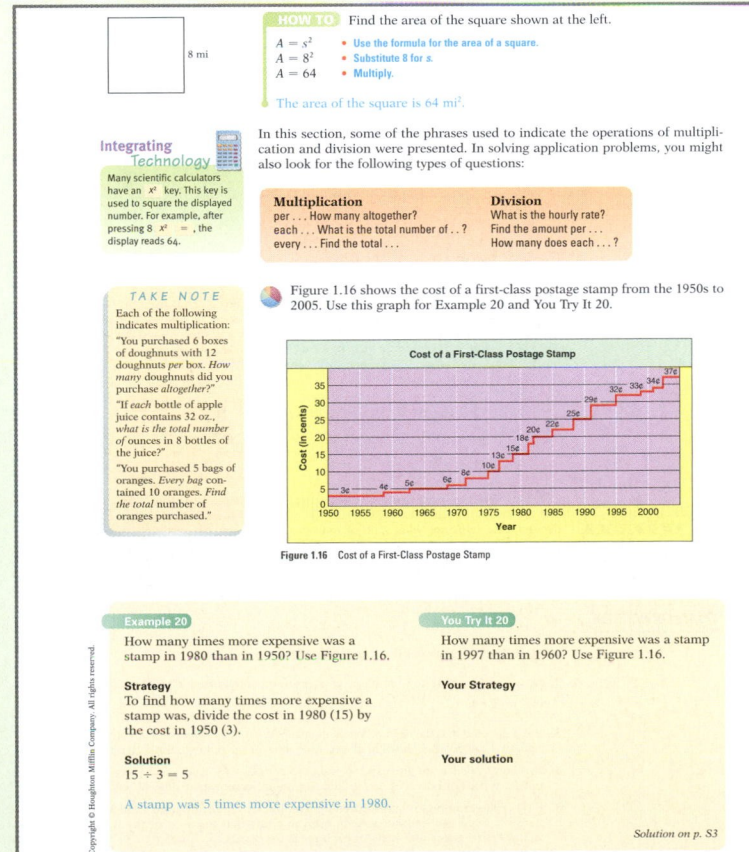

HOW TO Examples

HOW TO examples use annotations to explain what is happening in key steps of the complete, worked-out solutions.

Integrating Technology

These margin notes provide suggestions for using a scientific calculator.

Take Note

These margin notes alert students to a point requiring special attention or are used to amplify the concept under discussion.

Exercises and Projects

Exercises

The exercise sets of *Prealgebra and Introductory Algebra* emphasize skill building, skill maintenance, and applications. Concept-based writing or developmental exercises have been integrated within the exercise sets. Icons identify appropriate writing 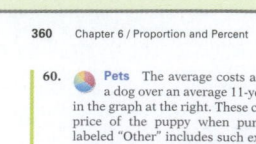, data analysis 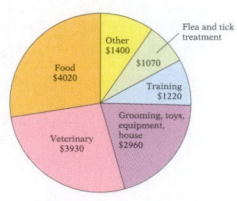, and calculator  exercises.

Included in each exercise set are **Applying the Concepts,** which present extensions of topics, require analysis, or offer challenge problems. The writing exercises ask students to explain answers, write about a topic in the section, or research and report on a related topic.

Page 360

360   Chapter 6 / Proportion and Percent

60. Pets   The average costs associated with owning a dog over an average 11-year life span are shown in the graph at the right. These costs do not include the price of the puppy when purchased. The category labeled "Other" includes such expenses as fencing and repairing furniture damaged by the pet. What percent of the total cost is spent on food? Round to the nearest tenth of a percent.

61. Manufacturing   During a quality control test, a manufacturer of computer boards found that 56 boards were defective. This was 0.7% of the total number of computer boards tested. How many of the tested computer boards were not defective?

62. Agriculture   Of the 572 million pounds of cranberries grown in the United States in a recent year, Wisconsin growers produced 291.72 million pounds. What percent of the total cranberry crop was produced in Wisconsin?

63. Politics   The results of a survey in which 32,840 full-time college and university faculty members were asked to describe their political views are shown at the right. How many more faculty members described their political views as liberal than described their views as far left?

64. Politics   The results of a survey in which 32,840 full-time college and university faculty members were asked to describe their political views are shown at the right. How many fewer faculty members described their political views as conservative than described their views as middle-of-the-road?

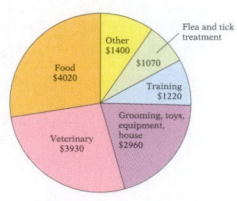

Cost of Owning a Dog
Source: American Kennel Club, *USA Today* research

Political View	Percent of Faculty Members Responding
Far left	5.3%
Liberal	42.3%
Middle of the road	34.3%
Conservative	17.7%
Far right	0.3%

Source: Higher Education Research Institute, UCLA

APPLYING THE CONCEPTS

65. Find 10% of a number and subtract it from the original number. Now take 10% of the new number and subtract it from the new number. Is this the same as taking 20% of the original number and subtracting it from the original number? Explain.

66. Increase a number by 10%. [...] result the original number? [...]

67. Compensation   Your emplo[...] on the job, a 6% raise the ne[...] your salary after the third y[...] would have been if you had [...]

68. Visit a savings and loa[...] write a report on the m[...]

69. Find five different use[...] used in those instances [...]

Page 360

Page 37

Objective C   To solve application problems and use formulas

113.    Mathematics   Find the sum of all the whole numbers less than 21.

114. Mathematics   Find the sum of all the natural numbers greater than 89 and less than 101.

115. Mathematics   Find the difference between the smallest four-digit number and the largest two-digit number.

116. Demography   The figure at the right shows the expected U.S. population aged 100 and over every two years from 2010 to 2020.
   a. Which two-year period has the smallest increase in the number of people aged 100 and over?
   b. Which two-year period has the greatest increase?

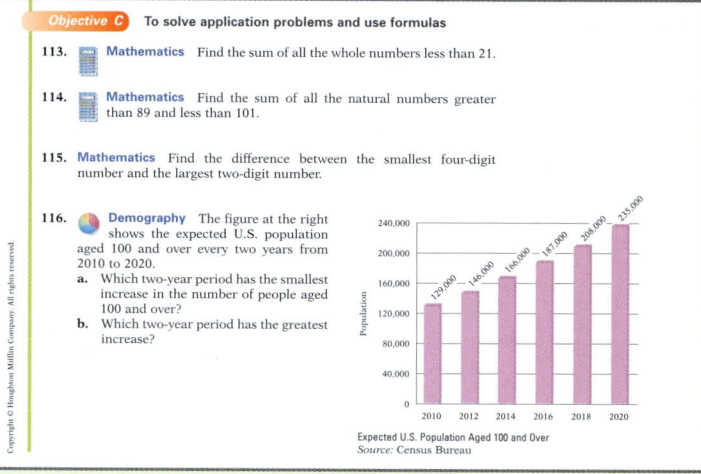

Expected U.S. Population Aged 100 and Over
Source: Census Bureau

Page 37

Page 164

Projects and Group Activities

The Projects and Group Activities feature at the end of each chapter can be used as extra credit or for cooperative learning activities. The projects cover various aspects of mathematics, including the use of calculators, collecting data from the Internet, data analysis, and extended applications.

164   Chapter 2 / Fractions and Decimals

Projects and Group Activities

Music   In musical notation, notes are printed on a **staff,** which is a set of five horizontal lines and the spaces between them. The notes of a musical composition are grouped into **measures,** or **bars.** Vertical lines separate measures on a staff. The shape of a note indicates how long it should be held. The whole note has the longest time value of any note. Each time value is divided by 2 in order to find the next smallest time value.

The **time signature** is a fraction that appears at the beginning of a piece of music. The numerator of the fraction indicates the number of beats in a measure. The denominator indicates what kind of note receives 1 beat. For example, music written in $\frac{2}{4}$ time has 2 beats to a measure, and a quarter note receives 1 beat. One measure in $\frac{2}{4}$ time may have 1 half note, 2 quarter notes, 4 eighth notes, or any other combination of notes totaling 2 beats. Other common time signatures are $\frac{4}{4}$, $\frac{3}{4}$, and $\frac{6}{8}$.

1. Explain the meaning of the 6 and the 8 in the time signature $\frac{6}{8}$.

2. Give some possible combinations of notes in one measure of a piece written in $\frac{4}{4}$ time.

3. What does a dot at the right of a note indicate? What is the effect of a dot at the right of a half note? At the right of a quarter note? At the right of an eighth note?

4. Symbols called rests are used to indicate periods of silence in a piece of music. What symbols are used to indicate the different time values of rests?

5. Find some examples of musical compositions written in different time signatures. Use a few measures from each to show that the sum of the time values of the notes and rests in each measure equals the numerator of the time signature.

Page 164

End of Chapter

Chapter Summary

At the end of each chapter there is a Chapter Summary that includes Key Words, Essential Rules and Procedures, and an example of each. Each entry includes an objective reference and a page reference indicating where the concept is introduced. These chapter summaries provide a single point of reference as the student prepares for a test.

Chapter 14 Summary

Key Words	Examples
A *set* is a collection of objects. The objects are called the *elements* of the set. The *roster method* of writing a set encloses a list of the elements in braces. [14.1A, p. 719]	Using the roster method, the set of the first three positive integers is written {1, 2, 3}.
The *empty set* or *null set*, written ∅, is the set that contains no elements. [14.1A, p. 719]	The set of cars that can travel faster than 1000 mph is an empty set.

Essential Rules and Procedures	Examples
Addition Property of Inequalities [14.2A, p. 725] The same term can be added to each side of an inequality without changing the solution set of the inequality. If $a > b$, then $a + c > b + c$. If $a < b$, then $a + c < b + c$.	$x - 3 < -7$ $x - 3 + 3 < -7 + 3$ $x < -4$
Multiplication Property of Inequalities [14.2B, p. 727] Each side of an inequality can be multiplied by the same positive number without changing the solution set of the inequality. If $a > b$ and $c > 0$, then $ac > bc$. If $a < b$ and $c > 0$, then $ac < bc$.	$4x > -8$ $\dfrac{4x}{4} > \dfrac{-8}{4}$ $x > -2$
If each side of an inequality is multiplied by the same negative number and the inequality symbol is reversed, then the solution set of the inequality is not changed. If $a > b$ and $c < 0$, then $ac < bc$. If $a < b$ and $c < 0$, then $ac > bc$.	$-2x < 6$ $\dfrac{-2x}{-2} > \dfrac{6}{-2}$ $x > -3$

Page 744

Chapter Review Exercises

Chapter Review Exercises are found at the end of each chapter. These exercises are selected to help the student integrate all of the topics presented in the chapter.

Page 745

Chapter 14 Review Exercises

1. Solve: $2x - 3 > x + 15$

2. Find $A \cap B$, given $A = \{0, 2, 4, 6, 8\}$ and $B = \{-2, -4\}$.

3. Use set-builder notation to write the set of odd integers greater than -8.

4. Find $A \cup B$, given $A = \{6, 8, 10\}$ and $B = \{2, 4, 6\}$.

Chapter Test

Each Chapter Test is designed to simulate a possible test of the material in the chapter.

Page 747

Chapter 14 Test

1. Graph: $\{x | x < 5\} \cap \{x | x > 0\}$

   $\begin{array}{ccccccccccc} \text{\textbardbl} & & & & & & & & & & \\ -5 & -4 & -3 & -2 & -1 & 0 & 1 & 2 & 3 & 4 & 5 \end{array}$

2. Use set-builder notation to write the set of positive integers less than 50.

3. Use the roster method to write the set of even positive integers between 3 and 9.

4. Solve: $3(2x - 5) \geq 8x - 9$

Cumulative Review Exercises

Cumulative Review Exercises, which appear at the end of each chapter (beginning with Chapter 2), help students maintain skills learned in previous chapters.

The answers to all Chapter Review Exercises, all Chapter Test exercises, and all Cumulative Review Exercises are given in the Answer Section. Along with the answer, there is a reference to the objective that pertains to the exercise.

Page 749

Cumulative Review Exercises

1. Simplify: $2[5a - 3(2 - 5a) - 8]$

2. Solve: $\dfrac{5}{8} - 4x = \dfrac{1}{8}$

3. Solve: $2x - 3[x - 2(x - 3)] = 2$

4. Simplify: $(-3a)(-2a^3b^2)^2$

Page A34

CUMULATIVE REVIEW EXERCISES

1. $40a - 28$ [4.2D]   2. $\dfrac{1}{8}$ [5.2A]   3. 4 [5.3B]   4. $-12a^7b^4$ [9.2B]   5. $-\dfrac{1}{b^4}$ [9.4A]

6. $4x - 2 - \dfrac{4}{4x - 1}$ [9.5B]   7. 0 [12.1D]   8. $3a^2(3x + 1)(3x - 1)$ [10.4B]   9. $\dfrac{1}{x + 2}$ [11.1C]

10. $\dfrac{18a}{(2a - 3)(a + 3)}$ [11.3B]   11. $-\dfrac{5}{9}$ [11.5A]   12. $C = S + Rt$ [11.6A]   13. $-\dfrac{7}{3}$ [12.3B]

Instructor's Annotated Edition—Format and Features

This edition offers a complete Instructor's Annotated Edition. Student pages are reduced, creating a margin for the following instructor-only features.

- **Instructor Notes**
  Instructor notes include teaching ideas, warnings about common student errors, and historical notes.

- **New Vocabulary**
  A list of new vocabulary introduced within a lesson is provided for appropriate objectives. There are similar lists for *New Symbols, Formulas, Rules, Properties,* and *Equations.*

- **Vocabulary to Review**
  A list of vocabulary terms introduced in previous objectives that students will need to recall in order to understand the material in the present lesson is provided for appropriate objectives. There are similar lists for *Symbols, Formulas, Rules, Properties,* and *Equations.*

- **In-Class Examples**
  For *every* objective, extra examples are offered that can be used during the presentation of the lesson. These extra examples do not duplicate the examples presented in the student textbook.

- **Discuss the Concepts**
  These questions, or requests for an explanation, can be used for class discussion or for writing exercises. They require students to verbalize the basic concepts presented in the lesson.

- **Concept Check**
  These questions or exercises can be used after the presentation of a lesson to test student understanding of the concepts developed.

- **Optional Student Activity**
  These exercises can be assigned at the conclusion of the lesson. They can serve as a class activity, individual work, or as cooperative learning projects. In general, activities were written to be accomplished within about a five-minute period.

- **Suggested Assignment**
  At the beginning of *every* exercise set, there is a suggested homework assignment that covers the essential topics of the section.

- **Quick Quiz**
  For *every* objective, there is a short quiz that can be given to students. These quizzes are designed to check basic concepts. These quizzes can also be downloaded from our website at **math.college.hmco.com/instructors** and are available on the Instructor *ClassPrep CD.*

- **Answers to Writing Exercises, Focus on Problem Solving, and Projects and Group Activities**
  Suggested answers to all the writing exercises are given in the Instructor's Annotated Edition. Also included, where appropriate, are answers to Focus on Problem Solving exercises and exercises in the Projects and Group Activities feature.

- **PowerPoints**
  Next to many of the graphs or tables in the text, there is a 🅿 that indicates that a PowerPoint slide of that graph is available. These PowerPoints can be downloaded from our website at **math.college.hmco.com/instructors** and are available on the Instructor *ClassPrep CD.*

- **Answers to Exercises**
  Answers to all the exercises are given in the Instructor's Annotated Edition. Only answers to the odd exercises are given in the Answer Section in the Student Edition. Answers to writing exercises are not included in the Student Edition.

Instructor Resources

Prealgebra and Introductory Algebra has a complete set of support materials for the instructor.

Instructor's Annotated Edition This edition contains a replica of the student text and additional resources just for the instructor. These include: *Instructor Notes, New Vocabulary/Symbols, etc., Vocabulary/Symbols, etc. to Review, In-Class Examples, Discuss the Concepts, Concept Checks, Optional Student Activities, Suggested Assignments, Quick Quizzes, Answers to Writing Exercises/Focus on Problem Solving/Projects and Group Activities,* and *PowerPoint* icons. Answers to all exercises are also provided.

Online Instructor's Solutions Manual The *Online Instructor's Solutions Manual* contains worked-out solutions for all exercises in the text.

Online Instructor's Resource Manual with Testing This resource includes four ready-to-use printed *Chapter Tests* per chapter, suggested *Course Sequences*, and a printout of the *AIM for Success* PowerPoint slide show. All resources are also available on the *ClassPrep CD.*

Online Teaching Center The free Houghton Mifflin Teaching Center on our website, **college.hmco.com/pic/aufmannPAIA1e,** contains an abundance of instructor resources.

HM ClassPrep™ with HM Testing (powered by Diploma™) *HM ClassPrep* offers a combination of two class-management tools including supplements and text-specific resources for the instructor. *HM Testing* (powered by *Diploma*) offers instructors a flexible and powerful tool for test generation and test management. Now supported by the Brownstone Research Group's market-leading *Diploma* software, this new version of *HM Testing* significantly improves on functionality and ease of use by offering all the tools needed to create, author, deliver, and customize multiple types of tests—including authoring and editing algorithmic questions. *Diploma* is currently in use at thousands of college and university campuses throughout the United States and Canada.

Blackboard®, WebCT®, and eCollege® Houghton Mifflin can provide you with valuable content to include in your existing Blackboard, WebCT, and eCollege systems. This text-specific content enables instructors to teach all or part of their course online. Contact your Houghton Mifflin sales representative for cartridge availability.

Eduspace® Powerful, customizable, and interactive, Eduspace, powered by Blackboard, is Houghton Mifflin's online learning tool. Eduspace provides instructors with online courses and content. By pairing the widely recognized tools of Blackboard with quality, text-specific content from Houghton Mifflin Company, Eduspace makes it easy for instructors to create all or part of a course online. Homework exercises, quizzes, tests, tutorials, and supplemental study materials all come ready-to-use. Instructors can choose to use the content as is or modify it, and they can even add their own. Visit **www.eduspace.com** for more information.

Student Resources

Student Solutions Manual The *Student Solutions Manual* contains complete solutions to all odd-numbered exercises in the text.

Math Study Skills Workbook by Paul D. Nolting This workbook is designed to reinforce skills and minimize frustration for students in any math class, lab, or study skills course. It offers a wealth of study tips and sound advice on note taking, time management, and reducing math anxiety. In addition, numerous opportunities for self-assessment enable students to track their own progress.

Eduspace Powerful, customizable, and interactive, Eduspace, powered by Blackboard, is Houghton Mifflin's online learning tool for instructors and students. Eduspace is a text-specific web-based learning environment that your instructor can use to offer students a combination of practice exercises, multimedia tutorials, video explanations, online algorithmic homework, and more. Specific content is available 24 hours a day to help you succeed in your course.

HM mathSpace® Student Tutorial CD-ROM For students who prefer the portability of a CD-ROM, this tutorial provides opportunities for self-paced review and practice with algorithmically generated exercises and step-by-step solutions.

Houghton Mifflin Instructional Videos and DVDs Text-specific videos and DVDs, hosted by Dana Mosely, cover all sections of the text and provide a valuable resource for further instruction and review.

Online Study Center The free Houghton Mifflin Study Center on our website, **college.hmco.com/pic/aufmannPAIA1e,** contains an abundance of student resources.

SMARTHINKING® Houghton Mifflin's unique partnership with SMARTHINKING brings students real-time, online tutorial support when they need it most. This partnership offers students a range of tutorial services exclusively for students using Houghton Mifflin texts. Using state-of-the-art whiteboard technology and feedback tools, students interact and communicate with "e-structors." These specially trained tutors guide students through the learning and problem-solving process without providing answers or rewriting a student's work.

SMARTHINKING offers three levels of service.*

- **Live Tutorial Help** provides real-time, one-on-one instruction.

- **Questions Any Time** allows students to e-mail questions to the tutor outside the scheduled tutorial sessions and receive a reply, usually within 24 hours.

- **Independent Study Resources** connect students with around-the-clock additional educational resources, ranging from interactive websites to Frequently Asked Questions.

Visit **smarthinking.com** for more information.
Limits apply; terms and hours of SMARTHINKING service are subject to change.

Acknowledgments

The authors would like to thank the people who reviewed this manuscript and provided many valuable suggestions.

Maria E. Bennett, *West Shore Community College*
Linda Clay, *Albuquerque Technical Vocational Institute*
Sharon J. Edgmon, *Bakersfield College*
Sally J. Keely, *University of Phoenix*
Lynette J. King, *Gadsden State Community College*
Richard Leedy, *Polk Community College*
Benjamin Moulton, *Utah Valley State College*
George Pasles
Kevin Yokoyama, *College of the Redwoods*

AIM for Success

Welcome to *Prealgebra and Introductory Algebra*. As you begin this course, we know two important facts: (1) We want you to succeed. (2) You want to succeed. To do that requires an effort from each of us. For the next few pages, we are going to show you what is required of you to achieve that success and how you can use the features of this text to be successful.

Motivation

One of the most important keys to success is motivation. We can try to motivate you by offering interesting or important ways mathematics can benefit you. But, in the end, the motivation must come from you. On the first day of class, it is easy to be motivated. Eight weeks into the term, it is harder to keep that motivation.

To stay motivated, there must be outcomes from this course that are worth your time, money, and energy.

List some reasons you are taking this course.

_____

_____

_____

TAKE NOTE

Motivation alone will not lead to success. For instance, suppose a person who cannot swim is placed in a boat, taken out to the middle of a lake, and then thrown overboard. That person has a lot of motivation but there is a high likelihood the person will drown without some help. Motivation gives us the desire to learn but is not the same as learning.

Although we hope that one of the reasons you listed was an interest in mathematics, we know that many of you are taking this course because it is required to graduate, it is a prerequisite for a course you must take, or because it is required for your major. Although you may not agree that this course is necessary, it is! If you are motivated to graduate or complete the requirements for your major, then use that motivation to succeed in this course. Do not become distracted from your goal to complete your education!

Commitment

To be successful, you must make a commitment to succeed. This means devoting time to math so that you achieve a better understanding of the subject.

List some activities (sports, hobbies, talents such as dance, art, or music) that you enjoy and at which you would like to become better.

ACTIVITY	TIME SPENT	TIME WISHED SPENT
_____	_____	_____
_____	_____	_____
_____	_____	_____

Thinking about these activities, put the number of hours that you spend each week practicing these activities next to the activity. Next to that number, indicate the number of hours per week you would like to spend on these activities.

Whether you listed surfing or sailing, aerobics or restoring cars, or any other activity you enjoy, note how many hours a week you spend doing it. To succeed in math, you must be willing to commit the same amount of time. Success requires some sacrifice.

The "I Can't Do Math" Syndrome

There may be things you cannot do, such as lift a two-ton boulder. You can, however, do math. It is much easier than lifting the two-ton boulder. When you first learned the activities you listed above, you probably could not do them well.

With practice, you got better. With practice, you will be better at math. Stay focused, motivated, and committed to success.

It is difficult for us to emphasize how important it is to overcome the "I Can't Do Math" Syndrome. If you listen to interviews of very successful athletes after a particularly bad performance, you will note that they focus on the positive aspect of what they did, not the negative. Sports psychologists encourage athletes to always be positive—to have a "Can Do" attitude. Develop this attitude toward math.

Strategies for Success

Textbook Review   Right now, do a 15-minute "textbook review" of this book. Here's how:

First, read the table of contents. Do it in three minutes or less. Next, look through the entire book, page by page. Move quickly. Scan titles, look at pictures, notice diagrams.

A textbook review shows you where a course is going. It gives you the big picture. That's useful because brains work best when going from the general to the specific. Getting the big picture before you start makes details easier to recall and understand later on.

Your textbook review will work even better if, as you scan, you look for ideas or topics that are interesting to you. List three facts, topics, or problems that you found interesting during your textbook review.

_____

_____

_____

The idea behind this technique is simple: It's easier to work at learning material if you know it's going to be useful to you.

Not all the topics in this book will be "interesting" to you. But that is true of any subject. Surfers find that on some days the waves are better than others, musicians find some music more appealing than other music, computer gamers find some computer games more interesting than others, car enthusiasts find some cars more exciting than others. Some car enthusiasts would rather have a completely restored 1957 Chevrolet than a new Ferrari.

Know the Course Requirements   To do your best in this course, you must know exactly what your instructor requires. Course requirements may be stated in a *syllabus*, which is a printed outline of the main topics of the course, or they may be presented orally. When they are listed in a syllabus or on other printed pages, keep them in a safe place. When they are presented orally, make sure to take complete notes. In either case, it is important that you understand them completely and follow them exactly. Be sure you know the answer to each of the following questions.

1. What is your instructor's name?
2. Where is your instructor's office?
3. At what times does your instructor hold office hours?
4. Besides the textbook, what other materials does your instructor require?
5. What is your instructor's attendance policy?
6. If you must be absent from a class meeting, what should you do before returning to class? What should you do when you return to class?

7. What is the instructor's policy regarding collection or grading of homework assignments?

8. What options are available if you are having difficulty with an assignment? Is there a math tutoring center?

9. If there is a math lab at your school, where is it located? What hours is it open?

10. What is the instructor's policy if you miss a quiz?

11. What is the instructor's policy if you miss an exam?

12. Where can you get help when studying for an exam?

Remember: Your instructor wants to see you succeed. If you need help, ask! Do not fall behind. If you are running a race and fall behind by 100 yards, you may be able to catch up but it will require more effort than had you not fallen behind.

TAKE NOTE

Besides time management, there must be realistic ideas of how much time is available. There are very few people who can *successfully* work full-time and go to school full-time. If you work 40 hours a week, take 15 units, spend the recommended study time given at the right, and sleep 8 hours a day, you will use over 80% of the available hours in a week. That leaves less than 20% of the hours in a week for family, friends, eating, recreation, and other activities.

Time Management   We know that there are demands on your time. Family, work, friends, and entertainment all compete for your time. We do not want to see you receive poor job evaluations because you are studying math. However, it is also true that we do not want to see you receive poor math test scores because you devoted too much time to work. When several competing and important tasks require your time and energy, the only way to manage the stress of being successful at both is to manage your time efficiently.

Instructors often advise students to spend twice the amount of time outside of class studying as they spend in the classroom. Time management is important if you are to accomplish this goal and succeed in school. The following activity is intended to help you structure your time more efficiently.

List the name of each course you are taking this term, the number of class hours each course meets, and the number of hours you should spend studying each subject outside of class. Then fill in a weekly schedule like the one printed below. Begin by writing in the hours spent in your classes, the hours spent at work (if you have a job), and any other commitments that are not flexible with respect to the time that you do them. Then begin to write down commitments that are more flexible, including hours spent studying. Remember to reserve time for activities such as meals and exercise. You should also schedule free time.

	Monday	Tuesday	Wednesday	Thursday	Friday	Saturday	Sunday
7–8 a.m.							
8–9 a.m.							
9–10 a.m.							
10–11 a.m.							
11–12 p.m.							
12–1 p.m.							
1–2 p.m.							
2–3 p.m.							
3–4 p.m.							
4–5 p.m.							
5–6 p.m.							
6–7 p.m.							
7–8 p.m.							
8–9 p.m.							
9–10 p.m.							
10–11 p.m.							
11–12 a.m.							

We know that many of you must work. If that is the case, realize that working 10 hours a week at a part-time job is equivalent to taking a three-unit class. If you must work, consider letting your education progress at a slower rate to allow you to be successful at both work and school. There is no rule that says you must finish school in a certain time frame.

Schedule Study Time   As we encouraged you to do by filling out the time management form above, schedule a certain time to study. You should think of this time the way you would the time for work or class—that is, reasons for missing study time should be as compelling as reasons for missing work or class. "I just didn't feel like it" is not a good reason to miss your scheduled study time.

Although this may seem like an obvious exercise, list a few reasons you might want to study.

_____

_____

_____

Of course, we have no way of knowing the reasons you listed, but from our experience one reason given quite frequently is "To pass the course." There is nothing wrong with that reason. If that is the most important reason for you to study, then use it to stay focused.

One method of keeping to a study schedule is to form a ***study group.*** Look for people who are committed to learning, who pay attention in class, and who are punctual. Ask them to join your group. Choose people with similar educational goals but different methods of learning. You can gain insight from seeing the material from a new perspective. Limit groups to four or five people; larger groups are unwieldy.

There are many ways to conduct a study group. Begin with the following suggestions and see what works best for your group.

1. Test each other by asking questions. Each group member might bring two or three sample test questions to each meeting.
2. Practice teaching each other. Many of us who are teachers learned a lot about our subject when we had to explain it to someone else.
3. Compare class notes. You might ask other students about material in your notes that is difficult for you to understand.
4. Brainstorm test questions.
5. Set an agenda for each meeting. Set approximate time limits for each agenda item and determine a quitting time.

And finally, probably the most important aspect of studying is that it should be done in relatively small chunks. If you can study only three hours a week for this course (probably not enough for most people), do it in blocks of one hour on three separate days, preferably after class. Three hours of studying on a Sunday is not as productive as three hours of paced study.

Text Features That Promote Success

There are 16 chapters in this text. Each chapter is divided into sections, and each section is subdivided into learning objectives. Each learning objective is labeled with a letter from A to E.

Preparing for a Chapter   Before you begin a new chapter, you should take some time to review previously learned skills. There are two ways to do this. The first is to complete the ***Cumulative Review Exercises,*** which occur after every chapter (except Chapter 1). For instance, turn to page 517. The questions in this review are taken from the previous chapters. The answers for all these exercises can be found on page A19. Turn to that page now and locate the answers for the Chapter 9 Cumulative Review Exercises. After the answer to the first exercise, which is $\frac{5}{144}$, you will see the objective reference [3.4A]. This means that this question was taken from Chapter 3, Section 4, Objective A. If you missed this question, you should return to that objective and restudy the material.

A second way of preparing for a new chapter is to complete the ***Prep Test.*** This test focuses on the particular skills that will be required for the new chapter. Turn to page 476 to see a Prep Test. The answers for the Prep Test are the first set of answers in the answer section for a chapter. Turn to page A17 to see the answers for the Chapter 9 Prep Test. Note that an objective reference is given for each question. If you answer a question incorrectly, restudy the objective from which the question was taken.

Before the class meeting in which your professor begins a new section, you should read each objective statement for that section. Next, browse through the objective material, being sure to note each word in bold type. These words indicate important concepts that you must know in order to learn the material. Do not worry about trying to understand all the material. Your professor is there to assist you with that endeavor. The purpose of browsing through the material is so that your brain will be prepared to accept and organize the new information when it is presented to you.

Turn to page 3. Write down the title of the first objective in Section 1.1. Write down the words under the title of the objective that are in bold print. It is not necessary for you to understand the meaning of these words. You are in this class to learn their meaning.

_____   _____

_____   _____

_____   _____

_____   _____

Math Is Not a Spectator Sport   To learn mathematics you must be an active participant. Listening and watching your professor do mathematics is not enough. Mathematics requires that you interact with the lesson you are studying. If you filled in the blanks above, you were being interactive. There are other ways this textbook has been designed to help you be an active learner.

Annotated Examples   The HOW TO feature indicates an example with explanatory remarks to the right of the work. Using paper and pencil, you should work along as you go through the example.

$$\frac{3}{4}x - 2 = -11$$

$$\frac{3}{4}x - 2 + 2 = -11 + 2$$

$$\frac{3}{4}x = -9$$

$$\frac{4}{3} \cdot \frac{3}{4}x = \frac{4}{3}(-9)$$

$$x = -12$$

HOW TO Solve: $\frac{3}{4}x - 2 = -11$

The goal is to write the equation in the form *variable = constant*.

$$\frac{3}{4}x - 2 = -11$$

$$\frac{3}{4}x - 2 + 2 = -11 + 2$$ • Add **2** to each side of the equation.

$$\frac{3}{4}x = -9$$ • Simplify.

$$\frac{4}{3} \cdot \frac{3}{4}x = \frac{4}{3}(-9)$$ • Multiply each side of the equation by $\frac{4}{3}$.

$$x = -12$$ • The equation is in the form *variable = constant*.

The solution is -12.

TAKE NOTE

Check: $\frac{3}{4}x - 2 = -11$

$$\frac{3}{4}(-12) - 2 \mid -11$$

$$-9 - 2 \mid -11$$

$$-11 = -11$$

A true equation

Page 292

When you complete the example, get a clean sheet of paper. Write down the problem and then try to complete the solution without referring to your notes or the book. When you can do that, move on to the next part of the objective.

Leaf through the book now and write down the page numbers of two other occurrences of a HOW TO example.

You Try Its   One of the key instructional features of this text is the paired examples. Notice that in each example box, the example on the left is completely worked out and the "You Try It" example on the right is not. Study the worked-out example carefully by working through each step. Then work the You Try It. If you get stuck, refer to the page number at the end of the example, which directs you to the place where the You Try It is solved—a complete worked-out solution is provided. Try to use the given solution to get a hint for the step you are stuck on. Then try to complete your solution.

Example 5

Solve: $2x + 4 - 5x = 10$

Solution

$$2x + 4 - 5x = 10$$
$$-3x + 4 = 10$$ • Combine like terms.
$$-3x + 4 - 4 = 10 - 4$$
$$-3x = 6$$
$$\frac{-3x}{-3} = \frac{6}{-3}$$
$$x = -2$$

The solution is -2.

You Try It 5

Solve: $x - 5 + 4x = 25$

Your solution

$$x - 5 + 4x = 25$$
$$5x - 5 = 25$$
$$5x - 5 + 5 = 25 + 5$$
$$5x = 30$$
$$\frac{5x}{5} = \frac{30}{5}$$
$$x = 6$$

The solution is 6.

Solution on p. S14

Page 295

When you have completed your solution, check your work against the solution we provided. (Turn to page S14 to see the solution of You Try It 5.) Be aware that frequently there is more than one way to solve a problem. Your answer, however, should be the same as the given answer. If you have any question as to whether your method will "always work," check with your instructor or with someone in the math center.

Browse through the textbook and write down the page numbers where two other paired example features occur.

Remember: Be an active participant in your learning process. When you are sitting in class watching and listening to an explanation, you may think that you understand. However, until you actually try to do it, you will have no confirmation of the new knowledge or skill. Most of us have had the experience of sitting in class thinking we knew how to do something only to get home and realize that we didn't.

TAKE NOTE

There is a strong connection between reading and being a successful student in math or any other subject. If you have difficulty reading, consider taking a reading course. Reading is much like other skills. There are certain things you can learn that will make you a better reader.

Word Problems   Word problems are difficult because we must read the problem, determine the quantity we must find, think of a method to do that, and then actually solve the problem. In short, we must formulate a *strategy* to solve the problem and then devise a *solution*.

Note in the paired example below that part of every word problem is a strategy and part is a solution. The strategy is a written description of how we will solve the problem. In the corresponding You Try It, you are asked to formulate a strategy. Do not skip this step, and be sure to write it out.

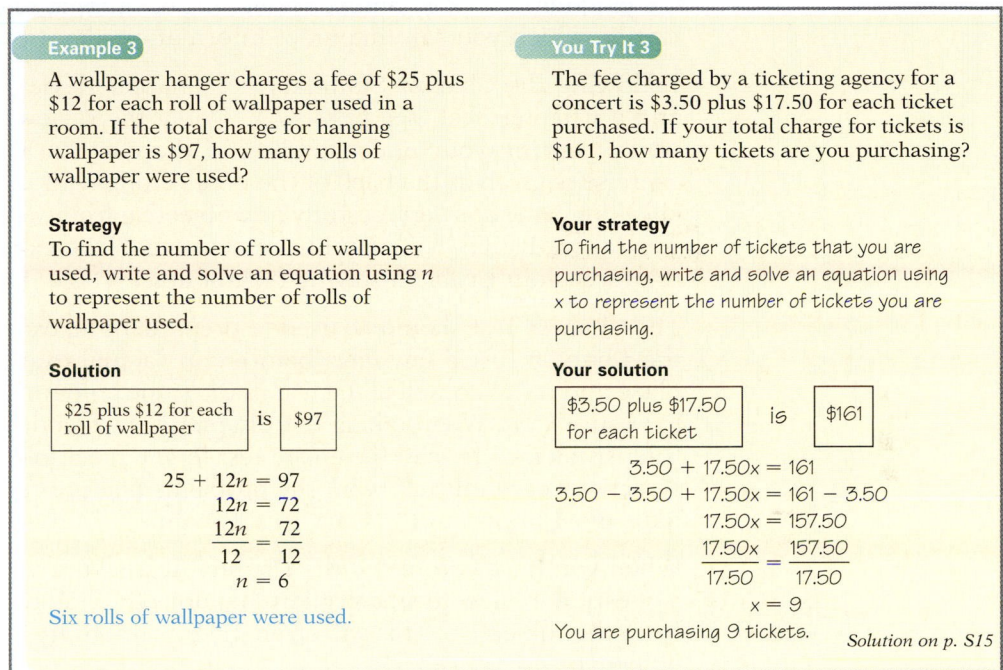

Example 3

A wallpaper hanger charges a fee of $25 plus $12 for each roll of wallpaper used in a room. If the total charge for hanging wallpaper is $97, how many rolls of wallpaper were used?

Strategy

To find the number of rolls of wallpaper used, write and solve an equation using n to represent the number of rolls of wallpaper used.

Solution

| $25 plus $12 for each roll of wallpaper | is | $97 |

$$25 + 12n = 97$$
$$12n = 72$$
$$\frac{12n}{12} = \frac{72}{12}$$
$$n = 6$$

Six rolls of wallpaper were used.

You Try It 3

The fee charged by a ticketing agency for a concert is $3.50 plus $17.50 for each ticket purchased. If your total charge for tickets is $161, how many tickets are you purchasing?

Your strategy

To find the number of tickets that you are purchasing, write and solve an equation using x to represent the number of tickets you are purchasing.

Your solution

| $3.50 plus $17.50 for each ticket | is | $161 |

$$3.50 + 17.50x = 161$$
$$3.50 - 3.50 + 17.50x = 161 - 3.50$$
$$17.50x = 157.50$$
$$\frac{17.50x}{17.50} = \frac{157.50}{17.50}$$
$$x = 9$$

You are purchasing 9 tickets.   *Solution on p. S15*

Page 310

Multiply each side of an equation by the same number. *Do not use zero.*

Rule Boxes   Pay special attention to rules placed in boxes. These rules give you the reasons certain types of problems are solved the way they are. When you see a rule, try to rewrite the rule in your own words.

The equations $2x = 6$, $10x = 30$, and $-8x = -24$ are equivalent equations; each equation has 3 as its solution. These examples suggest that multiplying each side of an equation by the same nonzero number produces an equivalent equation.

Multiplication Property of Equations

Each side of an equation can be multiplied by the same *nonzero* number without changing the solution of the equation. In symbols, if $c \neq 0$, then the equation $a = b$ has the same solutions as the equation $ac = bc$.

Page 283

Chapter Exercises   When you have completed studying an objective, do the exercises in the exercise set that correspond with that objective. The exercises are labeled with the same letter as the objective. Math is a subject that needs to be learned in small sections and practiced continually in order to be mastered. Doing all of the exercises in each exercise set will help you master the problem-solving techniques necessary for success. As you work through the exercises for an objective, check your answers to the odd-numbered exercises with those in the back of the book.

Preparing for a Test   There are important features of this text that can be used to prepare for a test.

- Chapter Summary
- Chapter Review Exercises
- Chapter Test

After completing a chapter, read the Chapter Summary. (See page 367 for the Chapter 6 Summary.) This summary highlights the important topics covered in the chapter. The page number following each topic refers you to the page in the text on which you can find more information about the concept.

Following the Chapter Summary are Chapter Review Exercises (see page 369) and a Chapter Test (see page 371). Doing the review exercises is an important way of testing your understanding of the chapter. The answer to each review exercise is given at the back of the book, along with its objective reference. After checking your answers, restudy any objective from which a question you missed was taken. It may be helpful to retry some of the exercises for that objective to reinforce your problem-solving techniques.

The Chapter Test should be used to prepare for an exam. We suggest that you try the Chapter Test a few days before your actual exam. Take the test in a quiet place and try to complete the test in the same amount of time you will be allowed for your exam. When taking the Chapter Test, practice the strategies of successful test takers: (1) scan the entire test to get a feel for the questions; (2) read the directions carefully; (3) work the problems that are easiest for you first; and perhaps most importantly, (4) try to stay calm.

When you have completed the Chapter Test, check your answers. If you missed a question, review the material in that objective and rework some of the exercises from that objective. This will strengthen your ability to perform the skills in that objective.

Is it difficult to be successful? YES! Successful music groups, artists, professional athletes, chefs, and <u>Write your major here</u> have to work very hard to achieve their goals. They focus on their goals and ignore distractions. The things we ask you to do to achieve success take time and commitment. We are confident that if you follow our suggestions, you will succeed.

Whole Numbers

How much would it cost to take a vacation that looks as relaxing as this one? Do you have enough money in the bank or would you need to earn more and save? Would you have to pay for travel, accommodations, and food? If you are considering taking a vacation, you will need to know the answers to these and other questions. The **Focus on Problem Solving on page 71** illustrates how you would use operations on whole numbers to determine costs associated with planning a vacation.

OBJECTIVES

Section 1.1

A To identify the order relation between two whole numbers

B To write whole numbers in words, in standard form, and in expanded form

C To round a whole number to a given place value

D To solve application problems and read statistical graphs

Section 1.2

A To add whole numbers

B To subtract whole numbers

C To solve application problems and use formulas

Section 1.3

A To multiply whole numbers

B To evaluate expressions that contain exponents

C To divide whole numbers

D To factor numbers and find the prime factorization of numbers

E To solve application problems and use formulas

Section 1.4

A To use the Order of Operations Agreement to simplify expressions

Need help? For online student resources, such as section quizzes, visit this textbook's website at **math.college.hmco.com/students**.

2

Features of the IAE
Among the features included in the margins of this Instructor's Annotated Edition are the following:

Instructor Note
These offer suggestions for presenting the material in the objective or information that may be helpful.

In-Class Examples
These exercises can be used during your presentation of the lesson. They do not duplicate examples presented in the textbook.

Discuss the Concepts
These questions or requests for an explanation can be used for class discussion or for writing exercises. They require students to verbalize the basic concepts presented in the lesson.

Concept Check
These questions or exercises can be used after the presentation of a lesson to test student understanding of the concepts developed.

Optional Student Activity
This exercise can be assigned at the conclusion of the lesson. It can serve as a class activity, as individual work, or as a cooperative learning project. In general, activities were written to be accomplished within about a five-minute period.

PREP TEST • • •

Do these exercises to prepare for Chapter 1.

1. Name the number of ♦s shown below.

   ♦ ♦ ♦ ♦ ♦ ♦ ♦ ♦
   8

2. Write the numbers from 1 to 10.

   1 ___ ___ ___ ___ ___ ___ ___ ___ 10
   1 2 3 4 5 6 7 8 9 10

3. Match the number with its word form.
   a. 4   A. five
   b. 2   B. one
   c. 5   C. zero
   d. 1   D. four
   e. 3   E. two
   f. 0   F. three
   a and D; b and E; c and A; d and B; e and F; f and C

4. How many American flags contain the color green?
   0

5. Write the number of states in the United States of America as a word, not a number.
   Fifty

GO FIGURE • • •

Five adults and two children want to cross a river in a rowboat. The boat can hold one adult or two children or one child. Everyone is able to row the boat. What is the minimum number of trips that will be necessary for everyone to get to the other side?   21 trips

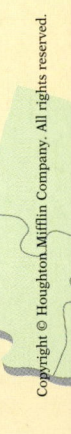

Instructor Note
Margin notes entitled *Point of Interest, Take Note, Study Tip,* and *Integrating Technology* are printed in the student text. The *Point of Interest* feature provides a historical note or mathematical fact of interest. The *Take Note* feature flags important information or provides assistance in understanding a concept. The *Study Tip* feature offers suggestions for using this text and approaches for creating good study habits. The *Integrating Technology* feature describes some of the functions of a calculator.

1.1 Introduction to Whole Numbers

Objective A **To identify the order relation between two whole numbers**

The **natural numbers** are 1, 2, 3, 4, 5, 6, 7, 8, 9, 10, 11, ...

The three dots mean that the list continues on and on and there is no largest natural number. The natural numbers are also called the **counting numbers.**

The **whole numbers** are 0, 1, 2, 3, 4, 5, 6, 7, 8, 9, 10, 11, ... Note that the whole numbers include the natural numbers and zero.

Just as distances are associated with markings on the edge of a ruler, the whole numbers can be associated with points on a line. This line is called the **number line** and is shown below.

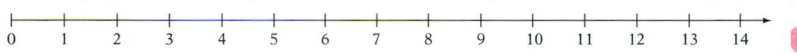

The arrowhead at the right indicates that the number line continues to the right.

The **graph of a whole number** is shown by placing a heavy dot on the number line directly above the number. Shown below is the graph of 6 on the number line.

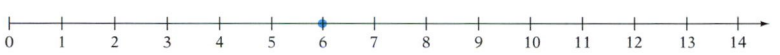

On the number line, the numbers get larger as we move from left to right. The numbers get smaller as we move from right to left. Therefore, the number line can be used to visualize the order relation between two whole numbers.

A number that appears to the right of a given number is **greater than** the given number. The symbol for *is greater than* is >.

8 is to the right of 3.
8 is greater than 3.
8 > 3

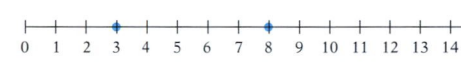

A number that appears to the left of a given number is **less than** the given number. The symbol for *is less than* is <.

5 is to the left of 12.
5 is less than 12.
5 < 12

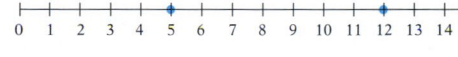

An **inequality** expresses the relative order of two mathematical expressions. 8 > 3 and 5 < 12 are inequalities.

Point of Interest

Among the slang words for zero are *zilch*, *zip*, and *goose egg*. The word *love* for zero in scoring a tennis game comes from the French for "the egg": *l'oeuf*.

TAKE NOTE

An inequality symbol, < or >, points to the smaller number. The symbol opens toward the larger number.

Objective 1.1A

New Vocabulary
natural numbers
counting numbers
whole numbers
number line
graph of a whole number
greater than (>)
less than (<)
inequality

Discuss the Concepts
Explain how you differentiate between the symbols < and >. (Some students will say that the symbol points toward the smaller number or opens toward the larger number.)

Concept Check
Determine which of the following are true and which are false. If a statement is false, correct it.
1. 27 < 32
   True
2. 82 > 87
   False. 82 < 87 or 87 > 82
3. 8050 < 8005
   False. 8050 > 8005 or 8005 < 8050

In-Class Examples (Objective 1.1A)

1. Graph 0 on the number line.

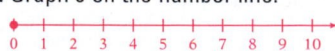

2. On the number line, which number is 4 units to the left of 9?   5

3. Place the correct symbol, < or >, between the two numbers.
6409   6490   6409 < 6490

4. Write the given numbers in order from smallest to largest.
483, 497, 492, 406, 438   406, 438, 483, 492, 497

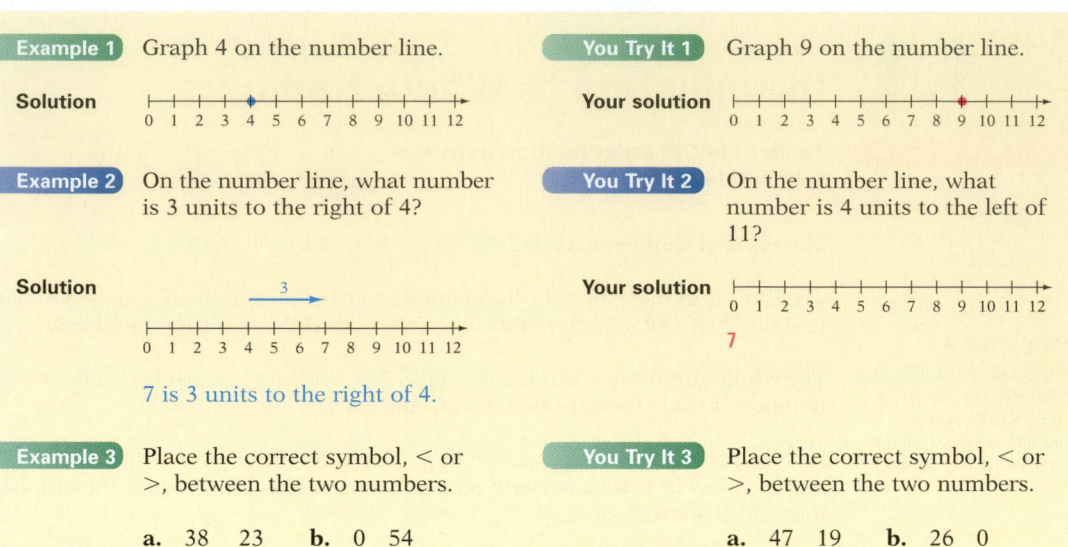

Example 1   Graph 4 on the number line.

Solution
0 1 2 3 4 5 6 7 8 9 10 11 12

You Try It 1   Graph 9 on the number line.

Your solution
0 1 2 3 4 5 6 7 8 9 10 11 12

Example 2   On the number line, what number is 3 units to the right of 4?

Solution
3 →

0 1 2 3 4 5 6 7 8 9 10 11 12

7 is 3 units to the right of 4.

You Try It 2   On the number line, what number is 4 units to the left of 11?

Your solution
0 1 2 3 4 5 6 7 8 9 10 11 12

7

Example 3   Place the correct symbol, < or >, between the two numbers.

a. 38   23   **b.** 0   54

Solution   **a.** 38 > 23   **b.** 0 < 54

You Try It 3   Place the correct symbol, < or >, between the two numbers.

a. 47   19   **b.** 26   0

Your solution   **a.** 47 > 19   **b.** 26 > 0

Example 4   Write the given numbers in order from smallest to largest.

16, 5, 47, 0, 83, 29

Solution   0, 5, 16, 29, 47, 83

You Try It 4   Write the given numbers in order from smallest to largest.

52, 17, 68, 0, 94, 3

Your solution   0, 3, 17, 52, 68, 94

Solutions on p. S1

Objective 1.1B

New Vocabulary

standard form
place value
place-value chart
period
expanded form

Point of Interest

The Romans represented numbers using M for 1000, D for 500, C for 100, L for 50, X for 10, V for 5, and I for 1. For example, MMDCCCLXXVI represented 2876. The Romans could represent any number up to the largest they would need for their everyday life, except zero.

Objective B   To write whole numbers in words, in standard form, and in expanded form

When a whole number is written using the digits 0, 1, 2, 3, 4, 5, 6, 7, 8, and 9, it is said to be in **standard form.** The position of each digit in the number determines the digit's **place value.** The diagram below shows a **place-value chart** naming the first twelve place values. The number 64,273 is in standard form and has been entered in the chart.

In the number 64,273, the position of the digit 6 determines that its place value is ten-thousands.

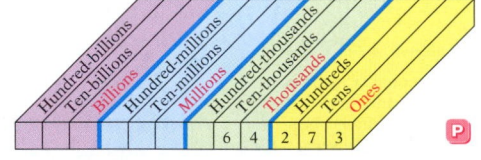

When a number is written in standard form, each group of digits separated by a comma is called a **period.** The number 5,316,709,842 has four periods. The period names are shown in color in the place-value chart above.

In-Class Examples (Objective 1.1B)

1. Write 3,007,609 in words.
   Three million seven thousand six hundred nine

2. Write two hundred forty-seven thousand sixty-three in standard form.
   247,063

3. Write 29,049 in expanded form.
   20,000 + 9000 + 40 + 9

Point of Interest

George Washington used a code to communicate with his men. He had a book in which each word or phrase was represented by a three-digit number. The numbers were arbitrarily assigned to each entry. Messages appeared as a string of numbers and thus could not be decoded by the enemy.

To write a number in words, start from the left. Name the number in each period. Then write the period name in place of the comma.

5,316,709,842 is read "five billion three hundred sixteen million seven hundred nine thousand eight hundred forty-two."

To write a whole number in standard form, write the number named in each period, and replace each period name with a comma.

Six million fifty-one thousand eight hundred seventy-four is written 6,051,874. The zero is used as a place holder for the hundred-thousands place.

The whole number 37,286 can be written in **expanded form** as

$$30,000 + 7000 + 200 + 80 + 6$$

The place-value chart can be used to find the expanded form of a number.

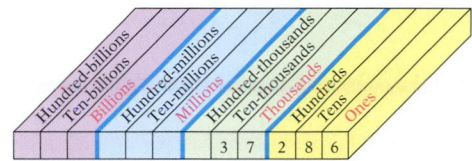

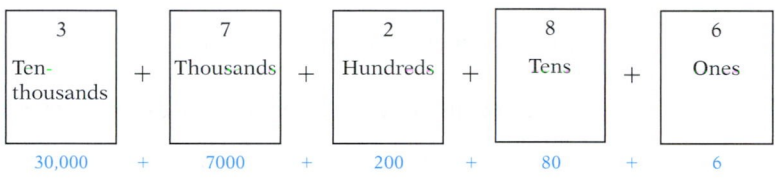

Write the number 510,409 in expanded form.

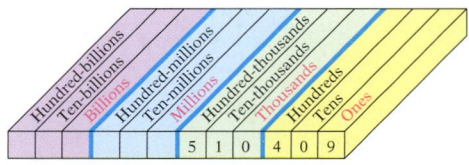

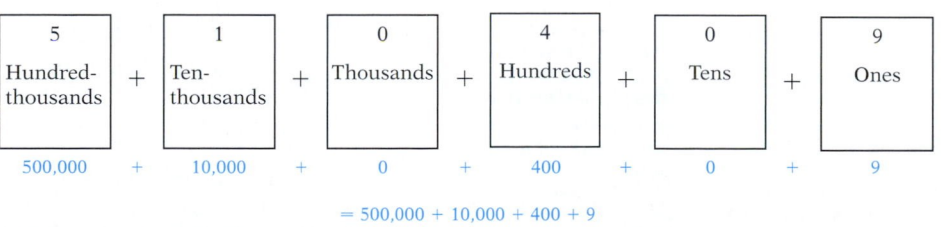

$$= 500,000 + 10,000 + 400 + 9$$

Discuss the Concepts

1. Explain how we use the words written in red in the place-value chart when reading a number.

2. How do the values of the following numbers differ?
   a. 3842
   b. Three thousand eight hundred forty-two
   c. 3000 + 800 + 40 + 2

In helping students understand that there is no difference among the values of the numbers in Exercise 2 above, and that they are merely different ways of writing the same number, you might discuss that, for example, the number 5 can be written as 5, five, or JHℝ.

Concept Check

Write the following numbers in standard form and then arrange them in order from smallest to largest.

1. Eighteen thousand nine hundred thirty-two
2. Three thousand four hundred seventy-five
3. Five million three thousand seven hundred ninety
4. Four hundred twenty-six thousand one hundred eight
   3475; 18,932; 426,108; 5,003,790

Optional Student Activity

Write each number in standard form.

1. 9000 + 200 + 40 + 8
   9248
2. 50,000 + 900 + 30 + 6
   50,936
3. 700,000 + 1000 + 80
   701,080
4. 6,000,000 + 40,000 + 500 + 20
   6,040,520

Example 5

Write 82,593,071 in words.

Solution

Eighty-two million five hundred ninety-three thousand seventy-one

You Try It 5

Write 46,032,715 in words.

Your solution

Forty-six million thirty-two thousand seven hundred fifteen

Example 6

Write four hundred six thousand nine in standard form.

Solution

406,009

You Try It 6

Write nine hundred twenty thousand eight in standard form.

Your solution

920,008

Example 7

Write 32,598 in expanded form.

Solution

30,000 + 2000 + 500 + 90 + 8

You Try It 7

Write 76,245 in expanded form.

Your solution

70,000 + 6000 + 200 + 40 + 5

Solutions on p. S1

Objective 1.1C

Vocabulary to Review

place value   [1.1B]

New Vocabulary

rounding

Discuss the Concepts

Explain how to round a four-digit number to the nearest hundred.

Objective C   **To round a whole number to a given place value**

When the distance to the sun is given as 93,000,000 mi, the number represents an approximation to the true distance. Giving an approximate value for an exact number is called **rounding.** A number is rounded to a given place value.

48 is closer to 50 than it is to 40. 48 rounded to the nearest ten is 50.

4872 rounded to the nearest ten is 4870.

4872 rounded to the nearest hundred is 4900.

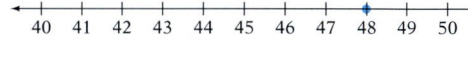

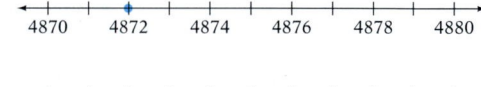

A number is rounded to a given place value without using the number line by looking at the first digit to the right of the given place value.

If the digit to the right of the given place value is less than 5, replace that digit and all digits to the right of it by zeros.

Round 12,743 to the nearest hundred.

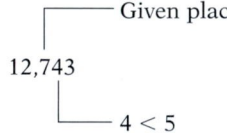

12,743 rounded to the nearest hundred is 12,700.

In-Class Examples (Objective 1.1C)

Round the number to the given place value.

1. 356; tens   360
2. 150; hundreds   200
3. 4060; hundreds   4100
4. 2369; thousands   2000
5. 35,099; thousands   35,000

6. 228,560; ten-thousands   230,000
7. 1,485,000; millions   1,000,000

If the digit to the right of the given place value is greater than or equal to 5, increase the digit in the given place value by 1, and replace all other digits to the right by zeros.

Round 46,738 to the nearest thousand.

Given place value

46,738

7 > 5

46,738 rounded to the nearest thousand is 47,000.

HOW TO Round 29,873 to the nearest thousand.

Given place value

29,873

8 > 5   Round up by adding 1 to the 9 (9 + 1 = 10). Carry the 1 to the ten-thousands place (2 + 1 = 3).

29,873 rounded to the nearest thousand is 30,000.

Example 8

Round 435,278 to the nearest ten-thousand.

Solution

Given place value

435,278

5 = 5

435,278 rounded to the nearest ten-thousand is 440,000.

You Try It 8

Round 529,374 to the nearest ten-thousand.

Your solution
530,000

Example 9

Round 1967 to the nearest hundred.

Solution

Given place value

1967

6 > 5

1967 rounded to the nearest hundred is 2000.

You Try It 9

Round 7985 to the nearest hundred.

Your solution
8000

Solutions on p. S1

Optional Student Activity

Determine whether the numbers are exact values or approximations.

1. The population of Dallas, Texas, is 1,100,000.
Approximation

2. In 1975 there were 19 female delegates to the U.S. House of Representatives.
Exact

3. At Sotheby's auction of Jackie Kennedy Onassis's estate, a bidder paid $700,000 for the woods in JFK's set of golf clubs.
Exact

4. The projected population of people aged 100 and over in the United States in 2020 is 235,000.
Approximation

5. In a recent year, General Motors spent $2,121,040 on advertising.
Exact

Objective 1.1D

New Vocabulary

pictograph
circle graph
bar graph
double-bar graph
broken-line graph

Discuss the Concepts

1. Provide some examples of data that would display well in a pictograph.

2. What characteristics of a bar graph make it easy to compare the data displayed in one?

Concept Check

1. In Figure 1.1, what does the dollar sign represent?
Ten billion dollars

2. How many sectors are in the circle graph in Figure 1.2?
8

3. What years are represented in the bar graph in Figure 1.3?
2005, 2010, 2015, 2020, 2025, and 2030

4. What do the numbers printed at the top of the bars in the bar graph in Figure 1.4 represent?
Fuel efficiency in miles per gallon

Optional Student Activity

A good source of graphs is *USA Today*. As an assignment, and to bring writing into the course, have students select a graph from that or another newspaper and write a paragraph that describes the data.

Objective D   **To solve application problems and read statistical graphs**

Graphs are displays that provide a pictorial representation of data. The advantage of graphs is that they present information in a way that is easily read.

A **pictograph** uses symbols to represent information. The symbol chosen usually has a connection to the data it represents.

 Figure 1.1 represents the net worth of America's richest billionaires. Each symbol represents ten billion dollars.

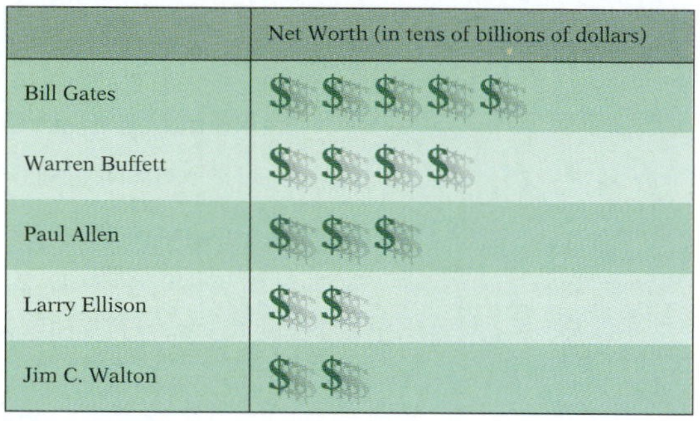

Bill Gates

	Net Worth (in tens of billions of dollars)
Bill Gates	$ $ $ $ $
Warren Buffett	$ $ $ $
Paul Allen	$ $ $
Larry Ellison	$ $
Jim C. Walton	$ $

Figure 1.1   Net Worth of America's Richest Billionaires
Source: www.Forbes.com

From the pictograph, we can see that Bill Gates has the greatest net worth. Warren Buffett's net worth is $10 billion more than Paul Allen's net worth.

 A typical household in the United States has an average after-tax income of $40,550. The **circle graph** in Figure 1.2 represents how this annual income is spent. The complete circle represents the total amount, $40,550. Each sector of the circle represents the amount spent on a particular expense.

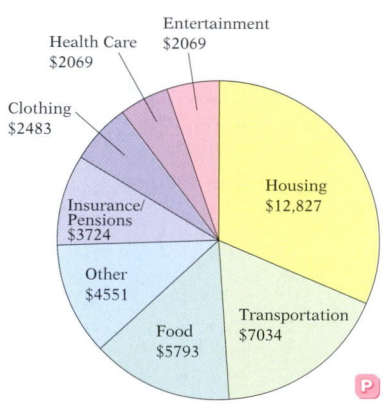

From the circle graph, we can see that the largest amount is spent on housing. We can see that the amount spent on food ($5793) is less than the amount spent on transportation ($7034).

Figure 1.2   Average Annual Expenses in a U.S. Household
Source: American Demographics

In-Class Examples (Objective 1.1D)

1. According to Figure 1.1, who has the greater net worth, Paul Allen or Larry Ellison?
Paul Allen

2. According to Figure 1.2, which is the greater expense, housing or food?
Housing

3. According to Figure 1.5, what is the value of the policy after 15 years?
$48,100

The **bar graph** in Figure 1.3 shows the expected U.S. population aged 100 and over.

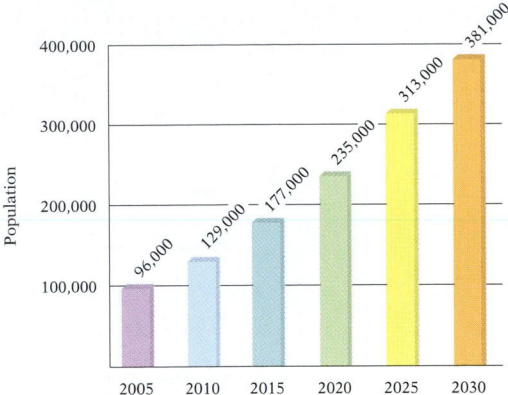

Figure 1.3   Expected U.S. Population Aged 100 and Over
Source: Census Bureau

In this bar graph, the horizontal axis is labeled with the years (2005, 2010, 2015, etc.) and the vertical axis is labeled with the numbers for the population. For each year, the height of the bar indicates the population for that year. For example, we can see that the expected population of those aged 100 and over in the year 2015 is 177,000. The graph indicates that the population of people aged 100 and over keeps increasing.

A **double-bar graph** is used to display data for purposes of comparison.

The double-bar graph in Figure 1.4 shows the fuel efficiency of four vehicles, as rated by the Environmental Protection Agency. These are among the most fuel-efficent 2005 model-year cars for city and highway mileage.

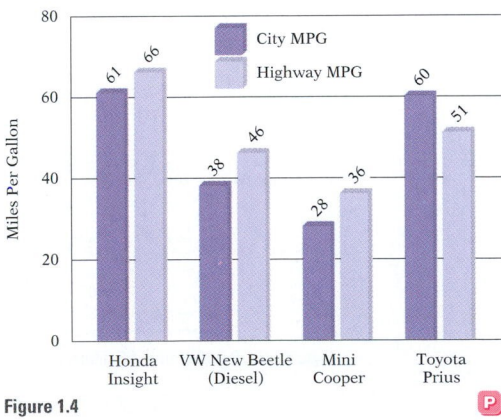

Figure 1.4

From the graph, we can see that the fuel efficiency of the Honda Insight is greater on the highway (66 mpg) than it is for city driving (61 mpg).

The **broken-line graph** in Figure 1.5 shows the effect of inflation on the value of a $100,000 life insurance policy. (An inflation rate of 5 percent is used here.)

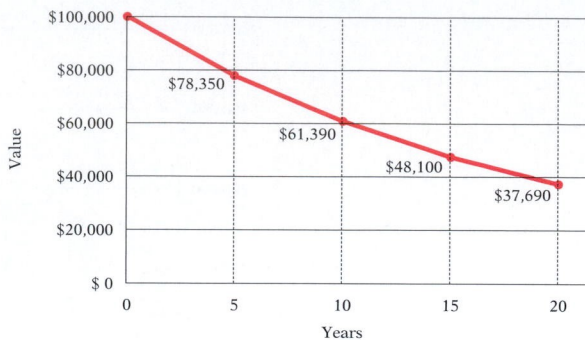

Figure 1.5   Effect of Inflation on the Value of a $100,000 Life Insurance Policy   Ⓟ

According to the line graph, after 5 years the purchasing power of the $100,000 has decreased to $78,350. We can see that the value of the $100,000 keeps decreasing over the 20-year period.

Two broken-line graphs are used so that data can be compared. Figure 1.6 shows the populations of California and Texas. The figures are those of the U.S. Census for the years 1900, 1925, 1950, 1975, and 2000. The numbers are rounded to the nearest thousand.

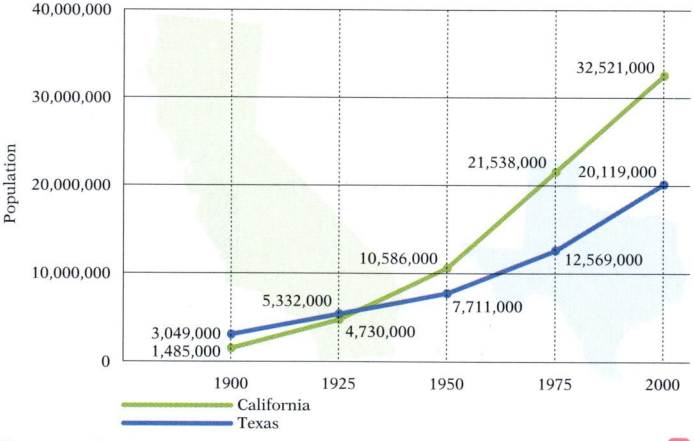

Figure 1.6   Populations of California and Texas   Ⓟ

From the graph, we can see that the population was greater in Texas in 1900 and 1925, while the population was greater in California in 1950, 1975, and 2000.

To solve an application problem, first read the problem carefully. The **Strategy** involves identifying the quantity to be found and planning the steps that are necessary to find that quantity. The **Solution** involves performing each operation stated in the Strategy and writing the answer.

The circle graph in Figure 1.7 shows the result of a survey of 300 people who were asked to name their favorite sport. Use this graph for Example 10 and You Try It 10.

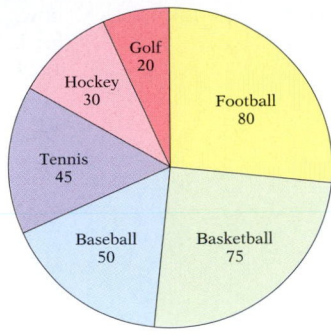

Figure 1.7   Distribution of Responses in a Survey

Example 10

According to Figure 1.7, which sport was named by the least number of people?

Strategy
To find the sport named by the least number of people, find the smallest number given in the circle graph.

Solution
The smallest number given in the graph is 20.

The sport named by the least number of people was golf.

You Try It 10

According to Figure 1.7, which sport was named by the greatest number of people?

Your Strategy

Your solution
Football

Example 11

The distance between St. Louis, Missouri, and Portland, Oregon, is 2057 mi. The distance between St. Louis, Missouri, and Seattle, Washington, is 2135 mi. Which distance is greater, St. Louis to Portland or St. Louis to Seattle?

Strategy
To find the greater distance, compare the numbers 2057 and 2135.

Solution
2135 > 2057

The greater distance is from St. Louis to Seattle.

You Try It 11

The distance between Los Angeles, California, and San Jose, California, is 347 mi. The distance between Los Angeles, California, and San Francisco, California, is 387 mi. Which distance is shorter, Los Angeles to San Jose or Los Angeles to San Francisco?

Your Strategy

Your solution
Los Angeles to San Jose

Solutions on p. S1

The bar graph in Figure 1.8 shows the states with the most sanctioned league bowlers. Use this graph for Example 12 and You Try It 12.

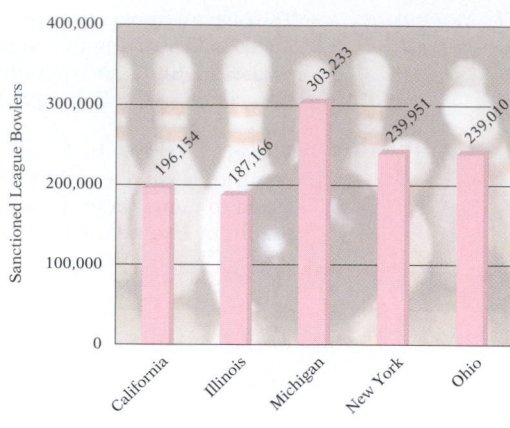

Figure 1.8 States with the Most Sanctioned League Bowlers
Sources: American Bowling Congress, Women's International Bowling Congress, Young American Bowling Alliance

Example 12

According to Figure 1.8, which state has the most sanctioned league bowlers?

Strategy
To determine which state has the most sanctioned league bowlers, locate the state that corresponds to the highest bar.

Solution
The highest bar corresponds to Michigan.

Michigan is the state with the most sanctioned league bowlers.

Example 13

The land area of the United States is 3,539,341 mi². What is the land area of the United States to the nearest ten-thousand square miles?

Strategy
To find the land area to the nearest ten-thousand square miles, round 3,539,341 to the nearest ten-thousand.

Solution
3,539,341 rounded to the nearest ten-thousand is 3,540,000.

To the nearest ten-thousand square miles, the land area of the United States is 3,540,000 mi².

You Try It 12

According to Figure 1.8, which state has fewer sanctioned league bowlers, New York or Ohio?

Your Strategy

Your solution
Ohio

You Try It 13

The land area of Canada is 3,851,809 mi². What is the land area of Canada to the nearest thousand square miles?

Your Strategy

Your solution
3,852,000 mi²

Solutions on p. S1

1.1 Exercises

Objective A   **To identify the order relation between two whole numbers**

1. ✎   How do the whole numbers differ from the natural numbers?

2. ✎   Explain how to round a four-digit number to the nearest hundred.

Graph the number on the number line.

3. 2

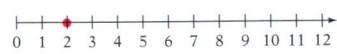

4. 7

5. 10

6. 1

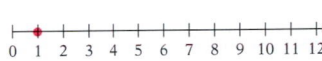

7. 5

8. 11

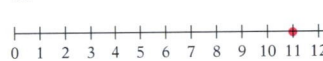

On the number line, which number is:

9. 4 units to the left of 9
   5

10. 5 units to the left of 8
   3

11. 3 units to the right of 2
   5

12. 4 units to the right of 6
   10

13. 7 units to the left of 7
   0

14. 8 units to the left of 11
   3

Place the correct symbol, < or >, between the two numbers.

15. 27   39
   <

16. 68   41
   >

17. 0   52
   <

18. 61   0
   >

19. 273   194
   >

20. 419   502
   <

21. 2761   3857
   <

22. 3827   6915
   <

23. 4610   4061
   >

24. 5600   56,000
   <

25. 8005   8050
   <

26. 92,010   92,001
   >

Write the given numbers in order from smallest to largest.

27. 21, 14, 32, 16, 11
   11, 14, 16, 21, 32

28. 18, 60, 35, 71, 27
   18, 27, 35, 60, 71

29. 72, 48, 84, 93, 13
   13, 48, 72, 84, 93

Section 1.1

Suggested Assignment
Exercises 3–107, odds
More challenging problems:
Exercises 108–110

Answers to Writing Exercises

1. Students should note that the whole numbers include 0 and the natural numbers do not.

2. To round a four-digit number to the nearest hundred, look at the digit in the tens place. If that digit is less than 5, the thousands digit and the hundreds digit remain unchanged and the digits in the tens and ones places are zeros. If the digit in the tens place is greater than or equal to 5, the hundreds digit is increased by 1 and the digits in the tens and ones places are zeros.

   Some students might note that if the digit in the tens place is greater than or equal to 5, and the digit in the hundreds place is 9, then the digit in the thousands place is increased by 1 and the digits in the hundreds, tens, and ones places are zeros.

Quick Quiz (Objective 1.1A)

1. On the number line, which number is 5 units to the right of 1?
   6

2. Place the correct symbol, < or >, between the two numbers.
   6857   8675
   6857 < 8675

3. Write the given numbers in order from smallest to largest.
   2516, 2615, 2165, 2651
   2165, 2516, 2615, 2651

30. 54, 45, 63, 28, 109
28, 45, 54, 63, 109

31. 26, 49, 106, 90, 77
26, 49, 77, 90, 106

32. 505, 496, 155, 358, 271
155, 271, 358, 496, 505

33. 736, 662, 204, 981, 399
204, 399, 662, 736, 981

34. 440, 404, 400, 444, 4000
400, 404, 440, 444, 4000

35. 377, 370, 307, 3700, 3077
307, 370, 377, 3077, 3700

Objective B   **To write whole numbers in words, in standard form, and in expanded form**

Write the number in words.

36. 704
Seven hundred four

37. 508
Five hundred eight

38. 374
Three hundred seventy-four

39. 635
Six hundred thirty-five

40. 2861
Two thousand
eight hundred sixty-one

41. 4790
Four thousand
seven hundred ninety

42. 48,297
Forty-eight thousand
two hundred ninety-seven

43. 53,614
Fifty-three thousnd
six hundred fourteen

44. 563,078
Five hundred sixty-three
thousand seventy-eight

45. 246,053
Two hundred forty-six
thousand fifty-three

46. 6,379,482
Six million three hundred
seventy-nine thousand four
hundred eighty-two

47. 3,842,905
Three million eight hundred
forty-two thousand nine
hundred five

Write the number in standard form.

48. Seventy-five
75

49. Four hundred ninety-six
496

50. Two thousand eight hundred fifty-one
2851

51. Fifty-three thousand three hundred forty
53,340

52. One hundred thirty thousand two hundred twelve
130,212

53. Five hundred two thousand one hundred forty
502,140

54. Eight thousand seventy-three
8073

55. Nine thousand seven hundred six
9706

Quick Quiz (Objective 1.1B)

1. Write 487,309 in words.
Four hundred eighty-seven thousand three hundred nine

2. Write six million eight thousand ninety-two in standard form.
6,008,092

3. Write 51,027 in expanded form.
$50,000 + 1000 + 20 + 7$

56. Six hundred three thousand one hundred thirty-two
603,132

57. Five million twelve thousand nine hundred seven
5,012,907

58. Three million four thousand eight
3,004,008

59. Eight million five thousand ten
8,005,010

Write the number in expanded form.

60. 6398
6000 + 300 + 90 + 8

61. 7245
7000 + 200 + 40 + 5

62. 46,182
40,000 + 6000 + 100 + 80 + 2

63. 532,791
500,000 + 30,000 + 2000 + 700 + 90 + 1

64. 328,476
300,000 + 20,000 + 8000 + 400 + 70 + 6

65. 5064
5000 + 60 + 4

66. 90,834
90,000 + 800 + 30 + 4

67. 20,397
20,000 + 300 + 90 + 7

68. 400,635
400,000 + 600 + 30 + 5

69. 402,708
400,000 + 2000 + 700 + 8

70. 504,603
500,000 + 4000 + 600 + 3

71. 8,000,316
8,000,000 + 300 + 10 + 6

Objective C   **To round a whole number to a given place value**

Round the number to the given place value.

72. 3049; tens
3050

73. 7108; tens
7110

74. 1638; hundreds
1600

75. 4962; hundreds
5000

76. 17,639; hundreds
17,600

77. 28,551; hundreds
28,600

78. 5326; thousands
5000

79. 6809; thousands
7000

80. 84,608; thousands
85,000

81. 93,825; thousands
94,000

82. 389,702; thousands
390,000

83. 629,513; thousands
630,000

84. 746,898; ten-thousands
750,000

85. 352,876; ten-thousands
350,000

86. 36,702,599; millions
37,000,000

87. 71,834,250; millions
72,000,000

Quick Quiz (Objective 1.1C)
Round the number to the given place value.
1. 4298; hundreds   4300
2. 29,074; tens   29,070
3. 67,524; thousands   68,000

Objective D   **To solve application problems and read statistical graphs**

88.   **Baseball**   During his baseball career, Eddie Collins had a record of 743 stolen bases. Max Carey had a record of 738 stolen bases during his baseball career. Who had more stolen bases, Eddie Collins or Max Carey?   Eddie Collins

89.   **Baseball**   During his baseball career, Ty Cobb had a record of 892 stolen bases. Billy Hamilton had a record of 937 stolen bases during his baseball career. Who had more stolen bases, Ty Cobb or Billy Hamilton?
Billy Hamilton

90.   **Nutrition**   The figure at the right shows the annual per capita turkey consumption in different countries.
   a.   What is the annual per capita turkey consumption in the United States?   a. 18 lb
   b.   In which country is the annual per capita turkey consumption the highest?   b. Israel

Britain	🦃🦃🦃🦃
Canada	🦃🦃🦃🦃
France	🦃🦃🦃🦃🦃
Ireland	🦃🦃🦃
Israel	🦃🦃🦃🦃🦃🦃🦃🦃🦃🦃
Italy	🦃🦃🦃🦃
U.S.	🦃🦃🦃🦃🦃🦃🦃🦃🦃

Each 🦃 represents 2 lb.

Per Capita Turkey Consumption
Source: National Turkey Federation   **P**

91.   **The Arts**   The play *Hello Dolly* was performed 2844 times on Broadway. The play *Fiddler on the Roof* was performed 3242 times on Broadway. Which play had the greater number of performances, *Hello Dolly* or *Fiddler on the Roof*?   Fiddler on the Roof

92.   **The Arts**   The play *Annie* was performed 2377 times on Broadway. The play *My Fair Lady* was performed 2717 times on Broadway. Which play had the greater number of performances, *Annie* or *My Fair Lady*?   My Fair Lady

93.   **Nutrition**   Two tablespoons of peanut butter contain 190 calories. Two tablespoons of grape jelly contain 114 calories. Which contains more calories, two tablespoons of peanut butter or two tablespoons of grape jelly?   Two tablespoons of peanut butter

94.   **History**   In 1892, the diesel engine was patented. In 1844, Samuel F. B. Morse patented the telegraph. Which was patented first, the diesel engine or the telegraph?   The telegraph

95.   **Geography**   The distance between St. Louis, Missouri, and Reno, Nevada, is 1892 mi. The distance between St. Louis, Missouri, and San Diego, California, is 1833 mi. Which is the shorter distance, St. Louis to Reno or St. Louis to San Diego?   St. Louis to San Diego

Samuel F. B. Morse

96.   **Movie Theaters**   The circle graph at the right shows the result of a survey of 150 people who were asked, "What bothers you most about movie theaters?"
   a.   Among the respondents, what was the most often mentioned complaint?   a. People talking
   b.   What was the least often mentioned complaint?   b. Uncomfortable seats

97.   **Astronomy**   As measured at the equator, the diameter of the planet Uranus is 32,200 mi and the diameter of the planet Neptune is 30,800 mi. Which planet is smaller, Uranus or Neptune?   Neptune

High Ticket Prices 33 | People Talking 42 | High Food Prices 31 | Dirty Floors 27 | Uncomfortable Seats 17   **P**

Distribution of Responses in a Survey

Quick Quiz (Objective 1.1D)

1.   The distance between Miami, Florida, and Dallas, Texas, is 1367 mi. The distance between Jacksonville, Florida, and Houston, Texas, is 884 mi. Which is the shorter distance, Miami to Dallas or Jacksonville to Houston?
Jacksonville to Houston

2.   The land area of Canada is 255,286 mi². What is the land area of Canada to the nearest ten thousand square miles?
260,000 mi²

98. 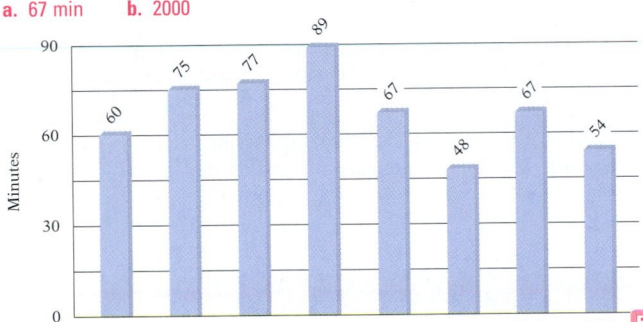 **Astronomy**   The diameter of Callisto, one of the moons orbiting Jupiter, is 4890 mi. The diameter of Ganymede, another of Jupiter's moons, is 5216 mi. Which is the larger moon, Callisto or Ganymede?
Ganymede

99. **Politics**   The figure below shows the length of the State of the Union Address for each of the years 1997 through 2004.
  a. What was the length of the State of the Union Address in 2001?
  b. In which of these years was the State of the Union Address the longest?
  a. 67 min   **b.** 2000

President Bush

Length of the State of the Union Address
Sources: USA Today; **www.whitehouse.gov; www.usconsulate.org**

100. **Geography**   The land area of Alaska is 570,833 mi². What is the land area of Alaska to the nearest thousand square miles?
571,000 mi²

101. **Geography**   The acreage of the Appalachian Trail is 161,546 acres. What is the acreage of the Appalachian Trail to the nearest ten-thousand acres?   160,000 acres

Alaska

102. **Automobile Accidents**   The figure below shows the number of crashes on U.S. roadways during each of the last six months of a recent year. Also shown is the number of vehicles involved in those crashes.
  a. Which was greater, the number of crashes in July or in October?
  b. Were there fewer vehicles involved in these crashes in July or in December?
  a. July   **b.** July

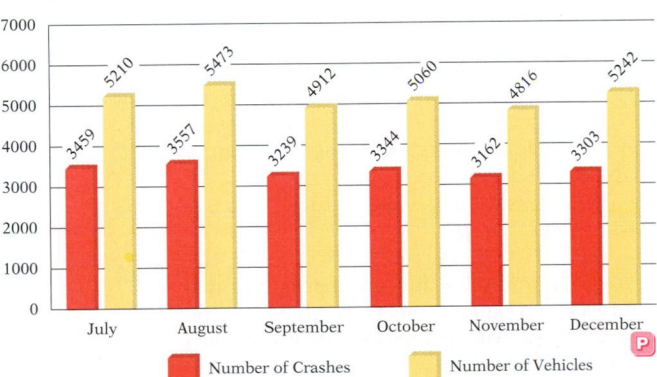

◼ Number of Crashes   ◼ Number of Vehicles

Accidents on U.S. Roadways
Source: National Highway Traffic Safety Administration

103. **School Enrollment** Actual and projected student enrollment in elementary and secondary schools in the United States is shown in the figure at the right. Enrollment figures are for the fall of each year. The jagged line at the bottom of the vertical axis indicates that this scale is missing the tens of millions from 0 to 30,000,000.

a. During which year was enrollment the lowest?

b. Did enrollment increase or decrease between 1975 and 1980?

a. 1985   b. Decrease

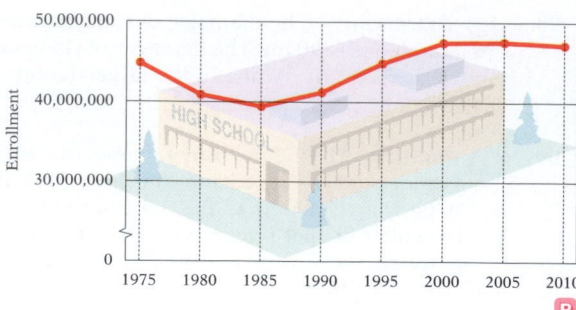

Enrollment in Elementary and Secondary Schools
Source: National Center for Education Statistics

104. **Aviation** The cruising speed of a Boeing 747 is 589 mph. What is the cruising speed of a Boeing 747 to the nearest ten miles per hour?

590 mph

105. **Physics** Light travels at a speed of 299,800 km/s. What is the speed of light to the nearest thousand kilometers per second?

300,000 km/s

APPLYING THE CONCEPTS

106. **Geography** Find the land area of the seven continents. List the continents in order from largest to smallest. List the oceans on Earth from largest to smallest.

Asia, Africa, North America, South America, Antarctica, Europe, Australia; Pacific, Atlantic, Indian, Arctic

107. **Mathematics** What is the largest three-digit number? What is the smallest five-digit number?

999; 10,000

108. What is the total enrollment of your school? To what place value would it be reasonable to round this number? Why? To what place value is the population of your town or city rounded? Why? To what place value is the population of your state rounded? To what place value is the population of the United States rounded?

For Exercise 109, answer true or false. If the answer is false, give an example to show that it is false.

109. a. If you are given two different whole numbers, then one of the numbers is always greater than the other number.   True

b. A rounded-off number is always less than its exact value.

False; 8270 rounded to the nearest hundred is 8300.

110. If 3846 is rounded to the nearest ten and then that number is rounded to the nearest hundred, is the result the same as what you get when you round 3846 to the nearest hundred? If not, which of the two methods is correct for rounding to the nearest hundred?

No. Round 3846 to the nearest hundred.

Answers to Writing Exercises

108. Answers to the question of the total enrollment of your school will vary. Generally, the population of a small school (less than 5000 students) is rounded to the nearest hundred; the population of a medium-size school (5000–20,000 students) or a large school (more than 20,000 students) is rounded to the nearest thousand.

The population of your state can be found in most annual almanacs. The place value to which the population is rounded will vary. For example, the population of New York might be rounded to the nearest million (18,000,000) while the population of Wyoming might be rounded to the nearest hundred-thousand (500,000).

Generally, the population of the United States is rounded to the nearest ten-million (290,000,000).

1.2 Addition and Subtraction of Whole Numbers

Objective A   To add whole numbers

Addition is the process of finding the total of two or more numbers.

On Arbor Day, a community group planted 3 trees along one street and 5 trees along another street. By counting, we can see that there were a total of 8 trees planted.

3   +   5   =   8

The 3 and 5 are called **addends.** The **sum** is 8.

The basic addition facts for adding one digit to one digit should be memorized. Addition of larger numbers requires the repeated use of the basic addition facts.

To add large numbers, begin by arranging the numbers vertically, keeping the digits of the same place value in the same column.

HOW TO   Add: 321 + 6472

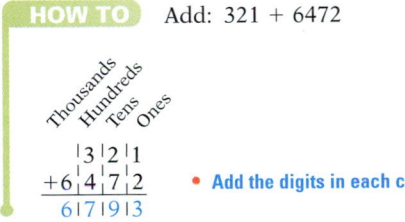

$$\begin{array}{r} 3\,|\,2\,|\,1 \\ +\,6\,|\,4\,|\,7\,|\,2 \\ \hline 6\,|\,7\,|\,9\,|\,3 \end{array}$$ • Add the digits in each column.

Study Tip

The HOW TO feature indicates an example with explanatory remarks. Using paper and pencil, you should work through the example. See *AIM for Success* page xxxi–xxxii.

HOW TO   Find the sum of 211, 45, 23, and 410.

$$\begin{array}{r} 211 \\ 45 \\ 23 \\ +\,410 \\ \hline 689 \end{array}$$

• Remember that a sum is the answer to an addition problem.
• Arrange the numbers vertically, keeping digits of the same place value in the same column.
• Add the numbers in each column.

The phrase *the sum of* was used in the example above to indicate the operation of addition. All of the phrases listed below indicate addition. An example of each is shown at the right of each phrase.

added to	6 added to 9	9 + 6
more than	3 more than 8	8 + 3
the sum of	the sum of 7 and 4	7 + 4
increased by	2 increased by 5	2 + 5
the total of	the total of 1 and 6	1 + 6
plus	8 plus 10	8 + 10

Objective 1.2A

New Vocabulary
addition
addend
sum
carrying
estimate
variable
variable expression
evaluating a variable expression
equation
left side of an equation
right side of an equation
solution of an equation

New Symbols
$\neq$ (is not equal to)

New Properties
Addition Property of Zero
Commutative Property of Addition
Associative Property of Addition

Discuss the Concepts
1. Provide at least three examples of situations in which we add numbers.
2. What is the difference between the Commutative Property of Addition and the Associative Property of Addition?

Concept Check
Which property justifies the statement?
1. $9 + 16 = 16 + 9$
Commutative Property of Addition
2. $23 + 0 = 23$
Addition Property of Zero
3. $6 + (8 + 5) = (6 + 8) + 5$
Associative Property of Addition

In-Class Examples (Objective 1.2A)

1. Estimate the sum of 347, 692, and 815.
1800

2. Identify the property that justifies the statement.
$x + 17 = 17 + x$
Commutative Property of Addition

3. Add: $29 + 6538 + 35{,}724 + 89$
42,380

4. Evaluate $a + b + c$ when $a = 9266$, $b = 8904$, and $c = 1795$.
19,965

5. Which number, 47, 37, 91, or 43, is a solution of the equation $64 = 27 + z$?
37

Optional Student Activity

Use a number line to illustrate the sum of 5, 3, and 6. Label the addends and the sum.

Optional Student Activity

Which numbers that are greater than 9 and less than 100 are increased by 9 when their digits are reversed?

12, 23, 34, 45, 56, 67, 78, 89

When the sum of the numbers in a column exceeds 9, addition involves **carrying**.

Integrating Technology

Most scientific calculators use *algebraic logic:* the add (+), subtract (−), multiply (×), and divide (÷) keys perform the indicated operation on the number in the display and the next number keyed in. For instance, for the example at the right, enter 359 + 478 = . The display reads 837.

HOW TO Add: 359 + 478

$$\begin{array}{r} \overset{1}{3}\;5\;9 \\ +\,4\;7\;8 \\ \hline 7 \end{array}$$

- Add the ones column.
  9 + 8 = **17** (1 ten + 7 ones).
  Write the 7 in the ones column and carry the 1 ten to the tens column.

$$\begin{array}{r} \overset{1\;1}{359} \\ +\,478 \\ \hline 37 \end{array}$$

- Add the tens column.
  1 + 5 + 7 = **13** (1 hundred + 3 tens).
  Write the 3 in the tens column and carry the 1 hundred to the hundreds column.

$$\begin{array}{r} \overset{1\;1}{359} \\ +\,478 \\ \hline 837 \end{array}$$

- Add the hundreds column.
  1 + 3 + 4 = **8** (8 hundreds).
  Write the 8 in the hundreds column.

HOW TO The bar graph in Figure 1.9 shows the population of each of the six New England states at the 2000 Census. What was the population of New England at the 2000 Census?

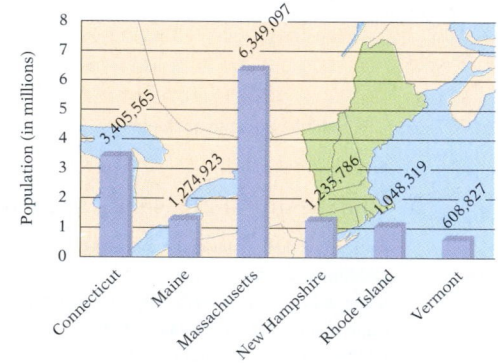

Figure 1.9   Population of the Six New England States

$$\begin{array}{r} 3,405,565 \\ 1,274,923 \\ 6,349,097 \\ 1,235,786 \\ 1,048,319 \\ +\quad 608,827 \\ \hline 13,922,517 \end{array}$$

The population of New England at the 2000 Census was 13,922,517 people.

Instructor Note
Estimation is an important skill.
Students should use this skill any
time a calculator is used.

An important skill in mathematics is the ability to determine whether an answer to a problem is reasonable. One method of determining whether an answer is reasonable is to use estimation. An **estimate** is an approximation.

Integrating Technology
Here is an example of why estimation is important when using a calculator.

Estimation is especially valuable when using a calculator. Suppose that you are adding 1497 and 2568 on a calculator. You enter the number 1497 correctly, but you inadvertently enter 256 instead of 2568 for the second addend. The sum reads 1753. If you quickly make an estimate of the answer, you can determine that the sum 1753 is not reasonable and that an error has been made.

$$\begin{array}{r} 1497 \\ +2568 \\ \hline 4065 \end{array} \qquad \begin{array}{r} 1497 \\ +\ 256 \\ \hline 1753 \end{array}$$

To estimate the answer to a calculation, round each number to the highest place value of the number; the first digit of each number will be nonzero and all other digits will be zero. Perform the calculation using the rounded numbers.

$$\begin{array}{r} 1497 \rightarrow\ \ 1000 \\ 2568 \rightarrow +3000 \\ \hline 4000 \end{array}$$

As shown above, the sum 4000 is an estimate of the sum of 1497 and 2568; it is very close to the actual sum, 4065. 4000 is not close to the incorrectly calculated sum, 1753.

HOW TO Estimate the sum of 35,498, 17,264, and 81,093.

$$\begin{array}{r} 35{,}498 \rightarrow\ \ \ 40{,}000 \\ 17{,}264 \rightarrow\ \ \ 20{,}000 \\ 81{,}093 \rightarrow +\ 80{,}000 \\ \hline 140{,}000 \end{array}$$

• Round each number to the nearest ten-thousand.

• Add the rounded numbers.

Note that 140,000 is close to the actual sum, 133,855.

Just as the word *it* is used in language to stand for an object, a letter of the alphabet can be used in mathematics to stand for a number. Such a letter is called a **variable.**

A mathematical expression that contains one or more variables is a **variable expression.** Replacing the variables in a variable expression with numbers and then simplifying the numerical expression is called **evaluating the variable expression.**

HOW TO Evaluate $a + b$ when $a = 678$ and $b = 294$.

$a + b$
$678 + 294$ • Replace *a* with 678 and *b* with 294.

$$\begin{array}{r} \overset{1\,1}{678} \\ +294 \\ \hline 972 \end{array}$$

• Arrange the numbers vertically.

• Add.

Variables are often used in algebra to describe mathematical relationships. Variables are used below to describe three properties, or rules, of addition. An example of each property is shown at the right.

> **The Addition Property of Zero**
>
> $a + 0 = a$ or $0 + a = a$

$5 + 0 = 5$

The Addition Property of Zero states that the sum of a number and zero is the number. The variable a is used here to represent any whole number. It can even represent the number zero because $0 + 0 = 0$.

> **The Commutative Property of Addition**
>
> $a + b = b + a$

$5 + 7 = 7 + 5$

$12 = 12$

The Commutative Property of Addition states that two numbers can be added in either order; the sum will be the same. Here the variables a and b represent any whole numbers. Therefore, if you know that the sum of 5 and 7 is 12, then you also know that the sum of 7 and 5 is 12, because $5 + 7 = 7 + 5$.

> **The Associative Property of Addition**
>
> $(a + b) + c = a + (b + c)$

$(2 + 3) + 4 = 2 + (3 + 4)$

$5 + 4 = 2 + 7$

$9 = 9$

The Associative Property of Addition states that when adding three or more numbers, we can group the numbers in any order; the sum will be the same. Note in the example at the right above that we can add the sum of 2 and 3 to 4, or we can add 2 to the sum of 3 and 4. In either case, the sum of the three numbers is 9.

HOW TO Rewrite the expression by using the Associative Property of Addition.

$(3 + x) + y$

$(3 + x) + y = 3 + (x + y)$ • **The Associative Property of Addition states that addends can be grouped in any order.**

Instructor Note
Emphasize the difference between an expression and an equation. You might put several examples on the board and ask students to label each. For example:
$3x + 7 = 9$
$3x + 7$
$4 - 6(y + 5)$
$a + b = 8$
$a + b - 8$
Explain to students that determining whether a number is a solution of an equation is a skill that will be used throughout the topic of solving equations. It is the way by which we check whether a solution is correct.

An **equation** expresses the equality of two numerical or variable expressions. In the example above, $(3 + x) + y$ is an expression; it does not contain an equals sign. $(3 + x) + y = 3 + (x + y)$ is an equation; it contains an equals sign.

Here is another example of an equation. The **left side** of the equation is the variable expression $n + 4$. The **right side** of the equation is the number 9.

$$n + 4 = 9$$

Just as a statement in English can be true or false, an equation may be true or false. The equation shown above is *true* if the variable is replaced by 5.

$$n + 4 = 9$$
$$5 + 4 = 9 \quad \text{True}$$

The equation is *false* if the variable is replaced by 8.

$$8 + 4 = 9 \quad \text{False}$$

A **solution** of an equation is a number that, when substituted for the variable, results in a true equation. The solution of the equation $n + 4 = 9$ is 5 because replacing n by 5 results in a true equation. When 8 is substituted for n, the result is a false equation; therefore, 8 is not a solution of the equation.

10 is a solution of $x + 5 = 15$ because $10 + 5 = 15$ is a true equation.

20 is not a solution of $x + 5 = 15$ because $20 + 5 = 15$ is a false equation.

HOW TO   Is 9 a solution of the equation $11 = 2 + x$?

$$11 = 2 + x$$
$$11 \mid 2 + 9 \qquad \bullet \text{ Replace } x \text{ by 9.}$$
$$11 = 11 \qquad \bullet \text{ Simplify the right side of the equation. Compare the results. If the results are equal, the given number is a solution of the equation. If the results are not equal, the given number is not a solution.}$$

Yes, 9 is a solution of the equation.

| **Example 1** | Estimate the sum of 379, 842, 693, and 518. | **You Try It 1** | Estimate the total of 6285, 3972, and 5140. |

Solution

$$
\begin{array}{rcr}
379 & \rightarrow & 400 \\
842 & \rightarrow & 800 \\
693 & \rightarrow & 700 \\
518 & \rightarrow & +\ 500 \\
\hline
& & 2400
\end{array}
$$

Your solution   15,000

| **Example 2** | Identify the property that justifies the statement. | **You Try It 2** | Identify the property that justifies the statement. |

$$7 + 2 = 2 + 7$$

$$33 + 0 = 33$$

Solution   The Commutative Property of Addition

Your solution   The Addition Property of Zero

Solutions on p. S1

The topic of the circle graph in Figure 1.10 is the eggs produced in the United States in a recent year. It shows where the eggs that were produced went or how they were used. Use this graph for Example 3 and You Try It 3.

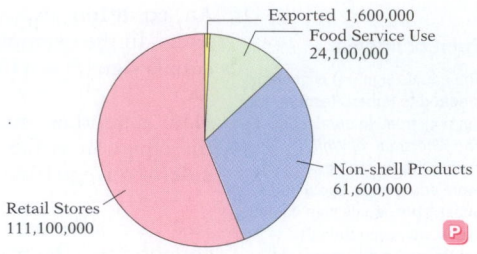

Exported 1,600,000
Food Service Use 24,100,000
Non-shell Products 61,600,000
Retail Stores 111,100,000

Figure 1.10 Distribution of Eggs Produced in the United States (in cases)

Source: American Egg Board. *USA Today.* Copyright © November 27, 2001.

Example 3 Use Figure 1.10 to determine the sum of the number of cases of eggs sold by retail stores and the number used for non-shell products.

Solution 111,100,000 cases of eggs were sold by retail stores. 61,600,000 cases of eggs were used for non-shell products.

```
  111,100,000
+  61,600,000
-------------
  172,700,000
```

172,700,000 cases of eggs were sold by retail stores and used for non-shell products.

You Try It 3 Use Figure 1.10 to determine the total number of cases of eggs produced during the year.

Your solution 198,400,000 cases of eggs

Example 4 Evaluate $x + y + z$ when $x = 8427$, $y = 3659$, and $z = 6281$.

Solution $x + y + z$
$8427 + 3659 + 6281$

```
   1 1 1
   8427
   3659
 + 6281
 ------
 18,367
```

You Try It 4 Evaluate $x + y + z$ when $x = 1692$, $y = 4783$, and $z = 5046$.

Your solution 11,521

Example 5 Is 6 a solution of the equation $9 + y = 14$?

Solution $9 + y = 14$
$\dfrac{9 + 6 \mid 14}{\quad 15 \neq 14}$ • The symbol $\neq$ is read "is not equal to."

No, 6 is not a solution of the equation $9 + y = 14$.

You Try It 5 Is 7 a solution of the equation $13 = b + 6$?

Your solution Yes

Solutions on p. S1

Objective B   **To subtract whole numbers**

Subtraction is the process of finding the difference between two numbers.

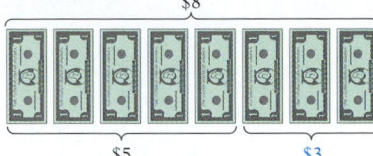

By counting, we see that the difference between $8 and $5 is $3.

$$\$8 \quad - \quad \$5 \quad = \quad \$3$$

$5 \qquad\qquad $3

Minuend − Subtrahend = Difference

Note that addition and subtraction are related.

	Subtrahend	5
+	Difference	+3
=	Minuend	8

The fact that the sum of the subtrahend and the difference equals the minuend can be used to check subtraction.

To subtract large numbers, begin by arranging the numbers vertically, keeping the digits of the same place value in the same column. Then subtract the numbers in each column.

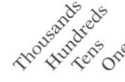 **HOW TO**   Find the difference between 8955 and 2432.

A *difference* is the answer to a subtraction problem.

```
      Thousands
        Hundreds
          Tens
            Ones
      8 | 9 | 5 | 5     Check:   Subtrahend    2432
        |   |   |              + Difference   +6523
    − 2 | 4 | 3 | 2            = Minuend       8955
      6 | 5 | 2 | 3
```

In the subtraction example above, the lower digit in each place value is smaller than the upper digit. When the lower digit is larger than the upper digit, subtraction involves **borrowing.**

HOW TO   Subtract: 692 − 378

```
  Hundreds            Hundreds            Hundreds            Hundreds
    Tens                Tens                Tens                Tens
      Ones                Ones                Ones                Ones

   8 + 1              8 + ①10              8  12               8  12
   6 | 9 | 2          6 | 9 | 2          6 | 9 | 2           6 | 9 | 2
     |   |              |   |              |   |               |   |
 − 3 | 7 | 8        − 3 | 7 | 8        − 3 | 7 | 8         − 3 | 7 | 8
                                                             3   1   4
```

8 > 2
Borrowing is necessary.
9 tens =
8 tens + 1 ten

Borrow 1 ten from the tens column and write 10 in the ones column.

Add the borrowed 10 to 2.

Subtract the numbers in each column.

In-Class Examples (Objective 1.2B)

1. Subtract and check: 35,021 − 9086
25,935

2. Estimate the difference between 65,271 and 29,403.
40,000

3. Evaluate $x − y$ when $x = 27{,}003$ and $y = 2905$.
24,098

4. Which number, 37, 89, 27, or 79, is a solution of the equation $53 = x − 26$?
79

New Vocabulary
subtraction
minuend
subtrahend
difference
borrowing

Discuss the Concepts

1. Provide at least three examples of situations in which we subtract numbers.

2. Explain why addition can be used to check the answer to a subtraction problem.

3. Explain the process of borrowing when subtracting 43 from 81.

Concept Check

For each subtraction problem, find the difference and then write a related addition problem.

1. 68 − 25
43; 25 + 43 = 68

2. 1539 − 722
817; 722 + 817 = 1539

Optional Student Activity

Use a number line to illustrate the difference between 12 and 7. Label the minuend, the subtrahend, and the difference.

Instructor Note

Borrowing can be related to money. For instance, if Kelly has $27 as 2 ten-dollar bills and 7 one-dollar bills and Chris wants to borrow $9, then Kelly can exchange a ten-dollar bill for 10 one-dollar bills. Kelly then has 1 ten-dollar bill and 17 one-dollar bills. Kelly now can give Chris 9 one-dollar bills. This leaves Kelly with 1 ten-dollar bill and 8 one-dollar bills.

Concept Check

Subtract

$$\begin{array}{r} 932 \\ -475 \\ \hline \end{array}$$

by first writing each number in expanded form.

$$\begin{array}{r} 120 \\ 800 \quad\; 20 \quad\; 12 \\ 900 + 30 + 2 \\ -400 - 70 - 5 \\ \hline 400 + 50 + 7 = 457 \end{array}$$

Note: Students will find this exercise difficult. They need to realize that they are borrowing 1 *ten* from 30 and 1 *hundred* from 900.

Subtraction may involve repeated borrowing.

HOW TO   Subtract: 7325 − 4698

$$\begin{array}{r} {\scriptstyle 1\;\;15} \\ 7\;3\;\cancel{2}\;\cancel{5} \\ -4\;6\;9\;8 \\ \hline 7 \end{array}$$

Borrow 1 ten (10 ones) from the tens column and add 10 to the 5 in the ones column. Subtract 15 − 8.

$$\begin{array}{r} {\scriptstyle 11} \\ {\scriptstyle 2\;\;1\;\;15} \\ 7\;\cancel{3}\;\cancel{2}\;\cancel{5} \\ -4\;6\;9\;8 \\ \hline 2\;7 \end{array}$$

Borrow 1 hundred (10 tens) from the hundreds column and add 10 to the 1 in the tens column. Subtract 11 − 9.

$$\begin{array}{r} {\scriptstyle 12\;\;11} \\ {\scriptstyle 6\;\;2\;\;1\;\;15} \\ \cancel{7}\;\cancel{3}\;\cancel{2}\;\cancel{5} \\ -4\;6\;9\;8 \\ \hline 2\;6\;2\;7 \end{array}$$

Borrow 1 thousand (10 hundreds) from the thousands column and add 10 to the 2 in the hundreds column. Subtract 12 − 6 and 6 − 4.

When there is a zero in the minuend, subtraction involves repeated borrowing.

HOW TO   Subtract: 3904 − 1775

$$\begin{array}{r} {\scriptstyle 8\;\;10} \\ 3\;\cancel{9}\;\cancel{0}\;4 \\ -1\;7\;7\;5 \\ \hline \end{array}$$

There is a 0 in the tens column. Borrow 1 hundred (10 tens) from the hundreds column and write 10 in the tens column.

$$\begin{array}{r} {\scriptstyle 9} \\ {\scriptstyle 8\;\;10\;\;14} \\ 3\;\cancel{9}\;\cancel{0}\;\cancel{4} \\ -1\;7\;7\;5 \\ \hline \end{array}$$

Borrow 1 ten from the tens column and add 10 to the 4 in the ones column.

$$\begin{array}{r} {\scriptstyle 9} \\ {\scriptstyle 8\;\;10\;\;14} \\ 3\;\cancel{9}\;\cancel{0}\;\cancel{4} \\ -1\;7\;7\;5 \\ \hline 2\;1\;2\;9 \end{array}$$

Subtract the numbers in each column.

Note that, for the preceding example, the borrowing could be performed as shown below.

Borrow 1 from 90. (90 − 1 = 89. The 8 is in the hundreds column. The 9 is in the tens column.) Add 10 to the 4 in the ones column. Then subtract the numbers in each column.

$$\begin{array}{r} {\scriptstyle 8\;\;9\;\;14} \\ 3\;\cancel{9}\;\cancel{0}\;\cancel{4} \\ -1\;7\;7\;5 \\ \hline 2\;1\;2\;9 \end{array}$$

HOW TO   Estimate the difference between 49,601 and 35,872.

$$\begin{array}{r} 49{,}601 \rightarrow\; 50{,}000 \\ 35{,}872 \rightarrow -40{,}000 \\ \hline 10{,}000 \end{array}$$

- **Round each number to the nearest ten-thousand.**

- **Subtract the rounded numbers.**

Note that 10,000 is close to the actual difference, 13,729.

TAKE NOTE

Note the order in which the numbers are subtracted when the phrase *less than* is used. Suppose that you have $10 and I have $6 *less than* you do; then I have $6 *less than* $10, or $10 − $6 = $4.

The phrase *the difference between* was used in the example above to indicate the operation of subtraction. All of the phrases listed below indicate subtraction. An example of each is shown at the right of each phrase.

minus	10 minus 3	$10 − 3$
less	8 less 4	$8 − 4$
less than	2 less than 9	$9 − 2$
the difference between	the difference between 6 and 1	$6 − 1$
decreased by	7 decreased by 5	$7 − 5$
subtract . . . from	subtract 11 from 20	$20 − 11$

HOW TO Evaluate $c − d$ when $c = 6183$ and $d = 2759$.

$$c − d$$
$$6183 − 2759$$ • Replace c with 6183 and d with 2759.

$$
\begin{array}{r}
\scriptstyle 5\ 11\ 7\ 13 \\
\cancel{6183} \\
-2759 \\
\hline
3424
\end{array}
$$ • Arrange the numbers vertically and then subtract.

Point of Interest

Someone who is our equal is our peer. Two make a pair. Both of the words *peer* and *pair* come from the Latin *par, paris,* meaning "equal."

HOW TO Is 23 a solution of the equation $41 − n = 17$?

$$41 − n = 17$$
$$41 − 23 \mid 17$$ • Replace n by 23.
$$18 \neq 17$$ • Simplify the left side of the equation. The results are not equal.

No, 23 is not a solution of the equation.

Example 6 Subtract and check:
57,004 − 26,189

Solution

$$
\begin{array}{r}
\scriptstyle 6\ \ \ 9\ \ 9\ 14 \\
5\,\cancel{7},\cancel{0}\,\cancel{0}\,\cancel{4} \\
-26{,}189 \\
\hline
30{,}815
\end{array}
$$

Check: $\begin{array}{r} 26{,}189 \\ +\,30{,}815 \\ \hline 57{,}004 \end{array}$

You Try It 6 Subtract and check:
49,002 − 31,865

Your solution 17,137

Example 7 Estimate the difference between 7261 and 4315. Then find the exact answer.

Solution

$$
\begin{array}{lll}
7261 \longrightarrow & 7000 & 7261 \\
4315 \longrightarrow & -4000 & -4315 \\
\hline
& 3000 & 2946
\end{array}
$$

You Try It 7 Estimate the difference between 8544 and 3621. Then find the exact answer.

Your solution 5000; 4923

Solutions on pp. S1–S2

Instructor Note

The phrases that indicate subtraction are more difficult for students, especially the phrase "2 less than 9," which means $9 − 2$.

Optional Student Activity

Write down any seven-digit number. Scramble the digits to make a different number. Subtract the smaller number from the larger number. Add the digits of the answer. If you end up with a two-digit number, add the digits to produce a one-digit number. What does the resulting number have to be?

9

The graph in Figure 1.11 shows the actual and projected world energy consumption in quadrillion British thermal units (Btu). Use this graph for Example 8 and You Try It 8.

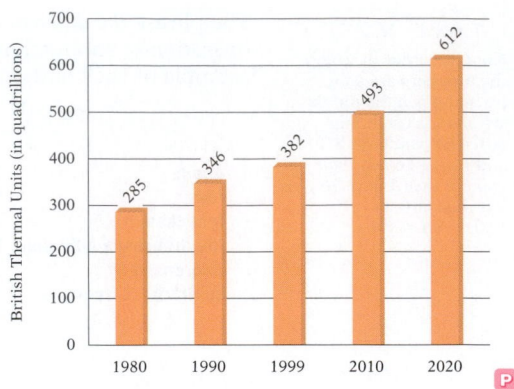

Figure 1.11   World Energy Consumption (in quadrillion British thermal units)

Sources: Energy Information Administration; Office of Energy Markets and End Use; *International Statistics Database and International Energy Annual;* World Energy Projection System

Example 8

Use Figure 1.11 to find the difference between the world energy consumption in 1980 and that projected for 2010.

Solution

2010: 493 quadrillion Btu
1980: 285 quadrillion Btu

$$
\begin{array}{r}
493 \\
-285 \\
\hline
208
\end{array}
$$

The difference between the world energy consumption in 1980 and that projected for 2010 is 208 quadrillion Btu.

You Try It 8

Use Figure 1.11 to find the difference between the world energy consumption in 1990 and that projected for 2020.

Your solution   266 quadrillion Btu

Example 9

Evaluate $x - y$ when $x = 3506$ and $y = 2477$.

Solution

$x - y$
$3506 - 2477$

$$
\begin{array}{r}
{\scriptstyle 4\ 9\ 16} \\
3\,5\,0\,6 \\
-\ 2\,4\,7\,7 \\
\hline
1\,0\,2\,9
\end{array}
$$

You Try It 9

Evaluate $x - y$ when $x = 7061$ and $y = 3229$.

Your solution   3832

Example 10

Is 39 a solution of the equation $24 = m - 15$?

Solution

$24 = m - 15$
$\overline{24 \mid 39 - 15}$ • **Replace _m_ by 39.**
$24 = 24$

Yes, 39 is a solution of the equation.

You Try It 10

Is 11 a solution of the equation $46 = 58 - p$?

Your solution   No

Solutions on p. S2

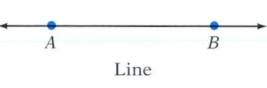

Objective C   **To solve application problems and use formulas**

One application of addition is calculating the perimeter of a figure. However, before defining perimeter, we will introduce some terms from geometry.

Two basic concepts in the study of geometry are point and line.

A **point** is symbolized by drawing a dot. A **line** is determined by two distinct points and extends indefinitely in both directions, as the arrows on the line shown at the right indicate. This line contains points *A* and *B*.

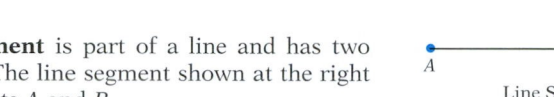

A **ray** starts at a point and extends indefinitely in *one* direction. The point at which a ray starts is called the **endpoint** of the ray. Point *A* is the endpoint of the ray shown at the right.

A **line segment** is part of a line and has two endpoints. The line segment shown at the right has endpoints *A* and *B*.

TAKE NOTE
The corner of a page of this book is a good model of a right angle.

An **angle** is formed by two rays with the same endpoint. An angle is measured in **degrees.** The symbol for degrees is a small raised circle, °. A **right angle** is an angle whose measure is 90°.

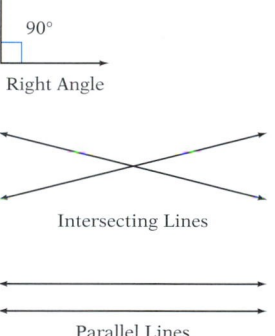

A **plane** is a flat surface and can be pictured as a floor or a wall. Figures that lie in a plane are called **plane figures.**

Lines in a plane can be intersecting or parallel. **Intersecting lines** cross at a point in the plane. **Parallel lines** never meet. The distance between them is always the same.

A **polygon** is a closed figure determined by three or more line segments that lie in a plane. The line segments that form the polygon are called its **sides.** The figures below are examples of polygons.

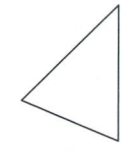  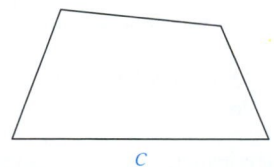 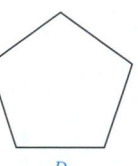 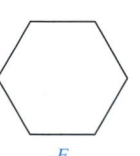

Objective 1.2C

New Vocabulary
point
line
ray
endpoint
line segment
angle
degrees
right angle
plane
plane figures
intersecting lines
parallel lines
polygon
sides of a polygon
triangle
quadrilateral
rectangle
perimeter

New Formulas
Perimeter of a triangle
$$P = a + b + c$$

Optional Student Activity
List five words or phrases that indicate the operation of addition. Use each one with numbers and then show the translation of your phrase into mathematical symbols.
See page 19 for a list of six such phrases, their use with numbers, and translations into mathematical symbols.

In-Class Examples (Objective 1.2C)
Note: Example 1 is a one-step problem. Example 2 is a two-step problem.

1. How much larger is Alaska than Texas? Alaska is 615,230 mi² in area and Texas is 276,277 mi² in area.
338,953 mi²

2. You drove a car 25,950 mi in a 3-year period. You drove 8070 mi the first year and 9759 mi the second year. How many miles did you drive the third year?
8121 mi

Rectangle

The name of a polygon is based on the number of its sides. A polygon with three sides is a **triangle.** Figure A on the previous page is a triangle. A polygon with four sides is a **quadrilateral.** Figures B and C are quadrilaterals.

Quadrilaterals are one of the most common types of polygons. Quadrilaterals are distinguished by their sides and angles. For example, a **rectangle** is a quadrilateral in which opposite sides are parallel, opposite sides are equal in length, and all four angles measure 90°.

The **perimeter** of a plane geometric figure is a measure of the distance around the figure.

The perimeter of a triangle is the sum of the lengths of the three sides.

Perimeter of a Triangle

The formula for the perimeter of a triangle is $P = a + b + c$, where P is the perimeter of the triangle and a, b, and c are the lengths of the sides of the triangle.

HOW TO   Find the perimeter of the triangle shown at the left.

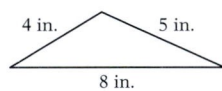

4 in.   5 in.

8 in.

$P = a + b + c$ • Use the formula for the perimeter of a triangle.
$P = 4 + 5 + 8$ • It does not matter which side you label a, b, or c.
$P = 17$ • Add.

The perimeter of the triangle is 17 in.

The perimeter of a quadrilateral is the sum of the lengths of its four sides.

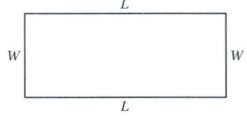

L

W   W

L

In a rectangle, opposite sides are equal in length. Usually the length, L, of a rectangle refers to the length of one of the longer sides of the rectangle, and the width, W, refers to the length of one of the shorter sides. The perimeter can then be represented as $P = L + W + L + W$.

HOW TO   Use the formula $P = L + W + L + W$ to find the perimeter of the rectangle shown at the left.

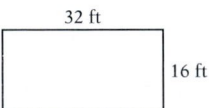

32 ft

16 ft

$P = L + W + L + W$ • Write the given formula for the perimeter of a rectangle.
$P = 32 + 16 + 32 + 16$ • Substitute 32 for L and 16 for W.
$P = 96$ • Add.

The perimeter of the rectangle is 96 ft.

In this section, some of the phrases used to indicate the operations of addition and subtraction were presented. In solving application problems, you might also look for the types of questions listed below.

Addition	**Subtraction**
How many . . . altogether?	How many more (or fewer) . . . ?
How many . . . in all?	How much is left?
How many . . . and . . . ?	How much larger (or smaller) . . . ?

The bar graph in Figure 1.12 shows the number of fatal accidents on amusement rides in the United States each year during the 1990s. Use this graph for Example 11 and You Try It 11.

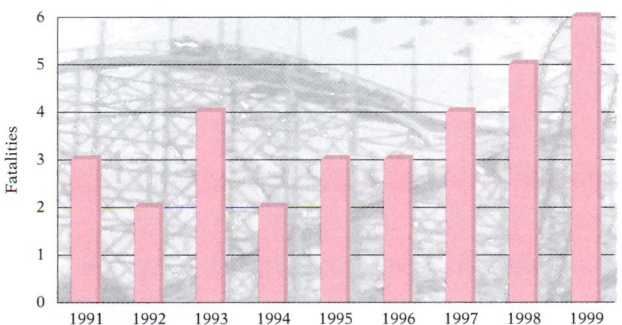

Figure 1.12   Number of Fatal Accidents on Amusement Rides
Source: USA Today, April 7, 2000

Example 11   Use Figure 1.12 to determine how many more fatal accidents occurred during the years 1995 through 1998 than occurred during the years 1991 through 1994.

You Try It 11   Use Figure 1.12 to find the total number of fatal accidents on amusement rides during 1991 through 1999.

Strategy   To find how many more fatalities occurred in 1995 through 1998 than occurred in 1991 through 1994:
• Find the total number of fatalities that occurred from 1995 through 1998 and the total number that occurred from 1991 through 1994.
• Subtract the smaller number from the larger.

Your Strategy

Solution   Fatalities during 1995–1998: 15
Fatalities during 1991–1994: 11

$15 - 11 = 4$

4 more fatalities occurred from 1995 to 1998 than occurred from 1991 to 1994.

Your solution   32 fatal accidents

Solution on p. S2

Another major pedagogical feature of this text is written *strategies* that accompany every application problem. For the paired You Try It, we ask students to provide their own written strategy. A suggested strategy, along with a complete solution to the problem, is given in the Solutions section at the back of the text.

Instructor Note

Optional Student Activity

1. What is the least number of plus signs that need to be put between the digits of the number 123,456,789 so that the sum is 126?
Five:
$12 + 34 + 56 + 7 + 8 + 9$

2. Now change three of the plus signs in the expression you wrote to minus signs so that the expression simplifies to 78.
$12 + 34 + 56 - 7 - 8 - 9$

Example 12

What is the price of a pair of skates that cost a business $109 and has a markup of $49? Use the formula $P = C + M$, where P is the price of a product paid by the consumer, C is the cost paid by the store for the product, and M is the markup.

Strategy
To find the price, replace C by 109 and M by 49 in the given formula and solve for P.

Solution

$P = C + M$

$P = 109 + 49$ • **Replace C by 109 and M by 49.**

$P = 158$

The price of the skates is $158.

You Try It 12

What is the price of a leather jacket that cost a business $148 and has a markup of $74? Use the formula $P = C + M$, where P is the price of a product paid by the consumer, C is the cost paid by the store for the product, and M is the markup.

Your Strategy

Your solution
$222

Example 13

Find the length of decorative molding needed to edge the tops of the walls in a rectangular room that is 12 ft long and 8 ft wide.

Strategy

Draw a diagram.

12 ft

8 ft

To find the length of molding needed, use the formula for the perimeter of a rectangle, $P = L + W + L + W$. $L = 12$ and $W = 8$.

Solution

$P = L + W + L + W$

$P = 12 + 8 + 12 + 8$ • **Replace L by 12 and W by 8.**

$P = 40$

40 ft of decorative molding is needed.

You Try It 13

Find the length of fencing needed to surround a rectangular corral that measures 60 ft on each side.

Your Strategy

Your solution
240 ft

Solutions on p. S2

1.2 Exercises

Objective A   **To add whole numbers**

1. 🖊 Provide at least three examples of situations in which we add numbers.

2. 🖊 Explain how to estimate the sum of 3287 and 4916.

Add.

| 3. | 732,453
+ 651,206
1,383,659 | 4. | 563,841
+ 726,053
1,289,894 | 5. | 2879
+ 3164
6043 | 6. | 9857
+ 1264
11,121 |

| 7. | 45,825
+ 66,327
112,152 | 8. | 56,442
+ 71,289
127,731 | 9. | 4037
3342
+ 5169
12,548 | 10. | 5242
7883
+ 4165
17,290 |

| 11. | 67,390
42,761
+ 89,405
199,556 | 12. | 34,801
97,302
+ 68,945
201,048 | 13. | 54,097
33,432
97,126
64,508
+ 78,310
327,473 | 14. | 23,086
44,697
67,302
83,441
+ 19,843
238,369 |

15. What is 88,123 increased by 80,451?
168,574

16. What is 44,765 more than 82,003?
126,768

17. What is 654 added to 7293?
7947

18. Find the sum of 658, 2709, and 10,935.
14,302

19. Find the total of 216, 8707, and 90,714.
99,637

20. Write the sum of x and y.
$x + y$

21. **College Enrollment** Use the figure at the right to find the total number of undergraduates enrolled at the college in 2005.
1872 students

22. **College Enrollment** Use the figure at the right to find the total number of undergraduates enrolled at the college in 2006.
1910 students

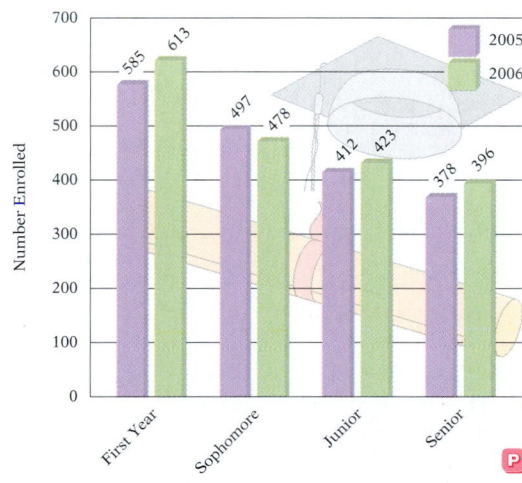

Undergraduates Enrolled in a Private College

Section 1.2

Suggested Assignment
Exercises 3–111, every other odd
Exercises 113–143, odds
More challenging problems:
   Exercises 145–147

Answers to Writing Exercises

1. Answers will vary.
2. To estimate the sum of 3287 and 4916, round each number to the nearest thousand.
3287 rounded to the nearest thousand is 3000.
4916 rounded to the nearest thousand is 5000.
Find the sum of the two rounded numbers.

$$3000 + 5000 = 8000$$

Quick Quiz (Objective 1.2A)

1. Estimate the sum of 1892, 367, and 405.
2800

2. Add: $47 + 8709 + 21,355 + 682$
30,793

3. Evaluate $a + b + c$ when $a = 257$, $b = 396$, and $c = 418$.
1071

4. Which number, 52, 16, 61, or 25, is a solution of the equation $43 = 18 + z$?
25

Estimate by rounding. Then find the exact answer.

23. 6742 + 8298
15,000; 15,040

24. 5426 + 1732
7000; 7158

25. 972,085 + 416,832
1,400,000; 1,388,917

26. 23,774 + 38,026
60,000; 61,800

27.
$$\begin{array}{r} 387 \\ 295 \\ 614 \\ + 702 \\ \hline \end{array}$$
2000; 1998

28.
$$\begin{array}{r} 528 \\ 163 \\ 947 \\ + 275 \\ \hline \end{array}$$
1900; 1913

29.
$$\begin{array}{r} 224{,}196 \\ 7{,}074 \\ + 98{,}531 \\ \hline \end{array}$$
307,000; 329,801

30.
$$\begin{array}{r} 1{,}607 \\ 873{,}925 \\ + 28{,}744 \\ \hline \end{array}$$
932,000; 904,276

Evaluate the variable expression $x + y$ for the given values of x and y.

31. $x = 574; y = 698$
1272

32. $x = 359; y = 884$
1243

33. $x = 4752; y = 7398$
12,150

34. $x = 6047; y = 9283$
15,330

35. $x = 38{,}229; y = 51{,}671$
89,900

36. $x = 74{,}376; y = 19{,}528$
93,904

Evaluate the variable expression $a + b + c$ for the given values of a, b, and c.

37. $a = 693; b = 508; c = 371$
1572

38. $a = 177; b = 892; c = 405$
1474

39. $a = 4938; b = 2615; c = 7038$
14,591

40. $a = 6059; b = 3774; c = 5136$
14,969

41. $a = 12{,}897; b = 36{,}075; c = 7038$
56,010

42. $a = 52{,}847; b = 3774; c = 5136$
61,757

Identify the property that justifies the statement.

43. $9 + 12 = 12 + 9$
The Commutative Property of Addition

44. $8 + 0 = 8$
The Addition Property of Zero

45. $11 + (13 + 5) = (11 + 13) + 5$
The Associative Property of Addition

46. $0 + 16 = 16 + 0$
The Commutative Property of Addition

47. $0 + 47 = 47$
The Addition Property of Zero

48. $(7 + 8) + 10 = 7 + (8 + 10)$
The Associative Property of Addition

Use the given property of addition to complete the statement.

49. The Addition Property of Zero
$28 + 0 = ?$
28

50. The Commutative Property of Addition
$16 + ? = 7 + 16$
7

51. The Associative Property of Addition
$9 + (? + 17) = (9 + 4) + 17$
4

52. The Addition Property of Zero
$0 + ? = 51$
51

53. The Commutative Property of Addition
$? + 34 = 34 + 15$
15

54. The Associative Property of Addition
$(6 + 18) + ? = 6 + (18 + 4)$
4

55. Is 38 a solution of the equation
$42 = n + 4?$
Yes

56. Is 17 a solution of the equation
$m + 6 = 13?$
No

57. Is 13 a solution of the equation
$2 + h = 16?$
No

58. Is 41 a solution of the equation
$n = 17 + 24?$
Yes

59. Is 30 a solution of the equation
$32 = x + 2?$
Yes

60. Is 29 a solution of the equation
$38 = 11 + z?$
No

Objective B   **To subtract whole numbers**

61. ✎ Provide at least three examples of situations in which we subtract numbers.

62. ✎ What is the difference between an expression and an equation?

Answers to Writing Exercises

61. Answers will vary.

62. Students should note that an equation contains an equals sign; an expression does not.

Subtract.

63. $\begin{array}{r} 883 \\ -467 \\ \hline 416 \end{array}$

64. $\begin{array}{r} 591 \\ -238 \\ \hline 353 \end{array}$

65. $\begin{array}{r} 360 \\ -172 \\ \hline 188 \end{array}$

66. $\begin{array}{r} 950 \\ -483 \\ \hline 467 \end{array}$

67. $\begin{array}{r} 657 \\ -193 \\ \hline 464 \end{array}$

68. $\begin{array}{r} 762 \\ -659 \\ \hline 103 \end{array}$

69. $\begin{array}{r} 407 \\ -199 \\ \hline 208 \end{array}$

70. $\begin{array}{r} 805 \\ -147 \\ \hline 658 \end{array}$

Quick Quiz (Objective 1.2B)

1. Subtract and check: $43,052 - 8976$
34,076

2. Estimate the difference between 492,317 and 280,459.
200,000

3. Evaluate $x - y$ when $x = 3624$ and $y = 891$.
2733

4. Which number, 97, 51, 79, or 5, is a solution of the equation $42 = x - 37?$
79

71. 6814
      − 3257
      3557

72. 7361
      − 4575
      2786

73. 5000
      − 2164
      2836

74. 4000
      − 1873
      2127

75. 3400
      − 1963
      1437

76. 7300
      − 2562
      4738

77. 30,004
      − 9,856
      20,148

78. 70,003
      − 8,246
      61,757

79. Find the difference between 2536 and 918.
1618

80. What is 1623 minus 287?
1336

81. What is 5426 less than 12,804?
7378

82. Find 14,801 less 3522.
11,279

83. Find 85,423 decreased by 67,875.
17,548

84. Write the difference between *x* and *y*.
$x - y$

85. 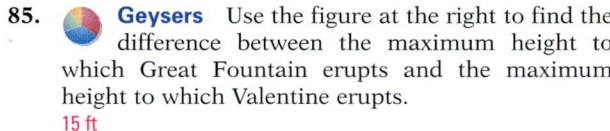 **Geysers**   Use the figure at the right to find the difference between the maximum height to which Great Fountain erupts and the maximum height to which Valentine erupts.
15 ft

86. 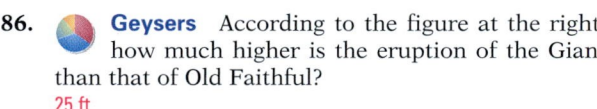 **Geysers**   According to the figure at the right, how much higher is the eruption of the Giant than that of Old Faithful?
25 ft

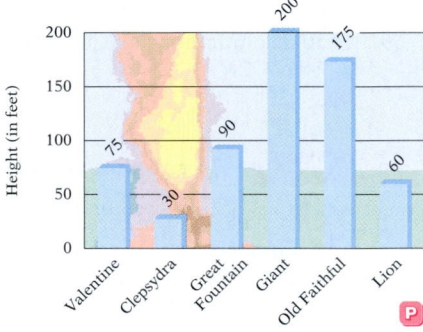

The Maximum Heights of the Eruptions of Six Geysers at Yellowstone National Park

Estimate by rounding. Then find the exact answer.

87. 7355 − 5219
2000; 2136

88. 8953 − 2217
7000; 6736

89. 59,126 − 20,843
40,000; 38,283

90. 63,051 − 29,478
30,000; 33,573

91. 36,287
      − 5,092
      35,000; 31,195

92. 58,316
      − 19,072
      40,000; 39,244

93. 224,196
      − 98,531
      100,000; 125,665

94. 873,925
      − 28,744
      870,000; 845,181

Evaluate the variable expression $x - y$ for the given values of x and y.

95. $x = 50; y = 37$
13

96. $x = 80; y = 33$
47

97. $x = 914; y = 271$
643

98. $x = 623; y = 197$
426

99. $x = 740; y = 385$
355

100. $x = 870; y = 243$
627

101. $x = 8672; y = 3461$
5211

102. $x = 7814; y = 3512$
4302

103. $x = 1605; y = 839$
766

104. $x = 1406; y = 968$
438

105. $x = 23,409; y = 5178$
18,231

106. $x = 56,397; y = 8249$
48,148

107. Is 24 a solution of the equation $29 = 53 - y$?
Yes

108. Is 31 a solution of the equation $48 - p = 17$?
Yes

109. Is 44 a solution of the equation $t - 16 = 60$?
No

110. Is 25 a solution of the equation $34 = x - 9$?
No

111. Is 27 a solution of the equation $82 - z = 55$?
Yes

112. Is 28 a solution of the equation $72 = 100 - d$?
Yes

Objective C   **To solve application problems and use formulas**

113. **Mathematics**   Find the sum of all the whole numbers less than 21.
210

114. **Mathematics**   Find the sum of all the natural numbers greater than 89 and less than 101.
1045

115. **Mathematics**   Find the difference between the smallest four-digit number and the largest two-digit number.
901

116. **Demography**   The figure at the right shows the expected U.S. population aged 100 and over every two years from 2010 to 2020.
a. Which two-year period has the smallest increase in the number of people aged 100 and over?
b. Which two-year period has the greatest increase?
a. 2010 to 2012   **b.** 2018 to 2020

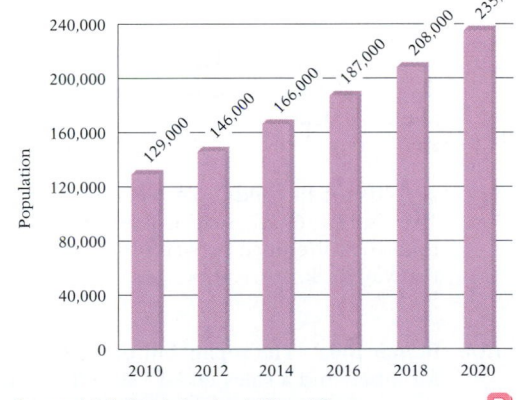

Expected U.S. Population Aged 100 and Over
Source: Census Bureau

Quick Quiz (Objective 1.2C)

Note: Example 1 is a one-step problem. Example 2 is a two-step problem.

1. After a trip of 728 mi, the odometer of your car read 65,412 mi. What was the odometer reading at the beginning of your trip?
64,684 mi

2. You had a bank balance of $843. You then wrote checks for $192, $65, and $19. Find the new bank balance.
$567

117. **Nutrition**   You eat an apple and one cup of cornflakes with one tablespoon of sugar and one cup of milk for breakfast. Find the total number
of calories consumed if one apple contains 80 calories, one cup of cornflakes has 95 calories, one tablespoon of sugar has 45 calories, and one
cup of milk has 150 calories.
370 calories

118. **Health**   You are on a diet to lose weight and are limited to 1500 calories
per day. If your breakfast and lunch contained 950 calories, how many
more calories can you consume during the rest of the day?
550 calories

119. **Geometry**   A rectangle has a length of 24 m and a width of 15 m. Find
the perimeter of the rectangle.
78 m

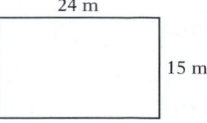

24 m

15 m

120. **Geometry**   Find the perimeter of a rectangle that has a length of 18 ft
and a width of 12 ft.
60 ft

121. **Geometry**   Find the perimeter of a triangle that has sides that measure
16 in., 12 in., and 15 in.
43 in.

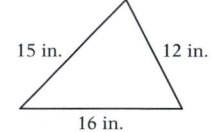

15 in.   12 in.

16 in.

122. **Geometry**   A triangle has sides of lengths 36 cm, 48 cm, and 60 cm.
Find the perimeter of the triangle.
144 cm

123. **Geometry**   A rectangular playground has a length of 160 ft and a width
of 120 ft. Find the length of hedge that surrounds the playground.
560 ft

124. **Geometry**   A rectangular vegetable garden has a length of 20 ft and a
width of 14 ft. How many feet of wire fence should be purchased to surround the garden?
68 ft

125. **Space Flights**   The Gemini-Titan 7 space flight made 206 orbits of
Earth. The Apollo-Saturn 7 space flight made 163 orbits of Earth.
How many more orbits did the Gemini-Titan 7 flight make than the
Apollo-Saturn 7 flight?
43 orbits

126. **Finances**   You had $1054 in your checking account before making a
deposit of $870. Find the amount in your checking account after you
made the deposit.
$1924

127. **Baseball Fields**   The seating capacity of SAFECO Field in Seattle
is 47,116. The seating capacity of Fenway Park in Boston is 36,298.
Find the difference between the seating capacity of SAFECO Field and
Fenway Park.
10,818 seats

128. **Repair Bills**   The repair bill on your car includes $358 for parts, $156
for labor, and a sales tax of $30. What is the total amount owed?
$544

Fenway Park

129. Purchasing   The computer system you would like to purchase includes an operating system priced at $830, a monitor that costs $245, an extended keyboard priced at $175, and a printer that sells for $395. What is the total cost of the computer system?
$1645

130.  **Geography**   The area of Lake Superior is 81,000 mi²; the area of Lake Michigan is 67,900 mi²; the area of Lake Huron is 74,000 mi²; the area of Lake Erie is 32,630 mi²; and the area of Lake Ontario is 34,850 mi². Estimate the total area of the five Great Lakes.
280,000 mi²

The Great Lakes

131. Travel   The odometer on your car read 58,376 this time last year. It now reads 77,912. Estimate the number of miles your car has been driven during the past year.
20,000 mi

Car Sales   The figure at the right shows the number of cars sold by a dealership for the first four months of 2005 and 2006. Use this graph for Exercises 132 to 134.

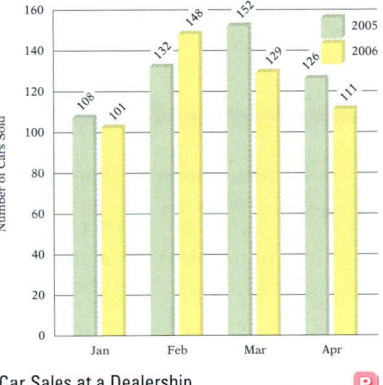

Car Sales at a Dealership

132. Between which two months did car sales decrease the most in 2006? What was the amount of decrease?
February to March; 19 cars

133. Between which two months did car sales increase the most in 2005? What was the amount of increase?
January to February; 24 cars

134. In which year were more cars sold during the four months shown?
2005

135. Investments   Use the formula $A = P + I$, where A is the value of an investment, P is the original investment, and I is the interest earned, to find the value of an investment that earned $775 in interest on an original investment of $12,500.
$13,275

136. Investments   Use the formula $A = P + I$, where A is the value of an investment, P is the original investment, and I is the interest earned, to find the value of an investment that earned $484 in interest on an original investment of $8800.
$9284

137. Mortgages   What is the mortgage loan amount on a home that sells for $290,000 with a down payment of $29,000? Use the formula $M = S - D$, where M is the mortgage loan amount, S is the selling price, and D is the down payment.
$261,000

138. Mortgages   What is the mortgage loan amount on a home that sells for $236,000 with a down payment of $47,200? Use the formula $M = S - D$, where M is the mortgage loan amount, S is the selling price, and D is the down payment.
$188,800

139. **Air Travel** What is the ground speed of an airplane traveling into a 25-mph head wind with an air speed of 375 mph? Use the formula $g = a - h$, where g is the ground speed, a is the air speed, and h is the speed of the head wind.
350 mph

140. **Air Travel** Find the ground speed of an airplane traveling into a 15-mph head wind with an air speed of 425 mph. Use the formula $g = a - h$, where g is the ground speed, a is the air speed, and h is the speed of the head wind.
410 mph

Speeds In some states, the speed limit on certain sections of highway is 70 mph. To test drivers' compliance with the speed limit, the highway patrol conducted a one-week study during which it recorded the speeds of motorists on one of these sections of highway. The results are recorded in the table at the right. Use this table for Exercises 141 to 144.

141. **a.** How many drivers were traveling at 70 mph or less?
b. How many drivers were traveling at 76 mph or more?
a. 9571 drivers   b. 4211 drivers

Speed	Number of Cars
> 80	1708
76 – 80	2503
71 – 75	3651
66 – 70	3717
61 – 65	2984
< 61	2870

142. Looking at the data in the table, is it possible to tell how many motorists were driving at 70 mph? Explain your answer.
No

143. Looking at the data in the table, is it possible to tell how many motorists were driving at less than 70 mph? Explain your answer.
No

144. Are more people driving at or below the posted speed limit, or are more people driving above the posted speed limit?
At or below the posted speed limit

APPLYING THE CONCEPTS

145. **Dice** If you roll two ordinary six-sided dice and add the two numbers that appear on top, how many different sums are possible?
11

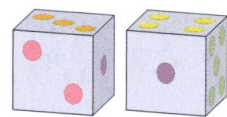

146. **Mathematics** How many two-digit numbers are there? How many three-digit numbers are there?
90; 900

147. Determine whether the statement is always true, sometimes true, or never true.
a. If a is any whole number, then $a - 0 = a$.
b. If a is any whole number, then $a - a = 0$.
a. Always true   b. Always true

148. ✏️ Find the circulation of your local newspaper and the population of the area served by that paper. What is the difference between the area's population and the newspaper's circulation? Why would this figure be of concern to the owner of the newspaper?

149. ✏️ What estimate is given for the size of the population of your state by the year 2025? What is the estimate of the size of the population of the United States by the year 2025? Estimates differ. On what basis was the estimate you recorded derived?

Answers to Writing Exercises

148. The difference between the population of the area served by the newspaper and the newspaper's circulation represents the number of people who live in the area but do not buy the newspaper. Therefore, it indicates the number of people in the target market who are not purchasing the paper. It could be interpreted as a measure of the popularity of the newspaper, or of the paper's success in meeting the needs of its market.

149. The U.S. Bureau of the Census publishes projections of the population of the United States and of the separate states in *Current Population Reports.* Estimates of U.S. population growth vary according to the projections used for the birth rate, the average life span of the population, and immigration figures. Population growth of states is further affected by internal migration.

1.3 Multiplication and Division of Whole Numbers

Objective A   To multiply whole numbers

A store manager orders six boxes of telephone answering machines. Each box contains eight answering machines. How many answering machines are ordered?

The answer can be calculated by adding six 8's.

$8 + 8 + 8 + 8 + 8 + 8 = 48$

This problem involves repeated addition of the same number. The answer can be calculated by a shorter process called multiplication. **Multiplication** is the repeated addition of the same number.

There is a total of 48 dots on the 6 dominoes.

The numbers that are multiplied are called **factors.** The answer is called the **product.**

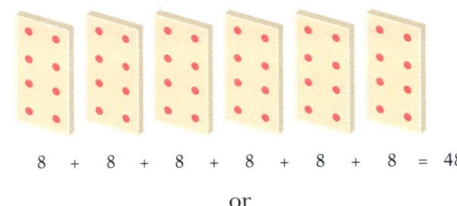

$8 + 8 + 8 + 8 + 8 + 8 = 48$

or

$\underset{\text{Factor}}{6} \times \underset{\text{Factor}}{8} = \underset{\text{Product}}{48}$

The times sign "×" is one symbol that is used to mean multiplication. Each of the expressions below also represents multiplication.

$$6 \cdot 8 \qquad 6(8) \qquad (6)(8) \qquad 6a \qquad 6(a) \qquad ab$$

The expression $6a$ means "6 times a." The expression ab means "a times b."

The basic facts for multiplying one-digit numbers should be memorized. Multiplication of larger numbers requires the repeated use of the basic multiplication facts.

Point of Interest

The cross X was first used as a symbol for multiplication in 1631 in a book titled *The Key to Mathematics*. Also in that year, another book, *Practice of the Analytical Art,* advocated the use of a dot to indicate multiplication.

HOW TO   Multiply: 37(4)

$$\begin{array}{r}\overset{2}{3}\,7 \\ \times \quad 4 \\ \hline 8\end{array}$$

- **Multiply 4 · 7.**
  4 · 7 = 28 (2 tens + 8 ones).
  Write the 8 in the ones column and carry the 2 to the tens column.

$$\begin{array}{r}\overset{2}{3}\,7 \\ \times \quad 4 \\ \hline 14\,8\end{array}$$

- **The 3 in 37 is 3 tens.**
  Multiply 4 · 3 tens = 12 tens.
  Add the carry digit: 12 tens + 2 tens = 14 tens.
  Write the 14.

New Vocabulary
multiplication
factors
product

New Properties
Multiplication Property of Zero
Multiplication Property of One
Commutative Property of Multiplication
Associative Property of Multiplication

Discuss the Concepts

1. Provide at least three examples of situations in which we multiply numbers.

2. What is the difference between the Multiplication Property of Zero and the Multiplication Property of One?

3. Explain how to multiply a number by 100.

Concept Check

Which property justifies the statement?

1. $8 \times 12 = 12 \times 8$
   Commutative Property of Multiplication

2. $39 \times 0 = 0$
   Multiplication Property of Zero

3. $2 \times (3 \times 9) = (2 \times 3) \times 9$
   Associative Property of Multiplication

4. $1 \times 45 = 45$
   Multiplication Property of One

In-Class Examples (Objective 1.3A)

1. Find the product of 100 and 27.   2700

2. Estimate the product of 5649 and 33.   180,000

3. Evaluate $2st$ when $s = 45$ and $t = 67$.   6030

4. What is twice 12,000?   24,000

5. Complete the statement by using the Multiplication Property of One. $1(?) = 63$
   63

6. Which number, 77, 12, 588, or 91, is a solution of the equation $84 = 7y$?
   12

Concept Check

What will be the last digit in the product?

1. 438×617 6

2. 9152×306 2

3. $32,480 \times 579$ 0

Optional Student Activity

Solve the multiplication problem below. Each letter stands for a single-digit number.

$$
\begin{array}{r}
S\,T\,R\,A\,W \\
\times \qquad 4 \\
\hline
W\,A\,R\,T\,S
\end{array}
$$

$21,978 \times 4 = 87,912$

Optional Student Activity

Assume the pattern continues, and find the next three numbers in the pattern.

1. 4, 8, 12, 16, . . .

20, 24, 28

2. 17, 170, 1700, . . .

17,000; 170,000; 1,700,000

3. 10, 200, 4000, . . .

80,000; 1,600,000; 32,000,000

In the preceding example, a number was multiplied by a one-digit number. The examples that follow illustrate multiplication by larger numbers.

HOW TO   Multiply: (47)(23)

Multiply by the ones digit.	Multiply by the tens digit.	Add.
$3 \cdot 47 = 141$	$2 \cdot 47 = 94$	

$$
\begin{array}{r}
47 \\
\times\ 23 \\
\hline
141
\end{array}
\qquad
\begin{array}{r}
47 \\
\times\ 23 \\
\hline
141 \\
94
\end{array}
\qquad
\begin{array}{r}
47 \\
\times\ 23 \\
\hline
141 \\
94 \\
\hline
1081
\end{array}
$$

The last digit is written in the ones column.

The last digit is written in the tens column.

The place-value chart illustrates the placement of the products.

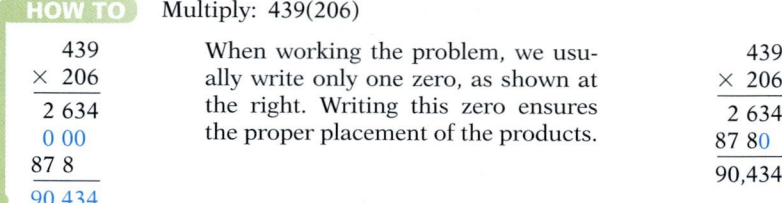

3 × 47
20 × 47
141 + 940

Note the placement of the products when we are multiplying by a factor that contains a zero.

HOW TO   Multiply: 439(206)

$$
\begin{array}{r}
439 \\
\times\ 206 \\
\hline
2\ 634 \\
0\ 00 \\
87\ 8 \\
\hline
90{,}434
\end{array}
$$

When working the problem, we usually write only one zero, as shown at the right. Writing this zero ensures the proper placement of the products.

$$
\begin{array}{r}
439 \\
\times\ 206 \\
\hline
2\ 634 \\
87\ 80 \\
\hline
90{,}434
\end{array}
$$

Note the pattern when the following numbers are multiplied.

Multiply the nonzero part of the factors. Attach the same number of zeros in the product as the total number of zeros in the factors.

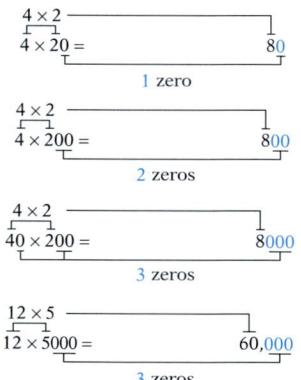

4×2
$4 \times 20 =$ 80
1 zero

4×2
$4 \times 200 =$ 800
2 zeros

4×2
$40 \times 200 =$ 8000
3 zeros

12×5
$12 \times 5000 =$ 60,000
3 zeros

HOW TO   Find the product of 600 and 70.

$600 \cdot 70 = 42{,}000$

• Remember that a *product* is the answer to a multiplication problem.

HOW TO Multiply: 3(20)(10)(4)

3(20)(10)4 = 60(10)(4) • **Multiply the first two numbers.**

= (600)(4) • **Multiply the product by the third number.**

= 2400 • **Continue multiplying until all the numbers have been multiplied.**

HOW TO   Figure 1.13 shows the average weekly earnings of full-time workers in the United States. Using these figures, calculate the earnings of a female full-time worker, age 22, for working for 4 weeks.

Multiply the number of weeks (4) times the amount earned for one week ($354).

4(354) = 1416

The average earnings of a 22-year-old, female, full-time worker for working for 4 weeks are $1416.

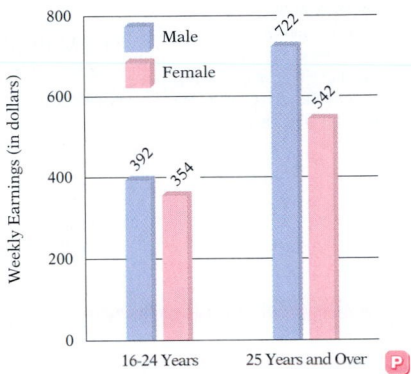

Figure 1.13 Average Weekly Earnings of Full-Time Workers
Source: Bureau of Labor Statistics

HOW TO Estimate the product of 345 and 92.

345 → 300
 92 → 90 • **Round each number to its highest place value.**

300 · 90 = 27,000 • **Multiply the rounded numbers.**

27,000 is an estimate of the product of 345 and 92.

The phrase *the product of* was used in the example above to indicate the operation of multiplication. All of the phrases below indicate multiplication. An example of each is shown at the right of each phrase.

times	8 times 4	8 · 4
the product of	the product of 9 and 5	9 · 5
multiplied by	7 multiplied by 3	3 · 7
twice	twice 6	2 · 6

HOW TO Evaluate *xyz* when *x* = 50, *y* = 2, and *z* = 7.

xyz • ***xyz* means *x* · *y* · *z*.**

50 · 2 · 7 • **Replace each variable by its value.**

= 100 · 7 • **Multiply the first two numbers.**

= 700 • **Multiply the product by the next number.**

Instructor Note

For reinforcement, the Addition and Multiplication Properties are presented again in Chapters 2 and 5.

As for addition, there are properties of multiplication.

> **The Multiplication Property of Zero**
>
> $a \cdot 0 = 0$ or $0 \cdot a = 0$

$8 \cdot 0 = 0$

The Multiplication Property of Zero states that the product of a number and zero is zero. The variable a is used here to represent any whole number. It can even represent the number zero because $0 \cdot 0 = 0$.

> **The Multiplication Property of One**
>
> $a \cdot 1 = a$ or $1 \cdot a = a$

$1 \cdot 9 = 9$

The Multiplication Property of One states that the product of a number and 1 is the number. Multiplying a number by 1 does not change the number.

> **The Commutative Property of Multiplication**
>
> $a \cdot b = b \cdot a$

$4 \cdot 9 = 9 \cdot 4$

$36 = 36$

The Commutative Property of Multiplication states that two numbers can be multiplied in either order; the product will be the same. Here the variables a and b represent any whole numbers. Therefore, for example, if you know that the product of 4 and 9 is 36, then you also know that the product of 9 and 4 is 36 because $4 \cdot 9 = 9 \cdot 4$.

> **The Associative Property of Multiplication**
>
> $(a \cdot b) \cdot c = a \cdot (b \cdot c)$

$(2 \cdot 3) \cdot 4 = 2 \cdot (3 \cdot 4)$

$6 \cdot 4 = 2 \cdot 12$

$24 = 24$

The Associative Property of Multiplication states that when multiplying three numbers, the numbers can be grouped in any order; the product will be the same. Note in the example at the right above that we can multiply the product of 2 and 3 by 4, or we can multiply 2 by the product of 3 and 4. In either case, the product of the three numbers is 24.

HOW TO   What is the solution of the equation $5x = 5$?

By the Multiplication Property of One, the product of a number and 1 is the number.

$$5x = 5$$
$$5(1) \mid 5$$
$$5 = 5$$

The solution is 1.

The check is shown at the right.

HOW TO   Is 7 a solution of the equation $3m = 21$?

$$3m = 21$$
$$3(7) \mid 21$$ • Replace *m* by 7.
$$21 = 21$$ • Simplify the left side of the equation. The results are equal.

Yes, 7 is a solution of the equation.

 Figure 1.14 shows the average monthly savings of individuals in seven different countries. Use this graph for Example 1 and You Try It 1.

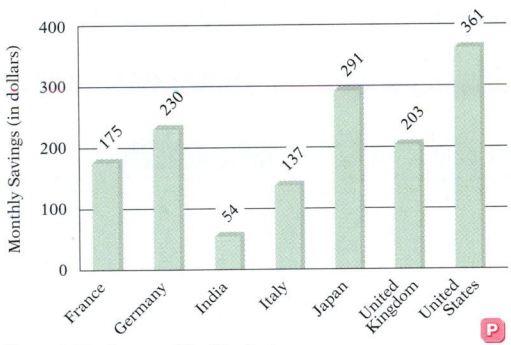

Figure 1.14   Average Monthly Savings
Source: Taylor Nelson - Sofres for American Express

Example 1   Use Figure 1.14 to determine the average annual savings of individuals in Japan.

Solution   The average monthly savings in Japan is $291. The number of months in one year is 12.

$$\begin{array}{r} 291 \\ \times\ 12 \\ \hline 582 \\ 291 \\ \hline 3492 \end{array}$$

The average annual savings of individuals in Japan is $3492.

You Try It 1   According to Figure 1.14, what is the average annual savings of individuals in France?

Your solution $2100

Solution on p. S2

Example 2 Estimate the product of 2871 and 49.

Solution

$$2871 \rightarrow 3000$$
$$49 \rightarrow 50$$

$$3000 \cdot 50 = 150,000$$

You Try It 2 Estimate the product of 8704 and 93.

Your solution 810,000

Example 3 Evaluate $3ab$ when $a = 10$ and $b = 40$.

Solution

$$3ab$$
$$3(10)(40) = 30(40)$$
$$= 1200$$

You Try It 3 Evaluate $5xy$ when $x = 20$ and $y = 60$.

Your solution 6000

Example 4 What is 800 times 300?

Solution

$$800 \cdot 300 = 240,000$$

You Try It 4 What is 90 multiplied by 7000?

Your solution 630,000

Example 5 Complete the statement by using the Associative Property of Multiplication.

$$(7 \cdot 8) \cdot 5 = 7 \cdot (? \cdot 5)$$

Solution $(7 \cdot 8) \cdot 5 = 7 \cdot (8 \cdot 5)$

You Try It 5 Complete the statement by using the Multiplication Property of Zero.

$$? \cdot 10 = 0$$

Your solution 0

Example 6 Is 9 a solution of the equation $82 = 9q$?

Solution

$$82 = 9q$$
$$\overline{82 \mid 9(9)}$$
$$82 \neq 81$$

No, 9 is not a solution of the equation.

You Try It 6 Is 11 a solution of the equation $7a = 77$?

Your solution Yes

Solutions on p. S2

Objective B   **To evaluate expressions that contain exponents**

Point of Interest

Lao-tzu, founder of Taoism, wrote: Counting gave birth to Addition, Addition gave birth to Multiplication, Multiplication gave birth to Exponentiation, Exponentiation gave birth to all the myriad operations.

Repeated multiplication of the same factor can be written in two ways:

$$4 \cdot 4 \cdot 4 \cdot 4 \cdot 4 \quad \text{or} \quad 4^5 \longleftarrow \text{exponent}$$
$$\quad\quad\quad\quad\quad\quad\quad\quad\quad\quad\uparrow\!\!\!\!\!\!\!\rule{1.5em}{0.4pt}\; \text{base}$$

The expression 4^5 is in **exponential form.** The **exponent,** 5, indicates how many times the **base,** 4, occurs as a factor in the multiplication.

It is important to be able to read numbers written in exponential form.

Point of Interest

One billion is too large a number for most of us to comprehend. If a computer were to start counting from 1 to 1 billion, writing to the screen one number every second of every day, it would take over 31 years for the computer to complete the task.

And if a billion is a large number, consider a googol. A googol is 1 with 100 zeros after it, or 10^{100}. Edward Kasner is the mathematician credited with thinking up this number, and his nine-year-old nephew is said to have thought up the name. The two then coined the word googolplex, which is 10^{googol}.

$$2 = 2^1$$ Read "two to the first power" or just "two." Usually the 1 is not written.
$$2 \cdot 2 = 2^2$$ Read "two squared" or "two to the second power."
$$2 \cdot 2 \cdot 2 = 2^3$$ Read "two cubed" or "two to the third power."
$$2 \cdot 2 \cdot 2 \cdot 2 = 2^4$$ Read "two to the fourth power."
$$2 \cdot 2 \cdot 2 \cdot 2 \cdot 2 = 2^5$$ Read "two to the fifth power."

Variable expressions can contain exponents.

$$x^1 = x$$ x to the first power is usually written simply as x.
$$x^2 = x \cdot x$$ x^2 means x times x.
$$x^3 = x \cdot x \cdot x$$ x^3 means x occurs as a factor 3 times.
$$x^4 = x \cdot x \cdot x \cdot x$$ x^4 means x occurs as a factor 4 times.

Each place value in the place-value chart can be expressed as a power of 10.

$$
\begin{aligned}
\text{Ten} &= 10 &= 10 &= 10^1 \\
\text{Hundred} &= 100 &= 10 \cdot 10 &= 10^2 \\
\text{Thousand} &= 1000 &= 10 \cdot 10 \cdot 10 &= 10^3 \\
\text{Ten-thousand} &= 10{,}000 &= 10 \cdot 10 \cdot 10 \cdot 10 &= 10^4 \\
\text{Hundred-thousand} &= 100{,}000 &= 10 \cdot 10 \cdot 10 \cdot 10 \cdot 10 &= 10^5 \\
\text{Million} &= 1{,}000{,}000 &= 10 \cdot 10 \cdot 10 \cdot 10 \cdot 10 \cdot 10 &= 10^6
\end{aligned}
$$

Note that the exponent on 10 when the number is written in exponential form is the same as the number of zeros in the number written in standard form. For example, $10^5 = 100{,}000$; the exponent on 10 is 5, and the number 100,000 has 5 zeros.

To evaluate a numerical expression containing exponents, write each factor as many times as indicated by the exponent and then multiply.

$$5^3 = 5 \cdot 5 \cdot 5 = 25 \cdot 5 = 125$$

$$2^3 \cdot 6^2 = (2 \cdot 2 \cdot 2) \cdot (6 \cdot 6) = 8 \cdot 36 = 288$$

Integrating Technology

A calculator can be used to evaluate an exponential expression. The y^x key (or on some calculators an x^y key or $\wedge$ key) is used to enter the exponent. For instance, for the example at the left, enter 4 y^x 3 = . The display reads 64.

HOW TO Evaluate the variable expression c^3 when $c = 4$.

$$c^3 = c \cdot c \cdot c$$
$$4^3 = 4 \cdot 4 \cdot 4$$
$$= 16 \cdot 4 = 64$$

• Replace c with 4 and then evaluate the exponential expression.

Example 7 Write $7 \cdot 7 \cdot 7 \cdot 4 \cdot 4$ in exponential form.

Solution $7 \cdot 7 \cdot 7 \cdot 4 \cdot 4 = 7^3 \cdot 4^2$

You Try It 7 Write $2 \cdot 2 \cdot 2 \cdot 3 \cdot 3 \cdot 3 \cdot 3$ in exponential form.

Your solution $2^3 \cdot 3^4$

Solution on p. S2

Solution on p. S2

Objective 1.3B

New Vocabulary
exponential form
exponent
base

Discuss the Concepts
Explain what an exponent is.

Concept Check
Simplify: $2^{40} \div 4^{20}$ 1

Optional Student Activity
Find the ones digit of 7^{97}. 7

Optional Student Activity
In which column will the number 1 million appear, column A, B, or C?

A	B	C
1	8	27
64	125	216
.	.	.
.	.	.
.	.	.

Column A; $1{,}000{,}000 = 100^3$

In-Class Examples (Objective 1.3B)

1. Write $a \cdot a \cdot a \cdot a \cdot b \cdot b \cdot b \cdot b \cdot b$ in exponential form.
$a^4 \cdot b^5$

2. Evaluate 7^3.
343

3. Evaluate 10^{11}.
100,000,000,000

4. Evaluate $3^3 \cdot 2^4$.
432

5. Evaluate $x^3 \cdot y^4$ when $x = 2$ and $y = 3$.
648

| **Example 8** | Evaluate 8^3. | **You Try It 8** | Evaluate 6^4. |

Example 8 Evaluate 8^3.

Solution $8^3 = 8 \cdot 8 \cdot 8 = 64 \cdot 8 = 512$

You Try It 8 Evaluate 6^4.

Your solution 1296

Example 9 Evaluate 10^7.

Solution $10^7 = 10,000,000$

(The exponent on 10 is 7. There are 7 zeroes in 10,000,000.)

You Try It 9 Evaluate 10^8.

Your solution 100,000,000

Example 10 Evaluate $3^3 \cdot 5^2$.

Solution $3^3 \cdot 5^2 = (3 \cdot 3 \cdot 3) \cdot (5 \cdot 5)$
$= 27 \cdot 25 = 675$

You Try It 10 Evaluate $2^4 \cdot 3^2$.

Your solution 144

Example 11 Evaluate $x^2 y^3$ when $x = 4$ and $y = 2$.

Solution $x^2 y^3$ ($x^2 y^3$ means x^2 times y^3.)

$4^2 \cdot 2^3 = (4 \cdot 4) \cdot (2 \cdot 2 \cdot 2)$
$= 16 \cdot 8$
$= 128$

You Try It 11 Evaluate $x^4 y^2$ when $x = 1$ and $y = 3$.

Your solution 9

Solutions on p. S2

Objective 1.3C

New Vocabulary

division
divisor
quotient
dividend
remainder

Point of Interest

The Chinese divided a day into 100 k'o, which was a unit equal to a little less than 15 min. Sundials were used to measure time during the daylight hours, and by A.D. 500, candles, water clocks, and incense sticks were used to measure time at night.

Objective C   **To divide whole numbers**

Division is used to separate objects into equal groups.

A store manager wants to display 24 new objects equally on 4 shelves. From the diagram, we see that the manager would place 6 objects on each shelf.

The manager's division problem can be written as follows:

Number of shelves — **Divisor**

Number on each shelf — **Quotient**

$4\overline{)24}$ with 6

Number of objects — **Dividend**

Note that the quotient multiplied by the divisor equals the dividend.

$4\overline{)24}^{\,6}$ because $\boxed{\underset{\text{Quotient}}{6}} \times \boxed{\underset{\text{Divisor}}{4}} = \boxed{\underset{\text{Dividend}}{24}}$

In-Class Examples (Objective 1.3C)

1. Find the quotient of 1519 and 7.   217
2. Estimate the quotient of 37,052 and 41.   1000
3. Divide: $487 \div 6$ 81 r1
4. Evaluate $\dfrac{x}{y}$ when $x = 10{,}890$ and $y = 9$.   1210

5. Which number, 3, 8, 16, or 48, is a solution of the equation $\dfrac{z}{12} = 4$?   48

Division is also represented by the symbol ÷ or by a fraction bar. Both are read "divided by."

$$9\overline{)54} \qquad 54 \div 9 = 6 \qquad \frac{54}{9} = 6$$

The fact that the quotient times the divisor equals the dividend can be used to illustrate properties of division.

$0 \div 4 = 0$ because $0 \cdot 4 = 0$.

$4 \div 4 = 1$ because $1 \cdot 4 = 4$.

$4 \div 1 = 4$ because $4 \cdot 1 = 4$.

$4 \div 0 = ?$ What number can be multiplied by 0 to get 4? There is no number whose product with 0 is 4 because the product of $? \cdot 0 = 4$ a number and zero is 0. **Division by zero is undefined.**

The properties of division are stated below. In these statements, the symbol ≠ is read "is not equal to."

Integrating Technology

Enter 4 ÷ 0 = . An error message is displayed because division by zero is undefined.

TAKE NOTE

Recall that the variable a represents any whole number. Therefore, for the first two properties, we must state that $a \neq 0$ in order to ensure that we are not dividing by zero.

> **Division Properties of Zero and One**
>
> If $a \neq 0$, $0 \div a = 0$.   Zero divided by any number other than zero is zero.
> If $a \neq 0$, $a \div a = 1$.   Any number other than zero divided by itself is one.
> $a \div 1 = a$ A number divided by one is the number.
> $a \div 0$ is undefined.   Division by zero is undefined.

The example below illustrates division of a larger whole number by a one-digit number.

HOW TO   Divide and check: $3192 \div 4$

$$
\begin{array}{r}
7 \\
4\overline{)3192} \\
-28 \\
\hline
39
\end{array}
$$

• Think $31 \div 4$.
• Subtract 7×4.
• Bring down the 9.

$$
\begin{array}{r}
79 \\
4\overline{)3192} \\
-28 \\
\hline
39 \\
-36 \\
\hline
32
\end{array}
$$

• Think $39 \div 4$.
• Subtract 9×4.
• Bring down the 2.

$$
\begin{array}{r}
798 \\
4\overline{)3192} \\
-28 \\
\hline
39 \\
-36 \\
\hline
32 \\
-32 \\
\hline
0
\end{array}
$$

• Think $32 \div 4$.
• Subtract 8×4.

Check:
$$
\begin{array}{r}
798 \\
\times \quad 4 \\
\hline
3192
\end{array}
$$

New Properties

Division Properties of Zero and One
For $a \neq 0$:
$0 \div a = 0$
$a \div a = 1$
$a \div 1 = a$
$a \div 0$ is undefined.

Discuss the Concepts

1. Explain the difference between $0 \div 9$ and $9 \div 0$.
2. Explain why multiplication can be used to check the answer to a division problem.
3. In what situation does a division problem have a remainder?
4. Explain how to check the answer to a division problem that has a remainder.
   Quotient × divisor + remainder = dividend

Concept Check

For each division problem, find the quotient and then write a related multiplication problem.

1. $98 \div 7$
   14; $14 \times 7 = 98$
2. $1968 \div 6$
   328; $328 \times 6 = 1968$

Instructor Note

To relate zero in division to a real situation, explain that $6 \div 3 = 2$ means that if $6 is divided equally among 3 people, each person receives $2. If $0 is divided equally among 3 people, each person receives $0 ($0 \div 3 = 0$). Now, how can $6 be divided equally among 0 people ($6 \div 0$)?

Concept Check

Which of the following numbers divide evenly into 144? (There is no remainder.)

1	10
2	11
3	12
4	13
5	14
6	15
7	16
8	17
9	18

1, 2, 3, 4, 6, 8, 9, 12, 16, 18

Instructor Note

Some students have difficulty with the concept of remainder. Have these students try to give 15 pennies to 4 students so that each student has the same number of pennies.

Optional Student Activity

Divide 111 by the sum of its digits (3). Divide 222 by the sum of its digits (6). Divide 333 by the sum of its digits (9). Continue this pattern through division of 999 by the sum of its digits (27). Describe the pattern. Why does this occur?

The quotient is always 37. Both the dividend and the divisor are multiplied by the same number.

The place-value chart can be used to show why this method works.

$$
\begin{array}{r}
7\ 9\ 8 \\
4)\overline{3\ 1\ 9\ 2} \\
-2\ 8\ 0\ 0 \quad \text{7 hundreds} \times 4 \\
\overline{3\ 9\ 2} \\
-3\ 6\ 0 \quad \text{9 tens} \times 4 \\
\overline{3\ 2} \\
-3\ 2 \quad \text{8 ones} \times 4 \\
\overline{0}
\end{array}
$$

Sometimes it is not possible to separate objects into a whole number of equal groups.

A packer at a bakery has 14 muffins to pack into 3 boxes. Each box will hold 4 muffins. From the diagram, we see that after the baker places 4 muffins in each box, there are 2 muffins left over. The 2 is called the **remainder.**

The packer's division problem can be written as follows:

Number of boxes ⟶ Divisor
Number in each box — Quotient
Total number of muffins — Dividend
Number left over — Remainder

$$
\begin{array}{r}
4 \\
3)\overline{14} \\
-12 \\
\overline{2}
\end{array}
\qquad \text{or} \qquad
\begin{array}{r}
4\ r2 \\
3)\overline{14}
\end{array}
$$

For any division problem, **(quotient · divisor) + remainder = dividend.** This result can be used to check a division problem.

> **HOW TO** Find the quotient of 389 and 24.
>
> $$
> \begin{array}{r}
> 16\ r5 \\
> 24)\overline{389} \\
> -24 \\
> \overline{149} \\
> -144 \\
> \overline{5}
> \end{array}
> $$
>
> *Check:* $(16 \cdot 24) + 5 = 384 + 5 = 389$

The phrase *the quotient of* was used in the example above to indicate the operation of division. The phrase *divided by* also indicates division.

the quotient of	the quotient of 8 and 4	$8 \div 4$
divided by	9 divided by 3	$9 \div 3$

HOW TO Estimate the result when 56,497 is divided by 28.

$56{,}497 \longrightarrow 60{,}000$ • **Round each number to its highest place value.**

$28 \quad\; \longrightarrow 30$

$60{,}000 \div 30 = 2000$ • **Divide the rounded numbers.**
2000 is an estimate of 56,497 ÷ 28.

HOW TO Evaluate $\dfrac{x}{y}$ when $x = 4284$ and $y = 18$.

$\dfrac{x}{y}$ • **Replace *x* with 4284 and *y* with 18.**

$\dfrac{4284}{18} = 238$ • $\dfrac{4284}{18}$ **means 4284 ÷ 18.**

HOW TO Is 42 a solution of the equation $\dfrac{x}{6} = 7$?

$\dfrac{x}{6} = 7$

$\dfrac{42}{6} \;\Big|\; 7$ • **Replace *x* by 42.**

$7 = 7$ • **Simplify the left side of the equation. The results are equal.**

42 is a solution of the equation.

Example 12 What is the quotient of 8856 and 42?

You Try It 12 What is 7694 divided by 24?

Solution

$$
\begin{array}{r}
210 \text{ r}36 \\
42\overline{)8856} \\
-\;84 \\
\hline
45 \\
-\;42 \\
\hline
36 \\
-\;\;0 \\
\hline
36
\end{array}
$$

• **Think** $42\overline{)36}$.
• **Subtract** $0 \cdot 42$.

Check: $(210 \cdot 42) + 36$
$= 8820 + 36 = 8856$

Your solution 320 r14

Solution on p. S3

Optional Student Activity

1. The number 10,981 is not divisible by 4. By rearranging the digits, find the largest possible number that is divisible by 4.
   91,180

2. Find a three-digit whole number such that, when it is divided by 11, the quotient is equal to the sum of its digits.
   198

Figure 1.15 shows a household's annual expenses of $44,000. Use this graph for Example 13 and You Try It 13.

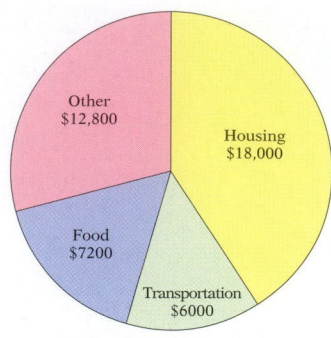

Figure 1.15   Annual Household Expenses

Example 13   Use Figure 1.15 to find the household's monthly expense for housing.

Solution   The annual expense for housing is $18,000.

$$18{,}000 \div 12 = 1500$$

The monthly expense is $1500.

You Try It 13   Use Figure 1.15 to find the household's monthly expense for food.

Your solution $600

Example 14   Estimate the quotient of 55,272 and 392.

Solution $55{,}272 \longrightarrow 60{,}000$
$392 \qquad \longrightarrow 400$

$$60{,}000 \div 400 = 150$$

You Try It 14   Estimate the quotient of 216,936 and 207.

Your solution   1000

Example 15   Evaluate $\dfrac{x}{y}$ when $x = 342$ and $y = 9$.

Solution $\dfrac{x}{y}$

$$\dfrac{342}{9} = 38$$

You Try It 15   Evaluate $\dfrac{x}{y}$ when $x = 672$ and $y = 8$.

Your solution   84

Example 16   Is 28 a solution of the equation $\dfrac{x}{7} = 4$?

Solution $\dfrac{x}{7} = 4$

$$\dfrac{28}{7} \;\Big|\; 4$$

$$4 = 4$$

Yes, 28 is a solution of the equation.

You Try It 16   Is 12 a solution of the equation $\dfrac{60}{y} = 2$?

Your solution   No

Solutions on p. S3

Objective D   **To factor numbers and find the prime factorization of numbers**

Natural number factors of a number divide that number evenly (there is no remainder).

1, 2, 3, and 6 are natural number factors of 6 because they divide 6 evenly.

Note that both the divisor and the quotient are factors of the dividend.

$$\begin{array}{cccc} 6 & 3 & 2 & 1 \\ 1\overline{)6} & 2\overline{)6} & 3\overline{)6} & 6\overline{)6} \end{array}$$

To find the factors of a number, try dividing the number by 1, 2, 3, 4, 5, ... Those numbers that divide the number evenly are its factors. Continue this process until the factors start to repeat.

Point of Interest

Twelve is the smallest abundant number, or number whose proper divisors add up to more than the number itself. The proper divisors of a number are all of its factors except the number itself. The proper divisors of 12 are 1, 2, 3, 4, and 6, which add up to 16, which is greater than 12. There are 246 abundant numbers between 1 and 1000.

A perfect number is one whose proper divisors add up to exactly that number. For example, the proper divisors of 6 are 1, 2, and 3, which add up to 6. There are only three perfect numbers less than 1000: 6, 28, and 496.

HOW TO   Find all the factors of 42.

$42 \div 1 = 42$	1 and 42 are factors.
$42 \div 2 = 21$	2 and 21 are factors.
$42 \div 3 = 14$	3 and 14 are factors.
$42 \div 4$	Will not divide evenly
$42 \div 5$	Will not divide evenly
$42 \div 6 = 7$	6 and 7 are factors.
$42 \div 7 = 6$	7 and 6 are factors.

The factors are repeating.
All the factors of 42 have been found.

The factors of 42 are 1, 2, 3, 6, 7, 14, 21, and 42.

The following rules are helpful in finding the factors of a number.

2 is a factor of a number if the digit in the ones place of the number is 0, 2, 4, 6, or 8.

436 ends in 6.
Therefore, 2 is a factor of 436 ($436 \div 2 = 218$).

3 is a factor of a number if the sum of the digits of the number is divisible by 3.

The sum of the digits of 489 is $4 + 8 + 9 = 21$. 21 is divisible by 3.
Therefore, 3 is a factor of 489 ($489 \div 3 = 163$).

4 is a factor of a number if the last two digits of the number are divisible by 4.

556 ends in 56.
56 is divisible by 4 ($56 \div 4 = 14$).
Therefore, 4 is a factor of 556 ($556 \div 4 = 139$).

5 is a factor of a number if the ones digit of the number is 0 or 5.

520 ends in 0.
Therefore, 5 is a factor of 520 ($520 \div 5 = 104$).

A **prime number** is a natural number greater than 1 that has exactly two natural number factors, 1 and the number itself. 7 is prime because its only factors are 1 and 7. If a number is not prime, it is a **composite** number. Because 6 has factors of 2 and 3, 6 is a composite number. The prime numbers less than 50 are

2, 3, 5, 7, 11, 13, 17, 19, 23, 29, 31, 37, 41, 43, 47

Objective 1.3D

New Vocabulary

natural number factors
prime number
composite number
prime factorization

Discuss the Concepts

1. What does it mean for a number to be divisible by another number?

2. How can you determine if 13 is a factor of 91? Is 13 a factor of 91?

3. What is the difference between a prime number and a composite number?

4. Is $4 \cdot 5$ the prime factorization of 20? Why or why not?

Instructor Note

Have students find other abundant numbers. (See the *Point of Interest* at the left.) Some examples are 18, 20, 24, and 30.

Concept Check

1. Find the largest factor of 111,111,111,111 that is less than 111,111,111,111. 37,037,037,037 because 111,111,111,111 is not divisible by 2 and 111,111,111,111 ÷ 3 = 37,037,037,037.

2. List all the numbers between 210 and 220 that do not have a factor of 2, 3, 4, 5, 6, 7, 8, or 9. 211

3. There are three different digits such that any two of them, written in any order, serve as the digits of a two-digit prime number. What are these three digits? 1, 3, 7

In-Class Examples (Objective 1.3D)

1. Find all the factors of 35. 1, 5, 7, 35
2. Find the prime factorization of 150. $2 \cdot 3 \cdot 5^2$
3. Find the prime factorization of 291. $3 \cdot 97$

Instructor Note

You may want to explain to students that it is not necessary to start with the smallest prime factor. As long as we divide by prime factors, we will obtain the same factorization. Starting with the smallest prime factor just provides a systematic method of attacking the problem.

Optional Student Activity

1. What is the ones digit in the product of all the factors of 100?
   0

2. Find the difference between the sum of all the factors of 66 and the sum of all the factors of 70.
   0. Both sums equal 144.

The **prime factorization** of a number is the expression of the number as a product of its prime factors. To find the prime factors of 90, begin with the smallest prime number as a trial divisor and continue with prime numbers as trial divisors until the final quotient is prime.

HOW TO Find the prime factorization of 90.

$$\begin{array}{r} 45 \\ 2\overline{)90} \end{array} \qquad \begin{array}{r} 15 \\ 3\overline{)45} \\ 2\overline{)90} \end{array} \qquad \begin{array}{r} 5 \\ 3\overline{)15} \\ 3\overline{)45} \\ 2\overline{)90} \end{array}$$

Divide 90 by 2. | 45 is not divisible by 2. | Divide 15 by 3.
 | Divide 45 by 3. | 5 is prime.

The prime factorization of 90 is $2 \cdot 3 \cdot 3 \cdot 5$, or $2 \cdot 3^2 \cdot 5$.

Finding the prime factorization of larger numbers can be more difficult. Try each prime number as a trial divisor. Stop when the square of the trial divisor is greater than the number being factored.

HOW TO Find the prime factorization of 201.

$$\begin{array}{r} 67 \\ 3\overline{)201} \end{array}$$

• 67 cannot be divided evenly by 2, 3, 5, 7, or 11. Prime numbers greater than 11 need not be tried because $11^2 = 121$ and $121 > 67$.

The prime factorization of 201 is $3 \cdot 67$.

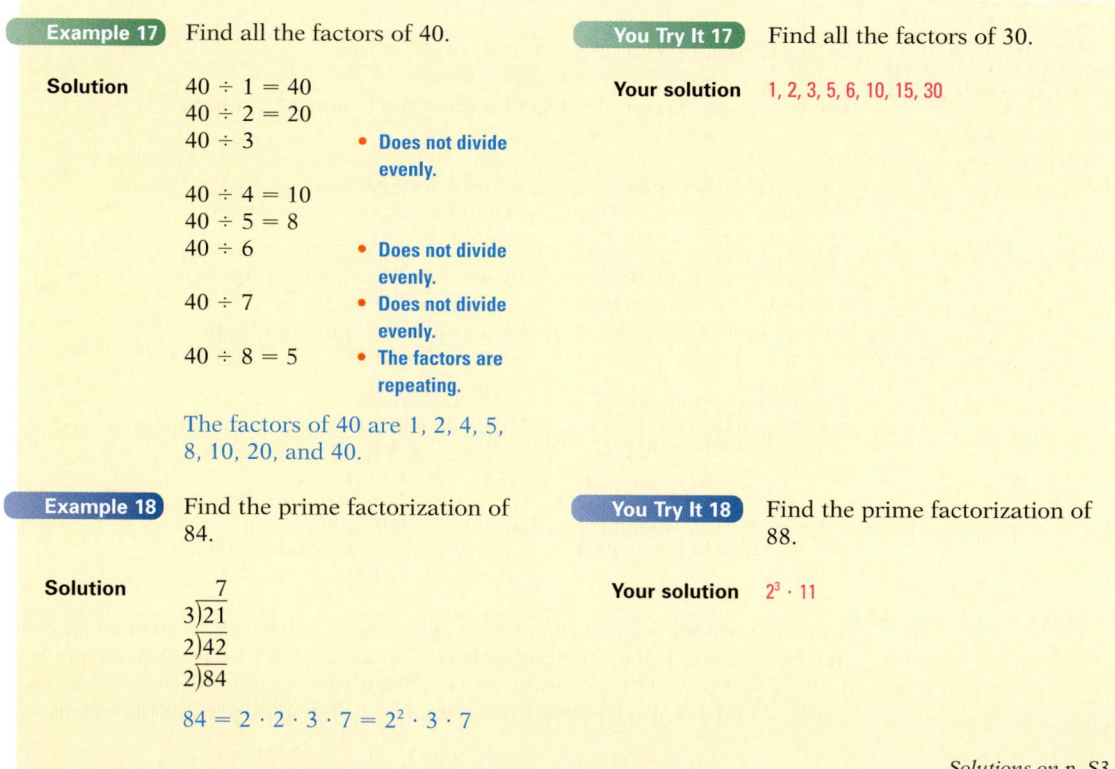

Example 17   Find all the factors of 40.

Solution

$40 \div 1 = 40$
$40 \div 2 = 20$
$40 \div 3$ • **Does not divide evenly.**
$40 \div 4 = 10$
$40 \div 5 = 8$
$40 \div 6$ • **Does not divide evenly.**
$40 \div 7$ • **Does not divide evenly.**
$40 \div 8 = 5$ • **The factors are repeating.**

The factors of 40 are 1, 2, 4, 5, 8, 10, 20, and 40.

You Try It 17   Find all the factors of 30.

Your solution   1, 2, 3, 5, 6, 10, 15, 30

Example 18   Find the prime factorization of 84.

Solution

$$\begin{array}{r} 7 \\ 3\overline{)21} \\ 2\overline{)42} \\ 2\overline{)84} \end{array}$$

$84 = 2 \cdot 2 \cdot 3 \cdot 7 = 2^2 \cdot 3 \cdot 7$

You Try It 18   Find the prime factorization of 88.

Your solution $2^3 \cdot 11$

Solutions on p. S3

Example 19 Find the prime factorization of 141.

Solution

$$\begin{array}{r} 47 \\ 3\overline{)141} \end{array}$$

- Try only 2, 3, 5, and 7 because $7^2 = 49$ and $49 > 47$.

$141 = 3 \cdot 47$

You Try It 19 Find the prime factorization of 295.

Your solution $5 \cdot 59$

Solution on p. S3

Objective E   **To solve application problems and use formulas**

In Section 1.2, we defined perimeter as the distance around a plane figure. The perimeter of a rectangle was given as $P = L + W + L + W$. This formula is commonly written as $P = 2L + 2W$.

> **TAKE NOTE**
> Remember that $2L$ means 2 times L, and $2W$ means 2 times W.

Perimeter of a Rectangle

The formula for the perimeter of a rectangle is $P = 2L + 2W$, where P is the perimeter of the rectangle, L is the length, and W is the width.

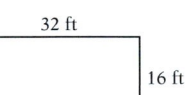

HOW TO Find the perimeter of the rectangle shown at the left.

$P = 2L + 2W$ • Use the formula for the perimeter of a rectangle.
$P = 2(32) + 2(16)$ • Substitute 32 for L and 16 for W.
$P = 64 + 32$ • Find the product of 2 and 32 and the product of 2 and 16.
$P = 96$ • Add.

The perimeter of the rectangle is 96 ft.

A **square** is a rectangle in which each side has the same length. Letting s represent the length of each side of a square, the perimeter of the square can be represented $P = s + s + s + s$. Note that we are adding *four s*'s. We can write the addition as multiplication: $P = 4s$.

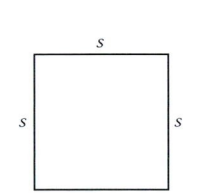

$P = s + s + s + s$
$P = 4s$

Perimeter of a Square

The formula for the perimeter of a square is $P = 4s$, where P is the perimeter and s is the length of a side of the square.

Objective 1.3E

Vocabulary to Review
perimeter   [1.2C]
rectangle   [1.2C]

New Formulas
Perimeter of a rectangle
   $P = 2L + 2W$
Perimeter of a square
   $P = 4s$
Area of a rectangle
   $A = LW$
Area of a square
   $A = s^2$

Instructor Note

The Order of Operations Agreement is presented in the next section. Instruct students using the formula $P = 2L + 2W$ to perform the multiplications first, and then the addition.

In-Class Examples (Objective 1.3E)

1. A lottery prize of $857,000 is divided equally among four winners. What amount does each winner receive? $214,250

2. A shipment of 9810 diodes requires testing. The diodes are divided equally among 15 employees. How many diodes must each employee test?
654 diodes

3. The Environmental Protection Agency estimates that a motorcycle gets 43 miles per gallon of gasoline. How many miles can it get on 8 gallons of gasoline?   344 mi

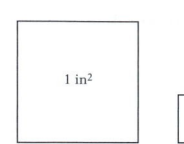

28 km

HOW TO   Find the perimeter of the square shown at the left.

$P = 4s$ • **Use the formula for the perimeter of a square.**
$P = 4(28)$ • **Substitute 28 for s.**
$P = 112$ • **Multiply.**

The perimeter of the square is 112 km.

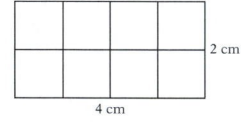

1 in²

1 cm²

Area is the amount of surface in a region. Area can be used to describe the size of a skating rink, the floor of a room, or a playground. Area is measured in square units.

A square that measures 1 inch on each side has an area of 1 square inch, which is written 1 in². A square that measures 1 centimeter on each side has an area of 1 square centimeter, which is written 1 cm².

Larger areas can be measured in square feet (ft²), square meters (m²), acres (43,560 ft²), square miles (mi²), or any other square unit.

2 cm

4 cm

The area of the rectangle is 8 cm².

The area of a geometric figure is the number of squares that are necessary to cover the figure. In the figure at the left, a rectangle has been drawn and covered with squares. Eight squares, each of area 1 cm², were used to cover the rectangle. The area of the rectangle is 8 cm². Note from this figure that the area of a rectangle can be found by multiplying the length of the rectangle by its width.

Area of a Rectangle

The formula for the area of a rectangle is $A = LW$, where A is the area, L is the length, and W is the width of the rectangle.

---

10 ft

25 ft

HOW TO   Find the area of the rectangle shown at the left.

$A = LW$ • **Use the formula for the area of a rectangle.**
$A = 25(10)$ • **Substitute 25 for L and 10 for W.**
$A = 250$ • **Multiply.**

The area of the rectangle is 250 ft².

s

$A = s \cdot s = s^2$

A square is a rectangle in which all sides are the same length. Therefore, both the length and the width of a square can be represented by s, and $A = LW = s \cdot s = s^2$.

Area of a Square

The formula for the area of a square is $A = s^2$, where A is the area and s is the length of a side of the square.

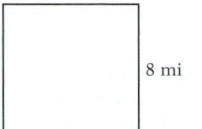

8 mi

HOW TO Find the area of the square shown at the left.

$A = s^2$ • Use the formula for the area of a square.
$A = 8^2$ • Substitute 8 for s.
$A = 64$ • Multiply.

The area of the square is 64 mi².

Integrating Technology

Many scientific calculators have an x^2 key. This key is used to square the displayed number. For example, after pressing 8 x^2 = , the display reads 64.

In this section, some of the phrases used to indicate the operations of multiplication and division were presented. In solving application problems, you might also look for the following types of questions:

Multiplication	**Division**
per . . . How many altogether?	What is the hourly rate?
each . . . What is the total number of . . ?	Find the amount per . . .
every . . . Find the total . . .	How many does each . . . ?

TAKE NOTE

Each of the following indicates multiplication:

"You purchased 6 boxes of doughnuts with 12 doughnuts *per* box. *How many* doughnuts did you purchase *altogether*?"

"If *each* bottle of apple juice contains 32 oz., *what is the total number of* ounces in 8 bottles of the juice?"

"You purchased 5 bags of oranges. *Every* bag contained 10 oranges. *Find the total* number of oranges purchased."

 Figure 1.16 shows the cost of a first-class postage stamp from the 1950s to 2005. Use this graph for Example 20 and You Try It 20.

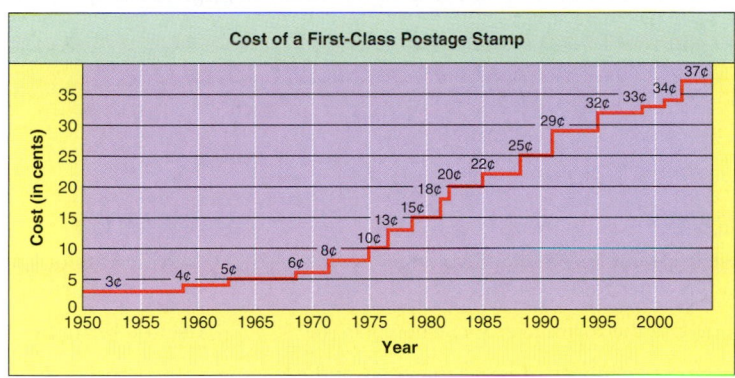

Figure 1.16 Cost of a First-Class Postage Stamp

Example 20

How many times more expensive was a stamp in 1980 than in 1950? Use Figure 1.16.

Strategy
To find how many times more expensive a stamp was, divide the cost in 1980 (15) by the cost in 1950 (3).

Solution
$15 \div 3 = 5$

A stamp was 5 times more expensive in 1980.

You Try It 20

How many times more expensive was a stamp in 1997 than in 1960? Use Figure 1.16.

Your Strategy

Your solution
8 times

Solution on p. S3

Optional Student Activity

A community organization is planning a bus trip to the city. Members have signed up either to go to the Museum of Fine Arts or to see *Phantom of the Opera.* The number attending each event is shown below. Buses will be ordered for the event, and each bus will carry 44 members. It is too expensive per person to have less than 20 people on a bus. Vans seating 11 people can be ordered through the bus company. How many buses are needed for the trip? Should any vans be ordered? If so, how many?

Museum of Fine Arts:
    82 people

Phantom of the Opera:
    109 people
4 buses; 2 vans (for 15 extra people)

Example 21

Find the amount of sod needed to cover a football field. A football field measures 120 yd by 50 yd.

Strategy

Draw a diagram

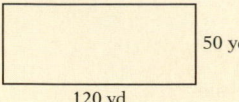

To find the amount of sod needed, use the formula for the area of a rectangle, $A = LW$. $L = 120$ and $W = 50$.

Solution

$A = LW$
$A = 120(50)$
$A = 6000$

6000 ft² of sod are needed.

You Try It 21

A homeowner wants to carpet the family room. The floor is square and measures 6 m on each side. How much carpet should be purchased?

Your Strategy

Your solution
36 m²

Example 22

At what rate of speed would you need to travel in order to drive a distance of 294 mi in 6 h? Use the formula $r = \dfrac{d}{t}$, where r is the average rate of speed, d is the distance, and t is the time.

Strategy

To find the rate of speed, replace d by 294 and t by 6 in the given formula and solve for r.

Solution

$r = \dfrac{d}{t}$

$r = \dfrac{294}{6} = 49$

You would need to travel at a speed of 49 mph.

You Try It 22

At what rate of speed would you need to travel in order to drive a distance of 486 mi in 9 h? Use the formula $r = \dfrac{d}{t}$, where r is the average rate of speed, d is the distance, and t is the time.

Your Strategy

Your solution
54 mph

Solutions on p. S3

1.3 Exercises

Objective A   To multiply whole numbers

1. ✎ Explain how to rewrite the addition $6 + 6 + 6 + 6 + 6$ as multiplication.

2. ✎ Provide at least three examples of situations in which we multiply numbers.

Multiply.

3. $(9)(127)$
1143

4. $(4)(623)$
2492

5. $(6709)(7)$
46,963

6. $(3608)(5)$
18,040

7. $8 \cdot 58,769$
470,152

8. $7 \cdot 60,047$
420,329

9. $683 \\ \times\ 71$
48,493

10. $591 \\ \times\ 92$
54,372

11. $7053 \\ \times\ \ 46$
324,438

12. $6704 \\ \times\ \ 58$
388,832

13. $3285 \\ \times\ 976$
3,206,160

14. $5327 \\ \times\ 624$
3,324,048

15. Find the product of 500 and 3.
1500

16. Find 30 multiplied by 80.
2400

17. What is 40 times 50?
2000

18. What is twice 700?
1400

19. What is the product of 400, 3, 20, and 0?
0

20. Write the product of f and g.
fg

21. Write the product of q, r, and s.
qrs

22. 🥧 **Physical Exercise** The figure at the right shows the number of calories burned on three different exercise machines during 1 h of a light, moderate, or vigorous workout. How many calories would you burn by (a) working out vigorously on a stair climber for a total of 6 h? (b) working out moderately on a treadmill for a total of 12 h?
(a) 2238 calories   **(b)** 4236 calories

Calories Burned on Exercise Machines
Source: Journal of American Medical Association

Estimate by rounding. Then find the exact answer.

23. $3467 \cdot 359$
1,200,000; 1,244,653

24. $8745(63)$
540,000; 550,935

25. $(39,246)(29)$
1,200,000; 1,138,134

26. $64,409 \cdot 67$
4,200,000; 4,315,403

Quick Quiz (Objective 1.3A)

1. Find the product of 78 and 4.   312
2. Estimate the product of 6291 and 59.   360,000
3. Evaluate $3cd$ when $c = 12$ and $d = 6$.   216
4. Multiply 1000 times 8.   8000

Section 1.3

Suggested Assignment

Exercises 3–165, every other odd
Exercises 167–187, odds
More challenging problems:
   Exercises 189, 190

Answers to Writing Exercises

1. Students should note that there are five 6's. Therefore, the addition $6 + 6 + 6 + 6 + 6$ can be written as 5 times 6: 5×6.
2. Answers will vary.

27. 745(63)
42,000; 46,935

28. 432 · 91
36,000; 39,312

29. (8941)(726)
6,300,000; 6,491,166

30. 2837(216)
600,000; 612,792

Evaluate the expression for the given values of the variables.

31. ab, when $a = 465$ and $b = 32$
14,880

32. cd, when $c = 381$ and $d = 25$
9525

33. $7a$, when $a = 465$
3255

34. $6n$, when $n = 382$
2292

35. xyz, when $x = 5$, $y = 12$, and $z = 30$
1800

36. abc, when $a = 4$, $b = 20$, and $c = 50$
4000

37. $2xy$, when $x = 67$ and $y = 23$
3082

38. $4ab$, when $a = 95$ and $b = 33$
12,540

Identify the property that justifies the statement.

39. $1 \cdot 29 = 29$
The Multiplication Property of One

40. $(10 \cdot 5) \cdot 8 = 10 \cdot (5 \cdot 8)$
The Associative Property of Multiplication

41. $43 \cdot 1 = 1 \cdot 43$
The Commutative Property of Multiplication

42. $0(76) = 0$
The Multiplication Property of Zero

Use the given property of multiplication to complete the statement.

43. The Commutative Property of Multiplication
$19 \cdot ? = 30 \cdot 19$
30

44. The Associative Property of Multiplication
$(? \cdot 6)100 = 5(6 \cdot 100)$
5

45. The Multiplication Property of Zero
$45 \cdot 0 = ?$
0

46. The Multiplication Property of One
$? \cdot 77 = 77$
1

47. Is 6 a solution of the equation $4x = 24$?
Yes

48. Is 0 a solution of the equation $4 = 4n$?
No

49. Is 23 a solution of the equation $96 = 3z$?
No

50. Is 14 a solution of the equation $56 = 4c$?
Yes

51. Is 19 a solution of the equation $2y = 38$?
Yes

52. Is 11 a solution of the equation $44 = 3a$?
No

Objective B   To evaluate expressions that contain exponents

Write in exponential form.

53. $2 \cdot 2 \cdot 2 \cdot 7 \cdot 7 \cdot 7 \cdot 7 \cdot 7$
$2^3 \cdot 7^5$

54. $3 \cdot 3 \cdot 3 \cdot 3 \cdot 3 \cdot 3 \cdot 5 \cdot 5 \cdot 5$
$3^6 \cdot 5^3$

55. $2 \cdot 2 \cdot 3 \cdot 3 \cdot 3 \cdot 5 \cdot 5 \cdot 5 \cdot 5$
$2^2 \cdot 3^3 \cdot 5^4$

56. $7 \cdot 7 \cdot 11 \cdot 11 \cdot 11 \cdot 19 \cdot 19 \cdot 19 \cdot 19$
$7^2 \cdot 11^3 \cdot 19^4$

57. $c \cdot c$
c^2

58. $d \cdot d \cdot d$
d^3

59. $x \cdot x \cdot x \cdot y \cdot y \cdot y$
$x^3 y^3$

60. $a \cdot a \cdot b \cdot b \cdot b \cdot b$
$a^2 b^4$

Evaluate.

61. 2^5
32

62. 2^6
64

63. 10^6
1,000,000

64. 10^9
1,000,000,000

65. $2^3 \cdot 5^2$
200

66. $2^4 \cdot 3^2$
144

67. $3^2 \cdot 10^3$
9000

68. $2^4 \cdot 10^2$
1600

69. $0^2 \cdot 6^2$
0

70. $4^3 \cdot 0^3$
0

71. $2^2 \cdot 5 \cdot 3^3$
540

72. $5^2 \cdot 2 \cdot 3^4$
4050

73. Find the square of 12.
144

74. What is the cube of 6?
216

75. Find the cube of 8.
512

76. What is the square of 11?
121

77. Write the fourth power of a.
a^4

78. Write the fifth power of t.
t^5

Evaluate the expression for the given values of the variables.

79. $x^3 y$, when $x = 2$ and $y = 3$
24

80. $x^2 y$, when $x = 3$ and $y = 4$
36

Quick Quiz (Objective 1.3B)

1. Write $2 \cdot 2 \cdot 3 \cdot 3 \cdot 3 \cdot 3$ in exponential form.
$2^2 \cdot 3^4$

2. Evaluate 10^9.
1,000,000,000

3. Evaluate $3^3 \cdot 7$.
189

4. Evaluate $x^2 \cdot y^5$ when $x = 6$ and $y = 2$.
1152

81. ab^6, when $a = 5$ and $b = 2$
320

82. ab^3, when $a = 7$ and $b = 4$
448

83. c^2d^2, when $c = 3$ and $d = 5$
225

84. m^3n^3, when $m = 5$ and $n = 10$
125,000

Objective C   **To divide whole numbers**

85. Provide at least three examples of situations in which we divide numbers.

86. In what situation does a division problem have a remainder?

Divide.

87. $9\overline{)2763}$
307

88. $4\overline{)2160}$
540

89. $5\overline{)1549}$
309 r4

90. $8\overline{)1636}$
204 r4

91. $15,300 \div 6$
2550

92. $43,500 \div 5$
8700

93. $681 \div 32$
21 r9

94. $879 \div 41$
21 r18

95. $9152 \div 62$
147 r38

96. $4161 \div 23$
180 r21

97. $7408 \div 37$
200 r8

98. $5207 \div 26$
200 r7

99. $31,546 \div 78$
404 r34

100. $38,976 \div 64$
609

101. $7713 \div 476$
16 r97

102. $8947 \div 223$
40 r27

103. Find the quotient of 7256 and 8.
907

104. What is the quotient of 8172 and 9?
908

105. What is 6168 divided by 7?
881 r1

106. Find 4153 divided by 9.
461 r4

107. Write the quotient of c and d.
$\dfrac{c}{d}$

108. Insurance Claims The table at the right shows the sources of laptop insurance claims in a recent year. Claims have been rounded to the nearest ten-thousand dollars.
a. What was the average monthly claim for theft?
b. For all sources combined, find the average claims per month.
a. $25,000   b. $95,000

Source	Claims (in dollars)
Accidents	560,000
Theft	300,000
Power Surge	80,000
Lightning	50,000
Transit	20,000
Water/flood	20,000
Other	110,000

Source: Safeware, The Insurance Company

Quick Quiz (Objective 1.3C)

1. Find the quotient of 711 and 9.
79

2. Estimate the quotient of 62,108 and 59.
1000

3. Divide: $6526 \div 96$
67 r94

4. Evaluate $\dfrac{x}{y}$ when $x = 6728$ and $y = 8$.
841

5. Which number, 3, 4, 30, or 144, is a solution of the equation $\dfrac{z}{24} = 6$?
144

Estimate by rounding. Then find the exact answer.

109. $36{,}472 \div 47$
800; 776

110. $62{,}176 \div 58$
1000; 1072

111. $389{,}804 \div 76$
5000; 5129

112. $637{,}072 \div 29$
20,000; 21,968

113. $79\overline{)38{,}984}$
500; 493 r37

114. $53\overline{)11{,}792}$
200; 222 r26

115. $219\overline{)332{,}004}$
1500; 1516

116. $324\overline{)632{,}124}$
2000; 1951

Evaluate the variable expression $\dfrac{x}{y}$ for the given values of x and y.

117. $x = 48; y = 1$
48

118. $x = 56; y = 56$
1

119. $x = 79; y = 0$
Undefined

120. $x = 0; y = 23$
0

121. $x = 39{,}200; y = 4$
9800

122. $x = 16{,}200; y = 3$
5400

123. Is 9 a solution of the equation $\dfrac{36}{z} = 4$?

Yes

124. Is 60 a solution of the equation $\dfrac{n}{12} = 5$?

Yes

125. Is 49 a solution of the equation $56 = \dfrac{x}{7}$?

No

126. Is 16 a solution of the equation $6 = \dfrac{48}{y}$?

No

Objective D   **To factor numbers and find the prime factorization of numbers**

Find all the factors of the number.

127. 10
1, 2, 5, 10

128. 20
1, 2, 4, 5, 10, 20

129. 12
1, 2, 3, 4, 6, 12

130. 9
1, 3, 9

131. 8
1, 2, 4, 8

132. 16
1, 2, 4, 8, 16

133. 13
1, 13

134. 17
1, 17

135. 18
1, 2, 3, 6, 9, 18

136. 24
1, 2, 3, 4, 6, 8, 12, 24

137. 25
1, 5, 25

138. 36
1, 2, 3, 4, 6, 9, 12, 18, 36

139. 56
1, 2, 4, 7, 8, 14, 28, 56

140. 45
1, 3, 5, 9, 15, 45

141. 28
1, 2, 4, 7, 14, 28

142. 32
1, 2, 4, 8, 16, 32

143. 48
1, 2, 3, 4, 6, 8, 12, 16, 24, 48

144. 64
1, 2, 4, 8, 16, 32, 64

145. 54
1, 2, 3, 6, 9, 18, 27, 54

146. 75
1, 3, 5, 15, 25, 75

Quick Quiz (Objective 1.3D)

1. Find all the factors of 108.
1, 2, 3, 4, 6, 9, 12, 18, 27, 36, 54, 108

2. Find the prime factorization of 110.
$2 \cdot 5 \cdot 11$

3. Find the prime factorization of 200.
$2^3 \cdot 5^2$

Find the prime factorization of the number.

147. 16
2^4

148. 24
$2^3 \cdot 3$

149. 12
$2^2 \cdot 3$

150. 27
3^3

151. 15
$3 \cdot 5$

152. 36
$2^2 \cdot 3^2$

153. 40
$2^3 \cdot 5$

154. 50
$2 \cdot 5^2$

155. 37
Prime

156. 83
Prime

157. 65
$5 \cdot 13$

158. 80
$2^4 \cdot 5$

159. 28
$2^2 \cdot 7$

160. 49
7^2

161. 42
$2 \cdot 3 \cdot 7$

162. 81
3^4

163. 51
$3 \cdot 17$

164. 89
Prime

165. 46
$2 \cdot 23$

166. 120
$2^3 \cdot 3 \cdot 5$

Objective E   **To solve application problems and use formulas**

167. Nutrition One ounce of cheddar cheese contains 115 calories. Find the number of calories in 4 oz of cheddar cheese.
460 calories

Nutrition Facts	Amount/Serving	% DV*	Amount/Serving	% DV*
Serv. Size 1 oz.	Total Fat 9g	14%	Total Carb. 1g	0%
Servings Per Package 12	Sat Fat 5g	25%	Fiber 0g	0%
Calories 115	Cholest. 30mg	10%	Sugars 0g	
Fat Cal. 80	Sodium 170mg	7%	Protein 7g	
*Percent Daily Values (DV) are based on a 2,000 calorie diet	Vitamin A 6% • Vitamin C 0% • Calcium 20% • Iron 0%			

168. 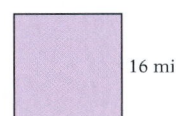 **Football** During his football career, John Riggins ran the ball 2916 times. He averaged about 4 yd per carry. About how many total yards did he gain during his career?
11,664 yd

169. Aviation A plane flying from Los Angeles to Boston uses 865 gal of jet fuel each hour. How many gallons of jet fuel are used on a 5-hour flight?
4325 gal

170. Geometry Find (a) the perimeter and (b) the area of a square that measures 16 mi on each side.
(a) 64 mi   (b) 256 mi²

16 mi

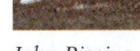
John Riggins

171. Geometry Find (a) the perimeter and (b) the area of a rectangle with a length of 24 m and a width of 15 m.
(a) 78 m   (b) 360 m²

172. Geometry Find the length of fencing needed to surround a square corral that measures 55 ft on each side.
220 ft

173. Geometry A homeowner plans to fence in the area around a swimming pool in the backyard. The area to be fenced in is a square measuring 24 ft on each side. How many feet of fencing should the homeowner purchase?
96 ft

Quick Quiz (Objective 1.3E)

Note: Question 1 is a one-step problem. Question 2 is a two-step problem.

1. A mechanic has a car payment of $197 each month. What is the total of the car payments over a 12-month period? $2364

2. A tannery produces and packages 320 briefcases each hour. Ten briefcases are put in each package for shipment. How many packages of briefcases can be produced in 8 h?
256 packages

174. Geometry   A solar panel is in the shape of a rectangle that has a width of 2 ft and a length of 3 ft. Find the area of the solar panel.
6 ft²

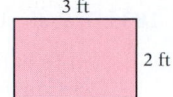

3 ft

2 ft

175. Geometry   What is the area of the floor of a two-car garage that is in the shape of a square that measures 24 ft on a side?
576 ft²

176. Geometry   A fieldstone patio is in the shape of a square that measures 9 ft on each side. What is the area of the patio?
81 ft²

177. Geometry   Find the amount of fabric needed for a rectangular flag that measures 308 cm by 192 cm.
59,136 cm²

178. Hourly Rate   A computer analyst doing consulting work received $5376 for working 168 h on a project. Find the hourly rate the consultant charged.
$32/h

179. Purchasing   A buyer for a department store purchased 215 suits at $83 each. Estimate the total cost of the order.
$16,000

180.   **Finances**   Financial advisors may predict how much money we should have saved for retirement by the ages of 35, 45, 55, and 65. One such prediction is included in the table below.
  a. A couple has earnings of $100,000 per year. According to the table, by how much should their savings grow per year from age 45 to 55?
  b. A couple has earnings of $50,000 per year. According to the table, by how much should their savings grow per year from age 55 to 65?
  a. $17,000   **b.** $8000

Minimum Levels of Savings Required for Married Couples to Be Prepared for Retirement				
Earnings	Savings Accumulation by Age			
	35	45	55	65
$50,000	8,000	23,000	90,000	170,000
$75,000	17,000	60,000	170,000	310,000
$100,000	34,000	110,000	280,000	480,000
$150,000	67,000	210,000	490,000	840,000

181. Loan Payments   Find the total amount paid on a loan when the monthly payment is $285 and the loan is paid off in 24 months. Use the formula $A = MN$, where A is the total amount paid, M is the monthly payment, and N is the number of payments.
$6840

182. Loan Payments   Find the total amount paid on a loan when the monthly payment is $187 and the loan is paid off in 36 months. Use the formula $A = MN$, where A is the total amount paid, M is the monthly payment, and N is the number of payments.
$6732

183. **Travel** Use the formula $t = \dfrac{d}{r}$, where t is the time, d is the distance, and r is the average rate of speed, to find the time it would take to drive 513 mi at an average speed of 57 mph.
9 h

184. **Travel** Use the formula $t = \dfrac{d}{r}$, where t is the time, d is the distance, and r is the average rate of speed, to find the time it would take to drive 432 mi at an average speed of 54 mph.
8 h

185. **Mutual Funds** The current value of the stocks in a mutual fund is $10,500,000. The number of shares outstanding is 500,000. Find the value per share of the fund. Use the formula $V = \dfrac{C}{S}$, where V is the value per share, C is the current value of the stocks in the fund, and S is the number of shares outstanding.
$21

186. **Mutual Funds** The current value of the stocks in a mutual fund is $4,500,000. The number of shares outstanding is 250,000. Find the value per share of the fund. Use the formula $V = \dfrac{C}{S}$, where V is the value per share, C is the current value of the stocks in the fund, and S is the number of shares outstanding.
$18

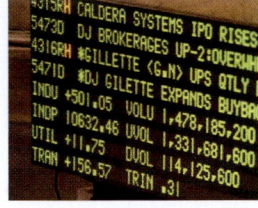

New York Stock Exchange

APPLYING THE CONCEPTS

187. **Time** There are 52 weeks in a year. Is this an exact figure or an approximation?
Approximation

188. **Mathematics** 13,827 is not divisible by 4. By rearranging the digits, find the largest possible number that is divisible by 4.
87,312

189. **Mathematics** A **palindromic number** is a whole number that remains unchanged when its digits are written in reverse order. For example, 818 is a palindromic number. Find the smallest three-digit multiple of 6 that is a palindromic number.
222

190. Determine whether the statement is always true, sometimes true, or never true.
a. Let a be any whole number. Then $a \cdot 0 = a$.
b. Let a be any whole number. Then $a \cdot 1 = 1$.
a. Sometimes true   **b.** Sometimes true

191. According to the National Safety Council, in a recent year, a death resulting from an accident occurred at the rate of one death every 5 min. At this rate, how many accidental deaths occurred each hour? each day? throughout the year? Explain how you arrived at your answers.

192. Prepare a monthly budget for a family of four. Explain how you arrived at the cost of each item. Annualize the budget you prepared.

Monthly Budget	
Rent	$975
Electricity	
Telephone	
Gas	
Food	

1.4 The Order of Operations Agreement

Objective A   To use the Order of Operations Agreement to simplify expressions

More than one operation may occur in a numerical expression. For example, the expression

$$4 + 3(5)$$

includes two arithmetic operations, addition and multiplication. The operations could be performed in different orders.

If we multiply first and then add, we have:		If we add first and then multiply, we have:	
	$4 + 3(5)$		$4 + 3(5)$
	$4 + 15$		$7(5)$
	19		35

To prevent more than one answer to the same problem, an Order of Operations Agreement is followed. By this agreement, 19 is the only correct answer.

Integrating Technology

Many calculators use the Order of Operations Agreement shown at the right. Enter 4 + 3 × 5 = into your calculator. If the answer is 19, your calculator uses the Order of Operations Agreement.

The Order of Operations Agreement

Step 1 Do all operations inside parentheses.

Step 2 Simplify any numerical expressions containing exponents.

Step 3 Do multiplication and division as they occur from left to right.

Step 4 Do addition and subtraction as they occur from left to right.

Integrating Technology

Here is an example of using the parentheses keys on a calculator. To evaluate $28(103 - 78)$, enter:

28 × (103 − 78)

= . Note that × is required on most calculators.

HOW TO Simplify: $2(4 + 1) - 2^3 + 6 \div 2$

$2(4 + 1) - 2^3 + 6 \div 2$
$= 2(5) - 2^3 + 6 \div 2$ • Perform operations in parentheses.
$= 2(5) - 8 + 6 \div 2$ • Simplify expressions with exponents.
$= 10 - 8 + 6 \div 2$ • Do multiplication and division as they
$= 10 - 8 + 3$ occur from left to right.
$= 2 + 3$ • Do addition and subtraction as they
$= 5$ occur from left to right.

One or more of the foregoing steps may not be needed to simplify an expression. In that case, proceed to the next step in the Order of Operations Agreement.

HOW TO Simplify: $8 + 9 \div 3$.

$8 + 9 \div 3$ • There are no parentheses (Step 1).
    There are no exponents (Step 2).
$= 8 + 3$ • Do the division (Step 3).
$= 11$ • Do the addition (Step 4).

New Procedures
Order of Operations Agreement

Discuss the Concepts

1. Why do we need an Order of Operations Agreement?

2. Describe the steps in the Order of Operations Agreement.

3. What operations are in the expression $12 + (9 - 5) \cdot 3$? In what order must these operations be performed?

Concept Check
Arrange the expressions in order from greatest value to least value.

$$12 \div 6 + 2$$
$$8 \cdot 4 + 5$$
$$7^2 - 14$$
$$15 + 3(7)$$
$$4(6 - 3) - 10$$
$$3(6 + 2) \div 8$$

$8 \cdot 4 + 5 > 15 + 3(7) >$
$7^2 - 14 > 12 \div 6 + 2 >$
$3(6 + 2) \div 8 > 4(6 - 3) - 10;$
$37 > 36 > 35 > 4 > 3 > 2$

Instructor Note

Have students try the *Projects and Group Activities* at the end of this chapter to determine if their calculators use the Order of Operations Agreement.

In-Class Examples (Objective 1.4A)

1. Simplify: $4^2 + 6(3 - 1)$ 28

2. Simplify: $9 - 6 + 6 \cdot 2 \div 3$ 7

3. Evaluate $2x + (x - y)^3$ when $x = 9$ and $y = 7$.
    26

Optional Student Activity

How can you place 12 sugar lumps in 3 coffee mugs so that there is an odd number of lumps in each mug?

Not possible

Instructor Note

An expression that uses the same digit three times and is equal to 30 is $3 + 3^3$.

Point of Interest

Try this: Use the same one-digit number three times to write an expression that is equal to 30.

HOW TO Evaluate $5a - (b + c)^2$ when $a = 6$, $b = 1$, and $c = 3$.

$$5a - (b + c)^2$$
$$5(6) - (1 + 3)^2$$
$$= 5(6) - (4)^2$$

- Replace a with 6, b with 1, and c with 3.
- Use the Order of Operations Agreement to simplify the resulting numerical expression. Perform operations inside parentheses.

$$= 5(6) - 16$$
$$= 30 - 16$$
$$= 14$$

- Simplify expressions with exponents.
- Do the multiplication.
- Do the subtraction.

Example 1

Simplify: $18 \div (6 + 3) \cdot 9 - 4^2$

Solution
$$18 \div (6 + 3) \cdot 9 - 4^2 = 18 \div 9 \cdot 9 - 4^2$$
$$= 18 \div 9 \cdot 9 - 16$$
$$= 2 \cdot 9 - 16$$
$$= 18 - 16$$
$$= 2$$

You Try It 1

Simplify: $4 \cdot (8 - 3) \div 5 - 2$

Your solution
2

Example 2

Simplify: $20 + 24(8 - 5) \div 2^2$

Solution
$$20 + 24(8 - 5) \div 2^2 = 20 + 24(3) \div 2^2$$
$$= 20 + 24(3) \div 4$$
$$= 20 + 72 \div 4$$
$$= 20 + 18$$
$$= 38$$

You Try It 2

Simplify: $16 + 3(6 - 1)^2 \div 5$

Your solution
31

Example 3

Evaluate $(a - b)^2 + 3c$ when $a = 6$, $b = 4$, and $c = 1$.

Solution
$$(a - b)^2 + 3c$$
$$(6 - 4)^2 + 3(1) = (2)^2 + 3(1)$$
$$= 4 + 3(1)$$
$$= 4 + 3$$
$$= 7$$

You Try It 3

Evaluate $(a - b)^2 + 5c$ when $a = 7$, $b = 2$, and $c = 4$.

Your solution
45

Solutions on p. S3

1.4 Exercises

Objective A   **To use the Order of Operations Agreement to simplify expressions**

1. ✏️ Why do we need an Order of Operations Agreement?

2. ✏️ What are the steps in the Order of Operations Agreement?

Simplify.

3. $8 \div 4 + 2$
    4

4. $12 - 9 \div 3$
    9

5. $6 \cdot 4 + 5$
    29

6. $5 \cdot 7 + 3$
    38

7. $4^2 - 3$
    13

8. $6^2 - 14$
    22

9. $5 \cdot (6 - 3) + 4$
    19

10. $8 + (6 + 2) \div 4$
    10

11. $9 + (7 + 5) \div 6$
    11

12. $14 \cdot (3 + 2) \div 10$
    7

13. $13 \cdot (1 + 5) \div 13$
    6

14. $14 - 2^3 + 9$
    15

15. $6 \cdot 3^2 + 7$
    61

16. $18 + 5 \cdot 3^2$
    63

17. $14 + 5 \cdot 2^3$
    54

18. $20 + (9 - 4) \cdot 2$
    30

19. $10 + (8 - 5) \cdot 3$
    19

20. $3^2 + 5 \cdot (6 - 2)$
    29

21. $2^3 + 4(10 - 6)$
    24

22. $3^2 \cdot 2^2 + 3 \cdot 2$
    42

23. $6(7) + 4^2 \cdot 3^2$
    186

24. $14 - 2(6)$
    2

25. $18 + 3(7)$
    39

26. $2(9 - 2) + 5$
    19

27. $6(8 - 3) - 12$
    18

28. $15 - (7 - 1) \div 3$
    13

29. $16 - (13 - 5) \div 4$
    14

30. $11 + 2 - 3 \cdot 4 \div 3$
    9

31. $17 + 1 - 8 \cdot 2 \div 4$
    14

32. $3(5 + 3) \div 8$
    3

Section 1.4

Suggested Assignment
Exercises 3–43, odds
More challenging problem:
    Exercise 46

Answers to Writing Exercises

1. We need an Order of Operations Agreement to ensure that there is only one correct answer to a problem that involves simplifying an arithmetic expression.

2. The steps in the Order of Operations Agreement are:
   1. Do all operations inside parentheses.
   2. Simplify any numerical expressions containing exponents.
   3. Do multiplication and division as they occur from left to right.
   4. Do addition and subtraction as they occur from left to right.

Quick Quiz (Objective 1.4A)

1. Simplify: $3^2 - 2(12 \div 6)$ 5
2. Simplify: $14 - (11 - 2) \div 3$ 11
3. Evaluate $x + 8y$ when $x = 7$ and $y = 10$.   87

Evaluate the expression for the given values of the variables.

33. $x - 2y$, where $x = 8$ and $y = 3$
2

34. $x + 6y$, where $x = 5$ and $y = 4$
29

35. $x^2 + 3y$, where $x = 6$ and $y = 7$
57

36. $3x^2 + y$, where $x = 2$ and $y = 9$
21

37. $x^2 + y \div x$, where $x = 2$ and $y = 8$
8

38. $x + y^2 \div x$, where $x = 4$ and $y = 8$
20

39. $4x + (x - y)^2$, where $x = 8$ and $y = 2$
68

40. $(x + y)^2 - 2y$, where $x = 3$ and $y = 6$
69

41. $x^2 + 3(x - y) + z^2$, where $x = 2$, $y = 1$, and $z = 3$
16

42. $x^2 + 4(x - y) \div z^2$, where $x = 8$, $y = 6$, and $z = 2$
66

43.   Use the inequality symbol $>$ to compare the expressions $11 + (8 + 4) \div 6$ and $12 + (9 - 5) \cdot 3$.
$12 + (9 - 5) \cdot 3 > 11 + (8 + 4) \div 6$ $[24 > 13]$

44.   Use the inequality symbol $<$ to compare the expressions $3^2 + 7(4 - 2)$ and $14 - 2^3 + 20$.
$3^2 + 7(4 - 2) < 14 - 2^3 + 20$ $[23 < 26]$

APPLYING THE CONCEPTS

45.   Arrange the expressions in order from greatest value to least value.

$27 \div 9 + 8$ $4 + 3 \cdot 12$
$81 - 8^2$ $50 - 6(8)$
$5(10 - 2) \div 4$ $2(1 + 4)^2 \div 10$
$4 + 3 \cdot 12 > 81 - 8^2 > 27 \div 9 + 8 > 5(10 - 2) \div 4 > 2(1 + 4)^2 \div 10 > 50 - 6(8)$
$[40 > 17 > 11 > 10 > 5 > 2]$

46.   What is the smallest prime number greater than $15 + (8 - 3)(2^4)$?
97

47.   ✏   Simplify $(47 + 48 + 49 + 51 + 52 + 53) \div 100$. What do you notice that will enable you to calculate the answer mentally?

Answers to Writing Exercises

47. The addends within the parentheses can be paired so that each pair equals 100 ($47 + 53 = 100$, $48 + 52 = 100$, $49 + 51 = 100$). The sum of the addends is $3 \cdot 100 = 300$. Therefore, $(47 + 48 + 49 + 51 + 52 + 53) \div 100 = 300 \div 100 = 3$.

Instructor Note
The feature entitled *Focus on Problem Solving* appears at the end of every chapter of the text. It provides optional material that can be used to enhance your students' problem-solving skills.

Focus on Problem Solving

Questions to Ask

You encounter problem-solving situations every day. Some problems are easy to solve, and you may mentally solve these problems without considering the steps you are taking in order to draw a conclusion. Others may be more challenging and may require more thought and consideration.

Suppose a friend suggests that you both take a trip over spring break. You'd like to go. What questions go through your mind? You might ask yourself some of the following questions:

How much will the trip cost? What will be the cost for travel, hotel rooms, meals, and so on?

Are some costs going to be shared by both me and my friend?

Can I afford it?

How much money do I have in the bank?

How much more money than I have now do I need?

How much time is there to earn that much money?

How much can I earn in that amount of time?

How much money must I keep in the bank in order to pay the next tuition bill (or some other expense)?

These questions require different mathematical skills. Determining the cost of the trip requires **estimation;** for example, you must use your knowledge of air fares or the cost of gasoline to arrive at an estimate of these costs. If some of the costs are going to be shared, you need to **divide** those costs by 2 in order to determine your share of the expense. The question regarding how much more money you need requires **subtraction:** the amount needed minus the amount currently in the bank. To determine how much money you can earn in the given amount of time requires **multiplication**—for example, the amount you earn per week times the number of weeks to be worked. To determine if the amount you can earn in the given amount of time is sufficient, you need to use your knowledge of **order relations** to compare the amount you can earn with the amount needed.

Facing the problem-solving situation described above may not seem difficult to you. The reason may be that you have faced similar situations before and, therefore, know how to work through this one. You may feel better prepared to deal with a circumstance such as this one because you know what questions to ask. An important aspect of learning to solve problems is learning what questions to ask. As you work through the application problems in this text, try to become more conscious of the mental process you are going through. You might begin the process by asking yourself the following questions whenever you are solving an application problem.

1. Have I read the problem enough times to be able to understand the situation being described?

2. Will restating the problem in different words help me to understand the problem situation better?

3. What facts are given? (You might make a list of the information contained in the problem.)

4. What information is being asked for?

5. What relationships exist among the given facts? What relationships exist among the given facts and the solution?

6. What mathematical operations are needed in order to solve the problem?

Try to focus on the problem-solving situation, not on the computation or on getting the answer quickly. And remember, the more problems you solve, the better able you will be to solve other problems in the future, partly because you are learning what questions to ask.

Projects and Group Activities

Order of Operations

Does your calculator use the Order of Operations Agreement? To find out, try this problem:

$$2 + 4 \cdot 7$$

If your answer is 30, then the calculator uses the Order of Operations Agreement. If your answer is 42, it does not use the agreement.

Even if your calculator does not use the Order of Operations Agreement, you can still correctly evaluate numerical expressions. The parentheses keys, (and), are used for this purpose.

Remember that $2 + 4 \cdot 7$ means $2 + (4 \cdot 7)$ because the multiplication must be completed before the addition. To evaluate this expression, enter the following:

Enter: 2   +   (4   ×   7   )   =

Display: 2   2   (4   4   7   28   30

When using your calculator to evaluate numerical expressions, insert parentheses around multiplications and around divisions. This has the effect of forcing the calculator to do the operations in the order you want.

For Exercises 1 to 10, evaluate.

1. $3 \cdot 8 - 5$

2. $6 + 8 \div 2$

3. $3 \cdot (8 - 2)^2$

4. $24 - (4 - 2)^2 \div 4$

5. $3 + (6 \div 2 + 4)^2 - 2$

6. $16 \div 2 + 4 \cdot (8 - 12 \div 4)^2 - 50$

7. $3 \cdot (15 - 2 \cdot 3) - 36 \div 3$

8. $4 \cdot 2^2 - (12 + 24 \div 6) + 5$

9. $16 \div 4 \cdot 3 + (3 \cdot 4 - 5) + 2$

10. $15 \cdot 3 \div 9 + (2 \cdot 6 - 3) + 4$

Surveys   On page 16 there is a circle graph showing the results of a survey of 150 people who were asked, "What bothers you most about movie theaters?" Note that the responses included (1) people talking in the theater, (2) high ticket prices, (3) high prices for food purchased in the theater, (4) dirty floors, and (5) uncomfortable seats.

Conduct a similar survey in your class. Ask each classmate which of the five conditions stated above is most irritating. Record the number of students who choose each one of the five possible responses. Prepare a bar graph to display the results of the survey. A model is provided below to help you get started.

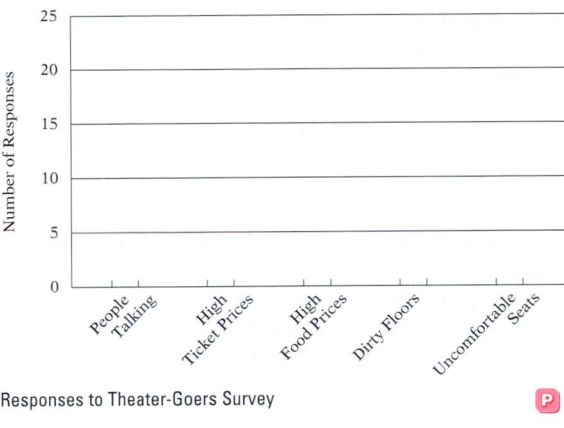

Responses to Theater-Goers Survey   P

Copyright © Houghton Mifflin Company. All rights reserved.

Patterns in Mathematics   For the circle at the left, use a straight line to connect each dot on the circle with every other dot on the circle. How many different straight lines are there?

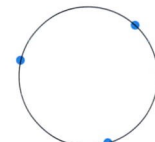

Follow the same procedure for each of the circles shown below. How many different straight lines are there in each?

Find a pattern to describe the number of dots on a circle and the corresponding number of different lines drawn. Use the pattern to determine the number of different lines that would be drawn in a circle with 7 dots and in a circle with 8 dots.

Now use the pattern to answer the following question. You are arranging a tennis tournament with 9 players. How many singles matches will be played among the 9 players if each player plays each of the other players only once?

Answers to Projects and Group Activities: Patterns in Mathematics

Number of straight lines in

a circle with 3 dots: 3

a circle with 4 dots: 6

a circle with 5 dots: 10

a circle with 6 dots: 15

The differences in the numbers of lines drawn for successive circles are

$$6 - 3 = 3$$
$$10 - 6 = 4$$
$$15 - 10 = 5$$

The difference increases by 1 each time.

For a circle with 7 dots, there will be

$$15 + 6 = 21$$

lines drawn.

For a circle with 8 dots, there will be

$$21 + 7 = 28$$

lines drawn.

For a circle with 9 dots, there will be

$$28 + 8 = 36$$

lines drawn. Therefore, 36 singles matches will be played among the 9 players.

Chapter 1 Summary

Key Words	Examples

The *natural numbers* or *counting numbers* are 1, 2, 3, 4, 5, 6, 7, 8, 9, 10, ... [1.1A, p. 3]

The *whole numbers* are 0, 1, 2, 3, 4, 5, 6, 7, 8, 9, 10, ... [1.1A, p. 3]

The symbol for "is less than" is $<$. The symbol for "is greater than" is $>$. A statement that uses the symbol $<$ or $>$ is an *inequality*. [1.1A, p. 3]

$3 < 7$
$9 > 2$

When a whole number is written using the digits 0, 1, 2, 3, 4, 5, 6, 7, 8, and 9, it is said to be in *standard form*. The position of each digit in the number determines the digit's *place value*. [1.1B, p. 4]

The number 598,317 is in standard form. The digit 8 is in the thousands place.

A *pictograph* represents data by using a symbol that is characteristic of the data. A *circle graph* represents data by the size of the sectors. A *bar graph* represents data by the height of the bars. A *broken-line graph* represents data by the position of the lines and shows trends or comparisons. [1.1D, pp. 8–10]

Addition is the process of finding the total of two or more numbers. The numbers being added are called *addends*. The answer is the *sum*. [1.2A, p. 19–20]

$$\begin{array}{r} {\scriptstyle 1\ 1\ 1} \\ 8762 \\ +\ 1359 \\ \hline 10,121 \end{array}$$

Subtraction is the process of finding the difference between two numbers. The *minuend* minus the *subtrahend* equals the *difference*. [1.2B, pp. 25–26]

$$\begin{array}{r} {\scriptstyle 4\ 11\ 11\ 6\ 13} \\ 5\,2,1\,7\,3 \\ -3\,4,9\,6\,8 \\ \hline 1\,7,2\,0\,5 \end{array}$$

Multiplication is the repeated addition of the same number. The numbers that are multiplied are called *factors*. The answer is the *product*. [1.3A, p. 41]

$$\begin{array}{r} {\scriptstyle 4\ 5} \\ 358 \\ \times\quad 7 \\ \hline 2506 \end{array}$$

The expression 3^5 is in *exponential form*. The *exponent*, 5, indicates how many times the *base*, 3, occurs as a factor in the multiplication. [1.3B, p. 46]

$5^4 = 5 \cdot 5 \cdot 5 \cdot 5 = 625$

Division is used to separate objects into equal groups. The *dividend* divided by the *divisor* equals the *quotient*. For any division problem, (*quotient* · *divisor*) + *remainder* = *dividend*. [1.3C, pp. 48–50]

$$\begin{array}{r} 93 \text{ r}3 \\ 7)\overline{654} \\ -63 \\ \hline 24 \\ -21 \\ \hline 3 \end{array}$$

Check: $(93 \cdot 7) + 3 = 651 + 3 = 654$

Natural number *factors* of a number divide that number evenly (there is no remainder). [1.3D, p. 53]

$18 \div 1 = 18$
$18 \div 2 = 9$
$18 \div 3 = 6$
$18 \div 4$ 4 does not divide 18 evenly.
$18 \div 5$ 5 does not divide 18 evenly.
$18 \div 6 = 3$ The factors are repeating.
The factors of 18 are 1, 2, 3, 6, 9, and 18.

A number greater than 1 is a *prime* number if its only whole number factors are 1 and itself. If a number is not prime, it is a *composite number.* [1.3D, p. 53]

The prime numbers less than 20 are 2, 3, 5, 7, 11, 13, 17, and 19.

The composite numbers less than 20 are 4, 6, 8, 9, 10, 12, 14, 15, 16, and 18.

The *prime factorization* of a number is the expression of the number as a product of its prime factors. [1.3D, p. 54]

$$\begin{array}{r} 7 \\ 3\overline{)21} \\ 2\overline{)42} \end{array}$$

The prime factorization of 42 is $2 \cdot 3 \cdot 7$.

A *variable* is a letter that is used to stand for a number. A mathematical expression that contains one or more variables is a *variable expression*. Replacing the variables in a variable expression with numbers and then simplifying the numerical expression is called *evaluating the variable expression.* [1.2A, p. 21]

To evaluate the variable expression $4ab$ when $a = 3$ and $b = 2$, replace a with 3 and b with 2. Simplify the resulting expression.

$4ab$
$4(3)(2) = 12(2) = 24$

An *equation* expresses the equality of two numerical or variable expressions. An equation contains an equals sign. A *solution* of an equation is a number that, when substituted for the variable, results in a true equation. [1.2A, p. 23]

6 is a solution of the equation $5 + x = 11$ because $5 + 6 = 11$ is a true equation.

Parallel lines never meet; the distance between them is always the same. [1.2C, p. 29]

Parallel Lines

An angle is measured in *degrees*. A 90° angle is a *right angle*. [1.2C, p. 29]

90°

Right Angle

A *polygon* is a closed figure determined by three or more line segments. The line segments that form the polygon are its *sides*. A *triangle* is a three-sided polygon. A *quadrilateral* is a four-sided polygon. A *rectangle* is a quadrilateral in which opposite sides are parallel, opposite sides are equal in length, and all four angles are right angles. A *square* is a rectangle in which all sides have the same length. The *perimeter* of a plane figure is a measure of the distance around the figure, and its area is the amount of surface in the region. [1.2C, pp. 29–30; 1.3E, pp. 55–56]

Triangle

Rectangle

Square

Essential Rules and Procedures	Examples
To round a number to a given place value: If the digit to the right of the given place value is less than 5, replace that digit and all digits to the right by zeros. If the digit to the right of the given place value is greater than or equal to 5, increase the digit in the given place value by 1, and replace all other digits to the right by zeros. [1.1C, pp. 6–7]	36,178 rounded to the nearest thousand is 36,000. 4592 rounded to the nearest thousand is 5000.

To estimate the answer to a calculation: Round each number to the highest place value of that number. Perform the calculation using the rounded numbers. [1.2A, p. 21]

$$39,471 \longrightarrow 40,000$$
$$12,586 \longrightarrow +10,000$$
$$\overline{50,000}$$

50,000 is an estimate of the sum of 39,471 and 12,586.

Properties of Addition [1.2A, p. 22]

Addition Property of Zero $a + 0 = a$ or $0 + a = a$	$7 + 0 = 7$
Commutative Property of Addition $a + b = b + a$	$8 + 3 = 3 + 8$
Associative Property of Addition $(a + b) + c = a + (b + c)$	$(2 + 4) + 6 = 2 + (4 + 6)$

Properties of Multiplication [1.3A, p. 44]

Multiplication Property of Zero $a \cdot 0 = 0$ or $0 \cdot a = 0$	$3 \cdot 0 = 0$
Multiplication Property of One $a \cdot 1 = a$ or $1 \cdot a = a$	$6 \cdot 1 = 6$
Commutative Property of Multiplication $a \cdot b = b \cdot a$	$2 \cdot 8 = 8 \cdot 2$
Associative Property of Multiplication $(a \cdot b) \cdot c = a \cdot (b \cdot c)$	$(2 \cdot 4) \cdot 6 = 2 \cdot (4 \cdot 6)$

Division Properties of Zero and One [1.3C, p. 49]

If $a \neq 0$, $0 \div a = 0$.	$0 \div 3 = 0$
If $a \neq 0$, $a \div a = 1$.	$3 \div 3 = 1$
$a \div 1 = a$	$3 \div 1 = 3$
$a \div 0$ is undefined.	$3 \div 0$ is undefined.

Order of Operations Agreement [1.4A, p. 67]

Step 1 Do all operations inside parentheses.	$5^2 - 3(2 + 4) = 5^2 - 3(6)$
Step 2 Simplify any numerical expressions containing exponents.	$= 25 - 3(6)$
Step 3 Do multiplication and division as they occur from left to right.	$= 25 - 18$
Step 4 Do addition and subtraction as they occur from left to right.	$= 7$

Geometric Formulas [1.2C, p. 30; 1.3E, pp. 55–56]

Perimeter of a Triangle	$P = a + b + c$
Perimeter of a Rectangle	$P = 2L + 2W$
Perimeter of a Square	$P = 4s$
Area of a Rectangle	$A = LW$
Area of a Square	$A = s^2$

Find the perimeter of a triangle with sides that measure 9 m, 6 m, and 5 m.

$$P = a + b + c$$
$$P = 9 + 6 + 5$$
$$P = 20$$

The perimeter of the triangle is 20 m.

Instructor Note
The notation [1.1A] following the answer to Exercise 1 indicates the objective that the student should review if that question is answered incorrectly. The notation [1.1A] means Chapter 1, Section 1, Objective A. This notation is used following every answer in all of the Prep Tests, Chapter Review Exercises, Chapter Tests, and Cumulative Review Exercises throughout the text.

Chapter 1 Review Exercises

1. Graph 8 on the number line.

    0 1 2 3 4 5 6 7 8 9 10 11 12

[1.1A]

2. Evaluate 10^4.
10,000 [1.3B]

3. Find the difference between 4207 and 1624.
2583 [1.2B]

4. Write $3 \cdot 3 \cdot 5 \cdot 5 \cdot 5 \cdot 5$ in exponential notation.
$3^2 \cdot 5^4$ [1.3B]

5. Add: $319 + 358 + 712$
1389 [1.2A]

6. Round 38,729 to the nearest hundred.
38,700 [1.1C]

7. Place the correct symbol, $<$ or $>$, between the two numbers.

247   163
$>$ [1.1A]

8. Write thirty-two thousand five hundred nine in standard form.
32,509 [1.1B]

9. Evaluate $2xy$ when $x = 50$ and $y = 7$.
700 [1.3A]

10. Find the quotient of 15,642 and 6.
2607 [1.3C]

11. Subtract: $6407 - 2359$
4048 [1.2B]

12. Estimate the sum of 482, 319, 570, and 146.
1500 [1.2A]

13. Find all the factors of 50.
1, 2, 5, 10, 25, 50 [1.3D]

14. Is 7 a solution of the equation $24 - y = 17$?
Yes [1.2B]

15. Simplify: $16 + 4(7 - 5)^2 \div 8$
18 [1.4A]

16. Identify the property that justifies the statement.

$10 + 33 = 33 + 10$
The Commutative Property of Addition   [1.2A]

17. Write 4,927,036 in words.
Four million nine hundred twenty-seven thousand thirty-six   [1.1B]

18. Evaluate x^3y^2 when $x = 3$ and $y = 5$.
675 [1.3B]

19. **Film Ratings** The circle graph at the right categorizes the 655 films released during a recent year by their ratings.
a. How many times more PG-13 films were released than NC-17 films?
b. How many times more R-rated films were released than NC-17 films?
a. 16 imes more   **b.** 61 times more   [1.3E]

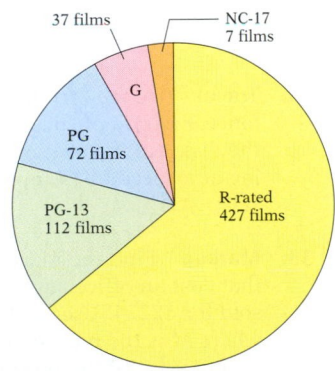

37 films
NC-17
7 films
G
PG
72 films
PG-13
112 films
R-rated
427 films

Ratings of Films Released
Source: MPA Worldwide Market Research

20. Divide: $6234 \div 92$
67 r70 [1.3C]

21. Find the product of 4 and 659.
2636 [1.3A]

22. Evaluate $x - y$ when $x = 270$ and $y = 133$.
137 [1.2B]

23. Find the prime factorization of 90.
$2 \cdot 3^2 \cdot 5$ [1.3D]

24. Evaluate $\dfrac{x}{y}$ when $x = 480$ and $y = 6$.
80 [1.3C]

25. Complete the statement by using the Multiplication Property of One.

$? \cdot 82 = 82$
1 [1.3A]

26. Simplify: $58 - 3 \cdot 4^2$
10 [1.4A]

27. Evaluate $x + y$ when $x = 683$ and $y = 249$.
932 [1.2A]

28. Multiply: $18 \cdot 24$
432 [1.3A]

29. Evaluate $(a + b)^2 - 2c$ when $a = 5$, $b = 3$, and $c = 4$.
56 [1.4A]

30. **Basketball** During his professional basketball career, Kareem Abdul-Jabbar had 17,440 rebounds. Elvin Hayes had 16,279 rebounds during his professional basketball career. Who had more rebounds, Abdul-Jabbar or Hayes?
Kareem Abdul-Jabbar [1.2C]

31. **Construction** A contractor quotes the cost of work on a new house, which is to have 2800 ft² of floor space, at $65 per square foot. Find the total cost of the contractor's work on the house.
$182,000 [1.3E]

32. **Geometry** A rectangle has a length of 25 m and a width of 12 m. Find (a) the perimeter and (b) the area of the rectangle.
(a) 74 m   (b) 300 m²   [1.2C, 1.3E]

Kareem Abdul-Jabbar

33. **College Enrollment** The line graph at the right shows the number of students enrolled in colleges.
a. During which decade did the student population increase the most?
b. What was the amount of increase?
a. 1960s   **b.** 4,792,000 students   [1.2C]

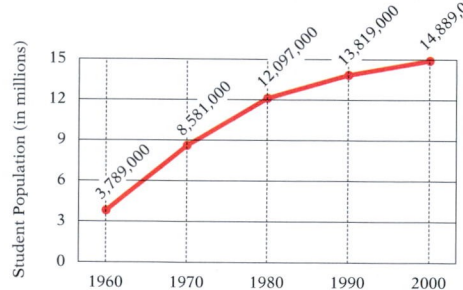

34. **Travel** Use the formula $d = rt$, where d is distance, r is rate of speed, and t is time, to find the distance traveled in 3 h by a cyclist traveling at a speed of 14 mph.
42 mi [1.3E]

Student Enrollment in Public and Private Colleges
Source: National Center for Educational Statistics

35. **Markup** Find the markup on a copy machine that cost an office supply business $1775 and sold for $2224. Use the formula $M = S - C$, where M is the markup on a product, S is the selling price of the product, and C is the cost of the product to the business.
$449 [1.2C]

Chapter 1 Test

1. Multiply: 3297×100
329,700 [1.3A]

2. Evaluate $2^4 \cdot 10^3$.
16,000 [1.3B]

3. Find the difference between 4902 and 873.
4029 [1.2B]

4. Write $x \cdot x \cdot x \cdot x \cdot y \cdot y \cdot y$ in exponential notation.
x^4y^3 [1.3B]

5. Is 7 a solution of the equation $23 = p + 16$?
Yes [1.2A]

6. Round 2961 to the nearest hundred.
3000 [1.1C]

7. Place the correct symbol, $<$ or $>$, between the two numbers.

7177   7717
$<$ [1.1A]

8. Write eight thousand four hundred ninety in standard form.
8490 [1.1B]

9. Write 382,904 in words.
Three hundred eighty-two thousand nine hundred four [1.1B]

10. Estimate the sum of 392, 477, 519, and 648.
2000 [1.2A]

11. Find the product of 8 and 1376.
11,008 [1.3A]

12. Estimate the product of 36,479 and 58.
2,400,000 [1.3A]

13. Find all the factors of 92.
1, 2, 4, 23, 46, 92 [1.3D]

14. Find the prime factorization of 240.
$2^4 \cdot 3 \cdot 5$ [1.3D]

15. Evaluate $x - y$ when $x = 39,241$ and $y = 8375$.
30,866 [1.2B]

16. Identify the property that justifies the statement.

$14 + y = y + 14$
The Commutative Property of Addition [1.2A]

17. Evaluate $\dfrac{x}{y}$ when $x = 3588$ and $y = 4$.
897 [1.3C]

18. Simplify: $27 - (12 - 3) \div 9$
26 [1.4A]

19. **Education** The table at the right shows the average annual earnings, based on level of education, for people aged 25 and older. What is the difference between average annual earnings for an individual with some college, but no degree, and for an individual with a bachelor's degree?
$13,900 [1.2C]

Educational Level	Average Annual Earnings
No high school diploma	21,400
High school diploma	28,800
Some college, no degree	32,400
Associate degree	35,400
Bachelor's degree	46,300
Master's degree	55,300

Source: Census Bureau; Bureau of Labor Statistics

20. Simplify: $5 + 2(4 - 3)^6$
7　[1.4A]

21. Write 3972 in expanded form.
3000 + 900 + 70 + 2　[1.1B]

22. Evaluate $5x + (x - y)^2$ when $x = 8$ and $y = 4$.
56　[1.4A]

23. Complete the statement by using the Associative Property of Addition.

$(3 + 7) + x = 3 + (? + x)$
7　[1.2A]

24. Mathematics What is the product of all the natural numbers less than 7?
720　[1.3E]

25. Purchasing You purchase a computer system that includes an operating system priced at $850, a monitor that cost $270, an extended keyboard priced at $175, and a printer for $425. You pay for the purchase by check. You had $2276 in your checking account before making the purchase. What was the balance in your account after making the purchase?
$556　[1.2C]

26. Geometry The length of each side of a square is 24 cm. Find (a) the perimeter and (b) the area of the square.
(a) 96 cm　　(b) 576 cm²　[1.3E]

27. Pay Deductions A data processor receives a total salary of $5690 per month. Deductions from the paycheck include $854 for taxes, $272 for retirement, and $108 for insurance. Find the data processor's monthly take-home pay.
$4456　[1.2C]

28. Navigation Systems The figure at the right shows the number of new vehicles sold with navigation systems.
a. Between which two years did the number of new vehicles sold with navigation systems increase the most?
b. What was the amount of that increase?
a. 2001–2002　　b. 125,000 vehicles　[1.2C]

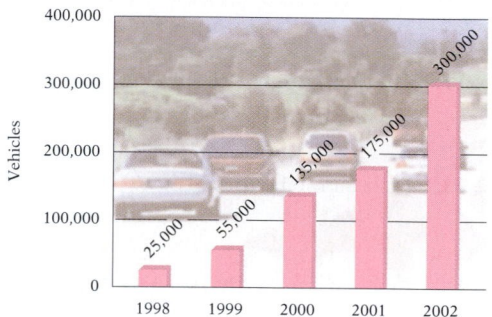

New Vehicles Sold with Navigation Systems
Source: J.D. Power and Associates

29. Commissions Use the formula $C = U \cdot R$, where C is the commission earned, U is the number of units sold, and R is the commission rate per unit, to find the commission earned from selling 480 boxes of greeting cards when the commission rate per box is $2.
$960　[1.3E]

30. Mutual Funds The current value of the stocks in a mutual fund is $5,500,000. The number of shares outstanding is 500,000. Find the value per share of the fund. Use the formula $V = \dfrac{C}{S}$, where V is the value per share, C is the current value of the stocks in the fund, and S is the number of shares outstanding.
$11　[1.3E]

2 Fractions and Decimals

Members of the Los Angeles Philharmonic follow conductor Esa-Pekka Salonen's movements to guarantee synchronized playing. By paying attention to both Salonen and the time signature for each musical piece, the musicians play in time with each other. The time signature appears as a fraction at the beginning of a piece of music and tells them how many beats to play per measure. The **project on page 164** demonstrates how to interpret the time signature.

OBJECTIVES

Section 2.1

A To find the least common multiple (LCM)
B To find the greatest common factor (GCF)

Section 2.2

A To write proper fractions, improper fractions, and mixed numbers
B To write equivalent fractions
C To identify the order relation between two fractions

Section 2.3

A To add fractions
B To subtract fractions
C To solve application problems and use formulas

Section 2.4

A To multiply fractions
B To divide fractions
C To simplify a complex fraction
D To solve application problems and use formulas

Section 2.5

A To read and write decimals
B To identify the order relation between two decimals
C To round a decimal to a given place value
D To solve application problems

Section 2.6

A To add and subtract decimals
B To multiply decimals
C To divide decimals
D To convert between decimals and fractions and identify the order relation between a decimal and a fraction
E To solve application problems and use formulas

Section 2.7

A To use the Order of Operations Agreement

Need help? For online student resources, such as section quizzes, visit this textbook's website at **math.college.hmco.com/students**.

Features of the IAE

For a description of the features included in the margins of this Instructor's Annotated Edition, see page 2.

PREP TEST • • •

Do these exercises to prepare for Chapter 2.

For Exercises 1 to 6, add, subtract, multiply, or divide.

1. 4×5
20 [1.3A]

2. $2 \cdot 2 \cdot 2 \cdot 3 \cdot 5$
120 [1.3A]

3. 9×1
9 [1.3A]

4. $6 + 4$
10 [1.2A]

5. $10 - 3$
7 [1.2B]

6. $63 \div 30$
2 r3 [1.3C]

7. Round 36,852 to the nearest hundred.
36,900 [1.1C]

8. Write 4791 in words.
Four thousand seven hundred ninety-one [1.1B]

9. Write six thousand eight hundred forty-two in standard form.
6842 [1.1B]

10. Which of the following numbers divide evenly into 12?

1 2 3 4 5 6 7 8 9 10 11 12
1, 2, 3, 4, 6, 12 [1.3C]

11. Simplify: $8 \times 7 + 3$
59 [1.4A]

12. Complete: $8 = ? + 1$
7 [1.2A]

13. Place the correct symbol, $<$ or $>$, between the two numbers.

44   48
$44 < 48$ [1.1A]

GO FIGURE • • •

You and a friend are swimming laps in a pool. You swim one lap every 4 minutes. Your friend swims one lap every 5 minutes. If you start at the same time from the same end of the pool, in how many minutes will both of you be at the starting point again? How many times will you have passed each other in the pool prior to that time?   20 min; 8 times

2.1 The Least Common Multiple and Greatest Common Factor

Objective A   To find the least common multiple (LCM)

S t u d y   T i p

Before you begin a new chapter, you should take some time to review previously learned skills. One way to do this is to complete the Prep Test. See page 82. This test focuses on the particular skills that will be required for the new chapter.

The **multiples of a number** are the products of that number and the numbers 1, 2, 3, 4, 5, ...

$3 \times 1 = 3$
$3 \times 2 = 6$
$3 \times 3 = 9$
$3 \times 4 = 12$ The multiples of 3 are 3, 6, 9, 12, 15, ...
$3 \times 5 = 15$

A number that is a multiple of two or more numbers is a **common multiple** of those numbers.

The multiples of 4 are 4, 8, 12, 16, 20, 24, 28, 32, 36, ...
The multiples of 6 are 6, 12, 18, 24, 30, 36, 42, ...
Some common multiples of 4 and 6 are 12, 24, and 36.

The **least common multiple (LCM)** is the smallest common multiple of two or more numbers.

The least common multiple of 4 and 6 is 12.

Listing the multiples of each number is one way to find the LCM. Another way to find the LCM uses the prime factorization of each number.

To find the LCM of 450 and 600, find the prime factorization of each number and write the factorization of each number in a table. Circle the greatest product in each column. The LCM is the product of the circled numbers.

	2	3	5
$450 =$	2	(3 · 3)	(5 · 5)
$600 =$	(2 · 2 · 2)	3	5 · 5

In the column headed by 5, the products are equal. Circle just one product.

The LCM is the product of the circled numbers.
The LCM = $2 \cdot 2 \cdot 2 \cdot 3 \cdot 3 \cdot 5 \cdot 5 = 1800$.

Example 1   Find the LCM of 24, 36, and 50.

Solution

	2	3	5
$24 =$	(2 · 2 · 2)	3	
$36 =$	2 · 2	(3 · 3)	
$50 =$	2		(5 · 5)

The LCM = $2 \cdot 2 \cdot 2 \cdot 3 \cdot 3 \cdot 5 \cdot 5$
$\qquad = 1800$.

You Try It 1   Find the LCM of 12, 27, and 50.

Your solution   2700

Solution on p. S4

New Vocabulary

multiples of a number
common multiple
least common multiple (LCM)

Discuss the Concepts

1. How can you find the multiples of 12?
2. How can you find some common multiples of 8 and 12?
3. Why is 24 the least common multiple of 8 and 12?

Concept Check

Find the LCM of 16, 20, and 40 by first listing the multiples of each number and then using the prime factorization method shown on page 83 of the textbook.   80

Optional Student Activity

The ancient Mayans used two calendars, a civil calendar of 365 days and a sacred calendar of 260 days. If a civil year and a sacred year begin on the same day, how many civil years and how many sacred years will pass before this situation occurs again?
The LCM of 365 and 260 is 18,980.
$18{,}980 \div 365 = 52$
$18{,}980 \div 260 = 73$
The situation occurs again after 52 civil years and 73 sacred years.

In-Class Examples (Objective 2.1A)

Find the LCM.
1. 14, 21   42
2. 2, 7, 14   14
3. 5, 12, 15   60

Objective 2.1B

Vocabulary to Review

factors of a number   [1.3D]

New Vocabulary

common factor
greatest common factor (GCF)

Discuss the Concepts

1. How can you find the factors of 24?
2. How can you find the common factors of 12 and 24?
3. Why is 12 the greatest common factor of 12 and 24?

Concept Check

Find the product of the GCF of 225 and 444 and the LCM of 225 and 444.
GCF = 3; LCM = 33,300;
3(33,300) = 99,900

Concept Check

Find the GCF of $2^5 \cdot 3^9 \cdot 5^7$ and $2^7 \cdot 3^2 \cdot 5^4$.
$2^5 \cdot 3^2 \cdot 5^4 = 180,000$

Instructor Note

The following model may help some students with the LCM and GCF.

The arrow indicates "divides into."

Optional Student Activity

What number must be multiplied by 200 so that the product has exactly 15 factors?
2; Factors: 1, 2, 4, 5, 8, 10, 16, 20, 25, 40, 50, 80, 100, 200, 400

Objective B   To find the greatest common factor (GCF)

Recall that a number that divides another number evenly is a factor of that number. The number 64 can be evenly divided by 1, 2, 4, 8, 16, 32, and 64, so the numbers 1, 2, 4, 8, 16, 32, and 64 are factors of 64.

A number that is a factor of two or more numbers is a **common factor** of those numbers.

The factors of 30 are 1, 2, 3, 5, 6, 10, 15, and 30.
The factors of 105 are 1, 3, 5, 7, 15, 21, 35, and 105.
The common factors of 30 and 105 are 1, 3, 5, and 15.

The **greatest common factor (GCF)** is the largest common factor of two or more numbers.

The greatest common factor of 30 and 105 is 15.

Listing the factors of each number is one way of finding the GCF. Another way to find the GCF uses the prime factorization of each number.

To find the GCF of 126 and 180, find the prime factorization of each number and write the factorization of each number in a table. Circle the least product in each column that does not have a blank. The GCF is the product of the circled numbers.

	2	3	5	7
126 =	②	(3 · 3)		7
180 =	2 · 2	3 · 3	5	

In the column headed by 3, the products are equal. Circle just one product.
Columns 5 and 7 have a blank, so 5 and 7 are not common factors of 126 and 180. Do not circle any number in these columns.

The GCF is the product of the circled numbers.
The GCF = $2 \cdot 3 \cdot 3 = 18$.

Example 2   Find the GCF of 90, 168, and 420.

Solution

	2	3	5	7
90 =	②	3 · 3	5	
168 =	2 · 2 · 2	③		7
420 =	2 · 2	3	5	7

The GCF = $2 \cdot 3 = 6$.

You Try It 2   Find the GCF of 36, 60, and 72.

Your solution   12

Example 3   Find the GCF of 7, 12, and 20.

Solution

	2	3	5	7
7 =				7
12 =	2 · 2	3		
20 =	2 · 2		5	

Because no numbers are circled, the GCF = 1.

You Try It 3   Find the GCF of 11, 24, and 30.

Your solution   1

Solutions on p. S4

In-Class Examples (Objective 2.1B)

Find the GCF.
1. 12, 18   6
2. 24, 64   8
3. 41, 67   1
4. 21, 27, 33   3

2.1 Exercises

Objective A   To find the least common multiple (LCM)

For Exercises 1 to 34, find the LCM.

1. 5, 8
40

2. 3, 6
6

3. 3, 8
24

4. 2, 5
10

5. 5, 6
30

6. 5, 7
35

7. 4, 6
12

8. 6, 8
24

9. 8, 12
24

10. 12, 16
48

11. 5, 12
60

12. 3, 16
48

13. 8, 14
56

14. 6, 18
18

15. 3, 9
9

16. 4, 10
20

17. 8, 32
32

18. 7, 21
21

19. 9, 36
36

20. 14, 42
42

21. 44, 60
660

22. 120, 160
480

23. 102, 184
9384

24. 123, 234
9594

25. 4, 8, 12
24

26. 5, 10, 15
30

27. 3, 5, 10
30

28. 2, 5, 8
40

29. 3, 8, 12
24

30. 5, 12, 18
180

31. 9, 36, 64
576

32. 18, 54, 63
378

33. 16, 30, 84
1680

34. 9, 12, 15
180

Objective B   To find the greatest common factor (GCF)

For Exercises 35 to 68, find the GCF.

35. 3, 5
1

36. 5, 7
1

37. 6, 9
3

38. 18, 24
6

39. 15, 25
5

40. 14, 49
7

41. 25, 100
25

42. 16, 80
16

43. 32, 51
1

44. 21, 44
1

45. 12, 80
4

46. 8, 36
4

47. 16, 140
4

48. 12, 76
4

Section 2.1

Suggested Assignment
Exercises 1–67, odds
Exercises 70–72, 74
More challenging problem:
    Exercise 73

Quick Quiz (Objective 2.1A)
Find the LCM.
1. 10, 25   50
2. 3, 6, 7   42
3. 2, 8, 64   64

49. 24, 30
6

50. 48, 144
48

51. 44, 96
4

52. 18, 32
2

53. 3, 5, 11
1

54. 6, 8, 10
2

55. 7, 14, 49
7

56. 6, 15, 36
3

57. 10, 15, 20
5

58. 12, 18, 20
2

59. 24, 40, 72
8

60. 3, 17, 51
1

61. 17, 31, 81
1

62. 14, 42, 84
14

63. 25, 125, 625
25

64. 12, 68, 92
4

65. 28, 35, 70
7

66. 1, 49, 153
1

67. 32, 56, 72
8

68. 24, 36, 48
12

Answers to Writing Exercises

71. The LCM of 2 and 3 is 6. The LCM of 5 and 7 is 35. The LCM of 11 and 19 is 209. The LCM of two prime numbers is the product of the two numbers. The LCM of three prime numbers is the product of the three numbers.

72. The GCF of 3 and 5 is 1. The GCF of 7 and 11 is 1. The GCF of 29 and 43 is 1. Because two prime numbers do not have a common factor other than 1, the GCF of two prime numbers is 1. Because three prime numbers do not have a common factor other than 1, the GCF of three prime numbers is 1.

73. Yes, the LCM of two numbers is always divisible by the GCF of the two numbers. The two numbers are factors of their LCM. The GCF of the two numbers is a factor of the LCM of the same numbers. That is, the LCM of two numbers always is divisible by the GCF of the two numbers. For example, the GCF of 4 and 6 is 2, and the LCM of 4 and 6 is 12. 12 is divisible by 2.

APPLYING THE CONCEPTS

69. Define the phrase *relatively prime numbers*. List three pairs of relatively prime numbers.
Relatively prime numbers are numbers with no common factors except the factor 1. Answers will vary. For example, 4 and 5, 8 and 9, 16 and 21.

70. **Work Schedules**   Joe Salvo, a lifeguard, works 3 days and then has a day off. A friend works 5 days and then has a day off. How many days after Joe and his friend have a day off together will they have another day off together?   12 days

71. Find the LCM of each of the following pairs of numbers: 2 and 3, 5 and 7, and 11 and 19. Can you draw a conclusion about the LCM of two prime numbers? Suggest a way of finding the LCM of three different prime numbers.

72. Find the GCF of each of the following pairs of numbers: 3 and 5, 7 and 11, and 29 and 43. Can you draw a conclusion about the GCF of two prime numbers? What is the GCF of three different prime numbers?

73. Is the LCM of two numbers always divisible by the GCF of the two numbers? If so, explain why. If not, give an example.

74. Using the pattern for the first two triangles at the right, determine the center number of the last triangle.
4

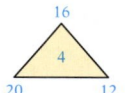
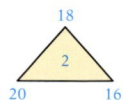
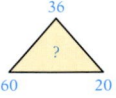

Quick Quiz (Objective 2.1B)

Find the GCF.
1. 6, 16   2
2. 4, 9   1
3. 26, 52   26
4. 12, 30, 60   6

2.2   Introduction to Fractions

Objective A   **To write proper fractions, improper fractions, and mixed numbers**

A recipe calls for $\frac{1}{2}$ cup of butter; a carpenter uses a $\frac{3}{8}$-inch screw; and a stockbroker might say that Sears closed down $\frac{3}{4}$. The numbers $\frac{1}{2}$, $\frac{3}{8}$, and $\frac{3}{4}$ are fractions.

A **fraction** can represent the number of equal parts of a whole. The circle at the right is divided into 8 equal parts. 3 of the 8 parts are shaded. The shaded portion of the circle is represented by the fraction $\frac{3}{8}$.

Each part of a fraction has a name.

Point of Interest
The fraction bar was first used in 1050 by al-Hassar. It is also called a vinculum.

$$\text{Fraction bar} \longrightarrow \frac{3 \leftarrow \textbf{Numerator}}{8 \leftarrow \textbf{Denominator}}$$

In a **proper fraction,** the numerator is smaller than the denominator. A proper fraction is less than 1.

$$\frac{1}{2} \qquad \frac{3}{8} \qquad \frac{3}{4}$$

Proper fractions

In an **improper fraction,** the numerator is greater than or equal to the denominator. An improper fraction is a number greater than or equal to 1.

$$\frac{7}{3} \qquad \frac{4}{4}$$

Improper fractions

The shaded portion of the circles at the right is represented by the improper fraction $\frac{7}{3}$.

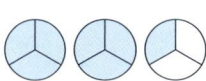

The shaded portion of the square at the right is represented by the improper fraction $\frac{4}{4}$.

A fraction bar can be read "divided by." Therefore, the fraction $\frac{4}{4}$ can be read "4 ÷ 4." Because a number divided by itself is equal to 1, 4 ÷ 4 = 1 and $\frac{4}{4} = 1$.

The shaded portion of the square above can be represented as $\frac{4}{4}$ or 1.

Since the fraction bar can be read as "divided by" and any number divided by 1 is the number, any whole number can be represented as an improper fraction. For example, $5 = \frac{5}{1}$ and $7 = \frac{7}{1}$.

Objective 2.2A

New Vocabulary
fraction
fraction bar
numerator
denominator
proper fraction
improper fraction
mixed number

Discuss the Concepts

1. Explain why the shaded portion of the diagram can be described as $\frac{2}{3}$.

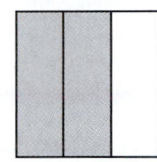

2. Explain the procedure for rewriting an improper fraction as a mixed number.
3. Explain the procedure for rewriting a mixed number as an improper fraction.

Concept Check

1. Write a proper fraction and explain why it is a proper fraction.
2. Write a mixed number and explain why it is a mixed number.
3. Write an improper fraction and explain why it is an improper fraction.

In-Class Examples (Objective 2.2A)

1. Express the shaded portion of the circles as an improper fraction and as a mixed number.

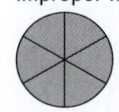 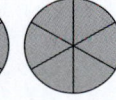 

$\frac{13}{6}$; $2\frac{1}{6}$

2. Write $\frac{15}{4}$ as a mixed number. $3\frac{3}{4}$

3. Write $\frac{24}{6}$ as a whole number.   4

4. Write $3\frac{4}{7}$ as an improper fraction. $\frac{25}{7}$

5. Write 12 as an improper fraction. $\frac{12}{1}$

Optional Student Activity

Use the symbol $<$, $>$, or $=$ to compare each number with the number 1.

1. $\dfrac{3}{7}$ $\dfrac{3}{7} < 1$

2. $\dfrac{9}{9}$ $\dfrac{9}{9} = 1$

3. $\dfrac{8}{5}$ $\dfrac{8}{5} > 1$

Instructor Note

As a classroom exercise, ask students to give real-world examples in which mixed numbers are used. Some possible answers: carpentry, sewing, recipes.

Because zero divided by any number other than zero is zero, **the numerator of a fraction can be zero.**

For example, $\dfrac{0}{6} = 0$ because $0 \div 6 = 0$.

Recall that division by zero is not defined. Therefore, **the denominator of a fraction cannot be zero.**

For example, $\dfrac{9}{0}$ is not defined because $\dfrac{9}{0} = 9 \div 0$, and division by zero is not defined.

A **mixed number** is a number greater than 1 with a whole number part and a fractional part.

The shaded portion of the circles at the right is represented by the mixed number $2\frac{1}{2}$.

Note from the diagram at the right that the improper fraction $\dfrac{5}{2}$ is equal to the mixed number $2\frac{1}{2}$.

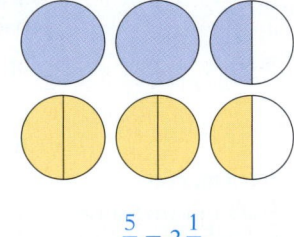

$$\dfrac{5}{2} = 2\dfrac{1}{2}$$

An improper fraction can be written as a mixed number.

To write $\dfrac{5}{2}$ as a mixed number, read the fraction bar as "divided by."

$\dfrac{5}{2}$ means $5 \div 2$.

Divide the numerator by the denominator.

$$\begin{array}{r} 2 \\ 2\overline{)5} \\ -4 \\ \hline 1 \end{array}$$

To write the fractional part of the mixed number, write the remainder over the divisor.

$$\begin{array}{r} 2\frac{1}{2} \\ 2\overline{)5} \\ -4 \\ \hline 1 \end{array}$$

Write the answer.

$$\dfrac{5}{2} = 2\dfrac{1}{2}$$

To write a mixed number as an improper fraction, multiply the denominator of the fractional part of the mixed number by the whole number part. The sum of this product and the numerator of the fractional part is the numerator of the improper fraction. The denominator remains the same.

HOW TO   Write $4\frac{5}{6}$ as an improper fraction.

$$4\dfrac{5}{6} = \dfrac{(6 \cdot 4) + 5}{6} = \dfrac{24 + 5}{6} = \dfrac{29}{6}$$

Example 1 Express the shaded portion of the circles as an improper fraction and as a mixed number.

Solution $\dfrac{19}{4}$; $4\dfrac{3}{4}$

You Try It 1 Express the shaded portion of the circles as an improper fraction and as a mixed number.

Your solution $\dfrac{19}{6}$; $3\dfrac{1}{6}$

Example 2 Write $\dfrac{14}{5}$ as a mixed number.

Solution

$$5\overline{)14}\;\;\;\;\dfrac{14}{5}=2\dfrac{4}{5}$$
$$\dfrac{-10}{4}$$

You Try It 2 Write $\dfrac{26}{3}$ as a mixed number.

Your solution $8\dfrac{2}{3}$

Example 3 Write $\dfrac{35}{7}$ as a whole number.

Solution

$$7\overline{)35}\;\;\;\;\dfrac{35}{7}=5$$
$$\dfrac{-35}{0}$$
• **Note:** The remainder is zero.

You Try It 3 Write $\dfrac{36}{4}$ as a whole number.

Your solution 9

Example 4 Write $12\dfrac{5}{8}$ as an improper fraction.

Solution

$$12\dfrac{5}{8}=\dfrac{(8\cdot12)+5}{8}=\dfrac{96+5}{8}$$
$$=\dfrac{101}{8}$$

You Try It 4 Write $9\dfrac{4}{7}$ as an improper fraction.

Your solution $\dfrac{67}{7}$

Example 5 Write 9 as an improper fraction.

Solution $9=\dfrac{9}{1}$

You Try It 5 Write 3 as an improper fraction.

Your solution $\dfrac{3}{1}$

Solutions on p. S4

Objective 2.2B

Vocabulary to Review

common factors [2.1B]

New Vocabulary

equivalent fractions
simplest form of a fraction

Discuss the Concepts

1. Explain the procedure for finding equivalent fractions.
2. Explain the procedure for simplifying fractions.

Concept Check

Name the equivalent fractions in the list below. Then name the fractions that are written in simplest form.

$$\frac{6}{8}, \frac{12}{16}, \frac{9}{16}, \frac{3}{4}, \frac{18}{24}$$

The equivalent fractions are $\frac{6}{8}, \frac{12}{16}, \frac{3}{4}$, and $\frac{18}{24}$. The fractions $\frac{3}{4}$ and $\frac{9}{16}$ are in simplest form.

Instructor Note

You may prefer to explain that a fraction can be simplified by dividing the numerator and denominator by the GCF of the numerator and denominator.

Instructor Note

To help some students understand equivalent fractions, use a pizza. By cutting the pizza into, say, eight pieces, students are able to see that

$$\frac{1}{2} = \frac{4}{8} \quad \text{and} \quad \frac{1}{4} = \frac{2}{8}$$

Objective B **To write equivalent fractions**

Point of Interest

Leonardo of Pisa, who was also called Fibonacci (c. 1175–1250), is credited with bringing the Hindu–Arabic number system to the Western world and promoting its use instead of the cumbersome Roman numeral system. He was also influential in promoting the idea of the fraction bar. His notation, however, was very different from what we use today. For instance, he wrote

$$\frac{3}{4} \frac{5}{7} \text{ to mean } \frac{5}{7} + \frac{3}{7 \cdot 4}.$$

Fractions can be graphed as points on a number line. The number lines at the right show thirds, sixths, and ninths graphed from 0 to 1.

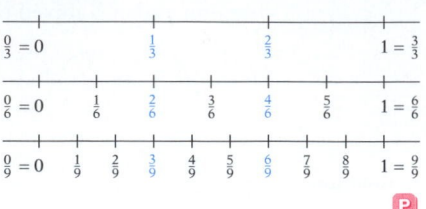

A particular point on the number line may be represented by different fractions, all of which are equal.

For example, $\frac{0}{3} = \frac{0}{6} = \frac{0}{9}, \frac{1}{3} = \frac{2}{6} = \frac{3}{9}, \frac{2}{3} = \frac{4}{6} = \frac{6}{9}$, and $\frac{3}{3} = \frac{6}{6} = \frac{9}{9}$.

Equal fractions with different denominators are called **equivalent fractions.** $\frac{1}{3}, \frac{2}{6}$, and $\frac{3}{9}$ are equivalent fractions. $\frac{2}{3}, \frac{4}{6}$, and $\frac{6}{9}$ are equivalent fractions.

Note that we can rewrite $\frac{2}{3}$ as $\frac{4}{6}$ by multiplying both the numerator and denominator of $\frac{2}{3}$ by 2.

$$\frac{2}{3} = \frac{2 \cdot 2}{3 \cdot 2} = \frac{4}{6}$$

Also, we can rewrite $\frac{4}{6}$ as $\frac{2}{3}$ by dividing both the numerator and denominator of $\frac{4}{6}$ by 2.

$$\frac{4}{6} = \frac{4 \div 2}{6 \div 2} = \frac{2}{3}$$

This suggests the following property of fractions.

> **Equivalent Fractions**
>
> The numerator and denominator of a fraction can be multiplied or divided by the same nonzero number. The resulting fraction is equivalent to the original fraction.
>
> $$\frac{a}{b} = \frac{a \cdot c}{b \cdot c}, \quad \frac{a}{b} = \frac{a \div c}{b \div c}, \quad \text{where } b \neq 0 \text{ and } c \neq 0$$

HOW TO Write an equivalent fraction with the given denominator.

$$\frac{3}{8} = \frac{}{40}$$

$40 \div 8 = 5$ • Divide the larger denominator by the smaller one.

$$\frac{3}{8} = \frac{3 \cdot 5}{8 \cdot 5} = \frac{15}{40}$$ • Multiply the numerator and denominator of the given fraction by the quotient (5).

A fraction is in **simplest form** when the numerator and denominator have no common factors other than 1. The fraction $\frac{3}{8}$ is in simplest form because 3 and 8 have no common factors other than 1. The fraction $\frac{15}{50}$ is not in simplest form because the numerator and denominator have a common factor of 5.

In-Class Examples (Objective 2.2B)

1. Write a fraction that is equivalent to $\frac{5}{9}$ and has a denominator of 27. $\frac{15}{27}$

2. Write a fraction that is equivalent to 6 and has a denominator of 5. $\frac{30}{5}$

3. Write $\frac{14}{21}$ in simplest form. $\frac{2}{3}$

4. Write $\frac{24}{16}$ in simplest form. $\frac{3}{2}$

5. Write $\frac{8d}{36}$ in simplest form. $\frac{2d}{9}$

To write a fraction in simplest form, divide the numerator and denominator of the fraction by their common factors.

> **HOW TO** Write $\frac{12}{15}$ in simplest form.
>
> $$\frac{12}{15} = \frac{12 \div 3}{15 \div 3} = \frac{4}{5}$$
>
> • **12 and 15 have a common factor of 3. Divide the numerator and denominator by 3.**

Simplifying a fraction requires that you recognize the common factors of the numerator and denominator. One way to do this is to write the prime factorizations of the numerator and denominator and then divide by the common prime factors.

> **HOW TO** Write $\frac{30}{42}$ in simplest form.
>
> $$\frac{30}{42} = \frac{\overset{1}{\cancel{2}} \cdot \overset{1}{\cancel{3}} \cdot 5}{\underset{1}{\cancel{2}} \cdot \underset{1}{\cancel{3}} \cdot 7} = \frac{5}{7}$$
>
> • **Write the prime factorization of the numerator and denominator. Divide by the common factors.**

> **HOW TO** Write $\frac{2x}{6}$ in simplest form.
>
> $$\frac{2x}{6} = \frac{\overset{1}{\cancel{2}} \cdot x}{\underset{1}{\cancel{2}} \cdot 3} = \frac{x}{3}$$
>
> • **Factor the numerator and denominator. Then divide by the common factors.**

Example 6 Write an equivalent fraction with the given denominator.

$$\frac{2}{5} = \frac{\ }{30}.$$

Solution $30 \div 5 = 6$

$$\frac{2}{5} = \frac{2 \cdot 6}{5 \cdot 6} = \frac{12}{30}$$

$\frac{12}{30}$ is equivalent to $\frac{2}{5}$.

You Try It 6 Write an equivalent fraction with the given denominator.

$$\frac{5}{8} = \frac{\ }{48}.$$

Your solution $\frac{30}{48}$

Example 7 Write an equivalent fraction with the given denominator.

$$3 = \frac{\ }{15}.$$

Solution $3 = \frac{3}{1}$ $15 \div 1 = 15$

$$3 = \frac{3}{1} = \frac{3 \cdot 15}{1 \cdot 15} = \frac{45}{15}$$

$\frac{45}{15}$ is equivalent to 3.

You Try It 7 Write an equivalent fraction with the given denominator.

$$8 = \frac{\ }{12}.$$

Your solution $\frac{96}{12}$

Solutions on p. S4

Optional Student Activity

1. Divide two-thirds of a circle into sixths and write the equivalent fraction.

$$\frac{4}{6}$$

2. Divide one-third of a circle into ninths and write the equivalent fraction.

$$\frac{3}{9}$$

3. Divide three-fourths of a circle into eighths and write the equivalent fraction.

$$\frac{6}{8}$$

4. Divide one-fourth of a circle into twelfths and write the equivalent fraction.

$$\frac{3}{12}$$

Instructor Note

As mentioned earlier, one of the main pedagogical features of this text is the paired examples. Using the model of the Example, students should work the You Try It. A *complete* solution can be found in the Appendix, so that students can check not only the answer but also their work.

Example 8   Write $\frac{18}{54}$ in simplest form.

Solution $\frac{18}{54} = \dfrac{\overset{1}{\cancel{2}} \cdot \overset{1}{\cancel{3}} \cdot \overset{1}{\cancel{3}}}{\underset{1}{\cancel{2}} \cdot \underset{1}{\cancel{3}} \cdot \underset{1}{\cancel{3}} \cdot 3} = \frac{1}{3}$

You Try It 8   Write $\frac{21}{84}$ in simplest form.

Your solution $\frac{1}{4}$

Example 9   Write $\frac{36}{20}$ in simplest form.

Solution $\frac{36}{20} = \dfrac{\overset{1}{\cancel{2}} \cdot \overset{1}{\cancel{2}} \cdot 3 \cdot 3}{\underset{1}{\cancel{2}} \cdot \underset{1}{\cancel{2}} \cdot 5} = \frac{9}{5}$

You Try It 9   Write $\frac{32}{12}$ in simplest form.

Your solution $\frac{8}{3}$

Example 10   Write $\frac{10m}{12}$ in simplest form.

Solution $\frac{10m}{12} = \dfrac{\overset{1}{\cancel{2}} \cdot 5 \cdot m}{\underset{1}{\cancel{2}} \cdot 2 \cdot 3} = \frac{5m}{6}$

You Try It 10   Write $\frac{11t}{11}$ in simplest form.

Your solution t

Solutions on p. S4

Objective 2.2C

New Vocabulary
least common denominator (LCD)

Discuss the Concepts
1. If two fractions have the same denominator, how can you determine which fraction is larger?
2. If two fractions have different denominators, how can you determine which fraction is larger?

Objective C

To identify the order relation between two fractions

The number line can be used to determine the order relation between two fractions.

A fraction that appears to the left of a given fraction on the number line is less than the given fraction.

$\frac{3}{8}$ is to the left of $\frac{5}{8}$.

$\frac{3}{8} < \frac{5}{8}$

A fraction that appears to the right of a given fraction on the number line is greater than the given fraction.

$\frac{7}{8}$ is to the right of $\frac{3}{8}$.

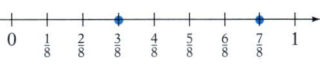

$\frac{7}{8} > \frac{3}{8}$

To find the order relation between two fractions with the *same* denominator, compare the numerators. The fraction with the smaller numerator is the smaller fraction. The larger fraction is the fraction with the larger numerator.

$\frac{3}{8}$ and $\frac{5}{8}$ have the same denominator. $\frac{3}{8} < \frac{5}{8}$ because $3 < 5$.

$\frac{7}{8}$ and $\frac{3}{8}$ have the same denominator. $\frac{7}{8} > \frac{3}{8}$ because $7 > 3$.

In-Class Examples (Objective 2.2C)

1. Place the correct symbol, $<$ or $>$, between the two numbers. $\frac{5}{8}$ $\frac{2}{3}$ $\frac{5}{8} < \frac{2}{3}$

2. Place the correct symbol, $<$ or $>$, between the two numbers. $\frac{5}{16}$ $\frac{3}{10}$ $\frac{5}{16} > \frac{3}{10}$

To compare two fractions with *different* denominators, rewrite the fractions with a common denominator. The common denominator is the least common multiple (LCM) of the denominators of the fractions. The LCM of the denominators is sometimes called the **least common denominator** or **LCD**.

> **HOW TO** Find the order relation between $\frac{5}{12}$ and $\frac{7}{18}$.
>
> The LCM of 12 and 18 is 36.   • **Find the LCM of the denominators.**
>
> $\frac{5}{12} = \frac{5 \cdot 3}{12 \cdot 3} = \frac{15}{36}$ ←——— Larger numerator   • **Write each fraction as an equivalent fraction with the LCM as the denominator.**
>
> $\frac{7}{18} = \frac{7 \cdot 2}{18 \cdot 2} = \frac{14}{36}$ ←——— Smaller numerator
>
> $\frac{15}{36} > \frac{14}{36}$ • **Compare the fractions.**
>
> $\frac{5}{12} > \frac{7}{18}$

Example 11 Place the correct symbol, < or >, between the two numbers.

$\frac{2}{3} \quad \frac{4}{7}$

Solution The LCM of 3 and 7 is 21.

$\frac{2}{3} = \frac{14}{21} \qquad \frac{4}{7} = \frac{12}{21}$

$\frac{14}{21} > \frac{12}{21}$

$\frac{2}{3} > \frac{4}{7}$

You Try It 11 Place the correct symbol, < or >, between the two numbers.

$\frac{4}{9} \quad \frac{8}{21}$

Your solution $\frac{4}{9} > \frac{8}{21}$

Example 12 Place the correct symbol, < or >, between the two numbers.

$\frac{7}{12} \quad \frac{11}{18}$

Solution The LCM of 12 and 18 is 36.

$\frac{7}{12} = \frac{21}{36} \qquad \frac{11}{18} = \frac{22}{36}$

$\frac{21}{36} < \frac{22}{36}$

$\frac{7}{12} < \frac{11}{18}$

You Try It 12 Place the correct symbol, < or >, between the two numbers.

$\frac{17}{24} \quad \frac{7}{9}$

Your solution $\frac{17}{24} < \frac{7}{9}$

Solutions on p. S4

Concept Check

Put the following fractions in order from smallest to largest.

$\frac{17}{30}, \frac{7}{12}, \frac{8}{15}, \frac{17}{25}, \frac{5}{9}, \frac{13}{24}$

$\frac{8}{15}, \frac{13}{24}, \frac{5}{9}, \frac{17}{30}, \frac{7}{12}, \frac{17}{25}$

Optional Student Activity

Use a diagram to show that $\frac{2}{3}$ is greater than $\frac{5}{8}$.

Section 2.2

Suggested Assignment

Exercises 1–117, odds
More challenging problems:
   Exercises 119–125, 128

2.2 Exercises

Objective A   To write proper fractions, improper fractions, and mixed numbers

Express the shaded portion of the circle as a fraction.

1.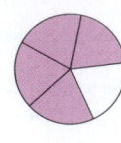

$\dfrac{4}{5}$

2.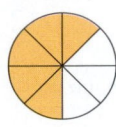

$\dfrac{5}{8}$

3.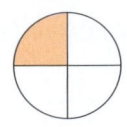

$\dfrac{1}{4}$

4.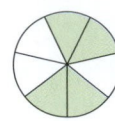

$\dfrac{4}{7}$

Express the shaded portion of the circles as an improper fraction and as a mixed number.

5.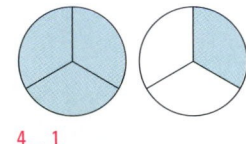

$\dfrac{4}{3}; 1\dfrac{1}{3}$

6.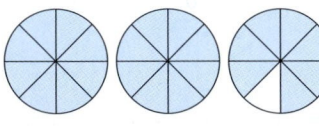

$\dfrac{23}{8}; 2\dfrac{7}{8}$

7.

$\dfrac{13}{5}; 2\dfrac{3}{5}$

8.

$\dfrac{15}{4}; 3\dfrac{3}{4}$

Write the improper fraction as a mixed number or a whole number.

9. $\dfrac{13}{4}$

$3\dfrac{1}{4}$

10. $\dfrac{14}{3}$

$4\dfrac{2}{3}$

11. $\dfrac{20}{5}$

4

12. $\dfrac{18}{6}$

3

13. $\dfrac{27}{10}$

$2\dfrac{7}{10}$

14. $\dfrac{31}{3}$

$10\dfrac{1}{3}$

15. $\dfrac{56}{8}$

7

16. $\dfrac{27}{9}$

3

17. $\dfrac{17}{9}$

$1\dfrac{8}{9}$

18. $\dfrac{8}{3}$

$2\dfrac{2}{3}$

19. $\dfrac{12}{5}$

$2\dfrac{2}{5}$

20. $\dfrac{19}{8}$

$2\dfrac{3}{8}$

21. $\dfrac{18}{1}$

18

22. $\dfrac{21}{1}$

21

23. $\dfrac{32}{15}$

$2\dfrac{2}{15}$

24. $\dfrac{39}{14}$

$2\dfrac{11}{14}$

25. $\dfrac{8}{8}$

1

26. $\dfrac{12}{12}$

1

27. $\dfrac{28}{3}$

$9\dfrac{1}{3}$

28. $\dfrac{43}{5}$

$8\dfrac{3}{5}$

Quick Quiz (Objective 2.2A)

1. Express the shaded portion of the circles as an improper fraction and as a mixed number.

$\dfrac{11}{6}; 1\dfrac{5}{6}$

2. Write $\dfrac{10}{3}$ as a mixed number. $3\dfrac{1}{3}$

3. Write $\dfrac{81}{9}$ as a whole number.   9

4. Write $4\dfrac{5}{9}$ as an improper fraction. $\dfrac{41}{9}$

5. Write 18 as an improper fraction. $\dfrac{18}{1}$

Write the mixed number or whole number as an improper fraction.

29. $2\frac{1}{4}$ **30.** $4\frac{2}{5}$ **31.** $5\frac{1}{2}$ **32.** $3\frac{2}{3}$ **33.** $2\frac{4}{5}$

$\frac{9}{4}$ $\frac{22}{5}$ $\frac{11}{2}$ $\frac{11}{3}$ $\frac{14}{5}$

34. $6\frac{3}{8}$ **35.** $7\frac{5}{6}$ **36.** $9\frac{1}{5}$ **37.** 7 **38.** 4

$\frac{51}{8}$ $\frac{47}{6}$ $\frac{46}{5}$ $\frac{7}{1}$ $\frac{4}{1}$

39. $8\frac{1}{4}$ **40.** $1\frac{7}{9}$ **41.** $10\frac{1}{3}$ **42.** $6\frac{3}{7}$ **43.** $4\frac{7}{12}$

$\frac{33}{4}$ $\frac{16}{9}$ $\frac{31}{3}$ $\frac{45}{7}$ $\frac{55}{12}$

44. $5\frac{4}{9}$ **45.** 8 **46.** 6 **47.** $12\frac{4}{5}$ **48.** $11\frac{5}{8}$

$\frac{49}{9}$ $\frac{8}{1}$ $\frac{6}{1}$ $\frac{64}{5}$ $\frac{93}{8}$

Objective B   **To write equivalent fractions**

Write an equivalent fraction with the given denominator.

49. $\frac{1}{2} = \frac{}{12}$ **50.** $\frac{1}{4} = \frac{}{20}$ **51.** $\frac{3}{8} = \frac{}{24}$ **52.** $\frac{9}{11} = \frac{}{44}$ **53.** $\frac{2}{17} = \frac{}{51}$

6 5 9 36 6

54. $\frac{9}{10} = \frac{}{80}$ **55.** $\frac{3}{4} = \frac{}{32}$ **56.** $\frac{5}{8} = \frac{}{32}$ **57.** $6 = \frac{}{18}$ **58.** $5 = \frac{}{35}$

72 24 20 108 175

59. $\frac{1}{3} = \frac{}{90}$ **60.** $\frac{3}{16} = \frac{}{48}$ **61.** $\frac{2}{3} = \frac{}{21}$ **62.** $\frac{4}{9} = \frac{}{36}$ **63.** $\frac{6}{7} = \frac{}{49}$

30 9 14 16 42

64. $\frac{7}{8} = \frac{}{40}$ **65.** $\frac{4}{9} = \frac{}{18}$ **66.** $\frac{11}{12} = \frac{}{48}$ **67.** $7 = \frac{}{4}$ **68.** $9 = \frac{}{6}$

35 8 44 28 54

Write the fraction in simplest form.

69. $\frac{3}{12}$ **70.** $\frac{10}{22}$ **71.** $\frac{33}{44}$ **72.** $\frac{6}{14}$ **73.** $\frac{4}{24}$

$\frac{1}{4}$ $\frac{5}{11}$ $\frac{3}{4}$ $\frac{3}{7}$ $\frac{1}{6}$

Quick Quiz (Objective 2.2B)

1. Write a fraction that is equivalent to $\frac{1}{2}$ and has a denominator of 32. $\frac{16}{32}$

2. Write a fraction that is equivalent to 8 and has a denominator of 11. $\frac{88}{11}$

3. Write $\frac{45}{81}$ in simplest form. $\frac{5}{9}$

4. Write $\frac{10b}{25}$ in simplest form. $\frac{2b}{5}$

74. $\dfrac{25}{75}$ **75.** $\dfrac{8}{33}$ **76.** $\dfrac{9}{25}$ **77.** $\dfrac{0}{8}$ **78.** $\dfrac{0}{11}$

$\dfrac{1}{3}$ $\dfrac{8}{33}$ $\dfrac{9}{25}$ 0 0

79. $\dfrac{42}{36}$ **80.** $\dfrac{30}{18}$ **81.** $\dfrac{16}{16}$ **82.** $\dfrac{24}{24}$ **83.** $\dfrac{21}{35}$

$\dfrac{7}{6}$ $\dfrac{5}{3}$ 1 1 $\dfrac{3}{5}$

84. $\dfrac{11}{55}$ **85.** $\dfrac{16}{60}$ **86.** $\dfrac{8}{84}$ **87.** $\dfrac{12}{20}$ **88.** $\dfrac{24}{36}$

$\dfrac{1}{5}$ $\dfrac{4}{15}$ $\dfrac{2}{21}$ $\dfrac{3}{5}$ $\dfrac{2}{3}$

89. $\dfrac{12m}{18}$ **90.** $\dfrac{20x}{25}$ **91.** $\dfrac{4y}{8}$ **92.** $\dfrac{14z}{28}$ **93.** $\dfrac{24a}{36}$

$\dfrac{2m}{3}$ $\dfrac{4x}{5}$ $\dfrac{y}{2}$ $\dfrac{z}{2}$ $\dfrac{2a}{3}$

94. $\dfrac{28z}{21}$ **95.** $\dfrac{8c}{8}$ **96.** $\dfrac{9w}{9}$ **97.** $\dfrac{18k}{3}$ **98.** $\dfrac{24t}{4}$

$\dfrac{4z}{3}$ c w $6k$ $6t$

Objective C   **To identify the order relation between two fractions**

Place the correct symbol, $<$ or $>$, between the two numbers.

99. $\dfrac{3}{8}$ $\dfrac{2}{5}$ **100.** $\dfrac{5}{7}$ $\dfrac{2}{3}$ **101.** $\dfrac{3}{4}$ $\dfrac{7}{9}$ **102.** $\dfrac{7}{12}$ $\dfrac{5}{8}$

$<$ $>$ $<$ $<$

103. $\dfrac{2}{3}$ $\dfrac{7}{11}$ **104.** $\dfrac{11}{14}$ $\dfrac{3}{4}$ **105.** $\dfrac{17}{24}$ $\dfrac{11}{16}$ **106.** $\dfrac{11}{12}$ $\dfrac{7}{9}$

$>$ $>$ $>$ $>$

107. $\dfrac{7}{15}$ $\dfrac{5}{12}$ **108.** $\dfrac{5}{8}$ $\dfrac{4}{7}$ **109.** $\dfrac{5}{9}$ $\dfrac{11}{21}$ **110.** $\dfrac{11}{30}$ $\dfrac{7}{24}$

$>$ $>$ $>$ $>$

111. $\dfrac{7}{12}$ $\dfrac{13}{18}$ **112.** $\dfrac{9}{11}$ $\dfrac{7}{8}$ **113.** $\dfrac{4}{5}$ $\dfrac{7}{9}$ **114.** $\dfrac{3}{4}$ $\dfrac{11}{13}$

$<$ $<$ $>$ $<$

115. $\dfrac{9}{16}$ $\dfrac{5}{9}$ **116.** $\dfrac{2}{3}$ $\dfrac{7}{10}$ **117.** $\dfrac{5}{8}$ $\dfrac{13}{20}$ **118.** $\dfrac{3}{10}$ $\dfrac{7}{25}$

$>$ $<$ $<$ $>$

Quick Quiz (Objective 2.2C)

1. Place the correct symbol, $<$ or $>$, between the two numbers. $\dfrac{1}{3}$ $\dfrac{5}{16}$ $\dfrac{1}{3} > \dfrac{5}{16}$

2. Place the correct symbol, $<$ or $>$, between the two numbers. $\dfrac{7}{9}$ $\dfrac{5}{6}$ $\dfrac{7}{9} < \dfrac{5}{6}$

APPLYING THE CONCEPTS

119. **Weight**   A ton is equal to 2000 lb. What fractional part of a ton is 250 lb?

   $\frac{1}{8}$

120. **Weight**   A pound is equal to 16 oz. What fractional part of a pound is 6 oz?

   $\frac{3}{8}$

121. **Time**   If a history class lasts 50 min, what fractional part of an hour is the history class?

   $\frac{5}{6}$

122. **Time**   If you sleep for 8 h one night, what fractional part of one day did you spend sleeping?

   $\frac{1}{3}$

123. **Jewelry**   Gold is designated by karats. Pure gold is 24 karats. What fractional part of an 18-karat gold bracelet is pure gold?

   $\frac{3}{4}$

The Food Industry   The table at the right shows the results of a survey that asked fast-food patrons their criteria for choosing where to go for fast food. Three out of every 25 people surveyed said that the speed of the service was most important. Use this table for Exercises 124 and 125.

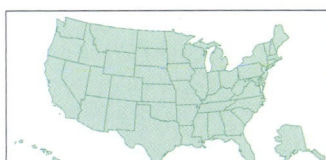

Fast-Food Patrons' Top Criteria for Fast-Food Restaurants	
Food Quality	$\frac{1}{4}$
Location	$\frac{13}{50}$
Menu	$\frac{4}{25}$
Price	$\frac{2}{25}$
Speed	$\frac{3}{25}$
Other	$\frac{3}{100}$

Source: Maritz Marketing Research, Inc.

124. According to the survey, do more people choose a fast-food restaurant on the basis of its location or on the basis of the quality of its food?
   Location

125. Which criterion was cited by most people?
   Location

126. **Card Games**   A standard deck of playing cards consists of 52 cards.
   a. What fractional part of a standard deck of cards is spades?
   b. What fractional part of a standard deck of cards is aces?

   a. $\frac{1}{4}$ **b.** $\frac{1}{13}$

127. **Geography**   What fraction of the states in the United States begin with the letter A?

   $\frac{2}{25}$

128. Is the expression $x < \frac{4}{9}$ true when $x = \frac{3}{8}$? Is it true when $x = \frac{5}{12}$?

   Yes; yes

Objective 2.3A

Vocabulary to Review

numerator   [2.2A]
denominator   [2.2A]
least common multiple (LCM)
    [2.1A]
least common denominator
    (LCD)   [2.2C]

New Procedures

Addition of fractions:

$$\frac{a}{b} + \frac{c}{b} = \frac{a + c}{b}$$

Instructor Note

We have chosen to present addition and subtraction of fractions prior to multiplication and division of fractions. If you prefer to present multiplication and division first, simply present Section 2.4 prior to Section 2.3.

Concept Check

Which of the following fractions, when added together, have a sum of 2?

$$\frac{1}{9}, \frac{2}{9}, \frac{4}{9}, \frac{5}{9}, \frac{7}{9}$$

$$\frac{2}{9}, \frac{4}{9}, \frac{5}{9}, \text{ and } \frac{7}{9}$$

Instructor Note

Ask your students what value x cannot be in the expression $\frac{4}{x} + \frac{8}{x}$.

2.3 Addition and Subtraction of Fractions

Objective A   **To add fractions**

Study Tip

Before the class meeting in which your professor begins a new section, you should read each objective statement for that section. Next, browse through the objective material. The purpose of browsing through the material is so that your brain will be prepared to accept and organize the new information when it is presented to you. See *AIM for Success*, page xxxi.

Suppose you and a friend order a pizza. The pizza has been cut into 8 equal pieces. If you eat 3 pieces of the pizza and your friend eats 2 pieces, then together you have eaten $\frac{5}{8}$ of the pizza.

Note that in adding the fractions $\frac{3}{8}$ and $\frac{2}{8}$, the numerators are added and the denominator remains the same.

$$\frac{3}{8} + \frac{2}{8} = \frac{3 + 2}{8}$$
$$= \frac{5}{8}$$

Addition of Fractions

To add fractions with the same denominator, add the numerators and place the sum over the common denominator.

$$\frac{a}{b} + \frac{c}{b} = \frac{a + c}{b}, \text{ where } b \neq 0$$

HOW TO   Add: $\frac{5}{16} + \frac{7}{16}$

$$\frac{5}{16} + \frac{7}{16} = \frac{5 + 7}{16}$$

- The denominators are the same. Add the numerators and place the sum over the common denominator.

$$= \frac{12}{16} = \frac{3}{4}$$

- Write the answer in simplest form.

HOW TO   Add: $\frac{4}{x} + \frac{8}{x}$

$$\frac{4}{x} + \frac{8}{x} = \frac{4 + 8}{x}$$

- The denominators are the same. Add the numerators and place the sum over the common denominator.

$$= \frac{12}{x}$$

Before two fractions can be added, the fractions must have the same denominator. To add fractions with different denominators, first rewrite the fractions as equivalent fractions with a common denominator. The common denominator is the least common multiple (LCM) of the denominators of the fractions. Recall that the LCM of denominators is sometimes called the least common denominator (LCD).

In-Class Examples (Objective 2.3A)

1. Add: $\frac{7}{12} + \frac{8}{15}$ $1\frac{7}{60}$

2. Add: $\frac{2}{3} + \frac{1}{8} + \frac{7}{12}$ $1\frac{3}{8}$

3. Find the total of 9 and $3\frac{7}{8}$. $12\frac{7}{8}$

4. Evaluate $x + y + z$ when $x = 3\frac{4}{15}$, $y = 2\frac{3}{5}$, and $z = 4\frac{7}{10}$. $10\frac{17}{30}$

Integrating Technology

Some scientific calculators have a fraction key, a^b/c. It is used to perform operations on fractions. To use this key to simplify the expression at the right, enter

5 a^b/c 6 + 3 a^b/c 8 =

$\dfrac{5}{6}$ $\dfrac{3}{8}$

HOW TO   Find the sum of $\dfrac{5}{6}$ and $\dfrac{3}{8}$.

The LCM of 6 and 8 is 24.   • The common denominator is the LCM of 6 and 8.

$$\dfrac{5}{6} + \dfrac{3}{8} = \dfrac{20}{24} + \dfrac{9}{24}$$

• Write the fractions as equivalent fractions with the common denominator.

$$= \dfrac{20 + 9}{24}$$

• Add the fractions.

$$= \dfrac{29}{24} = 1\dfrac{5}{24}$$

HOW TO   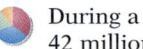   During a recent year, over 42 million Americans changed homes. Figure 2.1 shows what fractions of the people moved within the same county, moved to a different county in the same state, and moved to a different state. What fractional part of those who changed homes moved outside the county they had been living in?

Add the fraction of the people who moved to a different county in the same state and the fraction who moved to a different state.

$$\dfrac{4}{21} + \dfrac{1}{7} = \dfrac{4}{21} + \dfrac{3}{21} = \dfrac{7}{21} = \dfrac{1}{3}$$

$\dfrac{1}{3}$ of the Americans who changed homes moved outside of the county they had been living in.

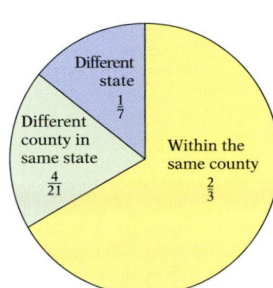

Figure 2.1   Where Americans Moved
Source: Census Bureau; *Geographical Mobility*

The mixed number $2\dfrac{1}{2}$ is the sum of 2 and $\dfrac{1}{2}$.

$$2\dfrac{1}{2} = 2 + \dfrac{1}{2}$$

Therefore, the sum of a whole number and a fraction is a mixed number.

$$2 + \dfrac{1}{2} = 2\dfrac{1}{2}$$

$$3 + \dfrac{4}{5} = 3\dfrac{4}{5}$$

$$8 + \dfrac{7}{9} = 8\dfrac{7}{9}$$

TAKE NOTE

$$5 + 4\dfrac{2}{7} = 5 + \left(4 + \dfrac{2}{7}\right)$$

$$= (5 + 4) + \dfrac{2}{7}$$

$$= 9 + \dfrac{2}{7} = 9\dfrac{2}{7}$$

The sum of a whole number and a mixed number is a mixed number.

HOW TO   Add: $5 + 4\dfrac{2}{7}$

$$5 + 4\dfrac{2}{7} = 9\dfrac{2}{7}$$

• Add the whole numbers (5 and 4). Write the fraction.

Optional Student Activity

Without calculating, decide which is larger. Explain your reasoning.

1. $\dfrac{1}{2} + \dfrac{2}{3}$ or $\dfrac{1}{4} + \dfrac{2}{5}$

2. $\dfrac{5}{6} + \dfrac{4}{9}$ or $\dfrac{3}{8} + \dfrac{3}{20}$

1. $\dfrac{1}{2} + \dfrac{2}{3}$

2. $\dfrac{5}{6} + \dfrac{4}{9}$

Explanations will vary. For example, for Exercise 1,

$\dfrac{1}{2} > \dfrac{1}{4}$ and $\dfrac{2}{3} > \dfrac{2}{5}$

so the first expression must be greater than the second.

Optional Student Activity

In a magic square, the sums across, down, and diagonally are the same. Determine whether these squares are magic squares.   **Yes**

1	$\dfrac{3}{8}$	$\dfrac{1}{2}$
$\dfrac{1}{8}$	$\dfrac{5}{8}$	$1\dfrac{1}{8}$
$\dfrac{3}{4}$	$\dfrac{7}{8}$	$\dfrac{1}{4}$

2	$\dfrac{3}{4}$	1
$\dfrac{1}{4}$	$1\dfrac{1}{4}$	$2\dfrac{1}{4}$
$1\dfrac{1}{2}$	$1\dfrac{3}{4}$	$\dfrac{1}{2}$

To add two mixed numbers, first write the fractional parts as equivalent fractions with a common denominator. Then add the fractional parts and add the whole numbers.

HOW TO   Add: $3\dfrac{5}{8} + 4\dfrac{7}{12}$

$3\dfrac{5}{8} + 4\dfrac{7}{12} = 3\dfrac{15}{24} + 4\dfrac{14}{24}$

- Write the fractions as equivalent fractions with a common denominator. The common denominator is the LCM of 8 and 12 (24).

$= 7\dfrac{29}{24}$

- Add the fractional parts and add the whole numbers.

$= 7 + \dfrac{29}{24}$

- Write the sum in simplest form.

$= 7 + 1\dfrac{5}{24}$

$= 8\dfrac{5}{24}$

HOW TO   Evaluate $x + y$ when $x = 2\dfrac{3}{4}$ and $y = 7\dfrac{5}{6}$.

$x + y$

$2\dfrac{3}{4} + 7\dfrac{5}{6}$

- Replace x with $2\dfrac{3}{4}$ and y with $7\dfrac{5}{6}$.

$= 2\dfrac{9}{12} + 7\dfrac{10}{12}$

- Write the fractions as equivalent fractions with a common denominator.

$= 9\dfrac{19}{12}$

- Add the fractional parts and add the whole numbers.

$= 10\dfrac{7}{12}$

- Write the sum in simplest form.

Example 1   Add: $\dfrac{9}{16} + \dfrac{5}{12}$

Solution $\dfrac{9}{16} + \dfrac{5}{12} = \dfrac{27}{48} + \dfrac{20}{48}$

$= \dfrac{27 + 20}{48} = \dfrac{47}{48}$

You Try It 1   Add: $\dfrac{7}{12} + \dfrac{3}{8}$

Your solution $\dfrac{23}{24}$

Solution on p. S4

Example 2

Add: $\dfrac{4}{5} + \dfrac{3}{4} + \dfrac{5}{8}$

Solution

$\dfrac{4}{5} + \dfrac{3}{4} + \dfrac{5}{8} = \dfrac{32}{40} + \dfrac{30}{40} + \dfrac{25}{40} = \dfrac{87}{40} = 2\dfrac{7}{40}$

You Try It 2

Add: $\dfrac{3}{5} + \dfrac{2}{3} + \dfrac{5}{6}$

Your solution

$2\dfrac{1}{10}$

Example 3

Find the sum of $12\dfrac{4}{7}$ and 19.

Solution

$12\dfrac{4}{7} + 19 = 31\dfrac{4}{7}$

You Try It 3

What is the sum of 16 and $8\dfrac{5}{9}$?

Your solution

$24\dfrac{5}{9}$

Example 4

Is $\dfrac{2}{3}$ a solution of $\dfrac{1}{4} + y = \dfrac{11}{12}$?

Solution

$$\dfrac{1}{4} + y = \dfrac{11}{12}$$

$$\begin{array}{c|c} \dfrac{1}{4} + \dfrac{2}{3} & \dfrac{11}{12} \\ \hline \dfrac{3}{12} + \dfrac{8}{12} & \dfrac{11}{12} \\ \dfrac{11}{12} & = \dfrac{11}{12} \end{array}$$

Yes, $\dfrac{2}{3}$ is a solution of $\dfrac{1}{4} + y = \dfrac{11}{12}$.

You Try It 4

Is $\dfrac{3}{8}$ a solution of $\dfrac{2}{3} + z = \dfrac{23}{24}$?

Your solution

No

Example 5

Evaluate $x + y + z$ when $x = 2\dfrac{1}{6}$, $y = 4\dfrac{3}{8}$, and $z = 7\dfrac{5}{9}$.

Solution

$x + y + z$

$2\dfrac{1}{6} + 4\dfrac{3}{8} + 7\dfrac{5}{9} = 2\dfrac{12}{72} + 4\dfrac{27}{72} + 7\dfrac{40}{72}$

$= 13\dfrac{79}{72} = 14\dfrac{7}{72}$

You Try It 5

Evaluate $x + y + z$ when $x = 3\dfrac{5}{6}$, $y = 2\dfrac{1}{9}$, and $z = 5\dfrac{5}{12}$.

Your solution

$11\dfrac{13}{36}$

Solutions on pp. S4–S5

Objective 2.3B

New Procedures

Subtraction of fractions:

$$\frac{a}{b} - \frac{c}{b} = \frac{a-c}{b}$$

Concept Check

Which of the following fractions, when subtracted, have a difference of $\frac{2}{5}$?

$$\frac{4}{5}, \frac{3}{5}, \frac{2}{5}, \frac{1}{5}$$

$\frac{4}{5}$ and $\frac{2}{5}$, or $\frac{3}{5}$ and $\frac{1}{5}$

Discuss the Concepts

1. Explain the procedure for subtracting two fractions with the same denominator.
2. What is the difference between the procedure for adding two fractions with the same denominator and the procedure for subtracting two fractions with the same denominator?
3. In subtraction of mixed numbers, when is borrowing necessary?

Concept Check

Which expression is larger?

1. $\frac{11}{12} - \frac{2}{3}$ or $\frac{2}{3} - \frac{1}{6}$

$\frac{2}{3} - \frac{1}{6}$

2. $\frac{3}{5} - \frac{1}{3}$ or $\frac{9}{10} - \frac{7}{8}$

$\frac{3}{5} - \frac{1}{3}$

Objective B   To subtract fractions

In the last objective, it was stated that in order for fractions to be added, the fractions must have the same denominator. The same is true for subtracting fractions: The two fractions must have the same denominator.

Point of Interest

The first woman mathematician for whom documented evidence exists is Hypatia (370–415). She lived in Alexandria, Egypt, and lectured at the Museum, the forerunner of our modern university. She made important contributions in mathematics, astronomy, and philosophy.

> **Subtraction of Fractions**
>
> To subtract fractions with the same denominator, subtract the numerators and place the difference over the common denominator.
>
> $$\frac{a}{b} - \frac{c}{b} = \frac{a-c}{b}, \qquad \text{where} \quad b \neq 0$$

HOW TO   Subtract: $\frac{5}{8} - \frac{3}{8}$

$$\frac{5}{8} - \frac{3}{8} = \frac{5-3}{8}$$

- The denominators are the same. Subtract the numerators and place the difference over the common denominator.

$$= \frac{2}{8} = \frac{1}{4}$$

- Write the answer in simplest form.

To subtract fractions with different denominators, first rewrite the fractions as equivalent fractions with a common denominator. The common denominator is the least common multiple (LCM) of the denominators of the fractions.

HOW TO   Subtract: $\frac{5}{12} - \frac{3}{8}$

The LCM of 12 and 8 is 24.

- The common denominator is the LCM of 12 and 8.

$$\frac{5}{12} - \frac{3}{8} = \frac{10}{24} - \frac{9}{24}$$

- Write the fractions as equivalent fractions with the common denominator.

$$= \frac{10-9}{24} = \frac{1}{24}$$

- Subtract the fractions.

To subtract mixed numbers when borrowing is not necessary, subtract the fractional parts and then subtract the whole numbers.

HOW TO   Find the difference between $5\frac{8}{9}$ and $2\frac{5}{6}$.

The LCM of 9 and 6 is 18.

$$5\frac{8}{9} - 2\frac{5}{6} = 5\frac{16}{18} - 2\frac{15}{18}$$

- Write the fractions as equivalent fractions with the LCM as the common denominator.

$$= 3\frac{1}{18}$$

- Subtract the fractional parts and subtract the whole numbers.

As in subtraction with whole numbers, subtraction of mixed numbers may involve borrowing.

In-Class Examples (Objective 2.3B)

1. Subtract: $\frac{3}{4} - \frac{3}{8}$ $\frac{3}{8}$

2. Subtract: $8 - 4\frac{3}{7}$ $3\frac{4}{7}$

3. What is $4\frac{1}{2}$ minus $1\frac{2}{5}$? $3\frac{1}{10}$

4. Is $\frac{7}{10}$ a solution of the equation $n - \frac{3}{5} = \frac{1}{10}$?

Yes

HOW TO   Subtract: $7 - 4\frac{2}{3}$

$7 - 4\frac{2}{3} = 6\frac{3}{3} - 4\frac{2}{3}$

- Borrow 1 from 7. Write the 1 as a fraction with the same denominator as the fractional part of the mixed number (3).

    Note: $7 = 6 + 1 = 6 + \frac{3}{3} = 6\frac{3}{3}$

$= 2\frac{1}{3}$

- Subtract the fractional parts and subtract the whole numbers.

HOW TO   Subtract: $9\frac{1}{8} - 2\frac{5}{6}$

$9\frac{1}{8} - 2\frac{5}{6} = 9\frac{3}{24} - 2\frac{20}{24}$

- Write the fractions as equivalent fractions with a common denominator.

$= 8\frac{27}{24} - 2\frac{20}{24}$

- $3 < 20$. Borrow 1 from 9. Add the 1 to $\frac{3}{24}$.

    Note: $9\frac{3}{24} = 9 + \frac{3}{24} = 8 + 1 + \frac{3}{24}$

    $= 8 + \frac{24}{24} + \frac{3}{24} = 8 + \frac{27}{24} = 8\frac{27}{24}$

$= 6\frac{7}{24}$

- Subtract.

HOW TO   Evaluate $x - y$ when $x = 7\frac{2}{9}$ and $y = 3\frac{5}{12}$.

$x - y$

$7\frac{2}{9} - 3\frac{5}{12}$

- Replace x with $7\frac{2}{9}$ and y with $3\frac{5}{12}$.

$= 7\frac{8}{36} - 3\frac{15}{36}$

- Write the fractions as equivalent fractions with a common denominator.

$= 6\frac{44}{36} - 3\frac{15}{36}$

- $8 < 15$. Borrow 1 from 7. Add the 1 to $\frac{8}{36}$.

    Note: $7\frac{8}{36} = 6 + \frac{36}{36} + \frac{8}{36} = 6\frac{44}{36}$

$= 3\frac{29}{36}$

- Subtract.

Instructor Note

Another money example that may reinforce the common denominator concept is: "Find 3 quarters minus 7 dimes." The concept of rewriting fractions as equivalent fractions with a common denominator is similar to exchanging all the coins for pennies. Three quarters equals 75 pennies, and 7 dimes equals 70 pennies.

$$\frac{3}{4} - \frac{7}{10} = \frac{75}{100} - \frac{70}{100}$$

$$= \frac{5}{100} = \frac{1}{20}$$

Optional Student Activity

Use a diagram to illustrate subtraction of two fractions with different denominators.

Example 6

Subtract: $\frac{5}{6} - \frac{3}{8}$

Solution

$\frac{5}{6} - \frac{3}{8} = \frac{20}{24} - \frac{9}{24} = \frac{11}{24}$

You Try It 6

Subtract: $\frac{5}{6} - \frac{7}{9}$

Your solution

$\frac{1}{18}$

Solution on p. S5

Example 7

Find the difference between $8\frac{5}{6}$ and $2\frac{3}{4}$.

Solution

$8\frac{5}{6} - 2\frac{3}{4} = 8\frac{10}{12} - 2\frac{9}{12} = 6\frac{1}{12}$

You Try It 7

Find the difference between $9\frac{7}{8}$ and $5\frac{2}{3}$.

Your solution

$4\frac{5}{24}$

Example 8

Subtract: $7 - 3\frac{5}{13}$

Solution

$7 - 3\frac{5}{13} = 6\frac{13}{13} - 3\frac{5}{13} = 3\frac{8}{13}$

You Try It 8

Subtract: $6 - 4\frac{2}{11}$

Your solution

$1\frac{9}{11}$

Solutions on p. S5

Objective C   **To solve application problems and use formulas**

Example 9

The length of a regulation NCAA football must be no less than $10\frac{7}{8}$ in. and no more than $11\frac{7}{16}$ in. What is the difference between the minimum and maximum lengths of an NCAA regulation football?

Strategy

To find the difference, subtract the minimum length $\left(10\frac{7}{8}\right)$ from the maximum length $\left(11\frac{7}{16}\right)$.

Solution

$11\frac{7}{16} - 10\frac{7}{8} = 11\frac{7}{16} - 10\frac{14}{16}$

$\qquad = 10\frac{23}{16} - 10\frac{14}{16} = \frac{9}{16}$

The difference is $\frac{9}{16}$ in.

You Try It 9

The Heller Research Group conducted a survey to determine favorite doughnut flavors. $\frac{2}{5}$ of the respondents named glazed doughnuts, $\frac{8}{25}$ named filled doughnuts, and $\frac{3}{20}$ named frosted doughnuts. What fraction of the respondents did not name glazed, filled, or frosted as their favorite type of doughnut?

Your Strategy

Your solution

$\frac{13}{100}$

Solution on p. S5

Objective 2.3C

Instructor Note

If students are distracted by fractions in an application problem, suggest that they reread the problem, substituting whole numbers for the fractions. This may help them determine how to solve the problem.

Optional Student Activity

The following are the average portions of each day that a person spends on each activity:

Sleeping, $\frac{1}{3}$

Working, $\frac{1}{3}$

Personal hygiene, $\frac{1}{24}$

Eating, $\frac{1}{8}$

Rest and relaxation, $\frac{1}{12}$

Do these five activities account for an entire day? Explain your answer.

No. These activities account for only 22 hours.

In-Class Examples (Objective 2.3C)

1. A carpenter built a header by nailing a $1\frac{1}{4}$-inch board to a $2\frac{5}{8}$-inch beam. Find the total thickness of the header. $3\frac{7}{8}$ in.

2. A flight from New York to Los Angeles takes $5\frac{1}{2}$ h. After the plane has been in the air for $2\frac{3}{4}$ h, how much flight time remains? $2\frac{3}{4}$ h

2.3 Exercises

Objective A   To add fractions

Suggested Assignment
Exercises 1–111, odds
More challenging problem:
  Exercise 113

Add.

1. $\dfrac{4}{11} + \dfrac{5}{11}$

$\dfrac{9}{11}$

2. $\dfrac{3}{7} + \dfrac{2}{7}$

$\dfrac{5}{7}$

3. $\dfrac{2}{3} + \dfrac{1}{3}$

1

4. $\dfrac{1}{2} + \dfrac{1}{2}$

1

5. $\dfrac{5}{6} + \dfrac{5}{6}$

$1\dfrac{2}{3}$

6. $\dfrac{3}{8} + \dfrac{7}{8}$

$1\dfrac{1}{4}$

7. $\dfrac{7}{18} + \dfrac{13}{18} + \dfrac{1}{18}$

$1\dfrac{1}{6}$

8. $\dfrac{8}{15} + \dfrac{2}{15} + \dfrac{11}{15}$

$1\dfrac{2}{5}$

9. $\dfrac{7}{b} + \dfrac{9}{b}$

$\dfrac{16}{b}$

10. $\dfrac{3}{y} + \dfrac{6}{y}$

$\dfrac{9}{y}$

11. $\dfrac{5}{c} + \dfrac{4}{c}$

$\dfrac{9}{c}$

12. $\dfrac{2}{a} + \dfrac{8}{a}$

$\dfrac{10}{a}$

13. $\dfrac{1}{x} + \dfrac{4}{x} + \dfrac{6}{x}$

$\dfrac{11}{x}$

14. $\dfrac{8}{n} + \dfrac{5}{n} + \dfrac{3}{n}$

$\dfrac{16}{n}$

15. $\dfrac{1}{4} + \dfrac{2}{3}$

$\dfrac{11}{12}$

16. $\dfrac{2}{3} + \dfrac{1}{2}$

$1\dfrac{1}{6}$

17. $\dfrac{7}{15} + \dfrac{9}{20}$

$\dfrac{11}{12}$

18. $\dfrac{4}{9} + \dfrac{1}{6}$

$\dfrac{11}{18}$

19. $\dfrac{2}{3} + \dfrac{1}{12} + \dfrac{5}{6}$

$1\dfrac{7}{12}$

20. $\dfrac{3}{8} + \dfrac{1}{2} + \dfrac{5}{12}$

$1\dfrac{7}{24}$

21. $\dfrac{7}{12} + \dfrac{3}{4} + \dfrac{4}{5}$

$2\dfrac{2}{15}$

22. $\dfrac{7}{11} + \dfrac{1}{2} + \dfrac{5}{6}$

$1\dfrac{32}{33}$

23. $8 + 7\dfrac{2}{3}$

$15\dfrac{2}{3}$

24. $6 + 9\dfrac{3}{5}$

$15\dfrac{3}{5}$

25. $2\dfrac{1}{6} + 3\dfrac{1}{2}$

$5\dfrac{2}{3}$

26. $1\dfrac{3}{10} + 4\dfrac{3}{5}$

$5\dfrac{9}{10}$

27. $8\dfrac{3}{5} + 6\dfrac{9}{20}$

$15\dfrac{1}{20}$

28. $7\dfrac{5}{12} + 3\dfrac{7}{9}$

$11\dfrac{7}{36}$

29. $5\dfrac{5}{12} + 4\dfrac{7}{9}$

$10\dfrac{7}{36}$

30. $2\dfrac{11}{12} + 3\dfrac{7}{15}$

$6\dfrac{23}{60}$

31. $2\dfrac{1}{4} + 3\dfrac{1}{2} + 1\dfrac{2}{3}$

$7\dfrac{5}{12}$

32. $1\dfrac{2}{3} + 2\dfrac{5}{6} + 4\dfrac{7}{9}$

$9\dfrac{5}{18}$

Quick Quiz (Objective 2.3A)

1. Add: $\dfrac{1}{3} + \dfrac{5}{8}$ $\dfrac{23}{24}$

2. Add: $\dfrac{3}{4} + \dfrac{1}{2} + \dfrac{5}{6}$ $2\dfrac{1}{12}$

3. Find the total of $4\dfrac{1}{2}$ and $8\dfrac{1}{5}$. $12\dfrac{7}{10}$

4. Evaluate $x + y$ when $x = 3\dfrac{4}{5}$ and $y = 9\dfrac{3}{7}$. $13\dfrac{8}{35}$

Solve.

33. Find the total of $\frac{2}{7}$, $\frac{3}{14}$, and $\frac{1}{4}$.

$\frac{3}{4}$

34. Find the total of $\frac{1}{3}$, $\frac{5}{18}$, and $\frac{2}{9}$.

$\frac{5}{6}$

35. Find $3\frac{7}{12}$ plus $2\frac{5}{8}$.

$6\frac{5}{24}$

36. Find $5\frac{4}{9}$ plus $6\frac{5}{6}$.

$12\frac{5}{18}$

37. Find $\frac{7}{8}$ increased by $1\frac{1}{3}$.

$2\frac{5}{24}$

38. Find the sum of $7\frac{11}{15}$, $2\frac{7}{10}$, and $5\frac{2}{5}$.

$15\frac{5}{6}$

Evaluate the variable expression $x + y$ for the given values of x and y.

39. $x = \frac{3}{5}, y = \frac{4}{5}$

$1\frac{2}{5}$

40. $x = \frac{5}{8}, y = \frac{3}{8}$

1

41. $x = \frac{5}{6}, y = \frac{8}{9}$

$1\frac{13}{18}$

42. $x = \frac{5}{8}, y = \frac{1}{6}$

$\frac{19}{24}$

Evaluate the variable expression $x + y + z$ for the given values of x, y, and z.

43. $x = \frac{3}{8}, y = \frac{1}{4}, z = \frac{7}{12}$

$1\frac{5}{24}$

44. $x = \frac{5}{6}, y = \frac{2}{3}, z = \frac{7}{24}$

$1\frac{19}{24}$

45. $x = 1\frac{1}{2}, y = 3\frac{3}{4}, z = 6\frac{5}{12}$

$11\frac{2}{3}$

46. $x = 7\frac{2}{3}, y = 2\frac{5}{6}, z = 5\frac{4}{9}$

$15\frac{17}{18}$

47. $x = 4\frac{3}{5}, y = 8\frac{7}{10}, z = 1\frac{9}{20}$

$14\frac{3}{4}$

48. $x = 2\frac{3}{14}, y = 5\frac{5}{7}, z = 3\frac{1}{2}$

$11\frac{3}{7}$

49. Is $\frac{3}{5}$ a solution of the equation $z + \frac{1}{4} = \frac{17}{20}$?

Yes

50. Is $\frac{3}{8}$ a solution of the equation $\frac{3}{4} = t + \frac{3}{8}$?

Yes

Objective B   To subtract fractions

Subtract.

51. $\frac{7}{12} - \frac{5}{12}$

$\frac{1}{6}$

52. $\frac{17}{20} - \frac{9}{20}$

$\frac{2}{5}$

53. $\frac{11}{24} - \frac{7}{24}$

$\frac{1}{6}$

54. $\frac{39}{48} - \frac{23}{48}$

$\frac{1}{3}$

55. $\dfrac{8}{d} - \dfrac{3}{d}$

$\dfrac{5}{d}$

56. $\dfrac{12}{y} - \dfrac{7}{y}$

$\dfrac{5}{y}$

57. $\dfrac{10}{n} - \dfrac{5}{n}$

$\dfrac{5}{n}$

58. $\dfrac{13}{c} - \dfrac{6}{c}$

$\dfrac{7}{c}$

59. $\dfrac{3}{7} - \dfrac{5}{14}$

$\dfrac{1}{14}$

60. $\dfrac{7}{8} - \dfrac{5}{16}$

$\dfrac{9}{16}$

61. $\dfrac{2}{3} - \dfrac{1}{6}$

$\dfrac{1}{2}$

62. $\dfrac{5}{21} - \dfrac{1}{6}$

$\dfrac{1}{14}$

63. $\dfrac{11}{12} - \dfrac{2}{3}$

$\dfrac{1}{4}$

64. $\dfrac{9}{20} - \dfrac{1}{30}$

$\dfrac{5}{12}$

65. $4\dfrac{11}{18} - 2\dfrac{5}{18}$

$2\dfrac{1}{3}$

66. $3\dfrac{7}{12} - 1\dfrac{1}{12}$

$2\dfrac{1}{2}$

67. $8\dfrac{3}{4} - 2$

$6\dfrac{3}{4}$

68. $6\dfrac{5}{9} - 4$

$2\dfrac{5}{9}$

69. $8\dfrac{5}{6} - 7\dfrac{3}{4}$

$1\dfrac{1}{12}$

70. $5\dfrac{7}{8} - 3\dfrac{2}{3}$

$2\dfrac{5}{24}$

71. $7 - 3\dfrac{5}{8}$

$3\dfrac{3}{8}$

72. $6 - 2\dfrac{4}{5}$

$3\dfrac{1}{5}$

73. $10 - 4\dfrac{8}{9}$

$5\dfrac{1}{9}$

74. $5 - 2\dfrac{7}{18}$

$2\dfrac{11}{18}$

75. $7\dfrac{3}{8} - 4\dfrac{5}{8}$

$2\dfrac{3}{4}$

76. $11\dfrac{1}{6} - 8\dfrac{5}{6}$

$2\dfrac{1}{3}$

77. $12\dfrac{5}{12} - 10\dfrac{17}{24}$

$1\dfrac{17}{24}$

78. $16\dfrac{1}{3} - 11\dfrac{5}{12}$

$4\dfrac{11}{12}$

79. $6\dfrac{2}{3} - 1\dfrac{7}{8}$

$4\dfrac{19}{24}$

80. $7\dfrac{7}{12} - 2\dfrac{5}{6}$

$4\dfrac{3}{4}$

81. $10\dfrac{2}{5} - 8\dfrac{7}{10}$

$1\dfrac{7}{10}$

82. $5\dfrac{5}{6} - 4\dfrac{7}{8}$

$\dfrac{23}{24}$

Solve.

83. What is $\dfrac{2}{3}$ less than $\dfrac{7}{8}$?

$\dfrac{5}{24}$

84. Find the difference between $\dfrac{8}{9}$ and $\dfrac{1}{6}$.

$\dfrac{13}{18}$

Quick Quiz (Objective 2.3B)

1. Subtract: $\dfrac{3}{4} - \dfrac{2}{5}$ $\dfrac{7}{20}$

2. Subtract: $11 - 8\dfrac{16}{17}$ $2\dfrac{1}{17}$

3. What is the difference between $\dfrac{5}{6}$ and $\dfrac{4}{15}$? $\dfrac{17}{30}$

4. Evaluate $x - y$ when $x = \dfrac{4}{5}$ and $y = \dfrac{1}{4}$. $\dfrac{11}{20}$

85. Find 8 less $1\frac{7}{12}$.

$6\frac{5}{12}$

86. Find 9 minus $5\frac{3}{20}$.

$3\frac{17}{20}$

Evaluate the variable expression $x - y$ for the given values of x and y.

87. $x = \frac{8}{9}, y = \frac{5}{9}$

$\frac{1}{3}$

88. $x = \frac{5}{6}, y = \frac{1}{6}$

$\frac{2}{3}$

89. $x = \frac{7}{15}, y = \frac{3}{10}$

$\frac{1}{6}$

90. $x = \frac{5}{6}, y = \frac{2}{15}$

$\frac{7}{10}$

91. $x = 5\frac{7}{9}, y = 4\frac{2}{3}$

$1\frac{1}{9}$

92. $x = 9\frac{5}{8}, y = 2\frac{3}{16}$

$7\frac{7}{16}$

93. $x = 5, y = 2\frac{7}{9}$

$2\frac{2}{9}$

94. $x = 8, y = 4\frac{5}{6}$

$3\frac{1}{6}$

95. Is $\frac{3}{4}$ a solution of the equation $\frac{4}{5} = \frac{31}{20} - y$?

Yes

96. Is $\frac{2}{3}$ a solution of the equation $\frac{2}{3} - x = 0$?

Yes

Objective C   **To solve application problems and use formulas**

97. Real Estate   You purchased $3\frac{1}{4}$ acres of land and then sold $1\frac{1}{2}$ acres of the property. How many acres of the property do you own now?

$1\frac{3}{4}$ acres

98. Carpentry   A $2\frac{3}{4}$-foot piece is cut from a 6-foot board. Find the length of the remaining piece of board.

$3\frac{1}{4}$ ft

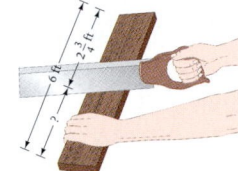

99. Community Service   You are required to contribute 20 h of community service to the town in which your college is located. After you have contributed $12\frac{1}{4}$ h, how many more hours of community service are still required of you?

$7\frac{3}{4}$ h

100. Horseracing   The 3-year-olds in the Kentucky Derby run $1\frac{1}{4}$ mi. The horses in the Belmont Stakes run $1\frac{1}{2}$ mi, and they run $1\frac{3}{16}$ mi in the Preakness Stakes. How much farther do the horses run in the Kentucky Derby than in the Preakness Stakes? How much farther do they run in the Belmont Stakes than in the Preakness Stakes?

$\frac{1}{16}$ mi; $\frac{5}{16}$ mi

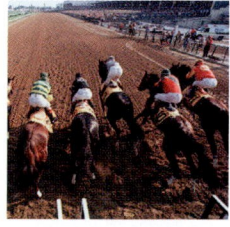

101. Boxing   A boxer is put on a diet to gain 15 lb in 4 weeks. The boxer gains $4\frac{1}{2}$ lb the first week and $3\frac{3}{4}$ lb the second week. How much weight must the boxer gain during the third and fourth weeks in order to gain a total of 15 lb?

$6\frac{3}{4}$ lb

102. **Construction**   A roofer and an apprentice are roofing a newly constructed house. In 1 day, the roofer completes $\frac{1}{3}$ of the job and the
apprentice completes $\frac{1}{4}$ of the job. How much of the job remains to be
done? Working at the same rate, can the roofer and the apprentice
complete the job in 1 more day?
$\frac{5}{12}$; Yes

103. **Sociology**   The table at the right shows the results of a survey in
which adults in the United States were asked how many evening
meals they cook at home during an average week.
  a. Which response was given most frequently?
  b. What fraction of the adult population cooks two or fewer dinners at
    home per week?
  c. What fraction of the adult population cooks five or more dinners
    at home per week? Is this less than half or more than half of the
    people?

  a. 5 meals   **b.** $\frac{23}{100}$ **c.** $\frac{49}{100}$; less than $\frac{1}{2}$

Responses to the question, "How many evening meals do you cook at home each week?"	
0	$\frac{2}{25}$
1	$\frac{1}{20}$
2	$\frac{1}{10}$
3	$\frac{13}{100}$
4	$\frac{3}{20}$
5	$\frac{21}{100}$
6	$\frac{9}{100}$
7	$\frac{19}{100}$

Source: Millward Brown for
Whirlpool

104. **Wages**   A student worked $4\frac{1}{3}$ h, 5 h, and $3\frac{2}{3}$ h this week at a part-time
job. The student is paid $9 an hour. How much did the student earn this
week?
$117

Golf   During the second half of the 1900s, greenskeepers mowed the grass
on golf putting surfaces progressively lower. The table at the right shows
the average grass height by decade. Use this table for Exercises 105 and 106.

105. What was the difference between the average height of the grass in the
1980s and the 1950s?
$\frac{3}{32}$ in.

106. Calculate the difference between the average grass height in the 1970s
and the 1960s.
$\frac{1}{32}$ in.

Average Height of Grass on Golf Putting Surfaces	
Decade	Height (in inches)
1950s	$\frac{1}{4}$
1960s	$\frac{7}{32}$
1970s	$\frac{3}{16}$
1980s	$\frac{5}{32}$
1990s	$\frac{1}{8}$

Source: Golf Course
Superintendents
Association of America

107. **Geometry**   You want to fence in the triangular plot of land shown at the
right. How many feet of fencing do you need? Use the formula
$P = a + b + c$.
$29\frac{1}{2}$ ft

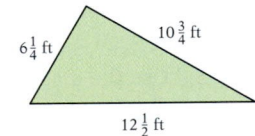

108. **Geometry**   The course of a yachting race is in the shape of a triangle
with sides that measure $4\frac{3}{10}$ mi, $3\frac{7}{10}$ mi, and $2\frac{1}{2}$ mi. Find the total length
of the course. Use the formula $P = a + b + c$.
$10\frac{1}{2}$ mi

Quick Quiz (Objective 2.3C)

1. A plumber worked $1\frac{1}{2}$ h of overtime on Monday,
$2\frac{1}{4}$ h of overtime on Tuesday, and $3\frac{1}{4}$ h of
overtime on Wednesday. Find the total number of
overtime hours worked during the 3 days.
7 h

2. A plane trip from Boston to San Francisco takes
$6\frac{1}{4}$ h. After the plane has been in the air for $3\frac{1}{2}$ h,
how much time remains before landing?
$2\frac{3}{4}$ h

109. Geometry   A flower garden in the yard of an historical home is in the shape of a triangle, as shown at the right. The wooden beams lining the edge of the garden need to be replaced. Find the total length of wood beams that must be purchased in order to replace the old beams. Use the formula $P = a + b + c$.

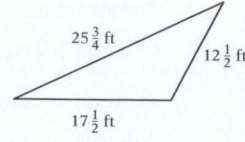

$55\dfrac{3}{4}$ ft

The Olympics   The table at the right shows the jump heights of three U.S. Olympic gold medalists in the high jump during the 1900s. Use this table for Exercises 110 and 111.

Olympic Gold Medalists in the High Jump		
Year	Athlete	Height of Jump (in feet)
1920	Richard Landon	$6\frac{1}{3}$
1924	Harold Osborn	$6\frac{1}{2}$
1996	Charles Austin	$7\frac{5}{6}$

Source: The World Almanac and Book of Facts

110. Find the difference between the height of Charles Austin's jump and the height of Richard Landon's jump.

$1\dfrac{1}{2}$ ft

111. How much higher was Charles Austin's jump than Harold Osborn's jump?

$1\dfrac{1}{3}$ ft

112. Demographics   Three-twentieths of the men in the United States are left-handed. (*Source:* Scripps Survey Research Center Poll) What fraction of the men in the United States are not left-handed?

$\dfrac{17}{20}$

Charles Austin

APPLYING THE CONCEPTS

113. The figure at the right is divided into five parts. Is each part of the figure $\dfrac{1}{5}$ of the figure? Why or why not?

No, because the parts are not equal in size.

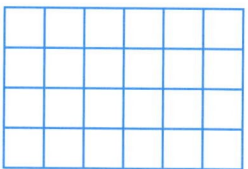

114. Draw a diagram that illustrates the addition of two fractions with the same denominator.

Answers will vary.

115. Use the diagram at the right to illustrate the sum of $\dfrac{1}{8}$ and $\dfrac{5}{6}$. Why does the figure contain 24 squares? Would it be possible to illustrate the sum of $\dfrac{1}{8}$ and $\dfrac{5}{6}$ if there were 48 squares in the figure? What if there were 16 squares? Make a list of the possible numbers of squares that could be used to illustrate the sum of $\dfrac{1}{8}$ and $\dfrac{5}{6}$.

The complete solution is in the *Solutions Manual.*

2.4 Multiplication and Division of Fractions

Objective A   To multiply fractions

To multiply two fractions, multiply the numerators and multiply the denominators.

> **Multiplication of Fractions**
>
> The product of two fractions is the product of the numerators over the product of the denominators.
>
> $$\frac{a}{b} \cdot \frac{c}{d} = \frac{ac}{bd},\quad \text{where}\quad b \neq 0\quad \text{and}\quad d \neq 0$$

Note that fractions do not need to have the same denominator in order to be multiplied.

HOW TO   Multiply: $\frac{2}{5} \cdot \frac{1}{3}$

$$\frac{2}{5} \cdot \frac{1}{3} = \frac{2 \cdot 1}{5 \cdot 3} = \frac{2}{15}$$

• **Multiply the numerators.**
Multiply the denominators.

The product $\frac{2}{5} \cdot \frac{1}{3}$ can be read "$\frac{2}{5}$ times $\frac{1}{3}$" or "$\frac{2}{5}$ of $\frac{1}{3}$."

Reading the times sign as "of" is useful in diagramming the product of two fractions.

$\frac{1}{3}$ of the bar at the right is shaded.

Shade $\frac{2}{5}$ of the $\frac{1}{3}$ already shaded.

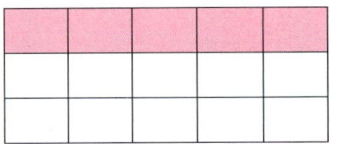

$\frac{2}{15}$ of the bar is now shaded.

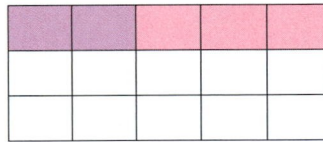

$$\frac{2}{5} \text{ of } \frac{1}{3} = \frac{2}{5} \cdot \frac{1}{3} = \frac{2}{15}$$

If a is a natural number, then $\frac{1}{a}$ is called the **reciprocal** or **multiplicative inverse** of a. Note that $a \cdot \frac{1}{a} = \frac{a}{1} \cdot \frac{1}{a} = \frac{a}{a} = 1$.

The product of a number and its multiplicative inverse is 1.

$$\frac{1}{8} \cdot 8 = 8 \cdot \frac{1}{8} = 1$$

Objective 2.4A

Vocabulary to Review
numerator   [2.2A]
denominator   [2.2A]
product   [1.3A]
exponent   [1.3B]
base   [1.3B]

New Vocabulary
reciprocal
multiplicative inverse

New Procedures
Multiplication of fractions:

$$\frac{a}{b} \cdot \frac{c}{d} = \frac{ac}{bd}$$

Discuss the Concepts
1. Describe the steps involved in multiplying $6\frac{1}{2}$ times 4.

2. How can you write the expression

$$\frac{2}{3} \cdot \frac{2}{3} \cdot \frac{2}{3} \cdot \frac{2}{3} \cdot \frac{2}{3}$$

as an exponential expression?

Concept Check
If two positive fractions, each less than 1, are multiplied, is the product always less than 1?
Yes

In-Class Examples (Objective 2.4A)

1. Multiply: $\frac{5}{6} \cdot \frac{18}{25} \cdot \frac{5}{9}$ $\frac{1}{3}$

2. Multiply: $\frac{c}{9} \cdot \frac{d}{7}$ $\frac{cd}{63}$

3. What is $\frac{5}{9}$ times 12? $6\frac{2}{3}$

4. Evaluate xy when $x = 2\frac{1}{4}$ and $y = 3\frac{1}{5}$. $7\frac{1}{5}$

5. Evaluate: $\left(\frac{2}{3}\right)^2\left(\frac{3}{4}\right)$ $\frac{1}{3}$

Concept Check

Which expression results in the largest product? Which results in the smallest product?

a. $5\frac{1}{3} \times 2\frac{5}{8}$

b. $6\frac{3}{7} \times 2\frac{4}{5}$

c. $4\frac{4}{5} \times 2\frac{1}{12}$

d. $6\frac{1}{4} \times 2\frac{6}{25}$

b is the largest product (18), and c is the smallest product (10).

Instructor Note

For the problem in the Point of Interest, one-third of a half-dozen is 2; one-fourth of the product of 2 and 8 is 4; $2 + 4 = 6$.

Point of Interest

Try this: What is the result if you take one-third of a half-dozen and add to it one-fourth of the product of the result and 8?

After multiplying two fractions, write the product in simplest form.

HOW TO   Multiply: $\dfrac{3}{8} \cdot \dfrac{4}{9}$

$$\frac{3}{8} \cdot \frac{4}{9} = \frac{3 \cdot 4}{8 \cdot 9}$$

- Multiply the numerators.
- Multiply the denominators.

$$= \frac{3 \cdot 2 \cdot 2}{2 \cdot 2 \cdot 2 \cdot 3 \cdot 3}$$

- Express the fraction in simplest form by first writing the prime factorization of each number.

$$= \frac{1}{6}$$

- Divide by the common factors and write the product in simplest form.

To multiply a whole number by a fraction or a mixed number, first write the whole number as a fraction with a denominator of 1.

HOW TO   Multiply: $3 \cdot \dfrac{5}{8}$

$$3 \cdot \frac{5}{8} = \frac{3}{1} \cdot \frac{5}{8}$$

- Write the whole number 3 as the fraction $\frac{3}{1}$.

$$= \frac{3 \cdot 5}{1 \cdot 8}$$

- Multiply the fractions. There are no common factors in the numerator and denominator.

$$= \frac{15}{8} = 1\frac{7}{8}$$

- Write the improper fraction as a mixed number.

HOW TO   Multiply: $\dfrac{x}{7} \cdot \dfrac{y}{5}$

$$\frac{x}{7} \cdot \frac{y}{5} = \frac{x \cdot y}{7 \cdot 5}$$

- Multiply the numerators.
- Multiply the denominators.

$$= \frac{xy}{35}$$

- Write the product in simplest form.

When a factor is a mixed number, first write the mixed number as an improper fraction. Then multiply.

HOW TO   Find the product of $4\frac{1}{6}$ and $2\frac{7}{10}$.

$$4\frac{1}{6} \cdot 2\frac{7}{10} = \frac{25}{6} \cdot \frac{27}{10}$$

- Write each mixed number as an improper fraction.

$$= \frac{25 \cdot 27}{6 \cdot 10}$$

- Multiply the fractions.

$$= \frac{5 \cdot 5 \cdot 3 \cdot 3 \cdot 3}{2 \cdot 3 \cdot 2 \cdot 5}$$

$$= \frac{45}{4} = 11\frac{1}{4}$$

- Write the product in simplest form.

HOW TO   Is $\frac{2}{3}$ a solution of the equation $\frac{3}{4}x = \frac{1}{2}$?

$$\frac{3}{4}x = \frac{1}{2}$$

$$\frac{3}{4}\left(\frac{2}{3}\right) \quad \bigg| \quad \frac{1}{2}$$

• Replace x by $\frac{2}{3}$ and then simplify.

$$\frac{3 \cdot 2}{4 \cdot 3} \quad \bigg| \quad \frac{1}{2}$$

$$\frac{3 \cdot 2}{2 \cdot 2 \cdot 3} \quad \bigg| \quad \frac{1}{2}$$

$$\frac{1}{2} = \frac{1}{2}$$

• The results are equal.

Yes, $\frac{2}{3}$ is a solution of the equation.

Recall that an exponent indicates the repeated multiplication of the same factor. For example,

$$3^5 = 3 \cdot 3 \cdot 3 \cdot 3 \cdot 3$$

The exponent, 5, indicates how many times the base, 3, occurs as a factor in the multiplication.

Point of Interest

René Descartes (1596–1650) was the first mathematician to extensively use exponential notation as it is used today. However, for some unknown reason, he always used xx for x^2.

The base of an exponential expression can be a fraction; for example, $\left(\frac{2}{3}\right)^4$. To evaluate this expression, write the factor as many times as indicated by the exponent and then multiply.

$$\left(\frac{2}{3}\right)^4 = \frac{2}{3} \cdot \frac{2}{3} \cdot \frac{2}{3} \cdot \frac{2}{3} = \frac{2 \cdot 2 \cdot 2 \cdot 2}{3 \cdot 3 \cdot 3 \cdot 3} = \frac{16}{81}$$

HOW TO   Evaluate $\left(\frac{3}{5}\right)^2 \cdot \left(\frac{5}{6}\right)^3$.

$$\left(\frac{3}{5}\right)^2 \cdot \left(\frac{5}{6}\right)^3$$

$$= \frac{3}{5} \cdot \frac{3}{5} \cdot \frac{5}{6} \cdot \frac{5}{6} \cdot \frac{5}{6}$$

• Write each factor as many times as indicated by the exponent.

$$= \frac{3 \cdot 3 \cdot 5 \cdot 5 \cdot 5}{5 \cdot 5 \cdot 6 \cdot 6 \cdot 6}$$

• Multiply.

$$= \frac{5}{24}$$

• Write the product in simplest form.

Example 1   Multiply: $\frac{6}{x} \cdot \frac{8}{y}$

Solution $\frac{6}{x} \cdot \frac{8}{y} = \frac{6 \cdot 8}{x \cdot y} = \frac{48}{xy}$

You Try It 1   Multiply: $\frac{y}{10} \cdot \frac{z}{7}$

Your solution $\frac{yz}{70}$

Solution on p. S5

Optional Student Activity

a. Shade $\frac{1}{3}$ of the diagram shown below.

b. Shade $\frac{2}{5}$ of the $\frac{1}{3}$ of the diagram you already shaded.

c. Then use the diagram to find the product of $\frac{2}{5}$ and $\frac{1}{3}$.

a. Five of the 15 parts should be shaded.

b. Two of the 15 parts should be shaded.

c. $\frac{2}{15}$

Example 2　Multiply: $\dfrac{7}{9} \cdot \dfrac{3}{14} \cdot \dfrac{2}{5}$

Solution

$$\dfrac{7}{9} \cdot \dfrac{3}{14} \cdot \dfrac{2}{5} = \dfrac{7 \cdot 3 \cdot 2}{9 \cdot 14 \cdot 5}$$

$$= \dfrac{7 \cdot 3 \cdot 2}{3 \cdot 3 \cdot 2 \cdot 7 \cdot 5} = \dfrac{1}{15}$$

You Try It 2　Multiply: $\dfrac{5}{12} \cdot \dfrac{9}{35} \cdot \dfrac{7}{8}$

Your solution　$\dfrac{3}{32}$

Example 3　What is the product of $\dfrac{7}{12}$ and 4?

Solution

$$\dfrac{7}{12} \cdot 4 = \dfrac{7}{12} \cdot \dfrac{4}{1}$$

$$= \dfrac{7 \cdot 4}{12 \cdot 1}$$

$$= \dfrac{7 \cdot 2 \cdot 2}{2 \cdot 2 \cdot 3 \cdot 1}$$

$$= \dfrac{7}{3} = 2\dfrac{1}{3}$$

You Try It 3　Find the product of $\dfrac{8}{9}$ and 6.

Your solution　$5\dfrac{1}{3}$

Example 4　Multiply: $7\dfrac{1}{2} \cdot 4\dfrac{2}{5}$

Solution

$$7\dfrac{1}{2} \cdot 4\dfrac{2}{5} = \dfrac{15}{2} \cdot \dfrac{22}{5} = \dfrac{15 \cdot 22}{2 \cdot 5}$$

$$= \dfrac{3 \cdot 5 \cdot 2 \cdot 11}{2 \cdot 5}$$

$$= \dfrac{33}{1} = 33$$

You Try It 4　Multiply: $3\dfrac{6}{7} \cdot 2\dfrac{4}{9}$

Your solution　$9\dfrac{3}{7}$

Example 5　Evaluate x^2y^2 when $x = 1\dfrac{1}{2}$ and $y = \dfrac{2}{3}$.

Solution　x^2y^2

$$\left(1\dfrac{1}{2}\right)^2 \cdot \left(\dfrac{2}{3}\right)^2 = \left(\dfrac{3}{2}\right)^2 \cdot \left(\dfrac{2}{3}\right)^2$$

$$= \dfrac{3}{2} \cdot \dfrac{3}{2} \cdot \dfrac{2}{3} \cdot \dfrac{2}{3}$$

$$= \dfrac{3 \cdot 3 \cdot 2 \cdot 2}{2 \cdot 2 \cdot 3 \cdot 3} = 1$$

You Try It 5　Evaluate x^4y^3 when $x = 2\dfrac{1}{3}$ and $y = \dfrac{3}{7}$.

Your solution　$2\dfrac{1}{3}$

Solutions on p. S5

Objective B   **To divide fractions**

Recall that the **reciprocal** of a fraction is that fraction with the numerator and denominator interchanged.

$$\text{The reciprocal of } \frac{3}{4} \text{ is } \frac{4}{3}.$$

$$\text{The reciprocal of } \frac{a}{b} \text{ is } \frac{b}{a}.$$

The process of interchanging the numerator and denominator of a fraction is called **inverting** the fraction.

To find the reciprocal of a whole number, first rewrite the whole number as a fraction with a denominator of 1. Then invert the fraction.

$$6 = \frac{6}{1}$$

The reciprocal of 6 is $\frac{1}{6}$.

Reciprocals are used to rewrite division problems as related multiplication problems. Look at the following two problems:

Point of Interest

Try this: What number when multiplied by its reciprocal is equal to 1?

$$6 \div 2 = 3 \qquad\qquad 6 \cdot \frac{1}{2} = 3$$

6 divided by 2 equals 3.   6 times the reciprocal of 2 equals 3.

Division is defined as multiplication by the reciprocal. Therefore, "divided by 2" is the same as "times $\frac{1}{2}$." Fractions are divided by making this substitution.

Division of Fractions

To divide two fractions, multiply by the reciprocal of the divisor.

$$\frac{a}{b} \div \frac{c}{d} = \frac{a}{b} \cdot \frac{d}{c}, \quad \text{where} \quad b \neq 0, \quad c \neq 0, \quad \text{and} \quad d \neq 0$$

HOW TO Divide: $\dfrac{2}{5} \div \dfrac{3}{4}$

$$\frac{2}{5} \div \frac{3}{4} = \frac{2}{5} \cdot \frac{4}{3}$$ • Rewrite the division as multiplication by the reciprocal.

$$= \frac{2 \cdot 4}{5 \cdot 3}$$ • Multiply the fractions.

$$= \frac{2 \cdot 2 \cdot 2}{5 \cdot 3} = \frac{8}{15}$$

Objective 2.4B

Vocabulary to Review
reciprocal of a fraction   [2.4A]

New Vocabulary
inverting a fraction

New Procedures
Division of fractions:

$$\frac{a}{b} \div \frac{c}{d} = \frac{a}{b} \cdot \frac{d}{c}$$

Discuss the Concepts
Explain why you "invert and multiply" when dividing a fraction by a fraction.

Concept Check
(*Note:* This is a classic problem that students frequently miss.)

What is 8 divided by $\frac{1}{2}$?   16

Instructor Note
The answer to the Point of Interest question is "any number except 0."

In-Class Examples (Objective 2.4B)

1. Divide: $\dfrac{5}{8} \div \dfrac{25}{42}$ $1\dfrac{1}{20}$

2. Divide: $\dfrac{b}{6} \div \dfrac{d}{8}$ $\dfrac{4b}{3d}$

3. What is 8 divided by $\dfrac{4}{5}$?   10

4. Evaluate $x \div y$ when $x = 4\dfrac{1}{2}$ and $y = 6$. $\dfrac{3}{4}$

Concept Check

Show by example that (1) the Commutative Property does not apply to division of fractions and (2) the Associative Property does not apply to division of fractions.

Answers will vary. For example:

(1) $\frac{1}{2} \div \frac{1}{4} = 2, \frac{1}{4} \div \frac{1}{2} = \frac{1}{2}$

(2) $\left(\frac{1}{2} \div \frac{1}{4}\right) \div \frac{1}{8} = 16$

$\frac{1}{2} \div \left(\frac{1}{4} \div \frac{1}{8}\right) = \frac{1}{4}$

Optional Student Activity

Find the sum of the reciprocals of all the whole number factors of 24.

$2\frac{1}{2}$

Instructor Note

Here is an extra-credit problem: One-quarter is the same part of one-third as one-half is of what number?   one-sixth

---

To divide a fraction and a whole number, first write the whole number as a fraction with a denominator of 1.

TAKE NOTE

$\frac{3}{4} \div 6 = \frac{1}{8}$ means that if $\frac{3}{4}$ is divided into 6 equal parts, each equal part is $\frac{1}{8}$ of the whole. For example, if 6 people share $\frac{3}{4}$ of a pizza, each person eats $\frac{1}{8}$ of the pizza.

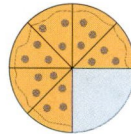

HOW TO Find the quotient of $\frac{3}{4}$ and 6.

$\frac{3}{4} \div 6 = \frac{3}{4} \div \frac{6}{1}$ • Write the whole number 6 as the fraction $\frac{6}{1}$.

$= \frac{3}{4} \cdot \frac{1}{6}$ • Rewrite the division as multiplication by the reciprocal.

$= \frac{3 \cdot 1}{4 \cdot 6}$ • Multiply the fractions.

$= \frac{3 \cdot 1}{2 \cdot 2 \cdot 2 \cdot 3}$

$= \frac{1}{8}$

When a number in a quotient is a mixed number, first write the mixed number as an improper fraction. Then divide the fractions.

HOW TO Divide: $\frac{2}{3} \div 1\frac{1}{4}$

$\frac{2}{3} \div 1\frac{1}{4} = \frac{2}{3} \div \frac{5}{4}$ • Write the mixed number $1\frac{1}{4}$ as an improper fraction.

$= \frac{2}{3} \cdot \frac{4}{5}$ • Rewrite the division as multiplication by the reciprocal.

$= \frac{2 \cdot 4}{3 \cdot 5} = \frac{8}{15}$ • Multiply the fractions.

Example 6 Divide: $\frac{4}{5} \div \frac{8}{15}$

Solution
$\frac{4}{5} \div \frac{8}{15} = \frac{4}{5} \cdot \frac{15}{8}$

$= \frac{4 \cdot 15}{5 \cdot 8}$

$= \frac{2 \cdot 2 \cdot 3 \cdot 5}{5 \cdot 2 \cdot 2 \cdot 2}$

$= \frac{3}{2} = 1\frac{1}{2}$

You Try It 6 Divide: $\frac{5}{6} \div \frac{10}{27}$

Your solution $2\frac{1}{4}$

Solution on p. S5

Example 7   Divide: $\dfrac{x}{2} \div \dfrac{y}{4}$

Solution

$$\dfrac{x}{2} \div \dfrac{y}{4} = \dfrac{x}{2} \cdot \dfrac{4}{y}$$

$$= \dfrac{x \cdot 4}{2 \cdot y}$$

$$= \dfrac{x \cdot 2 \cdot 2}{2 \cdot y} = \dfrac{2x}{y}$$

You Try It 7   Divide: $\dfrac{x}{8} \div \dfrac{y}{6}$

Your solution $\dfrac{3x}{4y}$

Example 8   Divide: $3\dfrac{4}{15} \div 2\dfrac{1}{10}$

Solution

$$3\dfrac{4}{15} \div 2\dfrac{1}{10} = \dfrac{49}{15} \div \dfrac{21}{10}$$

$$= \dfrac{49}{15} \cdot \dfrac{10}{21}$$

$$= \dfrac{49 \cdot 10}{15 \cdot 21}$$

$$= \dfrac{7 \cdot 7 \cdot 2 \cdot 5}{3 \cdot 5 \cdot 3 \cdot 7}$$

$$= \dfrac{14}{9} = 1\dfrac{5}{9}$$

You Try It 8   Divide: $4\dfrac{3}{8} \div 3\dfrac{1}{2}$

Your solution $1\dfrac{1}{4}$

Example 9   Evaluate $x \div y$ when $x = 3\dfrac{1}{8}$ and $y = 5$.

Solution

$x \div y$

$$3\dfrac{1}{8} \div 5 = \dfrac{25}{8} \div \dfrac{5}{1}$$

$$= \dfrac{25}{8} \cdot \dfrac{1}{5}$$

$$= \dfrac{25 \cdot 1}{8 \cdot 5}$$

$$= \dfrac{5 \cdot 5 \cdot 1}{2 \cdot 2 \cdot 2 \cdot 5} = \dfrac{5}{8}$$

You Try It 9   Evaluate $x \div y$ when $x = 2\dfrac{1}{4}$ and $y = 9$.

Your solution $\dfrac{1}{4}$

Solutions on p. S5

Objective 2.4C

New Vocabulary

complex fraction
main fraction bar

Discuss the Concepts

Explain the steps involved in simplifying the complex fraction

$$\dfrac{\dfrac{1}{4} + \dfrac{3}{8}}{\dfrac{1}{2}}.$$

Concept Check

The main fraction bar in a complex fraction can be read as what operation?

Division

Objective C   To simplify a complex fraction

A **complex fraction** is a fraction whose numerator or denominator contains one or more fractions. Examples of complex fractions are shown below.

Main fraction bar ⟶ $\dfrac{\dfrac{3}{4}}{\dfrac{7}{8}}$ $\dfrac{4}{3 - \dfrac{1}{2}}$ $\dfrac{\dfrac{9}{10} + \dfrac{3}{5}}{\dfrac{5}{6}}$ $\dfrac{3\frac{1}{2} \cdot 2\frac{5}{8}}{\left(4\frac{2}{3}\right) \div \left(3\frac{1}{5}\right)}$

Look at the first example given above and recall that the fraction bar can be read "divided by."

Therefore, $\dfrac{\dfrac{3}{4}}{\dfrac{7}{8}}$ can be read "$\dfrac{3}{4}$ divided by $\dfrac{7}{8}$" and can be written $\dfrac{3}{4} \div \dfrac{7}{8}$. This is the division of two fractions and can be simplified by multiplying by the reciprocal, as shown.

$$\dfrac{\dfrac{3}{4}}{\dfrac{7}{8}} = \dfrac{3}{4} \div \dfrac{7}{8} = \dfrac{3}{4} \cdot \dfrac{8}{7} = \dfrac{3 \cdot 8}{4 \cdot 7} = \dfrac{6}{7}$$

To simplify a complex fraction, first simplify the expression above the main fraction bar and the expression below the main fraction bar; the result is one number in the numerator and one number in the denominator. Then rewrite the complex fraction as a division problem by reading the main fraction bar as "divided by."

HOW TO   Simplify: $\dfrac{4}{3 - \dfrac{1}{2}}$

$\dfrac{4}{3 - \dfrac{1}{2}} = \dfrac{4}{\dfrac{5}{2}}$ • The numerator (4) is already simplified. Simplify the expression in the denominator.

Note: $3 - \dfrac{1}{2} = \dfrac{6}{2} - \dfrac{1}{2} = \dfrac{5}{2}$

$= 4 \div \dfrac{5}{2}$ • Rewrite the complex fraction as division.

$= \dfrac{4}{1} \div \dfrac{5}{2}$ • Divide.

$= \dfrac{4}{1} \cdot \dfrac{2}{5}$

$= \dfrac{8}{5} = 1\dfrac{3}{5}$ • Write the answer in simplest form.

In-Class Examples (Objective 2.4C)

1. Which number, $\dfrac{1}{6}$ or $\dfrac{1}{3}$, is a solution of the equation

$\dfrac{1\frac{5}{6} + x}{2\frac{1}{3} - x} = \dfrac{12}{13}$? $\dfrac{1}{6}$

2. Evaluate $\dfrac{2x - y}{4z}$ when $x = \dfrac{7}{8}$, $y = \dfrac{3}{4}$, and $z = \dfrac{1}{16}$.

4

HOW TO Evaluate $\dfrac{wx}{yz}$ when $w = 1\frac{1}{3}$, $x = 2\frac{5}{8}$, $y = 4\frac{1}{2}$, and $z = 3\frac{1}{3}$.

$$\dfrac{wx}{yz}$$

$$\dfrac{1\frac{1}{3} \cdot 2\frac{5}{8}}{4\frac{1}{2} \cdot 3\frac{1}{3}}$$

• Replace each variable with its given value.

$$= \dfrac{\dfrac{7}{2}}{15}$$

• Simplify the numerator.

Note: $1\frac{1}{3} \cdot 2\frac{5}{8} = \dfrac{4}{3} \cdot \dfrac{21}{8} = \dfrac{7}{2}$

Simplify the denominator.

Note: $4\frac{1}{2} \cdot 3\frac{1}{3} = \dfrac{9}{2} \cdot \dfrac{10}{3} = 15$

$$= \dfrac{7}{2} \div 15$$

• Rewrite the complex fraction as division.

$$= \dfrac{7}{2} \cdot \dfrac{1}{15} = \dfrac{7}{30}$$

• Divide by multiplying by the reciprocal.

Note: $15 = \dfrac{15}{1}$; the reciprocal of $\dfrac{15}{1}$ is $\dfrac{1}{15}$.

Example 10 Is $\dfrac{2}{3}$ a solution of $\dfrac{x + \dfrac{1}{2}}{x} = \dfrac{7}{4}$?

Solution

$$\dfrac{x + \dfrac{1}{2}}{x} = \dfrac{7}{4}$$

$$\dfrac{\dfrac{2}{3} + \dfrac{1}{2}}{\dfrac{2}{3}} \;\Big|\; \dfrac{7}{4}$$

$$\dfrac{\dfrac{7}{6}}{\dfrac{2}{3}} \;\Big|\; \dfrac{7}{4}$$

$$\dfrac{7}{6} \div \dfrac{2}{3} \;\Big|\; \dfrac{7}{4}$$

$$\dfrac{7}{6} \cdot \dfrac{3}{2} \;\Big|\; \dfrac{7}{4}$$

$$\dfrac{7}{4} = \dfrac{7}{4}$$

Yes, $\dfrac{2}{3}$ is a solution of the equation.

You Try It 10 Is $\dfrac{1}{2}$ a solution of $\dfrac{2y + 3}{y} = 2$?

Your solution No

Solution on p. S6

Example 11

Evaluate the variable expression $\frac{x-y}{z}$ when $x = 4\frac{1}{8}$, $y = 2\frac{5}{8}$, and $z = \frac{3}{4}$.

Solution

$$\frac{x-y}{z}$$

$$\frac{4\frac{1}{8} - 2\frac{5}{8}}{\frac{3}{4}} = \frac{\frac{2}{3}}{\frac{3}{4}}$$

$$= \frac{3}{2} \div \frac{3}{4}$$

$$= \frac{3}{2} \cdot \frac{4}{3} = 2$$

You Try It 11

Evaluate the variable expression $\frac{x}{y-z}$ when $x = 2\frac{4}{9}$, $y = 3$, and $z = 1\frac{1}{3}$.

Your solution

$1\frac{7}{15}$

Solution on p. S6

Objective 2.4D

New Vocabulary
base of a triangle
height of a triangle

New Formulas
Area of a triangle:

$$A = \frac{1}{2}bh$$

Objective D To solve application problems and use formulas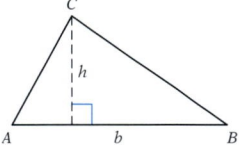

Figure *ABC* is a triangle. *AB* is the **base,** *b*, of the triangle. The line segment from *C* that forms a right angle with the base is the **height,** *h*, of the triangle. The formula for the area of a triangle is given below. Use this formula for Example 12 and You Try It 12.

Area of a Triangle

The formula for the area of a triangle is $A = \frac{1}{2}bh$, where *A* is the area of the triangle, *b* is the base, and *h* is the height.

In-Class Examples (Objective 2.4D)

1. A person can walk $3\frac{3}{4}$ mi in 1 h. How many miles can the person walk in $1\frac{1}{4}$ h? $4\frac{11}{16}$ mi

2. A building contractor bought $8\frac{1}{4}$ acres of land for $132,000. What was the cost per acre? $16,000

Example 12

A riveter uses metal plates that are in the shape of a triangle and have a base of 12 cm and a height of 6 cm. Find the area of one metal plate.

Strategy

To find the area, use the formula for the area of a triangle, $A = \frac{1}{2}bh$. $b = 12$ and $h = 6$.

Solution

$$A = \frac{1}{2}bh$$

$$A = \frac{1}{2}(12)(6)$$

$$A = 36$$

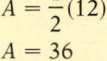

6 cm

12 cm

The area is 36 cm².

You Try It 12

Find the amount of felt needed to make a banner that is in the shape of a triangle with a base of 18 in. and a height of 9 in.

Your Strategy

Your solution
81 in²

Example 13

A 12-foot board is cut into pieces $2\frac{1}{2}$ ft long for use as bookshelves. What is the length of the remaining piece after as many shelves as possible are cut?

Strategy

To find the length of the remaining piece:
- Divide the total length (12) by the length of each shelf $\left(2\frac{1}{2}\right)$. The quotient is the number of shelves cut, with a certain fraction of a shelf left over.
- Multiply the fraction left over by the length of a shelf.

Solution

$$12 \div 2\frac{1}{2} = \frac{12}{1} \div \frac{5}{2} = \frac{12}{1} \cdot \frac{2}{5} = \frac{12 \cdot 2}{1 \cdot 5} = \frac{24}{5} = 4\frac{4}{5}$$

4 shelves, each $2\frac{1}{2}$ ft long, can be cut from the board. The piece remaining is $\frac{4}{5}$ of $2\frac{1}{2}$ ft long.

$$\frac{4}{5} \cdot 2\frac{1}{2} = \frac{4}{5} \cdot \frac{5}{2} = \frac{4 \cdot 5}{5 \cdot 2} = 2$$

The length of the remaining piece is 2 ft.

You Try It 13

The Booster Club is making 22 sashes for the high school band members. Each sash requires $1\frac{3}{8}$ yd of material at a cost of $12 per yard. Find the total cost of the material.

Your Strategy

Your solution
$363

Solutions on p. S6

Concept Check
(*Note:* Example 13 is difficult for students. Use the following problem if you have worked through this example with your students and want them to work another, similar problem either on their own or in small groups.)

A 14-foot piece of wood molding is cut into pieces $4\frac{1}{4}$ ft long.

What is the length of the remaining piece after as many pieces as possible have been cut?

$1\frac{1}{4}$ ft

Section 2.4

Suggested Assignment
Exercises 3–123, odds
More challenging problems:
Exercises 125, 126

Answers to Writing Exercises

1. The idea of needing a common denominator when adding two fractions is similar to adding objects in the "real world." The sum of 3 apples and 5 oranges is 3 apples and 5 oranges. We cannot add unlike objects. The sum of 3 apples and 5 apples is 8 apples. We can add like objects. The sum of 3 ninths and 5 ninths is 8 ninths; we must have the same denominator in order to add the fractions.

A common denominator is not required when multiplying two fractions. Multiplication does not require that we have like objects. One-half of 6 apples is 3 apples. One-half of $\frac{1}{4}$ of a pizza is $\frac{1}{8}$ of the pizza.

2. The problem can be restated as 1 times a number equals $\frac{3}{8}$. By the Multiplication Property of One, the product of 1 and a number is the number. Therefore, the unknown number is $\frac{3}{8}$.

$$1 \cdot \frac{3}{8} = \frac{3}{8}$$

2.4 Exercises

Objective A   To multiply fractions

1. ✏ Explain why you need a common denominator when adding or subtracting two fractions and why you don't need a common denominator when multiplying or dividing two fractions.

2. ✏ The product of 1 and a number is $\frac{3}{8}$. Find the number. Explain how you arrived at the answer.

Multiply.

3. $\frac{2}{3} \cdot \frac{9}{10}$

$\frac{3}{5}$

4. $\frac{3}{8} \cdot \frac{4}{5}$

$\frac{3}{10}$

5. $\frac{14}{15} \cdot \frac{6}{7}$

$\frac{4}{5}$

6. $\frac{15}{16} \cdot \frac{4}{9}$

$\frac{5}{12}$

7. $\frac{6}{7} \cdot \frac{0}{10}$

0

8. $\frac{5}{12} \cdot \frac{3}{0}$

Undefined

9. $\frac{9}{x} \cdot \frac{7}{y}$

$\frac{63}{xy}$

10. $\frac{4}{c} \cdot \frac{8}{d}$

$\frac{32}{cd}$

11. $\frac{2}{3} \cdot \frac{3}{8} \cdot \frac{4}{9}$

$\frac{1}{9}$

12. $\frac{5}{7} \cdot \frac{1}{6} \cdot \frac{14}{15}$

$\frac{1}{9}$

13. $6 \cdot \frac{1}{6}$

1

14. $\frac{1}{10} \cdot 10$

1

15. $\frac{3}{4} \cdot 8$

6

16. $\frac{5}{7} \cdot 14$

10

17. $\frac{6}{7} \cdot 0$

0

18. $0 \cdot \frac{9}{11}$

0

19. $\frac{5}{22} \cdot 2\frac{1}{5}$

$\frac{1}{2}$

20. $\frac{4}{15} \cdot 1\frac{7}{8}$

$\frac{1}{2}$

21. $3\frac{1}{2} \cdot 5\frac{3}{7}$

19

22. $2\frac{1}{4} \cdot 1\frac{1}{3}$

3

23. $8 \cdot 5\frac{1}{4}$

42

24. $3 \cdot 2\frac{1}{9}$

$6\frac{1}{3}$

25. $3\frac{1}{2} \cdot 1\frac{5}{7} \cdot \frac{11}{12}$

$5\frac{1}{2}$

26. $2\frac{2}{3} \cdot \frac{8}{9} \cdot 1\frac{5}{16}$

$3\frac{1}{9}$

27. Find the product of $\frac{3}{4}$ and $\frac{14}{15}$.

$\frac{7}{10}$

28. Find the product of $\frac{12}{25}$ and $\frac{5}{16}$.

$\frac{3}{20}$

Quick Quiz (Objective 2.4A)

1. Multiply: $\frac{5}{6} \cdot \frac{18}{25} \cdot \frac{5}{9}$ $\frac{1}{3}$

2. Multiply: $\frac{m}{8} \cdot \frac{n}{6}$ $\frac{mn}{48}$

3. What is $2\frac{1}{3}$ times 6? 14

4. Evaluate xy when $x = 2\frac{1}{2}$ and $y = 3\frac{2}{25}$. $7\frac{4}{5}$

29. What is $4\frac{4}{5}$ times $\frac{3}{8}$?

$1\frac{4}{5}$

30. What is $5\frac{1}{3}$ times $\frac{3}{16}$?

1

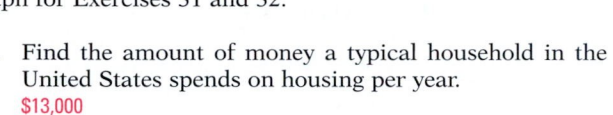 **Cost of Living**   A typical household in the United States has an average after-tax income of $45,000. The graph at the right represents how this annual income is spent. Use this graph for Exercises 31 and 32.

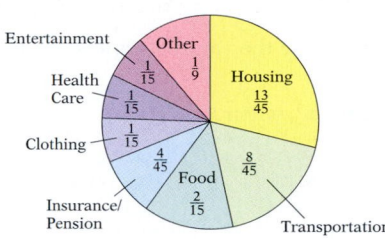

31. Find the amount of money a typical household in the United States spends on housing per year.

$13,000

How a Typical U.S. Household Spends Its Annual Income

Source: Based on data from American Demographics

32. How much money does a typical household in the United States spend annually on food?

$6000

Evaluate the variable expression xy for the given values of x and y.

33. $x = \frac{5}{16}, y = \frac{7}{15}$

$\frac{7}{48}$

34. $x = \frac{2}{5}, y = \frac{5}{6}$

$\frac{1}{3}$

35. $x = \frac{4}{7}, y = 6\frac{1}{8}$

$3\frac{1}{2}$

36. $x = 6\frac{3}{5}, y = 3\frac{1}{3}$

22

Evaluate the variable expression xyz for the given values of x, y, and z.

37. $x = \frac{3}{8}, y = \frac{2}{3}, z = \frac{4}{5}$

$\frac{1}{5}$

38. $x = 4, y = \frac{0}{8}, z = 1\frac{5}{9}$

0

39. $x = 2\frac{3}{8}, y = \frac{3}{19}, z = \frac{4}{9}$

$\frac{1}{6}$

40. $x = \frac{4}{5}, y = 15, z = \frac{7}{8}$

$10\frac{1}{2}$

41. $x = \frac{5}{6}, y = 3, z = 1\frac{7}{15}$

$3\frac{2}{3}$

42. $x = 4\frac{1}{2}, y = 3\frac{5}{9}, z = 1\frac{7}{8}$

30

43. Is $\frac{3}{4}$ a solution of the equation $\frac{4}{5}x = \frac{5}{3}$?

No

44. Is $\frac{1}{2}$ a solution of the equation $\frac{3}{4}p = \frac{3}{2}$?

No

Evaluate.

45. $\left(\frac{3}{4}\right)^2$

$\frac{9}{16}$

46. $\left(\frac{5}{8}\right)^2$

$\frac{25}{64}$

47. $\left(\frac{5}{8}\right)^3 \cdot \left(\frac{2}{5}\right)^2$

$\frac{5}{128}$

48. $\left(\frac{3}{5}\right)^3 \cdot \left(\frac{1}{3}\right)^2$

$\frac{3}{125}$

49. $\left(\dfrac{18}{25}\right)^2 \cdot \left(\dfrac{5}{9}\right)^3$

$\dfrac{4}{45}$

50. $\left(\dfrac{2}{3}\right)^3 \cdot \left(\dfrac{5}{6}\right)^2$

$\dfrac{50}{243}$

51. $7^2 \cdot \left(\dfrac{2}{7}\right)^3$

$1\dfrac{1}{7}$

52. $4^3 \cdot \left(\dfrac{5}{12}\right)^2$

$11\dfrac{1}{9}$

Evaluate the variable expression for the given values of x and y.

53. x^4, when $x = \dfrac{2}{3}$

$\dfrac{16}{81}$

54. y^3, when $y = \dfrac{3}{4}$

$\dfrac{27}{64}$

55. $x^3 y^2$, when $x = \dfrac{2}{3}$ and $y = 1\dfrac{1}{2}$

$\dfrac{2}{3}$

56. $x^2 y^4$, when $x = 2\dfrac{1}{3}$ and $y = \dfrac{3}{7}$

$\dfrac{9}{49}$

Objective B   To divide fractions

Divide.

57. $\dfrac{5}{7} \div \dfrac{2}{5}$

$1\dfrac{11}{14}$

58. $\dfrac{3}{8} \div \dfrac{2}{3}$

$\dfrac{9}{16}$

59. $0 \div \dfrac{7}{9}$

0

60. $0 \div \dfrac{4}{5}$

0

61. $6 \div \dfrac{3}{4}$

8

62. $8 \div \dfrac{2}{3}$

12

63. $\dfrac{3}{4} \div 6$

$\dfrac{1}{8}$

64. $\dfrac{2}{3} \div 8$

$\dfrac{1}{12}$

65. $\dfrac{9}{10} \div 0$

Undefined

66. $\dfrac{2}{11} \div 0$

Undefined

67. $\dfrac{b}{6} \div \dfrac{5}{d}$

$\dfrac{bd}{30}$

68. $\dfrac{y}{10} \div \dfrac{4}{z}$

$\dfrac{yz}{40}$

69. $3\dfrac{1}{3} \div \dfrac{5}{8}$

$5\dfrac{1}{3}$

70. $5\dfrac{1}{2} \div \dfrac{1}{4}$

22

71. $5\dfrac{1}{2} \div 11$

$\dfrac{1}{2}$

72. $4\dfrac{2}{3} \div 7$

$\dfrac{2}{3}$

73. $5\dfrac{2}{7} \div 1$

$5\dfrac{2}{7}$

74. $9\dfrac{5}{6} \div 1$

$9\dfrac{5}{6}$

75. $2\dfrac{4}{13} \div 1\dfrac{5}{26}$

$1\dfrac{29}{31}$

76. $3\dfrac{3}{8} \div 2\dfrac{7}{16}$

$1\dfrac{5}{13}$

Quick Quiz (Objective 2.4B)

1. Divide: $\dfrac{2}{9} \div \dfrac{1}{3}$ $\dfrac{2}{3}$

2. Divide: $\dfrac{p}{9} \div \dfrac{q}{6}$ $\dfrac{2p}{3q}$

3. What is 10 divided by $\dfrac{5}{8}$?   16

4. Evaluate $x \div y$ when $x = \dfrac{5}{6}$ and $y = 3\dfrac{3}{4}$. $\dfrac{2}{9}$

77. Find the quotient of $\frac{9}{10}$ and $\frac{3}{4}$.

$1\frac{1}{5}$

78. Find the quotient of $\frac{3}{5}$ and $\frac{12}{25}$.

$1\frac{1}{4}$

79. Find $\frac{7}{8}$ divided by $3\frac{1}{4}$.

$\frac{7}{26}$

80. Find $\frac{3}{8}$ divided by $2\frac{1}{4}$.

$\frac{1}{6}$

Evaluate the variable expression $x \div y$ for the given values of x and y.

81. $x = \frac{5}{8}, y = \frac{15}{2}$

$\frac{1}{12}$

82. $x = \frac{14}{3}, y = \frac{7}{9}$

6

83. $x = 18, y = \frac{3}{8}$

48

84. $x = 20, y = \frac{5}{6}$

24

The Food Industry The table at the right shows the net weight of four different boxes of cereal. Use this table for Exercises 85 and 86.

Cereal	Net Weight
Kellogg Honey Crunch Corn Flakes	24 oz
Nabisco Instant Cream of Wheat	28 oz
Post Shredded Wheat	18 oz
Quaker Oats	41 oz

85. Find the number of $\frac{3}{4}$-ounce servings in a box of Kellogg Honey Crunch Corn Flakes.

32 servings

86. Find the number of $1\frac{1}{4}$-ounce servings in a box of Post Shredded Wheat.

$14\frac{2}{5}$ servings

Objective C   To simplify a complex fraction

Simplify.

87. $\dfrac{\frac{9}{16}}{\frac{3}{4}}$

$\frac{3}{4}$

88. $\dfrac{\frac{7}{24}}{\frac{3}{8}}$

$\frac{7}{9}$

89. $\dfrac{\frac{2}{3} + \frac{1}{2}}{7}$

$\frac{1}{6}$

90. $\dfrac{5}{\frac{3}{8} - \frac{1}{4}}$

40

91. $\dfrac{2 + \frac{1}{4}}{\frac{3}{8}}$

6

92. $\dfrac{1 - \frac{3}{4}}{\frac{5}{12}}$

$\frac{3}{5}$

93. $\dfrac{\frac{9}{25}}{\frac{4}{5} - \frac{1}{10}}$

$\frac{18}{35}$

94. $\dfrac{\frac{9}{14} - \frac{1}{7}}{\frac{9}{14} + \frac{1}{7}}$

$\frac{7}{11}$

Quick Quiz (Objective 2.4C)

1. Simplify: $\dfrac{\frac{3}{5} + \frac{1}{3}}{\frac{4}{5}}$ $1\frac{1}{6}$

2. Evaluate $\dfrac{x}{y + z}$ when $x = \frac{3}{4}$, $y = \frac{1}{6}$, and $z = \frac{5}{12}$.

$1\frac{2}{7}$

95. $\dfrac{3 + 2\frac{1}{3}}{5\frac{1}{6} - 1}$

$1\frac{7}{25}$

96. $\dfrac{4 - 3\frac{5}{8}}{2\frac{1}{2} - \frac{3}{4}}$

$\frac{3}{14}$

97. $\dfrac{5\frac{2}{3} - 1\frac{1}{6}}{3\frac{5}{8} - 2\frac{1}{4}}$

$3\frac{3}{11}$

98. $\dfrac{3\frac{1}{4} - 2\frac{1}{2}}{4\frac{3}{4} + 1\frac{1}{2}}$

$\frac{3}{25}$

Evaluate the expression for the given values of the variables.

99. $\dfrac{x + y}{z}$, when $x = \frac{2}{3}$, $y = \frac{3}{4}$, and $z = \frac{1}{12}$

17

100. $\dfrac{x}{y + z}$, when $x = \frac{8}{15}$, $y = \frac{3}{5}$, and $z = \frac{2}{3}$

$\frac{8}{19}$

101. $\dfrac{x - y}{z}$, when $x = 2\frac{5}{8}$, $y = 1\frac{1}{4}$, and $z = 1\frac{3}{8}$

1

102. $\dfrac{x}{y - z}$, when $x = 2\frac{3}{10}$, $y = 3\frac{2}{5}$, and $z = 1\frac{4}{5}$

$1\frac{7}{16}$

103. Is $\frac{3}{4}$ a solution of the equation $\dfrac{4x}{x + 5} = \dfrac{4}{3}$?

No

104. Is $\frac{4}{5}$ a solution of the equation

$\dfrac{15y}{\frac{3}{10} + y} = 24$?

No

Objective D   **To solve application problems and use formulas**

Solve.

105. **Polo** A chukker is one period of play in a polo match. A chukker lasts $7\frac{1}{2}$ min. Find the length of time in four chukkers.

30 min

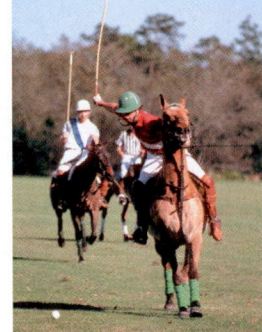

106. **Calendars** The Assyrian calendar was based on the phases of the moon. One lunation was $29\frac{1}{2}$ days long. There were 12 lunations in 1 year. Find the number of days in 1 year in the Assyrian calendar.

354 days

107. **Length** One rod is equal to $5\frac{1}{2}$ yd. How many feet are in one rod? How many inches are in one rod?

$16\frac{1}{2}$ ft; 198 in.

108. **Fuel Efficiency** A car used $12\frac{1}{2}$ gal of gasoline on a 275-mile trip. How many miles can this car travel on 1 gal of gasoline?

22 mi

109. **Housework** According to a national survey, the average couple spends $4\frac{1}{2}$ h cleaning house each week. How many hours does the average couple spend cleaning house each year?

234 h

Quick Quiz (Objective 2.4D)

1. A sports car gets 27 mi on each gallon of gasoline. How many miles can the car travel on $4\frac{2}{3}$ gal of gasoline?   126 mi

2. A station wagon used $15\frac{3}{10}$ gal of gasoline on a 459-mile trip. How many miles did this car travel on 1 gal of gasoline?   30 mi

110. **Manufacturing**   A factory worker can assemble a product in $7\frac{1}{2}$ min. How many products can the worker assemble in 1 h?
8 products

111. **Real Estate**   A developer purchases $25\frac{1}{2}$ acres of land and plans to set aside 3 acres for an entranceway to a housing development to be built on the property. Each house will be built on a $\frac{3}{4}$-acre plot of land. How many houses does the developer plan to build on the property?
30 houses

112. **Party Planning**   You are planning a barbecue for 25 people. You want to serve $\frac{1}{4}$-pound hamburger patties to your guests and you estimate each person will eat two hamburgers. How much hamburger meat should you buy for the barbecue?
$12\frac{1}{2}$ lb

113. **Board Games**   A wooden travel game board has hinges that allow the board to be folded in half. If the dimensions of the open board are 14 in. by 14 in. by $\frac{7}{8}$ in., what are the dimensions of the board when it is closed?
14 in. by 7 in. by $1\frac{3}{4}$ in.

114. **Carpentry**   A 16-foot board is cut into pieces $2\frac{1}{2}$ ft long for use as bookshelves. What is the length of the remaining piece after as many shelves as possible are cut?
1 ft

115.  **Nutrition**   According to the Center for Science in the Public Interest, the average teenage boy drinks $3\frac{1}{3}$ cans of soda per day. The average teenage girl drinks $2\frac{1}{3}$ cans of soda per day.
   a. The average teenage boy drinks how many cans of soda per week?
   b. If a can of soda contains 150 calories, how many calories does the average teenage boy consume each week in soda?
   c. How many more cans of soda per week does the average teenage boy drink than the average teenage girl?
   a. $23\frac{1}{3}$ cans   **b.** 3500 calories   **c.** 7 cans

116. **Wages**   Find the total wages of an employee who worked $26\frac{1}{2}$ h this week and who earns an hourly wage of $12.
$318

117. **Geometry**   A sail is in the shape of a triangle with a base of 12 m and a height of 16 m. How much canvas was needed to make the body of the sail?
96 m²

118. **Geometry**   A vegetable garden is in the shape of a triangle with a base of 21 ft and a height of 13 ft. Find the area of the vegetable garden.
$136\frac{1}{2}$ ft²

Answers to Writing Exercises

127. Our calendar year is 365 days. A complete orbit around the sun takes $365\frac{1}{4}$ days, so 1 solar year is $365\frac{1}{4}$ days. Therefore, after 365 days, Earth is $\frac{1}{4}$ day short of a complete orbit around the sun. After 2 years, Earth is $\frac{1}{4}$ day + $\frac{1}{4}$ day = $\frac{1}{2}$ day short of a complete orbit around the sun. After 4 years, Earth is $\frac{1}{4}$ day + $\frac{1}{4}$ day + $\frac{1}{4}$ day + $\frac{1}{4}$ day = 1 day short of a complete orbit around the sun. Therefore, every 4 years, our calendar incorporates a leap year, which is a 366-day year, in order to make up for the 1 extra day in the 4 solar years.

119. **Geometry** Find the area of a rectangle that has a length of $8\frac{1}{2}$ yd and a width of 5 yd. $42\frac{1}{2}$ yd²

120. **Geometry** What is the area of a rectangular recreational area that has a length of $3\frac{1}{4}$ mi and a width of $1\frac{1}{2}$ mi? $4\frac{7}{8}$ mi²

121. **Geometry** A city plans to plant grass seed in a public playground that has the shape of a triangle with a height of 24 m and a base of 20 m. Each bag of grass seed will seed 120 m². How many bags of seed should be purchased? 2 bags

122. **Oceanography** The pressure on a submerged object is given by $P = 15 + \frac{1}{2}D$, where D is the depth in feet and P is the pressure measured in pounds per square inch. Find the pressure on a diver who is at a depth of $12\frac{1}{2}$ ft. $21\frac{1}{4}$ lb/in²

123. **Hiking** Find the rate of a hiker who walked $4\frac{2}{3}$ mi in $1\frac{1}{3}$ h. Use the equation $r = \frac{d}{t}$, where r is the rate in miles per hour, d is the distance, and t is the time. $3\frac{1}{2}$ mph

124. **Physics** Find the amount of force necessary to push a 75-pound crate across a floor, where the coefficient of friction is $\frac{3}{8}$. Use the equation $F = \mu N$, where F is the force, μ is the coefficient of friction, and N is the weight of the crate. Force is measured in pounds. $28\frac{1}{8}$ lb

APPLYING THE CONCEPTS

125. **Cartography** On a map, two cities are $3\frac{1}{8}$ in. apart. If $\frac{1}{8}$ in. on the map represents 50 mi, what is the number of miles between the two cities? 1250 mi

126. Determine whether the statement is always true, sometimes true, or never true.

a. Let n be an even natural number. Then $\frac{1}{2}n$ is a whole number.

Sometimes true

b. Let n be an odd number. Then $\frac{1}{2}n$ is an improper fraction.

Sometimes true

127. On page 126, Exercise 106 describes the Assyrian calendar. Our calendar is based on the solar year. One solar year is $365\frac{1}{4}$ days. Use this fact to explain leap years.

2.5 Introduction to Decimals

Objective A

To read and write decimals

The price tag on a sweater reads $61.88. The number 61.88 is in **decimal notation.** A number written in decimal notation is often called simply a **decimal.**

A number written in decimal notation has three parts.

61	.	88
Whole-number part	**Decimal point**	**Decimal part**

The decimal part of the number represents a number less than one. For example, $.88 is less than one dollar. The decimal point (.) separates the whole-number part from the decimal part.

The position of a digit in a decimal determines the digit's place value. The place-value chart is extended to the right to show the place value of digits to the right of a decimal point.

In the decimal 458.302719, the position of the digit 7 determines that its place value is ten-thousandths.

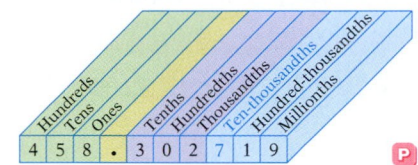

Note the relationship between fractions and numbers written in decimal notation.

seven tenths	seven hundredths	seven thousandths
$\frac{7}{10} = 0.7$	$\frac{7}{100} = 0.07$	$\frac{7}{1000} = 0.007$
1 zero in 10	2 zeros in 100	3 zeros in 1000
1 decimal place in 0.7	2 decimal places in 0.07	3 decimal places in 0.007

To write a decimal in words, write the decimal part of the number as though it were a whole number, and then name the place value of the last digit.

0.9684   nine thousand six hundred eighty-four ten-thousandths

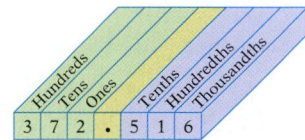

The decimal point in a decimal is read as "and."

372.516   three hundred seventy-two and five hundred sixteen thousandths

Point of Interest

The idea that all fractions should be represented in tenths, hundredths, and thousandths was presented in 1585 in Simon Stevin's publication *De Thiende* and its French translation, *La Disme*, which was widely read and accepted by the French. This may help to explain why the French accepted the metric system so easily 200 years later.

In *De Thiende*, Stevin argued in favor of his notation by including examples for astronomers, tapestry makers, surveyors, tailors, and the like. He stated that using decimals would enable calculations to be "performed . . . with as much ease as counterreckoning."

Objective 2.5A

Vocabulary to Review
standard form [1.1B]
place value [1.1B]

New Vocabulary
decimal notation
decimal
whole number part
decimal point
decimal part

Discuss the Concepts
1. Name the first seven place values to the left of the decimal point. Name the first five place values to the right of the decimal point.
2. What is the purpose of a decimal point in a number?
3. Name three situations in which decimals are used.

Instructor Note
Larger numbers are often written as 7.3 million or 2.3 billion. Have students write these numbers in standard form.

Concept Check
1. Why is the fraction $\frac{13}{100}$ equivalent to the decimal 0.13?
2. When writing a number in standard form, when is it necessary to use the digit 0?

In-Class Examples (Objective 2.5A)

1. Write $\frac{79}{100}$ as a decimal.   0.79

2. Write 0.281 as a fraction. $\frac{281}{1000}$

Write the decimal in words.
3. 6.053   Six and fifty-three thousandths
4. 4.3018
Four and three thousand eighteen ten-thousandths

Write the decimal in standard form.
5. One hundred thirty-four thousandths   0.134
6. Three and fifty-two millionths   3.000052

To write a decimal in standard form when it is written in words, write the whole number part, replace the word *and* with a decimal point, and write the decimal part so that the last digit is in the given place-value position.

four and twenty-three <u>hundredths</u>

3 is in the hundredths place.   4.2<u>3</u>

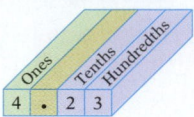

When writing a decimal in standard form, you may need to insert zeros after the decimal point so that the last digit is in the given place-value position.

ninety-one and eight <u>thousandths</u>

8 is in the thousandths place.
Insert two zeros so that the 8 is in   91.00<u>8</u>
the thousandths place.

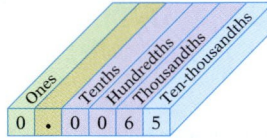

sixty-five <u>ten-thousandths</u>

5 is in the ten-thousandths place.
Insert two zeros so that the 5 is in   0.006<u>5</u>
the ten-thousandths place.

Example 1   Name the place value of the digit 8 in the number 45.687.

Solution   The digit 8 is in the hundredths place.

You Try It 1   Name the place value of the digit 4 in the number 907.1342.

Your solution   Thousandths

Example 2   Write $\frac{43}{100}$ as a decimal.

Solution $\frac{43}{100} = 0.43$ • Forty-three hundredths

You Try It 2   Write $\frac{501}{1000}$ as a decimal.

Your solution   0.501

Example 3   Write 0.289 as a fraction.

Solution $0.289 = \frac{289}{1000}$ • 289 thousandths

You Try It 3   Write 0.67 as a fraction.

Your solution $\frac{67}{100}$

Example 4   Write 293.50816 in words.

Solution   Two hundred ninety-three and fifty thousand eight hundred sixteen hundred-thousandths

You Try It 4   Write 55.6083 in words.

Your solution   Fifty-five and six thousand eighty-three ten-thousandths

Solutions on p. S6

Example 5 Write twenty-three and two hundred forty-seven millionths in standard form.	**You Try It 5** Write eight hundred six and four hundred ninety-one hundred-thousandths in standard form.
Solution 23.000247	**Your solution** 806.00491

Solution on p. S6

Objective B   **To identify the order relation between two decimals**

A whole number can be written as a decimal by writing a decimal point to the right of the last digit. For example,

$$62 = 62. \qquad 497 = 497.$$

You know that $62 and $62.00 both represent 62 dollars. Any number of zeros may be written to the right of the decimal point in a whole number without changing the value of the number.

$$62 = 62.00 = 62.0000 \qquad 497 = 497.0 = 497.000$$

Also, any number of zeros may be written to the right of the last digit in a decimal without changing the value of the number.

$$0.8 = 0.80 = 0.800 \qquad 1.35 = 1.350 = 1.3500 = 1.35000 = 1.350000$$

This fact is used to find the order relation between two decimals.

To compare two decimals, write the decimal part of each number so that each has the same number of decimal places. Then compare the two numbers.

HOW TO Place the correct symbol, $<$ or $>$, between the two numbers 0.693 and 0.71.

$0.71 = 0.710$	• 0.693 has 3 decimal places. 0.71 has 2 decimal places. Write 0.71 with 3 decimal places.
$0.693 < 0.710$	• Compare 0.693 and 0.710. 693 thousandths $<$ 710 thousandths
$0.693 < 0.71$	• Remove the zero written in 0.710.

HOW TO Place the correct symbol, $<$ or $>$, between the two numbers 5.8 and 5.493.

$5.8 = 5.800$	• Write 5.8 with 3 decimal places.
$5.800 > 5.493$	• Compare 5.800 and 5.493. The whole number part (5) is the same. 800 thousandths $>$ 493 thousandths
$5.8 > 5.493$	• Remove the extra zeros written in 5.800.

Point of Interest

The decimal point did not make its appearance until the early 1600s. Stevin's notation used subscripts with circles around them after each digit: 0 for ones, 1 for tenths (which he called "primes"), 2 for hundredths (called "seconds"), 3 for thousandths ("thirds"), and so on. For example, 1.375 would have been written

1 3 7 5
ⓞ ① ② ③

Objective 2.5B

Discuss the Concepts

1. Why do both $35 and $35.00 represent the same amount of money?
2. Explain how to determine the order relation between 0.52 and 0.483.

Concept Check

Write the following decimals in order from largest to smallest.
3.02, 3.20, 3.002, 3.22, 3.022
3.22, 3.20, 3.022, 3.02, 3.002

Instructor Note

When comparing decimals, have students write each number to the same place value by adding zeros if necessary. For example, it's easier to see that $0.693 < 0.71$ when you write it as $0.693 < 0.710$.

Optional Student Activity

Use a number line to show the order of the decimals 2.6 and 2.1.

In-Class Examples (Objective 2.5B)

Place the correct symbol, $<$, $>$, or $=$, between the two numbers.

1. 0.4500   0.45 $0.4500 = 0.45$
2. 0.0782   0.0728 $0.0782 > 0.0728$
3. 1.963   1.639 $1.963 > 1.639$

Example 6 Place the correct symbol, < or >, between the two numbers.

0.039   0.1001

Solution   0.039 = 0.0390

0.0390 < 0.1001

0.039 < 0.1001

You Try It 6 Place the correct symbol, < or >, between the two numbers.

0.065   0.0802

Your solution 0.065 < 0.0802

Example 7 Write the given numbers in order from smallest to largest.

1.01, 1.2, 1.002, 1.1, 1.12

Solution   1.010, 1.200, 1.002, 1.100, 1.120

1.002, 1.010, 1.100, 1.120, 1.200

1.002, 1.01, 1.1, 1.12, 1.2

You Try It 7 Write the given numbers in order from smallest to largest.

3.03, 0.33, 0.3, 3.3, 0.03

Your solution 0.03, 0.3, 0.33, 3.03, 3.3

Solutions on p. S6

Vocabulary to Review
rounding [1.1C]
place value [1.1B]

Discuss the Concepts
Explain how to round a decimal to the nearest hundredth.

Instructor Note
As a calculator activity, have students determine whether their calculators round or truncate. Using 2 ÷ 3 will serve as a good example.

Instructor Note
Explain to students that not all rounding is done as shown here. When sales tax is computed, the decimal is always rounded up to the nearest cent. Thus, a sales tax of $.132 would be rounded to $.14.

Objective C   To round a decimal to a given place value

In general, rounding decimals is similar to rounding whole numbers except that the digits to the right of the given place value are dropped instead of being replaced by zeros.

If the digit to the right of the given place value is less than 5, that digit and all digits to the right are dropped.

Round 6.9237 to the nearest hundredth.

┌── Given place value (hundredths)
6.9237
  └── 3 < 5 Drop the digits 3 and 7.

6.9237 rounded to the nearest hundredth is 6.92.

If the digit to the right of the given place value is greater than or equal to 5, increase the digit in the given place value by 1, and drop all digits to its right.

Round 12.385 to the nearest tenth.

┌── Given place value (tenths)
12.385
   └── 8 > 5 Increase 3 by 1 and drop all digits to the right of 3.

12.385 rounded to the nearest tenth is 12.4.

In-Class Examples (Objective 2.5C)
Round the decimal to the given place value.
1. 0.074; tenths   0.1
2. 840.156; hundredths   840.16
3. 5.60032; nearest whole number   6
4. 0.635457; hundred-thousandths   0.63546

5. The length of the marathon footrace in the Olympics is 42.195 km. What is the length of this race to the nearest tenth of a kilometer?
42.2 km

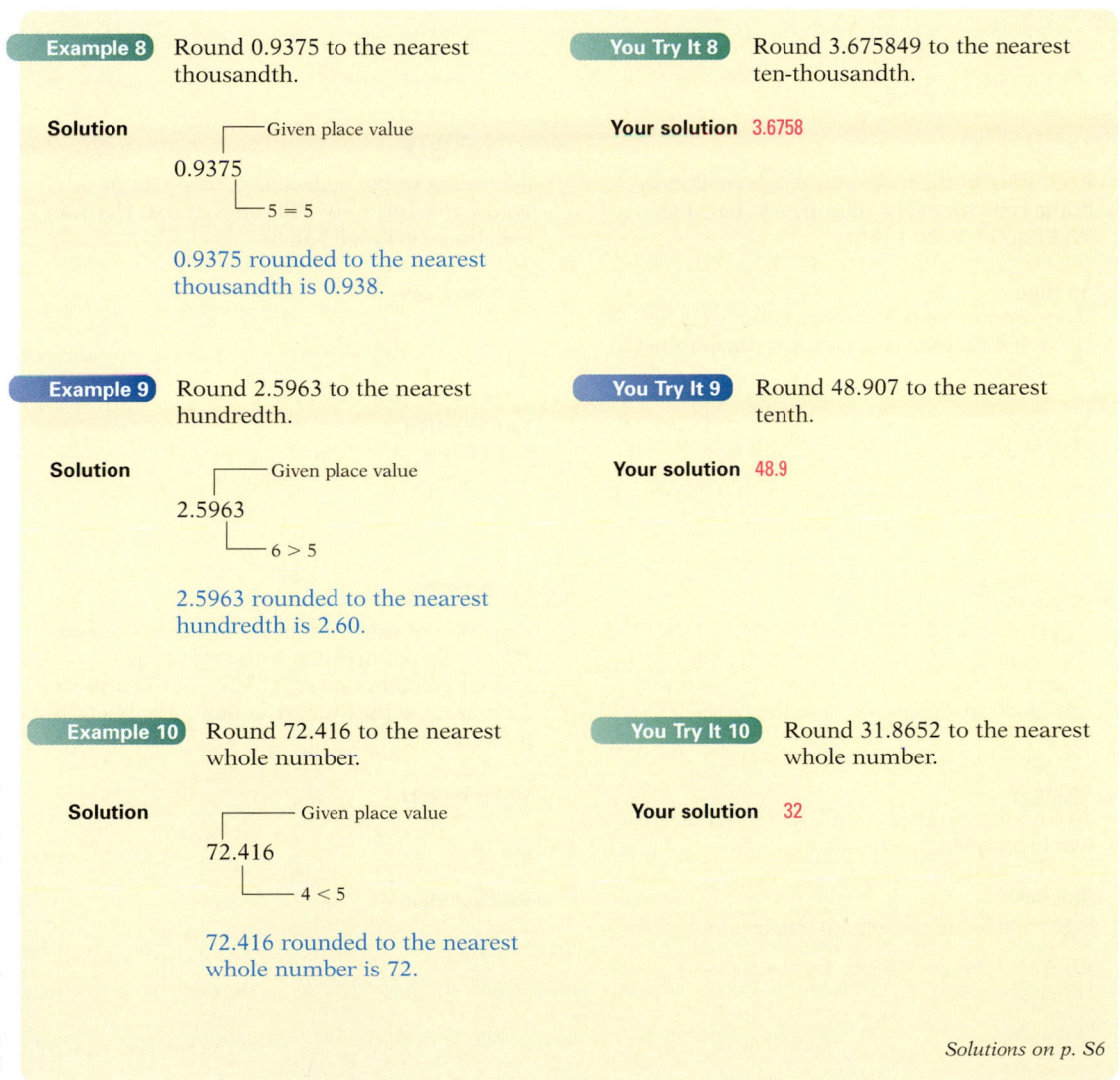

HOW TO Round 0.46972 to the nearest thousandth.

Given place value (thousandths)

0.46972

7 > 5   Round up by adding 1 to the 9 (9 + 1 = 10). Carry the 1 to the hundredths place (6 + 1 = 7).

0.46972 rounded to the nearest thousandth is 0.470.

Note that in this example, the zero in the given place value is not dropped. This indicates that the number is rounded to the nearest thousandth. If we dropped the zero and wrote 0.47, it would indicate that the number was rounded to the nearest hundredth.

Example 8 Round 0.9375 to the nearest thousandth.

Solution

Given place value

0.9375

5 = 5

0.9375 rounded to the nearest thousandth is 0.938.

You Try It 8 Round 3.675849 to the nearest ten-thousandth.

Your solution 3.6758

Example 9 Round 2.5963 to the nearest hundredth.

Solution

Given place value

2.5963

6 > 5

2.5963 rounded to the nearest hundredth is 2.60.

You Try It 9 Round 48.907 to the nearest tenth.

Your solution 48.9

Example 10 Round 72.416 to the nearest whole number.

Solution

Given place value

72.416

4 < 5

72.416 rounded to the nearest whole number is 72.

You Try It 10 Round 31.8652 to the nearest whole number.

Your solution 32

Solutions on p. S6

Concept Check

Round $\frac{49}{50}$ to the nearest tenth. Explain why it is incorrect to write the answer as 1.

1.0 indicates that the decimal has been rounded to the nearest tenth; 1 does not.

Optional Student Activity

Bring some newspapers to class. Arrange the students in small groups. Have students find and list situations in which decimals are used. Also ask them to determine whether the decimals they find are exact or approximations. For example, large numbers, such as 3.2 billion, used for describing the balance of trade or the national debt are approximations, whereas numbers used to describe business transactions, such as the exchange rate or stock dividends, are exact.

Objective 2.5D

Optional Student Activity

Bring to class pages from the business section of a newspaper. Arrange the students in small groups. Have each group select several companies from those listed in the newspaper. For each company, write down the price of a share of stock. List the prices in order from least to greatest. Then have the students round each price to the nearest dollar.

Objective D   **To solve application problems**

The table below shows the number of home runs hit, for every 100 times at bat, by four Major League baseball players. Use this table for Example 11 and You Try It 11.

Babe Ruth

Home Runs Hit for Every 100 At-Bats	
Harmon Killebrew	7.03
Ralph Kiner	7.09
Babe Ruth	8.05
Ted Williams	6.76

Source: Major League Baseball

Example 11

According to the table above, who had more home runs for every 100 times at bat, Ted Williams or Babe Ruth?

Strategy

To determine who had more home runs for every 100 times at bat, compare the numbers 6.76 and 8.05.

Solution

8.05 > 6.76

Babe Ruth had more home runs for every 100 at-bats.

You Try It 11

According to the table above, who had more home runs for every 100 times at bat, Harmon Killebrew or Ralph Kiner?

Your Strategy

Your solution

Ralph Kiner

Example 12

On average, an American goes to the movies 4.56 times per year. To the nearest whole number, how many times per year does an American go to the movies?

Strategy

To find the number, round 4.56 to the nearest whole number.

Solution

4.56 rounded to the nearest whole number is 5.

An American goes to the movies about 5 times per year.

You Try It 12

One of the driest cities in the Southwest is Yuma, Arizona, with an average annual precipitation of 2.65 in. To the nearest inch, what is the average annual precipitation in Yuma?

Your Strategy

Your solution

3 in.

Solutions on p. S7

In-Class Examples (Objective 2.5D)

1. The average rainfall in Seattle, Washington, in December is 5.6 in. In January, the average rainfall is 5.1 in.
   a. During which month, December or January, is the average rainfall greater?   December
   b. To the nearest inch, what is the average rainfall in December?   6 in.

2.5 Exercises

Objective A   To read and write decimals

Name the place value of the digit 5.

1. 76.31587
Thousandths

2. 291.508
Tenths

3. 432.09157
Ten-thousandths

4. 0.0006512
Hundred-thousandths

5. 38.2591
Hundredths

6. 0.0000853
Millionths

Write the fraction as a decimal.

7. $\frac{3}{10}$
0.3

8. $\frac{9}{10}$
0.9

9. $\frac{21}{100}$
0.21

10. $\frac{87}{100}$
0.87

11. $\frac{461}{1000}$
0.461

12. $\frac{853}{1000}$
0.853

13. $\frac{93}{1000}$
0.093

14. $\frac{61}{1000}$
0.061

Write the decimal as a fraction.

15. 0.1
$\frac{1}{10}$

16. 0.3
$\frac{3}{10}$

17. 0.47
$\frac{47}{100}$

18. 0.59
$\frac{59}{100}$

19. 0.289
$\frac{289}{1000}$

20. 0.601
$\frac{601}{1000}$

21. 0.09
$\frac{9}{100}$

22. 0.013
$\frac{13}{1000}$

Write the number in words.

23. 0.37
Thirty-seven hundredths

24. 25.6
Twenty-five and six tenths

25. 9.4
Nine and four tenths

26. 1.004
One and four thousandths

27. 0.0053
Fifty-three ten-thousandths

28. 41.108
Forty-one and one hundred
eight thousandths

29. 0.045
Forty-five thousandths

30. 3.157
Three and one hundred
fifty-seven thousandths

31. 26.04
Twenty-six and
four hundredths

Quick Quiz (Objective 2.5A)

1. Write $\frac{9}{1000}$ as a decimal.   0.009

Write the decimal in words.

2. 2.379
Two and three hundred seventy-nine thousandths

3. 0.00043   Forty-three hundred-thousandths

Write the decimal in standard form.

4. Nine and seven hundred four thousandths   9.704

5. Five and seventeen ten-thousandths   5.0017

Suggested Assignment
Exercises 1–81, odds
More challenging problems:
    Exercises 82, 83

Write the number in standard form.

32. Six hundred seventy-two thousandths
0.672

33. Three and eight hundred
six ten-thousandths
3.0806

34. Nine and four hundred seven
ten-thousandths
9.0407

35. Four hundred seven and three hundredths
407.03

36. Six hundred twelve and seven hundred
four thousandths
612.704

37. Two hundred forty-six and twenty-four
thousandths
246.024

38. Two thousand sixty-seven and nine
thousand two ten-thousandths
2067.9002

39. Seventy-three and two thousand six
hundred eighty-four hundred-thousandths
73.02684

Objective B   **To identify the order relation between two decimals**

Place the correct symbol, < or >, between the two numbers.

40. 0.16 0.6
<

41. 0.7 0.56
>

42. 5.54 5.45
>

43. 3.605 3.065
>

44. 0.047 0.407
<

45. 9.004 9.04
<

46. 1.0008 1.008
<

47. 9.31 9.031
>

48. 7.6005 7.605
<

49. 4.6 40.6
<

50. 0.31502 0.3152
<

51. 0.07046 0.07036
>

Write the given numbers in order from smallest to largest.

52. 0.39, 0.309, 0.399
0.309, 0.39, 0.399

53. 0.66, 0.699, 0.696, 0.609
0.609, 0.66, 0.696, 0.699

54. 0.24, 0.024, 0.204, 0.0024
0.0024, 0.024, 0.204, 0.24

55. 1.327, 1.237, 1.732, 1.372
1.237, 1.327, 1.372, 1.732

56. 0.06, 0.059, 0.061, 0.0061
0.0061, 0.059, 0.06, 0.061

57. 21.87, 21.875, 21.805, 21.78
21.78, 21.805, 21.87, 21.875

Quick Quiz (Objective 2.5B)

Place the correct symbol, <, >, or =, between the
two numbers.
1. 6.287 6.782   6.287 < 6.782
2. 1.035 1.035000   1.035 = 1.035000
3. 0.00923 0.156   0.00923 < 0.156

Objective C   **To round a decimal to a given place value**

Round the number to the given place value.

58. 6.249; tenths
6.2

59. 5.398; tenths
5.4

60. 21.007; tenths
21.0

61. 30.0092; tenths
30.0

62. 18.40937; hundredths
18.41

63. 413.5972; hundredths
413.60

64. 72.4983; hundredths
72.50

65. 6.061745; thousandths
6.062

66. 936.2905; thousandths
936.291

67. 96.8027; whole number
97

68. 47.3192; whole number
47

69. 5439.83; whole number
5440

70. 7014.96; whole number
7015

71. 0.023591; ten-thousandths
0.0236

72. 2.975268; hundred-thousandths
2.97527

Objective D   **To solve application problems**

73. **Weight**   A nickel weighs about 0.1763668 oz. Find the weight of a nickel to the nearest hundredth of an ounce.
0.18 oz

74. **Purchases**   The total cost of a parka, including sales tax, is $124.1093. Round the total cost to the nearest cent to find the amount a customer pays for the parka.
$124.11

75. **Boston Marathon**   Runners in the Boston Marathon run a distance of 26.21875 mi. To the nearest tenth of a mile, find the distance an entrant who completes the Boston Marathon runs.
26.2 miles

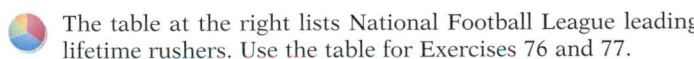 The table at the right lists National Football League leading lifetime rushers. Use the table for Exercises 76 and 77.

76. **Football**   Who had the greater average number of yards per carry, Tony Dorsett or Emmitt Smith?
Tony Dorsett

77. **Football**   Of all the players listed in the table, who has the greatest average number of yards per carry?
Barry Sanders

Football Player	Average Number of Yards per Carry
Eric Dickerson	4.43
Tony Dorsett	4.34
Walter Payton	4.36
Barry Sanders	4.99
Emmitt Smith	4.24

Source: Pro Football Hall of Fame

78. **Life Expectancy**   The average life expectancy in Great Britain is 75.3 years. The average life expectancy in Italy is 75.5 years (*Source:* U.S. Centers for Disease Control). In which country is the average life expectancy higher, Great Britain or Italy?
Italy

Quick Quiz (Objective 2.5C)
Round the decimal to the given place value.
1. 9.1384; tenths   9.1
2. 512.677; hundredths   512.68
3. 7.880102; nearest whole number   8
4. 3.4978205; hundred-thousandths   3.49782

Answers to Writing Exercises

84. Today, the timed events in the Olympics are usually recorded to the nearest hundredth. For example, Carl Lewis ran the 100-meter dash in 9.92 s and the 200-meter dash in 19.80 s. Florence Griffith-Joyner ran the 100-meter dash in 10.54 s. Bonnie Blair skated the 1000-meter event in 78.74 s.

85. A sales tax or meal tax is always rounded up to the nearest cent. Also, if a product is priced at "3 for $1.00," a customer is charged $.34 for one item.

Generally, interest paid by an institution is rounded down; for example, interest paid by a bank on a NOW account.

86. The Richter scale, presented by Beno Guttenberg and Charles Francis Richter in 1935, is used to quantitatively measure the magnitude of an earthquake. The weakest earthquakes that had been detected at that time were assigned values close to 0. Each increase of 1 unit on the Richter scale represents 10 times the magnitude. A microearthquake is a low-intensity earthquake of magnitude 2 or less on the Richter scale. An earthquake of magnitude 4.5 on the Richter scale will cause slight damage. An earthquake of magnitude 6 on the Richter scale is moderately destructive. No recorded earthquake has exceeded magnitude 9 on the Richter scale. The instrument used to measure earthquakes is the seismograph.

79. The Olympics The length of the marathon footrace in the Olympics is 42.195 km. What is the length of this race to the nearest tenth of a kilometer?
42.2 km

80. Credit Cards Credit card companies generally require a minimum payment on the balance of the account each month. Use the minimum payment schedule shown below to determine the minimum payment due on the given account balances.
a. $187.93 $20.00
b. $342.55 $35.00
c. $261.48 $30.00
d. $16.99 $16.99
e. $310.00 $35.00
f. $158.32 $20.00
g. $200.10 $25.00

If the New Balance Is:	The Minimum Required Payment Is:
Up to $20.00	The new balance
$20.01 to $200.00	$20.00
$200.01 to $250.00	$25.00
$250.01 to $300.00	$30.00
$300.01 to $350.00	$35.00
$350.01 to $400.00	$40.00

81. Shipping and Handling Shipping and handling charges when ordering online generally are based on the dollar amount of the order. Use the table shown below to determine the cost of shipping each order.
a. $12.42 $2.40
b. $23.56 $3.60
c. $47.80 $6.00
d. $66.91 $7.00
e. $35.75 $4.70
f. $20.00 $2.40
g. $18.25 $2.40

If the Amount Ordered Is:	The Shipping and Handling Charge Is:
$10.00 and under	$1.60
$10.01 to $20.00	$2.40
$20.01 to $30.00	$3.60
$30.01 to $40.00	$4.70
$40.01 to $50.00	$6.00
$50.01 and up	$7.00

APPLYING THE CONCEPTS

82. Indicate which digits of the number, if any, need not be entered on a calculator.
a. 1.500 **b.** 0.908 **c.** 60.07 **d.** 0.0032
a. 1.5~~00~~ **b.** ~~0~~.908 **c.** 60.07 **d.** ~~0~~.0032

83. Find a number between (a) 0.1 and 0.2, (b) 1 and 1.1, and (c) 0 and 0.005.
For example, (a) 0.15 (b) 1.05 (c) 0.001

84. To what place value are timed events in the Olympics recorded? Provide some specific examples of events and the winning times for each.

85. Provide an example of a situation in which a decimal is always rounded up, even if the digit to the right is less than 5. Provide an example of a situation in which a decimal is always rounded down, even if the digit to the right is 5 or greater than 5. (*Hint:* Think about situations in which money changes hands.)

86. Prepare a report on the Richter scale. Include in your report the magnitudes that classify an earthquake as strong or moderate, the magnitudes that classify an earthquake as a microearthquake, and the largest known recorded shocks.

Quick Quiz (Objective 2.5D)

1. The average annual precipitation in Albany, New York, is 38.60 in. The average annual precipitation in Columbus, Ohio, is 38.52 in.

a. In which city is the average annual precipitation greater? Albany

b. To the nearest inch, what is the average annual precipitation in Columbus? 39 in.

2.6 Operations on Decimals

Objective A To add and subtract decimals

To add decimals, write the numbers so that the decimal points are on a vertical line. Add as you would with whole numbers. Then write the decimal point in the sum directly below the decimal points in the addends.

HOW TO Add: $0.326 + 4.8 + 57.23$

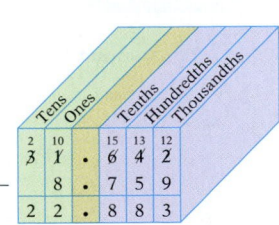

- Note that placing the decimal points on a vertical line ensures that digits of the same place value are added.

The sum is 62.356.

Point of Interest

Try this: Six different numbers are added together and their sum is 11. Four of the six numbers are 4, 3, 2, and 1. Find the other two numbers.

HOW TO Find the sum of 0.64, 8.731, 12, and 5.9

$$
\begin{array}{r}
\overset{1\;2}{}\\
0.64\\
8.731\\
12.\\
+\;5.9\\
\hline
27.271
\end{array}
$$

- Arrange the numbers vertically, placing the decimal points on a vertical line.

- Add the numbers in each column. Write the decimal point in the sum directly below the decimal points in the addends.

The sum is 27.271.

To subtract decimals, write the numbers so that the decimal points are on a vertical line. Subtract as you would with whole numbers. Then write the decimal point in the difference directly below the decimal point in the subtrahend.

HOW TO Subtract and check: $31.642 - 8.759$

- Note that placing the decimal points on a vertical line ensures that digits of the same place value are subtracted.

The difference is 22.883.

$$
\begin{array}{l}
\textit{Check:} \quad \text{Subtrahend} \qquad 8.759\\
\qquad\quad +\text{ Difference} \quad +22.883\\
\qquad\quad =\text{ Minuend} \qquad \overline{31.642}
\end{array}
$$

Vocabulary to Review

sum [1.2A]
difference [1.2B]

Discuss the Concepts

1. When adding decimals using a vertical format, why must the decimal points be aligned?
2. Explain how to subtract 2.6357 from 4.79. Why is 4.79 rewritten as 4.7900 before the subtraction is performed?

Instructor Note

The problem presented in the Point of Interest has a number of possible answers; for example, 0.2 and 0.8, 0.7 and 0.3, or 0.6 and 0.4.

Concept Check

Place plus signs between three of the numbers listed below so that the sum is 19.24.

$$0.12$$
$$3.45$$
$$6.78$$
$$9.01$$
$$2.34$$
$$3.45 + 6.78 + 9.01 = 19.24$$

In-Class Examples (Objective 2.6A)

1. Add: $3.514 + 22.6981 + 145.78$ 171.9921
2. What is 5.042 less than 12.36?   7.318
3. Subtract and check: $7.05 - 6.274$ 0.776
4. Estimate the difference between 8.769 and 3.515.
   5
5. Evaluate $x + y + z$ when $x = 3.5765$, $y = 35$, and $z = 11.08$.   49.6565

Optional Student Activity

Complete the following addition table. How is the Commutative Property of Multiplication represented in the table?

+	0.1	0.2	0.3	0.4	0.5
0.1					
0.2					
0.3					
0.4					
0.5					

Row 1: 0.2, 0.3, 0.4, 0.5, 0.6;
Row 2: 0.3, 0.4, 0.5, 0.6, 0.7;
Row 3: 0.4, 0.5, 0.6, 0.7, 0.8;
Row 4: 0.5, 0.6, 0.7, 0.8, 0.9;
Row 5: 0.6, 0.7, 0.8, 0.9, 1.0.
Answers will vary. For example: 0.4 + 0.3 = 0.3 + 0.4.

Instructor Note

The answers to the brain teaser in the Point of Interest is that the two coins are a half dollar and a nickel. One of these coins is not a nickel; it is a half dollar.

Point of Interest

Try this brain teaser. You have two U.S. coins that add up to $.55. One is not a nickel. What are these two coins?

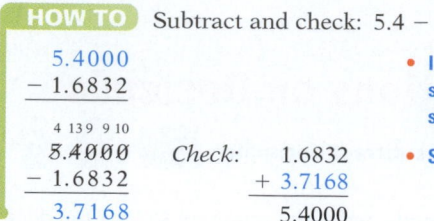

HOW TO   Subtract and check: $5.4 - 1.6832$

$$\begin{array}{r} 5.4000 \\ -\ 1.6832 \\ \hline \end{array}$$

$$\begin{array}{r} {}^{4\ 13\ 9\ 9\ 10} \\ \cancel{5}.4\cancel{0}\cancel{0}\cancel{0} \\ -\ 1.6832 \\ \hline 3.7168 \end{array}$$

Check: $\begin{array}{r} 1.6832 \\ +\ 3.7168 \\ \hline 5.4000 \end{array}$

- Insert zeros in the minuend so that it has the same number of decimal places as the subtrahend.
- Subtract and then check.

Recall that to estimate the answer to a calculation, round each number to the highest place value of the number; the first digit of each number will be nonzero and all other digits will be zero. Perform the calculation using the rounded numbers.

HOW TO   Estimate the sum of 23.037 and 16.7892.

$$\begin{array}{rcr} 23.037 & \longrightarrow & 20 \\ 16.7892 & \longrightarrow & +\ 20 \\ & & \hline \\ & & 40 \end{array}$$

$$\begin{array}{r} 23.037 \\ +\ 16.7892 \\ \hline 39.8262 \end{array}$$

- Round each number to the nearest ten.
- Add the rounded numbers.

40 is an estimate of the sum of 23.037 and 16.7892. Note that 40 is very close to the actual sum of 39.8262.

When a number in an estimation is a decimal less than 1, round the decimal so that there is one nonzero digit.

HOW TO   Estimate the difference between 4.895 and 0.6193.

$$\begin{array}{rcr} 4.895 & \longrightarrow & 5.0 \\ 0.6193 & \longrightarrow & -\ 0.6 \\ & & \hline \\ & & 4.4 \end{array}$$

$$\begin{array}{r} 4.8950 \\ -0.6193 \\ \hline 4.2757 \end{array}$$

- Round 4.895 to the nearest one. Round 0.6193 to the nearest tenth.
- Subtract the rounded numbers.

4.4 is an estimate of the difference between 4.895 and 0.6193. It is close to the actual difference of 4.2757.

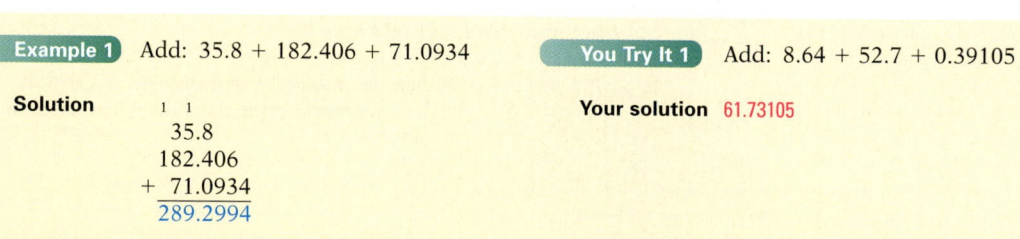

Example 1   Add: $35.8 + 182.406 + 71.0934$

Solution

$$\begin{array}{r} {}^{1\ \ \ 1} \\ 35.8 \\ 182.406 \\ +\ 71.0934 \\ \hline 289.2994 \end{array}$$

You Try It 1   Add: $8.64 + 52.7 + 0.39105$

Your solution   61.73105

Solution on p. S7

Example 2	Subtract and check: 73 − 8.16

Solution

    612 9 10
    73.00
 − 8.16
   64.84

Check:
  8.16
+64.84
 73.00

You Try It 2	Subtract and check: 25 − 4.91

Your solution 20.09

Example 3	Estimate the sum of 0.3927, 0.4856, and 0.2104.

Solution

0.3927 ⟶   0.4
0.4856 ⟶   0.5
0.2104 ⟶ +0.2
              1.1

You Try It 3	Estimate the sum of 6.514, 8.903, and 2.275.

Your solution 18

Example 4	Evaluate $x + y + z$ when $x = 1.6$, $y = 7.9$, and $z = 4.8$.

Solution

$x + y + z$
$1.6 + 7.9 + 4.8 = 9.5 + 4.8$
$\qquad\qquad\qquad\quad = 14.3$

You Try It 4	Evaluate $x + y + z$ when $x = 7.84$, $y = 3.05$, and $z = 2.19$.

Your solution 13.08

Solutions on p. S7

Objective B **To multiply decimals**

Decimals are multiplied as though they were whole numbers; then the decimal point is placed in the product. Writing the decimals as fractions shows where to write the decimal point in the product.

$$0.4 \cdot 2 = \frac{4}{10} \cdot \frac{2}{1} = \frac{8}{10} = 0.8$$

1 decimal place in 0.4   1 decimal place in 0.8

$$0.4 \cdot 0.2 = \frac{4}{10} \cdot \frac{2}{10} = \frac{8}{100} = 0.08$$

1 decimal place in 0.4
1 decimal place in 0.2   2 decimal places in 0.08

Optional Student Activity

Find the pattern in each of the following lists of numbers. Then name the next three numbers in each list.

1. 0.379, 0.329, 0.279, 0.229, . . .
   0.179, 0.129, 0.079
2. 1.684, 1.574, 1.464, 1.354, . . .
   1.244, 1.234, 1.024
3. 9.5, 8.75, 8, 7.25, . . .
   6.5, 5.75, 5

Objective 2.6B

Vocabulary to Review

product [1.3A]

In-Class Examples (Objective 2.6B)

1. Multiply: 0.97(0.632) 0.61304
2. Estimate the product of 0.842 and 5.7. 4.8
3. Find the product of 3.61 and 10^4. 36,100
4. Evaluate $30pq$ when $p = 5.4$ and $q = 0.2$. 32.4

Discuss the Concepts

1. Explain how to place the decimal point in a product when multiplying decimals.

2. How do you determine how many places to move the decimal point in a number being multiplied by a power of 10 written in exponential notation?

Concept Check

Light travels at a speed of 1.86×10^5 mi/s. Write this number in standard form.

186,000 mi/s

Optional Student Activity

Provide an example of multiplying tenths by hundredths using fractions. Rewrite the example using decimals in place of the fractions. Show that the results are the same.

Optional Student Activity

Each ▽ in the problem below represents one of the digits 2, 3, 4, 5, 6, 7, 8, or 9. The digits 0 and 1 are provided for you. Fill in the remaining digits.

$$\begin{array}{r} 0.\triangledown\triangledown\triangledown \\ \times \quad 0.\triangledown\triangledown \\ \hline 0.\triangledown\triangledown\triangledown\, 0\, 1 \end{array}$$

$0.927 \times 0.63 = 0.58401$

$$0.4 \cdot 0.02 = \frac{4}{10} \cdot \frac{2}{100} = \frac{8}{1000} = 0.008$$

1 decimal place in 0.4
2 decimal places in 0.02
3 decimal places in 0.008

To multiply decimals, multiply the numbers as you would whole numbers. Then write the decimal point in the product so that the number of decimal places in the product is the sum of the numbers of decimal places in the factors.

Integrating Technology

Scientific calculators have a floating decimal point. This means that the decimal point is automatically placed in the answer. For example, for the product at the right, enter

3 2 · 4 1 × 7 · 6 =

The display reads 246.316, with the decimal point in the correct position.

HOW TO Multiply: (32.41)(7.6)

$$\begin{array}{r} 32.41 \quad \text{2 decimal places} \\ \times \quad 7.6 \quad \text{1 decimal place} \\ \hline 19446 \\ 22687 \\ \hline 246.316 \quad \text{3 decimal places} \end{array}$$

Estimating the product of 32.41 and 7.6 shows that the decimal point has been correctly placed in the example above.

$$\begin{array}{rcl} 32.41 & \longrightarrow & 30 \\ 7.6 & \longrightarrow & \times\ 8 \\ & & \overline{240} \end{array}$$

• Round 32.41 to the nearest ten.
• Round 7.6 to the nearest one.
• Multiply the two numbers.

240 is an estimate of (32.41)(7.6). It is close to the actual product 246.316.

HOW TO Multiply: 0.061(0.08)

$$\begin{array}{r} 0.061 \quad \text{3 decimal places} \\ \times \quad 0.08 \quad \text{2 decimal places} \\ \hline 0.00488 \quad \text{5 decimal places} \end{array}$$

• Insert two zeros between the 4 and the decimal point so that there are 5 decimal places in the product.

To multiply a decimal by a power of 10 (10, 100, 1000, . . .), move the decimal point to the right the same number of places as there are zeros in the power of 10.

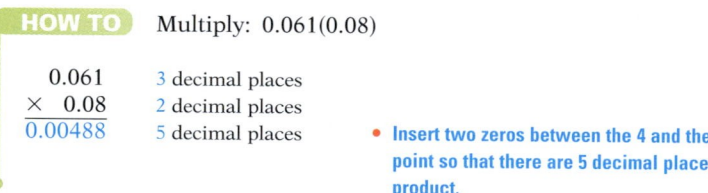

$2.7935 \cdot \underline{10}\qquad = 27.935$

1 zero — 1 decimal place

$2.7935 \cdot \underline{100}\qquad = 279.35$

2 zeros — 2 decimal places

$2.7935 \cdot \underline{1000}\qquad = 2793.5$

3 zeros — 3 decimal places

$2.7935 \cdot \underline{10,000}\qquad = 27,935.$

4 zeros — 4 decimal places

$2.7935 \cdot \underline{100,000} = 279,350.$

5 zeros — 5 decimal places

• A zero must be inserted before the decimal point.

Note that if the power of 10 is written in exponential notation, the exponent indicates how many places to move the decimal point.

$2.7935 \cdot 10^1 = 27.935$
1 decimal place

$2.7935 \cdot 10^2 = 279.35$
2 decimal places

$2.7935 \cdot 10^3 = 2793.5$
3 decimal places

$2.7935 \cdot 10^4 = 27,935.$
4 decimal places

$2.7935 \cdot 10^5 = 279,350.$
5 decimal places

HOW TO Find the product of 64.18 and 10^3.

$64.18 \cdot 10^3 = 64,180$ • The exponent on 10 is 3. Move the decimal point in 64.18 three places to the right.

Example 5 Multiply: 0.00073(0.052)

Solution
$$
\begin{array}{r}
0.00073 \\
\times\ \ \ \ 0.052 \\
\hline
146 \\
365 \\
\hline
0.00003796
\end{array}
$$

You Try It 5 Multiply: 0.000081(0.025)

Your solution 0.000002025

Example 6 Estimate the product of 0.7639 and 0.2188.

Solution
$$
\begin{array}{r}
0.7639 \longrightarrow\ \ \ 0.8 \\
0.2188 \longrightarrow \times\,0.2 \\
\hline
0.16
\end{array}
$$

You Try It 6 Estimate the product of 6.407 and 0.959.

Your solution 6

Example 7 What is 835.294 multiplied by 1000?

Solution $835.294 \cdot 1000 = 835,294$

You Try It 7 Find the product of 1.756 and 10^4.

Your solution 17,560

Solutions on p. S7

Example 8   Evaluate $50ab$ when $a = 0.9$ and $b = 0.2$.

Solution $50ab$
$50(0.9)(0.2) = 45(0.2)$
                $\hphantom{50(0.9)(0.2)} = 9$

You Try It 8   Evaluate $25xy$ when $x = 0.8$ and $y = 0.6$.

Your solution   12

Solution on p. S7

Objective 2.6C

Vocabulary to Review

divisor [1.3C]
dividend [1.3C]
quotient [1.3C]

New Symbol

≈ (is approximately equal to)

Discuss the Concepts

Explain how to divide a number by a power of 10.

Instructor Note

The paragraph at the right gives the justification for moving the decimal point before decimal numbers are divided.

Concept Check

Which of the following examples have the same quotient?

a. $15\overline{)180}$
b. $1.5\overline{)1.8}$
c. $1.5\overline{)18}$
d. $0.15\overline{)18}$
e. $0.15\overline{)1.8}$

a, c, and e have the same quotient.

Instructor Note

An alternative method for rounding quotients is

$$
\begin{array}{r}
1.86 \\
3\overline{)5.60} \\
-3 \\
\hline
2\,6 \\
-2\,4 \\
\hline
20 \\
-18 \\
\hline
2
\end{array}
$$

Since the last remainder is more than one-half of the divisor, 1.86 is rounded to 1.87.

Objective C   To divide decimals

Point of Interest

Benjamin Banneker (1731–1806) was the first African American to earn distinction as a mathematician and a scientist. He was on the survey team that determined the boundaries of Washington, D.C. The mathematics of surveying requires extensive use of decimals.

To divide decimals, move the decimal point in the divisor to the right so that the divisor is a whole number. Move the decimal point in the dividend the same number of places to the right. Place the decimal point in the quotient directly above the decimal point in the dividend. Then divide as you would with whole numbers.

HOW TO   Divide: $29.585 \div 4.85$

$$
4.85.\overline{)29.58.5}
$$

Move the decimal point 2 places to the right in the divisor. Move the decimal point 2 places to the right in the dividend. Place the decimal point in the quotient. Then divide as shown at the right.

$$
\begin{array}{r}
6.1 \\
485.\overline{)2958.5} \\
-2910 \\
\hline
48\,5 \\
-48\,5 \\
\hline
0
\end{array}
$$

Moving the decimal point the same number of places in the divisor and the dividend does not change the quotient because the process is the same as multiplying the numerator and denominator of a fraction by the same number. For the last example,

$$
4.85\overline{)29.585} = \frac{29.585}{4.85} = \frac{29.585 \cdot 100}{4.85 \cdot 100} = \frac{2958.5}{485} = 485\overline{)2958.5}
$$

In division of decimals, rather than writing the quotient with a remainder, we usually round the quotient to a specified place value. **The symbol ≈, which is read "is approximately equal to," is used to indicate that the quotient is an** approximate value after being rounded.

HOW TO   Divide and round to the nearest tenth: $0.86 \div 0.7$

$$
\begin{array}{r}
1.22 \approx 1.2 \\
0.7.\overline{)0.8.60} \\
-7 \\
\hline
1\,6 \\
-1\,4 \\
\hline
20 \\
-14 \\
\hline
6
\end{array}
$$

To round the quotient to the nearest tenth, the division must be carried to the hundredths place. Therefore, zeros must be inserted in the dividend so that the quotient has a digit in the hundredths place.

In-Class Examples (Objective 2.6C)

1. Divide: $83.08 \div 6.2$ 13.4
2. Estimate the quotient of 41.52 and 3.7.   10
3. Divide and round to the nearest hundredth:
    $32.087 \div 0.72$ 44.57
4. What is the quotient of 3812.5 and 1000?   3.8125

5. Evaluate $\dfrac{x}{y}$ when $x = 0.161$ and $y = 0.7$.   0.23

To divide a decimal by a power of 10 (10, 100, 1000, 10,000, . . .), move the decimal point to the left the same number of places as there are zeros in the power of 10.

$462.81 \div 1\underline{0}\qquad = 46.281$

    1 zero   1 decimal place

$462.81 \div 1\underline{00}\qquad = 4.6281$

    2 zeros   2 decimal places

$462.81 \div 1\underline{000}\qquad = 0.46281$

    3 zeros   3 decimal places

$462.81 \div 1\underline{0,000} = 0.046281$

    4 zeros   4 decimal places

- **A zero must be inserted between the decimal point and the 4.**

$462.81 \div 1\underline{00,000} = 0.0046281$

    5 zeros   5 decimal places

- **Two zeros must be inserted between the decimal point and the 4.**

If the power of 10 is written in exponential notation, the exponent indicates how many places to move the decimal point.

$462.81 \div 10^1 = 46.281$

                 1 decimal place

$462.81 \div 10^2 = 4.6281$

                 2 decimal places

$462.81 \div 10^3 = 0.46281$

                 3 decimal places

$462.81 \div 10^4 = 0.046281$

                 4 decimal places

$462.81 \div 10^5 = 0.0046281$

                 5 decimal places

HOW TO Find the quotient of 3.59 and 100.

$3.59 \div 100 = 0.0359$ • **There are two zeros in 100. Move the decimal point in 3.59 two places to the left.**

HOW TO What is the quotient of 64.79 and 10^4?

$64.79 \div 10^4 = 0.006479$ • **The exponent on 10 is 4. Move the decimal point in 64.79 four places to the left.**

Example 9   Divide: $431.97 \div 7.26$

Solution

$$
\begin{array}{r}
5\,9.5 \\
7.26\overline{)4\,3\,1.9\,7\,0} \\
-3\,6\,3\,0 \\
\hline
6\,8\,9\,7 \\
-6\,5\,3\,4 \\
\hline
3\,6\,3\,0 \\
-3\,6\,3\,0 \\
\hline
0
\end{array}
$$

You Try It 9   Divide: $314.746 \div 6.53$

Your solution   48.2

Example 10   Estimate the quotient of 8.37 and 0.219.

Solution

$8.37 \longrightarrow 8$
$0.219 \longrightarrow 0.2$

$8 \div 0.2 = 40$

You Try It 10   Estimate the quotient of 62.7 and 3.45.

Your solution   20

Example 11   Divide and round to the nearest hundredth: $448.2 \div 53$

Solution

$$
\begin{array}{r}
8.4\,5\,6 \approx 8.46 \\
53\overline{)4\,4\,8.2\,0\,0} \\
-4\,2\,4 \\
\hline
2\,4\,2 \\
-2\,1\,2 \\
\hline
3\,0\,0 \\
-2\,6\,5 \\
\hline
3\,5\,0 \\
-3\,1\,8 \\
\hline
3\,2
\end{array}
$$

You Try It 11   Divide and round to the nearest thousandth: $519.37 \div 86$

Your solution   6.039

Example 12   Find the quotient of 592.4 and 10^4.

Solution $592.4 \div 10^4 = 0.05924$

You Try It 12   What is 63.7 divided by 100?

Your solution   0.637

Example 13   Evaluate $\frac{x}{y}$ when $x = 76.8$ and $y = 0.8$.

Solution

$\dfrac{x}{y}$

$\dfrac{76.8}{0.8} = 96$

You Try It 13   Evaluate $\frac{x}{y}$ when $x = 40.6$ and $y = 0.7$.

Your solution   58

Solutions on p. S7

Objective D

To convert between decimals and fractions and identify the order relation between a decimal and a fraction

VIDEO & DVD / CD TUTOR / WEB / SSM

TAKE NOTE

The fraction bar can be read "divided by."

$$\frac{3}{4} = 3 \div 4$$

Dividing the numerator by the denominator results in a remainder of 0. The decimal 0.75 is a terminating decimal.

Because the fraction bar can be read "divided by," any fraction can be written as a decimal. To write a fraction as a decimal, divide the numerator of the fraction by the denominator.

HOW TO Convert $\frac{3}{4}$ to a decimal.

$$\begin{array}{r} 0.75 \\ 4\overline{)3.00} \\ -2\,8 \\ \hline 20 \\ -20 \\ \hline 0 \end{array}$$ ← This is a **terminating decimal.**

← The remainder is zero.

$$\frac{3}{4} = 0.75$$

HOW TO Convert $\frac{5}{11}$ to a decimal.

$$\begin{array}{r} 0.4545 \\ 11\overline{)5.0000} \\ -4\,4 \\ \hline 60 \\ -55 \\ \hline 50 \\ -44 \\ \hline 60 \\ -55 \\ \hline 5 \end{array}$$ ← This is a **repeating decimal.**

← The remainder is never zero.

TAKE NOTE

No matter how far we carry out the division, the remainder is never zero. The decimal $0.\overline{45}$ is a repeating decimal.

$$\frac{5}{11} = 0.\overline{45}$$ The bar over the digits 45 is used to show that these digits repeat.

HOW TO Convert $2\frac{4}{9}$ to a decimal.

$$\begin{array}{r} 0.444 = 0.\overline{4} \\ 9\overline{)4.000} \end{array}$$ • Write the fractional part of the mixed number as a decimal. Divide the numerator by the denominator.

$$2\frac{4}{9} = 2.\overline{4}$$ • The whole number part of the mixed number is the whole number part of the decimal.

To convert a decimal to a fraction, remove the decimal point and place the decimal part over a denominator equal to the place value of the last digit in the decimal.

$$0.57 = \overset{\text{hundredths}}{\frac{57}{100}}$$ $$7.65 = 7\overset{\text{hundredths}}{\frac{65}{100}} = 7\frac{13}{20}$$ $$8.6 = 8\overset{\text{tenths}}{\frac{6}{10}} = 8\frac{3}{5}$$

In-Class Examples (Objective 2.6D)

1. Convert $\frac{7}{18}$ to a decimal.   0.38$\overline{3}$

2. Convert $5\frac{5}{6}$ to a decimal.   5.8$\overline{3}$

3. Convert 4.96 to a fraction. $4\frac{24}{25}$

4. Place the correct symbol, $<$ or $>$, between the two numbers. $\frac{1}{6}$ 0.167 $\frac{1}{6} < 0.167$

Objective 2.6D

Vocabulary to Review

numerator [2.2A]
denominator [2.2A]
place value of decimals [2.5A]

New Vocabulary

terminating decimal
repeating decimal

Discuss the Concepts

1. Explain how to write a proper fraction as a decimal.
2. Explain how to write a mixed number as a decimal.
3. Explain how to write a decimal as a fraction.
4. Given both a fraction and a decimal, how can you determine which is the larger number?

Concept Check

Which of the following is true?

a. $\frac{137}{300} = 0.456666667$ F

b. $\frac{137}{300} < 0.45666666$ F

c. $\frac{137}{300} > 0.45666666$ T

Optional Student Activity

Do all proper fractions with a denominator of 6 have a decimal equivalent that contains repeating digits?

No. $\frac{3}{6} = 0.5$

HOW TO   Convert 4.375 to a fraction.

$$4.375 = 4\frac{375}{1000}$$ • The 5 in 4.375 is in the thousandths place. Write 0.375 as a fraction with a denominator of 1000.

$$= 4\frac{3}{8}$$ • Simplify the fraction.

Integrating Technology

Some calculators *truncate* a decimal number that exceeds the calculator display. This means that the digits beyond the calculator's display are not shown. For this type of calculator, $\frac{2}{3}$ would be shown as 0.66666666. Other calculators *round* a decimal number when the calculator display is exceeded. For this type of calculator, $\frac{2}{3}$ would be shown as 0.66666667.

To find the order relation between a fraction and a decimal, first rewrite the fraction as a decimal. Then compare the two decimals.

HOW TO   Find the order relation between $\frac{6}{7}$ and 0.855.

$$\frac{6}{7} \approx 0.8571$$ • Write the fraction as a decimal. Round to one more place value than the given decimal. (0.855 has 3 decimal places; round to 4 decimal places.)

$$0.8571 > 0.8550$$ • Compare the two decimals.

$$\frac{6}{7} > 0.855$$ • Replace the decimal approximation of $\frac{6}{7}$ with $\frac{6}{7}$.

Example 14   Convert $\frac{5}{8}$ to a decimal.

Solution

$$\begin{array}{r} 0.625 \\ 8)\overline{5.000} \end{array} \qquad \frac{5}{8} = 0.625$$

You Try It 14   Convert $\frac{4}{5}$ to a decimal.

Your solution   0.8

Example 15   Convert $3\frac{1}{3}$ to a decimal.

Solution   Write $\frac{1}{3}$ to a decimal.

$$\begin{array}{r} 0.333 \\ 3)\overline{1.000} \end{array} = 0.\overline{3}$$

$$3\frac{1}{3} = 3.\overline{3}$$

You Try It 15   Convert $1\frac{5}{6}$ to a decimal.

Your solution $1.8\overline{3}$

Example 16   Convert 7.25 to a fraction.

Solution

$$7.25 = 7\frac{25}{100} = 7\frac{1}{4}$$

You Try It 16   Convert 6.2 to a fraction.

Your solution $6\frac{1}{5}$

Solutions on p. S7

Example 17   Place the correct symbol, < or >, between the two numbers.

$$0.845 \qquad \frac{5}{6}$$

Solution $\frac{5}{6} \approx 0.8333$

$0.8450 > 0.8333$

$0.845 > \frac{5}{6}$

You Try It 17   Place the correct symbol, < or >, between the two numbers.

$$0.588 \qquad \frac{7}{12}$$

Your solution $0.588 > \frac{7}{12}$

Solution on p. S7

Objective E   **To solve application problems and use formulas**

Objective 2.6E

Optional Student Activity

Bring to class pages from the business section of a newspaper. Have students determine the difference between the high and low prices of several stocks over the past 52-week period. Ask them to determine which of the stocks has had the greatest range in price. Which stock is closest to its high price for the 52-week period? Which stock is the farthest below its high price for the 52-week period?

Then have students determine the cost of, say, 50 shares of Disney or 25 shares of Coca-Cola.

Students can also make comparisons. Ask them how much more it would cost to buy 20 shares of one automotive stock than 20 shares of another. (For example, you might use Ford and General Motors.) Other categories that might interest students include fast-food chains (such as McDonald's), retailers (such as Wal-Mart), sneaker manufacturers (such as Nike), or sports teams (such as the Boston Celtics).

Example 18

A 1-year subscription to a monthly magazine costs $93. The price of each issue at the newsstand is $9.80. How much would you save per issue by buying a year's subscription rather than buying each issue at the newsstand?

Strategy

To find the amount saved:

• Find the subscription price per issue by dividing the cost of the subscription (93) by the number of issues (12).
• Subtract the subscription price per issue from the newsstand price (9.80).

Solution

$$\begin{array}{r} 7.75 \\ 12\overline{)93.00} \\ -84 \\ \hline 9\,0 \\ -8\,4 \\ \hline 60 \\ -60 \\ \hline 0 \end{array} \qquad \begin{array}{r} 9.80 \\ -7.75 \\ \hline 2.05 \end{array}$$

The savings would be $2.05 per issue.

You Try It 18

You hand a postal clerk a ten-dollar bill to pay for the purchase of twelve 37¢ stamps. How much change do you receive?

Your Strategy

Your solution
$5.56

Solution on p. S8

In-Class Examples (Objective 2.6E)

1. A salesperson's commission checks for six months are $1649.52, $2731.18, $1711.98, $675.49, $2406.37, and $1986.06. Find the total commission income for the six months.   **$11,160.60**

2. You buy groceries for $57.92. How much change do you receive from a $100 bill?   **$42.08**

3. You bought a house with payments of $1024.50 per month for 20 years. Find the total amount of the payments.   **$245,880**

4. A jogger ran 7.4 mi in 45.88 min. What was the jogger's average time per mile?   **6.2 min**

Example 19

An overseas flight charges $12.80 for each kilogram or part of a kilogram over 50 kg of luggage weight. How much extra must be paid for three pieces of luggage weighing 21.4 kg, 19.3 kg, and 16.8 kg?

Strategy

To find the extra charge:
- Add the three weights (21.4, 19.3, and 16.8) to find the total weight of the luggage.
- Subtract 50 kg from the total weight of the luggage to find the excess weight.
- Round the difference up to the nearest whole number.
- Multiply the charge per kilogram of excess weight (12.80) by the excess weight.

Solution

$21.4 + 19.3 + 16.8 = 57.5$

$57.5 - 50 = 7.5$

7.5 rounded up to the nearest whole number is 8.

$12.80(8) = 102.40$

The extra charge for the luggage is $102.40.

You Try It 19

A health food store buys nuts in 100-pound containers and repackages the nuts in cellophane bags for resale. Each cellophane bag costs $.06, 2 lb of nuts are placed in each bag, and each bag of nuts is then sold for $12.50. Find the profit on a 100-pound container of nuts costing $475.

Your Strategy

Your solution

$147

Example 20

Use the formula $P = BF$, where P is the insurance premium, B is the base rate, and F is the rating factor, to find the insurance premium due on an insurance policy with a base rate of $342.50 and a rating factor of 2.2.

Strategy

To find the insurance premium due, replace B by 342.50 and F by 2.2 in the given formula and solve for P.

Solution

$P = BF$
$P = 342.50(2.2)$
$P = 753.50$

The insurance premium due is $753.50.

You Try It 20

Use the formula $P = BF$, where P is the insurance premium, B is the base rate, and F is the rating factor, to find the insurance premium due on an insurance policy with a base rate of $276.25 and a rating factor of 1.8.

Your Strategy

Your solution

$497.25

Solutions on p. S8

2.6 Exercises

Objective A   **To add and subtract decimals**

Add or subtract.

1. 1.864 + 39 + 25.0781
65.9421

2. 2.04 + 35.6 + 4.918
42.558

3. 35.9 + 8.217 + 146.74
190.857

Suggested Assignment
Exercises 1–117, every other odd
Exercises 119–149, odds
More challenging problem:
    Exercise 152

4. 12 + 73.59 + 6.482
92.072

5. 36.47 − 15.21
21.26

6. 85.69 − 2.13
83.56

7. 28 − 6.74
21.26

8. 5 − 1.386
3.614

9. 6.02 − 3.252
2.768

10. Find the sum of 2.536, 14.97, 8.014, and 21.67.
47.19

11. Find the total of 6.24, 8.573, 19.06, and 22.488.
56.361

12. What is 6.9217 decreased by 3.4501?
3.4716

13. What is 8.9 less than 62.57?
53.67

Estimate by rounding. Then find the exact answer.

14. 45.06 + 80.71
130; 125.77

15. 6.408 + 5.917
12; 12.325

16. 0.24 + 0.38 + 0.96
1.6; 1.58

17. 56.87 − 23.24
40; 33.63

18. 6.272 − 1.848
4; 4.424

19. 0.931 − 0.628
0.3; 0.303

20. 5.37 + 26.49
35; 31.86

21. 87.65 − 49.032
40; 38.618

22. 387.6 − 54.92
350; 332.68

Evaluate the variable expression $x + y + z$ for the given values of x, y, and z.

23. $x = 41.33; y = 26.095; z = 70.08$
137.505

24. $x = 6.059; y = 3.884; z = 15.71$
25.653

Evaluate the variable expression $x - y$ for the given values of x and y.

25. $x = 43.29; y = 18.76$
24.53

26. $x = 6.029; y = 4.708$
1.321

27. $x = 16.329; y = 4.54$
11.789

Quick Quiz (Objective 2.6A)

1. Add: 18.44 + 8.33091 + 25.7   52.47091

2. What is 18.9174 minus 8.82?   10.0974

3. Subtract and check: 29.843 − 12.76   17.083

4. Estimate the difference between 87.297 and 65.31.
  20

5. Evaluate $x + y + z$ when $x = 3.39$, $y = 4.5762$, and $z = 1.8$.   9.7662

Objective B   To multiply decimals

Multiply.

28. 0.9(0.3)
0.27

29. (3.4)(0.5)
1.70

30. (0.72)(3.7)
2.664

31. 8.29(0.004)
0.03316

32. What is the product of 5.92 and 100?
592

33. What is 1000 times 4.25?
4250

34. Find 0.82 times 10^2.
82

35. Find the product of 6.71 and 10^4.
67,100

Estimate by rounding. Then find the exact answer.

36. 86.4(4.2)
360; 362.88

37. (9.81)(0.77)
8.0; 7.5537

38. 0.238(8.2)
1.6; 1.9516

39. (6.88)(9.97)
70; 68.5936

40. (8.432)(0.043)
0.32; 0.362576

41. 28.45(1.13)
30; 32.1485

Evaluate the expression for the given values of the variables.

42. xy, when $x = 5.68$ and $y = 0.2$
1.136

43. ab, when $a = 6.27$ and $b = 8$
50.16

44. $40c$, when $c = 2.5$
100

45. $10t$, when $t = 4.8$
48

46. xy, when $x = 3.71$ and $y = 2.9$
10.759

Objective C   To divide decimals

Divide.

47. $16.15 \div 0.5$
32.3

48. $7.02 \div 3.6$
1.95

49. $27.08 \div 0.4$
67.7

50. $8.919 \div 0.9$
9.91

Divide. Round to the nearest tenth.

51. $55.63 \div 8.8$
6.3

52. $1.873 \div 1.4$
1.3

53. $52.8 \div 9.1$
5.8

54. $6.824 \div 0.053$
128.8

Quick Quiz (Objective 2.6B)

1. Multiply: 0.96(7)   6.72

2. Estimate the product of 0.918 and 6.2.   5.4

3. Find the product of 7.924 and 1000.   7924

4. Evaluate $2ab$ when $a = 3.6$ and $b = 9$.   64.8

Divide. Round to the nearest hundredth.

55. $6.457 \div 8$
0.81

56. $19.07 \div 0.54$
35.31

57. $0.0416 \div 0.53$
0.08

58. $31.792 \div 0.86$
36.97

59. Find the quotient of 52.78 and 10.
5.278

60. What is 37,942 divided by 1000?
37.942

61. What is the quotient of 48.05 and 10^2?
0.4805

62. Find 9.407 divided by 10^3.
0.009407

Estimate by rounding. Then divide and round to the nearest hundredth.

63. $42.43 \div 3.8$
10; 11.17

64. $678 \div 0.71$
1000; 954.93

65. $6.398 \div 5.5$
1; 1.16

66. $0.994 \div 0.456$
2; 2.18

67. $1.237 \div 0.021$
50; 58.90

68. $421.093 \div 4.087$
100; 103.03

69. $33.14 \div 4.6$
6; 7.20

70. $129.38 \div 4.47$
25; 28.94

Evaluate the variable expression $\frac{x}{y}$ for the given values of x and y.

71. $x = 52.8; y = 0.4$
132

72. $x = 3.542; y = 0.7$
5.06

73. $x = 2.436; y = 0.6$
4.06

74. $x = 0.648; y = 2.7$
0.24

75. $x = 26.22; y = 6.9$
3.8

76. $x = 8.034; y = 3.9$
2.06

Objective D   To convert between decimals and fractions and identify the order relation between a decimal and a fraction

Convert the fraction to a decimal. Place a bar over repeating digits of a repeating decimal.

77. $\dfrac{3}{8}$
0.375

78. $\dfrac{7}{15}$
$0.4\overline{6}$

79. $\dfrac{8}{11}$
$0.\overline{72}$

80. $\dfrac{9}{16}$
0.5625

81. $\dfrac{7}{12}$
$0.583\overline{3}$

82. $\dfrac{5}{3}$
$1.\overline{6}$

83. $\dfrac{7}{4}$
1.75

84. $2\dfrac{3}{4}$
2.75

85. $1\dfrac{1}{2}$
1.5

86. $3\dfrac{2}{9}$
$3.\overline{2}$

Quick Quiz (Objective 2.6C)

1. Divide: $36.608 \div 5.2$ 7.04

2. Estimate the quotient of 347.92 and 6.1.   50

3. Divide and round to the nearest hundredth:
$36.579 \div 5.2$ 7.03

4. What is the quotient of 14.08 and 10^3?   0.01408

5. Evaluate $\dfrac{x}{y}$ when $x = 257.93$ and $y = 100$.   2.5793

87. $4\frac{1}{6}$ **88.** $\frac{3}{25}$ **89.** $2\frac{1}{4}$ **90.** $6\frac{3}{5}$ **91.** $3\frac{8}{9}$

$4.1\overline{6}$ 0.12 2.25 6.6 $3.\overline{8}$

Convert the decimal to a fraction.

92. 0.6   **93.** 0.2   **94.** 0.25   **95.** 0.75   **96.** 0.48

$\frac{3}{5}$ $\frac{1}{5}$ $\frac{1}{4}$ $\frac{3}{4}$ $\frac{12}{25}$

97. 0.125   **98.** 0.325   **99.** 2.5   **100.** 3.4   **101.** 4.55

$\frac{1}{8}$ $\frac{13}{40}$ $2\frac{1}{2}$ $3\frac{2}{5}$ $4\frac{11}{20}$

102. 9.95   **103.** 1.72   **104.** 5.68   **105.** 0.045   **106.** 0.085

$9\frac{19}{20}$ $1\frac{18}{25}$ $5\frac{17}{25}$ $\frac{9}{200}$ $\frac{17}{200}$

Place the correct symbol, $<$ or $>$, between the two numbers.

107. $\frac{9}{10}$ 0.89   **108.** $\frac{7}{20}$ 0.34   **109.** $\frac{4}{5}$ 0.803   **110.** $\frac{3}{4}$ 0.706

$>$ $>$ $<$ $>$

111. 0.444 $\frac{4}{9}$ **112.** 0.72 $\frac{5}{7}$ **113.** 0.13 $\frac{3}{25}$ **114.** 0.25 $\frac{13}{50}$

$<$ $>$ $>$ $<$

115. $\frac{5}{16}$ 0.312   **116.** $\frac{7}{18}$ 0.39   **117.** $\frac{10}{11}$ 0.909   **118.** $\frac{8}{15}$ 0.543

$>$ $<$ $>$ $<$

Objective E   **To solve application problems and use formulas**

119. **Salaries** If you earn an annual salary of $47,619, what is your monthly salary?
$3968.25

120. **Car Insurance** You pay $947.60 a year in car insurance. The insurance is paid in four equal payments. Find the amount of each payment.
$236.90

121. **Unit Cost** A case of diet cola costs $8.89. If there are 24 cans in a case, find the cost per can. Round to the nearest cent.
$.37

Quick Quiz (Objective 2.6D)

1. Convert $\frac{1}{12}$ to a decimal. $0.08\overline{3}$

2. Convert $12\frac{1}{6}$ to a decimal. $12.1\overline{6}$

3. Convert 0.78 to a fraction. $\frac{39}{50}$

4. Place the correct symbol, $<$ or $>$, between the two numbers.

$\frac{5}{16}$ 0.313 $\frac{5}{16} < 0.313$

122. Purchasing   You make a down payment of $125 on a camcorder and agree to make payments of $34.17 a month for 9 months. Find the total cost of the camcorder.
$432.53

123. 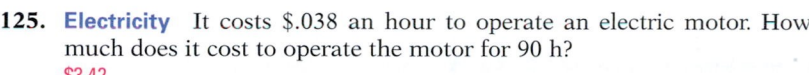 **Education**   The graph at the right shows where U.S. children in grades K–12 are being educated. Figures are in millions of children.
  a. Find the total number of children in grades K–12.
  b. How many more children are being educated in public school than in private school?
  a. 53.446 million children   **b.** 40.49 million children

Where Children in Grades K–12 are Being Educated in the United States
Source: U.S. Department of Education; Home School Legal Defense Association

124. Fuel Efficiency   You travel 295 mi on 12.5 gal of gasoline. How many miles can you travel on 1 gal of gasoline?
23.6 mi

125. Electricity   It costs $.038 an hour to operate an electric motor. How much does it cost to operate the motor for 90 h?
$3.42

126. Tolls   When the Massachusetts Turnpike opened, the toll for a passenger car that traveled the entire 136 mi of it was $5.60. Find the cost per mile. Round to the nearest cent.
$.04

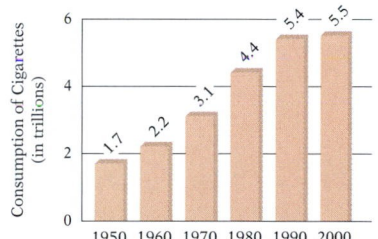

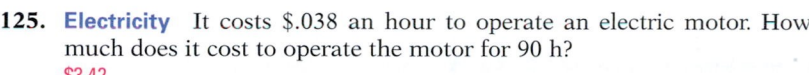 **Health**   The graph at the right shows that the worldwide consumption of cigarettes has been increasing. Use this table for Exercises 127 and 128.

127. a. Find the increase in cigarette consumption from 1950 to 1990.
  b. How many times greater was the cigarette consumption in 2000 than in 1960?
  a. 3.7 trillion cigarettes   **b.** 2.5 times

128. a. During which 10-year period was the increase in cigarette consumption greatest?
  b. During which 10-year period was the increase in cigarette consumption the least?
  a. 1970 to 1980   **b.** 1990 to 2000

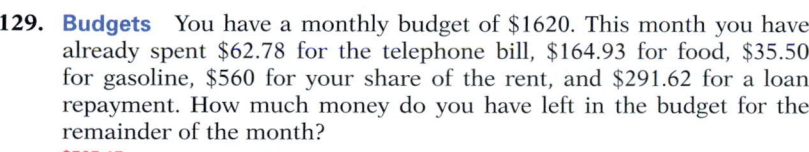

Worldwide Consumption of Cigarettes
Source: The Tobacco Atlas; U.S. Department of Agriculture

129. Budgets   You have a monthly budget of $1620. This month you have already spent $62.78 for the telephone bill, $164.93 for food, $35.50 for gasoline, $560 for your share of the rent, and $291.62 for a loan repayment. How much money do you have left in the budget for the remainder of the month?
$505.17

130. Banking You had a balance of $347.08 in your checking account. You then made a deposit of $189.53 and wrote a check for $62.89. Find the new balance in your checking account.
$473.72

131. Income A bookkeeper earns a salary of $660 for a 40-hour week. This week the bookkeeper worked 6 h of overtime at a rate of $24.75 for each hour of overtime worked. Find the bookkeeper's total income for the week.
$808.50

132. Computers The list below shows the average number of hours per week that students use a computer. On average, how many more hours per year does a second-grade student use a computer than a fifth-grade student?
36.4 h

Grade Level	Average Number of Hours of Computer Use Per Week
Pre Kindergarten – Kindergarten	3.9
1st – 3rd	4.9
4th – 6th	4.2
7th – 8th	6.9
9th – 12th	6.7

Source: Find/SVP American Learning Household Survey

133. Restaurant Meals Using the menu shown below, estimate the bill for the following order: 1 soup, 1 cheese sticks, 1 blackened swordfish, 1 chicken divan, and 1 carrot cake.
$69

Appetizers
Soup of the Day $5.75
Cheese Sticks $6.25
Potato Skins $6.50

Entrees
Roast Prime Rib $28.95
Blackened Swordfish $26.95
Chicken Divan $24.95

Desserts
Carrot Cake $7.25
Ice Cream Pie $8.50
Cheese Cake $9.75

134. Restaurant Meals Using the menu shown above, estimate the bill for the following order: 1 potato skins, 1 cheese sticks, 1 roast prime rib, 1 chicken divan, 1 ice cream pie, and 1 cheese cake.
$82

135. Business For $135, a druggist purchases 5 L of cough syrup and repackages it in 250-milliliter bottles. Each bottle costs the druggist $.55. Each bottle of cough syrup is sold for $11.89. Find the profit on the 5 L of cough syrup. (*Hint:* There are 1000 milliliters in 1 liter.)
$91.80

Quick Quiz (Objective 2.6E)

1. You have $655.12 in your checking account. You make deposits of $753.42, $49.90, $67.34, and $152.18. Find the amount in your checking account after making the deposits. $1677.96

2. A competitive swimmer beat the team's record time of 57.84 s in the 100-meter freestyle competition by 0.69 s. What is the new record time? 57.15 s

136. Shipping A confectioner ships holiday packs of candy and nuts anywhere in the United States. At the right is a price list for nuts and candy, and below that is a table of shipping charges to zones in the United States. Find the cost of sending the following orders to the given mail zones. For any fraction of a pound, use the next higher weight. Sixteen ounces is equal to one pound.
a. $52.90   **b.** $79.60   **c.** $61.05

Code	Description	Price
112	Almonds 16 oz	$4.75
116	Cashews 8 oz	$2.90
117	Cashews 16 oz	$5.50
130	Macadamias 7 oz	$5.25
131	Macadamias 16 oz	$9.95
149	Pecan halves 8 oz	$6.25
155	Mixed nuts 8 oz	$4.80
160	Cashew brittle 8 oz	$1.95
182	Pecan roll 8 oz	$3.70
199	Chocolate peanuts 8 oz	$1.90

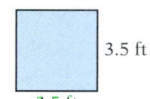

a. Code	Quantity	b. Code	Quantity	c. Code	Quantity
116	2	112	1	117	3
130	1	117	4	131	1
149	3	131	2	155	2
182	4	160	3	160	4
Mail to Zone 4.		182	5	182	1
		Mail to Zone 3.		199	3
				Mail to Zone 2.	

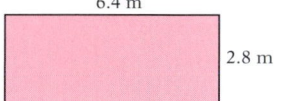

Pounds	Zone 1	Zone 2	Zone 3	Zone 4
1–3	$6.55	$6.85	$7.25	$7.75
4–6	$7.10	$7.40	$7.80	$8.30
7–9	$7.50	$7.80	$8.20	$8.70
10–12	$7.90	$8.20	$8.60	$9.10

137. Transportation A taxi costs $2.50 plus $.30 for each $\frac{1}{8}$ mi driven. Find the cost of hiring a taxi to get from the airport to your hotel, a distance of 4.5 mi.
$13.30

138. Geometry The length of each side of a square is 3.5 ft. Find the perimeter of the square. Use the formula $P = 4s$.
14 ft

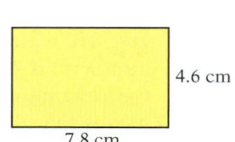

3.5 ft

3.5 ft

139. Geometry Find the perimeter of a rectangle that measures 4.5 in. by 3.25 in. Use the formula $P = 2L + 2W$.
15.5 in.

140. Geometry Find the perimeter of a rectangle that measures 2.8 m by 6.4 m. Use the formula $P = 2L + 2W$.
18.4 m

6.4 m

2.8 m

141. Geometry Find the area of a rectangle that measures 4.5 in. by 3.25 in. Use the formula $A = LW$.
14.625 in²

142. Geometry Find the area of a rectangle that has a length of 7.8 cm and a width of 4.6 cm. Use the formula $A = LW$.
35.88 cm²

4.6 cm

7.8 cm

143. Geometry Find the perimeter of a triangle with sides that measure 2.8 m, 4.75 m, and 6.4 m. Use the formula $P = a + b + c$.
13.95 m

144. Geometry The lengths of three sides of a triangle are 7.5 m, 6.1 m, and 4.9 m. Find the perimeter of the triangle. Use the formula $P = a + b + c$.
18.5 m

4.9 m   6.1 m

7.5 m

Quick Quiz (Objective 2.6E)
(*Continued*)

3. The cost of operating an electric lathe for 1 h is $.034. How much does it cost to operate the lathe for 25 h? $.85

4. You pay $1284.72 per year in life insurance premiums. You pay the premiums in 12 equal monthly payments. Find the amount of each monthly payment. $107.06

145. **Markup**   Use the formula $M = S - C$, where M is the markup on a consumer product, S is the selling price, and C is the cost of the product to the business, to find the markup on a product that cost a business $1653.19 and has a selling price of $2231.81.
$578.62

146. **Accounting**   The amount of an employee's earnings that is subject to federal withholding is called federal earnings. Find the federal earnings for an employee who earns $694.89 and has a withholding allowance of $132.69. Use the formula $F = E - W$, where F is the federal earnings, E is the employee's earnings, and W is the withholding allowance.
$562.20

147. **Consumerism**   Use the formula $M = \frac{C}{N}$, where M is the cost per mile for a rental car, C is the total cost, and N is the number of miles driven, to find the cost per mile when the total cost of renting a car is $260.16 and you drive the car 542 mi.
$.48

148. **Physics**   Find the force exerted on a falling object that has a mass of 4.25 kg. Use the formula $F = ma$, where F is the force exerted by gravity on a falling object, m is the mass of the object, and a is the acceleration due to gravity. The acceleration due to gravity is 9.8 m/s² (meters per second squared). The force is measured in newtons.
41.65 newtons

149. **Utilities**   Find the cost of operating an 1800-watt TV set for 5 h at a cost of $.06 per kilowatt-hour. Use the formula $c = 0.001wtk$, where c is the cost of operating an appliance, w is the number of watts, t is the time in hours, and k is the cost per kilowatt-hour.
$.54

150. **Home Equity**   Find the equity on a home that is valued at $225,000 when the homeowner has $167,853.25 in loans on the property. Use the formula $E = V - L$, where E is the equity, V is the value of the home, and L is the loan amount on the property.
$57,146.75

APPLYING THE CONCEPTS

151.   Find the product of 1.0035 and 1.00079 without using a calculator. Then find the product using a calculator and compare the two numbers. If your calculator has an eight-digit display, what number did the calculator display? Some calculators **truncate** the product, which means that the digits that cannot be displayed are discarded. Other calculators round the answer to the rightmost place value in the calculator's display. Determine which method your calculator uses to handle approximate answers.
1.004292765; truncate: 1.0042927, round: 1.0042928

152.   Convert $\frac{1}{9}$, $\frac{2}{9}$, $\frac{3}{9}$, and $\frac{4}{9}$ to decimals. Describe the pattern. Use the pattern to convert $\frac{5}{9}$, $\frac{7}{9}$, and $\frac{8}{9}$ to decimals.

2.7 The Order of Operations Agreement

Objective A   To use the Order of Operations Agreement

The Order of Operations Agreement applies when simplifying expressions containing fractions and decimals.

> **The Order of Operations Agreement**
>
> **Step 1.** Do all operations inside parentheses.
> **Step 2.** Simplify any numerical expressions containing exponents.
> **Step 3.** Do multiplication and division as they occur from left to right.
> **Step 4.** Do addition and subtraction as they occur from left to right.

HOW TO Simplify: $\left(\dfrac{1}{2}\right)^2 + \left(\dfrac{2}{3} \div \dfrac{5}{9}\right) \cdot \dfrac{5}{6}$

$\left(\dfrac{1}{2}\right)^2 + \left(\dfrac{2}{3} \div \dfrac{5}{9}\right) \cdot \dfrac{5}{6}$

$= \left(\dfrac{1}{2}\right)^2 + \left(\dfrac{6}{5}\right) \cdot \dfrac{5}{6}$ • Do the operation inside the parentheses (Step 1).

$= \dfrac{1}{4} + \left(\dfrac{6}{5}\right) \cdot \dfrac{5}{6}$ • Simplify the exponential expression (Step 2).

$= \dfrac{1}{4} + 1$ • Do the multiplication (Step 3).

$= 1\dfrac{1}{4}$ • Do the addition (Step 4).

A fraction bar acts like parentheses. Therefore, simplify the numerator and denominator of a fraction as part of Step 1 in the Order of Operations Agreement.

HOW TO Simplify: $6 - \dfrac{2+1}{15-8} \div \dfrac{3}{14}$

$6 - \dfrac{2+1}{15-8} \div \dfrac{3}{14}$

$= 6 - \dfrac{3}{7} \div \dfrac{3}{14}$ • Perform operations above and below the fraction bar.

$= 6 - \left(\dfrac{3}{7} \cdot \dfrac{14}{3}\right)$ • Do the division.

$= 6 - 2$

$= 4$ • Do the subtraction.

Objective 2.7A

Procedures to Review
Order of Operations Agreement
[1.4A]

Discuss the Concepts
In simplifying the expression

$$\frac{3}{4} + \frac{1}{2} \cdot \frac{8}{9}$$

why can't you begin by adding $\dfrac{3}{4}$ and $\dfrac{1}{2}$?

Optional Student Activity

1. Find the sum of the fraction halfway between $\dfrac{4}{5}$ and $\dfrac{2}{3}$ and the fraction halfway between $\dfrac{1}{2}$ and $\dfrac{1}{3}$. $1\dfrac{3}{20}$

2. Find the product of

$$\left(1 - \frac{1}{2^2}\right)\left(1 - \frac{1}{3^2}\right)\left(1 - \frac{1}{4^2}\right)$$
$$\cdots \left(1 - \frac{1}{9^2}\right)\left(1 - \frac{1}{10^2}\right)$$

The dots mean that the pattern continues. $\dfrac{11}{20}$

In-Class Examples (Objective 2.7A)
Simplify.

1. $\dfrac{7}{8} + \dfrac{1}{9} \div \dfrac{8}{9}$ 1

2. $(0.5)^2 + (4.9 - 2.4)$ 2.75

3. $\dfrac{3}{14} + \dfrac{6}{7} \cdot \left(\dfrac{2}{3} - \dfrac{1}{2}\right)^2$ $\dfrac{5}{21}$

HOW TO Evaluate $\dfrac{w + x}{y} - z$ when $w = \dfrac{3}{4}$, $x = \dfrac{1}{4}$, $y = 2$, and $z = \dfrac{1}{3}$.

$$\dfrac{w + x}{y} - z$$

$$\dfrac{\dfrac{3}{4} + \dfrac{1}{4}}{2} - \dfrac{1}{3}$$ • **Replace each variable with its given value.**

$$= \dfrac{1}{2} - \dfrac{1}{3}$$ • **Simplify the numerator of the complex fraction.**

$$= \dfrac{3}{6} - \dfrac{2}{6} = \dfrac{1}{6}$$ • **Do the subtraction.**

Example 1 Simplify: $0.2(5.6 - 2.5) + (1.4)^2$

Solution $0.2(5.6 - 2.5) + (1.4)^2$
$= 0.2(3.1) + (1.4)^2$ • **Parentheses**
$= 0.2(3.1) + 1.96$ • **Exponents**
$= 0.62 + 1.96$ • **Multiply.**
$= 2.58$ • **Add.**

You Try It 1 Simplify: $(1.2 - 0.8)^2 + (1.5)(6)$

Your solution 9.16

Example 2 Simplify: $\left(\dfrac{2}{3}\right)^2 \div \dfrac{7 - 2}{13 - 4} - \dfrac{1}{3}$

Solution
$$\left(\dfrac{2}{3}\right)^2 \div \dfrac{7 - 2}{13 - 4} - \dfrac{1}{3}$$
$$= \left(\dfrac{2}{3}\right)^2 \div \dfrac{5}{9} - \dfrac{1}{3}$$
$$= \dfrac{4}{9} \div \dfrac{5}{9} - \dfrac{1}{3}$$
$$= \dfrac{4}{9} \cdot \dfrac{9}{5} - \dfrac{1}{3}$$
$$= \dfrac{4}{5} - \dfrac{1}{3} = \dfrac{12}{15} - \dfrac{5}{15} = \dfrac{7}{15}$$

You Try It 2 Simplify: $\left(\dfrac{1}{2}\right)^3 \cdot \dfrac{7 - 3}{9 - 4} + \dfrac{4}{5}$

Your solution $\dfrac{9}{10}$

Solutions on p. S8

2.7 Exercises

Objective A   To use the Order of Operations Agreement

Simplify.

1. $\dfrac{3}{7} \cdot \dfrac{14}{15} + \dfrac{4}{5}$

$1\dfrac{1}{5}$

2. $\dfrac{3}{5} \div \dfrac{6}{7} + \dfrac{4}{5}$

$1\dfrac{1}{2}$

3. $\left(\dfrac{5}{6}\right)^2 - \dfrac{5}{9}$

$\dfrac{5}{36}$

4. $\left(\dfrac{3}{5}\right)^2 - \dfrac{3}{10}$

$\dfrac{3}{50}$

5. $\dfrac{3}{4} \cdot \left(\dfrac{11}{12} - \dfrac{7}{8}\right) + \dfrac{5}{16}$

$\dfrac{11}{32}$

6. $\dfrac{7}{18} + \dfrac{5}{6} \cdot \left(\dfrac{2}{3} - \dfrac{1}{6}\right)$

$\dfrac{29}{36}$

7. $\dfrac{11}{16} - \left(\dfrac{3}{4}\right)^2 + \dfrac{7}{8}$

1

8. $\left(\dfrac{2}{3}\right)^2 - \dfrac{7}{18} + \dfrac{5}{6}$

$\dfrac{8}{9}$

9. $\left(1\dfrac{1}{3} - \dfrac{5}{6}\right) + \dfrac{7}{8} \div \left(\dfrac{1}{2}\right)^2$

4

10. $\left(\dfrac{1}{4}\right)^2 \div \left(2\dfrac{1}{2} - \dfrac{3}{4}\right) + \dfrac{5}{7}$

$\dfrac{3}{4}$

11. $\left(\dfrac{2}{3}\right)^2 + \dfrac{8-7}{9-3} \div \dfrac{3}{8}$

$\dfrac{8}{9}$

12. $\left(\dfrac{1}{3}\right)^2 \cdot \dfrac{14-5}{10-6} + \dfrac{3}{4}$

1

13. $(0.5)(0.2)^2 + 1.7$

1.72

14. $0.3(4.8 - 1.7) + (1.2)^2$

2.37

15. $(1.8)^2 - 2.52 \div 1.8$

1.84

16. $(1.65 - 1.05)^2 \div 0.4 + 0.9$

1.8

17. $0.4(3 - 1.5) + (1.2)^2$

2.04

18. $(5 - 3.5)^2 + (0.75)(8)$

8.25

19. $\dfrac{1}{2} + \dfrac{\frac{13}{25}}{4 - \frac{3}{4}} \div \dfrac{1}{5}$

$1\dfrac{3}{10}$

20. $\dfrac{4}{5} + \dfrac{3 - \frac{7}{9}}{\frac{5}{6}} \cdot \dfrac{3}{8}$

$1\dfrac{4}{5}$

21. $\left(\dfrac{2}{3}\right)^2 + \dfrac{\frac{5}{8} - \frac{1}{4}}{\frac{2}{3} - \frac{1}{6}} \cdot \dfrac{8}{9}$

$1\dfrac{1}{9}$

Evaluate the expression for the given values of the variables.

22. $x^2 + \dfrac{y}{z}$, when $x = \dfrac{2}{3}$, $y = \dfrac{5}{8}$, and $z = \dfrac{3}{4}$

$1\dfrac{5}{18}$

23. $\dfrac{x}{y} - z^2$, when $x = \dfrac{5}{6}$, $y = \dfrac{1}{3}$, and $z = \dfrac{3}{4}$

$1\dfrac{15}{16}$

Quick Quiz (Objective 2.7A)

Simplify.

1. $(2.4)(5) + (4.1 - 3.9)^2$ 12.04

2. $\left(\dfrac{1}{3}\right)^2\left(\dfrac{4}{5} - \dfrac{1}{2}\right)$ $\dfrac{1}{30}$

3. $\left(\dfrac{1}{2}\right)^2 + \left(\dfrac{3}{5} - \dfrac{1}{2}\right) \div \dfrac{4}{15}$ $\dfrac{5}{8}$

Section 2.7

Suggested Assignment

Exercises 1–31, odds
More challenging problems:
    Exercises 32–34

Answers to Writing Exercises

35. The "puzzle" works as it does because the sum of the fractions $\frac{1}{2}$, $\frac{1}{3}$, and $\frac{1}{9}$ is $\frac{17}{18}$, not 1. As a result, the first child actually received $\frac{9}{17}$ of the horses, not $\frac{1}{2}$; the second child received $\frac{6}{17}$ of the horses, not $\frac{1}{3}$; and the third child got $\frac{2}{17}$ of the horses, not $\frac{1}{9}$.

24. $x - y^3z$, when $x = \frac{5}{6}$, $y = \frac{1}{2}$, and $z = \frac{8}{9}$

$\frac{13}{18}$

25. $xy^3 + z$, when $x = \frac{9}{10}$, $y = \frac{1}{3}$, and $z = \frac{7}{15}$

$\frac{1}{2}$

26. $\frac{wx}{y} + z$, when $w = \frac{4}{5}$, $x = \frac{5}{8}$, $y = \frac{3}{4}$, and $z = \frac{2}{3}$

$1\frac{1}{3}$

27. $\frac{w}{xy} - z$, when $w = 2\frac{1}{2}$, $x = 4$, $y = \frac{3}{8}$, and $z = \frac{2}{3}$ 1

28. $c^2 - ab$, when $a = 1.7$, $b = 0.6$, and $c = 2.8$
6.82

29. $(a + b)^2 - c$, when $a = 2.5$, $b = 1.8$, and $c = 0.4$
18.09

30. $\frac{b^2}{c} + 4a$, when $a = 1.5$, $b = 0.2$, and $c = 0.4$
6.1

31. $\frac{x}{y^2} + 3z$, when $x = 7.2$, $y = 0.6$, and $z = 3.5$
30.5

APPLYING THE CONCEPTS

32. Simplify: $\dfrac{\frac{3}{x} + \frac{2}{x}}{\frac{5}{6}}$

$\frac{6}{x}$

33. Given that x is a whole number, for what value of x will the expression $\left(\frac{3}{4}\right)^2 + x^5 \div \frac{7}{8}$ have a minimum value? What is the minimum value?

$0; \frac{9}{16}$

34. Which of the variables u, v, w, x, and y can be doubled so that $\dfrac{u + \frac{v}{w}}{\frac{x}{y}}$ is

(a) halved or (b) doubled?
(a) x (b) y

35. ✏️ A farmer died and left 17 horses to be divided among 3 children. The first child was to receive one-half of the horses, the second child one-third of the horses, and the third child one-ninth of the horses. The executor for the family's estate realized that 17 horses could not be divided by halves, thirds, or ninths and so added a neighbor's horse to the farmer's. With 18 horses, the executor gave 9 horses to the first child, 6 horses to the second child, and 2 horses to the third child. This accounted for the 17 horses, so the executor returned the borrowed horse to the neighbor. Explain why this worked.

Focus on Problem Solving

Common Knowledge   An application problem may not provide all the information that is needed to solve the problem. Sometimes, however, the necessary information is common knowledge.

> **HOW TO**   You are traveling by bus from Boston to New York. The trip is 4 h long. If the bus leaves Boston at 10 A.M., what time should you arrive in New York?
>
> What other information do you need to solve this problem?
>
> You need to know that, using a 12-hour clock, the hours run
>
> 10 A.M.
> 11 A.M.
> 12 P.M.
> 1 P.M.
> 2 P.M.
>
> Four hours after 10 A.M. is 2 P.M.
>
> You should arrive in New York at 2 P.M.

> **HOW TO**   You purchase a 37¢ stamp at the post office and hand the clerk a 1-dollar bill. How much change do you receive?
>
> What information do you need to solve this problem?
>
> You need to know that there are 100¢ in 1 dollar.
>
> Your change is 100¢ − 37¢.
>
> 100 − 37 = 63
>
>
>
> You receive 63¢ in change.

What information do you need to know to solve each of the following problems?

1. You sell a dozen tickets to a fundraiser. Each ticket costs $10. How much money do you collect?

2. The weekly lab period for your science course is 1 h and 20 min long. Find the length of the science lab period in minutes.

3. An employee's monthly salary is $3750. Find the employee's annual salary.

4. A survey revealed that eighth graders spend an average of 3 h each day watching television. Find the total time an eighth grader spends watching TV each week.

5. You want to buy a carpet for a room that is 15 ft wide and 18 ft long. Find the amount of carpet that you need.

Answers to Focus on Problem Solving: Common Knowledge

1. There are 12 in one dozen.
2. There are 60 min in 1 h.
3. There are 12 months in 1 year.
4. There are 7 days in 1 week.
5. You need to know the formula for the area of a rectangle.

Projects and Group Activities

Music In musical notation, notes are printed on a **staff,** which is a set of five horizontal lines and the spaces between them. The notes of a musical composition are grouped into **measures,** or **bars.** Vertical lines separate measures on a staff. The shape of a note indicates how long it should be held. The whole note has the longest time value of any note. Each time value is divided by 2 in order to find the next smallest time value.

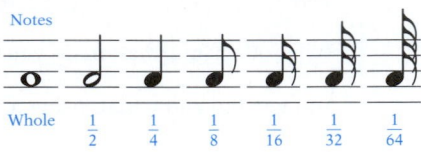

The **time signature** is a fraction that appears at the beginning of a piece of music. The numerator of the fraction indicates the number of beats in a measure. The denominator indicates what kind of note receives 1 beat. For example, music written in $\frac{2}{4}$ time has 2 beats to a measure, and a quarter note receives 1 beat. One measure in $\frac{2}{4}$ time may have 1 half note, 2 quarter notes, 4 eighth notes, or any other combination of notes totaling 2 beats. Other common time signatures are $\frac{4}{4}$, $\frac{3}{4}$, and $\frac{6}{8}$.

1. Explain the meaning of the 6 and the 8 in the time signature $\frac{6}{8}$.

2. Give some possible combinations of notes in one measure of a piece written in $\frac{4}{4}$ time.

3. What does a dot at the right of a note indicate? What is the effect of a dot at the right of a half note? At the right of a quarter note? At the right of an eighth note?

4. Symbols called rests are used to indicate periods of silence in a piece of music. What symbols are used to indicate the different time values of rests?

5. Find some examples of musical compositions written in different time signatures. Use a few measures from each to show that the sum of the time values of the notes and rests in each measure equals the numerator of the time signature.

Construction Suppose you are involved in building your own home. Design a stairway from the first floor of the house to the second floor. Some of the questions you will need to answer follow.

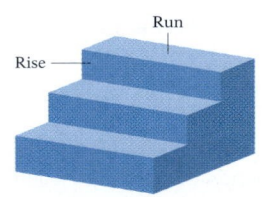

What is the distance from the floor of the first story to the floor of the second story?

Typically, what is the number of steps in a stairway?

What is a reasonable length for the run of each step?

What is the width of the wood being used to build the staircase?

In designing the stairway, remember that each riser should be the same height, that each run should be the same length, and that the width of the wood used for the steps will have to be incorporated into the calculation.

Answers to Projects and Group Activities: Music

1. There are 6 beats to a measure, and an eighth note receives 1 beat.

2. Answers will vary. Some examples include:
   1 whole note, 2 half notes, 4 quarter notes, 8 eighth notes, 1 half note and 2 quarter notes, 1 half note and 4 eighth notes, and 2 quarter notes and 4 eighth notes.

3. A dot at the right of a note increases its duration by one-half.
   A dotted half note equals a half note plus a quarter note.
   A dotted quarter note equals a quarter note plus an eighth note.
   A dotted eighth note equals an eighth note plus a sixteenth note.

4. Symbols used to indicate the different time values of rests are shown at the right below.

5. Examples will vary.

Answers to Projects and Group Activities: Construction

Designs will vary.

Symbols for Rests

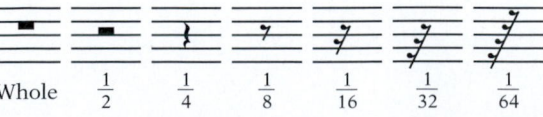

Chapter 2 Summary

Key Words	Examples
A number that is a multiple of two or more numbers is a *common multiple* of those numbers. The *least common multiple (LCM)* is the smallest common multiple of two or more numbers. [2.1A, p. 83]	12, 24, 36, 48, . . . are common multiples of 4 and 6. The LCM of 4 and 6 is 12.
A number that is a factor of two or more numbers is a *common factor* of those numbers. The *greatest common factor (GCF)* is the largest common factor of two or more numbers. [2.1B, p. 84]	The common factors of 12 and 16 are 1, 2, and 4. The GCF of 12 and 16 is 4.
A *fraction* can represent the number of equal parts of a whole. In a fraction, the *fraction bar* separates the *numerator* and the *denominator*. [2.2A, p. 87]	In the fraction $\frac{3}{4}$, the numerator is 3 and the denominator is 4.
In a *proper fraction*, the numerator is smaller than the denominator; a proper fraction is a number less than 1. In an *improper fraction*, the numerator is greater than or equal to the denominator; an improper fraction is a number greater than or equal to 1. A *mixed number* is a number greater than 1 with a whole-number part and a fractional part. [2.2A, pp. 87–88]	$\frac{2}{5}$ is a proper fraction. $\frac{7}{6}$ is an improper fraction. $4\frac{1}{10}$ is a mixed number; 4 is the whole-number part and $\frac{1}{10}$ is the fractional part.
Equal fractions with different denominators are called *equivalent fractions*. [2.2B, p. 90]	$\frac{3}{4}$ and $\frac{6}{8}$ are equivalent fractions.
A fraction is in *simplest form* when the numerator and denominator have no common factors other than 1. [2.2B, p. 90]	The fraction $\frac{11}{12}$ is in simplest form.
The *reciprocal* of a fraction is the fraction with the numerator and denominator interchanged. [2.4B, p. 115]	The reciprocal of $\frac{3}{8}$ is $\frac{8}{3}$. The reciprocal of 5 is $\frac{1}{5}$.
A *complex fraction* is a fraction whose numerator or denominator contains one or more fractions. [2.4C, p. 118]	$\dfrac{\frac{2}{3} - \frac{5}{8}}{\frac{1}{9}}$ is a complex fraction.
A number written in *decimal notation* has three parts: a whole number part, a decimal point, and a decimal part. The *decimal part* of a number represents a number less than 1. A number written in decimal notation is often simply called a *decimal*. [2.5A, p. 129]	For the decimal 31.25, 31 is the whole number part and 25 is the decimal part.

Essential Rules and Procedures	**Examples**

To find the LCM of two or more numbers, find the prime factorization of each number and write the factorization of each number in a table. Circle the greatest product in each column. The LCM is the product of the circled numbers. [2.1A, p. 83]

$$\begin{array}{c|c|c} & 2 & 3 \\ \hline 12 = & \boxed{2 \cdot 2} & 3 \\ \hline 18 = & 2 & \boxed{3 \cdot 3} \end{array}$$

The LCM of 12 and 18 is
$2 \cdot 2 \cdot 3 \cdot 3 = 36$.

To find the GCF of two or more numbers, find the prime factorization of each number and write the factorization of each number in a table. Circle the least product in each column that does not have a blank. The GCF is the product of the circled numbers. [2.1B, p. 84]

$$\begin{array}{c|c|c} & 2 & 3 \\ \hline 12 = & 2 \cdot 2 & ③ \\ \hline 18 = & ② & 3 \cdot 3 \end{array}$$

The GCF of 12 and 18 is $2 \cdot 3 = 6$.

To write an improper fraction as a mixed number, divide the numerator by the denominator. [2.2A, p. 88]

$\dfrac{29}{6} = 29 \div 6 = 4\dfrac{5}{6}$

To write a mixed number as an improper fraction, multiply the denominator of the fractional part of the mixed number by the whole number part. Add this product and the numerator of the fractional part. The sum is the numerator of the improper fraction. The denominator remains the same. [2.2A, p. 88]

$3\dfrac{2}{5} = \dfrac{5 \times 3 + 2}{5} = \dfrac{17}{5}$

To write a fraction in simplest form, divide the numerator and denominator of the fraction by their common factors. [2.2B, p. 91]

$\dfrac{30}{45} = \dfrac{2 \cdot \cancel{3} \cdot \cancel{5}}{\cancel{3} \cdot 3 \cdot \cancel{5}} = \dfrac{2}{3}$

To add fractions with the same denominators, add the numerators and place the sum over the common denominator.
$\dfrac{a}{b} + \dfrac{c}{b} = \dfrac{a + c}{b}$, where $b \neq 0$ [2.3A, p. 98]

$\dfrac{5}{12} + \dfrac{11}{12} = \dfrac{16}{12} = 1\dfrac{1}{3}$

To subtract fractions with the same denominators, subtract the numerators and place the difference over the common denominator.
$\dfrac{a}{b} - \dfrac{c}{b} = \dfrac{a - c}{b}$, where $b \neq 0$ [2.3B, p. 102]

$\dfrac{9}{16} - \dfrac{5}{16} = \dfrac{4}{16} = \dfrac{1}{4}$

To add or subtract fractions with different denominators, first rewrite the fractions as equivalent fractions with a common denominator. The common denominator is the least common multiple (LCM) of the denominators of the fractions. Then add or subtract the fractions. [2.3A/2.3B, pp. 98, 102]

$\dfrac{7}{8} + \dfrac{5}{6} = \dfrac{21}{24} + \dfrac{20}{24} = \dfrac{41}{24} = 1\dfrac{17}{24}$

$\dfrac{2}{3} - \dfrac{7}{16} = \dfrac{32}{48} - \dfrac{21}{48} = \dfrac{11}{48}$

To multiply two fractions, multiply the numerators; this is the numerator of the product. Multiply the denominators; this is the denominator of the product.
$\dfrac{a}{b} \cdot \dfrac{c}{d} = \dfrac{ac}{bd}$, where $b \neq 0$ and $d \neq 0$ [2.4A, p. 111]

$\dfrac{3}{4} \cdot \dfrac{2}{9} = \dfrac{3 \cdot 2}{4 \cdot 9} = \dfrac{3 \cdot 2}{2 \cdot 2 \cdot 3 \cdot 3} = \dfrac{1}{6}$

To divide two fractions, multiply the first fraction by the reciprocal of the second fraction.

$\dfrac{a}{b} \div \dfrac{c}{d} = \dfrac{a}{b} \cdot \dfrac{d}{c}$, where $b \neq 0$, $c \neq 0$, and $d \neq 0$ [2.4B, p. 115]

$$\dfrac{8}{15} \div \dfrac{4}{5} = \dfrac{8}{15} \cdot \dfrac{5}{4} = \dfrac{8 \cdot 5}{15 \cdot 4}$$
$$= \dfrac{2 \cdot 2 \cdot 2 \cdot 5}{3 \cdot 5 \cdot 2 \cdot 2} = \dfrac{2}{3}$$

To simplify a complex fraction, simplify the expression above the main fraction bar and simplify the expression below the main fraction bar. Then rewrite the complex fraction as a division problem by reading the main fraction bar as "divided by." [2.4C, p. 118]

$$\dfrac{\dfrac{8}{9} - \dfrac{2}{3}}{1\dfrac{1}{5}} = \dfrac{\dfrac{8}{9} - \dfrac{6}{9}}{\dfrac{6}{5}} = \dfrac{\dfrac{2}{9}}{\dfrac{6}{5}}$$
$$= \dfrac{2}{9} \div \dfrac{6}{5} = \dfrac{2}{9} \cdot \dfrac{5}{6} = \dfrac{5}{27}$$

The formula for the area of a triangle is $A = \dfrac{1}{2}bh$, where A is the area of the triangle, b is the base, and h is the height. [2.4D, p. 120]

Find the area of a triangle with a base measuring 6 ft and a height of 3 ft.

$$A = \dfrac{1}{2}bh = \dfrac{1}{2}(6)(3) = 9$$

There area is 9 ft².

To write a decimal in words, write the decimal part as though it were a whole number. Then name the place value of the last digit. The decimal point is read as "and." [2.5A, p. 129]

The decimal 12.875 is written in words as twelve and eight hundred seventy-five thousandths.

To write a decimal in standard form when it is written in words, write the whole number part, replace the word *and* with a decimal point, and write the decimal part so that the last digit is in the given place-value position. [2.5A, p. 130]

The decimal forty-nine and sixty-three thousandths is written in standard form as 49.063.

To compare two decimals, write the decimal part of each number so that each has the same number of decimal places. Then compare the two numbers. [2.5B, p. 131]

$1.790 > 1.789$

$0.8130 < 0.8315$

To round a decimal, use the same rules used with whole numbers, except drop the digits to the right of the given place value instead of replacing them with zeros. [2.5C, p. 132]

2.7134 rounded to the nearest tenth is 2.7.

0.4687 rounded to the nearest hundredth is 0.47.

To add or subtract decimals, write the decimals so that the decimal points are on a vertical line. Add or subtract as you would with whole numbers. Then write the decimal point in the answer directly below the decimal points in the given numbers. [2.6A, p. 139]

$$\begin{array}{r} {}^{1}\;{}^{1} \\ 1.35 \\ 20.8 \\ +\;0.76 \\ \hline 22.91 \end{array} \qquad \begin{array}{r} {}^{2}\;{}^{15}\;\;{}^{6}{}^{10} \\ 3\!\!\!/5.8\!\!\!/7\!\!\!/0 \\ -\;9.641 \\ \hline 26.229 \end{array}$$

To estimate the answer to a calculation, round each number to the highest place value of the number; the first digit of each number will be nonzero, and all other digits will be zero. If a number is a decimal less than 1, round the decimal so that there is one nonzero digit. Perform the calculation using the rounded numbers. [2.6A, p. 140]

$$
\begin{array}{ccc}
35.87 & \longrightarrow & 40 \\
61.09 & \longrightarrow & +60 \\
\hline
 & & 100
\end{array}
$$

$$
\begin{array}{ccc}
0.3876 & \longrightarrow & 0.4 \\
0.5472 & \longrightarrow & +0.5 \\
\hline
 & & 0.9
\end{array}
$$

To multiply decimals, multiply the numbers as you would whole numbers. Then write the decimal point in the product so that the number of decimal places in the product is the sum of the decimal places in the factors. [2.6B, p. 142]

$$
\begin{array}{lr}
26.83 & \text{2 decimal places} \\
\times \ 0.45 & \text{2 decimal places} \\
\hline
13415 & \\
10732 \ \ & \\
\hline
12.0735 & \text{4 decimal places}
\end{array}
$$

To multiply a decimal by a power of 10, move the decimal point to the right the same number of places as there are zeros in the power of 10. If the power of 10 is written in exponential notation, the exponent indicates how many places to move the decimal point. [2.6B, pp. 142–143]

$3.97 \cdot 10,000 = 39,700$

$0.641 \cdot 10^5 = 64,100$

To divide decimals, move the decimal point in the divisor to the right so that the divisor is a whole number. Move the decimal point in the dividend the same number of places to the right. Place the decimal point in the quotient directly above the decimal point in the dividend. Then divide as you would with whole numbers. [2.6C, p. 144]

$$
\begin{array}{r}
6.2 \\
0.39.\overline{)2.41.8} \\
-2\ 34 \\
\hline
7\ 8 \\
-7\ 8 \\
\hline
0
\end{array}
$$

To divide a decimal by a power of 10, move the decimal point to the left the same number of places as there are zeros in the power of 10. If the power of 10 is written in exponential notation, the exponent indicates how many places to move the decimal point. [2.6C, p. 145]

$972.8 \div 1000 = 0.9728$

$61.305 \div 10^4 = 0.0061305$

To write a fraction as a decimal, divide the numerator of the fraction by the denominator. [2.6D, p. 147]

$\dfrac{7}{8} = 7 \div 8 = 0.875$

To convert a decimal to a fraction, remove the decimal point and place the decimal part over a denominator equal to the place value of the last digit in the decimal. [2.6D, p. 147]

0.85 is eighty-five <u>hundredths</u>.

$0.85 = \dfrac{85}{100} = \dfrac{17}{20}$

To find the order relation between a decimal and a fraction, first rewrite the fraction as a decimal. Then compare the two decimals. [2.6D, p. 148]

Because $\dfrac{3}{11} \approx 0.273$, and

$0.273 > 0.260$, $\dfrac{3}{11} > 0.26$.

Order of Operations Agreement [2.7A, p. 159]

Step 1 Do all operations inside parentheses.
Step 2 Simplify any numerical expressions containing exponents.
Step 3 Do multiplication and division as they occur from left to right.
Step 4 Do addition and subtraction as they occur from left to right.

$$
\left(\dfrac{1}{3}\right)^2 + \left(\dfrac{11}{12} - \dfrac{5}{6}\right) \cdot 4
$$

$$
= \left(\dfrac{1}{3}\right)^2 + \dfrac{1}{12} \cdot 4
$$

$$
= \dfrac{1}{9} + \dfrac{1}{12} \cdot 4 = \dfrac{1}{9} + \dfrac{1}{3} = \dfrac{4}{9}
$$

Chapter 2 Review Exercises

1. Write $\dfrac{19}{2}$ as a mixed number. $9\dfrac{1}{2}$ [2.2A]

2. Subtract: $6\dfrac{2}{9} - 3\dfrac{7}{18}$ $2\dfrac{5}{6}$ [2.3B]

3. Evaluate $x \div y$ when $x = 2\dfrac{5}{8}$ and $y = 1\dfrac{3}{4}$.

   $1\dfrac{1}{2}$ [2.4B]

4. Write five and thirty-four thousandths in standard form.

   5.034   [2.5A]

5. Convert 0.28 to a fraction. $\dfrac{7}{25}$ [2.6D]

6. Find the product of 3 and $\dfrac{8}{9}$. $2\dfrac{2}{3}$ [2.4A]

7. Place the correct symbol, < or >, between the two numbers.

   8.039   8.31   <   [2.5B]

8. Place the correct symbol, < or >, between the two numbers.

   $\dfrac{3}{5}$ $\dfrac{7}{15}$ >   [2.2C]

9. Find the LCM of 50 and 75.
   150   [2.1A]

10. Find the product of 0.918 and 10^5.
    91,800   [2.6B]

11. Evaluate xy when $x = 8$ and $y = \dfrac{5}{12}$.

    $3\dfrac{1}{3}$ [2.4A]

12. Express the shaded portion of the circles as an improper fraction and as a mixed number.

     $\dfrac{10}{7}$; $1\dfrac{3}{7}$ [2.2A]

13. Place the correct symbol, < or >, between the two numbers.

    $\dfrac{3}{7}$ 0.429   <   [2.6D]

14. Simplify: $\dfrac{\dfrac{5}{8} - \dfrac{1}{4}}{\dfrac{1}{2} + \dfrac{1}{8}}$ $\dfrac{3}{5}$ [2.4C]

15. Write a fraction that is equivalent to $\dfrac{4}{9}$ and has a denominator of 72. $\dfrac{32}{72}$ [2.2B]

16. Evaluate x^2y^3 when $x = \dfrac{8}{9}$ and $y = \dfrac{3}{4}$.

    $\dfrac{1}{3}$ [2.4A]

17. Evaluate $ab^2 - c$ when $a = 4$, $b = \dfrac{1}{2}$, and $c = \dfrac{5}{7}$.

    $\dfrac{2}{7}$ [2.7A]

18. Find the GCF of 42 and 63.

    21   [2.1B]

19. Find the quotient of 14.2 and 10^3.
    0.0142   [2.6C]

20. Divide and round to the nearest tenth: $6.8 \div 47.92$ 0.1   [2.6C]

21. Find the quotient of $\dfrac{5}{9}$ and $\dfrac{2}{3}$.

    $\dfrac{5}{6}$ [2.4B]

22. Evaluate $\dfrac{x}{y}$ when $x = 0.396$ and $y = 3.6$.

    0.11   [2.6C]

23. Estimate the difference between 506.81 and 64.1.

440   [2.6A]

24. Multiply: (9.47)(0.26)

2.4622   [2.6B]

25. Evaluate $a - b$ when $a = 80.32$ and $b = 29.577$.

50.743   [2.6A]

26. Evaluate $\left(\dfrac{3}{8}\right)^2 \cdot 4^2$.

$2\dfrac{1}{4}$ [2.4A]

27. Find the sum of $3\dfrac{7}{12}$ and $5\dfrac{1}{2}$.

$9\dfrac{1}{12}$ [2.3A]

28. Write $\dfrac{30}{105}$ in simplest form.

$\dfrac{2}{7}$ [2.2B]

29. Evaluate $a - b$ when $a = 7$ and $b = 2\dfrac{3}{10}$.

$4\dfrac{7}{10}$ [2.3B]

30. 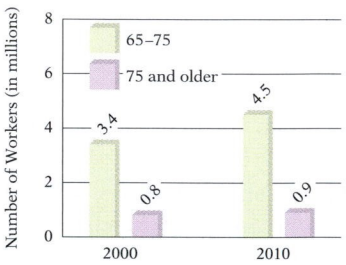 **Labor Force**   The graph at the right shows that the number of older workers is expected to increase during the first decade of the 21st century. Find the projected increase in the number of workers over the age of 65.

1.2 million workers   [2.6E]

Older Workers in the Labor Force   **P**
Source: Bureau of Labor Statistics

31. **Wrestling**   A wrestler is put on a diet to gain 12 lb in 4 weeks. The wrestler gains $3\dfrac{1}{2}$ lb the first week and $2\dfrac{1}{4}$ lb the second week. How much weight must the wrestler gain during the third and fourth weeks in order to gain a total of 12 lb?

$6\dfrac{1}{4}$ lb   [2.3C]

32. **Manufacturing**   An employee hired for piecework can assemble a unit in $2\dfrac{1}{2}$ min. How many units can this employee assemble during an 8-hour day?

192 units   [2.4D]

33. **Wages**   Find the overtime pay due an employee who worked $6\dfrac{1}{4}$ h of overtime this week. The employee's overtime rate is $24 an hour.

$150   [2.4D]

34. **Physics**   What is the final velocity, in feet per second, of an object dropped from a plane with a starting velocity of 0 ft/s and a fall of $15\dfrac{1}{2}$ s? Use the formula $V = S + 32t$, where V is the final velocity of a falling object, S is its starting velocity, and t is the time of the fall.

496 ft/s   [2.4D]

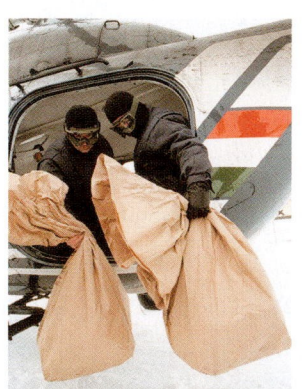

Chapter 2 Test

1. Write $\frac{18}{7}$ as a mixed number.

$2\frac{4}{7}$ [2.2A]

2. Subtract: $7\frac{3}{4} - 3\frac{5}{6}$

$3\frac{11}{12}$ [2.3B]

3. Evaluate xy when $x = 6\frac{3}{7}$ and $y = 3\frac{1}{2}$.

$22\frac{1}{2}$ [2.4A]

4. Find the product of $\frac{2}{3}$ and $\frac{7}{8}$.

$\frac{7}{12}$ [2.4A]

5. Find the LCM of 30 and 45.
90 [2.1A]

6. Write nine and thirty-three thousandths in standard form.
9.033 [2.5A]

7. Evaluate x^3y^2 when $x = 1\frac{1}{2}$ and $y = \frac{5}{6}$.

$2\frac{11}{32}$ [2.4A]

8. Write $3\frac{4}{5}$ as an improper fraction.

$\frac{19}{5}$ [2.2A]

9. What is $\frac{7}{12}$ divided by $\frac{3}{4}$?

$\frac{7}{9}$ [2.4B]

10. Place the correct symbol, < or >, between the two numbers.

4.003   4.009

< [2.5B]

11. Evaluate $\frac{x}{yz}$ when $x = \frac{7}{20}$, $y = \frac{2}{15}$, and $z = \frac{3}{8}$.
7 [2.4C]

12. Find the GCF of 18 and 54.

18 [2.1B]

13. How much larger is $\frac{13}{14}$ than $\frac{16}{21}$?

$\frac{1}{6}$ [2.3B]

14. Write $\frac{60}{75}$ in simplest form.

$\frac{4}{5}$ [2.2B]

15. Evaluate $x + y + z$ when $x = 1\frac{3}{8}$, $y = \frac{1}{2}$, and $z = \frac{5}{6}$.

$2\frac{17}{24}$ [2.3A]

16. Place the correct symbol, < or >, between the two numbers.

$\frac{5}{6}$ $\frac{11}{15}$

> [2.2C]

17. Evaluate $a^2b - c^2$ when $a = \frac{2}{3}$, $b = 9$, and $c = \frac{3}{5}$.

$3\frac{16}{25}$ [2.7A]

18. Place the correct symbol, < or >, between the two numbers.

0.22 $\frac{2}{9}$

< [2.6D]

19. Round 6.051367 to the nearest thousandth.

   6.051   [2.5C]

20. Evaluate $x \div y$ when $x = \frac{8}{9}$ and $y = \frac{16}{27}$.

   $1\frac{1}{2}$ [2.4B]

21. Find the difference between 30 and 7.247.
   22.753   [2.6A]

22. Estimate the difference between 92.34 and 17.95.

   70   [2.6A]

23. Find the total of 4.58, 3.9, and 6.017.
   14.497   [2.6A]

24. Evaluate $20cd$ when $c = 0.5$ and $d = 6.4$.

   64   [2.6B]

25. Find the quotient of 84.96 and 100.
   0.8496   [2.6C]

26. Write a fraction that is equivalent to $\frac{3}{7}$ and has a denominator of 28.

   $\frac{12}{28}$ [2.2B]

27. **The Film Industry**   The table at the right shows six James Bond films released between 1960 and 1970 and their gross box office income, in millions of dollars, in the United States. How much greater was the gross from *Thunderball* than the gross from *On Her Majesty's Secret Service*?
   $40.8 million   [2.6E]

Film	U.S. Box Office Gross
Dr. No	$16.1
On Her Majesty's Secret Service	$22.8
From Russia with Love	$24.8
You Only Live Twice	$43.1
Goldfinger	$51.1
Thunderball	$63.6

Source: **www.worldwideboxoffice.com**

28. **Community Service**   You are required to contribute 20 h of community service to the town in which your college is located. On one occasion you work $7\frac{1}{4}$ h, and on another occasion you work $2\frac{3}{4}$ h. How many more hours of community service are still required of you?
   10 h   [2.3C]

29. **Manufacturing**   An employee hired for piecework can assemble a unit in $4\frac{1}{2}$ min. How many units can this employee assemble in 6 h?
   80 units   [2.4C]

30. **Accounting**   The fundamental accounting equation is $A = L + S$, where A is the assets of the company, L is the liabilities of the company, and S is the stockholders' equity. Find the stockholders' equity in a company whose assets are $48.2 million and whose liabilities are $27.6 million.
   $20.6 million   [2.6E]

31. **Geometry**   The lengths of the three sides of a triangle are 8.75 m, 5.25 m, and 4.5 m. Find the perimeter of the triangle. Use the formula $P = a + b + c$.
   18.5 m   [2.6E]

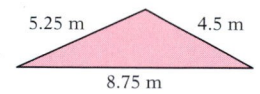

Cumulative Review Exercises

1. Find the quotient of 387.9 and 10^4.
0.03879 [2.6C]

2. Evaluate $(x + y)^2 - 2z$ when $x = 3$, $y = 2$, and $z = 5$.
15 [1.4A]

3. Find the prime factorization of 140.
$2^2 \cdot 5 \cdot 7$ [1.3D]

4. Write eight million seventy-two thousand ninety-two in standard form.
8,072,092 [1.1B]

5. Place the correct symbol, $<$ or $>$, between the two numbers.

$$\frac{7}{11} \quad \frac{4}{5}$$

$<$ [2.2C]

6. Find the GCF of 72 and 108.
36 [2.1B]

7. Find the difference between $\frac{5}{14}$ and $\frac{9}{42}$.

$\frac{1}{7}$ [2.3B]

8. Estimate the sum of 372, 541, 608, and 429.
1900 [1.2A]

9. Add: $6847 + 3501 + 924$

11,272 [1.2A]

10. Evaluate $x \div y$ when $x = 3\frac{2}{3}$ and $y = 2\frac{4}{9}$.

$1\frac{1}{2}$ [2.4B]

11. What is 36.92 increased by 18.5?
55.42 [2.6A]

12. Simplify: $\left(\frac{5}{9}\right)\left(\frac{3}{10}\right)\left(\frac{6}{7}\right)$

$\frac{1}{7}$ [2.4A]

13. Evaluate $x^4 y^2$ when $x = 2$ and $y = 10$.
1600 [1.3B]

14. Find the prime factorization of 260.
$2^2 \cdot 5 \cdot 13$ [1.3D]

15. Convert $\frac{19}{25}$ to a decimal.

0.76 [2.6D]

16. Estimate the difference between 89,357 and 66,042.
20,000 [1.2B]

17. **Vacation Days** The figure at the right shows the number of vacation days per year that are legally mandated in several countries.
  a. Which country mandates more vacation days, Ireland or Sweden?
  b. How many times more vacation days does Austria mandate than Switzerland?
  a. Sweden **b.** 1.5 times [2.6E]

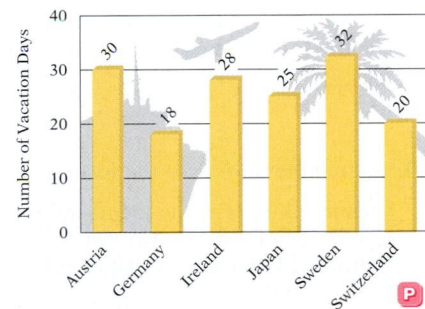

Number of Legally Mandated Vacation Days
Source: Economic Policy Institute; *World Almanac*

18. Divide: $\dfrac{8}{0}$

Undefined   [1.3C]

19. Simplify: $\dfrac{5}{7} + \dfrac{4}{21}$

$\dfrac{19}{21}$ [2.3A]

20. Subtract: $8\dfrac{3}{4} - 1\dfrac{5}{7}$

$7\dfrac{1}{28}$ [2.3B]

21. Evaluate $3a + (a - b)^3$ when $a = 4$ and $b = 1$.

39   [1.4A]

22. Simplify: $5(7 - 3) \div (4) + 6(2)$

17   [1.4A]

23. Evaluate $\dfrac{a}{b + c}$ when $a = \dfrac{3}{8}$, $b = \dfrac{1}{2}$, and $c = \dfrac{3}{4}$.

$\dfrac{3}{10}$ [2.4C]

24. Evaluate $x^3 y^4$ when $x = \dfrac{7}{12}$ and $y = \dfrac{6}{7}$.

$\dfrac{3}{28}$ [2.4A]

25. Divide and round to the nearest tenth: $2.617 \div 0.93$

2.8   [2.6C]

26. **Physical Fitness**   The chart at the right shows the calories burned per hour as a result of different aerobic activities. Suppose you weigh 150 lb. According to the chart, how many more calories would you burn by bicycling at 12 mph for 4 h than by walking at a rate of 3 mph for 5 h?
40 calories   [1.3E]

Activity	100 lb	150 lb
Bicycling, 6 mph	160	240
Bicycling, 12 mph	270	410
Jogging, 5 1/2 mph	440	660
Jogging, 7 mph	610	920
Jumping rope	500	750
Tennis, singles	265	400
Walking, 2 mph	160	240
Walking, 3 mph	210	320
Walking, 4 1/2 mph	295	440

27. **Demographics**   The Census Bureau projects that the population of New England will increase to 15,321,000 in 2020 from 13,581,000 in 2000. Find the projected increase in the population of New England during the 20-year period.
1,740,000 people   [1.2C]

28. **Sales**   The figure at the right shows how the average sales-person spends the workweek.
a. On average, how many hours per week does a salesperson work?
b. Does the average salesperson spend more time face-to-face selling or doing both administrative work and placing service calls?
a. 46.5 h   **b.** Face-to-face selling   [2.6E]

Average Salesperson's Workweek
Source: Dartnell's 28th Survey of Sales Force Compensation

29. **Travel**   A bicyclist rode for $\dfrac{3}{4}$ h at a rate of $5\dfrac{1}{2}$ mph. Use the equation $d = rt$, where d is the distance traveled, r is the rate of travel, and t is the time, to find the distance traveled by the bicyclist.

$4\dfrac{1}{8}$ mi   [2.4D]

30. **Consumerism**   Use the formula $C = \dfrac{M}{N}$, where C is the cost per visit at a health club, M is the membership fee, and N is the number of visits to the club, to find the cost per visit when your annual membership fee at a health club is $390 and you visit the club 125 times during the year.
$3.12   [2.6E]

3

Rational Numbers

Stock market reports involve signed numbers. Positive numbers indicate an increase in the price of a share of stock, and negative numbers indicate a decrease. Positive and negative numbers are also used to indicate whether a company has experienced a profit or loss over a specified period of time. An example is provided in **Exercise 148 on page 230.**

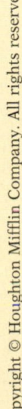

Need help? For online student resources, such as section quizzes, visit this textbook's website at **math.college.hmco.com/students.**

OBJECTIVES

Section 3.1

A To identify order relations between integers

B To find the opposite of a number

C To evaluate expressions that contain the absolute value symbol

D To solve application problems

Section 3.2

A To add integers

B To subtract integers

C To solve application problems

Section 3.3

A To multiply integers

B To divide integers

C To solve application problems

Section 3.4

A To add or subtract rational numbers

B To multiply or divide rational numbers

C To solve application problems

Section 3.5

A To use the Order of Operations Agreement to simplify expressions

Do these exercises to prepare for Chapter 3.

1. Place the correct symbol, $<$ or $>$, between the two numbers.

   54   45
   54 $>$ 45   [1.1A]

2. What is the distance from 4 to 8 on the number line?
   4 units   [1.1A]

For Exercises 3 to 14, add, subtract, multiply, or divide.

3. $7654 + 8193$
   15,847   [1.2A]

4. $6097 - 2318$
   3779   [1.2B]

5. 472×56
   26,432   [1.3A]

6. $\dfrac{144}{24}$
   6   [1.3C]

7. $\dfrac{2}{3} + \dfrac{3}{5}$
   $1\dfrac{4}{15}$ [2.3A]

8. $\dfrac{3}{4} - \dfrac{5}{16}$
   $\dfrac{7}{16}$ [2.3B]

9. $0.75 + 3.9 + 6.408$
   11.058   [2.6A]

10. $5.4 - 1.619$
    3.781   [2.6A]

11. $\dfrac{3}{4} \times \dfrac{8}{15}$
    $\dfrac{2}{5}$ [2.4A]

12. $\dfrac{5}{12} \div \dfrac{3}{4}$
    $\dfrac{5}{9}$ [2.4B]

13. 23.5×0.4
    9.4   [2.6B]

14. $0.96 \div 2.4$
    0.4   [2.6C]

15. Simplify: $(8 - 6)^2 + 12 \div 4 \cdot 3^2$
    31   [1.4A]

GO FIGURE ...

Super Yeast causes bread to double in volume each minute. If it takes one loaf of bread made with Super Yeast 30 min to fill the oven, how long does it take two loaves of bread made with Super Yeast to fill one-half the oven?   28 min

3.1 Introduction to Integers

Objective A   To identify order relations between integers

In Chapters 1 and 2, only zero and numbers greater than zero were discussed. In this chapter, numbers less than zero are introduced. Phrases such as "7 degrees below zero," "$50 in debt," and "20 feet below sea level" refer to numbers less than zero.

Numbers greater than zero are called **positive numbers.** Numbers less than zero are called **negative numbers.**

Point of Interest

Chinese manuscripts dating from about 250 B.C. contain the first recorded use of negative numbers. However, it was not until late in the 14th century that mathematicians generally accepted these numbers.

> **Positive and Negative Numbers**
>
> A number n is positive if $n > 0$.
> A number n is negative if $n < 0$.

A positive number can be indicated by placing the sign + in front of the number. For example, we can write +4 instead of 4. Both +4 and 4 represent "positive 4." Usually, however, the plus sign is omitted and it is understood that the number is a positive number.

A negative number is indicated by placing a negative sign (−) in front of the number. The number −1 is read "negative one," −2 is read "negative two," and so on.

The number line can be extended to the left of zero to show negative numbers.

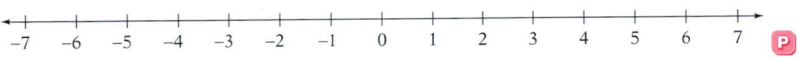

The **integers** are . . . , −4, −3, −2, −1, 0, 1, 2, 3, 4, ... The integers to the right of zero are the **positive integers.** The integers to the left of zero are the **negative integers.** Zero is an integer, but it is neither positive nor negative. The point corresponding to 0 on the number line is called the **origin.**

On a number line, the numbers get larger as we move from left to right. The numbers get smaller as we move from right to left. Therefore, a number line can be used to visualize the order relation between two integers.

A number that appears to the right of a given number on the number line is greater than (>) the given number. A number that appears to the left of a given number on the number line is less than (<) the given number.

2 is to the right of −3 on the number line.
2 is greater than −3.
2 > −3

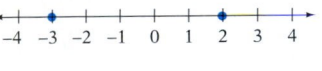

−4 is to the left of 1 on the number line.
−4 is less than 1.
−4 < 1

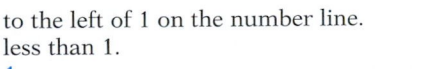

Objective 3.1A

New Vocabulary
positive numbers
negative numbers
integers
positive integers
negative integers

Symbols to Review
>
<

Discuss the Concepts
Classify each number as a positive integer, a negative integer, or neither.

$-10, 17, -8, 0, \dfrac{2}{3}, 465$

Instructor Note
Many of the graphs and tables in this text are available on Microsoft PowerPoint® slides. These selected art pieces are indicated by **P** . See, for example, the number line at the left.

Concept Check
A is a point on the number line halfway between −12 and 4. B is a point on the number line halfway between A and the graph of 2 on the number line. B is the graph of what number?
−1

In-Class Examples (Objective 3.1A)

1. On the number line, what number is 6 units to the left of 2?
−4

2. On the number line shown, if B is −3 and D is −1, what numbers are A and E?
A is −4 and E is 0.

A	B	C	D	E	F	

3. Place the correct symbol, < or >, between the two numbers.
−23   13   −23 < 13

4. Write the given numbers in order from smallest to largest.
5, −8, 0, −1, 9   −8, −1, 0, 5, 9

Optional Student Activity

Graph each of the following pairs of numbers on the number line. Then place the correct symbol, $<$ or $>$, between the two numbers.

1. -4 and 3 $-4 < 3$

2. 5 and -2 $5 > -2$

3. -1 and -3 $-1 > -3$

4. -5 and 0 $-5 < 0$

5. -7 and 1 $-7 < 1$

Order Relations

$a > b$ if a is to the right of b on the number line.

$a < b$ if a is to the left of b on the number line.

Example 1 On the number line, what number is 5 units to the right of -2?

Solution

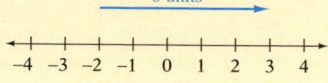

3 is 5 units to the right of -2.

You Try It 1 On the number line, what number is 4 units to the left of 1?

Your solution -3

Example 2 If G is 2 and I is 4, what numbers are B and D?

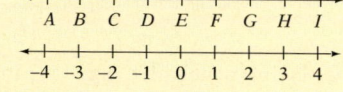

Solution

B is -3, and D is -1.

You Try It 2 If G is 1 and H is 2, what numbers are A and C?

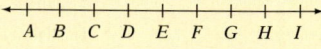

Your solution A is -5, and C is -3.

Example 3 Place the correct symbol, $<$ or $>$, between the two numbers.

a. -3 -1 **b.** 1 -2

Solution

a. -3 is to the left of -1 on the number line.

$-3 < -1$

b. 1 is to the right of -2 on the number line.

$1 > -2$

You Try It 3 Place the correct symbol, $<$ or $>$, between the two numbers.

a. 2 -5 **b.** -4 3

Your solution **a.** $2 > -5$ **b.** $-4 < 3$

Example 4 Write the given numbers in order from smallest to largest.

$5, -2, 3, 0, -6$

Solution $-6, -2, 0, 3, 5$

You Try It 4 Write the given numbers in order from smallest to largest.

$-7, 4, -1, 0, 8$

Your solution $-7, -1, 0, 4, 8$

Solutions on p. S8

Objective B   **To find the opposite of a number**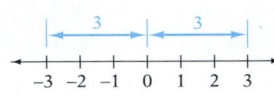

The distance from 0 to 3 on the number line is 3 units. The distance from 0 to −3 on the number line is 3 units. 3 and −3 are the same distance from 0 on the number line, but 3 is to the right of 0 and −3 is to the left of 0.

Two numbers that are the same distance from zero on the number line but are on opposite sides of zero are called **opposites.**

$$-3 \text{ is the opposite of } 3 \quad \text{and} \quad 3 \text{ is the opposite of } -3.$$

For any number n, the opposite of n is $-n$ and the opposite of $-n$ is n.

We can now define the **integers** as the whole numbers and their opposites.

A negative sign can be read as "the opposite of."

$$-(3) = -3 \qquad \text{The opposite of positive 3 is negative 3.}$$

$$-(-3) = 3 \qquad \text{The opposite of negative 3 is positive 3.}$$

Therefore, $-(a) = -a$ and $-(-a) = a$.

Note that with the introduction of negative integers and opposites, the symbols + and − can be read in different ways.

$6 + 2$	"six plus two"	+ is read "plus"
$+2$	"positive two"	+ is read "positive"
$6 - 2$	"six minus two"	− is read "minus"
-2	"negative two"	− is read "negative"
$-(-6)$	"the opposite of negative six"	− is read first as "the opposite of" and then as "negative"

When the symbols + and − indicate the operations of addition and subtraction, spaces are inserted before and after the symbol. When the symbols + and − indicate the sign of a number (positive or negative), there is no space between the symbol and the number.

Integrating Technology

The +/− key on your calculator is used to find the opposite of a number. The − key is used to perform the operation of subtraction.

Example 5   Find the opposite number.

    a. −8   **b.** 15   **c.** a

Solution   **a.** 8   **b.** −15   **c.** $-a$

You Try It 5   Find the opposite number.

    a. 24   **b.** −13   **c.** $-b$

Your solution   **a.** −24   **b.** 13   **c.** b

Solution on p. S8

Objective 3.1B

Vocabulary to Review
whole numbers   [1.1A]

New Vocabulary
opposites

Discuss the Concepts

1. Explain the different ways in which the symbol "+" can be read.

2. Explain the different ways in which the symbol "−" can be read.

3. How can you determine whether the symbol "−" in an expression indicates the sign of a number or the operation of subtraction?

Concept Check

1. What number is neither a negative nor a positive number?

2. The number 5 is how many units from zero on the number line?

3. The number −5 is how many units from zero on the number line?

Optional Student Activity

1. Write a mathematical expression in which the symbol "−" is used to mean subtraction.

2. Write a mathematical expression in which the symbol "−" is used to indicate a negative number.

In-Class Examples (Objective 3.1B)

1. Find the opposite number.   **a.** 32   **b.** −19
   a. −32   **b.** 19

2. Write the expression in words.
   −5 − (−10)
   Negative five minus negative ten

3. Simplify.   **a.** −(46)   **b.** −(−d)
   a. −46   **b.** d

Example 6 Write the expression in words.

a. $7 - (-9)$ b. $-4 + 10$

Solution a. Seven minus negative nine

b. Negative four plus ten

You Try It 6 Write the expression in words.

a. $-3 - 12$ b. $8 + (-5)$

Your solution a. Negative three minus twelve

b. Eight plus negative five

Example 7 Simplify.

a. $-(-27)$ b. $-(-c)$

Solution a. $-(-27) = 27$

b. $-(-c) = c$

You Try It 7 Simplify.

a. $-(-59)$ b. $-(y)$

Your solution a. 59   b. $-y$

Solutions on p. S9

Objective 3.1C

New Vocabulary
absolute value of a number

Discuss the Concepts
True or false:

1. Every negative number is to the left of every positive number on the number line.   T

2. Every positive number is to the right of every negative number on the number line.   T

3. -6 is to the left of -1 on the number line.   T

4. -8 is to the right of -12 on the number line.   T

Instructor Note
The important point for a student to understand about absolute value is magnitude. If a student runs 5 mi west or 5 mi east, the distance is the same. Only the direction is different.

Objective C   **To evaluate expressions that contain the absolute value symbol**

The **absolute value** of a number is the distance from zero to the number on the number line. Distance is never a negative number. Therefore, the absolute value of a number is a positive number or zero. The symbol for absolute value is "| |."

The distance from 0 to 3 is 3 units. Thus $|3| = 3$ (the absolute value of 3 is 3).

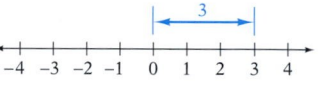

The distance from 0 to -3 is 3 units. Thus $|-3| = 3$ (the absolute value of -3 is 3).

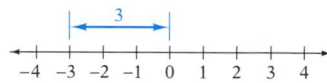

Because the distance from 0 to 3 and the distance from 0 to -3 are the same,

$$|3| = |-3| = 3$$

Absolute Value

The absolute value of a positive number is positive.	$|5| = 5$
The absolute value of a negative number is positive.	$|-5| = 5$
The absolute value of zero is zero.	$|0| = 0$

TAKE NOTE

It is important to be aware that the negative sign is *in front of the absolute value symbol*. This means $-|7| = -7$, but $|-7| = 7$.

HOW TO Evaluate $-|7|$.

The negative sign is *in front of* the absolute value symbol.

Recall that a negative sign can be read as "the opposite of."

Therefore, $-|7|$ can be read "the opposite of the absolute value of 7."

$-|7| = -7$

In-Class Examples (Objective 3.1C)

1. Find the absolute value of -67.
67

2. Evaluate.   a. $-|24|$ b. $-|-18|$
a. -24 b. -18

3. Evaluate $-|x|$ when $x = 6$.
-6

4. Write the given numbers in order from smallest to largest. $-|-6|, -8, -(-4), |3|$
$-8, -|-6|, |3|, -(-4)$

Example 8 Find the absolute value of **a.** 6 and **b.** −9.

Solution **a.** |6| = 6
b. |−9| = 9

You Try It 8 Find the absolute value of **a.** −8 and **b.** 12.

Your solution **a.** 8   **b.** 12

Example 9 Evaluate **a.** |−27| and **b.** −|−14|.

Solution **a.** |−27| = 27
b. −|−14| = −14

You Try It 9 Evaluate **a.** |0| and **b.** −|35|.

Your solution **a.** 0   **b.** −35

Example 10 Evaluate |−x|, where x = −4.

Solution |−x| = |−(−4)| = |4| = 4

You Try It 10 Evaluate |−y|, where y = 2.

Your solution 2

Example 11 Write the given numbers in order from smallest to largest.

|−7|, −5, |0|, −(−4), −|−3|

Solution |−7| = 7, |0| = 0,
−(−4) = 4, −|−3| = −3
−5, −|−3|, |0|, −(−4), |−7|

You Try It 11 Write the given numbers in order from smallest to largest.

|6|, |−2|, −(−1), −4, −|−8|

Your solution −|−8|, −4, −(−1), |−2|, |6|

Solutions on p. S9

Objective D To solve application problems

Example 12

Which is the colder temperature, −18°F or −15°F?

Strategy
To determine which is the colder temperature, compare the numbers −18 and −15. The lower number corresponds to the colder temperature.

Solution
−18 < −15

The colder temperature is −18°F.

You Try It 12

Which is closer to blastoff, −9 s and counting or −7 s and counting?

Your Strategy

Your solution
−7 s and counting

Solution on p. S9

Objective 3.1D

Optional Student Activity

The list below gives temperatures either on the surface or in the atmosphere of bodies in our solar system.

Mercury's dark side:
−346°F

Mercury's sunlit side:
950°F

Mars, daytime:
−17°F

Saturn's moon Titan:
−190°F

Neptune's atmosphere:
−360°F

Pluto, atmospheric high:
−163°F

Pluto, atmospheric low:
−397°F

a. List the temperatures given from coldest to warmest.
−397, −360, −346, −190, −163, −17, 950

b. How many of the temperatures listed are below −300°F?
Three of the temperatures listed are below −300°F (−397°F, −360°F, −346°F).

In-Class Examples (Objective 3.1D)

1. In a golf tournament, scores below par are recorded as negative numbers; scores above par are recorded as positive numbers. The winner of the tournament is the player who has the lowest score. If the four best finishers in a golf tournament had scores of −6, −9, −7, and −12, who won the tournament?
The player with a score of −12.

Section 3.1

Suggested Assignment

Exercises 1–119, odds
More challenging problems:
  Exercises 121–125, odd
  Exercise 126

3.1 Exercises

Objective A   To identify order relations between integers

Graph the number on the number line.

1.   −5

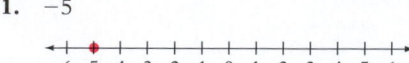

2.   −1

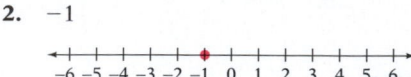

3.   −6

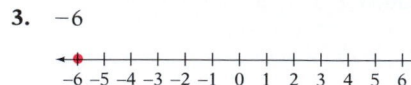

4.   −2

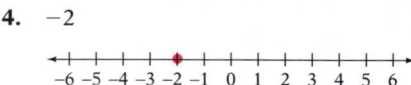

5. x, where $x = 5$

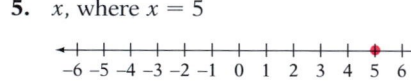

6. x, where $x = 0$

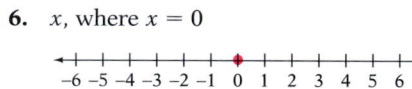

7. x, where $x = -4$

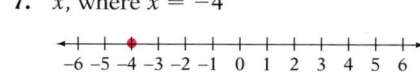

8. x, where $x = -3$

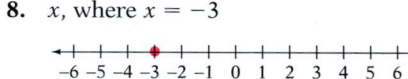

On the number line, which number is:

9.   3 units to the right of −2?
1

10.   5 units to the right of −3?
2

11.   4 units to the left of 3?
−1

12.   2 units to the left of −1?
−3

13.   6 units to the right of −3?
3

14.   4 units to the right of −4?
0

For Exercises 15–18, use the following number line.

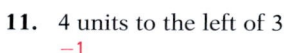

15.   If F is 1 and G is 2, what numbers are A and C?
A is −4 and C is −2.

16.   If G is 1 and H is 2, what numbers are B and D?
B is −4 and D is −2.

17.   If H is 0 and I is 1, what numbers are A and D?
A is −7 and D is −4.

18.   If G is 2 and I is 4, what numbers are B and E?
B is −3 and E is 0.

Quick Quiz (Objective 3.1A)

1. On the number line, what number is 5 units to the right of −8?
−3

2. On the number line shown, if B is −5 and D is −3, what numbers are A and F?
A is −6 and F is −1.

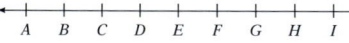

3. Place the correct symbol, < or >, between the two numbers.
34 −43 34 > −43

4. Write the given numbers in order from smallest to largest.
4, −10, 5, 0, −1 −10, −1, 0, 4, 5

Place the correct symbol, < or >, between the two numbers.

19. −2 −5
>

20. −6 −1
<

21. 3 −7
>

22. −11 −8
<

23. −42 −7
<

24. −21 −34
>

25. 53 −46
>

26. −27 −39
>

27. −51 −20
<

28. −136 0
<

29. −131 101
<

30. 127 −150
>

Write the given numbers in order from smallest to largest.

31. 3, −7, 0, −2
−7, −2, 0, 3

32. −4, 8, 6, −1
−4, −1, 6, 8

33. −3, 1, −5, 4
−5, −3, 1, 4

34. −6, 2, −8, 7
−8, −6, 2, 7

35. 9, −4, 5, 0
−4, 0, 5, 9

36. 6, −9, −12, 8
−12, −9, 6, 8

37. −10, 4, 12, −5, −7
−10, −7, −5, 4, 12

38. 11, −8, −1, 7, −6
−8, −6, −1, 7, 11

39. 10, −11, −2, 5, −7
−11, −7, −2, 5, 10

Objective B   **To find the opposite of a number**

Find the opposite of the number.

40. 22
−22

41. 45
−45

42. −31
31

43. −88
88

44. c
−c

45. n
−n

46. −w
w

47. −d
d

Write the expression in words.

48. −(−11)
The opposite of negative eleven

49. −(−13)
The opposite of negative thirteen

50. −(−d)
The opposite of negative d

51. −(−p)
The opposite of negative p

52. −2 + (−5)
Negative 2 plus negative five

53. 5 + (−10)
Five plus negative ten

54. 6 − (−7)
Six minus negative seven

55. −14 − (−3)
Negative fourteen minus negative three

56. 9 − 12
Nine minus twelve

57. −13 − 8
Negative thirteen minus eight

58. −a − b
Negative a minus b

59. m + (−n)
m plus negative n

Quick Quiz (Objective 3.1B)

1. Find the opposite number. **a.** 87 **b.** −22
a. −87 **b.** 22

2. Write the expression in words.
8 + (−3)
Eight plus negative three

3. Simplify. **a.** −(−93) **b.** −(b)
a. 93 **b.** −b

Simplify.

60. $-(-5)$
5

61. $-(-7)$
7

62. $-(-38)$
38

63. $-(-61)$
61

64. $-(29)$
-29

65. $-(46)$
-46

66. $-(-52)$
52

67. $-(-73)$
73

68. $-(-m)$
m

69. $-(-z)$
z

70. $-(b)$
$-b$

71. $-(p)$
$-p$

> **Objective C** To evaluate expressions that contain the absolute value symbol

Find the absolute value of the number.

72. 4
4

73. -4
4

74. -7
7

75. 9
9

76. -1
1

77. -11
11

78. 10
10

79. -12
12

Evaluate.

80. $|-15|$
15

81. $|-23|$
23

82. $-|33|$
-33

83. $-|27|$
-27

84. $|32|$
32

85. $|25|$
25

86. $-|-36|$
-36

87. $-|-41|$
-41

88. $-|-81|$
-81

89. $-|-93|$
-93

90. $|x|$, where $x = 7$
7

91. $|x|$, where $x = -10$
10

92. $|-x|$, where $x = 2$
2

93. $|-x|$, where $x = 8$
8

94. $|-y|$, where $y = -3$
3

95. $|-y|$, where $y = -6$
6

Place the correct symbol, $<$, $=$, or $>$, between the two numbers.

96. $|7|$ $|-9|$
$<$

97. $|-12|$ $|8|$
$>$

98. $|-5|$ $|-2|$
$>$

99. $|6|$ $|13|$
$<$

100. $|-8|$ $|3|$
$>$

101. $|-1|$ $|-17|$
$<$

102. $|-14|$ $|14|$
$=$

103. $|x|$ $|-x|$
$=$

Quick Quiz (Objective 3.1C)

1. Find the absolute value of -89.
89

2. Evaluate. **a.** $-|31|$ **b.** $-|-52|$
a. -31 **b.** -52

3. Evaluate $|-x|$ when $x = -9$.
9

4. Write the given numbers in order from smallest to largest.
$-5, -(-6), |8|, -|-9|$
$-|-9|, -5, -(-6), |8|$

Write the given numbers in order from smallest to largest.

104. $|-8|, -(-3), |2|, -|-5|$
$-|-5|, |2|, -(-3), |-8|$

105. $-|6|, -(4), |-7|, -(-9)$
$-|6|, -(4), |-7|, -(-9)$

106. $-(-1), |-6|, |0|, -|3|$
$-|3|, |0|, -(-1), |-6|$

107. $-|-7|, -9, -(5), |4|$
$-9, -|-7|, -(5), |4|$

108. $-|2|, -(-8), 6, |1|, -7$
$-7, -|2|, |1|, 6, -(-8)$

109. $-(-3), -|-8|, |5|, -|10|, -(-2)$
$-|10|, -|-8|, -(-2), -(-3), |5|$

Objective D   To solve application problems

Environmental Science   The table below gives equivalent temperatures for combinations of temperature and wind speed. For example, the combination of a temperature of 15°F and a wind blowing at 10 mph has a cooling power equal to 3°F. Use this table for Exercises 110 to 115.

Wind Chill Factors

Wind Speed (mph)	Thermometer Reading (degrees Fahrenheit)															
	25	20	15	10	5	0	-5	-10	-15	-20	-25	-30	-35	-40	-45
5	19	13	7	1	-5	-11	-16	-22	-28	-34	-40	-46	-52	-57	-63
10	15	9	3	-4	-10	-16	-22	-28	-35	-41	-47	-53	-59	-66	-72
15	13	6	0	-7	-13	-19	-26	-32	-39	-45	-51	-58	-64	-71	-77
20	11	4	-2	-9	-15	-22	-29	-35	-42	-48	-55	-61	-68	-74	-81
25	9	3	-4	-11	-17	-24	-31	-37	-44	-51	-58	-64	-71	-78	-84
30	8	1	-5	-12	-19	-26	-33	-39	-46	-53	-60	-67	-73	-80	-87
35	7	0	-7	-14	-21	-27	-34	-41	-48	-55	-62	-69	-76	-82	-89
40	6	-1	-8	-15	-22	-29	-36	-43	-50	-57	-64	-71	-78	-84	-91
45	5	-2	-9	-16	-23	-30	-37	-44	-51	-58	-65	-72	-79	-86	-93

110. Find the wind chill factor when the temperature is 5°F and the wind speed is 15 mph.
-13°F

111. Find the wind chill factor when the temperature is 10°F and the wind speed is 20 mph.
-9°F

112. Find the cooling power of a temperature of -10°F and a 5-mph wind.
-22°F

113. Find the cooling power of a temperature of -15°F and a 10-mph wind.
-35°F

114. Which feels colder, a temperature of 0°F with a 15-mph wind or a temperature of 10°F with a 25-mph wind?
0°F with a 15-mph wind

115. Which would feel colder, a temperature of -30°F with a 5-mph wind or a temperature of -20°F with a 10-mph wind?
-30°F with a 5-mph wind

Quick Quiz (Objective 3.1D)

1. The low temperature on Monday was -3°F. The low temperature on Tuesday was -8°F. The low temperature on Wednesday was -5°F. On which of the three days was the low temperature the coldest?   **Tuesday**

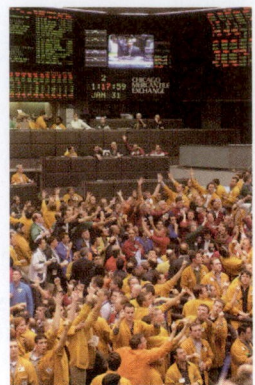

116. **Rocketry** Which is closer to blastoff, -12 min and counting or -17 min and counting?
−12 min and counting

117. **Stocks** In the stock market, the net change in the price of a share of stock is recorded as a positive or a negative number. If the price rises, the net change is positive. If the price falls, the net change is negative. If the net change for a share of Stock A is -2 and the net change for a share of Stock B is -1, which stock showed the least net change?
Stock B

118. **Business** Some businesses show a profit as a positive number and a loss as a negative number. During the first quarter of this year, the loss experienced by a company was recorded as $-12,575$. During the second quarter of this year, the loss experienced by the company was $-11,350$. During which quarter was the loss greater?
First quarter

119. **Business** Some businesses show a profit as a positive number and a loss as a negative number. During the third quarter of last year, the loss experienced by a company was recorded as $-26,800$. During the fourth quarter of last year, the loss experienced by the company was $-24,900$. During which quarter was the loss greater?
Third quarter

APPLYING THE CONCEPTS

120. Find the values of a for which $|a| = 7$.
7, −7

121. Find the values of y for which $|y| = 11$.
11, −11

122. Given that x is an integer, find all values of x for which $|x| < 5$.
−4, −3, −2, −1, 0, 1, 2, 3, 4

123. Given that c is an integer, find all values of c for which $|c| < 7$.
−6, −5, −4, −3, −2, −1, 0, 1, 2, 3, 4, 5, 6

124. **Mathematics** A is a point on the number line halfway between -9 and 3. B is a point halfway between A and the graph of 1 on the number line. B is the graph of what number?
−1

125. **a.** Name two numbers that are 4 units from 2 on the number line.
b. Name two numbers that are 5 units from 3 on the number line.
a. −2 and 6   **b.** −2 and 8

126. Determine whether the statement is always true, sometimes true, or never true.
a. The number $-n$ is a negative number.   Sometimes true
b. A number and its opposite are different numbers.   Sometimes true
c. $|x| > x$ Sometimes true
d. $|x| > -x$ Sometimes true
e. If n is a negative number, $-n$ is a positive number.   Always true
f. If n is a positive number, $-n$ is a negative number.   Always true

127. The $\boxed{+/-}$ key on a calculator changes the sign of the number in the calculator's display. In other words, it changes the number in the display to its opposite. Use the $\boxed{+/-}$ key on your calculator to display each of the following numbers:
a. -9 **b.** -20 **c.** -148 **d.** -573

3.2 Addition and Subtraction of Integers

Objective A   To add integers
VIDEO & DVD   CD TUTOR   WEB   SSM

Not only can an integer be graphed on a number line, an integer can be represented anywhere along a number line by an arrow. A positive number is represented by an arrow pointing to the right. A negative number is represented by an arrow pointing to the left. The absolute value of the number is represented by the length of the arrow. The integers 5 and −4 are shown on the number line in the figure below.

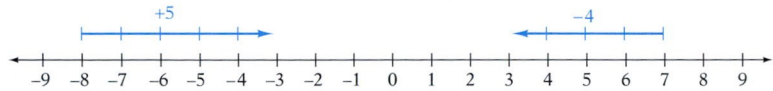

The sum of two integers can be shown on a number line. To add two integers, find the point on the number line corresponding to the first addend. At that point, draw an arrow representing the second addend. The sum is the number directly below the tip of the arrow.

$4 + 2 = 6$

$-4 + (-2) = -6$

$-4 + 2 = -2$

$4 + (-2) = 2$

The sums shown above can be categorized by the signs of the addends.

The addends have the same sign.

$4 + 2$	positive 4 plus positive 2
$-4 + (-2)$	negative 4 plus negative 2

The addends have different signs.

$-4 + 2$	negative 4 plus positive 2
$4 + (-2)$	positive 4 plus negative 2

The rule for adding two integers depends on whether the signs of the addends are the same or different.

In-Class Examples (Objective 3.2A)
1. Add: $-15 + (-27)$ −42
2. Add: $-30 + 16$ −14
3. Add: $-9 + (-21) + 18$ −12
4. What is 14 more than −6?   8
5. Evaluate $-x + y$ when $x = -9$ and $y = -7$.   2
6. Which number, 144, 30, −18, or 4, is a solution of the equation $6 = 24 + n$?   −18

Optional Student Activity

On the number line, illustrate each of the following sums.

1. $-3 + 5$

2. $-4 + 7$

3. $2 + (-6)$

4. $5 + (-9)$

5. $-1 + (-4)$

6. $-5 + (-2)$

Study Tip

Be sure to do all you need to do in order to be successful at adding and subtracting integers: Read through the introductory material, work through the HOW TO examples, study the paired Examples, do the You Try Its and check your solutions against those in the back of the book, and do the exercises in the 3.2 Exercise set. See *AIM for Success*, pages xxxi–xxxiii.

Integrating Technology

To add $-14 + (-47)$ with your calculator, enter the following:

14 [+/−] [+] 47 [+/−] [=]
└─ −14 ┘ └─ −47 ┘

Rule for Adding Two Integers

To add two integers with the same sign, add the absolute values of the numbers. Then attach the sign of the addends.

To add two integers with different signs, find the absolute values of the numbers. Subtract the smaller absolute value from the larger absolute value. Then attach the sign of the addend with the larger absolute value.

HOW TO Add: $(-4) + (-9)$

$|-4| = 4, |-9| = 9$
- The signs of the addends are the same. Find the absolute values of the numbers.

$4 + 9 = 13$
- Add the absolute values of the numbers.

$(-4) + (-9) = -13$
- Attach the sign of the addends. (Both addends are negative. The sum is negative.)

HOW TO Add: $-14 + (-47)$

$-14 + (-47) = -61$
- The signs are the same. Add the absolute values of the numbers. Attach the sign of the addends.

HOW TO Add: $6 + (-13)$

$|6| = 6, |-13| = 13$
- The signs of the addends are different. Find the absolute values of the numbers.

$13 - 6 = 7$
- Subtract the smaller absolute value from the larger absolute value.

$|-13| > |6|$
- Attach the sign of the number with the larger absolute value. Attach the negative sign.

$6 + (-13) = -7$

HOW TO Add: $162 + (-247)$

$247 - 162 = 85$
- The signs are different. Find the difference between the absolute values of the numbers.

$162 + (-247) = -85$
- Attach the sign of the number with the larger absolute value.

HOW TO Add: $-8 + 8$

$-8 + 8 = 0$
- The signs are different. Find the difference between the absolute values of the numbers.

$8 - 8 = 0$

Instructor Note
There are several ways to model the addition of integers. The model on the first page of this objective uses arrows. Another model uses money. For instance, if you are $8 in debt ($-8$) and you receive $5 ($+5$), then you are only $3 in debt ($-3$). Credit card debt is another model. If a student owes $100 ($-100$) and charges $25 more ($-25$), the student then owes $125 ($-125$).

   Examples such as these may help students to see that the rules are not arbitrary, but are designed to model everyday experience.

   Another model of addition of integers that is more manipulative in nature involves using chips: blue chips for positive numbers and red chips for negative numbers. One positive chip added to one negative chip gives zero. To add -8 and 5, place 8 red chips and 5 blue chips in a region. Pair as many red and blue chips as possible and remove the pairs from the region. The remaining chips give the answer—in this case, 3 red chips or -3.

   To model $(-8) + (-5)$, place 8 red chips in the region and then 5 more red chips in the region. There are no pairs of red and blue chips, so there are 13 red chips. Therefore, the answer is -13.

Note in this last example that we are adding a number and its opposite (-8 and 8), and the sum is 0. The opposite of a number is called its **additive inverse.** The opposite or additive inverse of -8 is 8, and the opposite or additive inverse of 8 is -8. **The sum of a number and its additive inverse is always zero.** This is known as the Inverse Property of Addition.

The properties of addition presented in Chapter 1 hold true for integers as well as for whole numbers. These properties are repeated below, along with the Inverse Property of Addition.

The Addition Property of Zero	$a + 0 = a$ or $0 + a = a$
The Commutative Property of Addition	$a + b = b + a$
The Associative Property of Addition	$(a + b) + c = a + (b + c)$
The Inverse Property of Addition	$a + (-a) = 0$ or $-a + a = 0$

HOW TO   Add: $(-4) + (-6) + (-8) + 9$

$(-4) + (-6) + (-8) + 9$
$= (-10) + (-8) + 9$ • **Add the first two numbers.**
$= (-18) + 9$ • **Add the sum to the third number.**
$= -9$ • **Continue until all the numbers have been added.**

HOW TO   The price of Byplex Corporation's stock fell each trading day of the first week of June 2005. Use Figure 3.1 to find the change in the price of Byplex stock over the week's time.

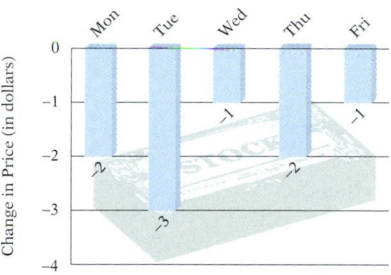

Figure 3.1   Change in Price of Byplex Corporation Stock

Add the five changes in price.

$-2 + (-3) + (-1) + (-2) + (-1)$
$= (-5) + (-1) + (-2) + (-1)$
$= -6 + (-2) + (-1)$
$= -8 + (-1) = -9$

The change in the price was -9.

This means that the price of the stock fell $9 per share.

HOW TO   Evaluate $-x + y$ when $x = -15$ and $y = -5$.

$-x + y$

$-(-15) + (-5)$ • Replace x with -15 and y with -5.

$= 15 + (-5)$ • Simplify $-(-15)$.

$= 10$ • Add.

HOW TO   Is -7 a solution of the equation $x + 4 = -3$?

> *TAKE NOTE*
> Recall that a solution of
> an equation is a number
> that, when substituted
> for the variable, results
> in a true equation.

$x + 4 = -3$

$-7 + 4 \;\big|\; -3$ • Replace x by -7 and then simplify.

$-3 = -3$ • The results are equal.

-7 is a solution of the equation.

Example 1   Add: $-97 + (-45)$

Solution $-97 + (-45)$ • The signs of the
$= -142$ addends are the same.

You Try It 1   Add: $-38 + (-62)$

Your solution -100

Example 2   Add: $81 + (-79)$

Solution $81 + (-79)$ • The signs of the
$= 2$ addends are different.

You Try It 2   Add: $47 + (-53)$

Your solution -6

Example 3   Add: $42 + (-12) + (-30)$

Solution $42 + (-12) + (-30)$
$= 30 + (-30)$
$= 0$

You Try It 3   Add: $-36 + 17 + (-21)$

Your solution -40

Example 4   What is -162 increased by 98?

Solution $-162 + 98 = -64$

You Try It 4   Find the sum of -154 and -37.

Your solution -191

Example 5   Evaluate $-x + y$ when $x = -11$ and $y = -2$.

Solution $-x + y$
$-(-11) + (-2) = 11 + (-2)$
$= 9$

You Try It 5   Evaluate $-x + y$ when $x = -3$ and $y = -10$.

Your solution -7

Solutions on p. S9

Example 6 Is -6 a solution of the equation $3 + y = -2$?

You Try It 6 Is -9 a solution of the equation $2 = 11 + a$?

Solution

$$3 + y = -2$$
$$\overline{3 + (-6)\ |\ -2}$$
$$-3 \neq -2$$

• **Replace y by -6.**

No, -6 is not a solution of the equation.

Your solution Yes

Solution on p. S9

Objective B   **To subtract integers**

Before the rules for subtracting two integers are explained, look at the translation into words of expressions that represent the difference of two integers.

$9 - 3$	positive 9 minus positive 3
$-9 - 3$	negative 9 minus positive 3
$9 - (-3)$	positive 9 minus negative 3
$-9 - (-3)$	negative 9 minus negative 3

Note that the sign $-$ is used in two different ways. One way is as a negative sign, as in -9 (negative 9). The second way is to indicate the operation of subtraction, as in $9 - 3$ (9 minus 3).

Look at the next four expressions and decide whether the second number in each expression is a positive number or a negative number.

1. $(-10) - 8$
2. $(-10) - (-8)$
3. $10 - (-8)$
4. $10 - 8$

In expressions 1 and 4, the second number is positive 8. In expressions 2 and 3, the second number is negative 8.

Opposites are used to rewrite subtraction problems as related addition problems. Notice below that the subtraction of whole numbers is the same as the addition of the opposite number.

Subtraction		Addition of the Opposite	
$8 - 4$	$=$	$8 + (-4)$	$= 4$
$7 - 5$	$=$	$7 + (-5)$	$= 2$
$9 - 2$	$=$	$9 + (-2)$	$= 7$

Objective 3.2B

Vocabulary to Review
minus [1.2B]
negative [3.1A]
opposite [3.1B]

New Rules
subtraction of integers

Discuss the Concepts
Explain why the absolute value of -16 is greater than the absolute value of 4.

Concept Check
State whether each "$-$" sign is a minus sign or a negative sign.
1. $7 - 6$
   Minus
2. $3 - (-9)$
   Minus; negative
3. $-4 - 1$
   Negative; minus
4. $-5 - (-2)$
   Negative; minus; negative
5. $(-8) - 10$
   Negative; minus

In-Class Examples (Objective 3.2B)

1. Subtract: $5 - 8$
   -3

2. Subtract: $-5 - (-3)$
   -2

3. What is 5 less than -9?
   -14

4. Simplify: $-4 - (-7) + (-3)$
   0

5. Evaluate $-x - (-y)$ when $x = -4$ and $y = -8$.
   -4

6. Which number, 48, -48, 36, or -36, is a solution of the equation $6 - y = -42$?
   48

Instructor Note

Although it is more complicated than the addition model, a subtraction model using blue and red chips can be used. Blue chips represent positive numbers, and red chips represent negative numbers. To model $5 - (-3)$, place 5 blue chips in a region. Subtraction requires removing 3 red chips, but since there are no red chips in the region, add 3 pairs of one red and one blue chip (essentially adding three zeros). Now the 3 red chips can be removed. The result is 8 blue chips.

Subtraction of integers can be written as the addition of the opposite number. To subtract two integers, rewrite the subtraction expression as the first number plus the opposite of the second number. Some examples are shown below.

First number	−	second number	=	First number	+	opposite of the second number
8	−	15	=	8	+	$(-15) = -7$
8	−	(-15)	=	8	+	$15 = 23$
−8	−	15	=	−8	+	$(-15) = -23$
−8	−	(-15)	=	−8	+	$15 = 7$

Rule for Subtracting Two Integers

To subtract two integers, add the opposite of the second integer to the first integer.

HOW TO   Subtract: $(-15) - 75$

$(-15) - 75$

$= (-15) + (-75)$
- Rewrite the subtraction operation as the sum of the first number and the opposite of the second number. The opposite of 75 is −75.

$= -90$
- Add.

HOW TO   Subtract: $6 - (-20)$

$6 - (-20)$

$= 6 + 20$
- Rewrite the subtraction operation as the sum of the first number and the opposite of the second number. The opposite of −20 is 20.

$= 26$

HOW TO   Subtract: $11 - 42$

$11 - 42$

$= 11 + (-42)$
- Rewrite the subtraction operation as the sum of the first number and the opposite of the second number. The opposite of 42 is −42.

$= -31$

TAKE NOTE

$42 - 11 = 31$
$11 - 42 = -31$
$42 - 11 \neq 11 - 42$

By the Commutative Property of Addition, the order in which two numbers are added does not affect the sum; $a + b = b + a$. However, note from this last example that the order in which two numbers are subtracted *does* affect the difference. The operation of subtraction is not commutative.

Integrating Technology

To subtract $-13 - 5 - (-8)$ with your calculator, enter the following:

13 +/− − 5 − 8 +/− =
$\underbrace{\quad}_{-13}$ $\underbrace{\quad}_{-8}$

When subtraction occurs several times in an expression, rewrite each subtraction as addition of the opposite and then add.

HOW TO Subtract: $-13 - 5 - (-8)$

$-13 - 5 - (-8)$
$= -13 + (-5) + 8$ • Rewrite each subtraction as addition of the opposite.
$= -18 + 8$ • Add.
$= -10$

HOW TO Simplify: $-14 + 6 - (-7)$

$-14 + 6 - (-7)$ • This problem involves both addition and subtraction. Rewrite the
$= -14 + 6 + 7$ subtraction as addition of the opposite.
$= -8 + 7$ • Add.
$= -1$

HOW TO Evaluate $a - b$ when $a = -2$ and $b = -9$.

$a - b$
$-2 - (-9)$ • Replace a with -2 and b with -9.
$= -2 + 9$ • Rewrite the subtraction as addition of the opposite.
$= 7$ • Add.

HOW TO Is -4 a solution of the equation $3 - a = 11 + a$?

$3 - a = 11 + a$
$3 - (-4) \mid 11 + (-4)$ • Replace a by -4 and then simplify.
$3 + 4 \mid 7$
$7 = 7$ • The results are equal.

Yes, -4 is a solution of the equation.

Instructor Note
Some students may have developed the habit of writing expressions such as

$3 + -4$

Explain to them that, to avoid confusion, the two symbols $+$ and $-$ should not be written in succession without using parentheses:

$3 + (-4)$

Example 7 Subtract: $-12 - (-17)$

Solution
$-12 - (-17)$
$= -12 + 17$ • Rewrite "$-$" as "$+$".
$= 5$ The opposite of -17 is 17.

You Try It 7 Subtract: $-35 - (-34)$

Your solution -1

Example 8 Subtract: $66 - (-90)$

Solution
$66 - (-90)$
$= 66 + 90$ • Rewrite "$-$" as "$+$".
$= 156$ The opposite of -90 is 90.

You Try It 8 Subtract: $83 - (-29)$

Your solution 112

Solutions on p. S9

Instructor Note

A calculator is one way to show students the difference between a minus sign and a negative sign. Have students try to calculate $5 - (-3)$ by pressing $5 \boxed{-} \boxed{-} 3 \boxed{=}$. The calculator will display an error message. Then have them calculate the difference correctly as $5 \boxed{-} 3 \boxed{+/-} \boxed{=}$.

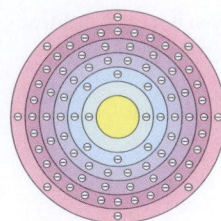

Radon

The table below shows the boiling point and the melting point in degrees Celsius of three chemical elements. Use this table for Example 9 and You Try It 9.

Chemical Element	Boiling Point	Melting Point
Mercury	357	−39
Radon	−62	−71
Xenon	−108	−112

Example 9 Use the table above to find the difference between the boiling point and the melting point of mercury.

Solution The boiling point of mercury is 357.

The melting point of mercury is −39.

$357 - (-39) = 357 + 39$
$\qquad\qquad\quad\; = 396$

The difference is 396°C.

You Try It 9 Use the table above to find the difference between the boiling point and the melting point of xenon.

Your solution 4°C

Example 10 What is −12 minus 8?

Solution
$-12 - 8$
$= -12 + (-8)$ • **Rewrite "−" as "+".**
$= -20$ **The opposite of 8 is −8.**

You Try It 10 What is 14 less than −8?

Your solution −22

Example 11 Subtract 91 from 43.

Solution
$43 - 91$
$= 43 + (-91)$ • **Rewrite "−" as "+".**
$= -48$ **The opposite of 91 is −91.**

You Try It 11 What is 25 decreased by 68?

Your solution −43

Example 12 Simplify:
$-8 - 30 - (-12) - 7 - (-14)$

Solution
$-8 - 30 - (-12) - 7 - (-14)$
$= -8 + (-30) + 12 + (-7) + 14$
$= -38 + 12 + (-7) + 14$
$= -26 + (-7) + 14$
$= -33 + 14$
$= -19$

You Try It 12 Simplify:
$-4 - (-3) + 12 - (-7) - 20$

Your solution −2

Solutions on p. S9

Example 13 Evaluate $-x - y$ when $x = -4$ and $y = -3$.

Solution
$$-x - y$$
$$-(-4) - (-3) = 4 - (-3)$$
$$= 4 + 3$$
$$= 7$$

You Try It 13 Evaluate $x - y$ when $x = -9$ and $y = 7$.

Your solution -16

Example 14 Is 8 a solution of the equation $-2 = 6 - x$?

Solution
$$\begin{array}{c|c} -2 = 6 - x \\ \hline -2 & 6 - 8 \\ -2 & 6 + (-8) \\ -2 = -2 \end{array}$$ • **Replace x by 8.**

Yes, 8 is a solution of the equation.

You Try It 14 Is -3 a solution of the equation $a - 5 = -8$?

Your solution   Yes

Solutions on p. S9

Objective C   **To solve application problems**

VIDEO & DVD   CD TUTOR   WEB   SSM

Figure 3.2 shows the melting points in degrees Celsius of six chemical elements. The abbreviations of the elements are:

F—Fluorine   H—Hydrogen
S—Sulfur   N—Nitrogen
O—Oxygen   Li—Lithium

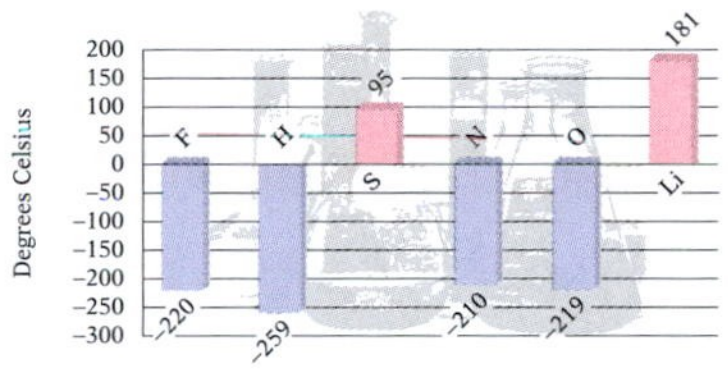

Use this graph for Example 15 and You Try It 15.

Figure 3.2   Melting Points of Chemical Elements

Example 15

Find the difference between the two lowest melting points shown in Figure 3.2.

Strategy
To find the difference, subtract the lowest melting point shown (-259) from the second lowest melting point shown (-220).

Solution
$$-220 - (-259) = -220 + 259 = 39$$

The difference is 39°C.

You Try It 15

Find the difference between the highest and lowest melting points shown in Figure 3.2.

Your Strategy

Your solution
440°C

Solution on p. S9

Objective 3.2C

Optional Student Activity

The list below gives temperatures either on the surface or in the atmosphere of bodies in our solar system. (This is the same list provided in Objective 3.1D.)

Mercury's dark side:
−346°F

Mercury's sunlit side:
950°F

Mars, daytime:
−17°F

Saturn's moon Titan:
−190°F

Neptune's atmosphere:
−360°F

Pluto, atmospheric high:
−163°F

Pluto, atmospheric low:
−397°F

a. Find the difference between the temperatures on the sunlit side and the dark side of Mercury.   1296°F

b. Find the difference between Pluto's atmospheric high and low temperatures.   234°F

Example 16

Find the temperature after an increase of 8°C from −5°C.

Strategy
To find the temperature, add the increase (8) to the previous temperature (−5).

Solution
−5 + 8 = 3

The temperature is 3°C.

Example 17

The average temperature on the sunlit side of the moon is approximately 215°F. The average temperature on the dark side is approximately −250°F. Find the difference between these average temperatures.

Strategy
To find the difference, subtract the average temperature on the dark side of the moon (−250) from the average temperature on the sunlit side (215).

Solution
$$215 - (-250) = 215 + 250$$
$$= 465$$

The difference is 465°F.

Example 18

The distance, d, between point a and point b on the number line is given by the formula $d = |a - b|$. Use the formula to find d when $a = 7$ and $b = -8$.

Strategy
To find d, replace a by 7 and b by −8 in the given formula and solve for d.

Solution
$$d = |a - b|$$
$$d = |7 - (-8)|$$
$$d = |7 + 8|$$
$$d = |15|$$
$$d = 15$$

The distance between the two points is 15 units.

You Try It 16

Find the temperature after an increase of 10°C from −3°C.

Your Strategy

Your solution
7°C

You Try It 17

The average temperature on Earth's surface is 57°F. The average temperature throughout Earth's stratosphere is −70°F. Find the difference between these average temperatures.

Your Strategy

Your solution
127°F

You Try It 18

The distance, d, between point a and point b on the number line is given by the formula $d = |a - b|$. Use the formula to find d when $a = -6$ and $b = 5$.

Your Strategy

Your solution
11 units

Solutions on pp. S9–S10

In-Class Examples (Objective 3.2C)

1. Find the temperature after a rise of 4°C from −2°C.   2°C

2. During a card game of Hearts, you had a score of 14 points before your opponent "shot the moon," subtracting a score of 26 from your total. What was your score after your opponent "shot the moon"?   −12 points

3.2 Exercises

Objective A   To add integers

1.   **a.** Explain the rule for adding two integers with the same sign.
   b. Explain the rule for adding two integers with different signs.

2.   Describe in words the Inverse Property of Addition. Provide an example of this property.

Add.

3. $-3 + (-8)$ -11	**4.** $-6 + (-9)$ -15	**5.** $-8 + 3$ -5	**6.** $-7 + 2$ -5
7. $-5 + 13$ 8	**8.** $-4 + 11$ 7	**9.** $6 + (-10)$ -4	**10.** $8 + (-12)$ -4
11. $3 + (-5)$ -2	**12.** $6 + (-7)$ -1	**13.** $-4 + (-5)$ -9	**14.** $-12 + (-12)$ -24
15. $-6 + 7$ 1	**16.** $-9 + 8$ -1	**17.** $(-5) + (-10)$ -15	**18.** $(-3) + (-17)$ -20
19. $-7 + 7$ 0	**20.** $-11 + 11$ 0	**21.** $(-15) + (-6)$ -21	**22.** $(-18) + (-3)$ -21
23. $0 + (-14)$ -14	**24.** $-19 + 0$ -19	**25.** $73 + (-54)$ 19	**26.** $-89 + 62$ -27

27. $2 + (-3) + (-4)$
-5

28. $7 + (-2) + (-8)$
-3

29. $-3 + (-12) + (-15)$
-30

30. $9 + (-6) + (-16)$
-13

31. $-17 + (-3) + 29$
9

32. $13 + 62 + (-38)$
37

33. $11 + (-22) + 4 + (-5)$
-12

34. $-14 + (-3) + 7 + (-6)$
-16

35. $-22 + 10 + 2 + (-18)$
-28

36. $-6 + (-8) + 13 + (-4)$
-5

37. $-25 + (-31) + 24 + 19$
-13

38. $10 + (-14) + (-21) + 8$
-17

39. What is 3 increased by -21?
-18

40. Find 12 plus -9.
3

41. What is 16 more than -5?
11

42. What is 17 added to -7?
10

43. Find the total of -3, -8, and 12.
1

44. Find the sum of 5, -16, and -13.
-24

Section 3.2

Suggested Assignment

Exercises 3–45, odds
Exercises 49–69, odds
Exercises 73–97, odds
Exercises 101–143, odds
More challenging problems:
 Exercises 144–146

Answers to Writing Exercises

1a. To add two integers with the same sign, add the absolute values of the integers. Then attach the sign of the integers added.

 b. To add two integers with different signs, find the absolute value of each integer. Then subtract the smaller of these absolute values from the larger. Finally, attach the sign of the integer with the larger absolute value.

 2. The Inverse Property of Addition states that the sum of a number and its opposite is zero. Examples will vary.

Quick Quiz (Objective 3.2A)

1. Add: $-23 + (-17)$ -40
2. Add: $-41 + 11$ -30
3. Add: $-6 + 9 + (-8)$ -5
4. What is the sum of -8 and 15?   7

5. Evaluate $-x + y$ when $x = -4$ and $y = -6$.
-2

6. Is -12 a solution of the equation $8 = 20 + n$?
Yes

45. Write the sum of x and -7.
$x + (-7)$

46. Write the total of $-a$ and b.
$-a + b$

47. **Balance of Trade** A nation's balance of trade is the difference between the values of its exports and imports. If the value of the exports is greater than the value of the imports, the result is a positive number and a *favorable balance of trade*. If the value of the exports is less than the value of the imports, the result is a negative number and an *unfavorable balance of trade*. The table at the right shows the unfavorable balance of trade in a recent year for the United States with four other countries. Find the total of the U.S. balance of trade with (a) Japan and Mexico, (b) Canada and Mexico, and (c) Japan and France.

U.S. Balance of Trade with Foreign Countries	
Japan	−57,931,000,000
Canada	−49,409,000,000
Mexico	−27,578,000,000
France	−10,140,000,000

Source: Bureau of Economic Analysis, U.S. Department of Commerce

a. $-\$85,509,000,000$ **b.** $-\$76,987,000,000$ **c.** $-\$68,071,000,000$

Evaluate the expression for the given values of the variables.

48. $x + y$, where $x = -5$ and $y = -7$
-12

49. $-a + b$, where $a = -8$ and $b = -3$
5

50. $a + b$, where $a = -8$ and $b = -3$
-11

51. $-x + y$, where $x = -5$ and $y = -7$
-2

52. $a + b + c$, where $a = -4$, $b = 6$, and $c = -9$
-7

53. $a + b + c$, where $a = -10$, $b = -6$, and $c = 5$
-11

54. $x + y + (-z)$, where $x = -3$, $y = 6$, and $z = -17$
20

55. $-x + (-y) + z$, where $x = -2$, $y = 8$, and $z = -11$
-17

Identify the property that justifies the statement.

56. $-12 + 5 = 5 + (-12)$
The Commutative Property of Addition

57. $-33 + 0 = -33$
The Addition Property of Zero

58. $-46 + 46 = 0$
The Inverse Property of Addition

59. $-7 + (3 + 2) = (-7 + 3) + 2$
The Associative Property of Addition

Use the given property of addition to complete the statement.

60. The Associative Property of Addition
$-11 + (6 + 9) = (? + 6) + 9$
-11

61. The Addition Property of Zero
$-13 + ? = -13$
0

62. The Commutative Property of Addition
$-2 + ? = -4 + (-2)$
-4

63. The Inverse Property of Addition
$? + (-18) = 0$
18

64. Is -3 a solution of the equation $x + 4 = 1$?
Yes

65. Is -8 a solution of the equation $6 = -3 + z$?
No

66. Is -6 a solution of the equation $6 = 12 + n$?
Yes

67. Is -8 a solution of the equation $-7 + m = -15$?
Yes

68. Is -2 a solution of the equation $3 + y = y + 3$?
Yes

69. Is -4 a solution of the equation $1 + z = z + 2$?
No

Objective B   **To subtract integers**

70. Explain how to rewrite the subtraction $8 - (-6)$ as addition of the opposite.

71. Explain the meanings of the words *minus* and *negative*.

Subtract.

72. $7 - 14$
-7

73. $6 - 9$
-3

74. $-7 - 2$
-9

75. $-9 - 4$
-13

76. $7 - (-2)$
9

77. $3 - (-4)$
7

78. $-6 - (-6)$
0

79. $-4 - (-4)$
0

80. $-12 - 16$
-28

81. $-10 - 7$
-17

82. $(-9) - (-3)$
-6

83. $(-7) - (-4)$
-3

84. $4 - (-14)$
18

85. $-4 - (-16)$
12

86. $(-14) - (-7)$
-7

87. $3 - (-24)$
27

88. $9 - (-9)$
18

89. $(-41) - 65$
-106

90. $57 - 86$
-29

91. $-95 - (-28)$
-67

92. How much larger is 5 than -11?
16

93. What is -10 decreased by -4?
-6

94. Find -13 minus -8.
-5

95. What is 6 less than -9?
-15

96. Write the difference between $-y$ and 5.
$-y - 5$

97. Write $-t$ decreased by r.
$-t - r$

Answers to Writing Exercises

70. In the subtraction $8 - (-6)$, the first "$-$" sign is read "minus." The second "$-$" is read "negative." To rewrite $8 - (-6)$ as addition of the opposite, we need to change subtraction to addition and change -6 to the opposite of -6. Therefore, we must change the minus sign to a plus sign and change the negative sign to a plus sign: $8 + (+6) = 8 + 6$.

71. The word *minus* indicates the operation of subtraction. The word *negative* indicates a number less than zero.

Quick Quiz (Objective 3.2B)

1. Subtract: $10 - 18$ -8

2. Subtract: $-6 - (-9)$ 3

3. Find 4 minus 22. -18

4. Simplify: $-6 - (-8) + (-5)$ -3

5. Evaluate $-x - y$ when $x = -7$ and $y = 4$. 3

6. Is -8 a solution of the equation $5 - y = 13$?
Yes

Temperature The figure at the right shows the highest and lowest temperatures ever recorded for selected regions of the world. Use this graph for Exercises 98 to 100.

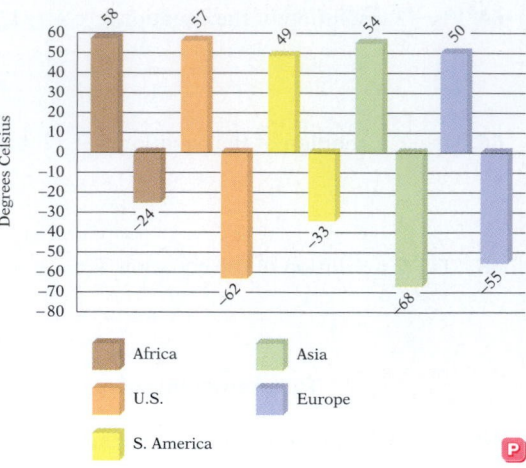

Highest and Lowest Temperatures Recorded (in degrees Celsius)

98. What is the difference between the highest and lowest temperatures ever recorded in Africa?
82°C

99. What is the difference between the highest and lowest temperatures ever recorded in South America?
82°C

100. What is the difference between the lowest temperature recorded in Europe and the lowest temperature recorded in Asia?
13°C

Simplify.

101. $-4 - 3 - 2$
-9

102. $4 - 5 - 12$
-13

103. $12 - (-7) - 8$
11

104. $-12 - (-3) - (-15)$
6

105. $4 - 12 - (-8)$
0

106. $-30 - (-65) - 29 - 4$
2

107. $-16 - 47 - 63 - 12$
-138

108. $42 - (-30) - 65 - (-11)$
18

109. $12 - (-6) + 8$
26

110. $-7 + 9 - (-3)$
5

111. $-8 - (-14) + 7$
13

112. $-4 + 6 - 8 - 2$
-8

113. $9 - 12 + 0 - 5$
-8

114. $11 - (-2) - 6 + 10$
17

115. $5 + 4 - (-3) - 7$
5

116. $-1 - 8 + 6 - (-2)$
-1

117. $-13 + 9 - (-10) - 4$
2

118. $6 - (-13) - 14 + 7$
12

Evaluate the expression for the given values of the variables.

119. $-x - y$, where $x = -3$ and $y = 9$
-6

120. $x - (-y)$, where $x = -3$ and $y = 9$
6

121. $-x - (-y)$, where $x = -3$ and $y = 9$
12

122. $a - (-b)$, where $a = -6$ and $b = 10$
4

123. $a - b - c$, where $a = 4$, $b = -2$, and $c = 9$
−3

124. $a - b - c$, where $a = -1$, $b = 7$, and $c = -15$
7

125. $x - y - (-z)$, where $x = -9$, $y = 3$, and $z = 30$
18

126. $-x - (-y) - z$, where $x = 8$, $y = 1$, and $z = -14$
7

127. Is −3 a solution of the equation $x - 7 = -10$?
Yes

128. Is −4 a solution of the equation $1 = 3 - y$?
No

129. Is −2 a solution of the equation $-5 - w = 7$?
No

130. Is −8 a solution of the equation $-12 = m - 4$?
Yes

131. Is −6 a solution of the equation $-t - 5 = 7 + t$?
Yes

132. Is −7 a solution of the equation $5 + a = -9 - a$?
Yes

Objective C **To solve application problems**

Geography The elevation, or height, of places on Earth is measured in relation to sea level, or the average level of the ocean's surface. The table below shows height above sea level as a positive number and depth below sea level as a negative number. Use the table below for Exercises 133 to 135.

Mt. Everest

Continent	Highest Elevation (in meters)		Lowest Elevation (in meters)	
Africa	Mt. Kilimanjaro	5895	Lake Assal	−156
Asia	Mt. Everest	8850	Dead Sea	−411
Europe	Mt. Elbrus	5642	Caspian Sea	−28
America	Mt. Aconcagua	6960	Death Valley	−86

133. What is the difference in elevation (a) between Mt. Aconcagua and Death Valley and (b) between Mt. Kilimanjaro and Lake Assal?
a. 7046 m   **b.** 6051 m

134. For which continent shown is the difference between the highest and lowest elevations greatest?
Asia

135. For which continent shown is the difference between the highest and lowest elevations smallest?
Europe

136. **Temperature** Find the temperature after a rise of 9°C from −6°C.
3°C

Quick Quiz (Objective 3.2C)

1. Find the temperature after a rise of 6°C from −10°C.   −4°C

2. In a card game of Hearts, you had a score of −15 points before you "shot the moon," entitling you to add 26 points to your score. What was your score after you "shot the moon"?   11 points

Aviation The table at the right shows the average temperatures at different cruising altitudes for airplanes. Use the table for Exercises 137 to 139.

Cruising Altitude	Average Temperature
12,000 ft	16°
20,000 ft	−12°
30,000 ft	−48°
40,000 ft	−70°
50,000 ft	−70°

137. What is the difference between the average temperatures at 12,000 ft and at 40,000 ft?
86°

138. What is the difference between the average temperatures at 40,000 ft and at 50,000 ft?
0°

139. How much colder is the average temperature at 30,000 ft than at 20,000 ft?
36°

140. **Golf** Use the equation $S = N - P$, where S is a golfer's score relative to par in a tournament, N is the number of strokes made by the golfer, and P is par, to find a golfer's score relative to par when the golfer made 196 strokes and par is 208.
−12

141. **Golf** Use the equation $S = N - P$, where S is a golfer's score relative to par in a tournament, N is the number of strokes made by the golfer, and P is par, to find a golfer's score relative to par when the golfer made 49 strokes and par is 52.
−3

142. **Mathematics** The distance, d, between point a and point b on the number line is given by the formula $d = |a - b|$. Find d when $a = 6$ and $b = -15$.
21

143. **Mathematics** The distance, d, between point a and point b on the number line is given by the formula $d = |a - b|$. Find d when $a = 7$ and $b = -12$.
19

APPLYING THE CONCEPTS

144. **Mathematics** Given the list of numbers at the right, find the largest difference that can be obtained by subtracting one number in the list from a different number in the list.
23

5, −2, −9, 11, 14

145. Determine whether the statement is always true, sometimes true, or never true.
 a. The difference between a number and its additive inverse is zero.
 b. The sum of a negative number and a negative number is a negative number.
 a. Sometimes true   **b.** Always true

146. The sum of two negative integers is −7. Find the two integers.
Answers will vary. Possible answers include −1 and −6, −2 and −5, −3 and −4.

147. Describe the steps involved in using a calculator to simplify $-17 - (-18) + (-5)$.

Answers to Writing Exercises

147. To simplify $-17 - (-18) + (-5)$ using a calculator, enter 17, press the +/− key, press the subtraction key (−), enter 18, press the +/− key, press the addition key (+), enter 5, press the +/− key, and press the = key. The result in the display should be −4.

3.3 Multiplication and Division of Integers

Objective A To multiply integers

When 5 is multiplied by a sequence of decreasing integers, each product decreases by 5.

$$5(3) = 15$$
$$5(2) = 10$$
$$5(1) = 5$$
$$5(0) = 0$$

The pattern developed can be continued so that 5 is multiplied by a sequence of negative numbers. To maintain the pattern of decreasing by 5, the resulting products must be negative.

$$5(-1) = -5$$
$$5(-2) = -10$$
$$5(-3) = -15$$
$$5(-4) = -20$$

This example illustrates that the product of a positive number and a negative number is negative.

When -5 is multiplied by a sequence of decreasing integers, each product increases by 5.

$$-5(3) = -15$$
$$-5(2) = -10$$
$$-5(1) = -5$$
$$-5(0) = 0$$

Point of Interest

Operations with negative numbers were not accepted until the late 13th century. One of the first attempts to prove that the product of two negative numbers is positive was made in the book *Ars Magna*, by Girolamo Cardan, in 1545.

The pattern developed can be continued so that -5 is multiplied by a sequence of negative numbers. To maintain the pattern of increasing by 5, the resulting products must be positive.

$$-5(-1) = 5$$
$$-5(-2) = 10$$
$$-5(-3) = 15$$
$$-5(-4) = 20$$

This example illustrates that the product of two negative numbers is positive.

The pattern for multiplication shown above is summarized in the following rule for multiplying integers.

> **Rule for Multiplying Two Integers**
>
> **To multiply two integers with the same sign,** multiply the absolute values of the factors. The product is **positive.**
>
> **To multiply two integers with different signs,** multiply the absolute values of the factors. The product is **negative.**

Integrating Technology

To multiply $(-6)(-15)$ with your calculator, enter the following:

6 +/− × 15 +/− =
  −6   −15

HOW TO Multiply: $-9(12)$

$-9(12) = -108$ • The signs are different. The product is negative.

HOW TO Multiply: $(-6)(-15)$

$(-6)(-15) = 90$ • The signs are the same. The product is positive.

Vocabulary to Review
product   [1.3A]

Properties to Review
Multiplication Property of Zero   [1.3A]
Multiplication Property of One   [1.3A]
Commutative Property of Multiplication   [1.3A]
Associative Property of Multiplication   [1.3A]

New Rules
multiplication of integers

Discuss the Concepts
Describe the rules for multiplying two integers.

Instructor Note
Another way to suggest that positive times negative equals negative is to use repeated addition. For instance, $(-5)(3)$ is $(-5) + (-5) + (-5) = -15$.
The idea that negative times negative equals positive seems arbitrary to students. You might relate it to using a double negative in English. For example, ask them the meaning of the sentence, "It is not impossible to run a 4-minute mile."

Concept Check
1. Which is greater: $(-2)(-3)$ or $(2356)(-5)$? $(-2)(-3)$
2. Is the product of two integers always greater than either of the integers? Explain.
   No. $(-2)(3) = -6$ and -6 is smaller than -2 and 3.

In-Class Examples (Objective 3.3A)
1. Find the product of -100 and 5. -500
2. Multiply: $-8(-4)(-2)$ -64
3. Evaluate $-xy$ when $x = -4$ and $y = -9$. -36
4. Which number, -48, -36, -7, or 7, is a solution of the equation $6x = -42$? -7

Concept Check

True or false:

1. The product of a nonzero number and its opposite is a positive number.   **F**

2. The product of an odd number of negative numbers is a negative number.   **T**

3. The product of an even number of negative numbers is a negative number.   **F**

4. To find the opposite of a number, multiply the number by −1.   **T**

Optional Student Activity

Illustrate each of the following products on the number line.

1. $3 \times (-3)$
2. $2 \times (-5)$
3. $3 \times (-2)$
4. $5 \times (-1)$

HOW TO 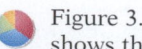 Figure 3.3 shows the melting points of bromine and mercury. The melting point of helium is seven times the melting point of mercury. Find the melting point of helium.

Multiply the melting point of mercury (−39°C) by 7.

$$-39(7) = -273$$

The melting point of helium is −273°C.

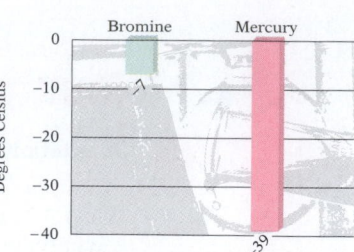

Figure 3.3   Melting Point of Chemical Elements (in degrees Celsius)

The properties of multiplication presented in Chapter 1 hold true for integers as well as whole numbers. These properties are repeated below.

The Multiplication Property of Zero	$a \cdot 0 = 0$ or $0 \cdot a = 0$
The Multiplication Property of One	$a \cdot 1 = a$ or $1 \cdot a = a$
The Commutative Property of Multiplication	$a \cdot b = b \cdot a$
The Associative Property of Multiplication	$(a \cdot b) \cdot c = a \cdot (b \cdot c)$

TAKE NOTE

For the example at the right, the product is the same if the numbers are multiplied in a different order. For instance,

$$2(-3)(-5)(-7)$$
$$= 2(-3)(35)$$
$$= 2(-105)$$
$$= -210$$

HOW TO Multiply: $2(-3)(-5)(-7)$

$2(-3)(-5)(-7)$

$= -6(-5)(-7)$ • Multiply the first two numbers.

$= 30(-7)$ • Then multiply the product by the third number.

$= -210$ • Continue until all the numbers have been multiplied.

By the Multiplication Property of One, $1 \cdot 6 = 6$ and $\mathbf{1} \cdot \mathbf{x} = \mathbf{x}$. Applying the rules for multiplication, we can extend this to $-1 \cdot 6 = -6$ and $\mathbf{-1} \cdot \mathbf{x} = \mathbf{-x}$.

TAKE NOTE

When variables are placed next to each other, it is understood that the operation is multiplication. $-ab$ means "the opposite of a times b."

HOW TO Evaluate $-ab$ when $a = -2$ and $b = -9$.

$-ab$

$-(-2)(-9)$ • Replace a with −2 and b with −9.

$= 2(-9)$ • Simplify $-(-2)$.

$= -18$ • Multiply.

HOW TO Is −4 a solution of the equation $5x = -20$?

$5x = -20$

$5(-4) \ \big| \ -20$ • Replace x by −4 and then simplify.

$-20 = -20$ • The results are equal.

Yes, −4 is a solution of the equation.

Example 1 Find −42 times 62.

Solution −42 · 62   • **The signs are different.**
            = −2604   **The product is negative.**

You Try It 1 What is −38 multiplied by 51?

Your solution −1938

Example 2 Multiply: −5(−4)(6)(−3)

Solution −5(−4)(6)(−3) = 20(6)(−3)
                      = 120(−3)
                      = −360

You Try It 2 Multiply: −7(−8)(9)(−2)

Your solution −1008

Example 3 Evaluate −5x when x = −11.

Solution −5x
            −5(−11) = 55

You Try It 3 Evaluate −9y when y = 20.

Your solution −180

Example 4 Is 5 a solution of the equation
            30 = −6z?

Solution 30 = −6z
            30 | −6(5)
            30 ≠ −30

No, 5 is not a solution of the
equation.

You Try It 4 Is −3 a solution of the equation
            12 = −4a?

Your solution Yes

Solutions on p. S10

Objective B   **To divide integers**

TAKE NOTE
Recall that the fraction
bar can be read "divided
by." Therefore,
$\frac{8}{2}$ can be read "8 divided
by 2."

For every division problem, there is a related multiplication problem.

$$\text{Division: } \frac{8}{2} = 4 \qquad \text{Related multiplication: } 4(2) = 8$$

This fact can be used to illustrate a rule for dividing integers.

$$\frac{12}{3} = 4 \quad \text{because} \quad 4(3) = 12 \quad \text{and} \quad \frac{-12}{-3} = 4 \quad \text{because} \quad 4(-3) = -12.$$

These two division examples suggest that the quotient of two numbers with the same sign is positive. Now consider these two examples.

$$\frac{12}{-3} = -4 \quad \text{because} \quad -4(-3) = 12$$

$$\frac{-12}{3} = -4 \quad \text{because} \quad -4(3) = -12$$

These two division examples suggest that the quotient of two numbers with different signs is negative. This property is summarized next.

Objective 3.3B

Concept Check

1. Is the quotient of two positive integers always smaller than either of the positive integers?

No. For example, $1 \div 1 = 1$.

2. Is the quotient of two nonzero integers always smaller than either integer?

No. $\dfrac{-6}{3} = -2$ but $-2 > -6$.

3. Can any two integers always be divided?

No. The divisor cannot be zero.

Rule for Dividing Two Integers

To divide two numbers with the same sign, divide the absolute values of the numbers. The quotient is **positive.**

To divide two numbers with different signs, divide the absolute values of the numbers. The quotient is **negative.**

Note from this rule that $\dfrac{12}{-3}$, $\dfrac{-12}{3}$, and $-\dfrac{12}{3}$ are all equal to -4.

If a and b are integers ($b \neq 0$), then $\dfrac{a}{-b} = \dfrac{-a}{b} = -\dfrac{a}{b}$.

HOW TO   Divide: $-36 \div 9$

$-36 \div 9 = -4$ • The signs are different. The quotient is negative.

Integrating Technology

To divide (-105) by (-5) with your calculator, enter the following:

105 [+/−] ÷ 5 [+/−] =

$\underbrace{}_{-105}$ $\underbrace{}_{-5}$

HOW TO   Divide: $(-105) \div (-5)$

$(-105) \div (-5) = 21$ • The signs are the same. The quotient is positive.

HOW TO   Figure 3.4 shows the record high and low temperatures in the United States for the first four months of the year. We can read from the graph that the record low temperature for April is $-36°F$. This is four times the record low temperature for September. What is the record low temperature for September?

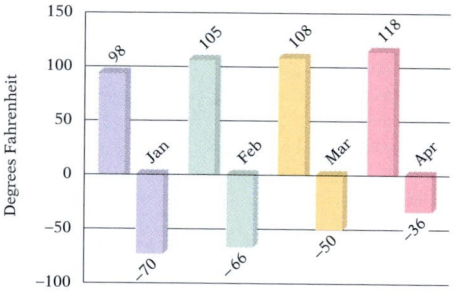

Figure 3.4   Record High and Low Temperatures, in Degrees Fahrenheit, in the United States for January, February, March, and April
Source: National Climatic Data Center, Asheville, NC, and Storm Phillips, STORMFAX, Inc.

To find the record low temperature for September, divide the record low for April (-36) by 4.

$-36 \div 4 = -9$

The record low temperature in the United States for the month of September is $-9°F$.

The division properties of zero and one, which were presented in Chapter 1, hold true for integers as well as whole numbers. These properties are repeated here.

Division Properties of Zero and One

If $a \neq 0$, $\dfrac{0}{a} = 0$.   If $a \neq 0$, $\dfrac{a}{a} = 1$.

$\dfrac{a}{1} = a$ $\dfrac{a}{0}$ is undefined.

Point of Interest

Historical manuscripts indicate that mathematics is at least 4000 years old. Yet it was only 400 years ago that mathematicians started using variables to stand for numbers. Before that time, mathematics was written in words.

HOW TO Evaluate $a \div (-b)$ when $a = -28$ and $b = -4$.

$a \div (-b)$
$-28 \div [-(-4)]$ • Replace a with -28 and b with -4.
$= -28 \div (4)$ • Simplify $-(-4)$.
$= -7$ • Divide.

HOW TO Is -4 a solution of the equation $\dfrac{-20}{x} = 5$?

$\dfrac{-20}{x} = 5$

$\dfrac{-20}{-4} \bigg| 5$ • Replace x by -4 and then simplify.

$5 = 5$ • The results are equal.

Yes, -4 is a solution of the equation.

Concept Check

What property of division is illustrated by each of the following?

1. $\dfrac{0}{-12} = 0$

Zero divided by any number other than zero is zero.

2. $\dfrac{-18}{1} = -18$

Any number divided by 1 is the number.

3. $\dfrac{-5}{0}$ is undefined.

Division by zero is undefined.

4. $\dfrac{-22}{-22} = 1$

Any number other than zero divided by itself is 1.

Example 5 Find the quotient of -23 and -23.

Solution $-23 \div (-23) = 1$ • If $a \neq 0$, $\dfrac{a}{a} = 1$.

You Try It 5 What is 0 divided by -17?

Your solution 0

Example 6 Divide: $\dfrac{95}{-5}$

Solution $\dfrac{95}{-5} = -19$ • The signs are different. The quotient is negative.

You Try It 6 Divide: $\dfrac{84}{-6}$

Your solution -14

Example 7 Divide: $x \div 0$

Solution Division by zero is not defined. $x \div 0$ is undefined.

You Try It 7 Divide: $x \div 1$

Your solution x

Solutions on p. S10

Example 8   Evaluate $\dfrac{-a}{b}$ when $a = -6$ and $b = -3$.

Solution

$$\dfrac{-a}{b}$$

$$\dfrac{-(-6)}{-3} = \dfrac{6}{-3} = -2$$

You Try It 8   Evaluate $\dfrac{a}{-b}$ when $a = -14$ and $b = -7$.

Your solution   −2

Example 9   Is −9 a solution of the equation $-3 = \dfrac{x}{3}$?

Solution $-3 = \dfrac{x}{3}$

$$-3 \;\Big|\; \dfrac{-9}{3}$$ • Replace x by −9.

$$-3 = -3$$

Yes, −9 is a solution of the equation.

You Try It 9   Is −3 a solution of the equation $\dfrac{-6}{y} = -2$?

Your solution   No

Solutions on p. S10

Objective 3.3C

New Vocabulary
average

Concept Check
The daily low temperatures (in degrees Celsius) during one week were recorded as −4°, −3°, 7°, 8°, 2°, −1°, and −2°. Find the average daily low temperature.
1°C

Objective C   **To solve application problems**

Example 10

The daily low temperatures during one week were recorded as follows: −10°, 2°, −1°, −9°, 1°, 0°, 3°. Find the average daily low temperature for the week.

Strategy
To find the average daily low temperature:
- Add the seven temperature readings.
- Divide by 7.

Solution
$$-10 + 2 + (-1) + (-9) + 1 + 0 + 3 = -14$$

$$-14 \div 7 = -2$$

The average daily low temperature was −2°.

You Try It 10

The daily high temperatures during one week were recorded as follows: −7°, −8°, 0°, −1°, −6°, −11°, −2°. Find the average daily high temperature for the week.

Your strategy

Your solution
−5°

Solution on p. S10

In-Class Examples (Objective 3.3C)

1. The combined scores, in relation to par, of the top seven golfers in a golf tournament equaled −98. What was the average score of the seven golfers?   −14

2. The daily low temperatures during one week were recorded as follows: −5°F, −8°F, 6°F, 8°F, 0°F, −6°F, −2°F. Find the average daily low temperature for the week.   −1°F

3.3 Exercises

Objective A   To multiply integers

1. 🖎 Give the rules for multiplying two integers.

2. 🖎 Name the operation in each expression and explain how you deter-mined that it was that operation.
   a. $8(-7)$ **b.** $8 - 7$ **c.** $8 - (-7)$ **d.** $-xy$ **e.** $x(-y)$ **f.** $-x - y$

Multiply.

3. $-4 \cdot 6$ -24	**4.** $-7 \cdot 3$ -21	**5.** $-2(-3)$ 6	**6.** $-5(-1)$ 5
7. $(9)(2)$ 18	**8.** $(3)(8)$ 24	**9.** $5(-4)$ -20	**10.** $4(-7)$ -28
11. $-8(2)$ -16	**12.** $-9(3)$ -27	**13.** $(-5)(-5)$ 25	**14.** $(-3)(-6)$ 18
15. $(-7)(0)$ 0	**16.** $-11(1)$ -11	**17.** $14(3)$ 42	**18.** $62(9)$ 558
19. $-32(4)$ -128	**20.** $-24(3)$ -72	**21.** $(-8)(-26)$ 208	**22.** $(-4)(-35)$ 140
23. $9(-27)$ -243	**24.** $8(-40)$ -320	**25.** $-5 \cdot (23)$ -115	**26.** $-6 \cdot (38)$ -228
27. $-7(-34)$ 238	**28.** $-4(-51)$ 204	**29.** $4 \cdot (-8) \cdot 3$ -96	**30.** $5 \cdot 7 \cdot (-2)$ -70
31. $(-6)(5)(7)$ -210	**32.** $(-9)(-9)(2)$ 162	**33.** $-8(-7)(-4)$ -224	**34.** $-1(4)(-9)$ 36

35. What is twice -20?
   -40

36. Find the product of 100 and -7.
   -700

37. What is -30 multiplied by -6?
   180

38. What is -9 times -40?
   360

Section 3.3

Suggested Assignment

Exercises 3–39, odds
Exercises 43–91, odds
Exercises 95–115, odds
More challenging problems:
   Exercises 120–123

Answers to Writing Exercises

1. To multiply two integers with the same sign, multiply the absolute values of the factors; the product is positive. To multiply two integers with different signs, multiply the absolute values of the factors; the product is negative.

2a. The operation in the expression $8(-7)$ is multiplication. There is no operation symbol between the 8 and the left parenthesis.

b. The operation in the expression $8 - 7$ is subtraction. There is a space before and after the minus sign. 7 is subtracted from 8.

c. The operation in the expression $8 - (-7)$ is subtraction. There is a space before and after the minus sign. -7 is subtracted from 8.

d. The operation in the expression $-xy$ is multiplication. The x and y are right next to each other with no sign in between.

e. The operation in the expression $x(-y)$ is multiplication. There is no operation symbol between the x and the left parenthesis.

f. The operation in the expression $-x - y$ is subtraction. There is a space before and after the minus sign. y is subtracted from $-x$.

Quick Quiz (Objective 3.3A)

1. Find the product of -80 and 7. -560
2. Multiply: $-5(-9)(-1)$ -45
3. Evaluate $-xy$ when $x = 6$ and $y = -4$. 24
4. Is -7 a solution of the equation $7x = -49$?   Yes

39. Write the product of $-q$ and r.
$-qr$

40. Write the product of $-f$, g, and h.
$-fgh$

41. 🔵 **Airlines** The table at the right shows the net income for the third quarter of 2004 for three companies. (*Note:* Negative net income indicates a loss.) If net income were to remain at this level throughout each quarter of the year 2005, what would be the annual net income for 2005 for (a) America West, (b) FLYi, and (c) Frontier Airlines?
a. $-\$188,400,000$ **b.** $-\$330,800,000$ **c.** $-\$8,400,000$

Company	Net Income 3rd Quarter of 2004
America West	−47,100,000
FLYi	−82,700,000
Frontier Airlines	−2,100,000

Identify the property that justifies the statement.

42. $0(-7) = 0$
The Multiplication Property of Zero

43. $1p = p$
The Multiplication Property of One

44. $-8(-5) = -5(-8)$
The Commutative Property of Multiplication

45. $-3(9 \cdot 4) = (-3 \cdot 9)4$
The Associative Property of Multiplication

Use the given property of multiplication to complete the statement.

46. The Commutative Property of Multiplication
$-3(-9) = -9(?)$
-3

47. The Associative Property of Multiplication
$?(5 \cdot 10) = (-6 \cdot 5)10$
-6

48. The Multiplication Property of Zero
$-81 \cdot ? = 0$
0

49. The Multiplication Property of One
$?(-14) = -14$
1

Evaluate the expression for the given values of the variables.

50. xy, when $x = -3$ and $y = -8$
24

51. $-xy$, when $x = -3$ and $y = -8$
-24

52. $x(-y)$, when $x = -3$ and $y = -8$
-24

53. $-xyz$, when $x = -6$, $y = 2$, and $z = -5$
-60

54. $-8a$, when $a = -24$
192

55. $-7n$, when $n = -51$
357

56. $5xy$, when $x = -9$ and $y = -2$
90

57. $8ab$, when $a = 7$ and $b = -1$
-56

58. $-4cd$, when $c = 25$ and $d = -8$
800

59. $-5st$, when $s = -40$ and $t = -8$
-1600

60. Is -4 a solution of the equation $6m = -24$?
Yes

61. Is -3 a solution of the equation $-5x = -15$?
No

62. Is -6 a solution of the equation $48 = -8y$?
Yes

63. Is 0 a solution of the equation $-8 = -8a$?
No

64. Is 7 a solution of the equation $-3c = 21$?
No

65. Is 9 a solution of the equation $-27 = -3c$?
Yes

Objective B   **To divide integers**

Divide.

66. $12 \div (-6)$
-2

67. $18 \div (-3)$
-6

68. $(-72) \div (-9)$
8

69. $(-64) \div (-8)$
8

70. $0 \div (-6)$
0

71. $-49 \div 1$
-49

72. $81 \div (-9)$
-9

73. $-40 \div (-5)$
8

74. $\dfrac{72}{-3}$
-24

75. $\dfrac{44}{-4}$
-11

76. $\dfrac{-93}{-3}$
31

77. $\dfrac{-98}{-7}$
14

78. $-114 \div (-6)$
19

79. $-91 \div (-7)$
13

80. $-53 \div 0$
Undefined

81. $(-162) \div (-162)$
1

82. $-128 \div 4$
-32

83. $-130 \div (-5)$
26

84. $(-200) \div 8$
-25

85. $(-92) \div (-4)$
23

Quick Quiz (Objective 3.3B)

1. What is the quotient of -50 and -10?   5

2. Divide: $\dfrac{-48}{4}$ -12

3. Divide: $y \div 0$ Undefined

4. Evaluate $(-c) \div (-d)$ when $c = -24$ and $d = 6$.
-4

5. Is 6 a solution of the equation $-18 = \dfrac{n}{-3}$?   No

86. Find the quotient of -700 and 70.
-10

87. Find 550 divided by -5.
-110

88. What is -670 divided by -10?
67

89. What is the quotient of -333 and -3?
111

90. Write the quotient of $-a$ and b.
$\dfrac{-a}{b}$

91. Write -9 divided by x.
$\dfrac{-9}{x}$

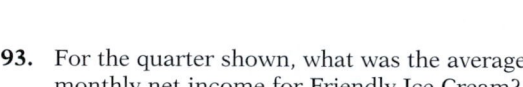 **Business**   The figure at the right shows the net income for the fourth quarter of a recent year for three companies. (*Note:* Negative income indicates a loss. One quarter of the year is 3 months.) Use this figure for Exercises 92 and 93.

92. For the quarter shown, what was the average monthly net income for Fresh Choice?
$-\$984{,}000$

93. For the quarter shown, what was the average monthly net income for Friendly Ice Cream?
$-\$236{,}000$

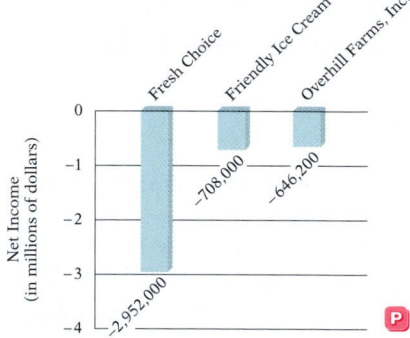

Net Income for Fourth Quarter
Source: **www.wsj.com**

Evaluate the expression for the given values of the variables.

94. $a \div b$, where $a = -36$ and $b = -4$
9

95. $-a \div b$, where $a = -36$ and $b = -4$
-9

96. $a \div (-b)$, where $a = -36$ and $b = -4$
-9

97. $(-a) \div (-b)$, where $a = -36$ and $b = -4$
9

98. $\dfrac{x}{y}$, where $x = -42$ and $y = -7$
6

99. $\dfrac{-x}{y}$, where $x = -42$ and $y = -7$
-6

100. $\dfrac{x}{-y}$, where $x = -42$ and $y = -7$
-6

101. $\dfrac{-x}{-y}$, where $x = -42$ and $y = -7$
6

102. Is 20 a solution of the equation $\dfrac{m}{-2} = -10$?

Yes

103. Is 18 a solution of the equation $6 = \dfrac{-c}{-3}$?

Yes

104. Is 0 a solution of the equation $0 = \dfrac{a}{-4}$?

Yes

105. Is -3 a solution of the equation $\dfrac{21}{n} = 7$?

No

106. Is -6 a solution of the equation $\dfrac{x}{2} = \dfrac{-18}{x}$?

No

107. Is 8 a solution of the equation $\dfrac{m}{-4} = \dfrac{-16}{m}$?

Yes

Objective C   **To solve application problems**

108. **Golf** The combined scores of the top five golfers in a tournament equaled -10 (10 under par). What was the average score of the five golfers?

-2

109. **Golf** The combined scores of the top four golfers in a tournament equaled -12 (12 under par). What was the average score of the four golfers?

-3

Temperature The following figure shows the record low temperatures, in degrees Fahrenheit, in the United States for each month. Use this figure for Exercises 110 to 112.

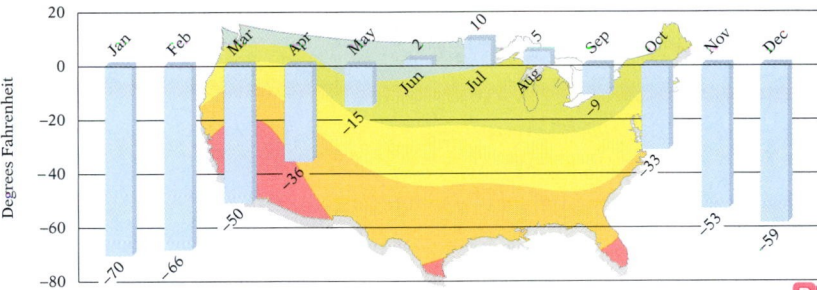

Record Low Temperatures, in Degrees Fahrenheit, in the United States
Source: National Climatic Data Center, Asheville, NC, and Storm Phillips, STORMFAX, Inc.

110. What is the average record low temperature for July, August, and September?

2°F

111. What is the average record low temperature for the first three months of the year?

-62°F

112. What is the average record low temperature for the four months with the lowest record low temperatures?

-62°F

Quick Quiz (Objective 3.3C)

1. The combined scores, in relation to par, of the top nine golfers in a golf tournament equaled -63. What was the average score of the nine golfers?
-7

2. The daily low temperatures during one week were recorded as follows: 4°F, -6°F, 8°F, -2°F, -9°F, -11°F, -5°F. Find the average daily low temperature for the week. -3°F

113.   Temperature The daily low temperatures during one week were recorded as follows: 4°, −5°, 8°, −1°, −12°, −14°, −8°. Find the average daily low temperature for the week.
−4°

114.   Temperature The daily high temperatures during one week were recorded as follows: −6°, −11°, 1°, 5°, −3°, −9°, −5°. Find the average daily high temperature for the week.
−4°

115.   **Environmental Science** The wind chill factor when the temperature is −20°F and the wind is blowing at 15 mph is five times the wind chill factor when the temperature is 10°F and the wind is blowing at 20 mph. If the wind chill factor at 10°F with a 20-mph wind is −9°, what is the wind chill factor at −20°F with a 15-mph wind?
−45°

Mathematics A geometric sequence is a list of numbers in which each number after the first is found by multiplying the preceding number in the list by the same number. For example, in the sequence 1, 3, 9, 27, 81, . . . , each number after the first is found by multiplying the preceding number in the list by 3. To find the multiplier in a geometric sequence, divide the second number in the sequence by the first number; for the example above, $3 \div 1 = 3$.

116.   Find the next three numbers in the geometric sequence −5, 15, −45, ...
135, −405, 1215

117.   Find the next three numbers in the geometric sequence 2, −4, 8, ...
−16, 32, −64

118.   Find the next three numbers in the geometric sequence −3, −12, −48, ...
−192, −768, −3072

119.   Find the next three numbers in the geometric sequence −1, −5, −25, ...
−125, −625, −3125

APPLYING THE CONCEPTS

120.   Use repeated addition to show that the product of two integers with different signs is a negative number.

121.   Mathematics
a. Find the largest possible product of two negative integers whose sum is −18.
b. Find the smallest possible sum of two negative integers whose product is 16.
a. 81   **b.** −17

122.   Determine whether the statement is always true, sometimes true, or never true.
a. The product of a number and its additive inverse is a negative number.
b. The product of an odd number of negative numbers is a negative number.
c. The square of a negative number is a positive number.
a. Sometimes true   **b.** Always true   **c.** Always true

123.   Find all negative integers x such that $1 - 3x < 12$.

124.   Describe the steps involved in using a calculator to simplify $(-2491) \div (-47)$.

125.   **a.** When is the product of two integers positive? When is it negative?
b. When is the quotient of two integers positive? When is it negative?

Answers to Writing Exercises

124. To simplify $(-2491) \div (-47)$ using a calculator, enter 2491, press the +/− key, press the division key (÷), enter 47, press the +/− key, and press the = key. The result in the display should be 53.

125a. The product of two integers is positive when the signs of the two integers are the same. The product is negative when the signs of the two integers are different.

b. The quotient of two integers is positive when the signs of the two integers are the same. The quotient is negative when the signs of the two integers are different.

3.4 Operations with Rational Numbers

Objective A   To add or subtract rational numbers

In this section, operations with rational numbers are discussed. A **rational number** is the quotient of two integers.

> **Rational Numbers**
>
> A rational number is a number that can be written in the form $\frac{a}{b}$, where a and b are integers and $b \neq 0$.

Each of the three numbers shown at the right is a rational number.

$$\frac{3}{4} \qquad \frac{-2}{9} \qquad \frac{13}{-5}$$

An integer can be written as the quotient of the integer and 1. Therefore, **every integer is a rational number.**

$$6 = \frac{6}{1} \qquad -8 = \frac{-8}{1}$$

A mixed number can be written as the quotient of two integers. Therefore, **every mixed number is a rational number.**

$$1\frac{4}{7} = \frac{11}{7} \qquad 3\frac{2}{5} = \frac{17}{5}$$

Recall that every fraction can be written as a decimal by dividing the numerator of the fraction by the denominator. The result is either a terminating decimal or a repeating decimal.

To write $\frac{5}{8}$ as a decimal, divide 5 by 8.

$$\underset{8)\overline{5.000}}{0.625} \longleftarrow \text{This is a \textbf{terminating decimal.}}$$

To write $\frac{4}{11}$ as a decimal, divide 4 by 11.

$$\underset{11)\overline{4.000}}{0.3636\ldots = 0.\overline{36}} \longleftarrow \text{This is a \textbf{repeating decimal.}}$$

Every rational number can be written as a terminating or a repeating decimal. Some numbers, for example, $\sqrt{7}$ and π, have decimal representations that never terminate or repeat. These numbers are called **irrational numbers.**

$$\sqrt{7} \approx 2.6457513\ldots \qquad \pi \approx 3.1415926\ldots$$

The rational numbers and the irrational numbers taken together are called the **real numbers.**

We begin the presentation of operations on rational numbers by adding rational numbers in fractional form. If an addend is a fraction containing a negative sign, rewrite the fraction with the negative sign in the numerator. Then add the numerators and place the sum over the common denominator.

TAKE NOTE

Rational numbers are fractions such as $-\frac{6}{7}$ and $\frac{10}{3}$, in which the numerator and denominator are integers. Rational numbers are also represented by repeating decimals such as 0.25767676 . . . or terminating decimals such as 1.73. An irrational number is neither a terminating decimal nor a repeating decimal. For example, 2.45445444544445 . . . is an irrational number.

TAKE NOTE

The three dots mean that the number continues without end.

Concept Check

Show that the number is a rational number by writing it as the quotient of two integers. (*Note:* Answers may vary. Possible answers are given.)

1. $-\dfrac{5}{6}$ $\dfrac{-5}{6}$

2. $-\dfrac{7}{25}$ $\dfrac{-7}{25}$

3. $3\dfrac{2}{9}$ $\dfrac{29}{9}$

4. -0.16 $\dfrac{-4}{25}$

5. -0.75 $\dfrac{-3}{4}$

6. 5.2 $\dfrac{26}{5}$

7. -2.4 $\dfrac{-12}{5}$

8. -6.25 $\dfrac{-25}{4}$

Optional Student Activity

Have students use calculators to write $\dfrac{5}{99}$, $\dfrac{17}{99}$, $\dfrac{42}{99}$, $\dfrac{63}{99}$, and $\dfrac{81}{99}$ as repeating decimals. Have them describe the pattern. Ask them to find fractions with other denominators that have interesting decimal patterns. (For example, ninths or elevenths.)

Study Tip

Have you considered joining a study group? Getting together regularly with other students in the class to go over material and quiz each other can be very beneficial. See *AIM for Success*, page xxx.

HOW TO   Add: $-\dfrac{5}{6} + \dfrac{3}{4}$

The LCM of 4 and 6 is 12.

$$-\dfrac{5}{6} + \dfrac{3}{4} = \dfrac{-5}{6} + \dfrac{3}{4}$$

$$= \dfrac{-10}{12} + \dfrac{9}{12}$$

$$= \dfrac{-10 + 9}{12}$$

$$= \dfrac{-1}{12} = -\dfrac{1}{12}$$

- The common denominator is the LCM of 4 and 6.
- Rewrite with the negative sign in the numerator.
- Rewrite each fraction in terms of the common denominator.
- Add the fractions.
- Simplify the numerator and write the negative sign in front of the fraction.

Although the sum in the last example could have been left as $\dfrac{-1}{12}$, **all answers in this text are written with the negative sign in front of the fraction.**

HOW TO   Add: $-\dfrac{2}{3} + \left(-\dfrac{4}{5}\right)$

$$-\dfrac{2}{3} + \left(-\dfrac{4}{5}\right) = \dfrac{-2}{3} + \dfrac{-4}{5}$$

$$= \dfrac{-10}{15} + \dfrac{-12}{15}$$

$$= \dfrac{-10 + (-12)}{15}$$

$$= \dfrac{-22}{15} = -1\dfrac{7}{15}$$

- Rewrite each negative fraction with the negative sign in the numerator.
- Rewrite each fraction as an equivalent fraction using the LCM as the denominator.
- Add the fractions.

HOW TO   Is $-\dfrac{2}{3}$ a solution of the equation $\dfrac{3}{4} + y = -\dfrac{1}{12}$?

$$\dfrac{3}{4} + y = -\dfrac{1}{12}$$

$$\begin{array}{c|c} \dfrac{3}{4} + \left(-\dfrac{2}{3}\right) & -\dfrac{1}{12} \\[2mm] \dfrac{9}{12} + \left(\dfrac{-8}{12}\right) & -\dfrac{1}{12} \\[2mm] \dfrac{9 + (-8)}{12} & -\dfrac{1}{12} \\[2mm] \dfrac{1}{12} \ne & -\dfrac{1}{12} \end{array}$$

- Replace y by $-\dfrac{2}{3}$. Then simplify.
- The common denominator is 12.
- The results are not equal.

No, $-\dfrac{2}{3}$ is not a solution of the equation.

In-Class Examples (Objective 3.4A)

(*Continued*)

6. Subtract: $6.2 - (-4.61)$ 10.81

7. Subtract: $3.8 - 7.4$ -3.6

8. Subtract: $-\dfrac{5}{12} - \left(-\dfrac{5}{6}\right)$. $\dfrac{5}{12}$

9. What is $4\dfrac{1}{2}$ less than $1\dfrac{3}{4}$? $-2\dfrac{3}{4}$

10. Evaluate $-x - (-y)$ when $x = -\dfrac{1}{4}$ and $y = -\dfrac{5}{8}$. $-\dfrac{3}{8}$

To subtract fractions with negative signs, first rewrite the fractions with the negative signs in the numerators.

HOW TO Simplify: $-\dfrac{2}{9} - \dfrac{5}{12}$

$$-\frac{2}{9} - \frac{5}{12} = \frac{-2}{9} - \frac{5}{12}$$

- Rewrite the negative fraction with the negative sign in the numerator.

$$= \frac{-8}{36} - \frac{15}{36}$$

- Write the fractions as equivalent fractions with a common denominator.

$$= \frac{-8 - 15}{36} = \frac{-23}{36}$$

- Subtract the numerators and place the difference over the common denominator.

$$= -\frac{23}{36}$$

- Write the negative sign in front of the fraction.

HOW TO Subtract: $\dfrac{2}{3} - \left(-\dfrac{4}{5}\right)$

$$\frac{2}{3} - \left(-\frac{4}{5}\right) = \frac{2}{3} + \frac{4}{5}$$

- Rewrite subtraction as addition of the opposite.

$$= \frac{10}{15} + \frac{12}{15}$$

- Write the fractions as equivalent fractions with a common denominator.

$$= \frac{10 + 12}{15}$$

- Add the fractions.

$$= \frac{22}{15} = 1\frac{7}{15}$$

HOW TO Evaluate $x - y$ when $x = -\dfrac{2}{5}$ and $y = -\dfrac{3}{10}$.

$x - y$

$$-\frac{2}{5} - \left(-\frac{3}{10}\right)$$

- Replace x with $-\dfrac{2}{5}$ and y with $-\dfrac{3}{10}$.

$$= -\frac{2}{5} + \frac{3}{10}$$

- Rewrite subtraction as addition of the opposite.

$$= \frac{-2}{5} + \frac{3}{10}$$

- Rewrite the negative fraction with the negative sign in the numerator.

$$= \frac{-4}{10} + \frac{3}{10}$$

- Write the fractions as equivalent fractions with a common denominator. *Note:* The LCM of 5 and 10 is 10.

$$= \frac{-4 + 3}{10} = \frac{-1}{10}$$

- Add the numerators and place the sum over the common denominator.

$$= -\frac{1}{10}$$

- Write the negative sign in front of the fraction.

Optional Student Activity

To determine whether a fraction can be written as a terminating decimal, first write the fraction in simplest form. Then look at the denominator of the fraction. If it contains prime factors of only 2's and/or 5's, then it can be expressed as a terminating decimal. If it contains prime factors other than 2's or 5's, it represents a repeating decimal. For example, if the denominator of a fraction in simplest form is 20, then the fraction can be written as a terminating decimal, because $20 = 2 \cdot 2 \cdot 5$ (only prime factors of 2 and 5). If the denominator of a fraction in simplest form is 6, the fraction represents a repeating decimal, because the denominator contains the prime factor 3 (a number other than 2 or 5).

Assume that each of the following numbers is the denominator of a fraction written in simplest form. Does the fraction represent a terminating or a repeating decimal? (*Continued on the next page*)

The sign rules for adding and subtracting decimals are the same rules used to add and subtract integers.

TAKE NOTE

Recall that the absolute value of a number is the distance from zero to the number on the number line. The absolute value of a number is a positive number or zero.
$|54.29| = 54.29$
$|-36.087| = 36.087$

HOW TO   Simplify: $-36.087 + 54.29$

$54.29 - 36.087 = 18.203$ • The signs of the addends are different. Subtract the smaller absolute value from the larger absolute value.

$|54.29| > |-36.087|$ • Attach the sign of the number with the larger absolute value.

$-36.087 + 54.29 = 18.203$ • The sum is positive.

Recall that the opposite of n is $-n$ and the opposite of $-n$ is n. To find the opposite of a number, change the sign of the number.

HOW TO   Simplify: $-2.86 - 10.3$

$-2.86 - 10.3$
$= -2.86 + (-10.3)$ • Rewrite subtraction as addition of the opposite. The opposite of 10.3 is -10.3.
$= -13.16$ • The signs of the addends are the same. Add the absolute values of the numbers. Attach the sign of the addends.

HOW TO   Evaluate $c - d$ when $c = 9.34$ and $d = -8.7$.

$c - d$
$9.34 - (-8.7)$ • Replace c with 9.34 and d with -8.7.
$= 9.34 + 8.7$ • Rewrite subtraction as addition of the opposite.
$= 18.04$ • Add.

Example 1

Add: $-\dfrac{3}{8} + \dfrac{3}{4} + \left(-\dfrac{5}{6}\right)$

Solution

$-\dfrac{3}{8} + \dfrac{3}{4} + \left(-\dfrac{5}{6}\right)$

$= \dfrac{-3}{8} + \dfrac{3}{4} + \dfrac{-5}{6}$ • Rewrite with negative signs in the numerators.

$= \dfrac{-9}{24} + \dfrac{18}{24} + \dfrac{-20}{24}$ • The LCD is 24.

$= \dfrac{-9 + 18 + (-20)}{24}$

$= \dfrac{-11}{24} = -\dfrac{11}{24}$ • Add the numerators.

You Try It 1

Add: $-\dfrac{5}{12} + \dfrac{5}{8} + \left(-\dfrac{1}{6}\right)$

Your solution

$\dfrac{1}{24}$

Solution on p. S10

(*Continued*)

a. 4	Terminating
b. 5	Terminating
c. 7	Repeating
d. 9	Repeating
e. 10	Terminating
f. 12	Repeating
g. 15	Repeating
h. 16	Terminating
i. 18	Repeating
j. 21	Repeating
k. 24	Repeating
l. 25	Terminating
m. 28	Repeating
n. 40	Terminating

Write another fraction in simplest form that represents a terminating decimal.
Answers will vary. For example, any fraction in simplest form with a denominator of 8 or 80.

Write another fraction in simplest form that represents a repeating decimal.
Answers will vary. For example, any fraction in simplest form with a denominator of 3 or 30.

Example 2

Subtract: $-\dfrac{5}{6} - \left(-\dfrac{3}{8}\right)$

Solution

$-\dfrac{5}{6} - \left(-\dfrac{3}{8}\right) = -\dfrac{5}{6} + \dfrac{3}{8}$ • **Rewrite "−" as "+". The opposite of $-\dfrac{3}{8}$ is $\dfrac{3}{8}$.**

$= \dfrac{-20}{24} + \dfrac{9}{24}$

$= \dfrac{-20 + 9}{24}$

$= \dfrac{-11}{24} = -\dfrac{11}{24}$

You Try It 2

Subtract: $-\dfrac{5}{6} - \dfrac{7}{9}$

Your solution

$-1\dfrac{11}{18}$

Example 3

What is -251.49 more than -638.7?

Solution

$-638.7 + (-251.49) = -890.19$

You Try It 3

What is 4.002 minus 9.378?

Your solution -5.376

Example 4

Evaluate $x + y + z$ when
$x = -1.6$, $y = 7.9$, and $z = -4.8$.

Solution

$x + y + z$
$-1.6 + 7.9 + (-4.8) = 6.3 + (-4.8)$
$\qquad\qquad\qquad\qquad = 1.5$

You Try It 4

Evaluate $x + y + z$ when
$x = -7.84$, $y = -3.05$, and $z = 2.19$.

Your solution -8.7

Example 5

Is $\dfrac{3}{8}$ a solution of the equation $\dfrac{2}{3} = w - \dfrac{5}{6}$?

Solution

$\dfrac{2}{3} = w - \dfrac{5}{6}$

$\begin{array}{c|c}
\dfrac{2}{3} & \dfrac{3}{8} - \dfrac{5}{6} \\
\end{array}$ • **Replace w by $\dfrac{3}{8}$.**

$\begin{array}{c|c}
\dfrac{2}{3} & \dfrac{9}{24} - \dfrac{20}{24} \\
\end{array}$

$\begin{array}{c|c}
\dfrac{2}{3} & \dfrac{-11}{24} \\
\end{array}$

$\dfrac{2}{3} \neq -\dfrac{11}{24}$

No, $\dfrac{3}{8}$ is not a solution of the equation.

You Try It 5

Is $-\dfrac{1}{4}$ a solution of the equation $\dfrac{2}{3} - v = \dfrac{11}{12}$?

Your solution
Yes

Solutions on p. S10

Objective 3.4B

Vocabulary to Review
reciprocal　[2.4A]

Rules to Review
multiplication of integers　[3.3A]
division of integers　[3.3B]

Discuss the Concepts
1. Explain how to multiply two fractions when one is a positive number and one is a negative number.
2. Explain how to divide two fractions when both fractions are negative numbers.
3. How can you determine whether the sum of two decimals, one a positive number and one a negative number, is positive or negative?

Concept Check
Find two fractions, one positive and one negative, with different denominators, whose sum is $\frac{1}{2}$.

Answers will vary. One example is
$\frac{4}{5} + \left(-\frac{3}{10}\right) = \frac{1}{2}$.

Optional Student Activity
Replace the question mark with the number that makes the equation true.

1. $4.3(?) = -2.58$　-0.6
2. $32.4 = -9(?)$　-3.6
3. $(?)(-0.2) = -1.6$　-8
4. $\dfrac{?}{-0.4} = -20$　8
5. $\dfrac{?}{-8} = -3.1$　24.8
6. $\dfrac{?}{0.48} = -12.5$　-6

Objective B　To multiply or divide rational numbers

The product of two rational numbers written in fractional form is the product of the numerators over the product of the denominators. The sign rules are the same rules used to multiply integers.

The product of two numbers with the same sign is positive.

The product of two numbers with different signs is negative.

HOW TO　Multiply: $-\dfrac{3}{4} \cdot \dfrac{8}{15}$

$-\dfrac{3}{4} \cdot \dfrac{8}{15} = -\left(\dfrac{3}{4} \cdot \dfrac{8}{15}\right)$　• The signs are different. The product is negative.

$= -\dfrac{3 \cdot 8}{4 \cdot 15}$　• Multiply the numerators. Multiply the denominators.

$= -\dfrac{3 \cdot 2 \cdot 2 \cdot 2}{2 \cdot 2 \cdot 3 \cdot 5}$　• Write the product in simplest form.

$= -\dfrac{2}{5}$

HOW TO　Multiply: $-\dfrac{3}{8}\left(-\dfrac{2}{5}\right)\left(-\dfrac{10}{21}\right)$

$-\dfrac{3}{8}\left(-\dfrac{2}{5}\right)\left(-\dfrac{10}{21}\right)$

$= \left(\dfrac{3}{8} \cdot \dfrac{2}{5}\right)\left(-\dfrac{10}{21}\right)$　• Multiply the first two fractions. The product is positive.

$= -\left(\dfrac{3}{8} \cdot \dfrac{2}{5} \cdot \dfrac{10}{21}\right)$　• The product of the first two fractions and the third fraction is negative.

$= -\dfrac{3 \cdot 2 \cdot 10}{8 \cdot 5 \cdot 21}$　• Multiply the numerators. Multiply the denominators.

$= -\dfrac{3 \cdot 2 \cdot 2 \cdot 5}{2 \cdot 2 \cdot 2 \cdot 5 \cdot 3 \cdot 7}$　• Write the product in simplest form.

$= -\dfrac{1}{14}$

Thus, the product of three negative fractions is negative. We can modify the rule for multiplying positive and negative fractions to say that **the product of an odd number of negative fractions is negative and the product of an even number of negative fractions is positive.**

The sign rules for dividing positive and negative fractions are the same rules used to divide integers.

The quotient of two numbers with the same sign is positive.

The quotient of two numbers with different signs is negative.

In-Class Examples (Objective 3.4B)

1. Multiply: $-\dfrac{5}{6}\left(\dfrac{3}{10}\right)$　$-\dfrac{1}{4}$

2. Multiply: $-6.8(-2.1)$　14.28

3. Evaluate $-xy$ when $x = -2$ and $y = -\dfrac{5}{12}$.　$-\dfrac{5}{6}$

4. Which number, 1.8, -1.8, -18, or -0.18, is a solution of the equation $-3x = 5.4$?　-1.8

HOW TO Simplify: $-\dfrac{7}{10} \div \left(-\dfrac{14}{15}\right)$

$-\dfrac{7}{10} \div \left(-\dfrac{14}{15}\right) = \dfrac{7}{10} \div \dfrac{14}{15}$ • The signs are the same. The quotient is positive.

$\quad = \dfrac{7}{10} \cdot \dfrac{15}{14}$ • Rewrite the division as multiplication by the reciprocal.

$\quad = \dfrac{7 \cdot 15}{10 \cdot 14}$ • Multiply the fractions.

$\quad = \dfrac{7 \cdot 3 \cdot 5}{2 \cdot 5 \cdot 2 \cdot 7}$

$\quad = \dfrac{3}{4}$

The sign rules for multiplying decimals are the same rules used to multiply integers.

The product of two numbers with the same sign is positive.
The product of two numbers with different signs is negative.

HOW TO Multiply: $(-3.2)(-0.008)$

$(-3.2)(-0.008) = 0.0256$ • The signs are the same. The product is positive. Multiply the absolute values of the numbers.

HOW TO Is -0.6 a solution of the equation $4.3a = -2.58$?

$4.3a = -2.58$

$\overline{4.3(-0.6)\ \big|\ -2.58}$ • Replace a by -0.6 and then simplify.

$\quad -2.58 = -2.58$ • The results are equal.

Yes, -0.6 is a solution of the equation.

The sign rules for dividing decimals are the same rules used to divide integers.

The quotient of two numbers with the same sign is positive.
The quotient of two numbers with different signs is negative.

HOW TO Divide: $-1.16 \div 2.9$

$-1.16 \div 2.9 = -0.4$ • The signs are different. The quotient is negative. Divide the absolute values of the numbers.

HOW TO Evaluate $c \div d$ when $c = -8.64$ and $d = -0.4$.

$c \div d$

$(-8.64) \div (-0.4)$ • Replace c with -8.64 and d with -0.4.

$\quad = 21.6$ • The signs are the same. The quotient is positive. Divide the absolute values of the numbers.

In-Class Examples (Objective 3.4B)

(*Continued*)

5. Divide: $-\dfrac{3}{4} \div \left(-\dfrac{9}{16}\right)$ $1\dfrac{1}{3}$

6. Divide and round to the nearest tenth:
$-13.97 \div 28.4$ -0.5

7. Find $-\dfrac{3}{8}$ divided by $2\dfrac{1}{4}$. $-\dfrac{1}{6}$

8. Evaluate $a \div (-b)$ when $a = -8.4$ and $b = -0.4$. -21

Example 6   Multiply: $-\dfrac{3}{4}\left(\dfrac{1}{2}\right)\left(-\dfrac{8}{9}\right)$

You Try It 6   Multiply: $-\dfrac{1}{3}\left(-\dfrac{5}{12}\right)\left(\dfrac{8}{15}\right)$

Solution $-\dfrac{3}{4}\left(\dfrac{1}{2}\right)\left(-\dfrac{8}{9}\right)$

Your solution $\dfrac{2}{27}$

$= \dfrac{3}{4} \cdot \dfrac{1}{2} \cdot \dfrac{8}{9}$ • **The product of two negative fractions is positive.**

$= \dfrac{3 \cdot 1 \cdot 8}{4 \cdot 2 \cdot 9}$

$= \dfrac{3 \cdot 1 \cdot 2 \cdot 2 \cdot 2}{2 \cdot 2 \cdot 2 \cdot 3 \cdot 3} = \dfrac{1}{3}$

Example 7   What is the product of $-\dfrac{1}{2}$ and $\dfrac{2}{5}$?

You Try It 7   Multiply $3\dfrac{6}{7}$ by $-\dfrac{4}{9}$.

Solution

$-\dfrac{1}{2} \cdot \dfrac{2}{5} = -\left(\dfrac{1}{2} \cdot \dfrac{2}{5}\right)$ • **The signs are different. The product is negative.**

Your solution $-1\dfrac{5}{7}$

$= -\dfrac{1 \cdot 2}{2 \cdot 5}$

$= -\dfrac{1}{5}$

Example 8   What is the quotient of 6 and $-\dfrac{3}{5}$?

You Try It 8   Find the quotient of 4 and $-\dfrac{6}{7}$.

Solution $6 \div \left(-\dfrac{3}{5}\right)$

Your solution $-4\dfrac{2}{3}$

$= -\left(\dfrac{6}{1} \div \dfrac{3}{5}\right)$ • **The signs are different. The quotient is negative.**

$= -\left(\dfrac{6}{1} \cdot \dfrac{5}{3}\right)$

$= -\dfrac{6 \cdot 5}{1 \cdot 3}$

$= -\dfrac{2 \cdot 3 \cdot 5}{1 \cdot 3}$

$= -\dfrac{10}{1} = -10$

Solutions on p. S11

Example 9   Multiply: $-3.42(6.1)$

Solution $-3.42(6.1)$ • **The signs are**
         $= -20.862$ **different. The**
                         product is negative.

You Try It 9   Multiply: $(-0.7)(-5.8)$

Your solution   4.06

Example 10   Divide and round to the nearest tenth: $-6.94 \div -1.5$

Solution $-6.94 \div (-1.5)$ • **The signs are the**
           ≈ 4.6 **same. The quotient**
                         is positive.

You Try It 10   Divide and round to the nearest tenth: $-25.7 \div 0.31$

Your solution -82.9

Example 11   Evaluate the variable expression xy when $x = 1\frac{4}{5}$ and $y = -\frac{5}{6}$.

Solution xy

$$1\frac{4}{5}\left(-\frac{5}{6}\right) = -\left(\frac{9}{5}\cdot\frac{5}{6}\right)$$

$$= -\frac{9\cdot5}{5\cdot6}$$

$$= -\frac{3\cdot3\cdot5}{5\cdot2\cdot3}$$

$$= -\frac{3}{2} = -1\frac{1}{2}$$

You Try It 11   Evaluate the variable expression xy when $x = -5\frac{1}{8}$ and $y = -\frac{2}{3}$.

Your solution $3\frac{5}{12}$

Example 12   Evaluate $\frac{x}{y}$ when $x = -76.8$ and $y = 0.8$.

Solution $\frac{x}{y}$

$$\frac{-76.8}{0.8} = -96$$

You Try It 12   Evaluate $\frac{x}{y}$ when $x = -40.6$ and $y = -0.7$.

Your solution   58

Solutions on p. S11

Example 13   Evaluate $50ab$ when $a = -0.9$ and $b = -0.2$.

Solution $50ab$
$50(-0.9)(-0.2) = -45(-0.2)$
$= 9$

You Try It 13   Evaluate $25xy$ when $x = -0.8$ and $y = 0.6$.

Your solution -12

Example 14   Is -0.4 a solution of the equation $\frac{8}{x} = -20$?

Solution

$$\frac{8}{x} = -20$$

$$\frac{8}{-0.4} \quad\bigg|\quad -20 \qquad \bullet \text{ Replace } x \text{ by } -0.4.$$

$$-20 = -20$$

Yes, -0.4 is a solution of the equation.

You Try It 14   Is -1.2 a solution of the equation $-2 = \frac{d}{-0.6}$?

Your solution   No

Solutions on p. S11

Objective 3.4C

Optional Student Activity

The table below shows the total trade balance, in billions of dollars, for the United States during the 1990s. (*Source:* U.S. Department of Commerce, Bureau of Economic Analysis)

Year	Trade Balance
1991	6.616
1992	−47.724
1993	−82.681
1994	−118.605
1995	−109.457
1996	−123.318
1997	−140.540
1998	−217.138
1999	−331.479

1. In which year was the trade balance the lowest? **1999**

2. Calculate the difference between the trade balance in 1991 and that in 1999.
$338.095 billion

3. Between which two consecutive years was the difference in the trade balance greatest?
1998−1999

4. How many times greater was the trade balance in 1999 than in 1992? Round to the nearest whole number.

7 times greater

5. Calculate the average trade balance per quarter in 1997.
−$35.135 billion

6. Examine the data. Would you expect the trade balance to have increased or decreased from 1999 to 2000? Why?
Answers will vary.

Objective C   To solve application problems

Example 15

In Fairbanks, Alaska, the average temperature during the month of July is 61.5°F. During the month of January, the average temperature in Fairbanks is −12.7°F. What is the difference between the average temperature in Fairbanks during July and the average temperature during January?

Strategy
To find the difference, subtract the average temperature in January (−12.7°F) from the average temperature in July (61.5°F).

Solution
$61.5 - (-12.7) = 61.5 + 12.7 = 74.2$

The difference between the average temperature during July and the average temperature during January in Fairbanks is 74.2°F.

You Try It 15

On January 10, 1911, in Rapid City, South Dakota, the temperature fell from 12.78°C at 7:00 A.M. to −13.33°C at 7:15 A.M. How many degrees did the temperature fall during the 15-minute period?

Your strategy

Your solution
26.11°C

Solution on p. S11

In-Class Examples (Objective 3.4C)

1. The lowest temperature recorded in Australia is −9.4°F. The highest temperature recorded is 128.0°F. (*Source:* National Climatic Data Center) Find the difference between these two extremes.
137.4°F

2. On November 3, 2004, the closing price of a share of Pepsi Bottling Group, Inc. stock was $27.93. The change in the closing price from the previous day was −$.21. Find the closing price on the previous day. **$28.14**

3.4 Exercises

Objective A   **To add or subtract rational numbers**

Simplify.

1. $\dfrac{5}{8} - \dfrac{5}{6}$

$-\dfrac{5}{24}$

2. $\dfrac{1}{9} - \dfrac{5}{27}$

$-\dfrac{2}{27}$

3. $-\dfrac{5}{12} - \dfrac{3}{8}$

$-\dfrac{19}{24}$

4. $-\dfrac{5}{6} - \dfrac{5}{9}$

$-1\dfrac{7}{18}$

5. $-\dfrac{6}{13} + \dfrac{17}{26}$

$\dfrac{5}{26}$

6. $-\dfrac{7}{12} + \dfrac{5}{8}$

$\dfrac{1}{24}$

7. $-\dfrac{5}{8} - \left(-\dfrac{11}{12}\right)$

$\dfrac{7}{24}$

8. $-\dfrac{7}{12} - \left(-\dfrac{7}{8}\right)$

$\dfrac{7}{24}$

9. $\dfrac{5}{12} - \dfrac{11}{15}$

$-\dfrac{19}{60}$

10. $\dfrac{2}{5} - \dfrac{14}{15}$

$-\dfrac{8}{15}$

11. $-\dfrac{3}{4} - \dfrac{5}{8}$

$-1\dfrac{3}{8}$

12. $-\dfrac{2}{3} - \dfrac{5}{8}$

$-1\dfrac{7}{24}$

13. $-\dfrac{5}{2} - \left(-\dfrac{13}{4}\right)$

$\dfrac{3}{4}$

14. $-\dfrac{7}{3} - \left(-\dfrac{3}{2}\right)$

$-\dfrac{5}{6}$

15. $-\dfrac{3}{8} - \dfrac{5}{12} - \dfrac{3}{16}$

$-\dfrac{47}{48}$

16. $-\dfrac{5}{16} + \dfrac{3}{4} - \dfrac{7}{8}$

$-\dfrac{7}{16}$

17. $\dfrac{1}{2} - \dfrac{3}{8} - \left(-\dfrac{1}{4}\right)$

$\dfrac{3}{8}$

18. $\dfrac{3}{4} - \left(-\dfrac{7}{12}\right) - \dfrac{7}{8}$

$\dfrac{11}{24}$

19. $\dfrac{1}{3} - \dfrac{1}{4} - \dfrac{1}{5}$

$-\dfrac{7}{60}$

20. $\dfrac{5}{16} + \dfrac{1}{8} - \dfrac{1}{2}$

$-\dfrac{1}{16}$

21. $\dfrac{1}{2} + \left(-\dfrac{3}{8}\right) + \dfrac{5}{12}$

$\dfrac{13}{24}$

22. $-\dfrac{3}{8} + \dfrac{3}{4} - \left(-\dfrac{3}{16}\right)$

$\dfrac{9}{16}$

23. $3.4 + (-6.8)$

-3.4

24. $-4.9 + 3.27$

-1.63

25. $-8.32 + (-0.57)$

-8.89

26. $-3.5 + 7$

3.5

27. $-4.8 + (-3.2)$

-8.0

Section 3.4

Suggested Assignment

Exercises 1–143, every other odd
Exercises 144–148
More challenging problems:
    Exercises 149, 150

Quick Quiz (Objective 3.4A)

1. Add: $-\dfrac{3}{8} + \left(-\dfrac{1}{4}\right)$ $-\dfrac{5}{8}$

2. Subtract: $-\dfrac{3}{5} - \dfrac{4}{15}$ $-\dfrac{13}{15}$

3. Add: $-4.68 + 2.97$ -1.71

4. Subtract: $-5.9 - (-6.31)$ 0.41

5. Evaluate $-x + y$ when $x = \dfrac{2}{3}$ and $y = \dfrac{5}{6}$. $\dfrac{1}{6}$

28. $6.2 + (-4.29)$
1.91

29. $-4.6 + 3.92$
-0.68

30. $7.2 + (-8.42)$
-1.22

31. $-45.71 + (-135.8)$
-181.51

32. $-35.274 + 12.47$
-22.804

33. $4.2 + (-6.8) + 5.3$
2.7

34. $6.7 + 3.2 + (-10.5)$
-0.6

35. $-4.5 + 3.2 + (-19.4)$
-20.7

36. $2.09 - 6.72 - 5.4$
-10.03

37. $-18.39 + 4.9 - 23.7$
-37.19

38. $19 - (-3.72) - 82.75$
-60.03

39. $-3.09 - 4.6 - 27.3$
-34.99

40. What is $-\dfrac{5}{6}$ added to $\dfrac{4}{9}$?

$-\dfrac{7}{18}$

41. What is $\dfrac{7}{12}$ added to $-\dfrac{11}{16}$?

$-\dfrac{5}{48}$

42. What is $-\dfrac{2}{3}$ more than $-\dfrac{5}{6}$?

$-1\dfrac{1}{2}$

43. What is $-\dfrac{7}{12}$ more than $-\dfrac{5}{9}$?

$-1\dfrac{5}{36}$

44. What is $-\dfrac{7}{12}$ minus $\dfrac{7}{9}$?

$-1\dfrac{13}{36}$

45. What is $\dfrac{3}{5}$ decreased by $-\dfrac{7}{10}$?

$1\dfrac{3}{10}$

46. What is the sum of -65.47 and -32.91?
-98.38

47. Find -138.72 minus 510.64.
-649.36

48. What is 4.793 less than -6.82?
-11.613

49. How much greater is -31 than -62.09?
31.09

Evaluate the variable expression $x + y$ for the given values of x and y.

50. $x = -\dfrac{3}{8}, y = \dfrac{2}{9}$

$-\dfrac{11}{72}$

51. $x = \dfrac{3}{10}, y = -\dfrac{7}{15}$

$-\dfrac{1}{6}$

52. $x = -\dfrac{5}{8}, y = -\dfrac{1}{6}$

$-\dfrac{19}{24}$

53. $x = -\dfrac{3}{8}, y = -\dfrac{5}{6}$

$-1\dfrac{5}{24}$

54. $x = 62.97, y = -43.85$
19.12

55. $x = 5.904, y = -7.063$
-1.159

56. $x = -125.41, y = 361.55$
236.14

57. $x = -6.175, y = -19.49$
-25.665

Evaluate the variable expression $x - y$ for the given values of x and y.

58. $x = -\dfrac{11}{12}, y = \dfrac{5}{12}$

$-1\dfrac{1}{3}$

59. $x = -\dfrac{15}{16}, y = \dfrac{5}{16}$

$-1\dfrac{1}{4}$

60. $x = -\dfrac{2}{3}, y = -\dfrac{3}{4}$

$\dfrac{1}{12}$

61. $x = -\dfrac{5}{12}, y = -\dfrac{5}{9}$

$\dfrac{5}{36}$

62. $x = -21.073, y = 6.48$

-27.553

63. $x = -3.69, y = -1.527$

-2.163

64. $x = -8.21, y = -6.798$

-1.412

65. Is $-\dfrac{5}{6}$ a solution of the equation $\dfrac{1}{4} + x = -\dfrac{7}{12}$?

Yes

66. Is $\dfrac{5}{8}$ a solution of the equation $-\dfrac{1}{4} = x - \dfrac{7}{8}$?

Yes

67. Is -1.2 a solution of the equation $6.4 = 5.2 + a$?

No

68. Is -2.8 a solution of the equation $0.8 - p = 3.6$?

Yes

69. Is -0.5 a solution of the equation $x - 0.5 = 1$?

No

70. Is 36.8 a solution of the equation $27.4 = y - 9.4$?

Yes

Objective B   **To multiply or divide rational numbers**

Simplify.

71. $\dfrac{1}{2}\left(-\dfrac{3}{4}\right)$

$-\dfrac{3}{8}$

72. $-\dfrac{2}{9}\left(-\dfrac{3}{14}\right)$

$\dfrac{1}{21}$

73. $\left(-\dfrac{3}{8}\right)\left(-\dfrac{4}{15}\right)$

$\dfrac{1}{10}$

74. $\left(-\dfrac{3}{4}\right)\left(-\dfrac{8}{27}\right)$

$\dfrac{2}{9}$

75. $-\dfrac{1}{2}\left(\dfrac{8}{9}\right)$

$-\dfrac{4}{9}$

76. $\dfrac{5}{12}\left(-\dfrac{8}{15}\right)$

$-\dfrac{2}{9}$

77. $\left(-\dfrac{5}{12}\right)\left(\dfrac{42}{65}\right)$

$-\dfrac{7}{26}$

78. $\left(\dfrac{3}{8}\right)\left(-\dfrac{15}{41}\right)$

$-\dfrac{45}{328}$

79. $\left(-\dfrac{15}{8}\right)\left(-\dfrac{16}{3}\right)$

10

80. $\left(-\dfrac{5}{7}\right)\left(-\dfrac{14}{15}\right)$

$\dfrac{2}{3}$

81. $\dfrac{5}{8}\left(-\dfrac{7}{12}\right)\left(\dfrac{16}{25}\right)$

$-\dfrac{7}{30}$

82. $\left(\dfrac{1}{2}\right)\left(-\dfrac{3}{4}\right)\left(-\dfrac{5}{8}\right)$

$\dfrac{15}{64}$

83. $\dfrac{1}{3} \div \left(-\dfrac{1}{2}\right)$

$-\dfrac{2}{3}$

84. $-\dfrac{3}{8} \div \dfrac{7}{8}$

$-\dfrac{3}{7}$

85. $\left(-\dfrac{3}{4}\right) \div \left(-\dfrac{7}{40}\right)$

$4\dfrac{2}{7}$

Quick Quiz (Objective 3.4B)

1. Multiply: $-\dfrac{7}{12} \cdot \dfrac{5}{8} \cdot \dfrac{16}{25}$ $-\dfrac{7}{30}$

2. Divide: $\left(-\dfrac{3}{8}\right) \div \dfrac{7}{8}$ $-\dfrac{3}{7}$

3. Multiply: $3.9(-1.4)$ -5.46

4. Divide: $-29.61 \div 4.7$ -6.3

5. Evaluate $-xy$ when $x = 15.33$ and $y = -7$.

107.31

86. $\dfrac{5}{6} \div \left(-\dfrac{3}{4}\right)$

$-1\dfrac{1}{9}$

87. $-\dfrac{5}{12} \div \dfrac{15}{32}$

$-\dfrac{8}{9}$

88. $-\dfrac{5}{16} \div \left(-\dfrac{3}{8}\right)$

$\dfrac{5}{6}$

89. $\left(-\dfrac{3}{8}\right) \div \left(-\dfrac{5}{12}\right)$

$\dfrac{9}{10}$

90. $\left(-\dfrac{8}{19}\right) \div \dfrac{7}{38}$

$-2\dfrac{2}{7}$

91. $\left(-\dfrac{2}{3}\right) \div 4$

$-\dfrac{1}{6}$

92. $-6 \div \dfrac{4}{9}$

$-13\dfrac{1}{2}$

93. $-6.7(-4.2)$

28.14

94. $-8.9(-3.5)$

31.15

95. $-1.6(4.9)$

-7.84

96. $-14.3(7.9)$

-112.97

97. $(-0.78)(-0.15)$

0.117

98. $(-1.21)(-0.03)$

0.0363

99. $(-8.919) \div (-0.9)$

9.91

100. $-77.6 \div (-0.8)$

97

101. $59.01 \div (-0.7)$

-84.3

102. $(-7.04) \div (-3.2)$

2.2

103. $(-84.66) \div 1.7$

-49.8

104. $-3.312 \div (0.8)$

-4.14

105. $1.003 \div (-0.59)$

-1.7

106. $26.22 \div (-6.9)$

-3.8

107. Find $-\dfrac{9}{16}$ multiplied by $\dfrac{4}{27}$.

$-\dfrac{1}{12}$

108. Find $\dfrac{3}{7}$ multiplied by $-\dfrac{14}{15}$.

$-\dfrac{2}{5}$

109. What is the product of $-\dfrac{7}{24}, \dfrac{8}{21},$ and $\dfrac{3}{7}$?

$-\dfrac{1}{21}$

110. What is the product of $-\dfrac{5}{13}, -\dfrac{26}{75},$ and $\dfrac{5}{8}$?

$\dfrac{1}{12}$

111. What is $-\dfrac{15}{24}$ divided by $\dfrac{3}{5}$?

$-1\dfrac{1}{24}$

112. What is $\dfrac{5}{6}$ divided by $-\dfrac{10}{21}$?

$-1\dfrac{3}{4}$

113. Find the product of 2.7, -16, and 3.04.

-131.328

114. What is the product of 0.06, -0.4, and -1.5?

0.036

In Exercises 115 to 118, round answers to the nearest tenth.

115. Find the quotient of -19.04 and 0.75.
-25.4

116. What is the quotient of -21.892 and -0.96?
22.8

117. Find 27.735 divided by -60.3.
-0.5

118. What is -13.97 divided by 28.4?
-0.5

Evaluate the variable expression xy for the given values of x and y.

119. $x = -\dfrac{5}{16}, y = \dfrac{7}{15}$

$-\dfrac{7}{48}$

120. $x = -\dfrac{2}{5}, y = -\dfrac{5}{6}$

$\dfrac{1}{3}$

121. $x = -49, y = \dfrac{5}{14}$

$-17\dfrac{1}{2}$

122. $x = -\dfrac{3}{10}, y = -35$

$10\dfrac{1}{2}$

Evaluate the variable expression xyz for the given values of x, y, and z.

123. $x = 2\dfrac{3}{8}, y = -\dfrac{3}{19}, z = -\dfrac{4}{9}$

$\dfrac{1}{6}$

124. $x = \dfrac{4}{5}, y = -15, z = \dfrac{7}{8}$

$-10\dfrac{1}{2}$

125. $x = \dfrac{5}{6}, y = -3, z = 1\dfrac{7}{15}$

$-3\dfrac{2}{3}$

Evaluate the expression for the given values of the variables.

126. $10t$, when $t = -4.8$
-48

127. ab, when $a = 452$ and $b = -0.86$
-388.72

128. cd, when $c = -2.537$ and $d = -9.1$
23.0867

Evaluate the variable expression $x \div y$ for the given values of x and y.

129. $x = -\dfrac{5}{8}, y = -\dfrac{15}{2}$

$\dfrac{1}{12}$

130. $x = -\dfrac{14}{3}, y = -\dfrac{7}{9}$

6

131. $x = -18, y = \dfrac{3}{8}$

-48

132. $x = 20, y = -\dfrac{5}{6}$

-24

133. $x = -64.05, y = -6.1$
10.5

134. $x = -2.501, y = 0.41$
-6.1

135. $x = 1.003, y = -0.59$
-1.7

136. Is $-\dfrac{1}{6}$ a solution of the equation

$6x = 1$?

No

137. Is $-\dfrac{4}{5}$ a solution of the equation

$\dfrac{5}{4}n = -1$?

Yes

138. Is -8 a solution of the equation $1.6 = -0.2z$?
Yes

139. Is -1 a solution of the equation $-7.9c = -7.9$?
No

140. Is $-\dfrac{1}{3}$ a solution of the equation

$\dfrac{3}{4}y = -\dfrac{1}{4}$?

Yes

141. Is $\dfrac{2}{5}$ a solution of the equation

$-\dfrac{5}{6}z = \dfrac{1}{3}$?

No

142. Is -8.4 a solution of the equation

$21 = \dfrac{t}{0.4}$?

No

143. Is -0.9 a solution of the equation

$\dfrac{-2.7}{a} = \dfrac{a}{-0.3}$?

Yes

Objective C   **To solve application problems**

144.   **Temperature**   On January 23, 1916, the temperature in Browing, Montana, was 6.67°C. On January 24, 1916, the temperature in Browing was −48.9°C. Find the difference between the temperatures in Browing on these two days.
55.57°C

145.   **Temperature**   On January 22, 1943, in Spearfish, South Dakota, the temperature fell from 12.22°C at 9 A.M. to −20°C at 9:27 A.M. How many degrees did the temperature fall during the 27-minute period?
32.22°C

146.   **Chemistry**   The boiling point of nitrogen is −195.8°C, and the melting point is −209.86°C. Find the difference between the boiling point and the melting point of nitrogen.
14.06°C

147.   **Chemistry**   The boiling point of oxygen is −182.962°C. Oxygen's melting point is −218.4°C. What is the difference between the boiling point and the melting point of oxygen?
35.438°C

148.   **Stocks**   The chart at the right shows the closing price of a share of stock on November 2, 2004, for each of three companies. Also shown is the change in the price from the previous day. To find the closing price on the previous day, subtract the change in price from the closing price on November 2.

Company	Closing Price	Change in Price
J.C. Penney Co.	34.33	−0.45
Sears, Roebuck and Co.	34.80	−0.13
Target Corp.	50.50	+0.03

  a. Find the closing price on the previous day for Sears. $34.93
  b. Find the closing price on the previous day for Target. $50.47
  c. Find the closing price on the previous day for JC Penney. $34.78

APPLYING THE CONCEPTS

149. Determine whether the statement is true or false.
  a. Every integer is a rational number.   True
  b. Every whole number is an integer.   True
  c. Every integer is a positive number.   False
  d. Every rational number is an integer.   False

150. **Number Problems**
  a. Find a rational number between 0.1 and 0.2.
  b. Find a rational number between 1 and 1.1.
  c. Find a rational number between 0 and 0.005.
  Answers will vary. For example, **a.** 0.15, **b.** 1.05, **c.** 0.001.

Quick Quiz (Objective 3.4C)

1. The lowest temperature recorded in North America is −81.4°F. The highest temperature recorded is 134.0°F. (*Source:* National Climatic Data Center) Find the difference between these two extremes.   215.4°F

2. On November 3, 2004, the closing price of a share of Cott Corp. stock was $24.73. The change in the closing price from the previous day was −$.24. Find the closing price on the previous day.
$24.97

3.5 The Order of Operations Agreement

Objective A To use the Order of Operations Agreement to simplify expressions

The Order of Operations Agreement, introduced in Chapter 1, is repeated here for your reference.

The Order of Operations Agreement

Step 1 Do all operations inside parentheses.
Step 2 Simplify any numerical expressions containing exponents.
Step 3 Do multiplication and division as they occur from left to right.
Step 4 Do addition and subtraction as they occur from left to right.

TAKE NOTE
The −3 is squared only when the negative sign is *inside* the parentheses. In $(-3)^2$, we are squaring −3; in -3^2, we are finding the opposite of 3^2.

Note how the following expressions containing exponents are simplified.

$(-3)^2 = (-3)(-3) = 9$ The (-3) is squared. Multiply −3 by −3.

$-(3^2) = -(3 \cdot 3) = -9$ Read $-(3^2)$ as "the opposite of three squared." 3^2 is 9. The opposite of 9 is −9.

$-3^2 = -(3^2) = -9$ The expression -3^2 is the same as $-(3^2)$.

HOW TO Simplify: $8 - 4 \div (-2)$

$8 - 4 \div (-2) = 8 - (-2)$ • There are no operations inside parentheses (Step 1).
There are no exponents (Step 2).
Do the division (Step 3).

$= 8 + 2 = 10$ • Do the subtraction (Step 4).

Integrating Technology
As shown above and at the right, the value of -3^2 is different from the value of $(-3)^2$. The keystrokes to evaluate each of these on your calculator are different. To evaluate -3^2, enter

3 x^2 +/−

To evaluate $(-3)^2$, enter

3 +/− x^2

HOW TO Simplify: $(-3)^2 - 2(8 - 3) + (-5)$

$(-3)^2 - 2(8 - 3) + (-5)$
$= (-3)^2 - 2(5) + (-5)$ • Perform operations inside parentheses.
$= 9 - 2(5) + (-5)$ • Simplify expressions with exponents.
$= 9 - 10 + (-5)$ • Do multiplication and division as they occur from left to right.
$= 9 + (-10) + (-5)$ • Do addition and subtraction as they occur from left to right.
$= -1 + (-5)$
$= -6$

Objective 3.5A

Procedures to Review
Order of Operations Agreement
[1.4A]

Instructor Note
The exercises in this section ask students to recall the Order of Operations Agreement and to practice a combination of operations with rational numbers.

Concept Check
Put the following expressions in order from smallest value to largest value.

a. $(-3) \div \left(-\dfrac{3}{2}\right) \div \left(-\dfrac{1}{8}\right)$

b. $3 - (-4 + 2) \div 2$

c. $\left(-\dfrac{1}{2}\right)^3 - \left(\dfrac{3}{4} \div \dfrac{9}{16}\right) + \dfrac{1}{6}$

d. $8 - (-7) \times (6 - 10) + 3^2$

a (-16); d (-11); c $\left(-1\dfrac{7}{24}\right)$;

b (4)

Optional Student Activity

1. What is the smallest prime number greater than $32 \div (1 - 9) + (-3)^2$?   7
2. What is the smallest integer greater than $-2^2 - (-3)^2 + 5(4) \div 10 - (-6)$?   −4
3. What is the smallest prime number that divides evenly into the sum of 3^9 and 5^{11}?   2

In-Class Examples (Objective 3.5A)
Simplify.
1. $(-4)^2 - 8 + 9(3)$ 35
2. $(-6)^2 \times (6 - 4)^2 - (-12) \div 4$ 147
3. $\left(\dfrac{1}{4} - \dfrac{1}{2}\right)^2 \div \dfrac{3}{8}$ $\dfrac{1}{6}$
4. $0.7(3.1 - 6.6) + 5.5$ 3.05

Optional Student Activity

Using the Order of Operations Agreement, explain how to evaluate Exercise 39 in the exercises for Section 3.5.

Optional Student Activity

Is -3 a solution of the equation $x^3 + 3x^2 - 5x - 15 = 0$?

Yes

HOW TO Evaluate $ab - b^2$ when $a = 2$ and $b = -6$.

$ab - b^2$

$2(-6) - (-6)^2$ • Replace a with 2 and each b with -6.

$\quad = 2(-6) - 36$ • Use the Order of Operations Agreement to simplify the resulting numerical expression. Simplify the exponential expression.

$\quad = -12 - 36$ • Do the multiplication.

$\quad = -12 + (-36)$ • Do the subtraction.

$\quad = -48$

Example 1 Simplify $12 \div (-2)^2 - 5$

Solution

$12 \div (-2)^2 - 5$

$= 12 \div 4 - 5$ • Exponents

$= 3 - 5$ • Division

$= 3 + (-5)$ • Subtraction

$= -2$

You Try It 1 Simplify: $8 \div 4 \cdot 4 - (-2)^2$

Your solution 4

Example 2 Simplify: $\left(-\dfrac{2}{3}\right)^2 \div \dfrac{7 - 2}{13 - 4} - \dfrac{1}{3}$

Solution

$\left(-\dfrac{2}{3}\right)^2 \div \dfrac{7 - 2}{13 - 4} - \dfrac{1}{3}$

$= \left(-\dfrac{2}{3}\right)^2 \div \dfrac{5}{9} - \dfrac{1}{3}$ • Simplify above and below the fraction bar.

$= \dfrac{4}{9} \div \dfrac{5}{9} - \dfrac{1}{3}$ • Exponents

$= \dfrac{4}{9} \cdot \dfrac{9}{5} - \dfrac{1}{3}$ • Division

$= \dfrac{4}{5} - \dfrac{1}{3} = \dfrac{7}{15}$ • Subtraction

You Try It 2 Simplify: $\left(-\dfrac{1}{2}\right)^3 \cdot \dfrac{7 - 3}{4 - 9} + \dfrac{4}{5}$

Your solution $\dfrac{9}{10}$

Example 3 Evaluate $6a \div (-b)$ when $a = -2$ and $b = -3$.

Solution

$6a \div (-b)$

$6(-2) \div (-(-3))$

$= 6(-2) \div (3)$

$= -12 \div 3$

$= -4$

You Try It 3 Evaluate $3a - 4b$ when $a = -2$ and $b = 5$.

Your solution -26

Solutions on p. S11

Copyright © Houghton Mifflin Company. All rights reserved.

3.5 Exercises

Objective A   To use the Order of Operations Agreement to simplify expressions

Suggested Assignment
Exercises 1–55, odds
More challenging problems:
    Exercises 58–60

Simplify.

1. $3 - 12 \div 2$
−3

2. $-16 \div 2 + 8$
0

3. $2(3 - 5) - 2$
−6

4. $2 - (8 - 10) \div 2$
3

5. $4 - (-3)^2$
−5

6. $(-2)^2 - 6$
−2

7. $4 \cdot (2 - 4) - 4$
−12

8. $6 - 2 \cdot (1 - 3)$
10

9. $4 - (-2)^2 + (-3)$
−3

10. $-3 + (-6)^2 - 1$
32

11. $3^3 - 4(2)$
19

12. $9 \div 3 - (-3)^2$
−6

13. $3 \cdot (6 - 2) \div 6$
2

14. $4 \cdot (2 - 7) \div 5$
−4

15. $2^3 - (-3)^2 + 2$
1

16. $6(8 - 2) \div 4$
9

17. $6 - 2(1 - 5)$
14

18. $(-2)^2 - (-3)^2 + 1$
−4

19. $6 - (-4)(-3)^2$
42

20. $4 - (-5)(-2)^2$
24

21. $4 \cdot 2 - 3 \cdot 7$
−13

22. $16 \div 2 - 9 \div 3$
5

23. $(-2)^2 - 5(3) - 1$
−12

24. $4 - 2 \cdot 7 - 3^2$
−19

25. $(-1) \cdot (4 - 7)^2 \div 9 + 6 - 3 - 4(2)$
−6

26. $(-3)^2 \cdot (5 - 7)^2 - (-9) \div 3$
39

27. $(1.2)^2 - 4.1(0.3)$
0.21

28. $2.4(-3) - 2.5$
−9.7

29. $1.6 - (-1.6)^2$
−0.96

30. $4.1(8) \div (-4.1)$
−8

31. $(4.1 - 3.9) - 0.7^2$
−0.29

32. $1.8(-2.3) - 2$
−6.14

33. $-\dfrac{1}{2} + \dfrac{3}{8} \div \left(-\dfrac{3}{4}\right)$
−1

34. $\left(\dfrac{3}{4}\right)^2 - \dfrac{3}{8}$
$\dfrac{3}{16}$

35. $\left(\dfrac{1}{2}\right)^2 - \left(-\dfrac{1}{2}\right)^2$
0

Quick Quiz (Objective 3.5A)

Simplify.

1. $3 - (-5)^2 - 6 + 2(4)$ −20

2. $24 \div (5 - 7) \div (-2)^2$ −3

3. $\dfrac{11}{32} - \left(\dfrac{3}{8}\right)\left(-\dfrac{1}{6}\right)$ $\dfrac{13}{32}$

4. $-5.5(4.2 - 4.6) \div 1.1$ 2

36. $-\dfrac{2}{3}\left(\dfrac{5}{8}\right) \div \dfrac{2}{7}$

$-1\dfrac{11}{24}$

37. $\dfrac{1}{2} - \left(\dfrac{3}{4} - \dfrac{3}{8}\right) \div \dfrac{1}{3}$

$-\dfrac{5}{8}$

38. $\dfrac{3}{8} \div \left(-\dfrac{1}{2}\right)^2 + 2$

$3\dfrac{1}{2}$

Evaluate the variable expression given $a = -2$, $b = 4$, $c = -1$, and $d = 3$.

39. $3a + 2b$

2

40. $6b \div (-a)$

12

41. $bc \div (2a)$

1

42. $a^2 - b^2$

−12

43. $b^2 - c^2$

15

44. $2a - (c + a)^2$

−13

45. $(b - a)^2 + 4c$

32

46. $\dfrac{b + c}{d}$

1

47. $\dfrac{d - b}{c}$

1

48. $\dfrac{2d + b}{-a}$

5

49. $\dfrac{b - d}{c - a}$

1

50. $\dfrac{bd}{a} \div c$

6

51. $(d - a)^2 \div 5$

5

52. $(b + c)^2 + (a + d)^2$

10

53. $(d - a)^2 - 3c$

28

54. $(b + d)^2 - 4a$

57

Evaluate the expression for the given values of the variables.

55. $x - y^3z$, when $x = \dfrac{5}{6}$, $y = \dfrac{1}{2}$, and $z = \dfrac{8}{9}$

$\dfrac{13}{18}$

56. $xy^3 + z$, when $x = \dfrac{9}{10}$, $y = \dfrac{1}{3}$, and $z = \dfrac{7}{15}$

$\dfrac{1}{2}$

APPLYING THE CONCEPTS

57. What is the smallest integer greater than
$-2^2 - (-3)^2 + 5(4) \div 10 - (-6)$?

−4

58. Evaluate.
  a. $1^3 + 2^3 + 3^3 + 4^3$
  b. $(-1)^3 + (-2)^3 + (-3)^3 + (-4)^3$
  c. $1^3 + 2^3 + 3^3 + 4^3 + 5^3$
  d. Based on your answers to parts (a), (b), and (c), evaluate
  $(-1)^3 + (-2)^3 + (-3)^3 + (-4)^3 + (-5)^3$.
  a. 100   **b.** −100   **c.** 225   **d.** −225

59. Is −4 a solution of the equation $x^2 - 2x - 8 = 0$?
No

60. Evaluate $a \div bc$ and $a \div (bc)$ when $a = 16$, $b = 2$, and $c = -4$.
Explain why the answers are not the same.

Answers to Writing Exercises

60. $a \div bc$ when $a = 16$, $b = 2$, and $c = -4$ is:

$16 \div (2)(-4) = 8(-4) = -32$

$a \div (bc)$ when $a = 16$, $b = 2$, and $c = -4$ is:

$16 \div [2(-4)]$
$= 16 \div [-8]$
$= -2$

By Step 3 of the Order of Operations Agreement, multiplication and division are performed as they occur from left to right. Therefore, in the first solution above, the division $16 \div 2$ must be performed first, then the multiplication. By Step 1 of the Order of Operations Agreement, operations in parentheses must be performed first. Therefore, in the second solution above, the operation of multiplication inside the parentheses (bc) must be performed first, then the division.

Focus on Problem Solving

Drawing Diagrams   How do you best remember something? Do you remember best what you hear? The word *aural* means "pertaining to the ear"; people with a strong aural memory remember best those things that they hear. The word *visual* means "pertaining to the sense of sight"; people with a strong visual memory remember best that which they see written down. Some people claim that their memory is in their writing hand—they remember something only if they write it down! The method by which you best remember something is probably also the method by which you can best learn something new.

In problem-solving situations, try to capitalize on your strengths. If you tend to understand the material better when you hear it spoken, read application problems aloud or have someone else read them to you. If writing helps you to organize ideas, rewrite application problems in your own words.

No matter what your main strength, visualizing a problem can be a valuable aid in problem solving. A drawing, sketch, diagram, or chart can be a useful tool in problem solving, just as calculators and computers are tools. A diagram can be helpful in gaining an understanding of the relationships inherent in a problem-solving situation. A sketch will help you to organize the given information and can lead to your being able to focus on the method by which the solution can be determined.

> **HOW TO**   A tour bus drives 5 mi south, then 4 mi west, then 3 mi north, then 4 mi east. How far is the tour bus from the starting point?
>
> Draw a diagram of the given information.
>
> From the diagram, we can see that the solution can be determined by subtracting 3 from 5: $5 - 3 = 2$.
>
> The bus is 2 mi from the starting point.

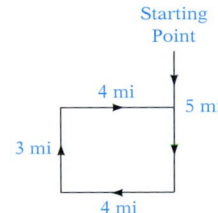

> **HOW TO**   If you roll two ordinary six-sided dice and multiply the two numbers that appear on top, how many different possible products are there?
>
> Make a chart of the possible products. In the chart below, repeated products are marked with an asterisk.

$1 \cdot 1 = 1$	$2 \cdot 1 = 2$ (*)	$3 \cdot 1 = 3$ (*)	$4 \cdot 1 = 4$ (*)	$5 \cdot 1 = 5$ (*)	$6 \cdot 1 = 6$ (*)
$1 \cdot 2 = 2$	$2 \cdot 2 = 4$ (*)	$3 \cdot 2 = 6$ (*)	$4 \cdot 2 = 8$ (*)	$5 \cdot 2 = 10$ (*)	$6 \cdot 2 = 12$ (*)
$1 \cdot 3 = 3$	$2 \cdot 3 = 6$ (*)	$3 \cdot 3 = 9$	$4 \cdot 3 = 12$ (*)	$5 \cdot 3 = 15$ (*)	$6 \cdot 3 = 18$ (*)
$1 \cdot 4 = 4$	$2 \cdot 4 = 8$	$3 \cdot 4 = 12$ (*)	$4 \cdot 4 = 16$	$5 \cdot 4 = 20$ (*)	$6 \cdot 4 = 24$ (*)
$1 \cdot 5 = 5$	$2 \cdot 5 = 10$	$3 \cdot 5 = 15$	$4 \cdot 5 = 20$	$5 \cdot 5 = 25$	$6 \cdot 5 = 30$ (*)
$1 \cdot 6 = 6$	$2 \cdot 6 = 12$	$3 \cdot 6 = 18$	$4 \cdot 6 = 24$	$5 \cdot 6 = 30$	$6 \cdot 6 = 36$

> By counting the products that are not repeats, we can see that there are 18 different possible products.

Look at Sections 1 and 2 of this chapter. You will notice that number lines are used to help you visualize the integers, as an aid in ordering integers, to help you understand the concepts of opposite and absolute value, and to illustrate addition of integers. As you begin your work with integers, you may find that sketching a number line proves helpful in coming to understand a problem or in working through a calculation that involves integers.

Projects and Group Activities

Answers to Projects and Group Activities: Multiplication of Integers

1. See the numbers in red on the grid.
2. +, −, +, −
3. Answers will vary.
4. Descriptions will vary.
5. A product divided by one of its factors equals the other factor. Examples will vary.

Multiplication of Integers

The grid at the left has four regions, or quadrants, numbered counterclockwise, starting at the upper right, with the Roman numerals I, II, III, IV.

Quadrant II   Quadrant I

−25	−20	−15	−10	−5	5	5	10	15	20	25
−20	−16	−12	−8	−4	4	4	8	12	16	20
−15	−12	−9	−6	−3	3	3	6	9	12	15
−10	−8	−6	−4	−2	2	2	4	6	8	10
−5	−4	−3	−2	−1	1	1	2	3	4	5
−5	−4	−3	−2	−1	0	1	2	3	4	5
5	4	3	2	1	−1	−1	−2	−3	−4	−5
10	8	6	4	2	−2	−2	−4	−6	−8	−10
15	12	9	6	3	−3	−3	−6	−9	−12	−15
20	16	12	8	4	−4	−4	−8	−12	−16	−20
25	20	15	10	5	−5	−5	−10	−15	−20	−25

Quadrant III   Quadrant IV

1. Complete Quadrant I by multiplying each of the horizontal numbers, 1 through 5, by each of the vertical numbers, 1 through 5. The product 4(3) has been filled in for you. Complete Quadrants II, III, and IV by again multiplying each horizontal number by each vertical number.
2. What is the sign of all the products in Quadrant I? Quadrant II? Quadrant III? Quadrant IV?
3. Describe at least three patterns that you observe in the completed table.
4. How does the table show that multiplication of integers is commutative?
5. How can you use the table to find the quotient of two integers? Provide at least two examples of division of integers.

Answers to Projects and Group Activities: Closure

Whole Numbers:
   Addition—Yes
   Subtraction—No
   Multiplication—Yes
   Division—No
Integers:
   Addition—Yes
   Subtraction—Yes
   Multiplication—Yes
   Division—No
Rational Numbers:
   Addition—Yes
   Subtraction—Yes
   Multiplication—Yes
   Division—Yes

Closure

Study Tip

Three important features of this text that can be used to prepare for a test are the:

- Chapter Summary
- Chapter Review Exercises
- Chapter Test

See *AIM for Success*, page xxxiv.

The whole numbers are said to be *closed* with respect to addition because when two whole numbers are added, the result is a whole number. The whole numbers are not closed with respect to subtraction because, for example, 4 and 7 are whole numbers but 4 − 7 = −3, and −3 is not a whole number. Complete the table below by entering a Y if the operation is closed for those numbers and an N if it is not closed. When we discuss whether multiplication and division are closed, zero is not included because division by zero is not defined.

	Add	Subtract	Multiply	Divide
Whole numbers	Y	N		
Integers				
Rational numbers				

Chapter 3 Summary

Key Words	**Examples**
A number n is a *positive number* if $n > 0$. A number n is a *negative number* if $n < 0$. [3.1A, p. 177]	Positive numbers are numbers greater than zero. 9, 87, and 603 are positive numbers. Negative numbers are numbers less than zero. -5, -41, and -729 are negative numbers.
The *integers* are . . . , -4, -3, -2, -1, 0, 1, 2, 3, 4, . . . The integers can be defined as the whole numbers and their opposites. *Positive integers* are to the right of zero on the number line. *Negative integers* are to the left of zero on the number line. [3.1A, p. 177]	-729, -41, -5, 9, 87, and 603 are integers. 0 is an integer, but it is neither a positive nor a negative integer.
Opposite numbers are two numbers that are the same distance from zero on the number line but are on opposite sides of zero. The opposite of a number is called its *additive inverse*. [3.1B, p. 179; 3.2A, p. 189]	8 is the opposite, or additive inverse, of -8. -2 is the opposite, or additive inverse, of 2.
The *absolute value* of a number is the distance from zero to the number on the number line. The absolute value of a number is a positive number or zero. The symbol for absolute value is "$|\,|$". [3.1C, p. 180]	$|9| = 9$ $|-9| = 9$ $-|9| = -9$
A *rational number* is a number that can be written in the form $\frac{a}{b}$, where a and b are integers and $b \neq 0$. [3.4A, p. 215]	$\frac{3}{7}$, $-\frac{5}{8}$, 9, -2, $4\frac{1}{2}$, 0.6, and $0.\overline{3}$ are rational numbers.

Essential Rules and Procedures	**Examples**
To add integers with the same sign, add the absolute values of the numbers. Then attach the sign of the addends. [3.2A, p. 188]	$6 + 4 = 10$ $-6 + (-4) = -10$
To add integers with different signs, find the absolute values of the numbers. Subtract the lesser absolute value from the greater absolute value. Then attach the sign of the addend with the greater absolute value. [3.2A, p. 188]	$-6 + 4 = -2$ $6 + (-4) = 2$
To subtract two integers, add the opposite of the second integer to the first integer. [3.2B, p. 192]	$6 - 4 = 6 + (-4) = 2$ $6 - (-4) = 6 + 4 = 10$ $-6 - 4 = -6 + (-4) = -10$ $-6 - (-4) = -6 + 4 = -2$

To multiply integers with the same sign, multiply the absolute values of the factors. The product is positive. [3.3A, p. 203]	$3 \cdot 5 = 15$ $-3(-5) = 15$
To multiply integers with different signs, multiply the absolute values of the factors. The product is negative. [3.3A, p. 203]	$-3(5) = -15$ $3(-5) = -15$
To divide two numbers with the same sign, divide the absolute values of the numbers. The quotient is positive. [3.3B, p. 206]	$15 \div 3 = 5$ $(-15) \div (-3) = 5$
To divide two numbers with different signs, divide the absolute values of the numbers. The quotient is negative. [3.3B, p. 206]	$-15 \div 3 = -5$ $15 \div (-3) = -5$

Order Relations	$a > b$ if a is to the right of b on the number line. $a < b$ if a is to the left of b on the number line. [3.1A, p. 178]	$-6 > -12$ $-8 < 4$

Properties of Addition [3.2A, p. 189]

Addition Property of Zero $a + 0 = a$ or $0 + a = a$	$-6 + 0 = -6$
Commutative Property of Addition $a + b = b + a$	$-8 + 4 = 4 + (-8)$
Associative Property of Addition $(a + b) + c = a + (b + c)$	$(-5 + 4) + 6 = -5 + (4 + 6)$
Inverse Property of Addition $a + (-a) = 0$ or $-a + a = 0$	$7 + (-7) = 0$

Properties of Multiplication [3.3A, p. 204]

Multiplication Property of Zero $a \cdot 0 = 0$ or $0 \cdot a = 0$	$-9(0) = 0$
Multiplication Property of One $a \cdot 1 = a$ or $1 \cdot a = a$	$-3(1) = -3$
Commutative Property of Multiplication $a \cdot b = b \cdot a$	$-2(6) = 6(-2)$
Associative Property of Multiplication $(a \cdot b) \cdot c = a \cdot (b \cdot c)$	$(-2 \cdot 4) \cdot 5 = -2 \cdot (4 \cdot 5)$

Division Properties of Zero and One [3.3B, p. 207]

If $a \neq 0$, $0 \div a = 0$.	$0 \div (-5) = 0$
If $a \neq 0$, $a \div a = 1$.	$-5 \div (-5) = 1$
$a \div 1 = a$	$-5 \div 1 = -5$
$a \div 0$ is undefined.	$-5 \div 0$ is undefined.

The Order of Operations Agreement [3.5A, p. 231]

Step 1 Do all operations inside parentheses.	$(-4)^2 - 3(1 - 5) = (-4)^2 - 3(-4)$
Step 2 Simplify any numerical expressions containing exponents.	$= 16 - 3(-4)$
	$= 16 - (-12)$
Step 3 Do multiplication and division as they occur from left to right.	$= 16 + 12$
	$= 28$
Step 4 Do addition and subtraction as they occur from left to right.	

Chapter 3 Review Exercises

1. Write the expression $8 - (-1)$ in words.
Eight minus negative one   [3.1B]

2. Evaluate $-|-36|$.
-36 [3.1C]

3. Find the product of -40 and -5.
200   [3.3A]

4. Evaluate $-a \div b$ when $a = -27$ and $b = -3$.
-9 [3.3B]

5. Add: $-28 + 14$
-14 [3.2A]

6. Simplify: $-(-13)$
13   [3.1B]

7. Graph -2 on the number line.

[3.1A]

8. What is the sum of -65.47 and -32.91?
-98.38 [3.4A]

9. Divide: $-51 \div (-3)$
17   [3.3B]

10. Find the quotient of 840 and -4.
-210 [3.3B]

11. Subtract: $-6 - (-7) - 15 - (-12)$
-2 [3.2B]

12. Evaluate $-ab$ when $a = -2$ and $b = -9$.
-18 [3.3A]

13. Find the sum of 18, -13, and -6.
-1 [3.2A]

14. Multiply: $-18(4)$
-72 [3.3A]

15. Simplify: $(-2)^2 - (-3)^2 \div (1 - 4)^2 \cdot 2 - 6$
-4 [3.5A]

16. Evaluate $-x - y$ when $x = -1$ and $y = 3$.
-2 [3.2B]

17. Find the difference between -15 and -28.
13   [3.2B]

18. What is the quotient of $-\frac{1}{5}$ and $-\frac{7}{10}$?
$\frac{2}{7}$ [3.4B]

19. Is -9 a solution of $-6 - t = 3$?
Yes   [3.2B]

20. Simplify: $-9 + 16 - (-7)$
14   [3.2B]

21. Divide: $\dfrac{0}{-17}$
0   [3.3B]

22. Multiply: $-5(2)(-6)(-1)$
-60 [3.3A]

23. Add: $3 + (-9) + 4 + (-10)$
-12 [3.2A]

24. Evaluate $(a - b)^2 - 2a$ when $a = -2$ and $b = -3$.
5   [3.5A]

25. Place the correct symbol, $<$ or $>$, between the two numbers.
$-8 > -10$ [3.1A]

26. Simplify: $3 \div \left(\dfrac{1}{2} - \dfrac{1}{4}\right) - 3$
9   [3.5A]

27. Find the absolute value of -27.
27   [3.1C]

28. Multiply: $-0.8(3.5)$
-2.8 [3.4B]

29. What is $\dfrac{7}{12}$ added to $-\dfrac{11}{16}$?
$-\dfrac{5}{48}$ [3.4A]

30. **Temperature**   Which is colder, a temperature of $-4°C$ or $-12°C$?
$-12°C$ [3.1D]

31. 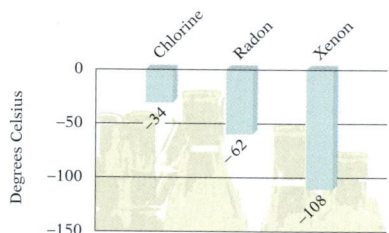 **Chemistry**   The figure at the right shows the boiling point in degrees Celsius of three chemical elements. The boiling point of neon is seven times the highest boiling point shown in the table. What is the boiling point of neon?
$-238°C$ [3.3C]

Boiling Points of Chemical Elements

32. **Temperature**   Find the temperature after an increase of $5.5°C$ from $-8.5°C$.
$-3°C$ [3.4C]

33. **Mathematics**   The distance, d, between point a and point b on the number line is given by the formula $d = |a - b|$. Find d when $a = 7$ and $b = -5$.
12   [3.2C]

Chapter 3 Test

1. Write the expression $-3 + (-5)$ in words.
   Negative three plus negative five   [3.1B]

2. Evaluate $-|-34|$.
   -34 [3.1C]

3. What is 3 minus -15?
   18   [3.2B]

4. Evaluate $a + b$ when $a = -11$ and $b = -9$.
   -20 [3.2A]

5. Evaluate $(-x)(-y)$ when $x = -4$ and $y = -6$.
   24   [3.3A]

6. What is $-\frac{5}{6}$ added to $\frac{4}{9}$?
   $-\frac{7}{18}$ [3.4A]

7. What is -360 divided by -30?
   12   [3.3B]

8. Find the sum of -3, -6, and 11.
   2   [3.2A]

9. Place the correct symbol between the two numbers.
   $16 > -19$ [3.1A]

10. Subtract: $7 - (-3) - 12$
    -2 [3.2B]

11. Evaluate $a - b - c$ when $a = 6$, $b = -2$, and $c = 11$.
    -3 [3.2B]

12. Simplify: $-(-49)$
    49   [3.1B]

13. Find the product of 50 and -5.
    -250 [3.3A]

14. Write the given numbers in order from smallest to largest.
    $-|5|, -(-11), |-9|, -(3)$
    $-|5|, -(3), |-9|, -(-11)$ [3.1C]

15. Is -9 a solution of the equation $17 - x = 8$?
    No   [3.2B]

16. On the number line, which number is 2 units to the right of -5?
    -3 [3.1A]

17. Divide: $\dfrac{0}{-16}$
    0   [3.3B]

18. Evaluate $2bc - (c + a)^3$ when $a = -2$, $b = 4$, and $c = -1$.
    19   [3.5A]

19. Find the opposite of 25.
−25   [3.1B]

20. What is 4.793 less than −6.82?
−11.613   [3.4A]

21. Subtract: 0 − 11
−11   [3.2B]

22. Divide: −96 ÷ (−4)
24   [3.3B]

23. Simplify: 16 ÷ 4 − 12 ÷ (−2)
10   [3.5A]

24. Evaluate $\dfrac{-x}{y}$ when $x = -56$ and $y = -8$.
−7   [3.3B]

25. Evaluate $3xy$ when $x = -2$ and $y = -10$.
60   [3.3A]

26. Simplify: $7 \div \left(\dfrac{1}{7} - \dfrac{3}{14} \right) - 9$
−107   [3.5A]

27. Divide: $-18 \div \dfrac{2}{3}$
−27   [3.4B]

28. Evaluate xy when $x = -0.3$ and $y = 5.1$.
−1.53   [3.4B]

29. What is 14 less than 4?
−10   [3.2B]

30. **Temperature**   Find the temperature after an increase of 11°C from −6.5°C.
4.5°C   [3.4C]

31. **Environmental Science**   The wind-chill factor when the temperature is −25°F and the wind is blowing at 40 mph is four times the wind-chill factor when the temperature is −5°F and the wind is blowing at 5 mph. If the wind-chill factor at −5°F with a 5-mph wind is −16°, what is the wind-chill factor at −25°F with a 40-mph wind?
−64°F   [3.3C]

32. **Temperature**   The high temperature today is 8° lower than the high temperature yesterday. The high temperature today is −13°C. What was the high temperature yesterday?
−5°C   [3.2C]

33. **Mathematics**   The distance, d, between point a and point b on the number line is given by the formula $d = |a - b|$. Find d when $a = 4$ and $b = -12$.
16 units   [3.2C]

Cumulative Review Exercises

1. Find the difference between -27 and -32.
   5   [3.2B]

2. Estimate the product of 439 and 28.
   12,000   [1.3A]

3. Divide: $16.15 \div 0.5$
   32.3   [2.6C]

4. Simplify: $16 \div (3 + 5) \cdot 9 - 2^4$
   2   [1.5A]

5. Evaluate $-|-82|$.
   -82 [3.1C]

6. Write three hundred nine thousand four hundred eighty in standard form.
   309,480   [1.1B]

7. Evaluate $5xy$ when $x = 80$ and $y = 6$.
   2400   [1.3A]

8. What is -294 divided by -14?
   21   [3.3B]

9. Subtract: $-28 - (-17)$
   -11 [3.2B]

10. Find the sum of -24, 16, and -32.
    -40 [3.2A]

11. Find all the factors of 44.
    1, 2, 4, 11, 22, 44   [1.3D]

12. Evaluate $x^4 y^2$ when $x = \frac{1}{2}$ and $y = 4$.
    1   [2.4A]

13. Round 629,874 to the nearest thousand.
    630,000   [1.1C]

14. Estimate the sum of 356, 481, 294, and 117.
    1300   [1.2A]

15. Evaluate $-a - b$ when $a = -4$ and $b = -5$.
    9   [3.2B]

16. Find the product of -100 and 25.
    -2500 [3.3A]

17. Find the sum of 3.97 and 4.8.
    8.77   [2.6A]

18. Add: $2\frac{1}{6} + 3\frac{1}{2}$
    $5\frac{2}{3}$ [2.3A]

19. Simplify: $(1 - 5)^2 \div (-6 + 4) + 8(-3)$
    -32 [3.5A]

20. Evaluate $-c \div d$ when $c = -32$ and $d = -8$.
    -4 [3.3B]

21. Find the quotient of $\frac{9}{10}$ and $\frac{3}{4}$.
    $1\frac{1}{5}$ [2.4B]

22. Place the correct symbol, $<$ or $>$, between the two numbers.
    $-62 < 26$ [3.1A]

23. What is -18 multiplied by -7?
126   [3.3A]

24. Divide: $(-3.312) \div (-0.8)$
4.14   [3.4B]

25. Write $2 \cdot 2 \cdot 2 \cdot 2 \cdot 2 \cdot 7 \cdot 7$ in exponential notation.
$2^5 \cdot 7^2$ [1.3B]

26. Evaluate $4a + (a - b)^3$ when $a = 5$ and $b = 2$.
47   [1.4A]

27. Add: $5971 + 482 + 3609$
10,062   [1.2A]

28. What is 5 less than -21?
-26 [3.2B]

29. Estimate the difference between 7352 and 1986.
5000   [1.2B]

30. Evaluate $3^4 \cdot 5^2$.
2025   [1.3B]

31. 🥧 **History** The land area of the United States prior to the Louisiana Purchase was 891,364 mi². The land area of the Louisiana Purchase, which was purchased from France in 1803, was 831,321 mi². What was the land area of the United States immediately after the Louisiana Purchase?
1,722,685 mi²   [1.2C]

32. 🥧 **History** Albert Einstein was born on March 14, 1879. He died on April 18, 1955. How old was Albert Einstein when he died?
76 years old   [1.2C]

33. **Finances** A customer makes a down payment of $3550 on a car costing $17,750. Find the amount that remains to be paid.
$14,200   [1.2C]

34. **Real Estate** A construction company is considering purchasing a 25-acre tract of land on which to build single-family homes. If the price is $3690 per acre, what is the total cost of the land?
$92,250   [1.3E]

Albert Einstein

35. **Temperature** Find the temperature after an increase of 7°C from -12°C.
-5°C   [3.2C]

36. 🥧 **Temperature** Record temperatures, in degrees Fahrenheit, for four states in the United States are shown at the right.
a. What is the difference between the record high and record low temperatures in Arizona?
b. For which state is the difference between the record high and record low temperatures greatest?
a. 168°F   **b.** Alaska   [3.2C]

Record Temperatures (in degrees Fahrenheit)		
State	Lowest	Highest
Alabama	-27	112
Alaska	-80	100
Arizona	-40	128
Arkansas	-29	120

Source: The World Almanac and Book of Facts 2003

37. **Sales** As a sales representative, your goal is to sell $120,000 in merchandise during the year. You sold $28,550 in merchandise during the first quarter of the year, $34,850 during the second quarter, and $31,700 during the third quarter. What must your sales for the fourth quarter be if you are to meet your goal for the year?
$24,900   [1.2C]

38. **Golf** Use the equation $S = N - P$, where S is a golfer's score relative to par in a tournament, N is the number of strokes made by the golfer, and P is par, to find a golfer's score relative to par when the golfer made 198 strokes and par is 206.
-8 [3.2C]

4 Variable Expressions

Have you ever purchased a product online? How satisfied were you with the experience? A recent survey conducted by Consumer Internet Barometer found that approximately one-fourth of people who had purchased a product online were extremely satisfied with their experience. See **Exercise 64 on page 268.** In this exercise, you are asked to express the number of people who were extremely satisfied with their online purchase in terms of the number of people who purchased a product online. To do this, you must use a variable expression. Variable expressions are the topic of this chapter.

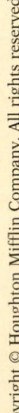

Need help? For online student resources, such as section quizzes, visit this textbook's website at **math.college.hmco.com/students.**

OBJECTIVES

Section 4.1

A To evaluate a variable expression

Section 4.2

A To simplify a variable expression using the Properties of Addition

B To simplify a variable expression using the Properties of Multiplication

C To simplify a variable expression using the Distributive Property

D To simplify general variable expressions

Section 4.3

A To translate a verbal expression into a variable expression, given the variable

B To translate a verbal expression into a variable expression and then simplify

C To translate application problems

PREP TEST • • •

Do these exercises to prepare for Chapter 4.

1. Subtract: $-12 - (-15)$
3 [3.2B]

2. Divide: $-36 \div (-9)$
4 [3.3B]

3. Add: $-\dfrac{3}{4} + \dfrac{5}{6}$
$\dfrac{1}{12}$ [3.4A]

4. What is the reciprocal of $-\dfrac{9}{4}$?
$-\dfrac{4}{9}$ [2.4A]

5. Divide: $-\dfrac{3}{4} \div \left(-\dfrac{5}{2}\right)$
$\dfrac{3}{10}$ [3.4B]

6. Evaluate: -2^4
-16 [3.5A]

7. Evaluate: $\left(\dfrac{2}{3}\right)^3$
$\dfrac{8}{27}$ [2.4A]

8. Evaluate: $3 \cdot 4^2$
48 [1.4A]

9. Evaluate: $7 - 2 \cdot 3$
1 [1.4A]

10. Evaluate: $5 - 7(3 - 2^2)$
12 [1.4A]

GO FIGURE • • •

Two fractions are inserted between $\dfrac{1}{4}$ and $\dfrac{1}{2}$ so that the difference between any two successive fractions is the same. Find the sum of the four fractions.
$\dfrac{3}{2}$

4.1 Evaluating Variable Expressions

Objective A   To evaluate a variable expression

Point of Interest

Historical manuscripts indicate that mathematics is at least 4000 years old. Yet it was only 400 years ago that mathematicians started using variables to stand for numbers. The idea that a letter can stand for some number was a critical turning point in mathematics. Today, x is used by most nations as the standard letter for a single unknown. In fact, x-rays were so named because the scientists who discovered them did not know what they were and thus labeled them the "unknown rays" or x-rays.

Often we discuss a quantity without knowing its exact value—for example, the price of gold next month, the cost of a new automobile next year, or the tuition cost for next semester. Recall that a letter of the alphabet, called a **variable,** is used to stand for a quantity that is unknown or that can change, or *vary*. An expression that contains one or more variables is called a **variable expression.**

A variable expression is shown at the right. The expression can be rewritten by writing subtraction as the addition of the opposite.

$$3x^2 - 5y + 2xy - x - 7$$

$$3x^2 + (-5y) + 2xy + (-x) + (-7)$$

Note that the expression has five addends. The **terms** of a variable expression are the addends of the expression. The expression has five terms.

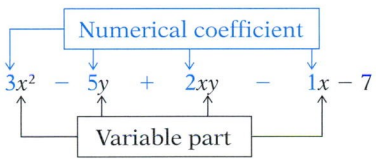

The terms $3x^2$, $-5y$, $2xy$, and $-x$ are **variable terms.**

The term -7 is a **constant term,** or simply a **constant.**

Each variable term is composed of a **numerical coefficient** and a **variable part** (the variable or variables and their exponents).

When the numerical coefficient is 1 or -1, the 1 is usually not written ($x = 1x$ and $-x = -1x$).

Variable expressions occur naturally in science. In a physics lab, a student may discover that a weight of 1 pound will stretch a spring $\frac{1}{2}$ inch. Two pounds will stretch the spring 1 inch. By experimenting, the student can discover that the distance the spring will stretch is found by multiplying the weight by $\frac{1}{2}$. By letting W represent the weight attached to the spring, the student can represent the distance the spring stretches by the variable expression $\frac{1}{2}W$.

With a weight of W pounds, the spring will stretch $\frac{1}{2} \cdot W = \frac{1}{2}W$ inches.

With a weight of 10 pounds, the spring will stretch $\frac{1}{2} \cdot 10 = 5$ inches. The number that is substituted for W, in this case 10, is called the **value of the variable** W.

With a weight of 3 pounds, the spring will stretch $\frac{1}{2} \cdot 3 = 1\frac{1}{2}$ inches.

Vocabulary to Review

variable   [1.2A]
variable expression   [1.2A]
evaluating a variable expression [1.2A]

New Vocabulary

terms
variable terms
constant term
constant
numerical coefficient
variable part
value of a variable

Concept Check

Consider the expression $4x^3 - x^2 + 3x - 9$.

1. Determine the number of terms in the expression.
   Four

2. List the numerical coefficients of the variable terms of the expression.
   $4, -1, 3$

3. Identify the constant term of the variable expression.
   -9

Optional Student Activity

Create a variable expression that has four terms, one of which is a constant. Then list the numerical coefficients of the variable terms.   Answers will vary.

Instructor Note

The students have evaluated variable expressions in Chapters 1, 2, and 3 of this text. However, this section is included here to reinforce the concepts already learned, as well as to introduce additional vocabulary associated with variable expressions. It also serves to prepare students for the next section on simplifying variable expressions.

In-Class Examples (Objective 4.1A)

1. Name the variable terms of the expression
   $3b^3 - 4b - 2$.
   $3b^3, -4b$

2. Evaluate $3a^2 - 4ab$ when $a = 5$ and $b = -4$.
   155

3. Evaluate $\dfrac{x^3 + y^3}{x + y}$ when $x = 2$ and $y = -3$.
   19

4. Evaluate $a^2 - 5(a - 2b) - c^2$ when $a = -3$, $b = 2$, and $c = -1$.   43

Concept Check

Is the statement $|x| = x$ sometimes true, always true, or never true?

Sometimes true; true when $x \geq 0$

Optional Student Activity

Evaluate $-x^2$ for the following values of x.

1. -3 -9
2. -2 -4
3. 0 0
4. 2 -4
5. 3 -9

Instructor Note

Emphasize that the result of evaluating a variable expression is one number.

Optional Student Activity

Evaluate y^y for the following values of y.

1. 1 1
2. 2 4
3. 3 27

Integrating Technology

See the appendix Keystroke Guide: *Evaluating Variable Expressions* for instructions on using a graphing calculator to evaluate variable expressions.

Recall that replacing each variable by its value and then simplifying the resulting numerical expression is called **evaluating a variable expression.**

HOW TO   Evaluate $ab - b^2$ when $a = 2$ and $b = -3$.

Replace each variable in the expression by its value. Then use the Order of Operations Agreement to simplify the resulting numerical expression.

$ab - b^2$

$2(-3) - (-3)^2 = -6 - 9$
$= -15$

When $a = 2$ and $b = -3$, the value of $ab - b^2$ is -15.

Example 1   Name the variable terms of the expression $2a^2 - 5a + 7$.

Solution $2a^2$ and $-5a$

You Try It 1   Name the constant term of the expression $6n^2 + 3n - 4$.

Your solution -4

Example 2   Evaluate $x^2 - 3xy$ when $x = 3$ and $y = -4$.

Solution
$x^2 - 3xy$
$3^2 - 3(3)(-4) = 9 - 3(3)(-4)$
$= 9 - 9(-4)$
$= 9 - (-36)$
$= 9 + 36 = 45$

You Try It 2   Evaluate $2xy + y^2$ when $x = -4$ and $y = 2$.

Your solution -12

Example 3   Evaluate $\dfrac{a^2 - b^2}{a - b}$ when $a = 3$ and $b = -4$.

Solution
$\dfrac{a^2 - b^2}{a - b}$

$\dfrac{3^2 - (-4)^2}{3 - (-4)} = \dfrac{9 - 16}{3 - (-4)}$

$= \dfrac{-7}{7} = -1$

You Try It 3   Evaluate $\dfrac{a^2 + b^2}{a + b}$ when $a = 5$ and $b = -3$.

Your solution 17

Example 4   Evaluate $x^2 - 3(x - y) - z^2$ when $x = 2$, $y = -1$, and $z = 3$.

Solution
$x^2 - 3(x - y) - z^2$
$2^2 - 3[2 - (-1)] - 3^2$
$= 2^2 - 3(3) - 3^2$
$= 4 - 3(3) - 9$
$= 4 - 9 - 9$
$= -5 - 9$
$= -14$

You Try It 4   Evaluate $x^3 - 2(x + y) + z^2$ when $x = 2$, $y = -4$, and $z = -3$.

Your solution 21

Solutions on p. S12

4.1 Exercises

Objective A   To evaluate a variable expression

For Exercises 1 to 3, name the terms of the variable expression. Then underline the constant term.

1. $2x^2 + 5x - 8$
   $2x^2, 5x, \underline{-8}$

2. $-3n^2 - 4n + 7$
   $-3n^2, -4n, \underline{7}$

3. $6 - a^4$
   $-a^4, \underline{6}$

For Exercises 4 to 6, name the variable terms of the expression. Then underline the variable part of each term.

4. $9b^2 - 4ab + a^2$
   $9\underline{b^2}, -4\underline{ab}, \underline{a^2}$

5. $7x^2y + 6xy^2 + 10$
   $7\underline{x^2y}, 6\underline{xy^2}$

6. $5 - 8n - 3n^2$
   $-8\underline{n}, -3\underline{n^2}$

For Exercises 7 to 9, name the coefficients of the variable terms.

7. $x^2 - 9x + 2$
   $1, -9$

8. $12a^2 - 8ab - b^2$
   $12, -8, -1$

9. $n^3 - 4n^2 - n + 9$
   $1, -4, -1$

10. ✏ What is the numerical coefficient of a variable term?

11. ✏ Explain the meaning of the phrase "evaluate a variable expression."

For Exercises 12 to 32, evaluate the variable expression when $a = 2$, $b = 3$, and $c = -4$.

12. $3a + 2b$
    12

13. $a - 2c$
    10

14. $-a^2$
    -4

15. $2c^2$
    32

16. $-3a + 4b$
    6

17. $3b - 3c$
    21

18. $b^2 - 3$
    6

19. $-3c + 4$
    16

20. $16 \div (2c)$
    -2

21. $6b \div (-a)$
    -9

22. $bc \div (2a)$
    -3

23. $b^2 - 4ac$
    41

24. $a^2 - b^2$
    -5

25. $b^2 - c^2$
    -7

26. $(a + b)^2$
    25

27. $a^2 + b^2$
    13

28. $2a - (c + a)^2$
    0

29. $(b - a)^2 + 4c$
    -15

30. $b^2 - \dfrac{ac}{8}$
    10

31. $\dfrac{5ab}{6} - 3cb$
    41

32. $(b - 2a)^2 + bc$
    -11

Section 4.1

Suggested Assignment

Exercises 1–49, odds
More challenging problems:
  Exercises 55, 57, 58

Answers to Writing Exercises

10. The numerical coefficient is the number factor of a variable term.

11. The phrase "evaluate a variable expression" means to replace each variable in the expression with its given value and then to simplify the resulting numerical expression.

Quick Quiz (Objective 4.1A)

1. Name the terms of the variable expression $3x^2 - 4x - 5$. Then underline the constant term.
   $3x^2, -4x, \underline{-5}$

2. Name the variable terms of the expression $4x^2y + 5xy^2 + 9$. Then underline the variable part of each term.
   $4\underline{x^2y}, 5\underline{xy^2}$

3. Name the coefficients of the variable terms of $5x^3 - x^2 + 3x - 4$.
   $5, -1, 3$

4. Evaluate $2x - (y + z)^2$ when $x = 5$, $y = -3$, and $z = 6$.
   1

For Exercises 33 to 50, evaluate the variable expression when $a = -2$, $b = 4$, $c = -1$, and $d = 3$.

33. $\dfrac{b + c}{d}$

1

34. $\dfrac{d - b}{c}$

1

35. $\dfrac{2d + b}{-a}$

5

36. $\dfrac{b + 2d}{b}$

$\dfrac{5}{2}$

37. $\dfrac{b - d}{c - a}$

1

38. $\dfrac{2c - d}{-ad}$

$-\dfrac{5}{6}$

39. $(b + d)^2 - 4a$

57

40. $(d - a)^2 - 3c$

28

41. $(d - a)^2 \div 5$

5

42. $3(b - a) - bc$

22

43. $\dfrac{b - 2a}{bc^2 - d}$

8

44. $\dfrac{b^2 - a}{ad + 3c}$

-2

45. $\dfrac{1}{3}d^2 - \dfrac{3}{8}b^2$

-3

46. $\dfrac{5}{8}a^4 - c^2$

9

47. $\dfrac{-4bc}{2a - b}$

-2

48. $-\dfrac{3}{4}b + \dfrac{1}{2}(ac + bd)$

4

49. $-\dfrac{2}{3}d - \dfrac{1}{5}(bd - ac)$

-4

50. $(b - a)^2 - (d - c)^2$

20

51. The value of z is the value of $a^2 - 2a$ when $a = -3$. Find the value of z^2.
225

52. The value of a is the value of $3x^2 - 4x - 5$ when $x = -2$. Find the value of $3a - 4$. 41

53. The value of c is the value of $a^2 + b^2$ when $a = 2$ and $b = -2$. Find the value of $c^2 - 4$. 60

APPLYING THE CONCEPTS

For Exercises 54 to 57, evaluate the following expressions for $x = 2$, $y = 3$, and $z = -2$.

54. $3^x - x^3$
1

55. z^x
4

56. $x^x - y^y$
-23

57. $y^{(x^2)}$
81

58. For each of the following, determine the first natural number x, greater than 1, for which the second expression is larger than the first.

a. $x^3, 3^x$

b. $x^4, 4^x$

c. $x^5, 5^x$

d. $x^6, 6^x$

2

5

6

7

59. On the basis of your answer to Exercise 58, make a conjecture that appears to be true about the two expressions x^n and n^x, where $n = 3, 4, 5, 6, 7, \ldots$ and x is a natural number greater than 1.
$n^x > x^n$ if $x \ge n + 1$

4.2 Simplifying Variable Expressions

Objective A To simplify a variable expression using the Properties of Addition

Like terms of a variable expression are terms with the same variable part. (Because $x^2 = x \cdot x$, x^2 and x are not like terms.)

Constant terms are like terms. 4 and 9 are like terms.

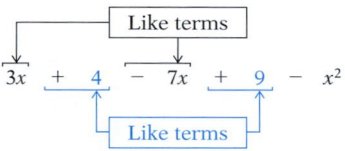

To simplify a variable expression, use the Distributive Property to combine like terms by adding the numerical coefficients. The variable part remains unchanged.

TAKE NOTE
Here is an example of the Distributive Property with just numbers.
$2(5 + 9) = 2(5) + 2(9)$
$= 10 + 18 = 28$
This is the same result we would obtain using the Order of Operations Agreement.
$2(5 + 9) = 2(14) = 28$
The usefulness of the Distributive Property will become more apparent as we explore variable expressions.

> **Distributive Property**
>
> If a, b, and c are real numbers, then $a(b + c) = ab + ac$.

The Distributive Property can also be written $ba + ca = (b + c)a$. This form is used to simplify a variable expression.

To simplify $2x + 3x$, use the Distributive Property to add the numerical coefficients of the like variable terms. This is called **combining like terms.**

$$2x + 3x = (2 + 3)x$$
$$= 5x$$

HOW TO Simplify: $5y - 11y$

$5y - 11y = (5 - 11)y$ • Use the **Distributive Property**.
$= -6y$

TAKE NOTE
Simplifying an expression means combining like terms. A constant term (5) and a variable term (7p) are not like terms and therefore cannot be combined.

HOW TO Simplify: $5 + 7p$

The terms 5 and $7p$ are not like terms.

The expression $5 + 7p$ is in simplest form.

> **The Associative Property of Addition**
>
> If a, b, and c are real numbers, then $(a + b) + c = a + (b + c)$.

When three or more terms are added, the terms can be grouped (with parentheses, for example) in any order. The sum is the same. For example,

$(5 + 7) + 15 = 5 + (7 + 15)$ $(3x + 5x) + 9x = 3x + (5x + 9x)$
$12 + 15 = 5 + 22$ $8x + 9x = 3x + 14x$
$27 = 27$ $17x = 17x$

In-Class Examples (Objective 4.2A)
1. Simplify: $4a - 5b - 3a + 2b$
   $a - 3b$
2. Simplify: $y^2 + 2 + 9y^2 - 14$
   $10y^2 - 12$
3. Simplify: $2z^2 - 3z - 5 - 3z^2 - 5$
   $-z^2 - 3z - 10$

Vocabulary to Review
additive inverse   [3.2A]

New Vocabulary
like terms
combining like terms

Properties to Review
Associative Property of Addition
   [1.2A]
Commutative Property of
   Addition   [1.2A]
Addition Property of Zero
   [1.2A]

New Properties
Distributive Property
Inverse Property of Addition

Instructor Note
Combining like terms can be related to many everyday experiences. For instance, 5 bricks plus 7 bricks is 12 bricks. But 5 bricks plus 7 nails is 5 bricks plus 7 nails. Students need to be constantly reminded that algebra is a reflection of our experiences, not some arbitrarily made-up system of rules.

Discuss the Concepts
1. Are $3x^2$ and $-4x^2$ like terms? Why or why not?
   Yes. They have the same variable raised to the same power.
2. Tell why b and b^2 are not like terms.
   Their exponents are different.

Optional Student Activity
1. Make up three sentences similar to "5 bricks plus 7 bricks is 12 bricks" that illustrate that like terms can be combined.
2. Make up three sentences similar to "3 cows plus 4 boys is NOT 7 cowboys" that illustrate that unlike terms cannot be combined.

Concept Check

Answer each of the following.
1. Is $0x$ the same as 0?   Yes
2. Is $1x$ the same as 1?
   No. $1x = x$

Concept Check

Determine the additive inverse of each of the following.
1. $4xy$ $-4xy$
2. $-6x^2$ $6x^2$
3. 0 0

Concept Check

True or false?
1. $x^2 + 7x^2 = 7x^2 + x^2$ True
2. $5y + 0 = 0$ False
3. $7x - 5 = 5 - 7x$ False
4. $8 - 8x = 0$ False

The Commutative Property of Addition

If a and b are real numbers, then $a + b = b + a$.

When two like terms are added, the terms can be added in either order. The sum is the same. For example,

$$15 + (-28) = (-28) + 15 \qquad 2x + (-4x) = -4x + 2x$$
$$-13 = -13 \qquad\qquad -2x = -2x$$

The Addition Property of Zero

If a is a real number, then $a + 0 = 0 + a = a$.

The sum of a term and zero is the term. For example,

$$-9 + 0 = 0 + (-9) = -9 \qquad 5x + 0 = 0 + 5x = 5x$$

The Inverse Property of Addition

If a is a real number, then $a + (-a) = (-a) + a = 0$.

The sum of a term and its opposite is zero. Recall that the opposite of a number is called its **additive inverse.**

$$12 + (-12) = (-12) + 12 = 0 \qquad 7x + (-7x) = -7x + 7x = 0$$

HOW TO   Simplify: $8x + 4y - 8x + y$

$8x + 4y - 8x + y$
$= (8x - 8x) + (4y + y)$ • Use the Commutative and Associative Properties of Addition to rearrange and group like terms.

$= 0 + 5y = 5y$ • Combine like terms.

HOW TO   Simplify: $4x^2 + 5x - 6x^2 - 2x + 1$

$4x^2 + 5x - 6x^2 - 2x + 1$
$= (4x^2 - 6x^2) + (5x - 2x) + 1$ • Use the Commutative and Associative Properties of Addition to rearrange and group like terms.

$= -2x^2 + 3x + 1$ • Combine like terms.

Example 1   Simplify: $3x + 4y - 10x + 7y$

Solution $3x + 4y - 10x + 7y = -7x + 11y$

You Try It 1   Simplify: $3a - 2b - 5a + 6b$

Your solution $-2a + 4b$

Example 2   Simplify: $x^2 - 7 + 4x^2 - 16$

Solution $x^2 - 7 + 4x^2 - 16 = 5x^2 - 23$

You Try It 2   Simplify: $-3y^2 + 7 + 8y^2 - 14$

Your solution $5y^2 - 7$

Solutions on p. S12

Objective B   To simplify a variable expression
using the Properties of Multiplication

In simplifying variable expressions, the following Properties of Multiplication are used.

> **TAKE NOTE**
> The Associative Property of Multiplication allows us to multiply a coefficient by a number. Without this property, the expression $2(3x)$ could not be changed.

> **The Associative Property of Multiplication**
>
> If a, b, and c are real numbers, then $(ab)c = a(bc)$.

When three or more factors are multiplied, the factors can be grouped in any order. The product is the same.

$$3(5 \cdot 6) = (3 \cdot 5)6 \qquad 2(3x) = (2 \cdot 3)x$$
$$3(30) = (15)6 \qquad\qquad\quad = 6x$$
$$90 = 90$$

> **TAKE NOTE**
> The Commutative Property of Multiplication allows us to rearrange factors. This property, along with the Associative Property of Multiplication, allows us to simplify some variable expressions.

> **The Commutative Property of Multiplication**
>
> If a and b are real numbers, then $ab = ba$.

Two factors can be multiplied in either order. The product is the same.

$$5(-7) = -7(5) \qquad (5x) \cdot 3 = 3 \cdot (5x) \qquad \bullet \text{ Commutative Property of Multiplication}$$
$$-35 = -35 \qquad\qquad\quad = (3 \cdot 5)x \qquad \bullet \text{ Associative Property of Multiplication}$$
$$= 15x$$

> **The Multiplication Property of One**
>
> If a is a real number, then $a \cdot 1 = 1 \cdot a = a$.

The product of a term and 1 is the term.

$$9 \cdot 1 = 1 \cdot 9 = 9 \qquad\qquad (8x) \cdot 1 = 1 \cdot (8x) = 8x$$

> **The Inverse Property of Multiplication**
>
> If a is a real number, and a is not equal to zero, then
>
> $$a \cdot \frac{1}{a} = \frac{1}{a} \cdot a = 1$$

> **TAKE NOTE**
> We must state that $x \neq 0$ because division by zero is undefined.

$\frac{1}{a}$ is called the **reciprocal** of a. $\frac{1}{a}$ is also called the **multiplicative inverse** of a. The product of a number and its reciprocal is 1.

$$7 \cdot \frac{1}{7} = \frac{1}{7} \cdot 7 = 1 \qquad\qquad x \cdot \frac{1}{x} = \frac{1}{x} \cdot x = 1, \quad x \neq 0$$

The multiplication properties just discussed are used to simplify variable expressions.

> **HOW TO**   Simplify: $2(-x)$
>
> $2(-x) = 2(-1 \cdot x)$ • Use the **Associative Property of**
> $\qquad = [2(-1)]x$ **Multiplication** to group factors.
> $\qquad = -2x$

Objective 4.2B

Vocabulary to Review
multiplicative inverse [2.4A]
reciprocal [2.4A]

Properties to Review
Associative Property of
  Multiplication [1.3A]
Commutative Property of
  Multiplication [1.3A]
Multiplication Property of One
  [1.3A]

New Properties
Inverse Property of
  Multiplication

Concept Check
Determine the reciprocal of 4.
$\frac{1}{4}$

Discuss the Concepts
Determine whether each of the following equations is true. In each case, tell which property supports your answer.
1. $-4y \cdot 1 = -4y$
True; the Multiplication Property of One
2. $(6x) \cdot 2 = 2 \cdot (6x)$
True; the Commutative Property of Multiplication
3. $5 \cdot \frac{1}{5} = 5$

False. The Inverse Property of Multiplication justifies the statement $5 \cdot \frac{1}{5} = 1$.
4. $3(4x) = (3 \cdot 4)x$
True; the Associative Property of Multiplication

In-Class Examples (Objective 4.2B)
1. Simplify: $-6(4y^2)$ $-24y^2$
2. Simplify: $-3(-12b)$ $36b$
3. Simplify: $(5a)(-6)$ $-30a$

Instructor Note

Simplifying expressions such as these prepares the students for solving equations.

Concept Check

Simplify.

1. $(2x) \div \dfrac{1}{2}$ $4x$

2. $\left(\dfrac{1}{2} \div 2\right)x$ $\dfrac{1}{4}x$

3. Replace the question mark to make a true statement.

$? \cdot (2x) = x$ $\dfrac{1}{2}$

HOW TO   Simplify: $\dfrac{3}{2}\left(\dfrac{2x}{3}\right)$

$\dfrac{3}{2}\left(\dfrac{2x}{3}\right) = \dfrac{3}{2}\left(\dfrac{2}{3}x\right)$ • Note that $\dfrac{2x}{3} = \dfrac{2}{3}x$.

$\qquad = \left(\dfrac{3}{2} \cdot \dfrac{2}{3}\right)x$ • Use the Associative Property of Multiplication to group factors.

$\qquad = 1 \cdot x$

$\qquad = x$

HOW TO   Simplify: $(16x)2$

$(16x)2 = 2(16x)$ • Use the Commutative and Associative Properties of Multiplication to rearrange and group factors.

$\qquad = (2 \cdot 16)x$

$\qquad = 32x$

Example 3   Simplify: $-2(3x^2)$

Solution $-2(3x^2) = -6x^2$

You Try It 3   Simplify: $-5(4y^2)$

Your solution $-20y^2$

Example 4   Simplify: $-5(-10x)$

Solution $-5(-10x) = 50x$

You Try It 4   Simplify: $-7(-2a)$

Your solution $14a$

Example 5   Simplify: $-\dfrac{3}{4}\left(\dfrac{2}{3}x\right)$

Solution $-\dfrac{3}{4}\left(\dfrac{2}{3}x\right) = -\dfrac{1}{2}x$

You Try It 5   Simplify: $-\dfrac{3}{5}\left(-\dfrac{7}{9}a\right)$

Your solution $\dfrac{7}{15}a$

Solutions on p. S12

Objective 4.2C

Properties to Review

Distributive Property   [4.2A]

Objective C   **To simplify a variable expression using the Distributive Property**

Recall that the Distributive Property states that if a, b, and c are real numbers, then

$$a(b + c) = ab + ac$$

The Distributive Property is used to remove parentheses from a variable expression.

HOW TO   Simplify: $3(2x + 7)$

$3(2x + 7) = 3(2x) + 3(7)$ • Use the **Distributive Property**. Multiply each term inside the parentheses by **3**.

$\qquad = 6x + 21$

In-Class Examples (Objective 4.2C)

1. Simplify: $4(3 - 9x)$ $12 - 36x$

2. Simplify: $(6 + 7y)8$ $48 + 56y$

3. Simplify: $-9(-5a + 2b)$ $45a - 18b$

4. Simplify: $4(y^2 - 3y + 7)$ $4y^2 - 12y + 28$

5. Simplify: $-5(a^2 + 8a - 6)$ $-5a^2 - 40a + 30$

Study Tip

To learn mathematics, you must be an active participant. Listening and watching your professor do mathematics is not enough. Take notes in class and mentally think through every question your instructor asks. Try to answer each question even if you are not called upon to answer it verbally. Ask questions when you have them. See *AIM for Success*, pages xxxi–xxxiii, for other ways to be an active learner.

HOW TO   Simplify: $-5(4x + 6)$

$$-5(4x + 6) = -5(4x) + (-5)(6)$$ • Use the **Distributive Property**.
$$= -20x - 30$$

HOW TO   Simplify: $-(2x - 4)$

$$-(2x - 4) = -1(2x - 4)$$ • Use the **Distributive Property**.
$$= -1(2x) - (-1)(4)$$
$$= -2x + 4$$

Note: When a negative sign immediately precedes the parentheses, the sign of each term inside the parentheses is changed.

HOW TO   Simplify: $-\dfrac{1}{2}(8x - 12y)$

$$-\dfrac{1}{2}(8x - 12y) = -\dfrac{1}{2}(8x) - \left(-\dfrac{1}{2}\right)(12y)$$ • Use the **Distributive Property**.

$$= -4x + 6y$$

An extension of the Distributive Property is used when an expression contains more than two terms.

HOW TO   Simplify: $3(4x - 2y - z)$

$$3(4x - 2y - z) = 3(4x) - 3(2y) - 3(z)$$ • Use the **Distributive Property**.
$$= 12x - 6y - 3z$$

Instructor Note

A problem such as $-(5x - 2)$ can be thought of either as the opposite of $5x - 2$ or as $-1(5x - 2)$.

Concept Check

Each of the following has an error. Correct the right-hand side of the statement.

1. $-(3x - 5) = 3x + 5$
   $-3x + 5$
2. $4(3 + 9y) = 12 + 9y$
   $12 + 36y$
3. $(8w - 3)7 = 56w + 21$
   $56w - 21$
4. $-6(3x - 2y + z) =$
   $-18x - 12y - 6z$
   $-18x + 12y - 6z$

Example 6

Simplify: $7(4 + 2x)$

Solution
$7(4 + 2x) = 28 + 14x$

You Try It 6

Simplify: $5(3 + 7b)$

Your solution
$15 + 35b$

Example 7

Simplify: $(2x - 6)2$

Solution
$(2x - 6)2 = 4x - 12$

You Try It 7

Simplify: $(3a - 1)5$

Your solution
$15a - 5$

Example 8

Simplify: $-3(-5a + 7b)$

Solution
$-3(-5a + 7b) = 15a - 21b$

You Try It 8

Simplify: $-8(-2a + 7b)$

Your solution
$16a - 56b$

Solutions on p. S12

Concept Check

Simplify.

1. $4(3x^2 - 2x + 5)$
$12x^2 - 8x + 20$

2. $-5(-x^2 + 3x - 2)$
$5x^2 - 15x + 10$

Example 9   Simplify: $3(x^2 - x - 5)$

Solution $3(x^2 - x - 5) = 3x^2 - 3x - 15$

You Try It 9   Simplify: $3(12x^2 - x + 8)$

Your solution $36x^2 - 3x + 24$

Example 10   Simplify: $-2(x^2 + 5x - 4)$

Solution $-2(x^2 + 5x - 4)$
$= -2x^2 - 10x + 8$

You Try It 10   Simplify: $3(-a^2 - 6a + 7)$

Your solution $-3a^2 - 18a + 21$

Solutions on p. S12

Objective 4.2D

Concept Check

Find the error in the following simplification.

$7 - 4(2x + 3) = 3(2x + 3)$
$= 6x + 9$

The Order of Operations Agreement was applied incorrectly. The Distributive Property must be applied first.

Discuss the Concepts

Is it appropriate to use the Distributive Property to simplify the expression $2(3 \cdot x)$?
No. The Associative Property of Multiplication is used.

Optional Student Activity

a. Evaluate $5 + 2(4x - 3)$ when $x = 5$.   39

b. Use the Distributive Property to simplify $5 + 2(4x - 3)$.
$8x - 1$

c. Evaluate the answer to part b when $x = 5$.   39

d. Should your answers to parts a and c have been the same? Why or why not?
Yes; $5 + 2(4x - 3)$ is equivalent to $8x - 1$, so evaluating either one for $x = 5$ should result in the same answer.

Objective D   To simplify general variable expressions

When simplifying variable expressions, use the Distributive Property to remove parentheses and brackets used as grouping symbols.

HOW TO   Simplify: $4(x - y) - 2(-3x + 6y)$

$4(x - y) - 2(-3x + 6y)$

$= 4x - 4y + 6x - 12y$ • Use the Distributive Property.

$= 10x - 16y$ • Combine like terms.

Example 11   Simplify: $2x - 3(2x - 7y)$

Solution $2x - 3(2x - 7y) = 2x - 6x + 21y$
$= -4x + 21y$

You Try It 11   Simplify: $3y - 2(y - 7x)$

Your solution $y + 14x$

Example 12   Simplify:
$7(x - 2y) - (-x - 2y)$

Solution $7(x - 2y) - (-x - 2y)$
$= 7x - 14y + x + 2y$
$= 8x - 12y$

You Try It 12   Simplify:
$-2(x - 2y) - (-x + 3y)$

Your solution $-x + y$

Example 13   Simplify: $2x - 3[2x - 3(x + 7)]$

Solution $2x - 3[2x - 3(x + 7)]$
$= 2x - 3[2x - 3x - 21]$
$= 2x - 3[-x - 21]$
$= 2x + 3x + 63$
$= 5x + 63$

You Try It 13   Simplify: $3y - 2[x - 4(2 - 3y)]$

Your solution $-2x - 21y + 16$

Solutions on p. S12

In-Class Examples (Objective 4.2D)

1. Simplify: $5a - 4(3a + 2b)$ $-7a - 8b$
2. Simplify: $-6(c - d) - (-c - 3d)$ $-5c + 9d$
3. Simplify: $9w - 4[3w - 5(w + 6)]$ $17w + 120$

4.2 Exercises

Objective A   **To simplify a variable expression using the Properties of Addition**

1. ✏️ What are *like terms*? Give an example of two like terms. Give an example of two terms that are not like terms.

2. ✏️ Explain the meaning of the phrase "simplify a variable expression."

For Exercises 3 to 38, simplify.

3. $6x + 8x$
$14x$

4. $12x + 13x$
$25x$

5. $9a - 4a$
$5a$

6. $12a - 3a$
$9a$

7. $4y - 10y$
$-6y$

8. $8y - 6y$
$2y$

9. $7 - 3b$
$7 - 3b$

10. $5 + 2a$
$5 + 2a$

11. $-12a + 17a$
$5a$

12. $-3a + 12a$
$9a$

13. $5ab - 7ab$
$-2ab$

14. $9ab - 3ab$
$6ab$

15. $-12xy + 17xy$
$5xy$

16. $-15xy + 3xy$
$-12xy$

17. $-3ab + 3ab$
0

18. $-7ab + 7ab$
0

19. $-\dfrac{1}{2}x - \dfrac{1}{3}x$
$-\dfrac{5}{6}x$

20. $-\dfrac{2}{5}y + \dfrac{3}{10}y$
$-\dfrac{1}{10}y$

21. $2.3x + 4.2x$
$6.5x$

22. $6.1y - 9.2y$
$-3.1y$

23. $x - 0.55x$
$0.45x$

24. $0.65A - A$
$-0.35A$

25. $5a - 3a + 5a$
$7a$

26. $10a - 17a + 3a$
$-4a$

27. $-5x^2 - 12x^2 + 3x^2$
$-14x^2$

28. $-y^2 - 8y^2 + 7y^2$
$-2y^2$

29. $\dfrac{3}{4}x - \dfrac{1}{3}x - \dfrac{7}{8}x$
$-\dfrac{11}{24}x$

30. $-\dfrac{2}{5}a - \left(-\dfrac{3}{10}a\right) - \dfrac{11}{15}a$
$-\dfrac{5}{6}a$

31. $7x - 3y + 10x$
$17x - 3y$

32. $8y + 8x - 8y$
$8x$

33. $3a + (-7b) - 5a + b$
$-2a - 6b$

34. $-5b + 7a - 7b + 12a$
$19a - 12b$

35. $3x + (-8y) - 10x + 4x$
$-3x - 8y$

36. $3y + (-12x) - 7y + 2y$
$-12x - 2y$

37. $x^2 - 7x + (-5x^2) + 5x$
$-4x^2 - 2x$

38. $3x^2 + 5x - 10x^2 - 10x$
$-7x^2 - 5x$

Section 4.2

Suggested Assignment

Exercises 1–137, every other odd
More challenging problems:
   Exercises 139, 140, 144

Answers to Writing Exercises

1. Like terms are variable terms with the same variable part. Constant terms are also like terms. Examples of like terms are $4x$ and $-9x$. Examples of terms that are not like terms are $4x^2$ and $-9x$. The terms 4 and 9 are also like terms; 4 and $4x$ are not.

2. To simplify a variable expression, combine the like terms.

Quick Quiz (Objective 4.2A)

Simplify.

1. $9b - 5b$ $4b$
2. $8x^2 - x^2 + 2x^2$ $9x^2$
3. $4y^2 + 3y - 6y^2 + 2y$ $-2y^2 + 5y$

Objective B   **To simplify a variable expression using the Properties of Multiplication**

For Exercises 39 to 78, simplify.

39. $4(3x)$
$12x$

40. $12(5x)$
$60x$

41. $-3(7a)$
$-21a$

42. $-2(5a)$
$-10a$

43. $-2(-3y)$
$6y$

44. $-5(-6y)$
$30y$

45. $(4x)2$
$8x$

46. $(6x)12$
$72x$

47. $(3a)(-2)$
$-6a$

48. $(7a)(-4)$
$-28a$

49. $(-3b)(-4)$
$12b$

50. $(-12b)(-9)$
$108b$

51. $-5(3x^2)$
$-15x^2$

52. $-8(7x^2)$
$-56x^2$

53. $\frac{1}{3}(3x^2)$
x^2

54. $\frac{1}{6}(6x^2)$
x^2

55. $\frac{1}{5}(5a)$
a

56. $\frac{1}{8}(8x)$
x

57. $-\frac{1}{2}(-2x)$
x

58. $-\frac{1}{4}(-4a)$
a

59. $-\frac{1}{7}(-7n)$
n

60. $-\frac{1}{9}(-9b)$
b

61. $(3x)\left(\frac{1}{3}\right)$
x

62. $(12x)\left(\frac{1}{12}\right)$
x

63. $(-6y)\left(-\frac{1}{6}\right)$
y

64. $(-10n)\left(-\frac{1}{10}\right)$
n

65. $\frac{1}{3}(9x)$
$3x$

66. $\frac{1}{7}(14x)$
$2x$

67. $-\frac{1}{5}(10x)$
$-2x$

68. $-\frac{1}{8}(16x)$
$-2x$

69. $-\frac{2}{3}(12a^2)$
$-8a^2$

70. $-\frac{5}{8}(24a^2)$
$-15a^2$

71. $-\frac{1}{2}(-16y)$
$8y$

72. $-\frac{3}{4}(-8y)$
$6y$

73. $(16y)\left(\frac{1}{4}\right)$
$4y$

74. $(33y)\left(\frac{1}{11}\right)$
$3y$

75. $(-6x)\left(\frac{1}{3}\right)$
$-2x$

76. $(-10x)\left(\frac{1}{5}\right)$
$-2x$

77. $(-8a)\left(-\frac{3}{4}\right)$
$6a$

78. $(21y)\left(-\frac{3}{7}\right)$
$-9y$

Objective C   **To simplify a variable expression using the Distributive Property**

For Exercises 79 to 117, simplify.

79. $-(x + 2)$
$-x - 2$

80. $-(x + 7)$
$-x - 7$

81. $2(4x - 3)$
$8x - 6$

82. $5(2x - 7)$
$10x - 35$

83. $-2(a + 7)$
$-2a - 14$

84. $-5(a + 16)$
$-5a - 80$

85. $-3(2y - 8)$
$-6y + 24$

86. $-5(3y - 7)$
$-15y + 35$

Quick Quiz (Objective 4.2B)

Simplify.

1. $5(6a)$ $30a$

2. $-\frac{1}{8}(-8b)$ b

3. $(-12c)\frac{2}{3}$ $-8c$

87. $(5 - 3b)7$

$35 - 21b$

88. $(10 - 7b)2$

$20 - 14b$

89. $\dfrac{1}{3}(6 - 15y)$

$2 - 5y$

90. $\dfrac{1}{2}(-8x + 4y)$

$-4x + 2y$

91. $3(5x^2 + 2x)$

$15x^2 + 6x$

92. $6(3x^2 + 2x)$

$18x^2 + 12x$

93. $-2(-y + 9)$

$2y - 18$

94. $-5(-2x + 7)$

$10x - 35$

95. $(-3x - 6)5$

$-15x - 30$

96. $(-2x + 7)7$

$-14x + 49$

97. $2(-3x^2 - 14)$

$-6x^2 - 28$

98. $5(-6x^2 - 3)$

$-30x^2 - 15$

99. $-3(2y^2 - 7)$

$-6y^2 + 21$

100. $-8(3y^2 - 12)$

$-24y^2 + 96$

101. $3(x^2 - y^2)$

$3x^2 - 3y^2$

102. $5(x^2 + y^2)$

$5x^2 + 5y^2$

103. $-\dfrac{2}{3}(6x - 18y)$

$-4x + 12y$

104. $-\dfrac{1}{2}(x - 4y)$

$-\dfrac{1}{2}x + 2y$

105. $-(6a^2 - 7b^2)$

$-6a^2 + 7b^2$

106. $3(x^2 + 2x - 6)$

$3x^2 + 6x - 18$

107. $4(x^2 - 3x + 5)$

$4x^2 - 12x + 20$

108. $-2(y^2 - 2y + 4)$

$-2y^2 + 4y - 8$

109. $\dfrac{3}{4}(2x - 6y + 8)$

$\dfrac{3}{2}x - \dfrac{9}{2}y + 6$

110. $-\dfrac{2}{3}(6x - 9y + 1)$

$-4x + 6y - \dfrac{2}{3}$

111. $4(-3a^2 - 5a + 7)$

$-12a^2 - 20a + 28$

112. $-5(-2x^2 - 3x + 7)$

$10x^2 + 15x - 35$

113. $-3(-4x^2 + 3x - 4)$

$12x^2 - 9x + 12$

114. $3(2x^2 + xy - 3y^2)$

$6x^2 + 3xy - 9y^2$

115. $5(2x^2 - 4xy - y^2)$

$10x^2 - 20xy - 5y^2$

116. $-(3a^2 + 5a - 4)$

$-3a^2 - 5a + 4$

117. $-(8b^2 - 6b + 9)$

$-8b^2 + 6b - 9$

Objective D   **To simplify general variable expressions**

For Exercises 118 to 138, simplify.

118. $4x - 2(3x + 8)$

$-2x - 16$

119. $6a - (5a + 7)$

$a - 7$

120. $9 - 3(4y + 6)$

$-12y - 9$

121. $10 - (11x - 3)$

$-11x + 13$

122. $5n - (7 - 2n)$

$7n - 7$

123. $8 - (12 + 4y)$

$-4y - 4$

Quick Quiz (Objective 4.2C)

Simplify.

1. $-(y - 4)$ $-y + 4$

2. $4(3a + 5)$ $12a + 20$

3. $-\dfrac{1}{2}(4x + 8y - 6z)$ $-2x - 4y + 3z$

124. $3(x + 2) - 5(x - 7)$
$-2x + 41$

125. $2(x - 4) - 4(x + 2)$
$-2x - 16$

126. $12(y - 2) + 3(7 - 3y)$
$3y - 3$

127. $6(2y - 7) - (3 - 2y)$
$14y - 45$

128. $3(a - b) - (a + b)$
$2a - 4b$

129. $2(a + 2b) - (a - 3b)$
$a + 7b$

130. $4[x - 2(x - 3)]$
$-4x + 24$

131. $2[x + 2(x + 7)]$
$6x + 28$

132. $-2[3x + 2(4 - x)]$
$-2x - 16$

133. $-5[2x + 3(5 - x)]$
$5x - 75$

134. $-3[2x - (x + 7)]$
$-3x + 21$

135. $-2[3x - (5x - 2)]$
$4x - 4$

136. $2x - 3[x - (4 - x)]$
$-4x + 12$

137. $-7x + 3[x - (3 - 2x)]$
$2x - 9$

138. $-5x - 2[2x - 4(x + 7)] - 6$
$-x + 50$

APPLYING THE CONCEPTS

139. Determine whether the statement is true or false. If the statement is false, give an example that illustrates that it is false.
  a. Division is a commutative operation.   False. For example, $8 \div 2 \neq 2 \div 8$
  b. Division is an associative operation.   False. For example, $(12 \div 4) \div 2 \neq 12 \div (4 \div 2)$
  c. Subtraction is an associative operation.   False. For example, $(9 - 2) - 3 \neq 9 - (2 - 3)$
  d. Subtraction is a commutative operation.   False. For example, $10 - 4 \neq 4 - 10$

140. Is the statement "any number divided by itself is 1" a true statement? If not, for what number or numbers is the statement not true?
No. 0

141. Does every number have a multiplicative inverse? If not, which real number or numbers do not have a multiplicative inverse?   No. 0

142. ✏ In your own words, explain the Distributive Property.

143. ✏ Give examples of two operations that occur in everyday experience that are not commutative (for example, putting on socks and then shoes).

144. Define an operation $\otimes$ as $a \otimes b = (a \cdot b) - (a + b)$. For example, $7 \otimes 5 = (7 \cdot 5) - (7 + 5) = 35 - 12 = 23$.
  a. Is $\otimes$ a commutative operation? Support your answer.
    Yes
  b. Is $\otimes$ an associative operation? Support your answer.
    No

Answers to Writing Exercises

142. Students might provide a description such as the following: The Distributive Property tells us to multiply each term inside parentheses by the factor outside the parentheses.

143. Examples of two operations that occur in everyday experience that are not commutative are (1) unlocking the car door and starting the car and (2) taking a shower and drying oneself off.

Quick Quiz (Objective 4.2D)

Simplify.
1. $8 - 5(3x + 2)$ $-15x - 2$
2. $7(3y - 4) - 2(4y - 3)$ $13y - 22$
3. $3a - 5[a - (6 - a)]$ $-7a + 30$

4.3 Translating Verbal Expressions into Variable Expressions

Objective A To translate a verbal expression into a variable expression, given the variable

One of the major skills required in applied mathematics is to translate a verbal expression into a variable expression. This requires recognizing the verbal phrases that translate into mathematical operations. A partial list of the verbal phrases used to indicate the different mathematical operations follows.

Addition	added to	6 added to y	$y + 6$
	more than	8 more than x	$x + 8$
	the sum of	the sum of x and z	$x + z$
	increased by	t increased by 9	$t + 9$
	the total of	the total of 5 and y	$5 + y$
Subtraction	minus	x minus 2	$x - 2$
	less than	7 less than t	$t - 7$
	decreased by	m decreased by 3	$m - 3$
	the difference between	the difference between y and 4	$y - 4$
	subtract...from...	subtract 9 from z	$z - 9$
Multiplication	times	10 times t	$10t$
	twice	twice w	$2w$
	of	one-half of x	$\frac{1}{2}x$
	the product of	the product of y and z	yz
	multiplied by	y multiplied by 11	$11y$
Division	divided by	x divided by 12	$\frac{x}{12}$
	the quotient of	the quotient of y and z	$\frac{y}{z}$
	the ratio of	the ratio of t to 9	$\frac{t}{9}$
Power	the square of	the square of x	x^2
	the cube of	the cube of a	a^3

Point of Interest

The way in which expressions are symbolized has changed over time. The following shows how some of the expressions at the right may have appeared in the early 16th century.

R p. 9 for $x + 9$. The symbol R was used for a variable to the first power. The symbol p. was used for plus.

R m. 3 for $x - 3$. The symbol R is still used for the variable. The symbol m. was used for minus.

The square of a variable was designated by Q and the cube was designated by C. The expression $x^2 + x^3$ was written **Q p. C.**

HOW TO Translate "14 less than the cube of x" into a variable expression.

14 <u>less than</u> the <u>cube</u> of x • Identify the words that indicate the mathematical operations.

$x^3 - 14$ • Use the identified operations to write the variable expression.

Objective 4.3A

Instructor Note

In all of these phrases, whatever is mentioned first may be written first, and whatever is mentioned second is written second—with the exception of the phrases *less than* and *subtracted from*. They are the only phrases in this list that require a reversal of the terms. For example, *5 added to x* may be written as $5 + x$, but *5 less than x* is written as $x - 5$.

Optional Student Activity

What do you think the word *thrice* means?
Thrice means "times three."

In-Class Examples (Objective 4.3A)

1. Translate "the sum of 4 times x and 7" into a variable expression. $4x + 7$

2. Translate "the product of 5 and the difference between w and 3" into a variable expression. $5(w - 3)$

3. Translate "3 less than the quotient of a number and 4" into a variable expression. $\frac{x}{4} - 3$

Copyright © Houghton Mifflin Company. All rights reserved.

Instructor Note

Students may need to be reminded that the word *and* is not in the list of clue words for addition. Although in common English usage the question "What is two and two?" means two *plus* two, in formal mathematics the word *and* is never used as a clue word for addition. It does, however, have a special usage; it is a connector. The word *and* indicates where the appropriate mathematical operator is placed.

Concept Check

In each of the following, two verbal expressions are given. Determine whether they represent the same algebraic expression.

1. "The sum of *x* and 4"; "4 more than *x*"   Same

2. "2 times the sum of *x* and 3"; "the sum of 2 times *x* and 3"   Different

3. "3*y* subtracted from 7"; "7 less than 3*y*"   Different

4. "The ratio of twice *x* to *y*"; "the quotient of the product of 2 and *x* and *y*"   Same

Optional Student Activity

What are some of the ways in which $3(x + 4)$ could be written as a verbal expression?
Answers will vary. One option is "the product of 3 and the sum of *x* and 4."

Translating a phrase that contains the word *sum, difference, product,* or *quotient* can sometimes cause a problem. In the examples at the right, note where the operation symbol is placed.

the *sum* of *x and y* $x + y$

the *difference* between *x and y* $x - y$

the *product* of *x and y* $x \cdot y$

the *quotient* of *x and y* $\dfrac{x}{y}$

HOW TO   Translate "the difference between the square of *x* and the sum of *y* and *z*" into a variable expression.

the difference between the square of *x* and the sum of *y* and *z*

$x^2 - (y + z)$

- Identify the words that indicate the mathematical operations.

- Use the identified operations to write the variable expression.

Example 1

Translate "the total of 3 times *n* and *n*" into a variable expression.

Solution
the total of 3 times *n* and *n*
$3n + n$

You Try It 1

Translate "the difference between twice *n* and one-third of *n*" into a variable expression.

Your solution
$2n - \dfrac{1}{3}n$

Example 2

Translate "*m* decreased by the sum of *n* and 12" into a variable expression.

Solution
m decreased by the sum of *n* and 12
$m - (n + 12)$

You Try It 2

Translate "the quotient of 7 less than *b* and 15" into a variable expression.

Your solution
$\dfrac{b - 7}{15}$

Solutions on p. S12

Objective B   **To translate a verbal expression into a variable expression and then simplify**

In most applications that involve translating phrases into variable expressions, the variable to be used is not given. To translate these phrases, a variable must be assigned to an unknown quantity before the variable expression can be written.

HOW TO Translate "a number multiplied by the total of six and the cube of the number" into a variable expression.

the unknown number: n
                             • Assign a variable to one of the unknown quantities.

the cube of the number: n^3
the total of six and the cube
  of the number: $6 + n^3$
                             • Use the assigned variable to write an expression for any other unknown quantity.

$n(6 + n^3)$
                             • Use the assigned variable to write the variable expression.

Example 3

Translate "a number added to the product of four and the square of the number" into a variable expression.

Solution
the unknown number: n
the square of the number: n^2
the product of four and the square of the
   number: $4n^2$
$4n^2 + n$

You Try It 3

Translate "negative four multiplied by the total of ten and the cube of a number" into a variable expression.

Your solution
$-4(10 + n^3)$

Example 4

Translate "four times the sum of one-half of a number and fourteen" into a variable expression. Then simplify.

Solution
the unknown number: n

one-half of the number: $\frac{1}{2}n$

the sum of one-half of the number and

  fourteen: $\frac{1}{2}n + 14$

$4\left(\frac{1}{2}n + 14\right)$

$2n + 56$

You Try It 4

Translate "five times the difference between a number and sixty" into a variable expression. Then simplify.

Your solution
$5(x - 60); 5x - 300$

Solutions on p. S12

Objective 4.3B

Concept Check
Translate. Then simplify.
1. "A number minus the product of eight and the number"
   $n - 8n; -7n$
2. "The product of three and the difference between a number and four" $3(n - 4); 3n - 12$

Optional Student Activity
Create five different translation problems that have $4n$ as the answer.
Answers will vary. One possibility is "the product of four and a number." Another is "the difference between six times a number and twice the number."

In-Class Examples (Objective 4.3B)

1. Translate "the product of three and the sum of four times the square of a number and one."
   $3(4x^2 + 1)$

2. Translate "six times the total of the quotient of a number and three and two" into a variable expression. Then simplify.

   $6\left(\dfrac{n}{3} + 2\right); 2n + 12$

264 Chapter 4 / Variable Expressions

Objective 4.3C

Concept Check

1. The length of a rectangular piece of paper is three times the width. If *w* represents the width of the paper, express the length of the paper in terms of the variable *w*. $3w$

2. A board 8 ft long is cut into two pieces. If *L* represents the length of the longer piece, express the length of the shorter piece in terms of the variable *L*. $8 - L$

Discuss the Concepts

A jar contained 2000 jelly beans. Someone then removed a handful of the jelly beans.

1. How does the quantity of jelly beans in the jar now compare to the number before the handful was removed?
There are now fewer than 2000 jelly beans in the jar.

2. What expression involving subtraction would indicate that the number of jelly beans is now less than 2000?
$2000 - n, n > 0$

3. Is there enough information to determine the actual number of jelly beans remaining in the jar? Explain.
No. The number of jelly beans in the handful is unknown.

Objective C   To translate application problems

Many of the applications of mathematics require that you identify the unknown quantity, assign a variable to that quantity, and then attempt to express other unknown quantities in terms of the variable.

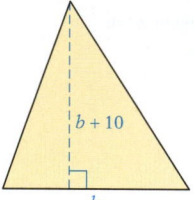

HOW TO   The height of a triangle is 10 ft longer than the base of the triangle. Express the height of the triangle in terms of the base of the triangle.

the base of the triangle: b

• Assign a variable to the base of the triangle.

the height is 10 more than the base: $b + 10$

• Express the height of the triangle in terms of b.

Example 5

The length of a swimming pool is 4 ft less than two times the width. Express the length of the pool in terms of the width.

Solution
the width of the pool: w
the length is 4 ft less than two times the width: $2w - 4$

You Try It 5

The speed of a new jet plane is twice the speed of an older model. Express the speed of the new model in terms of the speed of the older model.

Your solution
The speed of the older model: s
The speed of the new jet plane is twice the speed of the older model: $2s$.

Example 6

A banker divided $5000 between two accounts, one paying 10% annual interest and the second paying 8% annual interest. Express the amount invested in the 10% account in terms of the amount invested in the 8% account.

Solution
the amount invested at 8%: x
the amount invested at 10%: $5000 - x$

You Try It 6

A guitar string 6 ft long was cut into two pieces. Express the length of the shorter piece in terms of the length of the longer piece.

Your solution
The length of the longer piece: y
The length of the shorter piece: $6 - y$

Solutions on p. S12

In-Class Examples (Objective 4.3C)

1. The force of gravity on the moon is approximately one-sixth the force of gravity on Earth. Represent the force of gravity on the moon in terms of the force of gravity on Earth.
Force of gravity on Earth: F; force of gravity on the moon: $\dfrac{1}{6}F$

4.3 Exercises

Objective A   To translate a verbal expression into a
variable expression, given the variable

Suggested Assignment
Exercises 1–57, every other odd
Exercises 59–67, odds
More challenging problems:
    Exercises 68, 69

For Exercises 1 to 26, translate into a variable expression.

1. the sum of 8 and y
$8 + y$

2. a less than 16
$16 - a$

3. t increased by 10
$t + 10$

4. p decreased by 7
$p - 7$

5. z added to 14
$z + 14$

6. q multiplied by 13
$13q$

7. 20 less than the square of x
$x^2 - 20$

8. 6 times the difference between m and 7
$6(m - 7)$

9. the sum of three-fourths of n and 12
$\frac{3}{4}n + 12$

10. b decreased by the product of 2 and b
$b - 2b$

11. 8 increased by the quotient of n and 4
$8 + \frac{n}{4}$

12. the product of -8 and y
$-8y$

13. the product of 3 and the total of y and 7
$3(y + 7)$

14. 8 divided by the difference between x and 6
$\frac{8}{x - 6}$

15. the product of t and the sum of t and 16
$t(t + 16)$

16. the quotient of 6 less than n and twice n
$\frac{n - 6}{2n}$

17. 15 more than one-half of the square of x
$\frac{1}{2}x^2 + 15$

18. 19 less than the product of n and -2
$-2n - 19$

19. the total of 5 times the cube of n and the square of n
$5n^3 + n^2$

20. the ratio of 9 more than m to m
$\frac{m + 9}{m}$

21. r decreased by the quotient of r and 3
$r - \frac{r}{3}$

22. four-fifths of the sum of w and 10
$\frac{4}{5}(w + 10)$

23. the difference between the square of x and the total of x and 17
$x^2 - (x + 17)$

24. s increased by the quotient of 4 and s
$s + \frac{4}{s}$

25. the product of 9 and the total of z and 4
$9(z + 4)$

26. n increased by the difference between 10 times n and 9
$n + (10n - 9)$

Quick Quiz (Objective 4.3A)

Translate into a variable expression.

1. The product of 7 and the sum of n and 4 $7(n + 4)$

2. The quotient of 5 less than d and 3 $\frac{d - 5}{3}$

3. 5 less than the quotient of d and 3 $\frac{d}{3} - 5$

Objective B   **To translate a verbal expression into a variable expression and then simplify**

For Exercises 27 to 58, translate into a variable expression. Then simplify.

27. twelve minus a number
$12 - x$

28. a number divided by eighteen
$\dfrac{x}{18}$

29. two-thirds of a number
$\dfrac{2}{3}x$

30. twenty more than a number
$x + 20$

31. the quotient of twice a number and nine
$\dfrac{2x}{9}$

32. ten times the difference between a number and fifty
$10(x - 50); 10x - 500$

33. eight less than the product of eleven and a number
$11x - 8$

34. the sum of five-eighths of a number and six
$\dfrac{5}{8}x + 6$

35. nine less than the total of a number and two
$(x + 2) - 9; x - 7$

36. the difference between a number and three more than the number
$x - (x + 3); -3$

37. the quotient of seven and the total of five and a number
$\dfrac{7}{5 + x}$

38. four times the sum of a number and nineteen
$4(x + 19); 4x + 76$

39. five increased by one-half of the sum of a number and three
$5 + \dfrac{1}{2}(x + 3); \dfrac{1}{2}x + \dfrac{13}{2}$

40. the quotient of fifteen and the sum of a number and twelve
$\dfrac{15}{x + 12}$

41. a number added to the difference between twice the number and four
$(2x - 4) + x; 3x - 4$

42. the product of two-thirds and the sum of a number and seven
$\dfrac{2}{3}(x + 7); \dfrac{2}{3}x + \dfrac{14}{3}$

43. the product of five less than a number and seven
$(x - 5)7; 7x - 35$

44. the difference between forty and the quotient of a number and twenty
$40 - \dfrac{x}{20}$

45. the quotient of five more than twice a number and the number
$\dfrac{2x + 5}{x}$

46. the sum of the square of a number and twice the number
$x^2 + 2x$

47. a number decreased by the difference between three times the number and eight
$x - (3x - 8); -2x + 8$

48. the sum of eight more than a number and one-third of the number
$(x + 8) + \dfrac{1}{3}x; \dfrac{4}{3}x + 8$

Quick Quiz (Objective 4.3B)

Translate into a variable expression. Then simplify.

1. Five less than the total of a number and seven
$(x + 7) - 5; x + 2$

2. A number decreased by the difference between six and the product of the number and four
$x - (6 - 4x); 5x - 6$

3. The sum of a number divided by eight and the number $\dfrac{x}{8} + x; \dfrac{9}{8}x$

49. a number added to the product of three and the number
$3x + x; 4x$

50. a number increased by the total of the number and nine
$x + (x + 9); 2x + 9$

51. five more than the sum of a number and six
$(x + 6) + 5; x + 11$

52. a number decreased by the difference between eight and the number
$x - (8 - x); 2x - 8$

53. a number minus the sum of the number and ten
$x - (x + 10); -10$

54. the difference between one-third of a number and five-eighths of the number
$\frac{1}{3}x - \frac{5}{8}x; -\frac{7}{24}x$

55. the sum of one-sixth of a number and four-ninths of the number
$\frac{1}{6}x + \frac{4}{9}x; \frac{11}{18}x$

56. two more than the total of a number and five
$(x + 5) + 2; x + 7$

57. the sum of a number divided by three and the number
$\frac{x}{3} + x; \frac{4}{3}x$

58. twice the sum of six times a number and seven
$2(6x + 7); 12x + 14$

Objective C   **To translate application problems**

59. **E-mail**   According to Brightmail, Inc., one-half of all e-mail filtered by its e-mail program in a recent month was unsolicited e-mail (spam). Express the amount of spam in terms of the number of e-mails filtered.
Number of e-mails: A
Number of spam e-mails: $\frac{1}{2}A$

60. **Area Codes**   In 1951, phone companies began using area codes. According to information found at **www.area-code.com,** at the beginning of 2004 there were 207 more area codes than there were in 1951. Express the number of area codes in 2004 in terms of the number of area codes in 1951.
Number of area codes in 1951: A
Number of area codes in 2004: $A + 207$

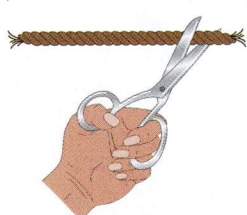

61. **Sailing**   A halyard 12 ft long was cut into two pieces of different lengths. Use one variable to express the lengths of the two pieces.
Length of one piece: S
Length of second piece: $12 - S$

62. **Natural Resources**   Twenty gallons of crude oil were poured into two containers of different sizes. Use one variable to express the amount of oil poured into each container.
Gallons of oil in first container: g
Gallons of oil in second container: $20 - g$

Quick Quiz (Objective 4.3C)

1. The number of grooves on the outer edge of a dime is one less than the number of grooves on the outer edge of a quarter. Express the number of grooves on the outer edge of a dime in terms of the number of grooves on the outer edge of a quarter.
Number of grooves on a quarter: g
Number of grooves on a dime: $g - 1$

2. A wire 50 ft long was cut into two pieces of different lengths. Use one variable to express the lengths of the two pieces.
Length of one piece: L
Length of second piece: $50 - L$

63. **Rates of Cars**   Two cars start at the same place and travel at different rates in opposite directions. Two hours later the cars are 200 mi apart. Express the distance traveled by the slower car in terms of the distance traveled by the faster car.
Distance traveled by faster car: x
Distance traveled by slower car: $200 - x$

64.  **Online Sales**   A recent survey conducted by Consumer Internet Barometer found that approximately 25% of those people who purchased a product online were extremely satisfied with their experience. Express the number of people who were extremely satisfied with their online purchases in terms of the number of people surveyed.
Number of people surveyed who purchased a product online: N
Number of people who were extremely pleased with their purchases: $0.25N$

65. **Medicine**   According to the American Podiatric Medical Association, the bones in your foot account for one-fourth of all the bones in your body. Express the number of bones in your foot in terms of the total number of bones in your body.
Number of bones in your body: N
Number of bones in your foot: $\frac{1}{4}N$

66. **Basketball**   The diameter of a basketball is approximately 4 times the diameter of a baseball. Express the diameter of a basketball in terms of the diameter of a baseball.
Diameter of a baseball: d
Diameter of a basketball: $4d$

67. **Cost of Living**   A cost-of-living calculator provided by **Realtor.com** shows that a person living in San Francisco, California, would need approximately twice the salary of a person living in Daytona Beach, Florida, to maintain the same standard of living. Express the salary needed in Daytona Beach in terms of the salary needed in San Francisco.
Salary needed in San Francisco: S
Salary needed in Daytona Beach: $\frac{1}{2}S$

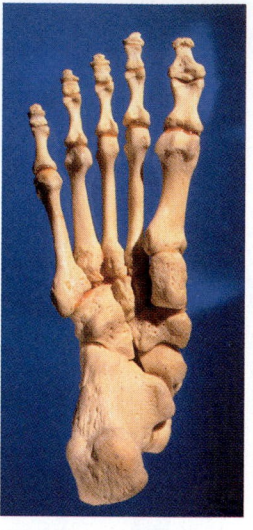

APPLYING THE CONCEPTS

68. **Metalwork**   A wire whose length is given as x inches is bent into a square. Express the length of a side of the square in terms of x.
$\frac{1}{4}x$

69. **Chemistry**   The chemical formula for glucose (sugar) is $C_6H_{12}O_6$. This formula means that there are 12 hydrogen atoms for every 6 carbon atoms and 6 oxygen atoms in each molecule of glucose (see the figure at the right). If x represents the number of oxygen atoms in a pound of sugar, express the number of hydrogen atoms in the pound of sugar in terms of x. $2x$

70. Translate the expressions $5x + 8$ and $5(x + 8)$ into phrases.

71. In your own words, explain how variables are used.

72. Explain the similarities and differences between the expressions "the difference between x and 5" and "5 less than x."

Answers to Writing Exercises

70. Two examples of the translation of $5x + 8$ are "eight more than the product of five and a number" and "the sum of five times a number and eight." Two examples of the translation of $5(x + 8)$ are "five times the sum of a number and eight" and "the product of five and eight more than a number."

71. Students should express the idea that variables are used to represent unknown quantities or quantities that can change according to different circumstances.

72. Both of the expressions "the difference between x and 5" and "5 less than x" translate into $x - 5$. However, in translating the two expressions from words, the word *difference* maintains the order in which the two quantities are subtracted, whereas the words *less than* reverse the order in which the two quantities are subtracted.

Focus on Problem Solving

From Concrete to Abstract   In your study of algebra, you will find that the problems are less concrete than those you studied in arithmetic. Problems that are concrete provide information pertaining to a specific instance. Algebra is more abstract. Abstract problems are theoretical; they are stated without reference to a specific instance. Let's look at an example of an abstract problem.

How many minutes are in h hours?

A strategy that can be used to solve this problem is to solve the same problem after substituting a number for the variable.

How many minutes are in 5 hours?

You know that there are 60 minutes in 1 hour. To find the number of minutes in 5 hours, multiply 5 by 60.

$60 \cdot 5 = 300$ There are 300 minutes in 5 hours.

Use the same procedure to find the number of minutes in h hours: multiply h by 60.

$60 \cdot h = 60h$ There are $60h$ minutes in h hours.

This problem might be taken a step further:

If you walk 1 mile in x minutes, how far can you walk in h hours?

Consider the same problem using numbers in place of the variables.

If you walk 1 mile in 20 minutes, how far can you walk in 3 hours?

To solve this problem, you need to calculate the number of minutes in 3 hours (multiply 3 by 60), and divide the result by the number of minutes it takes to walk 1 mile (20 minutes).

$$\frac{60 \cdot 3}{20} = \frac{180}{20} = 9$$ If you walk 1 mile in 20 minutes, you can walk 9 miles in 3 hours.

Use the same procedure to solve the related abstract problem. Calculate the number of minutes in h hours (multiply h by 60), and divide the result by the number of minutes it takes to walk 1 mile (x minutes).

$$\frac{60 \cdot h}{x} = \frac{60h}{x}$$ If you walk 1 mile in x minutes, you can walk $\frac{60h}{x}$ miles in h hours.

At the heart of the study of algebra is the use of variables. It is the variables in the problems above that make them abstract. But it is variables that allow us to generalize situations and state rules about mathematics.

Try each of the following problems.

1. How many hours are in d days?

Answers to Focus on Problem Solving: From Concrete to Abstract

1. $24d$
2. dh
3. $\frac{d}{p}$
4. $\frac{d}{s}$
5. $v - t$
6. $\frac{t}{g}$
7. $\frac{32q}{j}$
8. $\frac{60m}{s}$
9. $\frac{60hp}{m}$
10. $\frac{5q}{n}$

2. You earn d dollars an hour. What are your wages for working h hours?
3. If p is the price of one share of stock, how many shares can you purchase with d dollars?
4. A company pays a television station d dollars to air a commercial lasting s seconds. What is the cost per second?
5. After every v videotape rentals, you are entitled to one free rental. You have rented t tapes, where $t < v$. How many more do you need to rent before you are entitled to a free rental?
6. Your car gets g miles per gallon. How many gallons of gasoline does your car consume traveling t miles?
7. If you drink j ounces of juice each day, how many days will q quarts of the juice last?
8. A TV station has m minutes of commercials each hour. How many ads lasting s seconds each can be sold for each hour of programming?
9. A factory worker can assemble p products in m minutes. How many products can the factory worker assemble in h hours?
10. If one candy bar costs n nickels, how many candy bars can be purchased with q quarters?

Projects and Group Activities

Prime and Composite Numbers

Recall that a prime number is a natural number greater than 1 whose only natural-number factors are itself and 1. The number 11 is a prime number because the only natural-number factors of 11 are 11 and 1.

Eratosthenes, a Greek philosopher and astronomer who lived from 270 to 190 B.C., devised a method of identifying prime numbers. It is called the **Sieve of Eratosthenes.** The procedure is illustrated below.

~~1~~	②	③	~~4~~	⑤	~~6~~	⑦	~~8~~	~~9~~	~~10~~
⑪	~~12~~	⑬	~~14~~	~~15~~	~~16~~	⑰	~~18~~	⑲	~~20~~
~~21~~	~~22~~	㉓	~~24~~	~~25~~	~~26~~	~~27~~	~~28~~	㉙	~~30~~
㉛	~~32~~	~~33~~	~~34~~	~~35~~	~~36~~	㊲	~~38~~	~~39~~	~~40~~
㊶	~~42~~	㊸	~~44~~	~~45~~	~~46~~	㊼	~~48~~	~~49~~	~~50~~
~~51~~	~~52~~	(53)	~~54~~	~~55~~	~~56~~	~~57~~	~~58~~	(59)	~~60~~
(61)	~~62~~	~~63~~	~~64~~	~~65~~	~~66~~	(67)	~~68~~	~~69~~	~~70~~
(71)	~~72~~	(73)	~~74~~	~~75~~	~~76~~	~~77~~	~~78~~	(79)	~~80~~
~~81~~	~~82~~	(83)	~~84~~	~~85~~	~~86~~	~~87~~	~~88~~	(89)	~~90~~
~~91~~	~~92~~	~~93~~	~~94~~	~~95~~	~~96~~	(97)	~~98~~	~~99~~	~~100~~

List all the natural numbers from 1 to 100. Cross out the number 1, because it is not a prime number. The number 2 is prime; circle it. Cross out all the other multiples of 2 (4, 6, 8,…), because they are not prime. The number 3 is prime; circle it. Cross out all the other multiples of 3 (6, 9, 12,…) that are not already crossed out. The number 4, the next consecutive number in the list, has already been crossed out. The number 5 is prime; circle it. Cross out all the other multiples of 5 that are not already crossed out. Continue in this manner until all the prime numbers less than 100 are circled.

A composite number is a natural number greater than 1 that has a natural-number factor other than itself and 1. The number 21 is a composite number because it has factors of 3 and 7. All the numbers crossed out in the preceding table, except the number 1, are composite numbers.

1. Use the Sieve of Eratosthenes to find the prime numbers between 100 and 200.

2. How many prime numbers are even numbers?

3. Find the "twin primes" between 100 and 200. Twin primes are two prime numbers whose difference is 2. For instance, 3 and 5 are twin primes; 5 and 7 are also twin primes.

4. **a.** List two prime numbers that are consecutive natural numbers.
   b. Can there be any other pairs of prime numbers that are consecutive natural numbers?

5. Some primes are the sum of a square and 1. For example, $5 = 2^2 + 1$. Find another prime p such that $p = n^2 + 1$, where n is a natural number.

6. Find a prime number p such that $p = n^2 - 1$, where n is a natural number.

Chapter 4 Summary

Key Words	Examples
A *variable* is a letter that is used for a quantity that is unknown or that can change. A *variable expression* is an expression that contains one or more variables. [4.1A, p. 247]	$4x + 2y - 6z$ is a variable expression. It contains the variables x, y, and z.
The *terms* of a variable expression are the addends of the expression. Each term is a *variable term* or a *constant term*. [4.1A, p. 247]	The expression $2a^2 - 3b^3 + 7$ has three terms: $2a^2$, $-3b^3$, and 7. $2a^2$ and $-3b^3$ are variable terms. 7 is a constant term.
A variable term is composed of a *numerical coefficient* and a *variable part*. [4.1A, p. 247]	For the expression $-7x^3y^2$, -7 is the coefficient and x^3y^2 is the variable part.
In a variable expression, replacing each variable by its value and then simplifying the resulting numerical expression is called *evaluating the variable expression*. [4.1A, p. 248]	To evaluate $2ab - b^2$ when $a = 3$ and $b = -2$, replace a by 3 and b by -2 and then simplify the numerical expression. $2(3)(-2) - (-2)^2 = -16$

Like terms of a variable expression are terms with the same variable part. Constant terms are like terms. [4.2A, p. 251]	For the expressions $3a^2 + 2b - 3$ and $2a^2 - 3a + 4$, $3a^2$ and $2a^2$ are like terms; -3 and 4 are like terms.
To simplify the sum of like variable terms, use the Distributive Property to add the numerical coefficients. This is called *combining like terms*. [4.2A, p. 251]	$5y + 3y = (5 + 3)y$ $= 8y$
The *additive inverse* of a number is the opposite of the number. [4.2A, p. 252]	-4 is the additive inverse of 4. $\frac{2}{3}$ is the additive inverse of $-\frac{2}{3}$. 0 is the additive inverse of 0.
The *multiplicative inverse* of a number is the *reciprocal of the number*. [4.2B, p. 253]	$\frac{3}{4}$ is the multiplicative inverse of $\frac{4}{3}$. $-\frac{1}{4}$ is the multiplicative inverse of -4.

Essential Rules and Procedures

Examples

The Distributive Property [4.2A, p. 251]
If a, b, and c are real numbers, then $a(b + c) = ab + ac$.

$5(4 + 7) = 5 \cdot 4 + 5 \cdot 7$
$= 20 + 35 = 55$

The Associative Property of Addition [4.2A, p. 251]
If a, b, and c are real numbers, then $(a + b) + c = a + (b + c)$.

$-4 + (2 + 7) = -4 + 9 = 5$
$(-4 + 2) + 7 = -2 + 7 = 5$

The Commutative Property of Addition [4.2A, p. 252]
If a and b are real numbers, then $a + b = b + a$.

$2 + 5 = 7$ and $5 + 2 = 7$

The Addition Property of Zero [4.2A, p. 252]
If a is a real number, then $a + 0 = 0 + a = a$.

$-8 + 0 = -8$ and $0 + (-8) = -8$

The Inverse Property of Addition [4.2A, p. 252]
If a is a real number, then $a + (-a) = (-a) + a = 0$.

$5 + (-5) = 0$ and $(-5) + 5 = 0$

The Associative Property of Multiplication [4.2B, p. 253]
If a, b, and c are real numbers, then $(ab)c = a(bc)$.

$-3 \cdot (5 \cdot 4) = -3(20) = -60$
$(-3 \cdot 5) \cdot 4 = -15 \cdot 4 = -60$

The Commutative Property of Multiplication [4.2B, p. 253]
If a and b are real numbers, then $ab = ba$.

$-3(7) = -21$ and $7(-3) = -21$

The Multiplication Property of One [4.2B, p. 253]
If a is a real number, then $a \cdot 1 = 1 \cdot a = a$.

$-3(1) = -3$ and $1(-3) = -3$

The Inverse Property of Multiplication [4.2B, p. 253]
If a is a real number, and a is not equal to zero, then
$a \cdot \frac{1}{a} = \frac{1}{a} \cdot a = 1$.

$-3 \cdot -\frac{1}{3} = 1$ and $-\frac{1}{3} \cdot -3 = 1$

Chapter 4 Review Exercises

1. Simplify: $3(x^2 - 8x - 7)$
$3x^2 - 24x - 21$ [4.2C]

2. Simplify: $7x + 4x$
$11x$ [4.2A]

3. Simplify: $6a - 4b + 2a$
$8a - 4b$ [4.2A]

4. Simplify: $(-50n)\left(\dfrac{1}{10}\right)$

$-5n$ [4.2B]

5. Evaluate $(5c - 4a)^2 - b$ when $a = -1$, $b = 2$, and $c = 1$.
79   [4.1A]

6. Simplify: $5(2x - 7)$
$10x - 35$ [4.2C]

7. Simplify: $2(6y^2 + 4y - 5)$
$12y^2 + 8y - 10$ [4.2C]

8. Simplify: $\dfrac{1}{4}(-24a)$

$-6a$ [4.2B]

9. Simplify: $-6(7x^2)$
$-42x^2$ [4.2B]

10. Simplify: $-9(7 + 4x)$
$-63 - 36x$ [4.2C]

11. Simplify: $12y - 17y$
$-5y$ [4.2A]

12. Evaluate $2bc \div (a + 7)$ when $a = 3$, $b = -5$, and $c = 4$.
-4 [4.1A]

13. Simplify: $7 - 2(3x + 4)$
$-6x - 1$ [4.2D]

14. Simplify: $6 + 2[2 - 5(4a - 3)]$
$-40a + 40$ [4.2D]

15. Simplify: $6(8y - 3) - 8(3y - 6)$
$24y + 30$ [4.2D]

16. Simplify: $5c + (-2d) - 3d - (-4c)$
$9c - 5d$ [4.2A]

17. Simplify: $5(4x)$
$20x$ [4.2B]

18. Simplify: $-4(2x - 9) + 5(3x + 2)$
$7x + 46$ [4.2D]

19. Evaluate $(b - a)^2 + c$ when $a = -2$, $b = 3$, and $c = 4$.
29   [4.1A]

20. Simplify: $-9r + 2s - 6s + 12s$
$-9r + 8s$ [4.2A]

21. Evaluate $(2x - y)^2 + (2x + y)^2$ when $x = -2$ and $y = -3$.
50 [4.1A]

22. Evaluate $b^2 - 4ac$ when $b = -4$, $a = 1$, and $c = -3$.
28 [4.1A]

23. Simplify: $4x - 3x^2 + 2x - x^2$
$-4x^2 + 6x$ [4.2A]

24. Simplify: $5[2 - 3(6x - 1)]$
$-90x + 25$ [4.2D]

25. Simplify: $(7a^2 - 2a + 3)4$
$28a^2 - 8a + 12$ [4.2C]

26. Simplify: $18 - (4x - 2)$
$-4x + 20$ [4.2D]

27. Translate "two-thirds of the total of x and 10" into a variable expression.
$\frac{2}{3}(x + 10)$ [4.3A]

28. Translate "6 less than x" into a variable expression.
$x - 6$ [4.3A]

29. Translate "a number plus twice the number" into a variable expression. Then simplify.
$x + 2x$; $3x$ [4.3B]

30. Translate "three times a number plus the product of five and one less than the number" into a variable expression. Then simplify.
$3x + 5(x - 1)$; $8x - 5$ [4.3B]

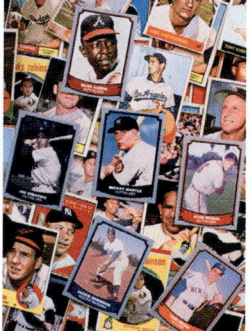

31. **Baseball Cards**   A baseball card collection contains five times as many National League players' cards as American League players' cards. Express the number of National League players' cards in the collection in terms of the number of American League players' cards.
Number of American League cards: A
Number of National League cards: $5A$ [4.3C]

32. **Nutrition**   A candy bar contains eight more calories than twice the number of calories in an apple. Express the number of calories in a candy bar in terms of the number of calories in an apple.
Number of calories in an apple: a
Number of calories in a candy bar: $2a + 8$ [4.3C]

Chapter 4 Test

1. Simplify: $3x - 5x + 7x$
  $5x$ [4.2A]

2. Simplify: $-3(2x^2 - 7y^2)$
  $-6x^2 + 21y^2$ [4.2C]

3. Simplify: $2x - 3(x - 2)$
  $-x + 6$ [4.2D]

4. Simplify: $2x + 3[4 - (3x - 7)]$
  $-7x + 33$ [4.2D]

5. Simplify: $3x - 7y - 12x$
  $-9x - 7y$ [4.2A]

6. Evaluate $b^2 - 3ab$ when $a = 3$ and $b = -2$.
  22 [4.1A]

7. Simplify: $\dfrac{1}{5}(10x)$
  $2x$ [4.2B]

8. Simplify: $5(2x + 4) - 3(x - 6)$
  $7x + 38$ [4.2D]

9. Simplify: $-5(2x^2 - 3x + 6)$
  $-10x^2 + 15x - 30$ [4.2C]

10. Simplify: $3x + (-12y) - 5x - (-7y)$
  $-2x - 5y$ [4.2A]

11. Evaluate $\dfrac{-2ab}{2b - a}$ when $a = -4$ and $b = 6$.
  3 [4.1A]

12. Simplify: $(12x)\left(\dfrac{1}{4}\right)$
  $3x$ [4.2B]

13. Simplify: $-7y^2 + 6y^2 - (-2y^2)$
  y^2 [4.2A]

14. Simplify: $-2(2x - 4)$
  $-4x + 8$ [4.2C]

15. Simplify: $\dfrac{2}{3}(-15a)$
  $-10a$ [4.2B]

16. Simplify: $-2[x - 2(x - y)] + 5y$
  $2x + y$ [4.2D]

17. Simplify: $(-3)(-12y)$
36y [4.2B]

18. Simplify: $5(3 - 7b)$
$15 - 35b$ [4.2C]

19. Translate "the difference between the squares of a and b" into a variable expression.
$a^2 - b^2$ [4.3A]

20. Translate "ten times the difference between a number and three" into a variable expression. Then simplify.
$10(x - 3)$; $10x - 30$ [4.3B]

21. Translate "the sum of a number and twice the square of the number" into a variable expression.
$x + 2x^2$ [4.3B]

22. Translate "three less than the quotient of six and a number" into a variable expression.
$\dfrac{6}{x} - 3$ [4.3B]

23. Translate "b decreased by the product of b and seven" into a variable expression.
$b - 7b$ [4.3A]

24. **Baseball**   The speed of a pitcher's fastball is twice the speed of the catcher's return throw. Express the speed of the fastball in terms of the speed of the return throw.
Speed of return throw: s; speed of fastball: 2s [4.3C]

25. **Metalwork**   A wire is cut into two lengths. The length of the longer piece is 3 in. less than four times the length of the shorter piece. Express the length of the longer piece in terms of the length of the shorter piece.
Shorter piece: x; longer piece: $4x - 3$ [4.3C]

Cumulative Review Exercises

1. Add: $-4 + 7 + (-10)$
-7 [3.2A]

2. Subtract: $-16 - (-25) - 4$
5 [3.2B]

3. Multiply: $(-2)(3)(-4)$
24 [3.3A]

4. Divide: $(-60) \div 12$
-5 [3.3B]

5. Find the prime factorization of 110.
$2 \cdot 5 \cdot 11$ [1.3D]

6. Simplify: $\dfrac{7}{12} - \dfrac{11}{16} - \left(-\dfrac{1}{3}\right)$

$\dfrac{11}{48}$ [3.4A]

7. Simplify: $-\dfrac{5}{12} \div \dfrac{5}{2}$

$-\dfrac{1}{6}$ [3.4B]

8. Simplify: $\left(-\dfrac{9}{16}\right) \cdot \left(\dfrac{8}{27}\right) \cdot \left(-\dfrac{3}{2}\right)$

$\dfrac{1}{4}$ [3.4B]

9. Estimate the sum of 397, 516, and 408.
1300 [1.2A]

10. Simplify: $-2^5 \div (3 - 5)^2 - (-3)$
-5 [3.5A]

11. Simplify: $\left(-\dfrac{3}{4}\right)^2 \div \left(\dfrac{3}{8} - \dfrac{11}{12}\right)$

$-\dfrac{27}{26}$ [3.5A]

12. Evaluate $a^2 - 3b$ when $a = 2$ and $b = -4$.
16 [4.1A]

13. Simplify: $-2x^2 - (-3x^2) + 4x^2$
$5x^2$ [4.2A]

14. Simplify: $5a - 10b - 12a$
$-7a - 10b$ [4.2A]

15. Write eight and three hundred fifty-seven thousandths in standard form.
8.357 [2.5A]

16. Place the correct symbol, $<$ or $>$, between the two numbers.
5.101 5.013
$5.101 > 5.013$ [2.5B]

17. Simplify: $3(8 - 2x)$
$24 - 6x$ [4.2C]

18. Simplify: $-2(-3y + 9)$
$6y - 18$ [4.2C]

19. Estimate the difference between 32.76 and 19.8.
10   [2.6A]

20. Round 8.667 to the nearest tenth.
8.7   [2.5C]

21. Simplify: $-4(2x^2 - 3y^2)$
$-8x^2 + 12y^2$ [4.2C]

22. Simplify: $-3(3y^2 - 3y - 7)$
$-9y^2 + 9y + 21$ [4.2C]

23. Simplify: $-3x - 2(2x - 7)$
$-7x + 14$ [4.2D]

24. Simplify: $4(3x - 2) - 7(x + 5)$
$5x - 43$ [4.2D]

25. Simplify: $2x + 3[x - 2(4 - 2x)]$
$17x - 24$ [4.2D]

26. Simplify: $3[2x - 3(x - 2y)] + 3y$
$-3x + 21y$ [4.2D]

27. Translate "the sum of one-half of b and b" into a variable expression.
$\frac{1}{2}b + b$ [4.3A]

28. Translate "10 divided by the difference between y and 2" into a variable expression.
$\frac{10}{y - 2}$ [4.3A]

29. Translate "the difference between eight and the quotient of a number and twelve" into a variable expression.
$8 - \frac{x}{12}$ [4.3B]

30. Translate "the sum of a number and two more than the number" into a variable expression. Then simplify.
$x + (x + 2); 2x + 2$ [4.3B]

31. **Geometry**   The length of each side of a square is 2.25 in. Find the perimeter of the square. Use the formula $P = 4s$.
9 in.   [2.6E]

32. **Internet Connections**   The speed of a DSL (Digital Subscriber Line) Internet connection is ten times faster than that of a dial-up connection. Express the speed of the DSL connection in terms of the speed of the dial-up connection.
Speed of dial-up connection: s
Speed of DSL connection: $10s$ [4.3C]

chapter

5

Solving Equations

Suppose that this bus is traveling at the posted speed limit of 45 mph and is making a 180-mile drive between two cities. How long will the bus ride last? To figure out the answer, you will need to use the formula $d = rt$, where d is the distance traveled, r is the rate of speed, and t is the time spent traveling. In this case, d is 180 and r is 45. When you solve for t, you get the correct answer of 4 h. The formula $d = rt$ is also used to solve **Exercises 34 and 35 on page 322**, which are more advanced motion problems involving bus travel.

OBJECTIVES

Section 5.1

A To determine whether a given number is a solution of an equation

B To solve an equation of the form $x + a = b$

C To solve an equation of the form $ax = b$

D To solve uniform motion problems

Section 5.2

A To solve an equation of the form $ax + b = c$

B To solve application problems using formulas

Section 5.3

A To solve an equation of the form $ax + b = cx + d$

B To solve an equation containing parentheses

C To solve application problems using formulas

Section 5.4

A To solve integer problems

B To translate a sentence into an equation and solve

Section 5.5

A To solve value mixture problems

B To solve uniform motion problems

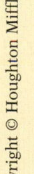

Need help? For online student resources, such as section quizzes, visit this textbook's website at **math.college.hmco.com/students.**

PREP TEST • • •

Do these exercises to prepare for Chapter 5.

1. Subtract: $8 - 12$

-4 [3.2B]

2. Multiply: $-\dfrac{3}{4}\left(-\dfrac{4}{3}\right)$

1 [3.4B]

3. Multiply: $-\dfrac{5}{8}(16)$

-10 [3.4B]

4. Simplify: $\dfrac{-3}{-3}$

1 [3.4B]

5. Simplify: $-16 + 7y + 16$

$7y$ [4.2A]

6. Simplify: $8x - 9 - 8x$

-9 [4.2A]

7. Evaluate $2x + 3$ when $x = -4$.

-5 [4.1A]

GO FIGURE • • •

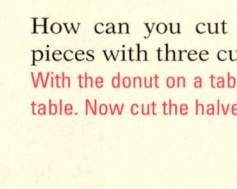

How can you cut a donut into eight equal pieces with three cuts of the knife?

With the donut on a table, slice the donut parallel to the table. Now cut the halved donut into quarters.

5.1 Introduction to Equations

Objective A | To determine whether a given number is a solution of an equation

Point of Interest

One of the most famous equations ever stated is $E = mc^2$. This equation, stated by Albert Einstein, shows that there is a relationship between mass m and energy E. As a side note, the chemical element einsteinium was named in honor of Einstein.

An **equation** expresses the equality of two mathematical expressions. The expressions can be either numerical or variable expressions.

$$\left.\begin{array}{l} 9 + 3 = 12 \\ 3x - 2 = 10 \\ y^2 + 4 = 2y - 1 \\ z = 2 \end{array}\right\} \text{Equations}$$

The equation at the right is true if the variable is replaced by 5.

$$\begin{array}{l} x + 8 = 13 \\ 5 + 8 = 13 \qquad \text{A true equation} \end{array}$$

The equation is false if the variable is replaced by 7.

$$7 + 8 = 13 \qquad \text{A false equation}$$

A **solution of an equation** is a number that, when substituted for the variable, results in a true equation. 5 is a solution of the equation $x + 8 = 13$. 7 is not a solution of the equation $x + 8 = 13$.

HOW TO Is -2 a solution of $2x + 5 = x^2 - 3$?

TAKE NOTE

The Order of Operations Agreement applies to evaluating $2(-2) + 5$ and $(-2)^2 - 3$.

$$\begin{array}{c|c} 2x + 5 & = x^2 - 3 \\ \hline 2(-2) + 5 & (-2)^2 - 3 \\ -4 + 5 & 4 - 3 \\ 1 & = 1 \end{array}$$

- Replace x with -2.
- Evaluate the numerical expressions.
- If the results are equal, -2 is a solution of the equation. If the results are not equal, -2 is not a solution of the equation.

Yes, -2 is a solution of the equation.

Example 1 Is -4 a solution of
$5x - 2 = 6x + 2$?

Solution
$$\begin{array}{c|c} 5x - 2 & = 6x + 2 \\ \hline 5(-4) - 2 & 6(-4) + 2 \\ -20 - 2 & -24 + 2 \\ -22 & = -22 \end{array}$$

• Replace x with -4.

Yes, -4 is a solution.

You Try It 1 Is $\frac{1}{4}$ a solution of
$5 - 4x = 8x + 2$?

Your solution Yes

Example 2 Is -4 a solution of
$4 + 5x = x^2 - 2x$?

Solution
$$\begin{array}{c|c} 4 + 5x & = x^2 - 2x \\ \hline 4 + 5(-4) & (-4)^2 - 2(-4) \\ 4 + (-20) & 16 - (-8) \\ -16 & \neq 24 \end{array}$$

($\neq$ means "is not equal to")

No, -4 is not a solution.

You Try It 2 Is 5 a solution of
$10x - x^2 = 3x - 10$?

Your solution No

Solutions on p. S13

Vocabulary to Review
equation [1.2A]
solution of an equation [1.2A]

Concept Check

1. Is 3 a solution of $4x + 1 = 11$?
No

2. Is -5 a solution of $x^2 = 25$?
Yes

Instructor Note

Remind students that determining whether a given number is a solution of an equation is a way to check an answer after the equation has been solved.

Instructor Note

Optical scanners, such as those used for reading ZIP codes and universal product codes, use an equation to determine whether the machine has correctly read the numbers in the code.

In-Class Examples (Objective 5.1A)

1. Is -6 a solution of $4x + 3 = 2x - 9$?   Yes

2. Is $-\frac{2}{3}$ a solution of $4 - 6x = 9x + 1$?   No

Objective 5.1B

New Vocabulary
solve an equation
equivalent equations

Properties to Review
Addition Property of Zero   [1.2A]

New Properties
Addition Property of Equations

Instructor Note
To motivate the need for the properties of equations, you might consider asking students to find the solutions of equations such as the following by guessing. By the time they reach the last equation, they may have more appreciation for the properties.

$$x + 5 = 7$$
$$7 - x = 9$$
$$2x - 3 = 7$$
$$\frac{x}{2} - 1 = 3$$
$$5 - 3x = 27$$

Instructor Note
The model of an equation as a balance scale will appeal to some students. If the same weight is not added to each side of the scale, the pans no longer balance.

Discuss the Concepts
What is the solution of the equation $x = 9$? Use the answer to explain why the goal in solving the equations in this objective is to get the variable alone on one side of the equation.

Objective B To solve an equation of the form $x + a = b$

To **solve an equation** means to find a solution of the equation. The simplest equation to solve is an equation of the form *variable = constant*, because the constant is the solution.

The solution of the equation $x = 5$ is 5 because $5 = 5$ is a true equation.

The solution of the equation at the right is 7 because $7 + 2 = 9$ is a true equation.

$$x + 2 = 9 \qquad\qquad 7 + 2 = 9$$

Note that if 4 is added to each side of the equation $x + 2 = 9$, the solution is still 7.

$$\begin{aligned} x + 2 &= 9 \\ x + 2 + 4 &= 9 + 4 \\ x + 6 &= 13 \end{aligned} \qquad 7 + 6 = 13$$

If -5 is added to each side of the equation $x + 2 = 9$, the solution is still 7.

$$\begin{aligned} x + 2 &= 9 \\ x + 2 + (-5) &= 9 + (-5) \\ x - 3 &= 4 \end{aligned} \qquad 7 - 3 = 4$$

Equations that have the same solution are **equivalent equations.** The equations $x + 2 = 9$, $x + 6 = 13$, and $x - 3 = 4$ are equivalent equations; each equation has 7 as its solution. These examples suggest that adding the same number to both sides of an equation produces an equivalent equation. This is called the *Addition Property of Equations.*

> **Addition Property of Equations**
>
> The same number can be added to each side of an equation without changing its solution. In symbols, the equation $a = b$ has the same solution as the equation $a + c = b + c$.

In solving an equation, the goal is to rewrite the given equation in the form *variable = constant*. The Addition Property of Equations is used to remove a *term* from one side of the equation by adding the opposite of that term to each side of the equation.

HOW TO   Solve: $x - 4 = 2$

$$x - 4 = 2$$ • The goal is to rewrite the equation as *variable = constant.*
$$x - 4 + 4 = 2 + 4$$ • Add 4 to each side of the equation.
$$x + 0 = 6$$ • Simplify.
$$x = 6$$ • The equation is in the form *variable = constant.*

Check:
$$\frac{x - 4 = 2}{6 - 4 \,\big|\, 2}$$
$$2 = 2$$ • A true equation

The solution is 6.

Because subtraction is defined in terms of addition, the Addition Property of Equations also makes it possible to subtract the same number from each side of an equation without changing the solution of the equation.

In-Class Examples (Objective 5.1B)
Solve.

1. $x - \dfrac{1}{4} = \dfrac{5}{6}$ $\dfrac{13}{12}$

2. $3 + x = 9$ 6

3. $5 = x + 5$ 0

HOW TO Solve: $y + \dfrac{3}{4} = \dfrac{1}{2}$

$y + \dfrac{3}{4} = \dfrac{1}{2}$ • The goal is to rewrite the equation in the form *variable = constant*.

$y + \dfrac{3}{4} - \dfrac{3}{4} = \dfrac{1}{2} - \dfrac{3}{4}$ • Subtract $\dfrac{3}{4}$ from each side of the equation.

$y + 0 = \dfrac{2}{4} - \dfrac{3}{4}$ • Simplify.

$y = -\dfrac{1}{4}$ • The equation is in the form *variable = constant*.

The solution is $-\dfrac{1}{4}$. You should check this solution.

Example 3 Solve: $x + \dfrac{2}{5} = \dfrac{1}{3}$

Solution

$x + \dfrac{2}{5} = \dfrac{1}{3}$

$x + \dfrac{2}{5} - \dfrac{2}{5} = \dfrac{1}{3} - \dfrac{2}{5}$ • Subtract $\dfrac{2}{5}$ from each side.

$x + 0 = \dfrac{5}{15} - \dfrac{6}{15}$

$x = -\dfrac{1}{15}$

The solution is $-\dfrac{1}{15}$.

You Try It 3 Solve: $\dfrac{5}{6} = y - \dfrac{3}{8}$

Your solution $\dfrac{29}{24}$

Solution on p. S13

Objective C **To solve an equation of the form $ax = b$**

VIDEO & DVD   CD TUTOR   WWW WEB   SSM

The solution of the equation at the right is 3 because $2 \cdot 3 = 6$ is a true equation.

Note that if each side of $2x = 6$ is multiplied by 5, the solution is still 3.

If each side of $2x = 6$ is multiplied by -4, the solution is still 3.

$2x = 6$ $2 \cdot 3 = 6$

$2x = 6$
$5(2x) = 5 \cdot 6$
$10x = 30$ $10 \cdot 3 = 30$

$2x = 6$
$(-4)(2x) = (-4)6$
$-8x = -24$ $-8 \cdot 3 = -24$

The equations $2x = 6$, $10x = 30$, and $-8x = -24$ are equivalent equations; each equation has 3 as its solution. These examples suggest that multiplying each side of an equation by the same nonzero number produces an equivalent equation.

> **Multiplication Property of Equations**
>
> Each side of an equation can be multiplied by the same *nonzero* number without changing the solution of the equation. In symbols, if $c \neq 0$, then the equation $a = b$ has the same solutions as the equation $ac = bc$.

Concept Check

Which of the following are equations of the form $x + a = b$?

a. $y + 7.8 = -9.2$
b. $0.3 = z + 1.4$
c. $-9 = 3d$
d. $-8 + n = -5.6$

a, b, and d are equations of the form $x + a = b$.

Optional Student Activity

Make equations by using the numbers 4, 8, and 12 to fill in the boxes.

$$x + \square = \square - \square$$

What is the largest number solution possible? 0 What is the smallest number solution possible? -16

Objective 5.1C

Instructor Note

The requirement that each side be multiplied by a *nonzero* number is important. Later in the text, students will solve some equations by multiplying each side by a variable expression whose value may be zero. This can yield extraneous solutions.

Properties to Review

Multiplication Property of One [1.3A]

New Properties

Multiplication Property of Equations

In-Class Examples (Objective 5.1C)

Solve.

1. $-\dfrac{5}{8}x = 25$ -40

2. $4y - 10y = -42$ 7

3. $8 = \dfrac{3x}{4}$ $\dfrac{32}{3}$

4. $2z = 0$ 0

Instructor Note

Solutions to equations such as $ax = b$ have appeared in algebra texts for a long time. The problem below is an adaptation from Fibonacci's text *Liber Abaci,* which dates from 1202.

A merchant purchased eggs at a price of 7 eggs for 1 denarius and sold them at a price of 5 eggs for 1 denarius. The merchant's profit was 18 denarii. How much did the merchant invest? The resulting equation is

$$\frac{7}{5}x - x = 18$$

Solving this equation can serve as a classroom exercise. The solution is 45 denarii.

Instructor Note

Some students may not realize that $\frac{3}{4}x$ is equivalent to $\frac{3x}{4}$.

Concept Check

Which are equivalent equations?

a. $5x = -20$

b. $-2x = 8$

c. $24 = 6x$

d. $\frac{3}{4}x = -3$

e. $-1 = -\frac{1}{4}x$

f. $x = 4$

g. $x = -4$

a, b, d, and g are equivalent equations. c, e, and f are equivalent equations.

Optional Student Activity

Match each numbered equation with a lettered question that can be used to solve the equation.

1. $x + 3 = 8$ d

2. $x - 5 = 20$ b

3. $4x = 16$ a

4. $\frac{x}{7} = 1$ e

5. $99 = -9x$ c

a. Four times what number is equal to 16?

b. What number minus 5 is equal to 20?

c. 99 is equal to -9 times what number?

d. What number plus 3 is equal to 8?

e. What number divided by 7 is equal to 1?

The Multiplication Property of Equations is used to remove a coefficient by multiplying each side of the equation by the reciprocal of the coefficient.

HOW TO   Solve: $\frac{3}{4}z = 9$

$\frac{3}{4}z = 9$ • The goal is to rewrite the equation in the form *variable = constant.*

$\frac{4}{3} \cdot \frac{3}{4}z = \frac{4}{3} \cdot 9$ • Multiply each side of the equation by $\frac{4}{3}$.

$1 \cdot z = 12$ • Simplify.

$z = 12$ • The equation is in the form *variable = constant.*

The solution is 12. You should check this solution.

Because division is defined in terms of multiplication, each side of an equation can be divided by the same nonzero number without changing the solution of the equation.

TAKE NOTE

Remember to check the solution.

Check: $6x = 14$

$6\left(\dfrac{7}{3}\right) \bigg| 14$

$14 = 14$

HOW TO   Solve: $6x = 14$

$6x = 14$ • The goal is to rewrite the equation in the form *variable = constant.*

$\frac{6x}{6} = \frac{14}{6}$ • Divide each side of the equation by 6.

$x = \frac{7}{3}$ • Simplify. The equation is in the form *variable = constant.*

The solution is $\frac{7}{3}$.

When using the Multiplication Property of Equations, multiply each side of the equation by the reciprocal of the coefficient when the coefficient is a fraction. Divide each side of the equation by the coefficient when the coefficient is an integer or a decimal.

Example 4   Solve: $\dfrac{3x}{4} = -9$

Solution

$\dfrac{3x}{4} = -9$

$\dfrac{4}{3} \cdot \dfrac{3}{4}x = \dfrac{4}{3}(-9)$ • $\dfrac{3x}{4} = \dfrac{3}{4}x$

$x = -12$

The solution is -12.

You Try It 4   Solve: $-\dfrac{2x}{5} = 6$

Your solution -15

Example 5   Solve: $5x - 9x = 12$

Solution

$5x - 9x = 12$

$-4x = 12$ • Combine like terms.

$\dfrac{-4x}{-4} = \dfrac{12}{-4}$

$x = -3$

The solution is -3.

You Try It 5   Solve: $4x - 8x = 16$

Your solution -4

Solutions on p. S13

Objective D **To solve uniform motion problems**

Objective 5.1D

> **TAKE NOTE**
> A car traveling in a *circle* at a constant speed of 45 mph is *not* in uniform motion because the direction of the car is always changing.

Any object that travels at a constant speed in a straight line is said to be in *uniform motion*. **Uniform motion** means that the speed and direction of an object do not change. For instance, a car traveling at a constant speed of 45 mph on a straight road is in uniform motion.

The solution of a uniform motion problem is based on the **uniform motion equation** $d = rt$, where d is the distance traveled, r is the rate of travel, and t is the time spent traveling. For instance, suppose a car travels at 50 mph for 3 h. Because the rate (50 mph) and time (3 h) are known, we can find the distance traveled by solving the equation $d = rt$ for d.

$$d = rt$$
$$d = 50(3) \qquad \bullet\ r = 50,\, t = 3$$
$$d = 150$$

The car travels a distance of 150 mi.

> **Study Tip**
> Note that solving a word problem includes stating a strategy and using the strategy to find a solution. If you have difficulty with a word problem, write down the known information. Be very specific. Write out a phrase or sentence that states what you are trying to find. See *AIM for Success,* page xxxiii.

HOW TO A jogger runs 3 mi in 45 min. What is the rate of the jogger in miles per hour?

Strategy • Because the answer must be in miles per *hour* and the given time is in *minutes,* convert 45 min to hours.

• To find the rate of the jogger, solve the equation $d = rt$ for r.

Solution $45\ \text{min} = \dfrac{45}{60}\ \text{h} = \dfrac{3}{4}\ \text{h}$

$$d = rt$$
$$3 = r\left(\frac{3}{4}\right) \qquad \bullet\ d = 3,\, t = \frac{3}{4}$$
$$3 = \frac{3}{4}r$$
$$\left(\frac{4}{3}\right)3 = \left(\frac{4}{3}\right)\frac{3}{4}r \qquad \bullet\ \text{Multiply each side of the equation}$$
$$\qquad\qquad\qquad \text{by the reciprocal of } \frac{3}{4}.$$
$$4 = r$$

The rate of the jogger is 4 mph.

If two objects are moving in opposite directions, then the rate at which the distance between them is increasing is the sum of the speeds of the two objects. For instance, in the diagram below, two cars start from the same point and travel in opposite directions. The distance between them is changing at 70 mph.

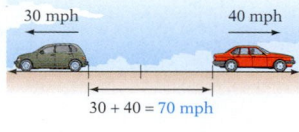

30 mph   40 mph

$30 + 40 = 70$ mph

New Vocabulary
uniform motion

New Equations
Uniform Motion Equation
$(d = rt)$

Instructor Note
The equation $d = rt$ is introduced in this objective, and students are asked to use this equation to solve problems of the form $ax = b$. This will give them exposure to the concepts involved in uniform motion problems prior to attempting the more difficult motion problems in Objective 5.5B.

Discuss the Concepts
1. Marvin and Luis, who runs faster than Marvin, are 100 yd apart and running toward each other.
   a. Who will have traveled the longer distance when they meet?   Luis
   b. When they meet, what is the total distance traveled by Marvin and Luis?
      100 yd
2. Fiona starts running on a straight race track. Two minutes later, Gabriella, who runs faster than Fiona, starts on the same course chasing Fiona. She catches up to Fiona 5 min later. Assume both women are running in a straight line.
   a. Are the distances run by Fiona and Gabriella the same, or does one woman run farther than the other?
      The distances are the same.
   b. Who was running the longer time?   Fiona

In-Class Examples (Objective 5.1D)

1. Ted leaves his house at 8:00 A.M. and arrives at work at 8:30 A.M. If the trip to work is 15 mi, determine Ted's average rate of speed.   30 mph

2. Joan leaves her house and travels at an average speed of 45 mph toward her cabin in the mountains. If the distance from her house to the cabin is 180 mi, how many hours will it take for Joan to arrive at her cabin if she stops one hour for lunch?   5 h

Concept Check

1. Suppose you are on a moving sidewalk that is moving at a rate of 4 mph. If you are walking in the direction of the sidewalk at a rate of 3 mph, what is your speed relative to a regular sidewalk that parallels the moving sidewalk?   7 mph

2. Suppose you are on a moving sidewalk that is moving at a rate of 4 mph. If you are walking in the opposite direction of the sidewalk at a rate of 3 mph, what is your speed relative to a stationary sidewalk that parallels the moving sidewalk?   1 mph

Similarly, if two objects are moving toward each other, the horizontal distance between them is decreasing at a rate that is equal to the sum of the speeds. The rate at which the two planes at the right are approaching one another is 800 mph.

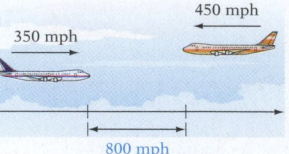

HOW TO   Two cars start from the same point and move in opposite directions. The car moving west is traveling 45 mph, and the car moving east is traveling 60 mph. In how many hours will the cars be 210 mi apart?

Strategy   The distance is 210 mi. Therefore, $d = 210$. The cars are moving in opposite directions, so the rate at which the distance between them is changing is the sum of the rates of the two cars. The rate is 45 mph + 60 mph = 105 mph. Therefore, $r = 105$. To find the time, solve the equation $d = rt$ for t.

Solution

$$d = rt$$
$$210 = 105t \qquad \bullet \ d = 210, \ r = 105$$
$$\frac{210}{105} = \frac{105t}{105} \qquad \bullet \ \text{Solve for } t.$$
$$2 = t$$

In 2 h, the cars will be 210 mi apart.

If a motorboat is on a river that is flowing at a rate of 4 mph, then the boat will float down the river at a speed of 4 mph when the motor is not on. Now suppose the motor is turned on and the power adjusted so that the boat will travel 10 mph without the aid of the current. Then, if the boat is moving with the current, its effective speed is the speed of the boat using power plus the speed of the current: 10 mph + 4 mph = 14 mph. (See the figure below.)

However, if the boat is moving against the current, the current slows the boat down, and the effective speed of the boat is the speed of the boat using power minus the speed of the current: 10 mph − 4 mph = 6 mph. (See the figure below.)

There are other situations in which the preceding concepts may be applied.

HOW TO An airline passenger is walking between two airline terminals and decides to get on a moving sidewalk that is 150 ft long. If the passenger walks at a rate of 7 ft/s and the moving sidewalk moves at a rate of 9 ft/s, how long, in seconds, will it take the passenger to walk from one end of the moving sidewalk to the other? Round to the nearest thousandth.

TAKE NOTE

The term ft/s is an abbreviation for "feet per second." Similarly, cm/s is "centimeters per second" and m/s is "meters per second."

Strategy The distance is 150 ft. Therefore, $d = 150$. The passenger is traveling at 7 ft/s and the moving sidewalk is traveling at 9 ft/s. The rate of the passenger is the sum of the two rates, or 16 ft/s. Therefore, $r = 16$. To find the time, solve the equation $d = rt$ for t.

Solution
$$d = rt$$
$$150 = 16t \quad \bullet\ d = 150,\ r = 16$$
$$\frac{150}{16} = \frac{16t}{16} \quad \bullet\ \text{Solve for } t.$$
$$9.375 = t$$

It will take 9.375 s for the passenger to travel the length of the moving sidewalk.

Example 6

Two cyclists start at the same time at opposite ends of an 80-mile course. One cyclist is traveling 18 mph, and the second cyclist is traveling 14 mph. How long after they begin will they meet?

Strategy
The distance is 80 mi. Therefore, $d = 80$. The cyclists are moving toward each other, so the rate at which the distance between them is changing is the sum of the rates of the two cyclists. The rate is 18 mph + 14 mph = 32 mph. Therefore, $r = 32$. To find the time, solve the equation $d = rt$ for t.

Solution
$$d = rt$$
$$80 = 32t \quad \bullet\ d = 80,\ r = 32$$
$$\frac{80}{32} = \frac{32t}{32} \quad \bullet\ \text{Solve for } t.$$
$$2.5 = t$$

The cyclists will meet in 2.5 h.

You Try It 6

A plane that can normally travel at 250 mph in calm air is flying into a headwind of 25 mph. How far can the plane fly in 3 h?

Your strategy

Your solution 675 mi

Solution on p. S13

Section 5.1

Suggested Assignment

Exercises 1–97, every other odd
Exercises 99–107, odds
More challenging problems:
  Exercises 109, 110

Answers to Writing Exercises

1. An equation contains an equals sign; an expression does not contain an equals sign.

2. A given number can be checked as a solution of an equation by replacing the variable with that number and then simplifying the resulting expressions. The given number is a solution if the left and right sides of the equation are equal.

21. Zero can be the solution of an equation. For instance, the solution of $x + 9 = 9$ is 0.

22. x is less than $-\frac{21}{43}$ because a positive constant is added to x to get $-\frac{21}{43}$.

5.1 Exercises

Objective A **To determine whether a given number is a solution of an equation**

1. What is the difference between an equation and an expression?

2. Explain how to determine whether a given number is a solution of an equation.

3. Is 4 a solution of $2x = 8$?
Yes

4. Is 3 a solution of $y + 4 = 7$?
Yes

5. Is -1 a solution of $2b - 1 = 3$?
No

6. Is -2 a solution of $3a - 4 = 10$?
No

7. Is 1 a solution of $4 - 2m = 3$?
No

8. Is 2 a solution of $7 - 3n = 2$?
No

9. Is 5 a solution of $2x + 5 = 3x$?
Yes

10. Is 4 a solution of $3y - 4 = 2y$?
Yes

11. Is -2 a solution of $3a + 2 = 2 - a$?
No

12. Is 3 a solution of $z^2 + 1 = 4 + 3z$?
No

13. Is 2 a solution of $2x^2 - 1 = 4x - 1$?
Yes

14. Is -1 a solution of $y^2 - 1 = 4y + 3$?
No

15. Is 4 a solution of $x(x + 1) = x^2 + 5$?
No

16. Is 3 a solution of $2a(a - 1) = 3a + 3$?
Yes

17. Is $-\frac{1}{4}$ a solution of $8t + 1 = -1$?
Yes

18. Is $\frac{1}{2}$ a solution of $4y + 1 = 3$?
Yes

19. Is $\frac{2}{5}$ a solution of $5m + 1 = 10m - 3$?
No

20. Is $\frac{3}{4}$ a solution of $8x - 1 = 12x + 3$?
No

Objective B **To solve an equation of the form $x + a = b$**

21. Can 0 ever be the solution of an equation? If so, give an example of an equation for which 0 is a solution.

22. Without solving $x + \frac{13}{15} = -\frac{21}{43}$, determine whether x is less than or greater than $-\frac{21}{43}$. Explain your answer.

For Exercises 23 to 58, solve and check.

23. $x + 5 = 7$
2

24. $y + 3 = 9$
6

25. $b - 4 = 11$
15

26. $z - 6 = 10$
16

27. $2 + a = 8$
6

28. $5 + x = 12$
7

29. $n - 5 = -2$
3

30. $x - 6 = -5$
1

31. $b + 7 = 7$
0

32. $y - 5 = -5$
0

33. $z + 9 = 2$
-7

34. $n + 11 = 1$
-10

35. $10 + m = 3$
-7

36. $8 + x = 5$
-3

37. $9 + x = -3$
-12

38. $10 + y = -4$
-14

Quick Quiz (Objective 5.1A)

1. Is -4 a solution of $3x - 5 = 7$?   No

2. Is $\frac{2}{3}$ a solution of $6x + 5 = 9$?   Yes

3. Is 6 a solution of $x(x + 2) = x^2 + 2$?   No

39. $2 = x + 7$
   −5

40. $-8 = n + 1$
   −9

41. $4 = m - 11$
   15

42. $-6 = y - 5$
   −1

43. $12 = 3 + w$
   9

44. $-9 = 5 + x$
   −14

45. $4 = -10 + b$
   14

46. $-7 = -2 + x$
   −5

47. $m + \dfrac{2}{3} = -\dfrac{1}{3}$
   −1

48. $c + \dfrac{3}{4} = -\dfrac{1}{4}$
   −1

49. $x - \dfrac{1}{2} = \dfrac{1}{2}$
   1

50. $x - \dfrac{2}{5} = \dfrac{3}{5}$
   1

51. $\dfrac{5}{8} + y = \dfrac{1}{8}$
   $-\dfrac{1}{2}$

52. $\dfrac{4}{9} + a = -\dfrac{2}{9}$
   $-\dfrac{2}{3}$

53. $m + \dfrac{1}{2} = -\dfrac{1}{4}$
   $-\dfrac{3}{4}$

54. $b + \dfrac{1}{6} = -\dfrac{1}{3}$
   $-\dfrac{1}{2}$

55. $x + \dfrac{2}{3} = \dfrac{3}{4}$
   $\dfrac{1}{12}$

56. $n + \dfrac{2}{5} = \dfrac{2}{3}$
   $\dfrac{4}{15}$

57. $-\dfrac{5}{6} = x - \dfrac{1}{4}$
   $-\dfrac{7}{12}$

58. $-\dfrac{1}{4} = c - \dfrac{2}{3}$
   $\dfrac{5}{12}$

Objective C   To solve an equation of the form $ax = b$

59. ✎ Without solving $-\dfrac{15}{41}x = -\dfrac{23}{25}$, determine whether x is less than or greater than 0. Explain your answer.

60. ✎ Explain why multiplying each side of an equation by the reciprocal of the coefficient of the variable is the same as dividing each side of the equation by the coefficient.

Answers to Writing Exercises

59. x is greater than 0 because a negative number times a positive number is a negative number.

60. Dividing by a number is the same as multiplying by the reciprocal of that number.

For Exercises 61 to 98, solve and check.

61. $5x = -15$
   −3

62. $4y = -28$
   −7

63. $3b = 0$
   0

64. $2a = 0$
   0

65. $-3x = 6$
   −2

66. $-5m = 20$
   −4

67. $-3x = -27$
   9

68. $-\dfrac{1}{6}n = -30$
   180

69. $20 = \dfrac{1}{4}c$
   80

70. $18 = 2t$
   9

71. $0 = -5x$
   0

72. $0 = -8a$
   0

73. $49 = -7t$
   −7

74. $\dfrac{x}{3} = 2$
   6

75. $\dfrac{x}{4} = 3$
   12

76. $-\dfrac{y}{2} = 5$
   −10

Quick Quiz (Objective 5.1B)
Solve.
1. $a + 5 = -8$ −13
2. $7 = b - 4$ 11
3. $c + \dfrac{5}{6} = \dfrac{1}{3}$ $-\dfrac{1}{2}$

77. $-\dfrac{b}{3} = 6$
-18

78. $\dfrac{3}{4}y = 9$
12

79. $\dfrac{2}{5}x = 6$
15

80. $-\dfrac{2}{3}d = 8$
-12

81. $-\dfrac{3}{5}m = 12$
-20

82. $\dfrac{2n}{3} = 0$
0

83. $\dfrac{5x}{6} = 0$
0

84. $\dfrac{-3z}{8} = 9$
-24

85. $\dfrac{3x}{4} = 2$
$\dfrac{8}{3}$

86. $\dfrac{3}{4}c = \dfrac{3}{5}$
$\dfrac{4}{5}$

87. $\dfrac{2}{9} = \dfrac{2}{3}y$
$\dfrac{1}{3}$

88. $-\dfrac{6}{7} = -\dfrac{3}{4}b$
$\dfrac{8}{7}$

89. $\dfrac{1}{5}x = -\dfrac{1}{10}$
$-\dfrac{1}{2}$

90. $-\dfrac{2}{3}y = -\dfrac{8}{9}$
$\dfrac{4}{3}$

91. $-1 = \dfrac{2n}{3}$
$-\dfrac{3}{2}$

92. $-\dfrac{3}{4} = \dfrac{a}{8}$
-6

93. $-\dfrac{2}{5}m = -\dfrac{6}{7}$
$\dfrac{15}{7}$

94. $5x + 2x = 14$
2

95. $3n + 2n = 20$
4

96. $7d - 4d = 9$
3

97. $10y - 3y = 21$
3

98. $2x - 5x = 9$
-3

Objective D   **To solve uniform motion problems**

99. As part of the training program for the Boston Marathon, a runner wants to build endurance by running at a rate of 9 mph for 20 min. How far will the runner travel in that time period?
3 mi

100. It takes a hospital dietician 40 min to drive from home to the hospital, a distance of 20 mi. What is the dietician's average rate of speed in miles per hour?
30 mph

101. Marcella leaves home at 9:00 A.M. and drives to school, arriving at 9:45 A.M. If the distance between home and school is 27 mi, what is Marcella's average rate of speed in miles per hour?
36 mph

Quick Quiz (Objective 5.1C)
Solve.
1. $3x = -21$ -7
2. $-12 = \dfrac{2}{3}x$ -18
3. $8x - 3x = 30$ 6

102. The Ride for Health Bicycle Club has chosen a 36-mile course for this Saturday's ride. If the riders plan on averaging 12 mph while they are riding, and they have a 1-hour lunch break planned, how long will it take them to complete the trip?
4 h

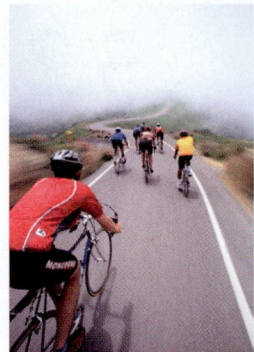

103. Palmer's average running speed is 3 kilometers per hour faster than his walking speed. If Palmer can run around a 30-kilometer course in 2 h, how many hours would it take for Palmer to walk the same course?
2.5 h

104. A shopping mall has a moving sidewalk that takes shoppers from the shopping area to the parking garage, a distance of 250 ft. If your normal walking rate is 5 ft/s and the moving sidewalk is traveling at 3 ft/s, how many seconds would it take for you to walk from one end of the moving sidewalk to the other end?
31.25 s

105. Two joggers start at the same time from opposite ends of an 8-mile jogging trail and begin running toward each other. One jogger is running at the rate of 5 mph, and the other jogger is running at a rate of 7 mph. How long, in minutes, after they start will the two joggers meet?
40 min

106. Two cyclists start from the same point at the same time and move in opposite directions. One cyclist is traveling at 8 mph, and the other cyclist is traveling at 9 mph. After 30 min, how far apart are the two cyclists?
8.5 mi

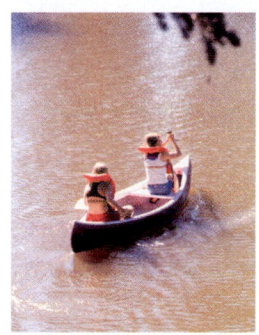

107. Petra and Celine can paddle their canoe at a rate of 10 mph in calm water. How long will it take them to travel 4 mi against the 2-mph current of the river?
0.5 h

108. At 8:00 A.M., a train leaves a station and travels at a rate of 45 mph. At 9:00 A.M., a second train leaves the same station on the same track and travels in the direction of the first train at a speed of 60 mph. At 10:00 A.M., how far apart are the two trains?
30 mi

APPLYING THE CONCEPTS

109. a. Make up an equation of the form $x + a = b$ that has 2 as a solution.
b. Make up an equation of the form $ax = b$ that has -1 as a solution.
a. Answers will vary.   **b.** Answers will vary.

110. ✏ Write out the steps for solving the equation $\frac{1}{2}x = -3$. Identify each Property of Real Numbers or Property of Equations as you use it.

111. ✏ In your own words, state the Addition Property of Equations and the Multiplication Property of Equations.

Answers to Writing Exercises

110. Look for the following explanations in your students' work.

$$\frac{1}{2}x = -3$$

$$2 \cdot \frac{1}{2}x = 2(-3)$$

By the Multiplication Property of Equations, each side of an equation can be multiplied by the same nonzero number without changing the solution of the equation. Multiply each side of the equation by the reciprocal of the coefficient of x.

$$\left(2 \cdot \frac{1}{2}\right)x = -6$$

Use the Associative Property of Multiplication to group 2 and $\frac{1}{2}$.

$$1x = -6$$

By the Inverse Property of Multiplication, the product of a nonzero number and its reciprocal is 1.

$$x = -6$$

By the Multiplication Property of One, the product of a number and 1 is the number.

111. Students should paraphrase the definitions of the properties given in the text.

The Addition Property of Equations states that the same number can be added to each side of an equation without changing the solution of the equation.

The Multiplication Property of Equations states that each side of an equation can be multiplied by the same nonzero number without changing the solution of the equation.

Quick Quiz (Objective 5.1D)

1. Chu Min runs 1 mph faster than Sasha. If Chu Min can run 6 mi in 45 min, determine Sasha's running speed.   7 mph

Objective 5.2A

Vocabulary to Review
least common multiple (LCM)
[2.1A]

New Vocabulary
clearing denominators

Optional Student Activity
The equation $ax + b = c$ is in the title of this objective. Give five examples of this type of equation.

Concept Check
1. Does the equation $13a + 17 = -54$ have the same solution as $-54 = 17 + 13a$?   Yes
2. Create an equation of the form $ax + b = c$ whose solution is -3.
   For instance, $3x + 5 = -4$

Optional Student Activity
Write a paragraph that you could e-mail to a friend who needs help solving the equation $3x - 7 = 5$.

Instructor Note
In the example at the right, the equation is solved. On the next page, we solve the same equation by first *clearing fractions*. Introducing this method now will prepare students to solve equations that contain fractions with variables in the denominator.

5.2   General Equations—Part I

Objective A   To solve an equation of the form $ax + b = c$

In solving an equation of the form $ax + b = c$, the goal is to rewrite the equation in the form *variable = constant*. This requires the application of both the Addition and the Multiplication Properties of Equations.

> **HOW TO**   Solve: $\frac{3}{4}x - 2 = -11$

The goal is to write the equation in the form *variable = constant*.

$$\frac{3}{4}x - 2 = -11$$

$$\frac{3}{4}x - 2 + 2 = -11 + 2$$ • **Add 2 to each side of the equation.**

$$\frac{3}{4}x = -9$$ • **Simplify.**

$$\frac{4}{3} \cdot \frac{3}{4}x = \frac{4}{3}(-9)$$ • **Multiply each side of the equation by $\frac{4}{3}$.**

$$x = -12$$ • **The equation is in the form *variable = constant*.**

The solution is -12.

> **TAKE NOTE**
> Check: $\frac{3}{4}x - 2 = -11$
> $$\frac{3}{4}(-12) - 2 \;\Big|\; -11$$
> $$-9 - 2 \;\Big|\; -11$$
> $$-11 = -11$$
> A true equation

Here is an example of solving an equation that contains more than one fraction.

> **HOW TO**   Solve: $\frac{2}{3}x + \frac{1}{2} = \frac{3}{4}$

$$\frac{2}{3}x + \frac{1}{2} = \frac{3}{4}$$

$$\frac{2}{3}x + \frac{1}{2} - \frac{1}{2} = \frac{3}{4} - \frac{1}{2}$$ • **Subtract $\frac{1}{2}$ from each side of the equation.**

$$\frac{2}{3}x = \frac{1}{4}$$ • **Simplify.**

$$\frac{3}{2}\left(\frac{2}{3}x\right) = \frac{3}{2}\left(\frac{1}{4}\right)$$ • **Multiply each side of the equation by $\frac{3}{2}$, the reciprocal of $\frac{2}{3}$.**

$$x = \frac{3}{8}$$

The solution is $\frac{3}{8}$.

It may be easier to solve an equation containing two or more fractions by multiplying each side of the equation by the least common multiple (LCM) of the denominators. For the equation above, the LCM of 3, 2, and 4 is 12. The LCM has the property that 3, 2, and 4 will divide evenly into it. Therefore, if both sides of the equation are multiplied by 12, the denominators will divide evenly into 12. The result is an equation that does not contain any fractions. Multiplying each side of an equation that contains fractions by the LCM of the denominators is called **clearing denominators.** It is an alternative method, as we show in the next example, of solving an equation that contains fractions.

In-Class Examples (Objective 5.2A)

Solve.
1. $8a + 3 = 10$ $\frac{7}{8}$
2. $7 = 12 + 5h$ -1
3. $\frac{2}{5}x - \frac{1}{4} = \frac{3}{2}$ $\frac{35}{8}$

4. $\frac{3}{8}x + \frac{2}{3} = \frac{5}{12}$ $-\frac{2}{3}$
5. $\frac{1}{3} - \frac{3}{5}x = \frac{1}{2}$ $-\frac{5}{18}$

HOW TO Solve: $\frac{2}{3}x + \frac{1}{2} = \frac{3}{4}$

$$\frac{2}{3}x + \frac{1}{2} = \frac{3}{4}$$

$$12\left(\frac{2}{3}x + \frac{1}{2}\right) = 12\left(\frac{3}{4}\right)$$ • Multiply each side of the equation by **12**, the LCM of 3, 2, and 4.

$$12\left(\frac{2}{3}x\right) + 12\left(\frac{1}{2}\right) = 12\left(\frac{3}{4}\right)$$ • Use the Distributive Property.

$$8x + 6 = 9$$ • Simplify.

$$8x + 6 - 6 = 9 - 6$$ • Subtract **6** from each side of the equation.

$$8x = 3$$

$$\frac{8x}{8} = \frac{3}{8}$$ • Divide each side of the equation by **8**.

$$x = \frac{3}{8}$$

The solution is $\frac{3}{8}$.

Note that both methods give exactly the same solution. You may use either method to solve an equation containing fractions.

Example 1 Solve: $3x - 7 = -5$

Solution

$$3x - 7 = -5$$
$$3x - 7 + 7 = -5 + 7$$ • Add **7** to each side.
$$3x = 2$$
$$\frac{3x}{3} = \frac{2}{3}$$ • Divide each side by **3**.
$$x = \frac{2}{3}$$

The solution is $\frac{2}{3}$.

You Try It 1 Solve: $5x + 7 = 10$

Your solution $\frac{3}{5}$

Example 2 Solve: $5 = 9 - 2x$

Solution

$$5 = 9 - 2x$$
$$5 - 9 = 9 - 9 - 2x$$ • Subtract **9** from each side.
$$-4 = -2x$$
$$\frac{-4}{-2} = \frac{-2x}{-2}$$ • Divide each side by **−2**.
$$2 = x$$

The solution is 2.

You Try It 2 Solve: $2 = 11 + 3x$

Your solution -3

Solutions on p. S13

Optional Student Activity

Solve the following equation first without clearing denominators and then by clearing denominators.

$$\frac{5}{6}x + \frac{3}{8} = \frac{3}{4}$$

What are the strengths and weaknesses of each method?

Instructor Note

One way to end this objective is to review the objective title, which is to solve equations of the form $ax + b = c$. If you ask students what the variables of the equation are, they may answer a, b, c, and x, and, in a sense, that is true. However, as written symbolic math evolved, it became customary to think of letters at the beginning of the alphabet as constants and those at the end of the alphabet as variables. This kind of implicit understanding is often lost on students.

For this objective, the goal was to solve for x given a, b, and c. In the next section, we solve $ax + b = cx + d$, again with the implicit understanding that a, b, c, and d are constants and coefficients.

Later in the text, we will introduce the equation $y = mx + b$, which also makes implicit assumptions about variables and constants. As students proceed through math courses, they will repeatedly be exposed to the same kinds of understandings.

Example 3 Solve: $\dfrac{2}{3} - \dfrac{x}{2} = \dfrac{3}{4}$

Solution

$$\frac{2}{3} - \frac{x}{2} = \frac{3}{4}$$

$$\frac{2}{3} - \frac{2}{3} - \frac{x}{2} = \frac{3}{4} - \frac{2}{3} \qquad \bullet \text{ Subtract } \frac{2}{3} \text{ from each side.}$$

$$-\frac{x}{2} = \frac{1}{12}$$

$$-2\left(-\frac{x}{2}\right) = -2\left(\frac{1}{12}\right) \qquad \bullet \text{ Multiply each side by } -2.$$

$$x = -\frac{1}{6}$$

The solution is $-\frac{1}{6}$.

You Try It 3 Solve: $\dfrac{5}{8} - \dfrac{2x}{3} = \dfrac{5}{4}$

Your solution $-\dfrac{15}{16}$

Example 4 Solve $\dfrac{4}{5}x - \dfrac{1}{2} = \dfrac{3}{4}$ by first clearing denominators.

Solution
The LCM of 5, 2, and 4 is 20.

$$\frac{4}{5}x - \frac{1}{2} = \frac{3}{4}$$

$$20\left(\frac{4}{5}x - \frac{1}{2}\right) = 20\left(\frac{3}{4}\right) \qquad \bullet \text{ Multiply each side by 20.}$$

$$20\left(\frac{4}{5}x\right) - 20\left(\frac{1}{2}\right) = 20\left(\frac{3}{4}\right) \qquad \bullet \text{ Use the Distributive Property.}$$

$$16x - 10 = 15$$

$$16x - 10 + 10 = 15 + 10 \qquad \bullet \text{ Add 10 to each side.}$$

$$16x = 25$$

$$\frac{16x}{16} = \frac{25}{16} \qquad \bullet \text{ Divide each side by 16.}$$

$$x = \frac{25}{16}$$

The solution is $\frac{25}{16}$.

You Try It 4 Solve $\dfrac{2}{3}x + 3 = \dfrac{7}{2}$ by first clearing denominators.

Your solution $\dfrac{3}{4}$

Solutions on p. S13

Example 5

Solve: $2x + 4 - 5x = 10$

Solution

$2x + 4 - 5x = 10$
$-3x + 4 = 10$ • Combine like terms.
$-3x + 4 - 4 = 10 - 4$
$-3x = 6$
$\dfrac{-3x}{-3} = \dfrac{6}{-3}$
$x = -2$

The solution is -2.

You Try It 5

Solve: $x - 5 + 4x = 25$

Your solution
6

Solution on p. S14

Objective B   **To solve application problems using formulas**

Example 6

To determine the total cost of production, an economist uses the equation $T = U \cdot N + F$, where T is the total cost, U is the unit cost, N is the number of units made, and F is the fixed cost. Use this equation to find the number of units made during a month when the total cost was $9000, the unit cost was $25, and the fixed cost was $3000.

Strategy

Given: $T = \$9000$
$U = \$25$
$F = \$3000$
Unknown: N

Solution

$T = U \cdot N + F$
$9000 = 25N + 3000$
$6000 = 25N$
$\dfrac{6000}{25} = \dfrac{25N}{25}$
$240 = N$

There were 240 units made.

You Try It 6

The pressure at a certain depth in the ocean can be approximated by the equation

$P = 15 + \dfrac{1}{2}D$, where P is the pressure in

pounds per square inch and D is the depth in feet. Use this equation to find the depth when the pressure is 45 pounds per square inch.

Your strategy

Your solution
60 ft

Solution on p. S14

Objective 5.2B

Optional Student Activity

In building highways, engineers provide for expansion joints so that the highway won't break when it expands on very hot days. For a particular 1-mile portion of a two-lane highway, an engineer used the equation $I = 0.06336T - 0.00084L$, where I is the expansion in feet, T is the temperature in degrees Fahrenheit, and L is the length of the highway in feet. If 1 mi of the highway expanded 2.5 ft, what was the temperature to the nearest tenth of a degree? (*Note:* 1 mi = 5280 ft)   109.5°F

In-Class Examples (Objective 5.2B)

1. The relationship between degrees Celsius and degrees Fahrenheit can be represented by the equation $F = 1.8C + 32$, where F is the temperature in degrees Fahrenheit and C is the temperature in degrees Celsius. Use this equation to determine the temperature in degrees Celsius when the temperature is 98.6°F.   37°C

Section 5.2

Suggested Assignment

Exercises 1–89, every other odd
Exercises 91–99, odds
More challenging problems:
    Exercises 101, 103–105

5.2 Exercises

Objective A   To solve an equation of the form $ax + b = c$

For Exercises 1 to 80, solve and check.

1. $3x + 1 = 10$
3

2. $4y + 3 = 11$
2

3. $2a - 5 = 7$
6

4. $5m - 6 = 9$
3

5. $5 = 4x + 9$
-1

6. $2 = 5b + 12$
-2

7. $2x - 5 = -11$
-3

8. $3n - 7 = -19$
-4

9. $4 - 3w = -2$
2

10. $5 - 6x = -13$
3

11. $8 - 3t = 2$
2

12. $12 - 5x = 7$
1

13. $4a - 20 = 0$
5

14. $3y - 9 = 0$
3

15. $6 + 2b = 0$
-3

16. $10 + 5m = 0$
-2

17. $-2x + 5 = -7$
6

18. $-5d + 3 = -12$
3

19. $-12x + 30 = -6$
3

20. $-13 = -11y + 9$
2

21. $2 = 7 - 5a$
1

22. $3 = 11 - 4n$
2

23. $-35 = -6b + 1$
6

24. $-8x + 3 = -29$
4

25. $-3m - 21 = 0$
-7

26. $-5x - 30 = 0$
-6

27. $-4y + 15 = 15$
0

28. $-3x + 19 = 19$
0

29. $9 - 4x = 6$
$\dfrac{3}{4}$

30. $3t - 2 = 0$
$\dfrac{2}{3}$

31. $9x - 4 = 0$
$\dfrac{4}{9}$

32. $7 - 8z = 0$
$\dfrac{7}{8}$

33. $1 - 3x = 0$
$\dfrac{1}{3}$

34. $9d + 10 = 7$
$-\dfrac{1}{3}$

35. $12w + 11 = 5$
$-\dfrac{1}{2}$

36. $6y - 5 = -7$
$-\dfrac{1}{3}$

37. $8b - 3 = -9$
$-\dfrac{3}{4}$

38. $5 - 6m = 2$
$\dfrac{1}{2}$

39. $7 - 9a = 4$
$\dfrac{1}{3}$

40. $9 = -12c + 5$
$-\dfrac{1}{3}$

Quick Quiz (Objective 5.2A)

Solve.

1. $7b + 5 = 61$ 8

2. $12 - 4c = 15$ $-\dfrac{3}{4}$

3. $\dfrac{4}{5}x + 3 = 7$ 5

4. $-3 = 6m + 4 + m$ -1

41. $10 = -18x + 7$

$-\dfrac{1}{6}$

42. $2y + \dfrac{1}{3} = \dfrac{7}{3}$

1

43. $4a + \dfrac{3}{4} = \dfrac{19}{4}$

1

44. $2n - \dfrac{3}{4} = \dfrac{13}{4}$

2

45. $3x - \dfrac{5}{6} = \dfrac{13}{6}$

1

46. $5y + \dfrac{3}{7} = \dfrac{3}{7}$

0

47. $9x + \dfrac{4}{5} = \dfrac{4}{5}$

0

48. $8 = 7d - 1$

$\dfrac{9}{7}$

49. $8 = 10x - 5$

$\dfrac{13}{10}$

50. $4 = 7 - 2w$

$\dfrac{3}{2}$

51. $7 = 9 - 5a$

$\dfrac{2}{5}$

52. $8t + 13 = 3$

$-\dfrac{5}{4}$

53. $12x + 19 = 3$

$-\dfrac{4}{3}$

54. $-6y + 5 = 13$

$-\dfrac{4}{3}$

55. $-4x + 3 = 9$

$-\dfrac{3}{2}$

56. $\dfrac{1}{2}a - 3 = 1$

8

57. $\dfrac{1}{3}m - 1 = 5$

18

58. $\dfrac{2}{5}y + 4 = 6$

5

59. $\dfrac{3}{4}n + 7 = 13$

8

60. $-\dfrac{2}{3}x + 1 = 7$

-9

61. $-\dfrac{3}{8}b + 4 = 10$

-16

62. $\dfrac{x}{4} - 6 = 1$

28

63. $\dfrac{y}{5} - 2 = 3$

25

64. $\dfrac{2x}{3} - 1 = 5$

9

65. $\dfrac{2}{3}x - \dfrac{5}{6} = -\dfrac{1}{3}$

$\dfrac{3}{4}$

66. $\dfrac{5}{4}x + \dfrac{2}{3} = \dfrac{1}{4}$

$-\dfrac{1}{3}$

67. $\dfrac{1}{2} - \dfrac{2}{3}x = \dfrac{1}{4}$

$\dfrac{3}{8}$

68. $\dfrac{3}{4} - \dfrac{3}{5}x = \dfrac{19}{20}$

$-\dfrac{1}{3}$

69. $\dfrac{3}{2} = \dfrac{5}{6} + \dfrac{3x}{8}$

$\dfrac{16}{9}$

70. $-\dfrac{1}{4} = \dfrac{5}{12} + \dfrac{5x}{6}$

$-\dfrac{4}{5}$

71. $\dfrac{11}{27} = \dfrac{4}{9} - \dfrac{2x}{3}$

$\dfrac{1}{18}$

72. $\dfrac{37}{24} = \dfrac{7}{8} - \dfrac{5x}{6}$

$-\dfrac{4}{5}$

73. $7 = \dfrac{2x}{5} + 4$

$\dfrac{15}{2}$

74. $5 - \dfrac{4c}{7} = 8$

$-\dfrac{21}{4}$

75. $7 - \dfrac{5}{9}y = 9$

$-\dfrac{18}{5}$

76. $6a + 3 + 2a = 11$

1

77. $5y + 9 + 2y = 23$

2

78. $7x - 4 - 2x = 6$

2

79. $11z - 3 - 7z = 9$

3

80. $2x - 6x + 1 = 9$

-2

81. Solve $3x + 4y = 13$ when $y = -2$.

$x = 7$

82. Solve $2x - 3y = 8$ when $y = 0$.

$x = 4$

83. Solve $-4x + 3y = 9$ when $x = 0$.

$y = 3$

84. Solve $5x - 2y = -3$ when $x = -3$.

$y = -6$

85. If $2x - 3 = 7$, evaluate $3x + 4$.
19

86. If $3x + 5 = -4$, evaluate $2x - 5$.
−11

87. If $4 - 5x = -1$, evaluate $x^2 - 3x + 1$.
−1

88. If $2 - 3x = 11$, evaluate $x^2 + 2x - 3$.
0

89. If $5x + 3 - 2x = 12$, evaluate $4 - 5x$.
−11

90. If $2x - 4 - 7x = 16$, evaluate $x^2 + 1$.
17

Objective B   **To solve application problems using formulas**

Physics   The distance s, in feet, that an object will fall in t seconds is given by $s = 16t^2 + vt$, where v is the initial velocity of the object in feet per second. Use this equation for Exercises 91 and 92.

91. Find the initial velocity of an object that falls 80 ft in 2 s.
8 ft/s

92. Find the initial velocity of an object that falls 144 ft in 3 s.
0 ft/s

Depreciation   A company uses the equation $V = C - 6000t$ to determine the depreciated value V, after t years, of a milling machine that originally cost C dollars. Equations like this are used in accounting for straight-line depreciation. Use this equation for Exercises 93 and 94.

93. A milling machine originally cost $50,000. In how many years will the depreciated value of the machine be $38,000?
2 years

94. A milling machine originally cost $78,000. In how many years will the depreciated value of the machine be $48,000?
5 years

Anthropology   Anthropologists approximate the height of a primate by the size of its humerus (the bone from the elbow to the shoulder) using the equation $H = 1.2L + 27.8$, where L is the length of the humerus and H is the height, in inches, of the primate. Use this equation for Exercises 95 and 96.

95. An anthropologist estimates the height of a primate to be 66 in. What is the approximate length of the humerus of this primate? Round to the nearest tenth of an inch.
31.8 in.

96. An anthropologist estimates the height of a primate to be 62 in. What is the approximate length of the humerus of this primate?
28.5 in.

Quick Quiz (Objective 5.2B)

1. A computer that costs a business $1200 has a selling price of $2000. Find the markup rate. Use the markup equation $S = C + rC$, where S is the selling price, C is the cost, and r is the markup rate.

$66\dfrac{2}{3}$%

2. The distance s, in feet, that an object will fall in t seconds is given by $s = 16t^2 + vt$, where v is the initial velocity of the object in feet per second. Determine the initial velocity of an object that falls 174 ft in 3 s.   10 ft/s

Car Safety   Black ice is an ice covering on roads that is especially difficult to see and therefore extremely dangerous for motorists. The distance that a car traveling 30 mph will slide on black ice after its brakes are applied is related to the outside temperature by the formula $C = \frac{1}{4}D - 45$, where C is the Celsius temperature and D is the distance in feet that the car will slide. Use this equation for Exercises 97 and 98.

97. Determine the distance a car will slide on black ice when the outside temperature is −3°C.
    168 ft

98. Determine the distance a car will slide on black ice when the outside temperature is −11°C.
    136 ft

Telecommunications   A telephone company estimates that the number N of phone calls per day between two cities of populations P_1 and P_2 that are d miles apart is given by the equation $N = \frac{2.51 P_1 P_2}{d^2}$. Use this equation for Exercises 99 and 100.

99. Estimate the population (P_1) of a city given that the population of a second city (P_2) is 48,000, the number of phone calls per day between the two cities is 1,100,000, and the distance between the cities is 75 mi. Round to the nearest thousand.
    51,000 people

100. Estimate the population (P_1) of a city given that the population of a second city (P_2) is 125,000, the number of phone calls per day between the two cities is 2,500,000, and the distance between the cities is 50 mi. Round to the nearest thousand.
     20,000 people

APPLYING THE CONCEPTS

101. If $2x + 1 = a$ and $3x - 2 = a$, find the value of a.
     $a = 7$

102. If $1 - 4x = y$ and $2x - 5 = y$, find the value of y.
     $y = -3$

103. Solve: $x \div 15 = 25$ remainder 10
     385

104. Does the sentence "Solve $3x - 4(x - 1)$" make sense? Why or why not?

105. The following problem does not contain enough information for you to find only one solution. Supply some additional information so that the problem has exactly one solution. Then write and solve an equation.
     The sum of two numbers is 15. Find the numbers.

Instructor Note

Here is an extension for Exercises 97 and 98.

1. For what temperatures is the formula valid?
   Temperatures less than or equal to 0°C

2. Will the car slide farther when the temperature is higher or when it is lower?   Higher

Answers to Writing Exercises

104. The sentence "Solve $3x - 4(x - 1)$" does not make sense because $3x - 4(x - 1)$ is not an equation; it is an expression. Equations can be solved. Expressions cannot be solved.

105. The student must supply any information that will ensure that the problem has only one solution. For example, the additional information "one number is twice the second number" restricts the number of solutions to one. Then the complete solution would be:

     Let x be one number. Then $2x$ is the other number.
     $x + 2x = 15$
     $3x = 15$
     $x = 5$
     $2x = 2(5) = 10$
     The two numbers are 5 and 10.

Objective 5.3A

Concept Check

If $ax + 8 = 2x - 6$ and $x = 2$, find the value of a.   **−5**

Concept Check

When solving an equation of the form $ax + b = cx + d$, would it always be correct to start by subtracting from both sides of the equation the variable term

a. on the right side of the equation?   **Yes**

b. on the left side of the equation?   **Yes**

c. with the smaller coefficient?   **Yes**

Optional Student Activity

1. Rick weighs 76 lb plus half his weight. How much does Rick weigh?   **152 lb**

2. At the local gym, you can either pay $75 for a year's membership and $5 per visit, or pay $150 for a year's membership plus $2 per visit. How many times during the year must you go to the gym for the two plans to be equal in price?   **25 visits**

3. The gauge on a water tank shows that the tank is $\frac{5}{8}$ full.

After 18 more gallons are drained from the tank, the gauge shows that it is $\frac{1}{4}$ full.

How many gallons of water were in the tank when it was $\frac{1}{2}$ full?   **24 gal**

Optional Student Activity

Write an explanation for solving $3x - 7 = 5x + 1$ that a friend who needs help solving this equation could use.

5.3 General Equations—Part II

Objective A   To solve an equation of the form $ax + b = cx + d$

In solving an equation of the form $ax + b = cx + d$, the goal is to rewrite the equation in the form *variable = constant*. Begin by rewriting the equation so that there is only one variable term in the equation. Then rewrite the equation so that there is only one constant term.

Study Tip

Always check the proposed solution of an equation. For the equation at the right:

$$\begin{array}{c|c} 2x + 3 = 5x - 9 \\ \hline 2(4) + 3 & 5(4) - 9 \\ 8 + 3 & 20 - 9 \\ 11 = 11 \end{array}$$

The solution checks.

HOW TO   Solve: $2x + 3 = 5x - 9$

$$2x + 3 = 5x - 9$$

$2x - 5x + 3 = 5x - 5x - 9$ • Subtract $5x$ from each side of the equation.

$-3x + 3 = -9$ • Simplify. There is only one variable term.

$-3x + 3 - 3 = -9 - 3$ • Subtract 3 from each side of the equation.

$-3x = -12$ • Simplify. There is only one constant term.

$\dfrac{-3x}{-3} = \dfrac{-12}{-3}$ • Divide each side of the equation by −3.

$x = 4$ • The equation is in the form *variable = constant*.

The solution is 4. You should verify this by checking this solution.

Example 1   Solve: $4x - 5 = 8x - 7$

Solution

$$4x - 5 = 8x - 7$$

$4x - 8x - 5 = 8x - 8x - 7$ • Subtract $8x$ from each side.

$-4x - 5 = -7$

$-4x - 5 + 5 = -7 + 5$ • Add 5 to each side.

$-4x = -2$

$\dfrac{-4x}{-4} = \dfrac{-2}{-4}$ • Divide each side by −4.

$x = \dfrac{1}{2}$

The solution is $\frac{1}{2}$.

You Try It 1   Solve: $5x + 4 = 6 + 10x$

Your solution $-\dfrac{2}{5}$

Solution on p. S14

Solution on p. S14

In-Class Examples (Objective 5.3A)

Solve.

1. $5x - 4 = 3x - 10$ **−3**

2. $8x + 3 - 4x = 5 + x$ $\dfrac{2}{3}$

3. $3x - 7 = 5x - 7$ **0**

Example 2   Solve: $3x + 4 - 5x = 2 - 4x$

Solution

$3x + 4 - 5x = 2 - 4x$

$-2x + 4 = 2 - 4x$ • **Combine like terms.**

$-2x + 4x + 4 = 2 - 4x + 4x$ • **Add 4x to each side.**

$2x + 4 = 2$

$2x + 4 - 4 = 2 - 4$ • **Subtract 4 from each side.**

$2x = -2$

$\dfrac{2x}{2} = \dfrac{-2}{2}$ • **Divide each side by 2.**

$x = -1$

The solution is -1.

You Try It 2   Solve: $5x - 10 - 3x = 6 - 4x$

Your solution $\dfrac{8}{3}$

Solution on p. S14

Objective B   **To solve an equation containing parentheses**

When an equation contains parentheses, one of the steps in solving the equation requires the use of the Distributive Property. The Distributive Property is used to remove parentheses from a variable expression.

Study Tip

An important element of success is practice. We cannot do anything well if we do not practice it repeatedly. Practice is crucial to success in mathematics. In this objective you are learning to solve equations containing parentheses. You will need to practice this procedure in order to be successful at it.

HOW TO   Solve: $4 + 5(2x - 3) = 3(4x - 1)$

$4 + 5(2x - 3) = 3(4x - 1)$

$4 + 10x - 15 = 12x - 3$ • **Use the Distributive Property.**

$10x - 11 = 12x - 3$ • **Simplify.**

$10x - 12x - 11 = 12x - 12x - 3$ • **Subtract 12x from each side of the equation.**

$-2x - 11 = -3$ • **Simplify.**

$-2x - 11 + 11 = -3 + 11$ • **Add 11 to each side of the equation.**

$-2x = 8$ • **Simplify.**

$\dfrac{-2x}{-2} = \dfrac{8}{-2}$ • **Divide each side of the equation by −2.**

$x = -4$ • **The equation is in the form variable = constant.**

The solution is -4. You should verify this by checking this solution.

In the next example, we solve an equation with parentheses and decimals.

In-Class Examples (Objective 5.3B)

1. Solve: $9x - 3(2x + 5) = 4(5x + 2) - 6$ -1

2. Solve: $5[6 - 2(5x + 1)] = 8x - 9$ $\dfrac{1}{2}$

3. If $5x = 3x + 10$, evaluate $4x^2 - 10$. 90

Optional Student Activity

Create an equation that contains parentheses and has a solution of 5.

Optional Student Activity

1. If $2x + 1 = 3a + 5 = 4x - 9$, find the value of a. 2
2. If $s = 4x - 3$ and $t = x + 5$, find the value of x for which $s = 2t - 1$. 6

Optional Student Activity

Solve.
1. $3(2x - 1) - (6x - 4) = -9$
   No solution
2. $2(5x - 6) - 3(x - 4)$
   $= 7x + 14$ No solution

HOW TO   Solve: $16 + 0.55x = 0.75(x + 20)$

$16 + 0.55x = 0.75(x + 20)$

$16 + 0.55x = 0.75x + 15$ • Use the Distributive Property.

$16 + 0.55x - 0.75x = 0.75x - 0.75x + 15$ • Subtract $0.75x$ from each side of the equation.

$16 - 0.20x = 15$ • Simplify.

$16 - 16 - 0.20x = 15 - 16$ • Subtract 16 from each side of the equation.

$-0.20x = -1$ • Simplify.

$\dfrac{-0.20x}{-0.20} = \dfrac{-1}{-0.20}$ • Divide each side of the equation by -0.20.

$x = 5$ • The equation is in the form *variable = constant*.

The solution is 5.

Example 3

Solve: $3x - 4(2 - x) = 3(x - 2) - 4$

Solution

$3x - 4(2 - x) = 3(x - 2) - 4$

$3x - 8 + 4x = 3x - 6 - 4$ • Distributive Property

$7x - 8 = 3x - 10$

$7x - 3x - 8 = 3x - 3x - 10$ • Subtract $3x$.

$4x - 8 = -10$

$4x - 8 + 8 = -10 + 8$ • Add 8.

$4x = -2$

$\dfrac{4x}{4} = \dfrac{-2}{4}$ • Divide by 4.

$x = -\dfrac{1}{2}$

The solution is $-\dfrac{1}{2}$.

You Try It 3

Solve: $5x - 4(3 - 2x) = 2(3x - 2) + 6$

Your solution 2

Example 4

Solve: $3[2 - 4(2x - 1)] = 4x - 10$

Solution

$3[2 - 4(2x - 1)] = 4x - 10$

$3[2 - 8x + 4] = 4x - 10$ • Distributive Property

$3[6 - 8x] = 4x - 10$

$18 - 24x = 4x - 10$ • Distributive Property

$18 - 24x - 4x = 4x - 4x - 10$ • Subtract $4x$.

$18 - 28x = -10$

$18 - 18 - 28x = -10 - 18$ • Subtract 18.

$-28x = -28$

$\dfrac{-28x}{-28} = \dfrac{-28}{-28}$ • Divide by -28.

$x = 1$

The solution is 1.

You Try It 4

Solve: $-2[3x - 5(2x - 3)] = 3x - 8$

Your solution 2

Solutions on p. S14

| **Objective C** | To solve application problems using formulas |

TAKE NOTE

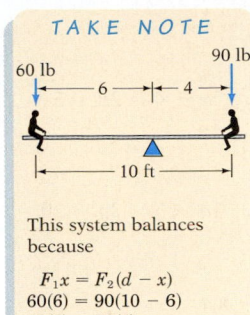

This system balances because

$$F_1x = F_2(d - x)$$
$$60(6) = 90(10 - 6)$$
$$60(6) = 90(4)$$
$$360 = 360$$

A lever system is shown at the right. It consists of a lever, or bar; a fulcrum; and two forces, F_1 and F_2. The distance d represents the length of the lever, x represents the distance from F_1 to the fulcrum, and $d - x$ represents the distance from F_2 to the fulcrum.

A principle of physics states that when the lever system balances, $F_1x = F_2(d - x)$.

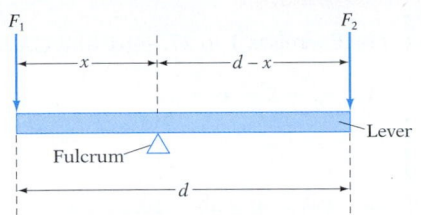

Discuss the Concepts

Consider the lever system equation $F_1x = F_2(d - x)$. Tell in words what each of the following represents.

1. F_1
2. F_2
3. x
4. d
5. $(d - x)$

Instructor Note

Although these problems are difficult for some students, there are many applications of lever systems. For instance, a bridge may be considered a lever system with the fulcrum at the center. When a truck or car is on the bridge, the lever equation is used to calculate the magnitude of the forces that are necessary, pushing upward at the ends of the bridge, so that the bridge will not move.

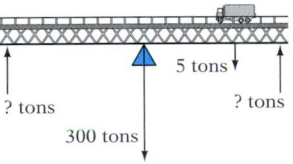

Example 5

A lever is 15 ft long. A force of 50 lb is applied to one end of the lever, and a force of 100 lb is applied to the other end. Where is the fulcrum located when the system balances?

Strategy
Make a drawing.

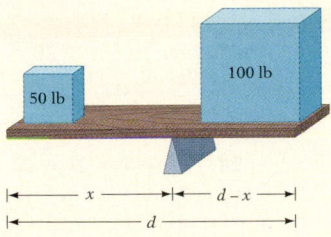

Given: $F_1 = 50$
       $F_2 = 100$
       $d = 15$
Unknown: x

Solution

$$F_1x = F_2(d - x)$$
$$50x = 100(15 - x)$$
$$50x = 1500 - 100x$$
$$50x + 100x = 1500 - 100x + 100x$$
$$150x = 1500$$
$$\frac{150x}{150} = \frac{1500}{150}$$
$$x = 10$$

The fulcrum is 10 ft from the 50-pound force.

You Try It 5

A lever is 25 ft long. A force of 45 lb is applied to one end of the lever, and a force of 80 lb is applied to the other end. Where is the location of the fulcrum when the system balances?

Your strategy

Your solution
16 ft from the 45-pound force

Solution on p. S14

In-Class Examples (Objective 5.3C)

1. A lever is 12 ft long. A force of 3 lb is applied to one end of the lever, and a force of 6 lb is applied to the other end. Where is the location of the fulcrum when the system balances?
8 ft from the 3-pound force

Section 5.3

Suggested Assignment

Exercises 1–57, every other odd
Exercises 59–71, odds
More challenging problems:
  Exercises 72, 73, 76

5.3 Exercises

Objective A   To solve an equation of the form $ax + b = cx + d$

For Exercises 1 to 27, solve and check.

1. $8x + 5 = 4x + 13$
2

2. $6y + 2 = y + 17$
3

3. $5x - 4 = 2x + 5$
3

4. $13b - 1 = 4b - 19$
-2

5. $15x - 2 = 4x - 13$
-1

6. $7a - 5 = 2a - 20$
-3

7. $3x + 1 = 11 - 2x$
2

8. $n - 2 = 6 - 3n$
2

9. $2x - 3 = -11 - 2x$
-2

10. $4y - 2 = -16 - 3y$
-2

11. $2b + 3 = 5b + 12$
-3

12. $m + 4 = 3m + 8$
-2

13. $4y - 8 = y - 8$
0

14. $5a + 7 = 2a + 7$
0

15. $6 - 5x = 8 - 3x$
-1

16. $10 - 4n = 16 - n$
-2

17. $5 + 7x = 11 + 9x$
-3

18. $3 - 2y = 15 + 4y$
-2

19. $2x - 4 = 6x$
-1

20. $2b - 10 = 7b$
-2

21. $8m = 3m + 20$
4

22. $9y = 5y + 16$
4

23. $8b + 5 = 5b + 7$
$\frac{2}{3}$

24. $6y - 1 = 2y + 2$
$\frac{3}{4}$

25. $7x - 8 = x - 3$
$\frac{5}{6}$

26. $2y - 7 = -1 - 2y$
$\frac{3}{2}$

27. $2m - 1 = -6m + 5$
$\frac{3}{4}$

28. If $5x = 3x - 8$, evaluate $4x + 2$.
-14

29. If $7x + 3 = 5x - 7$, evaluate $3x - 2$.
-17

30. If $2 - 6a = 5 - 3a$, evaluate $4a^2 - 2a + 1$.
7

31. If $1 - 5c = 4 - 4c$, evaluate $3c^2 - 4c + 2$.
41

32. If $2y + 3 = 5 - 4y$, evaluate $6y - 7$.
-5

33. If $3z + 1 = 1 - 5z$, evaluate $3z^2 - 7z + 8$.
8

Quick Quiz (Objective 5.3A)

1. Solve: $7x + 4 = 3x - 20$ -6

2. If $4 - 3x = 24 + 2x$, evaluate $3x^2 - 2x + 4$.   60

Objective B   **To solve an equation containing parentheses**

For Exercises 34 to 54, solve and check.

34. $5x + 2(x + 1) = 23$
3

35. $6y + 2(2y + 3) = 16$
1

36. $9n - 3(2n - 1) = 15$
4

37. $12x - 2(4x - 6) = 28$
4

38. $7a - (3a - 4) = 12$
2

39. $9m - 4(2m - 3) = 11$
−1

40. $5(3 - 2y) + 4y = 3$
2

41. $4(1 - 3x) + 7x = 9$
−1

42. $5y - 3 = 7 + 4(y - 2)$
2

43. $0.22(x + 6) = 0.2x + 1.8$
24

44. $0.05(4 - x) + 0.1x = 0.32$
2.4

45. $0.3x + 0.3(x + 10) = 300$
495

46. $2a - 5 = 4(3a + 1) - 2$
$-\dfrac{7}{10}$

47. $5 - (9 - 6x) = 2x - 2$
$\dfrac{1}{2}$

48. $7 - (5 - 8x) = 4x + 3$
$\dfrac{1}{4}$

49. $3[2 - 4(y - 1)] = 3(2y + 8)$
$-\dfrac{1}{3}$

50. $5[2 - (2x - 4)] = 2(5 - 3x)$
5

51. $3a + 2[2 + 3(a - 1)] = 2(3a + 4)$
$\dfrac{10}{3}$

52. $5 + 3[1 + 2(2x - 3)] = 6(x + 5)$
$\dfrac{20}{3}$

53. $-2[4 - (3b + 2)] = 5 - 2(3b + 6)$
$-\dfrac{1}{4}$

54. $-4[x - 2(2x - 3)] + 1 = 2x - 3$
2

55. If $4 - 3a = 7 - 2(2a + 5)$,
evaluate $a^2 + 7a$.
0

56. If $9 - 5x = 12 - (6x + 7)$,
evaluate $x^2 - 3x - 2$.
26

57. If $2z - 5 = 3(4z + 5)$, evaluate $\dfrac{z^2}{z - 2}$.
−1

58. If $3n - 7 = 5(2n + 7)$, evaluate $\dfrac{n^2}{2n - 6}$.
−2

Quick Quiz (Objective 5.3B)

1. Solve: $3 + 2[4x - 3(5 - x)] = 3(x - 20)$ −3

2. If $5x - 4 = 2(4x + 7)$, evaluate $3x^2 - 2x$.   120

Objective C   To solve application problems using formulas

Lever Systems   For Exercises 59 to 65, solve. Use the lever system equation $F_1x = F_2(d - x)$.

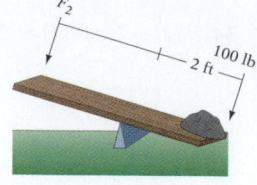

59. A lever 10 ft long is used to move a 100-pound rock. The fulcrum is placed 2 ft from the rock. What force must be applied to the other end of the lever to move the rock?
25 lb

60. An adult and a child are on a seesaw 14 ft long. The adult weighs 175 lb and the child weighs 70 lb. How many feet from the child must the fulcrum be placed so that the seesaw balances?
10 ft

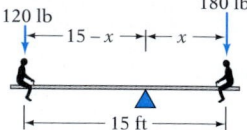

61. Two people are sitting 15 ft apart on a seesaw. One person weighs 180 lb. The second person weighs 120 lb. How far from the 180-pound person should the fulcrum be placed so that the seesaw balances?
6 ft

62. Two children are sitting on a seesaw that is 12 ft long. One child weighs 60 lb. The other child weighs 90 lb. How far from the 90-pound child should the fulcrum be placed so that the seesaw balances?
4.8 ft

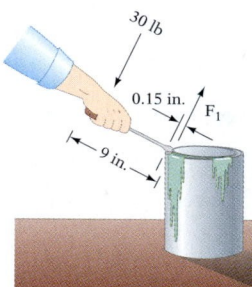

63. In preparation for a stunt, two acrobats are standing on a plank 18 ft long. One acrobat weighs 128 lb and the second acrobat weighs 160 lb. How far from the 128-pound acrobat must the fulcrum be placed so that the acrobats are balanced on the plank?
10 ft

64. A screwdriver 9 in. long is used as a lever to open a can of paint. The tip of the screwdriver is placed under the lip of the can with the fulcrum 0.15 in. from the lip. A force of 30 lb is applied to the other end of the screwdriver. Find the force on the lip of the can.
1770 lb

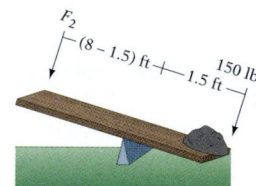

65. A metal bar 8 ft long is used to move a 150-pound rock. The fulcrum is placed 1.5 ft from the rock. What minimum force must be applied to the other end of the bar to move the rock? Round to the nearest tenth.
34.6 lb

Break-Even Point   To determine the break-even point, or the number of units that must be sold so that no profit or loss occurs, an economist uses the formula $Px = Cx + F$, where P is the selling price per unit, x is the number of units that must be sold to break even, C is the cost to make each unit, and F is the fixed cost. Use this equation for Exercises 66 to 71.

66. A business analyst has determined that the selling price per unit for a laser printer is $1600. The cost to make one laser printer is $950, and the fixed cost is $211,250. Find the break-even point.
325 laser printers

Quick Quiz (Objective 5.3C)

1. Two people are sitting on a seesaw that is 9 ft long. One person weighs 120 lb. The second person weighs 150 lb. How far from the 120-pound person should the fulcrum be placed so that the seesaw balances? Use the lever system equation $F_1x = F_2(d - x)$.   5 ft

67. A business analyst has determined that the selling price per unit for a gas barbecue is $325. The cost to make one gas barbecue is $175, and the fixed cost is $39,000. Find the break-even point.
260 barbecues

68. A manufacturer of thermostats determines that the cost per unit for a programmable thermostat is $38 and the fixed cost is $24,400. The selling price for the thermostat is $99. Find the break-even point.
400 thermostats

69. A manufacturing engineer determines that the cost per unit for a desk lamp is $12 and the fixed cost is $19,240. The selling price for the desk lamp is $49. Find the break-even point.
520 desk lamps

70. A manufacturing engineer determines the cost to make one compact disc to be $3.35 and the fixed cost to be $6180. The selling price for each compact disc is $8.50. Find the number of compact discs that must be sold to break even.
1200 compact discs

71. To manufacture a softball bat requires two steps. The first step is to cut a rough shape. The second step is to sand the bat to its final form. The cost to rough-shape a bat is $.45, and the cost to sand a bat to final form is $1.05. The total fixed cost for the two steps is $16,500. How many softball bats must be sold at a price of $7.00 each to break even?
3000 softball bats

APPLYING THE CONCEPTS

72. Write an equation of the form $ax + b = cx + d$ that has 4 as the solution.
Answers will vary.

For Exercises 73 to 76, solve. If the equation has no solution, write "no solution."

73. $3(2x - 1) - (6x - 4) = -9$ No solution

74. $7(3x + 6) - 4(3 + 5x) = 13 + x$ No solution

75. $\frac{1}{5}(25 - 10a) + 4 = \frac{1}{3}(12a - 15) + 14$ 0

76. $5[m + 2(3 - m)] = 3[2(4 - m) - 5]$ -21

77. ✏ The equation $x = x + 1$ has no solution, whereas the solution of the equation $2x + 3 = 3$ is zero. Is there a difference between no solution and a solution of zero? Explain your answer.

Answers to Writing Exercises

77. Many beginning algebra students do not differentiate between an equation that has no solution and an equation whose solution is zero. Students should explain that the solution of $2x + 3 = 3$ is the (real) number zero. However, there is no solution of $x = x + 1$ because there is no (real) number that is equal to itself plus 1.

Objective 5.4A

Vocabulary to Review
integers   [3.1A]

New Vocabulary
translate
even integer
odd integer
consecutive integers
consecutive even integers
consecutive odd integers

Concept Check

Translate "two more than a number is twelve" into an equation and solve.
$x + 2 = 12; x = 10$

Concept Check

1. If x represents the first of three consecutive integers, represent the second and third consecutive integers in terms of x. $x + 1, x + 2$

2. If x represents the first of three consecutive odd integers, represent the second and third consecutive odd integers in terms of x.
$x + 2, x + 4$

5.4 Translating Sentences into Equations

Objective A   To solve integer problems

An equation states that two mathematical expressions are equal. Therefore, to **translate** a sentence into an equation requires recognition of the words or phrases that mean "equals." Some of these phrases are listed below.

$$
\left.\begin{array}{l}
\text{equals} \\
\text{is} \\
\text{is equal to} \\
\text{amounts to} \\
\text{represents}
\end{array}\right\} \text{translate to } =
$$

Once the sentence is translated into an equation, the equation can be solved by rewriting the equation in the form *variable = constant*.

HOW TO   Translate "five less than a number is thirteen" into an equation and solve.

The unknown number: n • Assign a variable to the unknown number.

| Five less than a number | is | thirteen |

• Find two verbal expressions for the same value.

$$n - 5 \quad = \quad 13$$

• Write a mathematical expression for each verbal expression. Write the equals sign.

$$n - 5 + 5 = 13 + 5$$

• Solve the equation.

$$n = 18$$

The number is 18.

TAKE NOTE
You can check the solution to a translation problem.

Check:

5 less than 18 is 13
$18 - 5 \mid 13$
$13 = 13$

Recall that the integers are the numbers $\{. . . , -4, -3, -2, -1, 0, 1, 2, 3, 4, . . .\}$. An **even integer** is an integer that has no remainder when divided by 2. Examples of even integers are -8, 0, and 22. An **odd integer** is an integer that is not evenly divisible by 2. Examples of odd integers are -17, 1, and 39.

Consecutive integers are integers that follow one another in order. Examples of consecutive integers are shown at the right. (Assume that the variable n represents an integer.)

11, 12, 13
$-8, -7, -6$
$n, n + 1, n + 2$

TAKE NOTE
Both consecutive even and consecutive odd integers are represented using n, $n + 2, n + 4, ...$

Examples of **consecutive even integers** are shown at the right. (Assume that the variable n represents an even integer.)

24, 26, 28
$-10, -8, -6$
$n, n + 2, n + 4$

Examples of **consecutive odd integers** are shown at the right. (Assume that the variable n represents an odd integer.)

19, 21, 23
$-1, 1, 3$
$n, n + 2, n + 4$

In-Class Examples (Objective 5.4A)

Translate into an equation and solve.

1. The sum of two numbers is twenty-five. The total of four times the smaller number and two is six less than the product of two and the larger number. Find the two numbers.
$4x + 2 = 2(25 - x) - 6; 7, 18$

2. The sum of three consecutive odd integers is ninety-three. Find the three odd integers.
$n + (n + 2) + (n + 4) = 93; 29, 31, 33$

3. Three times the largest of three consecutive integers is ten more than the sum of the other two. Find the three integers.
$3(n + 2) = n + (n + 1) + 10; 5, 6, 7$

HOW TO The sum of three consecutive odd integers is forty-five. Find the integers.

Strategy

- First odd integer: n
  Second odd integer: $n + 2$
  Third odd integer: $n + 4$
- The sum of the three odd integers is 45.

• **Represent three consecutive odd integers.**

Solution

$$n + (n + 2) + (n + 4) = 45$$

• **Write an equation.**

$$3n + 6 = 45$$

• **Solve the equation.**

$$3n = 39$$

$$n = 13$$

• **The first odd integer is 13.**

$$n + 2 = 13 + 2 = 15$$

• **Find the second odd integer.**

$$n + 4 = 13 + 4 = 17$$

• **Find the third odd integer.**

The three consecutive odd integers are 13, 15, and 17.

Example 1

The sum of two numbers is sixteen. The difference between four times the smaller number and two is two more than twice the larger number. Find the two numbers.

Solution

The smaller number: n
The larger number: $16 - n$

The difference between four times the smaller number and two	is	two more than twice the larger number

$$4n - 2 = 2(16 - n) + 2$$
$$4n - 2 = 32 - 2n + 2$$
$$4n - 2 = 34 - 2n$$
$$4n + 2n - 2 = 34 - 2n + 2n$$
$$6n - 2 = 34$$
$$6n - 2 + 2 = 34 + 2$$
$$6n = 36$$
$$\frac{6n}{6} = \frac{36}{6}$$
$$n = 6$$

$$16 - n = 16 - 6 = 10$$

The smaller number is 6.
The larger number is 10.

You Try It 1

The sum of two numbers is twelve. The total of three times the smaller number and six amounts to seven less than the product of four and the larger number. Find the two numbers.

Your solution
5, 7

Solution on pp. S14–S15

Concept Check

The sum of three consecutive odd integers is twenty-seven. Find the integers. 7, 9, 11

Instructor Note

There is an opportunity, in problems that involve consecutive even or odd integers, to emphasize the importance of identifying the variable. Give students the problem "The sum of two consecutive odd integers is eleven. Find the integers." In this case $x = 4.5$, which is not an odd integer. Therefore, the problem has no solution.

Optional Student Activity

Twice the largest of three consecutive integers is three more than the sum of the other two. Find the integers. This statement is true for all consecutive integers.

Concept Check

Consider the problem "Find three consecutive integers such that three times the second equals ten more than the sum of the first and third." A correct equation to solve this problem is

$$3(x + 1) = x + (x + 2) + 10$$

Solving the equation yields the integers 9, 10, and 11. To check the solution, a student replaced x with 9, then 10, and then 11. 10 and 11 did not check as solutions, so the student concluded that the solution was not correct. However, 9, 10, and 11 are the correct integers. Explain the error in the student's conclusion.

Optional Student Activity

Create a consecutive-even-integers word problem that has the solution 4, 6, 8.

Objective 5.4B

Concept Check

1. A Chinese restaurant charges $11.95 for the adult buffet and $7.25 for the children's buffet. One family's bill came to $79.35. If there were three adults in the family, how many children were there?
6 children

2. A local feed store sells a 75-pound bag of feed for $10.90. If a customer buys more than one bag, each additional bag costs $10.50. A customer bought $84.40 worth of feed. How many 75-pound bags of feed did this customer purchase? **8 bags**

Optional Student Activity

1. One bleepet plus one-fifth of a bleepet equals twenty-one. What is the value of a bleepet?

$17\dfrac{1}{2}$

2. Aaron decided to quit smoking. He started by cutting back two cigarettes a day each day for a week. He smoked 119 cigarettes during the week. How many cigarettes did he smoke that day before he started to cut back?
25 cigarettes

Example 2

Find three consecutive even integers such that three times the second equals four more than the sum of the first and third.

Strategy
- First even integer: n
  Second even integer: $n + 2$
  Third even integer: $n + 4$
- Three times the second equals four more than the sum of the first and third.

Solution
$$3(n + 2) = n + (n + 4) + 4$$
$$3n + 6 = 2n + 8$$
$$3n - 2n + 6 = 2n - 2n + 8$$
$$n + 6 = 8$$
$$n = 2$$
$$n + 2 = 2 + 2 = 4$$
$$n + 4 = 2 + 4 = 6$$

The three integers are 2, 4, and 6.

You Try It 2

Find three consecutive integers whose sum is negative six.

Your strategy

Your solution
$-3, -2, -1$

Solution on p. S15

Objective B   To translate a sentence into an equation and solve

VIDEO & DVD   CD TUTOR   WWW WEB   SSM

Example 3

A wallpaper hanger charges a fee of $25 plus $12 for each roll of wallpaper used in a room. If the total charge for hanging wallpaper is $97, how many rolls of wallpaper were used?

Strategy
To find the number of rolls of wallpaper used, write and solve an equation using n to represent the number of rolls of wallpaper used.

Solution

$25 plus $12 for each roll of wallpaper	is	$97

$$25 + 12n = 97$$
$$12n = 72$$
$$\dfrac{12n}{12} = \dfrac{72}{12}$$
$$n = 6$$

Six rolls of wallpaper were used.

You Try It 3

The fee charged by a ticketing agency for a concert is $3.50 plus $17.50 for each ticket purchased. If your total charge for tickets is $161, how many tickets are you purchasing?

Your strategy

Your solution
9 tickets

Solution on p. S15

In-Class Examples (Objective 5.4B)

1. An electric company charges $.07 for each of the first 249 kWh (kilowatt-hours) and $.14 for each kilowatt-hour over 249 kWh. Find the number of kilowatt-hours used by a family whose electric bill was $28.77.   **330 kWh**

2. A piano wire 24 in. long is cut into two pieces. The length of the longer piece is 4 in. more than three times the length of the shorter piece. Find the length of each piece.
5 in., 19 in.

Example 4

A board 20 ft long is cut into two pieces. Five times the length of the shorter piece is 2 ft more than twice the length of the longer piece. Find the length of each piece.

Strategy

Let x represent the length of the shorter piece. Then $20 - x$ represents the length of the longer piece.

Make a drawing.

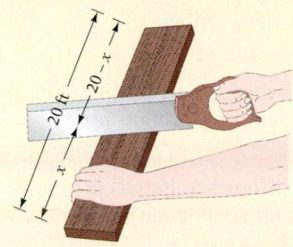

To find the lengths, write and solve an equation using x to represent the length of the shorter piece and $20 - x$ to represent the length of the longer piece.

Solution

Five times the length of the shorter piece	is	2 ft more than twice the length of the longer piece

$$5x = 2(20 - x) + 2$$
$$5x = 40 - 2x + 2$$
$$5x = 42 - 2x$$
$$5x + 2x = 42 - 2x + 2x$$
$$7x = 42$$
$$\frac{7x}{7} = \frac{42}{7}$$
$$x = 6$$

$20 - x = 20 - 6 = 14$

The length of the shorter piece is 6 ft.
The length of the longer piece is 14 ft.

You Try It 4

A wire 22 in. long is cut into two pieces. The length of the longer piece is 4 in. more than twice the length of the shorter piece. Find the length of each piece.

Your strategy

Your solution

6 in., 16 in.

Solution on p. S15

Instructor Note

Problems that begin with sentences such as "The sum of two numbers is twelve" and "A board 12 ft long is cut into two pieces" can have their two constituent parts represented identically, using n and $12 - n$.

Instructor Note

Some students have difficulty making the transition from the sentence "The sum of two numbers is sixteen" to the conclusion that if the first number is represented as n, then the second number may be represented as $16 - n$. The confusion is due in part to the use of the word *sum* and the fact that the representation of the two numbers does not include addition. It can be pointed out, however, that the sum of n and $(16 - n)$ is in fact 16.

Concept Check

A board 12 ft long is cut into two pieces.

1. If x represents the length of the shorter piece, represent the length of the longer piece in terms of x. $12 - x$

2. If x represents the length of the longer piece, represent the length of the shorter piece in terms of x. $12 - x$

Section 5.4

Suggested Assignment
Exercises 1–41, odds
More challenging problem:
Exercise 43

5.4 Exercises

Objective A   **To solve integer problems**

For Exercises 1 to 18, translate into an equation and solve.

1. The difference between a number and fifteen is seven. Find the number.
$x - 15 = 7; 22$

2. The sum of five and a number is three. Find the number.
$5 + x = 3; -2$

3. The product of seven and a number is negative twenty-one. Find the number.
$7x = -21; -3$

4. The quotient of a number and four is two. Find the number.
$\frac{x}{4} = 2; 8$

5. The difference between nine and a number is seven. Find the number.
$9 - x = 7; 2$

6. Three-fifths of a number is negative thirty. Find the number.
$\frac{3}{5}x = -30; -50$

7. The difference between five and twice a number is one. Find the number.
$5 - 2x = 1; 2$

8. Four more than three times a number is thirteen. Find the number.
$3x + 4 = 13; 3$

9. The sum of twice a number and five is fifteen. Find the number.
$2x + 5 = 15; 5$

10. The difference between nine times a number and six is twelve. Find the number.
$9x - 6 = 12; 2$

11. Six less than four times a number is twenty-two. Find the number.
$4x - 6 = 22; 7$

12. Four times the sum of twice a number and three is twelve. Find the number.
$4(2x + 3) = 12; 0$

13. Three times the difference between four times a number and seven is fifteen. Find the number. $3(4x - 7) = 15; 3$

14. Twice the difference between a number and twenty-five is three times the number. Find the number. $2(x - 25) = 3x; -50$

Instructor Note
For Exercises 15–18, the equations will be different if *x* is used to represent the larger number.
15. $3(20 - x) = 2x$
16. $3(15 - x) - 1 = x$
17. $2(14 - x) - x = 1$
18. $3(18 - x) + 2x = 44$

15. The sum of two numbers is twenty. Three times the smaller is equal to two times the larger. Find the two numbers.
$3x = 2(20 - x); 8, 12$

16. The sum of two numbers is fifteen. One less than three times the smaller is equal to the larger. Find the two numbers.
$3x - 1 = 15 - x; 4, 11$

17. The sum of two numbers is fourteen. The difference between two times the smaller and the larger is one. Find the two numbers.
$2x - (14 - x) = 1; 5, 9$

18. The sum of two numbers is eighteen. The total of three times the smaller and twice the larger is forty-four. Find the two numbers.
$3x + 2(18 - x) = 44; 8, 10$

19. The sum of three consecutive odd integers is fifty-one. Find the integers.
15, 17, 19

20. Find three consecutive even integers whose sum is negative eighteen.
$-8, -6, -4$

21. Find three consecutive odd integers such that three times the middle integer is one more than the sum of the first and third.
$-1, 1, 3$

22. Twice the smallest of three consecutive odd integers is seven more than the largest. Find the integers.   11, 13, 15

23. Find two consecutive even integers such that three times the first equals twice the second.   4, 6

24. Find two consecutive even integers such that four times the first is three times the second.   6, 8

25. Seven times the first of two consecutive odd integers is five times the second. Find the integers.   5, 7

26. Find three consecutive even integers such that three times the middle integer is four more than the sum of the first and third.
2, 4, 6

Quick Quiz (Objective 5.4A)
Translate into an equation and solve.
1. The product of five and a number is negative fifteen. Find the number. $5x = -15; -3$
2. Find three consecutive even integers such that four times the second is eight less than the third.
$4(x + 2) = (x + 4) - 8; -4, -2, 0$

> **Objective B**   **To translate a sentence into an equation and solve**

27. **Computer Science**   The processor speed of a personal computer is 3.2 gigahertz (GHz). This is three-fourths of the processor speed of a newer-model personal computer. Find the processor speed of the newer personal computer.
    4.2$\overline{6}$ GHz

28. **Computer Science**   The storage capacity of a hard-disk drive is 60 gigabytes. This is one-fourth of the storage capacity of a second hard-disk drive. Find the storage capacity of the second hard-disk drive.
    240 gigabytes

29. **Geometry**   An isosceles triangle has two sides of equal length. The length of the third side is 1 ft less than twice the length of an equal side. Find the length of each side when the perimeter is 23 ft. *Note:* The perimeter of a triangle is the sum of the lengths of the three sides.
    6 ft, 6 ft, 11 ft

30. **Geometry**   An isosceles triangle has two sides of equal length. The length of one of the equal sides is two more than three times the length of the third side. If the perimeter is 46 m, find the length of each side. *Note:* The perimeter of a triangle is the sum of the lengths of the three sides.
    20 m, 20 m, 6 m

31. **Union Dues**   A union charges monthly dues of $4.00 plus $.15 for each hour worked during the month. A union member's dues for March were $29.20. How many hours did the union member work during the month of March?
    168 h

32. **Technical Support**   A technical information hotline charges a customer $15.00 plus $2.00 per minute to answer questions about software. How many minutes did a customer who received a bill for $37 use this service?
    11 min

33. **Interior Decorating**   The total cost to paint the inside of a house was $1346. This cost included $125 for materials and $33 per hour for labor. How many hours of labor were required to paint the inside of the house?
    37 h

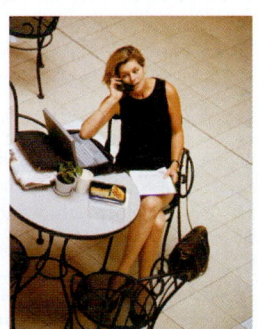

34. **Telecommunications**   The cellular phone service for a business executive is $35 per month plus $.40 per minute of phone use. In a month when the executive's cellular phone bill was $99.80, how many minutes did the executive use the phone?
    162 min

35. **Computer Science**   A computer screen consists of tiny dots of light called pixels. In a certain graphics mode, there are 1280 horizontal pixels. This is 768 less than twice the number of vertical pixels. Find the number of vertical pixels.
    1024 pixels

Quick Quiz (Objective 5.4B)

1. An investment of $8000 is divided into two accounts, one at a bank and one at a credit union. The value of the investment at the credit union is $1000 less than twice the value of the investment at the bank. Find the amount in each account.
   $3000 at the bank; $5000 at the credit union

Answers to Writing Exercises

43. A possible problem for $6x = 123$ is "A student worked 6 hours and earned $123. What was the student's hourly wage?" For the equation $8x + 100 = 300$, a possible problem is "A group of eight people spent $300 at an amusement park. This included $100 for lunch and the admission tickets for the eight people. Find the cost of each ticket."

44. There are many examples students may choose. Here are a few possibilities.

Accounting:

Assets = liabilities + stockholder equity

Finance: Interest = Prt

Auto Mechanics:

Gearbox efficiency = $\dfrac{\text{torque ratio}}{\text{gear ratio}}$

Chemistry:

Pressure = $\dfrac{k}{\text{volume}}$

45. The problem states that a 4-quart mixture of fruit juice is made from apple juice and cranberry juice. There are 6 more quarts of apple juice than of cranberry juice. If we let x = the number of quarts of cranberry juice, then $x + 6$ = the number of quarts of apple juice. The total number of quarts is 4. Therefore, we can write the equation

$$x + (x + 6) = 4$$
$$2x + 6 = 4$$
$$2x = -2$$
$$x = -1$$

Because x = the number of quarts of cranberry juice, there are -1 qt of cranberry juice in the mixture. We cannot add -1 qt to a mixture. The solution is not reasonable. We can see from the original problem that the answer will not be reasonable. If the total number of quarts in the mixture is 4, we cannot have more than 6 qt of apple juice in the mixture.

36. Electricity   The cost of electricity in a certain city is $.08 for each of the first 300 kWh (kilowatt-hours) and $.13 for each kilowatt-hour over 300 kWh. Find the number of kilowatt-hours used by a family with a $51.95 electric bill.
515 kWh

37. Contractors   Budget Plumbing charged $400 for a water softener and installation. The charge included $310 for the water softener and $30 per hour for labor. How many hours of labor were required for the job?
3 h

38. Purchasing   McPherson Cement sells cement for $75 plus $24 for each yard of cement. How many yards of cement can be purchased for $363?
12 yd

39. Carpentry   A 12-foot board is cut into two pieces. Twice the length of the shorter piece is 3 ft less than the length of the longer piece. Find the length of each piece.
3 ft, 9 ft

40. Fishing   A 14-yard fishing line is cut into two pieces. Three times the length of the longer piece is four times the length of the shorter piece. Find the length of each piece.
6 yd, 8 yd

41. Scholarships   Seven thousand dollars is divided into two scholarships. Twice the amount of the smaller scholarship is $1000 less than the larger scholarship. What is the amount of the larger scholarship?
$5000

42. Investing   An investment of $10,000 is divided into two accounts, one for stocks and one for mutual funds. The value of the stock account is $2000 less than twice the value of the mutual funds account. Find the amount in each account.
$6000 in stocks, $4000 in mutual funds

APPLYING THE CONCEPTS

43.   Make up two word problems, one that requires solving the equation $6x = 123$ and one that requires solving the equation $8x + 100 = 300$.

44.   A formula is an equation that relates variables in a known way. Find two examples of formulas that are used in your college major. Explain what each of the variables represents.

45.   It is always important to check the answer to an application problem to be sure the answer makes sense. Consider the following problem. A 4-quart mixture of fruit juices is made from apple juice and cranberry juice. There are 6 more quarts of apple juice than of cranberry juice. Write and solve an equation for the number of quarts of each juice used. Does the answer to this question make sense? Explain.

5.5 Mixture and Uniform Motion Problems

Objective A To solve value mixture problems

A value mixture problem involves combining ingredients that have different prices into a single blend. For example, a coffee merchant may blend two types of coffee into a single blend, or a candy manufacturer may combine two types of candy to sell as a variety pack.

The solution of a value mixture problem is based on the **value mixture equation** $AC = V$, where A is the amount of an ingredient, C is the cost per unit of the ingredient, and V is the value of the ingredient.

> **TAKE NOTE**
> The equation $AC = V$ is used to find the value of an ingredient. For example, the value of 4 lb of cashews costing $6 per pound is
> $$AC = V$$
> $$4 \cdot \$6 = V$$
> $$\$24 = V$$

HOW TO A coffee merchant wants to make 6 lb of a blend of coffee costing $5 per pound. The blend is made using a $6-per-pound grade and a $3-per-pound grade of coffee. How many pounds of each of these grades should be used?

> **Strategy for Solving a Value Mixture Problem**
>
> **1.** For each ingredient in the mixture, write a numerical or variable expression for the amount of the ingredient used, the unit cost of the ingredient, and the value of the amount used. For the blend, write a numerical or variable expression for the amount, the unit cost of the blend, and the value of the amount. The results can be recorded in a table.

> **TAKE NOTE**
> Use the information given in the problem to fill in the Amount and Unit Cost columns of the table. Fill in the Value column by multiplying the two expressions you wrote in each row. Use the expressions in the last column to write the equation.

The sum of the amounts is 6 lb.

Amount of $3 coffee: $6 - x$
Amount of $6 coffee: x

	Amount, A	·	Unit Cost, C	=	Value, V
$6 grade	x	·	6	=	$6x$
$3 grade	$6 - x$	·	3	=	$3(6 - x)$
$5 blend	6	·	5	=	$5(6)$

> **2.** Determine how the values of the ingredients are related. Use the fact that the sum of the values of all the ingredients is equal to the value of the blend.

The sum of the values of the $6 grade and the $3 grade is equal to the value of the $5 blend.

$$6x + 3(6 - x) = 5(6)$$
$$6x + 18 - 3x = 30$$
$$3x + 18 = 30$$
$$3x = 12$$
$$x = 4 \quad \bullet \text{ This is the amount of the \$6-grade coffee.}$$

$$6 - x = 6 - 4 = 2 \quad \bullet \text{ Find the amount of the \$3-grade coffee.}$$

The merchant must use 4 lb of the $6 coffee and 2 lb of the $3 coffee.

Objective 5.5A

New Equation
Value Mixture Equation
$(AC = V)$

Concept Check

1. A 15-pound bag contains coffee valued at $7 per pound. What is the total value of the 15-pound bag? $105

2. A 5-pound box of candy has a total value of $30. What is the cost per pound of the candy? $6 per pound

Discuss the Concepts

Why is the amount, A, of the mixture always larger than the amount of either ingredient?

In-Class Examples (Objective 5.5A)

1. A meatloaf mixture is made by combining ground turkey costing $1.49 per pound with ground beef costing $2.13 per pound. How many pounds of each should be used to make 8 lb of a meatloaf mixture that costs $1.89 per pound?
   Ground turkey: 3 lb; ground beef: 5 lb

Optional Student Activity

There are two basic types of value mixture problems. In what we may call type "a," the amount of at least one ingredient is known. In type "b," the amounts of both ingredients are unknown but the mixture amount is known. Type "a" problems are set up like the table in Example 1 on this page, with the Amount column filled in by entering the amounts of the two ingredients and adding them to fill in the amount of the mixture. Type "b" problems are set up in a manner similar to the table on the preceding page, with the bottom entry in the Amount column filled in with the known mixture amount. Read each of the value mixture problems in the exercises and determine whether each problem is a type "a" or a type "b" problem.

a: 1, 2, 3, 4, 5, 6, 9, 10, 11, 12, 17, 18
b: 7, 8, 13, 14, 15, 16

Example 1

How many ounces of a silver alloy that costs $4 an ounce must be mixed with 10 oz of an alloy that costs $6 an ounce to make a mixture that costs $4.32 an ounce?

Strategy

- Ounces of $4 alloy: x

	Amount	Cost	Value
$4 alloy	x	4	$4x$
$6 alloy	10	6	6(10)
$4.32 mixture	$10 + x$	4.32	$4.32(10 + x)$

- The sum of the values before mixing equals the value after mixing.

Solution

$$4x + 6(10) = 4.32(10 + x)$$
$$4x + 60 = 43.2 + 4.32x$$
$$-0.32x + 60 = 43.2$$
$$-0.32x = -16.8$$
$$x = 52.5$$

52.5 oz of the $4 silver alloy must be used.

You Try It 1

A gardener has 20 lb of a lawn fertilizer that costs $.80 per pound. How many pounds of a fertilizer that costs $.55 per pound should be mixed with this 20 lb of lawn fertilizer to produce a mixture that costs $.75 per pound?

Your strategy

Your solution

5 lb

Solution on p. S15

Objective B   To solve uniform motion problems

Recall from Section 5.1 that an object traveling at a constant speed in a straight line is in *uniform motion*. The solution of a uniform motion problem is based on the equation $rt = d$, where r is the rate of travel, t is the time spent traveling, and d is the distance traveled.

HOW TO   A car leaves a town traveling at 40 mph. Two hours later, a second car leaves the same town, on the same road, traveling at 60 mph. In how many hours will the second car pass the first car?

> **Strategy for Solving a Uniform Motion Problem**
>
> 1. For each object, write a numerical or variable expression for the rate, time, and distance. The results can be recorded in a table.

The first car traveled **2** h longer than the second car.

Unknown time for the second car: t
Time for the first car: $t + 2$

	Rate, r	·	Time, t	=	Distance, d
First car	40	·	$t + 2$	=	$40(t + 2)$
Second car	60	·	t	=	$60t$

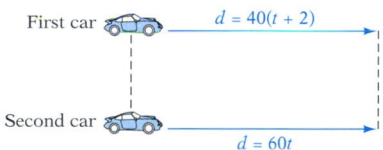

First car $d = 40(t + 2)$

Second car $d = 60t$

> 2. Determine how the distances traveled by the two objects are related. For example, the total distance traveled by both objects may be known, or it may be known that the two objects traveled the same distance.

The two cars travel the same distance.

$$40(t + 2) = 60t$$
$$40t + 80 = 60t$$
$$80 = 20t$$
$$4 = t$$

The second car will pass the first car in 4 h.

TAKE NOTE
Use the information given in the problem to fill in the Rate and Time columns of the table. Find the expression in the Distance column by multiplying the two expressions you wrote in each row.

Vocabulary to Review
uniform motion   [5.1D]

Equation to Review
Uniform Motion Equation
$d = rt$ [5.1D]

Instructor Note
One of the complications of distance-rate problems is that the variable may not directly represent the unknown. After doing a problem similar to the one at the left, show students a problem similar to Example 3 on the next page. The variable is time, but the unknown is distance.

In-Class Examples (Objective 5.5B)

1. Two planes leave an airport at the same time and fly in opposite directions. One of the planes is flying at 450 mph, and the other plane is flying at 550 mph. In how many hours will they be 2000 mi apart?   2 h

2. A cyclist starts on a course at 6 A.M. riding at 12 mph. An hour later, a second cyclist starts on the same course traveling at 18 mph. At what time will the second cyclist overtake the first cyclist?
9 A.M.

Optional Student Activity

Some students will find the following problem difficult.

Monica rides her bicycle to her friend Jorge's house and returns home by the same route. Monica rides her bike at constant speeds of 6 mph on level ground, 4 mph when going uphill, and 12 mph when going downhill. If her total riding time is 1 h, how far is it to Jorge's house?

Let x, y, and z be the distances on level ground, going uphill, and going downhill, respectively. Then

$$\frac{x}{6} + \frac{y}{4} + \frac{z}{12} + \frac{x}{6} + \frac{y}{12} +$$

$$\frac{z}{4} = 1 \text{ or}$$

$$4x + 4y + 4z = 12$$
$$4(x + y + z) = 12$$
$$x + y + z = 3$$

The distance is 3 mi.

Example 2

Two cars, one traveling 10 mph faster than the other, start at the same time from the same point and travel in opposite directions. In 3 h they are 300 mi apart. Find the rate of each car.

Strategy

- Rate of 1st car: r
  Rate of 2nd car: $r + 10$

	Rate	Time	Distance
1st car	r	3	$3r$
2nd car	$r + 10$	3	$3(r + 10)$

- The total distance traveled by the two cars is 300 mi.

Solution

$$3r + 3(r + 10) = 300$$
$$3r + 3r + 30 = 300$$
$$6r + 30 = 300$$
$$6r = 270$$
$$r = 45$$

$$r + 10 = 45 + 10 = 55$$

The first car is traveling 45 mph.
The second car is traveling 55 mph.

You Try It 2

Two trains, one traveling at twice the speed of the other, start at the same time on parallel tracks from stations that are 288 mi apart and travel toward each other. In 3 h, the trains pass each other. Find the rate of each train.

Your strategy

Your solution

32 mph; 64 mph

Example 3

How far can the members of a bicycling club ride out into the country at a speed of 12 mph and return over the same road at 8 mph if they travel a total of 10 h?

Strategy

- Time spent riding out: t
  Time spent riding back: $10 - t$

	Rate	Time	Distance
Out	12	t	$12t$
Back	8	$10 - t$	$8(10 - t)$

- The distance out equals the distance back.

Solution

$$12t = 8(10 - t)$$
$$12t = 80 - 8t$$
$$20t = 80$$
$$t = 4 \quad \text{(The time is 4 h.)}$$

The distance out $= 12t = 12(4) = 48$ mi.

The club can ride 48 mi into the country.

You Try It 3

A pilot flew out to a parcel of land and back in 5 h. The rate out was 150 mph, and the rate returning was 100 mph. How far away was the parcel of land?

Your strategy

Your solution

300 mi

Solutions on pp. S15–S16

5.5 Exercises

Objective A   **To solve value mixture problems**

Suggested Assignment
Exercises 1–35, odds
More challenging problems:
    Exercises 37, 38

1. An herbalist has 30 oz of herbs costing $2 per ounce. How many ounces of herbs costing $1 per ounce should be mixed with the 30 oz to produce a mixture costing $1.60 per ounce?
20 oz

2. The manager of a farmer's market has 500 lb of grain that costs $1.20 per pound. How many pounds of meal costing $.80 per pound should be mixed with the 500 lb of grain to produce a mixture that costs $1.05 per pound?
300 lb

3. Find the cost per pound of a meatloaf mixture made from 3 lb of ground beef costing $1.99 per pound and 1 lb of ground turkey costing $1.39 per pound.
$1.84

4. Find the cost per ounce of a sunscreen made from 100 oz of a lotion that costs $2.50 per ounce and 50 oz of a lotion that costs $4.00 per ounce.
$3

5. A snack food is made by mixing 5 lb of popcorn that costs $.80 per pound with caramel that costs $2.40 per pound. How much caramel is needed to make a mixture that costs $1.40 per pound?
3 lb

6. A wild birdseed mix is made by combining 100 lb of millet seed costing $.60 per pound with sunflower seeds costing $1.10 per pound. How many pounds of sunflower seeds are needed to make a mixture that costs $.70 per pound?
25 lb

7. Ten cups of a restaurant's house Italian dressing is made by blending olive oil costing $1.50 per cup with vinegar that costs $.25 per cup. How many cups of each are used if the cost of the blend is $.50 per cup?
Olive oil: 2 c; vinegar: 8 c

8. A high-protein diet supplement that costs $6.75 per pound is mixed with a vitamin supplement that costs $3.25 per pound. How many pounds of each should be used to make 5 lb of a mixture that costs $4.65 per pound?
Diet supplement: 2 lb; vitamin supplement: 3 lb

9. Find the cost per ounce of a mixture of 200 oz of a cologne that costs $5.50 per ounce and 500 oz of a cologne that costs $2.00 per ounce.
$3.00

10. Find the cost per pound of a trail mix made from 40 lb of raisins that cost $4.40 per pound and 100 lb of granola that costs $2.30 per pound.
$2.90

Quick Quiz (Objective 5.5A)

1. A trail mix is made by combining raisins that cost $4.20 per pound with granola that costs $2.20 per pound. How many pounds of each should be used to make 40 lb of trail mix that costs $2.75 per pound? Raisins: 11 lb; granola: 29 lb

11. A 20-ounce alloy of platinum that costs $220 per ounce is mixed with an alloy that costs $400 per ounce. How many ounces of the $400 alloy should be used to make an alloy that costs $300 per ounce?
16 oz

12. How many liters of a blue dye that costs $1.60 per liter must be mixed with 18 L of anil that costs $2.50 per liter to make a mixture that costs $1.90 per liter?
36 L

13. The manager of a specialty food store combined almonds that cost $4.50 per pound with walnuts that cost $2.50 per pound. How many pounds of each were used to make a 100-pound mixture that costs $3.24 per pound?
Almonds: 37 lb; walnuts: 63 lb

14. A goldsmith combined an alloy that cost $4.30 per ounce with an alloy that cost $1.80 per ounce. How many ounces of each were used to make a mixture of 200 oz costing $2.50 per ounce?
$4.30 alloy: 56 oz; $1.80 alloy: 144 oz

15. Adult tickets for a play cost $6.00, and children's tickets cost $2.50. For one performance, 370 tickets were sold. Receipts for the performance were $1723. Find the number of adult tickets sold.
228 adult tickets

16. Tickets for a piano concert sold for $4.50 for each adult. Student tickets sold for $2.00 each. The total receipts for 1720 tickets were $5980. Find the number of adult tickets sold.
1016 adult tickets

17. Find the cost per pound of sugar-coated breakfast cereal made from 40 lb of sugar that costs $1.00 per pound and 120 lb of corn flakes that cost $.60 per pound.
$.70

18. Find the cost per pound of a coffee mixture made from 8 lb of coffee that costs $9.20 per pound and 12 lb of coffee that costs $5.50 per pound.
$6.98

Objective B   **To solve uniform motion problems**

19. Two small planes start from the same point and fly in opposite directions. The first plane is flying 25 mph slower than the second plane. In 2 h, the planes are 470 mi apart. Find the rate of each plane.
105 mph, 130 mph

470 mi

20. Two cyclists start from the same point and ride in opposite directions. One cyclist rides twice as fast as the other. In 3 h, they are 81 mi apart. Find the rate of each cyclist.
9 mph, 18 mph

21. Two planes leave an airport at 8 A.M., one flying north at 480 km/h and the other flying south at 520 km/h. At what time will they be 3000 km apart?
11 A.M.

22. A long-distance runner started on a course running at an average speed of 6 mph. One-half hour later, a second runner began the same course running at an average speed of 7 mph. How long after the second runner started did the second runner overtake the first runner?
3 h

23. A motorboat leaves a harbor and travels at an average speed of 9 mph toward a small island. Two hours later, a cabin cruiser leaves the same harbor and travels at an average speed of 18 mph toward the same island. In how many hours after the cabin cruiser leaves the harbor will it be alongside the motorboat?
2 h

24. A 555-mile, 5-hour plane trip was flown at two speeds. For the first part of the trip, the average speed was 105 mph. For the remainder of the trip, the average speed was 115 mph. How long did the plane fly at each speed?
105 mph: 2 h; 115 mph: 3 h

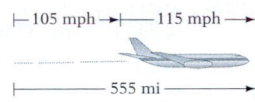

25. An executive drove from home at an average speed of 30 mph to an airport where a helicopter was waiting. The executive boarded the helicopter and flew to the corporate offices at an average speed of 60 mph. The entire distance was 150 mi. The entire trip took 3 h. Find the distance from the airport to the corporate offices.
120 mi

26. After a sailboat had been on the water for 3 h, a change in the wind direction reduced the average speed of the boat by 5 mph. The entire distance sailed was 57 mi. The total time spent sailing was 6 h. How far did the sailboat travel in the first 3 h?
36 mi

27. A car and a bus set out at 3 P.M. from the same point headed in the same direction. The average speed of the car is twice the average speed of the bus. In 2 h the car is 68 mi ahead of the bus. Find the rate of the car.
68 mph

28. A passenger train leaves a train depot 2 h after a freight train leaves the same depot. The freight train is traveling 20 mph slower than the passenger train. Find the rate of each train if the passenger train overtakes the freight train 3 h after the passenger train leaves the depot.
Passenger train: 50 mph; freight train: 30 mph

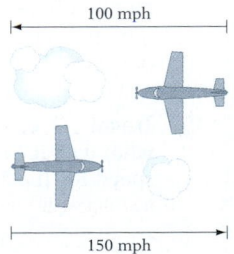

29. As part of flight training, a student pilot was required to fly to an airport and then return. The average speed on the way to the airport was 100 mph, and the average speed returning was 150 mph. Find the distance between the two airports if the total flying time was 5 h.
300 mi

Quick Quiz (Objective 5.5B)

1. A ship leaves the dock at 10 A.M. and travels south at 30 mph. One hour later, a second ship leaves the same dock and travels south at 50 mph. At what time does the second ship overtake the first ship? 12:30 P.M.

30. A ship traveling east at 25 mph is 10 mi from a harbor when another ship leaves the harbor traveling east at 35 mph. How long does it take the second ship to catch up to the first ship?
1 h

31. At 10 A.M., a plane leaves Boston, Massachusetts, for Seattle, Washington, a distance of 3000 mi. One hour later, a plane leaves Seattle for Boston. Both planes are traveling at a speed of 500 mph. How many hours after the plane leaves Seattle will the planes pass each other?
2.5 h

32. At noon, a train leaves Washington, D.C., headed for Charleston, South Carolina, a distance of 500 mi. The train travels at a speed of 60 mph. At 1 P.M., a second train leaves Charleston headed for Washington, D.C., traveling at 50 mph. How long after the train leaves Charleston will the two trains pass each other?
4 h

33. Two cyclists start at the same time from opposite ends of a course that is 51 mi long. One cyclist is riding at a rate of 16 mph, and the second cyclist is riding at a rate of 18 mph. How long after they begin will they meet?
1.5 h

34. A bus traveled on a level road for 2 h at an average speed that was 20 mph faster than its average speed on a winding road. The time spent on the winding road was 3 h. Find the average speed on the winding road if the total trip was 210 mi.
34 mph

35. A bus traveling at a rate of 60 mph overtakes a car traveling at a rate of 45 mph. If the car had a 1-hour head start, how far from the starting point does the bus overtake the car?
180 mi

36. A car traveling at 48 mph overtakes a cyclist who, riding at 12 mph, had a 3-hour head start. How far from the starting point does the car overtake the cyclist?
48 mi

APPLYING THE CONCEPTS

37. **Travel**   At 10 A.M., two campers left their campsite by canoe and paddled downstream at an average speed of 12 mph. They then turned around and paddled back upstream at an average rate of 4 mph. The total trip took 1 h. At what time did the campers turn around downstream?
10:15 A.M.

38. **Travel**   A car travels a 1-mile track at an average speed of 30 mph. At what average speed must the car travel the next mile so that the average speed for the 2 mi is 60 mph?
It is impossible to average 60 mph.

Focus on Problem Solving

Trial-and-Error Approach to Problem Solving

The questions below require an answer of always true, sometimes true, or never true. These problems are best solved by the trial-and-error method. The trial-and-error method of arriving at a solution to a problem involves repeated tests or experiments.

For example, consider the statement

> Both sides of an equation can be divided by the same number without changing the solution of the equation.

The solution of the equation $6x = 18$ is 3. If we divide both sides of the equation by 2, the result is $3x = 9$ and the solution is still 3. So the answer "never true" has been eliminated. We still need to determine whether there is a case for which the statement is not true. Is there a number that we could divide both sides of the equation by and the result would be an equation for which the solution is not 3?

If we divide both sides of the equation by 0, the result is $\frac{6x}{0} = \frac{18}{0}$; the solution of this equation is not 3 because the expressions on either side of the equals sign are undefined. Thus the statement is true for some numbers and not true for 0. The statement is sometimes true.

For Exercises 1 to 13, determine whether the statement is always true, sometimes true, or never true.

1. Both sides of an equation can be multiplied by the same number without changing the solution of the equation.

2. For an equation of the form $ax = b$, $a \neq 0$, multiplying both sides of the equation by the reciprocal of a will result in an equation of the form $x = constant$.

3. Adding -3 to each side of an equation yields the same result as subtracting 3 from each side of the equation.

4. An equation contains an equals sign.

5. The same variable term can be added to both sides of an equation without changing the solution of the equation.

6. An equation of the form $ax + b = c$ cannot be solved if a is a negative number.

7. The solution of the equation $\frac{x}{0} = 0$ is 0.

8. In solving an equation of the form $ax + b = cx + d$, subtracting cx from each side of the equation results in an equation with only one variable term in it.

9. If a rope 8 meters long is cut into two pieces and one of the pieces has a length of x meters, then the length of the other piece can be represented as $(x - 8)$ meters.

10. An even integer is a multiple of 2.

Answers to Focus on Problem Solving: Trial-and-Error Approach to Problem Solving

1. Sometimes true
2. Always true
3. Always true
4. Always true
5. Always true
6. Never true
7. Never true
8. Always true
9. Never true
10. Always true

Answers to Focus on Problem Solving
(continued)
11. Never true
12. Never true
13. Never true

Answers to Projects and Group Activities: Averages

1. Answers will vary.
2. Answers will vary.
3. Answers will vary.
4. 0
5. Steps 2 and 3: Answers will vary. Step 4: 0
6. The sums are the same.
7. If the activity were conducted again, the outcome would be the same.

Students should describe the average of a set of data as a number that represents the "middle" of the data. Therefore, the data values above the average and the data values below the average will "cancel" each other out.

11. If the first of three consecutive odd integers is n, then the second and third consecutive odd integers are represented by $n + 1$ and $n + 3$.

12. If we combine an alloy that costs \$8 an ounce with an alloy that costs \$5 an ounce, the cost of the resulting mixture will be greater than \$8 an ounce.

13. If the speed of one train is 20 mph slower than that of a second train, then the speeds of the two trains can be represented as r and $20 - r$.

Projects and Group Activities

Averages

We often discuss temperature in terms of average high or average low temperature. Temperatures collected over a period of time are analyzed to determine, for example, the average high temperature for a given month in your city or state. The following activity is planned to help you better understand the concept of "average."

1. Choose two cities in the United States. We will refer to them as City X and City Y. Over an 8-day period, record the daily high temperature for each city.

2. Determine the average high temperature for City X for the 8-day period. (Add the eight numbers, and then divide the sum by 8.) Do not round your answer.

3. Subtract the average high temperature for City X from each of the eight daily high temperatures for City X. You should have a list of eight numbers; the list should include positive numbers, negative numbers, and possibly zero.

4. Find the sum of the list of eight differences recorded in Step 3.

5. Repeat Steps 2 through 4 for City Y.

6. Compare the two sums found in Steps 4 and 5 for City X and City Y.

7. If you were to conduct this activity again, what would you expect the outcome to be? Use the results to explain what an average high temperature is. In your own words, explain what "average" means.

Addition and Multiplication Properties

The chart below is an addition table. Use it to answer Exercises 1 to 7.

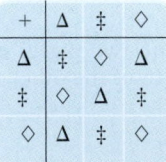

1. Find the sum of Δ and ‡.

2. What is ◇ plus ◇?

3. In our number system, 0 can be added to any number without changing that number; 0 is called the **additive identity.** What is the additive identity for the system in the chart on page 324? Explain your answer.

4. Does the Commutative Property of Addition apply to this system? Explain your answer.

5. What is −Δ (the opposite of Δ) equal to? Explain your answer.

6. What is −‡ (the opposite of ‡) equal to? Explain your answer.

7. Simplify −Δ + ‡ − ◇. Explain how you arrived at your answer.

The chart below is a multiplication table. Use it to answer Exercises 8 to 14.

×	£	¿	&
£	£	¿	&
¿	¿	¿	¿
&	&	¿	*

8. Find the product of £ and &.

9. What is ¿ times £?

10. Find the square of &.

11. Does the Commutative Property of Multiplication apply to this system? Explain your answer.

12. In our number system, the product of a number and 0 is 0. Is there an element in this system that corresponds to 0 in our system? Explain your answer.

13. In our number system, 1 can be multiplied by any number without changing that number; 1 is called the **multiplicative identity.** What is the multiplicative identity for the system in the chart above? Explain your answer.

14. Simplify & ÷ £ × ¿. Explain how you arrived at your answer.

Instructor Note

Before having students work on this activity, you might show them traditional addition and multiplication tables. Discuss the Commutative Properties of Addition and Multiplication, the Addition Property of Zero, the Multiplication Property of Zero, and the Multiplication Property of One.

Answers to Projects and Group Activities: Addition and Multiplication Properties

1. ◇
2. ◇
3. ◇
4. Yes. For example, ‡ + Δ = Δ + ‡. This is true for all elements in the table.
5. −Δ = ‡ because Δ + ‡ = 0.
6. −‡ = Δ because ‡ + Δ = 0.
7. −Δ + ‡ − ◇ = ‡ + ‡ − ◇
   = Δ − ◇ = Δ − 0 = Δ
8. &
9. ¿
10. *
11. Yes, For example, ¿ × £ = £ × ¿. This is true for all elements in the table.
12. Yes, ¿. The product of any symbol and ¿ is equal to ¿.
13. £. The product of £ and any symbol is equal to the symbol.
14. & ÷ £ × ¿ = & × ¿ = ¿

Chapter 5 Summary

Key Words	**Examples**
An *equation* expresses the equality of two mathematical expressions. [5.1A, p. 281]	$3 + 2(4x − 5) = x + 4$ is an equation.

A *solution of an equation* is a number that, when substituted for the variable, results in a true equation. [5.1A, p. 281]

-2 is a solution of $2 - 3x = 8$ because $2 - 3(-2) = 8$ is a true equation.

To *solve an equation* means to find a solution of the equation. The goal is to rewrite the equation in the form *variable = constant*, because the constant is the solution. [5.1B, p. 282]

The equation $x = -3$ is in the form *variable = constant*. The constant, -3, is the solution of the equation.

Consecutive integers follow one another in order. [5.4A, p. 308]

5, 6, 7 are consecutive integers.
$-9, -8, -7$ are consecutive integers.

Essential Rules and Procedures

Examples

Addition Property of Equations [5.1B, p. 282]
The same number can be added to each side of an equation without changing the solution of the equation.

If $a = b$, then $a + c = b + c$.

Multiplication Property of Equations [5.1C, p. 283]
Each side of an equation can be multiplied by the same *nonzero* number without changing the solution of the equation.

If $a = b$ and $c \neq 0$, then $ac = bc$.

Consecutive Integers [5.4A, p. 308]
$n, n + 1, n + 2, \ldots$

The sum of three consecutive integers is 33.

$n + (n + 1) + (n + 2) = 33$

Consecutive Even or Consecutive Odd Integers
$n, n + 2, n + 4$
[5.4A, p. 308]

The sum of three consecutive odd integers is 33.

$n + (n + 2) + (n + 4) = 33$

Value Mixture Equation [5.5A, p. 315]
Amount · Unit Cost = Value
$$AC = V$$

An herbalist has 30 oz of herbs costing \$4 per ounce. How many ounces of herbs costing \$2 per ounce should be mixed with the 30 oz to produce a mixture costing \$3.20 per ounce?

$30(4) + 2x = 3.20(30 + x)$

Uniform Motion Equation [5.1D, p. 285; 5.5B, p. 317]
Distance = Rate · Time
$$d = rt$$

A boat traveled from a harbor to an island at an average speed of 20 mph. The average speed on the return trip was 15 mph. The total trip took 3.5 h. How long did it take to travel to the island?

$20t = 15(3.5 - t)$

Chapter 5 Review Exercises

1. Solve: $x + 3 = 24$
21 [5.1B]

2. Solve: $x + 5(3x - 20) = 10(x - 4)$
10 [5.3B]

3. Solve: $5x - 6 = 29$
7 [5.2A]

4. Is 3 a solution of $5x - 2 = 4x + 5$?
No [5.1A]

5. Solve: $\dfrac{3}{5}a = 12$
20 [5.1C]

6. Solve: $6x + 3(2x - 1) = -27$
−2 [5.3B]

7. Solve: $x - 3 = -7$
−4 [5.1B]

8. Solve: $5x + 3 = 10x - 17$
4 [5.3A]

9. Solve: $7 - [4 + 2(x - 3)] = 11(x + 2)$
−1 [5.3B]

10. Solve: $-6x + 16 = -2x$
4 [5.3A]

11. Solve: $7 - 3x = 2 - 5x$

$-\dfrac{5}{2}$ [5.3A]

12. Solve: $-\dfrac{3}{8}x = -\dfrac{15}{32}$

$\dfrac{5}{4}$ [5.1C]

13. Solve: $35 - 3x = 5$

10 [5.2A]

14. Solve: $3x = 2(3x - 2)$

$\dfrac{4}{3}$ [5.3B]

15. **Lever Systems**   A lever is 12 ft long. At a distance of 2 ft from the fulcrum, a force of 120 lb is applied. How large a force must be applied to the other end so that the system will balance? Use the lever system equation $F_1 x = F_2(d - x)$.
24 lb [5.3C]

16. **Travel**   A bus traveled on a level road for 2 h at an average speed that was 20 mph faster than its speed on a winding road. The time spent on the winding road was 3 h. Find the average speed on the winding road if the total trip was 200 mi.
32 mph [5.5B]

17. **Integers**   The difference between nine and twice a number is five. Find the number.
     2   [5.4B]

18. **Integers**   The product of five and a number is fifty. Find the number.
     10   [5.4B]

19. **Juice Mixtures**   A health food store combined cranberry juice that cost $1.79 per quart with apple juice that cost $1.19 per quart. How many quarts of each were used to make 10 qt of cranapple juice costing $1.61 per quart?
     Cranberry juice: 7 qt; apple juice: 3 qt   [5.5A]

20. **Integers**   Four times the second of three consecutive integers equals the sum of the first and third integers. Find the integers.
     −1, 0, 1   [5.4A]

21. **Integers**   Translate "four less than the product of five and a number is sixteen" into an equation and solve.
     $5n - 4 = 16$; 4   [5.4A]

22.  **Buildings**   The Empire State Building is 1472 ft tall. This is 654 ft less than twice the height of the Eiffel Tower. Find the height of the Eiffel Tower.   1063 ft   [5.4B]

23. **Temperature**   Find the Celsius temperature when the Fahrenheit temperature is 100°. Use the formula $F = \frac{9}{5}C + 32$, where F is the Fahrenheit temperature and C is the Celsius temperature. Round to the nearest tenth.
     37.8°C   [5.2B]

24. **Travel**   A jet plane traveling at 600 mph overtakes a propeller-driven plane that had a 2-hour head start. The propeller-driven plane is traveling at 200 mph. How far from the starting point does the jet overtake the propeller-driven plane?
     600 mi   [5.5B]

25. **Integers**   The sum of two numbers is twenty-one. Three times the smaller number is two less than twice the larger number. Find the two numbers.
     8, 13   [5.4A]

26. **Farming**   A farmer harvested 28,336 bushels of corn. This amount represents an increase of 3036 bushels over last year's crop. How many bushels of corn did the farmer harvest last year?
     25,300 bushels   [5.4B]

Chapter 5 Test

1. Solve: $3x - 2 = 5x + 8$
   −5   [5.3A]

2. Solve: $x - 3 = -8$
   −5   [5.1B]

3. Solve: $3x - 5 = -14$
   −3   [5.2A]

4. Solve: $4 - 2(3 - 2x) = 2(5 - x)$
   2   [5.3B]

5. Is −2 a solution of $x^2 - 3x = 2x - 6$?
   No   [5.1A]

6. Solve: $7 - 4x = -13$
   5   [5.2A]

7. Solve: $5 = 3 - 4x$
   $-\dfrac{1}{2}$ [5.2A]

8. Solve: $5x - 2(4x - 3) = 6x + 9$
   $-\dfrac{1}{3}$ [5.3B]

9. Solve: $5x + 3 - 7x = 2x - 5$
   2   [5.3A]

10. Solve: $\dfrac{3}{4}x = -9$
    −12   [5.1C]

11. Solve: $\dfrac{x}{5} - 12 = 7$
    95   [5.2A]

12. Solve: $8 - 3x = 2x - 8$
    $\dfrac{16}{5}$ [5.3A]

13. Solve: $y - 4y + 3 = 12$
    −3   [5.2A]

14. Solve: $2x + 4(x - 3) = 5x - 1$
    11   [5.3B]

15. **Flour Mixtures**   A baker wants to make a 15-pound blend of flour that costs $.60 per pound. The blend is made using a rye flour that costs $.70 per pound and a wheat flour that costs $.40 per pound. How many pounds of each flour should be used?   Rye: 10 lb; wheat: 5 lb   [5.5A]

16. **Manufacturing**   A financial manager has determined that the cost per unit for a calculator is $15 and that the fixed cost per month is $2000. Find the number of calculators produced during a month in which the total cost was $5000. Use the equation $T = U \cdot N + F$, where T is the total cost, U is the cost per unit, N is the number of units produced, and F is the fixed cost.
    200 calculators   [5.2B]

17. **Integers**   Find three consecutive even integers whose sum is 36.
    10, 12, 14   [5.4A]

18. **Manufacturing**   A clock manufacturer's fixed costs per month are $5000. The unit cost for each clock is $15. Find the number of clocks made during a month in which the total cost was $65,000. Use the formula $T = U \cdot N + F$, where T is the total cost, U is the cost per unit, N is the number of units made, and F is the fixed costs.
    4000 clocks   [5.2B]

19. **Integers**   Translate "The difference between three times a number and fifteen is twenty-seven" into an equation and solve.
    $3x - 15 = 27$; 14   [5.4A]

20. **Travel**   A cross-country skier leaves a camp to explore a wilderness area. Two hours later a friend leaves the camp in a snowmobile, traveling 4 mph faster than the skier. This friend meets the skier 1 h later. Find the rate of the snowmobile.
    6 mph   [5.5B]

21. **Manufacturing**   A company makes 140 televisions per day. Three times the number of 15-inch TVs made equals 20 less than the number of 25-inch TVs made. Find the number of 25-inch TVs made each day.
    110 25-inch TVs   [5.4B]

22. **Integers**   The sum of two numbers is eighteen. The difference between four times the smaller number and seven is equal to the sum of two times the larger number and five. Find the two numbers.
    8, 10   [5.4A]

23. **Travel**   As part of flight training, a student pilot was required to fly to an airport and then return. The average speed to the airport was 90 mph, and the average speed returning was 120 mph. Find the distance between the two airports if the total flying time was 7 h.
    360 mi   [5.5B]

24. **Physics**   Find the time required for a falling object to increase in velocity from 24 ft/s to 392 ft/s. Use the formula $V = V_0 + 32t$, where V is the final velocity of a falling object, V_0 is the starting velocity of the falling object, and t is the time for the object to fall.
    11.5s   [5.2B]

25. **Chemistry**   A chemist mixes 100 g of water at 80°C with 50 g of water at 20°C. To find the final temperature of the water after mixing, use the equation $m_1(T_1 - T) = m_2(T - T_2)$, where m_1 is the quantity of water at the hotter temperature, T_1 is the temperature of the hotter water, m_2 is the quantity of water at the cooler temperature, T_2 is the temperature of the cooler water, and T is the final temperature of the water after mixing.
    60°C   [5.3C]

Cumulative Review Exercises

1. Subtract: $-6 - (-20) - 8$
   6   [3.2B]

2. Multiply: $(-2)(-6)(-4)$
   -48 [3.3A]

3. Subtract: $-\dfrac{5}{6} - \left(-\dfrac{7}{16}\right)$
   $-\dfrac{19}{48}$ [3.4A]

4. Divide: $-2\dfrac{1}{3} \div 1\dfrac{1}{6}$
   -2 [3.4B]

5. Simplify: $-4^2 \cdot \left(-\dfrac{3}{2}\right)^3$
   54   [3.5A]

6. Simplify: $25 - 3\dfrac{(5-2)^2}{2^3 + 1} - (-2)$
   24   [3.5A]

7. Evaluate $3(a - c) - 2ab$ when $a = 2$, $b = 3$, and $c = -4$.
   6   [4.1A]

8. Simplify: $3x - 8x + (-12x)$
   $-17x$ [4.2A]

9. Simplify: $2a - (-3b) - 7a - 5b$
   $-5a - 2b$ [4.2A]

10. Simplify: $(16x)\left(\dfrac{1}{8}\right)$
    $2x$ [4.2B]

11. Simplify: $-4(-9y)$
    $36y$ [4.2B]

12. Simplify: $-2(-x^2 - 3x + 2)$
    $2x^2 + 6x - 4$ [4.2C]

13. Simplify: $-2(x - 3) + 2(4 - x)$
    $-4x + 14$ [4.2D]

14. Simplify: $-3[2x - 4(x - 3)] + 2$
    $6x - 34$ [4.2D]

15. Is -3 a solution of $x^2 + 6x + 9 = x + 3$?
    Yes   [5.1A]

16. Is $\dfrac{1}{2}$ a solution of $3 - 8x = 12x - 2$?
    No   [5.1A]

17. Simplify: $\left(\dfrac{3}{8} - \dfrac{1}{4}\right) \div \dfrac{3}{4} + \dfrac{4}{9}$
    $\dfrac{11}{18}$ [3.5A]

18. Solve: $\dfrac{3}{5}x = -15$
    -25 [5.1C]

19. Solve: $7x - 8 = -29$
    -3 [5.2A]

20. Solve: $13 - 9x = -14$
    3   [5.2A]

21. Multiply: 9.67×0.0049
   0.047383   [2.6B]

22. Find 6 less than 13.
   7   [1.2B]

23. Solve: $8x - 3(4x - 5) = -2x - 11$
   13   [5.3B]

24. Solve: $6 - 2(5x - 8) = 3x - 4$
   2   [5.3B]

25. Solve: $5x - 8 = 12x + 13$
   −3   [5.3A]

26. Solve: $11 - 4x = 2x + 8$
   $\dfrac{1}{2}$ [5.3A]

27. Chemistry   A chemist mixes 300 g of water at 75°C with 100 g of water at 15°C. To find the final temperature of the water after mixing, use the equation $m_1(T_1 - T) = m_2(T - T_2)$, where m_1 is the quantity of water at the hotter temperature, T_1 is the temperature of the hotter water, m_2 is the quantity of water at the cooler temperature, T_2 is the temperature of the cooler water, and T is the final temperature of the water after mixing.   60°C   [5.3C]

28. Integers   Translate "The difference between twelve and the product of five and a number is negative eighteen" into an equation and solve.
   $12 - 5x = -18$; 6   [5.4A]

29. Construction   The area of a cement foundation of a house is 2000 ft². This is 200 ft² more than three times the area of the garage. Find the area of the garage.   600 ft²   [5.4B]

30. Flour Mixtures   How many pounds of an oat flour that costs $.80 per pound must be mixed with 40 lb of a wheat flour that costs $.50 per pound to make a blend that costs $.60 per pound?   20 lb   [5.5A]

31. Integers   Translate "the sum of three times a number and four" into a mathematical expression.
   $3n + 4$ [4.3B]

32. Integers   Three less than eight times a number is three more than five times the number. Find the number.
   2   [5.4B]

33. Sprinting   A sprinter ran to the end of a track at an average rate of 8 m/s and then jogged back to the starting point at an average rate of 3 m/s. The sprinter took 55 s to run to the end of the track and jog back. Find the length of the track.   120 m   [5.5B]

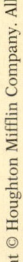

6

Proportion and Percent

Earned Run Average (ERA) is an important statistical measure of a baseball pitcher's performance. Mariano Rivera, a relief pitcher for the New York Yankees, is considered by many baseball fans to be the top relief pitcher in baseball. From 1995 to 2004, Rivera's career ERA was 2.43. In 1999, he won his third World Series with the Yankees and was named the Most Valuable Player of the series after pitching 12 scoreless innings and posting a 0.38 ERA (2 runs in $47\frac{1}{3}$ innings), which is ranked as the lowest post-season ERA in baseball history. A pitcher's ERA can be calculated by setting up a proportion, as seen in the Project on page 366.

OBJECTIVES

Section 6.1

A To write ratios and rates

Section 6.2

A To solve proportions
B To solve application problems

Section 6.3

A To write a percent as a fraction or a decimal
B To write a fraction or a decimal as a percent

Section 6.4

A To solve percent problems using the basic percent equation
B To solve percent problems using proportions
C To solve application problems

Section 6.5

A To solve problems involving simple interest

Need help? For online student resources, such as section quizzes, visit this textbook's website at **math.college.hmco.com/students.**

Do these exercises to prepare for Chapter 6.

1. Simplify: $\dfrac{8}{10}$

$\dfrac{4}{5}$ [2.2B]

2. Write as a decimal: $\dfrac{372}{15}$

24.8 [2.6C]

3. Which is greater, 4×33 or 62×2?

4×33 [1.1A/1.3A]

4. Multiply: $19 \times \dfrac{1}{100}$

$\dfrac{19}{100}$ [2.4A]

5. Multiply: 23×0.01

0.23 [2.6B]

6. Multiply: 0.47×100

47 [2.6B]

7. Multiply: $0.06 \times 47{,}500$

2850 [2.6B]

8. Divide: $60 \div 0.015$

4000 [2.6C]

9. Divide: $\dfrac{480}{0.06}$

8000 [2.6C]

10. Multiply $\dfrac{5}{8} \times 100$. Write the answer as a decimal.

62.5 [3.4B]

11. Write $\dfrac{200}{3}$ as a mixed number.

$66\dfrac{2}{3}$ [2.2A]

12. Divide $28 \div 16$. Write the answer as a decimal.

1.75 [2.6C]

GO FIGURE • • •

Suppose you threw six darts and all six hit the target shown. Which of the following could be your score?

4   15   58   28   29   31

28

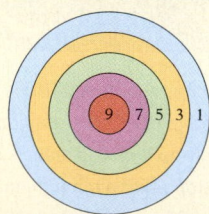

6.1 Ratios and Rates

Objective A To write ratios and rates

Point of Interest

It is believed that billiards was invented in France during the reign of Louis XI (1423–1483). In the United States, the standard billiard table is 4 ft 6 in. by 9 ft. This is a ratio of 1:2. The same ratio holds for carom and snooker tables, which are 5 ft by 10 ft.

In previous work, we have used quantities with units, such as 12 ft, 3 h, 2¢, and 15 acres. In these examples, the units are feet, hours, cents, and acres.

A **ratio** is the quotient or comparison of two quantities with the *same* unit. We can compare the measure of 3 ft to the measure of 8 ft by writing a quotient.

$$\frac{3 \text{ ft}}{8 \text{ ft}} = \frac{3}{8} \qquad 3 \text{ ft is } \frac{3}{8} \text{ of 8 ft.}$$

A ratio can be written in three ways:

1. As a fraction $\frac{3}{8}$

2. As two numbers separated by a colon   3:8

3. As two numbers separated by the word *to*   3 to 8

The ratio of 15 mi to 45 mi is written as

$$\frac{15 \text{ mi}}{45 \text{ mi}} = \frac{15}{45} = \frac{1}{3} \text{ or } 1{:}3 \text{ or } 1 \text{ to } 3$$

A ratio is in **simplest form** when the two numbers do not have a common factor. The units are not written in a ratio.

A **rate** is the comparison of two quantities with *different* units.

A catering company prepares 9 gal of coffee for every 50 people at a reception. This rate is written

$$\frac{9 \text{ gal}}{50 \text{ people}}$$

You traveled 200 mi in 6 h. The rate is written

$$\frac{200 \text{ mi}}{6 \text{ h}} = \frac{100 \text{ mi}}{3 \text{ h}}$$

A rate is in **simplest form** when the numbers have no common factors. The units are written as part of the rate.

Many rates are written as unit rates. A **unit rate** is a rate in which the number in the denominator is 1. The word *per* generally indicates a unit rate. It means "for each" or "for every." For example,

23 miles per gallon	• The unit rate is $\frac{23 \text{ mi}}{1 \text{ gal}}$.
65 miles per hour	• The unit rate is $\frac{65 \text{ mi}}{1 \text{ h}}$.
$4.78 per pound	• The unit rate is $\frac{\$4.78}{1 \text{ lb}}$.

Objective 6.1A

New Vocabulary

units
ratio
simplest form of a ratio
rate
simplest form of a rate
unit rate

Instructor Note

Ratios have applications to many disciplines. Investors talk of price–earnings ratios. Accountants use the current ratio, which is the ratio of current assets to current liabilities. Metallurgists use ratios to make various grades of steel.

   Unit rates are given in many situations. The EPA evaluates cars on the basis of miles driven per gallon of gas. A more difficult unit rate for students is the one used in the airline industry: cubic feet of fresh air per minute per person. Typical rates are: economy class, 7 ft³/min/person; first class: 50 ft³/min/person; cockpit: 150 ft³/min/person

Discuss the Concepts

1. Explain what a ratio is. Provide three examples of ratios.

2. What does it mean for a ratio to be in simplest form?

3. Describe three ways in which the ratio $10 to $2 can be written.

4. Explain what a rate is. Provide three examples of rates.

5. What is the difference between a ratio and a rate?

6. What does it mean for a rate to be in simplest form?

7. What is the student–faculty ratio at your school? What does this ratio mean?

In-Class Examples (Objective 6.1A)

1. Write the comparison 6 tons to 9 tons as a ratio in simplest form using a fraction, a colon (:), and the word *to*.

$\frac{2}{3}$, 2:3, 2 to 3

2. Write 84 ft in 9 s as a rate in simplest form.

$\frac{28 \text{ ft}}{3 \text{ s}}$

3. Write 297 mi on 9 gal as a unit rate.

33 mi/gal

Concept Check

State whether each of the following is a rate or a ratio. Explain your answer.

1. $\dfrac{3}{7}$ Ratio

2. $\dfrac{121 \text{ words}}{2 \text{ minutes}}$ Rate

3. $\dfrac{13 \text{ houses}}{20 \text{ acres}}$ Rate

4. $\dfrac{14}{5}$ Ratio

Optional Student Activity

Have students write ratios describing the class. For example: the ratio of male students to female students; the ratio of faculty to students; the ratio of students without children to students with children; the ratio of students wearing baseball caps to students not wearing baseball caps; the ratio of students using pens to students using pencils.

Unit rates make comparisons easier. For example, if you travel 37 mph and I travel 43 mph, we know that I am traveling faster than you are. It is more difficult to compare speeds if we are told that you are traveling $\dfrac{111 \text{ mi}}{3 \text{ h}}$ and I am traveling $\dfrac{172 \text{ mi}}{4 \text{ h}}$.

To find a unit rate, divide the number in the numerator of the rate by the number in the denominator of the rate. A unit rate is often written in decimal form.

HOW TO   A student received \$57 for working 6 h at the bookstore. Find the wage per hour (the unit rate).

$\dfrac{\$57}{6 \text{ h}}$ • Write the rate as a fraction.

$57 \div 6 = 9.5$ • Divide the number in the numerator of the rate (57) by the number in the denominator (6).

The unit rate is $\dfrac{\$9.50}{1 \text{ h}} = \$9.50/\text{h}$. This is read "\$9.50 per hour."

Example 1

Write the comparison of 12 to 8 as a ratio in simplest form using a fraction, a colon, and the word *to*.

Solution
$\dfrac{12}{8} = \dfrac{3}{2}$
$12:8 = 3:2$
$12 \text{ to } 8 = 3 \text{ to } 2$

You Try It 1

Write the comparison of 12 to 20 as a ratio in simplest form using a fraction, a colon, and the word *to*.

Your solution
$\dfrac{3}{5}$, 3:5, 3 to 5

Example 2

Write "12 hits in 26 times at bat" as a rate in simplest form.

Solution
$\dfrac{12 \text{ hits}}{26 \text{ at-bats}} = \dfrac{6 \text{ hits}}{13 \text{ at-bats}}$

You Try It 2

Write "20 bags of grass seed for 8 acres" as a rate in simplest form.

Your solution
$\dfrac{5 \text{ bags}}{2 \text{ acres}}$

Example 3

Write "285 mi in 5 h" as a unit rate.

Solution
$\dfrac{285 \text{ mi}}{5 \text{ h}}$
$285 \div 5 = 57$

The unit rate is 57 mph.

You Try It 3

Write "\$8.96 for 3.5 lb" as a unit rate.

Your solution
\$2.56/lb

Solutions on p. S16

6.1 Exercises

Objective A   To write ratios and rates

Section 6.1

Suggested Assignment
Exercises 1–27, odds
More challenging problems:
   Exercises 29, 30

Write the comparison as a ratio in simplest form using a fraction, a colon, and the word *to*.

1. 16 in. to 24 in.
$\frac{2}{3}$, 2:3, 2 to 3

2. 8 lb to 60 lb
$\frac{2}{15}$, 2:15, 2 to 15

3. 9 h to 24 h
$\frac{3}{8}$, 3:8, 3 to 8

4. $55 to $150
$\frac{11}{30}$, 11:30, 11 to 30

5. 9 ft to 2 ft
$\frac{9}{2}$, 9:2, 9 to 2

6. 50 min to 6 min
$\frac{25}{3}$, 25:3, 25 to 3

7. 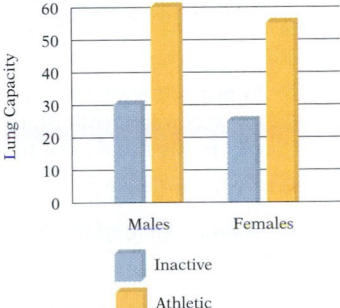 **Physical Fitness**   The figure at the right shows the lung capacity of inactive versus athletic 45-year-olds. Write the comparison of the lung capacity of an inactive male to that of an athletic male as a ratio in simplest form using a fraction, a colon, and the word *to*.
$\frac{1}{2}$, 1:2, 1 to 2

Lung Capacity (in milliliters of oxygen per kilogram of body weight per minute)

Write as a ratio in simplest form using a fraction.

8. **Construction**   The cost of building a patio cover was $1200 for labor and $3200 for materials. Find the ratio of the cost of materials to the cost of labor.
$\frac{8}{3}$

9. **Sports**   A baseball player had 3 errors in 42 fielding attempts. What is the ratio of the number of times the player did not make an error to the total number of attempts?
$\frac{13}{14}$

10. **Sports**   A basketball team won 18 games and lost 8 games during the season. What is the ratio of the number of games won to the total number of games?
$\frac{9}{13}$

11. **Mechanics**   Find the ratio of two meshed gears if one gear has 24 teeth and the other gear has 36 teeth.
$\frac{2}{3}$

Write as a rate in simplest form.

12. $85 for 3 shirts
$\frac{\$85}{3 \text{ shirts}}$

13. 150 mi in 6 h
$\frac{25 \text{ mi}}{1 \text{ h}}$

14. $76 for 8 h work
$\frac{\$19}{2 \text{ h}}$

15. $6.56 for 6 candy bars
$\frac{\$3.28}{3 \text{ bars}}$

16. 252 avocado trees on 6 acres
$\frac{42 \text{ trees}}{1 \text{ acre}}$

17. 9 children in 4 families
$\frac{9 \text{ children}}{4 \text{ families}}$

Quick Quiz (Objective 6.1A)

1. Write the comparison 20 days to 4 days as a ratio in simplest form using a fraction, a colon (:), and the word *to*.
$\frac{5}{1}$, 5:1, 5 to 1

2. Write 6 tablets in 4 h as a rate in simplest form.
$\frac{3 \text{ tablets}}{2 \text{ h}}$

3. Write 198 words in 4.5 min as a unit rate.
44 words/min

For Exercises 18–23, write as a unit rate.

18. $460 earned for 40 h of work
$11.50/h

19. $38,700 earned in 12 months
$3225/month

20. 387.8 mi in 7 h
55.4 mph

21. 364.8 mi on 9.5 gal of gas
38.4 mi/gal

22. $19.08 for 4.5 lb
$4.24/lb

23. $20.16 for 15 oz
$1.344/oz

24. **Sports** NCAA statistics show that for every 2800 college seniors playing college basketball, only 50 will play as rookies in the National Basketball Association. Write the ratio of the number of National Basketball Association rookies to the number of college seniors playing basketball.
$\frac{1}{56}$

25. **Energy** A transformer has 40 turns in the primary coil and 480 turns in the secondary coil. State the ratio of the number of turns in the primary coil to the number of turns in the secondary coil.
$\frac{1}{12}$

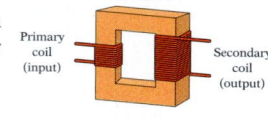

Transformer

Primary coil (input)   Secondary coil (output)

26. **Travel** An airplane flew 1155 mi in 2.5 h. Find the rate of travel.
462 mph

27. **Population Density** The table at the right shows the population and area of three countries. Find the population density (people per square mile) for each country. Round to the nearest tenth.
Australia: 6.6 people/mi²; India: 824.1 people/mi²; U.S.: 80.7 people/mi²

Country	Population	Area (in square miles)
Australia	19,547,000	2,968,000
India	1,045,845,000	1,269,000
United States	291,929,000	3,619,000

28. **Investments** An investor purchased 100 shares of stock for $2500. One year later the investor sold the stock for $3200. What was the investor's profit per share?
$7

APPLYING THE CONCEPTS

29. **Compensation** You have a choice of receiving a wage of $34,000 per year, $2840 per month, $650 per week, or $18 per hour. Which pay choice would you take? Assume a 40-hour week with 52 weeks per year.
$18/h

30. **Social Security** According to the Social Security Administration, the number of workers per retiree is expected to be as given in the table below.

Year	2010	2020	2030	2040
Number of workers per retiree	3.1	2.5	2.1	2.0

Why is the shrinking number of workers per retiree important to the Social Security Administration?

Answers to Writing Exercises

30. The fact that the number of workers per retiree is decreasing means that for each retiree drawing money out of Social Security, there are fewer and fewer workers paying into the Social Security system. In other words, fewer workers are supporting each retiree. Therefore, unless the amount paid into the system by each worker is increased, the funds to pay the Social Security benefits will be depleted.

6.2 Proportion

Objective A   To solve proportions

A **proportion** is the equality of two ratios or rates.

The equality $\dfrac{250 \text{ mi}}{5 \text{ h}} = \dfrac{50 \text{ mi}}{1 \text{ h}}$ is a proportion.

> **Definition of Proportion**
>
> If $\dfrac{a}{b}$ and $\dfrac{c}{d}$ are equal ratios or rates, then $\dfrac{a}{b} = \dfrac{c}{d}$ is a proportion.

Each of the four numbers in a proportion is called a **term.** Each term is numbered according to the following diagram.

$$\begin{array}{l}\text{first term} \longleftarrow \\ \text{second term} \longleftarrow\end{array}\ \dfrac{a}{b} = \dfrac{c}{d}\ \begin{array}{l}\longrightarrow \text{third term} \\ \longrightarrow \text{fourth term}\end{array}$$

The first and fourth terms of the proportion are called the **extremes,** and the second and third terms are called the **means.**

If we multiply the proportion by the least common multiple of the denominators, we obtain the following result:

$$\dfrac{a}{b} = \dfrac{c}{d}$$

$$bd\left(\dfrac{a}{b}\right) = bd\left(\dfrac{c}{d}\right)$$

$$ad = bc \qquad \textbf{\textit{ad}} \textbf{ is the product of the extremes.}$$
$$\textbf{\textit{bc}} \textbf{ is the product of the means.}$$

In any true proportion, **the product of the means equals the product of the extremes.** This is sometimes phrased as "the cross products are equal."

In the true proportion $\dfrac{3}{4} = \dfrac{9}{12}$, the cross products are equal.

$$\dfrac{3}{4}\ \diagdown\kern-0.9em\diagup\ \dfrac{9}{12} \quad \longrightarrow \quad 4 \cdot 9 = 36 \longleftarrow \text{ Product of the means}$$
$$\qquad\qquad\qquad\quad 3 \cdot 12 = 36 \longleftarrow \text{ Product of the extremes}$$

HOW TO Determine whether the proportion $\dfrac{47 \text{ mi}}{2 \text{ gal}} = \dfrac{304 \text{ mi}}{13 \text{ gal}}$ is a true proportion.

The product of the means: The product of the extremes:

$$2 \cdot 304 = 608 \qquad\qquad\qquad 47 \cdot 13 = 611$$

The proportion is not true because $608 \neq 611$.

Point of Interest

Proportions were studied by the earliest mathematicians. Clay tablets uncovered by archeologists show evidence of proportions in Egyptian and Babylonian cultures dating from 1800 B.C.

Study Tip

As you know, often in mathematics you learn one skill in order to perform another. This is true of this objective. You are learning to solve proportions. You will use this skill to solve application problems in the next objective and then to solve percent problems in Section 6.4.

Vocabulary to Review

ratio   [6.1A]
rate   [6.1A]

New Vocabulary

proportion
means
extremes
cross products
solve a proportion

Discuss the Concepts

1. What is a proportion?
2. What does the phrase "the cross products are equal" mean?

Optional Student Activity

On a calendar for any month, box in a few 2-by-2 number squares.

Su	Mo	Tu
	1	2
7	8	9
14	15	16
21	22	23
28	29	30

1. Compare the cross products in the number squares. What pattern do you notice?
   The products always differ by 7.

2. What pattern do you observe in the "cross sums"?
   The sums are always the same.

In-Class Examples (Objective 6.2A)

Determine if the proportion is true or not true.

1. $\dfrac{3}{7} = \dfrac{6}{14}$ True

2. $\dfrac{111 \text{ miles}}{3 \text{ hours}} = \dfrac{812 \text{ miles}}{22 \text{ hours}}$ Not true

Solve. Round to the nearest hundredth.

3. $\dfrac{60}{n} = \dfrac{24}{7}$ 17.5

4. $\dfrac{18}{20} = \dfrac{15}{n}$ 16.67

When three terms of a proportion are given, the fourth term can be found. To solve a proportion for an unknown term, use the fact that the product of the means equals the product of the extremes.

Integrating Technology

To use a calculator to solve the proportion at the right, multiply the second and third terms and divide by the fourth term. Enter

5 × 9 ÷ 16 =

The display reads 2.8125.

HOW TO Solve: $\dfrac{n}{5} = \dfrac{9}{16}$

$$\dfrac{n}{5} = \dfrac{9}{16}$$
$$5 \cdot 9 = n \cdot 16$$
$$45 = 16n$$
$$\dfrac{45}{16} = \dfrac{16n}{16}$$
$$2.8125 = n$$

- Find the number (n) that will make the proportion true.
- The product of the means equals the product of the extremes.
- Solve for n.

Example 1 Determine whether $\dfrac{15}{3} = \dfrac{90}{18}$ is a true proportion.

Solution
$$\dfrac{15}{3} \diagdown \dfrac{90}{18} \longrightarrow \begin{array}{l} 3 \cdot 90 = 270 \\ 15 \cdot 18 = 270 \end{array}$$

The product of the means equals the product of the extremes.

The proportion is true.

You Try It 1 Is $\dfrac{50 \text{ mi}}{3 \text{ gal}} = \dfrac{250 \text{ mi}}{12 \text{ gal}}$ a true proportion?

Your solution No

Example 2 Solve: $\dfrac{5}{9} = \dfrac{x}{45}$

Solution
$$\dfrac{5}{9} = \dfrac{x}{45}$$
$$9 \cdot x = 5 \cdot 45$$
$$9x = 225$$
$$\dfrac{9x}{9} = \dfrac{225}{9}$$
$$x = 25$$

- The cross products are equal.

You Try It 2 Solve: $\dfrac{7}{12} = \dfrac{42}{x}$

Your solution 72

Example 3 Solve $\dfrac{6}{n} = \dfrac{45}{124}$. Round to the nearest tenth.

Solution
$$\dfrac{6}{n} = \dfrac{45}{124}$$
$$n \cdot 45 = 6 \cdot 124$$
$$45n = 744$$
$$\dfrac{45n}{45} = \dfrac{744}{45}$$
$$n \approx 16.5$$

- The cross products are equal.

You Try It 3 Solve $\dfrac{5}{n} = \dfrac{3}{322}$. Round to the nearest hundredth.

Your solution 536.67

Solutions on p. S16

Example 4   Solve: $\dfrac{x+2}{3} = \dfrac{7}{8}$

You Try It 4   Solve: $\dfrac{4}{5} = \dfrac{3}{x-3}$

Solution $\dfrac{x+2}{3} = \dfrac{7}{8}$

$3 \cdot 7 = (x+2)8$ • **The cross products are equal.**

$21 = 8x + 16$

$5 = 8x$

$0.625 = x$

Your solution   6.75

Solution on p. S16

Objective B   To solve application problems

Objective 6.2B

Proportions are useful in many types of application problems. In recipes, proportions are used when a larger batch of ingredients is used than the recipe calls for. In mixing cement, the amounts of cement, sand, and rock are mixed in the same ratio. A map is drawn on a proportional basis, such as 1 in. representing 50 mi.

In setting up a proportion, keep the same units in the numerators and the same units in the denominators. For example, if *feet* is in the numerator on one side of the proportion, then *feet* must be in the numerator on the other side of the proportion.

Instructor Note

Students will have some difficulty setting up the proportions in this objective. Although there are a number of ways to set up a proportion correctly, you might tell them to write a proportion so that the units in the numerators are the same and the units in the denominators are the same.

TAKE NOTE

It is also correct to write the proportion with the costs in the numerators and the number of tires in the denominators:
$\dfrac{\$162.50}{2 \text{ tires}} = \dfrac{c}{5 \text{ tires}}$. The solution will be the same.

HOW TO   A customer sees an ad in a newspaper advertising 2 tires for $162.50. The customer wants to buy 5 tires and use one for the spare. How much will the 5 tires cost?

$\dfrac{2 \text{ tires}}{\$162.50} = \dfrac{5 \text{ tires}}{c}$ • **Write a proportion. Let c = the cost of the 5 tires.**

$162.50 \cdot 5 = 2 \cdot c$ • **The cross products are equal.**

$812.50 = 2c$

$\dfrac{812.50}{2} = \dfrac{2c}{2}$

$406.25 = c$

The 5 tires will cost $406.25.

Concept Check

Use the student–faculty ratio at your school and the number of students enrolled to estimate the number of faculty at the school.

In-Class Examples (Objective 6.2B)

1. A stock investment of 150 shares paid an annual dividend of $555. At this rate, what annual dividend would be paid on 180 shares of stock?
$666

2. A life insurance policy costs $8.52 for every $1000 of insurance. At this rate, what is the cost for $20,000 worth of life insurance? $170.40

Optional Student Activity

(*Note:* In this information age, we are bombarded by statistics. Rates are frequently quoted for everything from students per computer in public schools to state divorce rates per 1000 people. Using real data, such as in the problem below, might help students see meaning in the numbers they encounter.)

In the United States, the average annual number of deaths per million people ages 5 to 34 from asthma is 3.5. Approximately how many people ages 5 to 34 die from asthma each year in this country? Use a figure of 150,000,000 for the number of residents 5 to 34 years old. (*Source:* National Center for Health Statistics) 525 people

Optional Student Activity

(*Note:* This is a challenging problem for students.) If 5 bats eat 6 mosquitoes in 6 seconds, how many mosquitoes would be eaten by 30 bats in 30 seconds? 180 mosquitoes

Point of Interest

In late summer, the Congress Street Bridge in Austin, Texas, is home to about 1.5 million Mexican free-tail bats, who eat approximately 10,000 to 30,000 lb of insects per night.

Example 5

During a Friday, the ratio of stocks declining in price to those advancing was 5 to 3. If 450,000 shares advanced, how many shares declined on that day?

Strategy

To find the number of shares that declined in price, write and solve a proportion using n to represent the number of shares that declined in price.

Solution

$$\frac{5 \text{ (declining)}}{3 \text{ (advancing)}} = \frac{n \text{ shares declining}}{450{,}000 \text{ shares advancing}}$$

$$3n = 5 \cdot 450{,}000$$
$$3n = 2{,}250{,}000$$
$$\frac{3n}{3} = \frac{2{,}250{,}000}{3}$$
$$n = 750{,}000$$

750,000 shares declined in price.

Example 6

From previous experience, a manufacturer knows that in an average production run of 5000 calculators, 40 will be defective. What number of defective calculators can be expected from a run of 45,000 calculators?

Strategy

To find the number of defective calculators, write and solve a proportion using n to represent the number of defective calculators.

Solution

$$\frac{40 \text{ defective calculators}}{5000 \text{ calculators}} = \frac{n \text{ defective calculators}}{45{,}000 \text{ calculators}}$$

$$5000 \cdot n = 40 \cdot 45{,}000$$
$$5000n = 1{,}800{,}000$$
$$\frac{5000n}{5000} = \frac{1{,}800{,}000}{5000}$$
$$n = 360$$

The manufacturer can expect 360 defective calculators.

You Try It 5

An automobile can travel 396 mi on 11 gal of gas. At the same rate, how many gallons of gas would be necessary to travel 832 mi? Round to the nearest tenth.

Your Strategy

Your solution
23.1 gal

You Try It 6

An automobile recall was based on tests that showed 15 defective transmissions in 1200 cars. At this rate, how many defective transmissions will be found in 120,000 cars?

Your Strategy

Your solution
1500 defective transmissions

Solutions on p. S16

6.2 Exercises

Objective A   To solve proportions

Section 6.2

Suggested Assignment
Exercises 13–71, odds
More challenging problems:
  Exercises 75, 76

Determine whether the proportion is true or not true.

1. $\dfrac{27}{8} = \dfrac{9}{4}$ Not true

2. $\dfrac{3}{18} = \dfrac{4}{19}$ Not true

3. $\dfrac{45}{135} = \dfrac{3}{9}$ True

4. $\dfrac{3}{4} = \dfrac{54}{72}$ True

5. $\dfrac{16}{3} = \dfrac{48}{9}$ True

6. $\dfrac{15}{5} = \dfrac{3}{1}$ True

7. $\dfrac{6\ \text{min}}{5\ \text{cents}} = \dfrac{30\ \text{min}}{25\ \text{cents}}$ True

8. $\dfrac{7\ \text{tiles}}{4\ \text{ft}} = \dfrac{42\ \text{tiles}}{20\ \text{ft}}$ Not true

9. $\dfrac{15\ \text{ft}}{3\ \text{yd}} = \dfrac{90\ \text{ft}}{18\ \text{yd}}$ True

10. $\dfrac{\$65}{5\ \text{days}} = \dfrac{\$26}{2\ \text{days}}$ True

11. $\dfrac{1\ \text{gal}}{4\ \text{qt}} = \dfrac{7\ \text{gal}}{28\ \text{qt}}$ True

12. $\dfrac{300\ \text{ft}}{4\ \text{rolls}} = \dfrac{450\ \text{ft}}{7\ \text{rolls}}$ Not true

Solve. Round to the nearest hundredth.

13. $\dfrac{2}{3} = \dfrac{n}{15}$ 10

14. $\dfrac{7}{15} = \dfrac{n}{15}$ 7

15. $\dfrac{n}{5} = \dfrac{12}{25}$ 2.4

16. $\dfrac{n}{8} = \dfrac{7}{8}$ 7

17. $\dfrac{3}{8} = \dfrac{n}{12}$ 4.5

18. $\dfrac{5}{8} = \dfrac{40}{n}$ 64

19. $\dfrac{3}{n} = \dfrac{7}{40}$ 17.14

20. $\dfrac{7}{12} = \dfrac{25}{n}$ 42.86

21. $\dfrac{16}{n} = \dfrac{25}{40}$ 25.6

22. $\dfrac{15}{45} = \dfrac{72}{n}$ 216

23. $\dfrac{120}{n} = \dfrac{144}{25}$ 20.83

24. $\dfrac{65}{20} = \dfrac{14}{n}$ 4.31

25. $\dfrac{0.5}{2.3} = \dfrac{n}{20}$ 4.35

26. $\dfrac{1.2}{2.8} = \dfrac{n}{32}$ 13.71

27. $\dfrac{0.7}{1.2} = \dfrac{6.4}{n}$ 10.97

28. $\dfrac{2.5}{0.6} = \dfrac{165}{n}$ 39.6

29. $\dfrac{x}{6.25} = \dfrac{16}{87}$ 1.15

30. $\dfrac{x}{2.54} = \dfrac{132}{640}$ 0.52

31. $\dfrac{1.2}{0.44} = \dfrac{y}{14.2}$ 38.73

32. $\dfrac{12.5}{y} = \dfrac{102}{55}$ 6.74

33. $\dfrac{n+2}{5} = \dfrac{1}{2}$ 0.5

34. $\dfrac{5+n}{8} = \dfrac{3}{4}$ 1

35. $\dfrac{4}{3} = \dfrac{n-2}{6}$ 10

36. $\dfrac{3}{5} = \dfrac{n-7}{8}$ 11.8

Quick Quiz (Objective 6.2A)

Determine if the proportion is true or not true.

1. $\dfrac{4}{5} = \dfrac{13}{16}$ Not true

2. $\dfrac{\$300}{36\ \text{hours}} = \dfrac{\$200}{24\ \text{hours}}$ True

Solve. Round to the nearest hundredth.

3. $\dfrac{n}{14} = \dfrac{3}{7}$ 6

4. $\dfrac{4}{9} = \dfrac{n}{7}$ 3.11

37. $\dfrac{2}{n+3} = \dfrac{7}{12}$
0.43

38. $\dfrac{5}{n+1} = \dfrac{7}{3}$
1.14

39. $\dfrac{7}{10} = \dfrac{3+n}{2}$
−1.6

40. $\dfrac{3}{2} = \dfrac{5+n}{4}$
1

41. $\dfrac{x-4}{3} = \dfrac{3}{4}$
6.25

42. $\dfrac{x-1}{8} = \dfrac{5}{2}$
21

43. $\dfrac{6}{1} = \dfrac{x-2}{5}$
32

44. $\dfrac{7}{3} = \dfrac{x-4}{8}$
22.67

45. $\dfrac{5}{8} = \dfrac{2}{x-3}$
6.2

46. $\dfrac{5}{2} = \dfrac{1}{x-6}$
6.4

47. $\dfrac{3}{x-4} = \dfrac{5}{3}$
5.8

48. $\dfrac{8}{x-6} = \dfrac{5}{4}$
12.4

Objective B   **To solve application problems**

Solve.

49. Biology   In a drawing, the length of an amoeba is 2.6 in. The scale of the drawing is 1 in. on the drawing equals 0.002 in. on the amoeba. Find the actual length of the amoeba.
0.0052 in.

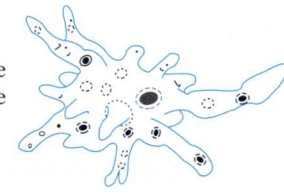

50. Insurance   A life insurance policy costs $15.22 for every $1000 of insurance. At this rate, what is the cost of $75,000 of insurance?
$1141.50

51. Sewing   Six children's robes can be made from 6.5 yd of material. How many robes can be made from 26 yd of material?
24 robes

52. Manufacturing   A computer manufacturer finds that an average of 3 defective hard drives are found in every 100 drives manufactured. How many defective drives are expected to be found in the production of 1200 hard drives?
36 defective drives

53. Taxes   The property tax on a $180,000 home is $4320. At this rate, what is the property tax on a home appraised at $280,000?
$6720

54. Medicine   The dosage of a certain medication is 2 mg for every 80 lb of body weight. How many milligrams of this medication are required for a person who weighs 220 lb?
5.5 mg

55. Travel   An automobile was driven 84 mi and used 3 gal of gasoline. At the same rate of consumption, how far would the car travel on 14.5 gal of gasoline?
406 mi

Quick Quiz (Objective 6.2B)

1. A liquid plant food is prepared by using 1 gal of water for each 1.5 teaspoons of plant food. At this rate, how many teaspoons of plant food are required for 5 gal of water?   7.5 teaspoons

2. For every 10 people who work in a city, 3 of them do not commute by public transportation. If 34,600 people work in the city, how many of them do not take public transportation?   10,380 people

56. **Nutrition**   If a 56-gram serving of pasta contains 7 g of protein, how many grams of protein are in a 454-gram box of the pasta?
56.75 g

57. **Consumerism**   If 4 grapefruit sell for $1.28, how much do 14 grapefruit cost?
$4.48

58. **Sports**   A halfback on a college football team has rushed for 435 yd in 5 games. At this rate, how many rushing yards will the halfback have in 12 games?
1044 yd

59. **Construction**   A building contractor estimates that 5 overhead lights are needed for every 400 ft² of office space. Using this estimate, how many light fixtures are necessary for an office building of 35,000 ft²?
438 lights

60. **Sports**   A softball player has hit 9 home runs in 32 games. At the same rate, how many home runs will the player hit in a 160-game schedule?
45 home runs

61. **Health**   A dieter has lost 3 lb in 5 weeks. At this rate, how long will it take the dieter to lose 36 lb?
60 weeks

62. **Consumerism**   Steak costs $25.20 for 3 lb. At this rate, how much does 8 lb of steak cost?
$67.20

63. **Business**   An automobile recall was based on engineering tests that showed 22 defects in 1000 cars. At this rate, how many defects would be found in 125,000 cars?
2750 defects

64. **Health**   Walking 5 mi in 2 h will use 650 calories. Walking at the same rate, how many miles would a person need to walk to lose 1 lb? (The burning of 3500 calories is equivalent to the loss of 1 lb.) Round to the nearest hundredth.
26.92 mi

65. **Travel**   An account executive bought a new car and drove 22,000 mi in the first 4 months. At the same rate, how many miles will the account executive drive in 3 years?
198,000 mi

66. **Investments**   An investment of $1500 earns $120 each year. At the same rate, how much additional money must be invested to earn $300 each year?
$2250

$1500	$1500 + x
earns	earns
$120	$300

67. **Investments**   A stock investment of $3500 earns a dividend of $280. At the same rate, how much additional money would have to be invested so that the total dividend is $400?
$1500

68.   Cartography   The scale on a map is $\frac{1}{2}$ in. equals 8 mi. What is the actual distance between two points that are $1\frac{1}{4}$ in. apart on the map?

20 mi

69.   Energy   A slow-burning candle will burn 1.5 in. in 40 min. How many inches of the candle will burn in 4 h?

9 in.

70.   Mixtures   A saltwater solution is made by dissolving $\frac{2}{3}$ lb of salt in 5 gal of water. At this rate, how many pounds of salt are required for 12 gal of water?

1.6 lb

2/3 lb of salt   x lb of salt

5 gal   12 gal

71.   Compensation   A management consulting firm recommends that the ratio of midmanagement salaries to junior management salaries be 7:5. Using this recommendation, find the yearly midmanagement salary when the junior management salary is $90,000.

$126,000

APPLYING THE CONCEPTS

72.   Determine whether the statement is true or false.

a.   A quotient $(a \div b)$ is a ratio.   **b.**   If $\frac{a}{b} = \frac{c}{d}$, then $\frac{b}{a} = \frac{d}{c}$.

True for $b \neq 0$ True

c.   If $\frac{a}{b} = \frac{c}{d}$, then $\frac{a}{c} = \frac{b}{d}$.   **d.**   If $\frac{a}{b} = \frac{c}{d}$, then $\frac{a}{d} = \frac{c}{b}$.

True   False

73.   If $\frac{a}{b} = \frac{c}{d}$, does $\frac{a}{b} = \frac{a+c}{b+d}$? Explain your answer.

Yes

74.   If $\frac{a}{b} = \frac{c}{d}$, show that $\frac{a+b}{b} = \frac{c+d}{d}$.

The complete solution is in the *Solutions Manual.*

75.   Elections   A survey of voters in a city claimed that 2 people of every 5 who voted cast a ballot in favor of city amendment A, and 3 people of every 4 who voted cast a ballot against amendment A. Is this possible? Explain your answer.

No

76.   🥧✏️   **Compensation**   In June 2002, *Time* magazine reported, "In 1980 the average CEO made 40 times the pay of the average factory worker; by 2000 the ratio had climbed to 531 to 1." What information would you need to know in order to determine the average pay of a CEO in 2000? With that information, how would you calculate the average pay of a CEO in 2000?

77.   ✏️   Write a paragraph describing how proportional representation is used to select the members of the U.S. House of Representatives.

Answers to Writing Exercises

76. To determine the average pay of a CEO in 2000, you would need to know the pay of the average factory worker in 2000. Suppose the average factory worker's pay in 2000 was $50,000. If we let n be the average pay of a CEO in 2000, then we can use the ratio $\frac{531}{1}$ and write the proportion $\frac{531}{1} = \frac{n}{50,000}$. Solve this proportion for n to determine the average pay of a CEO in 2000.

77. Your students can learn about proportional representation in the United States House of Representatives by consulting an information almanac or a history or government text. From a historical point of view, your students might find it interesting to read The Constitution of the United States of America, Article 1, Section 2, and Article XIV, Section 2, of the Amendments to the Constitution, both of which deal with proportional representation in the House of Representatives.

6.3 Percent

Objective A   **To write a percent as a fraction or a decimal**

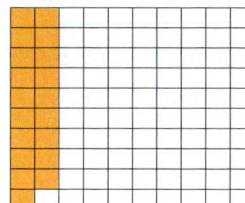

Percent means "parts of 100." The figure at the left has 100 parts. Because 19 of the 100 parts are shaded, 19% of the figure is shaded.

19 parts to 100 parts can be expressed as the ratio $\frac{19}{100}$. One percent can be expressed as 1 part to 100, or $\frac{1}{100}$. Thus 1% is $\frac{1}{100}$ or 0.01.

"A population growth rate of 5%," "a manufacturer's discount of 40%," and "an 8% increase in pay" are typical examples of the many ways in which percent is used in applied problems. When solving problems involving a percent, it is usually necessary either to rewrite the percent as a fraction or a decimal, or to rewrite a fraction or a decimal as a percent.

To write a percent as a fraction, remove the percent sign and multiply by $\frac{1}{100}$.

HOW TO   Write 67% as a fraction.

$$67\% = 67\left(\frac{1}{100}\right) = \frac{67}{100}$$
• Remove the percent sign and multiply by $\frac{1}{100}$.

To write a percent as a decimal, remove the percent sign and multiply by 0.01.

HOW TO   Write 19% as a decimal.

$$19\% \quad = \quad 19(0.01) \quad = \quad 0.19$$

Move the decimal point two places to the left. Then remove the percent sign.

• Remove the percent sign and multiply by 0.01. This is the same as moving the decimal point two places to the left.

Example 1   Write 150% as a fraction and as a decimal.

Solution $150\% = 150\left(\frac{1}{100}\right) = \frac{150}{100} = 1\frac{1}{2}$
$150\% = 150(0.01) = 1.50$

You Try It 1   Write 110% as a fraction and as a decimal.

Your solution $1\frac{1}{10}$, 1.10

Example 2   Write $66\frac{2}{3}\%$ as a fraction.

Solution $66\frac{2}{3}\% = 66\frac{2}{3}\left(\frac{1}{100}\right)$
$= \frac{200}{3}\left(\frac{1}{100}\right) = \frac{2}{3}$

You Try It 2   Write $16\frac{3}{8}\%$ as a fraction.

Your solution $\frac{131}{800}$

Solutions on p. S17

Objective 6.3A

New Vocabulary
percent
percent sign

New Symbols
%

Discuss the Concepts
Why do we multiply a percent by $\frac{1}{100}$ in order to rewrite it as a fraction and by 0.01 in order to rewrite it as a decimal?

Concept Check

1. Is 9.4% the same as $9\frac{2}{5}\%$? If not, what is the difference between the decimal equivalents of 9.4% and $9\frac{2}{5}\%$?
   Yes

2. Is $\frac{1}{2}\%$ the same as 0.5? If not, what is the difference between 0.5 and the decimal equivalent of $\frac{1}{2}\%$?   No. 0.495

Instructor Note
Example 2 and You Try It 2 are difficult for students. Here is an additional in-class example to use.

Write $12\frac{1}{2}\%$ as a fraction.

Solution:

$$12\frac{1}{2}\% = 12\frac{1}{2} \times \frac{1}{100}$$
$$= \frac{25}{2} \times \frac{1}{100}$$
$$= \frac{1}{8}$$

In-Class Examples (Objective 6.3A)

1. Write 72% as a fraction and as a decimal.
   $\frac{18}{25}$, 0.72

2. Write $15\frac{2}{3}\%$ as a fraction. $\frac{47}{300}$

3. Write 82.9% as a decimal.   0.829

Example 3   Write 0.35% as a decimal.

Solution $0.35\% = 0.35(0.01) = 0.0035$

You Try It 3   Write 0.8% as a decimal.

Your solution   0.008

Solution on p. S17

Objective B   To write a fraction or a decimal as a percent

A fraction or a decimal can be written as a percent by multiplying by 100%. Since 100% is $\frac{100}{100} = 1$, **multiplying by 100% is the same as multiplying by 1.**

HOW TO   Write $\frac{7}{8}$ as a percent.

$$\frac{7}{8} = \frac{7}{8}(100\%) = \frac{700}{8}\% = 87.5\%$$ • Multiply $\frac{7}{8}$ by 100%.

HOW TO   Write 0.64 as a percent.

$$0.64 \quad = \quad 0.64(100\%) \quad = \quad 64\%$$ • Multiply by 100%. This is the same as moving the decimal point two places to the right.

Move the decimal point two places to the right. Then write the percent sign.

Example 4   Write 1.78 as a percent.

Solution $1.78 = 1.78(100\%) = 178\%$

You Try It 4   Write 0.038 as a percent.

Your solution   3.8%

Example 5   Write $\frac{3}{11}$ as a percent. Write the remainder in fractional term.

Solution
$$\frac{3}{11} = \frac{3}{11}(100\%) = \frac{300}{11}\%$$
$$= 27\frac{3}{11}\%$$

You Try It 5   Write $\frac{9}{7}$ as a percent. Write the remainder in fractional term.

Your solution $128\frac{4}{7}\%$

Example 6   Write $1\frac{1}{7}$ as a percent. Round to the nearest tenth of a percent.

Solution
$$1\frac{1}{7} = \frac{8}{7} = \frac{8}{7}(100\%)$$
$$= \frac{800}{7}\% \approx 114.3\%$$

You Try It 6   Write $1\frac{5}{9}$ as a percent. Round to the nearest tenth of a percent.

Your solution   155.6%

Solutions on p. S17

In-Class Examples (Objective 6.3B)

1. Write 0.16 as a percent.   16%

2. Write $\frac{5}{12}$ as a percent. Round to the nearest tenth of a percent.   41.7%

3. Write $\frac{4}{9}$ as a percent. Write the remainder in fractional form. $44\frac{4}{9}\%$

6.3 Exercises

Objective A   To write a percent as a fraction or a decimal

1. ✏️   **a.** Explain how to convert a percent to a fraction.
        b. Explain how to convert a percent to a decimal.

2. ✏️   Explain why multiplying a number by 100% does not change the value of the number.

Write as a fraction and as a decimal.

3. 5%
$\frac{1}{20}$, 0.05

4. 60%
$\frac{3}{5}$, 0.60

5. 30%
$\frac{3}{10}$, 0.30

6. 90%
$\frac{9}{10}$, 0.90

7. 250%
$\frac{5}{2}$, 2.50

8. 140%
$\frac{7}{5}$, 1.40

9. 28%
$\frac{7}{25}$, 0.28

10. 66%
$\frac{33}{50}$, 0.66

11. 35%
$\frac{7}{20}$, 0.35

12. 8%
$\frac{2}{25}$, 0.08

13. 29%
$\frac{29}{100}$, 0.29

14. 83%
$\frac{83}{100}$, 0.83

Write as a fraction.

15. $11\frac{1}{9}\%$ $\frac{1}{9}$

16. $12\frac{1}{2}\%$ $\frac{1}{8}$

17. $37\frac{1}{2}\%$ $\frac{3}{8}$

18. $31\frac{1}{4}\%$ $\frac{5}{16}$

19. $66\frac{2}{3}\%$ $\frac{2}{3}$

20. $45\frac{5}{11}\%$ $\frac{5}{11}$

21. $6\frac{2}{3}\%$ $\frac{1}{15}$

22. $68\frac{3}{4}\%$ $\frac{11}{16}$

23. $\frac{1}{2}\%$ $\frac{1}{200}$

24. $83\frac{1}{3}\%$ $\frac{5}{6}$

25. $6\frac{1}{4}\%$ $\frac{1}{16}$

26. $3\frac{1}{3}\%$ $\frac{1}{30}$

Write as a decimal.

27. 7.3%
0.073

28. 9.1%
0.091

29. 15.8%
0.158

30. 16.7%
0.167

31. 0.3%
0.003

32. 0.9%
0.009

33. 121.2%
1.212

34. 18.23%
0.1823

35. 62.14%
0.6214

36. 0.15%
0.0015

37. 8.25%
0.0825

38. 5.05%
0.0505

39. 🥧 **Pets** The figure at the right shows some ways in which owners pamper their dogs. What fraction of the owners surveyed would buy a house or a car with their dog in mind?
$\frac{6}{25}$

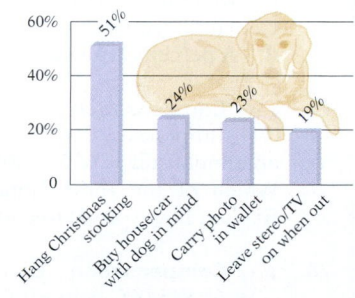

How Owners Pamper Their Dogs
Source: Purina Survey

Section 6.3

Suggested Assignment
Exercises 3–75, odds
More challenging problems:
    Exercises 77, 78

Answers to Writing Exercises

1a. Students should explain that to convert a percent to a fraction, drop the percent sign and multiply by $\frac{1}{100}$.

   b. Students should explain that to convert a percent to a decimal, drop the percent sign and multiply by 0.01.

2. Since 100% = 100 × 0.01 = 1, multiplying a number by 100% is the same as multiplying the number by 1. By the Multiplication Property of One, multiplying a number by 1 does not change the value of the number.

Quick Quiz (Objective 6.3A)

1. Write 65% as a fraction and as a decimal.

$\frac{13}{20}$, 0.65

2. Write $45\frac{1}{3}\%$ as a fraction. $\frac{34}{75}$

3. Write 34.27% as a decimal.   0.3427

Objective B **To write a fraction or a decimal as a percent**

Write as a percent.

40. 0.15
15%

41. 0.37
37%

42. 0.05
5%

43. 0.02
2%

44. 0.175
17.5%

45. 0.125
12.5%

46. 1.15
115%

47. 1.36
136%

48. 0.62
62%

49. 0.96
96%

50. 2.09
209%

51. 0.07
7%

Write as a percent. Round to the nearest tenth of a percent.

52. $\dfrac{27}{50}$
54%

53. $\dfrac{83}{100}$
83%

54. $\dfrac{37}{200}$
18.5%

55. $\dfrac{1}{3}$
33.3%

56. $\dfrac{5}{11}$
45.5%

57. $\dfrac{4}{9}$
44.4%

58. $\dfrac{7}{8}$
87.5%

59. $\dfrac{9}{20}$
45%

60. $1\dfrac{2}{3}$
166.7%

61. $2\dfrac{1}{2}$
250%

62. $\dfrac{2}{5}$
40%

63. $\dfrac{1}{6}$
16.7%

Write as a percent. Write the remainder in fractional form.

64. $\dfrac{17}{50}$
34%

65. $\dfrac{17}{25}$
68%

66. $\dfrac{3}{8}$
$37\dfrac{1}{2}\%$

67. $\dfrac{9}{16}$
$56\dfrac{1}{4}\%$

68. $1\dfrac{1}{4}$
125%

69. $2\dfrac{5}{8}$
$262\dfrac{1}{2}\%$

70. $1\dfrac{5}{9}$
$155\dfrac{5}{9}\%$

71. $2\dfrac{5}{6}$
$283\dfrac{1}{3}\%$

72. $\dfrac{12}{25}$
48%

73. $\dfrac{7}{30}$
$23\dfrac{1}{3}\%$

74. $\dfrac{3}{7}$
$42\dfrac{6}{7}\%$

75. $\dfrac{2}{9}$
$22\dfrac{2}{9}\%$

APPLYING THE CONCEPTS

76. Determine whether the statement is true or false. If the statement is false, give an example to show that it is false.
  a. Multiplying a number by a percent always decreases the number.
  b. Dividing by a percent always increases the number.
  c. The word *percent* means "per hundred."
  d. A percent is always less than one.
  a. False   **b.** False   **c.** True   **d.** False

77. ✏ **Compensation** Employee A had an annual salary of $42,000, Employee B had an annual salary of $48,000, and Employee C had an annual salary of $46,000 before each employee was given a 5% raise. Which of the three employees' annual salaries is now the highest? Explain how you arrived at your answer.

78. ✏ **Compensation** Each of three employees earned an annual salary of $45,000 before Employee A was given a 3% raise, Employee B was given a 6% raise, and Employee C was given a 4.5% raise. Which of the three employees now has the highest annual salary? Explain how you arrived at your answer.

Answers to Writing Exercises

77. Employee B's salary is the largest after the raise. You want students to explain that 5% of the largest original salary will be the largest raise.

78. Employee B's salary will be the largest after the raise. You want students to explain that, since all the original salaries are the same, the largest resulting salary will go to the employee with the largest percent raise.

Quick Quiz (Objective 6.3B)

1. Write 0.56 as a percent.   56%

2. Write $\dfrac{7}{45}$ as a percent. Round to the nearest tenth of a percent.   15.6%

3. Write $\dfrac{5}{6}$ as a percent. Write the remainder in fractional form. $83\dfrac{1}{3}\%$

6.4 The Basic Percent Equation

Objective A To solve percent problems using the basic percent equation

A real estate broker receives a payment that is 6% of a $275,000 sale. To find the amount the broker receives requires answering the question, *"6% of $275,000 is what?"* This sentence can be written using mathematical symbols and then solved for the unknown number. Recall that **of** is written as · (times), **is** is written as = (equals), and **what** is written as n (the unknown number).

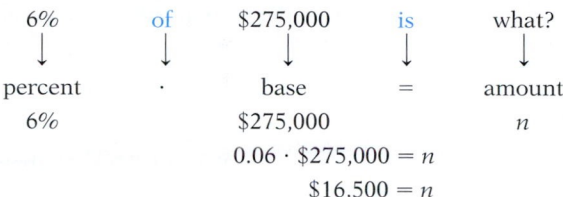

6%	of	$275,000	is	what?
↓	↓	↓	↓	↓
percent	·	base	=	amount
6%		$275,000		n

$$0.06 \cdot \$275,000 = n$$
$$\$16,500 = n$$

The broker receives a payment of $16,500.

The solution was found by solving the basic percent equation for amount.

> **The Basic Percent Equation**
>
> *Percent · base = amount*

Integrating Technology

The percent key % on a scientific calculator moves the decimal point to the left two places when pressed after a multiplication or division computation. For the example at the right, enter

800 × 2 · 5 % =

The display reads 20.

HOW TO Find 2.5% of 800.

$$Percent \cdot base = amount$$
$$0.025 \cdot 800 = n$$
$$20 = n$$

• Use the basic percent equation. Percent = 2.5% = 0.025, base = 800, amount = n

2.5% of 800 is 20.

A recent promotional game at a grocery store listed the probability of winning a prize as "1 chance in 2." A percent can be used to describe the chance of winning. This requires answering the question, *"What percent of 2 is 1?"*

The chance of winning can be found by solving the basic percent equation for percent.

What	percent	of	2	is	1?
	↓	↓	↓	↓	↓
	percent	·	base	=	amount
	n		2		1

$$n \cdot 2 = 1$$
$$n = \frac{1}{2}$$
$$n = \frac{1}{2}(100\%) = 50\%$$

• Write the fraction as a percent.

There is a 50% chance of winning a prize.

Objective 6.4A

New Vocabulary
base
amount

New Equations
Basic Percent Equation:
    Percent · base = amount

Concept Check
Create grids and ask the students to shade a given percent of a region. For example: Shade 40% of the region below.

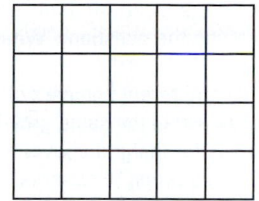

There are 20 squares.

40% × 20 = 0.40 × 20 = 8

8 squares should be shaded.

Shade different percents of the same grid (for example, for the grid above, shade 25% and 90%). Also create different grids; for example, shade 75% of a 5 × 8 grid or 37.5% of a 4 × 8 grid.

Instructor Note
Effective use of the basic percent equation is one of the most important skills a student can acquire. This section is devoted to solving the basic percent equation. The next objective gives you the option of teaching how to solve percent problems using proportions.

In-Class Examples (Objective 6.4A)

1. What is 45% of 80? 36
2. Find 12% of 425. 51
3. What percent of 80 is 25? 31.25%
4. 19 is what percent of 95? 20%
5. 10% of what is 20? 200
6. 7 is 14% of what? 50

Concept Check

Create grids and ask the students what percent of an entire region is shaded. For example, ask students what percent of the region below is shaded.

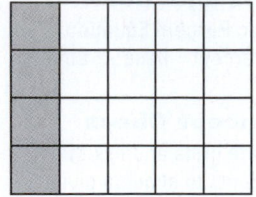

Since there are 20 squares, 4 of which are shaded, students are answering the question, "What percent of 20 is 4?" **20%**

Shade different numbers of squares within the same grid (for example, for the grid above, shade 6 squares, 10 squares, and 13 squares). Also create different grids. For example, shade 8 squares in a 4×6 grid, 9 squares in a 3×5 grid, or 7 squares in a 5×8 grid.

Discuss the Concepts

1. Multiplying a number by $33\frac{1}{3}\%$ is the same as multiplying it by what fraction?
2. Multiplying a number by 200% is the same as multiplying it by what whole number?
3. Multiplying a number by 150% is the same as multiplying it by what mixed number?

Instructor Note

In application problems involving percent, the basic percent equation frequently results in an equation of the form $ax = b$.

TAKE NOTE

We have written $n \cdot 20 = 32$ because that is the form of the basic percent equation. We could have written $20n = 32$.

HOW TO   32 is what percent of 20?

$$\text{Percent} \cdot \text{base} = \text{amount}$$
$$n \cdot 20 = 32$$
$$\frac{20n}{20} = \frac{32}{20}$$
$$n = 1.6$$
$$n = 160\%$$

• Use the basic percent equation.
  Percent = n, base = 20, amount = 32

• Write 1.6 as a percent.

32 is 160% of 20.

Each year an investor receives a payment that equals 8% of the value of an investment. This year that payment amounted to $640. To find the value of the investment this year, we must answer the question, "8% of what value is $640?"

The value of the investment can be found by solving the basic percent equation for the base.

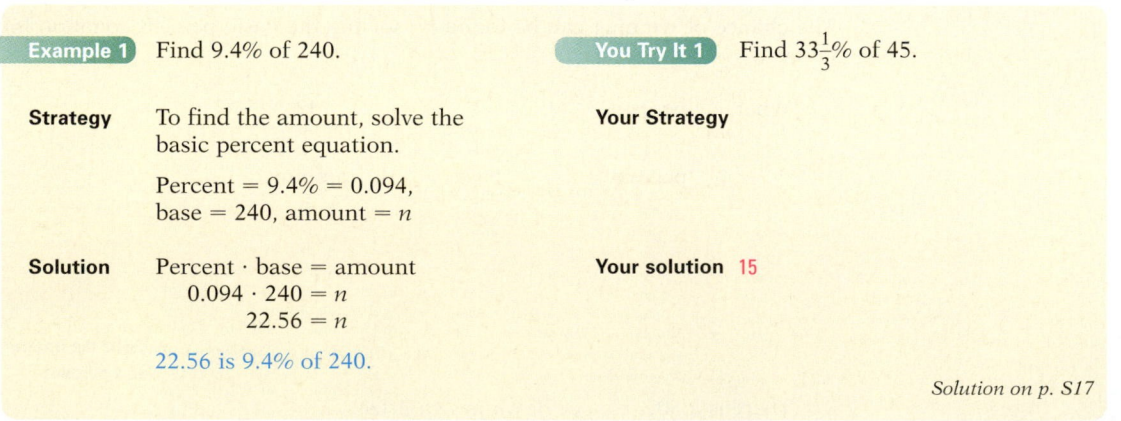

$$0.08 \cdot n = 640$$
$$\frac{0.08n}{0.08} = \frac{640}{0.08}$$
$$n = 8000$$

This year the investment is worth $8000.

TAKE NOTE

The base in the basic percent equation will usually follow the phrase *percent of*. Some percent problems may use the word *find*. In this case, we can substitute *what is* for *find*. See Example 1 below.

HOW TO   62% of what is 800? Round to the nearest tenth.

$$\text{Percent} \cdot \text{base} = \text{amount}$$
$$0.62 \cdot n = 800$$
$$\frac{0.62n}{0.62} = \frac{800}{0.62}$$
$$n \approx 1290.3$$

• Use the basic percent equation.
  Percent = 62% = 0.62, base = n, amount = 800

62% of 1290.3 is approximately 800.

Note from the previous three problems that if any two parts of the basic percent equation are given, the third part can be found.

Example 1   Find 9.4% of 240.

Strategy   To find the amount, solve the basic percent equation.

Percent = 9.4% = 0.094, base = 240, amount = n

Solution   Percent · base = amount
$$0.094 \cdot 240 = n$$
$$22.56 = n$$

22.56 is 9.4% of 240.

You Try It 1   Find $33\frac{1}{3}\%$ of 45.

Your Strategy

Your solution   **15**

Solution on p. S17

Example 2   What percent of 30 is 12?

Strategy   To find the percent, solve the basic percent equation.
Percent = n, base = 30, amount = 12

Solution   Percent · base = amount
$$n \cdot 30 = 12$$
$$\frac{30n}{30} = \frac{12}{30}$$
$$n = 0.4$$
$$n = 40\%$$

12 is 40% of 30.

You Try It 2   25 is what percent of 40?

Your Strategy

Your solution   62.5%

Example 3   60 is 2.5% of what?

Strategy   To find the percent, solve the basic percent equation.
Percent = 2.5% = 0.025,
base = n, amount = 60

Solution   Percent · base = amount
$$0.025 \cdot n = 60$$
$$\frac{0.025n}{0.025} = \frac{60}{0.025}$$
$$n = 2400$$

60 is 2.5% of 2400.

You Try It 3 $16\frac{2}{3}\%$ of what is 15?

Your Strategy

Your solution   90

Solutions on p. S17

Objective B   **To solve percent problems using proportions**

Percent problems can also be solved by using proportions. The proportion method is based on writing two ratios with quantities that can be found in the basic percent equation. One ratio is the percent ratio, written as $\frac{percent}{100}$. The second ratio is the amount-to-base ratio, written as $\frac{amount}{base}$. These two ratios form the proportion

$$\frac{\textbf{percent}}{\textbf{100}} = \frac{\textbf{amount}}{\textbf{base}}$$

The proportion method can be illustrated by a diagram. The rectangle at the left is divided into two parts. The whole rectangle is represented by 100 and the part by percent. On the other side, the whole rectangle is represented by the base and the part by amount. The ratio of the percent to 100 is equal to the ratio of the *amount* to the *base*.

Concept Check

For each of the following problems, do not do any calculations. Determine the answer by thinking about what the given percent does to the base in order to be equal to the given amount.

1. 25% of what is 800?
   a. 200
   b. 1600
   c. 3200   c

2. 20% of what is 16?
   a. 3.2
   b. 80
   c. 8   b

3. 90 is 150% of what?
   a. 60
   b. 135
   c. 100   a

4. 300 is $33\frac{1}{3}\%$ of what?
   a. 100
   b. 330
   c. 900   c

5. 200% of what is 50?
   a. 200
   b. 100
   c. 25   c

Objective 6.4B

Vocabulary to Review

ratio   [6.1A]
proportion   [6.2A]
base   [6.4A]
amount   [6.4A]

Instructor Note

This objective explains the proportion method of solving the basic percent equation. If you choose not to use this method, you can nonetheless use the exercises as a review of problems that involve percent. This will help students review the topic and recognize the different types of percent problems.

Discuss the Concepts

Do you prefer to solve percent problems using the basic percent equation or using proportions? Why?

Instructor Note

Some students find it easier to remember the proportion method by using the equation

$$\frac{\text{is}}{\text{of}} = \frac{n}{100}$$

Concept Check

The figure below shows the results of a survey of 1500 people who were asked how much they worry about privacy on the Internet.

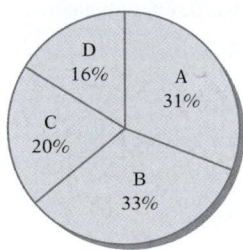

A—Worry a great deal
B—Worry somewhat
C—Don't worry very much
D—Don't worry at all

1. How many of the people said they worried somewhat about privacy on the Internet?
   495 people

2. How many of the people said they did not worry at all about privacy on the Internet?
   240 people

Study Tip

Remember that a vertical green bar indicates a worked-out example. Using paper and pencil, work through the example. See *AIM for Success*, page xxxi.

HOW TO   What is 32% of 85?

$$\frac{\text{percent}}{100} = \frac{\text{amount}}{\text{base}}$$

$$\frac{32}{100} = \frac{n}{85}$$

$$100 \cdot n = 32 \cdot 85$$

$$100n = 2720$$

$$\frac{100n}{100} = \frac{2720}{100}$$

$$n = 27.2$$

• Sketch a diagram.

• Percent = 32, base = 85, amount = n

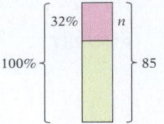

32% of 85 is 27.2.

Example 4   24% of what is 16? Round to the nearest hundredth.

Solution

$$\frac{\text{percent}}{100} = \frac{\text{amount}}{\text{base}}$$

$$\frac{24}{100} = \frac{16}{n}$$

• Percent = 24, amount = 16

$$24 \cdot n = 100 \cdot 16$$

$$24n = 1600$$

$$n = \frac{1600}{24} \approx 66.67$$

16 is approximately 24% of 66.67.

You Try It 4   8 is 25% of what?

Your solution   32

Example 5   Find 1.2% of 42.

Solution

$$\frac{\text{percent}}{100} = \frac{\text{amount}}{\text{base}}$$

$$\frac{1.2}{100} = \frac{n}{42}$$

• Percent = 1.2, base = 42

$$1.2 \cdot 42 = 100 \cdot n$$

$$50.4 = 100n$$

$$\frac{50.4}{100} = \frac{100n}{100}$$

$$0.504 = n$$

1.2% of 42 is 0.504.

You Try It 5   Find 0.74% of 1200.

Your solution   8.88

Example 6   What percent of 52 is 13?

Solution

$$\frac{\text{percent}}{100} = \frac{\text{amount}}{\text{base}}$$

$$\frac{n}{100} = \frac{13}{52}$$

• Amount = 13, base = 52

$$n \cdot 52 = 100 \cdot 13$$

$$52n = 1300$$

$$\frac{52n}{52} = \frac{1300}{52}$$

$$n = 25$$

25% of 52 is 13.

You Try It 6   What percent of 180 is 54?

Your solution   30%

Solutions on p. S17

Objective C   **To solve application problems**

HOW TO The circle graph at the right shows the causes of death for all police officers who died while on duty during a recent year. What percent of the deaths were due to traffic accidents? Round to the nearest tenth of a percent.

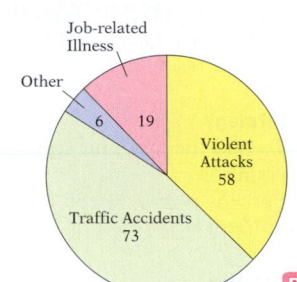

Figure 6.1 Causes of Death for Police Officers Killed in the Line of Duty
Source: International Union of Police Associations

Strategy   To find the percent:
- Find the total number of officers who died in the line of duty.
- Use the basic percent equation. Percent = *n*, base = total number who died, amount = number of deaths due to traffic accidents = 73

Solution $58 + 73 + 6 + 19 = 156$

$$\text{Percent} \cdot \text{base} = \text{amount}$$
$$n \cdot 156 = 73$$
$$\frac{156n}{156} = \frac{73}{156}$$
$$n \approx 0.468 = 46.8\%$$

Approximately 46.8% of the deaths were due to traffic accidents.

Example 7

During a recent year, 276 billion product coupons were issued by manufacturers. Shoppers redeemed 4.8 billion of these coupons. (*Source:* NCH NuWorld Consumer Behavior Study, America Coupon Council) What percent of the coupons issued were redeemed by customers? Round to the nearest tenth of a percent.

Strategy
To find the percent, use the basic percent equation. Percent = *n*, base = number of coupons issued = 276 billion, amount = number of coupons redeemed = 4.8 billion

Solution $\text{Percent} \cdot \text{base} = \text{amount}$
$$n \cdot 276 = 4.8$$
$$\frac{276n}{276} = \frac{4.8}{276}$$
$$n \approx 0.017$$
$$n \approx 1.7\%$$

Of the product coupons issued, approximately 1.7% were redeemed by customers.

You Try It 7

An instructor receives a monthly salary of $4330, and $649.50 is deducted for income tax. Find the percent of the instructor's salary deducted for income tax.

Your Strategy

Your solution
15%

Solution on p. S17

Objective 6.4C

Instructor Note
Although this is not foolproof, the base in the basic percent equation will generally follow the phrase "percent of" in application problems. Having students find this phrase will help them identify all the components of the percent equation.

Concept Check
A survey by the *Boston Globe* involved questioning elementary and middle-school students about television. Sixty-eight students, or 42.5% of those surveyed, said that they had a television in their bedroom at home. How many students were included in the *Globe* survey?
160 students

Optional Student Activity
1. At a department store, you purchase two items, one for $129 and the other for $99. The cash register automatically adds the sales tax and displays the total, $239.40. What is the sales tax rate expressed as a percent?   5%
2. A soccer team has won 25 of its first 30 games this season. The team has 20 more games to play. How many of the remaining games must the team win in order to win 80% of its games this season?
15 games

In-Class Examples (Objective 6.4C)
(*Note:* Solve for percent in Exercise 1, base in Exercise 2, and amount in Exercise 3.)

1. A soccer team won 42 out of the 56 games it played this season. What percent of the games played did the team win?   75%

2. A down payment of $2610 was paid on a new car. The down payment is 15% of the cost of the car. Find the cost of the car. $17,400

3. A growing company reinvested 54% of the $80,000 it earned into research and development. How much of the money earned was reinvested into research and development? $43,200

Optional Student Activity

(*Note:* This is a challenging problem.)

Five percent of the students enrolled at a state university were asked to name their favorite TV show. Of those surveyed, $\frac{1}{4}$ were first-year students, $\frac{1}{6}$ were sophomores, $\frac{1}{3}$ were juniors, $\frac{1}{5}$ were seniors, and 21 were graduate students. How many students are enrolled in the university?   **8400 students**

Example 8

A taxpayer pays a tax rate of 35% for state and federal taxes. The taxpayer has an income of $47,500. Find the amount of state and federal taxes paid by the taxpayer.

Strategy

To find the amount, solve the basic percent equation.
Percent = 35% = 0.35, base = 47,500, amount = n

Solution

Percent · base = amount
$$0.35 \cdot 47,500 = n$$
$$16,625 = n$$

The amount of taxes paid is $16,625.

You Try It 8

According to Board-Trac, approximately 19% of the country's 2.4 million surfers are women. Estimate the number of female surfers in this country. Write the number in standard form.

Your Strategy

Your solution
456,000 female surfers

Example 9

A department store has a blue blazer on sale for $114, which is 60% of the original price. What is the difference between the original price and the sale price?

Strategy

To find the difference between the original price and the sale price:
• Find the original price. Solve the basic percent equation.
  Percent = 60% = 0.60, amount = 114, base = n
• Subtract the sale price from the original price.

Solution

Percent · base = amount
$$0.60 \cdot n = 114$$
$$\frac{0.60n}{0.60} = \frac{114}{0.60}$$
$$n = 190$$

$$190 - 114 = 76$$

The difference in price is $76.

You Try It 9

An electrician's wage this year is $30.13 per hour, which is 115% of last year's wage. What was the increase in the hourly wage over last year?

Your Strategy

Your solution
$3.93

Solutions on p. S18

6.4 Exercises

Objective A   To solve percent problems using the basic percent equation

Solve. Use the basic percent equation.

1. 8% of 100 is what?
8

2. 16% of 50 is what?
8

3. 0.05% of 150 is what?
0.075

4. 0.075% of 625 is what?
0.46875

5. 15 is what percent of 90?
$16\frac{2}{3}\%$

6. 24 is what percent of 60?
40%

7. What percent of 16 is 6?
37.5%

8. What percent of 24 is 18?
75%

9. 10 is 10% of what?
100

10. 37 is 37% of what?
100

11. 2.5% of what is 30?
1200

12. 10.4% of what is 52?
500

13. Find 10.7% of 485.
51.895

14. Find 12.8% of 625.
80

15. 80% of 16.25 is what?
13

16. 26% of 19.5 is what?
5.07

17. 54 is what percent of 2000?
2.7%

18. 8 is what percent of 2500?
0.32%

19. 16.4 is what percent of 4.1?
400%

20. 5.3 is what percent of 50?
10.6%

21. 18 is 240% of what?
7.5

22. 24 is 320% of what?
7.5

Objective B   To solve percent problems using proportions

Solve. Use the proportion method.

23. 26% of 250 is what?
65

24. Find 18% of 150.
27

25. 37 is what percent of 148?
25%

26. What percent of 150 is 33?
22%

Suggested Assignment
Exercises 1–63, odds
More challenging problems:
Exercises 65–67

Section 6.4

Quick Quiz (Objective 6.4A)
1. What is 35% of 73?   25.55
2. Find 25% of 112.   28
3. 33 is what percent of 60?   55%
4. 56 is 70% of what?   80

27. 68% of what is 51?
75

28. 126 is 84% of what?
150

29. What percent of 344 is 43?
12.5%

30. 750 is what percent of 50?
1500%

31. 82 is 20.5% of what?
400

32. 2.4% of what is 21?
875

33. What is 6.5% of 300?
19.5

34. Find 96% of 75.
72

35. 7.4 is what percent of 50?
14.8%

36. What percent of 1500 is 693?
46.2%

37. Find 50.5% of 124.
62.62

38. What is 87.4% of 225?
196.65

39. 120% of what is 6?
5

40. 14 is 175% of what?
8

41. What is 250% of 18?
45

42. 325% of 4.4 is what?
14.3

43. 87 is what percent of 29?
300%

44. What percent of 38 is 95?
250%

Objective C **To solve application problems**

Solve.

45. Automotive Technology A mechanic estimates that the brakes of an RV still have 6000 mi of wear. This amount is 12% of the estimated safe-life use of the brakes. What is the estimated safe-life use of the brakes?
50,000 mi

46. Demographics Of the 281,422,000 people in the United States, 28.6% are under the age of 20. (*Source:* U.S. Census 2000) How many people in the United States are under the age of 20?
80,486,692 people

47. The Labor Force In Arkansas, Wal-Mart's home state, 41,000 workers are Wal-Mart employees. This is 3.3% of the Arkansas labor force. (*Source:* Wal-Mart Company Reports; Bureau of Labor Statistics) Find the number of workers in the Arkansas labor force.
1,242,424 workers

48. Astronomy The aphelion of Earth is its distance from the sun when it is farthest from the sun. The perihelion is its distance from the sun when it is nearest the sun, as shown in the figure at the right. What percent of the aphelion is the perihelion? Round to the nearest tenth of a percent.
96.8%

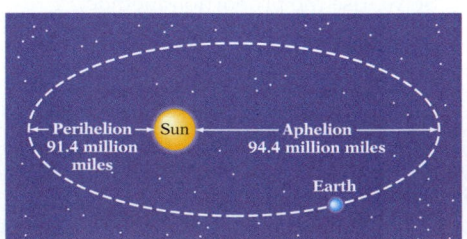

Quick Quiz (Objective 6.4B)

1. What is 14% of 250?   35
2. What percent of 140 is 49?   35%
3. 166 is 83% of what?   200

49. Fire Science A fire department received 24 false alarms out of a total of 200 alarms received. What percent of the alarms received were false alarms?
12%

50. Demographics The table at the right shows the projected increase in population from 2000 to 2040 for each of four counties in the Central Valley of California. What percent of the 2000 population of Sacramento County is the projected increase in population?
75%

County	2000 Population	Projected Increase by 2040
Sacramento	1,200,000	900,000
Kern	651,700	948,300
Fresno	794,200	705,800
San Joaquin	562,000	737,400

Source: California Department of Finance   P

51. Demographics The table at the right shows the projected increase in population from 2000 to 2040 for each of four counties in the Central Valley of California. What percent of the 2000 population of Kern County is the projected increase in population? Round to the nearest tenth of a percent.
145.5%

52. Business An antiques shop owner expects to receive $16\frac{2}{3}\%$ of the shop's sales as profit. What is the expected profit in a month when the total sales are $24,000?
$4000

53. Poultry In a recent year, North Carolina produced 1,300,000 lb of turkey. This was 18.6% of the U.S. total in that year. (*Source:* U.S. Census Bureau) Calculate the U.S. total turkey production for that year. Round to the nearest million.
7 million lb

54. Depreciation A used mobile home was purchased for $43,600. This amount was 64% of the cost of the mobile home when it was new. What did the mobile home cost when new?
$68,125

55. Agriculture A farmer is given an income tax credit of 15% of the cost of some farm machinery. What tax credit would the farmer receive on farm equipment that cost $85,000?
$12,750

56. Financing A used car is sold for $18,900. The buyer of the car makes a down payment of $3780. What percent of the selling price is the down payment?
20%

57. Medicine The active ingredient in a prescription skin cream is clobetasol propionate. It is 0.05% of the total ingredients. How many grams of clobetasol propionate are in a 30-gram tube of this cream?
0.015 g

58. Charitable Giving In a recent year, Americans gave $212 billion to charities. Use the figure at the right to determine how much of that amount came from individuals.
$160.696 billion

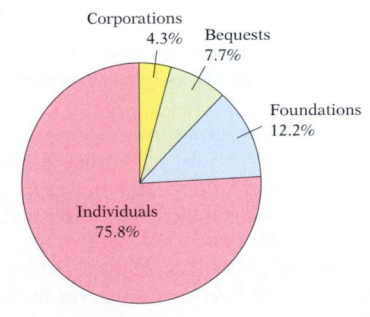

Corporations 4.3%   Bequests 7.7%
Foundations 12.2%
Individuals 75.8%

Charitable Giving   P
Sources: American Association of Fundraising Counsel; AP

59. Astronomy The diameter of Earth is approximately 8000 mi, and the diameter of the sun is approximately 870,000 mi. What percent of Earth's diameter is the sun's diameter?
10,875%

Quick Quiz (Objective 6.4C)

(*Note:* Solve for percent in Exercise 1, base in Exercise 2, and amount in Exercise 3.)

1. A down payment of $31,200 was paid on a new house costing $156,000. What percent of the purchase price is the down payment?   20%

2. A supermarket reduced the price of melon to $2.24 per pound, which is 80% of the original price. What was the original price?
$2.80 per pound

3. You bought a boat one year ago. Since then it has depreciated $2337, which is 12% of the price you paid for it. How much did you pay for the boat?
$19,475

60. **Pets** The average costs associated with owning a dog over an average 11-year life span are shown in the graph at the right. These costs do not include the price of the puppy when purchased. The category labeled "Other" includes such expenses as fencing and repairing furniture damaged by the pet. What percent of the total cost is spent on food? Round to the nearest tenth of a percent.
27.5%

61. **Manufacturing** During a quality control test, a manufacturer of computer boards found that 56 boards were defective. This was 0.7% of the total number of computer boards tested. How many of the tested computer boards were not defective?
7944 computer boards

62. **Agriculture** Of the 572 million pounds of cranberries grown in the United States in a recent year, Wisconsin growers produced 291.72 million pounds. What percent of the total cranberry crop was produced in Wisconsin?
51%

63. **Politics** The results of a survey in which 32,840 full-time college and university faculty members were asked to describe their political views are shown at the right. How many more faculty members described their political views as liberal than described their views as far left?
12,151 more faculty members

64. **Politics** The results of a survey in which 32,840 full-time college and university faculty members were asked to describe their political views are shown at the right. How many fewer faculty members described their political views as conservative than described their views as middle-of-the-road?
5451 fewer faculty members

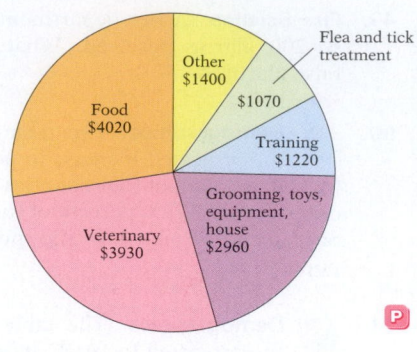

Cost of Owning a Dog
Source: American Kennel Club, *USA Today* research

Political View	Percent of Faculty Members Responding
Far left	5.3%
Liberal	42.3%
Middle of the road	34.3%
Conservative	17.7%
Far right	0.3%

Source: Higher Education Research Institute, UCLA

APPLYING THE CONCEPTS

65. Find 10% of a number and subtract it from the original number. Now take 10% of the new number and subtract it from the new number. Is this the same as taking 20% of the original number and subtracting it from the original number? Explain.
No

66. Increase a number by 10%. Now decrease the new number by 10%. Is the result the original number? Explain.
No

67. **Compensation** Your employer agrees to give you a 5% raise after 1 year on the job, a 6% raise the next year, and a 7% raise the following year. Is your salary after the third year greater than, less than, or the same as it would have been if you had received a 6% raise each year?
Less than

68. Visit a savings and loan institution or credit union to research and write a report on the meaning of *points* as they relate to a loan.

69. Find five different uses of percents and explain why percent was used in those instances.

Answers to Writing Exercises

68. A student's report should state that one of the fees that lenders charge is called points, and that 1 point is a term that lenders use for 1%. Points are paid before the loan is obtained and are charged on the amount of the loan. Therefore, a lender who charges 2 points for a $100,000 loan will charge 2% of $100,000, or $2000.

69. Student responses will vary. Their answers might include any of the following: a raise in wages; cost-of-living increases; commissions; exam grades; income tax brackets; sales tax; discounted merchandise; interest rates on car loans, certificates of deposit, or mortgages; a rate of return on an investment. If students are having difficulty finding examples, suggest that they look through a newspaper for ideas.

6.5 Simple Interest

Objective A   To solve problems involving simple interest

If you deposit money in a savings account at a bank, the bank will pay you for the privilege of using that money. The amount you deposit in the savings account is called the **principal.** The amount the bank pays you for the privilege of using the money is called **interest.**

If you borrow money from the bank in order to buy a car, the amount you borrow is called the **principal.** The additional amount of money you must pay the bank, above and beyond the amount borrowed, is called **interest.**

Whether you deposit money or borrow it, the amount of interest paid is usually computed as a percent of the principal. The percent used to determine the amount of interest to be paid is the **interest rate.** Interest rates are given for specific periods of time, such as months or years.

Interest computed on the original principal is called **simple interest.** Simple interest is the cost of a loan that is for a period of about 1 year or less.

> **The Simple Interest Formula**
>
> $I = Prt$,   where $I =$ simple interest earned, $P =$ principal,
> $r =$ annual simple interest rate, and $t =$ time (in years)

In the simple interest formula, t is the time in years. If a time period is given in days or months, it must be converted to years and then substituted in the formula for t. For example,

$$120 \text{ days is } \frac{120}{365} \text{ of a year.} \qquad 6 \text{ months is } \frac{6}{12} \text{ of a year.}$$

HOW TO Shannon O'Hara borrowed $5000 for 90 days at an annual simple interest rate of 7.5%. Find the simple interest due on the loan.

Strategy   To find the simple interest owed, use the simple interest formula.

$$P = 5000, r = 7.5\% = 0.075, t = \frac{90}{365}$$

Solution $I = Prt$

$$I = 5000(0.075)\left(\frac{90}{365}\right)$$

$$I \approx 92.47$$

The simple interest due on the loan is $92.47.

In the example above, we calculated that the simple interest due on Shannon O'Hara's 90-day, $5000 loan was $92.47. This means that at the end of the 90 days, Shannon owes $5000 + $92.47 = $5092.47. The principal plus the interest owed on a loan is called the **maturity value** of the loan.

Objective 6.5A

New Vocabulary
principal
interest
interest rate
simple interest
maturity value

New Formulas
Simple Interest Formula:
   $I = Prt$
Formula for the Maturity Value of a Simple Interest Loan:
   $M = P + I$

Instructor Note
Emphasize that the simple interest formula requires that the interest rate and the time have comparable units. If an *annual* interest rate is given, then time must be in *years*. If a *monthly* interest rate is given (as on most credit cards), then time must be in *months*.

Discuss the Concepts
1. Explain the difference between interest and interest rate.
2. Explain how to convert a number of months to a fractional part of a year.
3. Explain how to convert a number of days to a fractional part of a year.

Concept Check
Find the difference between (a) the maturity value of and (b) the monthly payment on a 6-month $10,000 loan at an annual interest rate of 9.75% and the same loan at an annual interest rate of 6.25%.
(a) $175; (b) $29.17

In-Class Examples (Objective 6.5A)

1. A rancher borrowed $120,000 for 180 days at an annual interest rate of 8.75%. What is the simple interest due on the loan? **$5178.08**

2. To finance the purchase of four new taxicabs, the owner of a fleet borrowed $84,000 for 8 months at an annual interest rate of 6.5%. Find the maturity value of the loan. **$87,640**

Instructor Note

In this objective, we have presented the *maturity value* of a loan. For an investment, such as a deposit in a bank savings account, the sum of the principal and the interest is called the *future value* of the investment. The formula is the same, but the name applied to the sum is different. You may want to introduce this term in class.

Optional Student Activity

Many traditional college students have little knowledge of finances but are very interested in learning about them. This objective may lend itself to some instruction on the use of credit.

Although the calculation of finance charges, given periodic payments and added purchases, is beyond the scope of this course, you may want to present a simplified version for the sake of illustration. For example, you might present a case in which the unpaid balance (say, $500) and the minimum payment due (perhaps $10) both remain constant. You might use a charge of 1.5% per month on the unpaid balance. Illustrate the total amount paid if only the minimum payment is paid each month. Calculate the total amount paid in finance charges. Using the same figures, illustrate the difference in these totals when the customer pays more than the minimum payment each month.

> **Formula for the Maturity Value of a Simple Interest Loan**
>
> $M = P + I$, where M = maturity value, P = principal, and I = simple interest

The example below illustrates solving the simple interest formula for the interest rate. The solution requires the Multiplication Property of Equations.

HOW TO Ed Pabas took out a 45-day, $12,000 loan. The simple interest on the loan was $168. To the nearest tenth of a percent, what is the simple interest rate?

Strategy   To find the simple interest rate, use the simple interest formula.
$$P = 12,000, \ t = \frac{45}{365}, \ I = 168$$

Solution
$$I = Prt$$
$$168 = 12,000r\left(\frac{45}{365}\right)$$
$$168 = \frac{540,000}{365}r$$
$$\frac{365}{540,000}(168) = \frac{365}{540,000} \cdot \frac{540,000}{365}r$$
$$0.114 \approx r$$

The simple interest rate on the loan is 11.4%.

Example 1

You arrange for a 9-month bank loan of $9000 at an annual simple interest rate of 8.5%. Find the total amount you must repay to the bank.

Strategy

To calculate the maturity value:
• Find the simple interest due on the loan by solving the simple interest formula for I.
$$t = \frac{9}{12}, P = 9000, r = 8.5\% = 0.085$$
• Use the formula for the maturity value of a simple interest loan, $M = P + I$.

Solution $I = Prt$
$$I = 9000(0.085)\left(\frac{9}{12}\right)$$
$$I = 573.73$$

$$M = P + I$$
$$M = 9000 + 573.75$$
$$M = 9573.75$$

The total amount owed to the bank is $9573.75.

You Try It 1

William Carey borrowed $12,500 for 8 months at an annual simple interest rate of 9.5%. Find the maturity value of the loan.

Your Strategy

Your solution
$13,291.67

Solution on p. S18

6.5 Exercises

Objective A   To solve problems involving simple interest

1. ✎ Explain what each variable in the simple interest formula represents.

2. ✎ Explain the difference between interest and interest rate.

3. **a.** In the table below, the interest rate is an annual simple interest rate. Complete the table by calculating the simple interest due on the loan.

Loan Amount	Interest Rate	Period	Interest	
$5000	6%	1 month	_____	$25
$5000	6%	2 months	_____	$50
$5000	6%	3 months	_____	$75
$5000	6%	4 months	_____	$100
$5000	6%	5 months	_____	$125

Use the pattern of your answers in the table to find the simple interest due on a $5000 loan that has an annual simple interest rate of 6% for a period of:
b. 6 months   **c.** 7 months   **d.** 8 months   **e.** 9 months
b. $150   c. $175   d. $200   e. $225

4. Use your solutions to Exercise 3 to answer the following questions:
   a. If you know the simple interest due on a 1-month loan, explain how you can use that figure to calculate the simple interest due on a 7-month loan for the same principal and the same interest rate.
   b. If the time period of a loan is doubled but the principal and interest rate remain the same, how many times greater is the simple interest due on the loan?
   a. Multiply the interest due by 7.   b. 2 times greater

5. Kristi Yang borrowed $15,000. The term of the loan was 90 days, and the annual simple interest rate was 7.4%. Find the simple interest due on the loan.
   $273.70

6. Hector Elizondo took out a 75-day loan of $7500 at an annual interest rate of 9.5%. Find the simple interest due on the loan.
   $146.40

7. To finance the purchase of 15 new cars, the Lincoln Car Rental Agency borrowed $100,000 for 9 months at an annual interest rate of 9%. What is the simple interest due on the loan?
   $6750

8. A home builder obtained a preconstruction loan of $50,000 for 8 months at an annual interest rate of 9.5%. What is the simple interest due on the loan?
   $3166.67

9. Assume that Visa charges Francesca 1.6% per month on her unpaid balance. Find the interest owed to Visa when her unpaid balance for the month is $1250.
   $20

Section 6.5

Suggested Assignment
Exercises 3–19, odds

Answers to Writing Exercises

1. In the simple interest formula $I = Prt$, I is the simple interest earned, P is the principal, r is the annual simple interest rate, and t is the time, in years.

2. Student descriptions must state that interest is an amount of money, while an interest rate is a percent.

Quick Quiz (Objective 6.5A)

1. A mechanic borrowed $15,000 for 90 days at an annual interest rate of 7.2%. What is the simple interest due on the loan?   **$266.30**

2. The owner of a convenience store borrowed $60,000 for 9 months at an annual interest rate of 8.6%. Find the maturity value of the loan.   **$63,870**

10. The Mission Valley Credit Union charges its customers an interest rate of 2% per month on money that is transferred into an account that is overdrawn. Find the interest owed to the credit union for 1 month when $800 is transferred into an overdrawn account.
$16

11. Find the simple interest Jacob Zucker owes on a 2-year loan of $8000 at an annual interest rate of 9%.
$1440

12. Find the simple interest Kara Tanamachi owes on a $1\frac{1}{2}$-year loan of $1500 at an annual interest rate of 7.5%.
$168.75

13. An auto parts dealer borrowed $150,000 at a 9.5% annual simple interest rate for 1 year. Find the maturity value of the loan.
$164,250

14. A corporate executive took out a $25,000 loan at an 8.2% annual simple interest rate for 1 year. Find the maturity value of the loan.
$27,050

15. Capitol City Bank approves a home-improvement loan application for $14,000 at an annual simple interest rate of 10.25% for 270 days. What is the maturity value of the loan?
$15,061.51

16. A credit union loans a member $5000 for the purchase of a used car. The loan is made for 18 months at an annual simple interest rate of 6.9%. What is the maturity value of the car loan?
$5517.50

17. A $12,000 investment earned $462 in interest in 6 months. Find the annual simple interest rate on the loan.
7.7%

18. Michele Gabrielle borrowed $3000 for 9 months and paid $168.75 in simple interest on the loan. Find the annual simple interest rate that Michele paid on the loan.
7.5%

19. An investor earned $937.50 on an investment of $50,000 in 75 days. Find the annual simple interest rate earned on the investment.
9.125%

20. Don Glover borrowed $18,000 for 210 days and paid $604.80 in simple interest on the loan. What annual simple interest rate did Don pay on the loan?
5.84%

APPLYING THE CONCEPTS

21. Interest has been described as a rental fee for money. Explain this description of interest.

22. Visit a savings and loan officer to collect information about the different kinds of home loans. Write a short essay describing the different kinds of loans available.

Answers to Writing Exercises

21. Explanations will vary. For example, students may describe interest as the amount of money charged for the privilege of borrowing money. In this sense, it could be considered a "rental fee."

22. There are a great variety of home loans available today. Students should certainly describe the difference between fixed-rate and variable-rate mortgages. They might discuss the difference between, for example, a 15-year fixed-rate mortgage and a 30-year fixed-rate mortgage, the relative monthly payments on each, and the different amounts of interest paid on each loan. The advantages and disadvantages of an adjustable-rate mortgage (ARM) might be mentioned—for example, lower initial interest rates but variable monthly payments during the life of the loan. Some students might describe biweekly mortgages or graduated-payment mortgages.

Focus on Problem Solving

Relevant Information   One of the challenges of problem solving is to separate the information that is relevant to the problem from other information. Following is an example from the physical sciences in which some relevant information was omitted.

Hooke's Law states that the distance a weight will stretch a spring is directly proportional to the weight on the spring. That is, $d = kF$, where d is the distance the spring is stretched and F is the force. In an experiment to verify this law, some physics students were continually getting inconsistent results. Finally, the instructor discovered that the heat produced when the lights were turned on was affecting the experiment. In this case, relevant information was omitted— namely, that the temperature of the spring can affect the distance it will stretch.

A lawyer drove 8 miles to the train station. After a 35-minute ride of 18 miles, the lawyer walked 10 minutes to the office. Find the total time it took the lawyer to get to work.

From this situation, answer the following questions before reading on.
a. What is asked for?
b. Is there enough information to answer the question?
c. Is information given that is not needed?

Here are the answers.
a. We want the total time for the lawyer to get to work.
b. No. We do not know the time it takes the lawyer to get to the train station.
c. Yes. Neither the distance to the train station nor the distance of the train ride is necessary to answer the question.

For each of the following problems, answer these questions:
a. What is asked for?
b. Is there enough information to answer the question?
c. Is information given that is not needed?

1. A customer bought six boxes of strawberries and paid with a $20 bill. What was the change?

2. A board is cut into two pieces. One piece is 3 feet longer than the other piece. What is the length of the original board?

3. A family rented a car for their vacation and drove 680 miles. The cost of the rental car was $21 per day with 150 free miles per day and $.15 for each mile driven above the number of free miles allowed. How many miles did the family drive per day?

4. An investor bought 8 acres of land for $80,000. One and one-half acres were set aside for a park, and the remainder of the land was developed into one-half-acre lots. How many lots were available for sale?

5. You wrote checks of $43.67, $122.88, and $432.22 after making a deposit of $768.55. How much do you have left in your checking account?

Answers to Focus on Problem Solving: Relevant Information

1a. The amount of change is asked for.
  b. No. We need to know the cost of the strawberries.
  c. No.
2a. The length of the original board is asked for.
  b. No. We need to know the length of the other piece.
  c. No.
3a. The number of miles driven per day is asked for.
  b. No. We are given only the total number of miles driven.
  c. Yes. None of the information given can be used to answer the question.
4a. We are asked for the number of lots available for sale.
  b. Yes.
  c. Yes. We do not need the price of the land.
5a. We are asked for the checking account balance.
  b. No. We don't know the balance before writing the checks and making the deposit.
  c. No.

Projects and Group Activities

Earned Run Average

One measure of a pitcher's success is earned run average. **Earned run average (ERA)** is the number of earned runs a pitcher gives up for every nine innings pitched. The definition of an earned run is somewhat complicated, but basically an earned run is a run that is scored as a result of hits and base running that involve no errors on the part of the pitcher's team. If the opposing team scores a run on an error (for example, a fly ball that should have been caught in the outfield but was fumbled), then that is not an earned run.

A proportion is used to calculate a pitcher's ERA. Remember that the statistic involves the number of earned runs *per nine innings*. The answer is always rounded to the nearest hundredth. Here is an example:

Earned Run Average Leaders

Year	Player, club	ERA
	National League	
1990	Danny Darwin, Houston	2.21
1991	Dennis Martinez, Montreal	2.39
1992	Bill Swift, San Francisco	2.08
1993	Greg Maddux, Atlanta	2.36
1994	Greg Maddux, Atlanta	1.56
1995	Greg Maddux, Atlanta	1.63
1996	Kevin Brown, Florida	1.89
1997	Pedro Martinez, Montreal	1.90
1998	Greg Maddux, Atlanta	2.22
1999	Randy Johnson, Arizona	2.48
2000	Kevin Brown, Los Angeles	2.58
2001	Randy Johnson, Arizona	2.49
2002	Randy Johnson, Arizona	2.32
2003	Jason Schmidt, San Francisco	2.34
2004	Jake Peavy, San Diego	2.27
	American League	
1990	Roger Clemens, Boston	1.93
1991	Roger Clemens, Boston	2.62
1992	Roger Clemens, Boston	2.41
1993	Kevin Appler, Kansas City	2.56
1994	Steve Ontiveros, Oakland	2.65
1995	Randy Johnson, Seattle	2.48
1996	Juan Guzman, Toronto	2.93
1997	Roger Clemens, Toronto	2.05
1998	Roger Clemens, Toronto	2.65
1999	Pedro Martinez, Boston	2.07
2000	Pedro Martinez, Boston	1.74
2001	Freddy Garcia, Seattle	3.05
2002	Pedro Martinez, Boston	2.26
2003	Pedro Martinez, Boston	2.22
2004	Johan Santana, Minnesota	2.61

HOW TO During the 2004 baseball season, Johan Santana gave up 66 earned runs and pitched 228 innings for the Minnesota Twins. Calculate Johan Santana's ERA.

Strategy

To calculate Johan Santana's ERA, let $x =$ the number of earned runs for every nine innings pitched. Then set up a proportion. Solve the proportion for x.

Solution

$$\frac{66 \text{ earned runs}}{228 \text{ innings}} = \frac{x}{9 \text{ innings}}$$
$$228 \cdot x = 66 \cdot 9$$
$$228x = 594$$
$$\frac{228x}{228} = \frac{594}{228}$$
$$x \approx 2.61$$

Johan Santana's ERA for the 2004 season was 2.61.

1. In 1979, Jeff Reardon's rookie year, he pitched 21 innings for the Mets and gave up 4 earned runs. Calculate Reardon's ERA for 1979.

2. Roger Clemens's first year with the Boston Red Sox was 1984. During that season, he pitched 133.1 innings and gave up 64 earned runs. Calculate Clemens's ERA for 1984.

3. During the 2003 baseball season, Ben Sheets of the Milwaukee Brewers pitched 220.2 innings and gave up 109 earned runs. During the 2004 season, he gave up 71 earned runs and pitched 237.0 innings. During which season was his ERA lower? How much lower?

4. In 1987, Nolan Ryan had the lowest ERA of any pitcher in the major leagues. He gave up 65 earned runs and pitched 211.2 innings for the Astros. Calculate Ryan's ERA for 1987.

Answers to Projects and Group Activities: Earned Run Average

1. 1.71
2. 4.33
3. 2004; 1.76
4. 2.77

Instructor Note

You might ask students to find the necessary statistics for a pitcher for their "home team," and calculate that pitcher's ERA.

Chapter 6 Summary

Key Words

Examples	

A *ratio* is the comparison of two quantities with the same unit. A ratio can be written in three ways: as a fraction, as two numbers separated by a colon, or as two numbers separated by the word *to*. A ratio is in simplest form when the two quantities do not have a common factor. [6.1A, p. 335]

The comparison 16 oz to 24 oz can be written as a ratio in simplest form:

$\frac{2}{3}$, 2:3, or 2 to 3

A *rate* is the comparison of two quantities with different units. A rate is in simplest form when the two quantities do not have a common factor. [6.1A, p. 335]

You earned \$63 for working 6 h. The rate is written $\frac{\$21}{2\,h}$.

A *unit rate* is a rate in which the denominator is 1. [6.1A, p. 335]

You traveled 144 mi in 3 h. The unit rate is 48 mph.

A *proportion* is the equality of two ratios or rates. Each of the four members in a proportion is called a *term*.

In the proportion $\frac{3}{5} = \frac{12}{20}$, 5 and 12 are the means; 3 and 20 are the extremes.

first term ⟵ $\frac{a}{b} = \frac{c}{d}$ ⟶ third term
second term ⟵ $\frac{a}{b} = \frac{c}{d}$ ⟶ fourth term

The second and third terms of the proportion are called the *means*, and the first and fourth terms are called the *extremes*. [6.2A, p. 339]

Percent means "parts of 100." [6.3A, p. 347]

23% means 23 of 100 equal parts.

Principal is the amount of money originally deposited or borrowed. *Interest* is the amount paid for the privilege of using someone else's money. The percent used to determine the amount of interest is the *interest rate*. Interest computed on the original amount is called *simple interest*. The principal plus the interest owed on a loan is called the *maturity value* of the loan. [6.5A, p. 361]

Consider a 1-year loan of \$5000 at an annual simple interest rate of 8%. The principal is \$5000. The interest rate is 8%. The interest paid on the loan is \$400. The maturity value is \$5000 + \$400 = \$5400.

Essential Rules and Procedures

Examples

To find a unit rate, divide the number in the numerator of the rate by the number in the denominator of the rate. [6.1A, p. 336]

You earned \$41 for working 4 h.

$$\frac{41}{4} = 41 \div 4 = 10.25$$

The unit rate is \$10.25 per hour.

To set up a proportion, keep the same units in the numerator and the same units in the denominator. [6.2B, p. 341]

Three machines fill 5 cereal boxes per minute. How many boxes can 8 machines fill per minute?

$$\frac{3\text{ machines}}{5\text{ cereal boxes}} = \frac{8\text{ machines}}{x\text{ cereal boxes}}$$

To solve a proportion, use the fact that the product of the means equals the product of the extremes. For the proportion $\dfrac{a}{b} = \dfrac{c}{d}$, $ad = bc$. [6.2A, p. 340]

$$\frac{6}{25} = \frac{9}{x}$$
$$25 \cdot 9 = 6 \cdot x$$
$$225 = 6x$$
$$\frac{225}{6} = \frac{6x}{6}$$
$$37.5 = x$$

To write a percent as a fraction, drop the percent sign and multiply by $\dfrac{1}{100}$. [6.3A, p. 347]

$$56\% = 56\left(\frac{1}{100}\right) = \frac{56}{100} = \frac{14}{25}$$

To write a percent as a decimal, drop the percent sign and multiply by 0.01. [6.3A, p. 347]

$$87\% = 87(0.01) = 0.87$$

To write a fraction as a percent, multiply by 100%. [6.3B, p. 348]

$$\frac{7}{20} = \frac{7}{20}(100\%) = \frac{700\%}{20} = 35\%$$

To write a decimal as a percent, multiply by 100%. [6.3B, p. 348]

$$0.325 = 0.325(100\%) = 32.5\%$$

The Basic Percent Equation [6.4A, p. 351]
Percent $\cdot$ base = amount

8% of 250 is what number?
Percent $\cdot$ base = amount
$$0.08 \cdot 250 = n$$
$$20 = n$$

Proportion Method of Solving a Percent Problem [6.4B, p. 353]
$$\frac{\text{percent}}{100} = \frac{\text{amount}}{\text{base}}$$

8% of 250 is what number?
$$\frac{\text{percent}}{100} = \frac{\text{amount}}{\text{base}}$$
$$\frac{8}{100} = \frac{n}{250}$$
$$100 \cdot n = 8 \cdot 250$$
$$100n = 2000$$
$$n = 20$$

Simple Interest Formula [6.5A, p. 361]
$I = Prt$, where I = simple interest earned, P = principal, r = annual simple interest rate, t = time (in years)

You borrow $10,000 for 180 days at an annual interest rate of 8%. Find the simple interest due on the loan.
$$I = Prt$$
$$I = 10,000(0.08)\left(\frac{180}{365}\right)$$
$$I \approx 394.52$$

Formula for the Maturity Value of a Simple Interest Loan [6.5A, p. 362]
$M = P + I$, where M = maturity value, P = principal, I = simple interest

Suppose you paid $400 in interest on a 1-year loan of $5000. The maturity value of the loan is $5000 + $400 = $5400.

Chapter 6 Review Exercises

1. Write the comparison 100 lb to 100 lb as a ratio in simplest form using a fraction, a colon, and the word *to*.

    $\dfrac{1}{1}$, 1:1, 1 to 1   [6.1A]

2. Write 18 roof supports for every 9 ft as a rate in simplest form.

    $\dfrac{\text{2 roof supports}}{\text{1 ft}}$ [6.1A]

3. Write $628 earned in 40 h as a unit rate.

    $15.70/h   [6.1A]

4. Write 8 h to 15 h as a ratio in simplest form using a fraction.

    $\dfrac{8}{15}$ [6.1A]

5. Solve: $\dfrac{n}{3} = \dfrac{8}{15}$

    1.6   [6.2A]

6. Write 15 lb of fertilizer for 12 trees as a rate in simplest form.

    $\dfrac{\text{5 lb}}{\text{4 trees}}$ [6.1A]

7. Write 171 mi driven in 3 h as a unit rate.

    57 mph   [6.1A]

8. Solve $\dfrac{2}{3.5} = \dfrac{n}{12}$. Round to the nearest hundredth.

    6.86   [6.2A]

9. Write 32% as a fraction.

    $\dfrac{8}{25}$ [6.3A]

10. Write 22% as a decimal.

    0.22   [6.3A]

11. Write 25% as a fraction and as a decimal.

    $\dfrac{1}{4}$, 0.25   [6.3A]

12. Write $3\dfrac{2}{5}$% as a fraction.

    $\dfrac{17}{500}$ [6.3A]

13. Write $\dfrac{7}{40}$ as a percent.

    17.5%   [6.3B]

14. Write $1\dfrac{2}{7}$ as a percent. Round to the nearest tenth of a percent.

    128.6%   [6.3B]

15. Write 2.8 as a percent.

    280%   [6.3B]

16. 42% of 50 is what?

    21   [6.4A/6.4B]

17. What percent of 3 is 15?

    500%   [6.4A/6.4B]

18. 12 is what percent of 18? Round to the nearest tenth of a percent.

    66.7%   [6.4A/6.4B]

19. 150% of 20 is what number?

    30   [6.4A/6.4B]

20. Find 18% of 85.

    15.3   [6.4A/6.4B]

21. 32% of what number is 180?

    562.5   [6.4A/6.4B]

22. 4.5 is what percent of 80?

    5.625%   [6.4A/6.4B]

23. **Technology**   In 3 years, the price of a graphing calculator went from $125 to $75. What is the ratio of the decrease in price to the original price?

$\dfrac{2}{5}$ [6.1A]

24. **Investments**   An investment of $8000 earns $520 in annual dividends. At the same rate, how much money must be invested to earn $780 in annual dividends?

$12,000   [6.2B]

25. **Lawn Care**   The directions on a bag of plant food recommend $\frac{1}{2}$ lb for every 50 ft² of lawn. How many pounds of plant food should be used on a lawn measuring 275 ft²?

2.75 lb   [6.2B]

26. **Profit Sharing**   Two attorneys share the profits of their firm in the ratio 3:2. If the attorney receiving the larger amount of this year's profits receives $96,000, what amount does the other attorney receive?

$64,000   [6.2B]

27. **Tourism**   The table at the right shows the countries with the highest projected numbers of tourists visiting in 2020. What percent of the tourists visiting these countries will be visiting China? Round to the nearest tenth of a percent.

34.0%   [6.4C]

Country	Projected Number of Tourists in 2020
China	137 million
France	93 million
Spain	71 million
USA	102 million

Source: The State of the World Atlas by Dan Smith

28. **Business**   A company spent 7% of its $120,000 budget for advertising. How much did the company spend for advertising?

$8400   [6.4C]

29. **Manufacturing**   A quality control inspector found that 1.2% of 4000 cellular telephones were defective. How many of the phones were not defective?

3952 telephones   [6.4C]

30. **Television**   According to the Cabletelevision Advertising Bureau, cable households watch an average of 61.35 h of television per week. On average, what percent of the week do cable households spend watching TV? Round to the nearest tenth of a percent.

36.5%   [6.4C]

31. **Finance**   Find the simple interest on a 45-day loan of $3000 at an annual simple interest rate of 8.6%.

$31.81   [6.5A]

32. **Finance**   A realtor took out a $10,000 loan at an 8.4% annual simple interest rate for 9 months. Find the maturity value of the loan.

$10,630   [6.5A]

33. **Travel**   An airline knowingly overbooks flights by selling 12% more tickets than there are seats available. How many tickets would this airline sell for an airplane that has 175 seats?

196 tickets   [6.4C]

Chapter 6 Test

1. Write the comparison 3 yd to 24 yd as a ratio in simplest form using a fraction, a colon, and the word *to*.

   $\frac{1}{8}$, 1:8, 1 to 8   [6.1A]

2. Write 16 oz of sugar for 64 cookies as a rate in simplest form.

   $\frac{1 \text{ oz}}{4 \text{ cookies}}$ [6.1A]

3. Write 120 mi driven in 200 min as a unit rate.
   0.6 mi/min   [6.1A]

4. Write 200 ft to 100 ft as a ratio in simplest form using a fraction.

   $\frac{2}{1}$ [6.1A]

5. Solve: $\frac{n}{5} = \frac{3}{20}$

   0.75   [6.2A]

6. Write 8 ft walked in 4 s as a unit rate.

   2 ft/s   [6.1A]

7. Write 2860 ft² mowed in 6 h as a unit rate. Round to the nearest hundredth.

   476.67 ft²/h   [6.1A]

8. Solve: $\frac{n}{4} = \frac{8}{9}$. Round to the nearest hundredth.
   3.56   [6.2A]

9. Write 86.4% as a decimal.
   0.864   [6.3A]

10. Write 0.4 as a percent.
    40%   [6.3B]

11. Write $\frac{5}{4}$ as a percent.

    125%   [6.3B]

12. Write $83\frac{1}{3}\%$ as a fraction.

    $\frac{5}{6}$ [6.3A]

13. Write 32% as a fraction.
    $\frac{8}{25}$ [6.3A]

14. Write 1.18 as a percent.

    118%   [6.3B]

15. 18 is 20% of what number?
    90   [6.4A/6.4B]

16. What is 68% of 73?
    49.64   [6.4A/6.4B]

17. What percent of 320 is 180?
    56.25%   [6.4A/6.4B]

18. 28 is 14% of what number?
    200   [6.4A/6.4B]

19. **Physical Fitness** A body builder who had been lifting weights for 2 years went from an original weight of 165 lb to 190 lb. What is the ratio of the original weight to the increased weight?

$\frac{33}{38}$ [6.1A]

20. **Taxes** The sales tax on a $95 purchase is $7.60. Find the sales tax on a car costing $39,200.

$3136   [6.2B]

21. **Elections** A preelection survey showed that 3 out of 4 registered voters would vote in a county election. At this rate, how many registered voters would vote in a county with 325,000 registered voters?

243,750 voters   [6.2B]

22. **Architecture** The scale on the architectural drawings for a new gymnasium is 1 in. equals 4 ft. How long is one of the rooms if it measures $12\frac{1}{2}$ in. on the drawing?

50 ft   [6.2B]

23. **Insurance** An insurance company expects that 2.2% of a company's employees will have an industrial accident. How many employees are expected to have an accident in a company that employs 1500 people?

33 accidents   [6.4C]

24. **Education** A student missed 16 questions on a history exam of 90 questions. What percent of the questions did the student answer correctly? Round to the nearest tenth.

82.2%   [6.4C]

25. **Compensation** An administrative assistant has a wage of $480 per week. This is 120% of last year's wage. What is the dollar increase in the assistant's weekly wage over last year?

$80   [6.4C]

26. **Nutrition** The table at the right shows the fat, saturated fat, cholesterol, and calorie content in a 90-gram ground-beef burger and in a 90-gram soy burger. The number of fat grams in the beef burger is what percent of the number of fat grams in the soy burger?

600%   [6.4C]

	Beef Burger	Soy Burger
Fat	24 g	4 g
Saturated Fat	10 g	1.5 g
Cholesterol	75 mg	0 mg
Calories	280	140

27. **Finance** You took out a 150-day, $40,000 business loan that had an annual simple interest rate of 9.25%. Find the maturity value of the loan.

$41,520.55   [6.5A]

Cumulative Review Exercises

1. Simplify: $18 \div \dfrac{6-3}{9} - (-3)$

   57 [3.5A]

2. Evaluate 5^4.

   625 [1.3B]

3. Subtract: $7\dfrac{5}{12} - 3\dfrac{5}{9}$

   $3\dfrac{31}{36}$ [2.3B]

4. Simplify: $\dfrac{4}{5} \div \dfrac{4}{5} + \dfrac{2}{3}$

   $1\dfrac{2}{3}$ [2.7A]

5. Find the quotient of 342 and -3.

   -114 [3.3B]

6. Evaluate $2a - 3ab$ when $a = 2$ and $b = -3$.

   22 [3.5A]

7. Solve: $5x - 20 = 0$

   4 [5.2A]

8. Solve: $3(x - 4) + 2x = 3$

   3 [5.3B]

9. Simplify: $-\dfrac{5}{8} - \left(-\dfrac{3}{4}\right) + \dfrac{5}{6}$

   $\dfrac{23}{24}$ [3.4A]

10. Find the product of 1.005 and 10^5.

    100,500 [2.6B]

11. Simplify: $(-5)^2 - (-8) \div (7 - 5)^2 \cdot 2 - 8$

    21 [3.5A]

12. Simplify: $\left(-\dfrac{2}{3}\right)\left(-\dfrac{3}{4}\right)^2$

    $-\dfrac{3}{8}$ [3.5A]

13. Simplify: $4 - (-3) + 5 - 8$

    4 [3.2B]

14. Simplify: $5 - 2(1 - 3a) + 2(a - 3)$

    $8a - 3$ [4.2D]

15. Solve: $\dfrac{3}{4}x = -9$

    -12 [5.1C]

16. Simplify: $-3y^2 + 3y - y^2 - 6y$

    $-4y^2 - 3y$ [4.2A]

17. Divide: $3\dfrac{5}{8} \div 2\dfrac{7}{12}$

    $1\dfrac{25}{62}$ [2.4B]

18. Write 30 cents to 1 dollar as a ratio in simplest form.

    $\dfrac{3}{10}$ [6.1A]

19. Write \$19,425 in 5 months as a unit rate.

    \$3885/month [6.1A]

20. Evaluate $a - b$ when $a = 102.5$ and $b = 77.546$.

    24.954 [2.6A]

21. Solve: $\dfrac{2}{3} = \dfrac{n}{48}$

    32 [6.2A]

22. Simplify: $\dfrac{\dfrac{1}{2} + \dfrac{3}{4}}{2 - \dfrac{5}{8}}$

    $\dfrac{10}{11}$ [2.4C]

23. 2.5 is what percent of 30? Round to the nearest tenth of a percent.
8.3% [6.4A/6.4B]

24. Find 42% of 60.
25.2 [6.4A/6.4B]

25. 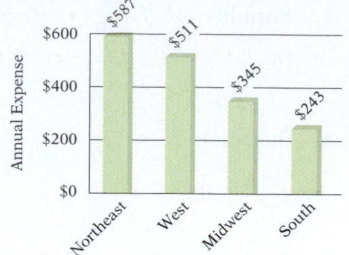 **Public Transportation** The figure at the right shows the average amount spent annually per household on public transportation, by region, in the United States. Find the difference between the average amount spent monthly per household in the Northeast and in the South. Round to the nearest cent.
$28.67 [2.6E]

Average Annual Expense per Household on
Public Transportation in the United States
Source: Bureau of Labor Statistics consumer expenditure survey

26. Five less than two-thirds of a number is three. Find the number.
12 [5.4A]

27. Translate "the difference between four times a number and three times the sum of the number and two" into a variable expression. Then simplify.
$4x - 3(x + 2); x - 6$ [4.3B]

28. **Travel** Your odometer reads 18,325 mi before you embark on a 125-mile trip. After you have driven $1\frac{1}{2}$ h, the odometer reads 18,386 mi. How many miles are left to drive?
64 mi [1.2C]

29. **Banking** You had a balance of $422.89 in your checking account. You then made a deposit of $122.35 and wrote a check for $279.76. Find the new balance in your checking account.
$265.48 [2.6E]

PAYMENT/ DEBIT (−)	√ T	FEE (IF ANY) (−)	DEPOSIT/ CREDIT (+)	BALANCE $
$		$	$ 122	35 422	89
279	76			

30. **Jobs** A data processor finished $\frac{2}{5}$ of a three-day job on the first day and $\frac{1}{3}$ on the second day. What part of the job is to be finished on the third day?
$\frac{4}{15}$ [2.3C]

31. **Health** According to the table at the right, what fraction of the population aged 75–84 is affected by Alzheimer's disease?
$\frac{1}{10}$ [6.3A]

Age Group	Percent Affected by Alzheimer's Disease
65 – 74	4%
75 – 84	10%
85 +	17%

Source: Mayo Clinic Family Health Book, Encyclopedia Americana, Associated Press

32. **Travel** A car is driven 402.5 mi on 11.5 gal of gas. Find the number of miles traveled per gallon of gas.
35 mi [6.1A]

33. **Mechanics** At a certain speed, the engine revolutions per minute (rpm) of a car in fourth gear is 2500. This is two-thirds of the rpm of the engine in third gear. Find the rpm of the engine in third gear.
3750 rpm [5.4B]

The best way to appreciate the different shapes and sizes of grassy areas in Buenos Aires' Rosedal Park is to view the garden from overhead. Each geometric shape, having its own set of dimensions, combines to form the entire park. **Example 4 on page 402** illustrates how to use a geometric formula to determine how large an area is and how much grass seed is needed for an area of that size.

OBJECTIVES

Section 7.1

A To solve problems involving lines and angles

B To solve problems involving angles formed by intersecting lines

C To solve problems involving the angles of a triangle

Section 7.2

A To solve problems involving the perimeter of geometric figures

B To solve problems involving the area of geometric figures

Section 7.3

A To solve problems using the Pythagorean Theorem

B To solve problems involving similar triangles

C To determine whether two triangles are congruent

Section 7.4

A To solve problems involving the volume of a solid

B To solve problems involving the surface area of a solid

Need help? For online student resources, such as section quizzes, visit this textbook's website at **math.college.hmco.com/students.**

Do these exercises to prepare for Chapter 7.

1. Simplify: $2(18) + 2(10)$

56   [1.4A]

2. Evaluate abc when $a = 2$, $b = 3.14$, and $c = 9$.

56.52   [4.1A]

3. Evaluate xyz^3 when $x = \frac{4}{3}$, $y = 3.14$, and $z = 3$.

113.04   [4.1A]

4. Solve: $x + 47 = 90$

43   [5.1B]

5. Solve: $32 + 97 + x = 180$

51   [5.1B]

6. Solve: $\dfrac{5}{12} = \dfrac{6}{x}$

14.4   [6.2A]

GO FIGURE • • •

Draw the figure that would come next.

 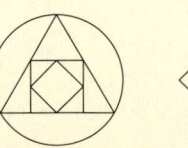 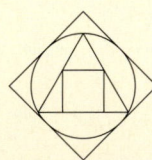 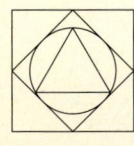

Circle inside diamond inside square inside triangle

7.1 Introduction to Geometry

Objective A   To solve problems involving lines and angles

Point of Interest

Geometry is one of the oldest branches of mathematics. Around 350 B.C., Euclid of Alexandria wrote *Elements*, which contained all of the known concepts of geometry. Euclid's contribution was to unify various concepts into a single deductive system that was based on a set of axioms.

The word *geometry* comes from the Greek words for *earth* and *measure*. In ancient Egypt, geometry was used by the Egyptians to measure land and to build structures such as the pyramids. Today geometry is used in many fields, such as physics, medicine, and geology. Geometry is also used in applied fields such as mechanical drawing and astronomy. Geometric forms are used in art and design.

Three basic concepts of geometry are point, line, and plane. A **point** is symbolized by drawing a dot. A **line** is determined by two distinct points and extends indefinitely in both directions, as the arrows on the line shown at the right indicate. This line contains points A and B and is represented by $\overleftrightarrow{AB}$. A line can also be represented by a single letter, such as ℓ.

A **ray** starts at a point and extends indefinitely in *one* direction. The point at which a ray starts is called the **endpoint** of the ray. The ray shown at the right is denoted by $\overrightarrow{AB}$. Point A is the endpoint of the ray.

A **line segment** is part of a line and has two endpoints. The line segment shown at the right is denoted by $\overline{AB}$.

The distance between the endpoints of $\overline{AC}$ is denoted by AC. If B is a point on $\overline{AC}$, then AC (the distance from A to C) is the sum of AB (the distance from A to B) and BC (the distance from B to C).

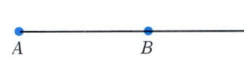

$AC = AB + BC$

HOW TO   Given the figure above and the fact that $AB = 22$ cm and $AC = 31$ cm, find BC.

$AC = AB + BC$ • Write an equation for the distances between points on the line segment.

$31 = 22 + BC$ • Substitute the given distances for AB and AC into the equation.

$9 = BC$ • Solve for BC.

$BC = 9$ cm

In this section we will be discussing figures that lie in a plane. A **plane** is a flat surface and can be pictured as a table top or blackboard that extends in all directions. Figures that lie in a plane are called **plane figures.**

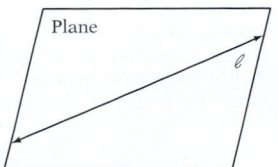

Plane

Objective 7.1A

Vocabulary to Review

point [1.2C]
line [1.2C]
ray [1.2C]
endpoint [1.2C]
line segment [1.2C]
plane [1.2C]
plane figures [1.2C]
intersecting lines [1.2C]
parallel lines [1.2C]
angle [1.2C]
degrees [1.2C]
right angle [1.2C]

New Vocabulary

vertex
side of an angle
protractor
perpendicular lines
complementary angles
straight angle
supplementary angles
acute angle
obtuse angle
adjacent angles

New Symbols

line, e.g., $\overleftrightarrow{AB}$

ray, e.g., $\overrightarrow{AB}$

line segment, e.g., $\overline{AB}$
∥ (is parallel to)
∠ (angle)
° (degrees)
∟ (right angle)
⊥ (is perpendicular to)

Instructor Note

Tell students that when no units (such as feet or meters) are given for lengths along a line segment, all the distances are assumed to be in the same unit of length.

In-Class Examples (Objective 7.1A)

1. How many degrees are in two-thirds of a revolution?   240°

2. Find the complement of a 38° angle.   52°

3. Find the supplement of a 57° angle.   123°

Discuss the Concepts

1. Describe each of the following: ray, line, and line segment.
2. Does the surface of Earth lie in a plane?
3. Using any objects in the classroom, provide examples of each of the following: a right angle, an acute angle, an obtuse angle, a plane, intersecting lines, parallel lines, and perpendicular lines. (If students need some assistance, suggest they consider the windows.)

Concept Check

1. Provide three names for the angle below.

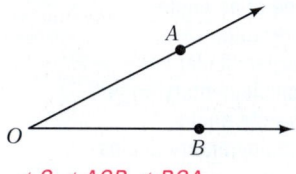

   ∠O, ∠AOB, ∠BOA

2. Name the number of degrees in a full circle, a straight angle, and a right angle.
   360°, 180°, 90°

3. How many dimensions does a point have? a line? a line segment? a ray? an angle?
   0; 1; 1; 1; 2

4. What is the name given to lines in a plane that do not intersect?
   Parallel lines

5. What is the name given to two lines that intersect at right angles?
   Perpendicular lines

Study Tip

A great many new vocabulary words are introduced in this chapter. There are six new terms on this page alone: intersecting lines, parallel lines, angle, vertex, sides, and degrees. All of these terms are in **bold type**. The bold type indicates that these are concepts you must know to learn the material. Be sure to study each new term as it is presented.

Point of Interest

The Babylonians knew that Earth is in approximately the same position in the sky every 365 days. Historians suggest that one complete revolution of a circle is called 360° because 360 is the closest number to 365 that is divisible by many natural numbers.

Lines in a plane can be intersecting or parallel. **Intersecting lines** cross at a point in the plane. **Parallel lines** never intersect. The distance between them is always the same.

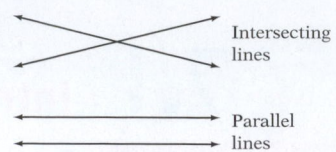

The symbol ∥ means "is parallel to." In the figure at the right, $j \parallel k$ and $\overline{AB} \parallel \overline{CD}$. Note that j contains $\overline{AB}$ and k contains $\overline{CD}$. Parallel lines contain parallel line segments.

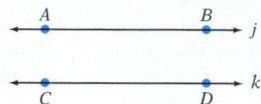

An **angle** is formed by two rays with the same endpoint. The **vertex** of the angle is the point at which the two rays meet. The rays are called the **sides** of the angle.

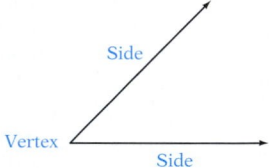

If A and C are points on rays r_1 and r_2, and B is the vertex, then the angle is called $\angle B$ or $\angle ABC$, where $\angle$ is the symbol for angle. Note that the angle is named by the vertex, or the vertex is the second point listed when the angle is named by giving three points. $\angle ABC$ could also be called $\angle CBA$.

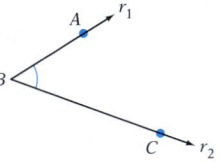

An angle can also be named by a variable written between the rays close to the vertex. In the figure at the right, $\angle x = \angle QRS$ and $\angle y = \angle SRT$. Note that in this figure, more than two rays meet at R. In this case, the vertex cannot be used to name an angle.

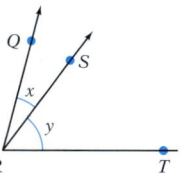

An angle is measured in **degrees.** The symbol for degrees is a small raised circle, °. Probably because early Babylonians believed that Earth revolves around the sun in approximately 360 days, the angle formed by a circle has a measure of 360° (360 degrees).

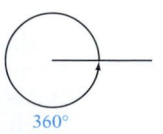

Point of Interest

The Leaning Tower of Pisa is the bell tower of the Cathedral in Pisa, Italy. Its construction began on August 9, 1173, and continued for about 200 years. The tower was designed to be vertical, but it started to lean during its construction. By 1350 it was 2.5° off from the vertical; by 1817, it was 5.1° off; and by 1990, it was 5.5° off. In 2001, work on the structure that returned its list to 5° was completed. (*Source: Time* magazine, June 25, 2001, pp. 34–35)

A **protractor** is used to measure an angle. Place the center of the line segment near the bottom edge of the protractor at the vertex of the angle and the line segment along a side of the angle. The angle shown in the figure below measures 58°.

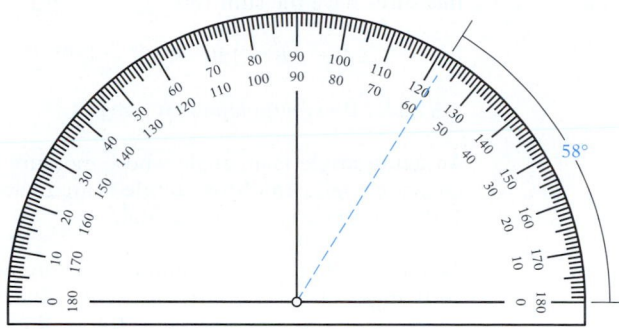

A 90° angle is called a **right angle.** The symbol ∟ represents a right angle.

Perpendicular lines are intersecting lines that form right angles.

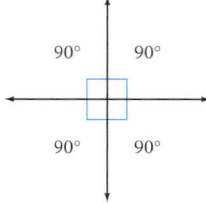

The symbol ⊥ means "is perpendicular to." In the figure at the right, $p \perp q$ and $\overline{AB} \perp \overline{CD}$. Note that line p contains $\overline{AB}$ and line q contains $\overline{CD}$. Perpendicular lines contain perpendicular line segments.

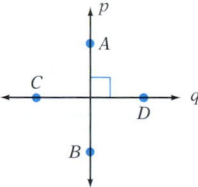

Complementary angles are two angles whose measures have the sum 90°.

$$\angle A + \angle B = 70° + 20° = 90°$$

$\angle A$ and $\angle B$ are complementary angles.

A 180° angle is called a **straight angle.**

∠AOB is a straight angle.

Supplementary angles are two angles whose measures have the sum 180°.

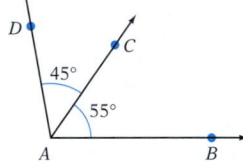

$$\angle A + \angle B = 130° + 50° = 180°$$

∠A and ∠B are supplementary angles.

An **acute angle** is an angle whose measure is between 0° and 90°. ∠B above is an acute angle. An **obtuse angle** is an angle whose measure is between 90° and 180°. ∠A above is an obtuse angle.

Two angles that share a common side are **adjacent angles.** In the figure at the right, ∠DAC and ∠CAB are adjacent angles. ∠DAC = 45° and ∠CAB = 55°.

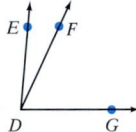

$$\angle DAB = \angle DAC + \angle CAB$$
$$= 45° + 55° = 100°$$

HOW TO   In the figure at the right, ∠EDG = 80°. ∠FDG is three times the measure of ∠EDF. Find the measure of ∠EDF.

$$\angle EDF + \angle FDG = \angle EDG$$
$$x + 3x = 80$$
$$4x = 80$$
$$x = 20$$

• Let x = the measure of ∠EDF. Then $3x$ = the measure of ∠FDG. Write an equation and solve for x, the measure of ∠EDF.

∠EDF = 20°

Example 1

Given $MN = 15$ mm, $NO = 18$ mm, and $MP = 48$ mm, find OP.

Solution

$$MN + NO + OP = MP$$
$$15 + 18 + OP = 48$$
$$33 + OP = 48$$
$$OP = 15$$

$OP = 15$ mm

You Try It 1

Given $QR = 24$ cm, $ST = 17$ cm, and $QT = 62$ cm, find RS.

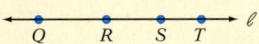

Your solution

21 cm

Solution on p. S18

Example 2

Given $XY = 9$ m and YZ is twice XY, find XZ.

$$\begin{array}{ccccc} \bullet & \bullet & & \bullet & \ell \\ X & Y & & Z & \end{array}$$

Solution

$XZ = XY + YZ$
$XZ = XY + 2(XY)$
$XZ = 9 + 2(9)$
$XZ = 9 + 18$
$XZ = 27$

$XZ = 27$ m

You Try It 2

Given $BC = 16$ ft and $AB = \frac{1}{4}(BC)$, find AC.

$$\begin{array}{ccccc} \bullet & \bullet & & \bullet & \ell \\ A & B & & C & \end{array}$$

Your solution
20 ft

Example 3

Find the complement of a 38° angle.

Strategy

Complementary angles are two angles whose sum is 90°. To find the complement, let x represent the complement of a 38° angle. Write an equation and solve for x.

Solution

$x + 38° = 90°$
$\quad\quad x = 52°$

The complement of a 38° angle is a 52° angle.

You Try It 3

Find the supplement of a 129° angle.

Your Strategy

Your solution
51°

Example 4

Find the measure of $\angle x$.

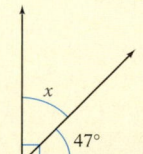

Strategy

To find the measure of $\angle x$, write an equation using the fact that the sum of the measure of $\angle x$ and 47° is 90°. Solve for $\angle x$.

Solution

$\angle x + 47° = 90°$
$\quad\quad \angle x = 43°$

The measure of $\angle x$ is 43°.

You Try It 4

Find the measure of $\angle a$.

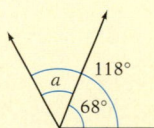

Your Strategy

Your solution
50°

Solutions on pp. S18–S19

Objective 7.1B

Vocabulary to Review

intersecting lines   [7.1A]
adjacent angles   [7.1A]

New Vocabulary

vertical angles
transversal
alternate interior angles
alternate exterior angles
corresponding angles

Discuss the Concepts

What is a transversal? Describe two different ways in which a transversal can intersect two other lines.

Concept Check

When a transversal intersects two parallel lines, which of the following are supplementary angles: vertical angles, adjacent angles, alternate interior angles, alternate exterior angles, corresponding angles?

Adjacent angles

Optional Student Activity

In the figure, $BD \parallel AC$ and $\angle BAC = \angle BCA$. Find the value of x.   65°

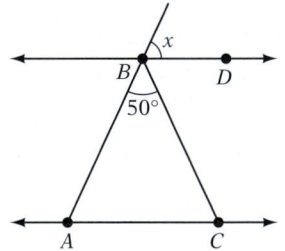

Objective B   To solve problems involving angles formed by intersecting lines

Point of Interest

Many cities in the New World, unlike those in Europe, were designed using rectangular street grids. Washington, D.C., was planned that way, except diagonal avenues were added, primarily for the purpose of enabling quick troop movement in the event that the city required defense. As an added precaution, monuments were constructed at major intersections so that attackers would not have a straight shot down a boulevard.

Four angles are formed by the intersection of two lines. If the two lines are perpendicular, each of the four angles is a right angle. If the two lines are not perpendicular, then two of the angles formed are acute angles and two of the angles are obtuse angles. The two acute angles are always opposite each other, and the two obtuse angles are always opposite each other.

In the figure at the right, $\angle w$ and $\angle y$ are acute angles. $\angle x$ and $\angle z$ are obtuse angles.

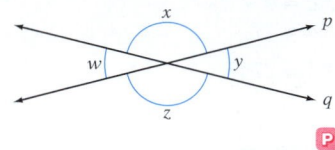

Two angles that are on opposite sides of the intersection of two lines are called **vertical angles.** Vertical angles have the same measure. $\angle w$ and $\angle y$ are vertical angles. $\angle x$ and $\angle z$ are vertical angles.

Vertical angles have the same measure.

$$\angle w = \angle y$$
$$\angle x = \angle z$$

Recall that two angles that share a common side are called **adjacent angles.** For the figure shown above, $\angle x$ and $\angle y$ are adjacent angles, as are $\angle y$ and $\angle z$, $\angle z$ and $\angle w$, and $\angle w$ and $\angle x$. Adjacent angles of intersecting lines are supplementary angles.

Adjacent angles of intersecting lines are supplementary angles.

$$\angle x + \angle y = 180°$$
$$\angle y + \angle z = 180°$$
$$\angle z + \angle w = 180°$$
$$\angle w + \angle x = 180°$$

HOW TO   Given that $\angle c = 65°$, find the measures of angles a, b, and d.

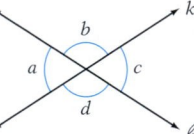

$\angle a = 65°$

$\angle b + \angle c = 180°$
$\angle b + 65° = 180°$
$\quad\quad \angle b = 115°$

$\angle d = 115°$

• $\angle a = \angle c$ because $\angle a$ and $\angle c$ are vertical angles.

• $\angle b$ is supplementary to $\angle c$ because $\angle b$ and $\angle c$ are adjacent angles of intersecting lines.

• $\angle d = \angle b$ because $\angle d$ and $\angle b$ are vertical angles.

In-Class Examples (Objective 7.1B)

1. In the figure, $\angle a = 62°$. Find $\angle b$.   118°

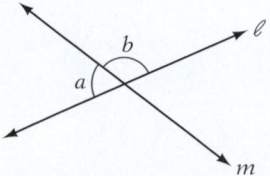

2. In the figure, $\ell_1 \parallel \ell_2$ and $\angle a = 103°$. Find $\angle b$.   77°

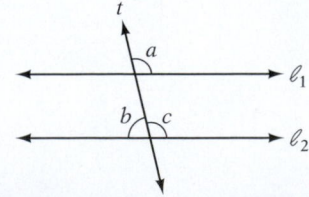

A line that intersects two other lines at different points is called a **transversal.**

If the lines cut by a transversal t are parallel lines and the transversal is perpendicular to the parallel lines, all eight angles formed are right angles.

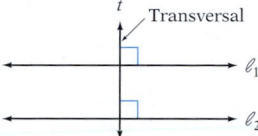

If the lines cut by a transversal t are parallel lines and the transversal is not perpendicular to the parallel lines, all four acute angles have the same measure and all four obtuse angles have the same measure. For the figure at the right,

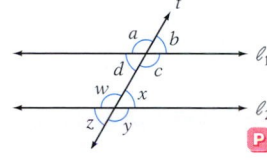

$$\angle b = \angle d = \angle x = \angle z$$

$$\angle a = \angle c = \angle w = \angle y$$

Alternate interior angles are two nonadjacent angles that are on opposite sides of the transversal and between the parallel lines. In the figure above, $\angle c$ and $\angle w$ are alternate interior angles; $\angle d$ and $\angle x$ are alternate interior angles. Alternate interior angles have the same measure.

Alternate interior angles have the same measure.

$$\angle c = \angle w$$
$$\angle d = \angle x$$

Alternate exterior angles are two nonadjacent angles that are on opposite sides of the transversal and outside the parallel lines. In the figure above, $\angle a$ and $\angle y$ are alternate exterior angles; $\angle b$ and $\angle z$ are alternate exterior angles. Alternate exterior angles have the same measure.

Alternate exterior angles have the same measure.

$$\angle a = \angle y$$
$$\angle b = \angle z$$

Corresponding angles are two angles that are on the same side of the transversal and are both acute angles or are both obtuse angles. For the figure above, the following pairs of angles are corresponding angles: $\angle a$ and $\angle w$, $\angle d$ and $\angle z$, $\angle b$ and $\angle x$, $\angle c$ and $\angle y$. Corresponding angles have the same measure.

Corresponding angles have the same measure.

$$\angle a = \angle w$$
$$\angle d = \angle z$$
$$\angle b = \angle x$$
$$\angle c = \angle y$$

HOW TO   Given that $\ell_1 \parallel \ell_2$ and $\angle c = 58°$, find the measures of $\angle f$, $\angle h$, and $\angle g$.

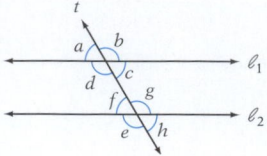

$\angle f = \angle c = 58°$ • $\angle c$ and $\angle f$ are alternate interior angles.
$\angle h = \angle c = 58°$ • $\angle c$ and $\angle h$ are corresponding angles.

$\angle g + \angle h = 180°$ • $\angle g$ is supplementary to $\angle h$.
$\angle g + 58° = 180°$
$\angle g = 122°$

Example 5

Find x.

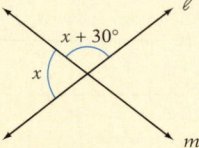

Strategy

The angles labeled are adjacent angles of intersecting lines and are therefore supplementary angles. To find x, write an equation and solve for x.

Solution

$x + (x + 30°) = 180°$
$2x + 30° = 180°$
$2x = 150°$
$x = 75°$

Example 6

Given $\ell_1 \parallel \ell_2$, find x.

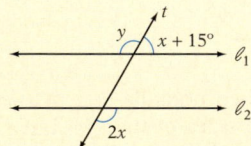

Strategy

$2x = y$ because alternate exterior angles have the same measure. $(x + 15°) + y = 180°$ because adjacent angles of intersecting lines are supplementary angles. Substitute $2x$ for y and solve for x.

Solution

$(x + 15°) + 2x = 180°$
$3x + 15° = 180°$
$3x = 165°$
$x = 55°$

You Try It 5

Find x.

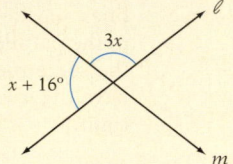

Your Strategy

Your solution
41°

You Try It 6

Given $\ell_1 \parallel \ell_2$, find x.

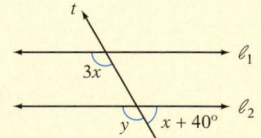

Your Strategy

Your solution
35°

Solutions on p. S19

Objective C

To solve problems involving the angles of a triangle

VIDEO & DVD   CD TUTOR   WEB   SSM

If the lines cut by a transversal are not parallel lines, the three lines will intersect at three points. In the figure at the right, the transversal *t* intersects lines *p* and *q*. The three lines intersect at points *A*, *B*, and *C*. These three points define three line segments, $\overline{AB}$, $\overline{BC}$, and $\overline{AC}$. The plane figure formed by these three line segments is called a **triangle.**

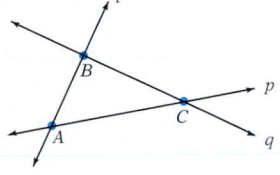

Each of the three points of intersection is the vertex of four angles. The angles within the region enclosed by the triangle are called **interior angles.** In the figure at the right, angles *a*, *b*, and *c* are interior angles. The sum of the measures of the interior angles of a triangle is 180°.

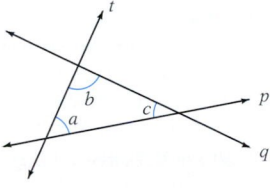

$$\angle a + \angle b + \angle c = 180°$$

> **The Sum of the Measures of the Interior Angles of a Triangle**
>
> The sum of the measures of the interior angles of a triangle is 180°.

An angle adjacent to an interior angle is an **exterior angle.** In the figure at the right, angles *m* and *n* are exterior angles for angle *a*. The sum of the measures of an interior and an exterior angle is 180°.

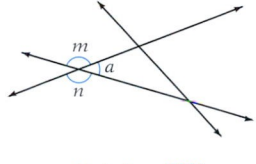

$$\angle a + \angle m = 180°$$
$$\angle a + \angle n = 180°$$

HOW TO   Given that $\angle c = 40°$ and $\angle d = 100°$, find the measure of $\angle e$.

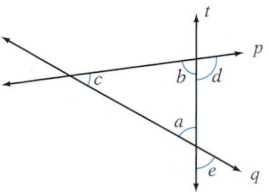

$\angle d$ and $\angle b$ are supplementary angles.

$$\angle d + \angle b = 180°$$
$$100° + \angle b = 180°$$
$$\angle b = 80°$$

The sum of the interior angles is 180°.

$$\angle c + \angle b + \angle a = 180°$$
$$40° + 80° + \angle a = 180°$$
$$120° + \angle a = 180°$$
$$\angle a = 60°$$

$\angle a$ and $\angle e$ are vertical angles.

$$\angle e = \angle a = 60°$$

Objective 7.1C

Vocabulary to Review
triangle   [1.2C]

New Vocabulary
interior angles
exterior angles

New Properties
The sum of the measures of the interior angles of a triangle is 180°.

Discuss the Concepts
In a right triangle, why are the two acute angles complementary?

Concept Check
1. The angles of a triangle are in the ratio 2:3:7. Find the number of degrees in the largest angle.   105°
2. The measures of the angles of a triangle are consecutive integers. Find the measure of each angle.   59°, 60°, 61°
3. Determine the measures of the angles of an equilateral triangle.   60°, 60°, 60°

In-Class Examples (Objective 7.1C)
1. A triangle has a 21° angle and a 64° angle. Find the measure of the third angle.   95°
2. A right triangle has a 52° angle. Find the measure of the other two angles.   90°, 38°
3. One angle of a right triangle is 2° less than three times the measure of the smallest angle. Find the measure of each angle.   23°, 67°, 90°

Discuss the Concepts

Use the figure in Example 7.

1. What does the square in the corner of the triangle mean?

2. If $\angle b = 30°$, what is the measure of $\angle d$? Why?

3. If $\angle b = 30°$, what is the measure of $\angle a$? Why?

4. If $\angle b = 30°$, what is the measure of $\angle y$? Why?

Optional Student Activity

Draw five triangles of different sizes. For each triangle, use a protractor to find the measure of each angle. (*Note:* You will want the triangles to be fairly large so that measuring the angles is not difficult.) Then find the sum of the angles of each triangle.

Example 7

Given that $\angle y = 55°$, find the measures of angles a, b, and d.

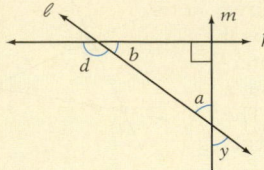

Strategy

• To find the measure of angle a, use the fact that $\angle a$ and $\angle y$ are vertical angles.
• To find the measure of angle b, use the fact that the sum of the measures of the interior angles of a triangle is 180°.
• To find the measure of angle d, use the fact that the sum of the measures of an interior and an exterior angle is 180°.

Solution

$\angle a = \angle y = 55°$

$\angle a + \angle b + 90° = 180°$
$55° + \angle b + 90° = 180°$
$\angle b + 145° = 180°$
$\angle b = 35°$

$\angle d + \angle b = 180°$
$\angle d + 35° = 180°$
$\angle d = 145°$

You Try It 7

Given that $\angle a = 45°$ and $\angle x = 100°$, find the measures of angles b, c, and y.

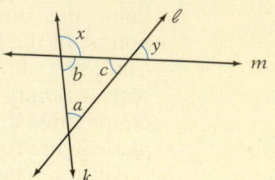

Your Strategy

Your solution
$\angle b = 80°, \angle c = 55°, \angle y = 55°$

Example 8

Two angles of a triangle measure 53° and 78°. Find the measure of the third angle.

Strategy

To find the measure of the third angle, use the fact that the sum of the measures of the interior angles of a triangle is 180°. Write an equation using x to represent the measure of the third angle. Solve the equation for x.

Solution

$x + 53° + 78° = 180°$
$x + 131° = 180°$
$x = 49°$

The measure of the third angle is 49°.

You Try It 8

One angle in a triangle is a right angle, and one angle measures 34°. Find the measure of the third angle.

Your Strategy

Your solution
56°

Solutions on p. S19

Section 7.1

7.1 Exercises

Objective A   **To solve problems involving lines and angles**

Suggested Assignment
Exercises 1–57, odds
More challenging problems:
  Exercises 62, 63

In Exercises 1–6, use a protractor to measure the angle. State whether the angle is acute, obtuse, or right.

1.
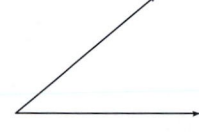

40°, acute

2.
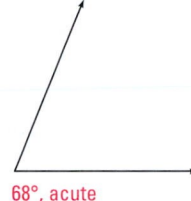

68°, acute

3.
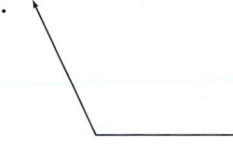

115°, obtuse

4.
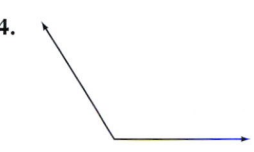

122°, obtuse

5.
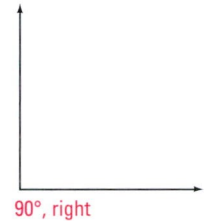

90°, right

6.
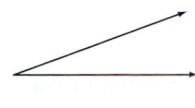

20°, acute

7. Find the complement of a 62° angle.
28°

8. Find the complement of a 31° angle.
59°

9. Find the supplement of a 162° angle.
18°

10. Find the supplement of a 72° angle.
108°

11. Given $AB = 12$ cm, $CD = 9$ cm, and $AD = 35$ cm, find the length of BC.
14 cm

12. Given $AB = 21$ mm, $BC = 14$ mm, and $AD = 54$ mm, find the length of CD.
19 mm

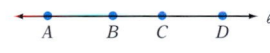

13. Given $QR = 7$ ft and RS is three times the length of QR, find the length of QS.
28 ft

14. Given $QR = 15$ in. and RS is twice the length of QR, find the length of QS.
45 in.

15. Given $EF = 20$ m and FG is one-half the length of EF, find the length of EG.
30 m

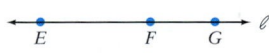

Quick Quiz (Objective 7.1A)

1. How many degrees are in three-fifths of a revolution?   216°

2. Find the complement of a 54° angle.   36°

3. Find the supplement of a 22° angle.   158°

16. Given *EF* = 18 cm and *FG* is one-third the length of *EF*, find the length of *EG*.

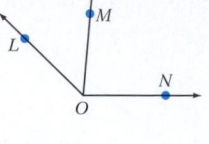

24 cm

17. Given ∠*LOM* = 53° and ∠*LON* = 139°, find the measure of ∠*MON*.

86°

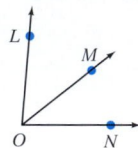

18. Given ∠*MON* = 38° and ∠*LON* = 85°, find the measure of ∠*LOM*.

47°

Find the measure of ∠*x*.

19.

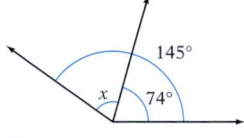

71°

20.

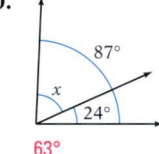

63°

Given that ∠*LON* is a right angle, find the measure of ∠*x*.

21.

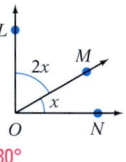

30°

22.

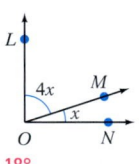

18°

23.

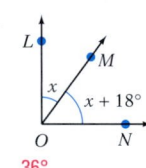

36°

24.

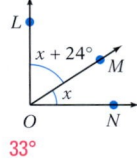

33°

Find the measure of ∠*a*.

25.

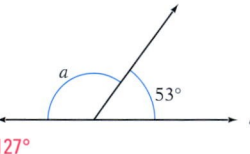

127°

26.

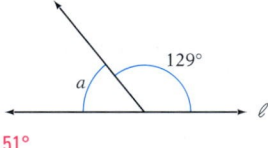

51°

27.

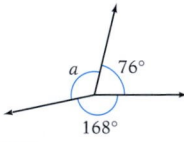

116°

28.

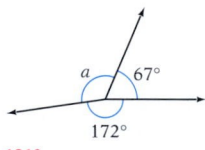

121°

In Exercises 29–34, find x.

29.

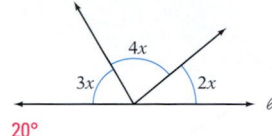

20°

30.

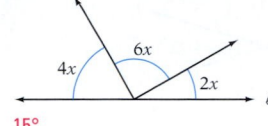

15°

31.

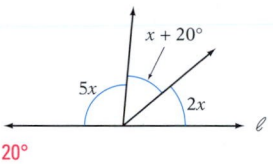

20°

32.

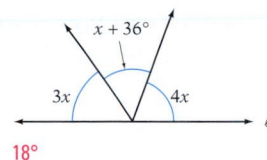

18°

33.

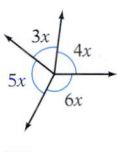

20°

34.

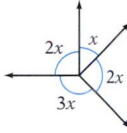

45°

35. Given $\angle a = 51°$, find the measure of $\angle b$.
141°

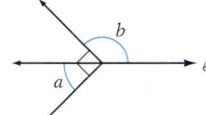

36. Given $\angle a = 38°$, find the measure of $\angle b$.
128°

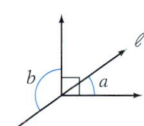

Objective B **To solve problems involving angles formed by intersecting lines**

Find the measure of $\angle x$.

37.

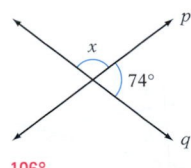

106°

38.

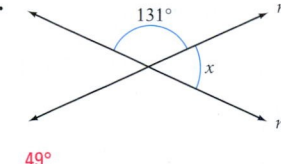

49°

Quick Quiz (Objective 7.1B)

1. In the figure, $\angle a = 118°$. Find $\angle b$.   62°

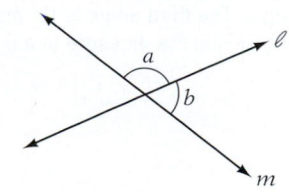

2. In the figure, $\ell_1 \parallel \ell_2$ and $\angle a = 67°$. Find $\angle b$.   113°

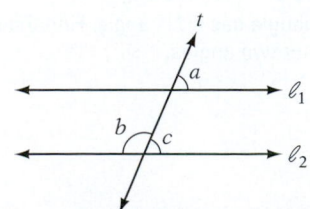

Find *x*.

39.

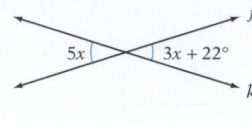

11°

40.

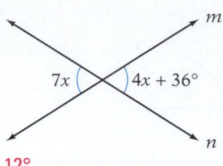

12°

Given that $\ell_1 \parallel \ell_2$, find the measures of angles *a* and *b*.

41.

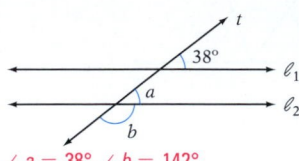

∠*a* = 38°, ∠*b* = 142°

42.

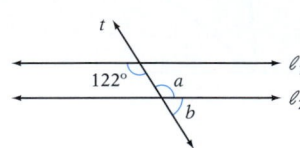

∠*a* = 122°, ∠*b* = 58°

43.

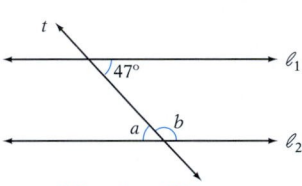

∠*a* = 47°, ∠*b* = 133°

44.

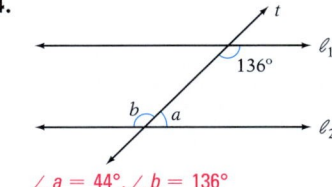

∠*a* = 44°, ∠*b* = 136°

Given that $\ell_1 \parallel \ell_2$, find *x*.

45.

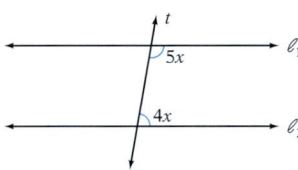

20°

46.

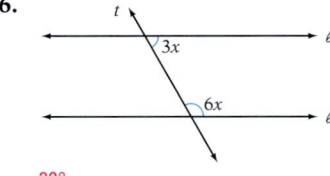

20°

47.

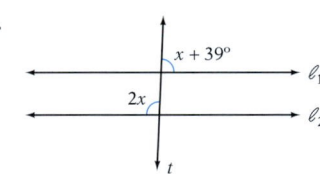

47°

48.

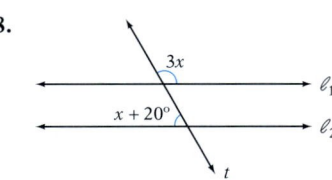

40°

Objective C   **To solve problems involving the angles of a triangle**

49. Given that ∠*a* = 95° and ∠*b* = 70°, find the measures of angles *x* and *y*.
∠*x* = 155°, ∠*y* = 70°

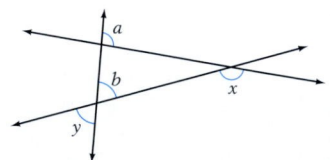

Quick Quiz (Objective 7.1C)

1. A triangle has a 110° angle and a 35° angle. Find the measure of the third angle.   35°

2. A right triangle has a 71° angle. Find the measure of the other two angles.   90°, 19°

3. In a triangle, one angle is twice the measure of a second angle. The third angle is 12° more than the second angle. Find the measure of each angle.
42°, 84°, 54°

50. Given that $\angle a = 35°$ and $\angle b = 55°$, find the measures of angles x and y.
$\angle x = 160°, \angle y = 145°$

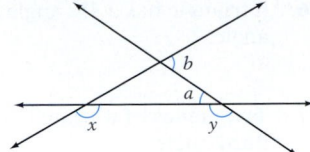

51. Given that $\angle y = 45°$, find the measures of angles a and b.
$\angle a = 45°, \angle b = 135°$

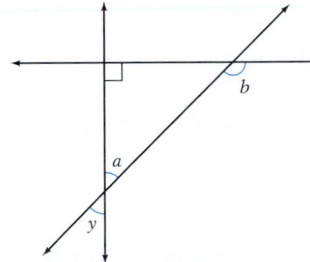

52. Given that $\angle y = 130°$, find the measures of angles a and b.
$\angle a = 40°, \angle b = 140°$

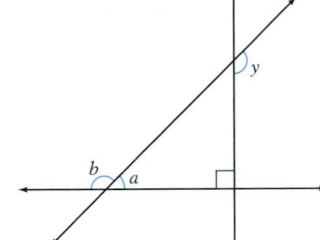

53. Given that $\overline{AO} \perp \overline{OB}$, express in terms of x the number of degrees in $\angle BOC$.
$90° - x$

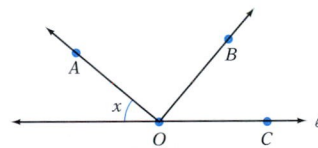

54. Given that $\overline{AO} \perp \overline{OB}$, express in terms of x the number of degrees in $\angle AOC$.
$75° - x$

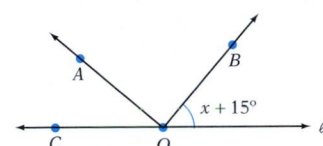

55. One angle in a triangle is a right angle, and one angle is equal to 30°. What is the measure of the third angle?
60°

56. A triangle has a 45° angle and a right angle. Find the measure of the third angle.
45°

57. Two angles of a triangle measure 42° and 103°. Find the measure of the third angle.
35°

58. A triangle has a 13° angle and a 65° angle. What is the measure of the third angle?
102°

APPLYING THE CONCEPTS

59. **a.** What is the smallest possible whole number of degrees in an angle of a triangle?
b. What is the largest possible whole number of degrees in an angle of a triangle?
a. 1°   **b.** 179°

60. Cut out a triangle and then tear off two of the angles, as shown at the right. Position the pieces you tore off so that angle *a* is adjacent to angle *b* and angle *c* is adjacent to angle *b*. Describe what you observe. What does this demonstrate?
See the *Solutions Manual.*

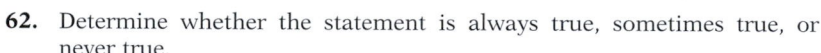

61. Construct a triangle with the given angle measures.
a. 45°, 45°, and 90°   **b.** 30°, 60°, and 90°   **c.** 40°, 40°, and 100°

62. Determine whether the statement is always true, sometimes true, or never true.
a. Two lines that are parallel to a third line are parallel to each other.
b. A triangle contains at least two acute angles.
c. Vertical angles are complementary angles.
a. Always true   **b.** Always true   **c.** Sometimes true

63. For the figure at the right, find the sum of the measures of angles *x*, *y*, and *z*.
360°

64. For the figure at the right, explain why $\angle a + \angle b = \angle x$. Write a rule that describes the relationship between an exterior angle of a triangle and the opposite interior angles. Use the rule to write an equation involving angles *a*, *c*, and *z*.

65. If $\overline{AB}$ and $\overline{CD}$ intersect at point *O*, and $\angle AOC = \angle BOC$, explain why $\overline{AB} \perp \overline{CD}$.

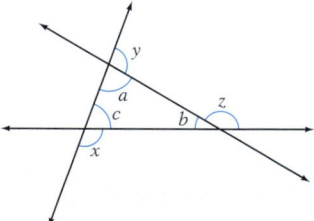

66. Do some research on the principle of reflection. Explain how this principle applies to the operation of a periscope and to the game of billiards.

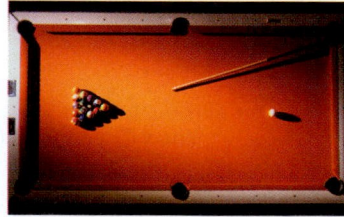

Answers to Writing Exercises

64. The sum of the interior angles of a triangle is 180°; therefore, $\angle a + \angle b + \angle c = 180°$. The sum of the measures of an interior and an exterior angle is 180°; therefore, $\angle c + \angle x = 180°$. Solving this equation for $\angle c$, $\angle c = 180° - \angle x$. Substitute $180° - \angle x$ for $\angle c$ in the equation $\angle a + \angle b + \angle c = 180°$: $\angle a + \angle b + (180° - \angle x) = 180°$. Add the measure of $\angle x$ to each side of the equation and subtract 180° from each side of the equation: $\angle a + \angle b = \angle x$. The measure of an exterior angle of a triangle is equal to the sum of the measures of the two opposite interior angles: $\angle a + \angle c = \angle z$.

65. $\angle AOC$ and $\angle BOC$ are supplementary angles. Therefore, $\angle AOC + \angle BOC = 180°$. Since $\angle AOC = \angle BOC$, by substitution, $\angle AOC + \angle AOC = 180°$. Therefore, $2(\angle AOC) = 180°$, and $\angle AOC = 90°$. So, $\overline{AB}$ is perpendicular to $\overline{CD}$.

66. The student should note that when a ray of light hits a surface, such as a mirror, the light is reflected at the same angle at which it hit the surface. In a periscope, light is reflected twice, with the result that light rays entering the periscope are parallel to light rays at eye level. A ball bouncing off the side of a billiard table will bounce off the side at the same angle at which it hit the side.

7.2 Plane Geometric Figures

Objective A   To solve problems involving the perimeter of geometric figures

A **polygon** is a closed figure determined by three or more line segments that lie in a plane. The line segments that form the polygon are called its **sides.** The figures below are examples of polygons.

 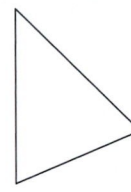 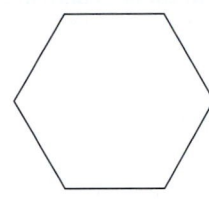 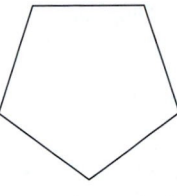 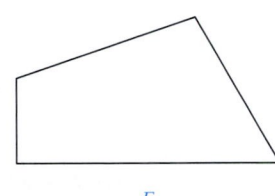

    A   *B*   *C*   *D*   *E*

A **regular polygon** is one in which all sides have the same length and all angles have the same measure. The polygons in Figures *A*, *C*, and *D* above are regular polygons.

The name of a polygon is based on the number of its sides. The table below lists the names of polygons that have from 3 to 10 sides.

Number of Sides	Name of the Polygon
3	Triangle
4	Quadrilateral
5	Pentagon
6	Hexagon
7	Heptagon
8	Octagon
9	Nonagon
10	Decagon

Point of Interest

Although a polygon is defined in terms of its sides, the word actually comes from the Latin word *polygonum*, meaning "many angles."

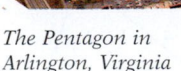

The Pentagon in Arlington, Virginia

Triangles and quadrilaterals are two of the most common types of polygons. Triangles are distinguished by the number of equal sides and also by the measures of their angles.

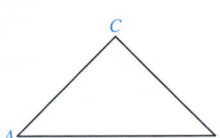   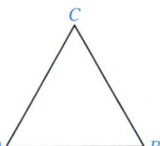   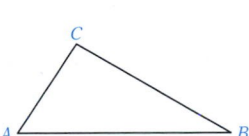

An **isosceles triangle** has two sides of equal length. The angles opposite the equal sides are of equal measure.

$AC = BC$
$\angle A = \angle B$

The three sides of an **equilateral triangle** are of equal length. The three angles are of equal measure.

$AB = BC = AC$
$\angle A = \angle B = \angle C$

A **scalene triangle** has no two sides of equal length. No two angles are of equal measure.

In-Class Examples (Objective 7.2A)

1. Find the perimeter of a square with sides equal to 15 m.   **60 m**

2. Find the perimeter of a rectangle with a length of 5 m and a width of 1.4 m.   **12.8 m**

3. Find the circumference of a circle with a radius of 11 cm. Round to the nearest hundreth.   **69.12 cm**

Objective 7.2A

Vocabulary to Review

plane figure [7.1A]
triangle [7.1C]
polygon [1.2C]
quadrilateral [1.2C]
rectangle [1.2C]
perimeter [1.2C]

New Vocabulary

sides of a polygon
regular polygon
pentagon
hexagon
heptagon
octagon
nonagon
decagon
isosceles triangle
equilateral triangle
scalene triangle
acute triangle
obtuse triangle
right triangle
parallelogram
square
rhombus
trapezoid
isosceles trapezoid
circle
center of a circle
diameter
radius
circumference

New Symbols

π (pi)

Formulas to Review

Perimeter of a triangle:
$P = a + b + c$ [1.2C]
Perimeter of a rectangle:
$P = 2L + 2W$ [1.3E]
Perimeter of a square:
$P = 4s$ [1.3E]

New Formulas

Diameter of a circle: $d = 2r$
Circumference of a circle:
$C = \pi d$ or $C = 2\pi r$

Discuss the Concepts

1. Is every square a rectangle? Is every rectangle a square?

2. In the definition of a polygon, what does the phrase "closed figure" mean? Draw a figure that is not closed.

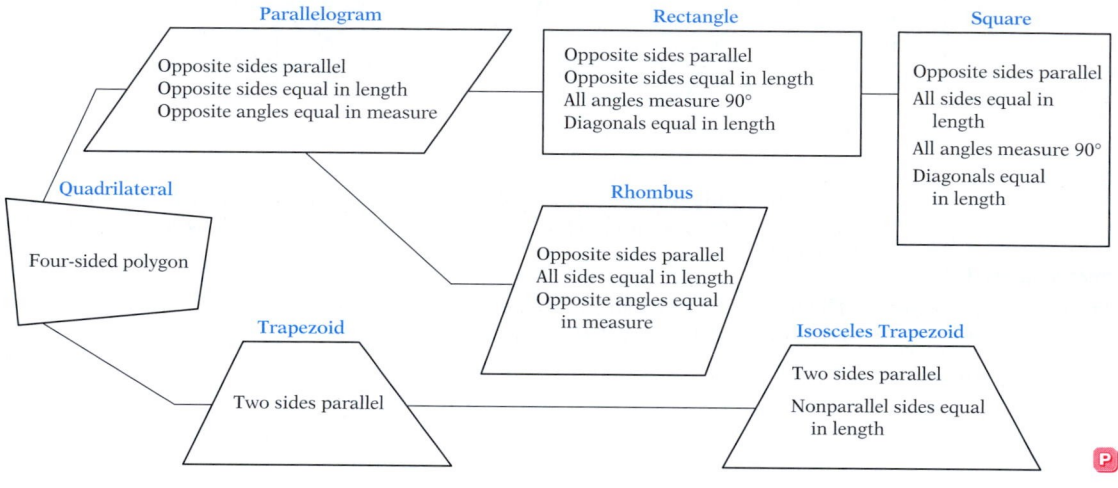

An **acute triangle** has three acute angles.

An **obtuse triangle** has an obtuse angle.

A **right triangle** has a right angle.

Quadrilaterals are also distinguished by their sides and angles, as shown below. Note that a rectangle, a square, and a rhombus are different forms of a parallelogram.

Parallelogram
Opposite sides parallel
Opposite sides equal in length
Opposite angles equal in measure

Rectangle
Opposite sides parallel
Opposite sides equal in length
All angles measure 90°
Diagonals equal in length

Square
Opposite sides parallel
All sides equal in length
All angles measure 90°
Diagonals equal in length

Quadrilateral
Four-sided polygon

Rhombus
Opposite sides parallel
All sides equal in length
Opposite angles equal in measure

Trapezoid
Two sides parallel

Isosceles Trapezoid
Two sides parallel
Nonparallel sides equal in length

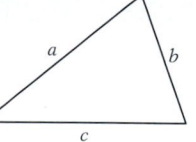

Instructor Note

The diagram at the right shows the relationships among quadrilaterals. A description of each quadrilateral is given within an example of that quadrilateral.

Concept Check

1. Figure A below is a rectangle. Label the length of the rectangle L and the width of the rectangle W.

Figure A

2. What is the name of a regular polygon that has three sides?
Equilateral triangle

3. What is the name of a parallelogram in which all angles are of equal measure?
Rectangle

The **perimeter** of a plane geometric figure is a measure of the distance around the figure. Perimeter is used in buying fencing for a lawn or determining how much baseboard is needed for a room.

The perimeter of a triangle is the sum of the lengths of the three sides.

Perimeter of a Triangle

Let a, b, and c be the lengths of the sides of a triangle. The perimeter, P, of the triangle is given by $P = a + b + c$.

$$P = a + b + c$$

HOW TO Find the perimeter of the triangle shown at the right.

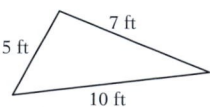

$$P = 5 + 7 + 10 = 22$$

The perimeter is 22 ft.

The perimeter of a quadrilateral is the sum of the lengths of its four sides.

A rectangle has four right angles and opposite sides of equal length. Usually the length, *L*, of a rectangle refers to the length of one of the longer sides of the rectangle, and the width, *W*, refers to the length of one of the shorter sides. The perimeter can then be represented by $P = L + W + L + W$.

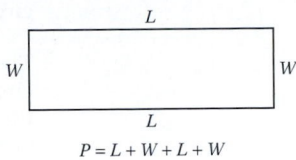

$$P = L + W + L + W$$

The formula for the perimeter of a rectangle is derived by combining like terms. $P = 2L + 2W$

> **Perimeter of a Rectangle**
>
> Let *L* represent the length and *W* the width of a rectangle. The perimeter, *P*, of the rectangle is given by $P = 2L + 2W$.

HOW TO   Find the perimeter of the rectangle shown at the right.

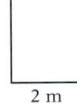

$P = 2L + 2W$
$P = 2(5) + 2(2)$ • **The length is 5 m. Substitute 5 for *L*.**
$P = 10 + 4$ **The width is 2 m. Substitute 2 for *W*.**
$P = 14$ **Solve for *P*.**

The perimeter is 14 m.

A square is a rectangle in which each side has the same length. If we let *s* represent the length of each side of a square, the perimeter of the square can be represented by $P = s + s + s + s$.

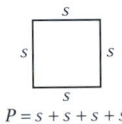

$$P = s + s + s + s$$

The formula for the perimeter of a square is derived by combining like terms. $P = 4s$

> **Perimeter of a Square**
>
> Let *s* represent the length of a side of a square. The perimeter, *P*, of the square is given by $P = 4s$.

HOW TO   Find the perimeter of the square shown at the right.

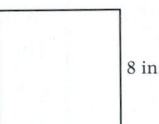

$P = 4s = 4(8) = 32$

The perimeter is 32 in.

Optional Student Activity

1. The base of isosceles triangle *ABC* (with $AB = BC$ and $\angle B = 42°$) and the base of equilateral triangle *CDE* lie on line segment *AE*. Find the measure of $\angle BCD$.   **51°**

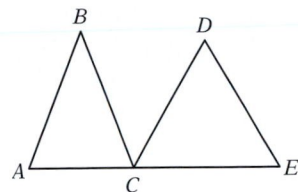

2. Three line segments are randomly chosen from line segments whose lengths are 1 cm, 2 cm, 3 cm, 4 cm, and 5 cm. What is the probability that a triangle can be formed from the line segments?

    $\dfrac{3}{10}$

3. Triangle *FJH* is an isosceles triangle in which $FJ = FH$, $FK = KJ$, and $FG = GH$. Find the perimeter of triangle *FJH* if $FK = 2x + 3$, $GH = 5x - 9$, and $JH = 4x$.   **60 units**

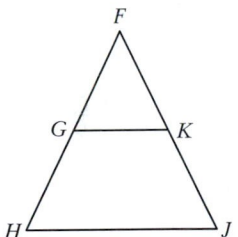

A **circle** is a plane figure in which all points are the same distance from point O, called the **center** of the circle.

The **diameter** of a circle is a line segment across the circle through point O. AB is a diameter of the circle at the right. The variable d is used to designate a diameter of a circle.

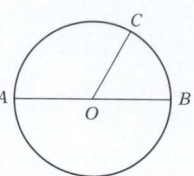

The **radius** of a circle is a line segment from the center of the circle to a point on the circle. OC is a radius of the circle at the right. The variable r is used to designate a radius of a circle.

The length of the diameter is twice the length of the radius.

$$d = 2r \text{ or } r = \frac{1}{2}d$$

Point of Interest

Archimedes (c. 287–212 B.C.) is the person who calculated that $\pi \approx 3\frac{1}{7}$. He actually showed that $3\frac{10}{71} < \pi < 3\frac{1}{7}$. The approximation $3\frac{10}{71}$ is a more accurate approximation of π than $3\frac{1}{7}$, but it is more difficult to use without a calculator.

The distance around a circle is called the **circumference.** The circumference, C, of a circle is equal to the product of π (pi) and the diameter.

$$C = \pi d$$

Because $d = 2r$, the formula for the circumference can be written in terms of r.

$$C = 2\pi r$$

Circumference of a Circle

The circumference, C, of a circle with diameter d and radius r is given by $C = \pi d$ or $C = 2\pi r$.

The formula for circumference uses the number π, which is an irrational number. The value of π can be approximated by a fraction or by a decimal.

$$\pi \approx \frac{22}{7} \text{ or } \pi \approx 3.14$$

The π key on a scientific calculator gives a closer approximation of π than 3.14. Use a scientific calculator to find approximate values in calculations involving π.

Integrating Technology

The π key on your calculator can be used to find decimal approximations to formulas that contain π. To perform the calculation at the right, enter
6 × π = .

HOW TO   Find the circumference of a circle with a diameter of 6 in.

$C = \pi d$ • **The diameter of the circle is given. Use the circumference**
$C = \pi(6)$ **formula that involves the diameter. $d = 6$**
$C = 6\pi$

The exact circumference of the circle is 6π in.

$C \approx 18.85$ • **An approximate measure is found by using the π key on a calculator.**

The approximate circumference is 18.85 in.

Example 1

A carpenter is designing a square patio with a perimeter of 44 ft. What is the length of each side?

Strategy

To find the length of each side, use the formula for the perimeter of a square. Substitute 44 for P and solve for s.

Solution

$P = 4s$
$44 = 4s$
$11 = s$

The length of each side of the patio is 11 ft.

You Try It 1

The infield of a softball field is a square with each side of length 60 ft. Find the perimeter of the infield.

Your Strategy

Your solution
240 ft

Example 2

The dimensions of a triangular sail are 18 ft, 11 ft, and 15 ft. What is the perimeter of the sail?

Strategy

To find the perimeter, use the formula for the perimeter of a triangle. Substitute 18 for a, 11 for b, and 15 for c. Solve for P.

Solution

$P = a + b + c$
$P = 18 + 11 + 15$
$P = 44$

The perimeter of the sail is 44 ft.

You Try It 2

What is the perimeter of a standard piece of typing paper that measures $8\frac{1}{2}$ in. by 11 in.?

Your Strategy

Your solution
39 in.

Example 3

Find the circumference of a circle with a radius of 15 cm. Round to the nearest hundredth.

Strategy

To find the circumference, use the circumference formula that involves the radius. An approximation is asked for; use the π key on a calculator. $r = 15$

Solution

$C = 2\pi r = 2\pi(15) = 30\pi \approx 94.25$

The circumference is approximately 94.25 cm.

You Try It 3

Find the circumference of a circle with a diameter of 9 in. Give the exact measure.

Your Strategy

Your solution
9π in.

Solutions on p. S19

Objective 7.2B

Vocabulary to Review
area   [1.3E]

New Vocabulary
base
height

Symbols to Review
in^2 (square inches)   [1.3E]
cm^2 (square centimeters)   [1.3E]
ft^2 (square feet)   [1.3E]
m^2 (square meters)   [1.3E]
mi^2 (square miles)   [1.3E]

Formulas to Review
Area of a rectangle:
  $A = LW$ [1.3E]
Area of a square:
  $A = s^2$ [1.3E]

New Formulas
Area of a parallelogram:
  $A = bh$
Area of a triangle:
  $A = \frac{1}{2}bh$
Area of a trapezoid:
  $A = \frac{1}{2}h(b_1 + b_2)$
Area of a circle:
  $A = \pi r^2$

Instructor Note
The first four square numbers are diagrammed in the Point of Interest. Ask students to find the next three square numbers. Have them form the square array, as well as represent the number in both standard form and exponential form.

Discuss the Concepts
What is wrong with the statement?
1. The perimeter is $40 \ m^2$.
2. The area is 120 ft.

Objective B   To solve problems involving the area of geometric figures

Area is the amount of surface in a region. Area can be used to describe the size of a rug, a parking lot, a farm, or a national park. Area is measured in square units.

Point of Interest

Figurate numbers are whole numbers that can be represented as regular geometric figures. For example, a square number is one that can be represented as a square array.

```
0      00     000    0000
       00     000    0000
              000    0000
                     0000
1      4      9      16
```

The square numbers are 1, 4, 9, 16, 25, ... They can be represented as $1^2, 2^2, 3^2, 4^2, 5^2, ...$

A square that measures 1 in. on each side has an area of 1 square inch, written $1 \ in^2$.

A square that measures 1 cm on each side has an area of 1 square centimeter, written $1 \ cm^2$.

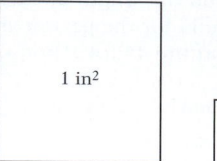

Larger areas can be measured in square feet (ft^2), square meters (m^2), acres (43,560 ft^2), square miles (mi^2), or any other square unit.

The area of a geometric figure is the number of squares that are necessary to cover the figure. In the figures below, two rectangles have been drawn and covered with squares. In the figure on the left, 12 squares, each of area 1 cm^2, were used to cover the rectangle. The area of the rectangle is 12 cm^2. In the figure on the right, 6 squares, each of area 1 in^2, were used to cover the rectangle. The area of the rectangle is 6 in^2.

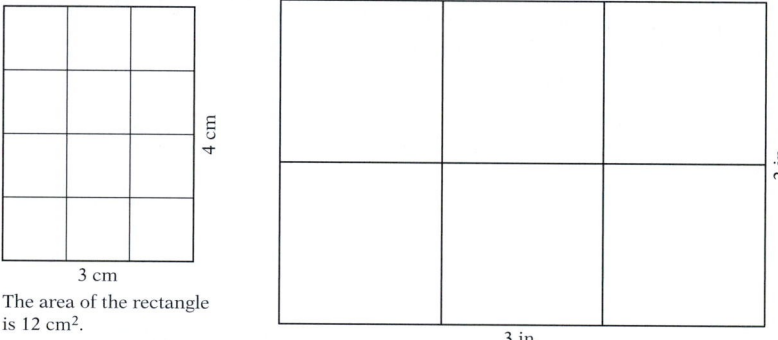

The area of the rectangle is 12 cm^2.

The area of the rectangle is 6 in^2.

Note from the above figures that the area of a rectangle can be found by multiplying the length of the rectangle by its width.

> **Area of a Rectangle**
>
> Let L represent the length and W the width of a rectangle. The area, A, of the rectangle is given by $A = LW$.

HOW TO   Find the area of the rectangle shown at the right.

$A = LW = 11(7) = 77$

● The area is 77 m^2.

In-Class Examples (Objective 7.2B)

1. Find the area of a triangle with a base of 5 cm and a height of 2.6 cm.   6.5 cm^2

2. Find the area of a square with a side of 8.5 ft.
72.25 ft^2

3. Find the area of a rectangle with a length of 37 in. and a width of 15 in.   555 in^2

4. Find the area of a circle with a diameter of 16 in. Round to the nearest hundredth.   201.06 in^2

A square is a rectangle in which all sides are the same length. Therefore, both the length and the width of a square can be represented by s, and $A = LW = s \cdot s = s^2$.

> **Area of a Square**
>
> Let s represent the length of a side of a square. The area, A, of the square is given by $A = s^2$.

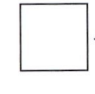

$A = s \cdot s = s^2$

HOW TO Find the area of the square shown at the right.

$A = s^2 = 9^2 = 81$

The area is 81 mi².

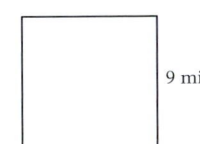

9 mi

Figure $ABCD$ is a parallelogram. BC is the **base,** b, of the parallelogram. AE, perpendicular to the base, is the **height,** h, of the parallelogram.

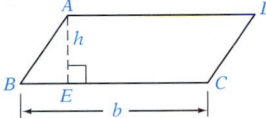

Any side of a parallelogram can be designated as the base. The corresponding height is found by drawing a line segment perpendicular to the base from the opposite side.

A rectangle can be formed from a parallelogram by cutting a right triangle from one end of the parallelogram and attaching it to the other end. The area of the resulting rectangle will equal the area of the original parallelogram.

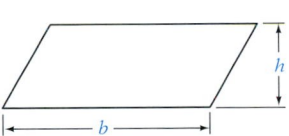

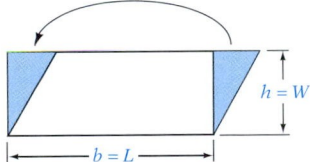

> **Area of a Parallelogram**
>
> Let b represent the length of the base and h the height of a parallelogram. The area, A, of the parallelogram is given by $A = bh$.

HOW TO Find the area of the parallelogram shown at the right.

$A = bh = 12 \cdot 6 = 72$

The area is 72 m².

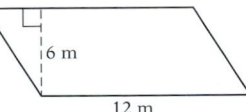

6 m

12 m

Concept Check

The concepts of square units and area are difficult for students. After introducing these ideas, ask students questions such as:

1. Would I measure the distance from Chicago to Boston in miles or square miles?   Miles

2. Is the amount of land cultivated by a gardener measured in feet or square feet?   Square feet

3. How is the size of a state park measured?
Acres or square miles

4. How is the length of a family room measured?
Feet or meters

Optional Student Activity

1. Use graph paper to draw different rectangles, each with a perimeter of 20 units. Investigate only whole number dimensions. What dimensions will result in a rectangle with the greatest possible area? **5 × 5**

2. Ancient Egyptians gave the formula for the area of a circle as $\left(\frac{8}{9}d\right)^2$, where d is the diameter. Does this formula give an area that is less than or greater than the correct formula? **Greater than**

Figure ABC is a triangle. AB is the **base, b,** of the triangle. CD, perpendicular to the base, is the **height, h,** of the triangle.

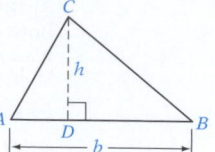

Any side of a triangle can be designated as the base. The corresponding height is found by drawing a line segment perpendicular to the base from the vertex opposite the base.

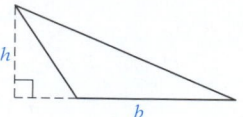

Consider the triangle with base b and height h shown at the right. By extending a line from C parallel to the base AB and equal in length to the base, a parallelogram is formed. The area of the parallelogram is bh and is twice the area of the triangle. Therefore, the area of the triangle is one-half the area of the parallelogram, or $\frac{1}{2}bh$.

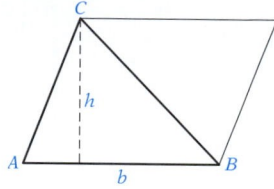

Area of a Triangle

Let b represent the length of the base and h the height of a triangle. The area, A, of the triangle is given by $A = \frac{1}{2}bh$.

HOW TO Find the area of a triangle with a base of 18 cm and a height of 6 cm.

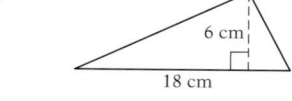

$$A = \frac{1}{2}bh = \frac{1}{2} \cdot 18 \cdot 6 = 54$$

The area is 54 cm².

Figure $ABCD$ is a trapezoid. AB is one **base, b_1,** of the trapezoid, and CD is the other base, b_2. AE, perpendicular to the two bases, is the **height, h.**

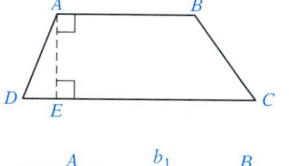

In the trapezoid at the right, the line segment BD divides the trapezoid into two triangles, ABD and BCD. In triangle ABD, b_1 is the base and h is the height. In triangle BCD, b_2 is the base and h is the height. The area of the trapezoid is the sum of the areas of the two triangles.

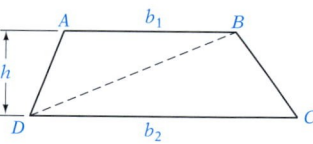

Area of trapezoid $ABCD$ = area of triangle ABD + area of triangle BCD

$$= \frac{1}{2}b_1h + \frac{1}{2}b_2h = \frac{1}{2}h(b_1 + b_2)$$

Area of a Trapezoid

Let b_1 and b_2 represent the lengths of the bases and h the height of a trapezoid. The area, A, of the trapezoid is given by $A = \frac{1}{2}h(b_1 + b_2)$.

HOW TO Find the area of a trapezoid that has bases measuring 15 in. and 5 in. and a height of 8 in.

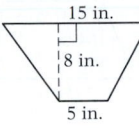

$$A = \frac{1}{2}h(b_1 + b_2)$$

$$= \frac{1}{2} \cdot 8(15 + 5) = 4(20) = 80$$

The area is 80 in^2.

The area of a circle is equal to the product of π and the square of the radius.

$A = \pi r^2$

Area of a Circle

The area, A, of a circle with radius r is given by $A = \pi r^2$.

HOW TO Find the area of a circle that has a radius of 6 cm.

$A = \pi r^2$ • Use the formula for the area of a circle.

$A = \pi(6)^2$ $r = 6$

$A = \pi(36)$

$A = 36\pi$

The exact area of the circle is 36π cm^2.

$A \approx 113.10$ • An approximate measure is found by using the π key on a calculator.

The approximate area of the circle is 113.10 cm^2.

Integrating Technology

To approximate 36π on your calculator, enter

36 × π = .

For your reference, all of the formulas for the perimeter and area of the geometric figures presented in this section are listed in the Chapter Summary, which begins on page 434.

Example 4

The Parks and Recreation Department of a city plans to plant grass seed in a playground that has the shape of a trapezoid, as shown below. Each bag of grass seed will seed 1500 ft². How many bags of grass seed should the department purchase?

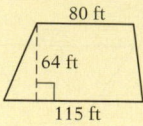

Strategy

To find the number of bags to be purchased:
• Use the formula for the area of a trapezoid to find the area of the playground.
• Divide the area of the playground by the area one bag will seed (1500).

Solution

$A = \dfrac{1}{2}h(b_1 + b_2)$

$A = \dfrac{1}{2} \cdot 64(80 + 115)$

$A = 6240$ • The area of the playground is 6240 ft².

$6240 \div 1500 = 4.16$

Because a portion of a fifth bag is needed, 5 bags of grass seed should be purchased.

Example 5

Find the area of a circle with a diameter of 5 ft. Give the exact measure.

Strategy

To find the area:
• Find the radius of the circle.
• Use the formula for the area of a circle. Leave the answer in terms of π.

Solution

$r = \dfrac{1}{2}d = \dfrac{1}{2}(5) = 2.5$

$A = \pi r^2 = \pi(2.5)^2 = \pi(6.25) = 6.25\pi$

The area of the circle is 6.25π ft².

You Try It 4

An interior designer decides to wallpaper two walls of a room. Each roll of wallpaper will cover 30 ft². Each wall measures 8 ft by 12 ft. How many rolls of wallpaper should be purchased?

Your Strategy

Your solution
7 rolls

You Try It 5

Find the area of a circle with a radius of 11 cm. Round to the nearest hundredth.

Your Strategy

Your solution
380.13 cm²

Solutions on p. S20

7.2 Exercises

Objective A To solve problems involving the perimeter of geometric figures

Section 7.2

Suggested Assignment
Exercises 13–99, odds
More challenging problems:
Exercises 101, 102, 105

Name each polygon.

1.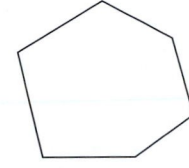

Hexagon

2.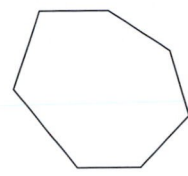

Heptagon

3.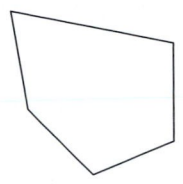

Pentagon

4.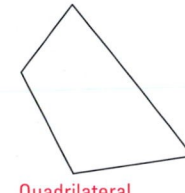

Quadrilateral

Classify the triangle as isosceles, equilateral, or scalene.

5.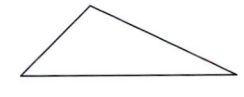

Scalene

6.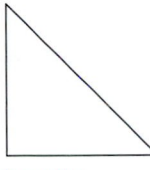

Isosceles

7.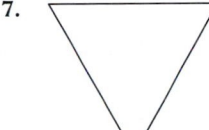

Equilateral

8.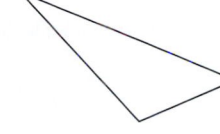

Scalene

Classify the triangle as acute, obtuse, or right.

9.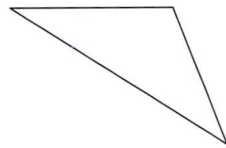

Obtuse

10.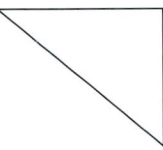

Right

11.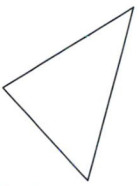

Acute

12.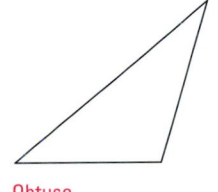

Obtuse

Find the perimeter of the figure.

13.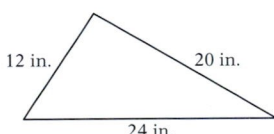
12 in.   20 in.   24 in.

56 in.

14.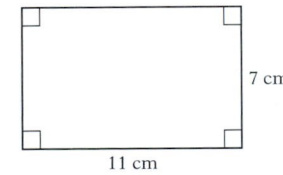
7 cm   11 cm

36 cm

15.
3.5 ft   3.5 ft

14 ft

16.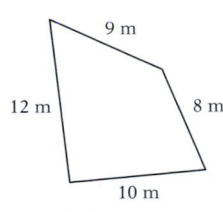
9 m   12 m   8 m   10 m

39 m

17.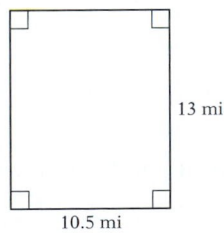
13 mi   10.5 mi

47 mi

18.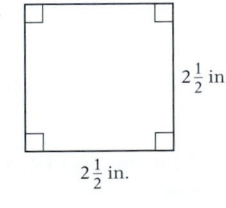
$2\frac{1}{2}$ in. $2\frac{1}{2}$ in.

10 in.

Quick Quiz (Objective 7.2A)

1. Find the perimeter of a square in which each side is 13.5 cm long. **54 cm**

2. Find the perimeter of a rectangle with a length of 3 m and a width of 0.75 m. **7.5 m**

3. Find the circumference of a circle with a radius of 14 in. Round to the nearest hundredth. **87.96 in.**

In Exercises 19–24, find the circumference of the figure.
Give both the exact value and an approximation to the nearest hundredth.

19.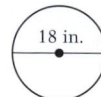

8π cm; 25.13 cm

20.

24π m; 75.40 m

21.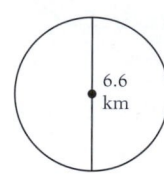

11π mi; 34.56 mi

22.

18π in.; 56.55 in.

23.

17π ft; 53.41 ft

24.

6.6π km; 20.73 km

25. The lengths of the three sides of a triangle are 3.8 cm, 5.2 cm, and 8.4 cm. Find the perimeter of the triangle.
17.4 cm

26. The lengths of the three sides of a triangle are 7.5 m, 6.1 m, and 4.9 m. Find the perimeter of the triangle.
18.5 m

27. The length of each of two sides of an isosceles triangle is $2\frac{1}{2}$ cm. The third side measures 3 cm. Find the perimeter of the triangle.
8 cm

28. The length of each side of an equilateral triangle is $4\frac{1}{2}$ in. Find the perimeter of the triangle.
$13\frac{1}{2}$ in.

29. A rectangle has a length of 8.5 m and a width of 3.5 m. Find the perimeter of the rectangle.
24 m

30. Find the perimeter of a rectangle that has a length of $5\frac{1}{2}$ ft and a width of 4 ft.
19 ft

31. The length of each side of a square is 12.2 cm. Find the perimeter of the square.
48.8 cm

32. Find the perimeter of a square that is 0.5 m on each side.
2 m

33. Find the perimeter of a regular pentagon that measures 3.5 in. on each side.
17.5 in.

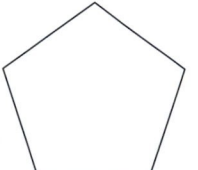

34. What is the perimeter of a regular hexagon that measures 8.5 cm on each side?
51 cm

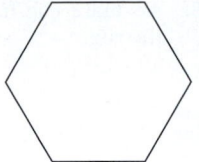

35. Find the circumference of a circle that has a diameter of 1.5 in. Give the exact value.
1.5π in.

36. 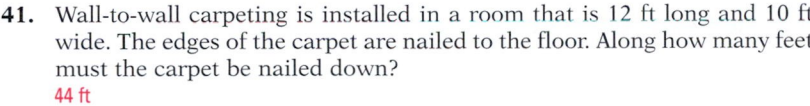 The diameter of a circle is 4.2 ft. Find the circumference of the circle. Round to the nearest hundredth.
13.19 ft

37. The radius of a circle is 36 cm. Find the circumference of the circle. Round to the nearest hundredth.
226.19 cm

38. Find the circumference of a circle that has a radius of 2.5 m. Give the exact value.
5π m

39. How many feet of fencing should be purchased for a rectangular garden that is 18 ft long and 12 ft wide?
60 ft

40. How many meters of binding are required to bind the edge of a rectangular quilt that measures 3.5 m by 8.5 m?
24 m

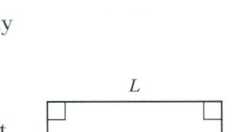

41. Wall-to-wall carpeting is installed in a room that is 12 ft long and 10 ft wide. The edges of the carpet are nailed to the floor. Along how many feet must the carpet be nailed down?
44 ft

42. The length of a rectangular park is 55 yd. The width is 47 yd. How many yards of fencing are needed to surround the park?
204 yd

43. The perimeter of a rectangular playground is 440 ft. If the width is 100 ft, what is the length of the playground?
120 ft

L

100 ft

44. A rectangular vegetable garden has a perimeter of 64 ft. The length of the garden is 20 ft. What is the width of the garden?
12 ft

45. Each of two sides of a triangular banner measures 18 in. If the perimeter of the banner is 46 in., what is the length of the third side of the banner?
10 in.

46. The perimeter of an equilateral triangle is 13.2 cm. What is the length of each side of the triangle?
4.4 cm

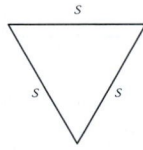

47. The perimeter of a square picture frame is 48 in. Find the length of each side of the frame.
12 in.

48. A square rug has a perimeter of 32 ft. Find the length of each edge of the rug.
8 ft

Solve. For Exercises 49 to 55, round to the nearest hundredth.

49. The circumference of a circle is 8 cm. Find the length of a diameter of the circle.
2.55 cm

50. The circumference of a circle is 15 in. Find the length of a radius of the circle.
2.39 in.

51. Find the length of molding needed to put around a circular table that is 4.2 ft in diameter.
13.19 ft

52. How much binding is needed to bind the edge of a circular rug that is 3 m in diameter?
9.42 m

53. A bicycle tire has a diameter of 24 in. How many feet does the bicycle travel when the wheel makes eight revolutions?
50.27 ft

54. A tricycle tire has a diameter of 12 in. How many feet does the tricycle travel when the wheel makes 12 revolutions?
37.70 ft

55. The distance from the surface of Earth to its center is 6356 km. What is the circumference of Earth?
39,935.93 km

56. Bias binding is to be sewed around the edge of a rectangular tablecloth measuring 72 in. by 45 in. If the bias binding comes in packages containing 15 ft of binding, how many packages of bias binding are needed for the tablecloth?
2 packages

Objective B   **To solve problems involving the area of geometric figures**

Find the area of the figure.

57.
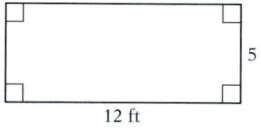
5 ft
12 ft

60 ft²

58.
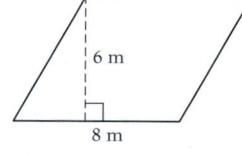
6 m
8 m

48 m²

59.
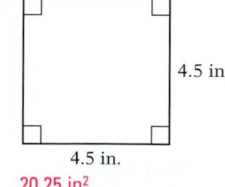
4.5 in.
4.5 in.

20.25 in²

Quick Quiz (Objective 7.2B)

1. Find the area of a triangle with a base of 10 ft and a height of 16 ft.   80 ft²

2. Find the area of a square with a side of 16 cm.   256 cm²

3. Find the area of a rectangle with a length of 64 cm and a width of 22 cm.   1408 cm²

4. Find the area of a circle with a diameter of 26 in. Round to the nearest hundredth.   530.93 in²

60.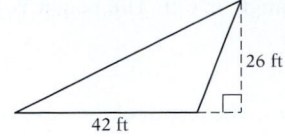

120 in²

61.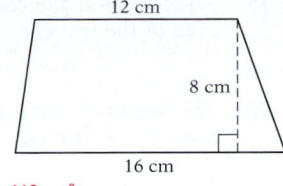

546 ft²

62.

112 cm²

 In Exercises 63–68, find the area of the figure.
Give both the exact value and an approximation to the nearest hundredth.

63.

16π cm²; 50.27 cm²

64.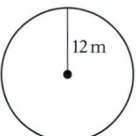

144π m²; 452.39 m²

65.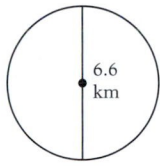

30.25π mi²; 95.03 mi²

66.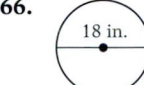

81π in²; 254.47 in²

67.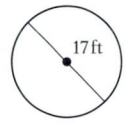

72.25π ft²; 226.98 ft²

68.

10.89π km²; 34.21 km²

 Solve.

69. The length of a side of a square is 12.5 cm. Find the area of the square.
156.25 cm²

70. Each side of a square measures $3\frac{1}{2}$ in. Find the area of the square.

12.25 in²

71. The length of a rectangle is 38 in., and the width is 15 in. Find the area of the rectangle.
570 in²

72. Find the area of a rectangle that has a length of 6.5 m and a width of 3.8 m.
24.7 m²

73. The length of the base of a parallelogram is 16 in., and the height is 12 in. Find the area of the parallelogram.
192 in²

74. The height of a parallelogram is 3.4 m, and the length of the base is 5.2 m. Find the area of the parallelogram.
17.68 m²

75. The length of the base of a triangle is 6 ft. The height is 4.5 ft. Find the area of the triangle.
13.5 ft²

76. The height of a triangle is 4.2 cm. The length of the base is 5 cm. Find the area of the triangle.
10.5 cm²

77. The length of one base of a trapezoid is 35 cm, and the length of the other base is 20 cm. If the height is 12 cm, what is the area of the trapezoid?
330 cm²

78. The height of a trapezoid is 5 in. The bases measure 16 in. and 18 in. Find the area of the trapezoid.
85 in²

79. The radius of a circle is 5 in. Find the area of the circle. Give the exact value.
25π in²

80. The diameter of a circle is 6.5 m. Find the area of the circle. Give the exact value.
10.5625π m²

81. The lens on the Hale telescope at Mount Palomar, California, has a diameter of 200 in. Find its area. Give the exact value.
$10{,}000\pi$ in²

82. An irrigation system waters a circular field that has a 50-foot radius. Find the area watered by the irrigation system. Give the exact value.
2500π ft²

83. Find the area of a rectangular flower garden that measures 14 ft by 9 ft.
126 ft²

84. What is the area of a square patio that measures 8.5 m on each side?
72.25 m²

85. Artificial turf is being used to cover a playing field. If the field is rectangular with a length of 100 yd and a width of 75 yd, how much artificial turf must be purchased to cover the field?
7500 yd²

86. A fabric wall hanging is to fill a space that measures 5 m by 3.5 m. Allowing for 0.1 m of the fabric to be folded back along each edge, how much fabric must be purchased for the wall hanging?
19.24 m²

87. The area of a rectangle is 300 in². If the length of the rectangle is 30 in., what is the width?
10 in.

88. The width of a rectangle is 12 ft. If the area is 312 ft², what is the length of the rectangle?
26 ft

89. The height of a triangle is 5 m. The area of the triangle is 50 m². Find the length of the base of the triangle.
20 m

90. The area of a parallelogram is 42 m². If the height of the parallelogram is 7 m, what is the length of the base?
6 m

91. You plan to stain the wooden deck attached to your house. The deck measures 10 ft by 8 ft. If a quart of stain will cover 50 ft², how many quarts of stain should you buy?
2 qt

92. You want to tile your kitchen floor. The floor measures 12 ft by 9 ft. How many tiles, each a square with side $1\frac{1}{2}$ ft, should you purchase for the job?
48 tiles

93. You are wallpapering two walls of a child's room, one measuring 9 ft by 8 ft and the other measuring 11 ft by 8 ft. The wallpaper costs $18.50 per roll, and each roll of the wallpaper will cover 40 ft². What is the cost to wallpaper the two walls?
$74

94. An urban renewal project involves reseeding a park that is in the shape of a square, 60 ft on each side. Each bag of grass seed costs $5.75 and will seed 1200 ft². How much money should be budgeted for buying grass seed for the park?
$17.25

95. A circle has a radius of 8 in. Find the increase in area when the radius is increased by 2 in. Round to the nearest hundredth.
113.10 in²

96. A circle has a radius of 6 cm. Find the increase in area when the radius is doubled. Round to the nearest hundredth.
339.29 cm²

97. You want to install wall-to-wall carpeting in your living room, which measures 15 ft by 24 ft. If the cost of the carpet you would like to purchase is $15.95 per square yard, what is the cost of the carpeting for your living room? (*Hint:* 9 ft² = 1 yd²)
$638

98. You want to paint the walls of your bedroom. Two walls measure 15 ft by 9 ft, and the other two walls measure 12 ft by 9 ft. The paint you wish to purchase costs $19.98 per gallon, and each gallon will cover 400 ft² of wall. Find the total amount you will spend on paint.
$39.96

99. A walkway 2 m wide surrounds a rectangular plot of grass. The plot is 30 m long and 20 m wide. What is the area of the walkway?
216 m²

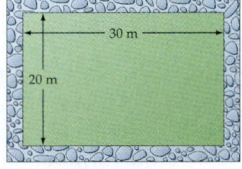

100. Pleated draperies for a window must be twice as wide as the width of the window. Draperies are being made for four windows, each 2 ft wide and 4 ft high. Since the drapes will fall slightly below the windowsill and extra fabric will be needed for hemming the drapes, 1 ft must be added to the height of the window. How much material must be purchased to make the drapes?
80 ft²

APPLYING THE CONCEPTS

101. Find the ratio of the areas of two squares if the ratio of the lengths of their sides is 2:3.
4:9

102. If both the length and the width of a rectangle are doubled, how many times larger is the area of the resulting rectangle?
Four times larger

103. If the formula $C = \pi d$ is solved for π, the resulting equation is $\pi = \dfrac{C}{d}$.
Therefore, π is the ratio of the circumference of a circle to the length of its diameter. Use several circular objects, such as coins, plates, tin cans, and wheels, to show that the ratio of the circumference of each object to its diameter is approximately equal to 3.14.

104. Derive a formula for the area of a circle in terms of the diameter of the circle.
$A = \dfrac{1}{4}\pi d^2$

105. Determine whether the statement is always true, sometimes true, or never true.
  a. Two triangles that have the same perimeter have the same area.
  b. Two rectangles that have the same area have the same perimeter.
  c. If two squares have the same area, then the sides of the squares have the same length.
  d. An equilateral triangle is also an isosceles triangle.
  e. All the radii (plural of radius) of a circle are equal.
  f. All the diameters of a circle are equal.
  a. Sometimes true   **b.** Sometimes true   **c.** Always true   **d.** Always true   **e.** Always true   **f.** Always true

106. Suppose a circle is cut into 16 equal pieces, which are then arranged as shown at the right. The figure formed resembles a parallelogram. What variable expression could describe the base of the parallelogram? What variable could describe its height? Explain how the formula for the area of a circle is derived from this approach.

107. Prepare a report on the history of quilts in the United States. Find examples of quilt patterns that incorporate regular polygons. Use pieces of cardboard to create the shapes needed for one block of one of the quilt patterns you learned about.

108. The **apothem** of a regular polygon is the perpendicular distance from the center of the polygon to a side. Explain how to derive a formula for the area of a regular polygon using the apothem. (*Hint:* Use the formula for the area of a triangle.)

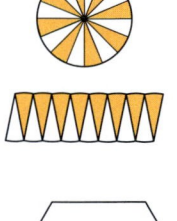
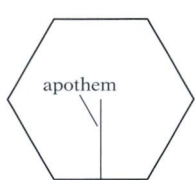
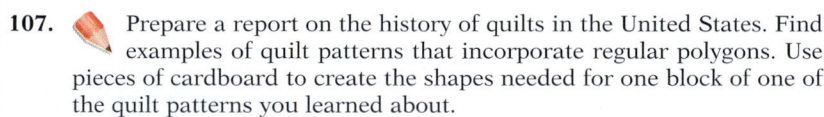

apothem

Answers to Writing Exercises

106. Students need to incorporate the following equations into their explanations: $b = \dfrac{1}{2}C$; $h = r$; $A = bh = \dfrac{1}{2}C \cdot r = \dfrac{1}{2}(\pi d)r = \dfrac{1}{2}\pi(2r)r = \pi r^2$.

107. There are a great number of quilt patterns that incorporate regular polygons—for example, Nine Patch Block, Grandmother's Flower Garden, Hour Glasses, Sunshine and Shadow, Field of Diamonds, and Trip Around the World.

108. Draw line segments from the center of the polygon to each of the vertices. The number of triangles formed is equal to the number of sides of the polygon. The area of the polygon is equal to the sum of the areas of the triangles. The area of each triangle is equal to one-half the product of the length of a side and the apothem. Multiply the area of one triangle by the number of triangles to find the area of the polygon. This can also be stated as "one-half the product of the apothem and the perimeter of the polygon."

7.3 Triangles

Objective A To solve problems using the Pythagorean Theorem

A **right triangle** contains one right angle. The side opposite the right angle is called the **hypotenuse.** The other two sides are called **legs.**

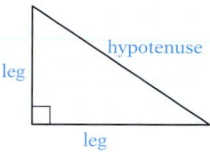

The angles in a right triangle are usually labeled with the capital letters A, B, and C, with C reserved for the right angle. The side opposite angle A is side a, the side opposite angle B is side b, and c is the hypotenuse.

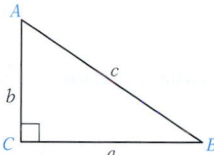

Point of Interest

The first known proof of the Pythagorean Theorem is in a Chinese textbook that dates from 150 B.C. The book is called *Nine Chapters on the Mathematical Art.* The diagram below is from that book and was used in the proof of the theorem.

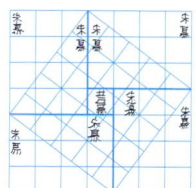

The Greek mathematician Pythagoras is generally credited with the discovery that the square of the hypotenuse of a right triangle is equal to the sum of the squares of the two legs. This is called the **Pythagorean Theorem.**

The figure at the right is a right triangle with legs measuring 3 units and 4 units and a hypotenuse measuring 5 units. Each side of the triangle is also the side of a square. The number of square units in the area of the largest square is equal to the sum of the numbers of square units in the areas of the smaller squares.

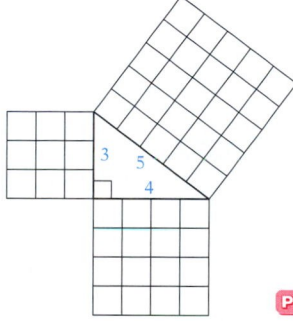

Square of the hypotenuse = sum of the squares of the two legs

$$5^2 = 3^2 + 4^2$$
$$25 = 9 + 16$$
$$25 = 25$$

> **Pythagorean Theorem**
>
> If a and b are the lengths of the legs of a right triangle and c is the length of the hypotenuse, then $c^2 = a^2 + b^2$.

If the lengths of two sides of a right triangle are known, the Pythagorean Theorem can be used to find the length of the third side.

Objective 7.3A

Vocabulary to Review
right triangle [7.2A]

New Vocabulary
hypotenuse
legs

New Theorems
Pythagorean Theorem:
$$c^2 = a^2 + b^2$$

Discuss the Concepts
How can you tell which side is the hypotenuse and which sides are the legs in a right triangle?

Concept Check
(*Note:* For Exercises 2 and 3 below, encourage students to draw a diagram.)

1. Label the right triangle below. Include the right angle symbol, the hypotenuse, and the legs.

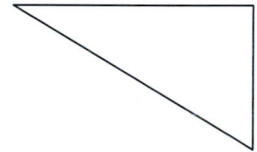

2. The diagonal of a rectangle is a line drawn from one vertex to the opposite vertex. Find the length of the diagonal of a rectangle that is 5 m wide and 11 m long. Round to the nearest tenth.   12.1 m

3. Marta Lightfoot leaves a dock in her sailboat and sails 2.5 mi due east. She then tacks and sails 4 mi due north. The walkie-talkie Marta has on board has a range of 5 mi. Will she be able to call a friend on the dock from her location using the walkie-talkie?   Yes

---

In-Class Examples (Objective 7.3A)

Find the unknown side of the triangle. Round to the nearest thousandth.

1. 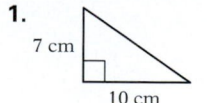   12.207 cm

2. 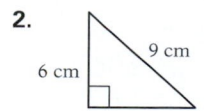   6.708 cm

Optional Student Activity

The numbers 3, 4, and 5 are called a **Pythagorean triple** because they are natural numbers that satisfy the equation of the Pythagorean Theorem.

In each exercise below, determine whether the numbers are a Pythagorean triple.

1. 5, 7, and 9   No

2. 8, 15, and 17   Yes

3. 11, 60, and 61   Yes

4. 28, 45, and 53   Yes

Mathematicians have investigated Pythagorean triples and have found formulas that will generate these triples. One such formula is

$$a = m^2 - n^2$$
$$b = 2mn$$
$$c = m^2 + n^2, \text{ where } m > n$$

For instance, let $m = 2$ and $n = 1$. Then $a = 2^2 - 1^2 = 3$, $b = 2(2)(1) = 4$, and $c = 2^2 + 1^2 = 5$. This is the Pythagorean triple 3, 4, 5.

Find the Pythagorean triple produced by each of the following.

5. $m = 3$ and $n = 1$ 6, 8, 10

6. $m = 5$ and $n = 2$ 20, 21, 29

7. $m = 4$ and $n = 2$ 12, 16, 20

8. $m = 6$ and $n = 1$ 12, 35, 37

Optional Student Activity

The radius of the circle shown below is 3 in. Find the length of a side of the square drawn inside the circle. Round to the nearest tenth.   4.2 in.

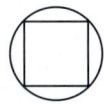

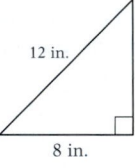

Integrating Technology

The way in which you evaluate the square root of a number depends on the type of calculator you have. Here are two possible keystroke combinations to find $\sqrt{35}$:

35 √ =

or

√ 35 ENTER

The first method is used on many scientific calculators. The second method is used on many graphing calculators.

Consider a right triangle with legs that measure 5 cm and 12 cm. Use the Pythagorean Theorem, with $a = 5$ and $b = 12$, to find the length of the hypotenuse. (If you let $a = 12$ and $b = 5$, the result is the same.)

$$c^2 = a^2 + b^2$$
$$c^2 = 5^2 + 12^2$$
$$c^2 = 25 + 144$$
$$c^2 = 169$$

This equation states that the square of c is 169. Since $13^2 = 169$, $c = 13$ and the length of the hypotenuse is 13 cm. We can find c by taking the square root of 169: $\sqrt{169} = 13$. This suggests the following property.

> **Principal Square Root Property**
>
> If $r^2 = s$, then $r = \sqrt{s}$, and r is called the **square root** of s.

The Principal Square Root Property and its application can be illustrated as follows: Because $5^2 = 25$, $5 = \sqrt{25}$. Therefore, if $c^2 = 25$, $c = \sqrt{25} = 5$.

Recall that numbers whose square roots are integers, such as 25, are perfect squares. If a number is not a perfect square, a calculator can be used to find an approximate square root when a decimal approximation is required.

HOW TO   The length of one leg of a right triangle is 8 in. The hypotenuse is 12 in. Find the length of the other leg. Round to the nearest hundredth.

$$a^2 + b^2 = c^2$$ • Use the Pythagorean Theorem.
$$8^2 + b^2 = 12^2$$ **$a = 8$, $c = 12$**
$$64 + b^2 = 144$$ • Solve for b^2.
$$b^2 = 80$$ **(If you let $b = 8$ and solve for a^2, the result is the same.)**
$$b = \sqrt{80}$$ • Use the Principal Square Root Property.
                     Since $b^2 = 80$, b is the square root of 80.
$$b \approx 8.94$$ • Use a calculator to approximate $\sqrt{80}$.

The length of the other leg is approximately 8.94 in.

Example 1

The two legs of a right triangle measure 12 ft and 9 ft. Find the hypotenuse of the right triangle.

Strategy

To find the hypotenuse, use the Pythagorean Theorem. $a = 12$, $b = 9$

Solution

$$c^2 = a^2 + b^2$$
$$c^2 = 12^2 + 9^2$$
$$c^2 = 144 + 81$$
$$c^2 = 225$$
$$c = \sqrt{225}$$
$$c = 15$$

The length of the hypotenuse is 15 ft.

You Try It 1

The hypotenuse of a right triangle measures 6 m, and one leg measures 2 m. Find the measure of the other leg. Round to the nearest hundredth.

Your Strategy

Your solution

5.66 m

Solution on p. S20

Objective B   **To solve problems involving similar triangles**

Similar objects have the same shape but not necessarily the same size. A tennis ball is similar to a basketball. A model ship is similar to an actual ship.

Similar objects have corresponding parts; for example, the rudder on the model ship corresponds to the rudder on the actual ship. The relationship between the sizes of each of the corresponding parts can be written as a ratio, and each ratio will be the same. If the rudder on the model ship is $\frac{1}{100}$ the size of the rudder on the actual ship, then the model wheelhouse is $\frac{1}{100}$ the size of the actual wheelhouse, the width of the model is $\frac{1}{100}$ the width of the actual ship, and so on.

The two triangles ABC and DEF shown at the right are similar. Side $\overline{AB}$ corresponds to side $\overline{DE}$, side $\overline{BC}$ corresponds to side $\overline{EF}$, and side $\overline{AC}$ corresponds to side $\overline{DF}$. The ratios of corresponding sides are equal.

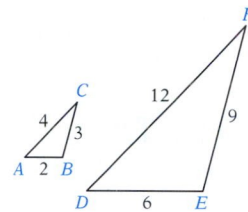

$$\frac{AB}{DE} = \frac{2}{6} = \frac{1}{3}, \frac{BC}{EF} = \frac{3}{9} = \frac{1}{3}, \text{ and } \frac{AC}{DF} = \frac{4}{12} = \frac{1}{3}$$

Since the ratios of corresponding sides are equal, three proportions can be formed.

$$\frac{AB}{DE} = \frac{BC}{EF}, \frac{AB}{DE} = \frac{AC}{DF}, \text{ and } \frac{BC}{EF} = \frac{AC}{DF}$$

The corresponding angles in similar triangles are equal. Therefore,

$$\angle A = \angle D, \angle B = \angle E, \text{ and } \angle C = \angle F$$

Triangles ABC and DEF at the right are similar triangles. AH and DK are the heights of the triangles. The ratio of the heights of similar triangles equals the ratio of corresponding sides.

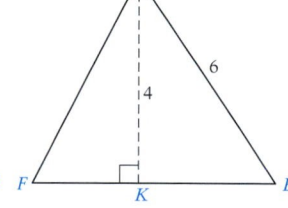

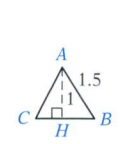

Ratio of corresponding sides $= \frac{1.5}{6} = \frac{1}{4}$

Ratio of heights $= \frac{1}{4}$

> **Properties of Similar Triangles**
>
> For similar triangles, the ratios of corresponding sides are equal. The ratio of corresponding heights is equal to the ratio of corresponding sides.

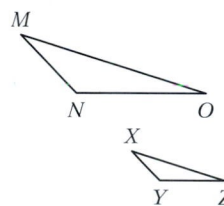

In-Class Examples (Objective 7.3B)

1. Find the ratio of corresponding sides for the similar triangles.

$\frac{2}{3}$

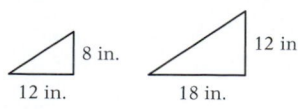

2. Triangles *ABC* and *DEF* are similar. Find side *DE*.
  6.75 m

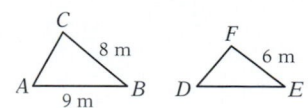

Optional Student Activity

1. For each pair of similar triangles, write the ratio of corresponding sides in simplest form.

   a. $\frac{1}{2}$

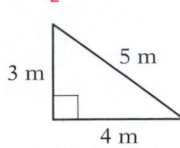
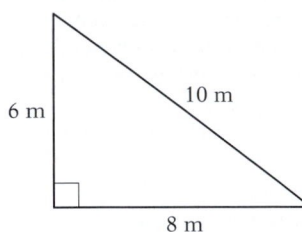

   b. $\frac{1}{3}$

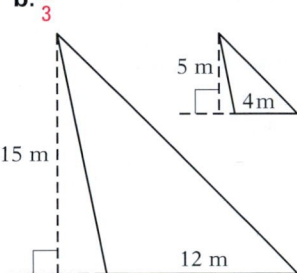

   c. $\frac{2}{3}$

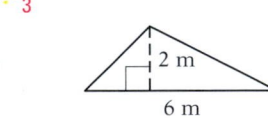
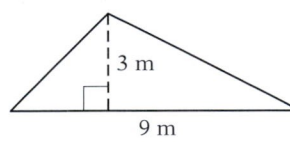

2. Find the area of each triangle in Activity 1.
   a. 6 m² and 24 m²; **b.** 10 m² and 90 m²; **c.** 6 m² and $\frac{27}{2}$ m²

3. For each pair of similar triangles in Activity 1, write the ratio of the areas in simplest form.
   a. $\frac{1}{4}$; **b.** $\frac{1}{9}$; **c.** $\frac{4}{9}$

4. Use your results from Activities 1 and 3 to make a conjecture about the ratio of the area of two similar triangles.
   The ratio of the areas is the square of the ratio of their corresponding sides.

Point of Interest

Many mathematicians have studied similar objects. Thales of Miletus (c. 624 B.C. – 547 B.C.) discovered that he could determine the heights of pyramids and other large objects by measuring a small object and the length of its shadow and then making use of similar triangles.

HOW TO The two triangles at the right are similar triangles. Find the length of side *EF*. Round to the nearest tenth.

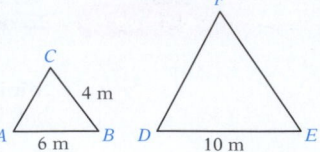

$$\frac{EF}{BC} = \frac{DE}{AB}$$

$$\frac{EF}{4} = \frac{10}{6}$$

$$6(EF) = 4(10)$$

$$6(EF) = 40$$

$$EF \approx 6.7$$

- The triangles are similar, so the ratios of corresponding sides are equal.

The length of side *EF* is approximately 6.7 m.

Example 2

Triangles *ABC* and *DEF* are similar. Find *FG*, the height of triangle *DEF*.

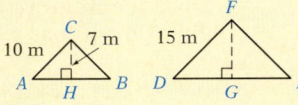

Strategy
To find *FG*, write a proportion using the fact that, in similar triangles, the ratio of corresponding sides equals the ratio of corresponding heights. Solve the proportion for *FG*.

Solution

$$\frac{AB}{DE} = \frac{CH}{FG}$$

$$\frac{8}{12} = \frac{4}{FG}$$

$$8(FG) = 12(4)$$

$$8(FG) = 48$$

$$FG = 6$$

The height *FG* of triangle *DEF* is 6 cm.

You Try It 2

Triangles *ABC* and *DEF* are similar. Find *FG*, the height of triangle *DEF*.

Your Strategy

Your solution
10.5 m

Solution on p. S20

Objective C To determine whether two triangles are congruent

Congruent objects have the same shape *and* the same size.

The two triangles at the right are congruent. They have the same size.

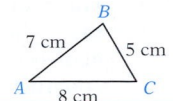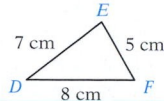

Congruent and similar triangles differ in that the corresponding sides and angles of congruent triangles must be equal, whereas for similar triangles, corresponding angles are equal, but corresponding sides are not necessarily the same length.

The three major rules used to determine whether two triangles are congruent are given below.

> **Side-Side-Side Rule (SSS)**
>
> Two triangles are congruent if the three sides of one triangle equal the corresponding three sides of a second triangle.

In the triangles at the right, $AC = DE$, $AB = EF$, and $BC = DF$. The corresponding sides of triangles ABC and DEF are equal. The triangles are congruent by the SSS Rule.

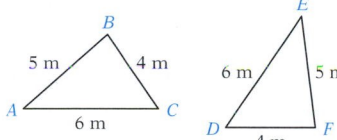

> **Side-Angle-Side Rule (SAS)**
>
> If two sides and the included angle of one triangle are equal to two sides and the included angle of a second triangle, the two triangles are congruent.

In the two triangles at the right, $AB = EF$, $AC = DE$, and $\angle BAC = \angle DEF$. The triangles are congruent by the SAS Rule.

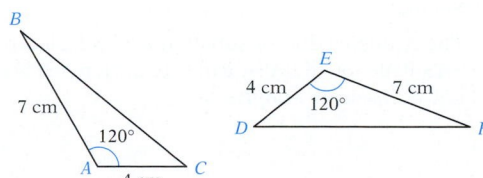

Objective 7.3C

Vocabulary to Review
similar objects   [7.3B]

New Vocabulary
congruent objects

New Rules
Side-Side-Side Rule: SSS
Side-Angle-Side Rule: SAS
Angle-Side-Angle Rule: ASA

Discuss the Concepts
1. What is the difference between similar objects and congruent objects?
2. The text discusses three rules to determine whether two triangles are congruent: the SSS Rule, the SAS Rule, and the ASA Rule. Are two triangles congruent if all three angles of one triangle are equal to all three angles of the second triangle?
3. What does the phrase "the included angle" mean?

In-Class Examples (Objective 7.3C)

1. Given triangle *ABC* and triangle *DEF*, do the conditions $\angle B = \angle F$, $AB = DF$, and $BC = EF$ guarantee that triangle *ABC* is congruent to triangle *DEF*? If they are congruent, by what rule are they congruent?   Yes, SAS Rule

2. Given triangle *ABC* and triangle *DEF*, do the conditions $\angle B = \angle F$, $\angle A = \angle E$, and $\angle C = \angle D$ guarantee that triangle *ABC* is congruent to triangle *DEF*? If they are congruent, by what rule are they congruent?   No

Concept Check

1. Is it possible for two circles not to be similar?   No

2. Is it possible for two circles not to be congruent?   Yes

3. Use the SSS Rule to draw two congruent triangles.
   Check that students have labeled all three sides of each triangle.

4. Use the SAS Rule to draw two congruent triangles.
   Check that students have labeled two sides and the included angle on each triangle.

5. Draw two triangles, each with a 90° angle and sides that measure 3 in. and 4 in., that are not congruent.

Optional Student Activity

In the figure below, triangles *ADB*, *CDA*, and *BDC* are congruent triangles. Find the measure of ∠*DAC*.   30°

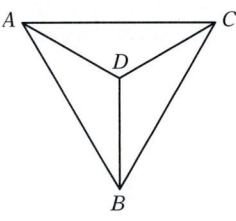

Angle-Side-Angle Rule (ASA)

If two angles and the included side of one triangle are equal to two angles and the included side of a second triangle, the two triangles are congruent.

For triangles *ABC* and *DEF* at the right, ∠*A* = ∠*F*, ∠*C* = ∠*E*, and *AC* = *EF*. The triangles are congruent by the ASA Rule.

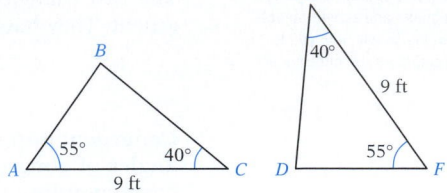

HOW TO   Given triangle *PQR* and triangle *MNO*, do the conditions ∠*P* = ∠*O*, ∠*Q* = ∠*M*, and *PQ* = *MO* guarantee that triangle *PQR* is congruent to triangle *MNO*?

Draw a sketch of the two triangles and determine whether one of the rules for congruence is satisfied.

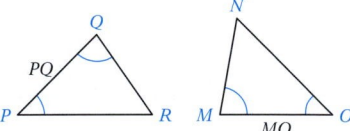

Because two angles and the included side of one triangle are equal to two angles and the included side of the second triangle, the triangles are congruent by the ASA Rule.

Example 3

In the figure below, is triangle *ABC* congruent to triangle *DEF*?

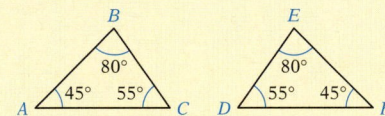

Strategy

To determine whether the triangles are congruent, determine whether one of the rules for congruence is satisfied.

Solution

The triangles do not satisfy the SSS Rule, the SAS Rule, or the ASA Rule. The triangles are not necessarily congruent.

You Try It 3

In the figure below, is triangle *PQR* congruent to triangle *MNO*?

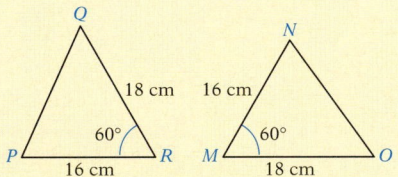

Your Strategy

Your solution
Yes, by the SAS Rule

Solution on p. S20

7.3 Exercises

Objective A   **To solve problems using the Pythagorean Theorem**

Suggested Assignment
Exercises 1–43, odds
More challenging problems:
    Exercises 44, 45

 Find the unknown side of the triangle. Round to the nearest tenth.

1.
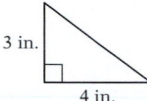
3 in.
4 in.

5 in.

2.
5 in.
12 in.

13 in.

3.
5 cm
7 cm

8.6 cm

4.
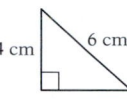
7 cm
9 cm

11.4 cm

5.
15 ft
10 ft

11.2 ft

6.
20 ft
18 ft

8.7 ft

7.
4 cm   6 cm

4.5 cm

8.
9 m   12 m

7.9 m

9.
9 yd
9 yd

12.7 yd

 Solve. Round to the nearest tenth.

10. A ladder 8 m long is leaning against a building. How high on the building will the ladder reach when the bottom of the ladder is 3 m from the building?
7.4 m

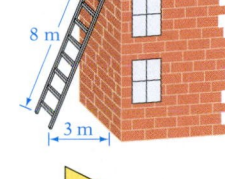
8 m
3 m

11. Find the distance between the centers of the holes in the metal plate.
8.5 cm

3 cm
8 cm

12. If you travel 18 mi east and then 12 mi north, how far are you from your starting point?
21.6 mi

13. Find the perimeter of a right triangle with legs that measure 5 cm and 9 cm.
24.3 cm

14. Find the perimeter of a right triangle with legs that measure 6 in. and 8 in.
24 in.

Quick Quiz (Objective 7.3A)

Find the unknown side of the triangle. Round to the nearest thousandth.

1.
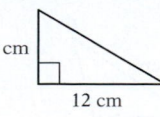
7 cm
12 cm

13.892 cm

2.
14 mi
11 mi

8.660 mi

Objective B **To solve problems involving similar triangles**

Find the ratio of corresponding sides for the similar triangles.

15.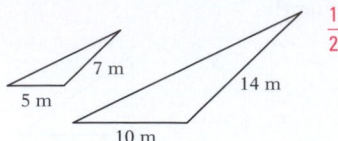
$\dfrac{1}{2}$

16.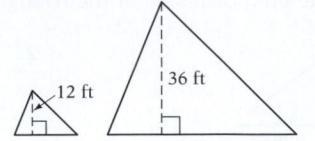
$\dfrac{1}{3}$

17.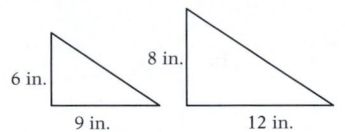
$\dfrac{3}{4}$

18.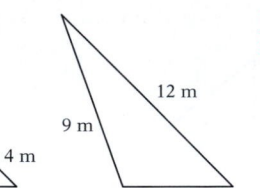
$\dfrac{1}{3}$

In Exercises 19–28, triangles *ABC* and *DEF* are similar triangles.
Solve and round to the nearest tenth.

19. Find side *DE*.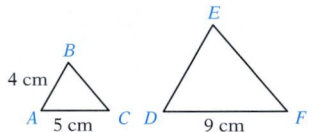
7.2 cm

20. Find side *DE*.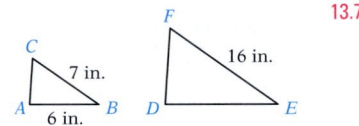
13.7 in.

21. Find the height of triangle *DEF*.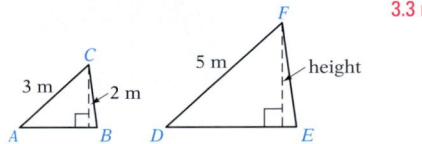
3.3 m

22. Find the height of triangle *ABC*.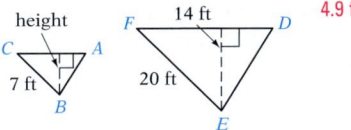
4.9 ft

23. Find the perimeter of triangle *ABC*.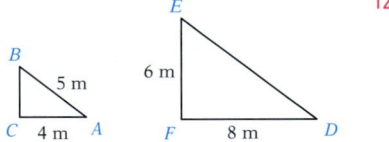
12 m

24. Find the perimeter of triangle *DEF*.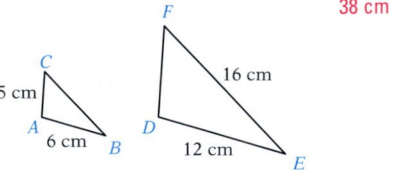
38 cm

25. Find the perimeter of triangle *ABC*.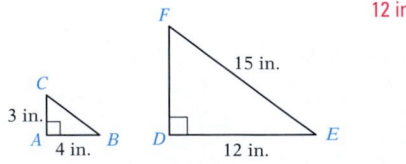
12 in.

26. Find the area of triangle *DEF*.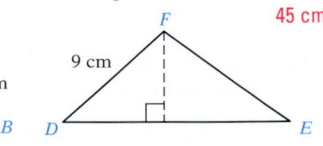
45 cm²

Quick Quiz (Objective 7.3B)

1. Find the ratio of corresponding sides for the
similar triangles.

$\dfrac{3}{4}$

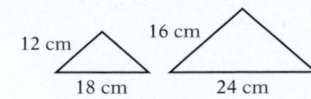

2. Triangles *ABC* and *DEF* are similar. Find side *AC*.
4 cm

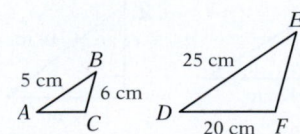

27. Find the area of triangle *ABC*.

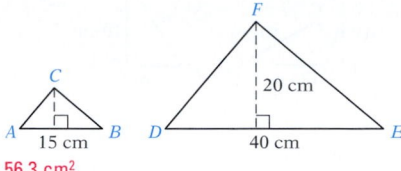

56.3 cm²

28. Find the area of triangle *DEF*.

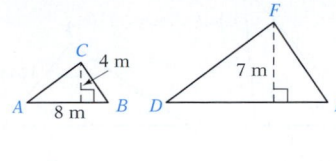

49 m²

The sun's rays, objects on Earth, and the shadows cast by them form similar triangles. Use this fact to solve Exercises 29–32.

29. Find the height of the flagpole.

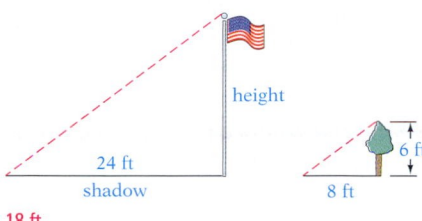

18 ft

30. Find the height of the flagpole.

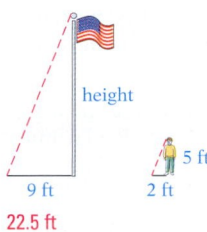

22.5 ft

31. Find the height of the building.

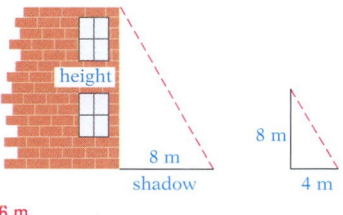

16 m

32. Find the height of the building.

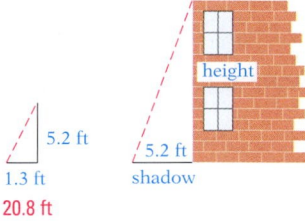

20.8 ft

Objective C   **To determine whether two triangles are congruent**

In Exercises 33–38, determine whether the two triangles are congruent. If they are congruent, state by what rule they are congruent.

33.

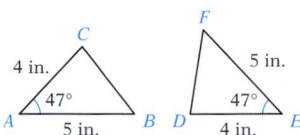

Yes, SAS Rule

34.

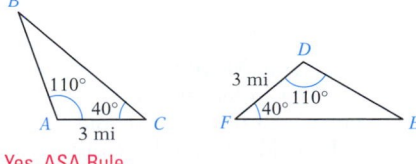

Yes, ASA Rule

35.

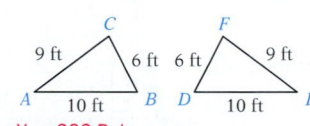

Yes, SSS Rule

36.

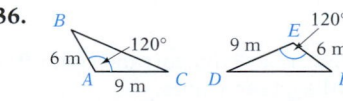

Yes, SAS Rule

Quick Quiz (Objective 7.3C)

1. Given triangle *ABC* and triangle *DEF*, do the conditions ∠*A* = ∠*F*, *AB* = *DF*, and ∠*B* = ∠*D* guarantee that triangle *ABC* is congruent to triangle *DEF*? If they are congruent, by what rule are they congruent?   **Yes, ASA Rule**

2. Given triangle *ABC* and triangle *DEF*, do the conditions *AB* = *EF*, *CA* = *DE*, and *BC* = *FD* guarantee that triangle *ABC* is congruent to triangle *DEF*? If they are congruent, by what rule are they congruent?   **Yes, SSS Rule**

37.

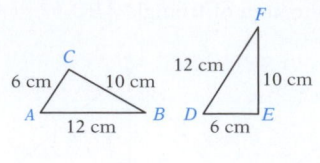

Yes, ASA Rule

38.

Yes, SSS Rule

39. Given triangle *ABC* and triangle *DEF*, do the conditions $\angle C = \angle E$, $AC = EF$, and $BC = DE$ guarantee that triangle *ABC* is congruent to triangle *DEF*? If they are congruent, by what rule are they congruent?
Yes, SAS Rule

40. Given triangle *PQR* and triangle *MNO*, do the conditions $PR = NO$, $PQ = MO$, and $QR = MN$ guarantee that triangle *PQR* is congruent to triangle *MNO*? If they are congruent, by what rule are they congruent?
Yes, SSS Rule

41. Given triangle *LMN* and triangle *QRS*, do the conditions $\angle M = \angle S$, $\angle N = \angle Q$, and $\angle L = \angle R$ guarantee that triangle *LMN* is congruent to triangle *QRS*? If they are congruent, by what rule are they congruent?
No

42. Given triangle *DEF* and triangle *JKL*, do the conditions $\angle D = \angle K$, $\angle E = \angle L$, and $DE = KL$ guarantee that triangle *DEF* is congruent to triangle *JKL*? If they are congruent, by what rule are they congruent?
Yes, ASA Rule

43. Given triangle *ABC* and triangle *PQR*, do the conditions $\angle B = \angle P$, $BC = PQ$, and $AC = QR$ guarantee that triangle *ABC* is congruent to triangle *PQR*? If they are congruent, by what rule are they congruent?
No

APPLYING THE CONCEPTS

44. Determine whether the statement is always true, sometimes true, or never true.
  a. If two angles of one triangle are equal to two angles of a second triangle, then the triangles are similar triangles.
  b. Two isosceles triangles are similar triangles.
  c. Two equilateral triangles are similar triangles.
  a. Always true   **b.** Sometimes true   **c.** Always true

45. **Home Maintenance** You need to clean the gutters of your home. The gutters are 24 ft above the ground. For safety, the distance a ladder reaches up a wall should be four times the distance from the bottom of the ladder to the base of the side of the house. Therefore, the ladder must be 6 ft from the base of the house. Will a 25-foot ladder be long enough to reach the gutters? Explain how you determined your answer.

46. What is a Pythagorean triple? Provide at least three examples of Pythagorean triples.

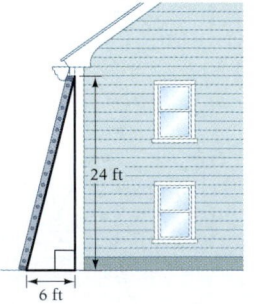

24 ft

6 ft

Answers to Writing Exercises

45. To determine if a 25-foot ladder is long enough to reach 24 ft up the side of the home when the bottom of the ladder is 6 ft from the base of the house, use the Pythagorean Theorem to find the hypotenuse of a right triangle with legs that measure 24 ft and 6 ft.

$$c^2 = a^2 + b^2$$
$$c^2 = 24^2 + 6^2$$
$$c^2 = 576 + 36$$
$$c^2 = 612$$
$$c \approx 24.74$$

Compare the length of the hypotenuse with 25. If the hypotenuse is shorter than 25 ft, the ladder will reach the gutters. If the hypotenuse is longer than 25 feet, the ladder will not reach the gutters.

$$24.74 < 25$$

The hypotenuse is shorter than 25 ft. The ladder will reach the gutters.

46. A Pythagorean triple is a set of three numbers, *a, b,* and *c,* that satisfies $a^2 + b^2 = c^2$. The three numbers 5, 12, and 13 form a Pythagorean triple because $5^2 + 12^2 = 13^2$. Other examples of Pythagorean triples include 3, 4, and 5; 6, 8, and 10; 8, 15, and 17; 20, 21, and 29; 9, 40, and 41.

7.4 Solids

Objective A — To solve problems involving the volume of a solid

Geometric solids are figures in space. Five common geometric solids are the rectangular solid, the sphere, the cylinder, the cone, and the pyramid.

A **rectangular solid** is one in which all six sides, called **faces**, are rectangles. The variable L is used to represent the length of a rectangular solid, W its width, and H its height.

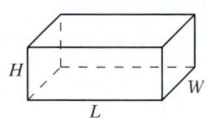

A **sphere** is a solid in which all points are the same distance from point O, called the **center** of the sphere. The **diameter**, d, of a sphere is a line across the sphere going through point O. The **radius**, r, is a line from the center to a point on the sphere. AB is a diameter and OC is a radius of the sphere shown at the right.

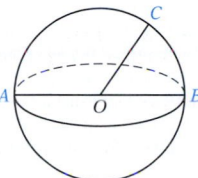

The most common cylinder, called a **right circular cylinder**, is one in which the bases are circles and are perpendicular to the height of the cylinder. The variable r is used to represent the radius of a base of a cylinder, and h represents the height. In this text, only right circular cylinders are discussed.

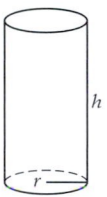

A **right circular cone** is obtained when one base of a right circular cylinder is shrunk to a point, called the **vertex**, V. The variable r is used to represent the radius of the base of the cone, and h represents the height. The variable l is used to represent the **slant height**, which is the distance from a point on the circumference of the base to the vertex. In this text, only right circular cones are discussed.

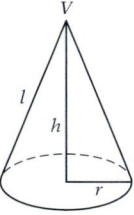

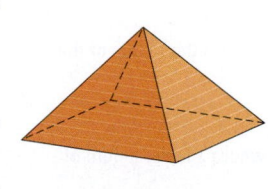

The base of a **regular pyramid** is a regular polygon, and the sides are isosceles triangles. The height, h, is the distance from the vertex, V, to the base and is perpendicular to the base. The variable l is used to represent the **slant height**, which is the height of one of the isosceles triangles on the face of the pyramid. The regular square pyramid at the right has a square base. This is the only type of pyramid discussed in this text.

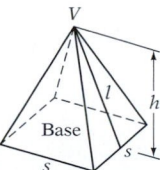

Objective 7.4A

New Vocabulary
geometric solids
rectangular solid
face
sphere
center of a sphere
diameter of a sphere
radius of a sphere
right circular cylinder
right circular cone
vertex
height
regular pyramid
slant height
cube
volume

New Symbols
in³ (cubic inches)
cm³ (cubic centimeters)
ft³ (cubic feet)
m³ (cubic meters)

New Formulas
Volume of a rectangular solid:
$V = LWH$
Volume of a cube:
$V = s^3$
Volume of a sphere:
$V = \frac{4}{3}\pi r^3$
Volume of a right circular cylinder:
$V = \pi r^2 h$
Volume of a right circular cone:
$V = \frac{1}{3}\pi r^2 h$
Volume of a regular square pyramid:
$V = \frac{1}{3}s^2 h$

In-Class Examples (Objective 7.4A)

Solve. Round to the nearest hundredth.

1. Find the volume of a rectangular solid with a length of 6 m, a width of 400 cm, and a height of 4.5 m.   108 m³

2. Find the volume of a cube with a side of 5 ft 3 in. Express your answer in terms of cubic feet.
144.70 ft³

3. Find the volume of a sphere with a radius of 6 mm.
904.78 mm³

4. Find the volume of a cylinder with a radius of 15 cm and a height of 14 cm.   9896.02 cm³

Discuss the Concepts

The difficulty students have distinguishing linear measure from square measure is compounded with cubic measure. Ask students to give examples of things that would be measured in, for instance, feet, square feet, and cubic feet—for example, the length of a room, the area of the floor, and the volume of air in the room. Here are two more examples.

a. The distance across a lake, the area of the surface of the lake, and the volume of water in the lake

b. The length of a driveway, the area of the driveway that needs to be snowplowed, and the volume of asphalt used to pave the driveway

Concept Check

1. Indicate which of the following are rectangular solids: a juice box, a milk carton, a can of soup, a compact disk, the plastic container a compact disk is packaged in.

   A juice box and the plastic container a compact disk is packaged in

2. Indicate which of the following units could *not* be used to measure the volume of a cylinder: ft³, m³, yd², cm³, mi.   yd², mi

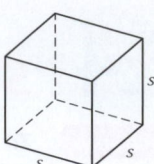

Point of Interest

Originally, the human body was used as the standard of measure. A mouthful was used as a unit of measure in ancient Egypt; it was later referred to as a *half jigger*. In French, the word for *inch* is *pouce*, which means thumb. A *span* was the distance from the tip of the outstretched thumb to the tip of the little finger. The *cubit* referred to the distance from the elbow to the end of the fingers. A *fathom* was the distance from the tip of the fingers on one hand to the tip of the fingers on the other hand when standing with arms fully extended out from the sides. The *hand*, where 1 hand = 4 inches, is still used today to measure horses.

A **cube** is a special type of rectangular solid. Each of the six faces of a cube is a square. The variable *s* is used to represent the length of one side of a cube.

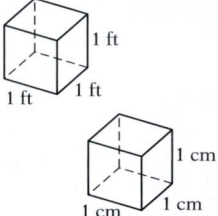

Volume is a measure of the amount of space occupied by a geometric solid. Volume can be used to describe the amount of trash in a landfill, the amount of concrete poured for the foundation of a house, or the amount of water in a town's reservoir.

A cube that is 1 ft on each side has a volume of 1 cubic foot, which is written 1 ft³. A cube that measures 1 cm on each side has a volume of 1 cubic centimeter, written 1 cm³.

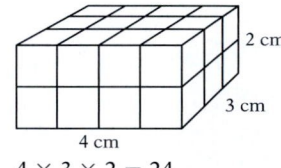

The volume of a solid is the number of cubes that are necessary to exactly fill the solid. The volume of the rectangular solid at the right is 24 cm³ because it will hold exactly 24 cubes, each 1 cm on a side. Note that the volume can be found by multiplying the length times the width times the height.

$4 \times 3 \times 2 = 24$

The formulas for the volumes of the geometric solids described above are given below.

Volumes of Geometric Solids

The volume, *V*, of a **rectangular solid** with length *L*, width *W*, and height *H* is given by $V = LWH$.

The volume, *V*, of a **cube** with side *s* is given by $V = s^3$.

The volume, *V*, of a **sphere** with radius *r* is given by $V = \frac{4}{3}\pi r^3$.

The volume, *V*, of a **right circular cylinder** is given by $V = \pi r^2 h$, where *r* is the radius of the base and *h* is the height.

The volume, *V*, of a **right circular cone** is given by $V = \frac{1}{3}\pi r^2 h$, where *r* is the radius of the circular base and *h* is the height.

The volume, *V*, of a **regular square pyramid** is given by $V = \frac{1}{3}s^2 h$, where *s* is the length of a side of the base and *h* is the height.

HOW TO   Find the volume of a sphere with a diameter of 6 in.

$r = \dfrac{1}{2}d = \dfrac{1}{2}(6) = 3$ • First find the radius of the sphere.

$V = \dfrac{4}{3}\pi r^3$ • Use the formula for the volume of a sphere.

$V = \dfrac{4}{3}\pi (3)^3$

$V = \dfrac{4}{3}\pi (27)$

$V = 36\pi$ • The exact volume of the sphere is 36π in³.

$V \approx 113.10$ • An approximate measure can be found by using the π key on a calculator.

Integrating Technology

To approximate 36π on your calculator, enter 36 × π = .

The approximate volume is 113.10 in³.

Example 1

The length of a rectangular solid is 5 m, the width is 3.2 m, and the height is 4 m. Find the volume of the solid.

Strategy

To find the volume, use the formula for the volume of a rectangular solid. $L = 5$, $W = 3.2$, $H = 4$

Solution

$V = LWH = 5(3.2)(4) = 64$

The volume of the rectangular solid is 64 m³.

Example 2

The radius of the base of a cone is 8 cm. The height is 12 cm. Find the volume of the cone. Round to the nearest hundredth.

Strategy

To find the volume, use the formula for the volume of a cone. An approximation is asked for; use the π key on a calculator. $r = 8$, $h = 12$

Solution

$V = \dfrac{1}{3}\pi r^2 h$

$V = \dfrac{1}{3}\pi (8)^2(12) = \dfrac{1}{3}\pi (64)(12) = 256\pi \approx 804.25$

The volume is approximately 804.25 cm³.

You Try It 1

Find the volume of a cube that measures 2.5 m on a side.

Your Strategy

Your solution
15.625 m³

You Try It 2

The diameter of the base of a cylinder is 8 ft. The height of the cylinder is 22 ft. Find the exact volume of the cylinder.

Your Strategy

Your solution
352π ft³

Solutions on p. S20

Optional Student Activity

1. A foot is what fraction of a yard?

   $\dfrac{1}{3}$

2. A square foot is what fraction of a square yard?

   $\dfrac{1}{9}$

3. A cubic foot is what fraction of a cubic yard?

   $\dfrac{1}{27}$

4. An inch is what fraction of a foot?

   $\dfrac{1}{12}$

5. A square inch is what fraction of a square foot?

   $\dfrac{1}{144}$

6. A cubic inch is what fraction of a cubic foot?

   $\dfrac{1}{1728}$

7. A centimeter is what fraction of a meter?

   $\dfrac{1}{100}$

8. A square centimeter is what fraction of a square meter?

   $\dfrac{1}{10,000}$

9. A cubic centimeter is what fraction of a cubic meter?

   $\dfrac{1}{1,000,000}$

Objective 7.4B

Vocabulary to Review
area   [7.2B]
rectangular solid   [7.4A]
cube   [7.4A]
sphere   [7.4A]
right circular cylinder   [7.4A]
right circular cone   [7.4A]
regular pyramid   [7.4A]

New Vocabulary
surface area

New Formulas
Surface area of a rectangular
   solid:
   $SA = 2LW + 2LH + 2WH$
Surface area of a cube:
   $SA = 6s^2$
Surface area of a sphere:
   $SA = 4\pi r^2$
Surface area of a right circular
   cylinder:
   $SA = 2\pi r^2 + 2\pi rh$
Surface area of a right circular
   cone:
   $SA = \pi r^2 + \pi rl$
Surface area of a regular
   pyramid:
   $SA = s^2 + 2sl$

Objective B   To solve problems involving the
surface area of a solid

The **surface area** of a solid is the total area on the surface of the solid. Suppose you want to cover a geometric solid with wallpaper. The amount of wallpaper needed is equal to the surface area of the figure.

When a rectangular solid is cut open and flattened out, each face is a rectangle. The surface area, SA, of the rectangular solid is the sum of the areas of the six rectangles:

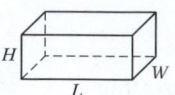

$$SA = LW + LH + WH + LW + WH + LH$$

which simplifies to

$$SA = 2LW + 2LH + 2WH$$

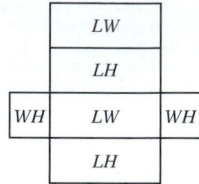

The surface area of a cube is the sum of the areas of the six faces of the cube. The area of each face is s^2. Therefore, the surface area, SA, of a cube is given by the formula $SA = 6s^2$.

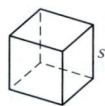

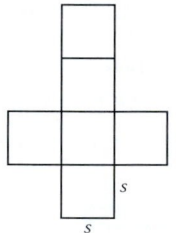

When a cylinder is cut open and flattened out, the top and bottom of the cylinder are circles. The side of the cylinder flattens out to a rectangle. The length of the rectangle is the circumference of the base, which is $2\pi r$; the width is h, the height of the cylinder. Therefore, the area of the rectangle is $2\pi rh$. The area of each circle is πr^2. The surface area, SA, of the cylinder is

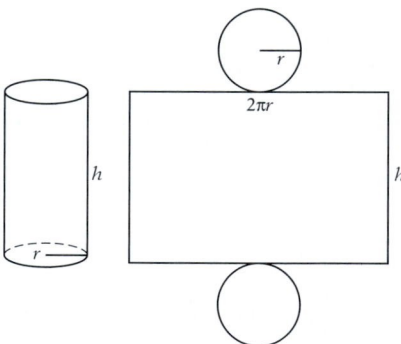

$$SA = \pi r^2 + 2\pi rh + \pi r^2$$

which simplifies to

$$SA = 2\pi r^2 + 2\pi rh$$

In-Class Examples (Objective 7.4B)

Solve. Round to the nearest hundredth.

1. Find the surface area of a cube that measures
2.5 in. on a side.
37.5 in²

2. Find the surface area of a sphere that has a
diameter of 8 cm.
201.06 cm²

3. Find the surface area of a cone with a slant height
of 3 in. The radius of the base is 2 in.
31.42 in²

The surface area of a pyramid is the area of the base plus the area of the four isosceles triangles. A side of the square base is s; therefore, the area of the base is s^2. The slant height, l, is the height of each triangle, and s is the base of each triangle. The surface area, SA, of a pyramid is

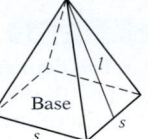

$$SA = s^2 + 4\left(\frac{1}{2}sl\right)$$

which simplifies to

$$SA = s^2 + 2sl$$

Formulas for the surface areas of geometric solids are given below.

Surface Areas of Geometric Solids

The surface area, SA, of a **rectangular solid** with length L, width W, and height H is given by $SA = 2LW + 2LH + 2WH$.

The surface area, SA, of a **cube** with side s is given by $SA = 6s^2$.

The surface area, SA, of a **sphere** with radius r is given by $SA = 4\pi r^2$.

The surface area, SA, of a **right circular cylinder** is given by $SA = 2\pi r^2 + 2\pi rh$, where r is the radius of the base and h is the height.

The surface area, SA, of a **right circular cone** is given by $SA = \pi r^2 + \pi rl$, where r is the radius of the circular base and l is the slant height.

The surface area, SA, of a **regular pyramid** is given by $SA = s^2 + 2sl$, where s is the length of a side of the base and l is the slant height.

Ⓟ

HOW TO   Find the surface area of a sphere with a diameter of 18 cm.

$r = \dfrac{1}{2}d = \dfrac{1}{2}(18) = 9$ • First find the radius of the sphere.

$SA = 4\pi r^2$ • Use the formula for the surface area of a sphere.
$SA = 4\pi(9)^2$
$SA = 4\pi(81)$
$SA = 324\pi$

The exact surface area of the sphere is 324π cm².

$SA \approx 1017.88$ • An approximate measure can be found by using the π key on a calculator.

The approximate surface area is 1017.88 cm².

Integrating
Technology

To approximate 324π on your calculator, enter
324 × π = .

Discuss the Concepts

1. Explain the difference between volume and surface area.

2. In what units is surface area measured? Why?

Concept Check

1. How much larger is the surface area of a cube with a side of length 8 cm than the surface area of a sphere with a radius of 9 cm? Round to the nearest hundredth.
633.88 cm²

2. How much larger is the surface area of a cone with a radius of 3 in. and a slant height of 4 in. than the surface area of a pyramid in which the length of a side of the base is 3 in. and the slant height is 4 in.? Round to the nearest hundredth.
32.97 in²

Optional Student Activity

Find the surface area of the solid. Round to the nearest hundredth.

1.

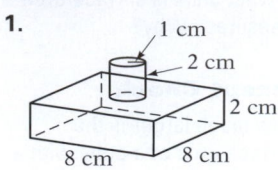

204.57 cm²

2.

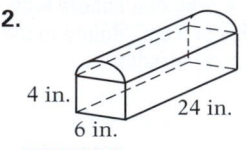

638.47 in²

Example 3

The diameter of the base of a cone is 5 m, and the slant height is 4 m. Find the surface area of the cone. Give the exact measure.

Strategy

To find the surface area of the cone:
• Find the radius of the base of the cone.
• Use the formula for the surface area of a cone. Leave the answer in terms of π.

Solution

$r = \dfrac{1}{2}d = \dfrac{1}{2}(5) = 2.5$

$SA = \pi r^2 + \pi r l$
$SA = \pi(2.5)^2 + \pi(2.5)(4)$
$SA = \pi(6.25) + \pi(2.5)(4)$
$SA = 6.25\pi + 10\pi$
$SA = 16.25\pi$

The surface area of the cone is 16.25π m².

You Try It 3

The diameter of the base of a cylinder is 6 ft, and the height is 8 ft. Find the surface area of the cylinder. Round to the nearest hundredth.

Your Strategy

Your solution
207.35 ft²

Example 4

Find the area of a label used to cover a soup can that has a radius of 4 cm and a height of 12 cm. Round to the nearest hundredth.

Strategy

To find the area of the label, use the fact that the surface area of the side of a cylinder is given by $2\pi r h$. An approximation is asked for; use the π key on a calculator. $r = 4$, $h = 12$

Solution

Area of the label $= 2\pi r h$
Area of the label $= 2\pi(4)(12) = 96\pi \approx 301.59$

The area is approximately 301.59 cm².

You Try It 4

Which has a larger surface area, a cube with a side measuring 10 cm or a sphere with a diameter measuring 8 cm?

Your Strategy

Your solution
The cube

Solutions on pp. S20–S21

7.4 Exercises

Objective A   To solve problems involving the volume of a solid

Suggested Assignment
Exercises 1–51, odds
More challenging problems:
Exercises 52–55

 In Exercises 1–6, find the volume of the figure. For calculations involving π, give both the exact value and an approximation to the nearest hundredth.

1.
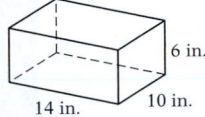
6 in.
14 in.   10 in.

840 in³

2.
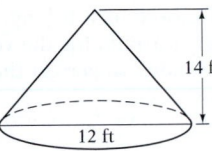
14 ft
12 ft

168π ft³, 527.79 ft³

3.
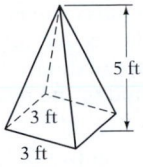
5 ft
3 ft
3 ft

15 ft³

4.
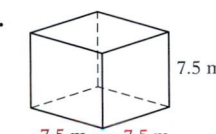
7.5 m
7.5 m   7.5 m

421.875 m³

5.
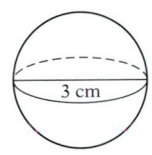
3 cm

4.5π cm³; 14.14 cm³

6.
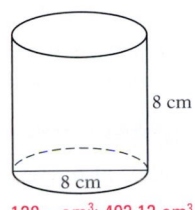
8 cm
8 cm

128π cm³; 402.12 cm³

 Solve.

7. A rectangular solid has a length of 6.8 m, a width of 2.5 m, and a height of 2 m. Find the volume of the solid.
34 m³

8. Find the volume of a rectangular solid that has a length of 4.5 ft, a width of 3 ft, and a height of 1.5 ft.
20.25 ft³

9. Find the volume of a cube whose side measures 2.5 in.
15.625 in³

10. The length of a side of a cube is 7 cm. Find the volume of the cube.
343 cm³

11. The diameter of a sphere is 6 ft. Find the volume of the sphere. Give the exact measure.
36π ft³

12. Find the volume of a sphere that has a radius of 1.2 m. Round to the nearest tenth.
7.2 m³

13. The diameter of the base of a cylinder is 24 cm. The height of the cylinder is 18 cm. Find the volume of the cylinder. Round to the nearest hundredth.
8143.01 cm³

14. The radius of the base of a cone is 5 in. The height of the cone is 9 in. Find the volume of the cone. Give the exact measure.
75π in³

15. The height of a cone is 15 cm. The diameter of the cone is 10 cm. Find the volume of the cone. Round to the nearest hundredth.
392.70 cm³

Quick Quiz (Objective 7.4A)

Solve. Round to the nearest hundredth.

1. Find the volume of a rectangular solid with a length of 3 m, a width of 0.9 m, and a height of 5 m.
13.5 m³

2. Find the volume of a cube with a side of 8 ft.
512 ft³

3. Find the volume of a sphere with a 5-foot diameter.   65.45 ft³

4. Find the volume of a cylinder with a radius of 30 cm and a height of 42 cm.   118,752.20 cm³

16. The length of a side of the base of a pyramid is 6 in., and the height is 10 in. Find the volume of the pyramid.
120 in³

17. The height of a pyramid is 8 m, and the length of a side of the base is 9 m. What is the volume of the pyramid?
216 m³

18. The index finger on the Statue of Liberty is 8 ft long. The circumference at the second joint is 3.5 ft. Use the formula for the volume of a cylinder to approximate the volume of the index finger on the Statue of Liberty. Round to the nearest hundredth.
7.80 ft³

19. The volume of a freezer with a length of 7 ft and a height of 3 ft is 52.5 ft³. Find the width of the freezer.
2.5 ft

20. The length of an aquarium is 18 in., and the width is 12 in. If the volume of the aquarium is 1836 in³, what is the height of the aquarium?
8.5 in.

21. The volume of a cylinder with a height of 10 in. is 502.4 in³. Find the radius of the base of the cylinder. Round to the nearest hundredth.
4.00 in.

22. The diameter of the base of a cylinder is 14 cm. If the volume of the cylinder is 2310 cm³, find the height of the cylinder. Round to the nearest hundredth.
15.01 cm

23. A rectangular solid has a square base and a height of 5 in. If the volume of the solid is 125 in³, find the length and the width.
Length: 5 in., width: 5 in.

24. The volume of a rectangular solid is 864 m³. The rectangular solid has a square base and a height of 6 m. Give the dimensions of the solid.
Height: 6 m, length: 12 m, width: 12 m

25. An oil storage tank, which is in the shape of a cylinder, is 4 m high and has a diameter of 6 m. The oil tank is two-thirds full. Find the number of cubic meters of oil in the tank. Round to the nearest hundredth.
75.40 m³

26. A silo, which is in the shape of a cylinder, is 16 ft in diameter and has a height of 30 ft. The silo is three-fourths full. Find the volume of the portion of the silo that is not being used for storage. Round to the nearest hundredth.
1507.96 ft³

Objective B **To solve problems involving the surface area of a solid**

 Find the surface area of the figure.

27.
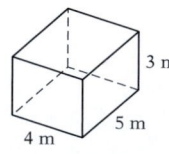
3 m
5 m
4 m
94 m²

28.
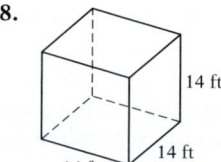
14 ft
14 ft
14 ft
1176 ft²

29.

5 m
4 m
4 m
56 m²

Quick Quiz (Objective 7.4B)
Solve. Round to the nearest hundredth.

1. Find the surface area of a rectangular solid that has a height of 4 ft, a length of 6 ft, and a width of 2 ft.   88 ft²

2. Find the surface area of a sphere that has a diameter of 3 in.   28.27 In²

3. Find the surface area of a cylinder with a height of 5 m. The radius of the base is 2 m.   87.96 m²

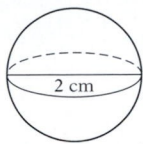 In Exercises 30–32, find the surface area of the figure. Give both the exact value and an approximation to the nearest hundredth.

30.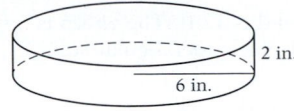

2 cm

4π cm²; 12.57 cm²

31.

2 in.

6 in.

96π in²; 301.59 in²

32.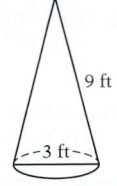

9 ft

3 ft

15.75π ft²; 49.48 ft²

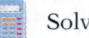 Solve.

33. The height of a rectangular solid is 5 ft. The length is 8 ft, and the width is 4 ft. Find the surface area of the solid.
184 ft²

34. The width of a rectangular solid is 32 cm. The length is 60 cm, and the height is 14 cm. What is the surface area of the solid?
6416 cm²

35. The side of a cube measures 3.4 m. Find the surface area of the cube.
69.36 m²

36. Find the surface area of a cube that has a side measuring 1.5 in.
13.5 in²

37. Find the surface area of a sphere with a diameter of 15 cm. Give the exact value.
225π cm²

38. The radius of a sphere is 2 in. Find the surface area of the sphere. Round to the nearest hundredth.
50.27 in²

39. The radius of the base of a cylinder is 4 in. The height of the cylinder is 12 in. Find the surface area of the cylinder. Round to the nearest hundredth.
402.12 in²

40. The diameter of the base of a cylinder is 1.8 m. The height of the cylinder is 0.7 m. Find the surface area of the cylinder. Give the exact value.
2.88π m²

41. The slant height of a cone is 2.5 ft. The radius of the base is 1.5 ft. Find the surface area of the cone. Give the exact value.
6π ft²

42. The diameter of the base of a cone is 21 in. The slant height is 16 in. What is the surface area of the cone? Round to the nearest hundredth.
874.15 in²

43. The length of a side of the base of a pyramid is 9 in., and the slant height is 12 in. Find the surface area of the pyramid.
297 in²

44. The slant height of a pyramid is 18 m, and the length of a side of the base is 16 m. What is the surface area of the pyramid?
832 m²

45. The surface area of a rectangular solid is 108 cm². The height of the solid is 4 cm, and the length is 6 cm. Find the width of the rectangular solid.
3 cm

46. The length of a rectangular solid is 12 ft. The width is 3 ft. If the surface area is 162 ft², find the height of the rectangular solid.
3 ft

47. A can of paint will cover 300 ft². How many cans of paint should be purchased in order to paint a cylinder that has a height of 30 ft and a radius of 12 ft?
11 cans

48. A hot air balloon is in the shape of a sphere. Approximately how much fabric was used to construct the balloon if its diameter is 32 ft? Round to the nearest whole number.
3217 ft²

49. How much glass is needed to make a fish tank that is 12 in. long, 8 in. wide, and 9 in. high? The fish tank is open at the top.
456 in²

50. Find the area of a label used to cover a can of juice that has a diameter of 16.5 cm and a height of 17 cm. Round to the nearest hundredth.
881.22 cm²

51. The length of a side of the base of a pyramid is 5 cm, and the slant height is 8 cm. How much larger is the surface area of this pyramid than the surface area of a cone with a diameter of 5 cm and a slant height of 8 cm? Round to the nearest hundredth.
22.53 cm²

Answers to Writing Exercises

55a. For example, cut perpendicular to the top and bottom faces and parallel to two of the sides.

  b. For example, beginning at an edge that is perpendicular to the bottom face, cut at an angle through to the bottom face.

  c. For example, beginning at the top face, at a distance d from the vertex, cut at an angle to the bottom face, ending at a distance greater than d from the vertex directly below the first chosen vertex.

  d. For example, beginning on the top face, at a distance d from a vertex, cut across the cube to a point just below the opposite vertex, intersecting the bottom face.

APPLYING THE CONCEPTS

52. Half of a sphere is called a **hemisphere.** Derive formulas for the volume and surface area of a hemisphere.
See the *Solutions Manual.*

53. Determine whether the statement is always true, sometimes true, or never true.
  a. The slant height of a regular pyramid is longer than the height.
  b. The slant height of a cone is shorter than the height.
  c. The four triangular faces of a regular pyramid are equilateral triangles.
  a. Always true   **b.** Never true   **c.** Sometimes true

54. a. What is the effect on the surface area of a rectangular solid if the width and height are doubled?
  b. What is the effect on the volume of a rectangular solid if both the length and the width are doubled?
  c. What is the effect on the volume of a cube if the length of each side of the cube is doubled?
  d. What is the effect on the surface area of a cylinder if the radius and height are doubled?
  a. Doubled + 4*WH*   **b.** Quadrupled   **c.** 8 times larger   **d.** 4 times larger

55. Explain how you could cut through a cube so that the face of the resulting solid is (a) a square, (b) an equilateral triangle, (c) a trapezoid, (d) a hexagon.

Focus on Problem Solving

Trial and Error Some problems in mathematics are solved by using **trial and error.** The trial-and-error method of arriving at a solution to a problem involves repeated tests or experiments until a satisfactory conclusion is reached.

Many of the Applying the Concepts exercises in this text require a trial-and-error method of solution. For example, an exercise on page 430 reads as follows:

Explain how you could cut through a cube so that the face of the resulting solid is (a) a square, (b) an equilateral triangle, (c) a trapezoid, (d) a hexagon.

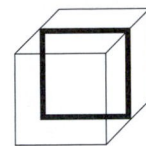

There is no formula to apply to this problem; there is no computation to perform. This problem requires picturing a cube and the results after it is cut through at different places on its surface and at different angles. For part (a), cutting perpendicular to the top and bottom of the cube and parallel to two of its sides will result in a square. The other shapes may prove more difficult.

When solving problems of this type, keep an open mind. Sometimes when using the trial-and-error method, we are hampered by narrowness of vision; we cannot expand our thinking to include other possibilities. Then, when we see someone else's solution, it appears so obvious to us! For example, for the question above, it is necessary to conceive of cutting through the cube at places other than the top surface; we need to be open to the idea of beginning the cut at one of the corner points of the cube.

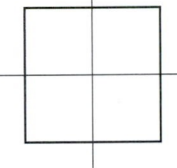

A topic of the Projects and Group Activities in this chapter is symmetry. Here again, the trial-and-error method is used to determine the lines of symmetry inherent in an object. For example, in determining lines of symmetry for a square, begin by drawing a square. The horizontal line of symmetry and the vertical line of symmetry may be immediately obvious to you.

But there are two others. Do you see that a line drawn through opposite corners of the square is also a line of symmetry?

Many of the questions in this text that require an answer of "always true," "sometimes true," or "never true" are best solved by the trial-and-error method. For example, consider the statement presented in Section 2 of this chapter:

Two rectangles that have the same area have the same perimeter.

Try some numbers. Each of two rectangles, one measuring 6 units by 2 units and another measuring 4 units by 3 units, has an area of 12 square units, but the perimeter of the first is 16 units and the perimeter of the second is 14 units. So, the answer "always true" has been eliminated. We still need to determine whether there is a case for which the statement *is* true. After experimenting with a lot of numbers, you may come to realize that we are trying to determine whether it is possible for two different pairs of factors of a number to have the same sum. Is it?

Don't be afraid to perform many experiments, and remember that *errors*, or tests that "don't work," are a part of the trial-and-*error* process.

Focus on Problem Solving: Trial and Error

No, it is not possible for two different pairs of factors of a number to have the same sum.

Projects and Group Activities: Lines of Symmetry

1. No

2. Besides A and H, the capital letters B, C, D, E, I, M, O, T, V, W, X, and Y have axes of symmetry. Note that the letters I, O, and X have more than one axis of symmetry. (For some letters, it depends on how the letter is written; an example is the letter U.)

3. The lowercase letters c, i, l, t, v, and w have one axis of symmetry. (For some letters, it depends on how the letter is written; an example is the letter t.)

4. The lowercase letters o and x have more than one axis of symmetry.

5. An isosceles triangle has one axis of symmetry. An equilateral triangle has three axes of symmetry. A rectangle has two axes of symmetry. A square has four axes of symmetry. A circle has an infinite number of axes of symmetry. A trapezoid has no axis of symmetry.

6. If a figure is unchanged after being rotated 180° about a point O, then the figure has point symmetry. A parallelogram, a rectangle, a square, and a circle have point symmetry.

   If a figure is unchanged after being rotated more than 0° and less than 360°, then the figure has rotational symmetry. An equilateral triangle, a rectangle, a square, a circle, and a parallelogram have rotational symmetry. It may be helpful to note that a regular pentagon has rotational symmetry but not point symmetry.

7. Examples of symmetry in nature include people and animals, snowflakes, starfish, and many types of flowers. Examples of symmetry in art and architecture can be found in history texts and in art and architectural magazines.

Projects and Group Activities

Lines of Symmetry

Look at the letter A printed at the left. If the letter were folded along line l, the two sides of the letter would match exactly. This letter has **symmetry** with respect to line l. Line l is called the **axis of symmetry.**

Now consider the letter H printed below at the left. Both lines l_1 and l_2 are axes of symmetry for this letter; the letter could be folded along either line and the two sides would match exactly.

1. Does the letter A have more than one axis of symmetry?

2. Find axes of symmetry for other capital letters of the alphabet.

3. Which lowercase letters have one axis of symmetry?

4. Do any of the lowercase letters have more than one axis of symmetry?

5. Find the numbers of axes of symmetry for the plane geometric figures presented in this chapter.

6. There are other types of symmetry. Look up the meaning of *point symmetry* and *rotational symmetry*. Which plane geometric figures provide examples of these types of symmetry?

7. Find examples of symmetry in nature, art, and architecture.

Preparing a Circle Graph

In Section 1 of this chapter, a protractor was used to measure angles. Preparing a circle graph requires the ability to use a protractor to draw angles.

To draw an angle of 142°, first draw a ray. Place a dot at the endpoint of the ray. This dot will be the vertex of the angle.

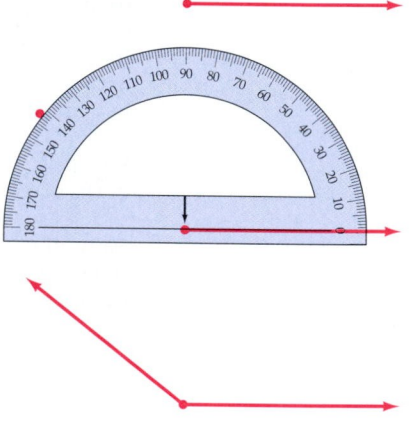

Place the line segment near the bottom edge of the protractor on the ray as shown in the figure at the right. Make sure the center of the line segment near the bottom edge of the protractor is located directly over the vertex point. Locate the position of the 142° mark. Place a dot next to the mark.

Remove the protractor and draw a ray from the vertex to the dot at the 142° mark.

An example of preparing a circle graph is given on the next page.

The revenues (in thousands of dollars) from four segments of a car dealership for the first quarter of a recent year were

| New car sales: | $2100 | Used car/truck sales: | $1500 |
| New truck sales: | $1200 | Parts/service: | $700 |

To draw a circle graph to represent the percent that each segment contributed to the total revenue from all four segments, proceed as follows.

Find the total revenue from all four segments.

$2100 + 1200 + 1500 + 700 = 5500$

Find what percent each segment is of the total revenue of $5500.

New car sales: $\dfrac{2100}{5500} \approx 38.2\%$

New truck sales: $\dfrac{1200}{5500} \approx 21.8\%$

Used car/truck sales: $\dfrac{1500}{5500} \approx 27.3\%$

Parts/service: $\dfrac{700}{5500} \approx 12.7\%$

Each percent represents a sector of the circle. Because the circle contains 360°, multiply each percent by 360° to find the measure of the angle for each sector. Round to the nearest whole number.

New car sales:

$0.382 \times 360° \approx 138°$

New truck sales:

$0.218 \times 360° \approx 78°$

Used car/truck sales:

$0.273 \times 360° \approx 98°$

Parts/service:

$0.127 \times 360° \approx 46°$

Draw a circle and use a protractor to draw the sectors representing the percents that each segment contributed to the total revenue.

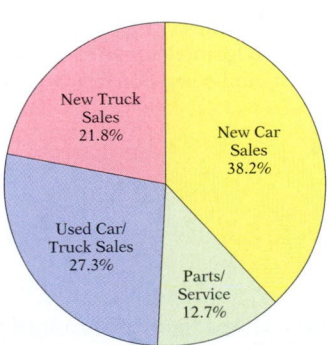

Collect data appropriate for display in a circle graph. [Some possibilities are last year's sales for the top three car manufacturers in the United States, votes cast in the last election for your state governor, the majors of the students in your math class, and the number of students enrolled in each class (senior, junior, etc.) at your college.] Then prepare the circle graph.

Projects and Group Activities: Preparing a Circle Graph

You might first have your students create a circle graph from data in which the percents are given. For example, shown below are U.S. adults' favorite pizza toppings. (*Source:* Market Facts for Bolla wines)

Pepperoni	43%
Sausage	19%
Mushrooms	14%
Vegetables	13%
Other	7%
Onions	4%

Then have students create a circle graph using data in which the percents must be calculated. For example, in a recent year, ten million tons of glass containers were produced. Given below is a list of the industries that use them. The category "Other" includes industries that manufacture items such as drugs and toiletries. (*Source:* Salomon Bros.)

Beer	4,600,000 tons
Food	3,500,000 tons
Wine and liquor	900,000 tons
Soft drinks	500,000 tons
Other	500,000 tons

Chapter 7 Summary

Key Words

Examples

A *line* is determined by two distinct points and extends indefinitely in both directions. A *line segment* is part of a line and has two endpoints. *Parallel lines* never meet; the distance between them is always the same. *Perpendicular lines* are intersecting lines that form right angles. [7.1A, pp. 377–379]

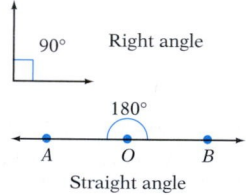

Parallel Lines

Perpendicular lines

A *ray* starts at a point and extends indefinitely in one direction. The point at which a ray starts is the *endpoint* of the ray. An *angle* is formed by two rays with the same endpoint. The *vertex* of an angle is the point at which the two rays meet. An angle is measured in *degrees*. A 90° angle is a *right angle*. A 180° angle is a *straight angle*. An *acute angle* is an angle whose measure is between 0° and 90°. An *obtuse angle* is an angle whose measure is between 90° and 180°. *Complementary angles* are two angles whose measures have the sum 90°. *Supplementary angles* are two angles whose measures have the sum 180°. [7.1A, pp. 377–380]

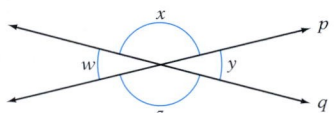

90°   Right angle

180°
A   O   B
Straight angle

Two angles that are on opposite sides of the intersection of two lines are *vertical angles;* vertical angles have the same measure. Two angles that share a common side are *adjacent angles;* adjacent angles of intersecting lines are supplementary angles. [7.1A, p. 380; 7.1B, p. 382]

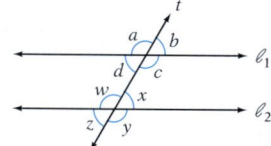

Angles w and y are vertical angles.
Angles x and y are adjacent angles.

A line that intersects two other lines at two different points is a *transversal*. If the lines cut by a transversal are parallel lines, equal angles are formed: *alternate interior angles, alternate exterior angles,* and *corresponding angles.* [7.1B, p. 383]

Parallel lines ℓ_1 and ℓ_2 are cut by transversal t. All four acute angles have the same measure. All four obtuse angles have the same measure.

A *polygon* is a closed figure determined by three or more line segments. The line segments that form the polygon are its *sides.* A *regular polygon* is one in which all sides have the same length and all angles have the same measure. Polygons are classified by the number of sides. A *quadrilateral* is a four-sided polygon. A parallelogram, a rectangle, a square, a rhombus, and a trapezoid are all quadrilaterals. [7.2A, pp. 393–394]

Number of Sides	Name of the Polygon
3	Triangle
4	Quadrilateral
5	Pentagon
6	Hexagon
7	Heptagon
8	Octagon
9	Nonagon
10	Decagon

A *triangle* is a plane figure formed by three line segments. An *isosceles triangle* has two sides of equal length. The three sides of an *equilateral triangle* are of equal length. A *scalene triangle* has no two sides of equal length. An *acute triangle* has three acute angles. An *obtuse triangle* has one obtuse angle. A *right triangle* has a right angle. [7.1C, p. 385; 7.2A, pp. 393–394]

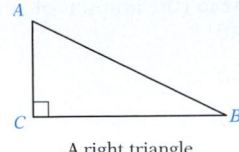

A right triangle

A *circle* is a plane figure in which all points are the same distance from the center of the circle. A *diameter* of a circle is a line segment across the circle through the center. A *radius* of a circle is a line segment from the center of the circle to a point on the circle. [7.2A, p. 396]

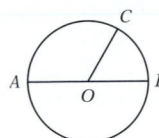

AB is a diameter of the circle.
OC is a radius.

Similar triangles have the same shape but not necessarily the same size. The ratios of corresponding sides are equal. The ratio of corresponding heights is equal to the ratio of corresponding sides. *Congruent triangles* have the same shape and the same size. [7.3B, p. 413; 7.3C, p. 415]

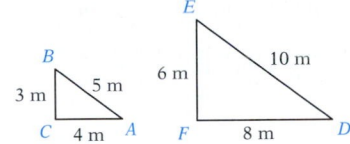

Triangles *ABC* and *DEF* are similar triangles. The ratio of corresponding sides is $\frac{1}{2}$.

Essential Rules and Procedures

Examples

Triangles [7.1C, p. 385, 7.3C, pp. 415–416]
Sum of the measures of the interior angles = 180°
Sum of an interior angle and the corresponding exterior angle = 180°
Rules to determine congruence: SSS Rule, SAS Rule, ASA Rule

In a right triangle, the measure of one acute angle is 12°. Find the measure of the other acute angle.

$$x + 12° + 90° = 180°$$
$$x + 102° = 180°$$
$$x = 78°$$

Formulas for Perimeter (the distance around a figure) [7.2A, pp. 394–396]
Triangle: $P = a + b + c$
Rectangle: $P = 2L + 2W$
Square: $P = 4s$
Circumference of a circle: $C = \pi d$ or $C = 2\pi r$

The length of a rectangle is 8 m. The width is 5.5 m. Find the perimeter of the rectangle.

$$P = 2L + 2W$$
$$P = 2(8) + 2(5.5)$$
$$P = 16 + 11$$
$$P = 27$$

The perimeter is 27 m.

Formulas for Area (the amount of surface in a region) [7.2B, pp. 398–401]

Triangle: $A = \dfrac{1}{2}bh$

Rectangle: $A = LW$
Square: $A = s^2$
Circle: $A = \pi r^2$
Parallelogram: $A = bh$

Trapezoid: $A = \dfrac{1}{2}h(b_1 + b_2)$

The length of the base of a parallelogram is 12 cm, and the height is 4 cm. Find the area of the parallelogram.

$A = bh$

$A = 12(4)$

$A = 48$

The area is 48 cm².

Formulas for Volume (the amount of space inside a figure in space) [7.4A, p. 422]

Rectangular solid: $V = LWH$
Cube: $V = s^3$

Sphere: $V = \dfrac{4}{3}\pi r^3$

Right circular cylinder: $V = \pi r^2 h$

Right circular cone: $V = \dfrac{1}{3}\pi r^2 h$

Regular pyramid: $V = \dfrac{1}{3}s^2 h$

Find the volume of a cube that measures 3 in. on a side.

$V = s^3$

$V = 3^3$

$V = 27$

The volume is 27 in³.

Formulas for Surface Area (the total area on the surface of a solid) [7.4B, p. 425]

Rectangular solid: $SA = 2LW + 2LH + 2WH$
Cube: $SA = 6s^2$
Sphere: $SA = 4\pi r^2$
Right circular cylinder: $SA = 2\pi r^2 + 2\pi rh$
Right circular cone: $SA = \pi r^2 + \pi rl$
Regular pyramid: $SA = s^2 + 2sl$

Find the surface area of a sphere with a diameter of 10 cm. Give the exact value.

$r = \dfrac{1}{2}d = \dfrac{1}{2}(10) = 5$

$SA = 4\pi r^2$

$SA = 4\pi(5^2)$

$SA = 4\pi(25)$

$SA = 100\pi$

The surface area is 100π cm².

Pythagorean Theorem [7.3A, p. 411]
If a and b are the lengths of the legs of a right triangle and c is the length of the hypotenuse, then $c^2 = a^2 + b^2$.

Two legs of a right triangle measure 6 ft and 8 ft. Find the hypotenuse of the right triangle.

$c^2 = a^2 + b^2$

$c^2 = 6^2 + 8^2$

$c^2 = 36 + 64$

$c^2 = 100$

$c = \sqrt{100}$

$c = 10$

The length of the hypotenuse is 10 ft.

Principal Square Root Property [7.3A, p. 412]
If $r^2 = s$, then $r = \sqrt{s}$, and r is called the **square root** of s.

If $c^2 = 16$, then $c = \sqrt{16} = 4$.

Chapter 7 Review Exercises

1. Given that $\angle a = 74°$ and $\angle b = 52°$, find the measures of angles x and y.

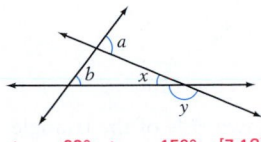

$\angle x = 22°$, $\angle y = 158°$ [7.1C]

2. Triangles ABC and DEF are similar. Find the perimeter of triangle ABC.

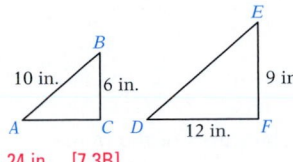

24 in.   [7.3B]

3. Given triangle ABC and triangle DEF, do the conditions $\angle B = \angle D$, $\angle A = \angle F$, and $\angle C = \angle E$ guarantee that triangle ABC is congruent to triangle DEF? If they are congruent, by what rule are they congruent?

No   [7.3C]

4. Find the measure of $\angle x$.

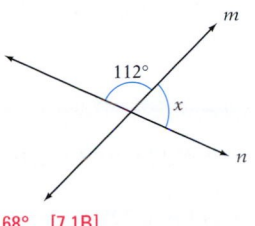

68°   [7.1B]

5. Determine whether the two triangles are congruent. If they are congruent, state by what rule they are congruent.

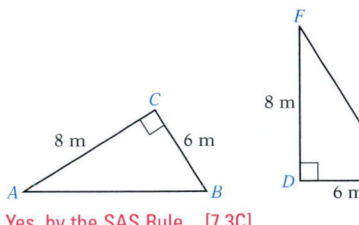

Yes, by the SAS Rule   [7.3C]

6. The two legs of a right triangle measure 4 in. and 6 in. Find the hypotenuse of the right triangle. Round to the nearest hundredth.

7.21 in.   [7.3A]

7. Given that $BC = 11$ cm and AB is three times the length of BC, find the length of AC.

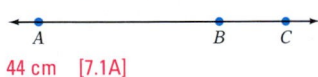

44 cm   [7.1A]

8. Find x.

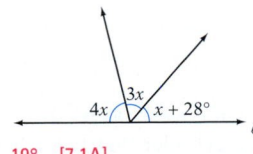

19°   [7.1A]

9. Find the area of the figure.

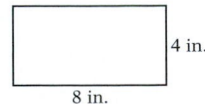

32 in²   [7.2B]

10. Find the volume of the figure.

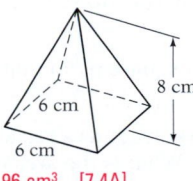

96 cm³   [7.4A]

11. Find the perimeter of the figure.

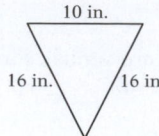

42 in.   [7.2A]

12. Given that $\ell_1 \parallel \ell_2$, find the measures of angles a and b.

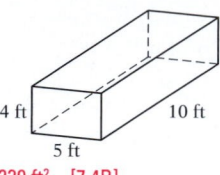

$\angle a = 138°, \angle b = 42°$ [7.1B]

13. Find the surface area of the figure.

220 ft²   [7.4B]

14. Find the unknown side of the triangle. Round to the nearest hundredth.

9.75 ft   [7.3A]

15. Find the volume of a cube whose side measures 3.5 in.
42.875 in³   [7.4A]

16. Find the supplement of a 32° angle.
148°   [7.1A]

17. Find the volume of a rectangular solid with a length of 6.5 ft, a width of 2 ft, and a height of 3 ft.
39 ft³   [7.4A]

18. Two angles of a triangle measure 37° and 48°. Find the measure of the third angle.
95°   [7.1C]

19. The height of a triangle is 7 cm. The area of the triangle is 28 cm². Find the length of the base of the triangle.
8 cm   [7.2B]

20. Find the volume of a sphere that has a diameter of 12 mm. Give the exact value.
288π mm³   [7.4A]

21. The perimeter of a square picture frame is 86 cm. Find the length of each side of the frame.
21.5 cm   [7.2A]

22. A can of paint will cover 200 ft². How many cans of paint should be purchased to paint a cylinder that has a height of 15 ft and a radius of 6 ft?
4 cans   [7.4B]

23. The length of a rectangular park is 56 yd. The width is 48 yd. How many yards of fencing are needed to surround the park?
208 yd   [7.2A]

24. What is the area of a square patio that measures 9.5 m on each side?
90.25 m²   [7.2B]

25. A walkway 2 m wide surrounds a rectangular plot of grass. The plot is 40 m long and 25 m wide. What is the area of the walkway?
276 m²   [7.2B]

Chapter 7 Test

1. For the right triangle shown below, determine the length of side *BC*. Round to the nearest hundredth.

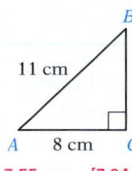

7.55 cm   [7.3A]

2. Determine whether the two triangles are congruent. If they are congruent, state by what rule they are congruent.

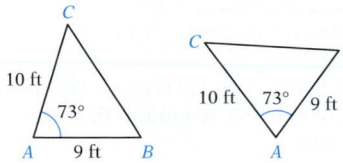

Congruent, SAS   [7.3C]

3. Determine the area of a rectangle with a length of 15 m and a width of 7.4 m.
111 m²   [7.2B]

4. Determine the area of a triangle whose base is 7 ft and whose height is 12 ft.
42 ft²   [7.2B]

5. Determine the exact volume of a right circular cone whose radius is 7 cm and whose height is 16 cm.
$\dfrac{784\pi}{3}$ cm³   [7.4A]

6. Determine the exact surface area of a pyramid whose square base is 3 m on each side and whose slant height is 11 m.

75 m²   [7.4B]

7. Determine the volume of the solid shown below. Round to the nearest hundredth.

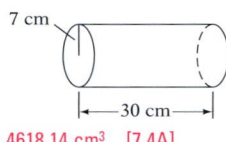

4618.14 cm³   [7.4A]

8. Determine the area of the trapezoid shown below.

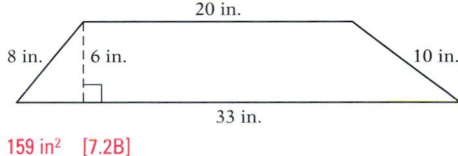

159 in²   [7.2B]

9. Determine the perimeter of the figure shown below.

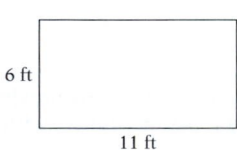

34 ft   [7.2A]

10. Two angles of a triangle measure 57° and 23°. Find the measure of the third angle.
100°   [7.1A]

11. Find *x*.

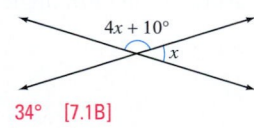

34°   [7.1B]

12. Name the figure shown below.

Octagon   [7.2A]

13. Determine whether the two triangles are congruent. If they are congruent, state by what rule they are congruent.

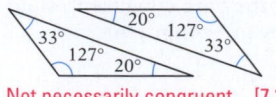

Not necessarily congruent   [7.3C]

14. Determine the volume of the rectangular solid shown below.

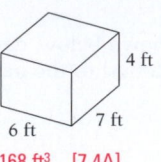

168 ft³   [7.4A]

15. Figure *ABC* is a right triangle. Determine the length of side *AB*. Round to the nearest hundredth.

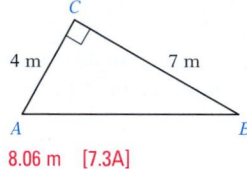

8.06 m   [7.3A]

16. Given that l_1 and l_2 are parallel lines, determine the measure of angle *a*.

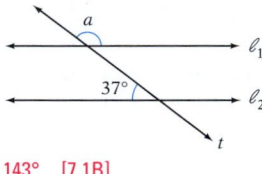

143°   [7.1B]

17. Determine the exact surface area of the right circular cylinder shown below.

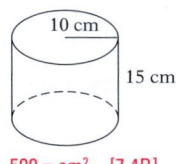

500π cm²   [7.4B]

18. Determine the measure of angle *a*.

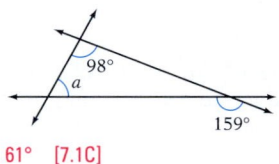

61°   [7.1C]

19. Triangles *ABC* and *DEF* are similar triangles. Determine the length of line segment *FG*. Round to the nearest hundredth.

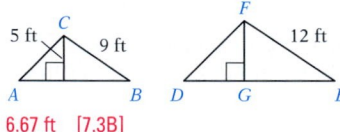

6.67 ft   [7.3B]

20. Triangles *ABC* and *DEF* are similar triangles. Determine the length of side *BC*. Round to the nearest hundredth.

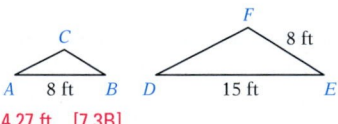

4.27 ft   [7.3B]

21. Determine the perimeter of a square whose side is 5 m.
20 m   [7.2A]

22. Determine the perimeter of a rectangle whose length is 8 cm and whose width is 5 cm.
26 cm   [7.2A]

23. Find the perimeter of a right triangle with legs that measure 12 ft and 18 ft. Round to the nearest tenth.
51.6 ft   [7.3A]

24. Two angles of a triangle measure 41° and 37°. Find the measure of the third angle.
102°   [7.1C]

25. Find the supplement of a 67° angle.
113°   [7.1A]

Cumulative Review Exercises

1. Find 8.5% of 2400.
204   [6.4A/6.4B]

2. Find all the factors of 78.
1, 2, 3, 6, 13, 26, 39, 78   [1.3D]

3. Divide: $4\dfrac{2}{3} \div 5\dfrac{3}{5}$

$\dfrac{5}{6}$ [2.4B]

4. Evaluate: $|-18|$

18   [3.1C]

5. Divide and round to the nearest tenth:
82.93 ÷ 6.5
12.8   [2.6C]

6. Subtract: $-6 - (-4)$
-2 [3.2B]

7. Find the measure of $\angle x$.

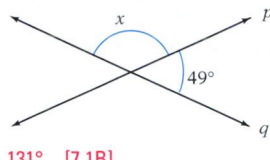

131°   [7.1B]

8. Find the unknown side of the triangle.

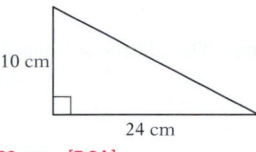

26 cm   [7.3A]

9. Add: $\dfrac{2}{3} + \dfrac{7}{10}$

$1\dfrac{11}{30}$ [2.3A]

10. Simplify: $-\dfrac{1}{4}(28b)$

$-7b$ [4.2B]

11. Simplify: $3(-4) - (-1 + 5)^2$

-28 [3.5A]

12. Solve: $3(2x + 5) = 18$

$\dfrac{1}{2}$ [5.3B]

13. Simplify: $3a + 5a - 7a + a$
$2a$ [4.2A]

14. Multiply: $-3(-25)$
75   [3.3A]

15. Simplify: $5(2x + 4) - (3x + 2)$
$7x + 18$ [4.2D]

16. Evaluate $2x + 3y^2z$ when $x = 5$, $y = -1$, and $z = -4$.
-2 [4.1A]

17. Evaluate $x^2y - 2z$ when $x = \dfrac{1}{2}$, $y = \dfrac{4}{5}$, and $z = -\dfrac{3}{10}$.

$\dfrac{4}{5}$ [4.1A]

18. Find the prime factorization of 78.
$2 \cdot 3 \cdot 13$ [1.3D]

19. Solve: $4x + 2 = 6x - 8$

5   [5.3A]

20. Write $\frac{3}{8}$ as a percent.

37.5%   [6.3B]

21. Translate "the product of eight and twice a number" into a variable expression. Then simplify.
$8(2n)$; $16n$ [4.3B]

22. **Uniform Motion**   Two cars, one traveling 5 mph faster than the other, start at the same time from the same point and travel in opposite directions. In 2 h they are 210 mi apart. Find the rate of each car.
50 mph; 55 mph   [5.5B]

23. Find the simple interest on a 270-day loan of $20,000 at an annual interest rate of 8.875%.
$1313.01   [6.5A]

24. **Mixtures**   How many ounces of a silver alloy that costs $3.50 per ounce must be mixed with 12 oz of an alloy that costs $5 per ounce to make a mixture that costs $4 per ounce?
24 oz   [5.5A]

25. **Cellular Phones**   The charge for cellular phone service for a business executive is $22 per month plus $.25 per minute of phone use. In a month when the executive's phone bill was $43.75, how many minutes did the executive use the cellular phone?
87 min   [5.4B]

26. **Taxes**   If the sales tax on a $12.50 purchase is $.75, what is the sales tax on a $75 purchase?
$4.50   [6.2B]

27. 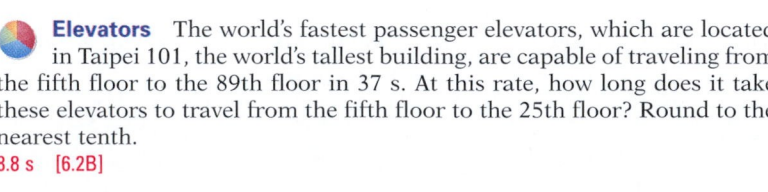 **Foreign Trade**   The figure at the right shows the values of the imports and exports during the first and second quarters of a recent year. Find the increase in the value of the imports from the first quarter to the second quarter.
$.08 trillion   [2.6E]

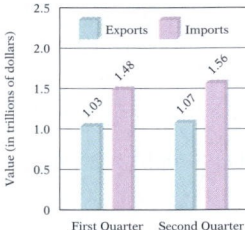

Values of Imports and Exports
Source: Bureau of Economic Analysis

28. **Geometry**   The volume of a box is 144 ft³. The length of the box is 12 ft, and the width is 4 ft. Find the height of the box.
3 ft   [7.4A]

29. **Oceanography**   The pressure, P, in pounds per square inch, at a certain depth in the ocean can be approximated by the equation $P = 15 + \frac{1}{2}D$, where D is the depth in feet. Use this equation to find the depth when the pressure is 35 lb/in².
40 ft   [5.2B]

30.  **Elevators**   The world's fastest passenger elevators, which are located in Taipei 101, the world's tallest building, are capable of traveling from the fifth floor to the 89th floor in 37 s. At this rate, how long does it take these elevators to travel from the fifth floor to the 25th floor? Round to the nearest tenth.
8.8 s   [6.2B]

chapter 8

Statistics and Probability

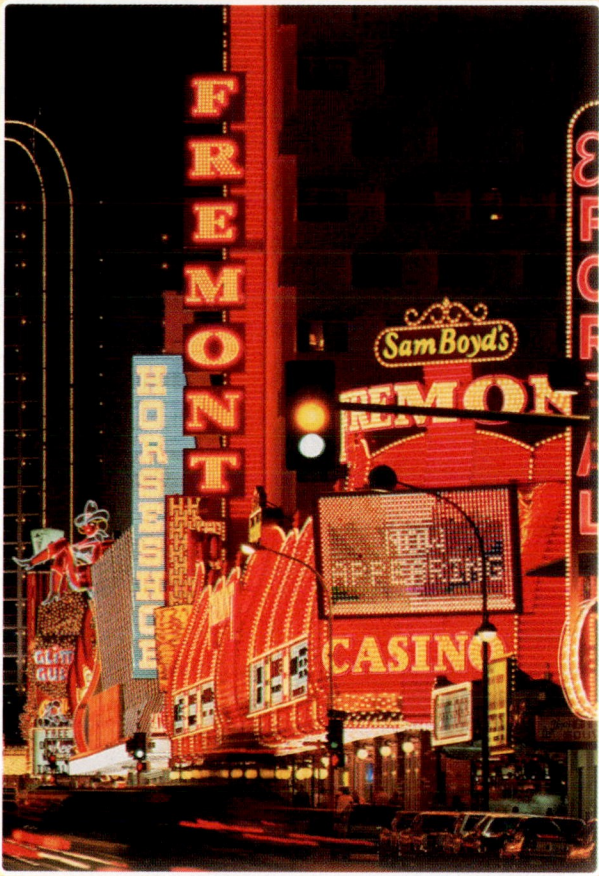

Las Vegas is home to some of the world's most famous casinos. The casinos rely on games of chance, such as Keno and Chuck-a-luck, and favorable probability to generate large cash revenues, some of which support the city. In the **exercises on pages 463 and 464,** you will calculate the probability of rolling certain combinations when tossing dice.

OBJECTIVES

Section 8.1

A To read and interpret graphs

Section 8.2

A To find the mean, median, and mode of a distribution
B To draw a box-and-whiskers plot

Section 8.3

A To calculate the probability of simple events

Need help? For online student resources, such as section quizzes, visit this textbook's website at **math.college.hmco.com/students.**

PREP TEST ● ● ●

Do these exercises to prepare for Chapter 8.

1.  **Mail**   Bill-related mail accounted for 49 billion of the 102 billion pieces of first-class mail handled by the U.S. Postal Service during a recent year. (*Source:* U.S. Postal Service) What percent of the pieces of first-class mail handled by the U.S. Postal Service was bill-related mail? Round to the nearest tenth of a percent.
   48.0%   [6.4A/6.4B]

2. **Education**   The table at the right shows the estimated cost of funding an education at a public college.

Enrollment Year	Cost of Public College
2005	$70,206
2006	$74,418
2007	$78,883
2008	$83,616
2009	$88,633
2010	$93,951

   Source: The College Board's Annual Survey of Colleges

   a. Between which two enrollment years is the increase in cost greatest?   Between 2009 and 2010

   b. What is the increase between these two years? $5318   [1.2C]

3. **Sports**   During the 1924 Summer Olympics in Paris, France, the United States won 45 gold medals, 27 silver medals, and 27 bronze medals. (*Source: The Ultimate Book of Sports Lists*)

   a. Find the ratio of gold medals won by the United States to silver medals won by the United States during the 1924 Summer Olympics. Write the ratio as a fraction in simplest form. $\frac{5}{3}$

   b. Find the ratio of silver medals won by the United States to bronze medals won by the United States during the 1924 Summer Olympics. Write the ratio using a colon.
   1:1   [6.1A]

4. **Television**   The table below shows the number of television viewers, in millions, who watch pay-cable channels, such as HBO and Showtime, each night of the week. (*Source:* Neilsen Media Research, analyzed by Initiative Media North America)

Mon	Tue	Wed	Thu	Fri	Sat	Sun
3.9	4.5	4.2	3.9	5.2	7.1	5.5

   a. Arrange the numbers in the table from least to greatest.   3.9, 3.9, 4.2, 4.5, 5.2, 5.5, 7.1   [2.5B]

   b. Find the average number of viewers per night.   4.9 million   [2.6E]

5. **The Military**   Approximately 90,000 women serve in the U.S. military. Five percent of these women serve in the Marine Corps. (*Source:* U.S. Department of Defense)

   a. Approximately how many women are in the Marine Corps?   4500 women   [6.4C]

   b. What fractional amount of women in the military are in the Marine Corps? $\frac{1}{20}$ [6.3A]

GO FIGURE ● ● ●

I have 2 brothers and 1 sister. My father's parents have 10 grandchildren. My mother's parents have 11 grandchildren. If no divorces or remarriages occurred, how many first cousins do I have?

13 first cousins

8.1 Statistical Graphs

Objective A To read and interpret graphs

 VIDEO & DVD CD TUTOR WEB SSM

Statistics is the branch of mathematics concerned with **data,** or numerical information. Graphs are used to display numerical information in a visual format that enables the reader to quickly see relationships and trends. Three of the most common types of graphs are the bar graph, the line graph, and the circle graph. Examples of each of these types of graphs are shown below.

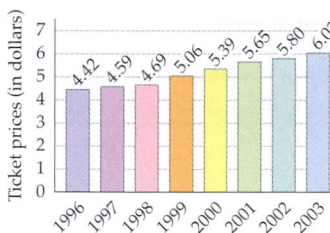

Figure 8.1 Average U.S. Movie Theater Ticket Prices (1996–2003)
Source: National Association of Theatre Owners (http://www.natoonline.org)

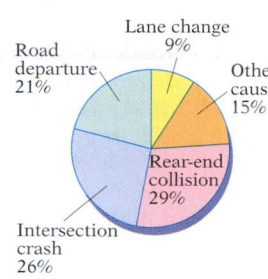

Figure 8.2 Types of Automobile Accidents, City of Twin Falls
Year: 2005
Total Number of Accidents: 4300

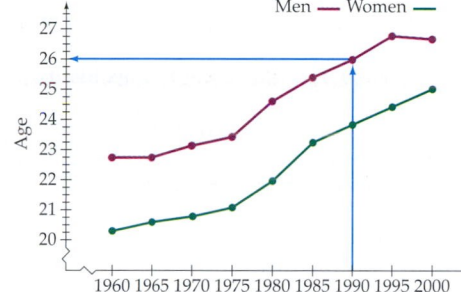

Figure 8.3 U.S. Average Age at First Marriage
Source: Bureau of the Census

The bar graph in Figure 8.1 displays the average U.S. movie theater ticket price for the years 1996 to 2003. Each vertical bar is used to display the average ticket price for a given year. The higher the bar, the greater the average ticket price for that year. We can see from the graph that the average movie theater ticket price has been increasing.

The circle graph in Figure 8.2 displays the percent of automobile accidents of a particular type that occurred in a given city for a given year. The largest sector of the circle corresponds to the largest percent of accidents of a given type, 29%.

Figure 8.3 shows two broken-line graphs. The upper broken-line graph displays the average age at first marriage for men for selected years from 1960 to 2000. The lower broken-line graph displays the average age at first marriage for women for selected years during the same time period. The line segments that connect the points on the graph indicate trends. Increasing trends are indicated by line segments that rise as they move to the right, and decreasing trends are indicated by line segments that fall as they move to the right. We can see from the graph that the average age at first marriage has been increasing for both men and women. The blue arrows in Figure 8.3 show that the average age at which men married for the first time in the year 1990 was about 26 years. The graph shows that, for the years shown, the average age at first marriage for men has always been greater than the average age at first marriage for women.

Objective 8.1A

New Vocabulary
statistics
data

Instructor Note
Bar graphs, circle graphs, and line graphs are introduced in Section 1.1. They are included here for completeness in discussing basic concepts from statistics. In this section, unlike Section 1.1, the students use their understanding of ratio and percent to answer questions about the data presented in the graphs. You may choose to review the material in Section 1.1 prior to presenting this material, or you may omit this section entirely.

Discuss the Concepts
1. Can the data in Figure 8.1 be displayed in a circle graph?
2. Can the data in Figure 8.2 be displayed in a bar graph?
3. What characteristics of a bar graph make it easy to compare the data that are displayed in one?

Instructor Note
You might draw the students' attention to the jagged portion of the vertical axis in Figure 8.3. Remind them that this indicates that some of the ages between 0 and 20 are not displayed, and that this break in the vertical axis enables us to display the graph in a compact form.

In-Class Examples (Objective 8.1A)

Use Figures 8.1, 8.2, and 8.3.

1. What percent of the 1996 average U.S. movie theater ticket price is the 2003 average U.S. movie theater ticket price? Round to the nearest percent.
   136%

2. Determine the number of lane-change accidents that occurred in Twin Falls in the year 2005.
   387 lane-change accidents

3. What was the average age at which women married for the first time in the year 1980?
   22 years

Concept Check

1. Why would the population of New York City in the years 1850, 1900, 1950, and 2000 be inappropriate data to display in a circle graph? Why would these data be appropriately represented by a line graph? (*Note:* You might discuss with your students the preference for presenting this type of data in a line graph rather than a bar graph by describing the population as continuous rather than discrete.)

2. Some circle graphs show data values and some show percents. Can we create a circle graph showing percents if we are given one showing the data values? Can we create a circle graph showing data values if we are given one showing the percents?

Optional Student Activity

Have small groups of students prepare three questions and answers based on the data shown in the circle graph below. Have the groups exchange questions and compare their answers to the other groups' answers.

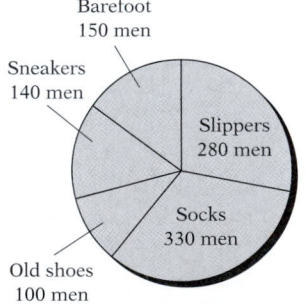

Barefoot
150 men

Sneakers
140 men

Slippers
280 men

Socks
330 men

Old shoes
100 men

Responses of 1000 men to the question "What type of footwear do you wear at home?"

Source: Based on data from L.B. Evans

At the right is the circle graph shown on the previous page. Use this graph for Example 1 and You Try It 1.

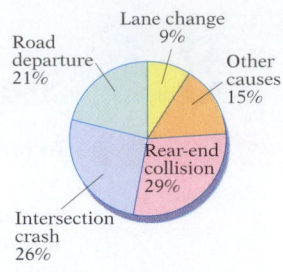

Figure 8.2 Types of Automobile Accidents, City of Twin Falls, Year: 2005
Total Number of Accidents: 4300

Example 1

a. Find the ratio, as a fraction in simplest form, of the percent of lane-change accidents to the percent of accidents resulting from "other causes."
b. Determine the number of rear-end collisions that occurred in Twin Falls in the year 2005.

Strategy

a. To find the ratio:
   - From the graph, find the percent of lane-change accidents and the percent of accidents resulting from "other causes."
   - Write the ratio in fractional form. Simplify.
b. To find the number of rear-end collisions:
   - From the graph, find the percent of accidents that were rear-end collisions.
   - Solve the basic percent equation for amount. The base is 4300.

Solution

a. Lane-change accidents: 9%

   Accidents resulting from "other causes:" 15%

   $$\frac{9\%}{15\%} = \frac{3}{5}$$

   The ratio is $\frac{3}{5}$.

b. Rear-end collisions: 29% = 0.29

   Percent · base = amount
   0.29 · 4300 = n
            1247 = n

   1247 rear-end collisions occurred in Twin Falls in 2005.

You Try It 1

a. Find the ratio, as a fraction in simplest form, of the percent of lane-change accidents to the percent of road-departure accidents.
b. Determine the number of accidents that occurred at intersections in Twin Falls in the year 2005.

Your strategy

Your solution

a. $\frac{3}{7}$

b. 1118 accidents

Solution on p. S21

8.1 Exercises

Objective A   To read and interpret graphs

Education   An accounting major recorded the number of units required in each discipline to graduate with a degree in accounting. The results are shown in the circle graph in Figure 8.4. Use this graph for Exercises 1 to 4.

1. How many units are required to graduate with a degree in accounting?
   128 units

2. What is the ratio of the number of units needed in finance to the number of units needed in accounting?
   $\frac{1}{3}$

3. What percent of the units required to graduate are taken in accounting? Round to the nearest tenth of a percent.
   35.2%

4. What percent of the units required to graduate are taken in mathematics? Round to the nearest tenth of a percent.
   9.4%

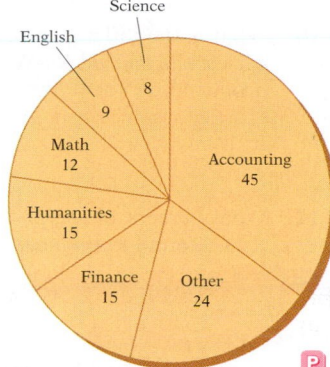

Figure 8.4 Number of units required to graduate with an accounting degree

Video Games   The circle graph in Figure 8.5 shows the breakdown of the approximately $3,100,000,000 that Americans spent on home video game equipment in one year. Use this graph for Exercises 5 to 8.

5. Find the amount of money spent on TV game machines.
   $1,085,000,000

6. Find the amount of money spent on portable game machines.
   $279,000,000

7. What fractional amount of the total money spent was spent on accessories?
   $\frac{2}{25}$

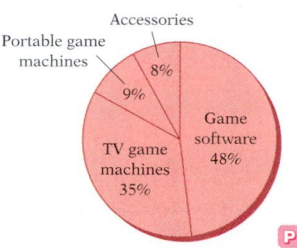

Figure 8.5 Percents of $3,100,000,000 spent annually on home video games
Source: The NPD Group, Toy Manufacturers of America

8. Is the amount spent for TV game machines more than three times the amount spent for portable game machines?
   Yes

Quick Quiz (Objective 8.1A)

1. According to the data in Figure 8.4, how many more units are required in accounting than in English to graduate with an accounting degree?
   36 units

2. According to the data in Figure 8.5, how much more money was spent on portable game machines than on accessories? $31,000,000

Automobile Production The bar graph in Figure 8.6 shows the regions in which all passenger cars were produced during a recent year. Use this graph for Exercises 9 to 11.

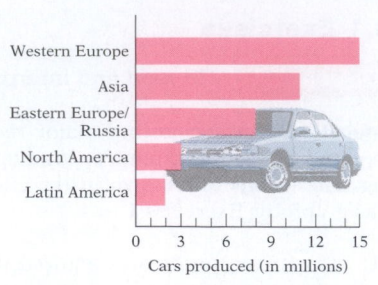

9. How many passenger cars were produced worldwide?
   39 million passenger cars

10. What is the difference between the number of passenger cars produced in Western Europe and the number produced in North America?
    12 million passenger cars

11. What percent of the passenger cars were produced in Asia? Round to the nearest percent.
    28%

Figure 8.6 Number of passenger cars produced in a recent year

Source: Copyright © 2000 by the *Los Angeles Times*. Reprinted with permission.

Health The double-broken-line graph in Figure 8.7 shows the number of Calories per day that should be consumed by women and men in various age groups. Use this graph for Exercises 12 to 14.

12. What is the difference between the number of Calories recommended for men and the number recommended for women 19 to 22 years of age?
    800 Calories

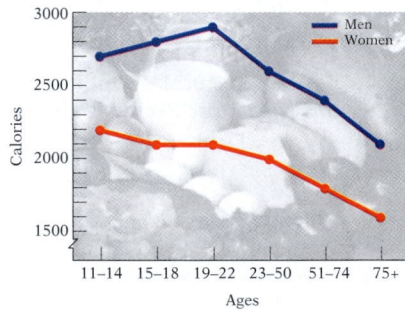

13. People of what gender and age have the lowest recommended number of Calories?
    Women, age 75+

Figure 8.7 Recommended number of Calories per day for women and men, by age

Source: Numbers, by Andrea Sutcliffe (HarperCollins)

14. Find the ratio of the number of Calories recommended for women 15 to 18 years old to the number recommended for women 51 to 74 years old.
    $\dfrac{7}{6}$

APPLYING THE CONCEPTS

15. The circle graph at the right shows a couple's expenditures last month. Write two observations about this couple's expenses.

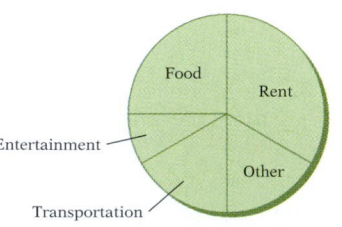

Answers to Writing Exercises

15. Answers will vary. For example:
  The couple's largest single expense was rent.
  Food represents approximately one-quarter of the month's expenditures.
  Rent represents approximately one-third of the month's expenditures.
  The expense for food is approximately the same as the combined expense for entertainment and transportation.
  The couple spent more for transportation than for entertainment.

8.2 Statistical Measures

Objective A To find the mean, median, and mode of a distribution

The average score on the math portion of the SAT was 432. The EPA estimates that a 2005 Ford Focus averages 35 miles per gallon on the highway. The average rainfall for portions of Kauai is 350 inches per year. Each of these statements uses one number to describe an entire collection of numbers. Such a number is called an *average*.

In statistics there are various ways to calculate an average. Three of the most common—*mean, median,* and *mode*—are discussed here.

An automotive engineer tests the miles-per-gallon ratings of 15 cars and records the results as follows:

Miles-per-Gallon Ratings of 15 Cars

25	22	21	27	25	35	29	31	25	26	21	39	34	32	28

The **mean** of the data is the sum of the measurements divided by the number of measurements. The symbol for the mean is $\bar{x}$.

> **Formula for the Mean**
>
> $$\bar{x} = \frac{\text{sum of the data values}}{\text{number of data values}}$$

To find the mean for the data above, add the numbers and then divide by 15.

$$\bar{x} = \frac{25 + 22 + 21 + 27 + 25 + 35 + 29 + 31 + 25 + 26 + 21 + 39 + 34 + 32 + 28}{15}$$

$$= \frac{420}{15} = 28$$

The mean number of miles per gallon for the 15 cars tested was 28 miles per gallon.

The mean is one of the most frequently computed averages. It is the one that is commonly used to calculate a student's performance in a class.

HOW TO The test scores for a student taking American history were 78, 82, 91, 87, and 93. What was the mean score for this student?

Strategy
To find the mean score, divide the sum of the test scores by 5, the number of scores.

Solution
$$\bar{x} = \frac{78 + 82 + 91 + 87 + 93}{5} = \frac{431}{5} = 86.2$$

The mean score for the history student was 86.2.

Integrating Technology
When using a calculator to calculate the mean, use parentheses to enclose the sum in the numerator.
(78 + 82 + 91 + 87 + 93) ÷ 5 =

Objective 8.2A

New Vocabulary
mean
median
mode

New Formulas
$$\bar{x} = \frac{\text{sum of the data values}}{\text{number of data values}}$$

Instructor Note
You might explain to your students that the mean is an appropriate measure when all the data values are relatively close. However, when the range of values is large compared to the values themselves, the mean may give an unrealistic picture of the data.

Discuss the Concepts
You may find it beneficial to have students work Exercise 1 on page 455 in small groups and then discuss their answers as a class.

Concept Check
1. In a typical college classroom, do you expect the mean age or the median age to be larger?
   The mean age
2. Provide a reason for the Bureau of Labor Statistics to cite median per capita income rather than mean per capita income.
   Only a few very large salaries would increase the mean to the point of distorting the "average" income.
(Continued on the next page)

In-Class Examples (Objective 8.2A)

1. A truck driver's records show the number of gallons of diesel fuel purchased each day of a 5-day trip.

Wed	Thu	Fri	Sat	Sun
74	86	93	79	88

   a. Find the mean number of gallons purchased.
   84 gal
   b. Find the median number of gallons purchased.
   86 gal

2. The ages of the six children in a small day-care center are 3, 4, 4, 5, 4, and 2 years. What is the mode of these data?   4 years

(*Continued*)

3. Suppose a real estate agent is interested in the cost of homes in the town of Wilmington. The realtor goes to City Hall and learns that the selling prices of the last 10 homes sold in the community were

$ 100,000
$ 100,000
$ 100,000
$ 100,000
$ 100,000
$ 100,000
$ 100,000
$ 100,000
$ 100,000
$1,100,000

Would a couple interested in buying a home in Wilmington be more interested in the mean or the median of these data? Why?
The mean is $200,000. The median is $100,000. The couple would be more interested in the median because it is more representative of the data. The couple should expect to pay $100,000 for a home in Wilmington. The $1,100,000 selling price is not typical of the prices of homes in that town, and it raised the mean by $100,000.

Optional Student Activity

1. On five tests on which scores could range from 0 to 100, you had an average of exactly 88. Find the lowest score you could have received on one of these tests.   40

2. If 51 is the mean of 50 consecutive even numbers, what is the largest of the numbers?   100

Study Tip

Word problems are difficult because we must read the problem, determine the quantity to be found, think of a method to find it, actually solve the problem, and then check the answer. In short, we must devise a *strategy* and then use that strategy to find the *solution*. See *AIM for Success*, page xxxiii.

TAKE NOTE

Some sets of numbers do not have a mode. For instance, 1, 6, 8, 10, 15, 32 has no mode. A set of numbers can have more than one mode. For instance, in the list 4, 2, 6, 2, 7, 9, 4, both 2 and 4 are modes for the data.

The **median** of a set of data is the number that separates the data into two equal parts when the numbers in the data are arranged from smallest to largest (or from largest to smallest). There is an equal number of values above the median and below the median.

To find the median of a set of numbers, first arrange the numbers from smallest to largest. The median is the number in the middle.

The result of arranging the miles-per-gallon ratings given on the previous page from smallest to largest is shown below.

21 21 22 25 25 25 26 **27** 28 29 31 32 34 35 39

7 values below the median   Middle number **Median**   7 values above the median

The median is 27 miles per gallon.

If data contain an *even* number of values, the median is the mean of the two middle numbers.

HOW TO The selling prices of the last six homes sold by a real estate agent were $175,000, $150,000, $250,000, $130,000, $245,000, and $190,000. Find the median selling price of these homes.

Strategy
To find the median selling price, arrange the numbers from smallest to largest. Because there is an even number of values, the median is the mean of the two middle numbers.

Solution
130,000 150,000 175,000 190,000 245,000 250,000

Middle 2 numbers

$$\text{Median} = \frac{175,000 + 190,000}{2} = 182,500$$

The median selling price was $182,500.

The **mode** of a set of numbers is the value that occurs most frequently. If a set of numbers has no number occurring more than once, then the data have no mode.

Here again are the data for the gasoline mileage ratings of 15 cars.

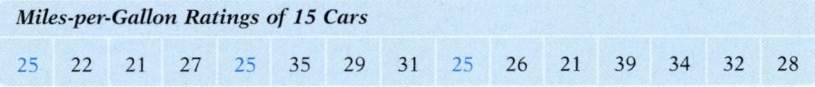

Miles-per-Gallon Ratings of 15 Cars														
25	22	21	27	25	35	29	31	25	26	21	39	34	32	28

25 is the number that occurs most frequently.

The mode is 25 miles per gallon.

Note from the miles-per-gallon example that the mean, median, and mode may be different.

Example 1

Twenty students were asked the number of units in which they were enrolled. The responses were as follows:

15	12	13	15	17	18	13	20	9	16
14	10	15	12	17	16	6	14	15	12

Find the mean number of units taken by these students.

Strategy

To find the mean number of units:

- Find the sum of the 20 numbers.
- Divide the sum by 20.

Solution

$15 + 12 + 13 + 15 + 17 + 18 + 13 + 20 + 9 +$
$\quad 16 + 14 + 10 + 15 + 12 + 17 + 16 + 6 +$
$\quad 14 + 15 + 12 = 279$

$$\bar{x} = \frac{279}{20} = 13.95$$

The mean is 13.95 units.

You Try It 1

The amounts spent by 12 customers at a McDonald's restaurant were as follows:

6.26	8.23	5.09	8.11	7.50	6.69
5.66	4.89	5.25	9.36	6.75	7.05

Find the mean amount spent by these customers. Round to the nearest cent.

Your strategy

Your solution

$6.74

Example 2

The starting hourly wages for an apprentice electrician for six different work locations are $10.90, $11.25, $10.10, $11.08, $11.56, and $10.55. Find the median starting hourly wage.

Strategy

To find the median starting hourly wage:

- Arrange the numbers from smallest to largest.
- Because there is an even number of values, the median is the mean of the two middle numbers.

Solution

10.10, 10.55, 10.90, 11.08, 11.25, 11.56

$$\text{Median} = \frac{10.90 + 11.08}{2} = 10.99$$

The median starting hourly wage is $10.99.

You Try It 2

The amounts of weight lost, in pounds, by 10 participants in a 6-month weight-reduction program were 22, 16, 31, 14, 27, 16, 29, 31, 40, and 10. Find the median weight loss for these participants.

Your strategy

Your solution

24.5 lb

Solutions on pp. S21–S22

Objective 8.2B

Vocabulary to Review

median   [8.2A]

New Vocabulary

quartile
first quartile
third quartile
range
interquartile range
box-and-whiskers plot

Instructor Note

Questions from statistics are included on many teachers' state competency exams. Hence the inclusion in this text of topics such as box-and-whiskers plots.

Discuss the Concepts

Suppose that the interquartile range on an exam was 10 points. What does this tell you about the test scores?

Instructor Note

You may want to explain to your students that the word *quartile* is used because a quartile divides the set of data into four sets of approximately equal size.

The second quartile, symbolized by Q_2, is the number that one-half of the data lie below and one-half of the data lie above. Therefore, it is the median of the data.

Concept Check

1. For each of the following, state whether the whiskers will be short or long relative to the box in a box-and-whiskers plot.

   a. The range is 50 and the interquartile range is 40.
   Short

   b. The range is 98 and the interquartile range is 31.
   Long

   c. The range is 60, Q_1 is 10, and Q_3 is 25.   Long

   d. The range is 100, Q_1 is 250, and Q_3 is 325.   Short

(*Continued on the next page*)

Objective B   To draw a box-and-whiskers plot

Recall from the last objective that an average is one number that is used to describe all the numbers in a set of data. For example, we know from the following statement that Erie gets a lot of snow each winter.

> The average annual snowfall in Erie, Pennsylvania, is 85 inches.

Now look at these two statements.

> The average annual temperature in San Francisco, California, is 57°F.
> The average annual temperature in St. Louis, Missouri, is 57°F.

San Francisco

The average annual temperature in both cities is the same. However, we do not expect the climate in St. Louis to be like San Francisco's climate. Although both cities have the same average annual temperature, their temperature ranges differ. In fact, the difference between the average monthly high temperatures in July and January in San Francisco is 14°F, whereas the difference between the average monthly high temperatures in July and January in St. Louis is 50°F.

St. Louis

Note that for this example, a single number (the average annual temperature) does not provide us with a very comprehensive picture of the climate of either of these two cities.

One method used to help us picture an entire set of data is a box-and-whiskers plot. To prepare a box-and-whiskers plot, we begin by separating a set of data into four parts, called **quartiles.** We will illustrate this by using the average monthly high temperatures for St. Louis, in degrees Fahrenheit. These are listed below for January through December.

39	47	58	72	81	88	89	89	85	76	49	47

Source: The Weather Channel

First list the numbers in order from smallest to largest and determine the median.

39	47	47	49	58	72	76	81	85	88	89	89

Median = 74

Now find the median of the data values below the median. The median of the data values below the median is called the **first quartile,** symbolized by Q_1. Also find the median of the data values above the median. The median of the data values above the median is called the **third quartile,** symbolized by Q_3.

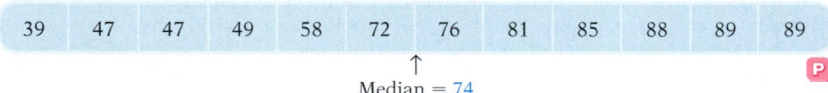

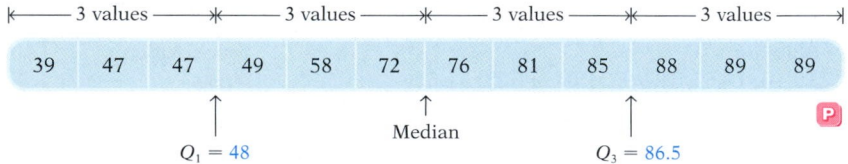

$Q_1 = 48$ Median $Q_3 = 86.5$

The first quartile, Q_1, is the number that one-quarter of the data lie below. This means that 25% of the data lie below the first quartile. The third quartile, Q_3, is the number that one-quarter of the data lie above. This means that 25% of the data lie above the third quartile.

In-Class Examples (Objective 8.2B)

1. The list below gives the numbers of years that U.S. presidents survived after leaving office.

0, 1, 2, 2, 3, 4, 6, 6, 7, 8, 8, 8, 8, 9, 11, 11, 12, 16, 17, 17, 19, 19, 19, 20, 21, 21, 25, 31

(*Note:* There are 28 values in this list. You may prefer to present your students with these numbers out of sequential order. Then the first step in organizing the data is to write the numbers from smallest to largest.) Find the lowest value, Q_1, the median, Q_3, and the highest value. (*Continued on the next page*)

The **range** of a set of numbers is the difference between the largest number and the smallest number in the set. The range describes the spread of the data. For the data above,

$$\text{Range} = \text{largest value} - \text{smallest value} = 89 - 39 = 50$$

The **interquartile range** is the difference between the third quartile, Q_3, and the first quartile, Q_1. For the data above,

$$\text{Interquartile range} = Q_3 - Q_1 = 86.5 - 48 = 38.5$$

The interquartile range is the distance that spans the "middle" 50% of the data values. Because it excludes the bottom fourth of the data values and the top fourth of the data values, it excludes any extremes in the numbers in the set.

A **box-and-whiskers plot**, or **boxplot**, is a graph that shows five numbers: the smallest value, the first quartile, the median, the third quartile, and the greatest value of a data set. Here are these five values for the data on St. Louis temperatures.

The smallest number	39
The first quartile, Q_1	48
The median	74
The third quartile, Q_3	86.5
The largest number	89

Think of a number line that includes the five values listed above. With this in mind, mark off the five values. Draw a box that spans the distance from Q_1 to Q_3. Draw a vertical line the height of the box at the median.

Listed below are the average monthly high temperatures for San Francisco.

57	60	61	64	68	71	71	73	74	73	60	59

Source: The Weather Channel

We can perform the same calculations on these data to determine the five values needed for the box-and-whiskers plot.

The smallest number	57
The first quartile, Q_1	60
The median	66
The third quartile, Q_3	72
The largest number	74

The box-and-whiskers plot is shown at the right, with the same scale used for the data on the St. Louis temperatures.

Note that by comparing the two boxplots, we can see that the range of temperatures in St. Louis is greater than the range of temperatures in San Francisco. For the St. Louis temperatures, there is a greater spread of the data below the median than above the median, whereas the spreads of the data above and below the median in the San Francisco boxplot are nearly equal.

TAKE NOTE
50% of the data in a distribution lie in the interquartile range.

TAKE NOTE
The number line above the boxplot at the right is shown as a reference only. It is not a part of the boxplot. It shows that the numbers labeled on the boxplot are plotted according to the distances on the number line above it.

TAKE NOTE
It is the "whiskers" on the box-and-whiskers plot that show the range of the data. The "box" on the box-and-whiskers plot shows the interquartile range of the data.

(*Continued*)

2. In the Concept Check in the last objective, we listed the selling prices of the last 10 homes sold in the town of Wilmington.

$ 100,000
$ 100,000
$ 100,000
$ 100,000
$ 100,000
$ 100,000
$ 100,000
$ 100,000
$ 100,000
$1,100,000

If data such as these are represented in a box-and-whiskers plot, will the plot have a long left whisker, a long box, or a long right whisker?
A long right whisker

Instructor Note
Emphasize that the boxplot at the left shows that there are as many months with average high temperatures between 48°F and 74°F as there are months with average high temperatures between 74°F and 86.5°F.

Optional Student Activity
Listed on the next page are the numbers of electoral college votes allotted to each of the 50 states and the District of Columbia. (*Source:* National Archives and Records Administration)

a. Draw a box-and-whiskers plot of the data.

b. What data value is responsible for the long whisker at the right?   55
c. What state does this correspond to?   California

(*Continued*)
a. Draw a box-and-whiskers plot of the data.
Lowest value = 0 years, Q_1 = 6 years, median = 10 years, Q_3 = 19 years, highest value = 31 years

b. Determine the range.   31 years
c. Determine the interquartile range.   13 years

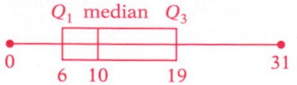

States and Electoral College Votes	
Alabama	9
Alaska	3
Arizona	10
Arkansas	6
California	55
Colorado	9
Connecticut	7
Delaware	3
D.C.	3
Florida	27
Georgia	15
Hawaii	4
Idaho	4
Illinois	21
Indiana	11
Iowa	7
Kansas	6
Kentucky	8
Louisiana	9
Maine	4
Maryland	10
Massachusetts	12
Michigan	17
Minnesota	10
Mississippi	6
Missouri	11
Montana	3
Nebraska	5
Nevada	5
New Hampshire	4
New Jersey	15
New Mexico	5
New York	31
North Carolina	15
North Dakota	3
Ohio	20
Oklahoma	7
Oregon	7
Pennsylvania	21
Rhode Island	4
South Carolina	8
South Dakota	3
Tennessee	11
Texas	34
Utah	5
Vermont	3
Virginia	13
Washington	11
West Virginia	5
Wisconsin	10
Wyoming	3

HOW TO The numbers of avalanche deaths in the United States during each of nine consecutive winters were 8, 24, 29, 13, 28, 30, 22, 26, and 32. (*Source:* Colorado Avalanche Information Center) Draw a box-and-whiskers plot of the data and determine the interquartile range.

Strategy

To draw the box-and-whiskers plot, arrange the data values from smallest to largest. Then find the median, Q_1, and Q_3. Use the smallest value, Q_1, the median, Q_3, and the largest value to draw the box-and-whiskers plot.

To find the interquartile range, find the difference between Q_3 and Q_1.

TAKE NOTE

Note that the left whisker in this box-and-whiskers plot is quite long. This indicates a set of data in which the median is closer to the largest data value. If a boxplot has a long whisker on the right, the median is closer to the smallest data value. If the two whiskers are approximately the same length, then the smallest and largest values are about the same distance from the median. See Example 3.

Solution

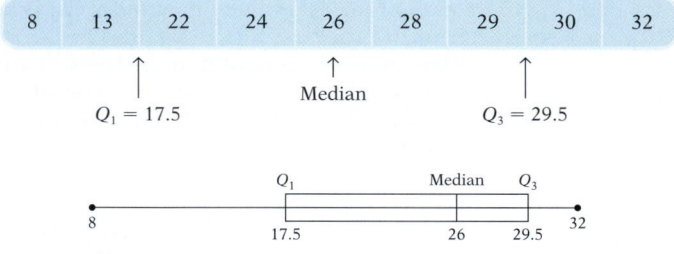

Interquartile range $= Q_3 - Q_1 = 29.5 - 17.5 = 12$

The interquartile range is 12 deaths.

Example 3

The average monthly snowfall amounts, in inches, for Buffalo, New York, from October through April are 1, 12, 24, 25, 18, 12, and 3. (*Source:* The Weather Channel) Draw a box-and-whiskers plot of the data.

Strategy

To draw the box-and-whiskers plot:

- Arrange the data values from smallest to largest.
- Find the median, Q_1, and Q_3.
- Use the smallest value, Q_1, the median, Q_3, and the largest value to draw the box-and-whiskers plot.

Solution

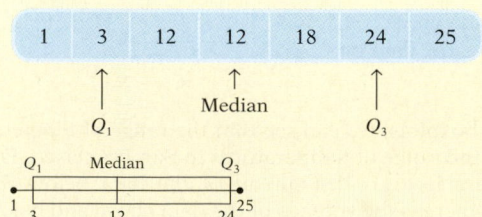

You Try It 3

The average monthly snowfall amounts, in inches, for Denver, Colorado, from October through April are 4, 7, 7, 8, 8, 9, and 13. (*Source:* The Weather Channel)
a. Draw a box-and-whiskers plot of the data.
b. How does the spread of the data within the interquartile range compare with the spread of the data in Example 3?

Your strategy

Your solution

a.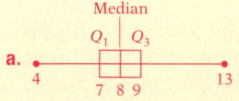

b. Answers about the spread of the data will vary.

Solution on p. S22

8.2 Exercises

Objective A   To find the mean, median, and mode of a distribution

Suggested Assignment
Exercises 1–21, odds
Exercises 24, 25
More challenging problems:
Exercises 23, 27, 28

1. State whether the mean, median, or mode is being used.
  a. Half of the houses in the new development are priced under $125,000.   Median
  b. The average bill for lunch at the college union is $7.95.   Mean
  c. The college bookstore sells more green college sweatshirts than any other color.   Mode
  d. In a recent year, there were as many people age 26 and younger in the world as there were people age 26 and older.   Median
  e. The majority of full-time students carry a load of 12 credit hours per semester.   Mode
  f. The average annual return on this investment is 6.5%.   Mean

2. Consumerism   The numbers of big-screen televisions sold each month for one year were recorded by an electronics store. The results were 15, 12, 20, 20, 19, 17, 22, 24, 17, 20, 15, and 27. Calculate the mean, the median, and the mode of the number of televisions sold per month.
Mean: 19 TVs; median: 19.5 TVs; mode: 20 TVs

3. The Airline Industry   The numbers of seats occupied on a jet for 16 trans–Atlantic flights were recorded. The numbers were 309, 422, 389, 412, 401, 352, 367, 319, 410, 391, 330, 408, 399, 387, 411, and 398. Calculate the mean, the median, and the mode of the number of seats occupied per flight.
Mean: 381.5625 seats; median: 394.5 seats; mode: no mode

4. Sports   The times, in seconds, for a 100-meter dash at a college track meet were 10.45, 10.23, 10.57, 11.01, 10.26, 10.90, 10.74, 10.64, 10.52, and 10.78.
  a. Calculate the mean time for the 100-meter dash.   10.61 s
  b. Calculate the median time for the 100-meter dash.   10.605 s

5. Consumerism   A consumer research group purchased identical items in eight grocery stores. The costs for the purchased items were $45.89, $52.12, $41.43, $40.67, $48.73, $42.45, $47.81, and $45.82. Calculate the mean and median costs of the purchased items.
Mean: $45.615; median: $45.855

6. Computers   One measure of a computer's hard-drive speed is called access time; this is measured in milliseconds (thousandths of a second). Find the mean and median for 11 hard drives whose access times were 5, 4.5, 4, 4.5, 5, 5.5, 6, 5.5, 3, 4.5, and 4.5. Round to the nearest tenth.
Mean: 4.7 milliseconds; median: 4.5 milliseconds

7. Health Plans   Eight health maintenance organizations (HMOs) presented group health insurance plans to a company. The monthly rates per employee were $423, $390, $405, $396, $426, $355, $404, and $430. Calculate the mean and median monthly rates for these eight HMOs.
Mean: $403.625; median: $404.50

8.  **Government**   The lengths of the terms, in years, of all the former Supreme Court chief justices are given in the table below. Find the mean and median length of term for the chief justices.   Mean: 12.4 years; median: 10 years

5	0	4	34	28	8	14	21
10	8	11	4	7	15	17	

Quick Quiz (Objective 8.2A)

1. A tourist center recorded the number of requests for information per day for a 5-day period.

Mon	Tue	Wed	Thu	Fri
124	130	127	126	148

  a. Find the mean number of requests.   131 requests
  b. Find the median number of requests.
    127 requests

2. The numbers of bags of flour used at a bakery during each of six days were 22, 24, 23, 25, 23, and 21. What is the mode of the data?

23 bags of flour

9.  **Life Expectancy**   The life expectancies, in years, in 10 selected Central and South American countries are given at the right.
   a. Find the mean life expectancy for this group of countries.
   b. Find the median life expectancy for this group of countries.
   Mean: 70.8 years; median: 72 years

Country	Life Expectancy
Brazil	62
Chile	75
Costa Rica	78
Ecuador	70
Guatemala	64
Panama	75
Peru	66
Trinidad and Tobago	71
Uruguay	74
Venezuela	73

10. **Education**   Your scores on six history tests were 78, 92, 95, 77, 94, and 88. If an "average score" of 90 receives an A for the course, which average, the mean or the median, would you prefer that the instructor use?
    Median

11. ✏ **Education**   One student received scores of 85, 92, 86, and 89 on four exams. A second student received scores of 90, 97, 91, and 94 (exactly 5 points more on each exam). Are the mean scores of the two students the same? If not, what is the relationship between the mean scores of the two students?

12.  **Defense Spending**   The table below shows the defense expenditures, in billions of dollars, by the federal government for 1965 through 1973, years during which the United States was actively involved in the Vietnam War.
    a. Calculate the mean annual defense expenditure. Round to the nearest tenth of a billion. $72.3 billion
    b. Find the median annual defense expenditure. $77.7 billion
    c. ✏ If the year 1965 were eliminated from the data, how would that affect the mean? The median?

Year	1965	1966	1967	1968	1969	1970	1971	1972	1973
Expenditures	$49.6	$56.8	$70.1	$80.5	$81.2	$80.3	$77.7	$78.3	$76.0

Source: Statistical Abstract of the United States

Objective B   **To draw a box-and-whiskers plot**

13. **a.** What percent of the data in a set of numbers lies above Q_3?   25%
    b. What percent of the data in a set of numbers lies above Q_1?   75%
    c. What percent of the data in a set of numbers lies below Q_3?   75%
    d. What percent of the data in a set of numbers lies below Q_1?   25%

14.  **U.S. Presidents**   The box-and-whiskers plot below shows the distribution of the ages of presidents of the United States at the time of their inauguration.
    a. What is the youngest age in the set of data?   42 years
    b. What is the oldest age?   69 years
    c. What is the first quartile?   51 years
    d. What is the third quartile?   58 years
    e. What is the median?   55 years
    f. Find the range.   27 years
    g. Find the interquartile range.   7 years

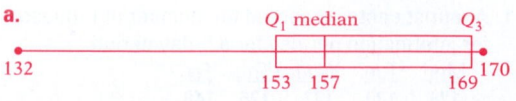
42   51   55   58   69

Quick Quiz (Objective 8.2B)

1. The numbers of federal, state, and local law enforcement officers killed in the line of duty during each of seven consecutive years were 157, 153, 169, 170, 132, 160, and 155.
   (*Source:* National Law Enforcement Officers Memorial Fund)
   a. Draw a box-and-whiskers plot of the data.
   b. Determine the range and the interquartile range.

a.

Q_1 median Q_3

132   170
         153   157   169

b. Range = 38 officers; interquartile range = 16 officers

15. 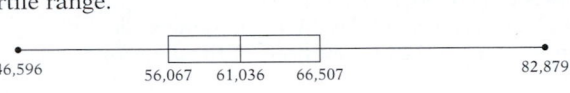 **Compensation** The box-and-whiskers plot below shows the distribution of median incomes for 50 states and the District of Columbia. What is the lowest value in the set of data? The highest value? The first quartile? The third quartile? The median? Find the range and the interquartile range.

Lowest = $46,596;
highest = $82,879;
Q_1 = $56,067;
Q_3 = $66,507;
median = $61,036;
range = $36,283;
interquartile range = $10,440

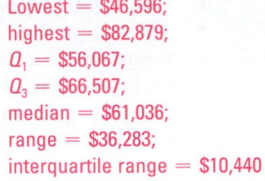

46,596   56,067 61,036 66,507   82,879

16. **Education** An aptitude test was taken by 200 students at the Fairfield Middle School. The box-and-whiskers plot at the right shows the distribution of their scores.

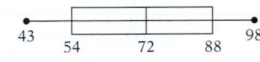

43   54   72   88   98

   a. How many students scored over 88?   50 students
   b. How many students scored below 72?   100 students
   c. How many scores are represented in each quartile?   50 scores
   d. What percent of the students had scores of at least 54?   75%

17. **Health** The cholesterol levels for 80 adults were recorded and then displayed in the box-and-whiskers plot shown at the right.

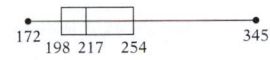

172 198 217 254   345

   a. How many adults had a cholesterol level above 217?   40 adults
   b. How many adults had a cholesterol level below 254?   60 adults
   c. How many cholesterol levels are represented in each quartile?   20 cholesterol levels
   d. What percent of the adults had a cholesterol level not more than 198?   25%

18. **Fuel Efficiency** The gasoline consumption of 19 cars was tested, and the results were recorded in the table below.

   a. Find the range, the first quartile, the third quartile, and the interquartile range.   Range = 17 mpg; Q_1 = 20 mpg; Q_3 = 30 mpg; interquartile range = 10 mpg
   b. Draw a box-and-whiskers plot of the data.
   c. Is the data value 21 in the interquartile range?   Yes

16   20   25   30   33

Miles per Gallon for 19 Cars									
33	21	30	32	20	31	25	20	16	24
22	31	30	28	26	19	21	17	26	

19. 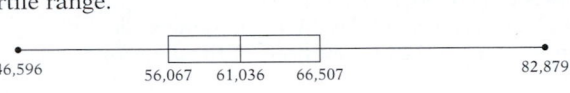 **Environment** Carbon dioxide is among the gases that contribute to global warming. The world's biggest emitters of carbon dioxide are listed below. The figures are emissions in millions of metric tons per year.

   a. Find the range, the first quartile, the third quartile, and the interquartile range.   In millions of metric tons per year, the range is 4.39 emissions; Q_1 = 0.56 emission; Q_3 = 2.10 emissions; the interquartile range is 1.54 emissions.
   b. Draw a box-and-whiskers plot of the data.
   c. What data value is responsible for the long whisker at the right?   4.80

Carbon Dioxide Emissions (in millions of metric tons per year)			
Canada	0.41	Japan	1.06
China	2.60	Russian Federation	2.10
Germany	0.87	Ukraine	0.61
India	0.76	United Kingdom	0.56
Italy	0.41	United States	4.80

Source: U.S. State Department

0.41   2.10   4.80
    0.56 0.815

Answers to Writing Exercises

(These exercises appear on page 458.)

20d. Answers will vary. For example, Seattle has months that average less rainfall than any month in Houston, and it has months that average more rainfall than any month in Houston. The range of the Seattle data is wider than the range of the Houston data. Almost all of the Houston data lie between the median and the third quartile of the Seattle data. The average amount of rain per month in Houston varies relatively little compared to the average amount of rain per month in Seattle.

21d. Answers will vary. For example, the distribution of the data is relatively similar for the two cities. However, the value of each of the five points on the boxplot for the Portland data is less than the corresponding value on the boxplot for the Orlando data. The average monthly rainfall in Portland is less than the average monthly rainfall in Orlando.

Answers to Writing Exercises

22a. To determine the mean, divide the sum of the values in the set of data by the number of values in the set.

b. To find the median, arrange the numbers from smallest to largest. If there is an *odd* number of values in the set of data, the median is the middle number. If there is an *even* number of values in the set of data, the median is the mean of the two middle numbers.

c. To find the mode, find the number in the set of data that occurs more often than any other number in the set.

24. The mode must be a value in the data set because it is the number that occurs most often in a set of data.

25a. Q_1 is the number that one-quarter of the data lie below.

b. Q_3 is the number that one-quarter of the data lie above.

c. $\overline{x}$ is the symbol for the mean of a set of data.

26. A box-and-whiskers plot displays the following values in a set of data: the lowest value, Q_1, the median, Q_3, and the highest value.
    The lowest value is shown at the far left of the plot (at the end of the left whisker). The highest value is shown at the far right of the plot (at the end of the right whisker). Q_1 is shown at the left end of the box, and Q_3 is shown at the right end of the box. The median is represented by the vertical line within the box.

27. The box does represent 50% of the data, but it provides a picture of the spread of the data. The box is not one-half the entire length of the box-and-whiskers plot because there is a greater spread of data in the interquartile range than in the first or fourth quarters of the data.

---

20. **Meteorology** The average monthly amounts of rainfall, in inches, from January through December for Seattle, Washington, and Houston, Texas, are listed below.

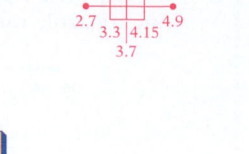

a. Is the difference between the means greater than 1 inch?   No
b. What is the difference between the medians?   0.8 inch
c. Draw a box-and-whiskers plot of each set of data. Use the same scale.
d. Describe the difference between the distributions of the data for Seattle and Houston.

Seattle	6.0	4.2	3.6	2.4	1.6	1.4	0.7	1.3	2.0	3.4	5.6	6.3
Houston	3.2	3.3	2.7	4.2	4.7	4.1	3.3	3.7	4.9	3.7	3.4	3.7

Source: The Weather Channel

21. **Meteorology** The average monthly amounts of rainfall, in inches, from January through December for Orlando, Florida, and Portland, Oregon, are listed below.

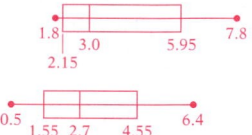

a. Is the difference between the means greater than 1 inch?   No
b. What is the difference between the medians?   0.3 inch
c. Draw a box-and-whiskers plot of each set of data. Use the same scale.
d. Describe the difference between the distributions of the data for Orlando and Portland.

Orlando	2.1	2.8	3.2	2.2	4.0	7.4	7.8	6.3	5.6	2.8	1.8	1.8
Portland	6.2	3.9	3.6	2.3	2.1	1.5	0.5	1.1	1.6	3.1	5.2	6.4

Source: The Weather Channel

APPLYING THE CONCEPTS

22.   **a.** Explain how to determine the mean of a set of data.
      b. Explain how to determine the median of a set of data.
c. Explain how to determine the mode of a set of data.

23. Write a set of data with five data values for which the mean, median, and mode are all 55.
Answers will vary. For example, 55, 55, 55, 55, 55, or 50, 55, 55, 55, 60

24.   A set of data has a mean of 16, a median of 15, and a mode of 14. Which of these numbers must be a value in the data set? Explain your answer.

25.   Explain each notation.
a. Q_1 **b.** Q_3 **c.** $\overline{x}$

26.   What values are shown on a box-and-whiskers plot? Explain how each is displayed.

27.   The box in a box-and-whiskers plot represents 50%, or one-half, of the data in a set. Why is the box in Example 3 of this section not one-half of the entire length of the box-and-whiskers plot?

28. Create a set of data containing 25 numbers that would correspond to the box-and-whiskers plot shown at the right.
Answers will vary. For example, 20, 21, 22, 24, 26, 27, 29, 31, 31, 32, 32, 33, 33, 36, 37, 37, 39, 40, 41, 43, 45, 46, 50, 54, 57

8.3 Introduction to Probability

Objective A   To calculate the probability of simple events

A weather forecaster estimates that there is a 75% chance of rain. A state lottery director claims that there is a $\frac{1}{9}$ chance of winning a prize in a new game offered by the lottery. Each of these statements involves uncertainty to some extent. The degree of uncertainty is called **probability.** For the statements above, the probability of rain is 75%, and the probability of winning a prize in the new lottery game is $\frac{1}{9}$.

A probability is determined from an **experiment,** which is any activity that has an observable outcome. Examples of experiments include

Tossing a coin and observing whether it lands heads up or tails up

Interviewing voters to determine their preference for a political candidate

Drawing a card from a standard deck of 52 cards

All the possible outcomes of an experiment are called the **sample space** of the experiment. The outcomes are listed between braces. For example:

The number cube shown at the left is rolled once. Any of the numbers from 1 to 6 could show on the top of the cube. The sample space is

$$\{1, 2, 3, 4, 5, 6\}$$

A fair coin is tossed once. (A fair coin is one for which heads and tails have an equal chance of landing face up.) If H represents "heads up" and T represents "tails up," then the sample space is

$$\{H, T\}$$

An **event** is one or more outcomes of an experiment. For the experiment of rolling the six-sided cube described above, some possible events are

The number is even: $\{2, 4, 6\}$
The number is a multiple of 3: $\{3, 6\}$
The number is less than 10: $\{1, 2, 3, 4, 5, 6\}$
(Note that in this case, the event is the entire sample space.)

HOW TO   The spinner at the left is spun once. Assume that the spinner does not come to rest on a line.

a. What is the sample space?

The arrow could come to rest on any one of the four sectors. The sample space is $\{1, 2, 3, 4\}$.

b. List the outcomes in the event that the spinner points to an odd number.

$\{1, 3\}$

In discussing experiments and events, it is convenient to refer to the **favorable outcomes** of an experiment. These are the outcomes of an experiment that satisfy the requirements of a particular event.

459

Objective 8.3A

New Vocabulary

probability
experiment
sample space
event
favorable outcomes
at random
theoretical probability
empirical probability

New Formulas

Theoretical Probability Formula
Empirical Probability Formula

Concept Check

1. For the situation of throwing one die, name an event that has a probability of $\frac{1}{2}$.

   Examples will vary. For example, throwing a 1, 2, or 3; throwing a 1, 3, or 5; throwing a 2, 4, or 6; throwing a 4, 5, or 6

2. For the experiment of throwing one die, what is the sum of the probabilities of throwing a 1, a 2, a 3, a 4, a 5, and a 6?   1

3. Find the probability of drawing an ace or a spade in one draw from a 52-card deck.
   $\frac{4}{13}$

4. A married couple wants to have three children. If the probability of having a boy is $\frac{1}{2}$ and the probability of having a girl is $\frac{1}{2}$, what is the probability that the couple will have exactly two girls?
   $\frac{3}{8}$

In-Class Examples (Objective 8.3A)

1. A coin is tossed three times. What is the probability that the outcome of the tosses is exactly $\{T, T, H\}$?
   $\frac{1}{8}$

2. Two dice are rolled. What is the probability that the sum of the dots on the upward faces is 4?
   $\frac{1}{12}$

Optional Student Activity

1. Have students work with the following: Each of three coins has a 1 on one side and a 2 on the other. All three coins are tossed. What is the probability of throwing a sum of 3? 4? 5? 6? 7?

$$\frac{1}{8}, \frac{3}{8}, \frac{3}{8}, \frac{1}{8}, 0$$

2. In how many different ways can a panel of four on-off switches be set if no two adjacent switches can be off?

8

3. What is the probability that a one-digit natural number selected at random is relatively prime to 12? ("Relatively prime to 12" means that the only common factor of the number and 12 is 1.)

$$\frac{1}{3}$$

4. The integers 1 through 300 are written on 300 identical pieces of paper. One of the pieces of paper is drawn randomly. What is the probability that the number drawn is divisible by 3? (When a number is divisible by 3, the sum of its digits is divisible by 3.)

$$\frac{1}{3}$$

For instance, consider the experiment of rolling a fair die once. The sample space is

$$\{1, 2, 3, 4, 5, 6\}$$

and one possible event would be rolling a number that is divisible by 3. The outcomes of the experiment that are favorable to the event are 3 and 6:

$$\{3, 6\}$$

The outcomes of the experiment of tossing a fair coin are *equally likely*. Either one of the outcomes is just as likely as the other. If a fair coin is tossed once, the probability of a head is $\frac{1}{2}$, and the probability of a tail is $\frac{1}{2}$. Both events are equally likely. The theoretical probability formula, given below, applies to experiments for which the outcomes are equally likely.

> **Study Tip**
>
> Remember to pay special attention to rule boxes such as the one at the right. They explain how certain types of problems are solved. See *AIM for Success*, page xxxiii.

> **Theoretical Probability Formula**
>
> The theoretical probability of an event is a fraction with the number of favorable outcomes of the experiment in the numerator and the total number of possible outcomes in the denominator.
>
> $$\text{Probability of an event} = \frac{\text{number of favorable outcomes}}{\text{number of possible outcomes}}$$

The probability of an event is a number from 0 to 1 that tells us how likely it is that a particular outcome will happen.

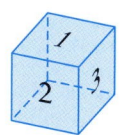

A probability of 0 means that the event is impossible.
The probability of getting a head when rolling the die shown at the left is 0.

A probability of 1 means that the event must happen.
The probability of getting either a head or a tail when tossing a coin is 1.

A probability of $\frac{1}{4}$ means that it is expected that the outcome will happen 1 in every 4 times the experiment is performed.

> **TAKE NOTE**
>
> The phrase **at random** means that each card has an equal chance of being drawn.

HOW TO   Each of the letters of the word *TENNESSEE* is written on a card, and the cards are placed in a hat. If one card is drawn at random from the hat, what is the probability that the card has the letter *E* on it?

Count the possible outcomes of the experiment.

There are 9 letters in *TENNESSEE*.
There are 9 possible outcomes of the experiment.

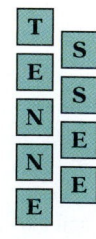

Count the number of outcomes of the experiment that are favorable to the event that a card with the letter *E* on it is drawn.

There are 4 cards with an *E* on them.

Use the probability formula.

$$\text{Probability of the event} = \frac{\text{number of favorable outcomes}}{\text{number of possible outcomes}} = \frac{4}{9}$$

The probability of drawing an *E* is $\frac{4}{9}$.

As shown above, calculating the probability of an event requires counting the number of possible outcomes of an experiment and the number of outcomes that are favorable to the event. One way to do this is to list the outcomes of the experiment in a systematic way. Using a table is often helpful.

When two dice are rolled, the sample space for the experiment can be recorded systematically, as in the following table.

Point of Interest

Romans called a die that was marked on four faces a *talus*, which meant "ankle bone." The ankle bone was considered an ideal die because it is roughly a rectangular solid and it has no marrow. Loose ankle bones from sheep were more likely than other bones to be left lying around after wolves had devoured their prey.

Possible Outcomes from Rolling Two Dice

HOW TO   Two dice are rolled once. Calculate the probability that the sum of the numbers on the two dice is 7.

Use the table above to count the number of possible outcomes of the experiment.

> There are 36 possible outcomes.

Count the number of outcomes of the experiment that are favorable to the event that a sum of 7 is rolled.

> There are 6 favorable outcomes: (1, 6), (2, 5), (3, 4), (4, 3), (5, 2), and (6, 1).

Use the probability formula.

$$\text{Probability of the event} = \frac{\text{number of favorable outcomes}}{\text{number of possible outcomes}} = \frac{6}{36} = \frac{1}{6}$$

The probability of a sum of 7 is $\frac{1}{6}$.

The probabilities calculated above are theoretical probabilities. The calculation of a **theoretical probability** is based on theory—for example, that either side of a coin is equally likely to land face up or that each of the six sides of a fair die is equally likely to land face up. Not all probabilities arise from such assumptions.

An **empirical probability** is based on observations of certain events. For instance, a weather forecast of a 75% chance of rain is an empirical probability. From historical records kept by the weather bureau, when a similar weather pattern existed, rain occurred 75% of the time. It is theoretically impossible to predict the weather, and only observations of past weather patterns can be used to predict future weather conditions.

> **Empirical Probability Formula**
>
> The empirical probability of an event is the ratio of the number of observations of the event to the total number of observations.
>
> $$\text{Probability of an event} = \frac{\text{number of observations of the event}}{\text{total number of observations}}$$

For example, suppose the records of an insurance company show that of 2549 claims for theft filed by policyholders, 927 were claims for more than $5000. The empirical probability that the next claim for theft that this company receives will be a claim for more than $5000 is the ratio of the number of claims for over $5000 to the total number of claims.

$$\frac{927}{2549} \approx 0.36$$

The probability is approximately 0.36.

Example 1

There are three choices, *a*, *b*, or *c*, for each of the two questions on a multiple-choice quiz. If the instructor randomly chooses a correct answer of *a*, *b*, or *c* for each question, what is the probability that the two correct answers on this quiz will be the same letter?

Strategy

To find the probability:

- List the outcomes of the experiment in a systematic way.
- Count the number of possible outcomes of the experiment.
- Count the number of outcomes of the experiment that are favorable to the event that the two correct answers on the quiz will be the same letter.
- Use the probability formula.

Solution

Possible outcomes: (a, a)   (b, a)   (c, a)
                                  (a, b)   (b, b)   (c, b)
                                  (a, c)   (b, c)   (c, c)

There are 9 possible outcomes.

There are 3 favorable outcomes:
(a, a), (b, b), (c, c)

$$\text{Probability} = \frac{\text{number of favorable outcomes}}{\text{number of possible outcomes}}$$

$$= \frac{3}{9} = \frac{1}{3}$$

The probability that the two correct answers will be the same letter is $\frac{1}{3}$.

You Try It 1

A professor writes three true/false questions for a quiz. If the professor randomly chooses which questions will have a "true" answer and which will have a "false" answer, what is the probability that two answers will be "true" and one will be "false"?

Your strategy

Your solution

$\dfrac{3}{8}$

Solution on p. S22

8.3 Exercises

Objective A   **To calculate the probability of simple events**

1. A coin is tossed four times. List the sample space of possible outcomes.
{(HHHH), (HHHT), (HHTT), (HTTT), (HTHH), (HTTH), (HTHT), (TTTT), (TTTH), (TTHH), (THHH), (TTHT), (THHT), (THTT), (THTH)}

2. Three cards—one red, one green, and one blue—are to be arranged in a stack. Using R for red, G for green, and B for blue, list all the different stacks that can be formed. (Some computer monitors are called RGB monitors for the colors red, green, and blue.)
RGB, RBG, GRB, GBR, BRG, BGR

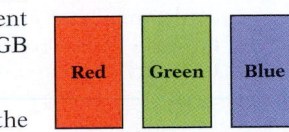

3. A tetrahedral die is one with four triangular sides. The sides show the numbers from 1 to 4. Suppose two tetrahedral dice are rolled. List the sample space of possible outcomes.
{(1, 1), (1, 2), (1, 3), (1, 4), (2, 1), (2, 2), (2, 3), (2, 4), (3, 1), (3, 2), (3, 3), (3, 4), (4, 1), (4, 2), (4, 3), (4, 4)}

4. A coin is tossed and then a die is rolled. List the sample space of possible outcomes. [To get you started, (H, 1) is one of the possible outcomes.]
{(H, 1), (H, 2), (H, 3), (H, 4), (H, 5), (H, 6), (T, 1), (T, 2), (T, 3), (T, 4), (T, 5), (T, 6)}

Tetrahedral die

5. The spinner at the right is spun once. Assume that the spinner does not come to rest on a line.
    a. What is the sample space?   {1, 2, 3, 4, 5, 6, 7, 8}
    b. List the outcomes in the event that the pointer comes to rest on a number less than 4.   {1, 2, 3}

6. A coin is tossed four times.
    a. What is the probability that the outcomes of the tosses are in the exact order HHTT? (See Exercise 1.) $\frac{1}{16}$
    b. What is the probability that the outcomes of the tosses consist of two heads and two tails? $\frac{3}{8}$
    c. What is the probability that the outcomes of the tosses consist of one head and three tails? $\frac{1}{4}$

7. Two dice are rolled.
    a. What is the probability that the sum of the dots on the upward faces is 5? $\frac{1}{9}$
    b. What is the probability that the sum of the dots on the upward faces is 15?   0
    c. What is the probability that the sum of the dots on the upward faces is less than 15?   1
    d. What is the probability that the sum of the dots on the upward faces is 2? $\frac{1}{36}$

8. A dodecahedral die has 12 sides numbered from 1 to 12. The die is rolled once.
    a. What is the probability that the upward face shows an 11? $\frac{1}{12}$
    b. What is the probability that the upward face shows a 5? $\frac{1}{12}$

9. A dodecahedral die has 12 sides numbered from 1 to 12. The die is rolled once.
    a. What is the probability that the upward face shows a number divisible by 4? $\frac{1}{4}$
    b. What is the probability that the upward face shows a number that is a multiple of 3? $\frac{1}{3}$

Dodecahedral die

10. Two tetrahedral dice are rolled (see Exercise 3).
    a. What is the probability that the sum on the upward faces is 4? $\frac{3}{16}$
    b. What is the probability that the sum on the upward faces is 6? $\frac{3}{16}$

Quick Quiz (Objective 8.3A)

1. A coin is tossed three times. What is the probability that the outcomes of the tosses consist of one tail and two heads?
$\frac{3}{8}$

2. Two dice are rolled. What is the probability that the sum of the dots on the upward faces is less than 4?
$\frac{1}{12}$

Suggested Assignment

Exercises 1–23, odds
More challenging problem:
    Exercise 22

11. Two dice are rolled. Which has the greater probability, throwing a sum of 10 or throwing a sum of 5?
    Throwing a sum of 5

12. Two dice are rolled once. Calculate the probability that the two numbers showing on the dice are equal. $\frac{1}{6}$

13. Each of the letters of the word *MISSISSIPPI* is written on a card, and the cards are placed in a hat. One card is drawn at random from the hat.
    a. What is the probability that the card has the letter *I* on it? $\frac{4}{11}$
    b. Which is greater, the probability of choosing an *S* or that of choosing a *P*?   **Choosing an *S***

14. There are five choices—choices *a* through *e*—for each question on a multiple-choice test. What is the probability of choosing the correct answer to a certain question by just guessing? $\frac{1}{5}$

15. Three blue marbles, four green marbles, and five red marbles are placed in a bag. One marble is chosen at random.
    a. What is the probability that the marble chosen is green? $\frac{1}{3}$
    b. Which is greater, the probability of choosing a blue marble or that of choosing a red marble?   **Choosing a red marble**

16. Which has the greater probability, drawing a jack, queen, or king from a deck of cards or drawing a spade?
    Drawing a spade

17. In a history class, a set of exams earned the following grades: 4 A's, 8 B's, 22 C's, 10 D's, and 3 F's. If a single student's exam is chosen from this class, what is the probability that it earned a B? $\frac{8}{47}$

18. A survey of 95 people showed that 37 preferred (to using a credit card) a cash discount of 2% if an item was purchased using cash or a check. Judging on the basis of this survey, what is the empirical probability that a person prefers a cash discount? Write the answer as a decimal rounded to the nearest hundredth.
    0.39

19. A survey of 725 people showed that 587 had a group health insurance plan where they worked. On the basis of this survey, what is the empirical probability that an employee has a group health insurance plan? Write the answer as a decimal rounded to the nearest hundredth.
    0.81

20. A television cable company surveyed some of its customers and asked them to rate the cable service as excellent, satisfactory, average, unsatisfactory, or poor. The results are recorded in the table at the right. What is the probability that a customer chosen at random from the survey rated the service as satisfactory or excellent? $\frac{185}{377}$

Quality of Service	Number Who Voted
Excellent	98
Satisfactory	87
Average	129
Unsatisfactory	42
Poor	21

APPLYING THE CONCEPTS

21. ✏️   If the spinner at the right is spun once, is each of the numbers 1 through 5 equally likely? Why or why not?

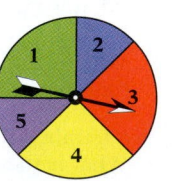

22. ✏️   The probability of tossing a coin and having it land heads up is $\frac{1}{2}$.

     Does this mean that if a coin is tossed 100 times, it will land heads up exactly 50 times? Explain your answer.

23. ✏️   Why can the probability of an event not be $\frac{5}{3}$?

Answers to Writing Exercises

21. No. The numbers 1 through 5 are not equally likely because the sizes of the sectors are different.

22. The probability of tossing a fair coin and having it land heads up is $\frac{1}{2}$. This does not mean that if the coin is tossed 100 times, it will land heads up exactly 50 times. Theoretical probabilities are obtained from logical reasoning, not from experimental data.

23. The mathematical probability of an event is a number from 0 to 1. Because $\frac{5}{3} = 1\frac{2}{3}$, $\frac{5}{3} > 1$. The probability of an event cannot be $\frac{5}{3}$ because a probability cannot be greater than 1.

Focus on Problem Solving

Inductive Reasoning Suppose that, beginning in January, you save $25 each month. The total amount you have saved at the end of each month can be described by a list of numbers.

25	50	75	100	125	150	175	
Jan	Feb	Mar	Apr	May	June	July	. . .

The list of numbers that indicates your total savings is an *ordered* list of numbers called a **sequence.** Each of the numbers in a sequence is called a **term** of the sequence. The list is ordered because the position of a number in the list indicates the month in which that total amount has been saved. For example, the 7th term of the sequence (indicating July) is 175. This number means that a total of $175 has been saved by the end of the 7th month.

Now consider a person who has a different savings plan. The total amount saved by this person for the first seven months is given by the sequence

$$20, 35, 50, 65, 80, 95, 110, . . .$$

The process you use to discover the next number in the above sequence is called *inductive reasoning.* **Inductive reasoning** involves making generalizations from specific examples; in other words, we reach a conclusion by making observations about particular facts or cases. In the case of the above sequence, the person saved $15 per month after the first month.

Here is another example of inductive reasoning. Find the next two letters of the sequence A, B, E, F, I, J, ...

By trying different patterns, we can determine that a pattern for this sequence is

$$A, B, C, D, E, F, G, H, I, J, . . .$$

That is, write the first two letters of the alphabet, skip the next two letters, write the next two letters, skip the next two letters, and so on. The next two letters are M, N.

Use inductive reasoning to solve the following problems.

1. What is the next term of the sequence: ban, ben, bin, bon, . . . ?

2. ▦ Using a calculator, determine the decimal representations of several proper fractions that have a denominator of 99. For instance, you may use $\frac{8}{99}, \frac{23}{99}$, and $\frac{75}{99}$. Now use inductive reasoning to explain the pattern, and use your reasoning to predict the decimal representation of $\frac{53}{99}$ without using a calculator.

3. Find the next number in the sequence 1, 1, 2, 3, 5, 8, 13, 21, ...

4. The decimal representation of a number begins 0.10100100010000100000 . . . . What are the next 10 digits in this number?

5. The first seven rows of a triangle of numbers called Pascal's triangle are given below. Find the next row.

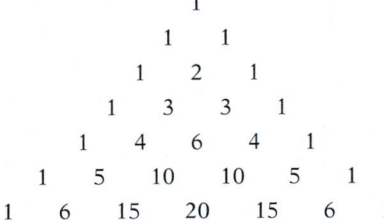

```
                1
              1   1
            1   2   1
          1   3   3   1
        1   4   6   4   1
      1   5   10  10  5   1
    1   6   15  20  15  6   1
```

Answers to Focus on Problem Solving: Inductive Reasoning

1. bun

2. Each decimal representation is a repeating decimal in which two digits repeat. If the numerator of the original fraction is one digit, then the repeating digits are 0 and the numerator. If the numerator of the original fraction is a two-digit number, then the repeating digits are the numerator.

$$\frac{53}{99} = 0.535353 . . .$$

3. 34

4. 1000000100

5. The next row is

1 7 21 35 35 21 7 1

Projects and Group Activities

Collecting, Organizing, Displaying, and Analyzing Data

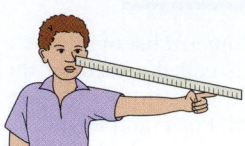

Projects and Group Activities: Collecting, Organizing, Displaying, and Analyzing Data

You will need centimeter tape measures for this project.

Before standardized units of measurement existed, measurements were made in terms of the human body. For example, the cubit was the distance from the end of the elbow to the tips of the fingers. The yard was the distance from the tip of the nose to the tips of the fingers on an outstretched arm.

For each student in the class, find the measure from the tip of the nose to the tips of the fingers on an outstretched arm. Round each measure to the nearest centimeter. Record all the measurements on the board.

1. From the data collected, determine each of the following.
   Mean _____
   Median _____
   Mode _____
   Range _____
   First quartile, Q_1 _____
   Third quartile, Q_3 _____
   Interquartile range _____

2. Prepare a box-and-whiskers plot of the data.

3. ✏ Write a description of the spread of the data.

4. ✏ Explain why we need standardized units of measurement.

Chapter 8 Summary

Key Words

Statistics is the branch of mathematics concerned with *data*, or numerical information. A *graph* is a pictorial representation of data. A *circle graph* represents data by the size of the sectors. [8.1A, p. 445]

Examples

The circle graph shows the results of a survey of 300 people who were asked to name their favorite sport.

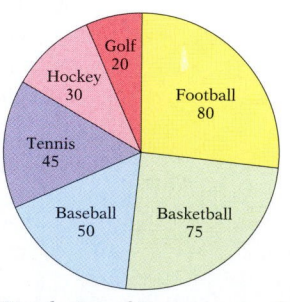

Distribution of Responses in a Survey

A *bar graph* represents data by the height of the bars.
[8.1A, p. 445]

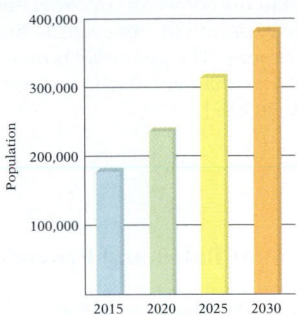 The bar graph shows the expected U.S. population aged 100 and over. *Source:* Census Bureau

Expected U.S. Population Aged 100 and Over

A *broken-line graph* represents data by the position of the lines and shows trends or comparisons. [8.1A, p. 455]

The line graph shows a recent graduate's cumulative debt in college loans at the end of each of the four years of college.

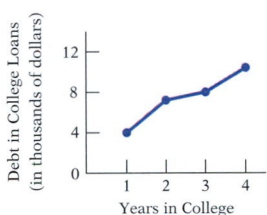

Cumulative Debt in College Loans

The *mean, median,* and *mode* are three types of averages used in statistics. The *mean* of a set of data is the sum of the data values divided by the number of values in the set. The *median* of a set of data is the number that separates the data into two equal parts when the data have been arranged from least to greatest (or greatest to least). There are an equal number of values above the median and below the median. The *mode* of a set of numbers is the value that occurs most frequently. [8.2A, pp. 449, 450]

Consider the following set of data.

24, 28, 33, 45, 45

The mean is 35.
The median is 33.
The mode is 45.

A *box-and-whiskers plot,* or *boxplot,* is a graph that shows five numbers: the least value, the first quartile, the median, the third quartile, and the greatest value of a data set. The *first quartile,* Q_1, is the number below which one-fourth of the data lie. The *third quartile,* Q_3, is the number above which one-fourth of the data lie. The box is placed around the values between the first quartile and the third quartile. The *range* is the difference between the largest number and the smallest number in the set. The range describes the spread of the data. The *interquartile range* is the difference between Q_3 and Q_1. [8.2B, pp. 452–453]

The box-and-whiskers plot for a set of test scores is shown below.

Range $= 96 - 45 = 51$
$Q_1 = 65$
$Q_3 = 86$
Interquartile range $= Q_3 - Q_1$
$= 86 - 65 = 21$

Probability is a number from 0 to 1 that tells us how likely it is that a certain outcome of an experiment will happen. An *experiment* is an activity with an observable outcome. All the possible outcomes of an experiment are called the *sample space* of the experiment. An *event* is one or more outcomes of an experiment. The *favorable outcomes* of an experiment are the outcomes that satisfy the requirements of a particular event. [8.3A, p. 459]

Tossing a single die is an example of an experiment. The sample space for this experiment is the set of possible outcomes:

$$\{1, 2, 3, 4, 5, 6\}$$

The event that the number landing face up is an odd number is represented by

$$\{1, 3, 5\}$$

Essential Rules and Procedures

Examples

To Find the Mean of a Set of Data [8.2A, p. 449]
Divide the sum of the values by the number of values in the set.
$$\bar{x} = \frac{\text{sum of the data values}}{\text{number of data values}}$$

Consider the following set of data.

24, 28, 33, 45, 45

$$\bar{x} = \frac{24 + 28 + 33 + 45 + 45}{5} = 35$$

To Find the Median [8.2A, p. 450]
1. Arrange the numbers from least to greatest.
2. If there is an *odd* number of values in the set of data, the median is the middle number. If there is an *even* number of values in the set of data, the median is the mean of the two middle numbers.

Consider the following set of data.

24, 28, 33, 35, 45, 45

The median is $\dfrac{33 + 35}{2} = 34$.

To Find Q_1 [8.2B, p. 452]
Arrange the numbers from least to greatest and locate the median. Q_1 is the median of the lower half of the data.

Consider the following data.

8	10	12	14	16	19	22
	↑		↑			
	Q_1		Median			

To Find Q_3 [8.2B, p. 452]
Arrange the numbers from least to greatest and locate the median. Q_3 is the median of the upper half of the data.

Consider the following data.

8	10	12	14	16	19	22
			↑		↑	
			Median		Q_3	

Theoretical Probability Formula [8.3A, p. 460]

$$\text{Probability of an event} = \frac{\text{number of favorable outcomes}}{\text{number of possible outcomes}}$$

A die is rolled. The probability of rolling a 2 or a 4 is $\dfrac{2}{6} = \dfrac{1}{3}$.

Empirical Probability Formula [8.3A, p. 462]

$$\text{Probability of an event} = \frac{\text{number of observations of the event}}{\text{total number of observations}}$$

A thumbtack is tossed 100 times. It lands point up 15 times and lands on its side 85 times. From this experiment, the empirical probability of "point up" is $\dfrac{15}{100} = \dfrac{3}{20}$.

Chapter 8 Review Exercises

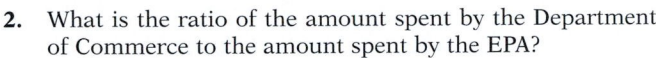 **Internet** The circle graph in Figure 8.8 shows the approximate amount of money that government agencies spent on maintaining Internet websites for a 3-year period. Use this graph for Exercises 1 to 3.

1. Find the total amount of money that these agencies spent on maintaining websites.
$349 million   [8.1A]

2. What is the ratio of the amount spent by the Department of Commerce to the amount spent by the EPA?
$\frac{9}{8}$ [8.1A]

3. What percent of the total money spent did NASA spend? Round to the nearest tenth of a percent.
8.9%   [8.1A]

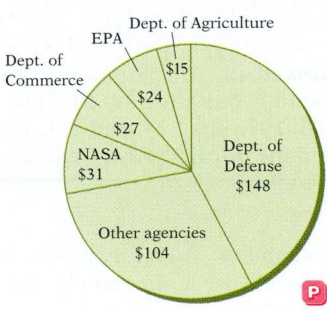

Figure 8.8 Amount federal agencies spent on websites (in millions of dollars)
Source: General Accounting Office

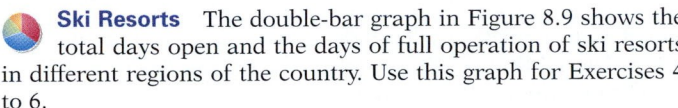 **Ski Resorts** The double-bar graph in Figure 8.9 shows the total days open and the days of full operation of ski resorts in different regions of the country. Use this graph for Exercises 4 to 6.

4. Find the difference between the total days open and the days of full operation for Midwest ski areas.
50 days   [8.1A]

5. What percent of the total days open were the days of full operation for the Rocky Mountain ski areas?
50%   [8.1A]

6. a. Which region had the lowest number of days of full operation?
b. How many days of full operation did this region have?
a. Southeast   **b.** 30 days   [8.1A]

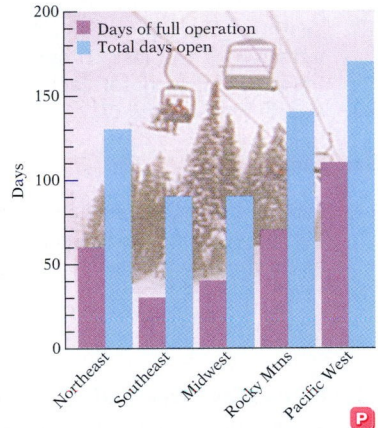

Figure 8.9
Source: Economic Analysis of United States Ski Areas

7. Health A health clinic administered a test for cholesterol to 11 people. The results were 180, 220, 160, 230, 280, 200, 210, 250, 190, 230, and 210. Find the mean and median of these data.
Mean: 214.$\overline{54}$; median: 210   [8.2A]

8. Newborns The weights, in pounds, of 10 babies born at a hospital were recorded as 6.3, 5.9, 8.1, 6.5, 7.2, 5.6, 8.9, 9.1, 6.9, and 7.2. Find the mean and median of these data.
Mean: 7.17 lb; median: 7.05 lb   [8.2A]

9. **Movies**   People leaving a new movie were asked to rate the movie as bad, good, very good, or excellent. The responses were bad, 28; good, 65; very good, 49; excellent, 28. What was the modal response for this survey?
    Good   [8.2A]

10. **Basketball**   The numbers of points scored by a basketball team for 15 games were 89, 102, 134, 110, 121, 124, 111, 116, 99, 120, 105, 109, 110, 124, and 131. Find the first quartile, the median, and the third quartile. Draw a box-and-whiskers plot.
    105, 111, 124   [8.2B]

    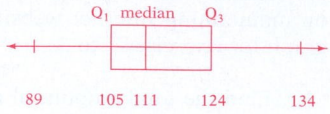

11. **Probability**   A coin is tossed and then a regular die is rolled. How many elements are in the sample space?
    12 elements   [8.3A]

12. **Probability**   A charity raffle sells 2500 raffle tickets for a big-screen television set. If you purchase 5 tickets, what is the probability that you will win the television?
    $\dfrac{1}{500}$ [8.3A]

13. **Testing**   An employee at a department of motor vehicles analyzed the written tests of the last 10 applicants for a drivers' license. The numbers of incorrect answers for each of these applicants were 2, 0, 3, 1, 0, 4, 5, 1, 3, and 1. Find Q_1 and Q_3.
    $Q_1 = 1$; $Q_3 = 3$ [8.2B]

14. **Probability**   A dodecahedral die has 12 sides numbered from 1 to 12. If this die is rolled once, what is the probability that a number divisible by 6 will be on the upward face?
    $\dfrac{1}{6}$ [8.3A]

15. **Probability**   One student is randomly selected from 3 first-year students, 4 sophomores, 5 juniors, and 2 seniors. What is the probability that the student is a junior?
    $\dfrac{5}{14}$ [8.3A]

16. **Physical Fitness**   The heart rates of 24 women tennis players were measured after each of them had run one-quarter of a mile. The results are listed in the table below.

80	82	99	91	93	87	103	94	73	96	86	80
97	94	108	81	100	109	91	84	78	96	96	100

    a. Find the mean, the median, and the mode for the data. Round to the nearest tenth.
    b. Find the range and the interquartile range for the data.
    a. Mean: 91.6 heartbeats per minute; median: 93.5 heartbeats per minute; mode: 96 heartbeats per minute   [8.2A]
    b. Range: 36 heartbeats per minute; interquartile range: 15 heartbeats per minute   [8.2B]

Chapter 8 Test

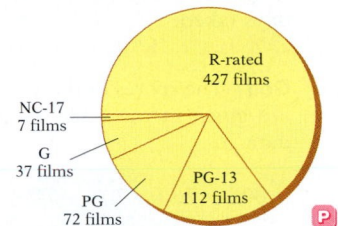

The Film Industry The circle graph in Figure 8.10 categorizes the 655 films released during a recent year by their ratings. Use this graph for Exercises 1 to 3.

1. How many more R-rated films were released than PG films?
   355 more   [8.1A]

2. The number of PG-13 films released was how many times the number of NC-17 films released?
   16 times   [8.1A]

3. What percent of the films released were rated G? Round to the nearest tenth of a percent.
   5.6%   [8.1A]

Figure 8.10 Ratings of films released
Source: MPA Worldwide Market Research

Education The broken-line graph in Figure 8.11 shows the number of students enrolled in colleges. Use this figure for Exercises 4 and 5.

4. During which decade did the student population increase the least?
   The 1990s   [8.1A]

5. Approximate the increase in college enrollment from 1960 to 2000.
   11 million students   [8.1A]

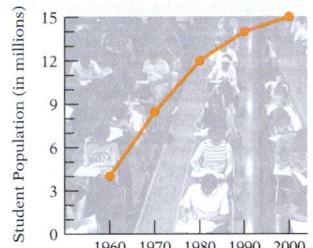

Figure 8.11 Student enrollment in public and private colleges
Source: National Center for Educational Statistics

6. **Bowling** The bowling scores for eight people were 138, 125, 162, 144, 129, 168, 184, and 173. What was the mean score for these eight people?
   152.875   [8.2A]

7. **Emergency Calls** The response times by an ambulance service to emergency calls were recorded by a public safety commission. The times (in minutes) were 17, 21, 11, 8, 22, 15, 11, 14, and 8. Determine the median response time for these calls.
   14 min   [8.2A]

8. **Education** Recent college graduates were asked to rate the quality of their education. The responses were 47, excellent; 86, very good; 32, good; 20, poor. What was the modal response?
   Very good   [8.2A]

9. **Manufacturing** The average time, in minutes, it takes a factory worker to assemble 14 different toys is given in the table. Determine the first quartile of the data.
   9.8   [8.2B]

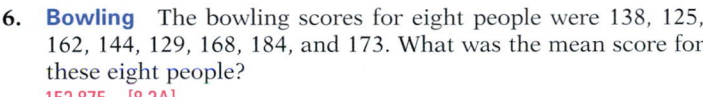

| 10.5 | 21.0 | 17.3 | 11.2 | 9.3 | 6.5 | 8.6 |
| 19.8 | 20.3 | 19.6 | 9.8 | 10.5 | 11.9 | 18.5 |

10. **Business** The number of vacation days taken last year by each of the employees of a firm was recorded. The box-and-whiskers plot at the right represents the data. **a.** Determine the range of the data. **b.** What was the median number of vacation days taken?
    a. 22 days   **b.** 14 vacation days   [8.2B]

Q_1 median Q_3

4   7   14   20   26

11. **Golf** The scores of the 14 leaders in a college golf tournament are given in the table. Draw a box-and-whiskers plot of the data.

Q_1 median Q_3

68   70   72.5   74   80

80	76	70	71	74	68	72	[8.2B]
74	70	70	73	75	69	73	

12. **Probability** Three coins—a nickel, a dime, and a quarter—are stacked. List the events in the sample space.
    {(N, D, Q), (N, Q, D), (D, N, Q), (D, Q, N), (Q, N, D), (Q, D, N)}   [8.3A]

13. **Probability** A cross-country flight has 14 passengers in first class, 32 passengers in business class, and 202 passengers in coach. If one passenger is selected at random, what is the probability that the person is in business class?
    $\frac{4}{31}$ [8.3A]

14. **Probability** Three playing cards—an ace, a king, and a queen—are randomly arranged and stacked. What is the probability that the ace is on top of the stack?
    $\frac{1}{3}$ [8.3A]

15. **Probability** A quiz contains three true/false questions. If a student attempts to answer the questions by just guessing, what is the probability that the student will answer all three questions correctly?
    $\frac{1}{8}$ [8.3A]

16. **Probability** A package of flower seeds contains 15 seeds for red flowers, 20 seeds for white flowers, and 10 seeds for pink flowers. If one seed is selected at random, what is the probability that it is not a seed for a red flower?
    $\frac{2}{3}$ [8.3A]

17. **Probability** A dodecahedral die has 12 sides. If this die is tossed once, what is the probability that the number on the upward face is less than 6?
    $\frac{5}{12}$ [8.3A]

18. **Quality Control** The length of time (in days) that various batteries operated a portable CD player continuously are given in the table below.

2.9	2.4	3.1	2.5	2.6	2.0	3.0	2.3	2.4	2.7
2.0	2.4	2.6	2.7	2.1	2.9	2.8	2.4	2.0	2.8

    a. Find the mean for the data.   2.53 days   [8.2A]
    b. Find the median for the data.   2.55 days   [8.2A]
    c. Draw a box-and-whiskers plot for the data.

2.0   2.35   2.55   2.8   3.1   [8.2B]

Cumulative Review Exercises

1. Simplify: $2^2 \cdot 3^3 \cdot 5$
540 [1.3B]

2. Simplify: $3^2 \cdot (5 - 2) \div 3 + 5$
14 [1.4A]

3. Find the LCM of 24 and 40.
120 [2.1A]

4. Write $\frac{60}{144}$ in simplest form.
$\frac{5}{12}$ [2.2B]

5. Find the total of $4\frac{1}{2}$, $2\frac{3}{8}$, and $5\frac{1}{5}$.
$12\frac{3}{40}$ [2.3A]

6. Subtract: $12\frac{5}{8} - 7\frac{11}{12}$
$4\frac{17}{24}$ [2.3B]

7. Multiply: $\frac{5}{8} \times 3\frac{1}{5}$
2 [2.4A]

8. Find the quotient of $3\frac{1}{5}$ and $4\frac{1}{4}$.
$\frac{64}{85}$ [2.4B]

9. Simplify: $\frac{5}{8} \div \left(\frac{3}{4} - \frac{2}{3} \right) + \frac{3}{4}$
$8\frac{1}{4}$ [2.7A]

10. Write two hundred nine and three hundred five thousandths in standard form.
209.305 [2.5A]

11. Find the product of 4.092 and 0.69.
2.82348 [2.6B]

12. Convert $16\frac{2}{3}$ to a decimal. Round to the nearest hundredth.
16.67 [2.6D]

13. Write "330 miles on 12.5 gallons of gas" as a unit rate.
26.4 mpg [6.1A]

14. Solve the proportion: $\frac{n}{5} = \frac{16}{25}$
3.2 [6.2A]

15. Write $\frac{4}{5}$ as a percent.
80% [6.3B]

16. 8 is 10% of what?
80 [6.4A/6.4B]

17. What is 38% of 43?
16.34 [6.4A/6.4B]

18. What percent of 75 is 30?
40% [6.4A/6.4B]

19. **Compensation** Tanim Kamal, a salesperson at a department store, receives $100 per week plus 2% commission on sales. Find the income for a week in which Tanim had $27,500 in sales.
$650   [6.4C]

20. **Insurance** A life insurance policy costs $8.15 for every $1000 of insurance. At this rate, what is the cost for $50,000 of life insurance?
$407.50   [6.2B]

21. **Simple Interest** A contractor borrowed $125,000 for 6 months at an annual simple interest rate of 6%. Find the interest due on the loan.
$3750   [6.5A]

22. **Business** A compact disc player that costs a retailer $180 is priced at 155% of the cost. Find the price of the compact disc player.
$279   [6.4C]

23. **Finance** The circle graph in Figure 8.12 shows how a family's monthly income of $3000 is budgeted. How much is budgeted for food?
$570   [8.1A]

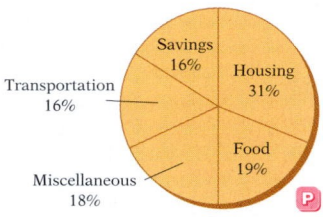

Figure 8.12 Budget for a monthly income of $3000

24. **Education** The double-broken-line graph in Figure 8.13 shows two students' scores on 5 math tests of 30 problems each. Find the difference between the scores of the two students on Test 1.
12 points   [8.1A]

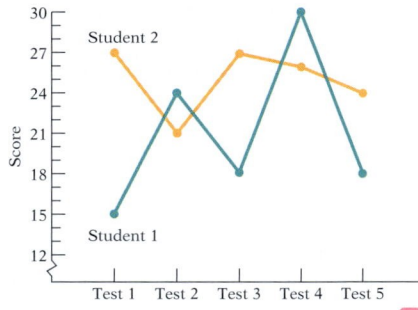

Figure 8.13

25. **Meteorology** The average daily high temperatures, in degrees Fahrenheit, for a week in Newtown were 56°, 72°, 80°, 75°, 68°, 62°, and 74°. Find the mean high temperature for the week. Round to the nearest tenth of a degree.
69.6°F   [8.2A]

26. **Probability** Two dice are rolled. What is the probability that the sum of the dots on the upward faces is 8?
$\frac{5}{36}$ [8.3A]

chapter

9

Polynomials

High-powered telescopes, such as the one shown here, allow scientists to look at objects that are extremely large but very far away. As an astronomer, this man uses scientific notation to describe distances in space. Scientific notation replaces very large or very small numbers with more concise expressions, making these numbers easier to read and write. **Exercises 115 to 122 on pages 501 and 502** describe some situations in which scientific notation is used.

OBJECTIVES

Section 9.1

A To add polynomials
B To subtract polynomials

Section 9.2

A To multiply monomials
B To simplify powers of monomials

Section 9.3

A To multiply a polynomial by a monomial
B To multiply two polynomials
C To multiply two binomials
D To multiply binomials that have special products
E To solve application problems

Section 9.4

A To divide monomials
B To write a number in scientific notation

Section 9.5

A To divide a polynomial by a monomial
B To divide polynomials

Need help? For online student resources, such as section quizzes, visit this textbook's website at **math.college.hmco.com/students.**

PREP TEST • • •

Do these exercises to prepare for Chapter 9.

1. Subtract: $-2 - (-3)$
1   [3.2B]

2. Multiply: $-3(6)$
-18 [3.3A]

3. Simplify: $-\dfrac{24}{-36}$
$\dfrac{2}{3}$ [3.3B]

4. Evaluate $3n^4$ when $n = -2$.
48   [4.1A]

5. If $\dfrac{a}{b}$ is a fraction in simplest form, what number is not a possible value of b?
0   [1.3C]

6. Are $2x^2$ and $2x$ like terms?
No   [4.1A]

7. Simplify: $3x^2 - 4x + 1 + 2x^2 - 5x - 7$
$5x^2 - 9x - 6$ [4.2A]

8. Simplify: $-4y + 4y$
0   [4.2A]

9. Simplify: $-3(2x - 8)$
$-6x + 24$ [4.2C]

10. Simplify: $3xy - 4y - 2(5xy - 7y)$
$-7xy + 10y$ [4.2D]

GO FIGURE • • •

If $x + y$, xy, and $\dfrac{x}{y}$ all equal the same number, find the values of x and y.

$x = \dfrac{1}{2}, y = -1$

9.1 Addition and Subtraction of Polynomials

Objective A   To add polynomials

> **TAKE NOTE**
> The expression $3\sqrt{x}$ is not a monomial because $\sqrt{x}$ cannot be written as a product of variables.
> The expression $\dfrac{2x}{y}$ is not a monomial because it is a *quotient* of variables.

A **monomial** is a number, a variable, or a product of numbers and variables. For instance,

7	b	$\dfrac{2}{3}a$	$12xy^2$
A number	A variable	A product of a number and a variable	A product of a number and variables

A **polynomial** is a variable expression in which the terms are monomials.

A polynomial of *one* term is a **monomial.** $-7x^2$ is a monomial.
A polynomial of *two* terms is a **binomial.** $4x + 2$ is a binomial.
A polynomial of *three* terms is a **trinomial.** $7x^2 + 5x - 7$ is a trinomial.

The **degree of a polynomial in one variable** is the greatest exponent on a variable. The degree of $4x^3 - 5x^2 + 7x - 8$ is 3; the degree of $2y^4 + y^2 - 1$ is 4. The degree of a nonzero constant is zero. For instance, the degree of 7 is zero. The number zero has no degree.

The terms of a polynomial in one variable are usually arranged so that the exponents of the variable decrease from left to right. This is called **descending order.**

$$5x^3 - 4x^2 + 6x - 1$$
$$7z^4 + 4z^3 + z - 6$$
$$2y^4 + y^3 - 2y^2 + 4y - 1$$

Polynomials can be added, using either a horizontal or a vertical format, by combining like terms.

> **HOW TO**   Add $(3x^3 - 7x + 2) + (7x^2 + 2x - 7)$. Use a horizontal format.
>
> $(3x^3 - 7x + 2) + (7x^2 + 2x - 7)$ • Use the Commutative and Associative Properties of Addition to rearrange and group like terms.
> $= 3x^3 + 7x^2 + (-7x + 2x) + (2 - 7)$
>
> $= 3x^3 + 7x^2 - 5x - 5$ • Then combine like terms.

> **HOW TO**   Add $(-4x^2 + 6x - 9) + (12 - 8x + 2x^3)$. Use a vertical format.
>
> $\begin{array}{r} -4x^2 + 6x - 9 \\ 2x^3 \phantom{{}+000} - 8x + 12 \\ \hline 2x^3 - 4x^2 - 2x + 3 \end{array}$
>
> • Arrange the terms of each polynomial in descending order with like terms in the same column.
> • Combine the terms in each column.

Example 1

Use a horizontal format to add
$(8x^2 - 4x - 9) + (2x^2 + 9x - 9)$.

Solution
$(8x^2 - 4x - 9) + (2x^2 + 9x - 9)$
$= (8x^2 + 2x^2) + (-4x + 9x) + (-9 - 9)$
$= 10x^2 + 5x - 18$

You Try It 1

Use a horizontal format to add
$(-4x^3 + 2x^2 - 8) + (4x^3 + 6x^2 - 7x + 5)$.

Your solution
$8x^2 - 7x - 3$

Solution on p. S22

In-Class Examples (Objective 9.1A)

1. Use a horizontal format to add.
   a. $(4x^2 + 5x - 7) + (8x^2 - 10x - 3)$
   $12x^2 - 5x - 10$
   b. $(7x^2 - 3x - 4) + (3x^2 - 5x + 6)$
   $10x^2 - 8x + 2$

2. Use a vertical format to add.
   a. $(-9x^2 - 6x + 8) + (4x^2 + 10x + 6)$
   $-5x^2 + 4x + 14$
   b. $(8x^2 + 9x - 3) + (-6x^2 - 12x - 7)$
   $2x^2 - 3x - 10$

New Vocabulary
monomial
polynomial
binomial
trinomial
descending order
degree of a polynomial in one variable

Instructor Note
An analogy may help students understand these terms. *Polynomial* is similar to the word *car*. Chevrolet and Ford are types of cars. Monomials and binomials are types of polynomials.
   As a class exercise, ask students to identify monomials. For instance, which of $\dfrac{1}{2}$, $\dfrac{2}{3}y$, $6x$, $6 + x$, $abxy$, and $\dfrac{3}{z}$ are monomials?
   Then ask students to identify polynomials. Here are some possible examples: $x + 7$, $\dfrac{x^2}{5} + \dfrac{x}{2} - 1$, $\dfrac{5}{x^2} + \dfrac{2}{x} - 1$, $\dfrac{x + 1}{x}$, and $\sqrt{2}x^4 - \pi x^2 - 7$.

Concept Check
Determine the degree of each of the following.
1. $9x^2 - 7x - 5$ 2
2. $3x^4 + x^2 - 5$ 4
3. $x + 6$ 1
4. -987 0

Discuss the Concepts
1. How is $3x^2 + 4x + 5$ similar to 345? *Hint:* Think of a value of x that would make these expressions equal.
   $345 = 3 \cdot 10^2 + 4 \cdot 10 + 5$
2. How is $4x^2 + 7$ similar to 407?
   $407 = 4 \cdot 10^2 + 0 \cdot 10 + 7$
3. How is addition of polynomials similar to addition of whole numbers?
   Like terms are added.

Fill in the blank. To add polynomials, add the _____ terms.

like

Objective 9.1B

New Vocabulary

opposite of a polynomial

Concept Check

1. Determine the opposite of $2x^2 - 12x + 5$.
   $-2x^2 + 12x - 5$

2. Replace the question mark to make a true statement.
   $(3x^2 - 4x + 5) + (?) = 0$
   $-3x^2 + 4x - 5$

Optional Student Activity

1. Make up two polynomials, each of degree 2, whose sum is also of degree 2.

2. Make up two polynomials, each of degree 2, whose sum is of degree 1.

3. Make up two polynomials, each of degree 2, whose sum is of degree 0.

1–3. Answers will vary. Possible answers include:
   1. $3x^2$ and $4x^2$
   2. $3x^2 + 5x$ and $-3x^2 + x$
   3. $3x^2$ and $-3x^2 - 2$

> **Example 2**
>
> Use a vertical format to add
> $(-5x^3 + 4x^2 - 7x + 9) + (2x^3 + 5x - 11)$.
>
> **Solution**
> $$\begin{array}{r} -5x^3 + 4x^2 - 7x + 9 \\ 2x^3 \quad\quad + 5x - 11 \\ \hline -3x^3 + 4x^2 - 2x - 2 \end{array}$$

> **You Try It 2**
>
> Use a vertical format to add
> $(6x^3 + 2x + 8) + (-9x^3 + 2x^2 - 12x - 8)$.
>
> **Your solution** $-3x^3 + 2x^2 - 10x$

Solution on p. S22

Objective B To subtract polynomials

The **opposite of the polynomial** $(3x^2 - 7x + 8)$ is $-(3x^2 - 7x + 8)$.

To simplify the opposite of a polynomial, change the sign of each term to its opposite.
$$-(3x^2 - 7x + 8) = -3x^2 + 7x - 8$$

> **TAKE NOTE**
> This is the same definition used for subtraction of integers: Subtraction is addition of the opposite.

Polynomials can be subtracted using either a horizontal or a vertical format. To subtract, add the opposite of the second polynomial to the first.

HOW TO Subtract $(4y^2 - 6y + 7) - (2y^3 - 5y - 4)$. Use a horizontal format.

$(4y^2 - 6y + 7) - (2y^3 - 5y - 4)$
$= (4y^2 - 6y + 7) + (-2y^3 + 5y + 4)$ • Add the opposite of the second polynomial to the first.
$= -2y^3 + 4y^2 + (-6y + 5y) + (7 + 4)$
$= -2y^3 + 4y^2 - y + 11$ • Combine like terms.

HOW TO Subtract $(9 + 4y + 3y^3) - (2y^2 + 4y - 21)$. Use a vertical format.

The opposite of $2y^2 + 4y - 21$ is $-2y^2 - 4y + 21$.

$$\begin{array}{r} 3y^3 \quad\quad + 4y + 9 \\ - 2y^2 - 4y + 21 \\ \hline 3y^3 - 2y^2 \quad\quad + 30 \end{array}$$

• Arrange the terms of each polynomial in descending order, with like terms in the same column.
• Note that $4y - 4y = 0$, but 0 is not written.

> **Example 3**
>
> Use a horizontal format to subtract
> $(7c^2 - 9c - 12) - (9c^2 + 5c - 8)$.
>
> **Solution**
> $(7c^2 - 9c - 12) - (9c^2 + 5c - 8)$
> $= (7c^2 - 9c - 12) + (-9c^2 - 5c + 8)$
> $= -2c^2 - 14c - 4$

> **You Try It 3**
>
> Use a horizontal format to subtract
> $(-4w^3 + 8w - 8) - (3w^3 - 4w^2 - 2w - 1)$.
>
> **Your solution**
> $-7w^3 + 4w^2 + 10w - 7$

> **Example 4**
>
> Use a vertical format to subtract
> $(3k^2 - 4k + 1) - (k^3 + 3k^2 - 6k - 8)$.
>
> **Solution**
> $$\begin{array}{r} 3k^2 - 4k + 1 \\ -k^3 - 3k^2 + 6k + 8 \\ \hline -k^3 \quad\quad + 2k + 9 \end{array}$$
> • Add the opposite of $(k^3 + 3k^2 - 6k - 8)$ to the first polynomial.

> **You Try It 4**
>
> Use a vertical format to subtract
> $(13y^3 - 6y - 7) - (4y^2 - 6y - 9)$.
>
> **Your solution**
> $13y^3 - 4y^2 + 2$

Solutions on p. S22

In-Class Examples (Objective 9.1B)

1. Use a horizontal format to subtract.
   a. $(6x^2 + 2x - 4) - (-8x^2 + x - 3)$
      $14x^2 + x - 1$
   b. $(-4x^2 + 7x + 6) - (7x^2 + 4x - 2)$
      $-11x^2 + 3x + 8$

2. Use a vertical format to subtract.
   a. $(7y^2 + 8y + 2) - (4y^3 + 9y^2 - 5y + 6)$
      $-4y^3 - 2y^2 + 13y - 4$
   b. $(-6y^2 - 3y - 5) - (-2y^3 - 3y^2 - 6y - 5)$
      $2y^3 - 3y^2 + 3y$

9.1 Exercises

Objective A **To add polynomials**

Suggested Assignment
Exercises 1–55, odds
More challenging problems:
Exercises 57, 58, 61

For Exercises 1 to 8, state whether the expression is a monomial.

1. 17 Yes

2. $3x^4$ Yes

3. $\dfrac{17}{\sqrt{x}}$ No

4. xyz Yes

5. $\dfrac{2}{3}y$ Yes

6. $\dfrac{xy}{z}$ No

7. $\sqrt{5}\,x$ Yes

8. πx Yes

For Exercises 9 to 16, state whether the expression is a monomial, a binomial, a trinomial, or none of these.

9. $3x + 5$
   Binomial

10. $2y - 3\sqrt{y}$
    None of these

11. $9x^2 - x - 1$
    Trinomial

12. $x^2 + y^2$
    Binomial

13. $\dfrac{2}{x} - 3$
    None of these

14. $\dfrac{ab}{4}$
    Monomial

15. $6x^2 + 7x$
    Binomial

16. $12a^4 - 3a + 2$
    Trinomial

For Exercises 17 to 26, add. Use a vertical format.

17. $(x^2 + 7x) + (-3x^2 - 4x)$
    $-2x^2 + 3x$

18. $(3y^2 - 2y) + (5y^2 + 6y)$
    $8y^2 + 4y$

19. $(y^2 + 4y) + (-4y - 8)$
    $y^2 - 8$

20. $(3x^2 + 9x) + (6x - 24)$
    $3x^2 + 15x - 24$

21. $(2x^2 + 6x + 12) + (3x^2 + x + 8)$
    $5x^2 + 7x + 20$

22. $(x^2 + x + 5) + (3x^2 - 10x + 4)$
    $4x^2 - 9x + 9$

23. $(-7x + x^3 + 4) + (2x^2 + x - 10)$
    $x^3 + 2x^2 - 6x - 6$

24. $(y^2 + 3y^3 + 1) + (-4y^3 - 6y - 3)$
    $-y^3 + y^2 - 6y - 2$

25. $(2a^3 - 7a + 1) + (1 - 4a - 3a^2)$
    $2a^3 - 3a^2 - 11a + 2$

26. $(5r^3 - 6r^2 + 3r) + (-3 - 2r + r^2)$
    $5r^3 - 5r^2 + r - 3$

For Exercises 27 to 36, add. Use a horizontal format.

27. $(4x^2 + 2x) + (x^2 + 6x)$
    $5x^2 + 8x$

28. $(-3y^2 + y) + (4y^2 + 6y)$
    $y^2 + 7y$

29. $(4x^2 - 5xy) + (3x^2 + 6xy - 4y^2)$
    $7x^2 + xy - 4y^2$

30. $(2x^2 - 4y^2) + (6x^2 - 2xy + 4y^2)$
    $8x^2 - 2xy$

31. $(2a^2 - 7a + 10) + (a^2 + 4a + 7)$
    $3a^2 - 3a + 17$

32. $(-6x^2 + 7x + 3) + (3x^2 + x + 3)$
    $-3x^2 + 8x + 6$

33. $(7x + 5x^3 - 7) + (10x^2 - 8x + 3)$
    $5x^3 + 10x^2 - x - 4$

34. $(4y + 3y^3 + 9) + (2y^2 + 4y - 21)$
    $3y^3 + 2y^2 + 8y - 12$

35. $(7 - 5r + 2r^2) + (3r^3 - 6r)$
    $3r^3 + 2r^2 - 11r + 7$

36. $(14 + 4y + 3y^3) + (-4y^2 + 21)$
    $3y^3 - 4y^2 + 4y + 35$

Quick Quiz (Objective 9.1A)

1. Is $\dfrac{x^2y}{3}$ a monomial? Yes

2. Is $3y^2 - 8$ a monomial, a binomial, a trinomial, or none of these? Binomial

3. Add: $(2x^2 - 3x + 4) + (5x^2 - 7x - 1)$
   $7x^2 - 10x + 3$

4. Add: $(3x^3 - 4x^2 + 7) + (5x^2 - 3x - 9)$
   $3x^3 + x^2 - 3x - 2$

Objective B   To subtract polynomials

For Exercises 37 to 46, subtract. Use a vertical format.

37. $(x^2 - 6x) - (x^2 - 10x)$
$4x$

38. $(y^2 + 4y) - (y^2 + 10y)$
$-6y$

39. $(2y^2 - 4y) - (-y^2 + 2)$
$3y^2 - 4y - 2$

40. $(-3a^2 - 2a) - (4a^2 - 4)$
$-7a^2 - 2a + 4$

41. $(x^2 - 2x + 1) - (x^2 + 5x + 8)$
$-7x - 7$

42. $(3x^2 + 2x - 2) - (5x^2 - 5x + 6)$
$-2x^2 + 7x - 8$

43. $(4x^3 + 5x + 2) - (1 + 2x - 3x^2)$
$4x^3 + 3x^2 + 3x + 1$

44. $(5y^2 - y + 2) - (-3 + 3y - 2y^3)$
$2y^3 + 5y^2 - 4y + 5$

45. $(-2y + 6y^2 + 2y^3) - (4 + y^2 + y^3)$
$y^3 + 5y^2 - 2y - 4$

46. $(4 - x - 2x^2) - (-2 + 3x - x^3)$
$x^3 - 2x^2 - 4x + 6$

For Exercises 47 to 56, subtract. Use a horizontal format.

47. $(y^2 - 10xy) - (2y^2 + 3xy)$
$-y^2 - 13xy$

48. $(x^2 - 3xy) - (-2x^2 + xy)$
$3x^2 - 4xy$

49. $(3x^2 + x - 3) - (4x + x^2 - 2)$
$2x^2 - 3x - 1$

50. $(5y^2 - 2y + 1) - (-y - 2 - 3y^2)$
$8y^2 - y + 3$

51. $(-2x^3 + x - 1) - (-x^2 + x - 3)$
$-2x^3 + x^2 + 2$

52. $(2x^2 + 5x - 3) - (3x^3 + 2x - 5)$
$-3x^3 + 2x^2 + 3x + 2$

53. $(1 - 2a + 4a^3) - (a^3 - 2a + 3)$
$3a^3 - 2$

54. $(7 - 8b + b^2) - (4b^3 - 7b - 8)$
$-4b^3 + b^2 - b + 15$

55. $(-1 - y + 4y^3) - (3 - 3y - 2y^2)$
$4y^3 + 2y^2 + 2y - 4$

56. $(-3 - 2x + 3x^2) - (4 - 2x^2 + 2x^3)$
$-2x^3 + 5x^2 - 2x - 7$

APPLYING THE CONCEPTS

57. What polynomial must be added to $3x^2 - 6x + 9$ so that the sum is $4x^2 + 3x - 2$? $x^2 + 9x - 11$

58. What polynomial must be subtracted from $2x^2 - x - 2$ so that the difference is $5x^2 + 3x + 1$? $-3x^2 - 4x - 3$

59. In your own words, explain the terms *monomial*, *binomial*, *trinomial*, and *polynomial*. Give an example of each.

60. Is it possible to subtract two polynomials, each of degree 3, and have the difference be a polynomial of degree 2? If so, give an example. If not, explain why not.

61. Is it possible to add two polynomials, each of degree 3, and have the sum be a polynomial of degree 2? If so, give an example. If not, explain why not.

Quick Quiz (Objective 9.1B)

1. Subtract using a vertical format.
$(4x^2 - 7x + 6) - (9x^2 - 12x + 8)$
$-5x^2 + 5x - 2$

2. Subtract using a horizontal format.
$(-7x^2 + 3x - 8) - (6x^2 - 12x + 8)$
$-13x^2 + 15x - 16$

9.2 Multiplication of Monomials

Objective A   **To multiply monomials**

Recall that in an exponential expression such as x^6, x is the base and 6 is the exponent. The exponent indicates the number of times the base occurs as a factor.

The product of exponential expressions with the *same* base can be simplified by writing each expression in factored form and writing the result with an exponent.

$$x^3 \cdot x^2 = \overbrace{(x \cdot x \cdot x)}^{3 \text{ factors}} \cdot \overbrace{(x \cdot x)}^{2 \text{ factors}}$$
$$\underbrace{}_{5 \text{ factors}}$$

$$= x^5$$

Note that adding the exponents results in the same product.

$$x^3 \cdot x^2 = x^{3+2} = x^5$$

Rule for Multiplying Exponential Expressions

If m and n are positive integers, then $x^m \cdot x^n = x^{m+n}$.

HOW TO   Simplify: $y^4 \cdot y \cdot y^3$

$y^4 \cdot y \cdot y^3 = y^{4+1+3}$ • The bases are the same. **Add the**
$\qquad\qquad\quad = y^8$ **exponents.** Recall that $y = y^1$.

HOW TO   Simplify: $(-3a^4b^3)(2ab^4)$

$(-3a^4b^3)(2ab^4) = (-3 \cdot 2)(a^4 \cdot a)(b^3 \cdot b^4)$ • Use the **Commutative** and
                                      Associative Properties of
                                      Multiplication to rearrange
                                      and group factors.

$\qquad\qquad\qquad\quad = -6(a^{4+1})(b^{3+4})$ • To multiply expressions
                                      with the same base, **add**
                                      the exponents.

$\qquad\qquad\qquad\quad = -6a^5b^7$ • **Simplify.**

TAKE NOTE

The Rule for Multiplying Exponential Expressions requires the bases to be the same. The expression a^5b^7 cannot be simplified.

Example 1   Simplify: $(-5ab^3)(4a^5)$

Solution
$(-5ab^3)(4a^5)$
$= (-5 \cdot 4)(a \cdot a^5)b^3$ • **Multiply coefficients. Add**
$= -20a^6b^3$ **exponents with same base.**

You Try It 1   Simplify: $(8m^3n)(-3n^5)$

Your solution $-24m^3n^6$

Example 2   Simplify: $(6x^3y^2)(4x^4y^5)$

Solution
$(6x^3y^2)(4x^4y^5)$
$= (6 \cdot 4)(x^3 \cdot x^4)(y^2 \cdot y^5)$ • **Multiply coefficients.**
$= 24x^7y^7$ **Add exponents with**
                                  same base.

You Try It 2   Simplify: $(12p^4q^3)(-3p^5q^2)$

Your solution $-36p^9q^5$

Solutions on p. S23

Objective 9.2A

Concept Check
Replace the ? to make a true statement.
1. $(x^3)(x) = x^{?+?} = x^?$ 3, 1, 4
2. $(x^4)(x^?) = x^7$ 3

Discuss the Concepts
Use the Rule for Multiplying Exponential Expressions to explain why $x \cdot x = x^2$.
x is the same as x^1;
$x \cdot x = x^1 \cdot x^1 = x^{1+1} = x^2$.

In-Class Examples (Objective 9.2A)
Simplify.
1. $(6x^4y)(-4y^6)$ $-24x^4y^7$
2. $(7x^6y^4)(-5x^2y^9)$ $-35x^8y^{13}$

Objective 9.2B

Discuss the Concepts

Are $6x^2$ and $(6x)^2$ the same? Explain.

No. In the expression $6x^2$, x is squared but 6 is not, whereas in the expression $(6x)^2$, both 6 and x are squared.

Concept Check

To which of the following does the rule $(x^m y^n)^p = x^{mp} y^{np}$ apply?

a. $(2x)^3$

b. $(xy^2 z^3)^4$

c. $(2 + x^2)^3$

d. $(x - y)^3$

e. $(-4xy^4 c^2)^5$

　　a, b, e

Objective B　To simplify powers of monomials

Point of Interest

One of the first symbolic representations of powers was given by Diophantus (c. 250 A.D.) in his book *Arithmetica*. He used Δ^Y for x^2 and κ^Y for x^3. The symbol Δ^Y was the first two letters of the Greek word *dunamis*, which means "power"; κ^Y was from the Greek word *kubos*, which means "cube." He also combined these symbols to denote higher powers. For instance, $\Delta\kappa^Y$ was the symbol for x^5.

The power of a monomial can be simplified by writing the power in factored form and then using the Rule for Multiplying Exponential Expressions.

$$(x^4)^3 = x^4 \cdot x^4 \cdot x^4 \qquad (a^2 b^3)^2 = (a^2 b^3)(a^2 b^3)$$

- Write in factored form.

$$= x^{4+4+4} \qquad\qquad = a^{2+2} b^{3+3}$$

- Use the Rule for Multiplying Exponential Expressions.

$$= x^{12} \qquad\qquad = a^4 b^6$$

Note that multiplying each exponent inside the parentheses by the exponent outside the parentheses results in the same product.

$$(x^4)^3 = x^{4 \cdot 3} = x^{12} \qquad (a^2 b^3)^2 = a^{2 \cdot 2} b^{3 \cdot 2} = a^4 b^6$$

- Multiply each exponent inside the parentheses by the exponent outside the parentheses.

> **Rule for Simplifying the Power of an Exponential Expression**
>
> If m and n are positive integers, then $(x^m)^n = x^{mn}$.

> **Rule for Simplifying the Power of a Product**
>
> If m, n, and p are positive integers, then $(x^m y^n)^p = x^{mp} y^{np}$.

HOW TO　Simplify: $(5x^2 y^3)^3$

$$(5x^2 y^3)^3 = 5^{1 \cdot 3} x^{2 \cdot 3} y^{3 \cdot 3}$$

- Use the **Rule for Simplifying the Power of a Product**. Note that $5 = 5^1$.

$$= 5^3 x^6 y^9$$

$$= 125 x^6 y^9$$

- Evaluate 5^3.

Example 3　Simplify: $(-2p^3 r)^4$

Solution

$$(-2p^3 r)^4 = (-2)^{1 \cdot 4} p^{3 \cdot 4} r^{1 \cdot 4}$$
$$= (-2)^4 p^{12} r^4 = 16 p^{12} r^4$$

- Use the Rule for Simplifying the Power of a Product.

You Try It 3　Simplify: $(-3a^4 bc^2)^3$

Your solution　$-27 a^{12} b^3 c^6$

Example 4　Simplify: $(2a^2 b)(2a^3 b^2)^3$

Solution

$$(2a^2 b)(2a^3 b^2)^3$$
$$= (2a^2 b)(2^{1 \cdot 3} a^{3 \cdot 3} b^{2 \cdot 3})$$
$$= (2a^2 b)(2^3 a^9 b^6)$$
$$= (2a^2 b)(8a^9 b^6)$$
$$= 16 a^{11} b^7$$

- Use the Rule for Simplifying the Power of a Product.

You Try It 4　Simplify: $(-xy^4)(-2x^3 y^2)^2$

Your solution　$-4x^7 y^8$

Solutions on p. S23

In-Class Examples (Objective 9.2B)

Simplify.

1. $(3x^4)^2$　$9x^8$

2. $(4x^2 y^3)(2xy^4)^3$　$32x^5 y^{15}$

3. $(-5xy^4 z^3)^3$　$-125 x^3 y^{12} z^9$

4. $(3ab^3)(-2a^2 b^4)^3$　$-24 a^7 b^{15}$

9.2 Exercises

Objective A   To multiply monomials

1. ✏ Explain how to multiply two monomials. Provide an example.

2. ✏ Explain how to simplify the power of a monomial. Provide an example.

For Exercises 3 to 35, simplify.

3. $(6x^2)(5x)$
$30x^3$

4. $(-4y^3)(2y)$
$-8y^4$

5. $(7c^2)(-6c^4)$
$-42c^6$

6. $(-8z^5)(5z^8)$
$-40z^{13}$

7. $(-3a^3)(-3a^4)$
$9a^7$

8. $(-5a^6)(-2a^5)$
$10a^{11}$

9. $(x^2)(xy^4)$
x^3y^4

10. $(x^2y^4)(xy^7)$
x^3y^{11}

11. $(-2x^4)(5x^5y)$
$-10x^9y$

12. $(-3a^3)(2a^2b^4)$
$-6a^5b^4$

13. $(-4x^2y^4)(-3x^5y^4)$
$12x^7y^8$

14. $(-6a^2b^4)(-4ab^3)$
$24a^3b^7$

15. $(2xy)(-3x^2y^4)$
$-6x^3y^5$

16. $(-3a^2b)(-2ab^3)$
$6a^3b^4$

17. $(x^2yz)(x^2y^4)$
x^4y^5z

18. $(-ab^2c)(a^2b^5)$
$-a^3b^7c$

19. $(-a^2b^3)(-ab^2c^4)$
$a^3b^5c^4$

20. $(-x^2y^3z)(-x^3y^4)$
x^5y^7z

21. $(-5a^2b^2)(6a^3b^6)$
$-30a^5b^8$

22. $(7xy^4)(-2xy^3)$
$-14x^2y^7$

23. $(-6a^3)(-a^2b)$
$6a^5b$

24. $(-2a^2b^3)(-4ab^2)$
$8a^3b^5$

25. $(-5y^4z)(-8y^6z^5)$
$40y^{10}z^6$

26. $(3x^2y)(-4xy^2)$
$-12x^3y^3$

27. $(x^2y)(yz)(xyz)$
$x^3y^3z^2$

28. $(xy^2z)(x^2y)(z^2y^2)$
$x^3y^5z^3$

29. $(3ab^2)(-2abc)(4ac^2)$
$-24a^3b^3c^3$

30. $(-2x^3y^2)(-3x^2z^2)(-5y^3z^3)$
$-30x^5y^5z^5$

31. $(4x^4z)(-yz^3)(-2x^3z^2)$
$8x^7yz^6$

32. $(-a^3b^4)(-3a^4c^2)(4b^3c^4)$
$12a^7b^7c^6$

33. $(-2x^2y^3)(3xy)(-5x^3y^4)$
$30x^6y^8$

34. $(4a^2b)(-3a^3b^4)(a^5b^2)$
$-12a^{10}b^7$

35. $(3a^2b)(-6bc)(2ac^2)$
$-36a^3b^2c^3$

Objective B   To simplify powers of monomials

For Exercises 36 to 66, simplify.

36. $(z^4)^3$
z^{12}

37. $(x^3)^5$
x^{15}

38. $(y^4)^2$
y^8

39. $(x^7)^2$
x^{14}

40. $(-y^5)^3$
$-y^{15}$

41. $(-x^2)^4$
x^8

42. $(-x^2)^3$
$-x^6$

43. $(-y^3)^4$
y^{12}

44. $(-3y)^3$
$-27y^3$

45. $(-2x^2)^3$
$-8x^6$

Section 9.2

Suggested Assignment

Exercises 1–65, odds
More challenging problems:
    Exercises 67–73, odds
    Exercises 79, 80

Answers to Writing Exercises

1. Answers will vary. Students should include the fact that when monomials are multiplied, the coefficients are multiplied and the exponents on the like variables are added. Example:
$(2x^3y^2)(3xy^4) = 6x^4y^6$

2. Answers will vary. Students should include the fact that when a monomial is raised to a power, the exponents are multiplied. Example:
$(3x^2y^4)^3 = 3^3x^6y^{12} = 27x^6y^{12}$

Quick Quiz (Objective 9.2A)

Simplify.

1. $(4a^3b^2)(5ab)$ $20a^4b^3$

2. $(-4x^2y^3z^2)(-3x^4yz^3)$ $12x^6y^4z^5$

3. $(2ab^2c)(3a^2b^3)(5a^3bc^2)$ $30a^6b^6c^3$

46. $(a^3b^4)^3$
a^9b^{12}

47. $(x^2y^3)^2$
x^4y^6

48. $(2x^3y^4)^5$
$32x^{15}y^{20}$

49. $(3x^2y)^2$
$9x^4y^2$

50. $(-2ab^3)^4$
$16a^4b^{12}$

51. $(-3x^3y^2)^5$
$-243x^{15}y^{10}$

52. $(3b^2)(2a^3)^4$
$48a^{12}b^2$

53. $(-2x)(2x^3)^2$
$-8x^7$

54. $(2y)(-3y^4)^3$
$-54y^{13}$

55. $(3x^2y)(2x^2y^2)^3$
$24x^8y^7$

56. $(a^3b)^2(ab)^3$
a^9b^5

57. $(ab^2)^2(ab)^2$
a^4b^6

58. $(-x^2y^3)^2(-2x^3y)^3$
$-8x^{13}y^9$

59. $(-2x)^3(-2x^3y)^3$
$64x^{12}y^3$

60. $(-3y)(-4x^2y^3)^3$
$192x^6y^{10}$

61. $(-2x)(-3xy^2)^2$
$-18x^3y^4$

62. $(-3y)(-2x^2y)^3$
$24x^6y^4$

63. $(ab^2)(-2a^2b)^3$
$-8a^7b^5$

64. $(a^2b^2)(-3ab^4)^2$
$9a^4b^{10}$

65. $(-2a^3)(3a^2b)^3$
$-54a^9b^3$

66. $(-3b^2)(2ab^2)^3$
$-24a^3b^8$

APPLYING THE CONCEPTS

For Exercises 67 to 74, simplify.

67. $3x^2 + (3x)^2$
$12x^2$

68. $4x^2 - (4x)^2$
$-12x^2$

69. $2x^6y^2 + (3x^2y)^2$
$2x^6y^2 + 9x^4y^2$

70. $(x^2y^2)^3 + (x^3y^3)^2$
$2x^6y^6$

71. $(2a^3b^2)^3 - 8a^9b^6$
0

72. $4y^2z^4 - (2yz^2)^2$
0

73. $(x^2y^4)^2 + (2xy^2)^4$
$17x^4y^8$

74. $(3a^3)^2 - 4a^6 + (2a^2)^3$
$13a^6$

For Exercises 75 to 78, answer true or false. If the answer is false, correct the right-hand side of the equation.

75. $(-a)^5 = -a^5$
True

76. $(-b)^8 = b^8$
True

77. $(x^2)^5 = x^{2+5} = x^7$
False. $(x^2)^5 = x^{2 \cdot 5} = x^{10}$

78. $x^3 + x^3 = 2x^{3+3} = 2x^6$
False. $x^3 + x^3 = 2x^3$

79. Evaluate $(2^3)^2$ and $2^{(3^2)}$. Are the results the same? If not, which expression has the larger value?
No. $2^{(3^2)}$ is larger. $(2^3)^2 = 2^6 = 64$; $2^{(3^2)} = 2^9 = 512$

80. ✏ If n is a positive integer and $x^n = y^n$, does $x = y$? Explain your answer.

81. ✏ The distance a rock will fall in t seconds is $16t^2$ (neglecting air resistance). Find other examples of quantities that can be expressed in terms of an exponential expression, and explain where the expression is used.

Answers to Writing Exercises

80. If n is a positive *odd* integer and $x^n = y^n$, then $x = y$. However, if n is a positive *even* integer and $x^n = y^n$, then it does not necessarily follow that $x = y$. For example, if $x = 3$, $y = -3$, and $n = 2$, then $x^n = y^n$ but $x \neq y$.

81. Answers will vary. Some examples include the area of a square, which can be expressed as s^2; the volume of a cube, which can be expressed as s^3; and the surface area of a cube, which can be expressed as $6s^2$.

Quick Quiz (Objective 9.2B)

Simplify.

1. $(x^4)^3$ x^{12}

2. $(3a^3bc^5)^2$ $9a^6b^2c^{10}$

3. $(5ab^3c^4)(-2a^2bc^3)^4$ $80a^9b^7c^{16}$

9.3 Multiplication of Polynomials

Objective A   To multiply a polynomial by a monomial

To multiply a polynomial by a monomial, use the Distributive Property and the Rule for Multiplying Exponential Expressions.

> **HOW TO**   Multiply: $-3a(4a^2 - 5a + 6)$
>
> $-3a(4a^2 - 5a + 6) = -3a(4a^2) - (-3a)(5a) + (-3a)(6)$ • **Use the Distributive Property.**
> $= -12a^3 + 15a^2 - 18a$

Example 1

Multiply: $(5x + 4)(-2x)$

Solution
$(5x + 4)(-2x) = -10x^2 - 8x$

You Try It 1

Multiply: $(-2y + 3)(-4y)$

Your solution
$8y^2 - 12y$

Example 2

Multiply: $2a^2b(4a^2 - 2ab + b^2)$

Solution
$2a^2b(4a^2 - 2ab + b^2)$
$= 8a^4b - 4a^3b^2 + 2a^2b^3$

You Try It 2

Multiply: $-a^2(3a^2 + 2a - 7)$

Your solution
$-3a^4 - 2a^3 + 7a^2$

Solutions on p. S23

Objective B   To multiply two polynomials

Multiplication of two polynomials requires the repeated application of the Distributive Property.

$$(y - 2)(y^2 + 3y + 1) = (y - 2)(y^2) + (y - 2)(3y) + (y - 2)(1)$$
$$= y^3 - 2y^2 + 3y^2 - 6y + y - 2$$
$$= y^3 + y^2 - 5y - 2$$

A convenient method of multiplying two polynomials is to use a vertical format similar to that used for multiplication of whole numbers.

$$
\begin{array}{r}
y^2 + 3y + 1 \\
y - 2 \\
\hline
\end{array}
$$

$-2y^2 - 6y - 2 = -2(y^2 + 3y + 1)$ • **Multiply the trinomial by −2.**
$y^3 + 3y^2 + y = y(y^2 + 3y + 1)$ • **Multiply the trinomial by y.**
$\overline{y^3 + y^2 - 5y - 2}$ • **Add the terms in each column.**

In-Class Examples (Objective 9.3A)
Multiply.
1. $(6a - 5)(-4a)$ $-24a^2 + 20a$
2. $3x^2y(5x^2 + 4xy - 2y^2)$ $15x^4y + 12x^3y^2 - 6x^2y^3$
3. $(2x + 4)3x$ $6x^2 + 12x$
4. $4xy(3x^2 - 2xy + 4y^2)$ $12x^3y - 8x^2y^2 + 16xy^3$

Concept Check

Replace the ? to make a true statement.

$(3x^2 + 2x + 5)(x + 3) =$
$3x^2(?) + 2x(?) + 5(?)$
$x + 3$

Optional Student Activity

Multiply: $(x^2 - xy + y^2)(x + y)$
$x^3 + y^3$

HOW TO Multiply: $(2a^3 + a - 3)(a + 5)$

$$
\begin{array}{r}
2a^3 \quad\quad + a - 3 \\
a + 5 \\
\hline
10a^3 \quad\quad + 5a - 15 \\
2a^4 \quad\quad + a^2 - 3a \\
\hline
2a^4 + 10a^3 + a^2 + 2a - 15
\end{array}
$$

• Note that spaces are provided so that like terms are in the same column.

• Add the terms in each column.

Example 3

Multiply: $(2b^3 - b + 1)(2b + 3)$

Solution

$$
\begin{array}{r}
2b^3 \quad\quad - b + 1 \\
2b + 3 \\
\hline
6b^3 \quad\quad - 3b + 3 = 3(2b^3 - b + 1) \\
4b^4 + \quad\quad - 2b^2 + 2b = 2b(2b^3 - b + 1) \\
\hline
4b^4 + 6b^3 - 2b^2 - b + 3
\end{array}
$$

You Try It 3

Multiply: $(2y^3 + 2y^2 - 3)(3y - 1)$

Your solution

$6y^4 + 4y^3 - 2y^2 - 9y + 3$

Solution on p. S23

Objective 9.3C

New Vocabulary
FOIL

Concept Check

Multiply.

1. $(x - 3)(x - 6)$
$x^2 - 9x + 18$

2. $(x + 4)(x + 5)$
$x^2 + 9x + 20$

3. $(x - 6)(x + 2)$
$x^2 - 4x - 12$

4. $(x + 4)(x - 7)$
$x^2 - 3x - 28$

Optional Student Activity

Make up a problem involving the product of two binomials. Write an explanation, using the FOIL method, that demonstrates how to simplify the product.

Objective C To multiply two binomials

It is frequently necessary to find the product of two binomials. The product can be found using a method called **FOIL,** which is based on the Distributive Property. The letters of FOIL stand for **F**irst, **O**uter, **I**nner, and **L**ast. To find the product of two binomials, add the products of the **F**irst terms, the **O**uter terms, the **I**nner terms, and the **L**ast terms.

TAKE NOTE

FOIL is not really a different way of multiplying. It is based on the Distributive Property.

$(2x + 3)(x + 5)$
$= 2x(x + 5) + 3(x + 5)$
   F   O   I   L
$= 2x^2 + 10x + 3x + 15$
$= 2x^2 + 13x + 15$

HOW TO Multiply: $(2x + 3)(x + 5)$

Multiply the **F**irst terms.	$(2x + 3)(x + 5)$	$2x \cdot x = 2x^2$
Multiply the **O**uter terms.	$(2x + 3)(x + 5)$	$2x \cdot 5 = 10x$
Multiply the **I**nner terms.	$(2x + 3)(x + 5)$	$3 \cdot x = 3x$
Multiply the **L**ast terms.	$(2x + 3)(x + 5)$	$3 \cdot 5 = 15$

$$
\begin{array}{c}
\quad\quad\quad\quad\quad\quad\quad F \quad\quad O \quad\quad I \quad\quad L \\
\text{Add the products.} \quad (2x + 3)(x + 5) = 2x^2 + 10x + 3x + 15 \\
\text{Combine like terms.} \quad\quad\quad\quad = 2x^2 + 13x + 15
\end{array}
$$

HOW TO Multiply: $(4x - 3)(3x - 2)$

$$(4x - 3)(3x - 2) = 4x(3x) + 4x(-2) + (-3)(3x) + (-3)(-2)$$
$$= 12x^2 - 8x - 9x + 6$$
$$= 12x^2 - 17x + 6$$

HOW TO Multiply: $(3x - 2y)(x + 4y)$

$$(3x - 2y)(x + 4y) = 3x(x) + 3x(4y) + (-2y)(x) + (-2y)(4y)$$
$$= 3x^2 + 12xy - 2xy - 8y^2$$
$$= 3x^2 + 10xy - 8y^2$$

In-Class Examples (Objective 9.3B)

Multiply.

1. $(2z^2 - 4z + 5)(4z - 2)$ $8z^3 - 20z^2 + 28z - 10$

2. $(6a^3 + 4a^2 - 3a)(3a - 2)$ $18a^4 - 17a^2 + 6a$

3. $(3x^3 - 2x + 5)(7x + 4)$
$21x^4 + 12x^3 - 14x^2 + 27x + 20$

Example 4

Multiply: $(2a - 1)(3a - 2)$

Solution
$(2a - 1)(3a - 2) = 6a^2 - 4a - 3a + 2$
$= 6a^2 - 7a + 2$

You Try It 4

Multiply: $(4y - 5)(2y - 3)$

Your solution
$8y^2 - 22y + 15$

Example 5

Multiply: $(3x - 2)(4x + 3)$

Solution
$(3x - 2)(4x + 3) = 12x^2 + 9x - 8x - 6$
$= 12x^2 + x - 6$

You Try It 5

Multiply: $(3b + 2)(3b - 5)$

Your solution
$9b^2 - 9b - 10$

Solutions on p. S23

Concept Check

Can FOIL be used to multiply $(x^2 - 5)(2x + 3)$? If so, what is the product?
Yes. $2x^3 + 3x^2 - 10x - 15$

Objective D   **To multiply binomials that have special products**

Objective 9.3D

Instructor Note

Remind students that the rule that applies to $(ab)^2$ is different from the rule that applies to $(a + b)^2$.

Using FOIL, it is possible to find a pattern for the product of the sum and difference of two terms and for the square of a binomial.

Product of the Sum and Difference of the Same Terms

$(a + b)(a - b) = a^2 - ab + ab - b^2$
$= a^2 - b^2$

Square of the first term
Square of the second term

Square of a Binomial

$(a + b)^2 = (a + b)(a + b) = a^2 + ab + ab + b^2$
$= a^2 + 2ab + b^2$

Square of the first term
Twice the product of the two terms
Square of the last term

Optional Student Activity

1. Demonstrate that $(a + b)^2 \neq a^2 + b^2$ by substituting numbers for a and b and then evaluating the expressions on each side of the equals sign.
2. Demonstrate that $(c - d)^2 \neq c^2 - d^2$ by substituting numbers for c and d and then evaluating the expressions on each side of the equals sign.

Concept Check

Multiply.
1. $(x + 3)(x - 3)$ $x^2 - 9$
2. $(x + 4)^2$ $x^2 + 8x + 16$

HOW TO   Multiply: $(2x + 3)(2x - 3)$

$(2x + 3)(2x - 3) = (2x)^2 - 3^2$ • This is the product of the sum and
$= 4x^2 - 9$ difference of the same terms.

TAKE NOTE
The word *expand* is used frequently to mean "multiply out a power."

HOW TO   Expand: $(3x - 2)^2$

$(3x - 2)^2 = (3x)^2 + 2(3x)(-2) + (-2)^2$ • This is the square of a
$= 9x^2 - 12x + 4$ binomial.

In-Class Examples (Objective 9.3C)

Multiply.
1. $(7x - 3)(5x - 6)$ $35x^2 - 57x + 18$
2. $(8y + 7)(3y - 4)$ $24y^2 - 11y - 28$
3. $(6w + 4)(5w - 2)$ $30w^2 + 8w - 8$

488   Chapter 9 / Polynomials

Concept Check

1. Multiply: $(5b - 7c)(5b + 7c)$
   $25b^2 - 49c^2$

2. Expand: $(4y - 5z)^2$
   $16y^2 - 40yz + 25z^2$

Instructor Note

You may want to tell students that binomials that are otherwise identical except that one is a sum and one is a difference are called *conjugates* of each other.

Optional Student Activity

Write an adaptation of the FOIL method that can be used to simplify the product of the sum and difference of the same terms. Demonstrate your method on the product $(3x + 8)(3x - 8)$.

Students may refer to the method as the "FL" method or the "$F^2 - L^2$" method.
$(3x + 8)(3x - 8) = (3x)(3x) + (8)(-8) = 9x^2 - 64$

Objective 9.3E

Concept Check

1. The length of a rectangle is 6 in. The width is 4 in. Find the area of the rectangle.
   24 in^2

2. The length of a rectangle is $(x + 3)$ in. The width is $(x - 2)$ in. Find the area of the rectangle in terms of the variable x.
   $(x^2 + x - 6)$ in^2

Example 6

Multiply: $(4z - 2w)(4z + 2w)$

Solution
$(4z - 2w)(4z + 2w) = 16z^2 - 4w^2$

You Try It 6

Multiply: $(2a + 5c)(2a - 5c)$

Your solution
$4a^2 - 25c^2$

Example 7

Expand: $(2r - 3s)^2$

Solution
$(2r - 3s)^2 = 4r^2 - 12rs + 9s^2$

You Try It 7

Expand: $(3x + 2y)^2$

Your solution
$9x^2 + 12xy + 4y^2$

Solutions on p. S23

Objective E   **To solve application problems**

Example 8

The length of a rectangle is $(x + 7)$ m. The width is $(x - 4)$ m. Find the area of the rectangle in terms of the variable x.

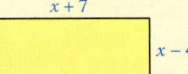
$x + 7$
$x - 4$

Strategy
To find the area, replace the variables L and W in the equation $A = L \cdot W$ by the given values and solve for A.

Solution
$A = L \cdot W$
$A = (x + 7)(x - 4)$
$A = x^2 - 4x + 7x - 28$
$A = x^2 + 3x - 28$

The area is $(x^2 + 3x - 28)$ m^2.

You Try It 8

The radius of a circle is $(x - 4)$ ft. Use the equation $A = \pi r^2$, where r is the radius, to find the area of the circle in terms of x. Leave the answer in terms of π.

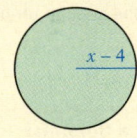
$x - 4$

Your strategy

Your solution
$(\pi x^2 - 8\pi x + 16\pi)$ ft^2

Solution on p. S23

In-Class Examples (Objective 9.3D)

1. Multiply: $(10x - 3)(10x + 3)$ $100x^2 - 9$
2. Expand: $(4x + 3y)^2$ $16x^2 + 24xy + 9y^2$

In-Class Examples (Objective 9.3E)

1. The radius of a circle is $(x + 5)$ ft. Use the equation $A = \pi r^2$, where r is the radius, to find the area of the circle in terms of x. Leave the answer in terms of π. $(\pi x^2 + 10\pi x + 25\pi)$ ft^2

9.3 Exercises

Objective A   To multiply a polynomial by a monomial

Section 9.3

Suggested Assignment
Exercises 1–83, every other odd
Exercises 85–109, odds
More challenging problems:
    Exercises 110–116

For Exercises 1 and 2, replace the question marks to make a true statement.

1. $3(4x - 5) = (?)(4x) - (?)(5) = ?$
$3, 3, 12x - 15$

2. $2x^2(3x + 7) = (?)(3x) + (?)(7) = ?$
$2x^2, 2x^2, 6x^3 + 14x^2$

For Exercises 3 to 34, multiply.

3. $x(x - 2)$
$x^2 - 2x$

4. $y(3 - y)$
$-y^2 + 3y$

5. $-x(x + 7)$
$-x^2 - 7x$

6. $-y(7 - y)$
$y^2 - 7y$

7. $3a^2(a - 2)$
$3a^3 - 6a^2$

8. $4b^2(b + 8)$
$4b^3 + 32b^2$

9. $-5x^2(x^2 - x)$
$-5x^4 + 5x^3$

10. $-6y^2(y + 2y^2)$
$-12y^4 - 6y^3$

11. $-x^3(3x^2 - 7)$
$-3x^5 + 7x^3$

12. $-y^4(2y^2 - y^6)$
$y^{10} - 2y^6$

13. $2x(6x^2 - 3x)$
$12x^3 - 6x^2$

14. $3y(4y - y^2)$
$-3y^3 + 12y^2$

15. $(2x - 4)3x$
$6x^2 - 12x$

16. $(3y - 2)y$
$3y^2 - 2y$

17. $(3x + 4)x$
$3x^2 + 4x$

18. $(2x + 1)2x$
$4x^2 + 2x$

19. $-xy(x^2 - y^2)$
$-x^3y + xy^3$

20. $-x^2y(2xy - y^2)$
$-2x^3y^2 + x^2y^3$

21. $x(2x^3 - 3x + 2)$
$2x^4 - 3x^2 + 2x$

22. $y(-3y^2 - 2y + 6)$
$-3y^3 - 2y^2 + 6y$

23. $-a(-2a^2 - 3a - 2)$
$2a^3 + 3a^2 + 2a$

24. $-b(5b^2 + 7b - 35)$
$-5b^3 - 7b^2 + 35b$

25. $x^2(3x^4 - 3x^2 - 2)$
$3x^6 - 3x^4 - 2x^2$

26. $y^3(-4y^3 - 6y + 7)$
$-4y^6 - 6y^4 + 7y^3$

27. $2y^2(-3y^2 - 6y + 7)$
$-6y^4 - 12y^3 + 14y^2$

28. $4x^2(3x^2 - 2x + 6)$
$12x^4 - 8x^3 + 24x^2$

29. $(a^2 + 3a - 4)(-2a)$
$-2a^3 - 6a^2 + 8a$

30. $(b^3 - 2b + 2)(-5b)$
$-5b^4 + 10b^2 - 10b$

31. $-3y^2(-2y^2 + y - 2)$
$6y^4 - 3y^3 + 6y^2$

32. $-5x^2(3x^2 - 3x - 7)$
$-15x^4 + 15x^3 + 35x^2$

33. $xy(x^2 - 3xy + y^2)$
$x^3y - 3x^2y^2 + xy^3$

34. $ab(2a^2 - 4ab - 6b^2)$
$2a^3b - 4a^2b^2 - 6ab^3$

Objective B   To multiply two polynomials

For Exercises 35 to 52, multiply.

35. $(x^2 + 3x + 2)(x + 1)$
$x^3 + 4x^2 + 5x + 2$

36. $(x^2 - 2x + 7)(x - 2)$
$x^3 - 4x^2 + 11x - 14$

37. $(a^2 - 3a + 4)(a - 3)$
$a^3 - 6a^2 + 13a - 12$

Quick Quiz (Objective 9.3A)
Multiply.
1. $y(y + 5)$ $y^2 + 5y$
2. $(a - 3)2a^2$ $2a^3 - 6a^2$
3. $-5b^3(2b^2 + 3b - 6)$ $-10b^5 - 15b^4 + 30b^3$

Quick Quiz (Objective 9.3B)
Multiply.
1. $(x^2 - 3x + 2)(x + 5)$ $x^3 + 2x^2 - 13x + 10$
2. $(y^2 + 4)(y - 1)$ $y^3 - y^2 + 4y - 4$
3. $(3a^3 - 2a^2 + a - 6)(2a - 5)$
$6a^4 - 19a^3 + 12a^2 - 17a + 30$

38. $(x^2 - 3x + 5)(2x - 3)$
$2x^3 - 9x^2 + 19x - 15$

39. $(-2b^2 - 3b + 4)(b - 5)$
$-2b^3 + 7b^2 + 19b - 20$

40. $(-a^2 + 3a - 2)(2a - 1)$
$-2a^3 + 7a^2 - 7a + 2$

41. $(-2x^2 + 7x - 2)(3x - 5)$
$-6x^3 + 31x^2 - 41x + 10$

42. $(-a^2 - 2a + 3)(2a - 1)$
$-2a^3 - 3a^2 + 8a - 3$

43. $(x^2 + 5)(x - 3)$
$x^3 - 3x^2 + 5x - 15$

44. $(y^2 - 2y)(2y + 5)$
$2y^3 + y^2 - 10y$

45. $(x^3 - 3x + 2)(x - 4)$
$x^4 - 4x^3 - 3x^2 + 14x - 8$

46. $(y^3 + 4y^2 - 8)(2y - 1)$
$2y^4 + 7y^3 - 4y^2 - 16y + 8$

47. $(5y^2 + 8y - 2)(3y - 8)$
$15y^3 - 16y^2 - 70y + 16$

48. $(3y^2 + 3y - 5)(4y - 3)$
$12y^3 + 3y^2 - 29y + 15$

49. $(5a^3 - 5a + 2)(a - 4)$
$5a^4 - 20a^3 - 5a^2 + 22a - 8$

50. $(3b^3 - 5b^2 + 7)(6b - 1)$
$18b^4 - 33b^3 + 5b^2 + 42b - 7$

51. $(y^3 + 2y^2 - 3y + 1)(y + 2)$
$y^4 + 4y^3 + y^2 - 5y + 2$

52. $(2a^3 - 3a^2 + 2a - 1)(2a - 3)$
$4a^4 - 12a^3 + 13a^2 - 8a + 3$

Objective C   **To multiply two binomials**

For Exercises 53 to 84, multiply.

53. $(x + 1)(x + 3)$
$x^2 + 4x + 3$

54. $(y + 2)(y + 5)$
$y^2 + 7y + 10$

55. $(a - 3)(a + 4)$
$a^2 + a - 12$

56. $(b - 6)(b + 3)$
$b^2 - 3b - 18$

57. $(y + 3)(y - 8)$
$y^2 - 5y - 24$

58. $(x + 10)(x - 5)$
$x^2 + 5x - 50$

59. $(y - 7)(y - 3)$
$y^2 - 10y + 21$

60. $(a - 8)(a - 9)$
$a^2 - 17a + 72$

61. $(2x + 1)(x + 7)$
$2x^2 + 15x + 7$

62. $(y + 2)(5y + 1)$
$5y^2 + 11y + 2$

63. $(3x - 1)(x + 4)$
$3x^2 + 11x - 4$

64. $(7x - 2)(x + 4)$
$7x^2 + 26x - 8$

65. $(4x - 3)(x - 7)$
$4x^2 - 31x + 21$

66. $(2x - 3)(4x - 7)$
$8x^2 - 26x + 21$

67. $(3y - 8)(y + 2)$
$3y^2 - 2y - 16$

68. $(5y - 9)(y + 5)$
$5y^2 + 16y - 45$

69. $(3x + 7)(3x + 11)$
$9x^2 + 54x + 77$

70. $(5a + 6)(6a + 5)$
$30a^2 + 61a + 30$

71. $(7a - 16)(3a - 5)$
$21a^2 - 83a + 80$

72. $(5a - 12)(3a - 7)$
$15a^2 - 71a + 84$

73. $(3a - 2b)(2a - 7b)$
$6a^2 - 25ab + 14b^2$

74. $(5a - b)(7a - b)$
$35a^2 - 12ab + b^2$

75. $(a - 9b)(2a + 7b)$
$2a^2 - 11ab - 63b^2$

Quick Quiz (Objective 9.3C)
Multiply.
1. $(x + 1)(x + 5)$ $x^2 + 6x + 5$
2. $(2x - 3)(3x - 4)$ $6x^2 - 17x + 12$
3. $(6x - 7)(4x + 3)$ $24x^2 - 10x - 21$

76. $(2a + 5b)(7a − 2b)$
$14a^2 + 31ab − 10b^2$

77. $(10a − 3b)(10a − 7b)$
$100a^2 − 100ab + 21b^2$

78. $(12a − 5b)(3a − 4b)$
$36a^2 − 63ab + 20b^2$

79. $(5x + 12y)(3x + 4y)$
$15x^2 + 56xy + 48y^2$

80. $(11x + 2y)(3x + 7y)$
$33x^2 + 83xy + 14y^2$

81. $(2x − 15y)(7x + 4y)$
$14x^2 − 97xy − 60y^2$

82. $(5x + 2y)(2x − 5y)$
$10x^2 − 21xy − 10y^2$

83. $(8x − 3y)(7x − 5y)$
$56x^2 − 61xy + 15y^2$

84. $(2x − 9y)(8x − 3y)$
$16x^2 − 78xy + 27y^2$

Objective D   **To multiply binomials that have special products**

For Exercises 85 to 92, multiply.

85. $(y − 5)(y + 5)$
$y^2 − 25$

86. $(y + 6)(y − 6)$
$y^2 − 36$

87. $(2x + 3)(2x − 3)$
$4x^2 − 9$

88. $(4x − 7)(4x + 7)$
$16x^2 − 49$

89. $(3x − 7)(3x + 7)$
$9x^2 − 49$

90. $(9x − 2)(9x + 2)$
$81x^2 − 4$

91. $(4 − 3y)(4 + 3y)$
$16 − 9y^2$

92. $(4x − 9y)(4x + 9y)$
$16x^2 − 81y^2$

For Exercises 93 to 102, expand.

93. $(x + 1)^2$
$x^2 + 2x + 1$

94. $(y − 3)^2$
$y^2 − 6y + 9$

95. $(3a − 5)^2$
$9a^2 − 30a + 25$

96. $(6x − 5)^2$
$36x^2 − 60x + 25$

97. $(2a + b)^2$
$4a^2 + 4ab + b^2$

98. $(x + 3y)^2$
$x^2 + 6xy + 9y^2$

99. $(x − 2y)^2$
$x^2 − 4xy + 4y^2$

100. $(2x − 3y)^2$
$4x^2 − 12xy + 9y^2$

101. $(5x + 2y)^2$
$25x^2 + 20xy + 4y^2$

102. $(2a − 9b)^2$
$4a^2 − 36ab + 81b^2$

Objective E   **To solve application problems**

103. **Geometry**   The length of a rectangle is $(5x)$ ft. The width is $(2x − 7)$ ft.
Find the area of the rectangle in terms of the variable x.
$(10x^2 − 35x)$ ft^2

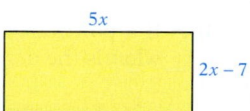
5x
2x − 7

104. **Geometry**   The width of a rectangle is $(3x + 1)$ in. The length of the
rectangle is twice the width. Find the area of the rectangle in terms of
the variable x.
$(18x^2 + 12x + 2)$ in^2

Quick Quiz (Objective 9.3D)

1. Multiply: $(5y + 3)(5y − 3)$ $25y^2 − 9$
2. Expand: $(x − 9)^2$ $x^2 − 18x + 81$
3. Expand: $(3x + 2y)^2$ $9x^2 + 12xy + 4y^2$

Quick Quiz (Objective 9.3E)

1. The width of a rectangle is $(3x + 2)$ ft. The length
of the rectangle is $(4x − 1)$ ft. Find the area of the
rectangle in terms of the variable x.
$(12x^2 + 5x − 2)$ ft^2

105. Geometry The length of a side of a square is $(2x + 1)$ km. Find the area of the square in terms of the variable x.
$(4x^2 + 4x + 1)$ km²

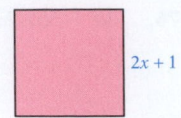

106. Geometry The radius of a circle is $(x + 4)$ cm. Find the area of the circle in terms of the variable x. Leave the answer in terms of π.
$(\pi x^2 + 8\pi x + 16\pi)$ cm²

107. Geometry The base of a triangle is $(4x)$ m and the height is $(2x + 5)$ m. Find the area of the triangle in terms of the variable x.
$(4x^2 + 10x)$ m²

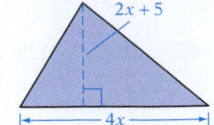

108. Sports A softball diamond has dimensions 45 ft by 45 ft. A base-path border x feet wide lies on both the first-base side and third-base side of the diamond. Express the combined area of the softball diamond and the base paths in terms of the variable x.
$(90x + 2025)$ ft²

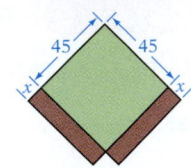

109. Sports An athletic field has dimensions 30 yd by 100 yd. An end zone that is w yards wide borders each end of the field. Express the combined area of the field and the end zones in terms of the variable w.
$(60w + 3000)$ yd²

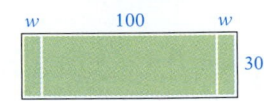

APPLYING THE CONCEPTS

110. Simplify: $(a + b)^2 - (a - b)^2$
$4ab$

111. Expand: $(x^2 + x - 3)^2$
$x^4 + 2x^3 - 5x^2 - 6x + 9$

112. Expand: $(a + 3)^3$
$a^3 + 9a^2 + 27a + 27$

113. What polynomial has quotient $3x - 4$ when divided by $4x + 5$?
$12x^2 - x - 20$

114. Add $x^2 + 2x - 3$ to the product of $2x - 5$ and $3x + 1$.
$7x^2 - 11x - 8$

115. Subtract $4x^2 - x - 5$ from the product of $x^2 + x + 3$ and $x - 4$.
$x^3 - 7x^2 - 7$

116. If a polynomial of degree 3 is multiplied by a polynomial of degree 2, what is the degree of the resulting polynomial?
5

117. Is it possible to multiply a polynomial of degree 2 by a polynomial of degree 2 and have the product be a polynomial of degree 3? If so, give an example. If not, explain why not.

Answers to Writing Exercises

117. No, it is not possible to multiply a polynomial of degree 2 by a polynomial of degree 2 and have the product be a polynomial of degree 3. A polynomial of degree 2 contains the term ax^2, $a \neq 0$. Two polynomials of degree 2 will have the terms ax^2 and bx^2, $a \neq 0$, $b \neq 0$. Multiplying these terms yields abx^4, where $ab \neq 0$. Therefore, the product will have an x^4 term and will be of degree 4.

9.4 Integer Exponents and Scientific Notation

Objective A   To divide monomials

The quotient of two exponential expressions with the same base can be simplified by writing each expression in factored form, dividing by the common factors, and then writing the result with an exponent.

$$\frac{x^5}{x^2} = \frac{\overset{1}{\cancel{x}} \cdot \overset{1}{\cancel{x}} \cdot x \cdot x \cdot x}{\underset{1}{\cancel{x}} \cdot \underset{1}{\cancel{x}}} = x^3$$

Note that subtracting the exponents gives the same result.

$$\frac{x^5}{x^2} = x^{5-2} = x^3$$

To divide two monomials with the same base, subtract the exponents of the like bases.

> **HOW TO**   Simplify: $\dfrac{a^7}{a^3}$
>
> $\dfrac{a^7}{a^3} = a^{7-3}$ • **The bases are the same. Subtract the exponents.**
>
> $\quad\ = a^4$

> **HOW TO**   Simplify: $\dfrac{r^8 t^6}{r^7 t}$
>
> $\dfrac{r^8 t^6}{r^7 t} = r^{8-7} t^{6-1}$ • **Subtract the exponents of the like bases.**
>
> $\quad\ = r t^5$

> **HOW TO**   Simplify: $\dfrac{p^7}{z^4}$
>
> Because the bases are not the same, $\dfrac{p^7}{z^4}$ is already in simplest form.

Consider the expression $\dfrac{x^4}{x^4}$, $x \neq 0$. This expression can be simplified, as shown below, by subtracting exponents or by dividing by common factors.

$$\frac{x^4}{x^4} = x^{4-4} = x^0 \qquad\qquad \frac{x^4}{x^4} = \frac{\overset{1}{\cancel{x}} \cdot \overset{1}{\cancel{x}} \cdot \overset{1}{\cancel{x}} \cdot \overset{1}{\cancel{x}}}{\underset{1}{\cancel{x}} \cdot \underset{1}{\cancel{x}} \cdot \underset{1}{\cancel{x}} \cdot \underset{1}{\cancel{x}}} = 1$$

The equations $\dfrac{x^4}{x^4} = x^0$ and $\dfrac{x^4}{x^4} = 1$ suggest the following definition of x^0.

> **Definition of Zero as an Exponent**
>
> If $x \neq 0$, then $x^0 = 1$. The expression 0^0 is not defined.

Objective 9.4A

Instructor Note

Here we are just verbalizing the rule for division of polynomials. The formal definition comes after we define negative exponents.

Have students write this rule, even just copy it onto a piece of paper, and then practice a few exercises, such as

$$\frac{a^4}{a^2} \quad \text{and} \quad \frac{y^8}{y}$$

It may also help to give them $\dfrac{a^9}{b^5}$

to emphasize that the bases must be the same.

Concept Check

Simplify.

1. $\dfrac{r^7}{r^4}$ r^3

2. $\dfrac{b^5}{b}$ b^4

3. $\dfrac{c^4}{c^5}$ $\dfrac{1}{c}$

Concept Check

Simplify.

1. $(-36)^0$ 1

2. $(14y^2)^0$, $y \neq 0$ 1

In-Class Examples (Objective 9.4A)

Simplify.

1. $(-3x^4 y^{-5})(-2x^{-3} y^{-1})$ $\dfrac{6x}{y^6}$

2. $\dfrac{3x^{-1} y^4}{6^{-1} x^3 y^{-2}}$ $\dfrac{18y^6}{x^4}$

3. $\left(\dfrac{8a^{-3} b^{-1} c^2}{12a^3 b^{-3} c^{-2}}\right)^{-2}$ $\dfrac{9a^{12}}{4b^4 c^8}$

Concept Check

Evaluate.

1. 3^{-1} $\dfrac{1}{3}$

2. -2^{-3} $-\dfrac{1}{8}$

Concept Check

Simplify.

1. $4b^{-3}$ $\dfrac{4}{b^3}$

2. a^3b^{-4} $\dfrac{a^3}{b^4}$

Discuss the Concepts

Explain why $2x^4y^{-3}$ is not in simplest form. What is the simplest form of this expression?

Only positive exponents may be used in the simplest form of an exponential expression. The simplest form of $2x^4y^{-3}$ is $\dfrac{2x^4}{y^3}$.

TAKE NOTE

In the example at the right, we indicated that $a \neq 0$. If we try to evaluate $(12a^3)^0$ when $a = 0$, we have

$$[12(0)^3]^0 = [12(0)]^0 = 0^0$$

However, 0^0 is not defined. Therefore, we must assume that $a \neq 0$. To avoid stating this for every example or exercise, we will assume that variables do not have values that result in the expression 0^0.

Point of Interest

In the 15th century, the expression $12^{2\overline{m}}$ was used to mean $12x^{-2}$. The use of $\overline{m}$ reflected an Italian influence. In Italy, m was used for minus and p was used for plus. It was understood that $2\overline{m}$ referred to an unnamed variable. Issac Newton, in the 17th century, advocated the negative exponent notation that we currently use.

TAKE NOTE

Note from the example at the right that 2^{-4} is a *positive* number. A negative exponent does not change the sign of a number.

TAKE NOTE

For the expression $3n^{-5}$, the exponent on n is -5 (*negative* 5). The n^{-5} is written in the denominator as n^5. The exponent on 3 is 1 (*positive* 1). The 3 remains in the numerator. Also, we indicated that $n \neq 0$. This is done because division by zero is not defined. In this textbook, we will assume that values of the variables are chosen so that division by zero does not occur.

HOW TO   Simplify: $(12a^3)^0$, $a \neq 0$

$(12a^3)^0 = 1$ • Any nonzero expression to the zero power is **1**.

HOW TO   Simplify: $-(4x^3y^7)^0$

$-(4x^3y^7)^0 = -(1) = -1$

Consider the expression $\dfrac{x^4}{x^6}$, $x \neq 0$. This expression can be simplified, as shown below, by subtracting exponents or by dividing by common factors.

$$\frac{x^4}{x^6} = x^{4-6} = x^{-2} \qquad \frac{x^4}{x^6} = \frac{\overset{1}{\cancel{x}} \cdot \overset{1}{\cancel{x}} \cdot \overset{1}{\cancel{x}} \cdot \overset{1}{\cancel{x}}}{\underset{1}{\cancel{x}} \cdot \underset{1}{\cancel{x}} \cdot \underset{1}{\cancel{x}} \cdot \underset{1}{\cancel{x}} \cdot x \cdot x} = \frac{1}{x^2}$$

The equations $\dfrac{x^4}{x^6} = x^{-2}$ and $\dfrac{x^4}{x^6} = \dfrac{1}{x^2}$ suggest that $x^{-2} = \dfrac{1}{x^2}$.

Definition of a Negative Exponent

If $x \neq 0$ and n is a positive integer, then

$$x^{-n} = \frac{1}{x^n} \qquad \text{and} \qquad \frac{1}{x^{-n}} = x^n$$

An exponential expression is in simplest form when it is written with only positive exponents.

HOW TO   Evaluate 2^{-4}.

$2^{-4} = \dfrac{1}{2^4}$ • Use the **Definition of a Negative Exponent**.

$= \dfrac{1}{16}$ • Evaluate the expression.

HOW TO   Simplify: $3n^{-5}$, $n \neq 0$

$3n^{-5} = 3 \cdot \dfrac{1}{n^5} = \dfrac{3}{n^5}$ • Use the **Definition of a Negative Exponent** to rewrite the expression with a positive exponent.

HOW TO   Simplify: $\dfrac{2}{5a^{-4}}$

$\dfrac{2}{5a^{-4}} = \dfrac{2}{5} \cdot \dfrac{1}{a^{-4}} = \dfrac{2}{5} \cdot a^4 = \dfrac{2a^4}{5}$ • Use the **Definition of a Negative Exponent** to rewrite the expression with a positive exponent.

The expression $\left(\dfrac{x^4}{y^3}\right)^2$, $y \neq 0$, can be simplified by squaring $\dfrac{x^4}{y^3}$ or by multiplying each exponent in the quotient by the exponent outside the parentheses.

$$\left(\frac{x^4}{y^3}\right)^2 = \left(\frac{x^4}{y^3}\right)\left(\frac{x^4}{y^3}\right) = \frac{x^4 \cdot x^4}{y^3 \cdot y^3} = \frac{x^{4+4}}{y^{3+3}} = \frac{x^8}{y^6} \qquad \left(\frac{x^4}{y^3}\right)^2 = \frac{x^{4 \cdot 2}}{y^{3 \cdot 2}} = \frac{x^8}{y^6}$$

> **Rule for Simplifying the Power of a Quotient**
>
> If m, n, and p are integers and $y \neq 0$, then $\left(\dfrac{x^m}{y^n}\right)^p = \dfrac{x^{mp}}{y^{np}}$.

TAKE NOTE

As a reminder, although it is not explicitly stated, we are assuming that $a \neq 0$ and $b \neq 0$. This is done to ensure that we do not have division by zero.

HOW TO   Simplify: $\left(\dfrac{a^3}{b^2}\right)^{-2}$

$\left(\dfrac{a^3}{b^2}\right)^{-2} = \dfrac{a^{3(-2)}}{b^{2(-2)}}$ • Use the **Rule for Simplifying the Power of a Quotient.**

$= \dfrac{a^{-6}}{b^{-4}} = \dfrac{b^4}{a^6}$ • Use the **Definition of a Negative Exponent** to write the expression with positive exponents.

The example above suggests the following rule.

> **Rule for Negative Exponents on Fractional Expressions**
>
> If $a \neq 0$, $b \neq 0$, and n is a positive integer, then
>
> $$\left(\frac{a}{b}\right)^{-n} = \left(\frac{b}{a}\right)^{n}$$

Now that zero as an exponent and negative exponents have been defined, a rule for dividing exponential expressions can be stated.

> **Rule for Dividing Exponential Expressions**
>
> If m and n are integers and $x \neq 0$, then $\dfrac{x^m}{x^n} = x^{m-n}$.

HOW TO   Evaluate $\dfrac{5^{-2}}{5}$.

$\dfrac{5^{-2}}{5} = 5^{-2-1} = 5^{-3}$ • Use the **Rule for Dividing Exponential Expressions.**

$= \dfrac{1}{5^3} = \dfrac{1}{125}$ • Use the **Definition of a Negative Exponent** to rewrite the expression with a positive exponent. Then evaluate.

Concept Check

Simplify.

1. $\left(\dfrac{x^3}{y^4}\right)^2$ $\dfrac{x^6}{y^8}$

2. $\left(\dfrac{x^4}{y^5}\right)^{-1}$ $\dfrac{y^5}{x^4}$

3. $\left(\dfrac{x}{y}\right)^{-3}$ $\dfrac{y^3}{x^3}$

Optional Student Activity

Evaluate $\dfrac{3^2}{3^{-2}}$.   81

Optional Student Activity

What is the simplest form of $(-2)^{-2}$? $\dfrac{1}{4}$

Optional Student Activity

By choosing values for m, n, and p, make up examples that demonstrate the simplification of $\left(\dfrac{x^m}{y^n}\right)^p$ for each of the following situations.

1. m, n, and p are all positive integers.

2. m and n are positive integers, and p is a negative integer.

3. m and n are negative integers, and p is a positive integer.

4. m and p are positive integers, and n is a negative integer.

5. n and p are positive integers, and m is a negative integer.

6. m, n, and p are all negative integers.

Instructor Note

Examples such as the one at the right are included to review the work on multiplying monomials and to demonstrate that negative exponents can be used for products of exponential expressions.

Instructor Note

There are a few different ways to simplify the expression at the right. Students can simplify the expression by starting as follows:

$$\left(\dfrac{6m^2n^3}{8m^7n^2}\right)^{-3} = \left(\dfrac{8m^7n^2}{6m^2n^3}\right)^3$$

This method uses the Rule for Negative Exponents on Fractional Expressions first.

HOW TO   Simplify: $\dfrac{x^4}{x^9}$

$\dfrac{x^4}{x^9} = x^{4-9}$ • Use the **Rule for Dividing Exponential Expressions**.

$\qquad = x^{-5}$ • Subtract the exponents.

$\qquad = \dfrac{1}{x^5}$ • Use the **Definition of a Negative Exponent** to rewrite the expression with a positive exponent.

The rules for simplifying exponential expressions and powers of exponential expressions are true for all integers. These rules are restated here, along with the rules for dividing exponential expressions.

Rules of Exponents

If m, n, and p are integers, then

$$x^m \cdot x^n = x^{m+n} \qquad (x^m)^n = x^{mn} \qquad (x^m y^n)^p = x^{mp} y^{np}$$

$$\dfrac{x^m}{x^n} = x^{m-n},\, x \neq 0 \qquad \left(\dfrac{x^m}{y^n}\right)^p = \dfrac{x^{mp}}{y^{np}},\, y \neq 0 \qquad x^{-n} = \dfrac{1}{x^n},\, x \neq 0$$

$$x^0 = 1,\, x \neq 0$$

HOW TO   Simplify: $(3ab^{-4})(-2a^{-3}b^7)$

$(3ab^{-4})(-2a^{-3}b^7) = [3 \cdot (-2)](a^{1+(-3)}b^{-4+7})$ • When multiplying expressions, add the exponents on like bases.

$\qquad = -6a^{-2}b^3$

$\qquad = -\dfrac{6b^3}{a^2}$

HOW TO   Simplify: $\left[\dfrac{6m^2n^3}{8m^7n^2}\right]^{-3}$

$\left[\dfrac{6m^2n^3}{8m^7n^2}\right]^{-3} = \left[\dfrac{3m^{2-7}n^{3-2}}{4}\right]^{-3}$ • Simplify inside the brackets.

$\qquad = \left[\dfrac{3m^{-5}n}{4}\right]^{-3}$ • Subtract the exponents.

$\qquad = \dfrac{3^{-3}m^{15}n^{-3}}{4^{-3}}$ • Use the Rule for Simplifying the Power of a Quotient.

$\qquad = \dfrac{4^3 m^{15}}{3^3 n^3} = \dfrac{64m^{15}}{27n^3}$ • Use the **Definition of a Negative Exponent** to rewrite the expression with positive exponents. Then simplify.

HOW TO   Simplify: $\dfrac{4a^{-2}b^5}{6a^5b^2}$

$\dfrac{4a^{-2}b^5}{6a^5b^2} = \dfrac{2a^{-2}b^5}{3a^5b^2}$ • **Divide the coefficients by their common factor.**

$= \dfrac{2a^{-2-5}b^{5-2}}{3}$ • **Use the Rule for Dividing Exponential Expressions.**

$= \dfrac{2a^{-7}b^3}{3} = \dfrac{2b^3}{3a^7}$ • **Use the Definition of a Negative Exponent to rewrite the expression with positive exponents.**

Concept Check

Simplify.

1. $(3a)(4a^{-3})^{-2}$ $\dfrac{3a^7}{16}$

2. $\dfrac{(3x^{-1})^{-3}}{x^{-2}}$ $\dfrac{x^5}{27}$

3. $\left(\dfrac{3x^{-1}}{6x^{-2}}\right)^{-1}$ $\dfrac{2}{x}$

Example 1   Simplify: $(-2x)(3x^{-2})^{-3}$

Solution

$(-2x)(3x^{-2})^{-3} = (-2x)(3^{-3}x^6)$ • **Rule for Simplifying the Power of a Product**

$= \dfrac{-2x^{1+6}}{3^3}$

$= -\dfrac{2x^7}{27}$

You Try It 1   Simplify: $(-2x^2)(x^{-3}y^{-4})^{-2}$

Your solution $-2x^8y^8$

Example 2   Simplify: $\dfrac{(2r^2t^{-1})^{-3}}{(r^{-3}t^4)^2}$

Solution

$\dfrac{(2r^2t^{-1})^{-3}}{(r^{-3}t^4)^2} = \dfrac{2^{-3}r^{-6}t^3}{r^{-6}t^8}$ • **Rule for Simplifying the Power of a Product**

$= 2^{-3}r^{-6-(-6)}t^{3-8}$ • **Rule for Dividing Exponential Expressions**

$= 2^{-3}r^0t^{-5}$

$= \dfrac{1}{2^3t^5}$

$= \dfrac{1}{8t^5}$ • **Write answer in simplest form.**

You Try It 2   Simplify: $\dfrac{(6a^{-2}b^3)^{-1}}{(4a^3b^{-2})^{-2}}$

Your solution $\dfrac{8a^8}{3b^7}$

Example 3   Simplify: $\left[\dfrac{4a^{-2}b^3}{6a^4b^{-2}}\right]^{-3}$

Solution

$\left[\dfrac{4a^{-2}b^3}{6a^4b^{-2}}\right]^{-3} = \left[\dfrac{2a^{-6}b^5}{3}\right]^{-3}$ • **Simplify inside brackets.**

$= \dfrac{2^{-3}a^{18}b^{-15}}{3^{-3}}$ • **Rule for Simplifying the Power of a Quotient**

$= \dfrac{27a^{18}}{8b^{15}}$ • **Write answer in simplest form.**

You Try It 3   Simplify: $\left[\dfrac{6r^3s^{-3}}{9r^3s^{-1}}\right]^{-2}$

Your solution $\dfrac{9s^4}{4}$

Solutions on p. S23

Objective 9.4B

New Vocabulary
scientific notation

Concept Check
Write the numbers in scientific notation.

1. 400 4×10^2

2. 0.03 3×10^{-2}

Concept Check
Write the numbers in decimal notation.

1. 4.3×10^6 4,300,000

2. 6.2×10^{-4} 0.00062

Optional Student Activity

The symbol for angstrom is Å. Research the definition of angstrom. Then convert 3.2 Å to millimeters, using scientific notation.

An angstrom is defined as a unit of length equal to one ten-millionth of a millimeter.

$$3.2 \text{ Å} = 3.2 \times 10^{-7} \text{ mm}$$

Objective B   **To write a number in scientific notation**

Integrating Technology
See the appendix Keystroke Guide: *Scientific Notation* for instructions on entering numbers in scientific notation on a calculator.

Point of Interest
An electron microscope uses wavelengths that are approximately 4×10^{-12} meter to make images of viruses.

The human eye can detect wavelengths between 4.3×10^{-7} meter and 6.9×10^{-7} meter. Although these are very short, they are approximately 10^5 times longer than the waves used in an electron microscope.

Very large and very small numbers abound in the natural sciences. For example, the mass of an electron is 0.00000000000000000000000000000911 kg. Numbers such as this are difficult to read, so a more convenient system called **scientific notation** is used. In scientific notation, a number is expressed as the product of two factors, one a number between 1 and 10, and the other a power of 10.

To express a number in scientific notation, write it in the form $a \times 10^n$, where a is a number between 1 and 10, and n is an integer.

For numbers greater than or equal to 10, move the decimal point to the right of the first digit. The exponent n is positive and equal to the number of places the decimal point has been moved.

$$240{,}000 = 2.4 \times 10^5$$
$$93{,}000{,}000 = 9.3 \times 10^7$$

For numbers less than 1, move the decimal point to the right of the first nonzero digit. The exponent n is negative. The absolute value of the exponent is equal to the number of places the decimal point has been moved.

$$0.0003 = 3 \times 10^{-4}$$
$$0.0000832 = 8.32 \times 10^{-5}$$

Changing a number written in scientific notation to decimal notation also requires moving the decimal point.

When the exponent is positive, move the decimal point to the right the same number of places as the exponent.

$$3.45 \times 10^6 = 3{,}450{,}000$$
$$2.3 \times 10^8 = 230{,}000{,}000$$

When the exponent is negative, move the decimal point to the left the same number of places as the absolute value of the exponent.

$$8.1 \times 10^{-3} = 0.0081$$
$$6.34 \times 10^{-7} = 0.000000634$$

Example 4   Write the number 824,300,000 in scientific notation.

Solution $824{,}300{,}000 = 8.243 \times 10^8$

You Try It 4   Write the number 0.000000961 in scientific notation.

Your solution 9.61×10^{-7}

Example 5   Write the number 6.8×10^{-10} in decimal notation.

Solution $6.8 \times 10^{-10} = 0.00000000068$

You Try It 5   Write the number 7.329×10^6 in decimal notation.

Your solution   7,329,000

Solutions on p. S23

In-Class Examples (Objective 9.4B)

1. Write the number 0.00394 in scientific notation.
3.94×10^{-3}

2. Write the number 3.8×10^4 in decimal notation.
38,000

9.4 Exercises

Objective A To divide monomials

For Exercises 1 and 2, replace the question marks to make a true statement.

1. $\dfrac{x^7}{x^5} = x^{?-?} = x^?$

7, 5, 2

2. $\dfrac{a^9}{a^{12}} = a^{?-?} = \dfrac{1}{a^?}$

9, 12, 3

For Exercises 3 to 10, evaluate.

3. 5^{-2} $\dfrac{1}{25}$

4. 3^{-3} $\dfrac{1}{27}$

5. $\dfrac{1}{8^{-2}}$ 64

6. $\dfrac{1}{12^{-1}}$ 12

7. $\dfrac{3^{-2}}{3}$ $\dfrac{1}{27}$

8. $\dfrac{5^{-3}}{5}$ $\dfrac{1}{625}$

9. $\dfrac{2^{-2}}{2^{-3}}$ 2

10. $\dfrac{3^2}{3^2}$ 1

For Exercises 11 to 94, simplify.

11. x^{-2} $\dfrac{1}{x^2}$

12. y^{-10} $\dfrac{1}{y^{10}}$

13. $\dfrac{1}{a^{-6}}$ a^6

14. $\dfrac{1}{b^{-4}}$ b^4

15. $4x^{-7}$ $\dfrac{4}{x^7}$

16. $-6y^{-1}$ $-\dfrac{6}{y}$

17. $\dfrac{2}{3}z^{-2}$ $\dfrac{2}{3z^2}$

18. $\dfrac{4}{5}a^{-4}$ $\dfrac{4}{5a^4}$

19. $\dfrac{5}{b^{-8}}$ $5b^8$

20. $\dfrac{-3}{v^{-3}}$ $-3v^3$

21. $\dfrac{1}{3x^{-2}}$ $\dfrac{x^2}{3}$

22. $\dfrac{2}{5c^{-6}}$ $\dfrac{2c^6}{5}$

23. $(ab^5)^0$ 1

24. $(32x^3y^4)^0$ 1

25. $-(3p^2q^5)^0$ -1

26. $-\left(\dfrac{2}{3}xy\right)^0$ -1

27. $\dfrac{y^7}{y^3}$ y^4

28. $\dfrac{z^9}{z^2}$ z^7

29. $\dfrac{a^8}{a^5}$ a^3

30. $\dfrac{c^{12}}{c^5}$ c^7

31. $\dfrac{p^5}{p}$ p^4

32. $\dfrac{w^9}{w}$ w^8

33. $\dfrac{4x^8}{2x^5}$ $2x^3$

34. $\dfrac{12z^7}{4z^3}$ $3z^4$

35. $\dfrac{22k^5}{11k^4}$ $2k$

36. $\dfrac{14m^{11}}{7m^{10}}$ $2m$

37. $\dfrac{m^9n^7}{m^4n^5}$ m^5n^2

38. $\dfrac{y^5z^6}{yz^3}$ y^4z^3

39. $\dfrac{6r^4}{4r^2}$ $\dfrac{3r^2}{2}$

40. $\dfrac{8x^9}{12x^6}$ $\dfrac{2x^3}{3}$

41. $\dfrac{-16a^7}{24a^6}$ $-\dfrac{2a}{3}$

42. $\dfrac{-18b^5}{27b^4}$ $-\dfrac{2b}{3}$

43. $\dfrac{y^3}{y^8}$ $\dfrac{1}{y^5}$

44. $\dfrac{z^4}{z^6}$ $\dfrac{1}{z^2}$

45. $\dfrac{a^5}{a^{11}}$ $\dfrac{1}{a^6}$

46. $\dfrac{m}{m^7}$ $\dfrac{1}{m^6}$

Quick Quiz (Objective 9.4A)

1. Evaluate: 5^{-2} $\dfrac{1}{25}$

2. Simplify: $-5x^{-2}$ $-\dfrac{5}{x^2}$

3. Simplify: $\dfrac{z^4}{z^6}$ $\dfrac{1}{z^2}$

4. Simplify: $\dfrac{(3^{-1}x^3y^2z^{-2})^{-3}}{(6^{-2}x^{-2}y^{-3}z^2)^{-2}}$ $\dfrac{z^{10}}{48x^{13}y^{12}}$

Section 9.4

Suggested Assignment
Exercises 1–93, every other odd
Exercises 97–121, odds
More challenging problems:
  Exercises 123, 125, 130, 131

47. $\dfrac{4x^2}{12x^5}$

$\dfrac{1}{3x^3}$

48. $\dfrac{6y^8}{8y^9}$

$\dfrac{3}{4y}$

49. $\dfrac{-12x}{-18x^6}$

$\dfrac{2}{3x^5}$

50. $\dfrac{-24c^2}{-36c^{11}}$

$\dfrac{2}{3c^9}$

51. $\dfrac{x^6y^5}{x^8y}$

$\dfrac{y^4}{x^2}$

52. $\dfrac{a^3b^2}{a^2b^3}$

$\dfrac{a}{b}$

53. $\dfrac{2m^6n^2}{5m^9n^{10}}$

$\dfrac{2}{5m^3n^8}$

54. $\dfrac{5r^3t^7}{6r^5t^7}$

$\dfrac{5}{6r^2}$

55. $\dfrac{pq^3}{p^4q^4}$

$\dfrac{1}{p^3q}$

56. $\dfrac{a^4b^5}{a^5b^6}$

$\dfrac{1}{ab}$

57. $\dfrac{3x^4y^5}{6x^4y^8}$

$\dfrac{1}{2y^3}$

58. $\dfrac{14a^3b^6}{21a^5b^6}$

$\dfrac{2}{3a^2}$

59. $\dfrac{14x^4y^6z^2}{16x^3y^9z}$

$\dfrac{7xz}{8y^3}$

60. $\dfrac{24a^2b^7c^9}{36a^7b^5c}$

$\dfrac{2b^2c^8}{3a^5}$

61. $\dfrac{15mn^9p^3}{30m^4n^9p}$

$\dfrac{p^2}{2m^3}$

62. $\dfrac{25x^4y^7z^2}{20x^5y^9z^{11}}$

$\dfrac{5}{4xy^2z^9}$

63. $(-2xy^{-2})^3$

$-\dfrac{8x^3}{y^6}$

64. $(-3x^{-1}y^2)^2$

$\dfrac{9y^4}{x^2}$

65. $(3x^{-1}y^{-2})^2$

$\dfrac{9}{x^2y^4}$

66. $(5xy^{-3})^{-2}$

$\dfrac{y^6}{25x^2}$

67. $(2x^{-1})(x^{-3})$

$\dfrac{2}{x^4}$

68. $(-2x^{-5})x^7$

$-2x^2$

69. $(-5a^2)(a^{-5})^2$

$-\dfrac{5}{a^8}$

70. $(2a^{-3})(a^7b^{-1})^3$

$\dfrac{2a^{18}}{b^3}$

71. $(-2ab^{-2})(4a^{-2}b)^{-2}$

$-\dfrac{a^5}{8b^4}$

72. $(3ab^{-2})(2a^{-1}b)^{-3}$

$\dfrac{3a^4}{8b^5}$

73. $(-5x^{-2}y)(-2x^{-2}y^2)$

$\dfrac{10y^3}{x^4}$

74. $\dfrac{a^{-3}b^{-4}}{a^2b^2}$

$\dfrac{1}{a^5b^6}$

75. $\dfrac{3x^{-2}y^2}{6xy^2}$

$\dfrac{1}{2x^3}$

76. $\dfrac{2x^{-2}y}{8xy}$

$\dfrac{1}{4x^3}$

77. $\dfrac{3x^{-2}y}{xy}$

$\dfrac{3}{x^3}$

78. $\dfrac{2x^{-1}y^4}{x^2y^3}$

$\dfrac{2y}{x^3}$

79. $\dfrac{2x^{-1}y^{-4}}{4xy^2}$

$\dfrac{1}{2x^2y^6}$

80. $\dfrac{(x^{-1}y)^2}{xy^2}$

$\dfrac{1}{x^3}$

81. $\dfrac{(x^{-2}y)^2}{x^2y^3}$

$\dfrac{1}{x^6y}$

82. $\dfrac{(x^{-3}y^{-2})^2}{x^6y^8}$

$\dfrac{1}{x^{12}y^{12}}$

83. $\dfrac{(a^{-2}y^3)^{-3}}{a^2y}$

$\dfrac{a^4}{y^{10}}$

84. $\dfrac{12a^2b^3}{-27a^2b^2}$

$-\dfrac{4b}{9}$

85. $\dfrac{-16xy^4}{96x^4y^4}$

$-\dfrac{1}{6x^3}$

86. $\dfrac{-8x^2y^4}{44y^2z^5}$

$-\dfrac{2x^2y^2}{11z^5}$

87. $\dfrac{22a^2b^4}{-132b^3c^2}$

$-\dfrac{a^2b}{6c^2}$

88. $\dfrac{-(8a^2b^4)^3}{64a^3b^8}$

$-8a^3b^4$

89. $\dfrac{-(14ab^4)^2}{28a^4b^2}$

$-\dfrac{7b^6}{a^2}$

90. $\dfrac{(2a^{-2}b^3)^{-2}}{(4a^2b^{-4})^{-1}}$

$\dfrac{a^6}{b^{10}}$

91. $\dfrac{(3^{-1}r^4s^{-3})^{-2}}{(6r^2s^{-1}t^{-2})^2}$

$\dfrac{s^8t^4}{4r^{12}}$

92. $\left(\dfrac{6x^{-4}yz^{-1}}{14xy^{-4}z^2}\right)^{-3}$

$\dfrac{343x^{15}z^9}{27y^{15}}$

93. $\left(\dfrac{15m^3n^{-2}p^{-1}}{25m^{-2}n^{-4}}\right)^{-3}$

$\dfrac{125p^3}{27m^{15}n^6}$

94. $\left(\dfrac{18a^4b^{-2}c^4}{12ab^{-3}d^2}\right)^{-2}$

$\dfrac{4d^4}{9a^6b^2c^8}$

Objective B   **To write a number in scientific notation**

95. ✏ Why might a number be written in scientific notation instead of decimal notation?

96. ✏ **a.** Explain how to write 0.00000076 in scientific notation.
  b. Explain how to write 4.3×10^8 in decimal notation.

For Exercises 97 to 105, write in scientific notation.

97. 0.00000000324
  3.24×10^{-9}

98. 0.00000012
  1.2×10^{-7}

99. 0.000000000000000003
  3×10^{-18}

100. 1,800,000,000
  1.8×10^9

101. 32,000,000,000,000,000
  3.2×10^{16}

102. 76,700,000,000,000
  7.67×10^{13}

103. 0.0000000000000000000122
  1.22×10^{-19}

104. 0.00137
  1.37×10^{-3}

105. 547,000,000
  5.47×10^8

For Exercises 106 to 114, write in decimal notation.

106. 2.3×10^{-12}
  0.0000000000023

107. 1.67×10^{-4}
  0.000167

108. 2×10^{15}
  2,000,000,000,000,000

109. 6.8×10^7
  68,000,000

110. 9×10^{-21}
  0.000000000000000000009

111. 3.05×10^{-5}
  0.0000305

112. 9.05×10^{11}
  905,000,000,000

113. 1.02×10^{-9}
  0.00000000102

114. 7.2×10^{-3}
  0.0072

115. 🔵 **Chemistry** Avogadro's number is used in chemistry, and its value is approximately 602,300,000,000,000,000,000,000. Express this number in scientific notation. 6.023×10^{23}

116. 🔵 **Geology** 5,980,000,000,000,000,000,000,000 kg is the approximate mass of the planet Earth. Write the mass of Earth in scientific notation. 5.98×10^{24} kg

Answers to Writing Exercises

95. Very large or very small numbers are written in scientific notation so that they are easier to read.

96a. To write 0.00000076 in scientific notation, move the decimal point 7 places to the right. Then multiply the resulting number by 10^{-7}.

  $0.00000076 = 7.6 \times 10^{-7}$

b. To write 4.3×10^8 in decimal notation, move the decimal point 8 places to the right.

  $4.3 \times 10^8 = 430,000,000$

Quick Quiz (Objective 9.4B)

1. Write in scientific notation.
  a. 41,300,000,000 4.13×10^{10}
  b. 0.000327 3.27×10^{-4}

2. Write in decimal notation.
  a. 2.4×10^{-5} 0.000024
  b. 5.76×10^6 5,760,000

502

117. **Physics** The length of an infrared light wave is approximately 0.0000037 m. Write this number in scientific notation.
3.7×10^{-6}

118. **Physics** Light travels approximately 16,000,000,000 mi in one day. Write this number in scientific notation.
1.6×10^{10}

119. **Computer Science** One unit used to measure the speed of a computer is the picosecond. One picosecond is 0.000000001 s. Write this number in scientific notation.
1×10^{-9}

120. **Astronomy** One light-year is the distance traveled by light in 1 year. One light-year is 5,880,000,000,000 mi. Write this number in scientific notation.
5.88×10^{12}

121. **Electricity** The electric charge on an electron is 0.00000000000000000016 coulomb. Write this number in scientific notation.
1.6×10^{-19}

122. **Chemistry** Approximately 35 teragrams (3.5×10^{13} g) of sulfur in the atmosphere is converted to sulfate each year. Write this number in decimal notation.
35,000,000,000,000

APPLYING THE CONCEPTS

123. Evaluate 2^x when $x = -2, -1, 0, 1$, and 2.
$\dfrac{1}{4}, \dfrac{1}{2}, 1, 2, 4$

124. Evaluate 3^x when $x = -2, -1, 0, 1$, and 2.
$\dfrac{1}{9}, \dfrac{1}{3}, 1, 3, 9$

125. Evaluate 2^{-x} when $x = -2, -1, 0, 1$, and 2.
$4, 2, 1, \dfrac{1}{2}, \dfrac{1}{4}$

126. Evaluate 3^{-x} when $x = -2, -1, 0, 1$, and 2.
$9, 3, 1, \dfrac{1}{3}, \dfrac{1}{9}$

For Exercises 127 to 129, determine whether each equation is true or false. If the equation is false, change the right-hand side of the equation to make a true equation.

127. $(2a)^{-3} = \dfrac{2}{a^3}$

False. $(2a)^{-3} = \dfrac{1}{8a^3}$

128. $\dfrac{x^{-3}}{y^{-3}} = \left(\dfrac{x}{y}\right)^{-3}$

True

129. $(2 + 3)^{-1} = 2^{-1} + 3^{-1}$

False. $(2 + 3)^{-1} = 5^{-1} = \dfrac{1}{5}$

130. Simplify: $\left(\dfrac{6x^4yz^3}{2x^2y^3}\right)\left(\dfrac{2x^2z^3}{4y^2z}\right) \div \left(\dfrac{6x^2y^3}{x^4y^2z}\right)$

$\dfrac{x^6z^6}{4y^5}$

131. If x is a nonzero real number, is x^{-2} always positive, always negative, or positive or negative depending on whether x is positive or negative? Explain your answer.

Answers to Writing Exercises

131. The expression x^{-2} is positive for all nonzero real numbers x. The reason is that $x^{-2} = 1/x^2$, and x^2 is positive for all nonzero values of x.

9.5 Division of Polynomials

Objective A   To divide a polynomial by a monomial

To divide a polynomial by a monomial, divide each term in the numerator by the denominator and write the sum of the quotients.

HOW TO   Divide: $\dfrac{6x^3 - 3x^2 + 9x}{3x}$

$$\dfrac{6x^3 - 3x^2 + 9x}{3x} = \dfrac{6x^3}{3x} - \dfrac{3x^2}{3x} + \dfrac{9x}{3x}$$

• Divide each term of the polynomial by the monomial.

$$= 2x^2 - x + 3$$

• Simplify each expression.

Example 1   Divide: $\dfrac{12x^2y - 6xy + 4x^2}{2xy}$

Solution

$$\dfrac{12x^2y - 6xy + 4x^2}{2xy} = \dfrac{12x^2y}{2xy} - \dfrac{6xy}{2xy} + \dfrac{4x^2}{2xy}$$

$$= 6x - 3 + \dfrac{2x}{y}$$

You Try It 1   Divide: $\dfrac{24x^2y^2 - 18xy + 6y}{6xy}$

Your solution $4xy - 3 + \dfrac{1}{x}$

Solution on p. S23

Objective B   To divide polynomials

Study Tip

An important element of success is practice. We cannot do anything well if we do not practice it repeatedly. Practice is crucial to success in mathematics. In this objective you are learning a new skill, how to divide polynomials. You will need to practice this skill over and over again in order to be successful at it.

The procedure for dividing two polynomials is similar to the procedure for dividing whole numbers. The same equation used to check division of whole numbers is used to check polynomial division.

(Quotient × divisor) + remainder = dividend

HOW TO   Divide: $(x^2 - 5x + 8) \div (x - 3)$

Step 1

$$\require{enclose}\begin{array}{r} x \\ x - 3 \enclose{longdiv}{x^2 - 5x + 8} \\ \underline{x^2 - 3x} \\ -2x + 8 \end{array}$$

• Think: $x\overline{)x^2} = \dfrac{x^2}{x} = x$

• Multiply: $x(x - 3) = x^2 - 3x$

• Subtract: $(x^2 - 5x) - (x^2 - 3x) = -2x$ Bring down the 8.

Step 2

$$\begin{array}{r} x - 2 \\ x - 3 \enclose{longdiv}{x^2 - 5x + 8} \\ \underline{x^2 - 3x} \\ -2x + 8 \\ \underline{-2x + 6} \\ 2 \end{array}$$

• Think: $x\overline{)-2x} = \dfrac{-2x}{x} = -2$

• Multiply: $-2(x - 3) = -2x + 6$

• Subtract: $(-2x + 8) - (-2x + 6) = 2$ The remainder is 2.

Check: $(x - 2)(x - 3) + 2 = x^2 - 5x + 6 + 2 = x^2 - 5x + 8$

$$(x^2 - 5x + 8) \div (x - 3) = x - 2 + \dfrac{2}{x - 3}$$

Objective 9.5A

Concept Check

Replace each ? to make a true statement.

1. $\dfrac{3x^2 + 6x}{3x} = \dfrac{3x^2}{?} + \dfrac{6x}{?} = ?$

   $3x, 3x, x + 2$

2. $\dfrac{4a^3 - 2a}{2a^2} = \dfrac{?}{2a^2} - \dfrac{?}{2a^2} = ?$

   $4a^3, 2a, 2a - \dfrac{1}{a}$

Objective 9.5B

Instructor Note

It may help some students if you start with the division algorithm for whole numbers and show them that a similar procedure is used to divide polynomials. $676 \div 21$ is an example you may consider using.

Instructor Note

Students are comfortable writing the answer to $15 \div 4$ as $3\dfrac{3}{4}$, which is $3 + \dfrac{3}{4}$. Tell students that this is the form in which the remainder for a quotient of polynomials is written.

In-Class Examples (Objective 9.5A)

Divide.

1. $\dfrac{18x^2y - 6xy + 12y}{6xy}$ $3x - 1 + \dfrac{2}{x}$

2. $\dfrac{24a^2b^2 + 16ab - 4b}{8ab}$ $3ab + 2 - \dfrac{1}{2a}$

Discuss the Concepts

In order for us to divide polynomials, how must the terms of the polynomials be arranged? Arrange the following division problem so that it is ready to be divided.

$$\frac{7x - 2 + x^3}{2 + x}$$

Both the dividend and the divisor must have their terms in descending order. Because there is no x^2 term, a zero can be inserted for that term. The problem may be arranged as follows:

$$x + 2\overline{)x^3 + 0 + 7x - 2}$$

or

$$x + 2\overline{)x^3 + 0x^2 + 7x - 2}$$

Concept Check

Replace each ? to make a true statement.

If $\frac{2x^2 - x - 1}{2x + 1} = x - 1$, then
$2x^2 - x - 1 = (?)\,(?).$
$x - 1, 2x + 1$

If a term is missing from the dividend, a zero can be inserted for that term. This helps keep like terms in the same column.

TAKE NOTE
Recall that a fraction bar means "divided by." Therefore, $6 \div 2$ can be written $\frac{6}{2}$, and $a \div b$ can be written $\frac{a}{b}$.

HOW TO Divide: $\dfrac{6x + 26 + 2x^3}{2 + x}$

$$\frac{2x^3 + 6x + 26}{x + 2}$$

- Arrange the terms of each polynomial in descending order.

$$
\begin{array}{r}
2x^2 - 4x + 14 \\
x + 2\overline{)2x^3 + 0 + 6x + 26} \\
\underline{2x^3 + 4x^2} \\
-4x^2 + 6x \\
\underline{-4x^2 - 8x} \\
14x + 26 \\
\underline{14x + 28} \\
-2
\end{array}
$$

- There is no x^2 term in $2x^3 + 6x + 26$. Insert a **zero** for the missing term.

Check:
$(2x^2 - 4x + 14)(x + 2) + (-2) = (2x^3 + 6x + 28) + (-2) = 2x^3 + 6x + 26$

$(2x^3 + 6x + 26) \div (x + 2) = 2x^2 - 4x + 14 - \dfrac{2}{x + 2}$

Example 2

Divide: $(8x^2 + 4x^3 + x - 4) \div (2x + 3)$

Solution

$$
\begin{array}{r}
2x^2 + x - 1 \\
2x + 3\overline{)4x^3 + 8x^2 + x - 4} \\
\underline{4x^3 + 6x^2} \\
2x^2 + x \\
\underline{2x^2 + 3x} \\
-2x - 4 \\
\underline{-2x - 3} \\
-1
\end{array}
$$

- Write the dividend in descending order.

$(4x^3 + 8x^2 + x - 4) \div (2x + 3)$

$= 2x^2 + x - 1 - \dfrac{1}{2x + 3}$

You Try It 2

Divide: $(2x^3 + x^2 - 8x - 3) \div (2x - 3)$

Your solution

$x^2 + 2x - 1 - \dfrac{6}{2x - 3}$

Example 3 Divide: $\dfrac{x^2 - 1}{x + 1}$

Solution

$$
\begin{array}{r}
x - 1 \\
x + 1\overline{)x^2 + 0 - 1} \\
\underline{x^2 + x} \\
-x - 1 \\
\underline{-x - 1} \\
0
\end{array}
$$

- Insert a zero for the missing term.

$(x^2 - 1) \div (x + 1) = x - 1$

You Try It 3 Divide: $\dfrac{x^3 - 2x + 1}{x - 1}$

Your solution

$x^2 + x - 1$

Solutions on p. S23

In-Class Examples (Objective 9.5B)

Divide.

1. $(6x^3 + x^2 - 18x + 10) \div (3x - 4)$

$2x^2 + 3x - 2 + \dfrac{2}{3x - 4}$

2. $\dfrac{x^3 - x - 6}{x - 2}$ $x^2 + 2x + 3$

9.5 Exercises

Objective A   To divide a polynomial by a monomial

For Exercises 1 to 24, divide.

1. $\dfrac{10a - 25}{5}$
$2a - 5$

2. $\dfrac{16b - 40}{8}$
$2b - 5$

3. $\dfrac{6y^2 + 4y}{y}$
$6y + 4$

4. $\dfrac{4b^3 - 3b}{b}$
$4b^2 - 3$

5. $\dfrac{3x^2 - 6x}{3x}$
$x - 2$

6. $\dfrac{10y^2 - 6y}{2y}$
$5y - 3$

7. $\dfrac{5x^2 - 10x}{-5x}$
$-x + 2$

8. $\dfrac{3y^2 - 27y}{-3y}$
$-y + 9$

9. $\dfrac{x^3 + 3x^2 - 5x}{x}$
$x^2 + 3x - 5$

10. $\dfrac{a^3 - 5a^2 + 7a}{a}$
$a^2 - 5a + 7$

11. $\dfrac{x^6 - 3x^4 - x^2}{x^2}$
$x^4 - 3x^2 - 1$

12. $\dfrac{a^8 - 5a^5 - 3a^3}{a^2}$
$a^6 - 5a^3 - 3a$

13. $\dfrac{5x^2y^2 + 10xy}{5xy}$
$xy + 2$

14. $\dfrac{8x^2y^2 - 24xy}{8xy}$
$xy - 3$

15. $\dfrac{9y^6 - 15y^3}{-3y^3}$
$-3y^3 + 5$

16. $\dfrac{4x^4 - 6x^2}{-2x^2}$
$-2x^2 + 3$

17. $\dfrac{3x^2 - 2x + 1}{x}$
$3x - 2 + \dfrac{1}{x}$

18. $\dfrac{8y^2 + 2y - 3}{y}$
$8y + 2 - \dfrac{3}{y}$

19. $\dfrac{-3x^2 + 7x - 6}{x}$
$-3x + 7 - \dfrac{6}{x}$

20. $\dfrac{2y^2 - 6y + 9}{y}$
$2y - 6 + \dfrac{9}{y}$

21. $\dfrac{16a^2b - 20ab + 24ab^2}{4ab}$
$4a - 5 + 6b$

22. $\dfrac{22a^2b - 11ab - 33ab^2}{11ab}$
$2a - 1 - 3b$

23. $\dfrac{9x^2y + 6xy - 3xy^2}{xy}$
$9x + 6 - 3y$

24. $\dfrac{5a^2b - 15ab + 30ab^2}{5ab}$
$a - 3 + 6b$

Objective B   To divide polynomials

For Exercises 25 and 26, replace the question marks to make a true statement.

25. If $\dfrac{x^2 - x - 6}{x - 3} = x + 2$, then
$x^2 - x - 6 = (?)(?).$ $(x + 2), (x - 3)$

26. If $\dfrac{x^2 + 2x - 3}{x - 2} = x + 4 + \dfrac{5}{x - 2}$, then
$x^2 + 2x - 3 = (?)(?) + ?.$ $(x + 4), (x - 2), 5$

For Exercises 27 to 53, divide.

27. $(b^2 - 14b + 49) \div (b - 7)$
$b - 7$

28. $(x^2 - x - 6) \div (x - 3)$
$x + 2$

29. $(y^2 + 2y - 35) \div (y + 7)$
$y - 5$

30. $(2x^2 + 5x + 2) \div (x + 2)$
$2x + 1$

31. $(2y^2 - 13y + 21) \div (y - 3)$
$2y - 7$

32. $(4x^2 - 16) \div (2x + 4)$
$2x - 4$

Quick Quiz (Objective 9.5A)

Divide.

1. $\dfrac{12x^3 - 15x^2 - 6x + 9}{3}$ $4x^3 - 5x^2 - 2x + 3$

2. $\dfrac{10x^3y - 15x^2y + 5xy - 10y}{5xy}$ $2x^2 - 3x + 1 - \dfrac{2}{x}$

33. $\dfrac{2y^2 + 7}{y - 3}$

$2y + 6 + \dfrac{25}{y - 3}$

34. $\dfrac{x^2 + 1}{x - 1}$

$x + 1 + \dfrac{2}{x - 1}$

35. $\dfrac{x^2 + 4}{x + 2}$

$x - 2 + \dfrac{8}{x + 2}$

36. $\dfrac{6x^2 - 7x}{3x - 2}$

$2x - 1 - \dfrac{2}{3x - 2}$

37. $\dfrac{6y^2 + 2y}{2y + 4}$

$3y - 5 + \dfrac{20}{2y + 4}$

38. $\dfrac{5x^2 + 7x}{x - 1}$

$5x + 12 + \dfrac{12}{x - 1}$

39. $(6x^2 - 5) \div (x + 2)$

$6x - 12 + \dfrac{19}{x + 2}$

40. $(a^2 + 5a + 10) \div (a + 2)$

$a + 3 + \dfrac{4}{a + 2}$

41. $(b^2 - 8b - 9) \div (b - 3)$

$b - 5 - \dfrac{24}{b - 3}$

42. $\dfrac{2y^2 - 9y + 8}{2y + 3}$

$y - 6 + \dfrac{26}{2y + 3}$

43. $\dfrac{3x^2 + 5x - 4}{x - 4}$

$3x + 17 + \dfrac{64}{x - 4}$

44. $(8x + 3 + 4x^2) \div (2x - 1)$

$2x + 5 + \dfrac{8}{2x - 1}$

45. $(10 + 21y + 10y^2) \div (2y + 3)$

$5y + 3 + \dfrac{1}{2y + 3}$

46. $\dfrac{15a^2 - 8a - 8}{3a + 2}$

$5a - 6 + \dfrac{4}{3a + 2}$

47. $\dfrac{12a^2 - 25a - 7}{3a - 7}$

$4a + 1$

48. $(5 - 23x + 12x^2) \div (4x - 1)$

$3x - 5$

49. $(24 + 6a^2 + 25a) \div (3a - 1)$

$2a + 9 + \dfrac{33}{3a - 1}$

50. $\dfrac{5x + 3x^2 + x^3 + 3}{x + 1}$

$x^2 + 2x + 3$

51. $\dfrac{7x + x^3 - 6x^2 - 2}{x - 1}$

$x^2 - 5x + 2$

52. $(x^4 - x^2 - 6) \div (x^2 + 2)$

$x^2 - 3$

53. $(x^4 + 3x^2 - 10) \div (x^2 - 2)$

$x^2 + 5$

APPLYING THE CONCEPTS

54. ✏️ In your own words, explain how to divide exponential expressions.

55. The product of a monomial and $4b$ is $12ab^2$. Find the monomial.
$3ab$

Answers to Writing Exercises

54. In their explanations of dividing exponential expressions, students should mention both dividing the coefficients and subtracting the exponents of like variables.

Quick Quiz (Objective 9.5B)

Divide.

1. $(3x^3 - 11x^2 + 10x - 12) \div (x - 3)$
$3x^2 - 2x + 4$

2. $\dfrac{16x^3 - 13x + 1}{4x - 1}$

$4x^2 + x - 3 - \dfrac{2}{4x - 1}$

Focus on Problem Solving

Dimensional Analysis

In solving application problems, it may be useful to include the units in order to organize the problem so that the answer is in the proper units. Using units to organize and check the correctness of an application is called **dimensional analysis.** We use the operations of multiplying units and dividing units in applying dimensional analysis to application problems.

The Rule for Multiplying Exponential Expressions states that we multiply two expressions with the same base by adding the exponents.

$$x^4 \cdot x^6 = x^{4+6} = x^{10}$$

In calculations that involve quantities, the units are operated on algebraically.

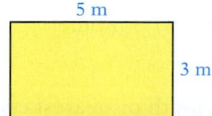

HOW TO   A rectangle measures 3 m by 5 m. Find the area of the rectangle.

$$A = LW = (3 \text{ m})(5 \text{ m}) = (3 \cdot 5)(\text{m} \cdot \text{m}) = 15 \text{ m}^2$$

The area of the rectangle is 15 m² (square meters).

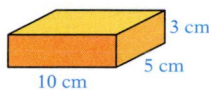

HOW TO   A box measures 10 cm by 5 cm by 3 cm. Find the volume of the box.

$$V = LWH = (10 \text{ cm})(5 \text{ cm})(3 \text{ cm}) = (10 \cdot 5 \cdot 3)(\text{cm} \cdot \text{cm} \cdot \text{cm}) = 150 \text{ cm}^3$$

The volume of the box is 150 cm³ (cubic centimeters).

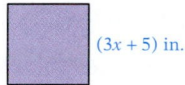

HOW TO   Find the area of a square whose side measures $(3x + 5)$ in.

$$A = s^2 = [(3x + 5) \text{ in.}]^2 = (3x + 5)^2 \text{ in}^2 = (9x^2 + 30x + 25) \text{ in}^2$$

The area of the square is $(9x^2 + 30x + 25)$ in² (square inches).

Dimensional analysis is used in the conversion of units.

The following example converts the unit miles to feet. The equivalent measures 1 mi = 5280 ft are used to form the following rates, which are called conversion

factors: $\dfrac{1 \text{ mi}}{5280 \text{ ft}}$ and $\dfrac{5280 \text{ ft}}{1 \text{ mi}}$. Because 1 mi = 5280 ft, both of the conversion factors

$\dfrac{1 \text{ mi}}{5280 \text{ ft}}$ and $\dfrac{5280 \text{ ft}}{1 \text{ mi}}$ are equal to 1.

To convert 3 mi to feet, multiply 3 mi by the conversion factor $\dfrac{5280 \text{ ft}}{1 \text{ mi}}$.

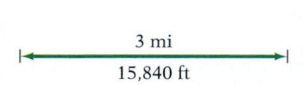

$$3 \text{ mi} = 3 \text{ mi} \cdot 1 = \frac{3 \text{ mi}}{1} \cdot \frac{5280 \text{ ft}}{1 \text{ mi}} = \frac{3 \text{ mi} \cdot 5280 \text{ ft}}{1 \text{ mi}} = 3 \cdot 5280 \text{ ft} = 15,840 \text{ ft}$$

There are two important points in the above illustration. First, you can think of dividing the numerator and denominator by the common unit "mile" just as you would divide the numerator and denominator of a fraction by a common factor. Second, the conversion factor $\dfrac{5280 \text{ ft}}{1 \text{ mi}}$ is equal to 1, and multiplying an expression by 1 does not change the value of the expression.

In the application problem that follows, the units are kept in the problem while the problem is worked.

 In 2003, a horse named Funny Cide ran a 1.25-mile race in 2.01 min. Find Funny Cide's average speed for the race in miles per hour. Round to the nearest tenth.

Strategy   To find the average speed, use the formula $r = \frac{d}{t}$, where r is the speed, d is the distance, and t is the time. Use the conversion factor $\frac{60 \text{ min}}{1 \text{ h}}$.

Solution $r = \dfrac{d}{t} = \dfrac{1.25 \text{ mi}}{2.01 \text{ min}} = \dfrac{1.25 \text{ mi}}{2.01 \text{ min}} \cdot \dfrac{60 \text{ min}}{1 \text{ h}}$

$= \dfrac{75 \text{ mi}}{2.01 \text{ h}} \approx 37.3 \text{ mph}$

Funny Cide's average speed was approximately 37.3 mph.

Try each of the following problems. Round to the nearest tenth or nearest cent.

1. Convert 88 ft/s to miles per hour.

2. Convert 8 m/s to kilometers per hour (1 km = 1000 m).

3. A carpet is to be placed in a meeting hall that is 36 ft wide and 80 ft long. At $21.50 per square yard, how much will it cost to carpet the meeting hall?

4. A carpet is to be placed in a room that is 20 ft wide and 30 ft long. At $22.25 per square yard, how much will it cost to carpet the area?

5. Find the number of gallons of water in a fish tank that is 36 in. long and 24 in. wide and is filled to a depth of 16 in. (1 gal = 231 in³).

6. Find the number of gallons of water in a fish tank that is 24 in. long and 18 in. wide and is filled to a depth of 12 in. (1 gal = 231 in³).

7. A $\frac{1}{4}$-acre commercial lot is on sale for $2.15 per square foot. Find the sale price of the commercial lot (1 acre = 43,560 ft²).

8. A 0.75-acre industrial parcel was sold for $98,010. Find the parcel's price per square foot (1 acre = 43,560 ft²).

9. A new driveway will require 800 ft³ of concrete. Concrete is ordered by the cubic yard. How much concrete should be ordered?

10. A piston-engined dragster traveled 440 yd in 4.936 s at Ennis, Texas, on October 9, 1988. Find the average speed of the dragster in miles per hour.

11. The Marianas Trench in the Pacific Ocean is the deepest part of the ocean. Its depth is 6.85 mi. Use an approximation of 4700 ft/s for the speed of sound under water to find the time it takes sound to travel from the surface to the bottom of the Marianas Trench and back.

Answers to Focus on Problem Solving: Dimensional Analysis

1. 60 mph
2. 28.8 km/h
3. $6880
4. $1483.33
5. 59.8 gal
6. 22.4 gal
7. $23,413.50
8. $3/ft²
9. 29.6 yd³
10. 182.3 mph
11. 15.4 s

Projects and Group Activities

Diagramming the Square of a Binomial

1. Explain why the diagram at the right represents $(a + b)^2 = a^2 + 2ab + b^2$.

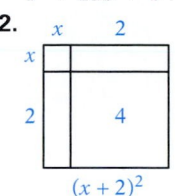

2. Draw similar diagrams representing each of the following.

$$(x + 2)^2$$

$$(x + 4)^2$$

Pascal's Triangle

Simplifying the power of a binomial is called *expanding the binomial*. The expansions of the first three powers of a binomial are shown below.

$$(a + b)^1 = a + b$$

$$(a + b)^2 = (a + b)(a + b) = a^2 + 2ab + b^2$$

$$(a + b)^3 = (a + b)^2(a + b) = (a^2 + 2ab + b^2)(a + b) = a^3 + 3a^2b + 3ab^2 + b^3$$

Find $(a + b)^4$. [*Hint:* $(a + b)^4 = (a + b)^3(a + b)$]

Find $(a + b)^5$. [*Hint:* $(a + b)^5 = (a + b)^4(a + b)$]

If we continue in this way, the results for $(a + b)^6$ are

$$(a + b)^6 = a^6 + 6a^5b + 15a^4b^2 + 20a^3b^3 + 15a^2b^4 + 6ab^5 + b^6$$

Now expand $(a + b)^8$. Before you begin, see whether you can find a pattern that will help you write the expansion of $(a + b)^8$ without having to multiply it out. Here are some hints.

1. Write out the variable terms of each binomial expansion from $(a + b)^1$ through $(a + b)^6$. Observe how the exponents on the variables change.

2. Write out the coefficients of all the terms without the variable parts. It will be helpful to make a triangular arrangement as shown at the left. Note that each row begins and ends with a 1. Also note (in the two shaded regions, for example) that any number in a row is the sum of the two closest numbers above it. For instance, $1 + 5 = 6$ and $6 + 4 = 10$.

```
      1   1
    1   2   1
  1   3   3   1
1   4   6   4   1
1   5  10  10   5   1
1   6  15  20  15   6   1
```

The triangle of numbers shown at the left is called Pascal's Triangle. To find the expansion of $(a + b)^8$, you need to find the eighth row of Pascal's Triangle. First find row 7. Then find row 8 and use the patterns you have observed to write the expansion $(a + b)^8$.

Pascal's Triangle has been the subject of extensive analysis, and many patterns have been found. See whether you can find some of them.

Answers to Projects and Group Activities: Diagramming the Square of a Binomial

1. There are two squares, one with area a^2 and one with area b^2. There are also two rectangles, with areas ab and ab. Adding these areas, we have $a^2 + b^2 + ab + ab = a^2 + 2ab + b^2$.

2.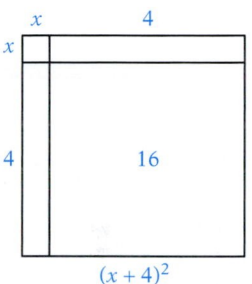

$$(x + 2)^2$$

$$(x + 4)^2$$

Answers to Projects and Group Activities: Pascal's Triangle

1. $(a + b)^1 = a + b$
$(a + b)^2 = a^2 + 2ab + b^2$
$(a + b)^3 = a^3 + 3a^2b + 3ab^2 + b^3$
$(a + b)^4 = a^4 + 4a^3b + 6a^2b^2 + 4ab^3 + b^4$
$(a + b)^5 = a^5 + 5a^4b + 10a^3b^2 + 10a^2b^3 + 5ab^4 + b^5$
$(a + b)^6 = a^6 + 6a^5b + 15a^4b^2 + 20a^3b^3 + 15a^2b^4 + 6ab^5 + b^6$

The exponents on a begin with the same exponent as that on the binomial and decrease by 1 each term.
  The exponents on b increase by 1 each term, beginning with zero up through the exponent on the binomial. The sum of the exponents in each term is the same as the exponent on the binomial.

2. Row 7:
1 7 21 35 35 21 7 1
Row 8:
1 8 28 56 70 56 28 8 1
$(a + b)^8 = a^8 + 8a^7b + 28a^6b^2 + 56a^5b^3 + 70a^4b^4 + 56a^3b^5 + 28a^2b^6 + 8ab^7 + b^8$

Chapter 9 Summary

Key Words	Examples
A *monomial* is a number, a variable, or a product of numbers and variables. [9.1A, p. 477]	5 is a number, y is a variable. $2a^3b^2$ is a product of numbers and variables. 5, y, and $2a^3b^2$ are monomials.
A *polynomial* is a variable expression in which the terms are monomials. [9.1A, p. 477]	$5x^2y - 3xy^2 + 2$ is a polynomial. Each term of this expression is a monomial.
A polynomial of two terms is a *binomial*. [9.1A, p. 477]	$x + 2$, $y^2 - 3$, and $6a + 5b$ are binomials.
A polynomial of three terms is a *trinomial*. [9.1A, p. 477]	$x^2 - 6x + 7$ is a trinomial.
The *degree of a polynomial in one variable* is the greatest exponent on a variable. [9.1A, p. 477]	The degree of $3x - 4x^3 + 17x^2 + 25$ is 3.
A polynomial in one variable is usually written in *descending order*, where the exponents of the variable terms decrease from left to right. [9.1A, p. 477]	The polynomial $2x^4 + 3x^2 - 4x - 7$ is written in descending order.
The *opposite of a polynomial* is the polynomial with the sign of every term changed to its opposite. [9.1B, p. 478]	The opposite of the polynomial $x^2 - 3x + 4$ is $-x^2 + 3x - 4$.

Essential Rules and Procedures	Examples
Addition of Polynomials [9.1A, p. 477] To add polynomials, add the coefficients of the like terms.	$(2x^2 + 3x - 4) + (3x^3 - 4x^2 + 2x - 5)$ $= 3x^3 + (2x^2 - 4x^2) + (3x + 2x)$ $\quad + (-4 - 5)$ $= 3x^3 - 2x^2 + 5x - 9$
Subtraction of Polynomials [9.1B, p. 478] To subtract polynomials, add the opposite of the second polynomial to the first.	$(3y^2 - 8y - 9) - (5y^2 - 10y + 3)$ $= (3y^2 - 8y - 9) + (-5y^2 + 10y - 3)$ $= (3y^2 - 5y^2) + (-8y + 10y)$ $\quad + (-9 - 3)$ $= -2y^2 + 2y - 12$
Rule for Multiplying Exponential Expressions [9.2A, p. 481] If m and n are integers, then $x^m \cdot x^n = x^{m+n}$.	$a^3 \cdot a^6 = a^{3+6} = a^9$

**Rule for Simplifying the Power of an
Exponential Expression** [9.2B, p. 482]
If m and n are integers, then $(x^m)^n = x^{mn}$.

$(c^3)^4 = c^{3 \cdot 4} = c^{12}$

Rule for Simplifying the Power of a Product [9.2B, p. 482]
If m, n, and p are integers, then $(x^m y^n)^p = x^{mp} y^{np}$.

$(a^3 b^2)^4 = a^{3 \cdot 4} b^{2 \cdot 4} = a^{12} b^8$

To multiply a polynomial by a monomial, use the
Distributive Property and the Rule for Multiplying Exponential
Expressions. [9.3A, p. 485]

$(-4y)(5y^2 + 3y - 8)$
$= (-4y)(5y^2) + (-4y)(3y) - (-4y)(8)$
$= -20y^3 - 12y^2 + 32y$

To multiply two polynomials, multiply each term of one
polynomial by each term of the other polynomial. [9.3B, p. 485]

$$\begin{array}{r} x^2 - 5x + 6 \\ x + 4 \\ \hline 4x^2 - 20x + 24 \\ x^3 - 5x^2 + 6x \\ \hline x^3 - x^2 - 14x + 24 \end{array}$$

FOIL Method [9.3C, p. 486]
To find the product of two binomials, add the products of
the **F**irst terms, the **O**uter terms, the **I**nner terms, and the
Last terms.

$(2x - 5)(3x + 4)$
$= (2x)(3x) + (2x)(4) + (-5)(3x)$
$\quad + (-5)(4)$
$= 6x^2 + 8x - 15x - 20$
$= 6x^2 - 7x - 20$

Product of the Sum and Difference of the Same Terms
[9.3D, p. 487]
$(a + b)(a - b) = a^2 - b^2$

$(3x + 4)(3x - 4) = (3x)^2 - 4^2$
$= 9x^2 - 16$

Square of a Binomial [9.3D, p. 487]
$(a + b)^2 = a^2 + 2ab + b^2$
$(a - b)^2 = a^2 - 2ab + b^2$

$(2x + 5)^2 = (2x)^2 + 2(2x)(5) + 5^2$
$= 4x^2 + 20x + 25$
$(3x - 4)^2 = (3x)^2 + 2(3x)(-4) + (-4)^2$
$= 9x^2 - 24x + 16$

Definition of Zero as an Exponent [9.4A, p. 493]
If $x \neq 0$, then $x^0 = 1$.

$17^0 = 1; (-6c)^0 = 1, c \neq 0$

Definition of a Negative Exponent [9.4A, p. 494]
If $x \neq 0$ and n is a positive integer, then $x^{-n} = \frac{1}{x^n}$ and $\frac{1}{x^{-n}} = x^n$.

$x^{-6} = \frac{1}{x^6}$ and $\frac{1}{x^{-6}} = x^6$

Rule for Simplifying the Power of a Quotient [9.4A, p. 495]

If m, n, and p are integers and $y \neq 0$, then $\left(\dfrac{x^m}{y^n}\right)^p = \dfrac{x^{mp}}{y^{np}}$.

$$\left(\dfrac{c^3}{a^5}\right)^2 = \dfrac{c^{3 \cdot 2}}{a^{5 \cdot 2}} = \dfrac{c^6}{a^{10}}$$

Rule for Negative Exponents on Fractional Expressions
[9.4A, p. 495]

If $a \neq 0$, $b \neq 0$, and n is a positive integer, then $\left(\dfrac{a}{b}\right)^{-n} = \left(\dfrac{b}{a}\right)^n$.

$$\left(\dfrac{x}{y}\right)^{-3} = \left(\dfrac{y}{x}\right)^3$$

Rule for Dividing Exponential Expressions [9.4A, p. 495]

If m and n are integers and $x \neq 0$, then $\dfrac{x^m}{x^n} = x^{m-n}$.

$$\dfrac{a^7}{a^2} = a^{7-2} = a^5$$

To Express a Number in Scientific Notation [9.4B, p. 498]
To express a number in scientific notation, write it in the form
$a \times 10^n$, where $1 \leq a < 10$ and n is an integer. If the number is
greater than 10, then n is a positive integer. If the number is
between 0 and 1, then n is a negative integer.

$367,000,000 = 3.67 \times 10^8$
$0.0000078 = 7.8 \times 10^{-6}$

**To Change a Number in Scientific Notation
to Decimal Notation** [9.4B, p. 498]
To change a number in scientific notation to decimal notation,
move the decimal point to the right if n is positive and to the
left if n is negative. Move the decimal point the same number
of places as the absolute value of the exponent on 10.

$2.418 \times 10^7 = 24,180,000$
$9.06 \times 10^{-5} = 0.0000906$

To divide a polynomial by a monomial, divide each term in
the numerator by the denominator and write the sum of the
quotients. [9.5A, p. 503]

$$\dfrac{8xy^3 - 4y^2 + 12y}{4y}$$
$$= \dfrac{8xy^3}{4y} - \dfrac{4y^2}{4y} + \dfrac{12y}{4y}$$
$$= 2xy^2 - y + 3$$

To check polynomial division, use the same equation used to
check division of whole numbers:

$$(\text{Quotient} \times \text{divisor}) + \text{remainder} = \text{dividend}$$

[9.5B, p. 503]

$$\begin{array}{r} x - 4 \\ x + 3 \overline{) x^2 - x - 10} \\ \underline{x^2 + 3x} \\ -4x - 10 \\ \underline{-4x - 12} \\ 2 \end{array}$$

Check:

$(x - 4)(x + 3) + 2 = x^2 - x - 12 + 2$
$ = x^2 - x - 10$

$(x^2 - x - 10) \div (x + 3) = x - 4 + \dfrac{2}{x + 3}$

Chapter 9 Review Exercises

1. Multiply: $(2b - 3)(4b + 5)$
$8b^2 - 2b - 15$ [9.3C]

2. Add: $(12y^2 + 17y - 4) + (9y^2 - 13y + 3)$
$21y^2 + 4y - 1$ [9.1A]

3. Simplify: $(xy^5z^3)(x^3y^3z)$
$x^4y^8z^4$ [9.2A]

4. Simplify: $\dfrac{8x^{12}}{12x^9}$
$\dfrac{2x^3}{3}$ [9.4A]

5. Multiply: $-2x(4x^2 + 7x - 9)$
$-8x^3 - 14x^2 + 18x$ [9.3A]

6. Simplify: $\dfrac{3ab^4}{-6a^2b^4}$
$-\dfrac{1}{2a}$ [9.4A]

7. Simplify: $(-2u^3v^4)^4$
$16u^{12}v^{16}$ [9.2B]

8. Evaluate: $(2^3)^2$
64 [9.2B]

9. Subtract: $(5x^2 - 2x - 1) - (3x^2 - 5x + 7)$
$2x^2 + 3x - 8$ [9.1B]

10. Simplify: $\dfrac{a^{-1}b^3}{a^3b^{-3}}$
$\dfrac{b^6}{a^4}$ [9.4A]

11. Simplify: $(-2x^3)^2(-3x^4)^3$
$-108x^{18}$ [9.2B]

12. Expand: $(5y - 7)^2$
$25y^2 - 70y + 49$ [9.3D]

13. Simplify: $(5a^7b^6)^2(4ab)$
$100a^{15}b^{13}$ [9.2B]

14. Divide: $\dfrac{12b^7 + 36b^5 - 3b^3}{3b^3}$
$4b^4 + 12b^2 - 1$ [9.5A]

15. Evaluate: -4^{-2}
$-\dfrac{1}{16}$ [9.4A]

16. Subtract: $(13y^3 - 7y - 2) - (12y^2 - 2y - 1)$
$13y^3 - 12y^2 - 5y - 1$ [9.1B]

17. Divide: $\dfrac{7 - x - x^2}{x + 3}$
$-x + 2 + \dfrac{1}{x + 3}$ [9.5B]

18. Multiply: $(2a - b)(x - 2y)$
$2ax - 4ay - bx + 2by$ [9.3C]

19. Multiply: $(3y^2 + 4y - 7)(2y + 3)$
$6y^3 + 17y^2 - 2y - 21$ [9.3B]

20. Divide: $(b^3 - 2b^2 - 33b - 7) \div (b - 7)$
$b^2 + 5b + 2 + \dfrac{7}{b - 7}$ [9.5B]

21. Multiply: $2ab^3(4a^2 - 2ab + 3b^2)$
$8a^3b^3 - 4a^2b^4 + 6ab^5$ [9.3A]

22. Multiply: $(2a - 5b)(2a + 5b)$
$4a^2 - 25b^2$ [9.3D]

23. Multiply: $(6b^3 - 2b^2 - 5)(2b^2 - 1)$
$12b^5 - 4b^4 - 6b^3 - 8b^2 + 5$ [9.3B]

24. Add: $(2x^3 + 7x^2 + x) + (2x^2 - 4x - 12)$
$2x^3 + 9x^2 - 3x - 12$ [9.1A]

25. Divide: $\dfrac{16y^2 - 32y}{-4y}$
$-4y + 8$ [9.5A]

26. Multiply: $(a + 7)(a - 7)$
$a^2 - 49$ [9.3D]

27. Write 37,560,000,000 in scientific notation.
3.756×10^{10} [9.4B]

28. Write 1.46×10^7 in decimal notation.
14,600,000   [9.4B]

29. Simplify: $(2a^{12}b^3)(-9b^2c^6)(3ac)$
$-54a^{13}b^5c^7$ [9.2A]

30. Divide: $(6y^2 - 35y + 36) \div (3y - 4)$
$2y - 9$ [9.5B]

31. Simplify: $(-3x^{-2}y^{-3})^{-2}$
$\dfrac{x^4y^6}{9}$ [9.4A]

32. Multiply: $(5a - 7)(2a + 9)$
$10a^2 + 31a - 63$ [9.3C]

33. Write 0.000000127 in scientific notation.
1.27×10^{-7} [9.4B]

34. Write 3.2×10^{-12} in decimal notation.
0.0000000000032   [9.4B]

35. **Geometry**   The length of a table-tennis table is 1 ft less than twice the width of the table. Let w represent the width of the table-tennis table. Express the area of the table in terms of the variable w.
$(2w^2 - w)$ ft^2 [9.3E]

36. **Geometry**   The side of a checkerboard is $(3x - 2)$ in. Express the area of the checkerboard in terms of the variable x.
$(9x^2 - 12x + 4)$ in^2 [9.3E]

Chapter 9 Test

1. Multiply: $2x(2x^2 - 3x)$
$4x^3 - 6x^2$ [9.3A]

2. Divide: $\dfrac{12x^3 - 3x^2 + 9}{3x^2}$
$4x - 1 + \dfrac{3}{x^2}$ [9.5A]

3. Simplify: $\dfrac{12x^2}{-3x^8}$
$-\dfrac{4}{x^6}$ [9.4A]

4. Simplify: $(-2xy^2)(3x^2y^4)$
$-6x^3y^6$ [9.2A]

5. Divide: $(x^2 + 1) \div (x + 1)$
$x - 1 + \dfrac{2}{x + 1}$ [9.5B]

6. Multiply: $(x - 3)(x^2 - 4x + 5)$
$x^3 - 7x^2 + 17x - 15$ [9.3B]

7. Simplify: $(-2a^2b)^3$
$-8a^6b^3$ [9.2B]

8. Simplify: $\dfrac{(3x^{-2}y^3)^3}{3x^4y^{-1}}$
$\dfrac{9y^{10}}{x^{10}}$ [9.4A]

9. Multiply: $(a - 2b)(a + 5b)$
$a^2 + 3ab - 10b^2$ [9.3C]

10. Divide: $\dfrac{16x^5 - 8x^3 + 20x}{4x}$
$4x^4 - 2x^2 + 5$ [9.5A]

11. Divide: $(x^2 + 6x - 7) \div (x - 1)$
$x + 7$ [9.5B]

12. Multiply: $-3y^2(-2y^2 + 3y - 6)$
$6y^4 - 9y^3 + 18y^2$ [9.3A]

13. Multiply: $(-2x^3 + x^2 - 7)(2x - 3)$
$-4x^4 + 8x^3 - 3x^2 - 14x + 21$ [9.3B]

14. Multiply: $(4y - 3)(4y + 3)$
$16y^2 - 9$ [9.3D]

15. Simplify: $(ab^2)(a^3b^5)$

a^4b^7 [9.2A]

16. Simplify: $\dfrac{2a^{-1}b}{2^{-2}a^{-2}b^{-3}}$

$8ab^4$ [9.4A]

17. Divide: $\dfrac{20a - 35}{5}$

$4a - 7$ [9.5A]

18. Subtract: $(3a^2 - 2a - 7) - (5a^3 + 2a - 10)$

$-5a^3 + 3a^2 - 4a + 3$ [9.1B]

19. Expand: $(2x - 5)^2$

$4x^2 - 20x + 25$ [9.3D]

20. Divide: $(4x^2 - 7) \div (2x - 3)$

$2x + 3 + \dfrac{2}{2x - 3}$ [9.5B]

21. Simplify: $\dfrac{-(2x^2y)^3}{4x^3y^3}$

$-2x^3$ [9.4A]

22. Multiply: $(2x - 7y)(5x - 4y)$

$10x^2 - 43xy + 28y^2$ [9.3C]

23. Add: $(3x^3 - 2x^2 - 4) + (8x^2 - 8x + 7)$

$3x^3 + 6x^2 - 8x + 3$ [9.1A]

24. Write 0.00000000302 in scientific notation.

3.02×10^{-9} [9.4B]

25. **Geometry**   The radius of a circle is $(x - 5)$ m. Use the equation $A = \pi r^2$, where r is the radius, to find the area of the circle in terms of the variable x. Leave the answer in terms of π.

$(\pi x^2 - 10\pi x + 25\pi)$ m^2 [9.3E]

$x - 5$

Cumulative Review Exercises

1. Simplify: $\dfrac{3}{16} - \left(-\dfrac{5}{8}\right) - \dfrac{7}{9}$

   $\dfrac{5}{144}$ [3.4A]

2. Evaluate $-3^2 \cdot \left(\dfrac{2}{3}\right)^3 \cdot \left(-\dfrac{5}{8}\right)$.

   $\dfrac{5}{3}$ [3.5A]

3. Simplify: $\left(-\dfrac{1}{2}\right)^3 \div \left(\dfrac{3}{8} - \dfrac{5}{6}\right) + 2$

   $\dfrac{25}{11}$ [3.5A]

4. Evaluate $\dfrac{b - (a - b)^2}{b^2}$ when $a = -2$ and $b = 3$.

   $-\dfrac{22}{9}$ [4.1A]

5. Simplify: $-2x - (-xy) + 7x - 4xy$

   $5x - 3xy$ [4.2A]

6. Simplify: $(12x)\left(-\dfrac{3}{4}\right)$

   $-9x$ [4.2B]

7. Simplify: $-2[3x - 2(4 - 3x) + 2]$

   $-18x + 12$ [4.2D]

8. Solve: $12 = -\dfrac{3}{4}x$

   -16 [5.1C]

9. Solve: $2x - 9 = 3x + 7$

   -16 [5.3A]

10. Solve: $2 - 3(4 - x) = 2x + 5$

    15 [5.3B]

11. 35.2 is what percent of 160?

    22% [6.4A/6.4B]

12. Add: $(4b^3 - 7b^2 - 7) + (3b^2 - 8b + 3)$

    $4b^3 - 4b^2 - 8b - 4$ [9.1A]

13. Subtract: $(3y^3 - 5y + 8) - (-2y^2 + 5y + 8)$

    $3y^3 + 2y^2 - 10y$ [9.1B]

14. Simplify: $(a^3b^5)^3$

    a^9b^{15} [9.2B]

15. Simplify: $(4xy^3)(-2x^2y^3)$

    $-8x^3y^6$ [9.2A]

16. Multiply: $-2y^2(-3y^2 - 4y + 8)$

    $6y^4 + 8y^3 - 16y^2$ [9.3A]

17. Multiply: $(2a - 7)(5a^2 - 2a + 3)$
$10a^3 - 39a^2 + 20a - 21$ [9.3B]

18. Multiply: $(3b - 2)(5b - 7)$
$15b^2 - 31b + 14$ [9.3C]

19. Simplify: $\dfrac{(-2a^2b^3)^2}{8a^4b^8}$

$\dfrac{1}{2b^2}$ [9.4A]

20. Divide: $(a^2 - 4a - 21) \div (a + 3)$
$a - 7$ [9.5B]

21. Write 6.09×10^{-5} in decimal notation.
0.0000609   [9.4B]

22. Translate "the difference between eight times a number and twice the number is eighteen" into an equation and solve.
$8x - 2x = 18; 3$ [5.4B]

23. **Juice Mixtures**   Find the cost per ounce of a fruit drink made from 200 oz of fruit juice that costs \$.25 per ounce and 300 oz of soda that costs \$.05 per ounce.
\$.13 per ounce   [5.5A]

24. **Transportation**   A car traveling at 50 mph overtakes a cyclist who, riding at 10 mph, has had a 2-hour head start. How far from the starting point does the car overtake the cyclist?
25 mi   [5.5B]

25. **Geometry**   The width of a rectangle is 40% of the length. The perimeter of the rectangle is 42 m. Find the length and width of the rectangle.
Length: 15 m; width: 6 m   [7.2A]

chapter 10

Factoring

A University of Florida receiver is being tackled during the play in the photo above. The University of Florida is part of the Southeastern conference of the NCAA. Imagine you have just gotten a job as the manager of this conference. One of your first tasks would be to organize the game schedule for the upcoming season. There are 12 teams in the Southeastern conference. How would you go about creating this schedule? **Exercises 71 and 72 on page 557** show you how to use a formula to determine the number of league games that must be scheduled if each team is to play every other team once.

OBJECTIVES

Section 10.1

A To factor a monomial from a polynomial
B To factor by grouping

Section 10.2

A To factor a trinomial of the form $x^2 + bx + c$
B To factor completely

Section 10.3

A To factor a trinomial of the form $ax^2 + bx + c$ by using trial factors
B To factor a trinomial of the form $ax^2 + bx + c$ by grouping

Section 10.4

A To factor the difference of two squares and perfect-square trinomials
B To factor completely

Section 10.5

A To solve equations by factoring
B To solve application problems

Need help? For online student resources, such as section quizzes, visit this textbook's website at **math.college.hmco.com/students**.

Do these exercises to prepare for Chapter 10.

1. Write 30 as a product of prime numbers.
$2 \cdot 3 \cdot 5$ [1.3D]

2. Simplify: $-3(4y - 5)$
$-12y + 15$ [4.2C]

3. Simplify: $-(a - b)$
$-a + b$ [4.2C]

4. Simplify: $2(a - b) - 5(a - b)$
$-3a + 3b$ [4.2D]

5. Solve: $4x = 0$
0 [5.1C]

6. Solve: $2x + 1 = 0$
$-\dfrac{1}{2}$ [5.2A]

7. Multiply: $(x + 4)(x - 6)$
$x^2 - 2x - 24$ [9.3C]

8. Multiply: $(2x - 5)(3x + 2)$
$6x^2 - 11x - 10$ [9.3C]

9. Simplify: $\dfrac{x^5}{x^2}$
x^3 [9.4A]

10. Simplify: $\dfrac{6x^4y^3}{2xy^2}$
$3x^3y$ [9.4A]

Without using a calculator (it probably wouldn't work anyway), how many digits are in the product $4^{54} \cdot 5^{100}$? **103**

10.1 Common Factors

Objective A   **To factor a monomial from a polynomial**

In Section 2.1B we discussed how to find the greatest common factor (GCF) of two or more integers. The **greatest common factor (GCF) of two or more monomials** is the product of the GCF of the coefficients and the common variable factors.

$$6x^3y = 2 \cdot 3 \cdot x \cdot x \cdot x \cdot y$$
$$8x^2y^2 = 2 \cdot 2 \cdot 2 \cdot x \cdot x \cdot y \cdot y$$
$$\text{GCF} = 2 \cdot x \cdot x \cdot y = 2x^2y$$

Note that the exponent of each variable in the GCF is the same as the *smallest* exponent of that variable in either of the monomials.

The GCF of $6x^3y$ and $8x^2y^2$ is $2x^2y$.

HOW TO   Find the GCF of $12a^4b$ and $18a^2b^2c$.

The common variable factors are a^2 and b; c is not a common variable factor.

$$12a^4b = 2 \cdot 2 \cdot 3 \cdot a^4 \cdot b$$
$$18a^2b^2c = 2 \cdot 3 \cdot 3 \cdot a^2 \cdot b^2 \cdot c$$
$$\text{GCF} = 2 \cdot 3 \cdot a^2 \cdot b = 6a^2b$$

To **factor a polynomial** means to write the polynomial as a product of other polynomials. In the example at the right, $2x$ is the GCF of the terms $2x^2$ and $10x$.

Multiply ←

Polynomial   =   **Factors**
$2x^2 + 10x$ $2x(x + 5)$

→ Factor

HOW TO   Factor: $5x^3 - 35x^2 + 10x$

Find the GCF of the terms of the polynomial.

$$5x^3 = 5 \cdot x^3$$
$$35x^2 = 5 \cdot 7 \cdot x^2$$
$$10x = 2 \cdot 5 \cdot x$$

The GCF is $5x$.

Rewrite the polynomial, expressing each term as a product with the GCF as one of the factors.

$$5x^3 - 35x^2 + 10x = 5x(x^2) + 5x(-7x) + 5x(2)$$
$$= 5x(x^2 - 7x + 2)$$

• Use the Distributive Property to write the polynomial as a product of factors.

TAKE NOTE

At the right, the factors in parentheses are determined by dividing each term of the trinomial by the GCF, $5x$.

$$\frac{5x^3}{5x} = x^2,$$

$$\frac{-35x^2}{5x} = -7x, \text{ and}$$

$$\frac{10x}{5x} = 2$$

See Objective 9.5A.

Objective 10.1A

Vocabulary to Review
greatest common factor (GCF)
  [2.1B]

New Vocabulary
greatest common factor (GCF) of two or more monomials
factor a polynomial

Concept Check
Identify the factors in the following products.
1. $3(x + 2)$ $3, (x + 2)$
2. $x(2x + 3)(x^2 + 4)$
  $x, 2x + 3, x^2 + 4$

Concept Check
Determine the GCF of the terms of the following polynomials.
1. $4x^2 + 6x - 8$ 2
2. $18x^3y - 24x^2y + 12xy$ $6xy$

Optional Student Activity
1. Determine the number of distinct factors of each monomial.
  a. x^4 5
  b. x^7 8
2. Determine the number of distinct factors of each monomial.
  a. x^2y^3 12
  b. x^3y^5 24
  c. x^4y^2 15
3. Find a formula for the number of distinct factors of x^ny^m, where n and m are positive integers. $mn + m + n + 1$

In-Class Examples (Objective 10.1A)

Factor.
1. $10y^2 - 15y^3z$ $5y^2(2 - 3yz)$
2. $12m^2 + 6m - 18$ $6(2m^2 + m - 3)$
3. $20x^4y^3 - 30x^3y^4 + 40x^2y^5$
  $10x^2y^3(2x^2 - 3xy + 4y^2)$

Discuss the Concepts

Is the factored form of $4x^2 - 36x$ equal to $4x(x - 9)$ or to $(x - 9)4x$? Explain.

Because multiplication is commutative, both answers are correct.

Optional Student Activity

Factor:
$12mn^2 - 18m^2n^3 + 36m^3n^4$

$6mn^2(2 - 3mn + 6m^2n^2)$

HOW TO   Factor: $21x^2y^3 - 6xy^5 + 15x^4y^2$

Find the GCF of the terms of the polynomial.

$$21x^2y^3 = 3 \cdot 7 \cdot x^2 \cdot y^3$$
$$6xy^5 = 2 \cdot 3 \cdot x \cdot y^5$$
$$15x^4y^2 = 3 \cdot 5 \cdot x^4 \cdot y^2$$

The GCF is $3xy^2$.

Rewrite the polynomial, expressing each term as a product with the GCF as one of the factors.

$$21x^2y^3 - 6xy^5 + 15x^4y^2 = 3xy^2(7xy) + 3xy^2(-2y^3) + 3xy^2(5x^3)$$
$$= 3xy^2(7xy - 2y^3 + 5x^3)$$

• Use the Distributive Property to write the polynomial as a product of factors.

Example 1

Factor: $8x^2 + 2xy$

Solution
The GCF is $2x$.

$$8x^2 + 2xy = 2x(4x) + 2x(y)$$
$$= 2x(4x + y)$$

You Try It 1

Factor: $14a^2 - 21a^4b$

Your solution
$7a^2(2 - 3a^2b)$

Example 2

Factor: $n^3 - 5n^2 + 2n$

Solution
The GCF is n.

$$n^3 - 5n^2 + 2n = n(n^2) + n(-5n) + n(2)$$
$$= n(n^2 - 5n + 2)$$

You Try It 2

Factor: $27b^2 + 18b + 9$

Your solution
$9(3b^2 + 2b + 1)$

Example 3

Factor: $16x^2y + 8x^4y^2 - 12x^4y^5$

Solution
The GCF is $4x^2y$.

$$16x^2y + 8x^4y^2 - 12x^4y^5$$
$$= 4x^2y(4) + 4x^2y(2x^2y) + 4x^2y(-3x^2y^4)$$
$$= 4x^2y(4 + 2x^2y - 3x^2y^4)$$

You Try It 3

Factor: $6x^4y^2 - 9x^3y^2 + 12x^2y^4$

Your solution
$3x^2y^2(2x^2 - 3x + 4y^2)$

Solutions on p. S24

Objective B   **To factor by grouping**

A factor that has two terms is called a **binomial factor.** In the examples at the right, the binomials in parentheses are binomial factors.

$$2a(a + b)^2$$
$$3xy(x - y)$$

The Distributive Property is used to factor a common binomial factor from an expression.

The common binomial factor of the expression $6x(x - 3) + y(x - 3)$ is $(x - 3)$. To factor that expression, use the Distributive Property to write the expression as a product of factors.

$$6x(x - 3) + y(x - 3) = (x - 3)(6x + y)$$

Consider the following simplification of $-(a - b)$.

$$-(a - b) = -1(a - b) = -a + b = b - a$$

Thus

$$b - a = -(a - b)$$

This equation is sometimes used to factor a common binomial from an expression.

> **HOW TO**   Factor: $2x(x - y) + 5(y - x)$
>
> $2x(x - y) + 5(y - x) = 2x(x - y) - 5(x - y)$ • $5(y - x) = 5[(-1)(x - y)]$
> $\qquad\qquad\qquad\qquad\quad = (x - y)(2x - 5)$ $\qquad\qquad = -5(x - y)$

A polynomial can be **factored by grouping** if its terms can be grouped and factored in such a way that a common binomial factor is found.

> **HOW TO**   Factor: $ax + bx - ay - by$
>
> $ax + bx - ay - by = (ax + bx) - (ay + by)$ • Group the first two terms and
> $\qquad\qquad\qquad\qquad\qquad\qquad\qquad\qquad$ the last two terms. Note that
> $\qquad\qquad\qquad\qquad\qquad\qquad\qquad\qquad$ $-ay - by = -(ay + by)$.
> $\qquad\qquad\qquad\qquad = x(a + b) - y(a + b)$ • Factor each group.
> $\qquad\qquad\qquad\qquad = (a + b)(x - y)$ • Factor the GCF, $(a + b)$,
> $\qquad\qquad\qquad\qquad\qquad\qquad\qquad\qquad$ from each group.

> **HOW TO**   Factor: $6x^2 - 9x - 4xy + 6y$
>
> $6x^2 - 9x - 4xy + 6y = (6x^2 - 9x) - (4xy - 6y)$ • Group the first two terms and
> $\qquad\qquad\qquad\qquad\qquad\qquad\qquad\qquad$ the last two terms. Note that
> $\qquad\qquad\qquad\qquad\qquad\qquad\qquad\qquad$ $-4xy + 6y = -(4xy - 6y)$.
> $\qquad\qquad\qquad\qquad\quad = 3x(2x - 3) - 2y(2x - 3)$ • Factor each group.
> $\qquad\qquad\qquad\qquad\quad = (2x - 3)(3x - 2y)$ • Factor the GCF, $(2x - 3)$,
> $\qquad\qquad\qquad\qquad\qquad\qquad\qquad\qquad$ from each group.

Objective 10.1B

New Vocabulary
binomial factor
factor by grouping

Instructor Note
It will help students if you give them some examples of inserting parentheses into expressions before they attempt the In-Class Examples below. Here are some suggestions.
$-a + 2b = -(a - 2b)$
$3x - 2y = -(-3x + 2y)$
$-4a - 3b = -(4a + 3b)$

Concept Check
Replace the ? to make a true statement.
1. $a - 3 = ?(3 - a)$ -1
2. $2 - (x - y) = 2 + (?)$
   $y - x$
3. $4x + (3a - b) = 4x - (?)$
   $b - 3a$

Optional Student Activity
Factor:
$4x(a - b) - 5(b - a) + 4(a - b)$
$(a - b)(4x + 9)$

In-Class Examples (Objective 10.1B)

Factor.
1. $6x(4x + 3) - 5(4x + 3)$ $(4x + 3)(6x - 5)$
2. $8x^2 - 12x - 6xy + 9y$ $(2x - 3)(4x - 3y)$
3. $7xy^2 - 3y + 14xy - 6$ $(7xy - 3)(y + 2)$
4. $5xy - 9y - 18 + 10x$ $(5x - 9)(y + 2)$

Concept Check

Factor.

1. $5x(a + 7b) + 3(7b + a)$
  $(a + 7b)(5x + 3)$

2. $x^2 - 4x + 2xy - 8y$
  $(x - 4)(x + 2y)$

3. $3x^2y - 5x - 9xy + 15$
  $(3xy - 5)(x - 3)$

4. $10mn - 4 + 2n - 5mn^2$
  $(5mn - 2)(2 - n)$

Optional Student Activity

Factor:
$x^2 - 2y - x + 2xy$
$(x + 2y)(x - 1)$

Example 4

Factor: $4x(3x - 2) - 7(3x - 2)$

Solution
$4x(3x - 2) - 7(3x - 2)$ • **$3x - 2$ is the common binomial factor.**

$= (3x - 2)(4x - 7)$

You Try It 4

Factor: $2y(5x - 2) - 3(2 - 5x)$

Your solution
$(5x - 2)(2y + 3)$

Example 5

Factor: $9x^2 - 15x - 6xy + 10y$

Solution
$9x^2 - 15x - 6xy + 10y$

$= (9x^2 - 15x) - (6xy - 10y)$ • $-6xy + 10y = -(6xy - 10y)$

$= 3x(3x - 5) - 2y(3x - 5)$ • **$3x - 5$ is the common factor.**

$= (3x - 5)(3x - 2y)$

You Try It 5

Factor: $a^2 - 3a + 2ab - 6b$

Your solution
$(a - 3)(a + 2b)$

Example 6

Factor: $3x^2y - 4x - 15xy + 20$

Solution
$3x^2y - 4x - 15xy + 20$

$= (3x^2y - 4x) - (15xy - 20)$ • $-15xy + 20 = -(15xy - 20)$

$= x(3xy - 4) - 5(3xy - 4)$ • **$3xy - 4$ is the common factor.**

$= (3xy - 4)(x - 5)$

You Try It 6

Factor: $2mn^2 - n + 8mn - 4$

Your solution
$(2mn - 1)(n + 4)$

Example 7

Factor: $4ab - 6 + 3b - 2ab^2$

Solution
$4ab - 6 + 3b - 2ab^2$

$= (4ab - 6) + (3b - 2ab^2)$

$= 2(2ab - 3) + b(3 - 2ab)$

$= 2(2ab - 3) - b(2ab - 3)$ • $3 - 2ab = -(2ab - 3)$

$= (2ab - 3)(2 - b)$ • **$2ab - 3$ is the common factor.**

You Try It 7

Factor: $3xy - 9y - 12 + 4x$

Your solution
$(x - 3)(3y + 4)$

Solutions on p. S24

10.1 Exercises

Objective A   To factor a monomial from a polynomial

1. Explain the meaning of "a common monomial factor of a polynomial."

2. Explain the meaning of "a factor" and the meaning of "to factor."

For Exercises 3 to 41, factor.

3. $5a + 5$
$5(a + 1)$

4. $7b - 7$
$7(b - 1)$

5. $16 - 8a^2$
$8(2 - a^2)$

6. $12 + 12y^2$
$12(1 + y^2)$

7. $8x + 12$
$4(2x + 3)$

8. $16a - 24$
$8(2a - 3)$

9. $30a - 6$
$6(5a - 1)$

10. $20b + 5$
$5(4b + 1)$

11. $7x^2 - 3x$
$x(7x - 3)$

12. $12y^2 - 5y$
$y(12y - 5)$

13. $3a^2 + 5a^5$
$a^2(3 + 5a^3)$

14. $9x - 5x^2$
$x(9 - 5x)$

15. $14y^2 + 11y$
$y(14y + 11)$

16. $6b^3 - 5b^2$
$b^2(6b - 5)$

17. $2x^4 - 4x$
$2x(x^3 - 2)$

18. $3y^4 - 9y$
$3y(y^3 - 3)$

19. $10x^4 - 12x^2$
$2x^2(5x^2 - 6)$

20. $12a^5 - 32a^2$
$4a^2(3a^3 - 8)$

21. $8a^8 - 4a^5$
$4a^5(2a^3 - 1)$

22. $16y^4 - 8y^7$
$8y^4(2 - y^3)$

23. $x^2y^2 - xy$
$xy(xy - 1)$

24. $a^2b^2 + ab$
$ab(ab + 1)$

25. $3x^2y^4 - 6xy$
$3xy(xy^3 - 2)$

26. $12a^2b^5 - 9ab$
$3ab(4ab^4 - 3)$

27. $x^2y - xy^3$
$xy(x - y^2)$

28. $3x^3 + 6x^2 + 9x$
$3x(x^2 + 2x + 3)$

29. $5y^3 - 20y^2 + 5y$
$5y(y^2 - 4y + 1)$

30. $2x^4 - 4x^3 + 6x^2$
$2x^2(x^2 - 2x + 3)$

31. $3y^4 - 9y^3 - 6y^2$
$3y^2(y^2 - 3y - 2)$

32. $2x^3 + 6x^2 - 14x$
$2x(x^2 + 3x - 7)$

33. $3y^3 - 9y^2 + 24y$
$3y(y^2 - 3y + 8)$

34. $2y^5 - 3y^4 + 7y^3$
$y^3(2y^2 - 3y + 7)$

35. $6a^5 - 3a^3 - 2a^2$
$a^2(6a^3 - 3a - 2)$

36. $x^3y - 3x^2y^2 + 7xy^3$
$xy(x^2 - 3xy + 7y^2)$

37. $2a^2b - 5a^2b^2 + 7ab^2$
$ab(2a - 5ab + 7b)$

38. $5y^3 + 10y^2 - 25y$
$5y(y^2 + 2y - 5)$

39. $4b^5 + 6b^3 - 12b$
$2b(2b^4 + 3b^2 - 6)$

40. $3a^2b^2 - 9ab^2 + 15b^2$
$3b^2(a^2 - 3a + 5)$

41. $8x^2y^2 - 4x^2y + x^2$
$x^2(8y^2 - 4y + 1)$

Objective B   To factor by grouping

For Exercises 42 to 68, factor.

42. $x(b + 4) + 3(b + 4)$
$(b + 4)(x + 3)$

43. $y(a + z) + 7(a + z)$
$(a + z)(y + 7)$

44. $a(y - x) - b(y - x)$
$(y - x)(a - b)$

45. $3r(a - b) + s(a - b)$
$(a - b)(3r + s)$

46. $x(x - 2) + y(2 - x)$
$(x - 2)(x - y)$

47. $t(m - 7) + 7(7 - m)$
$(m - 7)(t - 7)$

Quick Quiz (Objective 10.1A)

Factor.

1. $6ab + 9a$ $3a(2b + 3)$

2. $14x^2y^2 - 7xy$ $7xy(2xy - 1)$

3. $15r^2s + 20rs - 10rs^2$ $5rs(3r + 4 - 2s)$

Section 10.1

Suggested Assignment
Exercises 3–67, every other odd
More challenging problems:
   Exercises 69–71

Answers to Writing Exercises

1. A common monomial factor is a monomial that is a factor of each term of a polynomial.

2. The phrase *a factor* means one of the numbers or variables in a product. The phrase *to factor* means to write an expression as a product of factors.

48. $2x(7 + b) - y(b + 7)$
$(b + 7)(2x - y)$

49. $2y(4a - b) - (b - 4a)$
$(4a - b)(2y + 1)$

50. $8c(2m - 3n) + (3n - 2m)$
$(2m - 3n)(8c - 1)$

51. $x^2 + 2x + 2xy + 4y$
$(x + 2)(x + 2y)$

52. $x^2 - 3x + 4ax - 12a$
$(x - 3)(x + 4a)$

53. $p^2 - 2p - 3rp + 6r$
$(p - 2)(p - 3r)$

54. $t^2 + 4t - st - 4s$
$(t + 4)(t - s)$

55. $ab + 6b - 4a - 24$
$(a + 6)(b - 4)$

56. $xy - 5y - 2x + 10$
$(x - 5)(y - 2)$

57. $2z^2 - z + 2yz - y$
$(2z - 1)(z + y)$

58. $2y^2 - 10y + 7xy - 35x$
$(y - 5)(2y + 7x)$

59. $8v^2 - 12vy + 14v - 21y$
$(2v - 3y)(4v + 7)$

60. $21x^2 + 6xy - 49x - 14y$
$(7x + 2y)(3x - 7)$

61. $2x^2 - 5x - 6xy + 15y$
$(2x - 5)(x - 3y)$

62. $4a^2 + 5ab - 10b - 8a$
$(4a + 5b)(a - 2)$

63. $3y^2 - 6y - ay + 2a$
$(y - 2)(3y - a)$

64. $2ra + a^2 - 2r - a$
$(2r + a)(a - 1)$

65. $3xy - y^2 - y + 3x$
$(3x - y)(y + 1)$

66. $2ab - 3b^2 - 3b + 2a$
$(2a - 3b)(b + 1)$

67. $3st + t^2 - 2t - 6s$
$(3s + t)(t - 2)$

68. $4x^2 + 3xy - 12y - 16x$
$(4x + 3y)(x - 4)$

APPLYING THE CONCEPTS

69. Number Sense A natural number is a *perfect number* if it is the sum of all its factors less than itself. For example, 6 is a perfect number because all the factors of 6 that are less than 6 are 1, 2, and 3, and $1 + 2 + 3 = 6$.
  a. Find the one perfect number between 20 and 30.   28
  b. Find the one perfect number between 490 and 500.   496

70. Geometry In the equation $P = 2L + 2W$, what is the effect on P when the quantity $L + W$ doubles?
  P doubles.

71. Geometry Write an expression in factored form for the shaded portion in each of the following diagrams. Use the equation for the area of a rectangle ($A = LW$) and the equation for the area of a circle ($A = \pi r^2$).

a.
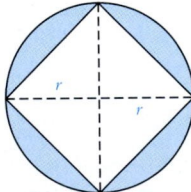

$r^2(\pi - 2)$

b.
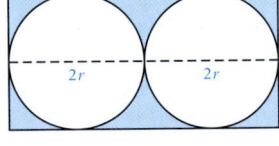

$2r^2(4 - \pi)$

c.
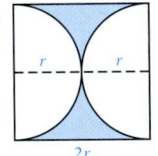

$r^2(4 - \pi)$

Quick Quiz (Objective 10.1B)
Factor.
1. $3xy - 9y + 2x - 6$ $(x - 3)(3y + 2)$
2. $2ay - a^2 - 3a + 6y$ $(2y - a)(a + 3)$

10.2 Factoring Polynomials of the Form $x^2 + bx + c$

Objective A   **To factor a trinomial of the form $x^2 + bx + c$**

Trinomials of the form $x^2 + bx + c$, where b and c are integers, are shown at the right.

$x^2 + 8x + 12;\ b = 8,\ c = 12$
$x^2 - 7x + 12;\ b = -7,\ c = 12$
$x^2 - 2x - 15;\ b = -2,\ c = -15$

To factor a trinomial of this form means to express the trinomial as the product of two binomials.

Trinomials expressed as the product of binomials are shown at the right.

$x^2 + 8x + 12 = (x + 6)(x + 2)$
$x^2 - 7x + 12 = (x - 3)(x - 4)$
$x^2 - 2x - 15 = (x + 3)(x - 5)$

The method by which factors of a trinomial are found is based on FOIL. Consider the following binomial products, noting the relationship between the constant terms of the binomials and the terms of the trinomials.

The signs in the binomials are the same.

$(x + 6)(x + 2) = x^2 + 2x + 6x + (6)(2) = x^2 + 8x + 12$
Sum of 6 and 2
Product of 6 and 2

$(x - 3)(x - 4) = x^2 - 4x - 3x + (-3)(-4) = x^2 - 7x + 12$
Sum of -3 and -4
Product of -3 and -4

The signs in the binomials are opposites.

$(x + 3)(x - 5) = x^2 - 5x + 3x + (3)(-5) = x^2 - 2x - 15$
Sum of 3 and -5
Product of 3 and -5

$(x - 4)(x + 6) = x^2 + 6x - 4x + (-4)(6) = x^2 + 2x - 24$
Sum of -4 and 6
Product of -4 and 6

Factoring $x^2 + bx + c$: IMPORTANT RELATIONSHIPS

1. When the constant term of the trinomial is positive, the constant terms of the binomials have the same sign. They are both positive when the coefficient of the x term in the trinomial is positive. They are both negative when the coefficient of the x term in the trinomial is negative.

2. When the constant term of the trinomial is negative, the constant terms of the binomials have opposite signs.

3. In the trinomial, the coefficient of x is the sum of the constant terms of the binomials.

4. In the trinomial, the constant term is the product of the constant terms of the binomials.

Ⓟ

Objective 10.2A

New Vocabulary
nonfactorable over the integers
prime polynomial

Instructor Note
Students sometimes see factoring as unrelated to multiplication. Remind students that the relationship between the binomial factors and the terms of the polynomial is based on multiplying binomials.

Concept Check
Fill in the blanks.
1. To factor $x^2 - 3x - 18$, find two numbers whose product is ___ and whose sum is ___.
   $-18, -3$
2. To factor $x^2 - 9x + 18$, find two numbers whose product is ___ and whose sum is ___.
   $18, -9$

Discuss the Concepts
How can you check the answer to a factoring problem?
Answers will vary. Students should include the concept that the product of the factors should be equivalent to the original expression.

Discuss the Concepts
What clues in $x^2 - 4x + 5$ tell which signs should be used in the factored form?
5 is positive, so both binomial constants have the same sign. -4 is negative, so the signs are both negative.

In-Class Examples (Objective 10.2A)
Factor.
1. $x^2 - 8x + 12$ $(x - 6)(x - 2)$
2. $x^2 + 14x + 33$ $(x + 11)(x + 3)$
3. $x^2 + 7x + 12$ $(x + 4)(x + 3)$
4. $x^2 - 4x - 12$ $(x - 6)(x + 2)$
5. $x^2 - 11x - 12$ $(x - 12)(x + 1)$

HOW TO   Factor: $x^2 - 7x + 10$

Because the constant term is positive and the coefficient of x is negative, the binomial constants will be negative. Find two negative factors of 10 whose sum is -7. The results can be recorded in a table.

Negative Factors of 10	Sum
$-1, -10$	-11
$-2, -5$	-7

• **These are the correct factors.**

$x^2 - 7x + 10 = (x - 2)(x - 5)$ • **Write the trinomial as a product of its factors.**

You can check the proposed factorization by multiplying the two binomials.

$$\text{Check: } (x - 2)(x - 5) = x^2 - 5x - 2x + 10$$
$$= x^2 - 7x + 10$$

HOW TO   Factor: $x^2 - 9x - 36$

The constant term is negative. The binomial constants will have opposite signs. Find two factors of -36 whose sum is -9.

Factors of -36	Sum
$+1, -36$	-35
$-1, +36$	35
$+2, -18$	-16
$-2, +18$	16
$+3, -12$	-9

• **Once the correct factors are found, it is not necessary to try the remaining factors.**

$x^2 - 9x - 36 = (x + 3)(x - 12)$ • **Write the trinomial as a product of its factors.**

For some trinomials, it is not possible to find integer factors of the constant term whose sum is the coefficient of the middle term. A polynomial that does not factor using only integers is **nonfactorable over the integers.**

HOW TO   Factor: $x^2 + 7x + 8$

The constant term is positive and the coefficient of x is positive. The binomial constants will be positive. Find two positive factors of 8 whose sum is 7.

Positive Factors of 8	Sum
$1, 8$	9
$2, 4$	6

• **There are no positive integer factors of 8 whose sum is 7.**

$x^2 + 7x + 8$ is nonfactorable over the integers.

Example 1   Factor: $x^2 - 8x + 15$

Solution
Find two negative factors of 15 whose sum is -8.

Factors	Sum
$-1, -15$	-16
$-3, -5$	-8

$x^2 - 8x + 15 = (x - 3)(x - 5)$

You Try It 1   Factor: $x^2 + 9x + 20$

Your solution
$(x + 4)(x + 5)$

Solution on p. S24

Example 2   Factor: $x^2 + 6x - 27$

You Try It 2   Factor: $x^2 + 7x - 18$

Solution
Find two factors of
-27 whose sum is 6.

Factors	Sum
$+1, -27$	-26
$-1, +27$	26
$+3, -9$	-6
$-3, +9$	6

Your solution
$(x + 9)(x - 2)$

$x^2 + 6x - 27 = (x - 3)(x + 9)$

Solution on p. S24

Objective B   **To factor completely**

A polynomial is **factored completely** when it is written as a product of factors that are nonfactorable over the integers.

TAKE NOTE
The first step in *any* factoring problem is to determine whether the terms of the polynomial have a *common factor*. If they do, factor it out first.

HOW TO   Factor: $4y^3 - 4y^2 - 24y$

$$4y^3 - 4y^2 - 24y = 4y(y^2) - 4y(y) - 4y(6)$$

   • The GCF is **4y**.

$$= 4y(y^2 - y - 6)$$

   • Use the Distributive Property to factor out the GCF.

$$= 4y(y + 2)(y - 3)$$

   • Factor $y^2 - y - 6$. The two factors of -6 whose sum is -1 are 2 and -3.

It is always possible to check the proposed factorization by multiplying the polynomials. Here is the check for the last example.

$$\begin{aligned} \text{Check: } 4y(y + 2)(y - 3) &= 4y(y^2 - 3y + 2y - 6) \\ &= 4y(y^2 - y - 6) \\ &= 4y^3 - 4y^2 - 24y \end{aligned}$$

   • This is the original polynomial.

HOW TO   Factor: $5x^2 + 60xy + 100y^2$

$$5x^2 + 60xy + 100y^2 = 5(x^2) + 5(12xy) + 5(20y^2)$$

   • The GCF is **5**.

$$= 5(x^2 + 12xy + 20y^2)$$

   • Use the Distributive Property to factor out the GCF.

$$= 5(x + 2y)(x + 10y)$$

   • Factor $x^2 + 12xy + 20y^2$. The two factors of 20 whose sum is 12 are 2 and 10.

TAKE NOTE
$2y$ and $10y$ are placed in the binomials. This is necessary so that the middle term contains xy and the last term contains y^2.

Note that $2y$ and $10y$ were placed in the binomials. The following check shows that this was necessary.

$$\begin{aligned} \text{Check: } 5(x + 2y)(x + 10y) &= 5(x^2 + 10xy + 2xy + 20y^2) \\ &= 5(x^2 + 12xy + 20y^2) \\ &= 5x^2 + 60xy + 100y^2 \end{aligned}$$

   • This is the original polynomial.

Concept Check
Factor.
1. $x^2 - 5x - 6$ $(x - 6)(x + 1)$
2. $x^2 - 5x + 6$ $(x - 2)(x - 3)$

Objective 10.2B

New Vocabulary
factor completely

Discuss the Concepts
Explain why $(4x + 2)(x - 5)$ is not factored completely. What is the correct factorization?
The terms of $4x + 2$ have a common factor, namely, 2. The complete factorization is $2(2x + 1)(x - 5)$.

Instructor Note
When the terms of a polynomial contain a common factor, some students will attempt to find the common factor and the two binomial factors in one step; they rarely obtain the correct final answer. Encourage the students to find only two factors at a time. Emphasize that the first step is to find the common factor and the resulting polynomial factor. After the first step is complete, the second step is to find the two binomial factors.

In-Class Examples (Objective 10.2B)
Factor.
1. $3x^2 - 6x - 72$ $3(x - 6)(x + 4)$
2. $2x^2 + 6x - 20$ $2(x + 5)(x - 2)$
3. $-4y^3 + 28y^2 - 48y$ $-4y(y - 4)(y - 3)$
4. $5x^2 - 10xy - 15y^2$ $5(x + y)(x - 3y)$

Concept Check

1. If $a(x + 3) = x^2 + 2x - 3$, find a. $x - 1$

2. If $-2x^3 - 6x^2 - 4x = a(x + 1)(x + 2)$, find a.
$-2x$

Optional Student Activity

Two binomials and one monomial are multiplied together. The resulting product is $6x^4y - 18x^3y^2 - 60x^2y^3$. Write an explanation telling how to determine the original three factors.

Answers will vary but should include determining the GCF first. The factored form is $6x^2y(x + 2y)(x - 5y)$.

HOW TO   Factor: $15 - 2x - x^2$

Because the coefficient of x^2 is -1, factor -1 from the trinomial and then write the resulting trinomial in descending order.

$15 - 2x - x^2 = -(x^2 + 2x - 15)$

$= -(x + 5)(x - 3)$

- $15 - 2x - x^2 = -1(-15 + 2x + x^2)$
  $= -(x^2 + 2x - 15)$
- Factor $x^2 + 2x - 15$. The two factors of -15 whose sum is 2 are 5 and -3.

Check: $-(x + 5)(x - 3) = -(x^2 + 2x - 15)$
$= -x^2 - 2x + 15$
$= 15 - 2x - x^2$

- This is the original polynomial.

TAKE NOTE

When the coefficient of the term with the highest power in a polynomial is negative, consider factoring out a negative GCF. Example 3 is another example of this technique.

Example 3

Factor: $-3x^3 + 9x^2 + 12x$

Solution
The GCF is $-3x$.
$-3x^3 + 9x^2 + 12x = -3x(x^2 - 3x - 4)$
Factor the trinomial $x^2 - 3x - 4$. Find two factors of -4 whose sum is -3.

Factors	Sum
$+1, -4$	-3

$-3x^3 + 9x^2 + 12x = -3x(x + 1)(x - 4)$

You Try It 3

Factor: $-2x^3 + 14x^2 - 12x$

Your solution
$-2x(x - 6)(x - 1)$

Example 4

Factor: $4x^2 - 40xy + 84y^2$

Solution
The GCF is 4.
$4x^2 - 40xy + 84y^2 = 4(x^2 - 10xy + 21y^2)$
Factor the trinomial $x^2 - 10xy + 21y^2$. Find two negative factors of 21 whose sum is -10.

Factors	Sum
$-1, -21$	-22
$-3, -7$	-10

$4x^2 - 40xy + 84y^2 = 4(x - 3y)(x - 7y)$

You Try It 4

Factor: $3x^2 - 9xy - 12y^2$

Your solution
$3(x + y)(x - 4y)$

Solutions on p. S24

10.2 Exercises

Objective A   **To factor a trinomial of the form $x^2 + bx + c$**

Section 10.2

Suggested Assignment
Exercises 1–133, every other odd
More challenging problems:
  Exercises 136–147, odds; 148

1. Fill in the blank. In factoring a trinomial, if the constant term is positive, then the signs in both binomial factors will be _____.   the same

2. Fill in the blanks. To factor $x^2 + 8x - 48$, we must find two numbers whose product is _____ and whose sum is _____. $-48, 8$

For Exercises 3 to 75, factor.

3. $x^2 + 3x + 2$
$(x + 1)(x + 2)$

4. $x^2 + 5x + 6$
$(x + 2)(x + 3)$

5. $x^2 - x - 2$
$(x + 1)(x - 2)$

6. $x^2 + x - 6$
$(x + 3)(x - 2)$

7. $a^2 + a - 12$
$(a + 4)(a - 3)$

8. $a^2 - 2a - 35$
$(a + 5)(a - 7)$

9. $a^2 - 3a + 2$
$(a - 1)(a - 2)$

10. $a^2 - 5a + 4$
$(a - 1)(a - 4)$

11. $a^2 + a - 2$
$(a + 2)(a - 1)$

12. $a^2 - 2a - 3$
$(a + 1)(a - 3)$

13. $b^2 - 6b + 9$
$(b - 3)(b - 3)$

14. $b^2 + 8b + 16$
$(b + 4)(b + 4)$

15. $b^2 + 7b - 8$
$(b + 8)(b - 1)$

16. $y^2 - y - 6$
$(y + 2)(y - 3)$

17. $y^2 + 6y - 55$
$(y + 11)(y - 5)$

18. $z^2 - 4z - 45$
$(z + 5)(z - 9)$

19. $y^2 - 5y + 6$
$(y - 2)(y - 3)$

20. $y^2 - 8y + 15$
$(y - 3)(y - 5)$

21. $z^2 - 14z + 45$
$(z - 5)(z - 9)$

22. $z^2 - 14z + 49$
$(z - 7)(z - 7)$

23. $z^2 - 12z - 160$
$(z + 8)(z - 20)$

24. $p^2 + 2p - 35$
$(p + 7)(p - 5)$

25. $p^2 + 12p + 27$
$(p + 3)(p + 9)$

26. $p^2 - 6p + 8$
$(p - 2)(p - 4)$

27. $x^2 + 20x + 100$
$(x + 10)(x + 10)$

28. $x^2 + 18x + 81$
$(x + 9)(x + 9)$

29. $b^2 + 9b + 20$
$(b + 4)(b + 5)$

30. $b^2 + 13b + 40$
$(b + 8)(b + 5)$

31. $x^2 - 11x - 42$
$(x + 3)(x - 14)$

32. $x^2 + 9x - 70$
$(x + 14)(x - 5)$

33. $b^2 - b - 20$
$(b + 4)(b - 5)$

34. $b^2 + 3b - 40$
$(b + 8)(b - 5)$

35. $y^2 - 14y - 51$
$(y + 3)(y - 17)$

36. $y^2 - y - 72$
$(y + 8)(y - 9)$

37. $p^2 - 4p - 21$
$(p + 3)(p - 7)$

38. $p^2 + 16p + 39$
$(p + 3)(p + 13)$

39. $y^2 - 8y + 32$
Nonfactorable over the integers

40. $y^2 - 9y + 81$
Nonfactorable over the integers

41. $x^2 - 20x + 75$
$(x - 5)(x - 15)$

42. $x^2 - 12x + 11$
$(x - 11)(x - 1)$

Quick Quiz (Objective 10.2A)

Factor.
1. $x^2 + 7x + 10$ $(x + 5)(x + 2)$
2. $x^2 + 5x - 24$ $(x + 8)(x - 3)$
3. $x^2 - 13x + 36$ $(x - 9)(x - 4)$

43. $p^2 + 24p + 63$
$(p + 3)(p + 21)$

44. $x^2 - 15x + 56$
$(x - 7)(x - 8)$

45. $x^2 + 21x + 38$
$(x + 2)(x + 19)$

46. $x^2 + x - 56$
$(x + 8)(x - 7)$

47. $x^2 + 5x - 36$
$(x + 9)(x - 4)$

48. $a^2 - 21a - 72$
$(a + 3)(a - 24)$

49. $a^2 - 7a - 44$
$(a + 4)(a - 11)$

50. $a^2 - 15a + 36$
$(a - 3)(a - 12)$

51. $a^2 - 21a + 54$
$(a - 3)(a - 18)$

52. $z^2 - 9z - 136$
$(z + 8)(z - 17)$

53. $z^2 + 14z - 147$
$(z + 21)(z - 7)$

54. $c^2 - c - 90$
$(c + 9)(c - 10)$

55. $c^2 - 3c - 180$
$(c + 12)(c - 15)$

56. $z^2 + 15z + 44$
$(z + 4)(z + 11)$

57. $p^2 + 24p + 135$
$(p + 9)(p + 15)$

58. $c^2 + 19c + 34$
$(c + 2)(c + 17)$

59. $c^2 + 11c + 18$
$(c + 2)(c + 9)$

60. $x^2 - 4x - 96$
$(x + 8)(x - 12)$

61. $x^2 + 10x - 75$
$(x + 15)(x - 5)$

62. $x^2 - 22x + 112$
$(x - 8)(x - 14)$

63. $x^2 + 21x - 100$
$(x + 25)(x - 4)$

64. $b^2 + 8b - 105$
$(b + 15)(b - 7)$

65. $b^2 - 22b + 72$
$(b - 4)(b - 18)$

66. $a^2 - 9a - 36$
$(a + 3)(a - 12)$

67. $a^2 + 42a - 135$
$(a + 45)(a - 3)$

68. $b^2 - 23b + 102$
$(b - 6)(b - 17)$

69. $b^2 - 25b + 126$
$(b - 7)(b - 18)$

70. $a^2 + 27a + 72$
$(a + 3)(a + 24)$

71. $z^2 + 24z + 144$
$(z + 12)(z + 12)$

72. $x^2 + 25x + 156$
$(x + 12)(x + 13)$

73. $x^2 - 29x + 100$
$(x - 4)(x - 25)$

74. $x^2 - 10x - 96$
$(x + 6)(x - 16)$

75. $x^2 + 9x - 112$
$(x + 16)(x - 7)$

Objective B   **To factor completely**

For Exercises 76 to 135, factor.

76. $2x^2 + 6x + 4$
$2(x + 1)(x + 2)$

77. $3x^2 + 15x + 18$
$3(x + 2)(x + 3)$

78. $18 + 7x - x^2$
$-(x - 9)(x + 2)$

79. $12 - 4x - x^2$
$-(x + 6)(x - 2)$

80. $ab^2 + 2ab - 15a$
$a(b + 5)(b - 3)$

81. $ab^2 + 7ab - 8a$
$a(b + 8)(b - 1)$

82. $xy^2 - 5xy + 6x$
$x(y - 2)(y - 3)$

83. $xy^2 + 8xy + 15x$
$x(y + 3)(y + 5)$

84. $z^3 - 7z^2 + 12z$
$z(z - 3)(z - 4)$

85. $-2a^3 - 6a^2 - 4a$
$-2a(a + 2)(a + 1)$

86. $-3y^3 + 15y^2 - 18y$
$-3y(y - 3)(y - 2)$

87. $4y^3 + 12y^2 - 72y$
$4y(y + 6)(y - 3)$

88. $3x^2 + 3x - 36$
$3(x + 4)(x - 3)$

89. $2x^3 - 2x^2 + 4x$
$2x(x^2 - x + 2)$

90. $5z^2 - 15z - 140$
$5(z + 4)(z - 7)$

91. $6z^2 + 12z - 90$
$6(z + 5)(z - 3)$

92. $2a^3 + 8a^2 - 64a$
$2a(a + 8)(a - 4)$

93. $3a^3 - 9a^2 - 54a$
$3a(a + 3)(a - 6)$

94. $x^2 - 5xy + 6y^2$
$(x - 2y)(x - 3y)$

95. $x^2 + 4xy - 21y^2$
$(x + 7y)(x - 3y)$

96. $a^2 - 9ab + 20b^2$
$(a - 4b)(a - 5b)$

97. $a^2 - 15ab + 50b^2$
$(a - 5b)(a - 10b)$

98. $x^2 - 3xy - 28y^2$
$(x + 4y)(x - 7y)$

99. $s^2 + 2st - 48t^2$
$(s + 8t)(s - 6t)$

100. $y^2 - 15yz - 41z^2$
Nonfactorable over
the integers

101. $x^2 + 85xy + 36y^2$
Nonfactorable over
the integers

102. $z^4 - 12z^3 + 35z^2$
$z^2(z - 5)(z - 7)$

103. $z^4 + 2z^3 - 80z^2$
$z^2(z + 10)(z - 8)$

104. $b^4 - 22b^3 + 120b^2$
$b^2(b - 10)(b - 12)$

105. $b^4 - 3b^3 - 10b^2$
$b^2(b + 2)(b - 5)$

106. $2y^4 - 26y^3 - 96y^2$
$2y^2(y + 3)(y - 16)$

107. $3y^4 + 54y^3 + 135y^2$
$3y^2(y + 3)(y + 15)$

108. $-x^4 - 7x^3 + 8x^2$
$-x^2(x + 8)(x - 1)$

109. $-x^4 + 11x^3 + 12x^2$
$-x^2(x - 12)(x + 1)$

110. $4x^2y + 20xy - 56y$
$4y(x + 7)(x - 2)$

111. $3x^2y - 6xy - 45y$
$3y(x + 3)(x - 5)$

Quick Quiz (Objective 10.2B)

Factor.

1. $3a^3 + 15a^2 + 18a$ $3a(a + 3)(a + 2)$

2. $4 - 3x - x^2$ $-(x + 4)(x - 1)$

3. $5x^3 - 15x^2y + 10xy^2$ $5x(x - y)(x - 2y)$

112. $c^3 + 18c^2 - 40c$
$c(c + 20)(c - 2)$

113. $-3x^3 + 36x^2 - 81x$
$-3x(x - 3)(x - 9)$

114. $-4x^3 - 4x^2 + 24x$
$-4x(x + 3)(x - 2)$

115. $x^2 - 8xy + 15y^2$
$(x - 3y)(x - 5y)$

116. $y^2 - 7xy - 8x^2$
$(y + x)(y - 8x)$

117. $a^2 - 13ab + 42b^2$
$(a - 6b)(a - 7b)$

118. $y^2 + 4yz - 21z^2$
$(y + 7z)(y - 3z)$

119. $y^2 + 8yz + 7z^2$
$(y + z)(y + 7z)$

120. $y^2 - 16yz + 15z^2$
$(y - z)(y - 15z)$

121. $3x^2y + 60xy - 63y$
$3y(x + 21)(x - 1)$

122. $4x^2y - 68xy - 72y$
$4y(x + 1)(x - 18)$

123. $3x^3 + 3x^2 - 36x$
$3x(x + 4)(x - 3)$

124. $4x^3 + 12x^2 - 160x$
$4x(x + 8)(x - 5)$

125. $4z^3 + 32z^2 - 132z$
$4z(z + 11)(z - 3)$

126. $5z^3 - 50z^2 - 120z$
$5z(z + 2)(z - 12)$

127. $4x^3 + 8x^2 - 12x$
$4x(x + 3)(x - 1)$

128. $5x^3 + 30x^2 + 40x$
$5x(x + 2)(x + 4)$

129. $5p^2 + 25p - 420$
$5(p + 12)(p - 7)$

130. $4p^2 - 28p - 480$
$4(p + 8)(p - 15)$

131. $p^4 + 9p^3 - 36p^2$
$p^2(p + 12)(p - 3)$

132. $p^4 + p^3 - 56p^2$
$p^2(p + 8)(p - 7)$

133. $t^2 - 12ts + 35s^2$
$(t - 5s)(t - 7s)$

134. $a^2 - 10ab + 25b^2$
$(a - 5b)(a - 5b)$

135. $a^2 - 8ab - 33b^2$
$(a + 3b)(a - 11b)$

APPLYING THE CONCEPTS

For Exercises 136 to 138, factor.

136. $2 + c^2 + 9c$
Nonfactorable over the integers

137. $x^2y - 54y - 3xy$
$y(x + 6)(x - 9)$

138. $45a^2 + a^2b^2 - 14a^2b$
$a^2(b - 9)(b - 5)$

For Exercises 139 to 141, find all integers k such that the trinomial can be factored over the integers.

139. $x^2 + kx + 35$
$-36, -12, 12, 36$

140. $x^2 + kx + 18$
$-19, -11, -9, 9, 11, 19$

141. $x^2 + kx + 21$
$-22, -10, 10, 22$

For Exercises 142 to 147, determine the positive integer values of k for which the following polynomials are factorable over the integers.

142. $y^2 + 4y + k$
3, 4

143. $z^2 + 7z + k$
6, 10, 12

144. $a^2 - 6a + k$
5, 8, 9

145. $c^2 - 7c + k$
6, 10, 12

146. $x^2 - 3x + k$
2

147. $y^2 + 5y + k$
4, 6

148. In Exercises 142 to 147, the requirement was given that $k > 0$. If k is allowed to be any integer, how many different values of k are possible for each polynomial?   An infinite number

10.3 Factoring Polynomials of the Form $ax^2 + bx + c$

Objective A

To factor a trinomial of the form $ax^2 + bx + c$ by using trial factors

Trinomials of the form $ax^2 + bx + c$, where a, b, and c are integers, are shown at the right.

$$3x^2 - x + 4; \ a = 3, b = -1, c = 4$$
$$6x^2 + 2x - 3; \ a = 6, b = 2, c = -3$$

These trinomials differ from those in the preceding section in that the coefficient of x^2 is not 1. There are various methods of factoring these trinomials. The method described in this objective is factoring polynomials by using trial factors.

To reduce the number of trial factors that must be considered, remember the following:

1. Use the signs of the constant term and the coefficient of x in the trinomial to determine the signs of the binomial factors. If the constant term is positive, the signs of the binomial factors will be the same as the sign of the coefficient of x in the trinomial. If the constant term is negative, the constant terms in the binomials will have opposite signs.

2. If the terms of the trinomial do not have a common factor, then the terms of each of the binomial factors will not have a common factor.

HOW TO Factor: $2x^2 - 7x + 3$

The terms have no common factor. The constant term is positive. The coefficient of x is negative. The binomial constants will be negative.

Positive Factors of 2 (coefficient of x^2)	Negative Factors of 3 (constant term)
1, 2	$-1, -3$

Write trial factors. Use the **O**uter and **I**nner products of FOIL to determine the middle term, $-7x$, of the trinomial.

Trial Factors	Middle Term
$(x - 1)(2x - 3)$	$-3x - 2x = -5x$
$(x - 3)(2x - 1)$	$-x - 6x = -7x$

Write the factors of the trinomial. $\quad 2x^2 - 7x + 3 = (x - 3)(2x - 1)$

HOW TO Factor: $3x^2 + 14x + 15$

The terms have no common factor. The constant term is positive. The coefficient of x is positive. The binomial constants will be positive.

Positive Factors of 3 (coefficient of x^2)	Positive Factors of 15 (constant term)
1, 3	1, 15
	3, 5

Write trial factors. Use the **O**uter and **I**nner products of FOIL to determine the middle term, $14x$, of the trinomial.

Trial Factors	Middle Term
$(x + 1)(3x + 15)$	Common factor
$(x + 15)(3x + 1)$	$x + 45x = 46x$
$(x + 3)(3x + 5)$	$5x + 9x = 14x$
$(x + 5)(3x + 3)$	Common factor

Write the factors of the trinomial. $\quad 3x^2 + 14x + 15 = (x + 3)(3x + 5)$

Objective 10.3A

Instructor Note
The first objective of this section is to factor by using trial factors. The second objective is to factor by grouping. You may skip one of these objectives or do both.

Discuss the Concepts
In multiplying two binomials, which portion of the FOIL method determines the middle term of the resulting trinomial?
The OI part of FOIL (that is, the sum of the Outer and Inner products) determines the middle term.

In-Class Examples (Objective 10.3A)
Factor.
1. $5x^2 - 2x - 3 \quad (5x + 3)(x - 1)$
2. $-12x^3 - 18x^2 + 30x \quad -6x(2x + 5)(x - 1)$

Copyright © Houghton Mifflin Company. All rights reserved.

Instructor Note

Remind students that eliminating the trial factors that have a common factor requires that the terms of the polynomial do not have a common factor.

Concept Check

Factor.

1. $2x^2 - 5x + 2$
   $(2x - 1)(x - 2)$

2. $-4x^3 - 10x^2 - 6x$
   $-2x(2x + 3)(x + 1)$

Instructor Note

Some trial factors may be eliminated by realizing that if both first terms or both last terms of the binomials are even, the middle term's coefficient will never be odd. For example, in factoring $12x^2 - 17x + 6$, do not choose $6x$ and $2x$ as the factors of $12x^2$ because 6 and 2 are even but -17, the coefficient of the middle term, is odd.

Discuss the Concepts

Consider factoring
$24x^2 - 29x - 4$.

1. Tell why you should *not* choose $12x$ and $2x$ as the factors of $24x^2$.
   Two even terms never make an odd middle term.

2. Tell why you should *not* choose $6x$ and $4x$ as the factors of $24x^2$.
   Two even terms never make an odd middle term.

3. Tell why you should *not* choose 2 and -2 as the factors of -4.
   Two even terms never make an odd middle term.

HOW TO Factor: $6x^3 + 14x^2 - 12x$

Factor the GCF, $2x$, from the terms.

$$6x^3 + 14x^2 - 12x = 2x(3x^2 + 7x - 6)$$

Factor the trinomial. The constant term is negative. The binomial constants will have opposite signs.

Positive Factors of 3	Factors of -6
1, 3	1, -6
	-1, 6
	2, -3
	-2, 3

Write trial factors. Use the **O**uter and **I**nner products of FOIL to determine the middle term, $7x$, of the trinomial.

It is not necessary to test trial factors that have a common factor.

Trial Factors	Middle Term
$(x + 1)(3x - 6)$	Common factor
$(x - 6)(3x + 1)$	$x - 18x = -17x$
$(x - 1)(3x + 6)$	Common factor
$(x + 6)(3x - 1)$	$-x + 18x = 17x$
$(x + 2)(3x - 3)$	Common factor
$(x - 3)(3x + 2)$	$2x - 9x = -7x$
$(x - 2)(3x + 3)$	Common factor
$(x + 3)(3x - 2)$	$-2x + 9x = 7x$

Write the factors of the trinomial.

$$6x^3 + 14x^2 - 12x = 2x(x + 3)(3x - 2)$$

For this example, all the trial factors were listed. Once the correct factors have been found, however, the remaining trial factors can be omitted. For the examples and solutions in this text, all trial factors except those that have a common factor will be listed.

Example 1 Factor: $3x^2 + x - 2$

Solution

| Positive factors of 3: 1, 3 | Factors of -2: 1, -2 |
| | -1, 2 |

Trial Factors	Middle Term
$(x + 1)(3x - 2)$	$-2x + 3x = x$
$(x - 2)(3x + 1)$	$x - 6x = -5x$
$(x - 1)(3x + 2)$	$2x - 3x = -x$
$(x + 2)(3x - 1)$	$-x + 6x = 5x$

$3x^2 + x - 2 = (x + 1)(3x - 2)$

You Try It 1 Factor: $2x^2 - x - 3$

Your solution
$(x + 1)(2x - 3)$

Example 2 Factor: $-12x^3 - 32x^2 + 12x$

Solution
The GCF is $-4x$.
$-12x^3 - 32x^2 + 12x = -4x(3x^2 + 8x - 3)$
Factor the trinomial.

| Positive factors of 3: 1, 3 | Factors of -3: 1, -3 |
| | -1, 3 |

Trial Factors	Middle Term
$(x - 3)(3x + 1)$	$x - 9x = -8x$
$(x + 3)(3x - 1)$	$-x + 9x = 8x$

$-12x^3 - 32x^2 + 12x = -4x(x + 3)(3x - 1)$

You Try It 2 Factor: $-45y^3 + 12y^2 + 12y$

Your solution
$-3y(3y - 2)(5y + 2)$

Solutions on pp. S24–S25

Objective B   **To factor a trinomial of the form $ax^2 + bx + c$ by grouping**

In the preceding objective, trinomials of the form $ax^2 + bx + c$ were factored by using trial factors. In this objective, these trinomials will be factored by grouping.

To factor $ax^2 + bx + c$, first find two factors of $a \cdot c$ whose sum is b. Then use factoring by grouping to write the factorization of the trinomial.

> **HOW TO**   Factor: $2x^2 + 13x + 15$
>
> Find two positive factors of 30 $(2 \cdot 15)$ whose sum is 13.
>
> | Positive Factors of 30 | Sum |
> |---|---|
> | 1, 30 | 31 |
> | 2, 15 | 17 |
> | 3, 10 | 13 |
>
> • Once the required sum has been found, the remaining factors need not be checked.
>
> $2x^2 + 13x + 15 = 2x^2 + 3x + 10x + 15$ • Use the factors of 30 whose sum is 13 to write $13x$ as $3x + 10x$.
>
> $\quad\quad = (2x^2 + 3x) + (10x + 15)$ • Factor by grouping.
> $\quad\quad = x(2x + 3) + 5(2x + 3)$
> $\quad\quad = (2x + 3)(x + 5)$
>
> Check: $(2x + 3)(x + 5) = 2x^2 + 10x + 3x + 15$
> $\quad\quad\quad\quad\quad\quad\quad = 2x^2 + 13x + 15$

> **HOW TO**   Factor: $6x^2 - 11x - 10$
>
> Find two factors of -60 $[6(-10)]$ whose sum is -11.
>
> | Factors of -60 | Sum |
> |---|---|
> | 1, -60 | -59 |
> | -1, 60 | 59 |
> | 2, -30 | -28 |
> | -2, 30 | 28 |
> | 3, -20 | -17 |
> | -3, 20 | 17 |
> | 4, -15 | -11 |
>
> $6x^2 - 11x - 10 = 6x^2 + 4x - 15x - 10$ • Use the factors of -60 whose sum is -11 to write $-11x$ as $4x - 15x$.
>
> $\quad\quad = (6x^2 + 4x) - (15x + 10)$ • Factor by grouping. Recall that $-15x - 10 = -(15x + 10)$.
> $\quad\quad = 2x(3x + 2) - 5(3x + 2)$
> $\quad\quad = (3x + 2)(2x - 5)$
>
> Check: $(3x + 2)(2x - 5) = 6x^2 - 15x + 4x - 10$
> $\quad\quad\quad\quad\quad\quad\quad = 6x^2 - 11x - 10$

Objective 10.3B

Concept Check

Fill in the blanks.

1. To factor $2x^2 - 5x + 2$ by grouping, find two numbers whose product is ____ and whose sum is ____.   4, -5

2. To factor $6x^2 + 7x - 3$ by grouping, $7x$ must be written as ____ − ____.   9x, 2x

---

In-Class Examples (Objective 10.3B)

Factor.

1. $3x^2 + 7x + 4$ $(3x + 4)(x + 1)$
2. $72x^3 - 42x^2 - 72x$ $6x(4x + 3)(3x - 4)$

HOW TO Factor: $3x^2 - 2x - 4$

Find two factors of -12 [$3(-4)$] whose sum is -2.

Factors of -12	Sum
1, -12	-11
-1, 12	11
2, -6	-4
-2, 6	4
3, -4	-1
-3, 4	1

TAKE NOTE
$3x^2 - 2x - 4$ is a prime polynomial.

Because no integer factors of -12 have a sum of -2, $3x^2 - 2x - 4$ is nonfactorable over the integers.

Example 3

Factor: $2x^2 + 19x - 10$

Solution

Factors of -20 [$2(-10)$]	Sum
-1, 20	19

$$2x^2 + 19x - 10 = 2x^2 - x + 20x - 10$$
$$= (2x^2 - x) + (20x - 10)$$
$$= x(2x - 1) + 10(2x - 1)$$
$$= (2x - 1)(x + 10)$$

You Try It 3

Factor: $2a^2 + 13a - 7$

Your solution
$(2a - 1)(a + 7)$

Example 4

Factor: $24x^2y - 76xy + 40y$

Solution
The GCF is $4y$.
$$24x^2y - 76xy + 40y = 4y(6x^2 - 19x + 10)$$

Negative Factors of 60 [$6(10)$]	Sum
-1, -60	-61
-2, -30	-32
-3, -20	-23
-4, -15	-19

$$6x^2 - 19x + 10 = 6x^2 - 4x - 15x + 10$$
$$= (6x^2 - 4x) - (15x - 10)$$
$$= 2x(3x - 2) - 5(3x - 2)$$
$$= (3x - 2)(2x - 5)$$

$$24x^2y - 76xy + 40y = 4y(6x^2 - 19x + 10)$$
$$= 4y(3x - 2)(2x - 5)$$

You Try It 4

Factor: $15x^3 + 40x^2 - 80x$

Your solution
$5x(3x - 4)(x + 4)$

Solutions on p. S25

10.3 Exercises

Objective A   To factor a trinomial of the form $ax^2 + bx + c$
by using trial factors

For Exercises 1 to 70, factor by using trial factors.

1. $2x^2 + 3x + 1$
$(x + 1)(2x + 1)$

2. $5x^2 + 6x + 1$
$(x + 1)(5x + 1)$

3. $2y^2 + 7y + 3$
$(y + 3)(2y + 1)$

4. $3y^2 + 7y + 2$
$(y + 2)(3y + 1)$

5. $2a^2 - 3a + 1$
$(a - 1)(2a - 1)$

6. $3a^2 - 4a + 1$
$(a - 1)(3a - 1)$

7. $2b^2 - 11b + 5$
$(b - 5)(2b - 1)$

8. $3b^2 - 13b + 4$
$(b - 4)(3b - 1)$

9. $2x^2 + x - 1$
$(x + 1)(2x - 1)$

10. $4x^2 - 3x - 1$
$(x - 1)(4x + 1)$

11. $2x^2 - 5x - 3$
$(x - 3)(2x + 1)$

12. $3x^2 + 5x - 2$
$(x + 2)(3x - 1)$

13. $2t^2 - t - 10$
$(t + 2)(2t - 5)$

14. $2t^2 + 5t - 12$
$(t + 4)(2t - 3)$

15. $3p^2 - 16p + 5$
$(p - 5)(3p - 1)$

16. $6p^2 + 5p + 1$
$(2p + 1)(3p + 1)$

17. $12y^2 - 7y + 1$
$(3y - 1)(4y - 1)$

18. $6y^2 - 5y + 1$
$(2y - 1)(3y - 1)$

19. $6z^2 - 7z + 3$
Nonfactorable
over the integers

20. $9z^2 + 3z + 2$
Nonfactorable
over the integers

21. $6t^2 - 11t + 4$
$(2t - 1)(3t - 4)$

22. $10t^2 + 11t + 3$
$(2t + 1)(5t + 3)$

23. $8x^2 + 33x + 4$
$(x + 4)(8x + 1)$

24. $7x^2 + 50x + 7$
$(x + 7)(7x + 1)$

25. $5x^2 - 62x - 7$
Nonfactorable
over the integers

26. $9x^2 - 13x - 4$
Nonfactorable
over the integers

27. $12y^2 + 19y + 5$
$(3y + 1)(4y + 5)$

28. $5y^2 - 22y + 8$
$(y - 4)(5y - 2)$

29. $7a^2 + 47a - 14$
$(a + 7)(7a - 2)$

30. $11a^2 - 54a - 5$
$(a - 5)(11a + 1)$

31. $3b^2 - 16b + 16$
$(b - 4)(3b - 4)$

32. $6b^2 - 19b + 15$
$(2b - 3)(3b - 5)$

33. $2z^2 - 27z - 14$
$(z - 14)(2z + 1)$

34. $4z^2 + 5z - 6$
$(z + 2)(4z - 3)$

35. $3p^2 + 22p - 16$
$(p + 8)(3p - 2)$

36. $7p^2 + 19p + 10$
$(p + 2)(7p + 5)$

Quick Quiz (Objective 10.3A)

Factor by using trial factors.

1. $2x^2 - 17x + 21$ $(2x - 3)(x - 7)$

2. $4x^2 + 11x - 20$ $(4x - 5)(x + 4)$

3. $12x^3y + x^2y - 6xy$ $xy(4x + 3)(3x - 2)$

Section 10.3

Suggested Assignment
Exercises 1–129, every other odd
More challenging problems:
Exercises 131–143, odds

37. $4x^2 + 6x + 2$
$2(x + 1)(2x + 1)$

38. $12x^2 + 33x - 9$
$3(x + 3)(4x - 1)$

39. $15y^2 - 50y + 35$
$5(y - 1)(3y - 7)$

40. $30y^2 + 10y - 20$
$10(y + 1)(3y - 2)$

41. $2x^3 - 11x^2 + 5x$
$x(x - 5)(2x - 1)$

42. $2x^3 - 3x^2 - 5x$
$x(x + 1)(2x - 5)$

43. $3a^2b - 16ab + 16b$
$b(a - 4)(3a - 4)$

44. $2a^2b - ab - 21b$
$b(a + 3)(2a - 7)$

45. $3z^2 + 95z + 10$
Nonfactorable over the integers

46. $8z^2 - 36z + 1$
Nonfactorable over the integers

47. $36x - 3x^2 - 3x^3$
$-3x(x + 4)(x - 3)$

48. $-2x^3 + 2x^2 + 4x$
$-2x(x - 2)(x + 1)$

49. $80y^2 - 36y + 4$
$4(4y - 1)(5y - 1)$

50. $24y^2 - 24y - 18$
$6(2y + 1)(2y - 3)$

51. $8z^3 + 14z^2 + 3z$
$z(2z + 3)(4z + 1)$

52. $6z^3 - 23z^2 + 20z$
$z(2z - 5)(3z - 4)$

53. $6x^2y - 11xy - 10y$
$y(2x - 5)(3x + 2)$

54. $8x^2y - 27xy + 9y$
$y(x - 3)(8x - 3)$

55. $10t^2 - 5t - 50$
$5(t + 2)(2t - 5)$

56. $16t^2 + 40t - 96$
$8(t + 4)(2t - 3)$

57. $3p^3 - 16p^2 + 5p$
$p(p - 5)(3p - 1)$

58. $6p^3 + 5p^2 + p$
$p(2p + 1)(3p + 1)$

59. $26z^2 + 98z - 24$
$2(z + 4)(13z - 3)$

60. $30z^2 - 87z + 30$
$3(2z - 5)(5z - 2)$

61. $10y^3 - 44y^2 + 16y$
$2y(y - 4)(5y - 2)$

62. $14y^3 + 94y^2 - 28y$
$2y(y + 7)(7y - 2)$

63. $4yz^3 + 5yz^2 - 6yz$
$yz(z + 2)(4z - 3)$

64. $12a^3 + 14a^2 - 48a$
$2a(2a - 3)(3a + 8)$

65. $42a^3 + 45a^2 - 27a$
$3a(2a + 3)(7a - 3)$

66. $36p^2 - 9p^3 - p^4$
$-p^2(p - 3)(p + 12)$

67. $9x^2y - 30xy^2 + 25y^3$
$y(3x - 5y)(3x - 5y)$

68. $8x^2y - 38xy^2 + 35y^3$
$y(2x - 7y)(4x - 5y)$

69. $9x^3y - 24x^2y^2 + 16xy^3$
$xy(3x - 4y)(3x - 4y)$

70. $9x^3y + 12x^2y + 4xy$
$xy(3x + 2)(3x + 2)$

Objective B   **To factor a trinomial of the form $ax^2 + bx + c$ by grouping**

For Exercises 71 to 130, factor by grouping.

71. $6x^2 - 17x + 12$
$(2x - 3)(3x - 4)$

72. $15x^2 - 19x + 6$
$(3x - 2)(5x - 3)$

73. $5b^2 + 33b - 14$
$(b + 7)(5b - 2)$

74. $8x^2 - 30x + 25$
$(2x - 5)(4x - 5)$

75. $6a^2 + 7a - 24$
$(2a - 3)(3a + 8)$

76. $14a^2 + 15a - 9$
$(2a + 3)(7a - 3)$

77. $4z^2 + 11z + 6$
$(z + 2)(4z + 3)$

78. $6z^2 - 25z + 14$
$(2z - 7)(3z - 2)$

79. $22p^2 + 51p - 10$
$(2p + 5)(11p - 2)$

80. $14p^2 - 41p + 15$
$(2p - 5)(7p - 3)$

81. $8y^2 + 17y + 9$
$(y + 1)(8y + 9)$

82. $12y^2 - 145y + 12$
$(y - 12)(12y - 1)$

83. $18t^2 - 9t - 5$
$(3t + 1)(6t - 5)$

84. $12t^2 + 28t - 5$
$(2t + 5)(6t - 1)$

85. $6b^2 + 71b - 12$
$(b + 12)(6b - 1)$

86. $8b^2 + 65b + 8$
$(b + 8)(8b + 1)$

87. $9x^2 + 12x + 4$
$(3x + 2)(3x + 2)$

88. $25x^2 - 30x + 9$
$(5x - 3)(5x - 3)$

89. $6b^2 - 13b + 6$
$(2b - 3)(3b - 2)$

90. $20b^2 + 37b + 15$
$(4b + 5)(5b + 3)$

91. $33b^2 + 34b - 35$
$(3b + 5)(11b - 7)$

92. $15b^2 - 43b + 22$
$(3b - 2)(5b - 11)$

93. $18y^2 - 39y + 20$
$(3y - 4)(6y - 5)$

94. $24y^2 + 41y + 12$
$(3y + 4)(8y + 3)$

95. $15a^2 + 26a - 21$
$(3a + 7)(5a - 3)$

96. $6a^2 + 23a + 21$
$(2a + 3)(3a + 7)$

97. $8y^2 - 26y + 15$
$(2y - 5)(4y - 3)$

98. $18y^2 - 27y + 4$
$(3y - 4)(6y - 1)$

99. $8z^2 + 2z - 15$
$(2z + 3)(4z - 5)$

100. $10z^2 + 3z - 4$
$(2z - 1)(5z + 4)$

101. $15x^2 - 82x + 24$
Nonfactorable over the integers

102. $13z^2 + 49z - 8$
Nonfactorable over the integers

103. $10z^2 - 29z + 10$
$(2z - 5)(5z - 2)$

104. $15z^2 - 44z + 32$
$(3z - 4)(5z - 8)$

105. $36z^2 + 72z + 35$
$(6z + 5)(6z + 7)$

106. $16z^2 + 8z - 35$
$(4z + 7)(4z - 5)$

107. $3x^2 + xy - 2y^2$
$(x + y)(3x - 2y)$

108. $6x^2 + 10xy + 4y^2$
$2(x + y)(3x + 2y)$

109. $3a^2 + 5ab - 2b^2$
$(a + 2b)(3a - b)$

110. $2a^2 - 9ab + 9b^2$
$(a - 3b)(2a - 3b)$

Quick Quiz (Objective 10.3B)

Factor by grouping.
1. $10x^2 + x - 2$ $(5x - 2)(2x + 1)$
2. $12x^2 + 31x + 9$ $(3x + 1)(4x + 9)$
3. $12x^3y + 10x^2y - 8xy$ $2xy(3x + 4)(2x - 1)$

111. $4y^2 - 11yz + 6z^2$
$(y - 2z)(4y - 3z)$

112. $2y^2 + 7yz + 5z^2$
$(y + z)(2y + 5z)$

113. $28 + 3z - z^2$
$-(z - 7)(z + 4)$

114. $15 - 2z - z^2$
$-(z - 3)(z + 5)$

115. $8 - 7x - x^2$
$-(x - 1)(x + 8)$

116. $12 + 11x - x^2$
$-(x - 12)(x + 1)$

117. $9x^2 + 33x - 60$
$3(x + 5)(3x - 4)$

118. $16x^2 - 16x - 12$
$4(2x + 1)(2x - 3)$

119. $24x^2 - 52x + 24$
$4(2x - 3)(3x - 2)$

120. $60x^2 + 95x + 20$
$5(3x + 4)(4x + 1)$

121. $35a^4 + 9a^3 - 2a^2$
$a^2(5a + 2)(7a - 1)$

122. $15a^4 + 26a^3 + 7a^2$
$a^2(3a + 1)(5a + 7)$

123. $15b^2 - 115b + 70$
$5(b - 7)(3b - 2)$

124. $25b^2 + 35b - 30$
$5(b + 2)(5b - 3)$

125. $3x^2 - 26xy + 35y^2$
$(x - 7y)(3x - 5y)$

126. $4x^2 + 16xy + 15y^2$
$(2x + 3y)(2x + 5y)$

127. $216y^2 - 3y - 3$
$3(8y - 1)(9y + 1)$

128. $360y^2 + 4y - 4$
$4(9y + 1)(10y - 1)$

129. $21 - 20x - x^2$
$-(x - 1)(x + 21)$

130. $18 + 17x - x^2$
$-(x - 18)(x + 1)$

APPLYING THE CONCEPTS

131. ✎ In your own words, explain how the signs of the last terms of the two binomial factors of a trinomial are determined.

For Exercises 132 to 137, factor.

132. $(x + 1)^2 - (x + 1) - 6$
$(x - 2)(x + 3)$

133. $(x - 2)^2 + 3(x - 2) + 2$
$x(x - 1)$

134. $(y + 3)^2 - 5(y + 3) + 6$
$y(y + 1)$

135. $2(y + 2)^2 - (y + 2) - 3$
$(2y + 1)(y + 3)$

136. $3(a + 2)^2 - (a + 2) - 4$
$(3a + 2)(a + 3)$

137. $4(y - 1)^2 - 7(y - 1) - 2$
$(4y - 3)(y - 3)$

For Exercises 138 to 143, find all integers k such that the trinomial can be factored over the integers.

138. $2x^2 + kx + 3$
$-5, 5, -7, 7$

139. $2x^2 + kx - 3$
$-1, 1, -5, 5$

140. $3x^2 + kx + 2$
$-5, 5, -7, 7$

141. $3x^2 + kx - 2$
$-1, 1, -5, 5$

142. $2x^2 + kx + 5$
$-7, 7, -11, 11$

143. $2x^2 + kx - 5$
$-3, 3, -9, 9$

Answers to Writing Exercises

131. Students should explain that the sign of the product of the last terms of the two binomial factors must be the same as the sign of the last term of the trinomial. Thus if the last term of the trinomial is positive, the last terms of the two binomial factors are either both positive or both negative, depending on the middle term of the trinomial. If the last term of the trinomial is negative, the last terms of the two binomial factors will have different signs.

10.4 Special Factoring

Objective A   To factor the difference of two squares and perfect-square trinomials

A polynomial of the form $a^2 - b^2$ is called a **difference of two squares.** Recall the following relationship from Objective 9.3D.

Product of the sum and difference of the same terms		Difference of two squares
$(a + b)(a - b)$	$=$	$a^2 - b^2$

> **TAKE NOTE**
> Note that the polynomial $x^2 + y^2$ is the *sum* of two squares. The sum of two squares is nonfactorable over the integers.

> **Factoring the Difference of Two Squares**
>
> The difference of two squares factors as the sum and difference of the same terms.
>
> $$a^2 - b^2 = (a + b)(a - b)$$

HOW TO   Factor: $x^2 - 16$

$x^2 - 16 = (x)^2 - (4)^2$ • $x^2 - 16$ is the difference of two squares.

$\qquad = (x + 4)(x - 4)$ • Factor the difference of squares.

Check: $(x + 4)(x - 4) = x^2 - 4x + 4x - 16$

$\qquad\qquad\qquad\qquad = x^2 - 16$

HOW TO   Factor: $8x^3 - 18x$

$8x^3 - 18x = 2x(4x^2 - 9)$ • The GCF is 2x.

$\qquad\quad = 2x[(2x)^2 - 3^2]$ • $4x^2 - 9$ is the difference of two squares.

$\qquad\quad = 2x(2x + 3)(2x - 3)$ • Factor the difference of squares.

You should check the factorization.

HOW TO   Factor: $x^2 - 10$

Because 10 cannot be written as the square of an integer, $x^2 - 10$ is nonfactorable over the integers.

A trinomial that can be written as the square of a binomial is called a **perfect-square trinomial.** Recall the pattern for finding the square of a binomial.

$$(a + b)^2 = a^2 + 2ab + b^2$$

Square of the first term ⟶ · ⟵ Square of the last term

Twice the product of the two terms

Objective 10.4A

New Vocabulary
difference of two squares
perfect-square trinomial

Concept Check
Tell whether or not the polynomial is a difference of two squares.
1. $x^2 - 25$ Yes
2. $4x^2 - 49$ Yes
3. $9x^2 - 27$ No
4. $x^2 + 36$ No
5. $16x^2 - y^2$ Yes
6. $100x^4 - 81$ Yes
7. $x^2 - 8$ No
8. $x^2y^2 - 1$ Yes

Concept Check
Tell whether or not the polynomial is a perfect-square trinomial.
1. $x^2 + 2xy + y^2$ Yes
2. $x^2 + 2x + 1$ Yes
3. $x^2 - 2x + 1$ Yes
4. $x^2 + 4x + 9$ No
5. $x^2 - 6x - 9$ No
6. $x^2 + y^2$ No
7. $x^2 - y^2$ No

In-Class Examples (Objective 10.4A)

Factor.
1. $36a^2 - 49b^2$ $(6a + 7b)(6a - 7b)$
2. $100x^2 - 180x + 81$ $(10x - 9)^2$
3. $16x^2 + 58x + 25$ $(8x + 25)(2x + 1)$
4. $(x^2 + 10x + 25) - 9y^2$
   $(x + 5 + 3y)(x + 5 - 3y)$

Concept Check

1. Replace each ? to make a true statement. $x^4 - 49 =$ $(?)^2 - (?)^2 = (?)(?)$
   $x^2, 7, x^2 + 7, x^2 - 7$

2. If $x^2 - a$ is the difference of two perfect squares, what are the possible values of a that are less than 10?   1, 4, 9

Concept Check

1. If $16x^2 + ax + 9$ is a perfect-square trinomial, what are the two possible values for a?
   $-24, 24$

2. If $49x^2 + 28x + b$ is a perfect-square trinomial, what is the value of b?   4

Instructor Note

Here is a geometric representation of the difference of squares.

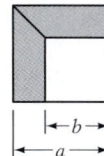

$a^2 - b^2$

The shaded area can be combined as shown below.

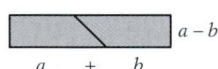

$a - b$

$(a + b)(a - b) = a^2 - b^2$

Factoring a Perfect-Square Trinomial

A perfect-square trinomial factors as the square of a binomial.

$$a^2 + 2ab + b^2 = (a + b)^2$$
$$a^2 - 2ab + b^2 = (a - b)^2$$

HOW TO   Factor: $4x^2 - 20x + 25$

Because the first and last terms are squares $[(2x)^2 = 4x^2; 5^2 = 25]$, try to factor this as the square of a binomial. Check the factorization.

$$4x^2 - 20x + 25 = (2x - 5)^2$$

Check: $(2x - 5)^2 = (2x)^2 + 2(2x)(-5) + 5^2$
$\qquad\qquad\quad = 4x^2 - 20x + 25$ • The factorization is correct.

$$4x^2 - 20x + 25 = (2x - 5)^2$$

HOW TO   Factor: $4x^2 + 37x + 9$

Because the first and last terms are squares $[(2x)^2 = 4x^2; 3^2 = 9]$, try to factor this as the square of a binomial. Check the proposed factorization.

$$4x^2 + 37x + 9 = (2x + 3)^2$$

Check: $(2x + 3)^2 = (2x)^2 + 2(2x)(3) + 3^2$
$\qquad\qquad\quad = 4x^2 + 12x + 9$

Because $4x^2 + 12x + 9 \neq 4x^2 + 37x + 9$, the proposed factorization is not correct. In this case, the polynomial is not a perfect-square trinomial. It may, however, still factor. In fact, $4x^2 + 37x + 9 = (4x + 1)(x + 9)$.

Example 1

Factor: $16x^2 - y^2$

Solution
$16x^2 - y^2 = (4x)^2 - y^2$ • The difference of two squares

$\qquad\qquad = (4x + y)(4x - y)$ • Factor.

You Try It 1

Factor: $25a^2 - b^2$

Your solution
$(5a + b)(5a - b)$

Example 2

Factor: $z^4 - 16$

Solution
$z^4 - 16 = (z^2)^2 - 4^2$ • The difference of two squares

$\qquad = (z^2 + 4)(z^2 - 4)$ • $z^2 - 4$ is the difference of two squares.
$\qquad = (z^2 + 4)(z^2 - 2^2)$

$\qquad = (z^2 + 4)(z + 2)(z - 2)$ • Factor.

You Try It 2

Factor: $n^4 - 81$

Your solution
$(n^2 + 9)(n + 3)(n - 3)$

Solutions on p. S25

Example 3

Factor: $9x^2 - 30x + 25$

Solution
$9x^2 = (3x)^2$, $25 = (5)^2$
$9x^2 - 30x + 25 = (3x - 5)^2$

Check: $(3x - 5)^2 = (3x)^2 + 2(3x)(-5) + 5^2$
$= 9x^2 - 30x + 25$

You Try It 3

Factor: $16y^2 + 8y + 1$

Your solution
$(4y + 1)^2$

Example 4

Factor: $9x^2 + 40x + 16$

Solution
Because $9x^2 = (3x)^2$, $16 = 4^2$, and
$40x \neq 2(3x)(4)$, the trinomial is not
a perfect-square trinomial.

Try to factor by another method.

$9x^2 + 40x + 16 = (9x + 4)(x + 4)$

You Try It 4

Factor: $x^2 + 15x + 36$

Your solution
$(x + 3)(x + 12)$

Solutions on p. S25

Concept Check

Suppose you wanted to make up a quiz that had as one of its problems to factor a perfect-square trinomial. How could you create the problem in such a way as to guarantee that you had a perfect-square trinomial?
Square a binomial.

Objective 10.4B

Objective B **To factor completely**
VIDEO & DVD   CD TUTOR   WEB   SSM

Study Tip

You have now learned to factor many different types of polynomials. You will need to be able to recognize each of the situations described in the box at the right. To test yourself, you might want to do the exercises in the Chapter Review.

General Factoring Strategy

1. Is there a common factor? If so, factor out the common factor.
2. Is the polynomial the difference of two perfect squares? If so, factor.
3. Is the polynomial a perfect-square trinomial? If so, factor.
4. Is the polynomial a trinomial that is the product of two binomials? If so, factor.
5. Does the polynomial contain four terms? If so, try factoring by grouping.
6. Is each binomial factor nonfactorable over the integers? If not, factor the binomial.

HOW TO   Factor: $z^3 + 4z^2 - 9z - 36$

$z^3 + 4z^2 - 9z - 36 = (z^3 + 4z^2) - (9z + 36)$ • Factor by grouping. Recall that $-9z - 36 = -(9z + 36)$.

$= z^2(z + 4) - 9(z + 4)$ • $z^3 + 4z^2 = z^2(z + 4)$
       $9z + 36 = 9(z + 4)$

$= (z + 4)(z^2 - 9)$ • Factor out the common binomial factor $(z + 4)$.

$= (z + 4)(z + 3)(z - 3)$ • Factor the difference of squares.

Discuss the Concepts

The first step in any factoring problem is to factor out any common factors. The second step depends on the number of terms the polynomial has. In each of the following cases, tell what could be a next step in factoring.

1. The polynomial has two terms.
Determine whether it is the difference of two squares; if so, factor.

2. The polynomial has three terms.
Answers will vary. Students may choose to determine whether the polynomial is a perfect-square trinomial, or they may try to "unFOIL" the trinomial.

3. The polynomial has four terms.
Try to factor by grouping.

In-Class Examples (Objective 10.4B)

Factor.
1. $9x^2 - 81$ $9(x + 3)(x - 3)$
2. $a^2b - 3a^2 - 16b + 48$ $(b - 3)(a + 4)(a - 4)$
3. $9x^2y - 48xy + 48y$ $3y(3x - 4)(x - 4)$

Concept Check

Explain how you could use the difference of two squares to find the products $42 \cdot 38$ and $84 \cdot 76$.
$(40 + 2)(40 - 2) = 1600 - 4 = 1596$, $(80 + 4)(80 - 4) = 6400 - 16 = 6384$

Optional Student Activity

List any three consecutive natural numbers. What is the relationship between the square of the middle number and the product of the first and third numbers? Is this always true? Try to prove your answer.
$(\text{middle})^2 - 1 = (\text{first})(\text{third})$
Let the numbers be $n - 1$, n, and $n + 1$. Then
$n^2 - 1 = (n - 1)(n + 1)$.

Example 5

Factor: $3x^2 - 48$

Solution
The GCF is 3.

$3x^2 - 48 = 3(x^2 - 16)$
$\qquad\qquad = 3(x + 4)(x - 4)$ • Factor the difference of two squares.

You Try It 5

Factor: $12x^3 - 75x$

Your solution
$3x(2x + 5)(2x - 5)$

Example 6

Factor: $x^3 - 3x^2 - 4x + 12$

Solution
Factor by grouping.

$x^3 - 3x^2 - 4x + 12$
$\quad = (x^3 - 3x^2) - (4x - 12)$ • Factor by grouping.
$\quad = x^2(x - 3) - 4(x - 3)$ • $x - 3$ is the common factor.
$\quad = (x - 3)(x^2 - 4)$ • $x^2 - 4$ is the difference of two squares.
$\quad = (x - 3)(x + 2)(x - 2)$ • Factor.

You Try It 6

Factor: $a^2b - 7a^2 - b + 7$

Your solution
$(b - 7)(a + 1)(a - 1)$

Example 7

Factor: $4x^2y^2 + 12xy^2 + 9y^2$

Solution
The GCF is y^2.

$4x^2y^2 + 12xy^2 + 9y^2$
$\quad = y^2(4x^2 + 12x + 9)$ • Factor the GCF, y^2.
$\quad = y^2(2x + 3)^2$ • Factor the perfect-square trinomial.

You Try It 7

Factor: $4x^3 + 28x^2 - 120x$

Your solution
$4x(x + 10)(x - 3)$

Solutions on p. S25

10.4 Exercises

Objective A   **To factor the difference of two squares and perfect-square trinomials**

Section 10.4

Suggested Assignment
Exercises 1–125, every other odd
More challenging problems:
    Exercises 127–131, odds

Answers to Writing Exercises

2. The sum of two squares is not factorable over the integers because there are no integers whose product is positive and whose sum is zero.

1. **a.** Provide an example of a binomial that is the difference of two squares.
    Answers will vary. For instance, $x^2 - 25$.
    b. Provide an example of a perfect-square trinomial.
    Answers will vary. For instance, $x^2 + 6x + 9$.

2. Explain why a binomial that is the sum of two squares is nonfactorable over the integers.

For Exercises 3 to 48, factor.

3. $x^2 - 4$
    $(x + 2)(x - 2)$

4. $x^2 - 9$
    $(x + 3)(x - 3)$

5. $a^2 - 81$
    $(a + 9)(a - 9)$

6. $a^2 - 49$
    $(a + 7)(a - 7)$

7. $y^2 + 2y + 1$
    $(y + 1)^2$

8. $y^2 + 14y + 49$
    $(y + 7)^2$

9. $a^2 - 2a + 1$
    $(a - 1)^2$

10. $x^2 - 12x + 36$
    $(x - 6)^2$

11. $4x^2 - 1$
    $(2x + 1)(2x - 1)$

12. $9x^2 - 16$
    $(3x + 4)(3x - 4)$

13. $x^6 - 9$
    $(x^3 + 3)(x^3 - 3)$

14. $y^{12} - 4$
    $(y^8 + 2)(y^6 - 2)$

15. $x^2 + 8x - 16$
    Nonfactorable over the integers

16. $z^2 - 18z - 81$
    Nonfactorable over the integers

17. $x^2 + 2xy + y^2$
    $(x + y)^2$

18. $x^2 + 6xy + 9y^2$
    $(x + 3y)^2$

19. $4a^2 + 4a + 1$
    $(2a + 1)^2$

20. $25x^2 + 10x + 1$
    $(5x + 1)^2$

21. $9x^2 - 1$
    $(3x + 1)(3x - 1)$

22. $1 - 49x^2$
    $(1 + 7x)(1 - 7x)$

23. $1 - 64x^2$
    $(1 + 8x)(1 - 8x)$

24. $t^2 + 36$
    Nonfactorable over the integers

25. $x^2 + 64$
    Nonfactorable over the integers

26. $64a^2 - 16a + 1$
    $(8a - 1)^2$

27. $9a^2 + 6a + 1$
    $(3a + 1)^2$

28. $x^4 - y^2$
    $(x^2 + y)(x^2 - y)$

29. $b^4 - 16a^2$
    $(b^2 + 4a)(b^2 - 4a)$

30. $16b^2 + 8b + 1$
    $(4b + 1)^2$

31. $4a^2 - 20a + 25$
    $(2a - 5)^2$

32. $4b^2 + 28b + 49$
    $(2b + 7)^2$

33. $9a^2 - 42a + 49$
    $(3a - 7)^2$

34. $9x^2 - 16y^2$
    $(3x + 4y)(3x - 4y)$

35. $25z^2 - y^2$
    $(5z + y)(5z - y)$

36. $x^2y^2 - 4$
    $(xy + 2)(xy - 2)$

37. $a^2b^2 - 25$
    $(ab + 5)(ab - 5)$

38. $16 - x^2y^2$
    $(4 + xy)(4 - xy)$

Quick Quiz (Objective 10.4A)

Factor.

1. $y^2 - 25$ $(y + 5)(y - 5)$

2. $16x^2 - 24xy + 9y^2$ $(4x - 3y)^2$

3. $(a + 2)^2 - 4$ $a(a + 4)$

39. $25x^2 - 1$
$(5x + 1)(5x - 1)$

40. $25a^2 + 30ab + 9b^2$
$(5a + 3b)^2$

41. $4a^2 - 12ab + 9b^2$
$(2a - 3b)^2$

42. $49x^2 + 28xy + 4y^2$
$(7x + 2y)^2$

43. $4y^2 - 36yz + 81z^2$
$(2y - 9z)^2$

44. $64y^2 - 48yz + 9z^2$
$(8y - 3z)^2$

45. $\dfrac{1}{x^2} - 4$
$\left(\dfrac{1}{x} + 2\right)\left(\dfrac{1}{x} - 2\right)$

46. $\dfrac{9}{a^2} - 16$
$\left(\dfrac{3}{a} + 4\right)\left(\dfrac{3}{a} - 4\right)$

47. $9a^2b^2 - 6ab + 1$
$(3ab - 1)^2$

48. $16x^2y^2 - 24xy + 9$
$(4xy - 3)^2$

Objective B **To factor completely**

For Exercises 49 to 126, factor.

49. $8y^2 - 2$
$2(2y + 1)(2y - 1)$

50. $12n^2 - 48$
$12(n + 2)(n - 2)$

51. $3a^3 + 6a^2 + 3a$
$3a(a + 1)^2$

52. $4rs^2 - 4rs + r$
$r(2s - 1)^2$

53. $m^4 - 256$
$(m^2 + 16)(m + 4)(m - 4)$

54. $81 - t^4$
$(9 + t^2)(3 + t)(3 - t)$

55. $9x^2 + 13x + 4$
$(9x + 4)(x + 1)$

56. $x^2 + 10x + 16$
$(x + 2)(x + 8)$

57. $16y^4 + 48y^3 + 36y^2$
$4y^2(2y + 3)^2$

58. $36c^4 - 48c^3 + 16c^2$
$4c^2(3c - 2)^2$

59. $y^8 - 81$
$(y^4 + 9)(y^2 + 3)(y^2 - 3)$

60. $32s^4 - 2$
$2(4s^2 + 1)(2s + 1)(2s - 1)$

61. $25 - 20p + 4p^2$
$(5 - 2p)^2$

62. $9 + 24a + 16a^2$
$(3 + 4a)^2$

63. $(4x - 3)^2 - y^2$
$(4x - 3 + y)(4x - 3 - y)$

64. $(2x + 5)^2 - 25$
$4x(x + 5)$

65. $(x^2 - 4x + 4) - y^2$
$(x - 2 + y)(x - 2 - y)$

66. $(4x^2 + 12x + 9) - 4y^2$
$(2x + 3 + 2y)(2x + 3 - 2y)$

Quick Quiz (Objective 10.4B)
Factor.
1. $8n^2 - 72$ $8(n + 3)(n - 3)$
2. $y^8 - 16$ $(y^4 + 4)(y^2 + 2)(y^2 - 2)$
3. $8 + 24x + 18x^2$ $2(2 + 3x)^2$

67. $5x^2 - 5$
$5(x + 1)(x - 1)$

68. $2x^2 - 18$
$2(x + 3)(x - 3)$

69. $x^3 + 4x^2 + 4x$
$x(x + 2)^2$

70. $y^3 - 10y^2 + 25y$
$y(y - 5)^2$

71. $x^4 + 2x^3 - 35x^2$
$x^2(x + 7)(x - 5)$

72. $a^4 - 11a^3 + 24a^2$
$a^2(a - 3)(a - 8)$

73. $5b^2 + 75b + 180$
$5(b + 3)(b + 12)$

74. $6y^2 - 48y + 72$
$6(y - 2)(y - 6)$

75. $3a^2 + 36a + 10$
Nonfactorable
over the integers

76. $5a^2 - 30a + 4$
Nonfactorable
over the integers

77. $2x^2y + 16xy - 66y$
$2y(x + 11)(x - 3)$

78. $3a^2b + 21ab - 54b$
$3b(a + 9)(a - 2)$

79. $x^3 - 6x^2 - 5x$
$x(x^2 - 6x - 5)$

80. $b^3 - 8b^2 - 7b$
$b(b^2 - 8b - 7)$

81. $3y^2 - 36$
$3(y^2 - 12)$

82. $3y^2 - 147$
$3(y + 7)(y - 7)$

83. $20a^2 + 12a + 1$
$(2a + 1)(10a + 1)$

84. $12a^2 - 36a + 27$
$3(2a - 3)^2$

85. $x^2y^2 - 7xy^2 - 8y^2$
$y^2(x + 1)(x - 8)$

86. $a^2b^2 + 3a^2b - 88a^2$
$a^2(b + 11)(b - 8)$

87. $10a^2 - 5ab - 15b^2$
$5(a + b)(2a - 3b)$

88. $16x^2 - 32xy + 12y^2$
$4(2x - y)(2x - 3y)$

89. $50 - 2x^2$
$-2(x + 5)(x - 5)$

90. $72 - 2x^2$
$-2(x + 6)(x - 6)$

91. $a^2b^2 - 10ab^2 + 25b^2$
$b^2(a - 5)^2$

92. $a^2b^2 + 6ab^2 + 9b^2$
$b^2(a + 3)^2$

93. $12a^3b - a^2b^2 - ab^3$
$ab(4a + b)(3a - b)$

94. $2x^3y - 7x^2y^2 + 6xy^3$
$xy(x - 2y)(2x - 3y)$

95. $12a^3 - 12a^2 + 3a$
$3a(2a - 1)^2$

96. $18a^3 + 24a^2 + 8a$
$2a(3a + 2)^2$

97. $243 + 3a^2$
$3(81 + a^2)$

98. $75 + 27y^2$
$3(25 + 9y^2)$

99. $12a^3 - 46a^2 + 40a$
$2a(2a - 5)(3a - 4)$

100. $24x^3 - 66x^2 + 15x$
$3x(2x - 5)(4x - 1)$

101. $4a^3 + 20a^2 + 25a$
$a(2a + 5)^2$

102. $2a^3 - 8a^2b + 8ab^2$
$2a(a - 2b)^2$

103. $27a^2b - 18ab + 3b$
$3b(3a - 1)^2$

104. $a^2b^2 - 6ab^2 + 9b^2$
$b^2(a - 3)^2$

105. $48 - 12x - 6x^2$
$-6(x - 2)(x + 4)$

106. $21x^2 - 11x^3 - 2x^4$
$-x^2(2x - 3)(x + 7)$

107. $x^4 - x^2y^2$
$x^2(x + y)(x - y)$

108. $b^4 - a^2b^2$
$b^2(b + a)(b - a)$

109. $18a^3 + 24a^2 + 8a$
$2a(3a + 2)^2$

110. $32xy^2 - 48xy + 18x$
$2x(4y - 3)^2$

111. $2b + ab - 6a^2b$
$-b(3a - 2)(2a + 1)$

112. $15y^2 - 2xy^2 - x^2y^2$
$-y^2(x - 3)(x + 5)$

113. $4x^4 - 38x^3 + 48x^2$
$2x^2(x - 8)(2x - 3)$

114. $3x^2 - 27y^2$
$3(x + 3y)(x - 3y)$

115. $x^4 - 25x^2$
$x^2(x + 5)(x - 5)$

116. $y^3 - 9y$
$y(y + 3)(y - 3)$

117. $a^4 - 16$
$(a^2 + 4)(a + 2)(a - 2)$

118. $15x^4y^2 - 13x^3y^3 - 20x^2y^4$
$x^2y^2(5x + 4y)(3x - 5y)$

119. $45y^2 - 42y^3 - 24y^4$
$-3y^2(2y + 5)(4y - 3)$

120. $a(2x - 2) + b(2x - 2)$
$2(x - 1)(a + b)$

121. $4a(x - 3) - 2b(x - 3)$
$2(x - 3)(2a - b)$

122. $x^2(x - 2) - (x - 2)$
$(x - 2)(x + 1)(x - 1)$

123. $y^2(a - b) - (a - b)$
$(a - b)(y + 1)(y - 1)$

124. $a(x^2 - 4) + b(x^2 - 4)$
$(x + 2)(x - 2)(a + b)$

125. $x(a^2 - b^2) - y(a^2 - b^2)$
$(a + b)(a - b)(x - y)$

126. $4(x - 5) - x^2(x - 5)$
$(x - 5)(2 + x)(2 - x)$

APPLYING THE CONCEPTS

For Exercises 127 to 132, find all integers k such that the trinomial is a perfect-square trinomial.

127. $4x^2 - kx + 9$
$-12, 12$

128. $x^2 + 6x + k$
9

129. $64x^2 + kxy + y^2$
$-16, 16$

130. $x^2 - 2x + k$
1

131. $25x^2 - kx + 1$
$-10, 10$

132. $x^2 + 10x + k$
25

Number Sense The following method can be used to determine the number of divisors of a composite number. First find the prime factorization (in exponential form) of the composite number. Add 1 to each exponent in the prime factorization and then compute the product of these exponents. This product is equal to the number of divisors of the composite number. For example, consider the composite number 12, which has six divisors, 1, 2, 3, 4, 6, and 12. The prime factorization of 12 is $2^2 \cdot 3^1$. Adding 1 to each of the exponents produces the numbers 3 and 2. The product of 3 and 2 is 6, the number of divisors of 12. In Exercises 133 to 136, determine the number of divisors of the composite number.

133. 60
12

134. 84
12

135. 288
18

136. 360
24

10.5 Solving Equations

Objective A   To solve equations by factoring

The Multiplication Property of Zero states that the product of a number and zero is zero. This property is stated below.

$$\text{If } a \text{ is a real number, then } a \cdot 0 = 0 \cdot a = 0.$$

Now consider $x \cdot y = 0$. For this to be a true equation, then either $x = 0$ or $y = 0$.

Principle of Zero Products

If the product of two factors is zero, then at least one of the factors must be zero.

If $a \cdot b = 0$, then $a = 0$ or $b = 0$.

The Principle of Zero Products is used to solve some equations.

HOW TO   Solve: $(x - 2)(x - 3) = 0$

$(x - 2)(x - 3) = 0$

$x - 2 = 0 \qquad x - 3 = 0$ • Let each factor equal zero (the Principle of Zero Products).

$\qquad x = 2 \qquad\qquad x = 3$ • Solve each equation for *x*.

Check:

$$\begin{array}{c|c} (x - 2)(x - 3) = 0 \\ \hline (2 - 2)(2 - 3) & 0 \\ 0(-1) & 0 \\ 0 = 0 \end{array} \qquad \begin{array}{c|c} (x - 2)(x - 3) = 0 \\ \hline (3 - 2)(3 - 3) & 0 \\ (1)(0) & 0 \\ 0 = 0 \end{array}$$

• A true equation   • A true equation

The solutions are 2 and 3.

An equation that can be written in the form $ax^2 + bx + c = 0$, $a \neq 0$, is a **quadratic equation.** A quadratic equation is in **standard form** when the polynomial is in descending order and equal to zero. The quadratic equations at the right are in standard form.

$3x^2 + 2x + 1 = 0$
$a = 3, b = 2, c = 1$

$4x^2 - 3x + 2 = 0$
$a = 4, b = -3, c = 2$

Objective 10.5A

New Vocabulary
quadratic equation
standard form

New Rules
Principle of Zero Products

Instructor Note
As an application of factoring, quadratic equations are introduced here. They can be omitted at this time. There is a complete discussion of the topic, including this material, later in the text.

Concept Check
Determine whether each equation below is a quadratic equation.
1. $2x^2 - 8 = 0$ Yes
2. $2x - 8 = 0$ No
3. $x^2 = 6x + 8$ Yes

Concept Check
Write each equation below in standard form.
1. $x^2 + 4 = 4x$
    $x^2 - 4x + 4 = 0$
2. $x + x^2 = 6$
    $x^2 + x - 6 = 0$

Concept Check
Consider the equation $(x + 4)(x - 5) = 0$.
1. Using the Principle of Zero Products, write the two equations that result by letting each factor equal zero.
    $x + 4 = 0; x - 5 = 0$
2. Rewrite each equation in part 1 in the form *variable* = *constant*.
    $x = -4; x = 5$

In-Class Examples (Objective 10.5A)
Solve.
1. $3x(x + 4) = 0$ $0, -4$
2. $16x^2 - 4 = 0$ $-\dfrac{1}{2}, \dfrac{1}{2}$
3. $(x + 3)(x - 5) = 9$ $6, -4$

Concept Check

Tell whether or not the equation can be solved by using the Principle of Zero Products without first rewriting the equation.

1. $4x(6x + 7) = 0$ Yes
2. $0 = (4x - 5)(3x + 8)$ Yes
3. $2x(x - 5) - 5 = 0$ No
4. $(x - 7)(y + 3) = 0$ Yes
5. $0 = (2x - 3)x + 3$ No
6. $0 = (2x - 3)(x + 3)$ Yes

Optional Student Activity

Examine the following solution of the equation $x^2 - 5x + 6 = 12$.

$(x - 3)(x - 2) = 3 \cdot 4$

Factor each side.

$x - 3 = 3$ or $x - 2 = 4$
$x = 6$ $x = 6$

Check:

$6^2 - 5(6) + 6 = 12$
$12 = 12$

The solution is 6.
Do you agree with this solution? If not, show a correct solution.

No. The correct solutions are −1 and 6.

Study Tip

Always check the solution of an equation. Here is a check of the solution −2 for the equation at the right:

$$\begin{array}{c|c} 2x^2 + x = 6 \\ \hline 2(-2)^2 + (-2) & 6 \\ 2(4) + (-2) & 6 \\ 8 + (-2) & 6 \\ 6 = 6 \end{array}$$

HOW TO   Solve: $2x^2 + x = 6$

$2x^2 + x = 6$
$2x^2 + x - 6 = 0$ • Write the equation in standard form.
$(2x - 3)(x + 2) = 0$ • Factor.
$2x - 3 = 0$ $x + 2 = 0$ • Use the Principle of Zero Products.
$2x = 3$ $x = -2$ • Solve each equation for x.
$x = \dfrac{3}{2}$

Check: $\dfrac{3}{2}$ and -2 check as solutions.

The solutions are $\dfrac{3}{2}$ and -2.

Example 1

Solve: $x(x - 3) = 0$

Solution
$x(x - 3) = 0$

$x = 0$ $x - 3 = 0$ • Use the Principle
                    $x = 3$ of Zero Products.

The solutions are 0 and 3.

You Try It 1

Solve: $2x(x + 7) = 0$

Your solution
0, −7

Example 2

Solve: $2x^2 - 50 = 0$

Solution
$2x^2 - 50 = 0$
$2(x^2 - 25) = 0$ • Factor the GCF, 2.
$2(x + 5)(x - 5) = 0$ • Factor the difference of two squares.

$x + 5 = 0$ $x - 5 = 0$ • Use the Principle of
$x = -5$ $x = 5$ Zero Products.

The solutions are −5 and 5.

You Try It 2

Solve: $4x^2 - 9 = 0$

Your solution
$\dfrac{3}{2}, -\dfrac{3}{2}$

Example 3

Solve: $(x - 3)(x - 10) = -10$

Solution
$(x - 3)(x - 10) = -10$
$x^2 - 13x + 30 = -10$ • Multiply $(x - 3)(x - 10)$.
$x^2 - 13x + 40 = 0$ • Add 10 to each side of
$(x - 8)(x - 5) = 0$ the equation. The equation
                            is now in standard form.

$x - 8 = 0$ $x - 5 = 0$
$x = 8$ $x = 5$

The solutions are 8 and 5.

You Try It 3

Solve: $(x + 2)(x - 7) = 52$

Your solution
−6, 11

Solutions on p. S26

Objective B   **To solve application problems**   VIDEO & DVD   CD TUTOR WWW WEB SSM

Example 4

The sum of the squares of two consecutive positive even integers is equal to 100. Find the two integers.

Strategy

First positive even integer: n
Second positive even integer: $n + 2$

The sum of the square of the first positive even integer and the square of the second positive even integer is 100.

Solution

$$n^2 + (n + 2)^2 = 100$$
$$n^2 + n^2 + 4n + 4 = 100$$
$$2n^2 + 4n + 4 = 100$$
$$2n^2 + 4n - 96 = 0$$
$$2(n^2 + 2n - 48) = 0$$
$$2(n - 6)(n + 8) = 0$$

$$n - 6 = 0 \qquad n + 8 = 0$$ • **Principle of**
$$n = 6 \qquad\qquad n = -8$$ **Zero Products**

Because -8 is not a positive even integer, it is not a solution.

$$n = 6$$
$$n + 2 = 6 + 2 = 8$$

The two integers are 6 and 8.

You Try It 4

The sum of the squares of two consecutive positive integers is 61. Find the two integers.

Your strategy

Your solution
5, 6

Solution on p. S26

In-Class Examples (Objective 10.5B)

1. The sum of the squares of two consecutive positive integers is 145. Find the two integers.
8, 9

2. The sum of the squares of two consecutive negative odd integers is 10. Find the two integers.
−3, −1

3. A garden measures 12 ft by 16 ft. A uniform border around the garden increases the total area to 357 ft². What is the width of the border?
2.5 ft

Example 5

A stone is thrown into a well with an initial speed of 4 ft/s. The well is 420 ft deep. How many seconds later will the stone hit the bottom of the well? Use the equation $d = vt + 16t^2$, where d is the distance in feet that the stone travels in t seconds when its initial speed is v feet per second.

Strategy

To find the time for the stone to drop to the bottom of the well, replace the variables d and v by their given values and solve for t.

Solution

$$d = vt + 16t^2$$
$$420 = 4t + 16t^2$$
$$0 = -420 + 4t + 16t^2$$
$$0 = 16t^2 + 4t - 420$$
$$0 = 4(4t^2 + t - 105)$$
$$0 = 4(4t + 21)(t - 5)$$

$$\begin{array}{ll} 4t + 21 = 0 & t - 5 = 0 \\ \quad\quad 4t = -21 & \quad\quad t = 5 \\ \quad\quad\ t = -\dfrac{21}{4} & \end{array}$$ • **Principle of Zero Products**

Because the time cannot be a negative number, $-\dfrac{21}{4}$ is not a solution.

The stone will hit the bottom of the well 5 s later.

You Try It 5

The length of a rectangle is 4 in. longer than twice the width. The area of the rectangle is 96 in². Find the length and width of the rectangle.

Your strategy

Your solution
Length: 16 in.; width: 6 in.

Solution on p. S26

10.5 Exercises

Objective A   To solve equations by factoring

1. ✏️ In your own words, explain the Principle of Zero Products.

2. Fill in the blanks. If $(x + 5)(2x - 7) = 0$, then _____ $= 0$ or _____ $= 0$. $x + 5, 2x - 7$

For Exercises 3 to 60, solve.

3. $(y + 3)(y + 2) = 0$ 4. $(y - 3)(y - 5) = 0$ 5. $(z - 7)(z - 3) = 0$ 6. $(z + 8)(z - 9) = 0$
   $-3, -2$ $3, 5$ $7, 3$ $-8, 9$

7. $x(x - 5) = 0$ 8. $x(x + 2) = 0$ 9. $a(a - 9) = 0$ 10. $a(a + 12) = 0$
   $0, 5$ $0, -2$ $0, 9$ $0, -12$

11. $y(2y + 3) = 0$ 12. $t(4t - 7) = 0$ 13. $2a(3a - 2) = 0$ 14. $4b(2b + 5) = 0$
   $0, -\dfrac{3}{2}$ $0, \dfrac{7}{4}$ $0, \dfrac{2}{3}$ $0, -\dfrac{5}{2}$

15. $(b + 2)(b - 5) = 0$ 16. $(b - 8)(b + 3) = 0$ 17. $x^2 - 81 = 0$ 18. $x^2 - 121 = 0$
   $-2, 5$ $8, -3$ $9, -9$ $11, -11$

19. $4x^2 - 49 = 0$ 20. $16x^2 - 1 = 0$ 21. $9x^2 - 1 = 0$ 22. $16x^2 - 49 = 0$
   $\dfrac{7}{2}, -\dfrac{7}{2}$ $\dfrac{1}{4}, -\dfrac{1}{4}$ $\dfrac{1}{3}, -\dfrac{1}{3}$ $\dfrac{7}{4}, -\dfrac{7}{4}$

23. $x^2 + 6x + 8 = 0$ 24. $x^2 - 8x + 15 = 0$ 25. $z^2 + 5z - 14 = 0$ 26. $z^2 + z - 72 = 0$
   $-4, -2$ $3, 5$ $2, -7$ $8, -9$

27. $2a^2 - 9a - 5 = 0$ 28. $3a^2 + 14a + 8 = 0$ 29. $6z^2 + 5z + 1 = 0$ 30. $6y^2 - 19y + 15 = 0$
   $-\dfrac{1}{2}, 5$ $-\dfrac{2}{3}, -4$ $-\dfrac{1}{3}, -\dfrac{1}{2}$ $\dfrac{5}{3}, \dfrac{3}{2}$

31. $x^2 - 3x = 0$ 32. $a^2 - 5a = 0$ 33. $x^2 - 7x = 0$ 34. $2a^2 - 8a = 0$
   $0, 3$ $0, 5$ $0, 7$ $0, 4$

35. $a^2 + 5a = -4$ 36. $a^2 - 5a = 24$ 37. $y^2 - 5y = -6$ 38. $y^2 - 7y = 8$
   $-1, -4$ $-3, 8$ $2, 3$ $-1, 8$

39. $2t^2 + 7t = 4$ 40. $3t^2 + t = 10$ 41. $3t^2 - 13t = -4$ 42. $5t^2 - 16t = -12$
   $\dfrac{1}{2}, -4$ $\dfrac{5}{3}, -2$ $\dfrac{1}{3}, 4$ $\dfrac{6}{5}, 2$

43. $x(x - 12) = -27$ 44. $x(x - 11) = 12$ 45. $y(y - 7) = 18$ 46. $y(y + 8) = -15$
   $3, 9$ $12, -1$ $9, -2$ $-3, -5$

Section 10.5

Suggested Assignment
Exercises 1–59, every other odd
Exercises 61–83, odds
More challenging problems:
   Exercises 85–91, odds

Answers to Writing Exercises

1. The Principle of Zero Products states that if the product of two numbers equals zero, then one or both of the numbers is (are) zero.

Quick Quiz (Objective 10.5A)

Solve.

1. $3x(2x - 1) = 0$ $0, \dfrac{1}{2}$

2. $x^2 - 6x - 27 = 0$ $-3, 9$

3. $(x + 2)(x - 9) = -24$ $1, 6$

47. $p(p + 3) = -2$ **48.** $p(p - 1) = 20$ **49.** $y(y + 4) = 45$ **50.** $y(y - 8) = -15$
   $-1, -2$ $5, -4$ $5, -9$ $3, 5$

51. $x(x + 3) = 28$ **52.** $p(p - 14) = 15$ **53.** $(x + 8)(x - 3) = -30$ **54.** $(x + 4)(x - 1) = 14$
   $4, -7$ $15, -1$ $-2, -3$ $-6, 3$

55. $(z - 5)(z + 4) = 52$ **56.** $(z - 8)(z + 4) = -35$ **57.** $(z - 6)(z + 1) = -10$
   $-8, 9$ $1, 3$ $1, 4$

58. $(a + 3)(a + 4) = 72$ **59.** $(a - 4)(a + 7) = -18$ **60.** $(2x + 5)(x + 1) = -1$
   $-12, 5$ $-5, 2$ $-\dfrac{3}{2}, -2$

Objective B   To solve application problems

61. Number Sense The square of a positive number is six more than five times the positive number. Find the number.
   6

62. Number Sense The square of a negative number is fifteen more than twice the negative number. Find the number.
   -3

63. Number Sense The sum of two numbers is six. The sum of the squares of the two numbers is twenty. Find the two numbers.
   2, 4

64. Number Sense The sum of two numbers is eight. The sum of the squares of the two numbers is thirty-four. Find the two numbers.
   3, 5

65. Number Sense The sum of the squares of two consecutive positive integers is forty-one. Find the two integers.
   4, 5

66. Number Sense The sum of the squares of two consecutive negative even integers is one hundred. Find the two integers.
   $-8, -6$

67. Number Sense The sum of two numbers is ten. The product of the two numbers is twenty-one. Find the two numbers.
   3, 7

68. Number Sense The sum of two numbers is thirteen. The product of the two numbers is forty. Find the two numbers.
   5, 8

Quick Quiz (Objective 10.5B)

1. The distance s, in feet, that an object will fall (neglecting air resistance) in t seconds is given by $s = vt + 16t^2$, where v is the initial velocity of the object in feet per second. An object is thrown downward from the top of a building 96 ft high. The initial velocity is 16 ft/s, and air resistance is neglected. How many seconds later will the object hit the ground?   2 s

Sum of Natural Numbers The formula $S = \frac{n^2 + n}{2}$ gives the sum, S, of the first n natural numbers. Use this formula for Exercises 69 and 70.

69. How many consecutive natural numbers beginning with 1 will give a sum of 78?
12

70. How many consecutive natural numbers beginning with 1 will give a sum of 171?
18

Sports The formula $N = \frac{t^2 - t}{2}$ gives the number, N, of football games that must be scheduled in a league with t teams if each team is to play every other team once. Use this formula for Exercises 71 and 72.

71. How many teams are in a league that schedules 15 games in such a way that each team plays every other team once?
6 teams

72. How many teams are in a league that schedules 45 games in such a way that each team plays every other team once?
10 teams

Physics The distance s, in feet, that an object will fall (neglecting air resistance) in t seconds is given by $s = vt + 16t^2$, where v is the initial velocity of the object in feet per second. Use this formula for Exercises 73 and 74.

73. An object is released from the top of a building 192 ft high. The initial velocity is 16 ft/s, and air resistance is neglected. How many seconds later will the object hit the ground?
3 s

74. In October 2003, the world's tallest building, Taipei 101, was completed. The top of the spire is 1667 ft above ground. If an object is released from this building at a point 640 ft above the ground at an initial velocity of 48 ft/s, assuming no air resistance, how many seconds later will the object reach the ground?
5 s

Sports The height h, in feet, that an object will attain (neglecting air resistance) in t seconds is given by $h = vt - 16t^2$, where v is the initial velocity of the object in feet per second. Use this formula for Exercises 75 and 76.

75. A golf ball is thrown onto a cement surface and rebounds straight up. The initial velocity of the rebound is 60 ft/s. How many seconds later will the golf ball return to the ground?
3.75 s

76. A foul ball leaves a bat, hits home plate, and travels straight up with an initial velocity of 64 ft/s. How many seconds later will the ball be 64 ft above the ground?
2 s

77. Geometry The length of a rectangle is 5 in. more than twice its width. Its area is 75 in². Find the length and width of the rectangle.
Length: 15 in.; width: 5 in.

78. **Geometry**   The width of a rectangle is 5 ft less than the length. The area of the rectangle is 176 ft². Find the length and width of the rectangle.
Length: 16 ft; width: 11 ft

79. **Geometry**   The height of a triangle is 4 m more than twice the length of the base. The area of the triangle is 35 m². Find the height of the triangle.
14 m

80. **Geometry**   The length of each side of a square is extended 5 in. The area of the resulting square is 64 in². Find the length of a side of the original square.
3 in.

81. **Publishing**   The page of a book measures 6 in. by 9 in. A uniform border around the page leaves 28 in² for type. What are the dimensions of the type area?
4 in. by 7 in.

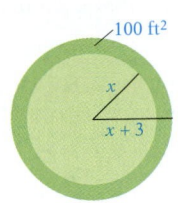

82. **Gardening**   A small garden measures 8 ft by 10 ft. A uniform border around the garden increases the total area to 143 ft². What is the width of the border?
1.5 ft

83. **Landscaping**   A landscape designer decides to increase the radius of a circular lawn by 3 ft. This increases the area of the lawn by 100 ft². Find the radius of the original circular lawn. Round to the nearest hundredth.
3.81 ft

100 ft²

x

$x + 3$

84.   **Geometry**   A circle has a radius of 10 in. Find the increase in area that occurs when the radius is increased by 2 in. Round to the nearest hundredth.
138.23 in²

APPLYING THE CONCEPTS

85.   Find $3n^2$ if $n(n + 5) = -4$.
3, 48

86.   Find $2n^2$ if $n(n + 3) = 4$.
2, 32

For Exercises 87 to 90, solve.

87. $2y(y + 4) = -5(y + 3)$ $-\dfrac{3}{2}, -5$

88. $(b + 5)^2 = 16$ $-9, -1$

89. $p^3 = 9p^2$ 0, 9

90. $(x + 3)(2x - 1) = (3 - x)(5 - 3x)$ 1, 18

91.   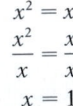 Explain the error made in solving the equation at the right. Solve the equation correctly.

$(x + 2)(x - 3) = 6$
$x + 2 = 6 \quad x - 3 = 6$
$\qquad x = 4 \qquad\quad x = 9$

92.   Explain the error made in solving the equation at the right. Solve the equation correctly.

$x^2 = x$
$\dfrac{x^2}{x} = \dfrac{x}{x}$
$x = 1$

Answers to Writing Exercises

91. The error in the solution
$(x + 2)(x - 3) = 6$
$x + 2 = 6 \quad x - 3 = 6$
$\quad x = 4 \qquad\quad x = 9$
occurs when the assumption is made that because
$(x + 2)(x - 3) = 6$,
$x + 2 = 6$ and/or
$x - 3 = 6$. In other words, it is an error to say that if the product of two numbers is 6, then at least one of the numbers must be 6. The correction solution is
$(x + 2)(x - 3) = 6$
$x^2 - x - 6 = 6$
$x^2 - x - 12 = 0$
$(x + 3)(x - 4) = 0$
$x = -3 \quad x = 4$

92. The error of accidentally dividing by zero takes place in the solution
$x^2 = x$
$\dfrac{x^2}{x} = \dfrac{x}{x}$
$x = 1$
Since one of the solutions of $x^2 = x$ is $x = 0$, an error occurs when the quotients in the equation $\dfrac{x^2}{x} = \dfrac{x}{x}$ are introduced. Here is the correct solution.
$x^2 = x$
$x^2 - x = 0$
$x(x - 1) = 0$
$x = 0 \quad x - 1 = 0$
$\qquad\qquad x = 1$

Focus on Problem Solving

Making a Table

There are six students using a gym. The wall on the gym has six lockers that are numbered 1, 2, 3, 4, 5, and 6. After a practice, the first student goes by and opens all the lockers. The second student shuts every second locker, the third student changes every third locker (opens a locker if it is shut, shuts a locker if it is open), the fourth student changes every fourth locker, the fifth student changes every fifth locker, and the sixth student changes every sixth locker. After the sixth student makes the changes, which lockers are open?

One method of solving this problem would be to create a table, as shown below.

Student / Locker	1	2	3	4	5	6
1	O	O	O	O	O	O
2	O	C	C	C	C	C
3	O	O	C	C	C	C
4	O	C	C	O	O	O
5	O	O	O	O	C	C
6	O	C	O	O	O	C

From this table, lockers 1 and 4 are open after the sixth student passes through.

Now extend this to more lockers and students. In each case, the nth student changes multiples of the nth locker. For instance, the 8th student would change the 8th, 16th, 24th, . . .

1. Suppose there were 10 lockers and 10 students. Which lockers would remain open?

2. Suppose there were 16 lockers and 16 students. Which lockers would remain open?

3. Suppose there were 25 lockers and 25 students. Which lockers would remain open?

4. Suppose there were 40 lockers and 40 students. Which lockers would remain open?

5. Suppose there were 50 lockers and 50 students. Which lockers would remain open?

6. Make a conjecture as to which lockers would be open if there were 100 lockers and 100 students.

7. Give a reason why your conjecture should be true. [*Hint:* Consider how many factors there are for the door numbers that remain open and those that remain closed. For instance, with 40 lockers and 40 students, locker 36 (which remains open) has factors 1, 2, 3, 4, 6, 9, 12, 18, 36—an odd number of factors. Locker 28, a closed locker, has factors 1, 2, 4, 7, 14, 28—an even number of factors.]

Answers to Focus on Problem Solving: Making a Table

1. 1, 4, 9
2. 1, 4, 9, 16
3. 1, 4, 9, 16, 25
4. 1, 4, 9, 16, 25, 36
5. 1, 4, 9, 16, 25, 36, 49
6. 1, 4, 9, 16, 25, 36, 49, 64, 81, 100.
7. If a locker number has an even number of factors, then it gets opened/closed as successive students whose position number is a factor of the locker number pass through the lockers. This leaves such a locker closed. The only lockers that remain open are those with an odd number of factors, namely, those whose numbers correspond to perfect squares.

Instructor Note

Question 7 of this Focus on Problem Solving is much more difficult than the other six questions.

Answers to Projects and Group Activities: Exploring Integers

1. One of two consecutive integers must be an even number. The product of an even number and an odd number is always even.

2. A product containing an even number is always an even number.

3. One added to an even number is an odd number.

4. Yes.

5. $(2n + 1)^2 - 1 = 4n(n + 1)$, which is divisible by 8.

6. No. One of two consecutive integers must be an even number and therefore not a prime number.

7. $n = 2$

8. Some possible answers are 3, 7, 127, 8191, and 131,071.

Projects and Group Activities

Exploring Integers *Number theory* is a branch of mathematics that focuses on integers and the relationships that exist among the integers. Some of the results from this field of study have important, practical applications for sending sensitive information such as credit card numbers over the Internet. In this project, you will be asked to discover some of those relationships.

1. If n is an integer, explain why the product $n(n + 1)$ is always an even number.

2. If n is an integer, explain why $2n$ is always an even integer.

3. If n is an integer, explain why $2n + 1$ is always an odd integer.

4. Select any odd integer greater than 1, square it, and then subtract 1. Try this for various odd integers greater than 1. Is the result always evenly divisible by 8?

5. Prove the assertion in Exercise 4. [*Suggestion:* From Exercise 2, an odd integer can be represented as $2n + 1$. Therefore, the assertion in Exercise 4 can be stated "$(2n + 1)^2 - 1$ is evenly divisible by 8." Expand this expression and explain why the result must be divisible by 8. You will need to use the result from Exercise 1.]

6. The integers 2 and 3 are consecutive prime numbers. Are there any other consecutive prime numbers? Why?

7. If n is a positive integer, for what values of n is $n^2 - 1$ a prime number?

8. A *Mersenne prime number* is a prime that can be written in the form $2^n - 1$, where n is also a prime number. For instance, $2^5 - 1 = 32 - 1 = 31$. Because 5 and 31 are prime numbers, 31 is a Mersenne prime number. On the other hand, $2^{11} - 1 = 2048 - 1 = 2047$. In this case, although 11 is a prime number, $2047 = 23 \cdot 89$, and so is not a prime number. Find two Mersenne prime numbers other than 31.

Chapter 10 Summary

Key Words	Examples
The *greatest common factor (GCF) of two or more monomials* is the product of the GCF of the coefficients and the common variable factors. [10.1A, p. 521]	The GCF of $8x^2y$ and $12xyz$ is $4xy$.
To *factor a polynomial* means to write the polynomial as a product of other polynomials. [10.1A, p. 521]	To factor $x^2 + 3x + 2$ means to write it as the product $(x + 1)(x + 2)$.

A factor that has two terms is called a *binomial factor*. [10.1B, p. 523]	$(x + 1)$ is a binomial factor of $3x(x + 1)$.
A polynomial that does not factor using only integers is *nonfactorable over the integers*. [10.2A, p. 528]	The trinomial $x^2 + x + 4$ is nonfactorable over the integers. There are no integers whose product is 4 and whose sum is 1.
A polynomial is *factored completely* if it is written as a product of factors that are nonfactorable over the integers. [10.2B, p. 529]	The polynomial $3y^3 + 9y^2 - 12y$ is factored completely as $3y(y + 4)(y - 1)$.
An equation that can be written in the form $ax^2 + bx + c = 0$, $a \neq 0$, is a *quadratic equation*. A quadratic equation is in *standard form* when the polynomial is written in descending order and equal to zero. [10.5A, p. 551]	The equation $2x^2 - 3x + 7 = 0$ is a quadratic equation in standard form.

Essential Rules and Procedures   **Examples**

Factoring by Grouping   [10.1B, p. 523]
A polynomial can be factored by grouping if its terms can be grouped and factored in such a way that a common binomial factor can be found.

$3a^2 - a - 15ab + 5b$
$= (3a^2 - a) - (15ab - 5b)$
$= a(3a - 1) - 5b(3a - 1)$
$= (3a - 1)(a - 5b)$

Factoring $x^2 + bx + c$: IMPORTANT RELATIONSHIPS
[10.2A, p. 527]

1. When the constant term of the trinomial is positive, the constant terms of the binomials have the same sign. They are both positive when the coefficient of the x term in the trinomial is positive. They are both negative when the coefficient of the x term in the trinomial is negative.

   $x^2 + 6x + 8 = (x + 4)(x + 2)$

   $x^2 - 6x + 5 = (x - 5)(x - 1)$

2. When the constant term of the trinomial is negative, the constant terms of the binomials have opposite signs.

   $x^2 - 4x - 21 = (x + 3)(x - 7)$

3. In the trinomial, the coefficient of x is the sum of the constant terms of the binomials.

   In the three examples above, note that $6 = 4 + 2$, $-6 = -5 + (-1)$, and $-4 = 3 + (-7)$.

4. In the trinomial, the constant term is the product of the constant terms of the binomials.

   In the three examples above, note that $8 = 4 \cdot 2$, $5 = -5(-1)$, and $-21 = 3(-7)$.

To factor $ax^2 + bx + c$ by grouping [10.3B, p. 537]

First find two factors of $a \cdot c$ whose sum is b. Then use factoring by grouping to write the factorization of the trinomial.

$3x^2 - 11x - 20$

$a \cdot c = 3(-20) = -60$

The product of 4 and -15 is -60.

The sum of 4 and -15 is -11.

$3x^2 + 4x - 15x - 20$
$$= (3x^2 + 4x) - (15x + 20)$$
$$= x(3x + 4) - 5(3x + 4)$$
$$= (3x + 4)(x - 5)$$

Factoring the Difference of Two Squares   [10.4A, p. 543]

The difference of two squares factors as the sum and difference of the same terms.

$a^2 - b^2 = (a + b)(a - b)$

$x^2 - 64 = (x + 8)(x - 8)$

$4x^2 - 81 = (2x)^2 - 9^2$
$$= (2x + 9)(2x - 9)$$

Factoring a Perfect-Square Trinomial   [10.4A, p. 544]

A perfect-square trinomial is the square of a binomial.

$a^2 + 2ab + b^2 = (a + b)^2$

$a^2 - 2ab + b^2 = (a - b)^2$

$x^2 + 14x + 49 = (x + 7)^2$

$x^2 - 10x + 25 = (x - 5)^2$

General Factoring Strategy [10.4B, p. 545]

1. Is there a common factor? Is so, factor out the common factor.

   $6x^2 - 8x = 2x(3x - 4)$

2. Is the polynomial the difference of two perfect squares? If so, factor.

   $9x^2 - 25 = (3x + 5)(3x - 5)$

3. Is the polynomial a perfect-square trinomial? If so, factor.

   $9x^2 + 6x + 1 = (3x + 1)^2$

4. Is the polynomial a trinomial that is the product of two binomials? If so, factor.

   $6x^2 + 5x - 6 = (3x - 2)(2x + 3)$

5. Does the polynomial contain four terms? If so, try factoring by grouping.

   $x^3 - 3x^2 + 2x - 6$
   $$= (x^3 - 3x^2) + (2x - 6)$$
   $$= x^2(x - 3) + 2(x - 3)$$
   $$= (x - 3)(x^2 + 2)$$

6. Is each binomial factor nonfactorable over the integers? If not, factor the binomial.

   $x^4 - 16 = (x^2 + 4)(x^2 - 4)$
   $$= (x^2 + 4)(x + 2)(x - 2)$$

Principle of Zero Products   [10.5A, p. 551]

If the product of two factors is zero, then at least one of the factors must be zero.

If $a \cdot b = 0$, then $a = 0$ or $b = 0$.

The Principle of Zero Products is used to solve a quadratic equation by factoring.

$$x^2 + x = 12$$
$$x^2 + x - 12 = 0$$
$$(x - 3)(x + 4) = 0$$

$x - 3 = 0 \qquad x + 4 = 0$

$\qquad x = 3 \qquad\qquad x = -4$

Chapter 10 Review Exercises

1. Factor: $b^2 - 13b + 30$
$(b - 3)(b - 10)$ [10.2A]

2. Factor: $4x(x - 3) - 5(3 - x)$
$(x - 3)(4x + 5)$ [10.1B]

3. Factor $2x^2 - 5x + 6$ by using trial factors.
Nonfactorable over the integers   [10.3A]

4. Factor: $5x^3 + 10x^2 + 35x$
$5x(x^2 + 2x + 7)$ [10.1A]

5. Factor: $14y^9 - 49y^6 + 7y^3$
$7y^3(2y^6 - 7y^3 + 1)$ [10.1A]

6. Factor: $y^2 + 5y - 36$
$(y - 4)(y + 9)$ [10.2A]

7. Factor $6x^2 - 29x + 28$ by using trial factors.
$(2x - 7)(3x - 4)$ [10.3A]

8. Factor: $12a^2b + 3ab^2$
$3ab(4a + b)$ [10.1A]

9. Factor: $a^6 - 100$
$(a^3 + 10)(a^3 - 10)$ [10.4A]

10. Factor: $n^4 - 2n^3 - 3n^2$
$n^2(n + 1)(n - 3)$ [10.2B]

11. Factor $12y^2 + 16y - 3$ by using trial factors.
$(6y - 1)(2y + 3)$ [10.3A]

12. Factor: $12b^3 - 58b^2 + 56b$
$2b(3b - 4)(2b - 7)$ [10.4B]

13. Factor: $9y^4 - 25z^2$
$(3y^2 + 5z)(3y^2 - 5z)$ [10.4A]

14. Factor: $c^2 + 8c + 12$
$(c + 6)(c + 2)$ [10.2A]

15. Factor $18a^2 - 3a - 10$ by grouping.
$(6a - 5)(3a + 2)$ [10.3B]

16. Solve: $4x^2 + 27x = 7$
$\frac{1}{4}, -7$ [10.5A]

17. Factor: $4x^3 - 20x^2 - 24x$
$4x(x - 6)(x + 1)$ [10.2B]

18. Factor: $3a^2 - 15a - 42$
$3(a + 2)(a - 7)$ [10.2B]

19. Factor $2a^2 - 19a - 60$ by grouping.
$(2a + 5)(a - 12)$ [10.3B]

20. Solve: $(x + 1)(x - 5) = 16$
$-3, 7$ [10.5A]

21. Factor: $21ax - 35bx - 10by + 6ay$
$(3a - 5b)(7x + 2y)$ [10.1B]

22. Factor: $a^2b^2 - 1$
$(ab + 1)(ab - 1)$ [10.4A]

23. Factor: $10x^2 + 25x + 4xy + 10y$
$(2x + 5)(5x + 2y)$ [10.1B]

24. Factor: $5x^2 - 5x - 30$
$5(x + 2)(x - 3)$ [10.2B]

25. Factor: $3x^2 + 36x + 108$
$3(x + 6)^2$ [10.4B]

26. Factor $3x^2 - 17x + 10$ by grouping.
$(3x - 2)(x - 5)$ [10.3B]

27. **Sports** The length of the field in field hockey is 20 yd less than twice the width of the field. The area of the field in field hockey is 6000 yd². Find the length and width of the field.
Length: 100 yd; width: 60 yd   [10.5B]

28. **Image Projection** The size, S, of an image from a slide projector depends on the distance, d, of the screen from the projector and is given by $S = d^2$. Find the distance between the projector and the screen when the size of the picture is 400 ft².
20 ft   [10.5B]

29. **Photography** A rectangular photograph has dimensions 15 in. by 12 in. A picture frame around the photograph increases the total area to 270 in². What is the width of the frame?
1.5 in. or $1\frac{1}{2}$ in.   [10.5B]

30. **Gardening** The length of each side of a square garden plot is extended 4 ft. The area of the resulting square is 576 ft². Find the length of a side of the original garden plot.
20 ft   [10.5B]

Chapter 10 Test

1. Factor: $ab + 6a - 3b - 18$
$(b + 6)(a - 3)$ [10.1B]

2. Factor: $2y^4 - 14y^3 - 16y^2$
$2y^2(y + 1)(y - 8)$ [10.2B]

3. Factor $8x^2 + 20x - 48$ by grouping.
$4(x + 4)(2x - 3)$ [10.3B]

4. Factor $6x^2 + 19x + 8$ by using trial factors.
$(2x + 1)(3x + 8)$ [10.3A]

5. Factor: $a^2 - 19a + 48$
$(a - 3)(a - 16)$ [10.2A]

6. Factor: $6x^3 - 8x^2 + 10x$
$2x(3x^2 - 4x + 5)$ [10.1A]

7. Factor: $x^2 + 2x - 15$
$(x + 5)(x - 3)$ [10.2A]

8. Solve: $4x^2 - 1 = 0$
$\dfrac{1}{2}, -\dfrac{1}{2}$ [10.5A]

9. Factor: $5x^2 - 45x - 15$
$5(x^2 - 9x - 3)$ [10.1A]

10. Factor: $p^2 + 12p + 36$
$(p + 6)^2$ [10.4A]

11. Solve: $x(x - 8) = -15$
$3, 5$ [10.5A]

12. Factor: $3x^2 + 12xy + 12y^2$
$3(x + 2y)^2$ [10.4B]

13. Factor: $b^2 - 16$
$(b + 4)(b - 4)$ [10.4A]

14. Factor $6x^2y^2 + 9xy^2 + 3y^2$ by grouping.
$3y^2(2x + 1)(x + 1)$ [10.3B]

15. Factor: $p^2 + 5p + 6$
$(p + 2)(p + 3)$ [10.2A]

16. Factor: $a(x - 2) + b(x - 2)$
$(x - 2)(a + b)$ [10.1B]

17. Factor: $x(p + 1) - (p + 1)$
$(p + 1)(x - 1)$ [10.1B]

18. Factor: $3a^2 - 75$
$3(a + 5)(a - 5)$ [10.4B]

19. Factor $2x^2 + 4x - 5$ by using trial factors.
Nonfactorable over the integers   [10.3A]

20. Factor: $x^2 - 9x - 36$
$(x + 3)(x - 12)$ [10.2A]

21. Factor: $4a^2 - 12ab + 9b^2$
$(2a - 3b)^2$ [10.4A]

22. Factor: $4x^2 - 49y^2$
$(2x + 7y)(2x - 7y)$ [10.4A]

23. Solve: $(2a - 3)(a + 7) = 0$
$\dfrac{3}{2}, -7$ [10.5A]

24. **Number Sense**   The sum of two numbers is ten. The sum of the squares of the two numbers is fifty-eight. Find the two numbers.
3, 7   [10.5B]

25. **Geometry**   The length of a rectangle is 3 cm longer than twice its width. The area of the rectangle is 90 cm². Find the length and width of the rectangle.
Length: 15 cm; width: 6 cm   [10.5B]

Cumulative Review Exercises

1. Subtract: $-2 - (-3) - 5 - (-11)$
   7   [3.2B]

2. Simplify: $(3 - 7)^2 \div (-2) - 3 \cdot (-4)$
   4   [3.5A]

3. Evaluate $-2a^2 \div (2b) - c$ when $a = -4$, $b = 2$, and $c = -1$.
   -7 [4.1A]

4. Simplify: $-\dfrac{3}{4}(-20x^2)$
   $15x^2$ [4.2B]

5. Simplify: $-2[4x - 2(3 - 2x) - 8x]$
   12   [4.2D]

6. Solve: $-\dfrac{5}{7}x = -\dfrac{10}{21}$
   $\dfrac{2}{3}$ [5.1C]

7. Solve: $3x - 2 = 12 - 5x$
   $\dfrac{7}{4}$ [5.3A]

8. Solve: $-2 + 4[3x - 2(4 - x) - 3] = 4x + 2$
   3   [5.3B]

9. 120% of what number is 54?
   45   [6.4A/6.4B]

10. Simplify: $(-3a^3b^2)^2$
   $9a^6b^4$ [9.2B]

11. Multiply: $(x + 2)(x^2 - 5x + 4)$
   $x^3 - 3x^2 - 6x + 8$ [9.3B]

12. Divide: $(8x^2 + 4x - 3) \div (2x - 3)$
   $4x + 8 + \dfrac{21}{2x - 3}$ [9.5B]

13. Simplify: $(x^{-4}y^3)^2$
   $\dfrac{y^6}{x^8}$ [9.4A]

14. Factor: $3a - 3b - ax + bx$
   $(a - b)(3 - x)$ [10.1B]

15. Factor: $15xy^2 - 20xy^4$
   $5xy^2(3 - 4y^2)$ [10.1A]

16. Factor: $x^2 - 5xy - 14y^2$
   $(x - 7y)(x + 2y)$ [10.2A]

17. Factor: $p^2 - 9p - 10$
   $(p - 10)(p + 1)$ [10.2A]

18. Factor: $18a^3 + 57a^2 + 30a$
   $3a(2a + 5)(3a + 2)$ [10.4B]

19. Factor: $36a^2 - 49b^2$
$(6a - 7b)(6a + 7b)$ [10.4A]

20. Factor: $4x^2 + 28xy + 49y^2$
$(2x + 7y)^2$ [10.4A]

21. Factor: $9x^2 + 15x - 14$
$(3x - 2)(3x + 7)$ [10.3A]

22. Factor: $18x^2 - 48xy + 32y^2$
$2(3x - 4y)^2$ [10.4B]

23. Factor: $3y(x - 3) - 2(x - 3)$
$(x - 3)(3y - 2)$ [10.1B]

24. Solve: $3x^2 + 19x - 14 = 0$
$\dfrac{2}{3}, -7$ [10.5A]

25. **Carpentry**   A board 10 ft long is cut into two pieces. Four times the length of the shorter piece is 2 ft less than three times the length of the longer piece. Find the length of each piece.
4 ft, 6 ft   [5.4B]

26. **Business**   A portable MP3 player that regularly sells for $165 is on sale for $99. Find the discount rate. Use the formula $S = R - rR$, where S is the sale price, R is the regular price, and r is the discount price. Write the answer as a percent.
40%   [5.2B/6.3B]

27. **Geometry**   Given that lines ℓ_1 and ℓ_2 are parallel, find the measures of angles a and b.
$\angle a = 72°$; $\angle b = 108°$ [7.1B]

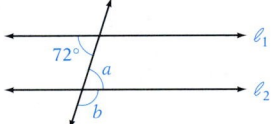

28. **Travel**   A family drove to a resort at an average speed of 42 mph and later returned over the same road at an average speed of 56 mph. Find the distance to the resort if the total driving time was 7 h.
168 mi   [5.5B]

29. **Consecutive Integers**   Find three consecutive even integers such that five times the middle integer is twelve more than twice the sum of the first and third integers.
10, 12, 14   [5.4A]

30. **Geometry**   The length of the base of a triangle is three times the height. The area of the triangle is 24 in². Find the length of the base of the triangle.
12 in.   [10.5B]

11

Rational Expressions

Deep-sea diving is not only a recreational sport enjoyed by tourists and adventure-seekers. Commercial deep-sea divers work on a variety of projects, including pipeline systems, offshore oilfield sites, and search-and-recovery missions. For any deep-sea diver, whether diving for recreation or for business, underwater survival depends on an oxygen tank. On land, the average adult at rest inhales and exhales about 1 cubic foot of air every 4 minutes. This air contains approximately 21% oxygen. The percent of oxygen needed by a person changes when he or she is underwater. **Exercise 82 on page 591** uses a variable expression that gives the recommended percent of oxygen for a diver as a function of the diver's depth.

Need help? For online student resources, such as section quizzes, visit this textbook's website at **math.college.hmco.com/students**.

OBJECTIVES

Section 11.1

A To simplify a rational expression
B To multiply rational expressions
C To divide rational expressions

Section 11.2

A To find the least common multiple (LCM) of two or more polynomials
B To express two fractions in terms of the LCM of their denominators

Section 11.3

A To add or subtract rational expressions with the same denominator
B To add or subtract rational expressions with different denominators

Section 11.4

A To simplify a complex fraction

Section 11.5

A To solve an equation containing fractions

Section 11.6

A To solve a literal equation for one of the variables

Section 11.7

A To solve work problems
B To use rational expressions to solve uniform motion problems

Do these exercises to prepare for Chapter 11.

1. Find the least common multiple (LCM) of 12 and 18.
    36   [2.1A]

2. Simplify: $\dfrac{9x^3y^4}{3x^2y^7}$

    $\dfrac{3x}{y^3}$ [9.4A]

3. Subtract: $\dfrac{3}{4} - \dfrac{8}{9}$

    $-\dfrac{5}{36}$ [3.4A]

4. Divide: $\left(-\dfrac{8}{11}\right) \div \dfrac{4}{5}$

    $-\dfrac{10}{11}$ [3.4B]

5. If a is a nonzero number, are the following two quantities equal: $\dfrac{0}{a}$ and $\dfrac{a}{0}$?

    No   [1.3C]

6. Solve: $\dfrac{2}{3}x - \dfrac{3}{4} = \dfrac{5}{6}$

    $\dfrac{19}{8}$ [5.2A]

7. Line l_1 is parallel to line l_2. Find the measure of angle a.
    130°   [7.1B]

    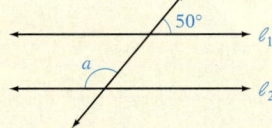

8. Factor: $x^2 - 4x - 12$
    $(x - 6)(x + 2)$ [10.2A]

9. Factor: $2x^2 - x - 3$
    $(2x - 3)(x + 1)$ [10.3A]

10. At 9:00 A.M., Anthony begins walking on a park trail at a rate of 9 m/min. Ten minutes later his sister Jean begins walking the same trail in pursuit of her brother at a rate of 12 m/min. At what time will Jean catch up to Anthony?
    9:40 A.M.   [5.5B]

GO FIGURE • • •

A mouse begins at corner A of a 12-foot-by-12-foot square maze traveling clockwise at a constant speed of 2 ft/s. Six seconds later, a second mouse starts from the same corner A and travels clockwise at a constant speed of 3 ft/s. How far apart are the two mice 18 s after the second mouse begins?
6 ft

11.1 Multiplication and Division of Rational Expressions

Objective A   To simplify a rational expression

A fraction in which the numerator and denominator are polynomials is called a **rational expression.** Examples of rational expressions are shown at the right.

$$\frac{5}{z}, \quad \frac{x^2 + 1}{2x - 1}, \quad \frac{y^2 + y - 1}{4y^2 + 1}$$

Care must be exercised with a rational expression to ensure that when the variables are replaced with numbers, the resulting denominator is not zero. Consider the rational expression at the right. The value of x cannot be 3 because the denominator would then be zero.

$$\frac{4x^2 - 9}{2x - 6}$$

$$\frac{4(3)^2 - 9}{2(3) - 6} = \frac{27}{0} \quad \text{Not a real number}$$

In the **simplest form of a rational expression,** the numerator and denominator have no common factors. The Multiplication Property of One is used to write a rational expression in simplest form.

HOW TO   Simplify: $\dfrac{x^2 - 4}{x^2 - 2x - 8}$

$$\frac{x^2 - 4}{x^2 - 2x - 8} = \frac{(x - 2)(x + 2)}{(x - 4)(x + 2)}$$ • **Factor the numerator and denominator.**

$$= \frac{x - 2}{x - 4} \cdot \boxed{\frac{x + 2}{x + 2}} = \frac{x - 2}{x - 4} \cdot 1$$

$$= \frac{x - 2}{x - 4}, x \neq -2, 4$$ • **The restriction $x \neq -2$ or 4 is necessary to prevent division by zero.**

This simplification is usually shown with slashes through the common factors:

$$\frac{x^2 - 4}{x^2 - 2x - 8} = \frac{(x - 2)\overset{1}{\cancel{(x + 2)}}}{(x - 4)\underset{1}{\cancel{(x + 2)}}}$$ • **Factor the numerator and denominator. Divide by the common factors.**

$$= \frac{x - 2}{x - 4}, x \neq -2, 4$$ • **The restriction $x \neq -2$ or 4 is necessary to prevent division by zero.**

In summary, to simplify a rational expression, factor the numerator and denominator. Then divide the numerator and denominator by the common factors.

HOW TO   Simplify: $\dfrac{10 + 3x - x^2}{x^2 - 4x - 5}$

$$\frac{10 + 3x - x^2}{x^2 - 4x - 5} = \frac{-(x^2 - 3x - 10)}{x^2 - 4x - 5}$$ • **Because the coefficient of x^2 in the numerator is −1, factor −1 from the numerator.**

$$= \frac{-\overset{1}{\cancel{(x - 5)}}(x + 2)}{\underset{1}{\cancel{(x - 5)}}(x + 1)}$$ • **Factor the numerator and denominator. Divide by the common factors.**

$$= -\frac{x + 2}{x + 1}, x \neq -1, 5$$

Objective 11.1A

New Vocabulary
rational expression
simplest form of a rational expression

Discuss the Concepts
Explain how to find the value of x that makes the denominator of the rational expression $\dfrac{5x}{3x + 6}$ equal to 0.

Instructor Note
Simplifying a rational expression is closely related to simplifying a rational number; the common factors are removed. Making this connection will help some students.

Optional Student Activity
Have students evaluate $\dfrac{x^2 - 4}{x^2 - 2x - 8}$ and $\dfrac{x - 2}{x - 4}$ for various values of x and determine that, except for at −2 and 4, the expressions are equal.

Concept Check
Explain why $x \neq 5$ in the example at the left, even though the rational expression $\dfrac{x + 2}{x + 1}$ is defined for $x = 5$.

In-Class Examples (Objective 11.1A)

Simplify.

1. $\dfrac{18x^5y^2}{12xy^3}$ $\dfrac{3x^4}{2y}$

2. $\dfrac{x^2 - 1}{x^2 + 4x - 5}$ $\dfrac{x + 1}{x + 5}$

3. $\dfrac{a^2 - 2a}{4 - 2a}$ $-\dfrac{a}{2}$

Concept Check

Simplify.

1. $\dfrac{(x+2)^3}{(x+2)^2}$ $x + 2$

2. $\dfrac{4x-3}{3-4x}$ -1

3. $\dfrac{4a^2(a-b)}{6a(b-a)}$ $-\dfrac{2a}{3}$

Concept Check

Write two different rational expressions that equal $\dfrac{3a}{4}$ when simplified.

Answers will vary. One example is $\dfrac{6a^2}{8a}$.

Instructor Note

It is important to emphasize that the numerator and denominator must be written in factored form before simplifying. This will help students avoid errors such as

$$\frac{x^2 - x}{x^2} = \frac{\cancel{x^2} - x}{\cancel{x^2}} = 1 - x$$

For the remaining examples, we will omit the restrictions on the variables that prevent division by zero and assume that the values of the variables are such that division by zero is not possible.

Example 1

Simplify: $\dfrac{4x^3y^4}{6x^4y}$

Solution

$\dfrac{4x^3y^4}{6x^4y} = \dfrac{2y^3}{3x}$ • Use rules of exponents.

You Try It 1

Simplify: $\dfrac{6x^5y}{12x^2y^3}$

Your solution

$\dfrac{x^3}{2y^2}$

Example 2

Simplify: $\dfrac{9 - x^2}{x^2 + x - 12}$

Solution

$$\dfrac{9 - x^2}{x^2 + x - 12} = \dfrac{\overset{-1}{\cancel{(3-x)}}(3+x)}{\underset{1}{\cancel{(x-3)}}(x+4)}$$ • $(3 - x)$
$= -1(x - 3)$

$$= -\dfrac{x+3}{x+4}$$

You Try It 2

Simplify: $\dfrac{x^2 + 2x - 24}{16 - x^2}$

Your solution

$-\dfrac{x+6}{x+4}$

Example 3

Simplify: $\dfrac{x^2 + 2x - 15}{x^2 - 7x + 12}$

Solution

$$\dfrac{x^2 + 2x - 15}{x^2 - 7x + 12} = \dfrac{(x+5)\overset{1}{\cancel{(x-3)}}}{\underset{1}{\cancel{(x-3)}}(x-4)} = \dfrac{x+5}{x-4}$$

You Try It 3

Simplify: $\dfrac{x^2 + 4x - 12}{x^2 - 3x + 2}$

Your solution

$\dfrac{x+6}{x-1}$

Solutions on p. S26

Objective 11.1B

Discuss the Concepts

Have students describe the steps involved in simplifying the product of two rational expressions.

Concept Check

Simplify.

1. $\dfrac{8x^2}{y} \cdot \dfrac{y}{12x}$ $\dfrac{2x}{3}$

2. $\dfrac{a-b}{a} \cdot \dfrac{3a}{b-a}$ -3

Objective B   **To multiply rational expressions**

The product of two fractions is a fraction whose numerator is the product of the numerators of the two fractions and whose denominator is the product of the denominators of the two fractions.

> **Multiplying Rational Expressions**
>
> Multiply the numerators. $\dfrac{a}{b} \cdot \dfrac{c}{d} = \dfrac{ac}{bd}$
> Multiply the denominators.

$\dfrac{2}{3} \cdot \dfrac{4}{5} = \dfrac{8}{15}$ $\dfrac{3x}{y} \cdot \dfrac{2}{z} = \dfrac{6x}{yz}$ $\dfrac{x+2}{x} \cdot \dfrac{3}{x-2} = \dfrac{3x+6}{x^2-2x}$

HOW TO Multiply: $\dfrac{x^2 + 3x}{x^2 - 3x - 4} \cdot \dfrac{x^2 - 5x + 4}{x^2 + 2x - 3}$

$$\dfrac{x^2 + 3x}{x^2 - 3x - 4} \cdot \dfrac{x^2 - 5x + 4}{x^2 + 2x - 3}$$

$$= \dfrac{x(x + 3)}{(x - 4)(x + 1)} \cdot \dfrac{(x - 4)(x - 1)}{(x + 3)(x - 1)}$$

• Factor the numerator and denominator of each fraction.

$$= \dfrac{x \overset{1}{(x + 3)} \overset{1}{(x - 4)} \overset{1}{(x - 1)}}{\underset{1}{(x - 4)}(x + 1) \underset{1}{(x + 3)} \underset{1}{(x - 1)}}$$

• Multiply. Then divide by the common factors.

$$= \dfrac{x}{x + 1}$$

• Write the answer in simplest form.

Instructor Note

Remind students that when they carry out the multiplication step, writing the product as a single fraction, they should leave the numerator and denominator in factored form. The simplified answer, too, may be left in factored form, as in Example 4. (Note, however, that in Objective 11.2B, students will need to multiply out the numerators when they apply the skill of multiplying rational expressions to the skill of writing fractions in terms of the LCM of their denominators.)

Example 4

Multiply: $\dfrac{10x^2 - 15x}{12x - 8} \cdot \dfrac{3x - 2}{20x - 25}$

Solution

$$\dfrac{10x^2 - 15x}{12x - 8} \cdot \dfrac{3x - 2}{20x - 25}$$

$$= \dfrac{5x(2x - 3)}{4(3x - 2)} \cdot \dfrac{(3x - 2)}{5(4x - 5)}$$

• Factor.

$$= \dfrac{\overset{1}{5}x(2x - 3)\overset{1}{(3x - 2)}}{4\underset{1}{(3x - 2)}\underset{1}{5}(4x - 5)}$$

• Divide by common factors.

$$= \dfrac{x(2x - 3)}{4(4x - 5)}$$

You Try It 4

Multiply: $\dfrac{12x^2 + 3x}{10x - 15} \cdot \dfrac{8x - 12}{9x + 18}$

Your solution

$\dfrac{4x(4x + 1)}{15(x + 2)}$

Optional Student Activity

1. Find two rational expressions whose product is 1.
   Answers will vary; for example, $\dfrac{x + 1}{5}$ and $\dfrac{5}{x + 1}$.

2. Find two rational expressions whose product is $\dfrac{12a}{5b^3}$.
   Answers will vary; for example, $\dfrac{3a}{5b}$ and $\dfrac{4}{b^2}$.

Example 5

Multiply: $\dfrac{x^2 + x - 6}{x^2 + 7x + 12} \cdot \dfrac{x^2 + 3x - 4}{4 - x^2}$

Solution

$$\dfrac{x^2 + x - 6}{x^2 + 7x + 12} \cdot \dfrac{x^2 + 3x - 4}{4 - x^2}$$

$$= \dfrac{(x + 3)(x - 2)}{(x + 3)(x + 4)} \cdot \dfrac{(x + 4)(x - 1)}{(2 - x)(2 + x)}$$

• Factor.

$$= \dfrac{\overset{1}{(x + 3)}\overset{-1}{(x - 2)}\overset{1}{(x + 4)}(x - 1)}{\underset{1}{(x + 3)}\underset{1}{(x + 4)}\underset{1}{(2 - x)}(2 + x)}$$

• Divide by common factors.

$$= -\dfrac{x - 1}{x + 2}$$

You Try It 5

Multiply: $\dfrac{x^2 + 2x - 15}{9 - x^2} \cdot \dfrac{x^2 - 3x - 18}{x^2 - 7x + 6}$

Your solution

$-\dfrac{x + 5}{x - 1}$

Solutions on pp. S26–S27

In-Class Examples (Objective 11.1B)

Multiply.

1. $\dfrac{28a^5b^7}{5x^4} \cdot \dfrac{15x^2}{14ab}$ $\dfrac{6a^4b^6}{x^2}$

2. $\dfrac{6x^2 - 10x}{3 - 3x} \cdot \dfrac{x^2 - 1}{12x - 20}$ $-\dfrac{x(x + 1)}{6}$

Objective 11.1C

New Vocabulary

reciprocal of a rational expression

Discuss the Concepts

Express each division problem as a multiplication problem.

1. $\dfrac{3}{a} \div \dfrac{b}{7}$ $\dfrac{3}{a} \cdot \dfrac{7}{b}$

2. $\dfrac{5x}{y} \div \dfrac{2x}{4y}$ $\dfrac{5x}{y} \cdot \dfrac{4y}{2x}$

3. $\dfrac{x-1}{x} \div \dfrac{x+3}{x}$

$\dfrac{x-1}{x} \cdot \dfrac{x}{x+3}$

Optional Student Activity

1. What should $\dfrac{5x^2}{3}$ be divided by to yield a quotient of 10?

   $\dfrac{x^2}{6}$

2. What should $\dfrac{2b}{a^3-a^2}$ be divided by to yield a quotient of $\dfrac{2}{a-1}$?

   $\dfrac{b}{a^2}$

Objective C   **To divide rational expressions**

The **reciprocal of a rational expression** is the rational expression with the numerator and denominator interchanged.

$$\text{Fraction} \left\{ \begin{array}{cc} \dfrac{a}{b} & \dfrac{b}{a} \\[6pt] x^2 = \dfrac{x^2}{1} & \dfrac{1}{x^2} \\[6pt] \dfrac{x+2}{x} & \dfrac{x}{x+2} \end{array} \right\} \text{Reciprocal}$$

> **Dividing Rational Expressions**
>
> Multiply the dividend by the reciprocal of the divisor. $\dfrac{a}{b} \div \dfrac{c}{d} = \dfrac{a}{b} \cdot \dfrac{d}{c} = \dfrac{ad}{bc}$

$$\dfrac{4}{x} \div \dfrac{y}{5} = \dfrac{4}{x} \cdot \dfrac{5}{y} = \dfrac{20}{xy}$$

$$\dfrac{x+4}{x} \div \dfrac{x-2}{4} = \dfrac{x+4}{x} \cdot \dfrac{4}{x-2} = \dfrac{4(x+4)}{x(x-2)}$$

The basis for the division rule is shown at the right.

$$\dfrac{a}{b} \div \dfrac{c}{d} = \dfrac{\frac{a}{b}}{\frac{c}{d}} = \dfrac{\frac{a}{b} \cdot \frac{d}{c}}{\frac{c}{d} \cdot \frac{d}{c}} = \dfrac{\frac{a}{b} \cdot \frac{d}{c}}{1} = \dfrac{a}{b} \cdot \dfrac{d}{c}$$

Example 6

Divide: $\dfrac{xy^2 - 3x^2y}{z^2} \div \dfrac{6x^2 - 2xy}{z^3}$

Solution

$\dfrac{xy^2 - 3x^2y}{z^2} \div \dfrac{6x^2 - 2xy}{z^3}$

$= \dfrac{xy^2 - 3x^2y}{z^2} \cdot \dfrac{z^3}{6x^2 - 2xy}$ • Multiply by the reciprocal.

$= \dfrac{xy(y - 3x) \cdot z^3}{z^2 \cdot 2x(3x - y)} = -\dfrac{yz}{2}$

You Try It 6

Divide: $\dfrac{a^2}{4bc^2 - 2b^2c} \div \dfrac{a}{6bc - 3b^2}$

Your solution

$\dfrac{3a}{2c}$

Example 7

Divide: $\dfrac{2x^2 + 5x + 2}{2x^2 + 3x - 2} \div \dfrac{3x^2 + 13x + 4}{2x^2 + 7x - 4}$

Solution

$\dfrac{2x^2 + 5x + 2}{2x^2 + 3x - 2} \div \dfrac{3x^2 + 13x + 4}{2x^2 + 7x - 4}$

$= \dfrac{2x^2 + 5x + 2}{2x^2 + 3x - 2} \cdot \dfrac{2x^2 + 7x - 4}{3x^2 + 13x + 4}$ • Multiply by the reciprocal.

$= \dfrac{(2x+1)(x+2) \cdot (2x-1)(x+4)}{(2x-1)(x+2) \cdot (3x+1)(x+4)} = \dfrac{2x+1}{3x+1}$

You Try It 7

Divide: $\dfrac{3x^2 + 26x + 16}{3x^2 - 7x - 6} \div \dfrac{2x^2 + 9x - 5}{x^2 + 2x - 15}$

Your solution

$\dfrac{x+8}{2x-1}$

Solutions on p. S27

In-Class Examples (Objective 11.1C)

Divide.

1. $\dfrac{12a^5b}{7xy^4} \div \dfrac{9ab^3}{35xy^2}$ $\dfrac{20a^4}{3b^2y^2}$

2. $\dfrac{4x - 8}{15 + 2x - x^2} \div \dfrac{x^2 - 4}{3x^2 - 15x}$ $-\dfrac{12x}{(x+3)(x+2)}$

11.1 Exercises

Objective A   To simplify a rational expression

1. Explain the procedure for writing a rational expression in simplest form.

2. Explain why the following simplification is incorrect.

$$\frac{x+3}{x} = \frac{\overset{1}{\cancel{x}}+3}{\underset{1}{\cancel{x}}} = 4$$

For Exercises 3 to 29, simplify.

3. $\dfrac{9x^3}{12x^4}$

$\dfrac{3}{4x}$

4. $\dfrac{16x^2y}{24xy^3}$

$\dfrac{2x}{3y^2}$

5. $\dfrac{(x+3)^2}{(x+3)^3}$

$\dfrac{1}{x+3}$

6. $\dfrac{(2x-1)^5}{(2x-1)^4}$

$2x-1$

7. $\dfrac{3n-4}{4-3n}$

-1

8. $\dfrac{5-2x}{2x-5}$

-1

9. $\dfrac{6y(y+2)}{9y^2(y+2)}$

$\dfrac{2}{3y}$

10. $\dfrac{12x^2(3-x)}{18x(3-x)}$

$\dfrac{2x}{3}$

11. $\dfrac{6x(x-5)}{8x^2(5-x)}$

$-\dfrac{3}{4x}$

12. $\dfrac{14x^3(7-3x)}{21x(3x-7)}$

$-\dfrac{2x^2}{3}$

13. $\dfrac{a^2+4a}{ab+4b}$

$\dfrac{a}{b}$

14. $\dfrac{x^2-3x}{2x-6}$

$\dfrac{x}{2}$

15. $\dfrac{4-6x}{3x^2-2x}$

$-\dfrac{2}{x}$

16. $\dfrac{5xy-3y}{9-15x}$

$-\dfrac{y}{3}$

17. $\dfrac{y^2-3y+2}{y^2-4y+3}$

$\dfrac{y-2}{y-3}$

18. $\dfrac{x^2+5x+6}{x^2+8x+15}$

$\dfrac{x+2}{x+5}$

19. $\dfrac{x^2+3x-10}{x^2+2x-8}$

$\dfrac{x+5}{x+4}$

20. $\dfrac{a^2+7a-8}{a^2+6a-7}$

$\dfrac{a+8}{a+7}$

21. $\dfrac{x^2+x-12}{x^2-6x+9}$

$\dfrac{x+4}{x-3}$

22. $\dfrac{x^2+8x+16}{x^2-2x-24}$

$\dfrac{x+4}{x-6}$

23. $\dfrac{x^2-3x-10}{25-x^2}$

$-\dfrac{x+2}{x+5}$

24. $\dfrac{4-y^2}{y^2-3y-10}$

$\dfrac{2-y}{y-5}$

25. $\dfrac{2x^3+2x^2-4x}{x^3+2x^2-3x}$

$\dfrac{2(x+2)}{x+3}$

26. $\dfrac{3x^3-12x}{6x^3-24x^2+24x}$

$\dfrac{x+2}{2(x-2)}$

27. $\dfrac{6x^2-7x+2}{6x^2+5x-6}$

$\dfrac{2x-1}{2x+3}$

28. $\dfrac{2n^2-9n+4}{2n^2-5n-12}$

$\dfrac{2n-1}{2n+3}$

29. $\dfrac{x^2+3x-28}{24-2x-x^2}$

$-\dfrac{x+7}{x+6}$

Section 11.1

Suggested Assignment

Exercises 1–77, odd
More challenging problems:
 Exercises 78–80, 83, 84

Answers to Writing Exercises

1. To write a rational expression in simplest form, factor the numerator and denominator. Then divide the numerator and denominator by their common factors.

2. The simplification is incorrect because the numerator and denominator were divided by a term rather than a factor.

Quick Quiz (Objective 11.1A)

Simplify.

1. $\dfrac{8ab-2a}{6-24b}$ $-\dfrac{a}{3}$

2. $\dfrac{x^2-x-12}{x^2+9x+18}$ $\dfrac{x-4}{x+6}$

3. $\dfrac{25-n^2}{n^2+n-30}$ $-\dfrac{n+5}{n+6}$

Objective B To multiply rational expressions

For Exercises 30 to 55, multiply.

30. $\dfrac{8x^2}{9y^3} \cdot \dfrac{3y^2}{4x^3}$

$\dfrac{2}{3xy}$

31. $\dfrac{14a^2b^3}{15x^5y^2} \cdot \dfrac{25x^3y}{16ab}$

$\dfrac{35ab^2}{24x^2y}$

32. $\dfrac{12x^3y^4}{7a^2b^3} \cdot \dfrac{14a^3b^4}{9x^2y^2}$

$\dfrac{8xy^2ab}{3}$

33. $\dfrac{18a^4b^2}{25x^2y^3} \cdot \dfrac{50x^5y^6}{27a^6b^2}$

$\dfrac{4x^3y^3}{3a^2}$

34. $\dfrac{3x-6}{5x-20} \cdot \dfrac{10x-40}{27x-54}$

$\dfrac{2}{9}$

35. $\dfrac{8x-12}{14x+7} \cdot \dfrac{42x+21}{32x-48}$

$\dfrac{3}{4}$

36. $\dfrac{3x^2+2x}{2xy-3y} \cdot \dfrac{2xy^3-3y^3}{3x^3+2x^2}$

$\dfrac{y^2}{x}$

37. $\dfrac{4a^2x-3a^2}{2by+5b} \cdot \dfrac{2b^3y+5b^3}{4ax-3a}$

ab^2

38. $\dfrac{x^2+5x+4}{x^3y^2} \cdot \dfrac{x^2y^3}{x^2+2x+1}$

$\dfrac{y(x+4)}{x(x+1)}$

39. $\dfrac{x^2+x-2}{xy^2} \cdot \dfrac{x^3y}{x^2+5x+6}$

$\dfrac{x^2(x-1)}{y(x+3)}$

40. $\dfrac{x^4y^2}{x^2+3x-28} \cdot \dfrac{x^2-49}{xy^4}$

$\dfrac{x^3(x-7)}{y^2(x-4)}$

41. $\dfrac{x^5y^3}{x^2+13x+30} \cdot \dfrac{x^2+2x-3}{x^7y^2}$

$\dfrac{y(x-1)}{x^2(x+10)}$

42. $\dfrac{2x^2-5x}{2xy+y} \cdot \dfrac{2xy^2+y^2}{5x^2-2x^3}$

$-\dfrac{y}{x}$

43. $\dfrac{3a^3+4a^2}{5ab-3b} \cdot \dfrac{3b^3-5ab^3}{3a^2+4a}$

$-ab^2$

44. $\dfrac{x^2-2x-24}{x^2-5x-6} \cdot \dfrac{x^2+5x+6}{x^2+6x+8}$

$\dfrac{x+3}{x+1}$

45. $\dfrac{x^2-8x+7}{x^2+3x-4} \cdot \dfrac{x^2+3x-10}{x^2-9x+14}$

$\dfrac{x+5}{x+4}$

46. $\dfrac{x^2+2x-35}{x^2+4x-21} \cdot \dfrac{x^2+3x-18}{x^2+9x+18}$

$\dfrac{x-5}{x+3}$

47. $\dfrac{y^2+y-20}{y^2+2y-15} \cdot \dfrac{y^2+4y-21}{y^2+3y-28}$

1

Quick Quiz (Objective 11.1B)

Multiply.

1. $\dfrac{x^2-5x+4}{2x^2+5x-3} \cdot \dfrac{x^2-9}{x^2-7x+12}$ $\dfrac{x-1}{2x-1}$

2. $\dfrac{10x^2-50x}{12x+24} \cdot \dfrac{2x+4}{5x-x^2}$ $-\dfrac{5}{3}$

48. $\dfrac{x^2 - 3x - 4}{x^2 + 6x + 5} \cdot \dfrac{x^2 + 5x + 6}{8 + 2x - x^2}$

$-\dfrac{x + 3}{x + 5}$

49. $\dfrac{25 - n^2}{n^2 - 2n - 35} \cdot \dfrac{n^2 - 8n - 20}{n^2 - 3n - 10}$

$-\dfrac{n - 10}{n - 7}$

50. $\dfrac{12x^2 - 6x}{x^2 + 6x + 5} \cdot \dfrac{2x^4 + 10x^3}{4x^2 - 1}$

$\dfrac{12x^4}{(x + 1)(2x + 1)}$

51. $\dfrac{8x^3 + 4x^2}{x^2 - 3x + 2} \cdot \dfrac{x^2 - 4}{16x^2 + 8x}$

$\dfrac{x(x + 2)}{2(x - 1)}$

52. $\dfrac{16 + 6x - x^2}{x^2 - 10x - 24} \cdot \dfrac{x^2 - 6x - 27}{x^2 - 17x + 72}$

$-\dfrac{x + 3}{x - 12}$

53. $\dfrac{x^2 - 11x + 28}{x^2 - 13x + 42} \cdot \dfrac{x^2 + 7x + 10}{20 - x - x^2}$

$-\dfrac{x + 2}{x - 6}$

54. $\dfrac{2x^2 + 5x + 2}{2x^2 + 7x + 3} \cdot \dfrac{x^2 - 7x - 30}{x^2 - 6x - 40}$

$\dfrac{x + 2}{x + 4}$

55. $\dfrac{x^2 - 4x - 32}{x^2 - 8x - 48} \cdot \dfrac{3x^2 + 17x + 10}{3x^2 - 22x - 16}$

$\dfrac{x + 5}{x - 12}$

Objective C   **To divide rational expressions**

56. ✏️ What is the reciprocal of a rational expression?

57. ✏️ Explain how to divide rational expressions.

For Exercises 58 to 77, divide.

58. $\dfrac{4x^2y^3}{15a^2b^3} \div \dfrac{6xy}{5a^3b^5}$

$\dfrac{2xy^2ab^2}{9}$

59. $\dfrac{9x^3y^4}{16a^4b^2} \div \dfrac{45x^4y^2}{14a^7b}$

$\dfrac{7a^3y^2}{40bx}$

60. $\dfrac{6x - 12}{8x + 32} \div \dfrac{18x - 36}{10x + 40}$

$\dfrac{5}{12}$

61. $\dfrac{28x + 14}{45x - 30} \div \dfrac{14x + 7}{30x - 20}$

$\dfrac{4}{3}$

62. $\dfrac{6x^3 + 7x^2}{12x - 3} \div \dfrac{6x^2 + 7x}{36x - 9}$

$3x$

63. $\dfrac{5a^2y + 3a^2}{2x^3 + 5x^2} \div \dfrac{10ay + 6a}{6x^3 + 15x^2}$

$\dfrac{3a}{2}$

64. $\dfrac{x^2 + 4x + 3}{x^2y} \div \dfrac{x^2 + 2x + 1}{xy^2}$

$\dfrac{y(x + 3)}{x(x + 1)}$

65. $\dfrac{x^3y^2}{x^2 - 3x - 10} \div \dfrac{xy^4}{x^2 - x - 20}$

$\dfrac{x^2(x + 4)}{y^2(x + 2)}$

66. $\dfrac{x^2 - 49}{x^4y^3} \div \dfrac{x^2 - 14x + 49}{x^4y^3}$

$\dfrac{x + 7}{x - 7}$

67. $\dfrac{x^2y^5}{x^2 - 11x + 30} \div \dfrac{xy^6}{x^2 - 7x + 10}$

$\dfrac{x(x - 2)}{y(x - 6)}$

Answers to Writing Exercises

56. The reciprocal of a rational expression is the expression with the numerator and denominator interchanged.

57. To divide rational expressions, multiply the dividend by the reciprocal of the divisor.

Quick Quiz (Objective 11.1C)

Divide.

1. $\dfrac{a^3b}{a^2 - 5a - 14} \div \dfrac{ab^6}{a^2 - 3a - 28}$ $\dfrac{a^2(a + 4)}{b^5(a + 2)}$

2. $\dfrac{10x + 10}{3x - 6} \div \dfrac{2x + 2}{3xy - 6y}$ $5y$

68. $\dfrac{4ax - 8a}{c^2} \div \dfrac{2y - xy}{c^3}$

$-\dfrac{4ac}{y}$

69. $\dfrac{3x^2y - 9xy}{a^2b} \div \dfrac{3x^2 - x^3}{ab^2}$

$-\dfrac{3by}{ax}$

70. $\dfrac{x^2 - 5x + 6}{x^2 - 9x + 18} \div \dfrac{x^2 - 6x + 8}{x^2 - 9x + 20}$

$\dfrac{x - 5}{x - 6}$

71. $\dfrac{x^2 + 3x - 40}{x^2 + 2x - 35} \div \dfrac{x^2 + 2x - 48}{x^2 + 3x - 18}$

$\dfrac{(x - 3)(x + 6)}{(x + 7)(x - 6)}$

72. $\dfrac{x^2 + 2x - 15}{x^2 - 4x - 45} \div \dfrac{x^2 + x - 12}{x^2 - 5x - 36}$

1

73. $\dfrac{y^2 - y - 56}{y^2 + 8y + 7} \div \dfrac{y^2 - 13y + 40}{y^2 - 4y - 5}$

1

74. $\dfrac{8 + 2x - x^2}{x^2 + 7x + 10} \div \dfrac{x^2 - 11x + 28}{x^2 - x - 42}$

$-\dfrac{x + 6}{x + 5}$

75. $\dfrac{x^2 - x - 2}{x^2 - 7x + 10} \div \dfrac{x^2 - 3x - 4}{40 - 3x - x^2}$

$\dfrac{x + 8}{x - 4}$

76. $\dfrac{2x^2 - 3x - 20}{2x^2 - 7x - 30} \div \dfrac{2x^2 - 5x - 12}{4x^2 + 12x + 9}$

$\dfrac{2x + 3}{x - 6}$

77. $\dfrac{6n^2 + 13n + 6}{4n^2 - 9} \div \dfrac{6n^2 + n - 2}{4n^2 - 1}$

$\dfrac{2n + 1}{2n - 3}$

Answers to Writing Exercises

78. It is not possible to choose a value of x such that $\dfrac{9}{x^2 + 1}$ is greater than 10 because $x^2 + 1$ is greater than or equal to 1 for all values of x.

79. Choosing a value of y very close to 3 makes $y - 3$ very close to 0. Dividing 1 by a number close to 0 produces a very large number. For instance, when

$y = 3.00000001$, $\dfrac{1}{y - 3}$ is greater than 10,000,000.

APPLYING THE CONCEPTS

78. Given the expression $\dfrac{9}{x^2 + 1}$, choose some values of x and evaluate the expression at those values. Is it possible to choose a value of x for which the value of the expression is greater than 10? If so, what is that value of x? If not, explain why it is not possible.

79. Given the expression $\dfrac{1}{y - 3}$, choose some values of y and evaluate the expression at those values. Is it possible to choose a value of y for which the value of the expression is greater than 10,000,000? If so, what is that value of y? If not, explain why it is not possible.

Geometry   For Exercises 80 and 81, write in simplest form the ratio of the shaded area of the figure to the total area of the figure.

80. 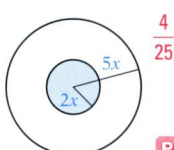 $\dfrac{4}{25}$

81. 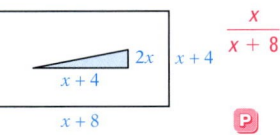 $\dfrac{x}{x + 8}$

For Exercises 82 to 84, complete the simplification.

82. $\dfrac{8x}{9y} \div \underline{\ ?\ } = \dfrac{10y}{3}$

$\dfrac{4x}{15y^2}$

83. $\dfrac{n}{n + 3} \div \underline{\ ?\ } = \dfrac{n}{n - 2}$

$\dfrac{n - 2}{n + 3}$

84. $\underline{\ ?\ } \div \dfrac{n - 1}{4n^3} = 2n^2(n + 1)$

$\dfrac{n^2 - 1}{2n}$

11.2 Expressing Fractions in Terms of the Least Common Multiple (LCM)

Objective A   To find the least common multiple (LCM) of two or more polynomials

Study Tip

The paragraph at the right begins with the word "Recall." This signals that the content refers to material presented earlier in the text. The ideas presented here will be more meaningful if you return to the introduction of least common multiple in Objective 2.1A and review the concepts presented there.

Recall that the least common multiple (LCM) of two or more numbers is the smallest number that contains the prime factorization of each number.

The LCM of 12 and 18 is 36 because 36 contains the prime factors of 12 and the prime factors of 18.

$$12 = 2 \cdot 2 \cdot 3$$
$$18 = 2 \cdot 3 \cdot 3$$

$$\text{LCM} = 36 = \overbrace{2 \cdot \underbrace{2 \cdot 3}_{\text{Factors of 18}} \cdot 3}^{\text{Factors of 12}}$$

The **least common multiple (LCM) of two or more polynomials** is the polynomial of least degree that contains all the factors of each polynomial.

To find the LCM of two or more polynomials, first factor each polynomial completely. The LCM is the product of each factor the greatest number of times it occurs in any one factorization.

TAKE NOTE

The LCM must contain the factors of each polynomial. As shown with the braces at the right, the LCM contains the factors of $4x^2 + 4x$ and the factors of $x^2 + 2x + 1$.

HOW TO   Find the LCM of $4x^2 + 4x$ and $x^2 + 2x + 1$.

The LCM of the polynomials is the product of the LCM of the numerical coefficients and each variable factor the greatest number of times it occurs in any one factorization.

$$4x^2 + 4x = 4x(x + 1) = 2 \cdot 2 \cdot x(x + 1)$$
$$x^2 + 2x + 1 = (x + 1)(x + 1)$$

$$\text{LCM} = \overbrace{2 \cdot 2 \cdot x}^{\text{Factors of } 4x^2 + 4x}(x + 1)\underbrace{(x + 1)}_{\text{Factors of } x^2 + 2x + 1} = 4x(x + 1)(x + 1)$$

Example 1

Find the LCM of $4x^2y$ and $6xy^2$.

Solution
$4x^2y = 2 \cdot 2 \cdot x \cdot x \cdot y$
$6xy^2 = 2 \cdot 3 \cdot x \cdot y \cdot y$
$\text{LCM} = 2 \cdot 2 \cdot 3 \cdot x \cdot x \cdot y \cdot y = 12x^2y^2$

You Try It 1

Find the LCM of $8uv^2$ and $12uw$.

Your solution
$24uv^2w$

Example 2

Find the LCM of $x^2 - x - 6$ and $9 - x^2$.

Solution
$x^2 - x - 6 = (x - 3)(x + 2)$
$9 - x^2 = -(x^2 - 9) = -(x + 3)(x - 3)$
$\text{LCM} = (x - 3)(x + 2)(x + 3)$

You Try It 2

Find the LCM of $m^2 - 6m + 9$ and $m^2 - 2m - 3$.

Your solution
$(m - 3)(m - 3)(m + 1)$

Solutions on p. S27

In-Class Examples (Objective 11.2A)
Find the LCM of the polynomials.
1. $24a^3b^2$ and $9ab^5$ $72a^3b^5$
2. $2x^2 - 4x$ and $x^2 - 4$ $2x(x - 2)(x + 2)$
3. $(3x + 1)^2$ and $(x - 3)(3x + 1)$ $(3x + 1)^2(x - 3)$
4. $x^2 - 3x - 40$ and $8 + 7x - x^2$
   $(x - 8)(x + 5)(x + 1)$

Objective 11.2B

Discuss the Concepts

Refer to the example at the right.

1. Explain why the LCM of the denominators is $12x^2(x-2)$.

2. Explain how to determine that the first fraction must be multiplied by $\frac{3(x-2)}{3(x-2)}$.

3. Explain how to determine that the second fraction must be multiplied by $\frac{2x}{2x}$.

Instructor Note

In Objective 11.1B, answers were left in factored form. Point out that now the numerators should be multiplied out (this is in preparation for actually adding fractions, as in Objective 11.3B).

Concept Check

Use the fractions $\frac{5}{a^2(a-4)}$ and $\frac{3}{a(a^2-16)}$.

1. Find the LCM of the denominators.
   $a^2(a-4)(a+4)$

2. What must the first fraction be multiplied by to change its denominator into the LCM?
   $\frac{a+4}{a+4}$

3. What must the second fraction be multiplied by to change its denominator into the LCM?
   $\frac{a}{a}$

Optional Student Activity

Write these three fractions in terms of the LCM of their denominators: $\frac{x-3}{x^2-1}$, $\frac{5x}{2-x-x^2}$, and $\frac{2x+1}{x^2+3x+2}$.

Objective B To express two fractions in terms of the LCM of their denominators

When adding and subtracting fractions, it is frequently necessary to express two or more fractions in terms of a common denominator. This common denominator is the LCM of the denominators of the fractions.

HOW TO Write the fractions $\frac{x+1}{4x^2}$ and $\frac{x-3}{6x^2-12x}$ in terms of the LCM of their denominators.

Find the LCM of the denominators.

The LCM is $12x^2(x-2)$.

For each fraction, multiply the numerator and the denominator by the factors whose product with the denominator is the LCM.

$$\frac{x+1}{4x^2} = \frac{x+1}{4x^2} \cdot \frac{3(x-2)}{3(x-2)} = \frac{3x^2-3x-6}{12x^2(x-2)}$$

$$\frac{x-3}{6x^2-12x} = \frac{x-3}{6x(x-2)} \cdot \frac{2x}{2x} = \frac{2x^2-6x}{12x^2(x-2)}$$ LCM

Example 3

Write the fractions $\frac{x+2}{3x^2}$ and $\frac{x-1}{8xy}$ in terms of the LCM of their denominators.

Solution

The LCM is $24x^2y$.

$$\frac{x+2}{3x^2} = \frac{x+2}{3x^2} \cdot \frac{8y}{8y} = \frac{8xy+16y}{24x^2y}$$

$$\frac{x-1}{8xy} = \frac{x-1}{8xy} \cdot \frac{3x}{3x} = \frac{3x^2-3x}{24x^2y}$$

You Try It 3

Write the fractions $\frac{x-3}{4xy^2}$ and $\frac{2x+1}{9y^2z}$ in terms of the LCM of their denominators.

Your solution

$$\frac{9xz-27z}{36xy^2z}, \frac{8x^2+4x}{36xy^2z}$$

Example 4

Write the fractions $\frac{2x-1}{2x-x^2}$ and $\frac{x}{x^2+x-6}$ in terms of the LCM of their denominators.

Solution

$$\frac{2x-1}{2x-x^2} = \frac{2x-1}{-(x^2-2x)} = -\frac{2x-1}{x^2-2x}$$

The LCM is $x(x-2)(x+3)$.

$$\frac{2x-1}{2x-x^2} = -\frac{2x-1}{x(x-2)} \cdot \frac{x+3}{x+3} = -\frac{2x^2+5x-3}{x(x-2)(x+3)}$$

$$\frac{x}{x^2+x-6} = \frac{x}{(x-2)(x+3)} \cdot \frac{x}{x} = \frac{x^2}{x(x-2)(x+3)}$$

You Try It 4

Write the fractions $\frac{x+4}{x^2-3x-10}$ and $\frac{2x}{25-x^2}$ in terms of the LCM of their denominators.

Your solution

$$\frac{x^2+9x+20}{(x+2)(x-5)(x+5)}, -\frac{2x^2+4x}{(x+2)(x-5)(x+5)}$$

Solutions on p. S27

In-Class Examples (Objective 11.2B)

Write the fractions in terms of the LCM of their denominators.

1. $\frac{8}{3x^2y}$ and $\frac{x}{6xy^3}$ $\frac{16y^2}{6x^2y^3}$ and $\frac{x^2}{6x^2y^3}$

2. $\frac{a^3}{a(a+1)^2}$ and $\frac{2}{a^2(a+1)}$ $\frac{a^4}{a^2(a+1)^2}$ and $\frac{2a+2}{a^2(a+1)^2}$

3. $\frac{x+2}{x^2-3x-4}$ and $\frac{1}{16-x^2}$

   $\frac{x^2+6x+8}{(x-4)(x+1)(x+4)}$ and $-\frac{x+1}{(x-4)(x+1)(x+4)}$

11.2 Exercises

Objective A   **To find the least common multiple (LCM) of two or more polynomials**

Suggested Assignment
Exercises 1–53, odds
More challenging problems:
   Exercises 54, 57, 60

For Exercises 1 to 33, find the LCM of the polynomials.

1. $8x^3y$
$12xy^2$
$24x^3y^2$

2. $6ab^2$
$18ab^3$
$18ab^3$

3. $10x^4y^2$
$15x^3y$
$30x^4y^2$

4. $12a^2b$
$18ab^3$
$36a^2b^3$

5. $8x^2$
$4x^2 + 8x$
$8x^2(x + 2)$

6. $6y^2$
$4y + 12$
$12y^2(y + 3)$

7. $2x^2y$
$3x^2 + 12x$
$6x^2y(x + 4)$

8. $4xy^2$
$6xy^2 + 12y^2$
$12xy^2(x + 2)$

9. $9x(x + 2)$
$12(x + 2)^2$
$36x(x + 2)^2$

10. $8x^2(x - 1)^2$
$10x^3(x - 1)$
$40x^3(x - 1)^2$

11. $3x + 3$
$2x^2 + 4x + 2$
$6(x + 1)^2$

12. $4x - 12$
$2x^2 - 12x + 18$
$4(x - 3)^2$

13. $(x - 1)(x + 2)$
$(x - 1)(x + 3)$
$(x - 1)(x + 2)(x + 3)$

14. $(2x - 1)(x + 4)$
$(2x + 1)(x + 4)$
$(2x - 1)(x + 4)(2x + 1)$

15. $(2x + 3)^2$
$(2x + 3)(x - 5)$
$(2x + 3)^2(x - 5)$

16. $(x - 7)(x + 2)$
$(x - 7)^2$
$(x - 7)^2(x + 2)$

17. $x - 1$
$x - 2$
$(x - 1)(x - 2)$

18. $(x + 4)(x - 3)$
$x + 4$
$x - 3$
$(x + 4)(x - 3)$

19. $x^2 - x - 6$
$x^2 + x - 12$
$(x - 3)(x + 2)(x + 4)$

20. $x^2 + 3x - 10$
$x^2 + 5x - 14$
$(x + 5)(x + 7)(x - 2)$

21. $x^2 + 5x + 4$
$x^2 - 3x - 28$
$(x + 4)(x + 1)(x - 7)$

22. $x^2 - 10x + 21$
$x^2 - 8x + 15$
$(x - 7)(x - 3)(x - 5)$

23. $x^2 - 2x - 24$
$x^2 - 36$
$(x - 6)(x + 6)(x + 4)$

24. $x^2 + 7x + 10$
$x^2 - 25$
$(x + 5)(x - 5)(x + 2)$

25. $x^2 - 7x - 30$
$x^2 - 5x - 24$
$(x - 10)(x - 8)(x + 3)$

26. $2x^2 - 7x + 3$
$2x^2 + x - 1$
$(2x - 1)(x - 3)(x + 1)$

27. $3x^2 - 11x + 6$
$3x^2 + 4x - 4$
$(3x - 2)(x - 3)(x + 2)$

28. $2x^2 - 9x + 10$
$2x^2 + x - 15$
$(2x - 5)(x - 2)(x + 3)$

29. $6 + x - x^2$
$x + 2$
$x - 3$
$(x + 2)(x - 3)$

30. $15 + 2x - x^2$
$x - 5$
$x + 3$
$(x + 3)(x - 5)$

31. $5 + 4x - x^2$
$x - 5$
$x + 1$
$(x - 5)(x + 1)$

32. $x^2 + 3x - 18$
$3 - x$
$x + 6$
$(x + 6)(x - 3)$

33. $x^2 - 5x + 6$
$1 - x$
$x - 6$
$(x - 3)(x - 2)(x - 1)(x - 6)$

Quick Quiz (Objective 11.2A)

Find the LCM of the polynomials.

1. $14a^4b^2$ and $21ab^5$ $42a^4b^5$

2. $3x^2y$ and $2x^2 - 10x$ $6x^2y(x - 5)$

3. $x^2 - 4x - 12$ and $36 - x^2$
$(x - 6)(x + 6)(x + 2)$

Objective B   To express two fractions in terms of the LCM of their denominators

For Exercises 34 to 53, write the fractions in terms of the LCM of their denominators.

34. $\dfrac{4}{x}, \dfrac{3}{x^2}$

$\dfrac{4x}{x^2}, \dfrac{3}{x^2}$

35. $\dfrac{5}{ab^2}, \dfrac{6}{ab}$

$\dfrac{5}{ab^2}, \dfrac{6b}{ab^2}$

36. $\dfrac{x}{3y^2}, \dfrac{z}{4y}$

$\dfrac{4x}{12y^2}, \dfrac{3yz}{12y^2}$

37. $\dfrac{5y}{6x^2}, \dfrac{7}{9xy}$

$\dfrac{15y^2}{18x^2y}, \dfrac{14x}{18x^2y}$

38. $\dfrac{y}{x(x-3)}, \dfrac{6}{x^2}$

$\dfrac{xy}{x^2(x-3)}, \dfrac{6x-18}{x^2(x-3)}$

39. $\dfrac{a}{y^2}, \dfrac{6}{y(y+5)}$

$\dfrac{ay+5a}{y^2(y+5)}, \dfrac{6y}{y^2(y+5)}$

40. $\dfrac{9}{(x-1)^2}, \dfrac{6}{x(x-1)}$

$\dfrac{9x}{x(x-1)^2}, \dfrac{6x-6}{x(x-1)^2}$

41. $\dfrac{a^2}{y(y+7)}, \dfrac{a}{(y+7)^2}$

$\dfrac{a^2y+7a^2}{y(y+7)^2}, \dfrac{ay}{y(y+7)^2}$

42. $\dfrac{3}{x-3}, \dfrac{5}{x(3-x)}$

$\dfrac{3x}{x(x-3)}, -\dfrac{5}{x(x-3)}$

43. $\dfrac{b}{y(y-4)}, \dfrac{b^2}{4-y}$

$\dfrac{b}{y(y-4)}, -\dfrac{b^2y}{y(y-4)}$

44. $\dfrac{3}{(x-5)^2}, \dfrac{2}{5-x}$

$\dfrac{3}{(x-5)^2}, -\dfrac{2x-10}{(x-5)^2}$

45. $\dfrac{3}{7-y}, \dfrac{2}{(y-7)^2}$

$\dfrac{3y-21}{(y-7)^2}, \dfrac{2}{(y-7)^2}$

46. $\dfrac{3}{x^2+2x}, \dfrac{4}{x^2}$

$\dfrac{3x}{x^2(x+2)}, \dfrac{4x+8}{x^2(x+2)}$

47. $\dfrac{2}{y-3}, \dfrac{3}{y^3-3y^2}$

$\dfrac{2y^2}{y^2(y-3)}, \dfrac{3}{y^2(y-3)}$

48. $\dfrac{x-2}{x+3}, \dfrac{x}{x-4}$

$\dfrac{x^2-6x+8}{(x+3)(x-4)}, \dfrac{x^2+3x}{(x+3)(x-4)}$

49. $\dfrac{x^2}{2x-1}, \dfrac{x+1}{x+4}$

$\dfrac{x^3+4x^2}{(2x-1)(x+4)}, \dfrac{2x^2+x-1}{(2x-1)(x+4)}$

50. $\dfrac{3}{x^2+x-2}, \dfrac{x}{x+2}$

$\dfrac{3}{(x+2)(x-1)}, \dfrac{x^2-x}{(x+2)(x-1)}$

51. $\dfrac{3x}{x-5}, \dfrac{4}{x^2-25}$

$\dfrac{3x^2+15x}{(x-5)(x+5)}, \dfrac{4}{(x-5)(x+5)}$

52. $\dfrac{x}{x^2+x-6}, \dfrac{2x}{x^2-9}$

$\dfrac{x^2-3x}{(x+3)(x-3)(x-2)}, \dfrac{2x^2-4x}{(x+3)(x-3)(x-2)}$

53. $\dfrac{x-1}{x^2+2x-15}, \dfrac{x}{x^2+6x+5}$

$\dfrac{x^2-1}{(x+5)(x-3)(x+1)}, \dfrac{x^2-3x}{(x+5)(x-3)(x+1)}$

Answers to Writing Exercises

54. The LCM of two polynomials is equal to the product of the two polynomials when the two polynomials have no common factors.

APPLYING THE CONCEPTS

54. When is the LCM of two polynomials equal to their product?

For Exercises 55 to 60, write the fractions in terms of the LCM of their denominators.

55. $\dfrac{8}{10^3}, \dfrac{9}{10^5}$

$\dfrac{800}{10^5}, \dfrac{9}{10^5}$

56. $3, \dfrac{2}{n}$

$\dfrac{3n}{n}, \dfrac{2}{n}$

57. $x, \dfrac{x}{x^2-1}$

$\dfrac{x^3-x}{x^2-1}, \dfrac{x}{x^2-1}$

58. $\dfrac{x^2+1}{(x-1)^3}, \dfrac{x+1}{(x-1)^2}, \dfrac{1}{x-1}$

$\dfrac{x^2+1}{(x-1)^3}, \dfrac{(x+1)(x-1)}{(x-1)^3}, \dfrac{(x-1)^2}{(x-1)^3}$

59. $\dfrac{c}{6c^2+7cd+d^2}, \dfrac{d}{3c^2-3d^2}$

$\dfrac{3c^2-3cd}{3(6c+d)(c-d)(c+d)}, \dfrac{6cd+d^2}{3(6c+d)(c-d)(c+d)}$

60. $\dfrac{1}{ab+3a-3b-b^2}, \dfrac{1}{ab+3a+3b+b^2}$

$\dfrac{a+b}{(a-b)(a+b)(b+3)}, \dfrac{a-b}{(a-b)(a+b)(b+3)}$

Quick Quiz (Objective 11.2B)

Write the fractions in terms of the LCM of their denominators.

1. $\dfrac{y}{10x^2}$ and $\dfrac{x}{15y^3}$ $\dfrac{3y^4}{30x^2y^3}$ and $\dfrac{2x^3}{30x^2y^3}$

2. $\dfrac{4}{y(2-y)}$ and $\dfrac{1}{(y-2)^2}$ $-\dfrac{4y-8}{y(y-2)^2}$ and $\dfrac{y}{y(y-2)^2}$

3. $\dfrac{x+1}{x^2-x}$ and $\dfrac{x}{x^2+6x-7}$

$\dfrac{x^2+8x+7}{x(x-1)(x+7)}$ and $\dfrac{x^2}{x(x-1)(x+7)}$

11.3 Addition and Subtraction of Rational Expressions

Objective A   **To add or subtract rational expressions with the same denominator**

VIDEO & DVD   CD TUTOR   WEB   SSM

When adding rational expressions in which the denominators are the same, add the numerators. The denominator of the sum is the common denominator.

$$\frac{5x}{18} + \frac{7x}{18} = \frac{5x + 7x}{18} = \frac{12x}{18} = \frac{2x}{3}$$

$$\frac{x}{x^2 - 1} + \frac{1}{x^2 - 1} = \frac{x + 1}{x^2 - 1} = \frac{\overset{1}{\cancel{(x + 1)}}}{(x - 1)\underset{1}{\cancel{(x + 1)}}} = \frac{1}{x - 1}$$

Note that the sum is written in simplest form.

When subtracting rational expressions with like denominators, subtract the numerators. The denominator of the difference is the common denominator. Write the answer in simplest form.

$$\frac{2x}{x - 2} - \frac{4}{x - 2} = \frac{2x - 4}{x - 2} = \frac{2\overset{1}{\cancel{(x - 2)}}}{\underset{1}{\cancel{x - 2}}} = 2$$

$$\frac{3x - 1}{x^2 - 5x + 4} - \frac{2x + 3}{x^2 - 5x + 4} = \frac{(3x - 1) - (2x + 3)}{x^2 - 5x + 4} = \frac{3x - 1 - 2x - 3}{x^2 - 5x + 4}$$

$$= \frac{x - 4}{x^2 - 5x + 4} = \frac{\overset{1}{\cancel{(x - 4)}}}{\underset{1}{\cancel{(x - 4)}}(x - 1)} = \frac{1}{x - 1}$$

> **Adding and Subtracting Rational Expressions with the Same Denominator**
>
> Add or subtract the numerators. Place the result over the common denominator.
> $$\frac{a}{b} + \frac{c}{b} = \frac{a + c}{b} \qquad \frac{a}{b} - \frac{c}{b} = \frac{a - c}{b}$$

Example 1

Subtract: $\dfrac{3x^2}{x^2 - 1} - \dfrac{x + 4}{x^2 - 1}$

Solution

$$\frac{3x^2}{x^2 - 1} - \frac{x + 4}{x^2 - 1} = \frac{3x^2 - (x + 4)}{x^2 - 1}$$

$$= \frac{3x^2 - x - 4}{x^2 - 1}$$

$$= \frac{(3x - 4)\overset{1}{\cancel{(x + 1)}}}{(x - 1)\underset{1}{\cancel{(x + 1)}}} = \frac{3x - 4}{x - 1}$$

You Try It 1

Subtract: $\dfrac{2x^2}{x^2 - x - 12} - \dfrac{7x + 4}{x^2 - x - 12}$

Your solution

$$\frac{2x + 1}{x + 3}$$

Solution on p. S27

In-Class Examples (Objective 11.3A)

1. Subtract: $\dfrac{2x^2}{x^2 + 7x - 8} - \dfrac{x + 1}{x^2 + 7x - 8}$ $\dfrac{2x + 1}{x + 8}$

2. Simplify: $\dfrac{3y^2 - 6}{y^2 + y - 20} + \dfrac{y - 9}{y^2 + y - 20} - \dfrac{2y^2 + y + 1}{y^2 + y - 20}$

$$\frac{y + 4}{y + 5}$$

Objective 11.3A

Discuss the Concepts

Have students describe the steps in the process of adding or subtracting rational expressions with the same denominator.

Concept Check

Find the sum.

1. $\dfrac{8a}{15b} + \dfrac{2a}{15b}$ $\dfrac{2a}{3b}$

2. $\dfrac{x + 4}{2x + 1} + \dfrac{x - 1}{2x + 1}$ $\dfrac{2x + 3}{2x + 1}$

Optional Student Activity

In the first subtraction example at the left, some students are surprised that $\dfrac{2x}{x - 2} - \dfrac{4}{x - 2} = 2$. Have students evaluate the expression $\dfrac{2x}{x - 2} - \dfrac{4}{x - 2}$ for several values of x to see that the calculation always yields 2 (except when $x = 2$).

Instructor Note

In the second subtraction example at the left, point out the importance of the parentheses around $2x + 3$. Emphasize that students must remember to subtract the whole numerator, which changes the sign of every term, not just the first term.

Concept Check

Find the difference.

1. $\dfrac{8a}{15b} - \dfrac{2a}{15b}$ $\dfrac{2a}{5b}$

2. $\dfrac{x + 4}{2x + 1} - \dfrac{x - 1}{2x + 1}$ $\dfrac{5}{2x + 1}$

Example 2

Simplify:

$$\frac{2x^2 + 5}{x^2 + 2x - 3} - \frac{x^2 - 3x}{x^2 + 2x - 3} + \frac{x - 2}{x^2 + 2x - 3}$$

Solution

$$\frac{2x^2 + 5}{x^2 + 2x - 3} - \frac{x^2 - 3x}{x^2 + 2x - 3} + \frac{x - 2}{x^2 + 2x - 3}$$

$$= \frac{(2x^2 + 5) - (x^2 - 3x) + (x - 2)}{x^2 + 2x - 3}$$

$$= \frac{2x^2 + 5 - x^2 + 3x + x - 2}{x^2 + 2x - 3}$$

$$= \frac{x^2 + 4x + 3}{x^2 + 2x - 3}$$

$$= \frac{\overset{1}{\cancel{(x + 3)}}(x + 1)}{\underset{1}{\cancel{(x + 3)}}(x - 1)} = \frac{x + 1}{x - 1}$$

You Try It 2

Simplify:

$$\frac{x^2 - 1}{x^2 - 8x + 12} - \frac{2x + 1}{x^2 - 8x + 12} + \frac{x}{x^2 - 8x + 12}$$

Your solution

$$\frac{x + 1}{x - 6}$$

Solution on p. S27

Objective 11.3B

Concept Check

Use the fractions $\frac{3}{10x^2}$ and $\frac{y + 1}{6x^2 y}$.

1. Find the least common denominator. $30x^2 y$

2. Rewrite each fraction with the common denominator.
$\frac{9y}{30x^2 y}$ and $\frac{5y + 5}{30x^2 y}$

3. Find the sum of the fractions.
$\frac{14y + 5}{30x^2 y}$

4. Find the difference of the fractions.
$\frac{4y - 5}{30x^2 y}$

Discuss the Concepts

Have students describe the steps involved in rewriting two or more rational expressions with a common denominator (this will review Objective 11.2B). The fractions in the example at the right can be used to illustrate the process.

Objective B

To add or subtract rational expressions with different denominators

Before two fractions with unlike denominators can be added or subtracted, each fraction must be expressed in terms of a common denominator. This common denominator is the LCM of the denominators of the fractions.

HOW TO Add: $\dfrac{x - 3}{x^2 - 2x} + \dfrac{6}{x^2 - 4}$

The LCM is $x(x - 2)(x + 2)$. • Find the LCM of the denominators.

$$\frac{x - 3}{x^2 - 2x} + \frac{6}{x^2 - 4}$$

$$= \frac{x - 3}{x(x - 2)} \cdot \frac{x + 2}{x + 2} + \frac{6}{(x - 2)(x + 2)} \cdot \frac{x}{x}$$
• Write each fraction in terms of the LCM.

$$= \frac{x^2 - x - 6}{x(x - 2)(x + 2)} + \frac{6x}{x(x - 2)(x + 2)}$$
• Multiply the factors in the numerators.

$$= \frac{(x^2 - x - 6) + 6x}{x(x - 2)(x + 2)}$$
• Add the fractions.

$$= \frac{x^2 + 5x - 6}{x(x - 2)(x + 2)}$$
• Simplify.

$$= \frac{(x + 6)(x - 1)}{x(x - 2)(x + 2)}$$
• Factor.

After combining the numerators over the common denominator, the last step is to factor the numerator to determine whether there are common factors in the numerator and denominator. For the previous example, there are no common factors, so the answer is in simplest form.

The process of adding and subtracting rational expressions is summarized below.

Instructor Note

The process of adding or subtracting rational expressions is a complex one, involving many steps and the application of several skills. It may help students to keep a list of the basic steps in front of them as they work.

Adding and Subtracting Rational Expressions

1. Find the LCM of the denominators.
2. Write each fraction as an equivalent fraction using the LCM as the denominator.
3. Add or subtract the numerators and place the result over the common denominator.
4. Write the answer in simplest form.

Ⓟ

Example 3

Simplify: $\dfrac{y}{x} - \dfrac{4y}{3x} + \dfrac{3y}{4x}$

Solution

The LCM of the denominators is $12x$.

$\dfrac{y}{x} - \dfrac{4y}{3x} + \dfrac{3y}{4x}$

$= \dfrac{y}{x} \cdot \dfrac{12}{12} - \dfrac{4y}{3x} \cdot \dfrac{4}{4} + \dfrac{3y}{4x} \cdot \dfrac{3}{3}$ • Write each fraction using the LCM as the denominator.

$= \dfrac{12y}{12x} - \dfrac{16y}{12x} + \dfrac{9y}{12x}$

$= \dfrac{12y - 16y + 9y}{12x} = \dfrac{5y}{12x}$ • Combine the numerators.

You Try It 3

Simplify: $\dfrac{z}{8y} - \dfrac{4z}{3y} + \dfrac{5z}{4y}$

Your solution

$\dfrac{z}{24y}$

Solution on p. S27

In-Class Examples (Objective 11.3B)

1. Simplify: $\dfrac{1}{6a} + \dfrac{4}{9b} - \dfrac{5}{2a}$ $\dfrac{4a - 21b}{9ab}$

2. Subtract: $\dfrac{x - 2}{10x} - \dfrac{x - 3}{15x}$ $\dfrac{1}{30}$

3. Add: $\dfrac{7}{x + 2} + 2$ $\dfrac{2x + 11}{x + 2}$

4. Subtract: $\dfrac{a - 1}{a^2 b} - \dfrac{a - 2}{ab}$ $\dfrac{-a^2 + 3a - 1}{a^2 b}$

5. Add: $\dfrac{2x}{3x - 2} + \dfrac{4}{x - 3}$ $\dfrac{2(x + 4)(x - 1)}{(3x - 2)(x - 3)}$

6. Add: $\dfrac{6y}{y^2 - 4} + \dfrac{3}{2 - y}$ $\dfrac{3}{y + 2}$

Optional Student Activity

Let x and y be positive integers. If $A = \dfrac{1}{x} + \dfrac{1}{y} + 1$ and $B = \dfrac{x+y}{xy}$, then which of the following is true? $A < B$, $A > B$, $A = B$

$A > B$

Example 4

Subtract: $\dfrac{2x}{x-3} - \dfrac{5}{3-x}$

Solution

Remember that $3 - x = -(x - 3)$.

Therefore, $\dfrac{5}{3-x} = \dfrac{5}{-(x-3)} = \dfrac{-5}{x-3}$.

$\dfrac{2x}{x-3} - \dfrac{5}{3-x}$

$= \dfrac{2x}{x-3} - \dfrac{-5}{x-3}$ • The LCM is $x - 3$.

$= \dfrac{2x - (-5)}{x-3} = \dfrac{2x+5}{x-3}$ • Combine the numerators.

You Try It 4

Add: $\dfrac{5x}{x-2} + \dfrac{3}{2-x}$

Your solution

$\dfrac{5x-3}{x-2}$

Example 5

Subtract: $\dfrac{2x}{2x-3} - \dfrac{1}{x+1}$

Solution

The LCM is $(2x - 3)(x + 1)$.

$\dfrac{2x}{2x-3} - \dfrac{1}{x+1}$

$= \dfrac{2x}{2x-3} \cdot \dfrac{x+1}{x+1} - \dfrac{1}{x+1} \cdot \dfrac{2x-3}{2x-3}$

$= \dfrac{2x^2 + 2x}{(2x-3)(x+1)} - \dfrac{2x-3}{(2x-3)(x+1)}$

$= \dfrac{(2x^2 + 2x) - (2x - 3)}{(2x-3)(x+1)}$

$= \dfrac{2x^2 + 2x - 2x + 3}{(2x-3)(x+1)} = \dfrac{2x^2 + 3}{(2x-3)(x+1)}$

You Try It 5

Add: $\dfrac{4x}{3x-1} + \dfrac{9}{x+4}$

Your solution

$\dfrac{4x^2 + 43x - 9}{(3x-1)(x+4)}$

Example 6

Add: $1 + \dfrac{3}{x^2}$

Solution

The LCM is x^2.

$1 + \dfrac{3}{x^2} = 1 \cdot \dfrac{x^2}{x^2} + \dfrac{3}{x^2} = \dfrac{x^2}{x^2} + \dfrac{3}{x^2} = \dfrac{x^2 + 3}{x^2}$

You Try It 6

Subtract: $2 - \dfrac{1}{x-3}$

Your solution

$\dfrac{2x-7}{x-3}$

Solutions on pp. S27–S28

Example 7

Add: $\dfrac{x+3}{x^2-2x-8} + \dfrac{3}{4-x}$

Solution

Recall that $\dfrac{3}{4-x} = \dfrac{-3}{x-4}$.

The LCM is $(x-4)(x+2)$.

$\dfrac{x+3}{x^2-2x-8} + \dfrac{3}{4-x}$

$= \dfrac{x+3}{(x-4)(x+2)} + \dfrac{(-3)}{x-4}$

$= \dfrac{x+3}{(x-4)(x+2)} + \dfrac{(-3)}{x-4} \cdot \dfrac{x+2}{x+2}$

$= \dfrac{x+3}{(x-4)(x+2)} + \dfrac{(-3)(x+2)}{(x-4)(x+2)}$

$= \dfrac{(x+3) + (-3)(x+2)}{(x-4)(x+2)}$

$= \dfrac{x+3-3x-6}{(x-4)(x+2)}$

$= \dfrac{-2x-3}{(x-4)(x+2)}$

You Try It 7

Add: $\dfrac{2x-1}{x^2-25} + \dfrac{2}{5-x}$

Your solution

$-\dfrac{11}{(x+5)(x-5)}$

Example 8

Simplify: $\dfrac{3x+2}{2x^2-x-1} - \dfrac{3}{2x+1} + \dfrac{4}{x-1}$

Solution

The LCM is $(2x+1)(x-1)$.

$\dfrac{3x+2}{2x^2-x-1} - \dfrac{3}{2x+1} + \dfrac{4}{x-1}$

$= \dfrac{3x+2}{(2x+1)(x-1)} - \dfrac{3}{2x+1} \cdot \dfrac{x-1}{x-1} + \dfrac{4}{x-1} \cdot \dfrac{2x+1}{2x+1}$

$= \dfrac{3x+2}{(2x+1)(x-1)} - \dfrac{3x-3}{(2x+1)(x-1)} + \dfrac{8x+4}{(2x+1)(x-1)}$

$= \dfrac{(3x+2) - (3x-3) + (8x+4)}{(2x+1)(x-1)}$

$= \dfrac{3x+2-3x+3+8x+4}{(2x+1)(x-1)}$

$= \dfrac{8x+9}{(2x+1)(x-1)}$

You Try It 8

Simplify:

$\dfrac{2x-3}{3x^2-x-2} + \dfrac{5}{3x+2} - \dfrac{1}{x-1}$

Your solution

$\dfrac{2(2x-5)}{(3x+2)(x-1)}$

Solutions on p. S28

588

Suggested Assignment

Exercises 1–79, odds
More challenging problems:
   Exercises 81–82

11.3 Exercises

Objective A   To add or subtract rational expressions with the same denominator

For Exercises 1 to 20, simplify.

1. $\dfrac{3}{y^2} + \dfrac{8}{y^2}$

$\dfrac{11}{y^2}$

2. $\dfrac{6}{ab} - \dfrac{2}{ab}$

$\dfrac{4}{ab}$

3. $\dfrac{3}{x+4} - \dfrac{10}{x+4}$

$-\dfrac{7}{x+4}$

4. $\dfrac{x}{x+6} - \dfrac{2}{x+6}$

$\dfrac{x-2}{x+6}$

5. $\dfrac{3x}{2x+3} + \dfrac{5x}{2x+3}$

$\dfrac{8x}{2x+3}$

6. $\dfrac{6y}{4y+1} - \dfrac{11y}{4y+1}$

$-\dfrac{5y}{4y+1}$

7. $\dfrac{2x+1}{x-3} + \dfrac{3x+6}{x-3}$

$\dfrac{5x+7}{x-3}$

8. $\dfrac{4x+3}{2x-7} + \dfrac{3x-8}{2x-7}$

$\dfrac{7x-5}{2x-7}$

9. $\dfrac{5x-1}{x+9} - \dfrac{3x+4}{x+9}$

$\dfrac{2x-5}{x+9}$

10. $\dfrac{6x-5}{x-10} - \dfrac{3x-4}{x-10}$

$\dfrac{3x-1}{x-10}$

11. $\dfrac{x-7}{2x+7} - \dfrac{4x-3}{2x+7}$

$\dfrac{-3x-4}{2x+7}$

12. $\dfrac{2n}{3n+4} - \dfrac{5n-3}{3n+4}$

$\dfrac{-3(n-1)}{3n+4}$

13. $\dfrac{x}{x^2+2x-15} - \dfrac{3}{x^2+2x-15}$

$\dfrac{1}{x+5}$

14. $\dfrac{3x}{x^2+3x-10} - \dfrac{6}{x^2+3x-10}$

$\dfrac{3}{x+5}$

15. $\dfrac{2x+3}{x^2-x-30} - \dfrac{x-2}{x^2-x-30}$

$\dfrac{1}{x-6}$

16. $\dfrac{3x-1}{x^2+5x-6} - \dfrac{2x-7}{x^2+5x-6}$

$\dfrac{1}{x-1}$

17. $\dfrac{4y+7}{2y^2+7y-4} - \dfrac{y-5}{2y^2+7y-4}$

$\dfrac{3}{2y-1}$

18. $\dfrac{x+1}{2x^2-5x-12} + \dfrac{x+2}{2x^2-5x-12}$

$\dfrac{1}{x-4}$

19. $\dfrac{2x^2+3x}{x^2-9x+20} + \dfrac{2x^2-3}{x^2-9x+20} - \dfrac{4x^2+2x+1}{x^2-9x+20}$

$\dfrac{1}{x-5}$

20. $\dfrac{2x^2+3x}{x^2-2x-63} - \dfrac{x^2-3x+21}{x^2-2x-63} - \dfrac{x-7}{x^2-2x-63}$

$\dfrac{x-2}{x-9}$

Quick Quiz (Objective 11.3A)

1. Subtract: $\dfrac{8a}{3a-1} - \dfrac{10a}{3a-1}$ $-\dfrac{2a}{3a-1}$

2. Subtract: $\dfrac{2x+3}{x-4} - \dfrac{x+5}{x-4}$ $\dfrac{x-2}{x-4}$

3. Add: $\dfrac{2x-8}{3x^2-2x-1} + \dfrac{x+9}{3x^2-2x-1}$ $\dfrac{1}{x-1}$

Objective B   **To add or subtract rational expressions with different denominators**

21. ✎ Explain the process of writing equivalent rational expressions using the LCM of the denominators of the rational expressions as the new denominator.

22. ✎ Explain the process of adding rational expressions with different denominators.

For Exercises 23 to 80, simplify.

23. $\dfrac{4}{x} + \dfrac{5}{y}$

$\dfrac{4y + 5x}{xy}$

24. $\dfrac{7}{a} + \dfrac{5}{b}$

$\dfrac{7b + 5a}{ab}$

25. $\dfrac{12}{x} - \dfrac{5}{2x}$

$\dfrac{19}{2x}$

26. $\dfrac{5}{3a} - \dfrac{3}{4a}$

$\dfrac{11}{12a}$

27. $\dfrac{1}{2x} - \dfrac{5}{4x} + \dfrac{7}{6x}$

$\dfrac{5}{12x}$

28. $\dfrac{7}{4y} + \dfrac{11}{6y} - \dfrac{8}{3y}$

$\dfrac{11}{12y}$

29. $\dfrac{5}{3x} - \dfrac{2}{x^2} + \dfrac{3}{2x}$

$\dfrac{19x - 12}{6x^2}$

30. $\dfrac{6}{y^2} + \dfrac{3}{4y} - \dfrac{2}{5y}$

$\dfrac{120 + 7y}{20y^2}$

31. $\dfrac{2}{x} - \dfrac{3}{2y} + \dfrac{3}{5x} - \dfrac{1}{4y}$

$\dfrac{52y - 35x}{20xy}$

32. $\dfrac{5}{2a} + \dfrac{7}{3b} - \dfrac{2}{b} - \dfrac{3}{4a}$

$\dfrac{21b + 4a}{12ab}$

33. $\dfrac{2x + 1}{3x} + \dfrac{x - 1}{5x}$

$\dfrac{13x + 2}{15x}$

34. $\dfrac{4x - 3}{6x} + \dfrac{2x + 3}{4x}$

$\dfrac{14x + 3}{12x}$

35. $\dfrac{x - 3}{6x} + \dfrac{x + 4}{8x}$

$\dfrac{7}{24}$

36. $\dfrac{2x - 3}{2x} + \dfrac{x + 3}{3x}$

$\dfrac{8x - 3}{6x}$

37. $\dfrac{2x + 9}{9x} - \dfrac{x - 5}{5x}$

$\dfrac{x + 90}{45x}$

38. $\dfrac{3y - 2}{12y} - \dfrac{y - 3}{18y}$

$\dfrac{7}{36}$

39. $\dfrac{x + 4}{2x} - \dfrac{x - 1}{x^2}$

$\dfrac{x^2 + 2x + 2}{2x^2}$

40. $\dfrac{x - 2}{3x^2} - \dfrac{x + 4}{x}$

$\dfrac{-3x^2 - 11x - 2}{3x^2}$

41. $\dfrac{x - 10}{4x^2} + \dfrac{x + 1}{2x}$

$\dfrac{2x^2 + 3x - 10}{4x^2}$

42. $\dfrac{x + 5}{3x^2} + \dfrac{2x + 1}{2x}$

$\dfrac{6x^2 + 5x + 10}{6x^2}$

43. $\dfrac{4}{x + 4} - x$

$\dfrac{-x^2 - 4x + 4}{x + 4}$

44. $2x + \dfrac{1}{x}$

$\dfrac{2x^2 + 1}{x}$

45. $5 - \dfrac{x - 2}{x + 1}$

$\dfrac{4x + 7}{x + 1}$

46. $3 + \dfrac{x - 1}{x + 1}$

$\dfrac{4x + 2}{x + 1}$

Answers to Writing Exercises

21. For each fraction, multiply the numerator and denominator by the factors whose product with the denominator is the LCM.

22. Students' explanations should include each of the following steps:
(1) Find the LCM of the denominators.
(2) Write each rational expression as an equivalent expression that has the LCM as the denominator.
(3) Add or subtract the numerators and place the result over the common denominator.
(4) Write the answer in factored form and express the answer in simplest form.

Quick Quiz (Objective 11.3B)

1. Subtract: $\dfrac{y - 1}{5y^2} - \dfrac{y + 3}{10y}$ $-\dfrac{y^2 + y + 2}{10y^2}$

2. Add: $\dfrac{4x}{x^2 - 36} + \dfrac{3}{6 - x}$ $\dfrac{x - 18}{(x - 6)(x + 6)}$

3. Subtract: $\dfrac{a - 1}{a - 2} - \dfrac{3a + 1}{a^2 + 3a - 10}$ $\dfrac{a + 3}{a + 5}$

47. $\dfrac{x+3}{6x} - \dfrac{x-3}{8x^2}$

$\dfrac{4x^2 + 9x + 9}{24x^2}$

48. $\dfrac{x+2}{xy} - \dfrac{3x-2}{x^2y}$

$\dfrac{x^2 - x + 2}{x^2y}$

49. $\dfrac{3x-1}{xy^2} - \dfrac{2x+3}{xy}$

$\dfrac{3x - 1 - 2xy - 3y}{xy^2}$

50. $\dfrac{4x-3}{3x^2y} + \dfrac{2x+1}{4xy^2}$

$\dfrac{16xy - 12y + 6x^2 + 3x}{12x^2y^2}$

51. $\dfrac{5x+7}{6xy^2} - \dfrac{4x-3}{8x^2y}$

$\dfrac{20x^2 + 28x - 12xy + 9y}{24x^2y^2}$

52. $\dfrac{x-2}{8x^2} - \dfrac{x+7}{12xy}$

$\dfrac{3xy - 6y - 2x^2 - 14x}{24x^2y}$

53. $\dfrac{3x-1}{6y^2} - \dfrac{x+5}{9xy}$

$\dfrac{9x^2 - 3x - 2xy - 10y}{18xy^2}$

54. $\dfrac{4}{x-2} + \dfrac{5}{x+3}$

$\dfrac{9x + 2}{(x-2)(x+3)}$

55. $\dfrac{2}{x-3} + \dfrac{5}{x-4}$

$\dfrac{7x - 23}{(x-3)(x-4)}$

56. $\dfrac{6}{x-7} - \dfrac{4}{x+3}$

$\dfrac{2(x+23)}{(x-7)(x+3)}$

57. $\dfrac{3}{y+6} - \dfrac{4}{y-3}$

$\dfrac{-y - 33}{(y+6)(y-3)}$

58. $\dfrac{2x}{x+1} + \dfrac{1}{x-3}$

$\dfrac{2x^2 - 5x + 1}{(x+1)(x-3)}$

59. $\dfrac{3x}{x-4} + \dfrac{2}{x+6}$

$\dfrac{3x^2 + 20x - 8}{(x-4)(x+6)}$

60. $\dfrac{4x}{2x-1} - \dfrac{5}{x-6}$

$\dfrac{4x^2 - 34x + 5}{(2x-1)(x-6)}$

61. $\dfrac{6x}{x+5} - \dfrac{3}{2x+3}$

$\dfrac{3(4x^2 + 5x - 5)}{(x+5)(2x+3)}$

62. $\dfrac{2a}{a-7} + \dfrac{5}{7-a}$

$\dfrac{2a - 5}{a - 7}$

63. $\dfrac{4x}{6-x} + \dfrac{5}{x-6}$

$\dfrac{-4x + 5}{x - 6}$

64. $\dfrac{x}{x^2-9} + \dfrac{3}{x-3}$

$\dfrac{4x + 9}{(x+3)(x-3)}$

65. $\dfrac{y}{y^2-16} + \dfrac{1}{y-4}$

$\dfrac{2(y+2)}{(y-4)(y+4)}$

66. $\dfrac{2x}{x^2-x-6} - \dfrac{3}{x+2}$

$\dfrac{-x + 9}{(x-3)(x+2)}$

67. $\dfrac{(x-1)^2}{(x+1)^2} - 1$

$-\dfrac{4x}{(x+1)^2}$

68. $1 - \dfrac{(y-2)^2}{(y+2)^2}$

$\dfrac{8y}{(y+2)^2}$

69. $\dfrac{x}{1-x^2} - 1 + \dfrac{x}{1+x}$

$\dfrac{2x - 1}{(1+x)(1-x)}$

70. $\dfrac{y}{x-y} + 2 - \dfrac{x}{y-x}$

$\dfrac{3x - y}{x - y}$

71. $\dfrac{3x - 1}{x^2 - 10x + 25} - \dfrac{3}{x - 5}$

$\dfrac{14}{(x - 5)^2}$

72. $\dfrac{2a + 3}{a^2 - 7a + 12} - \dfrac{2}{a - 3}$

$\dfrac{11}{(a - 4)(a - 3)}$

73. $\dfrac{x + 4}{x^2 - x - 42} + \dfrac{3}{7 - x}$

$\dfrac{-2(x + 7)}{(x + 6)(x - 7)}$

74. $\dfrac{x + 3}{x^2 - 3x - 10} + \dfrac{2}{5 - x}$

$\dfrac{-x - 1}{(x - 5)(x + 2)}$

75. $\dfrac{1}{x + 1} + \dfrac{x}{x - 6} - \dfrac{5x - 2}{x^2 - 5x - 6}$

$\dfrac{x - 4}{x - 6}$

76. $\dfrac{x}{x - 4} + \dfrac{5}{x + 5} - \dfrac{11x - 8}{x^2 + x - 20}$

$\dfrac{x + 3}{x + 5}$

77. $\dfrac{3x + 1}{x - 1} - \dfrac{x - 1}{x - 3} + \dfrac{x + 1}{x^2 - 4x + 3}$

$\dfrac{2x + 1}{x - 1}$

78. $\dfrac{4x + 1}{x - 8} - \dfrac{3x + 2}{x + 4} - \dfrac{49x + 4}{x^2 - 4x - 32}$

$\dfrac{x - 2}{x + 4}$

79. $\dfrac{2x + 9}{3 - x} + \dfrac{x + 5}{x + 7} - \dfrac{2x^2 + 3x - 3}{x^2 + 4x - 21}$

$\dfrac{-3(x^2 + 8x + 25)}{(x - 3)(x + 7)}$

80. $\dfrac{3x + 5}{x + 5} - \dfrac{x + 1}{2 - x} - \dfrac{4x^2 - 3x - 1}{x^2 + 3x - 10}$

$\dfrac{4(2x - 1)}{(x + 5)(x - 2)}$

APPLYING THE CONCEPTS

81. Transportation   Suppose that you drive about 12,000 mi per year and that the cost of gasoline averages $1.70 per gallon.

  a. Let x represent the number of miles per gallon your car gets. Write a variable expression for the amount you spend on gasoline in 1 year.

  b. Write and simplify a variable expression for the amount of money you will save each year if you can increase your gas mileage by 5 miles per gallon.

  c. If you currently get 25 miles per gallon and you increase your gas mileage by 5 miles per gallon, how much will you save in 1 year?

  a. $\dfrac{20,400}{x}$ dollars   **b.** $\dfrac{102,000}{x(x + 5)}$ dollars   **c.** $136

82. Deep-Sea Diving   A recommended percent of oxygen (by volume) in the air that a deep-sea diver breathes is given by $\dfrac{660}{d + 33}$, where d is the depth, in feet, at which the diver is working.

  a. What is the recommended percent of oxygen for a diver working at a depth of 50 ft? Round to the nearest percent.

  b. As the depth of the diver increases, does the recommended amount of oxygen increase or decrease?

  c. At sea level, the oxygen content of air is approximately 21%. Is this less than or more than the recommended amount of oxygen for a diver working at the water's surface?

  a. 8% **b.** Decrease **c.** More than

Objective 11.4A

Vocabulary to Review

complex fraction　[2.4C]

Concept Check

Refer to the three examples of complex fractions given in the text at the right. Tell what you would multiply each complex fraction by in order to simplify it.

$$\frac{2}{2}; \frac{x}{x}; \frac{(x-1)(x+4)}{(x-1)(x+4)}$$

Discuss the Concepts

Refer to the example at the bottom of the page.

1. What property allows you to multiply the numerator and denominator of the complex fraction by x^2?

   The Multiplication Property of One

2. Explain what happens in each step of the simplification process.

11.4　Complex Fractions

Objective A　To simplify a complex fraction

Point of Interest

There are many instances of complex fractions in application problems. The fraction $\dfrac{1}{\dfrac{1}{r_1}+\dfrac{1}{r_2}}$ is used to determine the total resistance in certain electric circuits.

Recall that a **complex fraction** is a fraction whose numerator or denominator contains one or more fractions. Examples of complex fractions are shown at the right.

$$\frac{3}{2-\dfrac{1}{2}},\quad \frac{4+\dfrac{1}{x}}{3+\dfrac{2}{x}},\quad \frac{\dfrac{1}{x-1}+x+3}{x-3+\dfrac{1}{x+4}}$$

To simplify a complex fraction, use one of the following methods.

TAKE NOTE

You may use either method to simplify a complex fraction. The result will be the same.

Simplifying Complex Fractions

Method 1: Multiply by 1 in the form $\dfrac{\text{LCM}}{\text{LCM}}$.

1. Determine the LCM of the denominators of the fractions in the numerator and denominator of the complex fraction.
2. Multiply the numerator and denominator of the complex fraction by the LCM.
3. Simplify.

Method 2: Multiply the numerator by the reciprocal of the denominator.

1. Simplify the numerator to a single fraction and simplify the denominator to a single fraction.
2. Using the definition for dividing fractions, multiply the numerator by the reciprocal of the denominator.
3. Simplify.

Here is an example of using Method 1.

HOW TO　Simplify: $\dfrac{9-\dfrac{4}{x^2}}{3+\dfrac{2}{x}}$

The LCM of x and x^2 is x^2.

- Find the **LCM** of the denominators of the fractions in the numerator and the denominator.

$$\frac{9-\dfrac{4}{x^2}}{3+\dfrac{2}{x}}=\frac{9-\dfrac{4}{x^2}}{3+\dfrac{2}{x}}\cdot\frac{x^2}{x^2}$$

- Multiply the numerator and denominator by the **LCM**.

$$=\frac{9\cdot x^2-\dfrac{4}{x^2}\cdot x^2}{3\cdot x^2+\dfrac{2}{x}\cdot x^2}=\frac{9x^2-4}{3x^2+2x}$$

- Use the **Distributive Property**.

$$=\frac{(3x-2)(3x+2)}{x(3x+2)}=\frac{3x-2}{x}$$

- Simplify.

In-Class Examples (Objective 11.4A)

1. Simplify: $\dfrac{a-\dfrac{10}{a-3}}{1+\dfrac{5}{a-3}}$　$a-5$

2. Simplify: $\dfrac{\dfrac{7}{x-3}-\dfrac{2}{3x}}{\dfrac{5}{3x}+\dfrac{1}{x-3}}$　$\dfrac{19x+6}{8x-15}$

Here is the same example using the second method.

HOW TO   Simplify: $\dfrac{9 - \dfrac{4}{x^2}}{3 + \dfrac{2}{x}}$

$\dfrac{9 - \dfrac{4}{x^2}}{3 + \dfrac{2}{x}} = \dfrac{\dfrac{9x^2}{x^2} - \dfrac{4}{x^2}}{\dfrac{3x}{x} + \dfrac{2}{x}} = \dfrac{\dfrac{9x^2 - 4}{x^2}}{\dfrac{3x + 2}{x}}$

- • Simplify the numerator to a single fraction and simplify the denominator to a single fraction.

$= \dfrac{9x^2 - 4}{x^2} \cdot \dfrac{x}{3x + 2}$

- • Multiply the numerator by the reciprocal of the denominator.

$= \dfrac{x(3x - 2)\overset{1}{\cancel{(3x + 2)}}}{x^{2}\underset{1}{\cancel{(3x + 2)}}}$

- • Simplify.

$= \dfrac{3x - 2}{x}$

For the examples below, we will use the first method.

Example 1

Simplify: $\dfrac{\dfrac{1}{x} + \dfrac{1}{2}}{\dfrac{1}{x^2} - \dfrac{1}{4}}$

Solution

The LCM of x, 2, x^2, and 4 is $4x^2$.

$\dfrac{\dfrac{1}{x} + \dfrac{1}{2}}{\dfrac{1}{x^2} - \dfrac{1}{4}} = \dfrac{\dfrac{1}{x} + \dfrac{1}{2}}{\dfrac{1}{x^2} - \dfrac{1}{4}} \cdot \dfrac{4x^2}{4x^2}$

- • Multiply by the LCM.

$= \dfrac{\dfrac{1}{x} \cdot 4x^2 + \dfrac{1}{2} \cdot 4x^2}{\dfrac{1}{x^2} \cdot 4x^2 - \dfrac{1}{4} \cdot 4x^2}$

- • Distributive Property

$= \dfrac{4x + 2x^2}{4 - x^2}$

- • Simplify.

$= \dfrac{2x\overset{1}{\cancel{(2 + x)}}}{(2 - x)\underset{1}{\cancel{(2 + x)}}}$

$= \dfrac{2x}{2 - x}$

You Try It 1

Simplify: $\dfrac{\dfrac{1}{3} - \dfrac{1}{x}}{\dfrac{1}{9} - \dfrac{1}{x^2}}$

Your solution

$\dfrac{3x}{x + 3}$

Solution on p. S28

Optional Student Activity

1. A salad dressing recipe calls for $\frac{1}{4}$ c vinegar, $\frac{1}{4}$ c sesame oil, and $\frac{1}{2}$ c canola oil.

   a. Write a complex fraction with the amount of vinegar in the numerator and the amount of oil in the denominator. $\dfrac{\dfrac{1}{4}}{\dfrac{1}{4} + \dfrac{1}{2}}$

   b. Write this complex fraction in simplest form. $\dfrac{1}{3}$

2. The approximation for π,
   $3 + \dfrac{1}{7 + \dfrac{1}{16}}$, was known to Zu Chongzhi, a Chinese mathematician who lived in the 5th century.

   a. Evaluate this expression as a fraction in lowest terms and as a decimal rounded to the nearest hundred-thousandth. $\dfrac{355}{113}$, 3.14159

   b. Is this decimal approximation larger or smaller than π? Smaller

Optional Student Activity

The following complex fraction is mentioned in the Point of Interest note on page 592:

$$\dfrac{1}{\dfrac{1}{r_1} + \dfrac{1}{r_2}}$$

The fraction gives the total resistance, measured in ohms, of an electrical circuit that contains two parallel resistors with resistances of r_1 and r_2.

1. Show that you can rewrite the resistance fraction in the following form:

$$\dfrac{r_1 r_2}{r_1 + r_2}$$

Multiply the complex fraction

by $\dfrac{r_1 r_2}{r_1 r_2}$ and then simplify.

2. Suppose an electrical circuit contains two parallel resistors with resistances of $r_1 = 2$ ohms and $r_2 = 3$ ohms. Calculate the total resistance in the circuit twice: once using the complex fraction shown above, and once using the fraction shown in part 1.

$\dfrac{6}{5}$ ohms

3. Repeat part 2 using $r_1 = 6$ ohms and $r_2 = 8$ ohms.

$\dfrac{24}{7}$ ohms

4. Which form of the resistance fraction do you find it easier to work with when doing the calculations in parts 2 and 3? Explain why.

Answers will vary.

Example 2

Simplify: $\dfrac{1 - \dfrac{2}{x} - \dfrac{15}{x^2}}{1 - \dfrac{11}{x} + \dfrac{30}{x^2}}$

Solution

The LCM of x and x^2 is x^2.

$$\dfrac{1 - \dfrac{2}{x} - \dfrac{15}{x^2}}{1 - \dfrac{11}{x} + \dfrac{30}{x^2}} = \dfrac{1 - \dfrac{2}{x} - \dfrac{15}{x^2}}{1 - \dfrac{11}{x} + \dfrac{30}{x^2}} \cdot \dfrac{x^2}{x^2}$$

• Multiply by the LCM.

$$= \dfrac{1 \cdot x^2 - \dfrac{2}{x} \cdot x^2 - \dfrac{15}{x^2} \cdot x^2}{1 \cdot x^2 - \dfrac{11}{x} \cdot x^2 + \dfrac{30}{x^2} \cdot x^2}$$

• Distributive Property

$$= \dfrac{x^2 - 2x - 15}{x^2 - 11x + 30}$$

$$= \dfrac{\overset{1}{\cancel{(x - 5)}}(x + 3)}{\underset{1}{\cancel{(x - 5)}}(x - 6)} = \dfrac{x + 3}{x - 6}$$

• Simplify.

You Try It 2

Simplify: $\dfrac{1 + \dfrac{4}{x} + \dfrac{3}{x^2}}{1 + \dfrac{10}{x} + \dfrac{21}{x^2}}$

Your solution

$\dfrac{x + 1}{x + 7}$

Example 3

Simplify: $\dfrac{x - 8 + \dfrac{20}{x + 4}}{x - 10 + \dfrac{24}{x + 4}}$

Solution

The LCM is $x + 4$.

$$\dfrac{x - 8 + \dfrac{20}{x + 4}}{x - 10 + \dfrac{24}{x + 4}}$$

$$= \dfrac{x - 8 + \dfrac{20}{x + 4}}{x - 10 + \dfrac{24}{x + 4}} \cdot \dfrac{x + 4}{x + 4}$$

• Multiply by the LCM.

$$= \dfrac{(x - 8)(x + 4) + \dfrac{20}{x + 4} \cdot (x + 4)}{(x - 10)(x + 4) + \dfrac{24}{x + 4} \cdot (x + 4)}$$

• Distributive Property

$$= \dfrac{x^2 - 4x - 32 + 20}{x^2 - 6x - 40 + 24} = \dfrac{x^2 - 4x - 12}{x^2 - 6x - 16}$$

• Simplify.

$$= \dfrac{(x - 6)\overset{1}{\cancel{(x + 2)}}}{(x - 8)\underset{1}{\cancel{(x + 2)}}} = \dfrac{x - 6}{x - 8}$$

You Try It 3

Simplify: $\dfrac{x + 3 - \dfrac{20}{x - 5}}{x + 8 + \dfrac{30}{x - 5}}$

Your solution

$\dfrac{x - 7}{x - 2}$

Solutions on p. S28

11.4 Exercises

Objective A   To simplify a complex fraction

For Exercises 1 to 30, simplify.

1. $\dfrac{1 + \dfrac{3}{x}}{1 - \dfrac{9}{x^2}}$

$\dfrac{x}{x - 3}$

2. $\dfrac{1 + \dfrac{4}{x}}{1 - \dfrac{16}{x^2}}$

$\dfrac{x}{x - 4}$

3. $\dfrac{2 - \dfrac{8}{x + 4}}{3 - \dfrac{12}{x + 4}}$

$\dfrac{2}{3}$

4. $\dfrac{5 - \dfrac{25}{x + 5}}{1 - \dfrac{3}{x + 5}}$

$\dfrac{5x}{x + 2}$

5. $\dfrac{1 + \dfrac{5}{y - 2}}{1 - \dfrac{2}{y - 2}}$

$\dfrac{y + 3}{y - 4}$

6. $\dfrac{2 - \dfrac{11}{2x - 1}}{3 - \dfrac{17}{2x - 1}}$

$\dfrac{4x - 13}{2(3x - 10)}$

7. $\dfrac{4 - \dfrac{2}{x + 7}}{5 + \dfrac{1}{x + 7}}$

$\dfrac{2(2x + 13)}{5x + 36}$

8. $\dfrac{5 + \dfrac{3}{x - 8}}{2 - \dfrac{1}{x - 8}}$

$\dfrac{5x - 37}{2x - 17}$

9. $\dfrac{1 - \dfrac{1}{x} - \dfrac{6}{x^2}}{1 - \dfrac{9}{x^2}}$

$\dfrac{x + 2}{x + 3}$

10. $\dfrac{1 + \dfrac{4}{x} + \dfrac{4}{x^2}}{1 - \dfrac{2}{x} - \dfrac{8}{x^2}}$

$\dfrac{x + 2}{x - 4}$

11. $\dfrac{1 - \dfrac{5}{x} - \dfrac{6}{x^2}}{1 + \dfrac{6}{x} + \dfrac{5}{x^2}}$

$\dfrac{x - 6}{x + 5}$

12. $\dfrac{1 - \dfrac{7}{a} + \dfrac{12}{a^2}}{1 + \dfrac{1}{a} - \dfrac{20}{a^2}}$

$\dfrac{a - 3}{a + 5}$

13. $\dfrac{1 - \dfrac{6}{x} + \dfrac{8}{x^2}}{\dfrac{4}{x^2} + \dfrac{3}{x} - 1}$

$\dfrac{-x + 2}{x + 1}$

14. $\dfrac{1 + \dfrac{3}{x} - \dfrac{18}{x^2}}{\dfrac{21}{x^2} - \dfrac{4}{x} - 1}$

$\dfrac{-x - 6}{x + 7}$

15. $\dfrac{x - \dfrac{4}{x + 3}}{1 + \dfrac{1}{x + 3}}$

$x - 1$

16. $\dfrac{y + \dfrac{1}{y - 2}}{1 + \dfrac{1}{y - 2}}$

$y - 1$

17. $\dfrac{1 - \dfrac{x}{2x + 1}}{x - \dfrac{1}{2x + 1}}$

$\dfrac{1}{2x - 1}$

18. $\dfrac{1 - \dfrac{2x - 2}{3x - 1}}{x - \dfrac{4}{3x - 1}}$

$\dfrac{1}{3x - 4}$

Quick Quiz (Objective 11.4A)

Simplify.

1. $\dfrac{1 - \dfrac{2}{y} - \dfrac{8}{y^2}}{1 + \dfrac{5}{y} + \dfrac{6}{y^2}}$ $\dfrac{y - 4}{y + 3}$

2. $\dfrac{x - 2 + \dfrac{3}{x + 2}}{3x - 2 + \dfrac{5}{x + 2}}$ $\dfrac{x - 1}{3x + 1}$

Suggested Assignment

Exercises 1–29, odds
More challenging problems:
   Exercises 32–35

19. $\dfrac{x - 5 + \dfrac{14}{x + 4}}{x + 3 - \dfrac{2}{x + 4}}$

$\dfrac{x - 3}{x + 5}$

20. $\dfrac{a + 4 + \dfrac{5}{a - 2}}{a + 6 + \dfrac{15}{a - 2}}$

$\dfrac{a - 1}{a + 1}$

21. $\dfrac{x + 3 - \dfrac{10}{x - 6}}{x + 2 - \dfrac{20}{x - 6}}$

$\dfrac{x - 7}{x - 8}$

22. $\dfrac{x - 7 + \dfrac{5}{x - 1}}{x - 3 + \dfrac{1}{x - 1}}$

$\dfrac{x - 6}{x - 2}$

23. $\dfrac{y - 6 + \dfrac{22}{2y + 3}}{y - 5 + \dfrac{11}{2y + 3}}$

$\dfrac{2y - 1}{2y + 1}$

24. $\dfrac{x + 2 - \dfrac{12}{2x - 1}}{x + 1 - \dfrac{9}{2x - 1}}$

$\dfrac{2x + 7}{2x + 5}$

25. $\dfrac{x - \dfrac{2}{2x - 3}}{2x - 1 - \dfrac{8}{2x - 3}}$

$\dfrac{x - 2}{2x - 5}$

26. $\dfrac{x + 3 - \dfrac{18}{2x + 1}}{x - \dfrac{6}{2x + 1}}$

$\dfrac{x + 5}{x + 2}$

27. $\dfrac{\dfrac{1}{x} - \dfrac{2}{x - 1}}{\dfrac{3}{x} + \dfrac{1}{x - 1}}$

$\dfrac{-x - 1}{4x - 3}$

28. $\dfrac{\dfrac{3}{n + 1} + \dfrac{1}{n}}{\dfrac{2}{n + 1} + \dfrac{3}{n}}$

$\dfrac{4n + 1}{5n + 3}$

29. $\dfrac{\dfrac{3}{2x - 1} - \dfrac{1}{x}}{\dfrac{4}{x} + \dfrac{2}{2x - 1}}$

$\dfrac{x + 1}{2(5x - 2)}$

30. $\dfrac{\dfrac{4}{3x + 1} + \dfrac{3}{x}}{\dfrac{6}{x} - \dfrac{2}{3x + 1}}$

$\dfrac{13x + 3}{2(8x + 3)}$

APPLYING THE CONCEPTS

For Exercises 31 to 39, simplify.

31. $1 + \dfrac{1}{1 + \dfrac{1}{2}}$

$\dfrac{5}{3}$

32. $1 + \dfrac{1}{1 + \dfrac{1}{1 + \dfrac{1}{2}}}$

$\dfrac{8}{5}$

33. $1 - \dfrac{1}{1 - \dfrac{1}{x}}$

$-\dfrac{1}{x - 1}$

34. $\dfrac{a^{-1} - b^{-1}}{a^{-2} - b^{-2}}$

$\dfrac{ab}{b + a}$

35. $\left(\dfrac{y}{4} - \dfrac{4}{y}\right) \div \left(\dfrac{4}{y} - 3 + \dfrac{y}{2}\right)$

$\dfrac{y + 4}{2(y - 2)}$

36. $\left(\dfrac{b}{8} - \dfrac{8}{b}\right) \div \left(\dfrac{8}{b} - 5 + \dfrac{b}{2}\right)$

$\dfrac{b + 8}{4(b - 2)}$

37. $\dfrac{1 + x^{-1}}{1 - x^{-1}}$

$\dfrac{x + 1}{x - 1}$

38. $\dfrac{x + x^{-1}}{x - x^{-1}}$

$\dfrac{x^2 + 1}{(x + 1)(x - 1)}$

39. $\dfrac{x^{-1}}{y^{-1}} + \dfrac{x}{y}$

$\dfrac{y^2 + x^2}{xy}$

11.5 Solving Equations Containing Fractions

Objective A To solve an equation containing fractions

Recall that to solve an equation containing fractions, clear denominators by multiplying each side of the equation by the LCM of the denominators. Then solve for the variable.

HOW TO Solve: $\dfrac{3x - 1}{4} + \dfrac{2}{3} = \dfrac{7}{6}$

$$\frac{3x - 1}{4} + \frac{2}{3} = \frac{7}{6}$$

$$12\left(\frac{3x - 1}{4} + \frac{2}{3}\right) = 12 \cdot \frac{7}{6}$$

• The LCM is **12**. To clear denominators, multiply each side of the equation by the LCM.

$$12\left(\frac{3x - 1}{4}\right) + 12 \cdot \frac{2}{3} = 12 \cdot \frac{7}{6}$$

• Simplify using the **Distributive Property** and the properties of fractions.

$$\overset{3}{\underset{1}{\cancel{12}}}\left(\frac{3x - 1}{\underset{1}{\cancel{4}}}\right) + \overset{4}{\underset{1}{\cancel{12}}} \cdot \frac{2}{\underset{1}{\cancel{3}}} = \overset{2}{\underset{1}{\cancel{12}}} \cdot \frac{7}{\underset{1}{\cancel{6}}}$$

$$9x - 3 + 8 = 14$$

• Solve for x.

$$9x + 5 = 14$$
$$9x = 9$$
$$x = 1$$

1 checks as a solution. The solution is 1.

Occasionally, a value of the variable that appears to be a solution of an equation will make one of the denominators zero. In this case, that value is not a solution of the equation.

HOW TO Solve: $\dfrac{2x}{x - 2} = 1 + \dfrac{4}{x - 2}$

$$\frac{2x}{x - 2} = 1 + \frac{4}{x - 2}$$

$$(x - 2)\frac{2x}{x - 2} = (x - 2)\left(1 + \frac{4}{x - 2}\right)$$

• The LCM is $x - 2$. Multiply each side of the equation by the LCM.

$$(x - 2)\frac{2x}{x - 2} = (x - 2) \cdot 1 + (x - 2)\frac{4}{x - 2}$$

• Simplify using the **Distributive Property** and the properties of fractions.

$$\frac{\overset{1}{\cancel{(x - 2)}}}{1} \cdot \frac{2x}{\underset{1}{\cancel{x - 2}}} = (x - 2) \cdot 1 + \frac{\overset{1}{\cancel{(x - 2)}}}{1} \cdot \frac{4}{\underset{1}{\cancel{x - 2}}}$$

$$2x = x - 2 + 4$$

• Solve for x.

$$2x = x + 2$$
$$x = 2$$

When x is replaced by 2, the denominators of $\dfrac{2x}{x - 2}$ and $\dfrac{4}{x - 2}$ are zero.

Therefore, the equation has no solution.

In-Class Examples (Objective 11.5A)

Solve.

1. $\dfrac{8}{4x - 3} = -4$ $\dfrac{1}{4}$

2. $\dfrac{3x}{2x + 1} + \dfrac{1}{x + 2} = \dfrac{4}{x + 2}$ $1, -1$

Objective 11.5A

Vocabulary to Review
clearing denominators   [5.2A]

Concept Check

Clear the denominators in each equation (rewrite each equation as an equation without fractions).

1. $\dfrac{x}{7} - \dfrac{2}{21} = \dfrac{x}{3}$ $3x - 2 = 7x$

2. $\dfrac{4}{y - 3} - 1 = \dfrac{1}{y - 3}$ $4 - (y - 3) = 1$

3. $\dfrac{5}{a - 1} = \dfrac{2}{a}$ $5a = 2(a - 1)$

Optional Student Activity

Find pairs of equivalent equations in the following list.

(a) $3x = 2(x + 1) - 1$

(b) $3 = 2 - \dfrac{x + 1}{x}$

(c) $\dfrac{3x}{x + 1} = 2 - \dfrac{1}{x + 1}$

(d) $3x = 2x - (x + 1)$

a and c, b and d

Instructor Note

The example at the left illustrates the importance of checking a solution of a rational equation when each side is multiplied by a variable expression.

You might emphasize that the Multiplication Property of Equations states that each side of an equation can be multiplied by the same *nonzero* number. If x is allowed to be 2 in this example, each side of the equation is multiplied by zero.

Copyright © Houghton Mifflin Company. All rights reserved.

Instructor Note

After learning to clear denominators in order to solve an equation containing fractions or to simplify a complex fraction, some students may try using the same technique when simply *adding* fractions. The following Discuss the Concepts question should help students learn to avoid making this error.

Discuss the Concepts

Look at the following two problems involving the fractions $\frac{x}{2}, \frac{1}{3}$, and $\frac{5}{2}$. Explain why you can clear the denominators in Problem 1 but not in Problem 2.

(1) Solve: $\frac{x}{2} + \frac{1}{3} = \frac{5}{2}$

(2) Simplify: $\frac{x}{2} + \frac{1}{3} + \frac{5}{2}$

Example 1

Solve: $\dfrac{x}{x+4} = \dfrac{2}{x}$

Solution

The LCM is $x(x + 4)$.

$$\frac{x}{x+4} = \frac{2}{x}$$

$$x(x+4)\left(\frac{x}{x+4}\right) = x(x+4)\left(\frac{2}{x}\right)$$ • Multiply by the LCM.

$$\frac{x\cancel{(x+4)}}{1} \cdot \frac{x}{\cancel{x+4}} = \frac{\cancel{x}(x+4)}{1} \cdot \frac{2}{\cancel{x}}$$ • Divide by common factors.

$$x^2 = (x+4)2$$ • Simplify.

$$x^2 = 2x + 8$$ • This is a quadratic equation.

$$x^2 - 2x - 8 = 0$$ • Write in standard form.

$$(x-4)(x+2) = 0$$ • Factor.

$$x - 4 = 0 \qquad x + 2 = 0$$ • Principle of Zero Products

$$x = 4 \qquad\qquad x = -2$$

Both 4 and −2 check as solutions.
The solutions are 4 and −2.

You Try It 1

Solve: $\dfrac{x}{x+6} = \dfrac{3}{x}$

Your solution
−3, 6

Example 2

Solve: $\dfrac{3x}{x-4} = 5 + \dfrac{12}{x-4}$

Solution

The LCM is $x - 4$.

$$\frac{3x}{x-4} = 5 + \frac{12}{x-4}$$

$$(x-4)\left(\frac{3x}{x-4}\right) = (x-4)\left(5 + \frac{12}{x-4}\right)$$ • Clear denominators.

$$\frac{\cancel{(x-4)}}{1} \cdot \frac{3x}{\cancel{x-4}} = (x-4)5 + \frac{\cancel{(x-4)}}{1} \cdot \frac{12}{\cancel{x-4}}$$

$$3x = (x-4)5 + 12$$ • Solve for x.

$$3x = 5x - 20 + 12$$

$$3x = 5x - 8$$

$$-2x = -8$$

$$x = 4$$

4 does not check as a solution.
The equation has no solution.

You Try It 2

Solve: $\dfrac{5x}{x+2} = 3 - \dfrac{10}{x+2}$

Your solution
No solution

Solutions on p. S29

11.5 Exercises

Objective A   **To solve an equation containing fractions**

1. ✎ Can 2 be a solution of the equation $\dfrac{6x}{x+1} - \dfrac{x}{x-2} = 4$? Explain your answer.

2. ✎ After multiplying each side of an equation by a variable expression, why must we check the solution?

For Exercises 3 to 35, solve.

3. $\dfrac{2x}{3} - \dfrac{5}{2} = -\dfrac{1}{2}$

3

4. $\dfrac{x}{3} - \dfrac{1}{4} = \dfrac{1}{12}$

1

5. $\dfrac{x}{3} - \dfrac{1}{4} = \dfrac{x}{4} - \dfrac{1}{6}$

1

6. $\dfrac{2y}{9} - \dfrac{1}{6} = \dfrac{y}{9} + \dfrac{1}{6}$

3

7. $\dfrac{2x-5}{8} + \dfrac{1}{4} = \dfrac{x}{8} + \dfrac{3}{4}$

9

8. $\dfrac{3x+4}{12} - \dfrac{1}{3} = \dfrac{5x+2}{12} - \dfrac{1}{2}$

2

9. $\dfrac{6}{2a+1} = 2$

1

10. $\dfrac{12}{3x-2} = 3$

2

11. $\dfrac{9}{2x-5} = -2$

$\dfrac{1}{4}$

12. $\dfrac{6}{4-3x} = 3$

$\dfrac{2}{3}$

13. $2 + \dfrac{5}{x} = 7$

1

14. $3 + \dfrac{8}{n} = 5$

4

15. $1 - \dfrac{9}{x} = 4$

-3

16. $3 - \dfrac{12}{x} = 7$

-3

17. $\dfrac{2}{y} + 5 = 9$

$\dfrac{1}{2}$

18. $\dfrac{6}{x} + 3 = 11$

$\dfrac{3}{4}$

19. $\dfrac{3}{x-2} = \dfrac{4}{x}$

8

20. $\dfrac{5}{x+3} = \dfrac{3}{x-1}$

7

21. $\dfrac{2}{3x-1} = \dfrac{3}{4x+1}$

5

22. $\dfrac{5}{3x-4} = \dfrac{-3}{1-2x}$

-7

23. $\dfrac{-3}{2x+5} = \dfrac{2}{x-1}$

-1

Section 11.5

Suggested Assignment
Exercises 1–35, odds
More challenging problems:
  Exercises 36–38

Answers to Writing Exercises

1. 2 cannot be a solution of
$\dfrac{6x}{x+1} - \dfrac{x}{x-2} = 4$ because
the denominator of $\dfrac{x}{x-2}$ is 0
when $x = 2$.

2. Multiplying each side of an equation by a variable expression may introduce an extraneous solution (a solution that is not a solution of the original equation). Therefore, the solutions must be checked.

Quick Quiz (Objective 11.5A)

Solve.

1. $5 - \dfrac{8}{x} = 1$ 2

2. $3 - \dfrac{4}{x+2} = \dfrac{2x}{x+2}$ No solution

24. $\dfrac{4}{5y-1}=\dfrac{2}{2y-1}$

-1

25. $\dfrac{4x}{x-4}+5=\dfrac{5x}{x-4}$

5

26. $\dfrac{2x}{x+2}-5=\dfrac{7x}{x+2}$

-1

27. $2+\dfrac{3}{a-3}=\dfrac{a}{a-3}$

No solution

28. $\dfrac{x}{x+4}=3-\dfrac{4}{x+4}$

No solution

29. $\dfrac{x}{x-1}=\dfrac{8}{x+2}$

$2, 4$

30. $\dfrac{x}{x+12}=\dfrac{1}{x+5}$

$2, -6$

31. $\dfrac{2x}{x+4}=\dfrac{3}{x-1}$

$-\dfrac{3}{2}, 4$

32. $\dfrac{5}{3n-8}=\dfrac{n}{n+2}$

$-\dfrac{2}{3}, 5$

33. $x+\dfrac{6}{x-2}=\dfrac{3x}{x-2}$

3

34. $x-\dfrac{6}{x-3}=\dfrac{2x}{x-3}$

$-1, 6$

35. $\dfrac{8}{y}=\dfrac{2}{y-2}+1$

4

APPLYING THE CONCEPTS

36. Explain the procedure for solving an equation containing fractions. Include in your discussion how the LCM of the denominators is used to eliminate fractions in the equation.

For Exercises 37 to 42, solve.

37. $\dfrac{3}{5}y-\dfrac{1}{3}(1-y)=\dfrac{2y-5}{15}$

0

38. $\dfrac{3}{4}a=\dfrac{1}{2}(3-a)+\dfrac{a-2}{4}$

1

39. $\dfrac{b+2}{5}=\dfrac{1}{4}b-\dfrac{3}{10}(b-1)$

$-\dfrac{2}{5}$

40. $\dfrac{x}{2x^2-x-1}=\dfrac{3}{x^2-1}+\dfrac{3}{2x+1}$

$0, -\dfrac{5}{2}$

41. $\dfrac{x+1}{x^2+x-2}=\dfrac{x+2}{x^2-1}+\dfrac{3}{x+2}$

$0, -\dfrac{2}{3}$

42. $\dfrac{y+2}{y^2-y-2}+\dfrac{y+1}{y^2-4}=\dfrac{1}{y+1}$

-3

Answers to Writing Exercises

36. Students' explanations should include each of the following steps:
(1) Find the LCM of the denominators of the fractions in the equation.
(2) Multiply each side of the equation by the LCM of the denominators.
(3) Simplify each side of the equation.
(4) Solve for the variable.

11.6 Literal Equations

Objective A To solve a literal equation for one of the variables

VIDEO & DVD   CD TUTOR   WEB   SSM

A **literal equation** is an equation that contains more than one variable. Examples of literal equations are shown at the right.

$$2x + 3y = 6$$
$$4w - 2x + z = 0$$

Formulas are used to express relationships among physical quantities. A **formula** is a literal equation that states a rule about measurements. Examples of formulas are shown at the right.

$$\frac{1}{R_1} + \frac{1}{R_2} = \frac{1}{R} \qquad \text{(Physics)}$$
$$s = a + (n - 1)d \qquad \text{(Mathematics)}$$
$$A = P + Prt \qquad \text{(Business)}$$

The Addition and Multiplication Properties can be used to solve a literal equation for one of the variables. The goal is to rewrite the equation so that the variable being solved for is alone on one side of the equation and all the other numbers and variables are on the other side.

HOW TO Solve $A = P(1 + i)$ for i.

The goal is to rewrite the equation so that i is on one side of the equation and all other variables are on the other side.

$$A = P(1 + i)$$
$$A = P + Pi \qquad \bullet \text{ Use the } \textbf{Distributive Property} \text{ to remove parentheses.}$$
$$A - P = P - P + Pi \qquad \bullet \text{ Subtract } \textbf{P} \text{ from each side of the equation.}$$
$$A - P = Pi$$
$$\frac{A - P}{P} = \frac{Pi}{P} \qquad \bullet \text{ Divide each side of the equation by } \textbf{P}.$$
$$\frac{A - P}{P} = i$$

Example 1

Solve $3x - 4y = 12$ for y.

Solution

$$3x - 4y = 12$$
$$3x - 3x - 4y = -3x + 12 \qquad \bullet \text{ Subtract } \textbf{3x}.$$
$$-4y = -3x + 12$$
$$\frac{-4y}{-4} = \frac{-3x + 12}{-4} \qquad \bullet \text{ Divide by } \textbf{-4}.$$
$$y = \frac{3}{4}x - 3$$

You Try It 1

Solve $5x - 2y = 10$ for y.

Your solution

$$y = \frac{5}{2}x - 5$$

Solution on p. S29

Objective 11.6A

New Vocabulary
literal equation
formula

Discuss the Concepts
Use the formula

$$A = P + Prt$$

which gives the value A of a principal P invested at a rate r for a given period of time t.

1. In what circumstances would it be helpful to have the formula solved for t? for P?
2. Describe the steps you would use to solve the equation for t.
3. If the formula is to be solved for P, P can appear only once in the formula. How can you rewrite the right side so that it has only one P? What would you do then to solve for P?

Instructor Note
Part 3 in the above Discuss the Concepts is difficult for students. This technique is also discussed in the Instructor Note on the next page.

Concept Check
Use the equation $5a = 7 - 2b$.

1. Solve for a. $a = \dfrac{7 - 2b}{5}$
2. Solve for b. $b = \dfrac{7 - 5a}{2}$

In-Class Examples (Objective 11.6A)

1. Solve for b: $A = \dfrac{1}{2}bh$ $b = \dfrac{2A}{h}$

2. Solve for x: $3x + 8y = 9$ $x = -\dfrac{8}{3}y + 3$

3. Solve for y: $7x - y = 12$ $y = 7x - 12$

Instructor Note

Example 4 will be difficult for students. Before doing that example or one similar to it, remind students that when solving $2x + 3x = 10$, they are using the Distributive Property to combine $2x$ and $3x$:

$$2x + 3x = (2 + 3)x = 5x$$

Now each side of the equation can be divided by 5.

For $ax + bx = 10$, the procedure is exactly the same, except that $a + b$ does not simplify further:

$$ax + bx = (a + b)x$$

Now each side of the equation can be divided by $(a + b)$.

Part 1 of the following Optional Student Activity can be used as another example illustrating this technique.

Optional Student Activity

1. Solve the equation
$cx - y = bx + 5$ for x.

$$x = \frac{y + 5}{c - b}$$

2. Solve this physics formula for R_1:

$$\frac{1}{R_1} + \frac{1}{R_2} = \frac{1}{R}$$

$$R_1 = \frac{RR_2}{R_2 - R}$$

Example 2

Solve $I = \dfrac{E}{R + r}$ for R.

Solution

$$I = \frac{E}{R + r}$$

$$(R + r)I = (R + r)\frac{E}{R + r} \quad \text{• Multiply by } (R + r).$$

$$RI + rI = E$$

$$RI + rI - rI = E - rI \quad \text{• Subtract } rI.$$

$$RI = E - rI$$

$$\frac{RI}{I} = \frac{E - rI}{I} \quad \text{• Divide by } I.$$

$$R = \frac{E - rI}{I}$$

You Try It 2

Solve $s = \dfrac{A + L}{2}$ for L.

Your solution
$$L = 2s - A$$

Example 3

Solve $L = a(1 + ct)$ for c.

Solution

$$L = a(1 + ct)$$

$$L = a + act \quad \text{• Distributive Property}$$

$$L - a = a - a + act \quad \text{• Subtract } a.$$

$$L - a = act$$

$$\frac{L - a}{at} = \frac{act}{at} \quad \text{• Divide by } at.$$

$$\frac{L - a}{at} = c$$

You Try It 3

Solve $S = a + (n - 1)d$ for n.

Your solution
$$n = \frac{S - a + d}{d}$$

Example 4

Solve $S = C - rC$ for C.

Solution

$$S = C - rC$$

$$S = C(1 - r) \quad \text{• Factor.}$$

$$\frac{S}{1 - r} = \frac{C(1 - r)}{1 - r} \quad \text{• Divide by } (1 - r).$$

$$\frac{S}{1 - r} = C$$

You Try It 4

Solve $S = rS + C$ for S.

Your solution
$$S = \frac{C}{1 - r}$$

Solutions on p. S29

11.6 Exercises

Objective A   To solve a literal equation for one of the variables

For Exercises 1 to 15, solve for y.

1. $3x + y = 10$
$y = -3x + 10$

2. $2x + y = 5$
$y = -2x + 5$

3. $4x - y = 3$
$y = 4x - 3$

4. $5x - y = 7$
$y = 5x - 7$

5. $3x + 2y = 6$
$y = -\dfrac{3}{2}x + 3$

6. $2x + 3y = 9$
$y = -\dfrac{2}{3}x + 3$

7. $2x - 5y = 10$
$y = \dfrac{2}{5}x - 2$

8. $5x - 2y = 4$
$y = \dfrac{5}{2}x - 2$

9. $2x + 7y = 14$
$y = -\dfrac{2}{7}x + 2$

10. $6x - 5y = 10$
$y = \dfrac{6}{5}x - 2$

11. $x + 3y = 6$
$y = -\dfrac{1}{3}x + 2$

12. $x + 2y = 8$
$y = -\dfrac{1}{2}x + 4$

13. $y - 2 = 3(x + 2)$

$y = 3x + 8$

14. $y + 4 = -2(x - 3)$

$y = -2x + 2$

15. $y - 1 = -\dfrac{2}{3}(x + 6)$

$y = -\dfrac{2}{3}x - 3$

For Exercises 16 to 24, solve for x.

16. $x + 3y = 6$
$x = -3y + 6$

17. $x + 6y = 10$
$x = -6y + 10$

18. $3x - y = 3$
$x = \dfrac{1}{3}y + 1$

19. $2x - y = 6$
$x = \dfrac{1}{2}y + 3$

20. $2x + 5y = 10$
$x = -\dfrac{5}{2}y + 5$

21. $4x + 3y = 12$
$x = -\dfrac{3}{4}y + 3$

22. $x - 2y + 1 = 0$
$x = 2y - 1$

23. $x - 4y - 3 = 0$
$x = 4y + 3$

24. $5x + 4y + 20 = 0$
$x = -\dfrac{4}{5}y - 4$

For Exercises 25 to 40, solve the formula for the given variable.

25. $d = rt;\ t$ (Physics)
$t = \dfrac{d}{r}$

26. $E = IR;\ R$ (Physics)
$R = \dfrac{E}{I}$

27. $PV = nRT;\ T$ (Chemistry)
$T = \dfrac{PV}{nR}$

28. $A = bh;\ h$ (Geometry)
$h = \dfrac{A}{b}$

Section 11.6

Suggested Assignment
Exercises 1–39, odds
More challenging problem:
Exercise 41

Quick Quiz (Objective 11.6A)

1. Use the equation $4x + y = 8$.
   a. Solve for y. $y = -4x + 8$

   b. Solve for x. $x = -\dfrac{1}{4}y + 2$

2. Use the formula $P = C + Cr$.
   a. Solve for r. $r = \dfrac{P - C}{C}$

   b. Solve for C. $C = \dfrac{P}{1 + r}$

29. $P = 2l + 2w; l$ (Geometry)

$l = \dfrac{P - 2w}{2}$

30. $F = \dfrac{9}{5}C + 32; C$ (Temperature conversion)

$C = \dfrac{5F - 160}{9}$

31. $A = \dfrac{1}{2}h(b_1 + b_2); b_1$ (Geometry)

$b_1 = \dfrac{2A - hb_2}{h}$

32. $C = \dfrac{5}{9}(F - 32); F$ (Temperature conversion)

$F = \dfrac{9C + 160}{5}$

33. $V = \dfrac{1}{3}Ah; h$ (Geometry)

$h = \dfrac{3V}{A}$

34. $P = R - C; C$ (Business)

$C = R - P$

35. $R = \dfrac{C - S}{t}; S$ (Business)

$S = C - Rt$

36. $P = \dfrac{R - C}{n}; R$ (Business)

$R = Pn + C$

37. $A = P + Prt; P$ (Business)

$P = \dfrac{A}{1 + rt}$

38. $T = fm - gm; m$ (Engineering)

$m = \dfrac{T}{f - g}$

39. $A = Sw + w; w$ (Physics)

$w = \dfrac{A}{S + 1}$

40. $a = S - Sr; S$ (Mathematics)

$S = \dfrac{a}{1 - r}$

APPLYING THE CONCEPTS

Business   Break-even analysis is a method used to determine the sales volume required for a company to break even, or experience neither a profit nor a loss on the sale of a product. The break-even point represents the number of units that must be made and sold for income from sales to equal the cost of the product. The break-even point can be calculated using the formula $B = \dfrac{F}{S - V}$, where F is the fixed costs, S is the selling price per unit, and V is the variable costs per unit. Use this information for Exercise 41.

41. a. Solve the formula $B = \dfrac{F}{S - V}$ for S. $S = \dfrac{F + BV}{B}$

b. Use your answer to part **a** to find the selling price per unit required for a company to break even. The fixed costs are $20,000, the variable costs per unit are $80, and the company plans to make and sell 200 desks.
$180

c. Use your answer to part **a** to find the selling price per unit required for a company to break even. The fixed costs are $15,000, the variable costs per unit are $50, and the company plans to make and sell 600 cameras.
$75

11.7 Application Problems

Objective A   To solve work problems

If a painter can paint a room in 4 h, then in 1 h the painter can paint $\frac{1}{4}$ of the room. The painter's rate of work is $\frac{1}{4}$ of the room each hour. The **rate of work** is the part of a task that is completed in 1 unit of time.

A pipe can fill a tank in 30 min. This pipe can fill $\frac{1}{30}$ of the tank in 1 min. The rate of work is $\frac{1}{30}$ of the tank each minute. If a second pipe can fill the tank in x min, the rate of work for the second pipe is $\frac{1}{x}$ of the tank each minute.

In solving a work problem, the goal is to determine the time it takes to complete a task. The basic equation that is used to solve work problems is

Rate of work × time worked = part of task completed

For example, if a faucet can fill a sink in 6 min, then in 5 min the faucet will fill $\frac{1}{6} \times 5 = \frac{5}{6}$ of the sink. In 5 min the faucet completes $\frac{5}{6}$ of the task.

Study Tip

Note in the examples in this section that solving a word problem includes stating a strategy and using the strategy to find a solution. If you have difficulty with a word problem, write down the known information. Be very specific. Write a phrase or sentence that states what you are trying to find. See *AIM for Success*, page xxxiii.

TAKE NOTE

Use the information given in the problem to fill in the "Rate of Work" and "Time Worked" columns of the table. Fill in the "Part of Task Completed" column by multiplying the two expressions you wrote in each row.

HOW TO   A painter can paint a wall in 20 min. The painter's apprentice can paint the same wall in 30 min. How long will it take them to paint the wall when they work together?

Strategy for Solving a Work Problem

1. For each person or machine, write a numerical or variable expression for the rate of work, the time worked, and the part of the task completed. The results can be recorded in a table.

Unknown time to paint the wall working together: t

	Rate of Work	·	Time Worked	=	Part of Task Completed
Painter	$\frac{1}{20}$	·	t	=	$\frac{t}{20}$
Apprentice	$\frac{1}{30}$	·	t	=	$\frac{t}{30}$

2. Determine how the parts of the task completed are related. Use the fact that the sum of the parts of the task completed must equal 1, the complete task.

$\frac{t}{20} + \frac{t}{30} = 1$ • The sum of the part of the task completed by the painter and the part of the task completed by the apprentice is 1.

$60\left(\frac{t}{20} + \frac{t}{30}\right) = 60 \cdot 1$ • Multiply by the **LCM of 20 and 30**.

$3t + 2t = 60$ • Distributive Property

$5t = 60$

$t = 12$

Working together, they will paint the wall in 12 min.

Objective 11.7A

New Vocabulary
rate of work

Concept Check
It takes Jenny 3 h to mow her lawn.
1. What is Jenny's rate of work?
$\frac{1}{3}$ of the lawn per hour

2. What fraction of the lawn can she mow in 2 h? $\frac{2}{3}$

3. What fraction of the lawn can she mow in $2\frac{1}{2}$ h? $\frac{5}{6}$

Discuss the Concepts

Refer to the equation $\frac{t}{20} + \frac{t}{30} = 1$ in the example at the left.

1. What does $\frac{t}{20}$ represent?
The part of the wall that the painter painted

2. What does $\frac{t}{30}$ represent?
The part of the wall that the apprentice painted

3. Why is the sum of $\frac{t}{20}$ and $\frac{t}{30}$ equal to 1?
The entire wall, or 100% of the wall, was painted by both people working together.

Optional Student Activity

Here is a challenge problem for students to try.

A construction project must be completed in 15 days. Twenty-five workers did one-half the job in 10 days. How many additional workers are necessary to complete the job in 5 days?

25 additional workers

Example 1

A small water pipe takes three times longer to fill a tank than does a large water pipe. With both pipes open, it takes 4 h to fill the tank. Find the time it would take the small pipe, working alone, to fill the tank.

Strategy

- Time for large pipe to fill the tank: t
  Time for small pipe to fill the tank: $3t$

Fills tank in $3t$ hours   Fills tank in t hours

Fills $\frac{4}{3t}$ of the tank in 4 hours   Fills $\frac{4}{t}$ of the tank in 4 hours

	Rate	Time	Part
Small pipe	$\frac{1}{3t}$	4	$\frac{4}{3t}$
Large pipe	$\frac{1}{t}$	4	$\frac{4}{t}$

- The sum of the parts of the task completed by each pipe must equal 1.

Solution

$$\frac{4}{3t} + \frac{4}{t} = 1$$

$$3t\left(\frac{4}{3t} + \frac{4}{t}\right) = 3t \cdot 1$$ • Multiply by the LCM of $3t$ and t.

$$4 + 12 = 3t$$ • Distributive Property

$$16 = 3t$$

$$\frac{16}{3} = t$$

$$3t = 3\left(\frac{16}{3}\right) = 16$$

The small pipe working alone takes 16 h to fill the tank.

You Try It 1

Two computer printers that work at the same rate are working together to print the payroll checks for a large corporation. After they work together for 2 h, one of the printers fails. The second printer requires 3 h more to complete the payroll checks. Find the time it would take one printer, working alone, to print the payroll.

Your strategy

Your solution

7 h

Solution on p. S29

Objective B   To use rational expressions to solve uniform motion problems

A car that travels constantly in a straight line at 30 mph is in uniform motion. **Uniform motion** means that the speed and direction of an object do not change.

The basic equation used to solve uniform motion problems is

$$\text{Distance} = \text{rate} \times \text{time}$$

An alternative form of this equation can be written by solving the equation for time.

$$\frac{\text{Distance}}{\text{Rate}} = \text{time}$$

This form of the equation is useful when the total time of travel for two objects or the time of travel between two points is known.

HOW TO   The speed of a boat in still water is 20 mph. The boat traveled 75 mi down a river in the same amount of time it took to travel 45 mi up the river. Find the rate of the river's current.

> **Strategy for Solving a Uniform Motion Problem**
>
> **1.** For each object, write a numerical or variable expression for the distance, rate, and time. The results can be recorded in a table.

The unknown rate of the river's current: r

	Distance	÷	Rate	=	Time
Down river	75	÷	$20 + r$	=	$\dfrac{75}{20 + r}$
Up river	45	÷	$20 - r$	=	$\dfrac{45}{20 - r}$

TAKE NOTE
Use the information given in the problem to fill in the "Distance" and "Rate" columns of the table. Fill in the "Time" column by dividing the two expressions you wrote in each row.

> **2.** Determine how the times traveled by each object are related. For example, it may be known that the times are equal, or the total time may be known.

$$\frac{75}{20 + r} = \frac{45}{20 - r}$$

• The time down the river is equal to the time up the river.

$$(20 + r)(20 - r)\frac{75}{20 + r} = (20 + r)(20 - r)\frac{45}{20 - r}$$

• Multiply by the **LCM**.

$$(20 - r)75 = (20 + r)45$$

$$1500 - 75r = 900 + 45r$$

• Distributive Property

$$-120r = -600$$

$$r = 5$$

The rate of the river's current is 5 mph.

Objective 11.7B

Vocabulary to Review
uniform motion   [5.1D]

Discuss the Concepts
In the example at the left, explain why the rate of the boat going down the river is represented by $20 + r$ and the rate of the boat going up the river is represented by $20 - r$. Going with the current, the boat is aided by the current. Going against the current, the boat is hindered by the current.

Concept Check
A trip involved 240 mi of highway driving and 120 mi of nonhighway driving. The driver's average highway speed was 20 mph faster than his average nonhighway speed. The total driving time was 7 h.

1. Write expressions to represent the driver's average nonhighway speed, his average highway speed, the time spent on highway driving, and the time spent on nonhighway driving.
$r, r + 20, \dfrac{240}{r + 20}, \dfrac{120}{r}$

2. Write an equation that expresses the fact that the total driving time was 7 h.
$\dfrac{120}{r} + \dfrac{240}{r + 20} = 7$

In-Class Examples (Objective 11.7B)

1. Lorenzo's bicycling rate is six times as fast as his walking rate. On his bicycle, he can complete a 9-mile route in $2\frac{1}{2}$ h less time than it takes him to walk the same route. Find Lorenzo's walking rate and his cycling rate.   3 mph, 18 mph

Optional Student Activity

Here is a challenge problem for students to try.

Two trains on the same track are heading toward each other. One is traveling at 40 mph; the other is traveling at 50 mph. How far apart are they 1 min before they would meet?   1.5 mi

Example 2

A cyclist rode the first 20 mi of a trip at a constant rate. For the next 16 mi, the cyclist reduced the speed by 2 mph. The total time for the 36 mi was 4 h. Find the rate of the cyclist for each leg of the trip.

Strategy

- Rate for the first 20 mi: r
  Rate for the next 16 mi: $r - 2$

r
20 mi
$r - 2$
16 mi

	Distance	Rate	Time
First 20 mi	20	r	$\dfrac{20}{r}$
Next 16 mi	16	$r - 2$	$\dfrac{16}{r - 2}$

- The total time for the trip was 4 h.

Solution

$$\frac{20}{r} + \frac{16}{r - 2} = 4$$ • The total time was 4 h.

$$r(r - 2)\left[\frac{20}{r} + \frac{16}{r - 2}\right] = r(r - 2) \cdot 4$$ • Multiply by the LCM.

$$(r - 2)20 + 16r = 4r^2 - 8r$$ • Distributive Property

$$20r - 40 + 16r = 4r^2 - 8r$$ • A quadratic equation

$$36r - 40 = 4r^2 - 8r$$

$$0 = 4r^2 - 44r + 40$$ • Standard form

$$0 = 4(r^2 - 11r + 10)$$

$$0 = 4(r - 10)(r - 1)$$ • Factor.

$$r - 10 = 0 \qquad r - 1 = 0$$ • Principle of Zero Products

$$r = 10 \qquad r = 1$$

The solution $r = 1$ mph is not possible, because the rate on the last 16 mi would then be -1 mph.

10 mph was the rate for the first 20 mi.
8 mph was the rate for the next 16 mi.

You Try It 2

The total time it took for a sailboat to sail back and forth across a lake 6 km wide was 2 h. The rate sailing back was three times the rate sailing across. Find the rate sailing out across the lake.

Your strategy

Your solution
4 km/h

Solution on pp. S29–S30

11.7 Exercises

Objective A   To solve work problems

1. ✎ Explain the meaning of the phrase "rate of work."

2. If $\frac{2}{5}$ of a room can be painted in 1 h, what is the rate of work? At the same rate, how long will it take to paint the entire room?
$\frac{2}{5}$ of the room each hour; $2\frac{1}{2}$ h

3. A park has two sprinklers that are used to fill a fountain. One sprinkler can fill the fountain in 3 h, whereas the second sprinkler can fill the fountain in 6 h. How long will it take to fill the fountain with both sprinklers operating?
2 h

4. One grocery clerk can stock a shelf in 20 min, whereas a second clerk requires 30 min to stock the same shelf. How long would it take to stock the shelf if the two clerks worked together?
12 min

5. One person with a skiploader requires 12 h to remove a large quantity of earth. A second, larger skiploader can remove the same amount of earth in 4 h. How long would it take to remove the earth with both skiploaders working together?
3 h

6. An experienced painter can paint a fence twice as fast as an inexperienced painter. Working together, the painters require 4 h to paint the fence. How long would it take the experienced painter, working alone, to paint the fence?
6 h

7. One computer can solve a complex prime factorization problem in 75 h. A second computer can solve the same problem in 50 h. How long would it take both computers, working together, to solve the problem?
30 h

8. A new machine can make 10,000 aluminum cans three times faster than an older machine. With both machines working, 10,000 cans can be made in 9 h. How long would it take the new machine, working alone, to make the 10,000 cans?
12 h

9. A small air conditioner can cool a room 5° in 75 min. A larger air conditioner can cool the room 5° in 50 min. How long would it take to cool the room 5° with both air conditioners working?
30 min

10. One printing press can print the first edition of a book in 55 min, whereas a second printing press requires 66 min to print the same number of copies. How long would it take to print the first edition with both presses operating?
30 min

11. Two oil pipelines can fill a small tank in 30 min. One of the pipelines requires 45 min to fill the tank. How long would it take the second pipeline, working alone, to fill the tank?
90 min

Suggested Assignment
Exercises 1–37, odds
More challenging problems:
  Exercises 38, 39

Answers to Writing Exercises

1. The rate of work is the amount of a task that is completed per unit of time.

Quick Quiz (Objective 11.7A)

1. One hose can fill a child's backyard pool in 24 min. A larger hose can fill the pool in 12 min. If the two hoses are used together, how long will it take to fill the pool?   8 min

12. Working together, two dock workers can load a crate in 6 min. One dock worker, working alone, can load the crate in 15 min. How long would it take the second dock worker, working alone, to load the crate?   10 min

13. A mason can construct a retaining wall in 10 h. With the mason's apprentice assisting, the task takes 6 h. How long would it take the apprentice, working alone, to construct the wall?   15 h

14. A mechanic requires 2 h to repair a transmission, whereas an apprentice requires 6 h to make the same repairs. The mechanic worked alone for 1 h and then stopped. How long will it take the apprentice, working alone, to complete the repairs?   3 h

15. One computer technician can wire a modem in 4 h, whereas it takes 6 h for a second technician to do the same job. After working alone for 2 h, the first technician quit. How long will it take the second technician to complete the wiring?   3 h

16. A wallpaper hanger requires 2 h to hang the wallpaper on one wall of a room. A second wallpaper hanger requires 4 h to hang the same amount of paper. The first wallpaper hanger worked alone for 1 h and then quit. How long will it take the second wallpaper hanger, working alone, to complete the wall?   2 h

17. Two welders who work at the same rate are welding the girders of a building. After they work together for 10 h, one of the welders quits. The second welder requires 20 more hours to complete the welds. Find the time it would have taken one of the welders, working alone, to complete the welds.   40 h

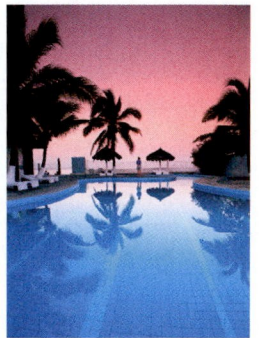

18. A large and a small heating unit are being used to heat the water of a pool. The larger unit, working alone, requires 8 h to heat the pool. After both units have been operating for 2 h, the larger unit is turned off. The small unit requires 9 h more to heat the pool. How long would it take the small unit, working alone, to heat the pool? $14\frac{2}{3}$ h

19. Two machines that fill cereal boxes work at the same rate. After they work together for 7 h, one machine breaks down. The second machine requires 14 h more to finish filling the boxes. How long would it have taken one of the machines, working alone, to fill the boxes?   28 h

20. A large and a small drain are opened to drain a pool. The large drain can empty the pool in 6 h. After both drains have been open for 1 h, the large drain becomes clogged and is closed. The smaller drain remains open and requires 9 h more to empty the pool. How long would it have taken the small drain, working alone, to empty the pool?   12 h

Objective B   **To use rational expressions to solve uniform motion problems**

21. Running at a constant speed, a jogger ran 24 mi in 3 h. How far did the jogger run in 2 h?
16 mi

22. For uniform motion, distance = rate · time. How is time related to distance and rate? How is rate related to distance and time?

Time = $\dfrac{\text{distance}}{\text{rate}}$; Rate = $\dfrac{\text{distance}}{\text{time}}$

23. Commuting from work to home, a lab technician traveled 10 mi at a constant rate through congested traffic. On reaching the expressway, the technician increased the speed by 20 mph. An additional 20 mi was traveled at the increased speed. The total time for the trip was 1 h. Find the rate of travel through the congested traffic.
20 mph

```
|--10 mi--|----20 mi----|  🚗
     r         r + 20
```

24. The president of a company traveled 1800 mi by jet and 300 mi on a prop plane. The rate of the jet was four times the rate of the prop plane. The entire trip took a total of 5 h. Find the rate of the jet plane.
600 mph

25. As part of a conditioning program, a jogger ran 8 mi in the same amount of time a cyclist rode 20 mi. The rate of the cyclist was 12 mph faster than the rate of the jogger. Find the rate of the jogger and that of the cyclist.
Jogger: 8 mph; cyclist: 20 mph

```
|--8 mi--| 🏃
    r

|-----20 mi-----| 🚴
      r + 12
```

26. An express train travels 600 mi in the same amount of time it takes a freight train to travel 360 mi. The rate of the express train is 20 mph faster than that of the freight train. Find the rate of each train.
Freight train: 30 mph; express train: 50 mph

27. To assess the damage done by a fire, a forest ranger traveled 1080 mi by jet and then an additional 180 mi by helicopter. The rate of the jet was four times the rate of the helicopter. The entire trip took a total of 5 h. Find the rate of the jet.
360 mph

28. A twin-engine plane can fly 800 mi in the same time it takes a single-engine plane to fly 600 mi. The rate of the twin-engine plane is 50 mph faster than that of the single-engine plane. Find the rate of the twin-engine plane.
200 mph

29. As part of an exercise plan, Camille Ellison walked for 40 min and then ran for 20 min. If Camille runs 3 mph faster than she walks and covered 5 mi during the 1-hour exercise period, what is her walking speed?
4 mph

30. A car and a bus leave a town at 1 P.M. and head for a town 300 mi away. The rate of the car is twice the rate of the bus. The car arrives 5 h ahead of the bus. Find the rate of the car.
60 mph

Quick Quiz (Objective 11.7B)

1. A plane flies 460 mph in calm air. Flying with the wind, the plane can travel 1560 mi in the same amount of time it takes to travel 1200 mi flying against the wind. Find the rate of the wind.
60 mph

31. A car is traveling at a rate that is 36 mph faster than the rate of a cyclist. The car travels 384 mi in the same time it takes the cyclist to travel 96 mi. Find the rate of the car.
48 mph

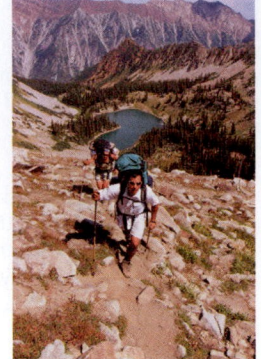

32. A backpacker hiking into a wilderness area walked 9 mi at a constant rate and then reduced this rate by 1 mph. Another 4 mi was hiked at this reduced rate. The time required to hike the 4 mi was 1 h less than the time required to walk the 9 mi. Find the rate at which the hiker walked the first 9 mi.   **3 mph**

33. A plane can fly 180 mph in calm air. Flying with the wind, the plane can fly 600 mi in the same amount of time it takes to fly 480 mi against the wind. Find the rate of the wind.   **20 mph**

34. A commercial jet can fly 550 mph in calm air. Traveling with the jet stream, the plane flew 2400 mi in the same amount of time it takes to fly 2000 mi against the jet stream. Find the rate of the jet stream.
50 mph

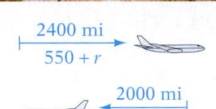

35. A cruise ship can sail at 28 mph in calm water. Sailing with the gulf current, the ship can sail 170 mi in the same amount of time it can sail 110 mi against the gulf current. Find the rate of the gulf current.
6 mph

36. Rowing with the current of a river, a rowing team can row 25 mi in the same amount of time it takes to row 15 mi against the current. The rate of the rowing team in calm water is 20 mph. Find the rate of the current.   **5 mph**

37. On a recent trip, a trucker traveled 330 mi at a constant rate. Because of road construction, the trucker then had to reduce the speed by 25 mph. An additional 30 mi was traveled at the reduced rate. The total time for the entire trip was 7 h. Find the rate of the trucker for the first 330 mi.   **55 mph**

APPLYING THE CONCEPTS

38. **Work**   One pipe can fill a tank in 2 h, a second pipe can fill the tank in 4 h, and a third pipe can fill the tank in 5 h. How long will it take to fill the tank with all three pipes working? $1\frac{1}{19}$ h

39. **Transportation**   Because of bad weather, a bus driver reduced the usual speed along a 150-mile bus route by 10 mph. The bus arrived only 30 min later than its usual arrival time. How fast does the bus usually travel?   **60 mph**

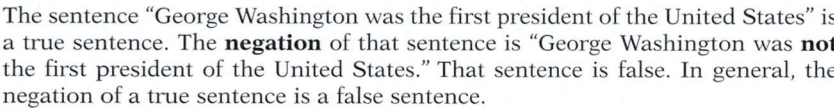

Focus on Problem Solving

Negations   The sentence "George Washington was the first president of the United States" is a true sentence. The **negation** of that sentence is "George Washington was **not** the first president of the United States." That sentence is false. In general, the negation of a true sentence is a false sentence.

The negation of a false sentence is a true sentence. For instance, the sentence "The moon is made of green cheese" is a false statement. The negation of that sentence, "The moon is **not** made of green cheese," is true.

The words *all, no* (or *none*), and *some* are called **quantifiers.** Writing the negation of a sentence that contains these words requires special attention. Consider the sentence "All pets are dogs." This sentence is not true because there are pets that are not dogs; cats, for example, are pets. Because the sentence is false, its negation must be true. You might be tempted to write "All pets are not dogs," but that sentence is not true because some pets are dogs. The correct negation of "All pets are dogs" is "Some pets are not dogs." Note the use of the word *some* in the negation.

Now consider the sentence "Some computers are portable." Because that sentence is true, its negation must be false. Writing "Some computers are not portable" as the negation is not correct, because that sentence is true. The negation of "Some computers are portable" is "No computers are portable."

The sentence "No flowers have red blooms" is false, because there is at least one flower (some roses, for example) that has red blooms. Because the sentence is false, its negation must be true. The negation is "Some flowers have red blooms."

Statement	*Negation*
All *A* are *B*.	Some *A* are not *B*.
No *A* are *B*.	Some *A* are *B*.
Some *A* are *B*.	No *A* are *B*.
Some *A* are not *B*.	All *A* are *B*.

Write the negation of the sentence.

1. All cats like milk.

2. All computers need people.

3. Some trees are tall.

4. No politicians are honest.

5. No houses have kitchens.

6. All police officers are tall.

7. All lakes are not polluted.

8. Some drivers are unsafe.

9. Some speeches are interesting.

10. All laws are good.

11. All businesses are not profitable.

12. All motorcycles are not large.

13. Some vegetables are good for you to eat.

14. Some banks are not open on Sunday.

Answers to Focus on Problem Solving: Negations

1. Some cats do not like milk.
2. Some computers do not need people.
3. No trees are tall.
4. Some politicians are honest.
5. Some houses have kitchens.
6. Some police officers are not tall.
7. Some lakes are polluted.
8. All drivers are safe.
9. No speeches are interesting.
10. Some laws are not good.
11. Some businesses are profitable.
12. Some motorcycles are large.
13. No vegetables are good for you to eat.
14. All banks are open on Sunday.

Projects and Group Activities

Body Mass Index   **Body mass index,** or **BMI,** expresses the relationship between a person's height and weight. It is a measurement for gauging a person's weight-related level of risk for high blood pressure, heart disease, and diabetes. A BMI value of 25 or less indicates a very low to low risk; a BMI value of 25 to 30 indicates low to moderate risk; a BMI of 30 or more indicates a moderate to very high risk.

The formula for body mass index is $B = \frac{705W}{H^2}$, where B is the BMI, W is the weight in pounds, and H is height in inches.

1. Amy is 140 lb and 5'8" tall. Calculate Amy's BMI. Round to the nearest tenth. Would you rank Amy as a low, moderate, or high risk for weight-related disease?

2. Roger is 5'11". How much should he weigh in order to have a BMI of 25? Round to the nearest whole number.

3. Brenda is 5'3". What should she weigh in order to have a BMI of 24? Round to the nearest whole number.

4. Carlos weighs 185 lb and is 5'9". How many pounds must Carlos lose in order to reach a BMI of 23? Round to the nearest whole number.

5. Pat is 6'3" and weighs 245 lb. Calculate the number of pounds Pat must lose in order to reach a BMI of 22. Round to the nearest whole number.

6. Zack weighs 205 lb and is 6'0". He would like to lower his BMI to 20. **a.** By how many points must Zack lower his BMI? Round to the nearest tenth. **b.** How many pounds must Zack lose in order to reach a BMI of 20? Round to the nearest whole number.

7. Felicia weighs 160 lb and is 5'7". She would like to lower her BMI to 20. **a.** By how many points must Felicia lower her BMI? Round to the nearest tenth. **b.** How many pounds must Felicia lose in order to reach a BMI of 20? Round to the nearest whole number.

Chapter 11 Summary

Key Words	Examples
A *rational expression* is a fraction in which the numerator and denominator are polynomials. A rational expression is in *simplest form* when the numerator and denominator have no common factors. [11.1A, p. 571]	$\frac{2x + 1}{x^2 + 4}$ is a rational expression in simplest form.
The *reciprocal of a rational expression* is the rational expression with the numerator and denominator interchanged. [11.1C, p. 574]	The reciprocal of $\frac{3x - y}{x + 4}$ is $\frac{x + 4}{3x - y}$.
The *least common multiple (LCM) of two or more polynomials* is the polynomial of least degree that contains all the factors of each polynomial. [11.2A, p. 579]	The LCM of $3x^2 - 6x$ and $x^2 - 4$ is $3x(x - 2)(x + 2)$, because it contains the factors of $3x^2 - 6x = 3x(x - 2)$ and the factors of $x^2 - 4 = (x - 2)(x + 2)$.

A *complex fraction* is a fraction whose numerator or denominator contains one or more fractions. [11.4A, p. 592]

$$\dfrac{x - \dfrac{2}{x+1}}{1 - \dfrac{4}{x}} \text{ is a complex fraction.}$$

A *literal equation* is an equation that contains more than one variable. A *formula* is a literal equation that states a rule about measurements. [11.6A, p. 601]

$3x - 4y = 12$ is a literal equation. $A = LW$ is a literal equation that is also the formula for the area of a rectangle.

Essential Rules and Procedures

Examples

Simplifying Rational Expressions [11.1A, p. 571]
Factor the numerator and denominator. Divide the numerator and denominator by the common factors.

$$\dfrac{x^2 - 3x - 10}{x^2 - 25} = \dfrac{(x+2)(x-5)}{(x+5)(x-5)}$$
$$= \dfrac{x+2}{x+5}$$

Multiplying Rational Expressions [11.1B, p. 572]
Multiply the numerators. Multiply the denominators. Write the answer in simplest form.

$$\dfrac{a}{b} \cdot \dfrac{c}{d} = \dfrac{ac}{bd}$$

$$\dfrac{x^2 - 3x}{x^2 + x} \cdot \dfrac{x^2 + 5x + 4}{x^2 - 4x + 3}$$
$$= \dfrac{x(x-3)}{x(x+1)} \cdot \dfrac{(x+1)(x+4)}{(x-3)(x-1)}$$
$$= \dfrac{x(x-3)(x+1)(x+4)}{x(x+1)(x-3)(x-1)}$$
$$= \dfrac{x+4}{x-1}$$

Dividing Rational Expressions [11.1C, p. 574]
Multiply the dividend by the reciprocal of the divisor. Write the answer in simplest form.

$$\dfrac{a}{b} \div \dfrac{c}{d} = \dfrac{a}{b} \cdot \dfrac{d}{c} = \dfrac{ad}{bc}$$

$$\dfrac{4x + 16}{3x - 6} \div \dfrac{x^2 + 6x + 8}{x^2 - 4}$$
$$= \dfrac{4x + 16}{3x - 6} \cdot \dfrac{x^2 - 4}{x^2 + 6x + 8}$$
$$= \dfrac{4(x+4)}{3(x-2)} \cdot \dfrac{(x-2)(x+2)}{(x+4)(x+2)} = \dfrac{4}{3}$$

Adding and Subtracting Rational Expressions [11.3B, p. 585]

1. Find the LCM of the denominators.

2. Write each fraction as an equivalent fraction using the LCM as the denominator.

3. Add or subtract the numerators and place the result over the common denominator.

4. Write the answer in simplest form.

$$\dfrac{a}{b} + \dfrac{c}{b} = \dfrac{a+c}{b} \qquad \dfrac{a}{b} - \dfrac{c}{b} = \dfrac{a-c}{b}$$

$$\dfrac{x}{x+1} - \dfrac{x+3}{x-2}$$
$$= \dfrac{x}{x+1} \cdot \dfrac{x-2}{x-2} - \dfrac{x+3}{x-2} \cdot \dfrac{x+1}{x+1}$$
$$= \dfrac{x(x-2)}{(x+1)(x-2)} - \dfrac{(x+3)(x+1)}{(x+1)(x-2)}$$
$$= \dfrac{x(x-2) - (x+3)(x+1)}{(x+1)(x-2)}$$
$$= \dfrac{(x^2 - 2x) - (x^2 + 4x + 3)}{(x+1)(x-2)}$$
$$= \dfrac{-6x - 3}{(x+1)(x-2)}$$

Simplifying Complex Fractions [11.4A, p. 592]

Method 1: Multiply by 1 in the form $\dfrac{\text{LCM}}{\text{LCM}}$.

1. Determine the LCM of the denominators of the fractions in the numerator and denominator of the complex fraction.
2. Multiply the numerator and denominator of the complex fraction by the LCM.
3. Simplify.

Method 1:

$$\frac{\dfrac{1}{x} + \dfrac{1}{y}}{\dfrac{1}{x} - \dfrac{1}{y}} = \frac{\dfrac{1}{x} + \dfrac{1}{y}}{\dfrac{1}{x} - \dfrac{1}{y}} \cdot \frac{xy}{xy}$$

$$= \frac{\dfrac{1}{x} \cdot xy + \dfrac{1}{y} \cdot xy}{\dfrac{1}{x} \cdot xy - \dfrac{1}{y} \cdot xy}$$

$$= \frac{y + x}{y - x}$$

Method 2: Multiply the numerator by the reciprocal of the denominator.

1. Simplify the numerator to a single fraction and simplify the denominator to a single fraction.
2. Using the definition for dividing fractions, multiply the numerator by the reciprocal of the denominator.
3. Simplify.

Method 2:

$$\frac{\dfrac{1}{x} + \dfrac{1}{y}}{\dfrac{1}{x} - \dfrac{1}{y}} = \frac{\dfrac{y + x}{xy}}{\dfrac{y - x}{xy}}$$

$$= \frac{y + x}{xy} \cdot \frac{xy}{y - x}$$

$$= \frac{y + x}{y - x}$$

Solving Equations Containing Fractions [11.5A, p. 597]

Clear denominators by multiplying each side of the equation by the LCM of the denominators. Then solve for the variable.

$$\frac{1}{2a} = \frac{2}{a} - \frac{3}{8}$$

$$8a\left(\frac{1}{2a}\right) = 8a\left(\frac{2}{a}\right) - 8a\left(\frac{3}{8}\right)$$

$$4 = 16 - 3a$$

$$-12 = -3a$$

$$4 = a$$

Solving Literal Equations [11.6A, p. 601]

Rewrite the equation so that the letter being solved for is alone on one side of the equation and all numbers and other variables are on the other side.

• Solve for *x*.

$$2x + ax = 5$$

$$x(2 + a) = 5$$

$$\frac{x(2 + a)}{2 + a} = \frac{5}{2 + a}$$

$$x = \frac{5}{2 + a}$$

Work Problems [11.7A, p. 605]

Rate of work × time worked = part of task completed

Pat can do a certain job in 3 h. Chris can do the same job in 5 h. How long would it take them, working together, to get the job done?

$$\frac{t}{3} + \frac{t}{5} = 1$$

Uniform Motion Problems with Rational Expressions [11.7B, p. 607]

$$\frac{\text{Distance}}{\text{Rate}} = \text{time}$$

Train A's speed is 15 mph faster than train B's speed. Train A travels 150 mi in the same amount of time it takes train B to travel 120 mi. Find the rate of train B.

$$\frac{120}{r} = \frac{150}{r + 15}$$

Chapter 11 Review Exercises

1. Divide: $\dfrac{6a^2b^7}{25x^3y} \div \dfrac{12a^3b^4}{5x^2y^2}$ $\dfrac{b^3y}{10ax}$ [11.1C]

2. Add: $\dfrac{x+7}{15x} + \dfrac{x-2}{20x}$ $\dfrac{7x+22}{60x}$ [11.3B]

3. Multiply: $\dfrac{3x^3+9x^2}{6xy^2-18y^2} \cdot \dfrac{4xy^3-12y^3}{5x^2+15x}$ $\dfrac{2xy}{5}$ [11.1B]

4. Divide: $\dfrac{2x(x-y)}{x^2y(x+y)} \div \dfrac{3(x-y)}{x^2y^2}$ $\dfrac{2xy}{3(x+y)}$ [11.1C]

5. Simplify: $\dfrac{x-\dfrac{16}{5x-2}}{3x-4-\dfrac{88}{5x-2}}$ $\dfrac{x-2}{3x-10}$ [11.4A]

6. Simplify: $\dfrac{x^2+x-30}{15+2x-x^2} - \dfrac{x+6}{x+3}$ [11.1A]

7. Simplify: $\dfrac{16x^5y^3}{24xy^{10}}$ $\dfrac{2x^4}{3y^7}$ [11.1A]

8. Solve: $\dfrac{20}{x+2} = \dfrac{5}{16}$ 62 [11.5A]

9. Divide: $\dfrac{10-23y+12y^2}{6y^2-y-5} \div \dfrac{4y^2-13y+10}{18y^2+3y-10}$ $\dfrac{(3y-2)^2}{(y-1)(y-2)}$ [11.1C]

10. Solve $3ax - x = 5$ for x. $x = \dfrac{5}{3a-1}$ [11.6A]

11. Solve: $\dfrac{2}{x} + \dfrac{3}{4} = 1$ 8 [11.5A]

12. Add: $\dfrac{x}{y} + \dfrac{3}{x}$ $\dfrac{x^2+3y}{xy}$ [11.3B]

13. Solve $5x + 4y = 20$ for y. $y = -\dfrac{5}{4}x + 5$ [11.6A]

14. Multiply: $\dfrac{8ab^2}{15x^3y} \cdot \dfrac{5xy^4}{16a^2b}$ $\dfrac{by^3}{6ax^2}$ [11.1B]

15. Simplify: $\dfrac{1-\dfrac{1}{x}}{1-\dfrac{8x-7}{x^2}}$ $\dfrac{x}{x-7}$ [11.4A]

16. Write each fraction in terms of the LCM of the denominators.

$\dfrac{x}{12x^2+16x-3}, \dfrac{4x^2}{6x^2+7x-3}$

$\dfrac{3x^2-x}{(2x+3)(6x-1)(3x-1)}, \dfrac{24x^3-4x^2}{(2x+3)(6x-1)(3x-1)}$ [11.2B]

17. Solve $T = 2(ab + bc + ca)$ for a.

$a = \dfrac{T-2bc}{2b+2c}$ [11.6A]

18. Solve: $\dfrac{5}{7} + \dfrac{x}{2} = 2 - \dfrac{x}{7}$ 2 [11.5A]

19. Simplify: $\dfrac{2+\dfrac{1}{x}}{3-\dfrac{2}{x}}$ $\dfrac{2x+1}{3x-2}$ [11.4A]

20. Subtract: $\dfrac{2x}{x-5} - \dfrac{x+1}{x-2}$ $\dfrac{x^2+5}{(x-5)(x-2)}$ [11.3B]

21. Solve $i = \dfrac{100m}{c}$ for c. $c = \dfrac{100m}{i}$ [11.6A]

22. Solve: $\dfrac{x+8}{x+4} = 1 + \dfrac{5}{x+4}$ No solution   [11.5A]

23. Divide: $\dfrac{20x^2 - 45x}{6x^3 + 4x^2} \div \dfrac{40x^3 - 90x^2}{12x^2 + 8x}$ $\dfrac{1}{x^2}$ [11.1C]

24. Add: $\dfrac{2y}{5y-7} + \dfrac{3}{7-5y}$ $\dfrac{2y-3}{5y-7}$ [11.3B]

25. Subtract: $\dfrac{5x+3}{2x^2+5x-3} - \dfrac{3x+4}{2x^2+5x-3}$

$\dfrac{1}{x+3}$ [11.3A]

26. Find the LCM of $10x^2 - 11x + 3$ and $20x^2 - 17x + 3$.

$(5x - 3)(2x - 1)(4x - 1)$ [11.2A]

27. Solve $4x + 9y = 18$ for y.

$y = -\dfrac{4}{9}x + 2$ [11.6A]

28. Multiply: $\dfrac{2x^2 - 5x - 3}{3x^2 - 7x - 6} \cdot \dfrac{3x^2 + 8x + 4}{x^2 + 4x + 4}$

$\dfrac{2x+1}{x+2}$ [11.1B]

29. Solve: $\dfrac{20}{2x+3} = \dfrac{17x}{2x+3} - 5$

5   [11.5A]

30. Add: $\dfrac{x-1}{x+2} + \dfrac{3x-2}{5-x} + \dfrac{5x^2 + 15x - 11}{x^2 - 3x - 10}$

$\dfrac{3x-1}{x-5}$ [11.3B]

31. Solve: $\dfrac{6}{x-7} = \dfrac{8}{x-6}$

10   [11.5A]

32. Solve: $\dfrac{3}{20} = \dfrac{x}{80}$

12   [11.5A]

33. **Work**   One hose can fill a pool in 15 h. The second hose can fill the pool in 10 h. How long would it take to fill the pool using both hoses?   6 h   [11.7A]

34. **Travel**   A car travels 315 mi in the same amount of time it takes a bus to travel 245 mi. The rate of the car is 10 mph faster than that of the bus. Find the rate of the car.   45 mph   [11.7B]

35. **Travel**   The rate of a jet is 400 mph in calm air. Traveling with the wind, the jet can fly 2100 mi in the same amount of time it takes to fly 1900 mi against the wind. Find the rate of the wind.   20 mph   [11.7B]

Chapter 11 Test

1. Subtract: $\dfrac{x}{x + 3} - \dfrac{2x - 5}{x^2 + x - 6}$

$\dfrac{x^2 - 4x + 5}{(x - 2)(x + 3)}$ [11.3B]

2. Solve: $\dfrac{3}{x + 4} = \dfrac{5}{x + 6}$

-1 [11.5A]

3. Multiply: $\dfrac{x^2 + 2x - 3}{x^2 + 6x + 9} \cdot \dfrac{2x^2 - 11x + 5}{2x^2 + 3x - 5}$

$\dfrac{(x - 5)(2x - 1)}{(x + 3)(2x + 5)}$ [11.1B]

4. Simplify: $\dfrac{16x^5 y}{24x^2 y^4}$

$\dfrac{2x^3}{3y^3}$ [11.1A]

5. Solve $d = s + rt$ for t.

$t = \dfrac{d - s}{r}$ [11.6A]

6. Solve: $\dfrac{6}{x} - 2 = 1$

2 [11.5A]

7. Simplify: $\dfrac{x^2 + 4x - 5}{1 - x^2}$

$-\dfrac{x + 5}{x + 1}$ [11.1A]

8. Find the LCM of $6x - 3$ and $2x^2 + x - 1$.

$3(2x - 1)(x + 1)$ [11.2A]

9. Subtract: $\dfrac{2}{2x - 1} - \dfrac{3}{3x + 1}$

$\dfrac{5}{(2x - 1)(3x + 1)}$ [11.3B]

10. Divide: $\dfrac{x^2 + 3x + 2}{x^2 + 5x + 4} \div \dfrac{x^2 - x - 6}{x^2 + 2x - 15}$

$\dfrac{x + 5}{x + 4}$ [11.1C]

11. Simplify: $\dfrac{1 + \dfrac{1}{x} - \dfrac{12}{x^2}}{1 + \dfrac{2}{x} - \dfrac{8}{x^2}}$

$\dfrac{x - 3}{x - 2}$ [11.4A]

12. Write each fraction in terms of the LCM of the denominators.

$\dfrac{3}{x^2 - 2x}, \dfrac{x}{x^2 - 4}$

$\dfrac{3x + 6}{x(x - 2)(x + 2)}, \dfrac{x^2}{x(x - 2)(x + 2)}$ [11.2B]

13. Subtract: $\dfrac{2x}{x^2 + 3x - 10} - \dfrac{4}{x^2 + 3x - 10}$

$\dfrac{2}{x + 5}$ [11.3A]

14. Solve $3x - 8y = 16$ for y.

$y = \dfrac{3}{8}x - 2$ [11.6A]

15. Solve: $\dfrac{2x}{x + 1} - 3 = \dfrac{-2}{x + 1}$

No solution [11.5A]

16. Multiply: $\dfrac{x^3 y^4}{x^2 - 4x + 4} \cdot \dfrac{x^2 - x - 2}{x^6 y^4}$

$\dfrac{x + 1}{x^3(x - 2)}$ [11.1B]

17. Divide: $\dfrac{8a^2 b^5}{3xy^4} \div \dfrac{4a^3 b}{9x^2 y}$

$\dfrac{6b^4 x}{ay^3}$ [11.1C]

18. Add: $\dfrac{4}{5x^2 y} + \dfrac{1}{5x^2 y}$

$\dfrac{1}{x^2 y}$ [11.3A]

19. **Work** A ski resort can manufacture enough machine-made snow to open its beginners' run in 4 h, whereas naturally falling snow would take 12 h to provide enough snow. If the resort makes snow at the same time it is snowing naturally, how long will it take until the run can be opened? 3 h [11.7A]

20. **Work** A pool can be filled with one pipe in 6 h, whereas a second pipe requires 12 h to fill the pool. How long would it take to fill the pool with both pipes turned on? 4 h [11.7A]

21. **Travel** A small plane can fly at 110 mph in calm air. Flying with the wind, the plane can fly 260 mi in the same amount of time it takes to fly 180 mi against the wind. Find the rate of the wind. 20 mph [11.7B]

22. **Travel** A jet ski can comfortably travel across calm water at 35 mph. If a rider traveled 4 mi down a river in the same amount of time it took to travel 3 mi back up the river, find the rate of the river's current. 5 mph [11.7B]

Cumulative Review Exercises

1. Evaluate: $\left(\dfrac{2}{3}\right)^2 \div \left(\dfrac{3}{2} - \dfrac{2}{3}\right) + \dfrac{1}{2}$

   $\dfrac{31}{30}$ [2.7A]

2. Evaluate $-a^2 + (a - b)^2$ when $a = -2$ and $b = 3$.

   21 [4.1A]

3. Simplify: $-2x - (-3y) + 7x - 5y$
   $5x - 2y$ [4.2A]

4. Simplify: $2[3x - 7(x - 3) - 8]$
   $-8x + 26$ [4.2D]

5. Solve: $4 - \dfrac{2}{3}x = 7$

   $-\dfrac{9}{2}$ [5.2A]

6. Solve: $3[x - 2(x - 3)] = 2(3 - 2x)$
   -12 [5.3B]

7. Find $16\dfrac{2}{3}\%$ of 60.

   10 [6.4A/6.4B]

8. Simplify: $(a^2 b^5)(ab^2)$
   $a^3 b^7$ [9.2A]

9. Multiply: $(a - 3b)(a + 4b)$
   $a^2 + ab - 12b^2$ [9.3C]

10. Divide: $\dfrac{15b^4 - 5b^2 + 10b}{5b}$

    $3b^3 - b + 2$ [9.5A]

11. Divide: $(x^3 - 8) \div (x - 2)$
    $x^2 + 2x + 4$ [9.5B]

12. Factor: $12x^2 - x - 1$
    $(4x + 1)(3x - 1)$ [10.3A]

13. Factor: $y^2 - 7y + 6$
    $(y - 6)(y - 1)$ [10.2A]

14. Factor: $2a^3 + 7a^2 - 15a$
    $a(2a - 3)(a + 5)$ [10.3A]

15. Factor: $4b^2 - 100$
    $4(b + 5)(b - 5)$ [10.4B]

16. Solve: $(x + 3)(2x - 5) = 0$
    $-3, \dfrac{5}{2}$ [10.5A]

17. Simplify: $\dfrac{12x^4 y^2}{18xy^7}$

    $\dfrac{2x^3}{3y^5}$ [11.1A]

18. Simplify: $\dfrac{x^2 - 7x + 10}{25 - x^2}$

    $-\dfrac{x - 2}{x + 5}$ [11.1A]

19. Divide: $\dfrac{x^2 - x - 56}{x^2 + 8x + 7} \div \dfrac{x^2 - 13x + 40}{x^2 - 4x - 5}$

1 [11.1C]

20. Subtract: $\dfrac{2}{2x - 1} - \dfrac{1}{x + 1}$

$\dfrac{3}{(2x - 1)(x + 1)}$ [11.3B]

21. Simplify: $\dfrac{1 - \dfrac{2}{x} - \dfrac{15}{x^2}}{1 - \dfrac{25}{x^2}}$

$\dfrac{x + 3}{x + 5}$ [11.4A]

22. Solve: $\dfrac{3x}{x - 3} - 2 = \dfrac{10}{x - 3}$

4 [11.5A]

23. Solve: $\dfrac{2}{x - 2} = \dfrac{12}{x + 3}$

3 [11.5A]

24. Solve $f = v + at$ for t.

$t = \dfrac{f - v}{a}$ [11.6A]

25. **Number Sense**   Translate "the difference between five times a number and thirteen is the opposite of eight" into an equation and solve.
$5x - 13 = -8$; $x = 1$ [5.4A]

26. **Home-Schooling**   According to the National Center for Education Statistics, 1.1 million students are home-schooled. This number is 2.2% of the school-age population in the United States. What is the school-age population in the United States?   50 million people [6.4C]

27. **Geometry**   The length of the base of a triangle is 2 in. less than twice the height. The area of the triangle is 30 in². Find the base and height of the triangle.   Base: 10 in.; height: 6 in. [10.5B]

28. **Insurance**   A life insurance policy costs $16 for every $1000 of coverage. At this rate, how much money would a policy of $5000 cost? $80 [6.2B]

29. **Work**   One water pipe can fill a tank in 9 min, whereas a second pipe requires 18 min to fill the tank. How long would it take both pipes, working together, to fill the tank?   6 min [11.7A]

30. **Travel**   The rower of a boat can row at a rate of 5 mph in calm water. Rowing with the current, the boat travels 14 mi in the same amount of time it takes to travel 6 mi against the current. Find the rate of the current.
2 mph [11.7B]

12 Linear Equations in Two Variables

This tennis player gets the energy for his workout from carbohydrates. Carbohydrates are the body's primary source of fuel for exercise. They can be released quickly and easily to fulfill the demands that exercise puts on the body. Since carbohydrates also fuel most of our muscular contractions, it is important to eat enough carbohydrates before any rigorous exercise. **Exercise 27 on page 662** presents data on the number of grams of carbohydrates burned as a strenuous tennis workout progresses.

OBJECTIVES

Section 12.1

A To graph points in a rectangular coordinate system

B To determine ordered-pair solutions of an equation in two variables

C To determine whether a set of ordered pairs is a function

D To evaluate a function written in functional notation

Section 12.2

A To graph an equation of the form $y = mx + b$

B To graph an equation of the form $Ax + By = C$

C To solve application problems

Section 12.3

A To find the x- and y-intercepts of a straight line

B To find the slope of a straight line

C To graph a line using the slope and the y-intercept

Section 12.4

A To find the equation of a line given a point and the slope

B To find the equation of a line given two points

C To solve application problems

Need help? For online student resources, such as section quizzes, visit this textbook's website at **math.college.hmco.com/students.**

PREP TEST • • •

Do these exercises to prepare for Chapter 12.

1. Simplify: $-\dfrac{5 - (-7)}{4 - 8}$

3 [3.5A]

2. Evaluate $\dfrac{a - b}{c - d}$ when $a = 3$, $b = -2$, $c = -3$, and $d = 2$.

-1 [4.1A]

3. Simplify: $-3(x - 4)$

$-3x + 12$ [4.2C]

4. Solve: $3x + 6 = 0$

-2 [5.2A]

5. Solve $4x + 5y = 20$ when $y = 0$.

$x = 5$ [5.2A]

6. Solve $3x - 7y = 11$ when $x = -1$.

$y = -2$ [5.2A]

7. Divide: $\dfrac{12x - 15}{-3}$

$-4x + 5$ [9.5A]

8. Solve: $\dfrac{2x + 1}{3} = \dfrac{3x}{4}$

4 [6.2A]

9. Solve $3x - 5y = 15$ for y.

$y = \dfrac{3}{5}x - 3$ [11.6A]

10. Solve $y + 3 = -\dfrac{1}{2}(x + 4)$ for y.

$y = -\dfrac{1}{2}x - 5$ [11.6A]

GO FIGURE • • •

Points A, B, C, and D lie on the same line and in that order. The ratio of AB to AC is $\dfrac{1}{4}$ and the ratio of BC to CD is $\dfrac{1}{2}$. Find the ratio of AB to CD.

$\dfrac{1}{6}$

12.1 The Rectangular Coordinate System

Objective A   To graph points in a rectangular coordinate system

Before the 15th century, geometry and algebra were considered separate branches of mathematics. That all changed when René Descartes, a French mathematician who lived from 1596 to 1650, founded **analytic geometry.** In this geometry, a *coordinate system* is used to study relationships between variables.

A **rectangular coordinate system** is formed by two number lines, one horizontal and one vertical, that intersect at the zero point of each line. The point of intersection is called the **origin.** The two lines are called **coordinate axes,** or simply **axes.** The axes determine a **plane,** which can be thought of as a large, flat sheet of paper. The two axes divide the plane into four regions called **quadrants,** which are numbered counterclockwise from I to IV.

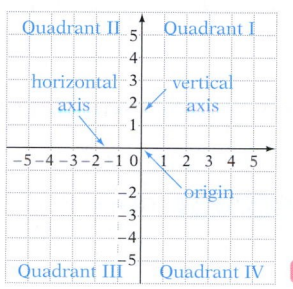

Each point in the plane can be identified by a pair of numbers called an **ordered pair.** The first number of the pair measures a horizontal distance and is called the **abscissa.** The second number of the pair measures a vertical distance and is called the **ordinate.** The **coordinates of a point** are the numbers in the ordered pair associated with the point. The abscissa is also called the **first coordinate** of the ordered pair, and the ordinate is also called the **second coordinate** of the ordered pair.

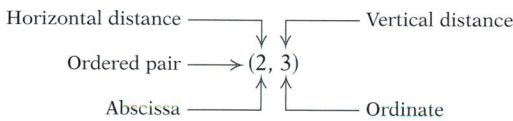

When drawing a rectangular coordinate system, we often label the horizontal axis x and the vertical axis y. In this case, the coordinate system is called an **xy-coordinate system.** The coordinates of the points are given by ordered pairs (x, y), where the abscissa is called the **x-coordinate** and the ordinate is called the **y-coordinate.**

To **graph or plot a point in the plane,** place a dot at the location given by the ordered pair. The **graph of an ordered pair** (x, y) is the dot drawn at the coordinates of the point in the plane. The points whose coordinates are $(3, 4)$ and $(-2.5, -3)$ are graphed in the figures below.

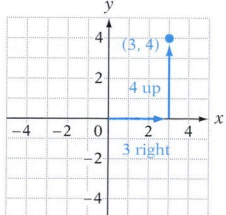 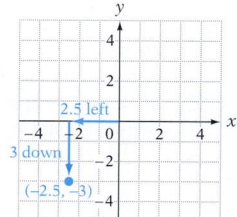

In-Class Examples (Objective 12.1A)

1. Graph the ordered pairs $(5, 0)$, $(-4, 1)$, $(-1, 2)$, and $(-2, -4)$.

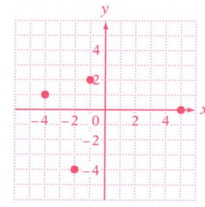

2. Use the graph at the right. Give the coordinates of the points labeled A and B. Give the abscissa of point C and the ordinate of point D.

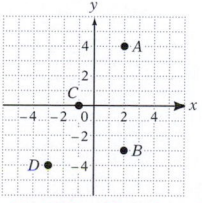

$A\,(2, 4)$, $B\,(2, -3)$, -1, -4

Objective 12.1A

Vocabulary to Review
plane   [1.2C]

New Vocabulary
analytic geometry
rectangular coordinate system
origin
coordinate axes (axes)
quadrants
ordered pair
abscissa
ordinate
coordinates of a point
first coordinate
second coordinate
xy-coordinate system
x-coordinate
y-coordinate
graph (plot) a point in the plane
graph of an ordered pair

Instructor Note
Although Descartes is given credit for introducing analytic geometry, there were others working on the same concept, most notably Pierre Fermat. Nowhere in Descartes's work is there a coordinate system as we draw it with two axes. Descartes did not use the word *coordinate* in his work. This word was introduced by Gottfried Leibnitz, who is also responsible for the use of the words *abscissa* and *ordinate*.

Instructor Note
Within the Microsoft PowerPoint® slides available with this text is a blank coordinate grid. It can be used to create a transparency on which to plot points, equations, and so on.

Concept Check
1. Plot five different points that have an *x*-coordinate of 3.

2. If the points in part 1 were connected with a line, would the line be horizontal or vertical?   Vertical

3. Plot five different points that have a *y*-coordinate of −2.

4. If the points in part 3 were connected with a line, would the line be horizontal or vertical?   Horizontal

Concept Check

1. In which quadrant is the graph of $(-3, 4)$?   II

2. On which axis is the graph of $(0, -4)$?   *y*-axis

Instructor Note

It may help students to think of an ordered pair as the address (location) of a point in the plane.

Optional Student Activity

1. Graph the ordered pairs $(-4, 0)$, $(-2, 1)$, $(0, 2)$, $(2, 3)$, and $(4, 4)$. What do you notice about these five points?

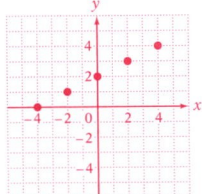

They all lie on the same line.

2. As you move from left to right, from one point to the next, what is the change in the *x*-coordinate? What is the change in the *y*-coordinate?
2, 1

3. Use the pattern that you observed in part 2 to locate two more points on the same line. Write the coordinates of these points.
For example, $(6, 5)$ and $(8, 6)$

TAKE NOTE
This is very important. An **ordered pair** is a *pair* of coordinates, and the *order* in which the coordinates are listed is crucial.

The points whose coordinates are $(3, -1)$ and $(-1, 3)$ are graphed at the right. Note that the graphed points are in different locations. *The order of the coordinates of an ordered pair is important.*

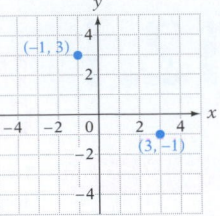

Each point in the plane is associated with an ordered pair, and each ordered pair is associated with a point in the plane. Although only the labels for integers are given on a coordinate grid, the graph of any ordered pair can be approximated. For example, the points whose coordinates are $(-2.3, 4.1)$ and $(\pi, 1)$ are shown on the graph at the right.

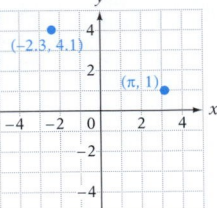

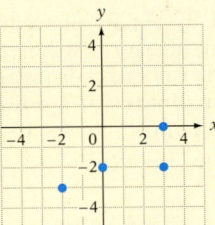
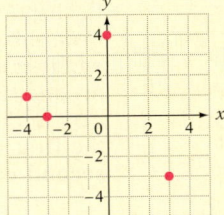
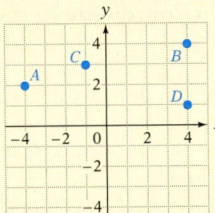
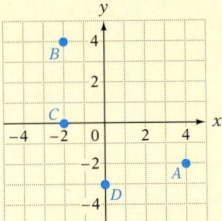

Objective B   **To determine ordered-pair solutions of an equation in two variables**

An xy-coordinate system is used to study the relationship between two variables. Frequently this relationship is given by an equation. Examples of equations in two variables include

$$y = 2x - 3 \qquad 3x + 2y = 6 \qquad x^2 - y = 0$$

A **solution of an equation in two variables** is an ordered pair (x, y) whose coordinates make the equation a true statement.

HOW TO   Is $(-3, 7)$ a solution of $y = -2x + 1$?

$$y = -2x + 1$$

7	$-2(-3) + 1$
	$6 + 1$

$7 = 7$

• Replace x with -3; replace y with 7.

• A true statement

$(-3, 7)$ is a solution of the equation $y = -2x + 1$.

Besides $(-3, 7)$, there are many other ordered-pair solutions of $y = -2x + 1$. For example, $(0, 1)$, $\left(-\frac{3}{2}, 4\right)$, and $(4, -7)$ are also solutions. In general, an equation in two variables has an infinite number of solutions. By choosing any value of x and substituting that value into the equation, we can calculate a corresponding value of y.

HOW TO   Find the ordered-pair solution of $y = \frac{2}{3}x - 3$ that corresponds to $x = 6$.

$$y = \frac{2}{3}x - 3$$

$$= \frac{2}{3}(6) - 3 \qquad \text{• Replace } x \text{ with 6.}$$

$$= 4 - 3 = 1 \qquad \text{• Simplify.}$$

The ordered-pair solution is $(6, 1)$.

The solutions of an equation in two variables can be graphed in an xy-coordinate system.

HOW TO   Graph the ordered-pair solutions of $y = -2x + 1$ when $x = -2, -1, 0, 1,$ and 2.

Use the values of x to determine ordered-pair solutions of the equation. It is convenient to record these in a table.

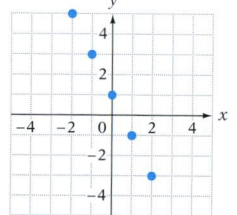

x	$y = -2x + 1$	y	(x, y)
-2	$-2(-2) + 1$	5	$(-2, 5)$
-1	$-2(-1) + 1$	3	$(-1, 3)$
0	$-2(0) + 1$	1	$(0, 1)$
1	$-2(1) + 1$	-1	$(1, -1)$
2	$-2(2) + 1$	-3	$(2, -3)$

Objective 12.1B

New Vocabulary

solution of an equation in two variables

Discuss the Concepts

Explain how to check whether or not the ordered pair $(5, 1)$ is a solution of the equation $y = 2x - 9$.

Concept Check

Find three ordered-pair solutions of the equation $y = x - 1$.

Instructor Note

Problems such as the one at the left are given to prepare the student to graph straight lines, which is the subject of the next section.
   We want to impress on the student the relationship between a solution of an equation and the pictorial representation of that solution, its graph.

In-Class Examples (Objective 12.1B)

1. Is $(0, -1)$ a solution of $2x - y = 1$?   Yes
2. Is $(2, -2)$ a solution of $y = 3x + 8$?   No
3. Graph the ordered-pair solutions of $y = -3x$ when $x = -1, 0,$ and 1.
4. Graph the ordered-pair solutions of $3x - 4y = 4$ when $x = -4, 0,$ and 4.

3.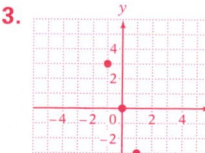

4.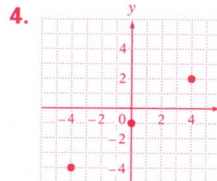

Discuss the Concepts

In Example 4, explain why it is helpful to solve the equation for y before substituting the four values of x.

Optional Student Activity

The example at the bottom of page 627 and Example 4 both show points lying on a straight line. This is not always the case. Graph the ordered-pair solutions of $y = x^2$ when $x = -3, -2, -1, 0, 1, 2,$ and 3. How would you describe the shape on which these points seem to lie?

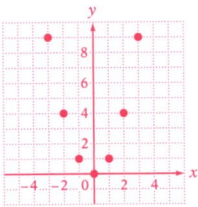

U-shaped

Example 3

Is $(3, -2)$ a solution of $3x - 4y = 15$?

Solution

$$3x - 4y = 15$$

$$\begin{array}{c|c} 3(3) - 4(-2) & 15 \\ 9 + 8 & \\ & 17 \neq 15 \end{array}$$

• Replace x with **3** and y with **-2**.

No. $(3, -2)$ is not a solution of $3x - 4y = 15$.

Example 4

Graph the ordered-pair solutions of $2x - 3y = 6$ when $x = -3, 0, 3,$ and 6.

Solution

$$2x - 3y = 6 \qquad \bullet \text{ Solve } 2x - 3y = 6 \text{ for } y.$$
$$-3y = -2x + 6$$
$$y = \frac{2}{3}x - 2$$

Replace x in $y = \frac{2}{3}x - 2$ with $-3, 0, 3,$ and 6. For each value of x, determine the value of y.

x	$y = \dfrac{2}{3}x - 2$	y	(x, y)
-3	$\dfrac{2}{3}(-3) - 2$	-4	$(-3, -4)$
0	$\dfrac{2}{3}(0) - 2$	-2	$(0, -2)$
3	$\dfrac{2}{3}(3) - 2$	0	$(3, 0)$
6	$\dfrac{2}{3}(6) - 2$	2	$(6, 2)$

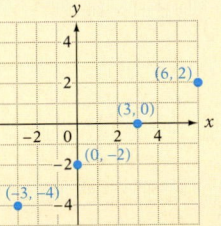

You Try It 3

Is $(-2, 4)$ a solution of $x - 3y = -14$?

Your solution

Yes

You Try It 4

Graph the ordered-pair solutions of $x + 2y = 4$ when $x = -4, -2, 0,$ and 2.

Your solution

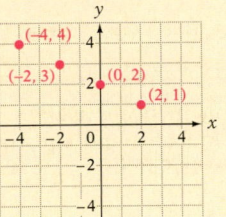

Solutions on p. S30

Objective C   **To determine whether a set of ordered pairs is a function**

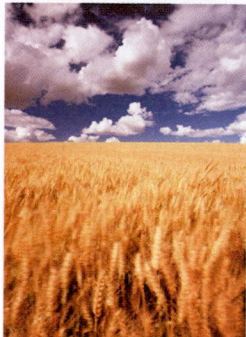

Discovering a relationship between two variables is an important task in the application of mathematics. Here are some examples.

- Botanists study the relationship between the number of bushels of wheat yielded per acre and the amount of watering per acre.
- Environmental scientists study the relationship between the incidents of skin cancer and the amount of ozone in the atmosphere.
- Business analysts study the relationship between the price of a product and the number of products that are sold at that price.

Each of these relationships can be described by a set of ordered pairs.

> **Definition of a Relation**
>
> A **relation** is any set of ordered pairs.

The following table shows the number of hours that each of nine students spent studying for a midterm exam and the grade that each of these nine students received.

Hours	3	3.5	2.75	2	4	4.5	3	2.5	5
Grade	78	75	70	65	85	85	80	75	90

This information can be written as the relation

{(3, 78), (3.5, 75), (2.75, 70), (2, 65), (4, 85), (4.5, 85), (3, 80), (2.5, 75), (5, 90)}

where the first coordinate of the ordered pair is the hours spent studying and the second coordinate is the score on the midterm.

The **domain** of a relation is the set of first coordinates of the ordered pairs; the **range** is the set of second coordinates. For the relation above,

Domain = {2, 2.5, 2.75, 3, 3.5, 4, 4.5, 5}   Range = {65, 70, 75, 78, 80, 85, 90}

The **graph of a relation** is the graph of the ordered pairs that belong to the relation. The graph of the relation given above is shown at the right. The horizontal axis represents the hours spent studying (the domain); the vertical axis represents the test score (the range). The axes could be labeled H for hours studied and S for test score.

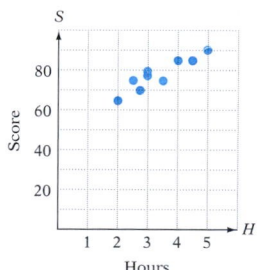

A *function* is a special type of relation in which no two ordered pairs have the same first coordinate.

> **Definition of a Function**
>
> A **function** is a relation in which no two ordered pairs have the same first coordinate.

Objective 12.1C

New Vocabulary
relation
domain
range
graph of a relation
function

Concept Check
Use this relation:
{(−2, 6), (−1, 3), (0, 0), (1, 3), (2, 6)}
1. What is the domain of the relation?   {−2, −1, 0, 1, 2}
2. What is the range of the relation?   {0, 3, 6}

Discuss the Concepts
Is every relation a function? Is every function a relation? Give examples to support your position.

In-Class Examples (Objective 12.1C)

1. The table at the right shows the one-way fares for the number of zones traveled on a commuter-rail train in Boston. Write a relation in which the first coordinate is the number of zones traveled and the second coordinate is the one-way fare. Is the relation a function?

{(1, 3.25), (2, 3.50), (3, 3.75), (4, 4.50), (5, 5.00), (6, 5.25)}; Yes

2. Does $|y| = 2x$ define y as a function of x?   No

Zones traveled	One-way fare (dollars)
1	3.25
2	3.50
3	3.75
4	4.50
5	5.00
6	5.25

Concept Check

1. Write a relation that pairs each element of the domain $\{-2, -1, 0, 2, 4\}$ with one-half of that element.

$$\left\{ (-2, -1), \left(-1, -\frac{1}{2}\right), (0, 0), (2, 1), (4, 2) \right\}$$

2. What is the range of the relation in part 1?

$$\left\{ -1, -\frac{1}{2}, 0, 1, 2 \right\}$$

3. Is the relation in part 1 a function?   Yes

Instructor Note

The concept of a function was beginning to form with the work of Fermat and Descartes. However, it was Euler who gave us the word and its first definition: "A function of a variable quantity is an analytic expression composed in any way whatsoever of the variable quantity and numbers or constant quantities."

The sense of function in Euler's definition is that of "*y* is a function of *x*," but it does not encompass the more general notion of a function as a certain set of ordered pairs.

The table at the right is the grading scale for a 100-point test. This table defines a relationship between the *score* on the test and a *letter grade*. Some of the ordered pairs of this function are (78, C), (97, A), (84, B), and (82, B).

Score	Grade
90–100	A
80–89	B
70–79	C
60–69	D
0–59	F

The grading-scale table defines a function, because no two ordered pairs can have the *same* first coordinate and *different* second coordinates. For instance, it is not possible to have the ordered pairs (72, C), and (72, B)—same first coordinate (test score) but different second coordinates (test grade). The domain of this function is {0, 1, 2, . . . , 99, 100}. The range is {A, B, C, D, F}.

The example of hours spent studying and test score given earlier is *not* a function, because (3, 78) and (3, 80) are ordered pairs of the relation that have the *same* first coordinate but *different* second coordinates.

Consider again the grading-scale example. Note that (84, B) and (82, B) are ordered pairs of the function. Ordered pairs of a function may have the same *second* coordinates but not the same first coordinates.

Although relations and functions can be given by tables, they are frequently given by an equation in two variables.

The equation $y = 2x$ expresses the relationship between a number, x, and twice the number, y. For instance, if $x = 3$, then $y = 6$, which is twice 3. To indicate exactly which ordered pairs are determined by the equation, the domain (values of x) is specified. If $x \in \{-2, -1, 0, 1, 2\}$, then the ordered pairs determined by the equation are $\{(-2, -4), (-1, -2), (0, 0), (1, 2), (2, 4)\}$. This relation is a function because no two ordered pairs have the same first coordinate.

The graph of the function $y = 2x$ with domain $\{-2, -1, 0, 1, 2\}$ is shown at the right. The horizontal axis (domain) is labeled x; the vertical axis (range) is labeled y.

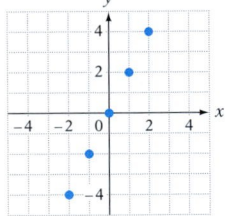

The domain $\{-2, -1, 0, 1, 2\}$ was chosen arbitrarily. Other domains could have been selected. The type of application usually influences the choice of the domain.

For the equation $y = 2x$, we say that "y is a function of x" because the set of ordered pairs is a function.

Not all equations, however, define a function. For instance, the equation $|y| = x + 2$ does not define y as a function of x. The ordered pairs (2, 4) and (2, -4) both satisfy the equation. Thus there are two ordered pairs with the same first coordinate but different second coordinates.

Example 5

The number of tournaments and the total earnings of the top five Ladies Professional Golf Association (LPGA) players are given in the following table.

Player	Tournaments	Winnings
A. Sorenstam	16	$1,914,506
Se Ri Pak	25	$1,561,928
Grace Park	25	$1,374,702
Hee-Won Han	26	$1,101,060
Juli Inkster	20	$1,012,455

Write a relation in which the first coordinate is the number of tournaments played and the second coordinate is the winnings per tournament rounded to the nearest dollar. Is the relation a function?

Solution

Find the winnings per tournament for each player by dividing the player's winnings by the number of tournaments played.

Sorenstam: $1,914,506 \div 16 \approx 119,657$
Pak: $1,561,928 \div 25 \approx 62,477$
Park: $1,374,702 \div 25 \approx 54,988$
Han: $1,101,060 \div 26 \approx 42,348$
Inkster: $1,012,455 \div 20 \approx 50,623$

The relation is $\{(16, 119{,}657), (25, 62{,}477), (25, 54{,}988), (26, 42{,}348), (20, 50{,}623)\}$. The relation is not a function. Two ordered pairs have the same first coordinate but different second coordinates.

You Try It 5

Six students decided to go on a diet and fitness program over the summer. Their weights (in pounds) at the beginning and end of the program are given in the table below.

Beginning	End
145	140
140	125
150	130
165	150
140	130
165	160

Write a relation wherein the first coordinate is the weight at the beginning of the summer and the second coordinate is the weight at the end of the summer. Is the relation a function?

Your solution

$\{(145, 140), (140, 125), (150, 130), (165, 150),$
$(140, 130), (165, 160)\}$
No. The relation is not a function.

Example 6

Does $y = x^2 + 3$, where $x \in \{-2, -1, 1, 3\}$, define y as a function of x?

Solution

Determine the ordered pairs defined by the equation. Replace x in $y = x^2 + 3$ by the given values and solve for y.

$\{(-2, 7), (-1, 4), (1, 4), (3, 12)\}$

No two ordered pairs have the same first coordinate. Therefore, the relation is a function and the equation $y = x^2 + 3$ defines y as a function of x.

Note that $(-1, 4)$ and $(1, 4)$ are ordered pairs that belong to this function. Ordered pairs of a function may have the same *second* coordinate but not the same *first* coordinate.

You Try It 6

Does $y = \frac{1}{2}x + 1$, where $x \in \{-4, 0, 2\}$, define y as a function of x?

Your solution

$\{(-4, -1), (0, 1), (2, 2)\}$
Yes, y is a function of x.

Solutions on p. S30

Optional Student Activity

An extra-credit problem for interested students is to define three situations that are represented by functions and three that are represented by relations but not functions. Here's an example of a relation that is not a function: The ordered pairs whose first coordinates are the runs scored by a baseball team and whose second coordinates are either *W* for win or *L* for lose.

Objective 12.1D

New Vocabulary

functional notation
value of a function at x
evaluating a function
dependent variable
independent variable

Concept Check

1. Evaluate $f(x) = x^2 + 4x + 6$ for $x = -5$ and $x = 1$.
11, 11

2. Based on the results of part 1, if $f(a) = f(b)$, does it follow that $a = b$?
No

Instructor Note

One way to assist students with evaluating a function is to use open parentheses. For instance,

$$f(\;\;) = (\;\;)^2 + (\;\;) - 3$$
$$\downarrow \qquad \downarrow \qquad \downarrow$$
$$f(-2) = (-2)^2 + (-2) - 3$$

Optional Student Activity

As noted at the right, $(-2, -1)$ is one ordered pair of the function given by $f(x) = x^2 + x - 3$. Find four more ordered pairs of this function. Graph the five ordered pairs.
For example, $(-1, -3)$, $(0, -3)$, $(1, -1)$, and $(2, 3)$

Discuss the Concepts

Have students describe how the independent and dependent variables of a function are related to the domain and range of the function.
The domain of a function is the set of all values of the independent variable. The range of a function is the set of all values of the dependent variable.

Objective D   To evaluate a function written in functional notation

When an equation defines y as a function of x, **functional notation** is frequently used to emphasize that the relation is a function. In this case, it is common to replace y in the function's equation with the symbol $f(x)$, where

$$f(x) \text{ is read "} f \text{ of } x \text{" or "the value of } f \text{ at } x \text{."}$$

For instance, the equation $y = x^2 + 3$ from Example 6 defined y as a function of x. The equation can also be written in functional notation as

$$f(x) = x^2 + 3$$

where y has been replaced with $f(x)$.

The symbol $f(x)$ is called the **value of a function at x** because it is the result of evaluating a variable expression. For instance, $f(4)$ means to replace x with 4 and then simplify the resulting numerical expression.

$$f(x) = x^2 + 3$$
$$f(4) = 4^2 + 3 \qquad \bullet \text{ Replace } x \text{ with } 4.$$
$$= 16 + 3 = 19$$

This process is called **evaluating a function.**

> **HOW TO**   Given $f(x) = x^2 + x - 3$, find $f(-2)$.
>
> $$f(x) = x^2 + x - 3$$
> $$f(-2) = (-2)^2 + (-2) - 3 \qquad \bullet \text{ Replace } x \text{ with } -2.$$
> $$= 4 - 2 - 3 = -1$$
> $$f(-2) = -1$$

In this example, $f(-2)$ is the second coordinate of an ordered pair of the function; the first coordinate is -2. Therefore, an ordered pair of this function is $(-2, f(-2))$, or, because $f(-2) = -1$, $(-2, -1)$.

For the function given by $y = f(x) = x^2 + x - 3$, y is called the **dependent variable** because its value depends on the value of x. The **independent variable** is x.

Functions can be written using other letters or even combinations of letters. For instance, some calculators use $ABS(x)$ for the absolute-value function. Thus the equation $y = |x|$ would be written $ABS(x) = |x|$, where $ABS(x)$ replaces y.

Example 7

Given $G(t) = \dfrac{3t}{t+4}$, find $G(1)$.

Solution

$$G(t) = \frac{3t}{t+4}$$

$$G(1) = \frac{3(1)}{1+4} \qquad \bullet \text{ Replace } t \text{ with } 1. \text{ Then simplify.}$$

$$G(1) = \frac{3}{5}$$

You Try It 7

Given $H(x) = \dfrac{x}{x-4}$, find $H(8)$.

Your solution
2

Solution on p. S30

In-Class Examples (Objective 12.1D)

1. Given $f(x) = -5 - 2x$, find $f(-1)$. -3

2. Given $g(t) = t^2 - t$, find $g(4)$.   12

3. Given $P(r) = \dfrac{2r^2}{r+1}$, find $P(-2)$. -8

12.1 Exercises

Objective A   To graph points in a rectangular coordinate system

Section 12.1

Suggested Assignment
Exercises 1–51, odds
More challenging problems:
  Exercises 54, 55

1. Graph $(-2, 1)$, $(3, -5)$, $(-2, 4)$, and $(0, 3)$.

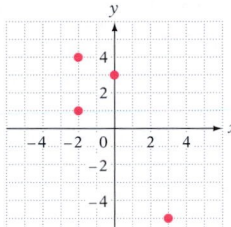

2. Graph $(5, -1)$, $(-3, -3)$, $(-1, 0)$, and $(1, -1)$.

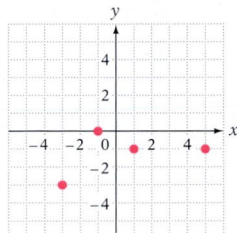

3. Graph $(0, 0)$, $(0, -5)$, $(-3, 0)$, and $(0, 2)$.

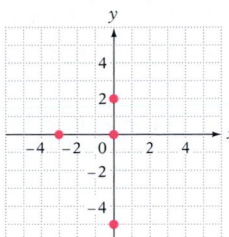

4. Graph $(-4, 5)$, $(-3, 1)$, $(3, -4)$, and $(5, 0)$.

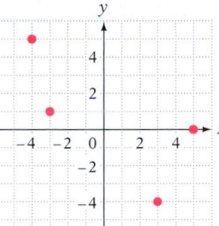

5. Graph $(-1, 4)$, $(-2, -3)$, $(0, 2)$, and $(4, 0)$.

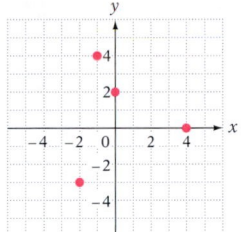

6. Graph $(5, 2)$, $(-4, -1)$, $(0, 0)$, and $(0, 3)$.

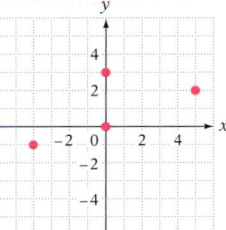

7. Find the coordinates of each of the points.

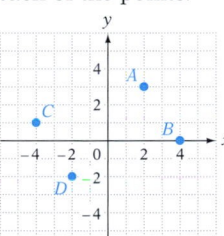

$A(2, 3)$, $B(4, 0)$, $C(-4, 1)$, $D(-2, -2)$

8. Find the coordinates of each of the points.

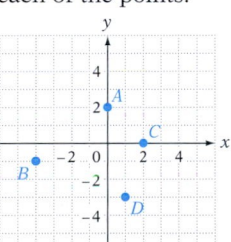

$A(0, 2)$, $B(-4, -1)$, $C(2, 0)$, $D(1, -3)$

9. Find the coordinates of each of the points.

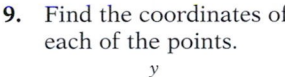
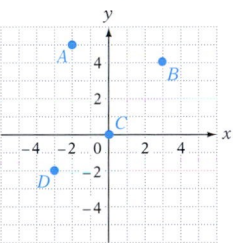

$A(-2, 5)$, $B(3, 4)$, $C(0, 0)$, $D(-3, -2)$

10. Find the coordinates of each of the points.

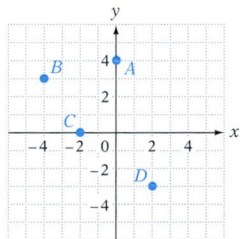

$A(0, 4)$, $B(-4, 3)$, $C(-2, 0)$, $D(2, -3)$

11. **a.** Name the abscissas of points A and C.
   b. Name the ordinates of points B and D.

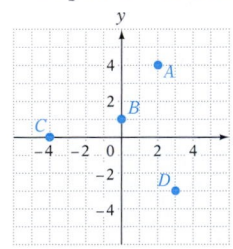

a. $2, -4$ **b.** $1, -3$

12. **a.** Name the abscissas of points A and C.
   b. Name the ordinates of points B and D.

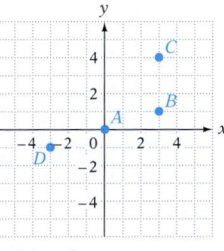

a. $0, 3$ **b.** $1, -1$

Quick Quiz (Objective 12.1A)

1. Graph the ordered pairs $(5, -1)$, $(-4, -2)$, $(1, 0)$, and $(3, 4)$.

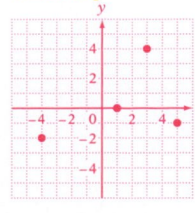

2. Use the graph at the right.
   a. Name the abscissas of A and C. $0, -3$
   b. Name the ordinates of B and D. $4, -2$

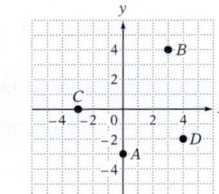

13. ✏ Suppose you are helping a student who is having trouble graphing ordered pairs. The work of the student is at the right. What can you say to this student to correct the error that is being made?

14. **a.** What are the signs of the coordinates of a point in the third quadrant?

b. What are the signs of the coordinates of a point in the fourth quadrant?

c. On an xy-coordinate system, what is the name of the axis for which all the x-coordinates are zero?

d. On an xy-coordinate system, what is the name of the axis for which all the y-coordinates are zero?

a. x-coordinate is negative, y-coordinate is negative; **b.** x-coordinate is positive, y-coordinate is negative; **c.** y-axis; **d.** x-axis

Objective B To determine ordered-pair solutions of an equation in two variables

15. Is $(3, 4)$ a solution of $y = -x + 7$?   Yes

16. Is $(2, -3)$ a solution of $y = x + 5$?   No

17. Is $(-1, 2)$ a solution of $y = \frac{1}{2}x - 1$?   No

18. Is $(1, -3)$ a solution of $y = -2x - 1$?   Yes

19. Is $(4, 1)$ a solution of $2x - 5y = 4$?   No

20. Is $(-5, 3)$ a solution of $3x - 2y = 9$?   No

21. Is $(0, 4)$ a solution of $3x - 4y = -4$?   No

22. Is $(-2, 0)$ a solution of $x + 2y = -1$?   No

For Exercises 23 to 28, graph the ordered-pair solutions of the equation for the given values of x.

23. $y = 2x$; $x = -2, -1, 0, 2$

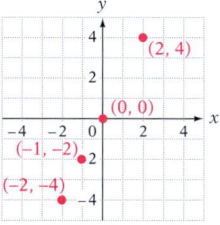

24. $y = -2x$; $x = -2, -1, 0, 2$

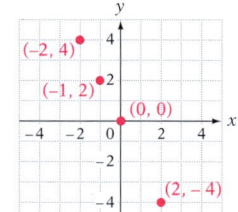

25. $y = \frac{2}{3}x + 1$; $x = -3, 0, 3$

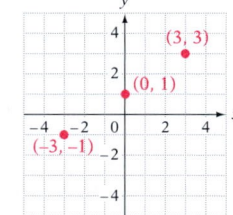

26. $y = -\frac{1}{3}x - 2$; $x = -3, 0, 3$

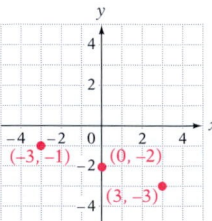

27. $2x + 3y = 6$; $x = -3, 0, 3$

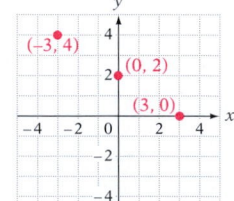

28. $x - 2y = 4$; $x = -2, 0, 2$

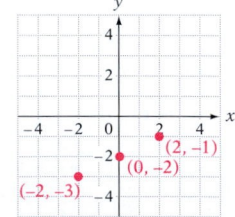

Quick Quiz (Objective 12.1B)

1. Is $(-3, 0)$ a solution of $y = \frac{1}{3}x - 1$?   No

2. Graph the ordered-pair solutions of $y = -\frac{2}{3}x - 2$ when $x = -3, 0,$ and 3.

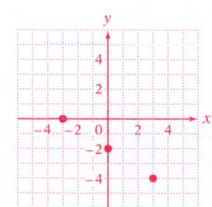

Objective C   **To determine whether a set of ordered pairs is a function**

29. ● **Biology**   The table below shows the length, in centimeters, of the humerus (the long bone of the forelimb, from shoulder to elbow) and the total wingspan, in centimeters, of several pterosaurs, which are extinct flying reptiles of the order Pterosauria. Write a relation in which the first coordinate is the length of the humerus and the second is the wingspan. Is the relation a function?

Humerus (in centimeters)	24	32	22	15	4.4	17	15	4.4
Wingspan (in centimeters)	600	750	430	300	68	370	310	55

{(24, 600), (32, 750), (22, 430), (15, 300), (4.4, 68), (17, 370), (15, 310), (4.4, 55)}; No

30. ● **Environmental Science**   The table below, based in part on data from the National Oceanic and Atmospheric Administration, shows the average annual concentration of atmospheric carbon dioxide (in parts per million) and the average sea surface temperature (in degrees Celsius) for eight consecutive years. Write a relation in which the first coordinate is the carbon dioxide concentration and the second coordinate is the average sea surface temperature. Is the relation a function?

Carbon dioxide concentration (in parts per million)	352	353	354	355	356	358	360	361
Sea surface temperature (in degrees Celsius)	15.4	15.4	15.1	15.1	15.2	15.4	15.3	15.5

{(352, 15.4), (353, 15.4), (354, 15.1), (355, 15.1), (356, 15.2), (358, 15.4), (360, 15.3), (361, 15.5)}; Yes

31. ● **Baseball**   The table at the right shows the number of home runs and the number of at-bats for the top five home-run leaders in the National League for the 2003 season. Write a relation in which the first coordinate is the number of at-bats and the second coordinate is the number of home runs per at-bats rounded to the nearest thousand. Is the relation a function?

{(390, 0.115), (591, 0.073), (517, 0.077), (576, 0.068), (605, 0.064)}; Yes

Player	At-bats	Home runs
Barry Bonds	390	45
Albert Pujois	591	43
Sammy Sosa	517	40
Gary Sheffield	576	39
Jeff Bagwell	605	39

32. ● **Nielsen Ratings**   The ratings (each rating point is 1,055,000 households) and share (the percentage of television sets in use tuned to a specific program) for selected television shows for a week in November 2003 are shown in the table at the right. Write a relation in which the first coordinate is the ratings and the second coordinate is the share. Is the relation a function?

{(18.1, 27.0), (13.6, 22.0), (13.4, 21.0), (13.2, 21.0), (11.3, 18.0)}; Yes

Television Show	Rating	Share
CSI	18.1	27.0
E.R.	13.6	22.0
Friends	13.4	21.0
CSI: Miami	13.2	21.0
60 Minutes	11.3	18.0

33. Does $y = -2x - 3$, where $x \in \{-2, -1, 0, 3\}$, define y as a function of x?
Yes

34. Does $y = 2x + 3$, where $x \in \{-2, -1, 1, 4\}$, define y as a function of x?
Yes

35. Does $|y| = x - 1$, where $x \in \{1, 2, 3, 4\}$, define y as a function of x?
No

36. Does $|y| = x + 2$, where $x \in \{-2, -1, 0, 3\}$, define y as a function of x?
No

37. Does $y = x^2$, where $x \in \{-2, -1, 0, 1, 2\}$, define y as a function of x?
Yes

38. Does $y = x^2 - 1$, where $x \in \{-2, -1, 0, 1, 2\}$, define y as a function of x?
Yes

Quick Quiz (Objective 12.1C)

1. The table at the right shows the ages of six children and their grades in school. Write a relation in which the first coordinate is the age of a child and the second coordinate is the child's grade in school. Is the relation a function?
{(6, 1), (7, 2), (8, 3), (7, 1)}; No

2. Does $y = x^2 - 4$, where $x \in \{-3, -1, 0, 1, 3\}$, define y as a function of x?   Yes

Age (years)	Grade in school
6	1
7	2
6	1
8	3
6	1
7	1

Objective D   To evaluate a function written in functional notation

39. Given $f(x) = 3x - 4$, find $f(4)$.
8

40. Given $f(x) = 5x + 1$, find $f(2)$.
11

41. Given $f(x) = x^2$, find $f(3)$.
9

42. Given $f(x) = x^2 - 1$, find $f(1)$.
0

43. Given $G(x) = x^2 + x$, find $G(-2)$.
2

44. Given $H(x) = x^2 - x$, find $H(-2)$.
6

45. Given $s(t) = \dfrac{3}{t - 1}$, find $s(-2)$.
-1

46. Given $P(x) = \dfrac{4}{2x + 1}$, find $P(-2)$.
$-\dfrac{4}{3}$

47. Given $h(x) = 3x^2 - 2x + 1$, find $h(3)$.
22

48. Given $Q(r) = 4r^2 - r - 3$, find $Q(2)$.
11

49. Given $f(x) = \dfrac{x}{x + 5}$, find $f(-3)$.
$-\dfrac{3}{2}$

50. Given $v(t) = \dfrac{2t}{2t + 1}$, find $v(3)$.
$\dfrac{6}{7}$

51. Given $g(x) = x^3 - x^2 + 2x - 7$, find $g(0)$.
-7

52. Given $F(z) = \dfrac{z}{z^2 + 1}$, find $F(0)$.
0

APPLYING THE CONCEPTS

53. Write a few sentences that describe the similarities and differences between relations and functions.

54. The graph of $y^2 = x$, where $x \in \{0, 1, 4, 9\}$, is shown at the right. Is this the graph of a function? Explain your answer.

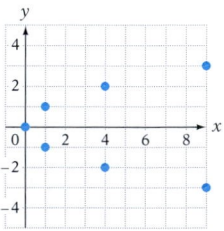

55. Is it possible to evaluate $f(x) = \dfrac{5}{x - 1}$ when $x = 1$? If so, what is $f(1)$? If not, explain why not.

Quick Quiz (Objective 12.1D)

1. Given $f(x) = -2x + 5$, find $f(-2)$.   9
2. Given $h(s) = s^2 + 3s$, find $h(-3)$.   0
3. Given $Q(t) = \dfrac{t^2}{t - 3}$, find $Q(1)$. $-\dfrac{1}{2}$

12.2 Linear Equations in Two Variables

Objective A   To graph an equation of the form $y = mx + b$

The **graph of an equation in two variables** is a graph of the ordered-pair solutions of the equation.

Consider $y = 2x + 1$. Choosing $x = -2, -1, 0, 1,$ and 2 and determining the corresponding values of y produces some of the ordered pairs of the equation. These are recorded in the table at the right. See the graph of the ordered pairs in Figure 1.

x	$y = 2x + 1$	y	(x, y)
-2	$2(-2) + 1$	-3	$(-2, -3)$
-1	$2(-1) + 1$	-1	$(-1, -1)$
0	$2(0) + 1$	1	$(0, 1)$
1	$2(1) + 1$	3	$(1, 3)$
2	$2(2) + 1$	5	$(2, 5)$

Choosing values of x that are not integers produces more ordered pairs to graph, such as $\left(-\frac{5}{2}, -4\right)$ and $\left(\frac{3}{2}, 4\right)$, as shown in Figure 2. Choosing still other values of x would result in more and more ordered pairs being graphed. The result would be so many dots that the graph would appear as the straight line shown in Figure 3, which is the graph of $y = 2x + 1$.

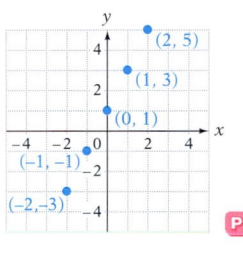

Figure 1

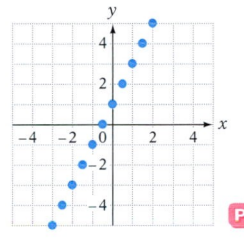

Figure 2

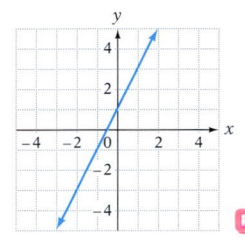

Figure 3

Equations in two variables have characteristic graphs. The equation $y = 2x + 1$ is an example of a *linear equation*, or *linear function*, because its graph is a straight line. It is also called a *first-degree equation in two variables* because the exponent on each variable is the first power.

> **Linear Equation in Two Variables**
>
> Any equation of the form $y = mx + b$, where m is the coefficient of x and b is a constant, is a **linear equation in two variables**, or **first-degree equation in two variables**, or a **linear function**. The graph of a linear equation in two variables is a straight line.

Examples of linear equations are shown at the right. These equations represent linear functions because there is only one possible y for each x. Note that for $y = 3 - 2x$, m is the coefficient of x and b is the constant.

$y = 2x + 1 \qquad (m = 2, b = 1)$

$y = x - 4 \qquad (m = 1, b = -4)$

$y = -\frac{3}{4}x \qquad \left(m = -\frac{3}{4}, b = 0\right)$

$y = 3 - 2x \qquad (m = -2, b = 3)$

The equation $y = x^2 + 4x + 3$ is not a linear equation in two variables because there is a term with a variable squared. The equation $y = \frac{3}{x - 4}$ is not a linear equation because a variable occurs in the denominator of a fraction.

Objective 12.2A

New Vocabulary

graph of an equation in two variables

linear equation in two variables

first-degree equation in two variables

linear function

Instructor Note

It is important for students to associate "graph is a straight line" with "$y = mx + b$." The Concept Check below gives exercises for students who may not see the connection.

Concept Check

1. Which of the equations below are linear equations in two variables?   b and d
   (a) $y = x^2 + 1$
   (b) $y = -x$
   (c) $y = \dfrac{1}{x}$
   (d) $y = 2 - \dfrac{1}{2}x$
   (e) $y = \sqrt{x} - 1$

2. Which of the equations above have graphs that are straight lines?   b and d

In-Class Examples (Objective 12.2A)

Graph.

1. $y = x - 2$

2. $y = -3x + 1$

3. $y = -\dfrac{4}{3}x$

1.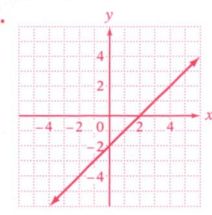

2.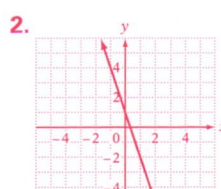

3.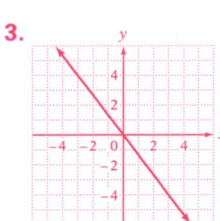

Discuss the Concepts

In the example at the right, we have chosen three values of x to draw the graph, even though only two values are necessary. Have students explain why choosing three values of x and then plotting the resulting points helps ensure accuracy.

Discuss the Concepts

As noted in the example at the right, when m is a fraction, it is helpful to choose values of x that simplify the calculation by eliminating the fraction. Describe the values of x that you would choose in order to find integer solutions to each of the following equations.

1. $y = \dfrac{1}{3}x - 1$

Multiples of 3, such as -3, 0, 3, and 6

2. $y = -\dfrac{2}{5}x + 2$

Multiples of 5, such as -5, 0, 5, and 10

Concept Check

Is $(2, -3)$ a solution of the equation whose graph is shown below?   No

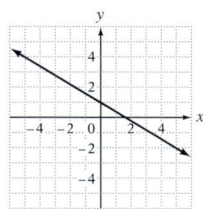

Integrating Technology

The Projects and Group Activities section at the end of this chapter contains information on using calculators to graph an equation.

To graph a linear equation, choose some values of x and then find the corresponding values of y. Because a straight line is determined by two points, it is sufficient to find only two ordered-pair solutions. However, it is recommended that at least three ordered-pair solutions be used to ensure accuracy.

HOW TO   Graph $y = -\dfrac{3}{2}x + 2$.

This is a linear equation with $m = -\dfrac{3}{2}$ and $b = 2$. Find at least three solutions. Because m is a fraction, choose values of x that will simplify the calculations. We have chosen $-2, 0,$ and 4 for x. (Any values of x could have been selected.)

x	$y = -\dfrac{3}{2}x + 2$	y	(x, y)
-2	$-\dfrac{3}{2}(-2) + 2$	5	$(-2, 5)$
0	$-\dfrac{3}{2}(0) + 2$	2	$(0, 2)$
4	$-\dfrac{3}{2}(4) + 2$	-4	$(4, -4)$

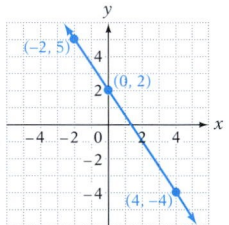

The graph of $y = -\dfrac{3}{2}x + 2$ is shown at the right.

Remember that a graph is a drawing of the ordered-pair solutions of an equation. Therefore, every point on the graph is a solution of the equation, and every solution of the equation is a point on the graph.

The graph at the right is the graph of $y = x + 2$. Note that $(-4, -2)$ and $(1, 3)$ are points on the graph and that these points are solutions of $y = x + 2$. The point whose coordinates are $(4, 1)$ is not a point on the graph and is not a solution of the equation.

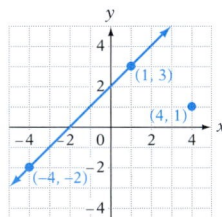

Example 1   Graph $y = 3x - 2$.

Solution

x	y
0	-2
-1	-5
2	4

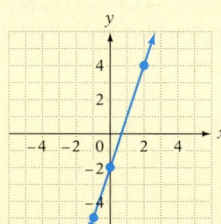

You Try It 1   Graph $y = 3x + 1$.

Your solution

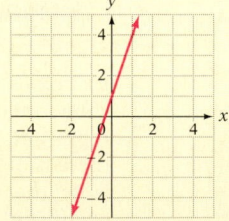

Solution on p. S30

Example 2   Graph $y = 2x$.

Solution

x	y
0	0
2	4
-2	-4

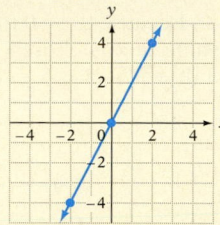

You Try It 2   Graph $y = -2x$.

Your solution

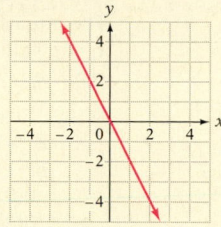

Example 3   Graph $y = \frac{1}{2}x - 1$.

Solution

x	y
0	-1
2	0
-2	-2

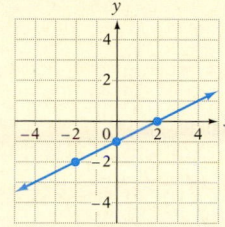

You Try It 3   Graph $y = \frac{1}{3}x - 3$.

Your solution

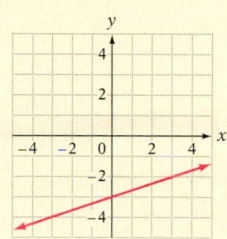

Solutions on p. S30

Objective B   To graph an equation of the form $Ax + By = C$

The equation $Ax + By = C$, where A and B are coefficients and C is a constant, is called the **standard form of a linear equation in two variables.** Examples are shown at the right.

$2x + 3y = 6$ $(A = 2, B = 3, C = 6)$
$x - 2y = -4$ $(A = 1, B = -2, C = -4)$
$2x + y = 0$ $(A = 2, B = 1, C = 0)$
$4x - 5y = 2$ $(A = 4, B = -5, C = 2)$

To graph an equation of the form $Ax + By = C$, first solve the equation for y. Then follow the same procedure used to graph $y = mx + b$.

Study Tip

Remember that a How To example indicates a worked-out example. Using paper and pencil, work through the example. See *AIM for Success,* pages xxxi–xxxii.

HOW TO   Graph $3x + 4y = 12$.

$3x + 4y = 12$
$4y = -3x + 12$

$y = -\frac{3}{4}x + 3$

x	y
0	3
4	0
-4	6

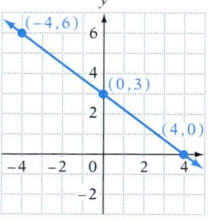

- Solve for y.
- Subtract $3x$ from each side of the equation.
- Divide each side of the equation by 4.
- Find three ordered-pair solutions of the equation.
- Graph the ordered pairs and then draw a line through the points.

Instructor Note

It is important for students to understand that each point on the graph is a solution of the equation and that each solution of the equation is a point on the graph. The following Optional Student Activity will reinforce this idea.

Optional Student Activity

1. Graph $y = 2x - 3$ on graph paper.

2. Use your graph to find the coordinates of several points on the graph. (Be sure to choose points other than the ones you plotted to draw the graph in part 1.)

3. Verify that the points whose coordinates you found in part 2 are solutions of the equation in part 1.

Objective 12.2B

New Vocabulary

standard form of a linear equation in two variables

Concept Check

Tell whether the given equation is in the form $Ax + By = C$, the form $y = mx + b$, or neither. If the equation is not in the form $y = mx + b$, rewrite it in this form.

1. $6x - 3y = 6$
   $Ax + By = C; y = 2x - 2$

2. $y = x - 1$ $y = mx + b$

3. $8 - 4y = x$
   Neither; $y = -\frac{1}{4}x + 2$

4. $5x + 4y = 4$
   $Ax + By = C; y = -\frac{5}{4}x + 1$

Discuss the Concepts

Have students describe the steps involved in changing an equation of the form $Ax + By = C$ into the form $y = mx + b$.

In-Class Examples (Objective 12.2B)

Graph.

1. $-2x + y = 1$
2. $x - 4y = 8$
3. $y = -3$
4. $x = 1$

1.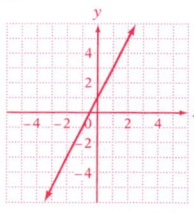

2.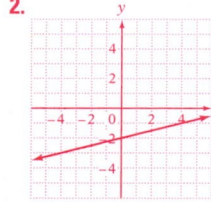

3.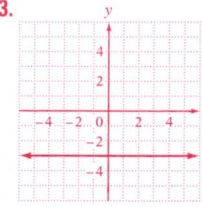

4.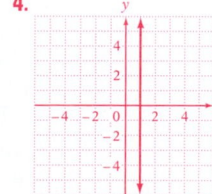

Concept Check

What is the equation of the
x-axis? $y = 0$

Concept Check

What is the equation of the
y-axis? $x = 0$

Optional Student Activity

On one set of axes, draw two vertical and two horizontal lines that intersect to form a square. Write the equations of your four lines.

The graph of a linear equation with one of the variables missing is either a horizontal or a vertical line.

The equation $y = 2$ could be written $0x + y = 2$. Because $0x = 0$ for any value of x, the value of y is always 2 no matter what value of x is chosen. For instance, replace x with -4, with -1, with 0, and with 3. In each case, $y = 2$.

$$0x + y = 2$$
$$0(-4) + y = 2 \qquad (-4, 2) \text{ is a solution.}$$
$$0(-1) + y = 2 \qquad (-1, 2) \text{ is a solution.}$$
$$0(0) + y = 2 \qquad (0, 2) \text{ is a solution.}$$
$$0(3) + y = 2 \qquad (3, 2) \text{ is a solution.}$$

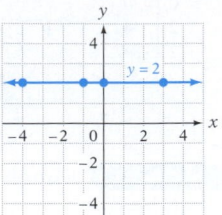

The solutions are plotted in the graph at the right, and a line is drawn through the plotted points. Note that the line is horizontal.

> **Graph of a Horizontal Line**
>
> The graph of $y = b$ is a horizontal line passing through $(0, b)$.

The equation $x = -2$ could be written $x + 0y = -2$. Because $0y = 0$ for any value of y, the value of x is always -2 no matter what value of y is chosen. For instance, replace y with -2, with 0, with 2, and with 3. In each case, $x = -2$.

$$x + 0y = -2$$
$$x + 0(-2) = -2 \qquad (-2, -2) \text{ is a solution.}$$
$$x + 0(0) = -2 \qquad (-2, 0) \text{ is a solution.}$$
$$x + 0(2) = -2 \qquad (-2, 2) \text{ is a solution.}$$
$$x + 0(3) = -2 \qquad (-2, 3) \text{ is a solution.}$$

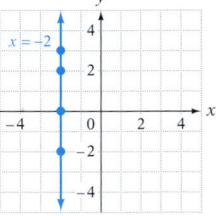

The solutions are plotted in the graph at the right, and a line is drawn through the plotted points. Note that the line is vertical.

> **Graph of a Vertical Line**
>
> The graph of $x = a$ is a vertical line passing through $(a, 0)$.

HOW TO   Graph $x = -3$ and $y = 1$ on the same coordinate grid.

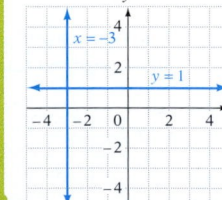

- **The graph of $x = -3$ is a vertical line passing through $(-3, 0)$.**

- **The graph of $y = 1$ is a horizontal line passing through $(0, 1)$.**

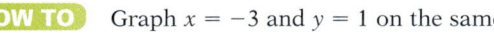

Example 4 Graph $2x - 5y = 10$.

Solution Solve $2x - 5y = 10$ for y.

$2x - 5y = 10$
$-5y = -2x + 10$
$y = \dfrac{2}{5}x - 2$

x	y
0	−2
5	0
−5	−4

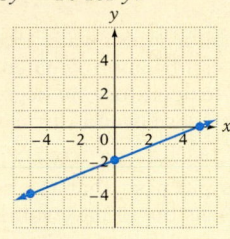

You Try It 4 Graph $5x - 2y = 10$.

Your solution

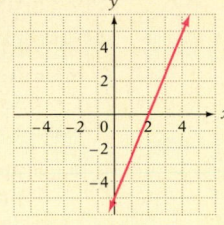

Example 5 Graph $x + 2y = 6$.

Solution Solve $x + 2y = 6$ for y.

$x + 2y = 6$
$2y = -x + 6$
$y = -\dfrac{1}{2}x + 3$

x	y
0	3
−2	4
4	1

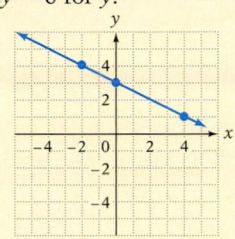

You Try It 5 Graph $x - 3y = 9$.

Your solution

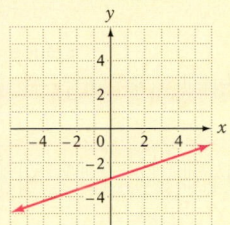

Example 6 Graph $y = -2$.

Solution
The graph of an equation of the form $y = b$ is a horizontal line passing through the point $(0, b)$.

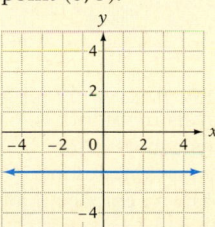

You Try It 6 Graph $y = 3$.

Your solution

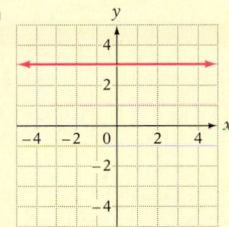

Example 7 Graph $x = 3$.

Solution
The graph of an equation of the form $x = a$ is a vertical line passing through the point $(a, 0)$.

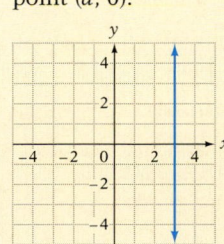

You Try It 7 Graph $x = -4$.

Your solution

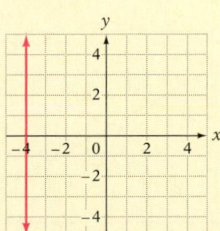

Solutions on p. S31

Instructor Note
In the next section, students will graph $Ax + By = C$ by using x- and y-intercepts. The goal of this objective is to have students first solve $Ax + By = C$ for y. Solving an equation for y is a skill that is used in a variety of situations. For instance, most graphing calculators require the form $y = mx + b$ when the equation of a line is to be graphed.

Objective 12.2C

Objective C   To solve application problems

There are a variety of applications of linear functions.

HOW TO   The temperature of a cup of water that has been placed in a microwave oven to be heated can be approximated by the equation $T = 0.7s + 65$, where T is the temperature (in degrees Fahrenheit) of the water s seconds after the microwave oven is turned on.

a. Graph this equation for $0 \le s \le 220$. (*Note:* In many applications, the domain of the variable is given so that the equation makes sense. For instance, it would not be sensible to have values of s that are less than 0. This would correspond to negative time. The choice of 220 is somewhat arbitrary and was chosen so that the water would not boil over.)

b. The point whose coordinates are (120, 149) is on the graph of this equation. Write a sentence that describes the meaning of this ordered pair.

Solution

a.

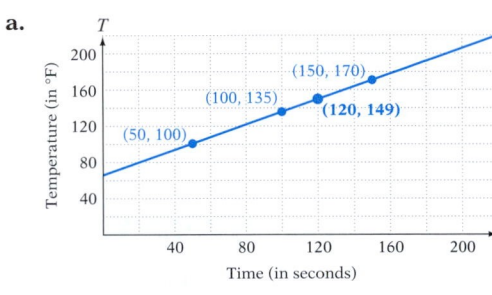

• Choosing $s = 50, 100,$ and 150, you can find the corresponding ordered pairs (50, 100), (100, 135), and (150, 170). Plot these points and draw a line through the points.

b. The point whose coordinates are (120, 149) means that 120 s (2 min) after the oven is turned on, the water temperature is 149°F.

Instructor Note

Part b of the example at the right asks the student to write a sentence that explains the meaning of an ordered pair. Questions like this require that the student do more than just manipulate symbols. They require an understanding of the ordered pair in the context of the application.

Concept Check

Refer to the example at the right.

1. Find the water temperature 40 s after the oven is turned on.   **93°F**

2. Verify that the point whose coordinates are (120, 149) is on the graph.

Discuss the Concepts

In Example 8, have students explain why the domain is $0 \le t \le 170$. t represents the time to download a portion of the file.
Time begins with 0 and ends when the file is completely downloaded 170 s later.

Optional Student Activity

1. Verify the graph shown in the solution of Example 8 by using the equation to find the coordinates of two more points on the graph.
For example, (0, 935) and (100, 385)

2. Using a different Internet connection, the number of kilobytes, K, of the MP3 file of Example 8 that remain to be downloaded t seconds after starting the download is given by $K = 935 - 11t$. Will this Internet connection complete the download in more or less time than the connection in Example 8?   **Less**

3. What is the difference between the download times for the Internet connection in Example 8 and the Internet connection in part 2 above?
85 s

Example 8

The number of kilobytes, K, of an MP3 file that remain to be downloaded t seconds after starting the download is given by $K = 935 - 5.5t$. Graph this equation for $0 \le t \le 170$. The point whose coordinates are (50, 660) is on this graph. Write a sentence that describes the meaning of this ordered pair.

Solution K

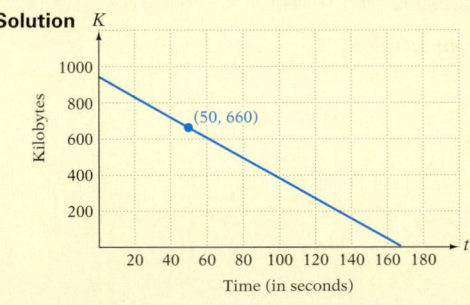

The ordered pair (50, 660) means that after 50 s, there are 660 K remaining to be downloaded.

You Try It 8

A car is traveling at a uniform speed of 40 mph. The distance, d, the car travels in t hours is given by $d = 40t$. Graph this equation for $0 \le t \le 5$. The point whose coordinates are (3, 120) is on the graph. Write a sentence that describes the meaning of this ordered pair.

Your solution d

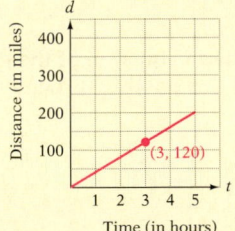

The ordered pair (3, 120) means that in 3 h the car will have traveled 120 mi.

Solution on p. S31

In-Class Examples (Objective 12.2C)

1. A long-distance telephone service costs $5.00 per month plus $.10 per minute of use. The equation that describes the monthly cost C, in dollars, for m minutes of long-distance calls is $C = 0.10m + 5.00$. Graph this equation for $0 \le m \le 300$. The point whose coordinates are (120, 17.00) is on the graph. Write a sentence that describes the meaning of this ordered pair.

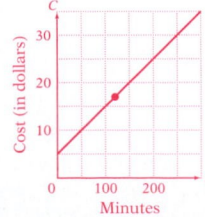

120 min of long-distance calls during 1 month costs $17.00.

12.2 Exercises

Objective A   To graph an equation of the form $y = mx + b$

For Exercises 1 to 18, graph.

1. $y = 2x - 3$

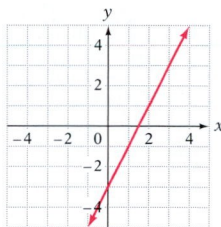

2. $y = -2x + 2$

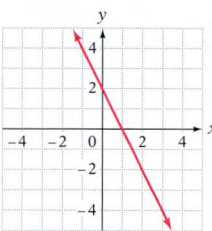

3. $y = \dfrac{1}{3}x$

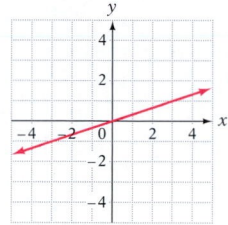

4. $y = -3x$

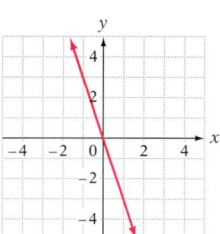

5. $y = \dfrac{2}{3}x - 1$

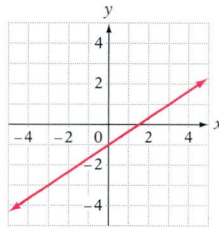

6. $y = \dfrac{3}{4}x + 2$

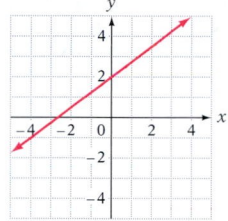

7. $y = -\dfrac{1}{4}x + 2$

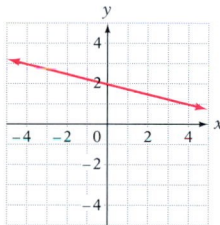

8. $y = -\dfrac{1}{3}x + 1$

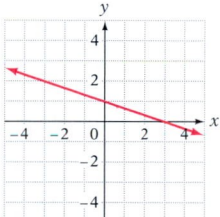

9. $y = -\dfrac{2}{5}x + 1$

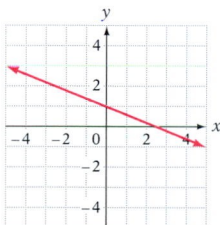

10. $y = -\dfrac{1}{2}x + 3$

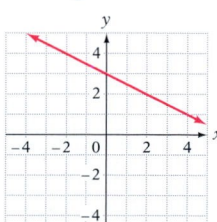

11. $y = 2x - 4$

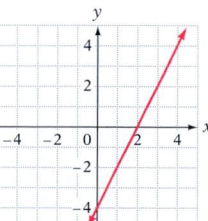

12. $y = 3x - 4$

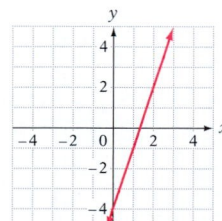

Suggested Assignment
Exercises 1–39, odds
More challenging problems:
   Exercises 43, 44

Quick Quiz (Objective 12.2A)

Graph.

1. $y = -x + 3$

2. $y = 4x$

3. $y = \dfrac{3}{4}x + 1$

1.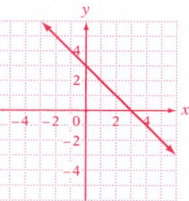

2.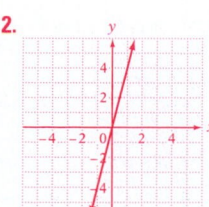

3.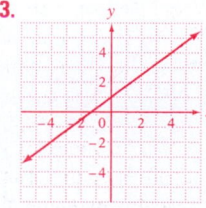

13. $y = x - 3$

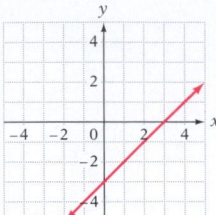

14. $y = x + 2$

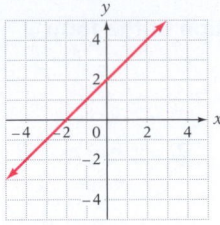

15. $y = -x + 2$

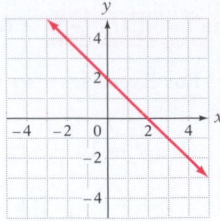

16. $y = -x - 1$

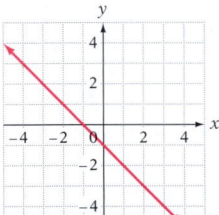

17. $y = -\dfrac{2}{3}x + 1$

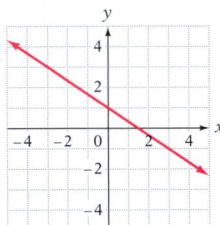

18. $y = 5x - 4$

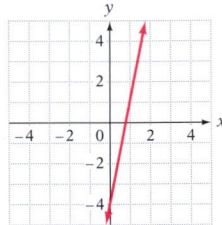

Objective B   To graph an equation of the form $Ax + By = C$

For Exercises 19 to 36, graph.

19. $3x + y = 3$

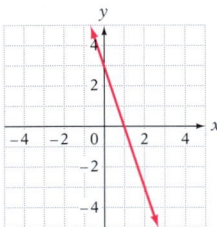

20. $2x + y = 4$

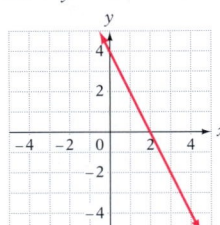

21. $2x + 3y = 6$

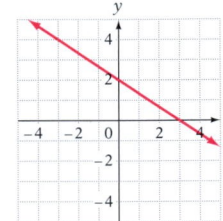

22. $3x + 2y = 4$

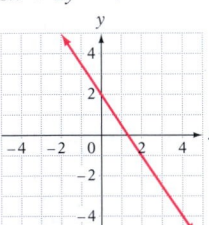

23. $x - 2y = 4$

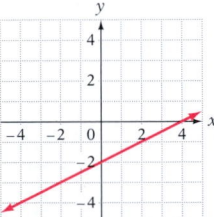

24. $x - 3y = 6$

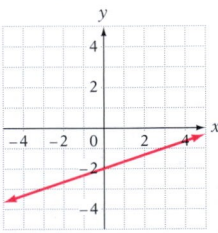

Quick Quiz
(Objective 12.2B)

Graph.

1. $x + y = 3$

2. $5x - 2y = 4$

3. $x = 2$

4. $y = -1$

1.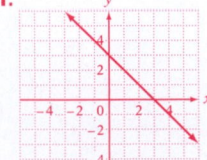

2.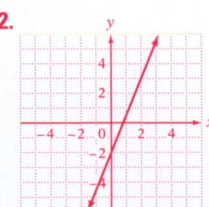

3.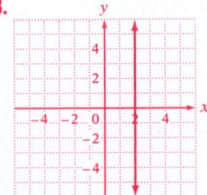

4.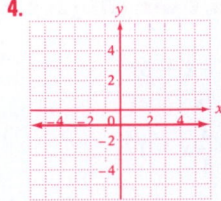

25. $2x - 3y = 6$

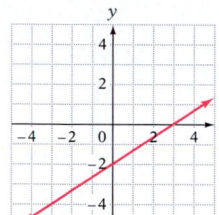

26. $3x - 2y = 8$

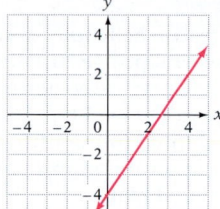

27. $2x + 5y = 10$

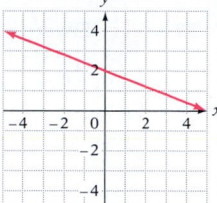

28. $3x + 4y = 12$

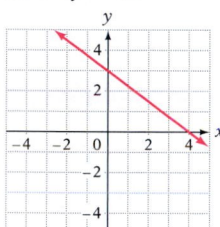

29. $x = 3$

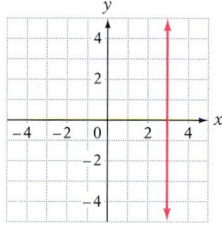

30. $y = -4$

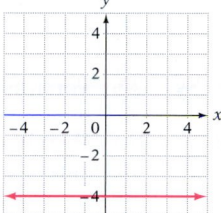

31. $x + 4y = 4$

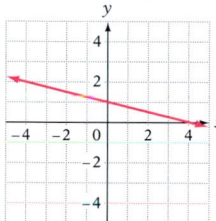

32. $4x - 3y = 12$

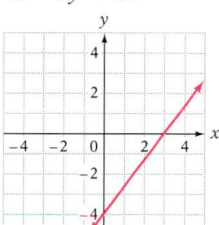

33. $y = 4$

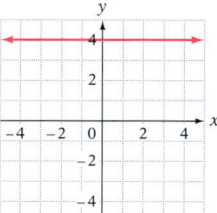

34. $x = -2$

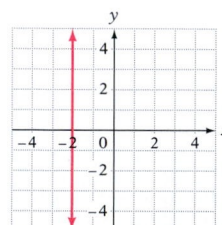

35. $\dfrac{x}{5} + \dfrac{y}{4} = 1$

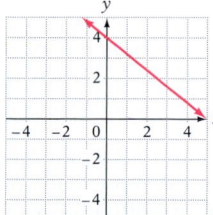

36. $\dfrac{x}{4} - \dfrac{y}{3} = 1$

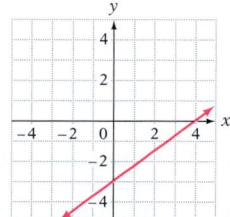

Objective C   To solve application problems

37. ✏️ **Emergency Response**   A rescue helicopter is rushing at a constant speed of 150 mph to reach several people stranded in the ocean 11 mi away after their boat sank. The rescuers can determine how far they are from the victims by using the equation $D = 11 - 2.5t$, where D is the distance in miles and t is the time elapsed in minutes. Graph this equation for $0 \le t \le 4$. The point $(3, 3.5)$ is on the graph. Write a sentence that describes the meaning of this ordered pair.
After flying for 3 min, the helicopter is 3.5 mi away from the victims.

38. ✏️ **Business**   A custom-illustrated sign or banner can be commissioned for a cost of $25 for the material and $10.50 per square foot for the artwork. The equation that represents this cost is given by $y = 10.50x + 25$, where y is the cost and x is the number of square feet in the sign. Graph this equation for $0 \le x \le 20$. The point $(15, 182.5)$ is on the graph. Write a sentence that describes the meaning of this ordered pair.
It costs $182.50 for a custom-illustrated sign 15 ft² in area.

39. ✏️ **Veterinary Science**   According to some veterinarians, the age, x, of a dog can be translated to "human years" by using the equation $H = 4x + 16$, where H is the human-equivalent age of the dog. Graph this equation for $2 \le x \le 21$. The point whose coordinates are $(6, 40)$ is on this graph. Write a sentence that explains the meaning of this ordered pair.
A dog 6 years old is equivalent in age to a human 40 years old.

40. 🥧 **HDTVs**   According to data from the Consumer Electronics Association, the projected number N (in millions) of sales of high-definition televisions (HDTVs) can be approximated by $N = 3t + 4$, where $0 \le t \le 4$ and $t = 0$ corresponds to the year 2003. Graph this equation. The point whose coordinates are $(3, 13)$ is on this graph. Write a sentence that explains the meaning of this ordered pair in the context of the problem.
In the year 2006, 13 million HDTVs will be sold.

APPLYING THE CONCEPTS

41. ✏️ Graph $y = 2x - 2$, $y = 2x$, and $y = 2x + 3$. What observation can you make about the graphs?

42. ✏️ Graph $y = x + 3$, $y = 2x + 3$, and $y = -\frac{1}{2}x + 3$. What observation can you make about the graphs?

43. For the equation $y = 3x + 2$, when the value of x changes from 1 to 2, does the value of y increase or decrease? What is the change in y? Suppose the value of x changes from 13 to 14. What is the change in y?
Increases; 3; 3

44. For the equation $y = -2x + 1$, when the value of x changes from 1 to 2, does the value of y increase or decrease? What is the change in y? Suppose the value of x changes from 13 to 14. What is the change in y?
Decreases; −2; −2

45. Telecommunications   A long-distance telephone company offers a flat rate of $.99 for the first 15 min of a phone call and then $.15 for each additional minute. The graph of this situation is a combination of the graphs of two linear equations: $C = 0.99$ when $0 < t \le 15$ and $C = 0.15(t - 15) + 0.99$ when $t > 15$. The graph is shown at the right.
a. What is the cost of a telephone call that lasts 5 min? $.99
b. What is the cost of a telephone call that lasts 20 min? $1.74

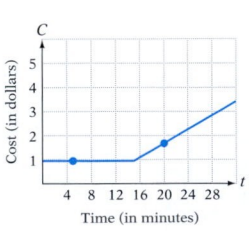

Quick Quiz (Objective 12.2C)

1. The temperature F, in degrees Fahrenheit, that corresponds to a temperature C in degrees Celsius is given by the equation $F = \frac{9}{5}C + 32$.

Graph this equation for $0 \le x \le 100$. The point whose coordinates are $(45, 113)$ is on the graph. Write a sentence that describes the meaning of this ordered pair.

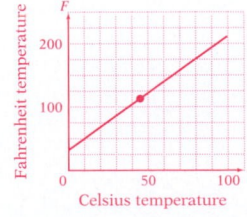

A temperature of 45 degrees Celsius is equivalent to a temperature of 113 degrees Fahrenheit.

12.3 Intercepts and Slopes of Straight Lines

Objective A   **To find the *x*- and *y*-intercepts of a straight line**

The graph of the equation $2x + 3y = 6$ is shown at the right. The graph crosses the *x*-axis at the point (3, 0) and crosses the *y*-axis at the point (0, 2). The point at which a graph crosses the *x*-axis is called the **x-intercept**. At the *x*-intercept, the *y*-coordinate is 0. The point at which a graph crosses the *y*-axis is called the **y-intercept**. At the *y*-intercept, the *x*-coordinate is 0.

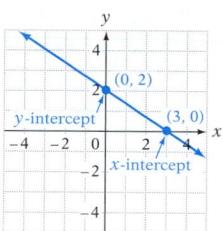

HOW TO   Find the *x*- and *y*-intercepts of the graph of the equation $2x - 3y = 12$.

TAKE NOTE
To find the *x*-intercept, let $y = 0$ and solve for *x*. To find the *y*-intercept, let $x = 0$ and solve for *y*.

To find the *x*-intercept, let $y = 0$. (Any point on the *x*-axis has *y*-coordinate 0.)

$$2x - 3y = 12$$
$$2x - 3(0) = 12$$
$$2x = 12$$
$$x = 6$$

The *x*-intercept is (6, 0).

To find the *y*-intercept, let $x = 0$. (Any point on the *y*-axis has *x*-coordinate 0.)

$$2x - 3y = 12$$
$$2(0) - 3y = 12$$
$$-3y = 12$$
$$y = -4$$

The *y*-intercept is (0, −4).

HOW TO   Find the *y*-intercept of $y = 3x + 4$.

$$y = 3x + 4 = 3(0) + 4 = 4$$ • Let $x = 0$.

The *y*-intercept is (0, 4).

For any equation of the form $y = mx + b$, the *y*-intercept is (0, *b*).

Some linear equations can be graphed by finding the *x*- and *y*-intercepts and then drawing a line through these two points.

Example 1   Find the *x*- and *y*-intercepts for $x - 2y = 4$. Graph the line.

Solution
To find the *x*-intercept, let $y = 0$ and solve for *x*.
$$x - 2y = 4$$
$$x - 2(0) = 4$$
$$x = 4 \qquad (4, 0)$$
To find the *y*-intercept, let $x = 0$ and solve for *y*.
$$x - 2y = 4$$
$$0 - 2y = 4$$
$$-2y = 4$$
$$y = -2 \qquad (0, -2)$$

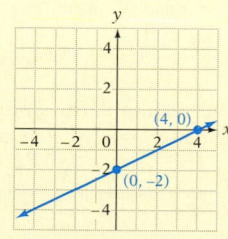

You Try It 1   Find the *x*- and *y*-intercepts for $y = 2x - 4$. Graph the line.

Your solution   *x*-intercept: *y*-intercept:
(2, 0)   (0, −4)

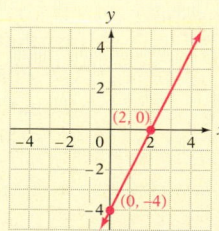

Solution on p. S31

Objective 12.3A

New Vocabulary
x-intercept
y-intercept

Discuss the Concepts
Have students describe how to find the *x*- and *y*-intercepts of a line from its equation.

Concept Check
1. Which coordinate of an *x*-intercept is 0?   *y*
2. Which coordinate of a *y*-intercept is 0?   *x*
3. Describe the line that has an *x*-intercept but no *y*-intercept. Vertical
4. Describe the line that has a *y*-intercept but no *x*-intercept. Horizontal

Instructor Note
The following Optional Student Activity can be used to have students "discover" that the *y*-intercept of the graph of $y = mx + b$ is (0, *b*).

Optional Student Activity
1. Find the *y*-intercept for each of the following.
   a. $y = -2x + 1$
   b. $y = x - 4$
   c. $y = \dfrac{2}{3}x + 2$
   d. $y = 3x$
   e. $y = 3 - \dfrac{x}{4}$
2. How is the *y*-coordinate of the *y*-intercept related to the constant term of the equation?
3. Make a conjecture about the relationship between the *y*-coordinate of the *y*-intercept and the constant term of the equation.

In-Class Examples (Objective 12.3A)

Find the *x*- and *y*-intercepts.

1. $x - 2y = 0$ (0, 0), (0, 0)

2. $y = \dfrac{1}{3}x - 3$ (9, 0), (0, −3)

3. $4x + 3y = -12$ (−3, 0), (0, −4)

4. $2x - 5y = 5$ $\left(\dfrac{5}{2}, 0\right)$, (0, −1)

Objective 12.3B

Vocabulary to Review

parallel lines [1.2C]
perpendicular lines [7.1A]

New Vocabulary

slope
positive slope
negative slope
zero slope
undefined slope

New Equation

Slope Formula, $m = \dfrac{y_2 - y_1}{x_2 - x_1}$

Instructor Note

Students need to understand that the slope of a line is the same regardless of which two points are used to calculate the slope. The following Optional Student Activity can be used to explore this idea, as well as to give students practice with calculating slope.

Optional Student Activity

From the graph in Figure 2, you can see that the line that passes through the points $(-2, -3)$ and $(6, 1)$ also passes through the points $(0, -2)$, $(2, -1)$, and $(4, 0)$. In parts 1–5, find the change in the y values, the change in the x values, and the ratio $\dfrac{\text{change in } y}{\text{change in } x}$ as you move from one point to the other.

1. From $(-2, -3)$ to $(0, -2)$

$1, 2, \dfrac{1}{2}$

2. From $(0, -2)$ to $(4, 0)$

$2, 4, \dfrac{1}{2}$

3. From $(6, 1)$ to $(4, 0)$

$-1, -2, \dfrac{1}{2}$

4. From $(6, 1)$ to $(2, -1)$

$-2, -4, \dfrac{1}{2}$

5. From $(-2, -3)$ to $(4, 0)$

$3, 6, \dfrac{1}{2}$

6. Make an observation based on the results of parts 1–5.

Any two points on a line can be used to find the slope of the line.

Objective B   To find the slope of a straight line

The graphs of $y = \frac{2}{3}x + 1$ and $y = 2x + 1$ are shown in Figure 1. Each graph crosses the y-axis at the point $(0, 1)$, but the graphs have different slants. The **slope** of a line is a measure of the slant of the line. The symbol for slope is m.

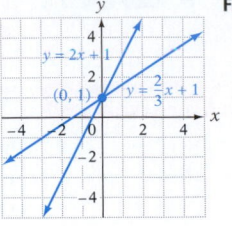

Figure 1

TAKE NOTE

The change in the y values can be thought of as the *rise* of the line, and the change in the x values can be thought of as the *run*. Then

$$\text{Slope} = m = \frac{\text{rise}}{\text{run}}$$

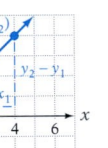

$$m = \frac{\text{rise}}{\text{run}}$$

The slope of a line containing two points is the ratio of the change in the y values of the two points to the change in the x values. The line containing the points $(-2, -3)$ and $(6, 1)$ is graphed in Figure 2. The change in the y values is the difference between the two ordinates.

$$\text{Change in } y = 1 - (-3) = 4$$

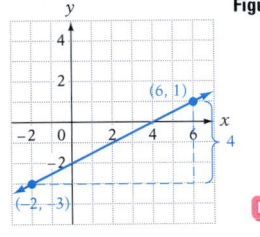

Figure 2

The change in the x values is the difference between the two abscissas (Figure 3).

$$\text{Change in } x = 6 - (-2) = 8$$

$$\text{Slope} = m = \frac{\text{change in } y}{\text{change in } x} = \frac{4}{8} = \frac{1}{2}$$

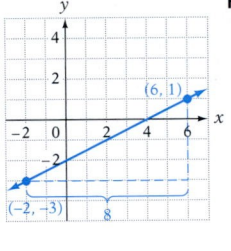

Figure 3

Slope Formula

If $P_1(x_1, y_1)$ and $P_2(x_2, y_2)$ are two points on a line and $x_1 \neq x_2$, then $m = \dfrac{y_2 - y_1}{x_2 - x_1}$ (Figure 4). If $x_1 = x_2$, the slope is undefined.

Figure 4

HOW TO   Find the slope of the line containing the points $(-1, 1)$ and $(2, 3)$.

Let P_1 be $(-1, 1)$ and P_2 be $(2, 3)$. Then $x_1 = -1$, $y_1 = 1$, $x_2 = 2$, and $y_2 = 3$.

$$m = \frac{y_2 - y_1}{x_2 - x_1} = \frac{3 - 1}{2 - (-1)} = \frac{2}{3}$$

The slope is $\frac{2}{3}$.

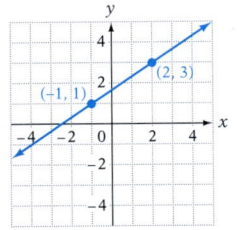

Positive slope

TAKE NOTE

Positive slope means that the value of y increases as the value of x increases.

A line that slants upward to the right always has a **positive slope**.

Note that you obtain the same results if the points are named oppositely. Let P_1 be $(2, 3)$ and P_2 be $(-1, 1)$. Then $x_1 = 2$, $y_1 = 3$, $x_2 = -1$, and $y_2 = 1$.

$$m = \frac{y_2 - y_1}{x_2 - x_1} = \frac{1 - 3}{-1 - 2} = \frac{-2}{-3} = \frac{2}{3}$$

The slope is $\frac{2}{3}$. Therefore, it does not matter which point is named P_1 and which is named P_2; the slope remains the same.

In-Class Examples (Objective 12.3B)

1. Find the slope of the line containing the points $(5, 3)$ and $(-2, 5)$. $-\dfrac{2}{7}$

2. The graph at the right shows the distance traveled by a car and the amount of gas used by the car. Find the slope of the line. Write a sentence that explains the meaning of the slope.

$m = 32$. The car gets 32 mpg.

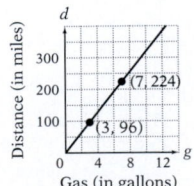

HOW TO Find the slope of the line containing the points $(-3, 4)$ and $(2, -2)$.

Let P_1 be $(-3, 4)$ and P_2 be $(2, -2)$.

$$m = \frac{y_2 - y_1}{x_2 - x_1} = \frac{-2 - 4}{2 - (-3)} = \frac{-6}{5} = -\frac{6}{5}$$

The slope is $-\frac{6}{5}$.

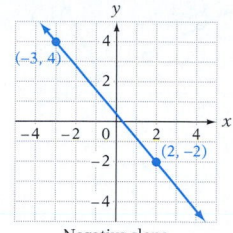

Negative slope

A line that slants downward to the right always has a **negative slope**.

HOW TO Find the slope of the line containing the points $(-1, 3)$ and $(4, 3)$.

Let P_1 be $(-1, 3)$ and P_2 be $(4, 3)$.

$$m = \frac{y_2 - y_1}{x_2 - x_1} = \frac{3 - 3}{4 - (-1)} = \frac{0}{5} = 0$$

The slope is 0.

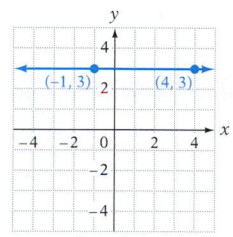

Zero slope

A horizontal line has **zero slope**.

HOW TO Find the slope of the line containing the points $(2, -2)$ and $(2, 4)$.

Let P_1 be $(2, -2)$ and P_2 be $(2, 4)$.

$$m = \frac{y_2 - y_1}{x_2 - x_1} = \frac{4 - (-2)}{2 - 2} = \frac{6}{0}$$ Division by zero is not defined.

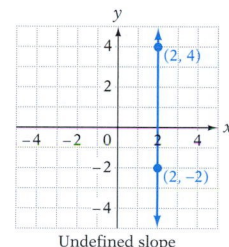

Undefined slope

A vertical line has **undefined slope**.

Two lines in the plane that never intersect are called parallel lines. The lines l_1 and l_2 in the figure at the right are parallel. Calculating the slope of each line, we have

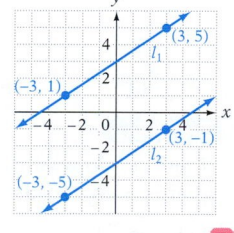

Slope of l_1: $m_1 = \frac{y_2 - y_1}{x_2 - x_1} = \frac{5 - 1}{3 - (-3)} = \frac{4}{6} = \frac{2}{3}$

Slope of l_2: $m_2 = \frac{y_2 - y_1}{x_2 - x_1} = \frac{-1 - (-5)}{3 - (-3)} = \frac{4}{6} = \frac{2}{3}$

Note that these parallel lines have the same slope. This is always true.

Parallel Lines

Two nonvertical lines in the plane are parallel if and only if they have the same slope. All vertical lines in the plane are parallel.

Concept Check

1. Graph the line that passes through the given pair of points and find its slope. Put parts a–c on one set of axes. Put parts d–f on another set of axes.
   a. $(0, -1)$, $(3, 0)$ $\frac{1}{3}$
   b. $(-1, -2)$, $(2, 1)$ 1
   c. $(-1, -4)$, $(1, 2)$ 3

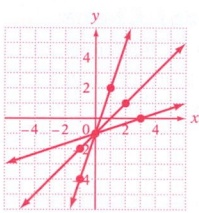

   d. $(0, 1)$, $(3, 0)$ $-\frac{1}{3}$
   e. $(0, 3)$, $(2, 1)$ -1
   f. $(2, 3)$, $(4, -3)$ -3

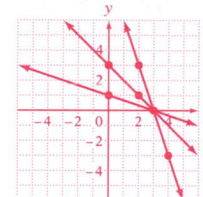

2. Describe how the sign (positive or negative) of a line's slope is related to the graph of the line.
   A line with positive slope rises from left to right. A line with negative slope falls from left to right.

3. Describe how the slope of a line is related to the steepness of the line.
   The larger the absolute value of the slope, the steeper the line.

Discuss the Concepts

Just as the phrase *the sum of* means to add, any expression that contains the word *per* translates mathematically as slope. Have students suggest some everyday situations that involve slope. Some examples are situations that include miles per gallon or feet per second.

Concept Check

Refer to the example about Florence Griffith-Joyner. Use the slope of the line to find how far Florence Griffith-Joyner had traveled after 8 s.   76 m

Optional Student Activity

Suppose a photocopy machine makes copies at a rate of 45 copies per minute.

1. Write an equation for the number of copies that can be printed in *x* minutes.
   $y = 45x$

2. What is the slope of the equation you wrote in part 1?
   45

3. How is the slope from part 2 related to the number of copies produced per minute?
   They are the same number.

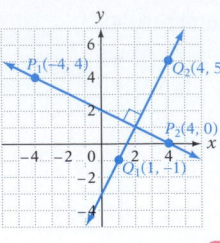

Two lines that intersect at a 90° angle (right angle) are perpendicular lines. The lines at the left are perpendicular.

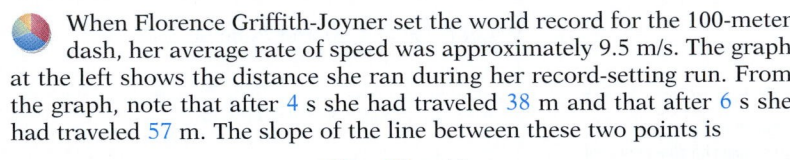

Perpendicular Lines

Two nonvertical lines in the plane are perpendicular if and only if the product of their slopes is −1. A vertical and a horizontal line are perpendicular.

The slope of the line between P_1 and P_2 is $m_1 = \frac{0 - 4}{4 - (-4)} = -\frac{4}{8} = -\frac{1}{2}$. The slope of the line between Q_1 and Q_2 is $m_2 = \frac{5 - (-1)}{4 - 1} = \frac{6}{3} = 2$. The product of the slopes is $\left(-\frac{1}{2}\right)2 = -1$. Because the product of the slopes is −1, the graphs are perpendicular.

There are many applications of the concept of slope. Here is an example.

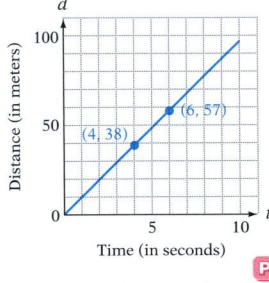

When Florence Griffith-Joyner set the world record for the 100-meter dash, her average rate of speed was approximately 9.5 m/s. The graph at the left shows the distance she ran during her record-setting run. From the graph, note that after 4 s she had traveled 38 m and that after 6 s she had traveled 57 m. The slope of the line between these two points is

$$m = \frac{57 - 38}{6 - 4} = \frac{19}{2} = 9.5$$

Note that the slope of the line is the same as the rate at which she was running, 9.5 m/s. The average speed of an object is related to slope.

Example 2

Find the slope of the line containing the points with coordinates $(-2, -3)$ and $(3, 4)$.

Solution
Let $P_1 = (-2, -3)$ and $P_2 = (3, 4)$.

$m = \frac{y_2 - y_1}{x_2 - x_1} = \frac{4 - (-3)}{3 - (-2)}$ • $y_2 = 4, y_1 = -3,$
$x_2 = 3, x_1 = -2$

$= \frac{7}{5}$

The slope is $\frac{7}{5}$.

You Try It 2

Find the slope of the line containing the points with coordinates $(1, 4)$ and $(-3, 8)$.

Your solution
−1

Example 3

Find the slope of the line containing the points with coordinates $(-1, 4)$ and $(-1, 0)$.

Solution
Let $P_1 = (-1, 4)$ and $P_2 = (-1, 0)$.

$m = \frac{y_2 - y_1}{x_2 - x_1} = \frac{0 - 4}{-1 - (-1)}$ • $y_2 = 0, y_1 = 4,$
$x_2 = -1, x_1 = -1$

$= \frac{-4}{0}$

The slope is undefined.

You Try It 3

Find the slope of the line containing the points with coordinates $(-1, 2)$ and $(4, 2)$.

Your solution
0

Solutions on p. S31

Example 4

The graph below shows the height of a plane above an airport during its 30-minute descent from cruising altitude to landing. Find the slope of the line. Write a sentence that explains the meaning of the slope.

Solution

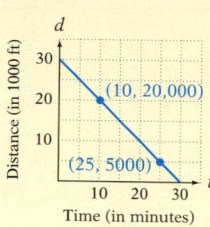

$$m = \frac{5000 - 20{,}000}{25 - 10} = \frac{-15{,}000}{15}$$

$$= -1000$$

A slope of -1000 means that the height of the plane is *decreasing* at the rate of 1000 ft/min.

You Try It 4

The graph below shows the approximate decline in the value of a used car over a 5-year period. Find the slope of the line. Write a sentence that states the meaning of the slope.

Your solution

$m = -850$

A slope of -850 means that the value of the car is decreasing at a rate of $850 per year.

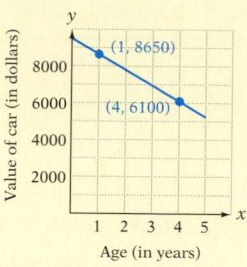

Solution on p. S31

Objective C   **To graph a line using the slope and the *y*-intercept**

The graph of the equation $y = \frac{2}{3}x + 1$ is shown at the right. The points $(-3, -1)$ and $(3, 3)$ are on the graph. The slope of the line between the two points is

$$m = \frac{3 - (-1)}{3 - (-3)} = \frac{4}{6} = \frac{2}{3}$$

Observe that the slope of the line is the coefficient of x in the equation $y = \frac{2}{3}x + 1$. Also recall that the y-intercept is $(0, 1)$, where 1 is the constant term of the equation.

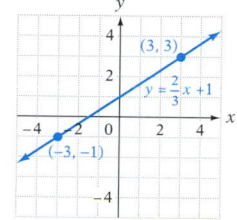

TAKE NOTE

Here are some equations in slope-intercept form.

$y = 2x - 3$: Slope is 2; y-intercept is $(0, -3)$.

$y = -x + 2$: Slope is -1 (recall that $-x = -1x$); y-intercept is $(0, 2)$.

$y = \frac{x}{2}$: Because $\frac{x}{2} = \frac{1}{2}x$, slope is $\frac{1}{2}$; y-intercept is $(0, 0)$.

> **Slope-Intercept Form of a Linear Equation**
>
> An equation of the form $y = mx + b$ is called the **slope-intercept form** of a straight line. The slope of the line is m, the coefficient of x. The y-intercept is $(0, b)$, where b is the constant term of the equation.

If a linear equation is not in slope-intercept form, solve the equation for y.

HOW TO   Find the slope and the y-intercept of the graph of $3x + 2y = 12$.

$$3x + 2y = 12$$
$$2y = -3x + 12 \quad \bullet \text{ Write the equation in slope-intercept}$$
$$y = -\frac{3}{2}x + 6 \qquad \text{ form by solving for } y.$$

The slope is $-\frac{3}{2}$; the y-intercept is $(0, 6)$.

When an equation of a line is in slope-intercept form, its graph can be drawn using the slope and the y-intercept. First locate the y-intercept. Use the slope to find a second point on the line. Then draw a line through the two points.

Objective 12.3C

New Vocabulary
slope-intercept form

New Equation
Slope-Intercept Form,
$y = mx + b$

Concept Check
Identify the slope and the y-intercept of each line from its equation.

1. $y = \frac{1}{2}x - 2$ $\frac{1}{2}$; $(0, -2)$

2. $y = x + 4$ 1; $(0, 4)$

3. $2x + y = 1$ -2; $(0, 1)$

4. $4y = -x + 8$ $-\frac{1}{4}$; $(0, 2)$

5. $2x + 7y = 14$ $-\frac{2}{7}$; $(0, 2)$

Discuss the Concepts
For each line in the above Concept Check, describe how to move from the y-intercept to a second point on the line.

In-Class Examples (Objective 12.3C)

Graph by using the slope and the y-intercept.

1. $y = x - 3$

2. $y = \frac{4}{3}x + 1$

3. $x + 2y = 4$

1.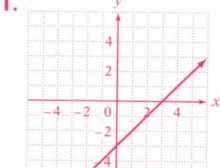

2.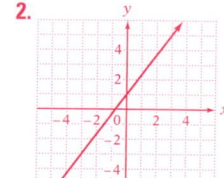

3.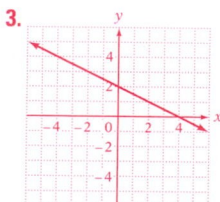

Discuss the Concepts

How can you check your work when graphing a line by using the slope and the y-intercept? That is, how can you verify that the second point you find using the slope is on the line of the given equation?

Substitute the coordinates of the point into the equation and determine whether it checks as a solution.

Optional Student Activity

1. Look at the form of each equation.

$$4x - 3y = 12$$
$$y = -\frac{1}{2}x + 2$$

Which equation would be easier to graph by using the slope and the y-intercept, and which equation would be easier to graph by finding the x- and y-intercepts?

Slope and y-intercept:

$$y = -\frac{1}{2}x + 2$$

x- and y-intercepts:

$$4x - 3y = 12$$

2. Graph the two lines.

HOW TO   Graph $y = 2x - 3$.

y-intercept $= (0, b) = (0, -3)$

$$m = 2 = \frac{2}{1} = \frac{\text{change in } y}{\text{change in } x}$$

Beginning at the y-intercept, move right 1 unit (change in x) and then up 2 units (change in y).

$(1, -1)$ is a second point on the graph.

Draw a line through the two points $(0, -3)$ and $(1, -1)$.

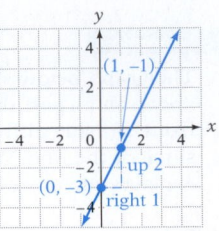

Example 5   Graph $y = -\frac{2}{3}x + 1$ by using the slope and the y-intercept.

Solution y-intercept $= (0, b) = (0, 1)$

$$m = -\frac{2}{3} = \frac{-2}{3} = \frac{\text{change in } y}{\text{change in } x}$$

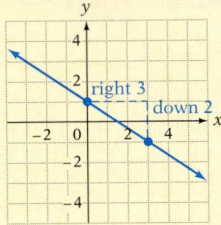

You Try It 5   Graph $y = -\frac{1}{4}x - 1$ by using the slope and the y-intercept.

Your solution

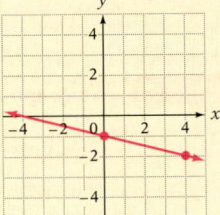

Example 6   Graph $2x - 3y = 6$ by using the slope and the y-intercept.

Solution   Solve the equation for y.

$$2x - 3y = 6$$
$$-3y = -2x + 6$$
$$y = \frac{2}{3}x - 2$$

y-intercept $= (0, -2);\ m = \frac{2}{3}$

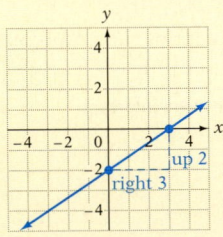

You Try It 6   Graph $x - 2y = 4$ by using the slope and the y-intercept.

Your solution

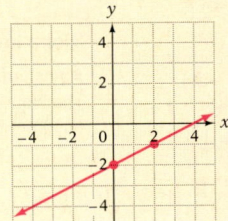

Solutions on pp. S31–S32

12.3 Exercises

Objective A   To find the *x*- and *y*-intercepts of a straight line

For Exercises 1 to 12, find the *x*- and *y*-intercepts.

1. $x - y = 3$
(3, 0), (0, −3)

2. $3x + 4y = 12$
(4, 0), (0, 3)

3. $y = 3x − 6$
(2, 0), (0, −6)

4. $y = 2x + 10$
(−5, 0), (0, 10)

5. $x − 5y = 10$
(10, 0), (0, −2)

6. $3x + 2y = 12$
(4, 0), (0, 6)

7. $y = 3x + 12$
(−4, 0), (0, 12)

8. $y = 5x + 10$
(−2, 0), (0, 10)

9. $2x − 3y = 0$
(0, 0), (0, 0)

10. $3x + 4y = 0$
(0, 0), (0, 0)

11. $y = -\dfrac{1}{2}x + 3$
(6, 0), (0, 3)

12. $y = \dfrac{2}{3}x − 4$
(6, 0), (0, −4)

For Exercises 13 to 18, find the *x*- and *y*-intercepts and then graph.

13. $5x + 2y = 10$

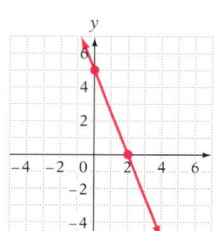

14. $x − 3y = 6$

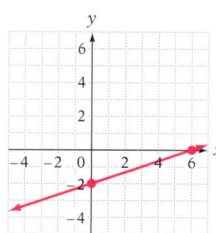

15. $y = \dfrac{3}{4}x − 3$

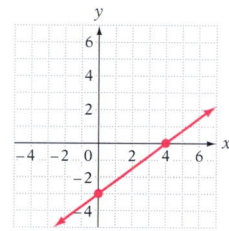

16. $y = \dfrac{2}{5}x − 2$

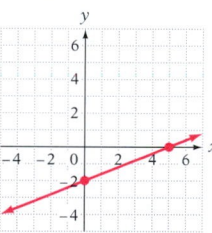

17. $5y − 3x = 15$

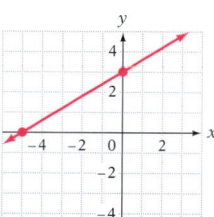

18. $9y − 4x = 18$

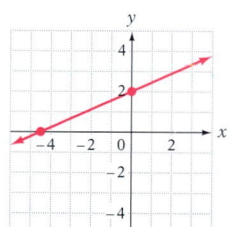

Objective B   To find the slope of a straight line

19.   Explain how to find the slope of a line given two points on the line.

20. What is the difference between a line that has zero slope and one that has undefined slope?

Section 12.3

Suggested Assignment
Exercises 1–65, odds
More challenging problems:
Exercises 66–68

Answers to Writing Exercises

19. The slope of a nonvertical line is calculated by selecting two points on the line, finding the difference between the *y*-coordinates, finding the difference between the *x*-coordinates, and then writing the ratio of the difference in the *y*-coordinates to the difference in the *x*-coordinates. For a vertical line, the slope is undefined.

20. A line that has zero slope is a horizontal line; a line whose slope is undefined is a vertical line.

Quick Quiz (Objective 12.3A)

Find the *x*- and *y*-intercepts and then graph.

1. $x − y = 2$ (2, 0), (0, −2)

2. $−x + 4y = 4$ (−4, 0), (0, 1)

3. $y = \dfrac{4}{3}x − 4$ (3, 0), (0, −4)

4. $y = −4x − 4$ (−1, 0), (0, −4)

1.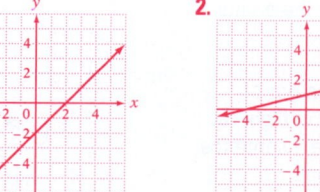

2.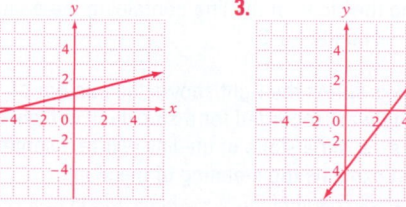

3.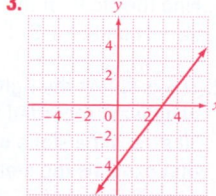

4.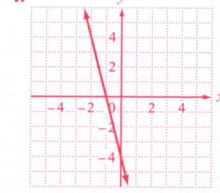

For Exercises 21 to 32, find the slope of the line containing the given points.

21. $P_1(4, 2)$, $P_2(3, 4)$ **22.** $P_1(2, 1)$, $P_2(3, 4)$ **23.** $P_1(-1, 3)$, $P_2(2, 4)$ **24.** $P_1(-2, 1)$, $P_2(2, 2)$

-2 3 $\dfrac{1}{3}$ $\dfrac{1}{4}$

25. $P_1(2, 4)$, $P_2(4, -1)$ **26.** $P_1(1, 3)$, $P_2(5, -3)$ **27.** $P_1(3, -4)$, $P_2(3, 5)$ **28.** $P_1(-1, 2)$, $P_2(-1, 3)$

$-\dfrac{5}{2}$ $-\dfrac{3}{2}$ Undefined   Undefined

29. $P_1(4, -2)$, $P_2(3, -2)$ **30.** $P_1(5, 1)$, $P_2(-2, 1)$ **31.** $P_1(0, -1)$, $P_2(3, -2)$ **32.** $P_1(3, 0)$, $P_2(2, -1)$

Zero   Zero $-\dfrac{1}{3}$ 1

For Exercises 33 to 40, determine whether the line through P_1 and P_2 is parallel, perpendicular, or neither parallel nor perpendicular to the line through Q_1 and Q_2.

33. $P_1(-3, 4)$, $P_2(2, -5)$; $Q_1(3, 6)$, $Q_2(-2, -3)$ **34.** $P_1(4, -5)$, $P_2(6, -9)$; $Q_1(5, -4)$, $Q_2(1, 4)$
Neither   Parallel

35. $P_1(0, 1)$, $P_2(2, 4)$; $Q_1(-4, -7)$, $Q_2(2, 5)$ **36.** $P_1(5, 1)$, $P_2(3, -2)$; $Q_1(0, -2)$, $Q_2(3, -4)$
Neither   Perpendicular

37. $P_1(-2, 4)$, $P_2(2, 4)$; $Q_1(-3, 6)$, $Q_2(4, 6)$ **38.** $P_1(1, -1)$, $P_2(3, -2)$; $Q_1(-4, 1)$, $Q_2(2, -5)$
Parallel   Neither

39. $P_1(7, -1)$, $P_2(-4, 6)$; $Q_1(3, 0)$, $Q_2(-5, 3)$ **40.** $P_1(5, -2)$, $P_2(-1, 3)$; $Q_1(3, 4)$, $Q_2(-2, -2)$
Neither   Perpendicular

41. 🟠 **Camera-Phone Sales**   The graph at the right is based on data from *InfoSync World*. It shows the projected camera-phone sales worldwide through 2008. Find the slope of the line. Write a sentence that states the meaning of the slope in the context of this problem.
$m = 33$. The worldwide sales of camera-phones are increasing by 33 million units per year.

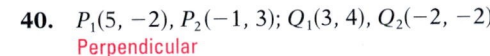

42. 🔺 **Deep-Sea Diving**   The pressure, in pounds per square inch, on a diver is shown in the graph at the right. Find the slope of the line. Write a sentence that explains the meaning of the slope.
$m = 0.5$. For each additional foot a diver descends below the surface of the water, the pressure on the diver increases by 0.5 lb/in².

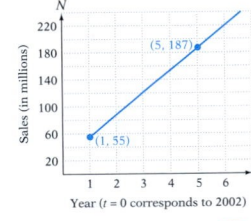

43. 🟠 **Depreciation**   The graph at the right, based on data from *Kelley Blue Book*, shows the decline in value of a 2002 Porsche Boxster as the number of miles the car is driven increases. Find the slope of the line. Write a sentence that states the meaning of the slope in the context of this problem.
$m = -180$. The value of the car is decreasing $180 for each additional 1000 mi the car is driven.

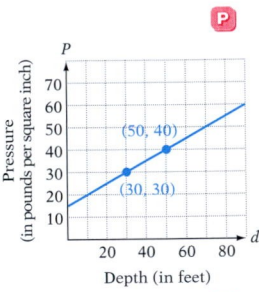

Quick Quiz (Objective 12.3B)

1. Find the slope of the line containing the points $(4, -1)$ and $(-1, 3)$. $-\dfrac{4}{5}$

2. The graph at the right shows the number of wingbeats recorded for a shellduck in flight for 30 s. Find the slope of the line. Write a sentence that explains the meaning of the slope.
$m = 3$. The shellduck beats its wings at a rate of 3 beats/s.

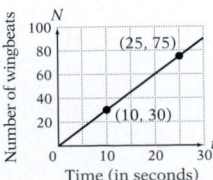

44. 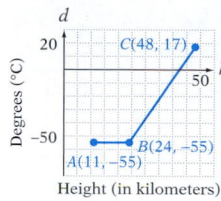 **Environmental Science**   The stratosphere extends from approximately 11 km to 50 km above Earth. The graph at the right shows how the temperature, in degrees Celsius, changes in the stratosphere. Explain the meaning of the horizontal line segment from *A* to *B*. Find the slope of the line from *B* to *C* and explain its meaning.

The temperature remains constant from A to B. m = 3. The temperature is increasing at a rate of 3°C/km.

Objective C   **To graph a line using the slope and the *y*-intercept**

For Exercises 45 to 50, find the slope and the *y*-intercept for the graph of the equation.

45. $2x - 3y = 6$

$m = \dfrac{2}{3}, (0, -2)$

46. $4x + 3y = 12$

$m = -\dfrac{4}{3}, (0, 4)$

47. $2x + 5y = 10$

$m = -\dfrac{2}{5}, (0, 2)$

48. $2x + y = 0$

$m = -2, (0, 0)$

49. $x - 4y = 0$

$m = \dfrac{1}{4}, (0, 0)$

50. $2x + 3y = 8$

$m = -\dfrac{2}{3}, \left(0, \dfrac{8}{3}\right)$

For Exercises 51 to 65, graph by using the slope and the *y*-intercept.

51. $y = 3x + 1$

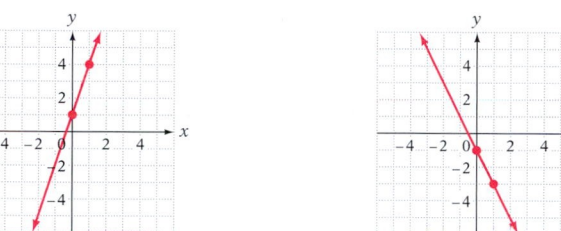

52. $y = -2x - 1$

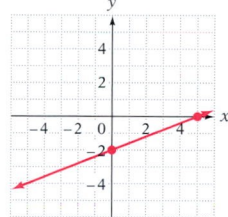

53. $y = \dfrac{2}{5}x - 2$

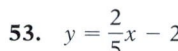

54. $y = \dfrac{3}{4}x + 1$

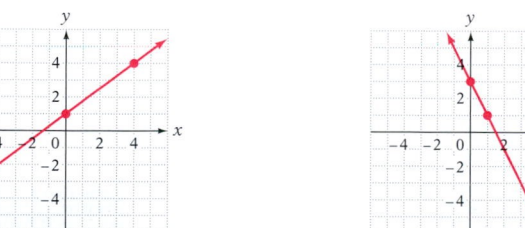

55. $2x + y = 3$

56. $3x - y = 1$

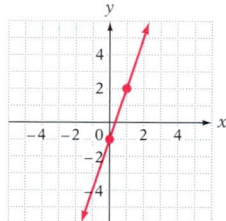

57. $x - 2y = 4$

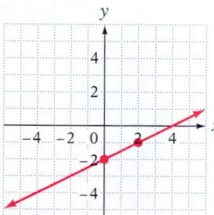

58. $x + 3y = 6$

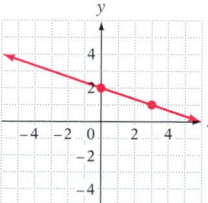

59. $y = \dfrac{2}{3}x$

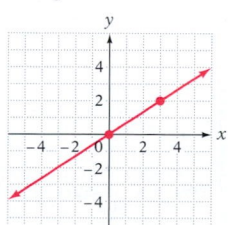

Quick Quiz (Objective 12.3C)

Graph by using the slope and *y*-intercept.

1. $4x - y = 2$

2. $y = \dfrac{1}{5}x$

3. $x + y = 3$

1.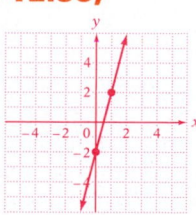

2.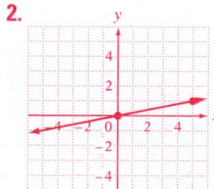

3.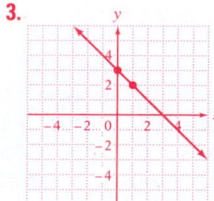

60. $y = \frac{1}{2}x$

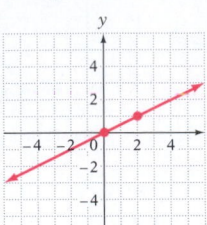

61. $y = -x + 1$

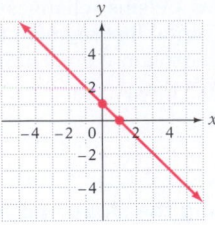

62. $y = -x - 3$

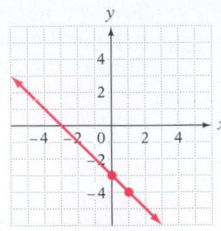

63. $3x - 4y = 12$

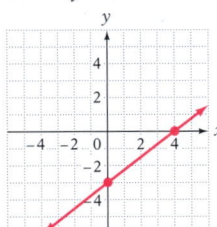

64. $5x - 2y = 10$

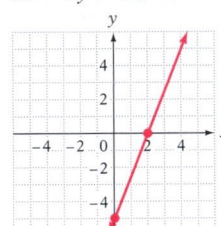

65. $y = -4x + 2$

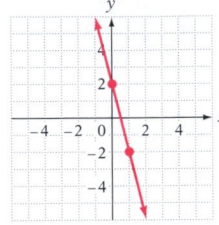

APPLYING THE CONCEPTS

66. Do all straight lines have a y-intercept? If not, give an example of one that does not.   No. For instance, the graph of $x = 3$.

67. If two lines have the same slope and the same y-intercept, must the graphs of the lines be the same? If not, give an example.   Yes

68. a. Graph: $\frac{x}{3} + \frac{y}{4} = 1$ **b.** Graph: $\frac{x}{2} - \frac{y}{3} = 1$ **c.** Graph: $-\frac{x}{4} + \frac{y}{2} = 1$

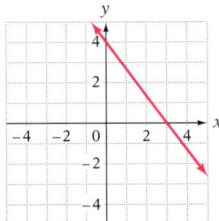
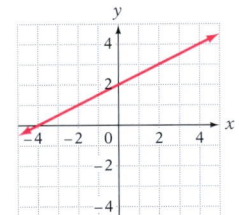

d. What observation can you make about the x- and y-intercepts of these graphs and the coefficients of x and y? Use this observation to draw the graph of $\frac{x}{4} - \frac{y}{3} = 1$.

The reciprocals of the coefficients of x and y are the x- and y-intercepts, respectively, of the graphs of the lines.

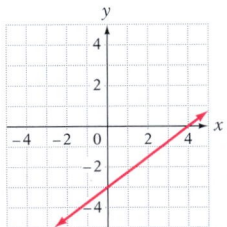

69. ✏️ **Safety**   What does the highway sign at the right have to do with slope?

Answers to Writing Exercises

69. The sign indicates a 6% downgrade, which means that the average slope of the road is 0.06. In other words, for every 100 ft in the horizontal direction, the road drops 6 ft in the vertical direction.

12.4 Equations of Straight Lines

Objective A To find the equation of a line given a point and the slope

In earlier sections, the equation of a line was given and you were asked to determine some properties of the line, such as its intercepts and slope. Here, the process is reversed. Given properties of a line, you will determine its equation.

If the slope and the y-intercept of a line are known, the equation of the line can be determined by using the slope-intercept form of a straight line.

HOW TO Find the equation of the line with slope $-\frac{1}{2}$ and y-intercept $(0, 3)$.

$y = mx + b$ • Use the slope-intercept formula.

$y = -\frac{1}{2}x + 3$ • $m = -\frac{1}{2}$; $(0, b) = (0, 3)$, so $b = 3$.

The equation of the line is $y = -\frac{1}{2}x + 3$.

When the slope and the coordinates of a point other than the y-intercept are known, the equation of the line can be found by using the formula for slope.

Suppose a line passes through the point $(3, 1)$ and has a slope of $\frac{2}{3}$. The equation of the line with these properties is determined by letting (x, y) be the coordinates of an unknown point on the line. Because the slope of the line is known, use the slope formula to write an equation. Then solve for y.

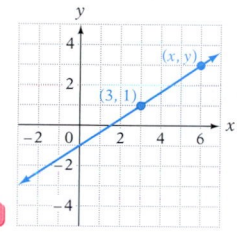

$\dfrac{y - 1}{x - 3} = \dfrac{2}{3}$ • $\dfrac{y_2 - y_1}{x_2 - x_1} = m; m = \dfrac{2}{3}; (x_2, y_2) = (x, y); (x_1, y_1) = (3, 1)$

$\dfrac{y - 1}{x - 3}(x - 3) = \dfrac{2}{3}(x - 3)$ • Multiply each side by $(x - 3)$.

$y - 1 = \dfrac{2}{3}x - 2$ • Simplify.

$y = \dfrac{2}{3}x - 1$ • Solve for y. Add 1 to each side.

The equation of the line is $y = \frac{2}{3}x - 1$.

The same procedure that was used above is used to derive the *point-slope formula*. We use this formula to determine the equation of a line when we are given the coordinates of a point on the line and the slope of the line.

Let (x_1, y_1) be the given coordinates of a point on a line, m the given slope of the line, and (x, y) the coordinates of an unknown point on the line. Then

$\dfrac{y - y_1}{x - x_1} = m$ • Formula for slope

$\dfrac{y - y_1}{x - x_1}(x - x_1) = m(x - x_1)$ • Multiply each side by $(x - x_1)$.

$y - y_1 = m(x - x_1)$ • Simplify.

In-Class Examples (Objective 12.4A)

Find the equation of the line with the given characteristics.

1. Has slope $-\frac{3}{4}$ and y-intercept $(0, 4)$

$y = -\frac{3}{4}x + 4$

2. Contains the point $(-6, -4)$ and has slope $\frac{1}{2}$

$y = \frac{1}{2}x - 1$

3. Contains the point $(2, 1)$ and has slope $\frac{2}{5}$

$y = \frac{2}{5}x + \frac{1}{5}$

Instructor Note

A model of the point-slope formula with open parentheses may help some students substitute correctly.

$$y - y_1 = m(x - x_1)$$
$$y - (\ ) = (\ )[x - (\ )]$$

Discuss the Concepts

Can the point-slope formula be used to find the equation of a line with zero slope? Can it be used to find the equation of a line whose slope is undefined?
Yes, No

Discuss the Concepts

1. Is it possible to find the equation of a line if only the slope is given? Why or why not?
   No. There are an infinite number of lines with the same slope. Each of them will have a different y-intercept.

2. Is it possible to find the equation of a line if only a point is given? Why or why not?
   No. There are an infinite number of lines that pass through a single point. Each of them will have a different slope.

Point-Slope Formula

If (x_1, y_1) is a point on a line with slope m, then $y - y_1 = m(x - x_1)$.

HOW TO Find the equation of the line that passes through the point $(2, 3)$ and has slope -2.

$$y - y_1 = m(x - x_1)$$ • Use the point-slope formula.
$$y - 3 = -2(x - 2)$$ • $m = -2$; $(x_1, y_1) = (2, 3)$
$$y - 3 = -2x + 4$$ • Solve for y.
$$y = -2x + 7$$

The equation of the line is $y = -2x + 7$.

Example 1

Find the equation of the line whose slope is $-\dfrac{2}{3}$ and whose y-intercept is $(0, -1)$.

Solution

Because the slope and the y-intercept are known, use the slope-intercept formula, $y = mx + b$.

$$y = -\frac{2}{3}x - 1$$ • $m = -\dfrac{2}{3}; b = -1$

You Try It 1

Find the equation of the line whose slope is $\dfrac{5}{3}$ and whose y-intercept is $(0, 2)$.

Your solution

$$y = \frac{5}{3}x + 2$$

Example 2

Use the point-slope formula to find the equation of the line that passes through the point $(-2, -1)$ and has slope $\dfrac{3}{2}$.

Solution

$$y - y_1 = m(x - x_1)$$
$$y - (-1) = \frac{3}{2}[x - (-2)]$$ • $m = \dfrac{3}{2};$
$$(x_1, y_1) = (-2, -1)$$
$$y + 1 = \frac{3}{2}(x + 2)$$
$$y + 1 = \frac{3}{2}x + 3$$
$$y = \frac{3}{2}x + 2$$

You Try It 2

Use the point-slope formula to find the equation of the line that passes through the point $(4, -2)$ and has slope $\dfrac{3}{4}$.

Your solution

$$y = \frac{3}{4}x - 5$$

Solutions on p. S32

Objective B   **To find the equation of a line given two points**

The point-slope formula is used to find the equation of a line when a point on the line and the slope of the line are known. But this formula can also be used to find the equation of a line given two points on the line. In this case,

1. Use the slope formula to determine the slope of the line between the points.

2. Use the point-slope formula, the slope you just calculated, and one of the given points to find the equation of the line.

HOW TO   Find the equation of the line that passes through the points whose coordinates are $(-3, -1)$ and $(3, 3)$.

Use the slope formula to determine the slope of the line between the points.

$$m = \frac{y_2 - y_1}{x_2 - x_1} = \frac{3 - (-1)}{3 - (-3)} = \frac{4}{6} = \frac{2}{3}$$ • $(x_1, y_1) = (-3, -1); (x_2, y_2) = (3, 3)$

Use the point-slope formula, the slope you just calculated, and one of the given points to find the equation of the line.

$$y - y_1 = m(x - x_1)$$ • **Point-slope formula**

$$y - (-1) = \frac{2}{3}[x - (-3)]$$ • $m = \frac{2}{3}; (x_1, y_1) = (-3, -1)$

$$y + 1 = \frac{2}{3}(x + 3)$$

$$y + 1 = \frac{2}{3}x + 2$$

$$y = \frac{2}{3}x + 1$$

Check:

$$y = \frac{2}{3}x + 1$$

$$\begin{array}{c|c} -1 & \frac{2}{3}(-3) + 1 \\ \hline -1 & -2 + 1 \\ -1 & = -1 \end{array}$$ • $(x, y) = (-3, -1)$

$$y = \frac{2}{3}x + 1$$

$$\begin{array}{c|c} 3 & \frac{2}{3}(3) + 1 \\ \hline 3 & 2 + 1 \\ 3 & = 3 \end{array}$$ • $(x, y) = (3, 3)$

The equation of the line that passes through the two points is $y = \frac{2}{3}x + 1$.

> **TAKE NOTE**
> You can verify that the equation $y = \frac{2}{3}x + 1$ passes through the points $(-3, -1)$ and $(3, 3)$ by substituting the coordinates of these points into the equation.

Example 3

Find the equation of the line that passes through the points $(-4, 0)$ and $(2, -3)$.

Solution

Find the slope of the line between the two points.

$$m = \frac{y_2 - y_1}{x_2 - x_1} = \frac{-3 - 0}{2 - (-4)} = \frac{-3}{6} = -\frac{1}{2}$$

Use the point-slope formula.

$$y - y_1 = m(x - x_1)$$ • **Point-slope formula**

$$y - 0 = -\frac{1}{2}[x - (-4)]$$ • $m = -\frac{1}{2}; (x_1, y_1) = (-4, 0)$

$$y = -\frac{1}{2}(x + 4)$$

$$y = -\frac{1}{2}x - 2$$

The equation of the line is $y = -\frac{1}{2}x - 2$.

You Try It 3

Find the equation of the line that passes through the points $(-6, -2)$ and $(3, 1)$.

Your solution

$$y = \frac{1}{3}x$$

Solution on p. S32

Objective 12.4B

Discuss the Concepts

Have students describe the steps involved in writing the equation of a line given the coordinates of two points on the line.

Concept Check

Repeat the example at the left, substituting the point $(3, 3)$ into the point-slope formula instead of the point $(-1, -3)$. Verify that the result is the same.

Optional Student Activity

Work with a partner.

1. Each person should write an equation of a line in $y = mx + b$ form (you will each select your own values for m and b). Do not show your equation to your partner.

2. Find the coordinates of two points on your line. If you chose a fractional slope, think about how you can find points with integer coordinates. Write your points on a separate piece of paper. Exchange papers with your partner.

3. Find the equation of the line that passes through the two points you have been given by your partner.

4. Check each other's work.

In-Class Examples (Objective 12.4B)

Find the equation of the line that goes through the given points.

1. $(4, -11)$ and $(-2, 1)$ $y = -2x - 3$

2. $(0, 0)$ and $(-8, -2)$ $y = \frac{1}{4}x$

3. $(1, -1)$ and $(-5, 7)$ $y = -\frac{4}{3}x + \frac{1}{3}$

Objective 12.4C

New Vocabulary
linear model
scatter diagram
line of best fit

Discuss the Concepts
Refer to the example at the right. Explain the meaning of the slope of the line of best fit.

For each additional 25.6 lb of weight, the spring will stretch 1 in. more.

Concept Check
Refer to the example at the right. Use the equation of the line of best fit to find the following.

1. The weight required to stretch the spring 3 in.   **75.5 lb**

2. The distance a 150-pound weight would stretch the spring   **About 5.9 in.**

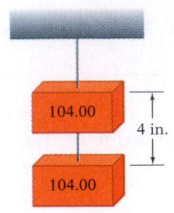

A **linear model** is a first-degree equation that is used to describe a relationship between quantities. In many cases, a linear model is used to approximate collected data. The data are graphed as points in a coordinate system, and then a line is drawn that approximates the data. The graph of the points is called a **scatter diagram**; the line is called the **line of best fit.**

Consider an experiment to determine the weight required to stretch a spring a certain distance. Data from such an experiment are shown in the table below.

Distance (in inches)	2.5	4	2	3.5	1	4.5
Weight (in pounds)	63	104	47	85	27	115

The accompanying graph shows the scatter diagram, which is the plotted points, and the line of best fit, which is the line that approximately goes through the plotted points. The equation of the line of best fit is $y = 25.6x - 1.3$, where x is the number of inches the spring is stretched and y is the weight in pounds.

The table below shows the values that the model would predict to the nearest tenth. Good linear models should predict values that are close to the actual values. A more thorough analysis of lines of best fit is undertaken in statistics courses.

Distance, x	2.5	4	2	3.5	1	4.5
Weight predicted using $y = 25.6x - 1.3$	62.7	101.1	49.9	88.3	24.3	113.9

Example 4

The data in the table below show the percent of people in the United States who purchase their music from the Internet. (*Source:* The Recording Industry Association of America) The line of best fit is $y = 0.62x + 0.667$, where x is the year (with 1997 corresponding to $x = 0$) and y is the percent of people in the United States purchasing music from the Internet.

Year	0	1	2	3	4	5
Percent	0.3	1.1	2.4	3.2	2.9	3.4

Graph the data and the line of best fit on the coordinate system below. Write a sentence that describes the meaning of the slope of the line.

Solution

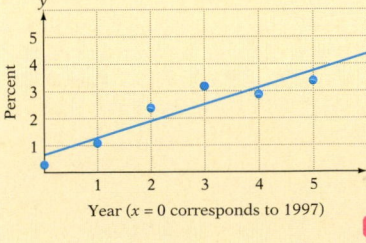

Year ($x = 0$ corresponds to 1997)

The slope of the line means that the percent of people purchasing music from the Internet is increasing 0.62% per year.

You Try It 4

The data in the table below show a reading test grade and a history test grade in a history class. The line of best fit is $y = 8.3x - 7.8$, where x is the reading test score and y is the history test score.

Reading	8.5	9.4	10.0	11.4	12.0
History	64	68	76	87	92

Graph the data and the line of best fit on the coordinate system below. Write a sentence that describes the meaning of the slope of the line of best fit.

Your solution

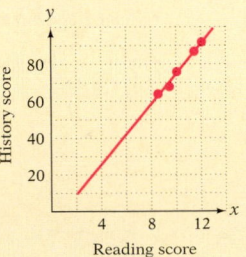

Reading score

The slope of the line means that the grade on the history test increases 8.3 points for each 1-point increase in the grade on the reading test.

Solution on p. S32

In-Class Examples (Objective 12.4C)

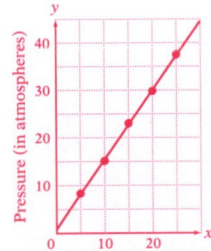

Depth (in meters)

1. The data in the table show the water pressure at various depths below the surface of the water. The line of best fit is $y = 1.46x + 0.81$. Graph the data and the line of best fit. Write a sentence that describes the meaning of the slope of the line of best fit.

For each additional meter of depth, the water pressure increases by 1.46 atmospheres.

Depth (in meters), x	5	10	15	20	25
Water pressure (in atmospheres), y	8.3	15.1	23	29.8	37.5

12.4 Exercises

Objective A   To find the equation of a line given a point and the slope

1. What is the point-slope formula and how is it used?

2. Can the point-slope formula be used to find the equation of any line? If not, equations for which types of lines cannot be found using this formula?

3. Find the equation of the line that contains the point $(0, 2)$ and has slope 2.
   $y = 2x + 2$

4. Find the equation of the line that contains the point $(0, -1)$ and has slope -2.
   $y = -2x - 1$

5. Find the equation of the line that contains the point $(-1, 2)$ and has slope -3.
   $y = -3x - 1$

6. Find the equation of the line that contains the point $(2, -3)$ and has slope 3.
   $y = 3x - 9$

7. Find the equation of the line that contains the point $(3, 1)$ and has slope $\frac{1}{3}$.
   $y = \frac{1}{3}x$

8. Find the equation of the line that contains the point $(-2, 3)$ and has slope $\frac{1}{2}$.
   $y = \frac{1}{2}x + 4$

9. Find the equation of the line that contains the point $(4, -2)$ and has slope $\frac{3}{4}$.
   $y = \frac{3}{4}x - 5$

10. Find the equation of the line that contains the point $(2, 3)$ and has slope $-\frac{1}{2}$.
    $y = -\frac{1}{2}x + 4$

11. Find the equation of the line that contains the point $(5, -3)$ and has slope $-\frac{3}{5}$.
    $y = -\frac{3}{5}x$

12. Find the equation of the line that contains the point $(5, -1)$ and has slope $\frac{1}{5}$.
    $y = \frac{1}{5}x - 2$

13. Find the equation of the line that contains the point $(2, 3)$ and has slope $\frac{1}{4}$.
    $y = \frac{1}{4}x + \frac{5}{2}$

14. Find the equation of the line that contains the point $(-1, 2)$ and has slope $-\frac{1}{2}$.
    $y = -\frac{1}{2}x + \frac{3}{2}$

Objective B   To find the equation of a line given two points

15. Find the equation of the line that passes through the points $(1, -1)$ and $(-2, -7)$.
    $y = 2x - 3$

16. Find the equation of the line that passes through the points $(2, 3)$ and $(3, 2)$.
    $y = -x + 5$

17. Find the equation of the line that passes through the points $(-2, 1)$ and $(1, -5)$.
    $y = -2x - 3$

18. Find the equation of the line that passes through the points $(-1, -3)$ and $(2, -12)$.
    $y = -3x - 6$

19. Find the equation of the line that passes through the points $(0, 0)$ and $(-3, -2)$.
    $y = \frac{2}{3}x$

20. Find the equation of the line that passes through the points $(0, 0)$ and $(-5, 1)$.
    $y = -\frac{1}{5}x$

Section 12.4

Suggested Assignment

Exercises 1–29, odds;
More challenging problems:
Exercises 31, 33, 35, 37

Answers to Writing Exercises

1. The point-slope formula is $y - y_1 = m(x - x_1)$. This formula is used to find the equation of a line when its slope and a point on the line are known.

2. No. The point-slope formula cannot be used to find the equation of a vertical line.

Quick Quiz (Objective 12.4A)

Find the equation of the line with the given characteristics.

1. Has slope $-\frac{2}{5}$ and y-intercept $(0, 3)$
   $y = -\frac{2}{5}x + 3$

2. Contains the point $(-3, -4)$ and has slope $\frac{2}{3}$
   $y = \frac{2}{3}x - 2$

3. Contains the point $(0, -5)$ and has slope -3
   $y = -3x - 5$

21. Find the equation of the line that passes through the points (2, 3) and (−4, 0).

$y = \frac{1}{2}x + 2$

22. Find the equation of the line that passes through the points (3, −1) and (0, −3).

$y = \frac{2}{3}x - 3$

23. Find the equation of the line that passes through the points (−4, 1) and (4, −5).

$y = -\frac{3}{4}x - 2$

24. Find the equation of the line that passes through the points (−5, 0) and (10, −3).

$y = -\frac{1}{5}x - 1$

25. Find the equation of the line that passes through the points (−2, 1) and (2, 4).

$y = \frac{3}{4}x + \frac{5}{2}$

26. Find the equation of the line that passes through the points (3, −2) and (−3, −3).

$y = \frac{1}{6}x - \frac{5}{2}$

Objective C   **To solve application problems**

27. **Carbohydrates** The data in the table below show the number of carbohydrates used for various amounts of time during a strenuous tennis workout. The line of best fit is $y = 1.55x + 1.45$, where x is the time of the workout in minutes and y is the number of carbohydrates used in grams.

Time of workout, x (in minutes)	5	10	20	30	60
Carbohydrates used, y (in grams)	10	15	33	49	94

Graph the data and the line of best fit on the coordinate system at the right. Write a sentence that describes the meaning of the slope of the line of best fit in the context of this problem.
The tennis player is using 1.55 g of carbohydrates/min.

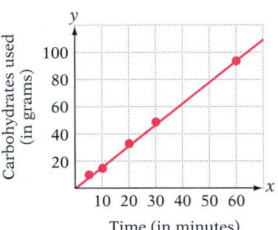

28. **Hydration** The data in the table below show the amount of water a professional tennis player loses for various times during a tennis match. The line of best fit is $y = 34.6x + 207$, where x is the time of the workout in minutes and y is the milliliters of water lost during the match.

Time of workout, x (in minutes)	10	20	30	40	50	60
Water lost, y (in milliliters)	600	900	1200	1500	2000	2300

Graph the data and the line of best fit on the coordinate system at the right. Write a sentence that describes the meaning of the slope of the line of best fit in the context of this problem.
The tennis player is losing 34.6 ml of water/min.

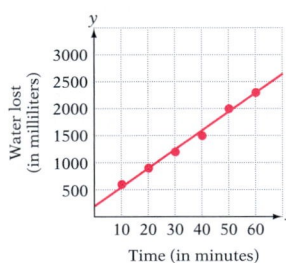

29. **Entertainment** The data in the table below show the decline in the percent of music purchased in stores in the United States. (*Source:* RIAA) The line of best fit is $y = -3x + 55$, where x is the year (with $x = 0$ corresponding to 1997) and y is the percent of music purchased in stores in the United States.

Year, x	1	2	3	4	5	6
Percent, y	52	51	45	42	42	37

Graph the data and the line of best fit on the coordinate system at the right. Write a sentence that describes the meaning of the slope of the line of best fit in the context of this problem.
The percent of music purchased in stores is decreasing 3% per year.

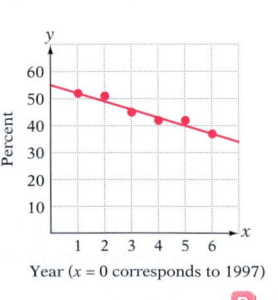

Quick Quiz (Objective 12.4B)

Find the equation of the line that goes through the given points.

1. (2, −4) and (−3, −9) $y = x - 6$

2. (0, 0) and (−2, 3) $y = -\frac{3}{2}x$

3. (−3, 2) and (6, 5) $y = \frac{1}{3}x + 3$

30. ✏️ **Evaporation** The data in the table below show the amount of water that evaporates from swimming pools of various surface areas. The line of best fit is $y = 0.17x - 1$, where x is the surface area of the swimming pool in square feet and y is the number of gallons of water that evaporate in one day.

Surface area, x (in square feet)	100	200	300	400	600	1000
Water evaporated, y (in gallons)	25	30	45	60	100	170

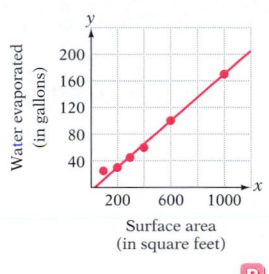

Water evaporated (in gallons)

Surface area (in square feet)

Graph the data and the line of best fit on the coordinate system at the right. Write a sentence that describes the meaning of the slope of the line of best fit in the context of this problem.

The amount of water that evaporates per day from a pool increases 0.17 gal for each additional square foot of surface area.

APPLYING THE CONCEPTS

In Exercises 31 to 34, the first two given points are on a line. Determine whether the third point is on the line.

31. $(-3, 2)$, $(4, 1)$; $(-1, 0)$
No

32. $(2, -2)$, $(3, 4)$; $(-1, 5)$
No

33. $(-3, -5)$, $(1, 3)$; $(4, 9)$
Yes

34. $(-3, 7)$, $(0, -2)$; $(1, -5)$
Yes

35. If $(-2, 4)$ are the coordinates of a point on the line whose equation is $y = mx + 1$, what is the slope of the line?
$-\dfrac{3}{2}$

36. If $(3, 1)$ are the coordinates of a point on the line whose equation is $y = mx - 3$, what is the slope of the line?
$\dfrac{4}{3}$

37. If $(0, -3)$, $(6, -7)$, and $(3, n)$ are coordinates of points on the same line, determine n.
-5

38. If $(-4, 11)$, $(2, -4)$, and $(6, n)$ are coordinates of points on the same line, determine n.
-14

The formula $y - y_1 = \dfrac{y_2 - y_1}{x_2 - x_1}(x - x_1)$, where $x_1 \neq x_2$, is called the **two-point formula** for a straight line. This formula can be used to find the equation of a line given two points. Use this formula for Exercises 39 and 40.

39. Find the equation of the line passing through $(-2, 3)$ and $(4, -1)$.
$y = -\dfrac{2}{3}x + \dfrac{5}{3}$

40. Find the equation of the line passing through $(3, -1)$ and $(4, -3)$.
$y = -2x + 5$

41. ✏️ Explain why the condition $x_1 \neq x_2$ is placed on the two-point formula given above.

42. ✏️ Explain how the two-point formula given above can be derived from the point-slope formula.

Answers to Writing Exercises

41. The condition $x_1 \neq x_2$ is placed on the two-point formula because if $x_1 = x_2$, the denominator would equal zero, and division by zero is not defined.

42. To derive the two-point formula from the point-slope formula, replace m in the point-slope formula by the formula for slope.

Quick Quiz (Objective 12.4C)

1. The data in the table show the tuition fees at a college for different years. The line of best fit is $y = 3230x + 5020$, where x is the year (with $x = 0$ corresponding to 2001) and y is the fee in dollars. Graph the data and the line of best fit. Write a sentence that describes the meaning of the slope of the line of best fit.

Year, x	0	1	2	3	4
Tuition (in dollars), y	5000	8200	11,700	14,500	18,000

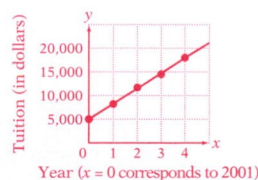

Tuition (in dollars)

Year ($x = 0$ corresponds to 2001)

The tuition fees are increasing at a rate of $3230 per year.

Focus on Problem Solving

Counterexamples   Some of the exercises in this text ask you to determine whether a statement is true or false. For instance, the statement "Every real number has a reciprocal" is false because 0 is a real number and 0 does not have a reciprocal.

Finding an example, such as "0 has no reciprocal," to show that a statement is not always true is called finding a *counterexample*. A **counterexample** is an example that shows that a statement is not always true.

Here are some counterexamples to the statement "The square of a number is always larger than the number."

$$\left(\frac{1}{2}\right)^2 = \frac{1}{4} \quad \text{but} \quad \frac{1}{4} < \frac{1}{2} \qquad 1^2 = 1 \quad \text{but} \quad 1 = 1$$

For Exercises 1 to 9, answer true if the statement is always true. If there is an instance when the statement is false, give a counterexample.

1. The product of two integers is always a positive number.

2. The sum of two prime numbers is never a prime number.

3. For all real numbers, $|x + y| = |x| + |y|$.

4. If x and y are nonzero real numbers and $x > y$, then $x^2 > y^2$.

5. The quotient of any two nonzero real numbers is less than either one of the numbers.

6. The reciprocal of a positive number is always smaller than the number.

7. If $x < 0$, then $|x| = -x$.

8. For any two real numbers x and y, $x + y > x - y$.

9. The list of numbers 1, 11, 111, 1111, 11111, . . . contains infinitely many composite numbers. (*Hint:* A number is divisible by 3 if the sum of the digits of the number is divisible by 3.)

Answers to Focus on Problem Solving: Counterexamples

The following are suggested counterexamples. Other answers are possible.

1. False. $(-3)4 = -12$

2. False. $2 + 3 = 5$

3. False. Let $x = -3$ and $y = 1$.

4. False. Let $x = 4$ and $y = -5$.

5. False. $2 \div \frac{1}{2} = 4$

6. False. The reciprocal of $\frac{1}{2}$ is 2.

7. True

8. False. Let $x = 3$ and $y = -4$.

9. True

Projects and Group Activities

Graphing Linear Equations with a Graphing Utility

A computer or graphing calculator screen is divided into *pixels*. There are approximately 6000 to 790,000 pixels available on the screen (depending on the computer or calculator). The greater the number of pixels, the smoother a graph will appear. A portion of a screen is shown at the left. Each little rectangle represents one pixel.

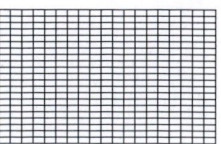

The graphing utilities that are used by computers or calculators to graph an equation do basically what we have shown in the text: They choose values of x and, for each, calculate the corresponding value of y. The pixel corresponding to the ordered pair is then turned on. The graph is jagged because pixels are much larger than the dots we draw on paper.

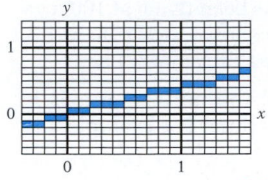

The graph of $y = 0.45x$ is shown at the left as a calculator drew it (jagged). The x- and y-axes have been chosen so that each pixel represents $\frac{1}{10}$ of a unit. Consider the regions of the graph where $x = 1$, 1.1, and 1.2.

The corresponding values of y are 0.45, 0.495, and 0.54. Because the y-axis is in tenths, the numbers 0.45, 0.495, and 0.54 are rounded to the nearest tenth before being plotted. Rounding 0.45, 0.495, and 0.54 to the nearest tenth results in 0.5 for each number. Thus the ordered pairs (1, 0.45), (1.1, 0.495), and (1.2, 0.54) are graphed as (1, 0.5), (1.1, 0.5), and (1.2, 0.5). These points appear as three illuminated horizontal pixels. However, if you use the TRACE feature of the calculator (see the Appendix), the actual y-coordinate for each value of x is displayed.

TAKE NOTE

Xmin and Xmax are the smallest and largest values of x that will be shown on the screen. Ymin and Ymax are the smallest and largest values of y that will be shown on the screen.

Here are the keystrokes for a TI-83 calculator to graph $y = \frac{2}{3}x + 1$. First the equation is entered. Then the domain (Xmin to Xmax) and the range (Ymin to Ymax) are entered. This is called the **viewing window.**

Integrating Technology

See the appendix Keystroke Guide: **Y=** and **WINDOW** for assistance.

By changing the keystrokes 2 [X,T,θ,n] [÷] 3 [+] 1, you can graph different equations.

For Exercises 1 to 4, graph on a graphing calculator.

1. $y = 2x + 1$ **2.** $y = -\frac{1}{2}x - 2$ **3.** $3x + 2y = 6$ **4.** $4x + 3y = 75$

Graphs of Motion

A graph can be useful in analyzing the motion of a body. For example, consider an airplane in uniform motion traveling at 100 m/s. The table at the right shows the distance, in meters, traveled by the plane at the end of each of five 1-second intervals.

Time (in seconds)	Distance (in meters)
0	0
1	100
2	200
3	300
4	400
5	500

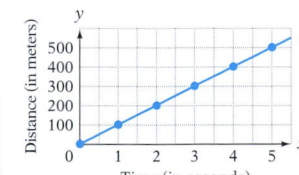

These data can be graphed on a rectangular coordinate system and a straight line drawn through the points plotted. The travel time is shown along the horizontal axis, and the distance traveled by the plane is shown along the vertical axis. (Note that the units along the two axes are not the same length.)

To write the equation for the line just graphed, use the coordinates of any two points on the line to find the slope. The y-intercept is (0, 0).

Let $(x_1, y_1) = (1, 100)$ and $(x_2, y_2) = (2, 200)$.

$$m = \frac{y_2 - y_1}{x_2 - x_1} = \frac{200 - 100}{2 - 1} = 100$$

$$y = mx + b$$
$$y = 100x + 0$$
$$y = 100x$$

Note that the slope of the line, 100, is equal to the speed, 100 m/s. *The slope of a distance-time graph represents the speed of the object.*

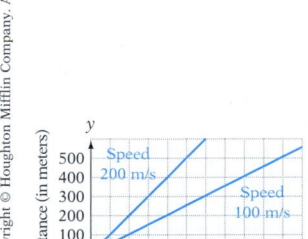

The distance-time graphs for two planes are shown at the left. One plane is traveling at 100 m/s, and the other is traveling at 200 m/s. The slope of the line representing the faster plane is greater than the slope of the line representing the slower plane.

Answers to Projects and Group Activities: Graphing Linear Equations with a Graphing Utility

1.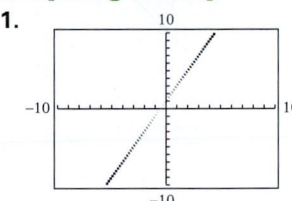

2.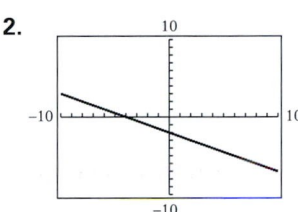

3.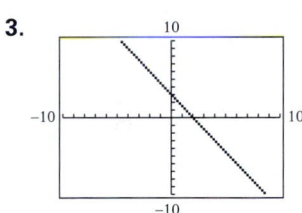

4.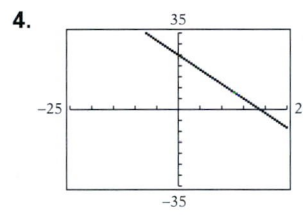

Answers to Projects and Group Activities: Graphs of Motion

1a.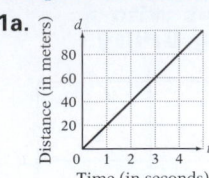

b. 20

c. $d = 20t$

d.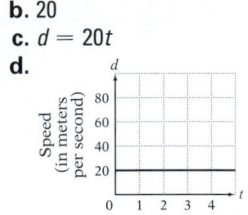

e. 60 m

2a.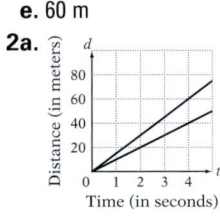

b. 10, 15

c. $d = 10t$, $d = 15t$

d. 25 m

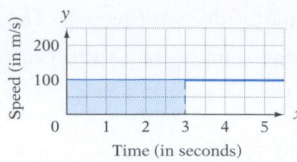

In the speed-time graph at the left, the time a plane has been flying at 100 m/s is shown along the horizontal axis, and its speed is shown along the vertical axis. Because the speed is constant, the graph is a horizontal line.

The area between the horizontal line graphed and the horizontal axis is equal to the distance traveled by the plane up to that time. For example, the area of the shaded region on the graph is

$$\text{Length} \cdot \text{width} = (3 \text{ s})(100 \text{ m/s}) = 300 \text{ m}$$

The distance traveled by the plane in 3 s is equal to 300 m.

1. A car in uniform motion is traveling at 20 m/s.
   a. Prepare a distance-time graph for the car for 0 s to 5 s.
   b. Find the slope of the line.
   c. Find the equation of the line.
   d. Prepare a speed-time graph for the car for 0 s to 5 s.
   e. Find the distance traveled by the car after 3 s.

2. One car in uniform motion is traveling at 10 m/s. A second car in uniform motion is traveling at 15 m/s.
   a. Prepare one distance-time graph for both cars for 0 s to 5 s.
   b. Find the slope of each line.
   c. Find the equation of each line graphed.
   d. Assuming the cars started at the same point at 0 s, find the distance between the cars at the end of 5 s.

Chapter 12 Summary

Key Words

A *rectangular coordinate system* is formed by two number lines, one horizontal and one vertical, that intersect at the zero point of each line. The number lines that make up a rectangular coordinate system are called the *coordinate axes,* or simply *axes.* The *origin* is the point of intersection of the two coordinate axes. Generally, the horizontal axis is labeled the *x*-axis and the vertical axis is labeled the *y*-axis. The coordinate system divides the plane into four regions called *quadrants.* The *coordinates of a point* in the plane are given by an *ordered pair* (x, y). The first number in the ordered pair is called the *abscissa* or *x-coordinate.* The second number in the ordered pair is the *ordinate* or *y-coordinate.* The *graph of an ordered pair* (x, y) is the dot drawn at the coordinates of the point in the plane. [12.1A, p. 625]

A *solution of an equation in two variables* is an ordered pair (x, y) that makes the equation a true statement. [12.1B, p. 627]

A *relation* is any set of ordered pairs. The *domain* of a relation is the set of first coordinates of the ordered pairs. The *range* is the set of second coordinates of the ordered pairs. [12.1C, p. 629]

Examples

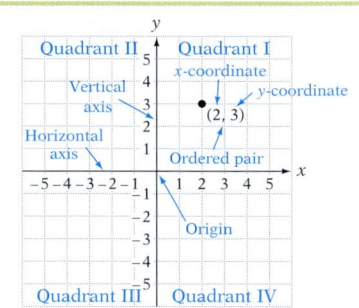

The ordered pair $(-1, 1)$ is a solution of the equation $y = 2x + 3$ because when -1 is substituted for x and 1 is substituted for y, the result is a true equation.

For the relation $\{(-1, 2), (2, 4), (3, 5), (3, 7)\}$, the domain is $\{-1, 2, 3\}$; the range is $\{2, 4, 5, 7\}$.

A *function* is a relation in which no two ordered pairs have the same first coordinate. [12.1C, p. 629]

The relation {(−2, −3), (0, 4), (1, 5)} is a function. No two ordered pairs have the same first coordinate.

The *graph of an equation in two variables* is a graph of the ordered-pair solutions of the equation. An equation of the form $y = mx + b$ is a *linear equation in two variables*. [12.2A, p. 637]

$y = 2x + 3$ is a linear equation in two variables. Its graph is shown at the right.

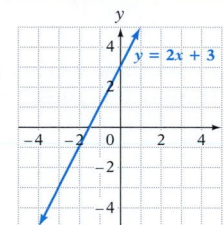

An equation written in the form $Ax + By = C$ is the *standard form of a linear equation in two variables*. [12.2B, p. 639]

$2x + 7y = 10$ is an example of a linear equation in two variables written in standard form.

The point at which a graph crosses the *x*-axis is called the *x-intercept*. At the *x*-intercept, the *y*-coordinate is 0. The point at which a graph crosses the *y*-axis is called the *y-intercept*. At the *y*-intercept, the *x*-coordinate is 0. [12.3A, p. 647]

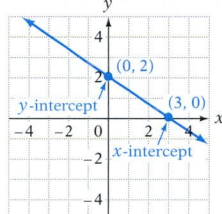

The *slope* of a line is a measure of the slant of the line. The symbol for slope is m. A line with *positive slope* slants upward to the right. A line with *negative slope* slants downward to the right. A horizontal line has *zero slope*. A vertical line has an *undefined slope*. [12.3A, pp. 648–649]

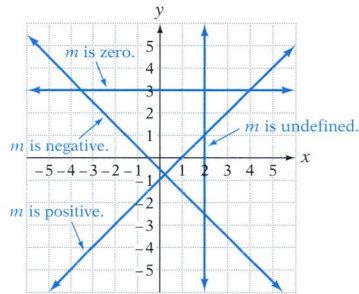

When data are graphed as points in a coordinate system, the graph is called a *scatter diagram*. A line drawn to approximate the data is called the *line of best fit*. [12.4C, p. 660]

The graph shown at the right is the scatter diagram and line of best fit for the spring data on page 660.

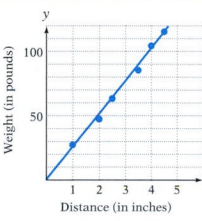

Essential Rules and Procedures

Examples

Functional Notation [12.1D, p. 632]

The equation of a function is written in functional notation when y is replaced by the symbol $f(x)$, where $f(x)$ is read "f of x" or "the value of f at x." To evaluate a function at a given value of x, replace x by the given value and then simplify the resulting numerical expression to find the value of $f(x)$.

$y = x^2 + 2x − 1$ is written in functional notation as $f(x) = x^2 + 2x − 1$. To evaluate $f(x) = x^2 + 2x − 1$ at $x = −3$, find $f(−3)$.

$$f(−3) = (−3)^2 + 2(−3) − 1$$
$$= 9 − 6 − 1 = 2$$

Horizontal and Vertical Lines [12.2B, p. 640]

The graph of $y = b$ is a horizontal line passing through $(0, b)$.
The graph of $x = a$ is a vertical line passing through $(a, 0)$.

The graph of $y = -2$ is a horizontal line passing through $(0, -2)$. The graph of $x = 3$ is a vertical line passing through $(3, 0)$.

To find the x-intercept, let $y = 0$ and solve for x.
To find the y-intercept, let $x = 0$ and solve for y.
[12.3A, p. 647]

To find the x-intercept of $4x - 5y = 20$, let $y = 0$ and solve for x. To find the y-intercept, let $x = 0$ and solve for y.

$$4x - 5y = 20 \qquad\qquad 4x - 5y = 20$$
$$4x - 5(0) = 20 \qquad\quad 4(0) - 5y = 20$$
$$4x = 20 \qquad\qquad\quad -5y = 20$$
$$x = 5 \qquad\qquad\qquad y = -4$$

The x-intercept is $(5, 0)$.   The y-intercept is $(0, -4)$.

Slope Formula [12.3B, p. 648]

If $P_1(x_1, y_1)$ and $P_2(x_2, y_2)$ are two points on a line and $x_1 \neq x_2$, then

$$m = \frac{y_2 - y_1}{x_2 - x_1}$$

To find the slope of the line between the points $(1, -2)$ and $(-3, -1)$, let $P_1 = (1, -2)$ and $P_2 = (-3, -1)$. Then

$$m = \frac{y_2 - y_1}{x_2 - x_1} = \frac{-1 - (-2)}{-3 - 1} = \frac{1}{-4} = -\frac{1}{4}.$$

Parallel Lines [12.3B, p. 649]

Two nonvertical lines in the plane are parallel if and only if they have the same slope. All vertical lines in the plane are parallel.

The slope of the line through $P_1(3, -6)$ and $P_2(5, -10)$ is $m_1 = \frac{-10 - (-6)}{5 - 3} = -2$.

The slope of the line through $Q_1(4, -5)$ and $Q_2(0, 3)$ is $m_2 = \frac{3 - (-5)}{0 - 4} = -2$.

Because $m_1 = m_2$, the lines are parallel.

Perpendicular Lines [12.3B, p. 650]

Two nonvertical lines in the plane are perpendicular if and only if the product of their slopes is -1. A vertical and a horizontal line are perpendicular.

The slope of the line through $P_1(5, -3)$ and $P_2(2, -1)$ is $m_1 = \frac{-1 - (-3)}{2 - 5} = -\frac{2}{3}$.

The slope of the line through $Q_1(1, -4)$ and $Q_2(3, -1)$ is $m_2 = \frac{-1 - (-4)}{3 - 1} = \frac{3}{2}$.

Because $m_1 m_2 = \left(-\frac{2}{3}\right)\left(\frac{3}{2}\right) = -1$, the lines are perpendicular.

Slope-Intercept Form of a Linear Equation [12.3C, p. 651]

An equation of the form $y = mx + b$ is called the slope-intercept form of a straight line. The slope of the line is m, the coefficient of x. The y-intercept is $(0, b)$, where b is the constant term of the equation.

For the line with equation $y = -3x + 2$, the slope is -3 and the y-intercept is $(0, 2)$.

Point-Slope Formula [12.4A, p. 658]

If (x_1, y_1) is a point on a line with slope m, then

$$y - y_1 = m(x - x_1)$$

The equation of the line that passes through the point $(5, -3)$ and has slope -2 is:

$$y - y_1 = m(x - x_1)$$
$$y - (-3) = -2(x - 5)$$
$$y + 3 = -2x + 10$$
$$y = -2x + 7$$

Chapter 12 Review Exercises

1. **a.** Graph the ordered pairs $(-2, 4)$ and $(3, -2)$.
   b. Name the abscissa of point A. -2
   c. Name the ordinate of point B. -4

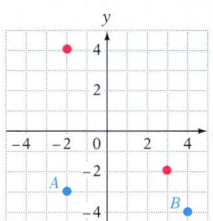

[12.1A]

2. Graph the ordered-pair solutions of $y = -\frac{1}{2}x - 2$ when $x \in \{-4, -2, 0, 2\}$.

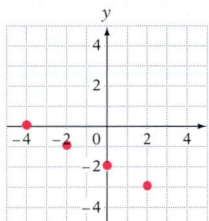

[12.1B]

3. Determine the equation of the line that passes through the points $(-1, 3)$ and $(2, -5)$.

$y = -\frac{8}{3}x + \frac{1}{3}$ [12.4B]

4. Determine the equation of the line that passes through the point $(6, 1)$ and has slope $-\frac{5}{2}$.

$y = -\frac{5}{2}x + 16$ [12.4A]

5. Graph $y = \frac{1}{4}x + 3$.

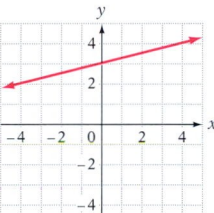

[12.2A]

6. Graph $5x + 3y = 15$.

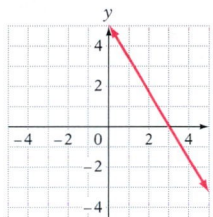

[12.2B]

7. Is the line that passes through $(7, -5)$ and $(6, -1)$ parallel, perpendicular, or neither parallel nor perpendicular to the line that passes through $(4, 5)$ and $(2, -3)$?
Neither [12.3B]

8. Given $f(x) = x^2 - 2$, find $f(-1)$.
-1 [12.1D]

9. Determine the equation of the line that passes through the points $(-2, 5)$ and $(4, 1)$.

$y = -\frac{2}{3}x + \frac{11}{3}$ [12.4B]

10. Does $y = -x + 3$, where $x \in \{-2, 0, 3, 5\}$, define y as a function of x?
Yes [12.1C]

11. Find the slope of the line containing the points $(9, 8)$ and $(-2, 1)$.
$\frac{7}{11}$ [12.3B]

12. Find the x- and y-intercepts of $3x - 2y = 24$.
$(8, 0), (0, -12)$ [12.3A]

13. Find the slope of the line containing the points $(-2, -3)$ and $(4, -3)$.
0 [12.3B]

14. Graph the line that has slope $\frac{1}{2}$ and y-intercept $(0, -1)$.

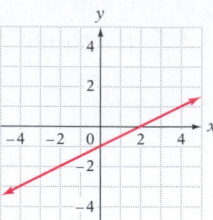

[12.3C]

15. Graph $x = -3$.

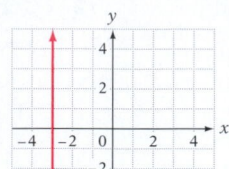

[12.2B]

16. Graph the line that has slope $-\frac{2}{3}$ and y-intercept $(0, 2)$.

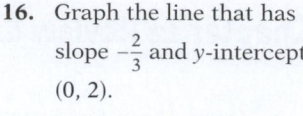

[12.3C]

17. Graph $y = -2x - 1$.

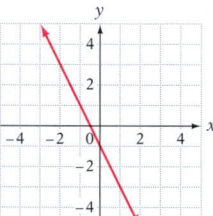

[12.2A]

18. Graph the line that has slope 2 and y-intercept $(0, -4)$.

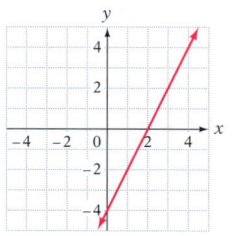

[12.3C]

19. Graph $3x - 2y = -6$.

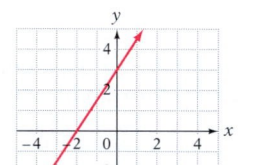

[12.2B]

20. Health The height and weight of eight seventh-grade students are shown in the following table. Write a relation in which the first coordinate is height, in inches, and the second coordinate is weight, in pounds. Is the relation a function?

Height (in inches)	55	57	53	57	60	61	58	54
Weight (in pounds)	95	101	94	98	100	105	97	95

{(55, 95), (57, 101), (53, 94), (57, 98), (60, 100), (61, 105), (58, 97), (54, 95)}; No   [12.1C]

21. 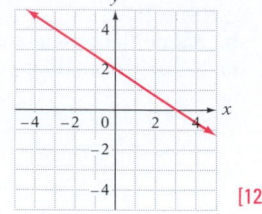 **Online Service** An online research service charges a monthly access fee of $75 plus $.45 per minute to use the service. An equation that represents the monthly cost to use this service is $C = 0.45x + 75$, where C is the monthly cost and x is the number of minutes of access used. Graph this equation for $0 \leq x \leq 100$. The point (50, 97.5) is on the graph. Write a sentence that describes the meaning of this ordered pair.
The cost of 50 min of access time for 1 month is $97.50.   [12.2C]

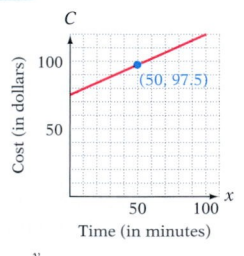

22. **Telecommunications** The data in the table below are estimates of the projected annual increase in average telephone bills for a family. The line of best fit is $y = 34x + 657$, where x is the year (with $x = 0$ corresponding to 1999) and y is the annual cost, in dollars, of telephone bills.

Year, x	0	1	2	3	4	5	6
Telephone bills, y (in dollars)	658	690	708	772	809	830	849

Graph the data and the line of best fit on the coordinate system at the right. Write a sentence that describes the meaning of the slope of the line of best fit.
The average annual telephone bill for a family is increasing by $34 per year.   [12.4C]

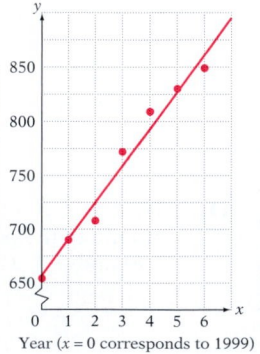

Chapter 12 Test

1. Find the ordered-pair solution of
$2x - 3y = 15$ corresponding to $x = 3$.
(3, −3)   [12.1B]

2. Graph the ordered-pair solutions of
$y = -\frac{3}{2}x + 1$ for $x \in \{-2, 0, 4\}$.   [12.1B]

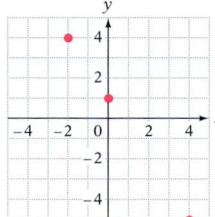

3. Does $y = \frac{1}{2}x - 3$ define y as a function of x
for $x \in \{-2, 0, 4\}$?   Yes   [12.1C]

4. Given $f(t) = t^2 + t$, find $f(2)$.
6   [12.1D]

5. Given $f(x) = x^2 - 2x$, find $f(-1)$.
3   [12.1D]

6. **Emergency Response**   The distance a house is from a fire station and the
amount of damage that the house sustained in a fire are given in the following table. Write a relation in which the first coordinate of the ordered
pair is the distance, in miles, from the fire station and the second coordinate
is the amount of damage in thousands of dollars. Is the relation a function?

Distance (in miles)	3.5	4.0	5.2	5.0	4.0	6.3	5.4
Damage (in thousands of dollars)	25	30	45	38	42	12	34

{(3.5, 25), (4.0, 30), (5.2, 45), (5.0, 38), (4.0, 42), (6.3, 12), (5.4, 34)}; No   [12.1C]

7. Graph $y = 3x + 1$.

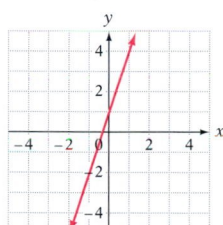

[12.2A]

8. Graph $y = -\frac{3}{4}x + 3$.

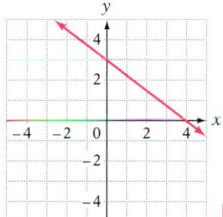

[12.2A]

9. Graph $3x - 2y = 6$.

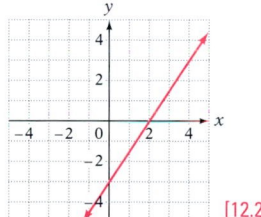

[12.2B]

10. Graph $x + 3 = 0$.

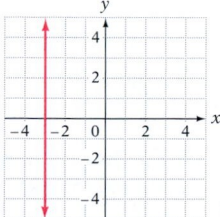

[12.2B]

11. Graph the line that has
slope $-\frac{2}{3}$ and y-intercept
(0, 4).

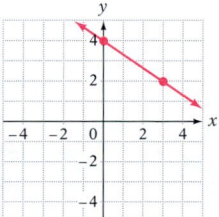

[12.3C]

12. Graph the line that has
slope 2 and y-intercept -2.

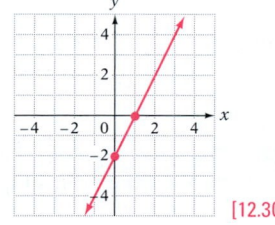

[12.3C]

13. 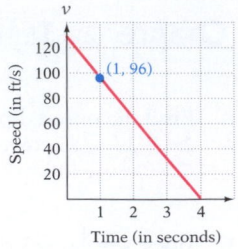 **Sports** The equation for the speed of a ball that is thrown straight up with an initial speed of 128 ft/s is $v = 128 - 32t$, where v is the speed of the ball after t seconds. Graph this equation for $0 \le t \le 4$. The point whose coordinates are (1, 96) is on the graph. Write a sentence that describes the meaning of this ordered pair.
After 1 s, the ball is traveling 96 ft/s.   [12.2C]

14. 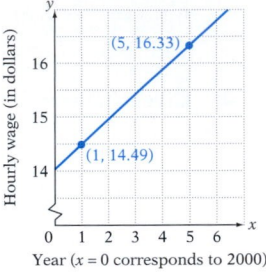 **Wages** The graph at the right shows the projected increase in the average hourly wage of a U.S. worker for the years 2000 through 2006 (with $x = 0$ corresponding to 2000). Find the slope of the line. Write a sentence that states the meaning of the slope.
$m = 0.46$. The average hourly wage is increasing by $.46 per year.   [12.3B]

15. **Tuition** The data in the table below are the average annual tuition costs for private 4-year colleges in the United States. The line of best fit is $y = 809x + 12{,}195$, where x is the year (with $x = 0$ corresponding to 1995) and y is the annual tuition cost in dollars rounded to the nearest 100.

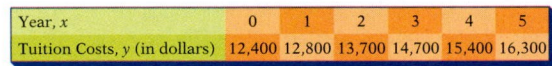

Year, x	0	1	2	3	4	5
Tuition Costs, y (in dollars)	12,400	12,800	13,700	14,700	15,400	16,300

Graph the data and the line of best fit on the coordinate system at the right. Write a sentence that describes the meaning of the slope of the line of best fit.
The average annual tuition for a private 4-year college is increasing $809 per year.   [12.4C]

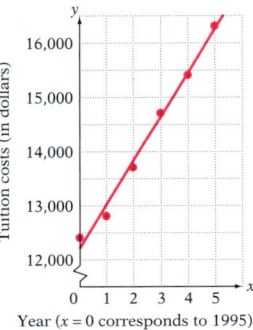

16. Find the x- and y-intercepts for $6x - 4y = 12$.
(2, 0), (0, −3)   [12.3A]

17. Find the x- and y-intercepts for $y = \frac{1}{2}x + 1$.
(−2, 0), (0, 1)   [12.3A]

18. Find the slope of the line containing the points (2, −3) and (4, 1).
2   [12.3B]

19. Is the line that passes through (2, 5) and (−1, 1) parallel, perpendicular, or neither parallel nor perpendicular to the line that passes through (−2, 3) and (4, 11)?
Parallel   [12.3B]

20. Find the slope of the line containing the points (−5, 2) and (−5, 7).
Undefined   [12.3B]

21. Find the slope of the line whose equation is $2x + 3y = 6$.
$-\frac{2}{3}$ [12.3B]

22. Find the equation of the line that contains the point (0, −1) and has slope 3.
$y = 3x - 1$ [12.4A]

23. Find the equation of the line that contains the point (−3, 1) and has slope $\frac{2}{3}$.
$y = \frac{2}{3}x + 3$ [12.4A]

24. Find the equation of the line that passes through the points (5, −4) and (−3, 1).
$y = -\frac{5}{8}x - \frac{7}{8}$ [12.4B]

25. Find the equation of the line that passes through the points (−2, 0) and (5, −2).
$y = -\frac{2}{7}x - \frac{4}{7}$ [12.4B]

Cumulative Review Exercises

1. Simplify: $12 - 18 \div 3 \cdot (-2)^2$

   -12 [3.5A]

2. Evaluate $\dfrac{a - b}{a^2 - c}$ when $a = -2$, $b = 3$, and $c = -4$.

   $-\dfrac{5}{8}$ [4.1A]

3. Given $f(x) = \dfrac{2}{x - 1}$, find $f(-2)$.

   $f(-2) = -\dfrac{2}{3}$ [12.1D]

4. Solve: $2x - \dfrac{2}{3} = \dfrac{7}{3}$

   $\dfrac{3}{2}$ [5.2A]

5. Solve: $3x - 2[x - 3(2 - 3x)] = x - 7$

   $\dfrac{19}{18}$ [5.3B]

6. Write $6\dfrac{2}{3}\%$ as a fraction.

   $\dfrac{1}{15}$ [6.3A]

7. Simplify: $(-2x^2y)^3(2xy^2)^2$
   $-32x^8y^7$ [9.2B]

8. Simplify: $\dfrac{-15x^7}{5x^5}$

   $-3x^2$ [9.4A]

9. Divide: $(x^2 - 4x - 21) \div (x - 7)$
   $x + 3$ [9.5B]

10. Factor: $5x^2 + 15x + 10$
    $5(x + 2)(x + 1)$ [10.2B]

11. Factor: $x(a + 2) + y(a + 2)$
    $(a + 2)(x + y)$ [10.1B]

12. Solve: $x(x - 2) = 8$
    4 and -2 [10.5A]

13. Multiply: $\dfrac{x^5y^3}{x^2 - x - 6} \cdot \dfrac{x^2 - 9}{x^2y^4}$

    $\dfrac{x^3(x + 3)}{y(x + 2)}$ [11.1B]

14. Subtract: $\dfrac{3x}{x^2 + 5x - 24} - \dfrac{9}{x^2 + 5x - 24}$

    $\dfrac{3}{x + 8}$ [11.3A]

15. Solve: $3 - \dfrac{1}{x} = \dfrac{5}{x}$

    2 [11.5A]

16. Solve $4x - 5y = 15$ for y.

    $y = \dfrac{4}{5}x - 3$ [11.6A]

17. Find the ordered-pair solution of $y = 2x - 1$ corresponding to $x = -2$.
(−2, −5)   [12.1B]

18. Find the slope of the line that contains the points (2, 3) and (−2, 3).
0   [12.3B]

19. Find the equation of the line that contains the point (2, −1) and has slope $\frac{1}{2}$.

$y = \frac{1}{2}x - 2$ [12.4A]

20. Find the equation of the line that contains the point (0, 2) and has slope −3.
$y = -3x + 2$ [12.4A]

21. Find the equation of the line that contains the point (−1, 0) and has slope 2.
$y = 2x + 2$ [12.4A]

22. Find the equation of the line that contains the point (6, 1) and has slope $\frac{2}{3}$.

$y = \frac{2}{3}x - 3$ [12.4A]

23. **Probability**   Four blue marbles, three red marbles, and two green marbles are placed in a bag. One marble is chosen at random. What is the probability that the marble chosen is not red?
$\frac{2}{3}$ [8.3A]

24. **Geometry**   The measure of the first angle of a triangle is 3° more than the measure of the second angle. The measure of the third angle is 5° more than twice the measure of the second angle. Find the measure of each angle.
46°, 43°, 91°   [7.1C]

25. **Taxes**   The real estate tax for a home that costs $50,000 is $625. At this rate, what is the value of a home for which the real estate tax is $1375?
$110,000   [6.2B]

26. **Work**   An electrician requires 6 h to wire a garage. An apprentice can do the same job in 10 h. How long would it take to wire the garage if both the electrician and the apprentice worked together?
$3\frac{3}{4}$ h   [11.7A]

27. Graph $y = \frac{1}{2}x - 1$.

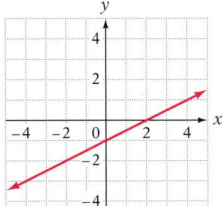

[12.2A]

28. Graph the line that has slope $-\frac{2}{3}$ and y-intercept 2.

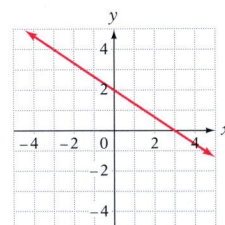

[12.3C]

chapter 13

Systems of Linear Equations

Recall from Section 5.1 that if a boat is traveling with the current, the effective speed of the boat is increased. If the boat is traveling against the current, the current slows the boat down. The same is true for airplanes and the wind. If the plane in the photo above is flying with a tailwind, its effective speed is increased. If the plane is flying against a headwind, the wind will slow the plane down. In **Exercises 2, 5, 6, and 8 on page 705,** you will use systems of equations to determine the rates of planes in calm air and rates of the wind.

OBJECTIVES

Section 13.1

A To solve a system of linear equations by graphing

Section 13.2

A To solve a system of linear equations by the substitution method
B To solve investment problems

Section 13.3

A To solve a system of linear equations by the addition method

Section 13.4

A To solve rate-of-wind or rate-of-current problems
B To solve application problems using two variables

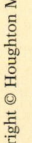

Need help? For online student resources, such as section quizzes, visit this textbook's website at
math.college.hmco.com/students.

PREP TEST ● ● ●

Do these exercises to prepare for Chapter 13.

1. Solve $3x - 4y = 24$ for y.
   $y = \dfrac{3}{4}x - 6$ [11.6A]

2. Solve: $50 + 0.07x = 0.05(x + 1400)$
   1000   [5.3B]

3. Simplify: $-3(2x - 7y) + 3(2x + 4y)$
   $33y$ [4.2D]

4. Simplify: $4x + 2(3x - 5)$
   $10x - 10$ [4.2D]

5. Is $(-4, 2)$ a solution of $3x - 5y = -22$?
   Yes   [12.1B]

6. Find the x- and y-intercepts for $3x - 4y = 12$.
   $(4, 0), (0, -3)$ [12.3A]

7. Are the graphs of $3x + y = 6$ and $y = -3x - 4$ parallel?
   Yes   [12.3B]

8. Graph: $y = \dfrac{5}{4}x - 2$
   [12.3C]

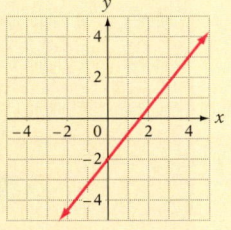

9. **Hiking**   One hiker starts along a trail walking at 3 mph. One-half hour later, another hiker starts on the same trail walking at a speed of 4 mph. How long after the second hiker starts will the two hikers be side-by-side?
   1.5 h   [5.5B]

GO FIGURE ● ● ●

Two children are running from school to home. Carla runs half the time and walks half the time. James runs half the distance and walks half the distance. Assuming the children run and walk at the same rate, which child arrives home first?
Carla

13.1 Solving Systems of Linear Equations by Graphing

Objective A   To solve a system of linear equations by graphing

Two or more equations considered together are called a **system of equations.**
Three examples of *linear* systems of equations in *two* variables are shown below, along with the graphs of the equations of each system.

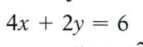

System I	System II	System III
$x - 2y = -8$	$4x + 2y = 6$	$4x + 6y = 12$
$2x + 5y = 11$	$y = -2x + 3$	$6x + 9y = -9$

For system I, the two lines intersect at a single point, $(-2, 3)$. Because this point lies on both lines, it is a solution of each equation of the system of equations. We can check this by replacing x with -2 and y with 3. The check is shown below.

$$\begin{array}{c|c}
x - 2y = -8 & \\
\hline
-2 - 2(3) & -8 \\
-2 - 6 & -8 \\
-8 = -8 \ \checkmark &
\end{array}
\qquad
\begin{array}{c|c}
2x + 5y = 11 & \\
\hline
2(-2) + 5(3) & 11 \\
-4 + 15 & 11 \\
11 = 11 \ \checkmark &
\end{array}$$

• Replace x with -2 and replace y with 3.

A **solution of a system of equations in two variables** is an ordered pair that is a solution of each equation of the system. The ordered pair $(-2, 3)$ is a solution of system I.

HOW TO   Is $(-1, 4)$ a solution of the system of equations? $7x + 3y = 5$
$3x - 2y = 12$

$$\begin{array}{c|c}
7x + 3y = 5 & \\
\hline
7(-1) + 3(4) & 5 \\
-7 + 12 & 5 \\
5 = 5 \ \checkmark &
\end{array}
\qquad
\begin{array}{c|c}
3x - 2y = 12 & \\
\hline
3(-1) - 2(4) & 12 \\
-3 - 8 & 12 \\
-11 \ne 12 &
\end{array}$$

• Replace x with -1 and replace y with 4.
• Does not check

Because $(-1, 4)$ is not a solution of both equations, $(-1, 4)$ is not a solution of the system of equations.

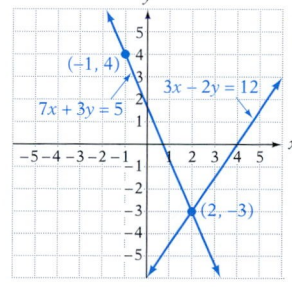

Using the system of equations above and the graph at the right, note that the graph of the ordered pair $(-1, 4)$ lies on the graph of $7x + 3y = 5$ but not on *both* lines. The ordered pair $(-1, 4)$ is *not* a solution of the system of equations. The graph of the ordered pair $(2, -3)$ does lie on both lines and therefore the ordered pair $(2, -3)$ is a solution of the system of equations.

TAKE NOTE

The systems of equations above are *linear systems of equations* because each of the equations in the system has a graph that is a line. Also, each equation has two variables. In future math courses, you will study equations that contain more than two variables.

Objective 13.1A

New Vocabulary
system of equations
solution of a system of equations
  in two variables
independent system
dependent system
inconsistent system

Instructor Note
One way to illustrate that sometimes more than one equation is necessary to arrive at a unique answer is to say to students, "Find two numbers whose sum is 10." They will soon realize that there are an infinite number of possible pairs of numbers. (You may have to suggest fractional answers or negative numbers.)
  Now say, "Find two numbers whose sum is 10 and whose difference is 6." Only the ordered pair $(8, 2)$ satisfies both conditions. Ask students to create this system of equations.

Optional Student Activity
Have students verify that $(2, -3)$ is a solution of the system of equations

$$7x + 3y = 5$$
$$3x - 2y = 12$$

shown at the left.

In-Class Examples (Objective 13.1A)

1. Is $(3, -4)$ a solution of the system of equations?
$2x - 3y = 18$
$x - 4y = 19$
Yes

2. Solve by graphing: $x - 2y = 0$
$2x - y = -3$

$(-2, -1)$

3. Solve by graphing: $x + 2y = 8$
$2x + 4y = 16$

Dependent. The solutions are the ordered pairs that satisfy $x + 2y = 8$.

4. Solve by graphing: $4x - 2y = 8$
$y = 2x + 1$

No solution

Optional Student Activity

Have students solve $4x + 2y = 6$ for y and verify that the two equations of the system of equations at the right are identical.

Concept Check

Ask students whether $(1, -2)$ is a solution of the system of equations at the right. Because the system has no solution, students should say no without doing any work.

Discuss the Concepts

Have students explain how independent, dependent, and inconsistent systems of equations differ.

Optional Student Activity

Ask students to draw the graphs of an independent, a dependent, and an inconsistent system of equations. These do not have to be graphs of particular equations, just lines that depict the given situations.

TAKE NOTE

The fact that there are an infinite number of ordered pairs that are solutions of the system at the right does not mean *every* ordered pair is a solution. For instance, $(0, 3)$, $(-2, 7)$, and $(2, -1)$ are solutions. However, $(3, 1)$, $(-1, 4)$, and $(1, 6)$ are not solutions. You should verify these statements.

System II from the preceding page and the graph of the equations of that system are shown again at the right. Note that the graph of $y = -2x + 3$ lies directly on top of the graph of $4x + 2y = 6$. Thus the two lines intersect at an infinite number of points. The graphs intersect at an infinite number of points, so there are an infinite number of solutions of this system of equations. Because each equation represents the same set of points, the solutions of the system of equations can be stated by using the ordered pairs of either one of the equations. Therefore, we can say, "The solutions are the ordered pairs that satisfy $4x + 2y = 6$," or we can say, "The solutions are the ordered pairs that satisfy $y = -2x + 3$."

$$4x + 2y = 6$$
$$y = -2x + 3$$

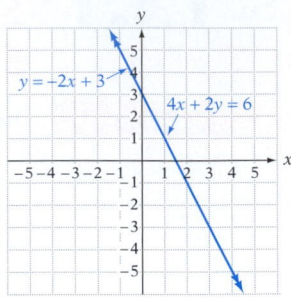

System III from the preceding page and the graph of the equations of that system are shown again at the right. Note that in this case, the graphs of the lines are parallel and do not intersect. Because the graphs do not intersect, there is no point that is on both lines. Therefore, the system of equations has no solution.

$$4x + 6y = 12$$
$$6x + 9y = -9$$

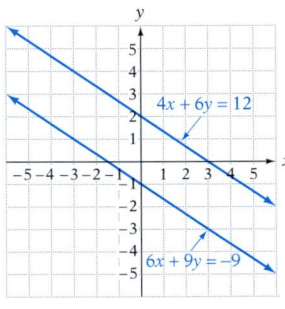

The preceding examples illustrate three types of systems of linear equations. An **independent system** has exactly one solution—the graphs intersect at one point. A **dependent system** has an infinite number of solutions—the graphs are the same line. An **inconsistent system** has no solution—the graphs are parallel lines.

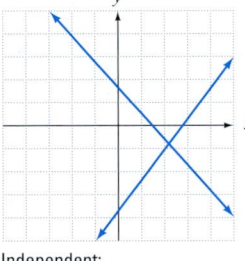

Independent:
one solution

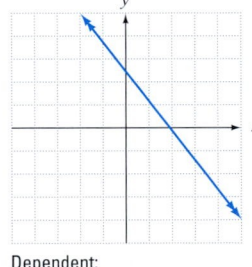

Dependent:
infinitely many solutions

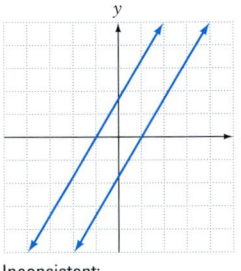

Inconsistent:
no solutions

HOW TO The graphs of the equations for the system of equations below are shown at the right. What is the solution of the system of equations?

$$2x + 3y = 6$$
$$2x + y = -2$$

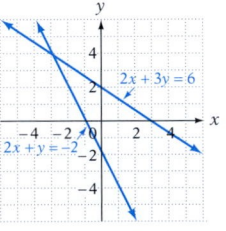

The graphs intersect at $(-3, 4)$. This is an *independent* system of equations. The solution of the system of equations is $(-3, 4)$.

HOW TO   The graphs of the equations of the system of equations at the right are shown below. What is the solution of the system of equations?

$$y = 2x - 2$$
$$x = \frac{1}{2}y + 1$$

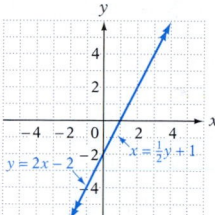

The two graphs lie directly on top of one another. Thus the two lines intersect at an infinite number of points, and the system of equations has an infinite number of solutions. This is a *dependent* system of equations. The solutions of the system of equations are the ordered pairs that satisfy $y = 2x - 2$.

Integrating Technology

The Projects and Group Activities at the end of this chapter discusses using a calculator to approximate the solution of an independent system of equations. Also see the appendix Keystroke Guide: *Intersect*.

Solving a system of equations means finding the ordered-pair solutions of the system. One way to do this is to draw the graphs of the equations in the system of equations and determine where the graphs intersect.

To solve a system of linear equations in two variables by graphing, graph each equation on the same coordinate system, and then determine the points of intersection.

HOW TO   Solve by graphing: $2x - y = -1$
$x + 2y = 7$

Graph each line.

The point of intersection of the two graphs lies on both lines and is therefore the solution of the system of equations.

The system of equations is independent. (1, 3) is a solution of each equation.

The solution is (1, 3).

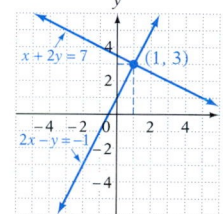

HOW TO   Solve by graphing: $y = 2x + 2$
$4x - 2y = 4$

Graph each line.

The graphs do not intersect. The system of equations is inconsistent.

The system of equations has no solution.

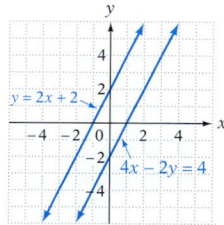

Concept Check

Can the solution of the system of equations

$$y = 2x - 2$$
$$x = \frac{1}{2}y + 1$$

be written as $2x - y = 2$?
Yes

Concept Check

Fill in the blank.

1. If a system of two linear equations in two variables is independent, then the slopes of the lines must be
_____.
different (not equal)

2. If a system of two linear equations in two variables is dependent, then the slopes of the lines must be _____ and the y-intercepts must be _____. equal, equal

3. If a system of two linear equations in two variables is inconsistent, then the slopes of the lines must be _____ and the y-intercepts must be _____. equal, different

Example 1

Is $(1, -3)$ a solution of the following system?
$3x + 2y = -3$
$x - 3y = 6$

Solution
Replace x with 1 and y with -3.

$$\begin{array}{r|r} 3x + 2y = -3 \\ \hline 3 \cdot 1 + 2(-3) & -3 \\ 3 + (-6) & -3 \\ -3 = -3 \end{array} \qquad \begin{array}{r|r} x - 3y = 6 \\ \hline 1 - 3(-3) & 6 \\ 1 - (-9) & 6 \\ 10 \neq 6 \end{array}$$

No, $(1, -3)$ is not a solution of the system of equations.

Example 2

Solve by graphing:
$x - 2y = 2$
$x + y = 5$

Solution

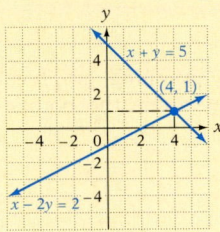

The solution is $(4, 1)$.

Example 3

Solve by graphing:
$4x - 2y = 6$
$y = 2x - 3$

Solution

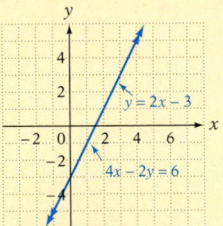

The solutions are the ordered pairs that satisfy the equation $y = 2x - 3$.

You Try It 1

Is $(-1, -2)$ a solution of the following system?
$2x - 5y = 8$
$-x + 3y = -5$

Your solution
Yes

You Try It 2

Solve by graphing:
$x + 3y = 3$
$-x + y = 5$

Your solution

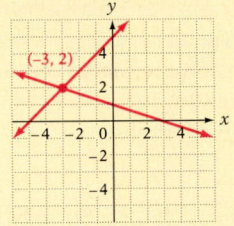

$(-3, 2)$

You Try It 3

Solve by graphing:
$y = 3x - 1$
$6x - 2y = -6$

Your solution

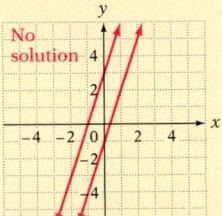

The system of equations is inconsistent and does not have a solution.

Solutions on p. S32

13.1 Exercises

Objective A   To solve a system of linear equations by graphing

Suggested Assignment

Exercises 1–41, odds
More challenging problems:
    Exercises 42–44

For Exercises 1 and 2, identify each system of equations (systems I, II, and III)
as (a) independent, (b) dependent, or (c) inconsistent.

1.   I 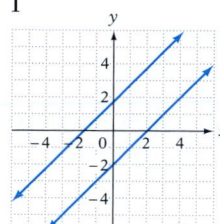   II 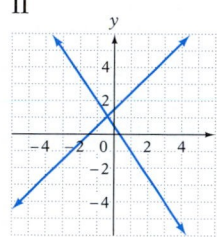   III 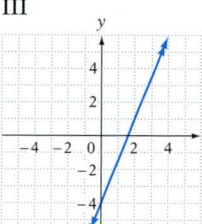   I: c; II: a; III: b

2.   I 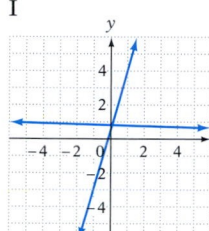   II 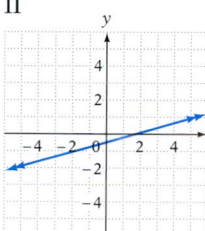   III 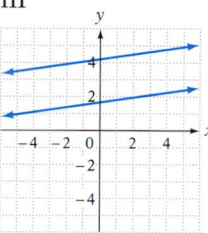   I: a; II: b; III: c

For Exercises 3 to 10, use the graphs of the equations of the system of equations to find the solution of the system of equations.

3. 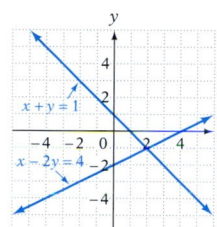 $(2, -1)$

4. 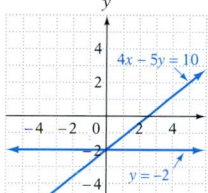 $(0, -2)$

5.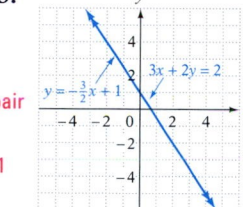

The ordered-pair solutions of

$$y = -\frac{3}{2}x + 1$$

6.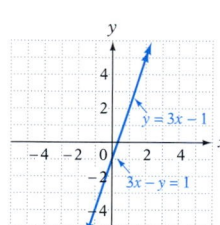

7.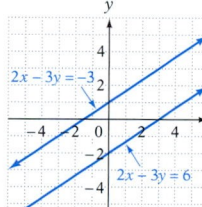

The ordered-pair solutions of $y = 3x - 1$

No solution

8.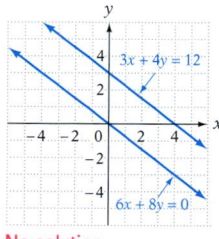

No solution

9. 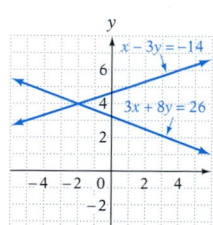 $(-2, 4)$

10.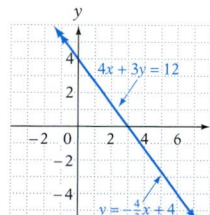

The ordered-pair solutions of

$$y = -\frac{4}{3}x + 4$$

Quick Quiz (Objective 13.1A)

1. Is $(3, -4)$ a solution of the system of equations?
    $2x - 3y = 18$
    $-x + 4y = -19$
    Yes

2. Solve by graphing: $x + 2y = 6$
                                $2x - y = 7$
    (4, 1)

3. Solve by graphing: $4x - 2y = 6$
                                $y = 2x - 3$
    Dependent. The solutions are the ordered-pair solutions of $y = 2x - 3$.

4. Solve by graphing: $x + 2y = 6$
                                $x = -2y + 2$
    No solution

11. Is (2, 3) a solution of $\begin{array}{l} 3x + 4y = 18 \\ 2x - y = 1 \end{array}$?

Yes

12. Is (2, −1) a solution of $\begin{array}{l} x - 2y = 4 \\ 2x + y = 3 \end{array}$?

Yes

13. Is (4, 3) a solution of $\begin{array}{l} 5x - 2y = 14 \\ x + y = 8 \end{array}$?

No

14. Is (2, 5) a solution of $\begin{array}{l} 3x + 2y = 16 \\ 2x - 3y = 4 \end{array}$?

No

15. Is (2, −3) a solution of $\begin{array}{l} y = 2x - 7 \\ 3x - y = 9 \end{array}$?

Yes

16. Is (−1, −2) a solution of $\begin{array}{l} 3x - 4y = 5 \\ y = x - 1 \end{array}$?

Yes

17. Is (0, 0) a solution of $\begin{array}{l} 3x + 4y = 0 \\ y = x \end{array}$?

Yes

18. Is (3, −4) a solution of $\begin{array}{l} 5x - 2y = 23 \\ 2x - 5y = 25 \end{array}$?

No

For Exercises 19 to 38, solve by graphing.

19. $x - y = 3$
$x + y = 5$

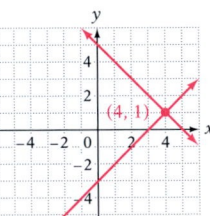

20. $2x - y = 4$
$x + y = 5$

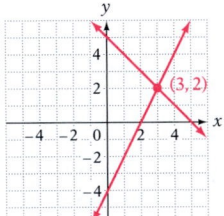

21. $x + 2y = 6$
$x - y = 3$

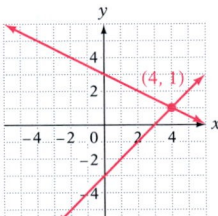

22. $3x - y = 3$
$2x + y = 2$

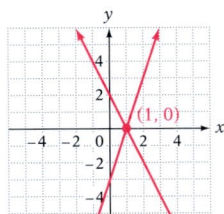

23. $3x - 2y = 6$
$y = 3$

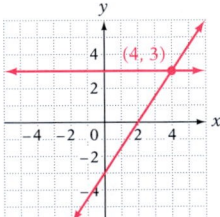

24. $x = 2$
$3x + 2y = 4$

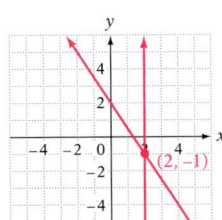

25. $x = 3$
$y = -2$

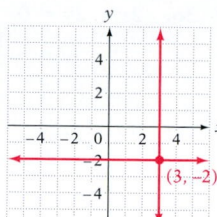

26. $x + 1 = 0$
$y - 3 = 0$

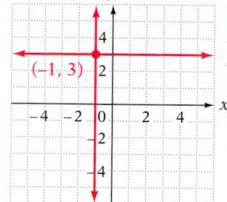

27. $y = 2x - 6$
$x + y = 0$

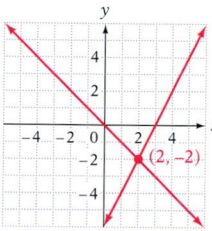

28. $5x - 2y = 11$
$y = 2x - 5$

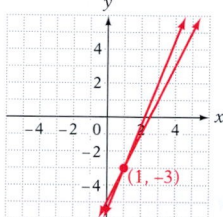

29. $2x + y = -2$
$6x + 3y = 6$

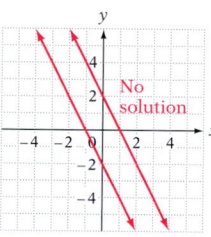

30. $x + y = 5$
$3x + 3y = 6$

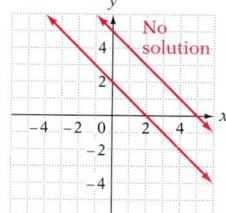

31. $y = 2x - 2$
$4x - 2y = 4$
The ordered-pair solutions
of $y = 2x - 2$

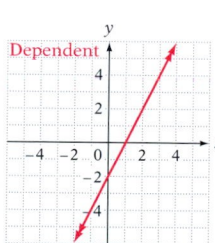

32. $y = -\dfrac{1}{3}x + 1$
$2x + 6y = 6$
The ordered-pair solutions
of $y = -\dfrac{1}{3}x + 1$

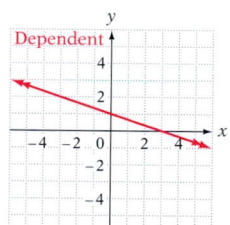

33. $x - y = 5$
$2x - y = 6$

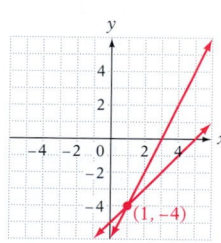

34. $5x - 2y = 10$
$3x + 2y = 6$

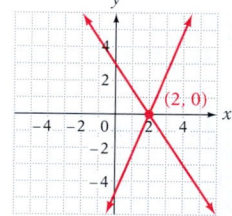

35. $3x + 4y = 0$
$2x - 5y = 0$

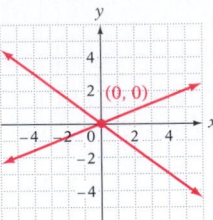

36. $2x - 3y = 0$
$y = -\dfrac{1}{3}x$

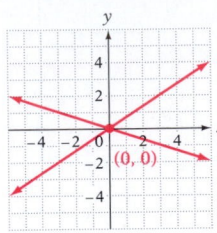

37. $x - 3y = 3$
$2x - 6y = 12$

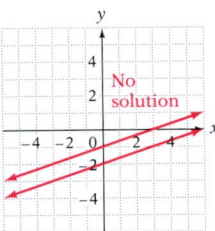

38. $4x + 6y = 12$
$6x + 9y = 18$

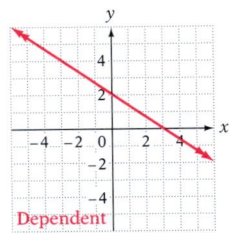

The ordered-pair solutions of $4x + 6y = 12$

APPLYING THE CONCEPTS

39. Determine whether the statement is always true, sometimes true, or never true.
  a. A solution of a system of two equations in two variables is a point in the plane.   Sometimes true
  b. Two parallel lines have the same slope.   Always true
  c. Two different lines with the same y-intercept are parallel.   Never true
  d. Two different lines with the same slope are parallel.   Always true

40. ✏ Explain how you can determine from the graph of a system of two linear equations in two variables whether it is an independent system of equations.

41. ✏ Explain how you can determine from the graph of a system of two linear equations in two variables whether it is an inconsistent system of equations.

42. Write a system of equations that has $(-2, 4)$ as its only solution.
  Answers will vary.

43. Write a system of equations for which there is no solution.
  Answers will vary.

44. Write a system of equations that is a dependent system of equations.
  Answers will vary.

Answers to Writing Exercises

40. If the system of equations is independent, the graphs intersect at exactly one point.

41. If the system of equations is inconsistent, the graphs of the equations are parallel and therefore do not intersect.

13.2 Solving Systems of Linear Equations by the Substitution Method

Objective A To solve a system of linear equations by the substitution method

A graphical solution of a system of equations is based on approximating the coordinates of a point of intersection. Algebraic methods can be used to find an exact solution of a system of equations. To solve a system of equations by the **substitution method,** one variable must be written in terms of the other variable.

HOW TO Solve by the substitution method: (1) $2x + 5y = -11$
                                               (2) $y = 3x - 9$

Equation (2) states that $y = 3x - 9$. Substitute $3x - 9$ for y in Equation (1). Then solve for x.

$$2x + 5y = -11$$ • This is Equation (1).
$$2x + 5(3x - 9) = -11$$ • From Equation (2), substitute $3x - 9$ for y.
$$2x + 15x - 45 = -11$$ • Solve for x.
$$17x - 45 = -11$$
$$17x = 34$$
$$x = 2$$

Now substitute the value of x into Equation (2) and solve for y.

$$y = 3x - 9$$ • This is Equation (2).
$$y = 3(2) - 9$$ • Substitute 2 for x.
$$y = 6 - 9 = -3$$

The solution is the ordered pair $(2, -3)$.

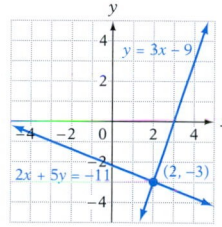

The graph of the equations in this system of equations is shown at the right. Note that the lines intersect at the point whose coordinates are $(2, -3)$, which is the algebraic solution we determined by the substitution method.

To solve a system of equations by the substitution method, we may need to solve one of the equations in the system of equations for one of its variables. For instance, the first step in solving the system of equations

(1) $x + 2y = -3$
(2) $2x - 3y = 5$

is to solve an equation of the system for one of its variables. Either equation can be used.

Solving Equation (1) for x:

$$x + 2y = -3$$
$$x = -2y - 3$$

Solving Equation (2) for x:

$$2x - 3y = 5$$
$$2x = 3y + 5$$
$$x = \frac{3y + 5}{2} = \frac{3}{2}y + \frac{5}{2}$$

Because solving Equation (1) for x does not result in fractions, it is the easier of the two equations to use.

Objective 13.2A

New Vocabulary
substitution method

Instructor Note
When students evaluate a variable expression, they replace a variable with a constant. Here the student is replacing a variable with a variable expression. Mentioning this connection will help some students master the substitution method of solving a system of equations.

Concept Check
Ask students whether the system of equations solved here is independent, dependent, or inconsistent.

In-Class Examples (Objective 13.2A)

1. Solve by substitution: $4x - 3y = 11$
                                $3x - y = 7$

$(2, -1)$

2. Solve by substitution: $6x + 2y = 8$
                                $y = -3x + 1$

No solution

3. Solve by substitution: $x - 2y = 8$
                                $y = \frac{1}{2}x - 4$

Dependent. The solutions are the ordered-pair solutions of $y = \frac{1}{2}x - 4$.

Here is the solution of the system of equations given on the preceding page.

HOW TO   Solve by the substitution method: (1) $x + 2y = -3$
(2) $2x - 3y = 5$

To use the substitution method, we must solve an equation for one of its variables. Equation (1) is used here because solving it for x does not result in fractions.

$x + 2y = -3$

(3) $x = -2y - 3$ • **Solve for x. This is Equation (3).**

Now substitute $-2y - 3$ for x in Equation (2) and solve for y.

$2x - 3y = 5$ • **This is Equation (2).**

$2(-2y - 3) - 3y = 5$ • **From Equation (3), substitute $-2y - 3$ for x.**

$-4y - 6 - 3y = 5$ • **Solve for y.**

$-7y - 6 = 5$

$-7y = 11$

$y = -\dfrac{11}{7}$

Substitute the value of y into Equation (3) and solve for x.

$x = -2y - 3$ • **This is Equation (3).**

$= -2\left(-\dfrac{11}{7}\right) - 3$ • **Substitute $-\dfrac{11}{7}$ for y.**

$= \dfrac{22}{7} - 3 = \dfrac{22}{7} - \dfrac{21}{7} = \dfrac{1}{7}$

The solution is $\left(\dfrac{1}{7}, -\dfrac{11}{7}\right)$.

The graph of the system of equations given above is shown at the right. It would be difficult to determine the exact solution of this system of equations from the graphs of the equations.

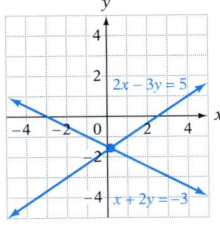

HOW TO   Solve by the substitution method: (1) $y = 3x - 1$
(2) $y = -2x - 6$

$y = -2x - 6$

$3x - 1 = -2x - 6$ • **Substitute $3x - 1$ for y in Equation (2).**

$5x = -5$ • **Solve for x.**

$x = -1$

Substitute this value of x into Equation (1) or Equation (2) and solve for y. Equation (1) is used here.

$y = 3x - 1$

$y = 3(-1) - 1 = -4$ • **Substitute -1 for x.**

The solution is $(-1, -4)$.

The substitution method can be used to analyze inconsistent and dependent systems of equations. If, when solving a system of equations algebraically, the variable is eliminated and the result is a false equation, such as $0 = 4$, the system of equations is inconsistent. If the variable is eliminated and the result is a true equation, such as $12 = 12$, the system of equations is dependent.

HOW TO   Solve by the substitution method: (1) $2x + 3y = 3$

(2) $y = -\dfrac{2}{3}x + 3$

$2x + 3y = 3$	• This is Equation (1).
$2x + 3\left(-\dfrac{2}{3}x + 3\right) = 3$	• From Equation (2), replace y with $-\dfrac{2}{3}x + 3$.
$2x - 2x + 9 = 3$	• Solve for x.
$9 = 3$	• This is not a true equation.

Because $9 = 3$ is not a true equation, the system of equations has no solution.

Solving Equation (1) above for y, we have $y = -\dfrac{2}{3}x + 1$.

Comparing this with Equation (2) reveals that the slopes are equal and the y-intercepts are different. The graphs of the equations that make up this system of equations are parallel and thus never intersect. Because the graphs do not intersect, there are no solutions of the system of equations. The system of equations is inconsistent.

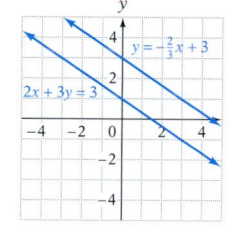

HOW TO   Solve by the substitution method: (1) $x = 2y + 3$

(2) $4x - 8y = 12$

$4x - 8y = 12$	• This is Equation (2).
$4(2y + 3) - 8y = 12$	• From Equation (1), replace x with $2y + 3$.
$8y + 12 - 8y = 12$	• Solve for y.
$12 = 12$	• This is a true equation.

The true equation $12 = 12$ indicates that any ordered pair (x, y) that satisfies one equation of the system satisfies the other equation. Therefore, the system of equations has an infinite number of solutions. The solutions are the ordered pairs (x, y) that are solutions of $x = 2y + 3$.

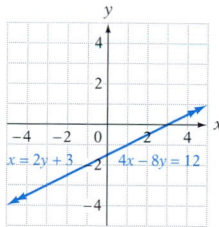

TAKE NOTE

As we mentioned in the previous section, when the system of equations is dependent, either equation can be used to write the ordered-pair solutions. Thus we could have said, "The solutions are the ordered pairs (x, y) that are solutions of $4x - 8y = 12$." Also note that, as we show at the right, if we solve each equation for y, the equations have the same slope-intercept form. This means we could also say, "The solutions are the ordered pairs (x, y) that are solutions of $y = \dfrac{1}{2}x - \dfrac{3}{2}$." When a system of equations is dependent, there are many ways in which the solutions can be stated.

If we write Equation (1) and Equation (2) in slope-intercept form, we have

$$x = 2y + 3 \qquad\qquad 4x - 8y = 12$$
$$-2y = -x + 3 \qquad\qquad -8y = -4x + 12$$
$$y = \frac{1}{2}x - \frac{3}{2} \qquad\qquad y = \frac{1}{2}x - \frac{3}{2}$$

The slope-intercept forms of the equations are the same, and therefore the graphs are the same. If we graph these two equations, we essentially graph one over the other. Accordingly, the graphs intersect at an infinite number of points.

Instructor Note

To prepare students for the inconsistent system of equations at the left, try asking whether 5 is a solution of $x + 2 = 8$. Most students will say no, because $7 \neq 8$. Now present students with the system of equations at the left and ask whether $(3, 1)$ is a solution. They should determine that it is a solution of Equation (2) but not of Equation (1), and therefore is not a solution of the system. Now try a few more ordered pairs that are solutions of Equation (2), such as $(-3, 5)$, $(0, 3)$, and $(6, -1)$, and have students verify that they are not solutions of the system. Using the substitution method for this system of equations shows that *any* ordered pair that is a solution of Equation (2) is not a solution of Equation (1).

A similar strategy can be used for dependent systems of equations. In this case, however, every solution of Equation (1) *is* a solution of Equation (2). Using the substitution method for this system of equations shows that *any* ordered pair that is a solution of Equation (1) is a solution of Equation (2).

Optional Student Activity

Find the time t between successive alignments of the hour and minute hands on a clock. [*Hint:* Begin with the hands aligned at 12:00. Let $d°$ be the angle at which the hands next align. The time t it takes the hour hand to rotate $d°$ equals the time it takes the minute hand to rotate $(d + 360)°$. The hour hand rotates at $30°/h$, and the minute hand rotates at $360°/h$.

Thus $t = \dfrac{(d + 360)°}{360}$ and $t = \dfrac{d°}{30}$.

Now solve the system of equations by substitution.]
The time is approximately 65.5 min.

Example 1   Solve by substitution:
(1) $3x + 4y = -2$
(2) $-x + 2y = 4$

Solution
$-x + 2y = 4$ • Solve Equation (2) for x.
$\quad -x = -2y + 4$
$\quad\ x = 2y - 4$

Substitute in Equation (1).
(1) $3x + 4y = -2$
$\quad 3(2y - 4) + 4y = -2$ • $x = 2y - 4$
$\quad 6y - 12 + 4y = -2$ • Solve for y.
$\quad\ 10y - 12 = -2$
$\quad\quad\ 10y = 10$
$\quad\quad\quad\ y = 1$

Substitute in $x = 2y - 4$.
$\quad x = 2y - 4$
$\quad x = 2(1) - 4$ • $y = 1$
$\quad x = 2 - 4$
$\quad x = -2$

The solution is $(-2, 1)$.

You Try It 1   Solve by substitution:
(1) $7x - y = 4$
(2) $3x + 2y = 9$

Your solution   (1, 3)

Example 2   Solve by substitution:
$4x + 2y = 5$
$\quad\quad y = -2x + 1$

Solution
$\quad\quad 4x + 2y = 5$
$4x + 2(-2x + 1) = 5$ • $y = -2x + 1$
$\quad 4x - 4x + 2 = 5$ • Solve for x.
$\quad\quad\quad\quad 2 = 5$ • Not a true equation

The system of equations is inconsistent and therefore does not have a solution.

You Try It 2   Solve by substitution:
$3x - y = 4$
$\quad y = 3x + 2$

Your solution   No solution

Example 3   Solve by substitution:
$\quad\quad y = 3x - 2$
$6x - 2y = 4$

Solution
$\quad\quad\quad 6x - 2y = 4$
$6x - 2(3x - 2) = 4$ • $y = 3x - 2$
$\quad 6x - 6x + 4 = 4$ • Solve for x.
$\quad\quad\quad\quad 4 = 4$ • A true equation

The system of equations is dependent. The solutions are the ordered pairs that satisfy the equation $y = 3x - 2$.

You Try It 3   Solve by substitution:
$\quad\quad y = -2x + 1$
$6x + 3y = 3$

Your solution   The system of equations is dependent. The solutions are the ordered pairs that satisfy the equation $y = -2x + 1$.

Solutions on pp. S32–S33

Objective B   **To solve investment problems**

The annual simple interest that an investment earns is given by the equation $Pr = I$, where P is the principal, or the amount invested, r is the simple interest rate, and I is the simple interest.

For instance, if you invest $750 at a simple interest rate of 6%, then the interest earned after 1 year is calculated as follows:

$$Pr = I$$
$$750(0.06) = I \qquad \bullet \text{ Replace } P \text{ with } 750 \text{ and } r \text{ with } 0.06 \text{ (6\%).}$$
$$45 = I \qquad \bullet \text{ Simplify.}$$

The amount of interest earned is $45.

HOW TO   A medical lab technician decided to open an Individual Retirement Account (IRA) by placing $2000 in two simple interest accounts. On one account, a corporate bond fund, the annual simple interest rate is 7.5%. On the second account, a real estate investment trust, the annual simple interest rate is 9%. If the technician wants to have annual earnings of $168 from these two investments, how much must be invested in each account?

Strategy for Solving Simple-Interest Investment Problems

1. For each amount invested, use the equation $Pr = I$. Write a numerical or variable expression for the principal, the interest rate, and the interest earned.

Amount invested at 7.5%: x
Amount invested at 9%: y

	Principal, P	·	Interest rate, r	=	Interest earned, I
Amount at 7.5%	x	·	0.075	=	$0.075x$
Amount at 9%	y	·	0.09	=	$0.09y$

2. Write a system of equations. One equation will express the relationships among the amounts invested. The second equation will express the relationships among the amounts of interest earned by the investments.

The total amount invested is $2000: $x + y = 2000$
The total annual interest earned is $168: $0.075x + 0.09y = 168$

Solve the system of equations.

(1) $x + y = 2000$
(2) $0.075x + 0.09y = 168$

Solve Equation (1) for y and substitute into Equation (2).

(3) $y = -x + 2000$

$$0.075x + 0.09(-x + 2000) = 168 \qquad \bullet \text{ Substitute } -x + 2000 \text{ for } y.$$
$$0.075x - 0.09x + 180 = 168$$
$$-0.015x = -12$$
$$x = 800$$

Substitute the value of x into Equation (3) and solve for y.

$$y = -x + 2000$$
$$y = -800 + 2000 = 1200 \qquad \bullet \text{ Substitute } -800 \text{ for } x.$$

The amount invested at 7.5% is $800. The amount invested at 9% is $1200.

Objective 13.2B

New Equations

Simple Interest Equation,
$I = Prt$

Instructor Note

Some students may recall the simple interest equation as $I = Prt$. Because we are discussing *annual* interest, $t = 1$. Thus we just write $I = Pr$.

Instructor Note

Students may not realize that investors do not always choose to put all their money into the account with the greatest interest rate because that account usually has the most risk. Placing money in different accounts allows the investor to diversify.

Optional Student Activity

Have students use the simple interest equation to check the solution of this problem. They should show that the interest earned on $800 plus the interest earned on $1200 is $168 and that $800 + 1200 = 2000$.

In-Class Examples (Objective 13.2B)

1. A web page designer invested a total of $10,000 in two accounts, a money market account and a high-yield corporate bond fund. The annual interest rate on the money market account was 3.5%, and the annual interest rate on the high-yield corporate bond fund was 9.25%. If the designer received annual interest income of $723.75, how much was invested in each account?
$3500 at 3.5%; $6500 at 9.25%

690

Example 4

A hair stylist invested some money at an annual simple interest rate of 5.2%. A second investment, $1000 more than the first, was invested at an annual simple interest rate of 7.2% so that the total interest earned was $320. How much was invested in each account?

Strategy

• Amount invested at 5.2%: x
  Amount invested at 7.2%: y

	Principal	Rate	Interest
Amount at 5.2%	x	0.052	$0.052x$
Amount at 7.2%	y	0.072	$0.072y$

• The second investment is $1000 more than the first investment:

$y = x + 1000$

The sum of the interest earned at 5.2% and the interest earned at 7.2% equals $320.

$0.052x + 0.072y = 320$

Solution

(1) $y = x + 1000$
(2) $0.052x + 0.072y = 320$

Replace y in Equation (2) with $x + 1000$ from Equation (1). Then solve for x.

$$0.052x + 0.072y = 320$$
$$0.052x + 0.072(x + 1000) = 320 \quad \bullet \; y = x + 1000$$
$$0.052x + 0.072x + 72 = 320 \quad \bullet \text{ Solve for } x.$$
$$0.124x + 72 = 320$$
$$0.124x = 248$$
$$x = 2000$$

$y = x + 1000$
$ = 2000 + 1000 \quad \bullet \; x = 2000$
$ = 3000$

$2000 was invested at an annual simple interest rate of 5.2%; $3000 was invested at 7.2%.

You Try It 4

The manager of a city's investment income wishes to place $330,000 in two simple interest accounts. The first account earns 6.5% annual interest and the second account earns 4.5% annual interest. How much should be invested in each account so that both accounts earn the same annual interest?

Your strategy

Your solution

$135,000 at 6.5%; $195,000 at 4.5%

Solution on p. S33

13.2 Exercises

Objective A   **To solve a system of linear equations by the substitution method**

1. ✎ Describe in your own words the process of solving a system of equations by the substitution method.

2. ✎ When you solve a system of equations by the substitution method, how do you determine whether the system of equations is dependent?

For Exercises 3 to 32, solve by substitution.

3. $2x + 3y = 7$
   $x = 2$
   (2, 1)

4. $y = 3$
   $3x - 2y = 6$
   (4, 3)

5. $y = x - 3$
   $x + y = 5$
   (4, 1)

6. $y = x + 2$
   $x + y = 6$
   (2, 4)

7. $x = y - 2$
   $x + 3y = 2$
   (−1, 1)

8. $x = y + 1$
   $x + 2y = 7$
   (3, 2)

9. $y = 4 - 3x$
   $3x + y = 5$
   No solution

10. $y = 2 - 3x$
    $6x + 2y = 7$
    No solution

11. $x = 3y + 3$
    $2x - 6y = 12$
    No solution

12. $x = 2 - y$
    $3x + 3y = 6$
    Dependent. The solutions satisfy the equation $x = 2 - y$.

13. $3x + 5y = -6$
    $x = 5y + 3$
    $\left(-\dfrac{3}{4}, -\dfrac{3}{4}\right)$

14. $y = 2x + 3$
    $4x - 3y = 1$
    (−5, −7)

15. $3x + y = 4$
    $4x - 3y = 1$
    (1, 1)

16. $x - 4y = 9$
    $2x - 3y = 11$
    $\left(\dfrac{17}{5}, -\dfrac{7}{5}\right)$

17. $3x - y = 6$
    $x + 3y = 2$
    (2, 0)

18. $4x - y = -5$
    $2x + 5y = 13$
    $\left(-\dfrac{6}{11}, \dfrac{31}{11}\right)$

19. $3x - y = 5$
    $2x + 5y = -8$
    (1, −2)

20. $3x + 4y = 18$
    $2x - y = 1$
    (2, 3)

21. $4x + 3y = 0$
    $2x - y = 0$
    (0, 0)

22. $5x + 2y = 0$
    $x - 3y = 0$
    (0, 0)

23. $2x - y = 2$
    $6x - 3y = 6$
    Dependent. The solutions satisfy the equation $2x - y = 2$.

Quick Quiz (Objective 13.2A)

1. Solve by the substitution method:
   $3x - 4y = 9$
   $y = 2x - 1$ (−1, −3)

2. Solve by the substitution method:
   $2x + 7y = -5$
   $x + 4y = -3$ (1, −1)

3. Solve by the substitution method:
   $2x + 3y = 3$
   $y = -\dfrac{2}{3}x + 1$
   Dependent. The solutions are the ordered pairs that satisfy $2x + 3y = 3$.

Section 13.2

Suggested Assignment
Exercises 1–45, odds
More challenging problems:
Exercises 46–49, 53

Answers to Writing Exercises

1. Students should include the following:
   (i) Solve, if necessary, one of the equations for x or y (or whatever variables are used in the system).
   (ii) Assuming Equation (1) was solved for y, replace y in Equation (2) with the expression to which it is equal.
   (iii) Solve for x in Equation (2).
   (iv) Use the value of x to determine y.
   (v) Write the answer as an ordered pair.
   (vi) Check the result.

2. The system of equations is dependent if the equation reduces to a true equation, such as $4 = 4$. In this case (as in the case of an inconsistent system of equations), the variable drops out and we are left with only the constants.

24. $3x + y = 4$
$9x + 3y = 12$
Dependent. The solutions
satisfy the equation $3x + y = 4$.

25. $x = 3y + 2$
$y = 2x + 6$
$(-4, -2)$

26. $x = 4 - 2y$
$y = 2x - 13$
$(6, -1)$

27. $y = 2x + 11$
$y = 5x - 19$
$(10, 31)$

28. $y = 2x - 8$
$y = 3x - 13$
$(5, 2)$

29. $y = -4x + 2$
$y = -3x - 1$
$(3, -10)$

30. $x = 3y + 7$
$x = 2y - 1$
$(-17, -8)$

31. $x = 4y - 2$
$x = 6y + 8$
$(-22, -5)$

32. $x = 3 - 2y$
$x = 5y - 10$
$\left(-\dfrac{5}{7}, \dfrac{13}{7}\right)$

Objective B   **To solve investment problems**

33.   An investment of $3500 is divided between two simple interest accounts. On one account, the annual simple interest rate is 5%, and on the second account, the annual simple interest rate is 7.5%. How much should be invested in each account so that the total interest earned from the two accounts is $215?
$1900 at 5%; $1600 at 7.5%

34.   A mortgage broker purchased two trust deeds for a total of $250,000. One trust deed earns 7% simple annual interest, and the second one earns 8% simple annual interest. If the total annual interest earned from the two trust deeds is $18,500, what was the purchase price of each trust deed?
7% trust deed: $150,000; 8% trust deed: $100,000

35.   When Sara Whitehorse changed jobs, she rolled over the $6000 in her retirement account into two simple interest accounts. On one account, the annual simple interest rate is 9%; on the second account, the annual simple interest rate is 6%. How much must be invested in each account if the accounts earn the same amount of annual interest?
$3600 at 6%; $2400 at 9%

36.   An animal trainer decided to take the $15,000 won on a game show and deposit it in two simple interest accounts. Part of the winnings were placed in an account paying 7% annual simple interest, and the remaining money was used to purchase a government bond that earns 6.5% annual simple interest. The amount of interest earned for 1 year was $1020. How much was invested in each account?
$9000 at 7%; $6000 at 6.5%

37.   A police officer has chosen a high-yield stock fund that earns 8% annual simple interest for part of a $6000 investment. The remaining portion is used to purchase a preferred stock that earns 11% annual simple interest. How much should be invested in each account so that the amount earned on the 8% account is twice the amount earned on the 11% account?
$4400 at 8%; $1600 at 11%

Quick Quiz (Objective 13.2B)

1. A landscape architect invested a total of $20,000 in two accounts, a municipal bond fund and a real estate investment trust. The annual interest rate on the municipal bond fund was 4.25%, and the annual interest rate on the real estate investment trust was 9%. If the architect received annual interest income of $1230, how much was invested in each account? $12,000 at 4.25%; $8000 at 9%

38. To plan for the purchase of a new car, a deposit was made into an account that earns 7% annual simple interest. Another deposit, $1500 less than the first deposit, was placed in a certificate of deposit earning 9% annual simple interest. The total interest earned on both accounts for 1 year was $505. How much money was deposited in the certificate of deposit?
$2500

39. The Pacific Investment Group invested some money in a certificate of deposit (CD) that earns 6.5% annual simple interest. Twice the amount invested at 6.5% was invested in a second CD that earns 8.5% annual simple interest. If the total annual interest earned from the two investments was $4935, how much was invested at 6.5%? $21,000

40. A corporation gave a university $300,000 to support product safety research. The university deposited some of the money in a 10% simple interest account and the remainder in an 8.5% simple interest account. How much should be deposited in each account so that the annual interest earned is $28,500? $200,000 at 10%; $100,000 at 8.5%

41. Ten co-workers formed an investment club, and each deposited $2000 in the club's account. They decided to take the total amount and invest some of it in preferred stock that pays 8% annual simple interest and the remainder in a municipal bond that pays 7% annual simple interest. The amount of interest earned each year from the investments was $1520. How much was invested in each? $12,000 at 8%; $8000 at 7%

42. A financial consultant advises a client to invest part of $30,000 in municipal bonds that earn 6.5% annual simple interest and the remainder of the money in 8.5% corporate bonds. How much should be invested in each so that the total interest earned each year is $2190?
$18,000 at 6.5%; $12,000 at 8.5%

43. Alisa Rhodes placed some money in a real estate investment trust that earns 7.5% annual simple interest. A second investment, which was one-half the amount placed in the real estate investment trust, was used to purchase a trust deed that earns 9% annual simple interest. If the total annual interest earned from the two investments was $900, how much was invested in the trust deed? $3750

APPLYING THE CONCEPTS

44. Consumerism   Suppose a breadmaker costs $180 and the ingredients and electricity needed to make one loaf of bread cost $.95. If a comparable loaf of bread at a grocery store costs $1.55, how many loaves of bread must you make before the breadmaker pays for itself?   300 loaves

45. Consumerism   Suppose a natural gas clothes dryer costs $240 and uses $.45 worth of gas to dry a load of clothes for 1 hour. If a laundromat charges $1.75 to use a dryer for 1 hour, how many loads of clothes must you dry before the purchase of the gas dryer is more economical?
185 loads

For Exercises 46 to 48, find the value of k for which the system of equations has no solution.

46. $2x - 3y = 7$
$kx - 3y = 4$
2

47. $8x - 4y = 1$
$2x - ky = 3$
1

48. $x = 4y + 4$
$kx - 8y = 4$
2

49. The following was offered as a solution of the given system of equations.

(1) $y = \dfrac{1}{2}x + 2$

(2) $2x + 5y = 10$

$$2x + 5y = 10 \qquad \bullet \text{ Equation (2)}$$

$$2x + 5\left(\dfrac{1}{2}x + 2\right) = 10 \qquad \bullet \text{ Substitute } \dfrac{1}{2}x + 2 \text{ for } y.$$

$$2x + \dfrac{5}{2}x + 10 = 10 \qquad \bullet \text{ Solve for } x.$$

$$\dfrac{9}{2}x = 0$$

$$x = 0$$

At this point the student stated that because $x = 0$, the system of equations has no solution. If this assertion is correct, is the system of equations independent, dependent, or inconsistent? If the assertion is not correct, what is the correct solution?

The assertion is not correct. The system of equations is independent. The solution is $(0, 2)$.

50. ✎ When you solve a system of equations by the substitution method, how do you determine whether the system of equations is inconsistent?

51. Investments   A sales representative invests in a stock paying a 9% dividend. A research consultant invests $5000 more than the sales representative in bonds paying 8% annual simple interest. The research consultant's income from the investment is equal to the sales representative's. Find the amount of the research consultant's investment.
$45,000

52. Investments   A plant manager invested $3000 more in stocks than in bonds. The stocks paid 8% annual simple interest, and the bonds paid 9.5% annual simple interest. Both investments yielded the same income. Find the total annual interest received on both investments.
$3040

53. Compound Interest   The exercises in this objective were based on annual *simple* interest, r. For simple interest, the amount of interest earned after 1 year is given by $I = Pr$. For **compound interest,** the interest earned for a certain period of time (usually daily or monthly) is added to the principal before the interest for the next period is calculated. The compound interest earned in 1 year is given by the formula $I = P\left[\left(1 + \dfrac{r}{n}\right)^{n} - 1\right]$, where n is the number of times per year that interest is compounded. For instance, if interest is compounded daily, then $n = 365$; if interest is compounded monthly, then $n = 12$. Suppose an investment of $5000 is deposited into three different accounts. The first account earns 8% annual simple interest, the second earns 8% compounded monthly ($n = 12$), and the third earns 8% compounded daily ($n = 365$). Find the amount of interest earned annually from each account.
Simple interest: $400; compounded monthly: $415.00; compounded daily: $416.39

Answers to Writing Exercises

50. The system of equations is inconsistent if the equation reduces to a false equation, such as $0 = 2$. In this case (as in the case of a dependent system of equations), the variable drops out and we are left with only the constants.

13.3 Solving Systems of Equations by the Addition Method

Objective A   To solve a system of linear equations by the addition method

Another method of solving a system of equations is called the **addition method.** This method is based on the Addition Property of Equations.

Note, for the system of equations at the right, the effect of adding Equation (2) to Equation (1). Because $2y$ and $-2y$ are opposites, adding the equations results in an equation with only one variable.

$$
\begin{array}{ll}
(1) & 5x + 2y = 11 \\
(2) & 3x - 2y = 13 \\
\hline
& 8x + 0y = 24 \\
& 8x = 24
\end{array}
$$

Solving $8x = 24$ for x gives the first coordinate of the ordered-pair solution of the system of equations.

$$\frac{8x}{8} = \frac{24}{8}$$
$$x = 3$$

The second coordinate is found by substituting the value of x into Equation (1) or Equation (2) and then solving for y. Equation (1) is used here.

$$
\begin{array}{ll}
(1) & 5x + 2y = 11 \\
& 5(3) + 2y = 11 \\
& 15 + 2y = 11 \\
& 2y = -4 \\
& y = -2
\end{array}
$$

The solution is $(3, -2)$.

Sometimes adding the two equations does not eliminate one of the variables. In this case, you can use the Multiplication Property of Equations to rewrite one or both of the equations so that the coefficients of one variable are opposites. Then add the equations and solve for the variables.

> **HOW TO**   Solve by the addition method:
> $$
> \begin{array}{ll}
> (1) & 4x + y = 5 \\
> (2) & 2x - 5y = 19
> \end{array}
> $$
>
> Multiply Equation (2) by -2. The coefficients of x will then be opposites.
>
> $$
> \begin{array}{ll}
> & -2(2x - 5y) = -2 \cdot 19 \\
> (3) & -4x + 10y = -38
> \end{array}
> $$
> • Multiply Equation (2) by -2.
> • Simplify. This is Equation (3).
>
> Add Equation (1) to Equation (3). Then solve for y.
>
> $$
> \begin{array}{ll}
> (1) & 4x + y = 5 \\
> (3) & -4x + 10y = -38 \\
> \hline
> & 11y = -33 \\
> & y = -3
> \end{array}
> $$
> • Note that the coefficients of x are opposites.
> • Add the two equations.
> • Solve for y.
>
> Substitute the value of y into Equation (1) or Equation (2) and solve for x. Equation (1) is used here.
>
> $$
> \begin{array}{ll}
> (1) & 4x + y = 5 \\
> & 4x + (-3) = 5 \\
> & 4x - 3 = 5 \\
> & 4x = 8 \\
> & x = 2
> \end{array}
> $$
> • Substitute -3 for y.
> • Solve for x.
>
> The solution is $(2, -3)$.

Objective 13.3A

New Vocabulary
addition method

Instructor Note
It may help some students to review the Addition Property of Equations before starting this section. For instance, give them $3x - 5 = 7$, add 5 to each side of the equation $(3x - 5 + 5 = 7 + 5)$, and simplify.
Now present this as

$$3x - 5 = 7$$
$$5 = 5$$

Now add the two equations and simplify.
For $\begin{array}{l} 5x + 2y = 11 \\ 3x - 2y = 13 \end{array}$, the situation is similar. Add $3x - 2y$ to each side of Equation (1).

$$5x + 2y + (3x - 2y)$$
$$= 11 + (3x - 2y)$$

Now use substitution to replace $3x - 2y$ on the right side of the equation with 13. After simplifying, we have $8x = 24$.

Optional Student Activity
Have students solve the system of equations

$$4x + y = 5$$
$$2x - 5y = 19$$

shown at the left by multiplying Equation (1) by 5 and then adding. This will not only give students some extra practice, but will also illustrate that there is more than one way to use the Multiplication Property of Equations to solve a system of equations.

In-Class Examples (Objective 13.3A)

1. Solve by the addition method: $3x + 4y = -5$
$9x - 2y = 13$

$(1, -2)$

2. Solve by the addition method: $2x + 5y = 1$
$3x + 1 = 2y + 12$

$(3, -1)$

3. Solve by the addition method: $x + 3y = 3$
$3x + 9y = 9$

Dependent. The solutions are the ordered pairs that satisfy $x + 3y = 3$.

4. Solve by the addition method: $3x = 2y + 4$
$6x - 4y = 5$

No solution

Concept Check

For the systems of equations in parts 1 to 3, do the following:

a. Determine the numbers that Equation (1) and Equation (2) must be multiplied by to eliminate x.

b. Determine the numbers that Equation (1) and Equation (2) must be multiplied by to eliminate y.

1. (1) $5x - 3y = 8$
(2) $2x + 7y = -5$
a. $-2, 5$ or $2, -5$ b. $7, 3$

2. (1) $x + 5y = -8$
(2) $3x - 2y = 10$
a. $-3, 1$ or $3, -1$ b. $2, 5$

3. (1) $4x + 5y = 3$
(2) $6x - 15y = 37$
a. $-6, 4$ or $6, -4$ ($-3, 2$ or $3, -2$ is also acceptable.)
b. $3, 1$

Discuss the Concepts

Explain how to choose the numbers to multiply an equation by to eliminate y from a system of equations.

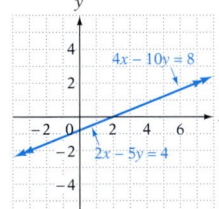

Sometimes each equation of a system of equations must be multiplied by a constant so that the coefficients of one of the variable terms are opposites.

HOW TO   Solve by the addition method:
(1) $3x + 7y = 2$
(2) $5x - 3y = -26$

To eliminate x, multiply Equation (1) by 5 and Equation (2) by -3. Note at the right how the constants are chosen.

$$5(3x + 7y) = 5 \cdot 2$$
$$-3(5x - 3y) = -3(-26)$$

• The negative is used so that the coefficients will be opposites.

$$\begin{array}{ll} 15x + 35y = 10 & \text{• 5 times Equation (1)} \\ -15x + 9y = 78 & \text{• } -3 \text{ times Equation (2)} \\ \hline 44y = 88 & \text{• Add the equations.} \\ y = 2 & \text{• Solve for } y. \end{array}$$

Substitute the value of y into Equation (1) or Equation (2) and solve for x. Equation (1) is used here.

(1) $3x + 7y = 2$
$$3x + 7(2) = 2 \qquad \text{• Substitute 2 for } y.$$
$$3x + 14 = 2 \qquad \text{• Solve for } x.$$
$$3x = -12$$
$$x = -4$$

The solution is $(-4, 2)$.

For the above system of equations, the value of x was determined by substitution. This value can also be determined by eliminating y from the system.

$$\begin{array}{ll} 9x + 21y = 6 & \text{• 3 times Equation (1)} \\ 35x - 21y = -182 & \text{• 7 times Equation (2)} \\ \hline 44x = -176 & \text{• Add the equations.} \\ x = -4 & \text{• Solve for } x. \end{array}$$

Note that this is the same value of x we determined by using substitution.

HOW TO   Solve by the addition method:
(1) $2x - 5y = 4$
(2) $4x - 10y = 8$

Eliminate x. Multiply Equation (1) by -2.

$$-2(2x - 5y) = -2(4) \qquad \text{• } -2 \text{ times Equation (1)}$$
(3) $-4x + 10y = -8 \qquad \text{• This is Equation (3).}$

Add Equation (3) to Equation (2) and solve for y.

(2) $4x - 10y = 8$
(3) $-4x + 10y = -8$
$$\overline{0x + 0y = 0}$$
$$0 = 0$$

The equation $0 = 0$ means that the system of equations is dependent. Therefore, the solutions of the system of equations are the ordered pairs that satisfy $2x - 5y = 4$.

The graphs of the two equations in the system of equations above are shown at the left. Note that one line is on top of the other line, and therefore the lines intersect infinitely often. The system of equations is dependent and the solutions are the ordered pairs that belong to either one of the equations.

HOW TO   Solve by the addition method: (1) $2x + y = 2$
(2) $4x + 2y = -5$

Eliminate y. Multiply Equation (1) by -2.

(1) $-2(2x + y) = -2 \cdot 2$ • **−2** times Equation **(1)**
(3) $-4x - 2y = -4$ • This is Equation **(3)**.

Add Equation (2) to Equation (3) and solve for x.

(3) $-4x - 2y = -4$
(2) $\underline{4x + 2y = -5}$
    $0x + 0y = -9$ • Add Equation **(2)** to Equation **(3)**.
    $0 = -9$ • This is not a true equation.

The system of equations is inconsistent and therefore does not have a solution.

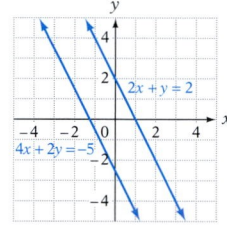

The graphs of the two equations in the system of equations above are shown at the left. Note that the graphs are parallel and therefore do not intersect. Thus the system of equations has no solution.

Example 1

Solve by the addition method:
(1) $2x + 4y = 7$
(2) $5x - 3y = -2$

Solution
Eliminate x.
$5(2x + 4y) = 5 \cdot 7$ • **5** times Equation **(1)**
$-2(5x - 3y) = -2(-2)$ • **−2** times Equation **(2)**

$\begin{aligned} 10x + 20y &= 35 \\ \underline{-10x + 6y} &= \underline{4} \\ 26y &= 39 \end{aligned}$ • Add the equations.

$y = \dfrac{39}{26} = \dfrac{3}{2}$ • Solve for y.

Substitute $\frac{3}{2}$ for y in Equation (1).

(1) $2x + 4y = 7$

$2x + 4\left(\dfrac{3}{2}\right) = 7$ • Replace y with $\dfrac{3}{2}$.

$2x + 6 = 7$ • Solve for x.

$2x = 1$

$x = \dfrac{1}{2}$

The solution is $\left(\dfrac{1}{2}, \dfrac{3}{2}\right)$.

You Try It 1

Solve by the addition method:
(1) $x - 2y = 1$
(2) $2x + 4y = 0$

Your solution

$\left(\dfrac{1}{2}, -\dfrac{1}{4}\right)$

Solution on p. S33

Instructor Note

Show students that if the two equations at the left are both solved for y, then we have

$$y = -2x + 2 \text{ and } y = -2x - \dfrac{5}{2}.$$

Because the graphs have the same slope but different y-intercepts, the graphs are parallel and do not intersect.

Discuss the Concepts

Have students discuss how it is possible to tell whether a system of equations is independent, dependent, or inconsistent if each equation of the system is in the form $y = mx + b$.

Give students the following systems of equations and ask them to identify the systems as independent, dependent, or inconsistent.

1. $y = -\dfrac{2}{3}x + 3$

$y = -\dfrac{2}{3}x - 3$

Inconsistent

2. $y = -x - 1$

$y = x + 1$

Independent

3. $y = 3x + 2$

$y = 3x + 2$

Dependent

4. $y = -2x + 3$

$y = 2x + 3$

Independent

Example 2

Solve by the addition method:
(1) $6x + 9y = 15$
(2) $4x + 6y = 10$

Solution

Eliminate x.

$\begin{aligned} 4(6x + 9y) &= 4 \cdot 15 \\ -6(4x + 6y) &= -6 \cdot 10 \end{aligned}$ • **4 times Equation (1)**
 • **−6 times Equation (2)**

$\begin{aligned} 24x + 36y &= 60 \\ -24x - 36y &= -60 \end{aligned}$

$\begin{aligned} 0x + 0y &= 0 \\ 0 &= 0 \end{aligned}$ • **Add the equations.**

The system of equations is dependent. The solutions are the ordered pairs that satisfy the equation $6x + 9y = 15$.

You Try It 2

Solve by the addition method:
$\begin{aligned} 2x - 3y &= 4 \\ -4x + 6y &= -8 \end{aligned}$

Your solution

The system of equations is dependent. The solutions are the ordered pairs that satisfy the equation $2x - 3y = 4$.

Example 3

Solve by the addition method:
(1) $2x = y + 8$
(2) $3x + 2y = 5$

Solution

Write Equation (1) in the form $Ax + By = C$.

$\begin{aligned} 2x &= y + 8 \end{aligned}$
(3) $2x - y = 8$ • **This is Equation (3).**

Eliminate y.

$\begin{aligned} 2(2x - y) &= 2 \cdot 8 \\ 3x + 2y &= 5 \end{aligned}$ • **2 times Equation (3)**
 • **This is Equation (2).**

$\begin{aligned} 4x - 2y &= 16 \\ 3x + 2y &= 5 \end{aligned}$

$\begin{aligned} 7x &= 21 \\ x &= 3 \end{aligned}$ • **Add the equations.**

Replace x in Equation (1).
(1) $2x = y + 8$
$\begin{aligned} 2 \cdot 3 &= y + 8 \\ 6 &= y + 8 \\ -2 &= y \end{aligned}$ • **Replace x with 3.**

The solution is $(3, -2)$.

You Try It 3

Solve by the addition method:
$\begin{aligned} 4x + 5y &= 11 \\ 3y &= x + 10 \end{aligned}$

Your solution

$(-1, 3)$

Solutions on pp. S33–S34

13.3 Exercises

Objective A To solve a system of linear equations by the addition method

For Exercises 1 to 36, solve by the addition method.

1. $x + y = 4$
$x - y = 6$
(5, −1)

2. $2x + y = 3$
$x - y = 3$
(2, −1)

3. $x + y = 4$
$2x + y = 5$
(1, 3)

4. $x - 3y = 2$
$x + 2y = -3$
(−1, −1)

5. $2x - y = 1$
$x + 3y = 4$
(1, 1)

6. $x - 2y = 4$
$3x + 4y = 2$
(2, −1)

7. $4x - 5y = 22$
$x + 2y = -1$
(3, −2)

8. $3x - y = 11$
$2x + 5y = 13$
(4, 1)

9. $2x - y = 1$
$4x - 2y = 2$
Dependent. The solutions
satisfy the equation
$2x - y = 1.$

10. $x + 3y = 2$
$3x + 9y = 6$
Dependent. The solutions
satisfy the equation
$x + 3y = 2.$

11. $4x + 3y = 15$
$2x - 5y = 1$
(3, 1)

12. $3x - 7y = 13$
$6x + 5y = 7$
(2, −1)

13. $2x - 3y = 1$
$4x - 6y = 2$
Dependent. The solutions
satisfy the equation
$2x - 3y = 1.$

14. $2x + 4y = 6$
$3x + 6y = 9$
Dependent. The solutions
satisfy the equation
$2x + 4y = 6.$

15. $3x - 6y = -1$
$6x - 4y = 2$
$\left(\dfrac{2}{3}, \dfrac{1}{2}\right)$

16. $5x + 2y = 3$
$3x - 10y = -1$
$\left(\dfrac{1}{2}, \dfrac{1}{4}\right)$

17. $5x + 7y = 10$
$3x - 14y = 6$
(2, 0)

18. $7x + 10y = 13$
$4x + 5y = 6$
(−1, 2)

19. $3x - 2y = 0$
$6x + 5y = 0$
(0, 0)

20. $5x + 2y = 0$
$3x + 5y = 0$
(0, 0)

21. $2x - 3y = 16$
$3x + 4y = 7$
(5, −2)

22. $3x + 4y = 10$
$4x + 3y = 11$
(2, 1)

23. $5x + 3y = 7$
$2x + 5y = 1$
$\left(\dfrac{32}{19}, -\dfrac{9}{19}\right)$

24. $-2x + 7y = 9$
$3x + 2y = -1$
(−1, 1)

Quick Quiz (Objective 13.3A)

1. Solve by the addition method: $2x - 5y = 19$
$\qquad\qquad\qquad\qquad\qquad\quad 3x + 4y = -6$

(2, −3)

2. Solve by the addition method: $3x + 5y = 1$
$\qquad\qquad\qquad\qquad\qquad\quad 3x + 4 = -5y + 12$

No solution

3. Solve by the addition method: $4x + 5y = 20$
$\qquad\qquad\qquad\qquad\qquad\qquad 8x + 10y = 40$

Dependent. The solutions are the ordered pairs
that satisfy the equation $4x + 5y = 20$.

Suggested Assignment
Exercises 1–37, odds
More challenging problems:
  Exercises 39, 41

25. $3x + 4y = 4$
$5x + 12y = 5$
$\left(\dfrac{7}{4}, -\dfrac{5}{16}\right)$

26. $2x + 5y = 2$
$3x + 3y = 1$
$\left(-\dfrac{1}{9}, \dfrac{4}{9}\right)$

27. $8x - 3y = 11$
$6x - 5y = 11$
$(1, -1)$

28. $4x - 8y = 36$
$3x - 6y = 15$
No solution

29. $5x + 15y = 20$
$2x + 6y = 12$
No solution

30. $y = 2x - 3$
$3x + 4y = -1$
$(1, -1)$

31. $3x = 2y + 7$
$5x - 2y = 13$
$(3, 1)$

32. $2y = 4 - 9x$
$9x - y = 25$
$(2, -7)$

33. $2x + 9y = 16$
$5x = 1 - 3y$
$(-1, 2)$

34. $3x - 4 = y + 18$
$4x + 5y = -21$
$\left(\dfrac{89}{19}, -\dfrac{151}{19}\right)$

35. $2x + 3y = 7 - 2x$
$7x + 2y = 9$
$(1, 1)$

36. $5x - 3y = 3y + 4$
$4x + 3y = 11$
$(2, 1)$

Answers to Writing Exercises

37. Student descriptions should include the following steps: (1) If necessary, multiply one or both of the equations by a constant so that the coefficients of one variable will be opposites. (2) Add the two equations and solve for the variable. (3) Substitute the value of the variable into either equation in the system and solve for the second variable. (4) Write the ordered-pair solution. (5) Check the solution.

APPLYING THE CONCEPTS

37. ✎ Describe in your own words the process of solving a system of equations by the addition method.

38. The point of intersection of the graphs of the equations $Ax + 2y = 2$ and $2x + By = 10$ is $(2, -2)$. Find A and B. $A = 3; B = -3$

39. The point of intersection of the graphs of the equations $Ax - 4y = 9$ and $4x + By = -1$ is $(-1, -3)$. Find A and B. $A = 3; B = -1$

40. For what value of k is the system of equations dependent?

a. $2x + 3y = 7$
$4x + 6y = k$
$k = 14$

b. $y = \dfrac{2}{3}x - 3$
$y = kx - 3$
$k = \dfrac{2}{3}$

c. $x = ky - 1$
$y = 2x + 2$
$k = \dfrac{1}{2}$

41. For what value of k is the system of equations inconsistent?

a. $x + y = 7$
$kx + y = 3$
$k = 1$

b. $x + 2y = 4$
$kx + 3y = 2$
$k = \dfrac{3}{2}$

c. $2x + ky = 1$
$x + 2y = 2$
$k = 4$

13.4 Application Problems in Two Variables

Objective A   To solve rate-of-wind or rate-of-current problems

We normally need two variables to solve motion problems that involve an object moving with or against a wind or current.

HOW TO   Flying with the wind, a small plane can fly 600 mi in 3 h. Against the wind, the plane can fly the same distance in 4 h. Find the rate of the plane in calm air and the rate of the wind.

> **Strategy for Solving Rate-of-Wind or Rate-of-Current Problems**
>
> Choose one variable to represent the rate of the object in calm conditions and a second variable to represent the rate of the wind or current. Using these variables, express the rate of the object with and against the wind or current. Use the equation $rt = d$ to write expressions for the distance traveled by the object. The results can be recorded in a table.

Rate of plane in calm air: p
Rate of wind: w

	Rate	·	Time	=	Distance
With the wind	$p + w$	·	3	=	$3(p + w)$
Against the wind	$p - w$	·	4	=	$4(p - w)$

> Determine how the expressions for distance are related.

The distance traveled with the wind is 600 mi. $3(p + w) = 600$
The distance traveled against the wind is 600 mi. $4(p - w) = 600$

Solve the system of equations.

$3(p + w) = 600$ $\quad\to\quad$ $\dfrac{1}{3} \cdot 3(p + w) = \dfrac{1}{3} \cdot 600$ $\quad\to\quad$ $p + w = 200$

$4(p - w) = 600$ $\quad$ $\dfrac{1}{4} \cdot 4(p - w) = \dfrac{1}{4} \cdot 600$ $\quad$ $\underline{p - w = 150}$

$\qquad\qquad\qquad\qquad\qquad\qquad\qquad\qquad\qquad\qquad\qquad 2p = 350$
$\qquad\qquad\qquad\qquad\qquad\qquad\qquad\qquad\qquad\qquad\quad p = 175$

$p + w = 200$
$175 + w = 200$ • $p = 175$
$\quad w = 25$

The rate of the plane in calm air is 175 mph.
The rate of the wind is 25 mph.

Objective 13.4A

Equation to Review
Uniform Motion Equation,
$d = rt$ [5.1D]

Instructor Note
The concepts presented here are not intuitive to some students. Having these students answer the Concept Check questions below may help them understand.

Concept Check

Suppose that a person has a powerboat with the throttle set to move the boat at 8 mph in calm water and that the rate of the current of a river is 4 mph.

1. What is the speed of the boat with the current?   12 mph
2. What is the speed of the boat against the current?   4 mph
3. If the rate of the current increases to 6 mph, what is the speed of the boat as it proceeds downstream?   14 mph
4. If the rate of the current increases to 6 mph, what is the speed of the boat upstream?   2 mph
5. If the rate of the current is 6 mph and the throttle is set so that the speed of the boat in calm water is 3 mph, what is the speed of the boat with the current?   9 mph
6. If the throttle is set so that the speed of the boat in calm water is 3 mph and the rate of the current is 6 mph, what is the speed of the boat against the current? Is the boat moving upstream or downstream?   −3 mph, downstream
7. If the rate of the current is 5 mph, what is the minimum throttle setting so that the boat will not move downstream?   5 mph

---

In-Class Examples (Objective 13.4A)

1. Traveling with the current, a boat went 22 mi in 2 h. Traveling against the current, the boat traveled 10 mi in 2 h. Find the rate of the boat in calm water and the rate of the current.
Rate of boat in calm water: 8 mph; rate of current: 3 mph

Example 1

A 450-mile trip from one city to another takes 3 h when a plane is flying with the wind. The return trip, against the wind, takes 5 h. Find the rate of the plane in still air and the rate of the wind.

Strategy

- Rate of the plane in still air: p
  Rate of the wind: w

	Rate	*Time*	*Distance*
With wind	$p + w$	3	$3(p + w)$
Against wind	$p - w$	5	$5(p - w)$

- The distance traveled with the wind is 450 mi. The distance traveled against the wind is 450 mi.

Solution

$3(p + w) = 450 \qquad \dfrac{1}{3} \cdot 3(p + w) = \dfrac{1}{3} \cdot 450$

$5(p - w) = 450 \qquad \dfrac{1}{5} \cdot 5(p - w) = \dfrac{1}{5} \cdot 450$

$$p + w = 150$$
$$\underline{p - w = 90}$$
$$2p = 240$$
$$p = 120$$

$p + w = 150$
$120 + w = 150 \qquad \bullet\; \boldsymbol{p = 120}$
$w = 30$

The rate of the plane in still air is 120 mph.
The rate of the wind is 30 mph.

You Try It 1

A canoeist paddling with the current can travel 15 mi in 3 h. Against the current, it takes the canoeist 5 h to travel the same distance. Find the rate of the current and the rate of the canoeist in calm water.

Your strategy

Your solution

Rate of current: 1 mph;
rate of canoeist in calm water: 4 mph

Solution on p. S34

Objective B

To solve application problems using two variables

The application problems in this section are varieties of those problems solved earlier in the text. Each of the strategies for the problems in this section will result in a system of equations.

HOW TO   A jeweler purchased 5 oz of a gold alloy and 20 oz of a silver alloy for a total cost of $540. The next day, at the same prices per ounce, the jeweler purchased 4 oz of the gold alloy and 25 oz of the silver alloy for a total cost of $450. Find the cost per ounce of the gold and silver alloys.

> **Strategy for Solving an Application Problem in Two Variables**
>
> Choose one variable to represent one of the unknown quantities and a second variable to represent the other unknown quantity. Write numerical or variable expressions for all of the remaining quantities. These results can be recorded in two tables, one for each of the conditions.

Cost per ounce of gold: g
Cost per ounce of silver: s

First day:

	Amount	·	Unit Cost	=	Value
Gold	5	·	g	=	$5g$
Silver	20	·	s	=	$20s$

Second day:

	Amount	·	Unit Cost	=	Value
Gold	4	·	g	=	$4g$
Silver	25	·	s	=	$25s$

> Determine a system of equations. Each table will give one equation of the system.

The total value of the purchase on the first day was $540. $5g + 20s = 540$

The total value of the purchase on the second day was $450. $4g + 25s = 450$

Solve the system of equations.

$5g + 20s = 540$ $4(5g + 20s) = 4 \cdot 540$ $20g + 80s = 2160$
$4g + 25s = 450$ $-5(4g + 25s) = -5 \cdot 450$ $\underline{-20g - 125s = -2250}$
                                                               $-45s = -90$
                                                               $s = 2$

$5g + 20s = 540$
$5g + 20(2) = 540$ • $s = 2$
$5g + 40 = 540$
$5g = 500$
$g = 100$

The cost per ounce of the gold alloy was $100.
The cost per ounce of the silver alloy was $2.

Point of Interest

The Babylonians had a method for solving a system of equations. Here is an adaptation of a problem from an ancient (around 1500 B.C.) Babylonian text. "There are two silver blocks. The sum of $\frac{1}{7}$ of the first block and $\frac{1}{11}$ of the second block is one sheqel (a weight). The first block diminished by $\frac{1}{7}$ of its weight equals the second diminished by $\frac{1}{11}$ of its weight. What are the weights of the two blocks?"

In-Class Examples (Objective 13.4B)

1. A jeweler purchased 5 oz of a gold alloy and 10 oz of a platinum alloy for a total cost of $2375. In a second purchase at the same prices, the jeweler bought 8 oz of the gold alloy and 7 oz of the platinum alloy for a total cost of $2225. What is the cost per ounce of the gold alloy and of the platinum alloy?
Gold: $125 per ounce; platinum: $175 per ounce

Optional Student Activity

1. **Fuel Mixtures** The octane number of 87 on gasoline means that it will fight engine "knock" as effectively as a reference fuel that is 87% isooctane, a type of gasoline. Suppose you want to fill an empty 18-gallon tank with some 87-octane gasoline and some 93-octane fuel to produce a mixture that is 89-octane. How much of each type of gasoline must you use?
   87-octane: 12 gal; 93-octaine: 6 gal

2. Have students find the current prices of 87-, 89-, and 93-octane gasoline. Is it cheaper to blend your own gasoline to achieve an octane rating of 89 or to buy it already blended? See Exercise 1 above.

3. Suppose 87-octane gas costs $1.60 per gallon and 93-octane gas costs $1.90 per gallon.
   a. By blending 87- and 93-octane gasoline, can a blend be produced that costs less than $1.60 per gallon?   No
   b. By blending 87- and 93-octane gasoline, can a blend be produced that costs more than $1.90 per gallon?   No
   c. By blending 87- and 93-octane gasoline, can a blend be produced that costs at least $1.60 per gallon and at most $1.90 per gallon?   Yes
   d. Prove your answer to part **c**.   Answers will vary.

Example 2

A store owner purchased 20 incandescent light bulbs and 30 fluorescent bulbs for a total cost of $40. A second purchase, at the same prices, included 30 incandescent bulbs and 10 fluorescent bulbs for a total cost of $25. Find the cost of an incandescent bulb and of a fluorescent bulb.

Strategy
Cost of an incandescent bulb: b
Cost of a fluorescent bulb: f

First purchase:

	Amount	Unit Cost	Value
Incandescent	20	b	$20b$
Fluorescent	30	f	$30f$

Second purchase:

	Amount	Unit Cost	Value
Incandescent	30	b	$30b$
Fluorescent	10	f	$10f$

The total cost of the first purchase was $40.
The total cost of the second purchase was $25.

Solution

$$20b + 30f = 40 \qquad 3(20b + 30f) = 3 \cdot 40$$
$$30b + 10f = 25 \qquad -2(30b + 10f) = -2 \cdot 25$$

$$\begin{array}{r} 60b + 90f = 120 \\ -60b - 20f = -50 \\ \hline 70f = 70 \\ f = 1 \end{array}$$

$$20b + 30f = 40$$
$$20b + 30(1) = 40 \qquad \bullet\ f = 1$$
$$20b = 10$$
$$b = \frac{1}{2}$$

The cost of an incandescent bulb is $.50.
The cost of a fluorescent bulb is $1.00.

You Try It 2

A citrus grower purchased 25 orange trees and 20 grapefruit trees for $290. The next week, at the same prices, the grower bought 20 orange trees and 30 grapefruit trees for $330. Find the cost of an orange tree and the cost of a grapefruit tree.

Your strategy

Your solution
Orange tree: $6;
Grapefruit tree: $7

Solution on p. S34

13.4 Exercises

Objective A   To solve rate-of-wind or rate-of-current problems

Suggested Assignment
Exercises 1–15, odds
More challenging problems:
  Exercises 17, 19, 20

1. A whale swimming against an ocean current traveled 60 mi in 2 h. Swimming in the opposite direction, with the current, the whale was able to travel the same distance in 1.5 h. Find the speed of the whale in calm water and the rate of the ocean current.   Whale: 35 mph; current: 5 mph

2. A plane flying with the jet stream flew from Los Angeles to Chicago, a distance of 2250 mi, in 5 h. Flying against the jet stream, the plane could fly only 1750 mi in the same amount of time. Find the rate of the plane in calm air and the rate of the wind.   Plane: 400 mph; wind: 50 mph

3. A rowing team rowing with the current traveled 40 km in 2 h. Rowing against the current, the team could travel only 16 km in 2 h. Find the rowing rate in calm water and the rate of the current.
    Rowing team: 14 km/h; current: 6 km/h

4. The bird capable of the fastest flying speed is the swift. A swift flying with the wind to a favorite feeding spot traveled 26 mi in 0.2 h. On returning, now flying against the wind, the swift was able to travel only 16 mi in the same amount of time. Find the rate of the swift in calm air and the rate of the wind.   Swift: 105 mph; wind: 25 mph

5. A private Learjet 31A was flying with a tailwind and traveled 1120 mi in 2 h. Flying against the wind on the return trip, the jet was able to travel only 980 mi in 2 h. Find the speed of the jet in calm air and the rate of the wind.   Jet: 525 mph; wind: 35 mph

6. A plane flying with a tailwind flew 300 mi in 2 h. Against the wind, it took 3 h to travel the same distance. Find the rate of the plane in calm air and the rate of the wind.   Plane: 125 mph; wind: 25 mph

7. A Boeing Apache Longbow military helicopter traveling directly into a strong headwind was able to travel 450 mi in 2.5 h. The return trip, now with a tailwind, took 1 h 40 min. Find the speed of the helicopter in calm air and the rate of the wind.   Helicopter: 225 mph; wind: 45 mph

8. A seaplane pilot flying with the wind flew from an ocean port to a lake, a distance of 240 mi, in 2 h. Flying against the wind, the pilot flew from the lake to the ocean port in 2 h 40 min. Find the rate of the plane in calm air and the rate of the wind.   Plane: 105 mph; wind: 15 mph

9. Rowing with the current, a canoeist paddled 14 mi in 2 h. Against the current, the canoeist could paddle only 10 mi in the same amount of time. Find the rate of the canoeist in calm water and the rate of the current.
    Canoeist: 6 mph; current: 1 mph

Objective B   To solve application problems using two variables

10. **Internet Services**   A computer online service charges one hourly price for regular use but a higher hourly rate for designated "premium" areas. One customer was charged $28 after spending 2 h in premium areas and 9 regular hours; another spent 3 h in premium areas and 6 regular hours and was charged $27. What does the online service charge per hour for regular and premium services?   Regular: $2; premium: $5

11. **Flour Mixtures**   A baker purchased 12 lb of wheat flour and 15 lb of rye flour for a total cost of $18.30. A second purchase, at the same prices, included 15 lb of wheat flour and 10 lb of rye flour. The cost of the second purchase was $16.75. Find the cost per pound of the wheat flour and of the rye flour.   Wheat: $.65; rye: $.70

Quick Quiz (Objective 13.4A)

1. Flying with the wind, a pilot was able to travel 600 mi in 2 h. Against the wind, it took the pilot 3 h to fly the same distance. Find the rate of the wind and the rate of the plane in calm air.
   Rate of wind: 50 mph; rate of plane in calm air: 250 mph

12. **Investments**   An investor owned 300 shares of an oil company and 200 shares of a movie company. The quarterly dividend from the two stocks was $165. After the investor sold 100 shares of the oil company and bought an additional 100 shares of the movie company, the quarterly dividend became $185. Find the dividend per share for each stock.
Oil: $.25; movie: $.45

13. **Purchasing**   A carpenter purchased 50 ft of redwood and 80 ft of pine for a total cost of $275. A second purchase, at the same prices, included 60 ft of redwood and 40 ft of pine for a total cost of $260. Find the cost per foot of redwood and of pine.
Redwood: $3.50; pine: $1.25

14. **Finance**   During one month, a homeowner used 400 units of electricity and 100 units of gas for a total cost of $310. The next month, 500 units of electricity and 150 units of gas were used for a total cost of $395. Find the cost per unit of gas.
$.30

15. **Food Mixtures**   A merchant mixed 10 lb of cinnamon tea with 8 lb of spice tea. The 18-pound mixture cost $58. A second mixture included 5 lb of the cinnamon tea and 10 lb of the spice tea. The 15-pound mixture cost $50. Find the cost per pound of the cinnamon tea and of the spice tea.
Cinnamon: $3; spice: $3.50

APPLYING THE CONCEPTS

16. **Geometry**   Two angles are supplementary. The measure of the larger angle is 15° more than twice the measure of the smaller angle. Find the measures of the two angles. (Recall that supplementary angles are two angles whose sum is 180°.)   55°, 125°

17. **Coin Problem**   The value of the nickels and dimes in a coin bank is $.25. If the number of nickels and the number of dimes were doubled, the value of the coins would be $.50. How many nickels and how many dimes are in the bank?
1 nickel and 2 dimes, or 3 nickels and 1 dime

18. **Investments**   An investor has $5000 to invest in two accounts. The first account earns 8% annual simple interest, and the second account earns 10% annual simple interest. How much money should be invested in each account so that the total annual simple interest earned is $600?
It is impossible to earn $600 in interest.

19. **Ancient Problem**   Solve the following problem, which dates from a Chinese manuscript called the Jinzhang that is approximately 2100 years old. "The price of 1 acre of good land is 300 pieces of gold; the price of 7 acres of bad land is 500 pieces of gold. One has purchased altogether 100 acres. The price was 10,000 pieces of gold. How much good land and how much bad land was bought?" Adapted from Victor J. Katz, *A History of Mathematics, An Introduction* (New York: Harper-Collins, 1993), p. 15.
12.5 acres of good land, 87.5 acres of bad land

20. **Coin Problem**   A coin bank contains only nickels or dimes, but there are no more than 27 coins. The value of the coins is $2.10. How many different combinations of nickels and dimes could be in the coin bank?   Seven

Quick Quiz (Objective 13.4B)

1. A chemist mixed 20 ml of reagent I with 30 ml of reagent II. The result was a solution that was 13% acetic acid. Using the same reagents, the chemist mixed 30 ml of reagent I with 20 ml of reagent II to make a solution that was 12% acetic acid. What percent acetic acid was in reagent I?
10%

Focus on Problem Solving

Using a Table and Searching for a Pattern

Consider the numbers 10, 12, and 28 and the sum of the proper factors (the natural number factors less than the number) of those numbers.

$$10: 1 + 2 + 5 = 8 \qquad 12: 1 + 2 + 3 + 4 + 6 = 16 \qquad 28: 1 + 2 + 4 + 7 + 14 = 28$$

10 is called a **deficient number** because the sum of its proper factors is less than the number ($8 < 10$). 12 is called an **abundant number** because the sum of its proper factors is greater than the number ($16 > 12$). 28 is called a **perfect number** because the sum of its proper factors equals the number ($28 = 28$).

Our goal for this Focus on Problem Solving is to try to find a method that will determine whether a number is deficient, abundant, or perfect without having to first find all the factors and then add them up. We will use a table and search for a pattern.

Before we begin, recall that a prime number is a number greater than 1 whose only factors are itself and 1, and that each natural number greater than 1 has a unique prime factorization. For instance, the prime factorization of 36 is given by $36 = 2^2 \cdot 3^2$. Note that the proper factors of 36 (1, 2, 3, 4, 6, 9, 12, 18) can be represented in terms of the same prime numbers.

$$1 = 2^0, \quad 2 = 2^1, \quad 3 = 3^1, \quad 4 = 2^2, \quad 6 = 2 \cdot 3, \quad 9 = 3^2, \quad 12 = 2^2 \cdot 3, \quad 18 = 2 \cdot 3^2$$

Now let us consider a trial problem of determining whether 432 is deficient, abundant, or perfect.

TAKE NOTE

This table contains the factor $432 = 2^4 \cdot 3^3$, which is not a proper factor.

We write the prime factorization of 432 as $2^4 \cdot 3^3$ and place the factors of 432 in a table, as shown at the right. This table contains *all* the factors of 432 represented in terms of the prime number factors. The sum of each column is shown at the bottom.

	1	$1 \cdot 3$	$1 \cdot 3^2$	$1 \cdot 3^3$
	2	$2 \cdot 3$	$2 \cdot 3^2$	$2 \cdot 3^3$
	2^2	$2^2 \cdot 3$	$2^2 \cdot 3^2$	$2^2 \cdot 3^3$
	2^3	$2^3 \cdot 3$	$2^3 \cdot 3^2$	$2^3 \cdot 3^3$
	2^4	$2^4 \cdot 3$	$2^4 \cdot 3^2$	$2^4 \cdot 3^3$
Sum	31	$31 \cdot 3$	$31 \cdot 3^2$	$31 \cdot 3^3$

Here is the calculation of the sum for the column headed by $1 \cdot 3$.

$$1 \cdot 3 + 2 \cdot 3 + 2^2 \cdot 3 + 2^3 \cdot 3 + 2^4 \cdot 3 = (1 + 2 + 2^2 + 2^3 + 2^4)3 = 31(3)$$

For the column headed by $1 \cdot 3^2$ there is a similar situation.

$$1 \cdot 3^2 + 2 \cdot 3^2 + 2^2 \cdot 3^2 + 2^3 \cdot 3^2 + 2^4 \cdot 3^2 = (1 + 2 + 2^2 + 2^3 + 2^4)3^2 = 31(3^2)$$

The sum of *all* the factors (including 432) is the sum of the last row.

$$\text{Sum of all factors} = 31 + 31 \cdot 3 + 31 \cdot 3^2 + 31 \cdot 3^3 = 31(1 + 3 + 3^2 + 3^3) = 31(40) = 1240$$

To find the sum of the *proper* factors, we must subtract 432 from 1240; we get 808 (see Take Note). Thus 432 is abundant.

We now look for some pattern for the sum of all the factors. Note that

Sum of left column $\quad 1 + 2 + 2^2 + 2^3 + 2^4 = $ ⌐————————⌐ $ = 1 + 3 + 3^2 + 3^3 \quad$ Sum of top row

$$31(40) = 1240$$

This suggests that the sum of the proper factors can be found by finding the sum of all the prime power factors for each prime, multiplying those numbers, and then subtracting the original number. Although we have not proved this for all cases, it is a true statement.

For instance, to find the sum of the proper factors of 3240, first find the prime factorization.

$$3240 = 2^3 \cdot 3^4 \cdot 5$$

Now find the following sums:

$$1 + 2 + 2^2 + 2^3 = 15 \qquad 1 + 3 + 3^2 + 3^3 + 3^4 = 121 \qquad 1 + 5 = 6$$

The sum of the proper factors = $(15)(121)6 - 3240 = 7650$. Thus 3240 is abundant.

Determine whether the number is deficient, abundant, or perfect.

1. 200   **2.** 3125   **3.** 8128   **4.** 10,000

5. Is a prime number deficient, abundant, or perfect?

Projects and Group Activities

Finding a Pattern

The Focus on Problem Solving involved finding the sum of $1 + 3 + 3^2$ and $1 + 2 + 2^2 + 2^3$. For sums that contain a larger number of terms, it may be difficult or time-consuming to try to evaluate the sum. Perhaps there is a pattern for these sums that we can use to calculate them without having to evaluate each exponential expression and then add the results.

Look at the following from the calculation of the sum of the factors of 3240.

$$1 + 2 + 2^2 + 2^3 = 1 + 2 + 4 + 8 = 15$$

$$\frac{2^{3+1} - 1}{2 - 1} = \frac{2^4 - 1}{1} = 16 - 1 = 15$$

$$1 + 3 + 3^2 + 3^3 + 3^4 = 1 + 3 + 9 + 27 + 81 = 121$$

$$\frac{3^{4+1} - 1}{3 - 1} = \frac{3^5 - 1}{2} = \frac{243 - 1}{2} = \frac{242}{2} = 121$$

Consider another sum of this type.

$$1 + 7 + 7^2 + 7^3 + 7^4 + 7^5 = 1 + 7 + 49 + 343 + 2401 + 16,807 = 19,608$$

$$\frac{7^{5+1} - 1}{7 - 1} = \frac{7^6 - 1}{6} = \frac{117,649 - 1}{6} = \frac{117,648}{6} = 19,608$$

On the basis of the examples shown above, make a conjecture about the value of $1 + n + n^2 + n^3 + n^4 + \cdots + n^k$, where n and k are natural numbers greater than 1.

Solving a System of Equations with a Graphing Calulator

A graphing calculator can be used to approximate the solution of a system of equations in two variables. Graph each equation of the system of equations, and then approximate the coordinates of the point of intersection. The process by which you approximate the solution depends on what model of calculator you have. In all cases, however, you must first solve each equation in the system of equations for y.

Solve: $2x - 5y = 9$
$\quad 4x + 3y = 2$

$2x - 5y = 9$ $\qquad$ $4x + 3y = 2$ $\qquad$ • **Solve each equation for *y*.**

$-5y = -2x + 9$ $\qquad$ $3y = -4x + 2$

$y = \dfrac{2}{5}x - \dfrac{9}{5}$ $\qquad$ $y = -\dfrac{4}{3}x + \dfrac{2}{3}$

For the TI-83, TI-83 Plus, or TI-84 Plus, press **Y=**. Enter one equation as Y1 and the other as Y2. The result should be similar to the screen at the left below. Press **GRAPH**. The graphs of the two equations should appear on the screen, as shown at the right below. If the point of intersection is not on the screen, adjust the viewing window by pressing the **WINDOW** key.

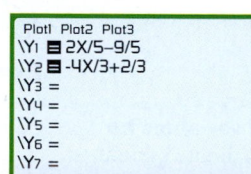

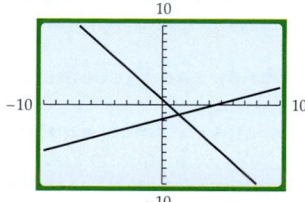

Integrating Technology

See the appendix Keystroke Guide: *Intersect* for instructions on using a graphing calculator to solve systems of equations.

Press **2nd** CALC 5 **ENTER** **ENTER** **ENTER**. After a few seconds, the point of intersection will show on the bottom of the screen as X = 1.4230769, Y = −1.230769.

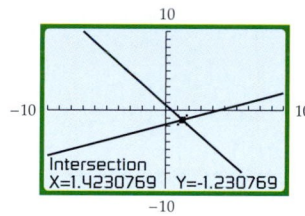

Answers to Projects and Group Activities: Solving a System of Equations with a Graphing Calculator

1. (1.716981132, −0.2264150943)

2. (2.875, 1.1875)

3. (0.3636363636, −0.5454545455)

4. (−19, −7)

For Exercises 1 to 4, solve by using a graphing calculator.

1. $4x - 5y = 8$
$\ 5x + 7y = 7$

2. $3x + 2y = 11$
$\ 7x - 6y = 13$

3. $x = 3y + 2$
$\ y = 4x - 2$

4. $x = 2y - 5$
$\ x = 3y + 2$

Chapter 13 Summary

Key Words

Examples

Two or more equations considered together are called a *system of equations*. [13.1A, p. 677]

An example of a system of equations is
$$2x - 3y = 9$$
$$3x + 4y = 5$$

A *solution of a system of equations in two variables* is an ordered pair that is a solution of each equation of the system. [13.1A, p. 677]

The solution of the system of equations shown above is the ordered pair (3, −1) because it is a solution of each equation of the system of equations.

An *independent system* of linear equations has exactly one solution. The graphs of the equations in an independent system of linear equations intersect at one point. [13.1A, p. 678]

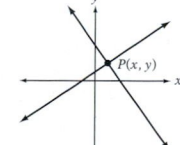

A *dependent system* of linear equations has an infinite number of solutions. The graphs of the equations in a dependent system of linear equations are the same line. [13.1A, p. 678]

If, when solving a system of equations algebraically, the variable is eliminated and the result is a true equation, such as $5 = 5$, then the system of equations is dependent. [13.2A, p. 687]

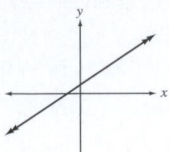

An *inconsistent system* of linear equations has no solution. The graphs of the equations of an inconsistent system of linear equations are parallel lines. [13.1A, p. 678]

If, when solving a system of equations algebraically, the variable is eliminated and the result is a false equation, such as $0 = 4$, then the system of equations is inconsistent. [13.2A, p. 687]

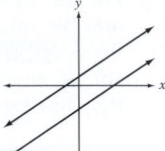

Essential Rules and Procedures

Examples

To solve a system of linear equations in two variables by graphing, graph each equation on the same coordinate system, and then determine the point of intersection. [13.1A, p. 679]

Solve by graphing: $x + 2y = 4$
$\qquad\qquad\qquad\quad 2x + y = -1$

The solution is $(-2, 3)$.

To solve a system of equations by the substitution method, one variable must be written in terms of the other variable. [13.2A, p. 685]

Solve by substitution: $2x + y = 5$ (1)
$\qquad\qquad\qquad\qquad\quad 3x - 2y = 11$ (2)
$2x + y = 5$
$\qquad y = -2x + 5$ • **Solve Equation (1) for *y*.**

$\qquad\quad 3x - 2y = 11$ • **Equation (2)**
$3x - 2(-2x + 5) = 11$ • **Substitute for *y***
$\qquad 3x + 4x - 10 = 11$ **in Equation (2).**
$\qquad\qquad 7x - 10 = 11$
$\qquad\qquad\qquad 7x = 21$
$\qquad\qquad\qquad\ x = 3$
$y = -2x + 5$
$y = -2(3) + 5$ • ***x* = 3**
$y = -1$
The solution is $(3, -1)$.

To solve a system of linear equations by the addition method, use the Multiplication Property of Equations to rewrite one or both of the equations so that the coefficients of one variable are opposites. Then add the equations and solve for the variables. [13.3A, p. 695]

Solve by the addition method:
$2x + 5y = 8$ (1)
$3x - 4y = -11$ (2)
$\ \ 6x + 15y = 24$ • **3 times Equation (1)**
$-6x + 8y = 22$ • **−2 times Equation (2)**
$\qquad\quad 23y = 46$ • **Add the equations.**
$\qquad\qquad y = 2$ • **Solve for *y*.**
$2x + 5y = 8$
$2x + 5(2) = 8$ • **Replace *y* with 2 in Eq. (1).**
$2x + 10 = 8$ • **Solve for *x*.**
$\qquad 2x = -2$
$\qquad\ x = -1$
The solution is $(-1, 2)$.

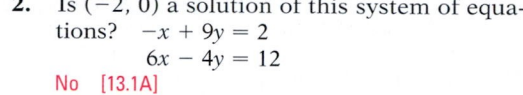

Chapter 13 Review Exercises

1. Is $(-1, -3)$ a solution of this system of equations? $5x + 4y = -17$
$2x - y = 1$
Yes   [13.1A]

2. Is $(-2, 0)$ a solution of this system of equations? $-x + 9y = 2$
$6x - 4y = 12$
No   [13.1A]

3. Solve by graphing:
$3x - y = 6$
$y = -3$

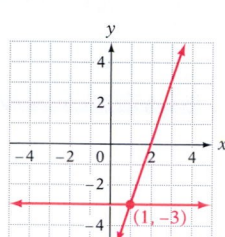

$(1, -3)$ [13.1A]

4. Solve by graphing:
$4x - 2y = 8$
$y = 2x - 4$

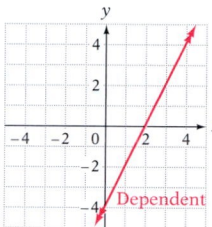

Dependent

The solutions are the ordered pairs that satisfy the equation $y = 2x - 4$.   [13.1A]

5. Solve by graphing:
$x + 2y = 3$
$y = -\dfrac{1}{2}x + 1$

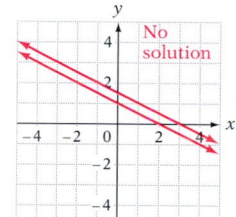

No solution

[13.1A]

6. Solve by substitution:
$4x + 7y = 3$
$x = y - 2$
$(-1, 1)$ [13.2A]

7. Solve by substitution:
$6x - y = 0$
$7x - y = 1$
$(1, 6)$ [13.2A]

8. Solve by the addition method:
$3x + 8y = -1$
$x - 2y = -5$
$(-3, 1)$ [13.3A]

9. Solve by the addition method:
$6x + 4y = -3$
$12x - 10y = -15$
$\left(-\dfrac{5}{6}, \dfrac{1}{2}\right)$ [13.3A]

10. Solve by substitution:
$12x - 9y = 18$
$y = \dfrac{4}{3}x - 3$
No solution   [13.2A]

11. Solve by substitution:
$8x - y = 2$
$y = 5x + 1$
$(1, 6)$ [13.2A]

12. Solve by the addition method:
$4x - y = 9$
$2x + 3y = -13$
$(1, -5)$ [13.3A]

13. Solve by the addition method:
$5x + 7y = 21$
$20x + 28y = 63$
No solution   [13.3A]

14. Solve by substitution:
$4x + 3y = 12$
$y = -\dfrac{4}{3}x + 4$
Dependent. The solutions satisfy the equation $y = -\dfrac{4}{3}x + 4$.
[13.2A]

15. Solve by substitution:
$7x + 3y = -16$
$x - 2y = 5$
$(-1, -3)$ [13.2A]

16. Solve by the addition method:
$3x + y = -2$
$-9x - 3y = 6$
Dependent. The solutions satisfy the equation $3x + y = -2$.
[13.3A]

17. Solve by the addition method:
$6x - 18y = 7$
$9x + 24y = 2$
$\left(\dfrac{2}{3}, -\dfrac{1}{6}\right)$ [13.3A]

18. Sculling   A sculling team rowing with the current went 24 mi in 2 h. Rowing against the current, the sculling team went 18 mi in 3 h. Find the rate of the sculling team in calm water and the rate of the current.
Sculling team: 9 mph; current: 3 mph   [13.4A]

19. Investments   An investor bought 1500 shares of stock, some at $6 per share and the rest at $25 per share. If $12,800 worth of stock was purchased, how many shares of each kind did the investor buy?
1300 shares at $6; 200 shares at $25   [13.4B]

20. Travel   A flight crew flew 420 km in 3 h with a tailwind. Flying against the wind, the flight crew flew 440 km in 4 h. Find the rate of the flight crew in calm air and the rate of the wind.
Flight crew: 125 km/h; wind: 15 km/h   [13.4A]

21. Travel   A small plane flying with the wind flew 360 mi in 3 h. Against a headwind, it took the plane 4 h to fly the same distance. Find the rate of the plane in calm air and the rate of the wind.
Plane: 105 mph; wind: 15 mph   [13.4A]

22. Postage   A small wood-carving company mailed 190 advertisements, some requiring $.25 postage and others requiring $.45. The total cost for mailing was $59.50. Find the number of advertisements mailed at each rate.
Number of ads requiring $.25: 130; number of ads requiring $.45: 60   [13.4B]

23. Investments   Terra Cotta Art Center receives an annual interest income of $915 from two simple interest investments. One investment, in a corporate bond fund, earns 8.5% annual simple interest. The second investment, in a real estate investment trust, earns 7% annual simple interest. If the total amount invested in the two accounts is $12,000, how much is invested in each account?
$7000 at 7%; $5000 at 8.5%   [13.2B]

24. Grain Mixtures   A silo contains a mixture of lentils and corn. If 50 bushels of lentils were added, there would be twice as many bushels of lentils as of corn. If 150 bushels of corn were added instead, there would be the same amount of corn as of lentils. How many bushels of each were originally in the silo?
350 bushels of lentils; 200 bushels of corn   [13.4B]

25. Investments   Mosher Children's Hospital received a $300,000 donation that it invested in two simple interest accounts, one earning 5.4% and the other earning 6.6%. If each account earned the same amount of annual interest, how much was invested in each account?
$165,000 at 5.4%; $135,000 at 6.6%   [13.2B]

Chapter 13 Test

1. Is $(-2, 3)$ a solution of this system?
$$2x + 5y = 11$$
$$x + 3y = 7$$
Yes   [13.1A]

2. Is $(1, -3)$ a solution of this system?
$$3x - 2y = 9$$
$$4x + y = 1$$
Yes   [13.1A]

3. Solve by graphing: $3x + 2y = 6$
                              $5x + 2y = 2$

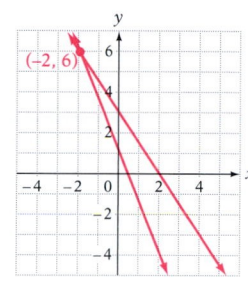

[13.1A]

4. Solve by substitution:
$$4x - y = 11$$
$$y = 2x - 5$$
$(3, 1)$ [13.2A]

5. Solve by substitution:
$$x = 2y + 3$$
$$3x - 2y = 5$$
$(1, -1)$ [13.2A]

6. Solve by substitution:
$$3x + 5y = 1$$
$$2x - y = 5$$
$(2, -1)$ [13.2A]

7. Solve by substitution:
$$3x - 5y = 13$$
$$x + 3y = 1$$
$\left(\dfrac{22}{7}, -\dfrac{5}{7}\right)$ [13.2A]

8. Solve by substitution:
$$2x - 4y = 1$$
$$y = \dfrac{1}{2}x + 3$$
No solution   [13.2A]

9. Solve by the addition method:
$$4x + 3y = 11$$
$$5x - 3y = 7$$
$(2, 1)$ [13.3A]

10. Solve by the addition method:
$$2x - 5y = 6$$
$$4x + 3y = -1$$
$\left(\dfrac{1}{2}, -1\right)$ [13.3A]

11. Solve by the addition method:
$x + 2y = 8$
$3x + 6y = 24$
Dependent. The solutions satisfy the equation
$x + 2y = 8$.   [13.3A]

12. Solve by the addition method:
$7x + 3y = 11$
$2x - 5y = 9$
$(2, -1)$ [13.3A]

13. Solve by the addition method:
$5x + 6y = -7$
$3x + 4y = -5$
$(1, -2)$ [13.3A]

14. **Travel**   With the wind, a plane flies 240 mi in 2 h. Against the wind, the plane requires 3 h to fly the same distance. Find the rate of the plane in calm air and the rate of the wind.
Plane: 100 mph; wind: 20 mph   [13.4A]

15. **Entertainment**   For the first performance of a play in a community theater, 50 reserved-seat tickets and 80 general-admission tickets were sold. The total receipts were $980. For the second performance, 60 reserved-seat tickets and 90 general-admission tickets were sold. The total receipts were $1140. Find the price of a reserved-seat ticket and the price of a general-admission ticket.
Reserved seat: $10; general admission: $6   [13.4B]

16. **Investments**   Bernardo Community Library received a $28,000 donation that it invested in two accounts, one earning 7.6% simple interest and the other earning 6.4% simple interest. If both accounts earned the same amount of annual interest, how much was invested in each account?
$15,200 at 6.4%; $12,800 at 7.6%   [13.2B]

Cumulative Review Exercises

1. Evaluate $\dfrac{a^2 - b^2}{2a}$ when $a = 4$ and $b = -2$.

$\dfrac{3}{2}$ [4.1A]

2. Solve: $-\dfrac{3}{4}x = \dfrac{9}{8}$

$-\dfrac{3}{2}$ [5.1C]

3. Given $f(x) = x^2 + 2x - 1$, find $f(2)$.

7 [12.1D]

4. Multiply: $(2a^2 - 3a + 1)(2 - 3a)$

$-6a^3 + 13a^2 - 9a + 2$ [9.3B]

5. Simplify: $\dfrac{(-2x^2y)^4}{-8x^3y^2}$

$-2x^5y^2$ [9.4A]

6. Divide: $(4b^2 - 8b + 4) \div (2b - 3)$

$2b - 1 + \dfrac{1}{2b - 3}$ [9.5B]

7. Simplify: $\dfrac{8x^{-2}y^5}{-2xy^4}$

$-\dfrac{4y}{x^3}$ [9.4A]

8. Factor: $4x^2y^4 - 64y^2$

$4y^2(xy - 4)(xy + 4)$ [10.4B]

9. Solve: $(x - 5)(x + 2) = -6$

$4, -1$ [10.5A]

10. Divide: $\dfrac{x^2 - 6x + 8}{2x^3 + 6x^2} \div \dfrac{2x - 8}{4x^3 + 12x^2}$

$x - 2$ [11.1C]

11. Add: $\dfrac{x - 1}{x + 2} + \dfrac{2x + 1}{x^2 + x - 2}$

$\dfrac{x^2 + 2}{(x + 2)(x - 1)}$ [11.3B]

12. Simplify: $\dfrac{x + 4 - \dfrac{7}{x - 2}}{x + 8 + \dfrac{21}{x - 2}}$

$\dfrac{x - 3}{x + 1}$ [11.4A]

13. Solve: $\dfrac{x}{2x - 3} + 2 = \dfrac{-7}{2x - 3}$

$-\dfrac{1}{5}$ [11.5A]

14. Solve $A = P + Prt$ for r.

$r = \dfrac{A - P}{Pt}$ [11.6A]

15. Find the x- and y-intercepts for $2x - 3y = 12$.

x-intercept: $(6, 0)$; y-intercept: $(0, -4)$ [12.3A]

16. Find the slope of the line that passes through the points $(2, -3)$ and $(-3, 4)$.

$-\dfrac{7}{5}$ [12.3B]

17. Find the equation of the line that passes through the point $(-2, 3)$ and has slope $-\dfrac{3}{2}$.

$y = -\dfrac{3}{2}x$ [12.4A]

18. Is $(2, 0)$ a solution of this system?
$$5x - 3y = 10$$
$$4x + 7y = 8$$

Yes [13.1A]

19. Solve by substitution:
$$3x - 5y = -23$$
$$x + 2y = -4$$
(−6, 1)   [13.2A]

20. Solve by the addition method:
$$5x - 3y = 29$$
$$4x + 7y = -5$$
(4, −3)   [13.3A]

21. **Investments**   A total of $8750 is invested in two accounts. On one account, the annual simple interest rate is 9.6%; on the second account, the annual simple interest rate is 7.2%. How much should be invested in each account so that both accounts earn the same interest?
$3750 at 9.6%; $5000 at 7.2%   [13.2B]

22. **Travel**   A passenger train leaves a train depot $\frac{1}{2}$ h after a freight train leaves the same depot. The freight train is traveling 8 mph slower than the passenger train. Find the rate of each train if the passenger train overtakes the freight train in 3 h.
Freight train: 48 mph; passenger train: 56 mph   [5.5B]

23. **Geometry**   The length of each side of a square is extended 4 in. The area of the resulting square is 144 in². Find the length of a side of the original square.
8 in.   [10.5B]

24. **Travel**   A plane can travel 160 mph in calm air. Flying with the wind, the plane can fly 570 mi in the same amount of time it takes to fly 390 mi against the wind. Find the rate of the wind.
30 mph   [13.4A]

25. Graph $2x - 3y = 6$.

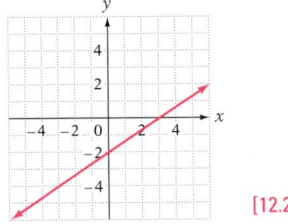

[12.2B]

26. Solve by graphing: $3x + 2y = 6$
$$3x - 2y = 6$$

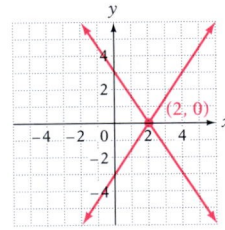

(2, 0)

[13.1A]

27. **Travel**   With the current, a motorboat can travel 48 mi in 3 h. Against the current, the boat requires 4 h to travel the same distance. Find the rate of the boat in calm water.
14 mph   [13.4A]

28. 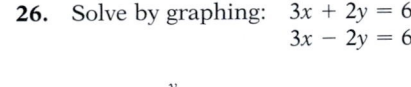 **Registered Voters**   In a recent year, the U.S. voting-age population was 205 million people, but only 156 million Americans were registered to vote. (*Source:* The Election Center) What percent of the voting-age population was registered to vote? Round to the nearest percent.
76%   [6.4C]

14

Inequalities

You know how to determine the average of four exam scores when each exam has the same weight: add the four scores and divide the sum by 4. But do you know how to determine what score you must receive on a fifth exam to earn a specific average for the five exams? In this chapter you will learn how to use an inequality to determine that fifth and final score. **Exercises 86 and 87 on page 732** give you different scores for the first three or four tests, and ask you to determine the score needed on the fourth or fifth test in order to achieve a specific average.

OBJECTIVES

Section 14.1

A To write a set using the roster method
B To write a set using set-builder notation
C To graph an inequality on the number line

Section 14.2

A To solve an inequality using the Addition Property of Inequalities
B To solve an inequality using the Multiplication Property of Inequalities
C To solve application problems

Section 14.3

A To solve general inequalities
B To solve application problems

Section 14.4

A To graph an inequality in two variables

Need help? For online student resources, such as section quizzes, visit this textbook's website at **math.college.hmco.com/students**.

PREP TEST • • •

Do these exercises to prepare for Chapter 14.

1. Place the correct symbol, $<$ or $>$, between the two numbers.
   $-45 < -27$ [3.1A]

2. Simplify: $3x - 5(2x - 3)$
   $-7x + 15$ [4.2D]

3. State the Addition Property of Equations.
   The same number can be added to each side of an equation without changing the solution of the equation. [5.1B]

4. State the Multiplication Property of Equations.
   Each side of an equation can be multiplied by the same nonzero number without changing the solution of the equation. [5.1C]

5. **Nutrition** A certain grade of hamburger contains 15% fat. How many pounds of fat are in 3 lb of this hamburger?
   0.45 lb [6.4C]

6. Solve: $4x - 5 = -7$
   $-\dfrac{1}{2}$ [5.2A]

7. Solve: $4 = 2 - \dfrac{3}{4}x$
   $-\dfrac{8}{3}$ [5.2A]

8. Solve: $7 - 2(2x - 3) = 3x - 1$
   2 [5.3B]

9. Graph: $y = \dfrac{2}{3}x - 3$
   [12.2A]

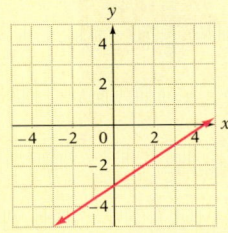

10. Graph: $3x + 4y = 12$
    [12.2B]

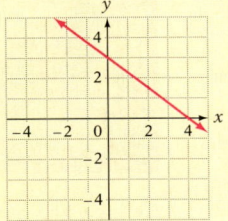

GO FIGURE • • •

Without using a calculator, which is the largest number: 2^{150}, 3^{100}, or 5^{50}?
3^{100}

14.1 Sets

Objective A   To write a set using the roster method

A **set** is a collection of objects, which are called the **elements** of the set. The **roster method** of writing a set encloses a list of the elements in braces.

The set of the last three letters of the alphabet is written {x, y, z}.

The set of the positive integers less than 5 is written {1, 2, 3, 4}.

> **HOW TO**   Use the roster method to write the set of integers between 0 and 10.
>
> $A = \{1, 2, 3, 4, 5, 6, 7, 8, 9\}$ • A set can be designated by a capital letter. Note that 0 and 10 are not elements of the set.

> **HOW TO**   Use the roster method to write the set of natural numbers.
>
> $A = \{1, 2, 3, 4, \ldots\}$ • The three dots mean that the pattern of numbers continues without end.

The **empty set,** or **null set,** is the set that contains no elements. The symbol $\varnothing$ or { } is used to represent the empty set.

The set of people who have run a 2-minute mile is the empty set.

> **Union of Two Sets**
>
> The **union** of two sets, written $A \cup B$, is the set that contains the elements of A and the elements of B.

> **HOW TO**   Find $A \cup B$, given $A = \{1, 2, 3, 4\}$ and $B = \{3, 4, 5, 6\}$.
>
> $A \cup B = \{1, 2, 3, 4, 5, 6\}$ • The union of A and B contains all the elements of A and all the elements of B. Elements in both sets are listed only once.

> **Intersection of Two Sets**
>
> The **intersection** of two sets, written $A \cap B$, is the set that contains the elements that are common to both A and B.

> **HOW TO**   Find $A \cap B$, given $A = \{1, 2, 3, 4\}$ and $B = \{3, 4, 5, 6\}$.
>
> $A \cap B = \{3, 4\}$ • The intersection of A and B contains the elements common to A and B.

Example 1

Use the roster method to write the set of the odd positive integers less than 12.

Solution
$A = \{1, 3, 5, 7, 9, 11\}$

You Try It 1

Use the roster method to write the set of the odd negative integers greater than -10.

Your solution
$A = \{-9, -7, -5, -3, -1\}$

Solution on p. S35

Objective 14.1A

New Vocabulary
set
element
roster method
empty set
null set
union of sets A and B
intersection of sets A and B

Instructor Note
Students want to write the empty set as {$\varnothing$}. It is very difficult to convince them that this is incorrect. Students, in general, have a difficult time with the idea that a set can be an element of another set.

Concept Check
Let N be the set of all positive even integers and let O be the set of all positive odd integers.
1. What is $N \cup O$?
   The positive integers
2. What is $N \cap O$? $\varnothing$

Discuss the Concepts
Decide whether each statement is true or false. Explain your reasoning.
1. For any set A, $A \cup \varnothing = A$.
   True
2. For any set A, $A \cap \varnothing = A$.
   False

In-Class Examples (Objective 14.1A)

1. Use the roster method to write the set of even integers between 20 and 30. $A = \{22, 24, 26, 28\}$
2. Find $A \cup B$ and $A \cap B$, given that $A = \{a, e, i, o, u\}$ and $B = \{a, b, c, d, e\}$.
   $\{a, b, c, d, e, i, o, u\}$; $\{a, e\}$
3. Find $C \cup D$ and $C \cap D$, given that $C = \{10, 20, 30, 40\}$ and $D = \{5, 15, 25, 35\}$.
   $\{5, 10, 15, 20, 25, 30, 35, 40\}$; $\varnothing$

Optional Student Activity

1. Find two sets A and B such that $A \cup B = \{-4, -2, 0, 1, 2, 3, 4, 5, 6\}$ and $A \cap B = \{2, 4\}$.
   For example,
   $A = \{1, 2, 3, 4, 5, 6\}$ and
   $B = \{-4, -2, 0, 2, 4\}$

2. Find two sets C and D such that $C \cup D = \{10, 20, 30, 40, 50, 60, 70, 80\}$ and $C \cap D = \varnothing$.
   For example,
   $C = \{10, 20, 30, 40\}$ and
   $D = \{50, 60, 70, 80\}$

Optional Student Activity

Use these sets:
$A = \{1, 2, 3, 4, 5, 6\}$
$B = \{2, 4, 6, 8, 10\}$
$C = \{1, 3, 5, 7, 9\}$

1. Find $(A \cup B) \cap C$. $\{1, 3, 5\}$
2. Find $A \cup (B \cap C)$. A
3. Find $(A \cap C) \cup B$.
   $\{1, 2, 3, 4, 5, 6, 8, 10\}$
4. Find $A \cap (C \cup B)$. A

Example 2

Use the roster method to write the set of the even positive integers.

Solution
$A = \{2, 4, 6, \ldots\}$

You Try It 2

Use the roster method to write the set of the odd positive integers.

Your solution
$A = \{1, 3, 5, \ldots\}$

Example 3

Find $D \cup E$, given $D = \{6, 8, 10, 12\}$ and $E = \{-8, -6, 10, 12\}$.

Solution
$D \cup E = \{-8, -6, 6, 8, 10, 12\}$

You Try It 3

Find $A \cup B$, given $A = \{-2, -1, 0, 1, 2\}$ and $B = \{0, 1, 2, 3, 4\}$.

Your solution
$A \cup B = \{-2, -1, 0, 1, 2, 3, 4\}$

Example 4

Find $A \cap B$, given $A = \{5, 6, 9, 11\}$ and $B = \{5, 9, 13, 15\}$.

Solution
$A \cap B = \{5, 9\}$

You Try It 4

Find $C \cap D$, given $C = \{10, 12, 14, 16\}$ and $D = \{10, 16, 20, 26\}$.

Your solution
$C \cap D = \{10, 16\}$

Example 5

Find $A \cap B$, given $A = \{1, 2, 3, 4\}$ and $B = \{8, 9, 10, 11\}$.

Solution
$A \cap B = \varnothing$

You Try It 5

Find $A \cap B$, given $A = \{-5, -4, -3, -2\}$ and $B = \{2, 3, 4, 5\}$.

Your solution
$A \cap B = \varnothing$

Solutions on p. S35

Objective 14.1B

New Vocabulary

set-builder notation

Concept Check

Use these two sets, which are written in set-builder notation.
$A = \{x \,|\, x > -2, x \in \text{integers}\}$
$B = \{x \,|\, x < 4, x \in \text{integers}\}$

1. Read each set out loud.
2. Find $A \cup B$. Write your answer using set-builder notation. $\{x \,|\, x \in \text{integers}\}$
3. Find $A \cap B$. Write your answer using the roster method. $\{-1, 0, 1, 2, 3\}$

Objective B   To write a set using set-builder notation

Point of Interest

The symbol $\in$ was first used in the book *Arithmeticae Principia*, published in 1889. It is the first letter of the Greek word $\varepsilon\sigma\tau\iota$, which means "is." The symbols for union and intersection were also introduced at that time.

Another method of representing sets is called **set-builder notation.** This method of writing a set uses a rule to describe the elements of the set. Using set-builder notation, we represent the set of all positive integers less than 10 as

$$\{x \,|\, x < 10, x \in \text{positive integers}\},$$ which is read "the set of all x such that x is less than 10 and x is an element of the positive integers."

HOW TO   Use set-builder notation to write the set of real numbers greater than 4.

$$\{x \,|\, x > 4, x \in \text{real numbers}\}$$

• "$x \in$ real numbers" is read "x is an element of the real numbers."

In-Class Examples (Objective 14.1B)

Use set-builder notation to write each set.

1. The even integers greater than 10
   $\{x \,|\, x > 10, x \in \text{even integers}\}$
2. The positive real numbers less than 25
   $\{x \,|\, x < 25, x \in \text{positive real numbers}\}$
3. The real numbers greater than -12
   $\{x \,|\, x > -12, x \in \text{real numbers}\}$
4. The integers less than -100
   $\{x \,|\, x < -100, x \in \text{integers}\}$

Example 6

Use set-builder notation to write the set of negative integers greater than −100.

Solution
$\{x \mid x > -100, x \in \text{negative integers}\}$

You Try It 6

Use set-builder notation to write the set of positive even integers less than 59.

Your solution
$\{x \mid x < 59, x \in \text{positive even integers}\}$

Example 7

Use set-builder notation to write the set of real numbers less than 60.

Solution
$\{x \mid x < 60, x \in \text{real numbers}\}$

You Try It 7

Use set-builder notation to write the set of real numbers greater than −3.

Your solution
$\{x \mid x > -3, x \in \text{real numbers}\}$

Solutions on p. S35

Objective C   To graph an inequality on the number line

An expression that contains the symbol $>, <, \geq$ (is greater than or equal to), or $\leq$ (is less than or equal to) is called an **inequality.** An inequality expresses the relative order of two mathematical expressions. The expressions can be either numerical or variable.

$$4 > 2$$
$$3x \leq 7$$
$$x^2 - 2x > y + 4$$

Inequalities

The **solution set of an inequality** is the set of numbers each element of which, when substituted for the variable, results in a true inequality. The solution set of an inequality can be graphed on the number line.

TAKE NOTE

For the remainder of this section, all variables will represent real numbers. Given this convention, the expression $\{x \mid x > 1, x \in \text{real numbers}\}$ will be written as $\{x \mid x > 1\}$, as shown in the example at the right.

HOW TO   Graph: $\{x \mid x > 1\}$

The graph is the real numbers greater than 1. The parenthesis at 1 indicates that 1 is not included in the graph.

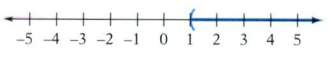

HOW TO   Graph: $\{x \mid x \geq 1\}$

The bracket at 1 indicates that 1 is included in the solution set.

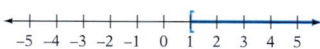

HOW TO   Graph: $\{x \mid -1 > x\}$

$-1 > x$ is equivalent to $x < -1$. The numbers less than −1 are to the left of −1 on the number line.

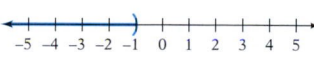

The union of two sets is the set that contains all the elements of each set.

HOW TO   Graph: $\{x \mid x > 4\} \cup \{x \mid x < 1\}$

The graph is the numbers greater than 4 and the numbers less than 1.

Objective 14.1C

Vocabulary to Review
inequality   [1.1A]

New Vocabulary
solution set of an inequality

Discuss the Concepts
Refer to the example at the left, in which the inequality $\{x \mid -1 > x\}$ is graphed. Discuss why $-1 > x$ is equivalent to $x < -1$.

Concept Check
Rewrite each of the following inequalities so that the variable appears on the left.
1. $2 < x$ $x > 2$
2. $-3 \geq a$ $a \leq -3$
3. $6 \leq n$ $n \geq 6$
4. State a rule for switching the sides of an inequality.
When the two sides of an inequality are switched, the inequality symbol must be reversed; for example, if $a > b$, then $b < a$.

Instructor Note
Students need to distinguish between $>$ and $\geq$ and between $<$ and $\leq$. Use the Concept Check and Discuss the Concepts questions on the next page to help students understand the important differences between these pairs of symbols.

In-Class Examples (Objective 14.1C)
Graph.
1. $\{x \mid x \geq -3\}$

3. $\{x \mid x < -1\} \cup \{x \mid x > 3\}$

2. $\{x \mid x \geq 0\} \cap \{x \mid x \leq 5\}$

Concept Check

Which of the numbers 3, 2.5, and 2 satisfy the given inequality?

1. $x > 2$ 3, 2.5

2. $x \le 3$ 3, 2.5, 2

Discuss the Concepts

Have students describe when to use a parenthesis and when to use a bracket when graphing an inequality.

Optional Student Activity

1. Write an inequality that describes the graph.

a.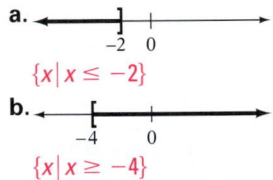

$\{x \mid x \le -2\}$

b.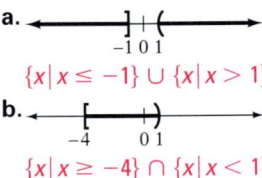

$\{x \mid x \ge -4\}$

2. Use set-builder notation to write a union or intersection of two inequalities that describes the graph.

a.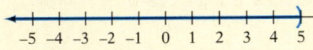

$\{x \mid x \le -1\} \cup \{x \mid x > 1\}$

b.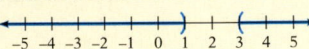

$\{x \mid x \ge -4\} \cap \{x \mid x < 1\}$

Optional Student Activity

1. Write an inequality that describes the given situation.

a. To avoid monthly fees, one must maintain a minimum balance, b, of $1000.

$b \ge 1000$

b. The temperature, t, never got above freezing (32°F).

$t \le 32$

2. Use set-builder notation to write a union or intersection of two inequalities that describes this situation: The daily high temperatures in March ranged from 18°F to 57°F.

$\{t \mid t \ge 18\} \cap \{t \mid t \le 57\}$

3. Make up your own examples like the ones in parts 1 and 2.

The intersection of two sets is the set that contains the elements common to both sets.

HOW TO   Graph: $\{x \mid x > -1\} \cap \{x \mid x < 2\}$

The graphs of $\{x \mid x > -1\}$ and $\{x \mid x < 2\}$ are shown at the right.

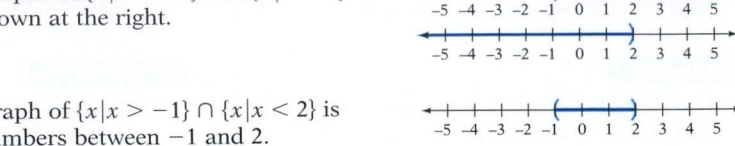

The graph of $\{x \mid x > -1\} \cap \{x \mid x < 2\}$ is the numbers between -1 and 2.

Example 8

Graph: $\{x \mid x < 5\}$

Solution
The solution set is the numbers less than 5.

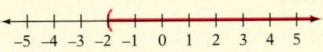

You Try It 8

Graph: $\{x \mid -2 < x\}$

Your solution

Example 9

Graph: $\{x \mid x > -2\} \cap \{x \mid x < 1\}$

Solution
The solution set is the numbers between -2 and 1.

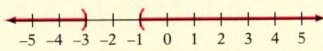

You Try It 9

Graph: $\{x \mid x > -1\} \cup \{x \mid x < -3\}$

Your solution

Example 10

Graph: $\{x \mid x > 3\} \cup \{x \mid x < 1\}$

Solution
The solution set is the numbers greater than 3 and the numbers less than 1.

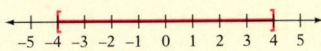

You Try It 10

Graph: $\{x \mid x \le 4\} \cap \{x \mid x \ge -4\}$

Your solution

Example 11

Graph: $\{x \mid x \le 5\} \cup \{x \mid x \ge -3\}$

Solution
The solution set is the real numbers.

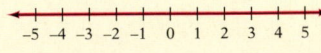

You Try It 11

Graph: $\{x \mid x < 2\} \cup \{x \mid x \ge -2\}$

Your solution

Solutions on p. S35

14.1 Exercises

Objective A   To write a set using the roster method

1. ✏️ Explain how to find the union of two sets.

2. ✏️ Explain how to find the intersection of two sets.

For Exercises 3 to 8, use the roster method to write the set.

3. The integers between 15 and 22
   $A = \{16, 17, 18, 19, 20, 21\}$

4. The integers between -10 and -4
   $A = \{-9, -8, -7, -6, -5\}$

5. The odd integers between 8 and 18
   $A = \{9, 11, 13, 15, 17\}$

6. The even integers between -11 and -1
   $A = \{-10, -8, -6, -4, -2\}$

7. The letters of the alphabet between a and d
   $A = \{b, c\}$

8. The letters of the alphabet between p and v
   $A = \{q, r, s, t, u\}$

For Exercises 9 to 16, find $A \cup B$.

9. $A = \{3, 4, 5\}$ $B = \{4, 5, 6\}$
   $A \cup B = \{3, 4, 5, 6\}$

10. $A = \{-3, -2, -1\}$ $B = \{-2, -1, 0\}$
    $A \cup B = \{-3, -2, -1, 0\}$

11. $A = \{-10, -9, -8\}$ $B = \{8, 9, 10\}$
    $A \cup B = \{-10, -9, -8, 8, 9, 10\}$

12. $A = \{a, b, c\}$ $B = \{x, y, z\}$
    $A \cup B = \{a, b, c, x, y, z\}$

13. $A = \{a, b, d, e\}$ $B = \{c, d, e, f\}$
    $A \cup B = \{a, b, c, d, e, f\}$

14. $A = \{m, n, p, q\}$ $B = \{m, n, o\}$
    $A \cup B = \{m, n, o, p, q\}$

15. $A = \{1, 3, 7, 9\}$ $B = \{7, 9, 11, 13\}$
    $A \cup B = \{1, 3, 7, 9, 11, 13\}$

16. $A = \{-3, -2, -1\}$ $B = \{-1, 1, 2\}$
    $A \cup B = \{-3, -2, -1, 1, 2\}$

For Exercises 17 to 22, find $A \cap B$.

17. $A = \{3, 4, 5\}$ $B = \{4, 5, 6\}$
    $A \cap B = \{4, 5\}$

18. $A = \{-4, -3, -2\}$ $B = \{-6, -5, -4\}$
    $A \cap B = \{-4\}$

19. $A = \{-4, -3, -2\}$ $B = \{2, 3, 4\}$
    $A \cap B = \varnothing$

20. $A = \{1, 2, 3, 4\}$ $B = \{1, 2, 3, 4\}$
    $A \cap B = \{1, 2, 3, 4\}$

21. $A = \{a, b, c, d, e\}$ $B = \{c, d, e, f, g\}$
    $A \cap B = \{c, d, e\}$

22. $A = \{m, n, o, p\}$ $B = \{k, l, m, n\}$
    $A \cap B = \{m, n\}$

Objective B   To write a set using set-builder notation

For Exercises 23 to 30, use set-builder notation to write the set.

23. The negative integers greater than -5
    $\{x \mid x > -5, x \in \text{negative integers}\}$

24. The positive integers less than 5
    $\{x \mid x < 5, x \in \text{positive integers}\}$

25. The integers greater than 30
    $\{x \mid x > 30, x \in \text{integers}\}$

26. The integers less than -70
    $\{x \mid x < -70, x \in \text{integers}\}$

27. The even integers greater than 5
    $\{x \mid x > 5, x \in \text{even integers}\}$

28. The odd integers less than -2
    $\{x \mid x < -2, x \in \text{odd integers}\}$

29. The real numbers greater than 8
    $\{x \mid x > 8, x \in \text{real numbers}\}$

30. The real numbers less than 57
    $\{x \mid x < 57, x \in \text{real numbers}\}$

Section 14.1

Suggested Assignment

Exercises 1–39, odds
More challenging problems:
Exercises 41–43

Answers to Writing Exercises

1. Student explanations should include the idea that to find the union of two sets, we list all the elements of the first set and then list all the elements of the second set that are not elements of the first set.

2. Student explanations should include the idea that to find the intersection of two sets, we list only those elements that are elements of both sets.

Quick Quiz (Objective 14.1A)

1. Use the roster method to write the set of integers between -6 and 0. $\{-5, -4, -3, -2, -1\}$

2. Find $A \cup B$ and $A \cap B$, given that $A = \{p, q, r\}$ and $B = \{m, n, o\}$. $\{m, n, o, p, q, r\}; \varnothing$

3. Find $C \cup D$ and $C \cap D$, given that $C = \{-3, -2, -1, 0\}$ and $D = \{0, 1, 2, 3\}$.
   $\{-3, -2, -1, 0, 1, 2, 3\}; \{0\}$

Quick Quiz (Objective 14.1B)

Use set-builder notation to write each set.

1. The odd integers less than 20
   $\{x \mid x < 20, x \in \text{odd integers}\}$

2. The real numbers greater than -15
   $\{x \mid x > -15, x \in \text{real numbers}\}$

3. The positive integers less than 32
   $\{x \mid x < 32, x \in \text{positive integers}\}$

4. The even integers greater than 50
   $\{x \mid x > 50, x \in \text{even integers}\}$

Objective C   **To graph an inequality on the number line**

For Exercises 31 to 40, graph.

31. $\{x \,|\, x > 2\}$

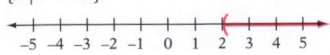

32. $\{x \,|\, x \geq -1\}$

33. $\{x \,|\, x \leq 0\}$

34. $\{x \,|\, x < 4\}$

35. $\{x \,|\, x > -2\} \cup \{x \,|\, x < -4\}$

36. $\{x \,|\, x > 4\} \cup \{x \,|\, x < -2\}$

37. $\{x \,|\, x > -2\} \cap \{x \,|\, x < 4\}$

38. $\{x \,|\, x > -3\} \cap \{x \,|\, x < 3\}$

39. $\{x \,|\, x \geq -2\} \cup \{x \,|\, x < 4\}$

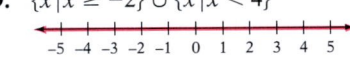

40. $\{x \,|\, x > 0\} \cup \{x \,|\, x \leq 4\}$

APPLYING THE CONCEPTS

41. Determine whether the statement is always true, sometimes true, or never true.
   a. Given that $a > 0$ and $b < 0$, then $ab > 0$.   Never true
   b. Given that $a < 0$, then $a^2 > 0$.   Always true
   c. Given that $a > 0$ and $b < 0$, then $a^2 > b$.   Always true

42. **a.** By trying various sets, make a conjecture about whether or not finding the union of two sets is a commutative operation.   Yes
   b. By trying various sets, make a conjecture about whether or not finding the union of two sets is an associative operation.   Yes

43. **a.** By trying various sets, make a conjecture about whether or not finding the intersection of two sets is a commutative operation.   Yes
   b. By trying various sets, make a conjecture about whether or not finding the intersection of two sets is an associative operation.   Yes

Quick Quiz (Objective 14.1C)

Graph.

1. $\{x \,|\, x < 4\}$

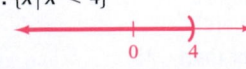

2. $\{x \,|\, x \geq 0\} \cup \{x \,|\, x \leq -3\}$

3. $\{x \,|\, x < 2\} \cap \{x \,|\, x > -4\}$

14.2 The Addition and Multiplication Properties of Inequalities

Objective A   To solve an inequality using the Addition Property of Inequalities

Recall that the solution set of an inequality is the set of numbers each element of which, when substituted for the variable, results in a true inequality.

The inequality at the right is true if the variable is replaced by 7, 9.3, or $\frac{15}{2}$.

$$x + 5 > 8$$

$$\left.\begin{array}{c} 7 + 5 > 8 \\ 9.3 + 5 > 8 \\ \frac{15}{2} + 5 > 8 \end{array}\right\} \text{True inequalities}$$

The inequality $x + 5 > 8$ is false if the variable is replaced by 2, 1.5, or $-\frac{1}{2}$.

$$\left.\begin{array}{c} 2 + 5 > 8 \\ 1.5 + 5 > 8 \\ -\frac{1}{2} + 5 > 8 \end{array}\right\} \text{False inequalities}$$

There are many values of the variable x that will make the inequality $x + 5 > 8$ true. The solution set of $x + 5 > 8$ is any number greater than 3.

At the right is the graph of the solution set of $x + 5 > 8$.

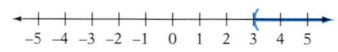

In solving an inequality, the goal is to rewrite the given inequality in the form *variable > constant* or *variable < constant*. The Addition Property of Inequalities is used to rewrite an inequality in this form.

> **Addition Property of Inequalities**
>
> The same term can be added to each side of an inequality without changing the solution set of the inequality.
>
> If $a > b$, then $a + c > b + c$.
> If $a < b$, then $a + c < b + c$.

The Addition Property of Inequalities also holds true for an inequality containing the symbol $\geq$ or $\leq$.

The Addition Property of Inequalities is used when, in order to rewrite an inequality in the form *variable > constant* or *variable < constant*, we must remove a term from one side of the inequality. Add the opposite of that term to each side of the inequality.

HOW TO   Solve: $x - 4 < -3$

$$x - 4 < -3$$
$$x - 4 + 4 < -3 + 4 \qquad \text{• Add 4 to each side of the inequality.}$$
$$x < 1 \qquad \text{• Simplify.}$$

At the right is the graph of the solution set of $x - 4 < -3$.

Objective 14.2A

Vocabulary to Review
solution set of an inequality [14.1C]

Optional Student Activity
This activity will prepare students for the Addition Property of Inequalities. Use the true inequality $3 < 7$. Have students add 4, 8, -3, -5, and -9 to each side of the inequality and determine whether the inequality remains true.
  Repeat the procedure, beginning with $5 > 2$.
  Ask students to state a property of inequalities that is similar to the Addition Property of Equations.

Concept Check
Refer to the example at the left.
1. Which of the following numbers is (are) a part of the solution of $x < 1$?

$$5, \ -4, \ \frac{1}{2}, \ 0, \ 1, \ \frac{4}{3}$$

$$-4, \frac{1}{2}, 0$$

2. Verify that your answers to part 1 are solutions of $x - 4 < -3$ by substituting each one into the inequality.

In-Class Examples (Objective 14.2A)

1. Solve and graph the solution set of $-6 \geq a - 3$.

$a \leq -3$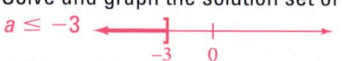

2. Solve.
  a. $10x + 6 < 4 + 9x$ $x < -2$
  b. $3y - \frac{2}{5} > \frac{1}{10} + 2y$ $y > \frac{1}{2}$
  c. $5.3x \leq 8 + 4.3x$ $x \leq 8$

Discuss the Concepts

Compare the Addition Property of Inequalities to the Addition Property of Equations.

Because subtraction is defined in terms of addition, the Addition Property of Inequalities allows the same term to be subtracted from each side of an inequality.

> **HOW TO**   Solve: $5x - 6 \leq 4x - 4$
>
> $$5x - 6 \leq 4x - 4$$
> $$5x - 4x - 6 \leq 4x - 4x - 4 \qquad \bullet \text{ Subtract } 4x \text{ from each side of the inequality.}$$
> $$x - 6 \leq -4 \qquad \bullet \text{ Simplify.}$$
> $$x - 6 + 6 \leq -4 + 6 \qquad \bullet \text{ Add 6 to each side of the inequality.}$$
> $$x \leq 2 \qquad \bullet \text{ Simplify.}$$

Instructor Note

In Example 1, remind students that the final solution can be written as $x > -2$ (see Discuss the Concepts on p. 721). Many people prefer to write solutions to equations and inequalities with the variable on the left side. Students must remember that although an equation can be "turned around" (for example, $2 = x$ can be written as $x = 2$), an inequality can be "turned around" only if the inequality symbol is reversed.

Example 1

Solve $3 < x + 5$ and graph the solution set.

Solution
$$3 < x + 5$$
$$3 - 5 < x + 5 - 5 \qquad \bullet \text{ Subtract 5.}$$
$$-2 < x$$

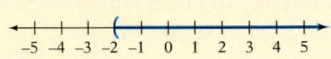

You Try It 1

Solve $x + 2 < -2$ and graph the solution set.

Your solution
$x < -4$

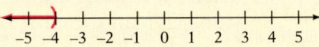

Example 2

Solve: $7x - 14 \leq 6x - 16$

Solution
$$7x - 14 \leq 6x - 16$$
$$7x - 6x - 14 \leq 6x - 6x - 16 \qquad \bullet \text{ Subtract } 6x.$$
$$x - 14 \leq -16$$
$$x - 14 + 14 \leq -16 + 14 \qquad \bullet \text{ Add 14.}$$
$$x \leq -2$$

You Try It 2

Solve: $5x + 3 > 4x + 5$

Your solution
$x > 2$

Solutions on p. S35

Objective B   To solve an inequality using the Multiplication Property of Inequalities

In solving an inequality, the goal is to rewrite the given inequality in the form *variable* > *constant* or *variable* < *constant*. The Multiplication Property of Inequalities is used when, in order to rewrite an inequality in this form, we must remove a coefficient from one side of the inequality.

> **Multiplication Property of Inequalities**
>
> Each side of an inequality can be multiplied by the same positive number without changing the solution set of the inequality.
>
> $$\text{If } a > b \text{ and } c > 0, \text{ then } ac > bc.$$
> $$\text{If } a < b \text{ and } c > 0, \text{ then } ac < bc.$$
>
> If each side of an inequality is multiplied by the same negative number and the inequality symbol is reversed, then the solution set of the inequality is not changed.
>
> $$\text{If } a > b \text{ and } c < 0, \text{ then } ac < bc.$$
> $$\text{If } a < b \text{ and } c < 0, \text{ then } ac > bc.$$

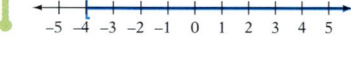

TAKE NOTE

Any time an inequality is multiplied or divided by a negative number, the inequality symbol must be reversed. Compare the next two examples.

$2x < -4$	Divide each side by *positive* 2.
$\dfrac{2x}{2} < \dfrac{-4}{2}$	
$x < -2$	Inequality *is not* reversed.

$-2x < 4$	Divide each side by *negative* 2.
$\dfrac{-2x}{-2} > \dfrac{4}{-2}$	
$x > -2$	Inequality *is* reversed.

$5 > 4$	• **A true inequality**
$5(2) > 4(2)$	• **Multiply by *positive* 2.**
$10 > 8$	• **Still a true inequality**

$6 < 9$	• **A true inequality**
$6(-3) > 9(-3)$	• **Multiply by *negative* 3 and *reverse* the inequality symbol.**
$-18 > -27$	• **Still a true inequality**

The Multiplication Property of Inequalities also holds true for an inequality containing the symbol $\geq$ or $\leq$.

HOW TO   Solve $-\dfrac{3}{2}x \leq 6$ and graph the solution set.

$$-\frac{3}{2}x \leq 6$$

• **Multiply each side of the inequality by $-\dfrac{2}{3}$. Because $-\dfrac{2}{3}$ is a negative number, the inequality symbol must be reversed.**

$$-\frac{2}{3}\left(-\frac{3}{2}x\right) \geq -\frac{2}{3}(6)$$

$$x \geq -4$$

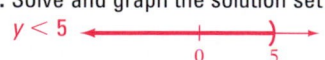

• **Graph $\{x \mid x \geq -4\}$.**

Because division is defined in terms of multiplication, the Multiplication Property of Inequalities allows each side of an inequality to be divided by a nonzero constant.

TAKE NOTE

As shown in the example at the right, the goal in solving an inequality can be *constant < variable* or *constant > variable*. We could have written the answer to this example as $x > -\dfrac{2}{3}$.

HOW TO   Solve: $-4 < 6x$

$$-4 < 6x$$

$$\frac{-4}{6} < \frac{6x}{6}$$

• **Divide each side of the inequality by 6.**

$$-\frac{2}{3} < x$$

• **Simplify: $\dfrac{-4}{6} = -\dfrac{2}{3}$**

Discuss the Concepts

In Example 3, have students explain why we reverse the inequality symbol when *dividing* both sides by −7.

Dividing by −7 is the same as multiplying by $-\frac{1}{7}$, and we must reverse the inequality symbol when multiplying both sides of an inequality by a negative number.

Example 3 Solve $-7x > 14$ and graph the solution set.

Solution
$$-7x > 14$$
$$\frac{-7x}{-7} < \frac{14}{-7} \quad \bullet \text{ Divide by } -7.$$
$$x < -2$$

You Try It 3 Solve $-3x > -9$ and graph the solution set.

Your solution $x < 3$

Example 4 Solve: $-\frac{5}{8}x \leq \frac{5}{12}$

Solution
$$-\frac{5}{8}x \leq \frac{5}{12}$$
$$-\frac{8}{5}\left(-\frac{5}{8}x\right) \geq -\frac{8}{5}\left(\frac{5}{12}\right) \quad \bullet \text{ Multiply by } -\frac{8}{5}.$$
$$x \geq -\frac{2}{3}$$

You Try It 4 Solve: $-\frac{3}{4}x \geq 18$

Your solution $x \leq -24$

Solutions on p. S35

Objective 14.2C

Objective C   To solve application problems

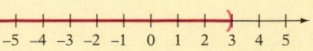

Concept Check

1. Refer to the situation described in Example 5. Another student's scores were 90, 91, 83, and 84. Can this student receive an A? Explain.

   No; a final test score of 100 will give the student only 448 of the necessary 450 points.

2. To merit a B in a course, a student's four test scores must average at least 80. One student's first three test scores were 75, 86, and 77. Can this student receive a B? Explain.

   Yes, by scoring 82 or higher

Optional Student Activity

A geometry theorem called the *Triangle Inequality Theorem* states that the sum of the measures of two sides of a triangle must be greater than the measure of the third side. Suppose two sides of a triangle measure 10 in. and 18 in. Let x be the length of the third side.

1. What are the possible values of x?

   Between 8 in. and 28 in.

2. Write your answer to part 1 using the intersection of two sets and set-builder notation.

   $\{x \mid x > 8\} \cap \{x \mid x < 28\}$

Example 5

A student must have at least 450 points out of 500 points on five tests to receive an A in a course. One student's results on the first four tests were 94, 87, 77, and 95. What scores on the last test will enable this student to receive an A in the course?

Strategy
To find the scores, write and solve an inequality using N to represent the possible scores on the last test.

Solution

Total number of points on the five tests	is greater than or equal to	450

$$94 + 87 + 77 + 95 + N \geq 450$$
$$353 + N \geq 450 \quad \bullet \text{ Simplify.}$$
$$353 - 353 + N \geq 450 - 353 \quad \bullet \text{ Subtract 353.}$$
$$N \geq 97$$

The student's score on the last test must be greater than or equal to 97.

You Try It 5

An appliance dealer will make a profit on the sale of a television set if the cost of the new set to the dealer is less than 70% of the selling price to the customer. What selling prices will enable the dealer to make a profit on a television set that costs the dealer $314?

Your strategy

Your solution
Any price greater than or equal to $448.58

Solution on p. S35

In-Class Examples (Objective 14.2C)

1. One-fourth of a number is less than five-sixths. Find the largest integer that satisfies this inequality.   3

2. In a person's daily diet, fat intake should be at most 30% of calorie intake. If a person who eats 2000 calories a day consumes 270 calories of fat at breakfast, how many more fat calories can this person consume during the rest of the day?
   ≤ 330 fat calories

14.2 Exercises

Objective A   To solve an inequality using the
Addition Property of Inequalities

Section 14.2

Suggested Assignment
Exercises 1–77, every other odd;
  81–87, odds
More challenging problems:
  Exercises 90, 92, 94

For Exercises 1 to 8, solve the inequality and graph the solution set.

1. $x + 1 < 3$ $x < 2$

2. $y + 2 < 2$ $y < 0$

3. $x - 5 > -2$ $x > 3$

4. $x - 3 > -2$ $x > 1$

5. $7 \le n + 4$ $n \ge 3$

6. $3 \le 5 + x$ $x \ge -2$

7. $x - 6 \le -10$ $x \le -4$

8. $y - 8 \le -11$ $y \le -3$

For Exercises 9 to 12, write an inequality that represents the set of numbers shown in the graph.

9. 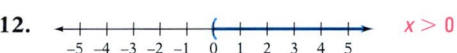 $x < 1$

10. $x \ge -2$

11. $x \le -3$

12. $x > 0$

For Exercises 13 to 42, solve.

13. $y - 3 \ge -12$
$y \ge -9$

14. $x + 8 \ge -14$
$x \ge -22$

15. $3x - 5 < 2x + 7$
$x < 12$

16. $5x + 4 < 4x - 10$
$x < -14$

17. $8x - 7 \ge 7x - 2$
$x \ge 5$

18. $3n - 9 \ge 2n - 8$
$n \ge 1$

19. $2x + 4 < x - 7$
$x < -11$

20. $9x + 7 < 8x - 7$
$x < -14$

21. $4x - 8 \le 2 + 3x$
$x \le 10$

22. $5b - 9 < 3 + 4b$
$b < 12$

23. $6x + 4 \ge 5x - 2$
$x \ge -6$

24. $7x - 3 \ge 6x - 2$
$x \ge 1$

25. $2x - 12 > x - 10$

$x > 2$

26. $3x + 9 > 2x + 7$

$x > -2$

27. $d + \dfrac{1}{2} < \dfrac{1}{3}$

$d < -\dfrac{1}{6}$

Quick Quiz (Objective 14.2A)

1. Solve and graph the solution set of $-5 \ge x - 4$.
$x \le -1$

2. Solve.
  a. $7n - 3 \ge 2 + 6n$ $n \ge 5$
  b. $b + \dfrac{1}{6} < -\dfrac{3}{4}$ $b < -\dfrac{11}{12}$
  c. $x + 4.2 \le 2.7$ $x \le -1.5$

28. $x - \dfrac{3}{8} < \dfrac{5}{6}$

$x < \dfrac{29}{24}$

29. $x + \dfrac{5}{8} \geq -\dfrac{2}{3}$

$x \geq -\dfrac{31}{24}$

30. $y + \dfrac{5}{12} \geq -\dfrac{3}{4}$

$y \geq -\dfrac{7}{6}$

31. $x - \dfrac{3}{8} < \dfrac{1}{4}$

$x < \dfrac{5}{8}$

32. $y + \dfrac{5}{9} \leq \dfrac{5}{6}$

$y \leq \dfrac{5}{18}$

33. $2x - \dfrac{1}{2} < x + \dfrac{3}{4}$

$x < \dfrac{5}{4}$

34. $6x - \dfrac{1}{3} \leq 5x - \dfrac{1}{2}$

$x \leq -\dfrac{1}{6}$

35. $3x + \dfrac{5}{8} > 2x + \dfrac{5}{6}$

$x > \dfrac{5}{24}$

36. $4b - \dfrac{7}{12} \geq 3b - \dfrac{9}{16}$

$b \geq \dfrac{1}{48}$

37. $3.8x < 2.8x - 3.8$
$x < -3.8$

38. $1.2x < 0.2x - 7.3$
$x < -7.3$

39. $x + 5.8 \leq 4.6$
$x \leq -1.2$

40. $n - 3.82 \leq 3.95$
$n \leq 7.77$

41. $x - 3.5 < 2.1$
$x < 5.6$

42. $x - 0.23 \leq 0.47$
$x \leq 0.70$

Objective B **To solve an inequality using the Multiplication Property of Inequalities**

For Exercises 43 to 52, solve the inequality and graph the solution set.

43. $3x < 12$ $x < 4$

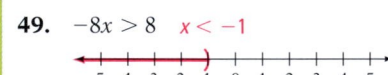

44. $8x \leq -24$ $x \leq -3$

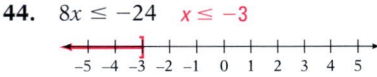

45. $15 \leq 5y$ $y \geq 3$

46. $-48 < 24x$ $x > -2$

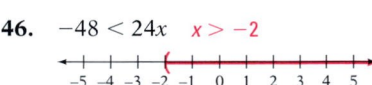

47. $16x \leq 16$ $x \leq 1$

48. $3x > 0$ $x > 0$

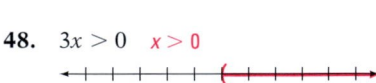

49. $-8x > 8$ $x < -1$

50. $-2n \leq -8$ $n \geq 4$

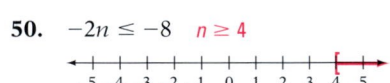

51. $-6b > 24$ $b < -4$

52. $-4x < 8$ $x > -2$

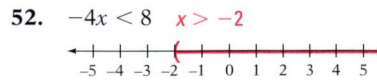

Quick Quiz (Objective 14.2B)

1. Solve and graph the solution set of $-2 \leq 2x$.

$x \geq -1$

2. Solve.

a. $\dfrac{3}{7} < \dfrac{9}{14}x$ $x > \dfrac{2}{3}$

b. $-5a > 0$ $a < 0$

c. $-4.3x \geq -8.6$ $x \leq 2$

For Exercises 53 to 79, solve.

53. $-5y \geq 0$

$y \leq 0$

54. $-3z < 0$

$z > 0$

55. $7x > 2$

$x > \dfrac{2}{7}$

56. $6x \leq -1$

$x \leq -\dfrac{1}{6}$

57. $2x \leq -5$

$x \leq -\dfrac{5}{2}$

58. $\dfrac{5}{6}n < 15$

$n < 18$

59. $\dfrac{3}{4}x < 12$

$x < 16$

60. $\dfrac{2}{3}y \geq 4$

$y \geq 6$

61. $10 \leq \dfrac{5}{8}x$

$x \geq 16$

62. $4 \geq \dfrac{2}{3}x$

$x \leq 6$

63. $-\dfrac{3}{7}x \leq 6$

$x \geq -14$

64. $-\dfrac{2}{11}b \geq -6$

$b \leq 33$

65. $-\dfrac{4}{7}x \geq -12$

$x \leq 21$

66. $\dfrac{2}{3}n < \dfrac{1}{2}$

$n < \dfrac{3}{4}$

67. $-\dfrac{3}{5}x < 0$

$x > 0$

68. $-\dfrac{2}{3}x \geq 0$

$x \leq 0$

69. $-\dfrac{3}{8}x \geq \dfrac{9}{14}$

$x \leq -\dfrac{12}{7}$

70. $-\dfrac{3}{5}x < -\dfrac{6}{7}$

$x > \dfrac{10}{7}$

71. $-\dfrac{4}{5}x < -\dfrac{8}{15}$

$x > \dfrac{2}{3}$

72. $-\dfrac{3}{4}y \geq -\dfrac{5}{8}$

$y \leq \dfrac{5}{6}$

73. $-\dfrac{8}{9}x \geq -\dfrac{16}{27}$

$x \leq \dfrac{2}{3}$

74. $1.5x \leq 6.30$

$x \leq 4.2$

75. $2.3x \leq 5.29$

$x \leq 2.3$

76. $-3.5d > 7.35$

$d < -2.1$

77. $-0.24x > 0.768$

$x < -3.2$

78. $4.25m > -34$

$m > -8$

79. $-3.9x \geq -19.5$

$x \leq 5$

Objective C **To solve application problems**

80. Number Sense Three-fifths of a number is greater than two-thirds. Find the smallest integer that satisfies this inequality.

2

81. Sports To be eligible for a basketball tournament, a basketball team must win at least 60% of its remaining games. If the team has 17 games remaining, how many games must the team win to qualify for the tournament?

$\geq$ 11 games

82. Taxes To avoid a tax penalty, at least 90% of a self-employed person's total annual income tax liability must be paid in estimated tax payments during the year. What amount of income tax must a person with an annual income tax liability of $3500 pay in estimated tax payments?

$\geq$ $3150

Quick Quiz (Objective 14.2C)

1. To earn an A in a course, a student must have an average of 90 or more on five tests. One student's first four test scores were 78, 88, 92, and 95. What scores can the student receive on the fifth test to earn an A? $\geq$ 97

83. **Recycling**   A service organization will receive a bonus of $200 for collecting more than 1850 lb of aluminum cans during its four collection drives. On the first three drives, the organization collected 505 lb, 493 lb, and 412 lb. How many pounds of cans must the organization collect on the fourth drive to receive the bonus?
> 440 lb

84. **Software Development**   Computer software engineers are fond of saying that software takes at least twice as long to develop as they think it will. According to that saying, how many hours will it take to develop a software product that an engineer thinks can be finished in 50 h?
≥ 100 h

85. **Health**   A government agency recommends a minimum daily allowance of vitamin C of 60 mg. How many additional milligrams of vitamin C does a person who has already drunk a glass of orange juice with 10 mg of vitamin C need in order to satisfy the recommended daily allowance?
≥ 50 mg

86. **Grading**   To pass a course with a B grade, a student must have an average of 80 points on five tests. The student's grades on the first four tests were 75, 83, 86, and 78. What scores can the student receive on the fifth test to earn a B grade?
≥ 78

87. **Grading**   A professor scores all tests with a maximum of 100 points. To earn an A grade in this course, a student must have an average of 92 on four tests. One student's grades on the first three tests were 89, 86, and 90. Can this student earn an A grade?
No

88. **Health**   A health official recommends a maximum cholesterol level of 200 units. By how many units must a patient with a cholesterol level of 275 units reduce her cholesterol level to satisfy the recommended maximum level?
≥ 75 units

APPLYING THE CONCEPTS

For Exercises 89 to 94, given that $a > b$ and that a and b are real numbers, determine for which real numbers c the statement is true. Use set-builder notation to write the answer.

89. $ac > bc$ $\{c \mid c > 0\}$

90. $ac < bc$ $\{c \mid c < 0\}$

91. $a + c > b + c$ $\{c \mid c \in \text{real numbers}\}$

92. $a + c < b + c$ $\varnothing$

93. $\dfrac{a}{c} > \dfrac{b}{c}$ $\{c \mid c > 0\}$

94. $\dfrac{a}{c} < \dfrac{b}{c}$ $\{c \mid c < 0\}$

95.   In your own words, state the Addition Property of Inequalities.

96.   In your own words, state the Multiplication Property of Inequalities.

Answers to Writing Exercises

95. Students should paraphrase the definition given in the text: The Addition Property of Inequalities states that the same term can be added to each side of an inequality without changing the solution set of the inequality.

96. Students should paraphrase the definition given in the text: The Multiplication Property of Inequalities states that we can multiply both sides of an inequality by the same positive number without changing the solution set of the inequality. Multiplying both sides of an inequality by the same negative number changes the direction of the inequality symbol.

14.3 General Inequalities

Objective A   To solve general inequalities

Solving an inequality frequently requires application of both the Addition and Multiplication Properties of Inequalities.

Study Tip

Be sure to verbalize the similarities and differences between solving the equations in Chapter 5 and solving these inequalities.

> **HOW TO**   Solve: $4y - 3 \geq 6y + 5$
>
> $4y - 3 \geq 6y + 5$
>
> $4y - 6y - 3 \geq 6y - 6y + 5$ • Subtract **6y** from each side of the inequality.
>
> $-2y - 3 \geq 5$ • Simplify.
>
> $-2y - 3 + 3 \geq 5 + 3$ • Add **3** to each side of the inequality.
>
> $-2y \geq 8$ • Simplify.
>
> $\dfrac{-2y}{-2} \leq \dfrac{8}{-2}$ • Divide each side of the inequality by **−2**. Because **−2** is a negative number, the inequality symbol must be reversed.
>
> $y \leq -4$

When an inequality contains parentheses, one of the steps in solving the inequality requires the use of the Distributive Property.

> **HOW TO**   Solve: $-2(x - 7) > 3 - 4(2x - 3)$
>
> $-2(x - 7) > 3 - 4(2x - 3)$
>
> $-2x + 14 > 3 - 8x + 12$ • Use the **Distributive Property** to remove parentheses.
>
> $-2x + 14 > -8x + 15$ • Simplify.
>
> $-2x + 8x + 14 > -8x + 8x + 15$ • Add **8x** to each side of the inequality.
>
> $6x + 14 > 15$ • Simplify.
>
> $6x + 14 - 14 > 15 - 14$ • Subtract **14** from each side of the inequality.
>
> $6x > 1$ • Simplify.
>
> $\dfrac{6x}{6} > \dfrac{1}{6}$ • Divide each side of the inequality by **6**.
>
> $x > \dfrac{1}{6}$

Example 1   Solve: $7x - 3 \leq 3x + 17$

Solution

$7x - 3 \leq 3x + 17$

$7x - 3x - 3 \leq 3x - 3x + 17$ • Subtract **3x**.

$4x - 3 \leq 17$

$4x - 3 + 3 \leq 17 + 3$ • Add **3**.

$4x \leq 20$

$\dfrac{4x}{4} \leq \dfrac{20}{4}$ • Divide by **4**.

$x \leq 5$

You Try It 1   Solve: $5 - 4x > 9 - 8x$

Your solution $x > 1$

Solution on p. S35

Discuss the Concepts

Have students compare the process of solving a first-degree inequality to the process of solving a first-degree equation. Be sure they note that the only difference is the requirement of reversing the inequality symbol when multiplying or dividing by a negative number.

Concept Check

Refer to the two examples worked out at the left.

1. Check the solution of $4y - 3 \geq 6y + 5$ by substituting a y-value of -4 and a y-value less than -4 into the inequality.

2. Check the solution of $-2(x - 7) > 3 - 4(2x - 3)$ by substituting an x-value greater than $\dfrac{1}{6}$ into the inequality.

Optional Student Activity

1. Write an inequality that can be solved using only the Addition Property of Inequalities.
   For example, $x + 4 < 1$

2. Write an inequality that can be solved using only the Multiplication Property of Inequalities.
   For example, $3x > 6$

3. Write an inequality that can be solved using both the Addition and Multiplication Properties of Inequalities.
   For example, $3x - 5 < x + 3$

In-Class Examples (Objective 14.3A)

Solve.

1. $0.3(70 + x) \leq x$ $x \geq 30$

2. $9 - 5(1 - 2x) > 6(x + 4)$ $x > 5$

3. $-3(8x + 2) < 5(8 - 4x)$ $x > -\dfrac{23}{2}$

Optional Student Activity

The following are linear inequalities in two variables (the subject of Objective 14.4A). Solve each inequality for y.

1. $5x - y \geq 1$ $y \leq 5x - 1$

2. $x + 6y < 12$ $y < -\dfrac{1}{6}x + 2$

3. $4 - 2(x - y) \leq y + x$
   $y \leq 3x - 4$

Discuss the Concepts

Use the Multiplication and Addition Properties of Inequalities to explain why, if $x < 8$, then $2x + 4 < 20$.

First, multiply both sides of $x < 8$ by 2, which gives $2x < 16$. Then add 4 to each side of $2x < 16$, which gives $2x + 4 < 20$.

Objective 14.3B

Concept Check

A pizza costs $10 plus $.75 per topping. You and your friend have $35 and want to save $20 to go to a movie. Write an inequality that you can solve for x, the number of toppings you can have on your pizza.

For example, $0.75x \leq 5$ or $10 + 0.75x \leq 35 - 20$

Example 2

Solve: $3(3 - 2x) \geq -5x - 2(3 - x)$

Solution

$$3(3 - 2x) \geq -5x - 2(3 - x)$$
$$9 - 6x \geq -5x - 6 + 2x \quad \bullet \text{ Distributive Property}$$
$$9 - 6x \geq -3x - 6$$
$$9 - 6x + 3x \geq -3x + 3x - 6 \quad \bullet \text{ Add } 3x.$$
$$9 - 3x \geq -6$$
$$9 - 9 - 3x \geq -6 - 9 \quad \bullet \text{ Subtract 9.}$$
$$-3x \geq -15$$
$$\dfrac{-3x}{-3} \leq \dfrac{-15}{-3} \quad \bullet \text{ Divide by } -3.$$
$$x \leq 5$$

You Try It 2

Solve: $8 - 4(3x + 5) \leq 6(x - 8)$

Your solution
$x \geq 2$

Solution on p. S35

Objective B   To solve application problems

Example 3

A rectangle is 10 ft wide and $(2x + 4)$ ft long. Express as an integer the maximum length of the rectangle when the area is less than 200 ft². (The area of a rectangle is equal to its length times its width.)

Strategy

To find the maximum length:

- Replace the variables in the area formula by the given values and solve for x.
- Replace the variable in the expression $2x + 4$ with the value found for x.

Solution

Length times width	is less than	200 ft²

$$10(2x + 4) < 200$$
$$20x + 40 < 200 \quad \bullet \text{ Distributive Property}$$
$$20x + 40 - 40 < 200 - 40 \quad \bullet \text{ Subtract 40.}$$
$$20x < 160$$
$$\dfrac{20x}{20} < \dfrac{160}{20} \quad \bullet \text{ Divide by 20.}$$
$$x < 8$$

The length is $(2x + 4)$ ft. Because $x < 8$, $2x + 4 < 2(8) + 4 = 20$. Therefore, the length is less than 20 ft. The maximum length is 19 ft.

You Try It 3

Company A rents cars for $8 a day and $.10 for every mile driven. Company B rents cars for $10 a day and $.08 per mile driven. You want to rent a car for 1 week. What is the maximum number of miles you can drive a Company A car if it is to cost you less than a Company B car?

Your strategy

Your solution
699 mi

Solution on p. S36

In-Class Examples (Objective 14.3B)

1. The label on a bottle of juice says that the drink is at least 15% real fruit juice. If the manufacturer begins with 450 oz of real juice, how many ounces of water and other ingredients can be added while still keeping the final mixture at least 15% real juice?   ≤ 2550 oz

14.3 Exercises

Section 14.3

Suggested Assignment
Exercises 1–27, odds
More challenging problems:
   Exercises 28, 31–34

Objective A   To solve general inequalities

For Exercises 1 to 20, solve.

1. $4x - 8 < 2x$
$x < 4$

2. $7x - 4 < 3x$
$x < 1$

3. $2x - 8 > 4x$
$x < -4$

4. $3y + 2 > 7y$
$y < \dfrac{1}{2}$

5. $8 - 3x \le 5x$
$x \ge 1$

6. $10 - 3x \le 7x$
$x \ge 1$

7. $3x + 2 > 5x - 8$
$x < 5$

8. $2n - 9 \ge 5n + 4$
$n \le -\dfrac{13}{3}$

9. $5x - 2 < 3x - 2$
$x < 0$

10. $8x - 9 > 3x - 9$
$x > 0$

11. $0.1(180 + x) > x$
$x < 20$

12. $x > 0.2(50 + x)$
$x > 12.5$

13. $2(2y - 5) \le 3(5 - 2y)$
$y \le \dfrac{5}{2}$

14. $2(5x - 8) \le 7(x - 3)$
$x \le -\dfrac{5}{3}$

15. $5(2 - x) > 3(2x - 5)$
$x < \dfrac{25}{11}$

16. $4(3d - 1) > 3(2 - 5d)$
$d > \dfrac{10}{27}$

17. $4 - 3(3 - n) \le 3(2 - 5n)$
$n \le \dfrac{11}{18}$

18. $15 - 5(3 - 2x) \le 4(x - 3)$
$x \le -2$

19. $2x - 3(x - 4) \ge 4 - 2(x - 7)$
$x \ge 6$

20. $4 + 2(3 - 2y) \le 4(3y - 5) - 6y$
$y \ge 3$

Objective B   To solve application problems

21. **Compensation**   The sales agent for a jewelry company is offered a flat monthly salary of $3200 or a salary of $1000 plus an 11% commission on the selling price of each item sold by the agent. If the agent chooses the $3200, what dollar amount does the agent expect to sell in 1 month?
$\le$ $20,000

22. **Compensation**   A baseball player is offered an annual salary of $200,000 or a base salary of $100,000 plus a bonus of $1000 for each hit over 100 hits. How many hits must the baseball player make to earn more than $200,000?
$\ge$ 201 hits

Quick Quiz (Objective 14.3A)

Solve.

1. $3x - 7 \le 7x + 11$ $x \ge -\dfrac{9}{2}$

2. $4(5 - x) > 3(2x + 10)$ $x < -1$

3. $12 - 3(4x - 1) \ge 6(4 - 3x)$ $x \ge \dfrac{3}{2}$

23. **Comparing Services**   A computer bulletin board service charges a flat fee of $10 per month or a fee of $4 per month plus $.10 for each minute the service is used. How many minutes must a person use this service to exceed $10 in fees?   > 60 min

24. **Comparing Services**   A site licensing fee for a computer program is $1500. Paying this fee allows the company to use the program at any computer terminal within the company. Alternatively, the company can choose to pay $200 for each individual computer it has. How many individual computers must a company have for the site license to be more economical for the company?   ≥ 8 computers

25. **Health**   For a product to be labeled orange juice, a state agency requires that at least 80% of the drink be real orange juice. How many ounces of artificial flavors can be added to 32 oz of real orange juice and have it still be legal to label the drink orange juice?   ≤ 8 oz

26. **Health**   Grade A hamburger cannot contain more than 20% fat. How much fat can a butcher mix with 300 lb of lean meat to meet the 20% requirement?   ≤ 75 lb

27. **Transportation**   A shuttle service taking skiers to a ski area charges $8 per person each way. Four skiers are debating whether to take the shuttle bus or rent a car for $45 plus $.25 per mile. Assuming that the skiers will share the cost of the car and that they want the least expensive method of transportation, determine how far away the ski area is if they choose the shuttle service.   > 38 mi

APPLYING THE CONCEPTS

28. Determine whether the statement is always true, sometimes true, or never true, given that a, b, and c are real numbers.
    a. If $a > b$, then $-a > -b$.   Never true
    b. If $a < b$, then $ac < bc$.   Sometimes true
    c. If $a > b$, then $a + c > b + c$.   Always true
    d. If $a \neq 0$, $b \neq 0$, and $a > b$, then $\frac{1}{a} > \frac{1}{b}$.   Sometimes true

For Exercises 29 and 30, use the roster method to list the set of positive integers that are solutions of the inequality.

29. $7 - 2b \leq 15 - 5b$ {1, 2}

30. $-6(2 - d) \geq 4d - 9$ {2, 3, 4, ...}

For Exercises 31 and 32, use the roster method to list the set of integers that are common to the solution sets of the two inequalities.

31. $5x - 12 \leq x + 8$
     $3x - 4 \geq 2 + x$ {3, 4, 5}

32. $3(x + 2) > 9x - 2$
     $4(x + 5) > 3(x + 6)$ {−1, 0, 1}

33. Determine the solution set of $2 - 3(x + 4) < 5 - 3x$.   {$x \mid x \in$ real numbers}

34. Determine the solution set of $3x + 2(x - 1) > 5(x + 1)$.   ∅

Quick Quiz (Objective 14.3B)

1. A telephone company offers two plans: a flat monthly fee of $36, or a monthly fee of $6 with a charge of $.15 per minute of calling time. How many minutes of calling time per month must a person use to make the flat fee of $36 the less expensive option?   >200 min

14.4 Graphing Linear Inequalities

Objective A   To graph an inequality in two variables

Point of Interest

Linear inequalities play an important role in applied mathematics. They are used in a branch of mathematics called *linear programming*, which was developed during World War II to solve problems in supplying the Air Force with the machine parts necessary to keep planes flying. Today, linear programming's applications extend to many other disciplines.

The graph of the linear equation $y = x - 2$ separates a plane into three sets:

The set of points on the line
The set of points above the line
The set of points below the line

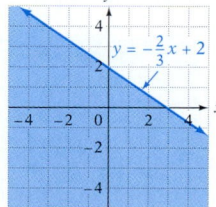

The point $(3, 1)$ is a solution of $y = x - 2$.

$$\begin{array}{c|c} y = x - 2 \\ \hline 1 & 3 - 2 \\ 1 = 1 \end{array}$$

The point $(3, 3)$ is a solution of $y > x - 2$.

$$\begin{array}{c|c} y > x - 2 \\ \hline 3 & 3 - 2 \\ 3 > 1 \end{array}$$

Any point **above** the line is a solution of $y > x - 2$.

Study Tip

Be sure to do all you need to do in order to be successful at graphing linear inequalities: Read through the introductory material, work through the How To examples, study the paired examples, do the You Try Its, and check your solutions against those in the back of the book. See *AIM for Success*, pages xxxi–xxxii.

The point $(3, -1)$ is a solution of $y < x - 2$.

$$\begin{array}{c|c} y < x - 2 \\ \hline -1 & 3 - 2 \\ -1 < 1 \end{array}$$

Any point **below** the line is a solution of $y < x - 2$.

The solution set of $y = x - 2$ is all points on the line. The solution set of $y > x - 2$ is all points above the line. The solution set of $y < x - 2$ is all points below the line. The solution set of an inequality in two variables is a **half-plane.**

The following illustrates the procedure for graphing a linear inequality.

HOW TO   Graph the solution set of $2x + 3y \leq 6$.

Solve the inequality for y.

$$2x + 3y \leq 6$$
$$2x - 2x + 3y \leq -2x + 6 \qquad \bullet \text{ Subtract } 2x \text{ from each side.}$$
$$3y \leq -2x + 6 \qquad \bullet \text{ Simplify.}$$
$$\frac{3y}{3} \leq \frac{-2x + 6}{3} \qquad \bullet \text{ Divide each side by 3.}$$
$$y \leq -\frac{2}{3}x + 2 \qquad \bullet \text{ Simplify.}$$

Change the inequality to an equality and graph $y = -\frac{2}{3}x + 2$. If the inequality is $\geq$ or $\leq$, the line is part of the solution set and is shown by a solid line. If the inequality is $>$ or $<$, the line is not part of the solution set and is shown by a dotted line.

If the inequality is $>$ or $\geq$, shade the upper half-plane. If the inequality is $<$ or $\leq$, shade the lower half-plane.

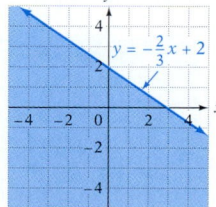

Objective 14.4A

New Vocabulary

half-plane

Instructor Note

This objective may be difficult for students because it combines skills from several other objectives. If necessary, review the process of graphing lines using the slope and y-intercept (see Objective 12.3C).

Instructor Note

Show students that as long as the graph of an equation does not pass through the origin, the ordered pair $(0, 0)$ affords an easy check of the correct region to shade. If $(0, 0)$ satisfies the linear inequality, then the point whose coordinates are $(0, 0)$ should be included in the shaded region. If $(0, 0)$ does not satisfy the inequality, the point $(0, 0)$ should not be in the shaded region.

Discuss the Concepts

Have students explain why the line is included in the graph of a linear inequality when the inequality symbol is $\geq$ or $\leq$.

Concept Check

Refer to the example at the left. How would the graph change if the inequality symbol were $>$ instead of $\leq$?
The line would be dotted instead of solid. The shading would be above the line instead of below.

Concept Check

Look at the graph at the top of the page. Write the inequality that describes the graph.
$y \geq x - 2$

In-Class Examples (Objective 14.4A)

Graph the solution set.

1. $x - y < -2$
2. $3x + 4y \leq -12$
3. $x \geq -1$

1.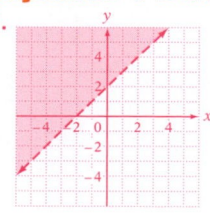

2.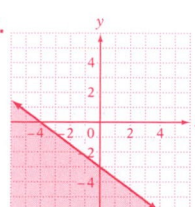

3.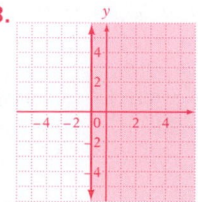

Concept Check

For a line with a steep slope, it can be difficult for some students to identify which half-plane is "above" the line and which is "below." For each line below, tell whether the point $(-3, 2)$ is above the line or below the line. Would $(-3, 2)$ be included in the solution of $y < mx + b$ or in the solution of $y > mx + b$?

1.

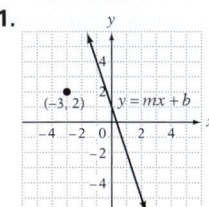

Below, $y < mx + b$

2.

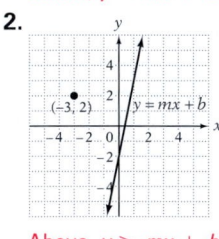

Above, $y > mx + b$

Concept Check

1. Which side of the line $x = -3$ is shaded to represent the solution set of $x \geq -3$?
Right

2. Which side of the line $x = 2$ is shaded to represent the solution set of $2 \leq x$?
Right

Discuss the Concepts

Have students describe the steps involved in graphing a linear inequality.

Example 1

Graph the solution set of $3x + y > -2$.

Solution

$$3x + y > -2$$
$$3x - 3x + y > -3x - 2 \quad \bullet \text{ Subtract } 3x.$$
$$y > -3x - 2$$

Graph $y = -3x - 2$ as a dotted line.
Shade the upper half-plane.

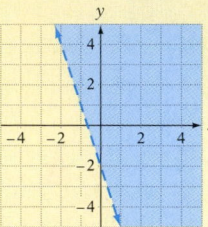

Example 2

Graph the solution set of $2x - y \geq 2$.

Solution

$$2x - y \geq 2$$
$$2x - 2x - y \geq -2x + 2 \quad \bullet \text{ Subtract } 2x.$$
$$-y \geq -2x + 2$$
$$-1(-y) \leq -1(-2x + 2) \quad \bullet \text{ Multiply by } -1.$$
$$y \leq 2x - 2$$

Graph $y = 2x - 2$ as a solid line.
Shade the lower half-plane.

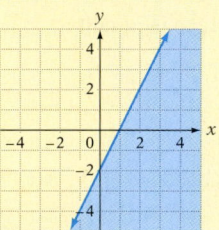

Example 3

Graph the solution set of $y > -1$.

Solution
Graph $y = -1$ as a dotted line.
Shade the upper half-plane.

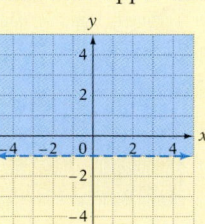

You Try It 1

Graph the solution set of $x - 3y < 2$.

Your solution

$y > \dfrac{1}{3}x - \dfrac{2}{3}$

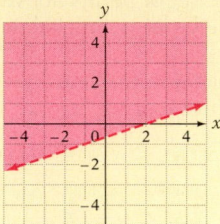

You Try It 2

Graph the solution set of $2x - 4y \leq 8$.

Your solution

$y \geq \dfrac{1}{2}x - 2$

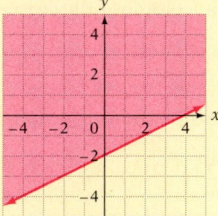

You Try It 3

Graph the solution set of $x < 3$.

Your solution

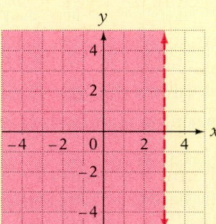

Solutions on p. S36

14.4 Exercises

Objective A   To graph an inequality in two variables

For Exercises 1 to 18, graph the solution set of the inequality.

1. $x + y > 4$

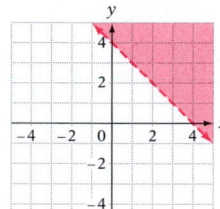

2. $x - y > -3$

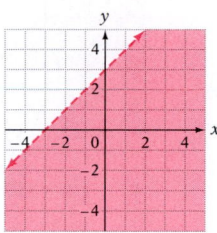

3. $2x - y < -3$

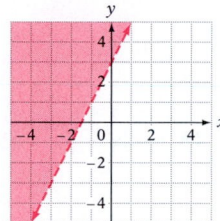

4. $3x - y < 9$

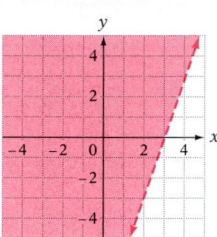

5. $2x + y \geq 4$

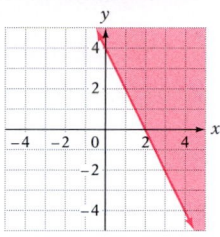

6. $3x + y \geq 6$

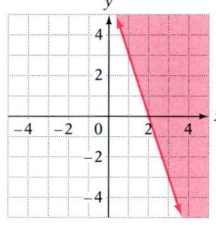

7. $y \leq -2$

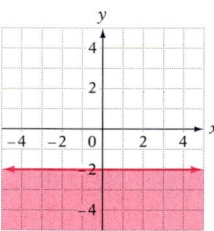

8. $y > 3$

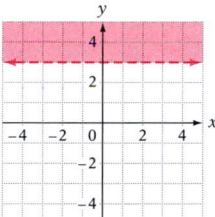

9. $3x - 2y < 8$

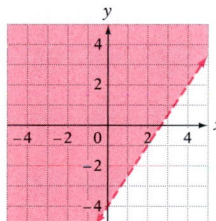

10. $5x + 4y > 4$

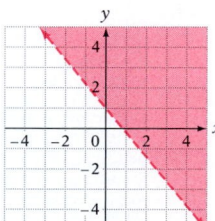

11. $-3x - 4y \geq 4$

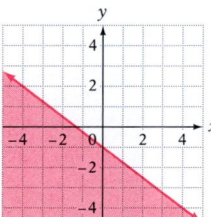

12. $-5x - 2y \geq 8$

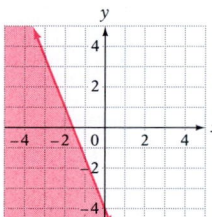

Section 14.4

Suggested Assignment
Exercises 1–17, odds
More challenging problems:
   Exercises 19, 22–24

Quick Quiz (Objective 14.4A)

Graph the solution set.

1. $x - 2y \geq 2$

2. $x + y > 3$

3. $y < 1$

1.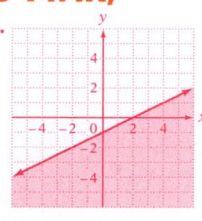

2.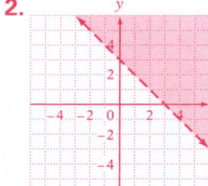

3.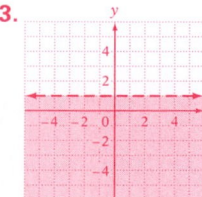

13. $6x + 5y \leq -10$

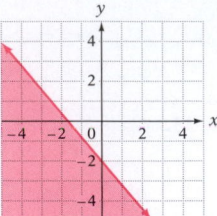

14. $2x + 2y \leq -4$

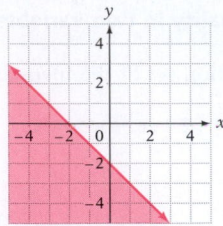

15. $-4x + 3y < -12$

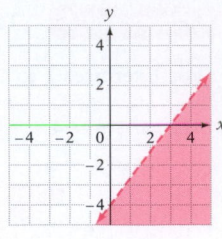

16. $-4x + 5y < 15$

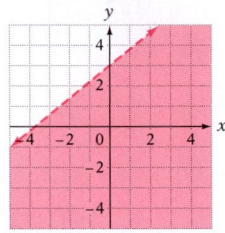

17. $-2x + 3y \leq 6$

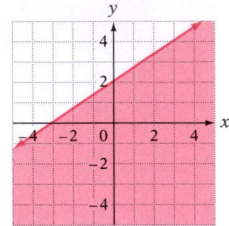

18. $3x - 4y > 12$

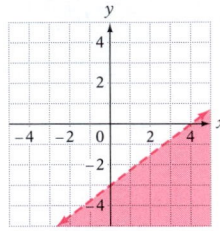

APPLYING THE CONCEPTS

For Exercises 19 to 21, graph the solution set of the inequality.

19. $\dfrac{x}{4} + \dfrac{y}{2} > 1$

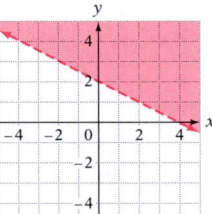

20. $2x - 3(y + 1) > y - (4 - x)$

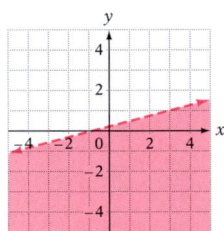

21. $4y - 2(x + 1) \geq 3(y - 1) + 3$

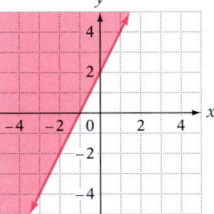

For Exercises 22 to 24, write the inequality given its graph.

22.

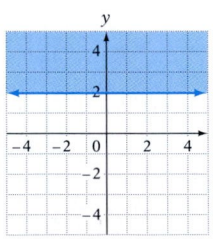

$y \geq 2$

23.

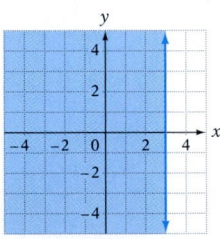

$x \leq 3$

24.

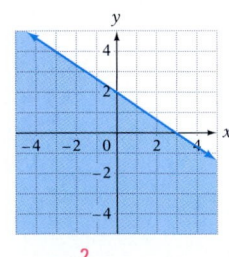

$y \leq -\dfrac{2}{3}x + 2$

Focus on Problem Solving

Graphing Data

Graphs are very useful in displaying data. By studying a graph, we can reach various conclusions about the data.

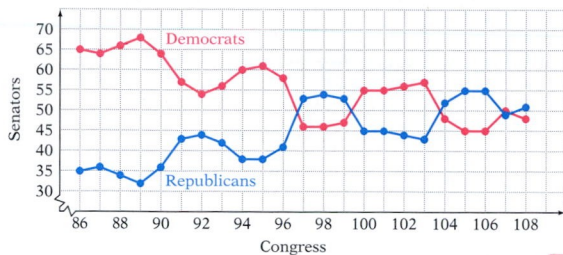

The double-line graph at the left shows the number of Democrats and the number of Republicans in the U.S. Senate for the 86th Congress (1959–1961) through the 108th Congress (2003–2005).

1. How many Democratic and how many Republican senators were in the 90th Congress?

2. In which Congress was the difference between the numbers of Democrats and Republicans the greatest?

3. In which Congress did the majority first change from Democratic to Republican?

4. Between which two Congresses did the number of Republican senators increase but the number of Democratic senators remain the same?

5. In what percent of the Congresses did the number of Democrats exceed the number of Republicans? Round to the nearest tenth.

6. In which Congresses were there a greater number of Republican senators than Democratic senators?

Year	DVD Recorders	DVD Players
2003	1	59
2004	2	62
2005	9	60
2006	18	54
2007	28	44
2008	44	34

The table at the left, based on data from Allied Business Intelligence, Inc., shows the actual and projected numbers (in millions) of worldwide shipments of DVD players and DVD recorders for the years 2003 through 2008.

7. Make a double-line graph of the data.

8. In which year will the number of DVD recorders shipped first exceed the number of DVD players shipped?

9. Based on these data, for the year 2009, would you expect the difference between the number of DVD recorders shipped and the number of DVD players shipped to be less than or greater than the difference in 2008?

Answers to Focus on Problem Solving: Graphing Data

1. Democrats: 64, Republicans: 36
2. 89th Congress
3. 97th Congress
4. 97th to 98th Congresses
5. 69.6%
6. 97th, 98th, 99th, 104th, 105th, 106th, 108th
7.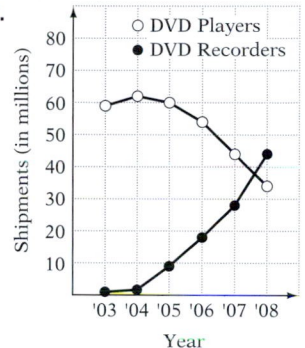
8. 2008
9. Greater. DVD player shipments are decreasing while DVD recorder shipments are increasing.

Projects and Group Activities

Standard Deviation and GPA

An automotive engineer tests the miles-per-gallon ratings of 15 cars and records the results as follows:

25 22 21 27 25 35 29 31 25 26 21 39 34 32 28

Recall that the **mean** of the data is the sum of the measurements divided by the number of measurements. The symbol for the mean is $\bar{x}$.

$$\text{Mean} = \bar{x} = \frac{\text{sum of all data values}}{\text{number of data values}}$$

To find the mean for the data on page 741, add the numbers and then divide by 15.

$$\bar{x} = \frac{25 + 22 + 21 + 27 + 25 + 35 + 29 + 31 + 25 + 26 + 21 + 39 + 34 + 32 + 28}{15}$$

$$= \frac{420}{15} = 28$$

The mean number of miles per gallon for the 15 cars tested is 28 mpg.

Consider two students, each of whom has taken five exams.

Scores for student A

| 84 | 86 | 83 | 85 | 87 |

Scores for student B

| 90 | 75 | 94 | 68 | 98 |

$$\bar{x} = \frac{84 + 86 + 83 + 85 + 87}{5} = \frac{425}{5} = 85 \qquad \bar{x} = \frac{90 + 75 + 94 + 68 + 98}{5} = \frac{425}{5} = 85$$

The mean score for student A is 85.　　　The mean score for student B is 85.

For each of these students, the mean (average) for the five exams is 85. However, student A has a more consistent record of scores than student B. One way to measure the consistency, or "clustering" near the mean, of data is to use the **standard deviation.**

To calculate the standard deviation:

Step 1. Sum the squares of the differences between each value of the data and the mean.

Step 2. Divide the result in Step 1 by the number of items in the data set.

Step 3. Take the square root of the result in Step 2.

The calculation for student A is shown at the right.

Step 1:

x	$x - \bar{x}$	$(x - \bar{x})^2$
84	$84 - 85$	$(-1)^2 = 1$
86	$86 - 85$	$1^2 = 1$
83	$83 - 85$	$(-2)^2 = 4$
85	$85 - 85$	$0^2 = 0$
87	$87 - 85$	$2^2 = 4$
		Total = 10

The symbol for standard deviation is the lowercase Greek letter *sigma*, σ.

Step 2: $\frac{10}{5} = 2$

Step 3: $\sigma = \sqrt{2} \approx 1.414$

The standard deviation for student A's scores is approximately 1.414.

Following a similar procedure for student B shows that the standard deviation for student B's scores is approximately 11.524. Because the standard deviation of student B's scores is greater than that of student A's scores (11.524 > 1.414), student B's scores are not as consistent as those of student A.

Another type of average is grade point average (GPA). It is calculated by multiplying the units for each class by the grade for that class, adding the results, and dividing by the total number of units taken. Here is an example using the grading scale A = 4, B = 3, C = 2, D = 1, and F = 0.

Class	Units	Grade
Math	4	B (= 3)
English	3	A (= 4)
French	5	C (= 2)
Biology	3	B (= 3)

$$\text{GPA} = \frac{4 \cdot 3 + 3 \cdot 4 + 5 \cdot 2 + 3 \cdot 3}{4 + 3 + 5 + 3} = \frac{43}{15} \approx 2.87$$

1. The weights in ounces of six newborn infants were recorded by a hospital. The weights were 96, 105, 84, 90, 102, and 99. Find the standard deviation of the weights. Round to the nearest hundredth.

2. The numbers of rooms occupied in a hotel on six consecutive days were 234, 321, 222, 246, 312, and 396. Find the standard deviation of the number of rooms occupied. Round to the nearest hundredth.

3. Seven coins were tossed 100 times. The numbers of heads recorded for each coin were 56, 63, 49, 50, 48, 53, and 52. Find the standard deviation of the number of heads. Round to the nearest hundredth.

4. The temperatures, in degrees Fahrenheit, for 11 consecutive days at a desert resort were 95°, 98°, 98°, 104°, 97°, 100°, 96°, 97°, 108°, 93°, and 104°. For the same days, temperatures in Antarctica were 27°, 28°, 28°, 30°, 28°, 27°, 30°, 25°, 24°, 26°, and 21°. Which location has the greater standard deviation of temperatures?

5. The scores for five college basketball games were 56, 68, 60, 72, and 64. The scores for five professional basketball games were 106, 118, 110, 122, and 114. Which scores have the greater standard deviation?

6. The weights in pounds of the players making up the five-man front line of a college football team are 210, 245, 220, 230, and 225. Find the standard deviation of the weights. Round to the nearest hundredth.

7. One student received test scores of 85, 92, 86, and 89. A second student received scores of 90, 97, 91, and 94 (exactly 5 points more on each test). Are the means of the two students the same? If not, what is the relationship between the means of the two students? Are the standard deviations of the scores of the two students the same? If not, what is the relationship between the standard deviations of the scores of the two students?

8. A grading scale that provides for plus or minus grades uses A = 4, A− = 3.7, B+ = 3.3, B = 3, B− = 2.7, C+ = 2.3, C = 2, C− = 1.7, D+ = 1.3, D = 1, D− = 0.7, and F = 0. Calculate the GPA of the student whose grades are given in the accompanying table.

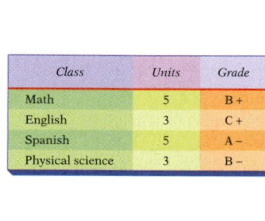

9. If you average 40 mph for 1 h and then 50 mph for 1 h, is your average speed $\frac{40 + 50}{2} = 45$ mph? Why or why not?

10. A company is negotiating with its employees the terms of a raise in salary. One proposal would add $500 a year to each employee's salary. The second proposal would give each employee a 4% raise. Explain how each of these proposals would affect the current mean and standard deviation of salaries for the company.

Chapter 14 Summary

Key Words	Examples

A *set* is a collection of objects. The objects are called the *elements* of the set. The *roster method* of writing a set encloses a list of the elements in braces. [14.1A, p. 719]

Using the roster method, the set of the first three positive integers is written {1, 2, 3}.

The *empty set* or *null set*, written ∅, is the set that contains no elements. [14.1A, p. 719]

The set of cars that can travel faster than 1000 mph is an empty set.

The *union* of two sets, written $A \cup B$, is the set that contains the elements of A and the elements of B. The *intersection* of two sets, written $A \cap B$, is the set that contains the elements that are common to both A and B. [14.1A, p. 719]

Let $A = \{2, 4, 6, 8\}$ and $B = \{0, 1, 2, 3, 4\}$. Then $A \cup B = \{0, 1, 2, 3, 4, 6, 8\}$ and $A \cap B = \{2, 4\}$.

Set-builder notation uses a rule to describe the elements of a set. [14.1B, p. 720]

Using set-builder notation, the set of real numbers greater than 2 is written $\{x \mid x > 2, x \in \text{real numbers}\}$.

An *inequality* is an expression that contains the symbol $<$, $>$, $\leq$, or $\geq$. The *solution set of an inequality* is a set of numbers each element of which, when substituted for the variable, results in a true inequality. The solution set of an inequality can be graphed on a number line. [14.1C, p. 721]

$3x - 1 < 5$ is an inequality. The solution set of $3x - 1 < 5$ is $\{x \mid x < 2\}$. The graph of the solution set is

−5 −4 −3 −2 −1 0 1 2 3 4 5

The solution set of a linear inequality in two variables is a *half-plane*. [14.4A, p. 737]

The solution set of $3x + 4y \geq 12$ is the half-plane shown at the right.

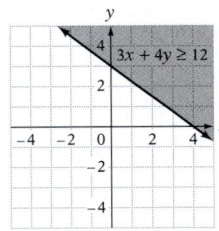

Essential Rules and Procedures

	Examples

Addition Property of Inequalities [14.2A, p. 725]
The same term can be added to each side of an inequality without changing the solution set of the inequality.

If $a > b$, then $a + c > b + c$.
If $a < b$, then $a + c < b + c$.

$$x - 3 < -7$$
$$x - 3 + 3 < -7 + 3$$
$$x < -4$$

Multiplication Property of Inequalities [14.2B, p. 727]
Each side of an inequality can be multiplied by the same positive number without changing the solution set of the inequality.

If $a > b$ and $c > 0$, then $ac > bc$.
If $a < b$ and $c > 0$, then $ac < bc$.

$$4x > -8$$
$$\frac{4x}{4} > \frac{-8}{4}$$
$$x > -2$$

If each side of an inequality is multiplied by the same negative number and the inequality symbol is reversed, then the solution set of the inequality is not changed.

If $a > b$ and $c < 0$, then $ac < bc$.
If $a < b$ and $c < 0$, then $ac > bc$.

$$-2x < 6$$
$$\frac{-2x}{-2} > \frac{6}{-2}$$
$$x > -3$$

Chapter 14 Review Exercises

1. Solve: $2x - 3 > x + 15$

$x > 18$ [14.2A]

2. Find $A \cap B$, given $A = \{0, 2, 4, 6, 8\}$ and $B = \{-2, -4\}$.

$A \cap B = \varnothing$ [14.1A]

3. Use set-builder notation to write the set of odd integers greater than -8.

$\{x \mid x > -8, x \in \text{odd integers}\}$ [14.1B]

4. Find $A \cup B$, given $A = \{6, 8, 10\}$ and $B = \{2, 4, 6\}$.

$A \cup B = \{2, 4, 6, 8, 10\}$ [14.1A]

5. Use the roster method to write the set of odd positive integers less than 8.

$A = \{1, 3, 5, 7\}$ [14.1A]

6. Solve: $12 - 4(x - 1) \le 5(x - 4)$

$x \ge 4$ [14.3A]

7. Graph: $\{x \mid x > 3\}$

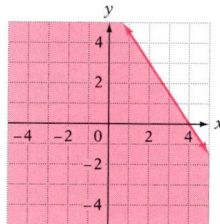   [14.1C]

8. Solve: $3x + 4 \ge -8$

$x \ge -4$ [14.3A]

9. Graph: $3x + 2y \le 12$

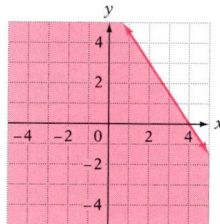   [14.4A]

10. Graph: $5x + 2y < 6$

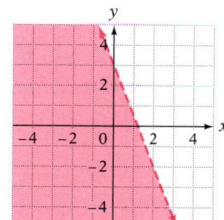   [14.4A]

11. Use set-builder notation to write the set of real numbers greater than 3.

$\{x \mid x > 3, x \in \text{real numbers}\}$ [14.1B]

12. Solve and graph the solution set of $x - 3 > -1$. $x > 2$

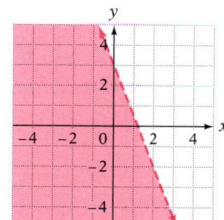   [14.2A]

13. Find $A \cap B$, given $A = \{1, 5, 9, 13\}$ and $B = \{1, 3, 5, 7, 9\}$.

$A \cap B = \{1, 5, 9\}$ [14.1A]

14. Graph: $\{x \mid x < 2\} \cup \{x \mid x > 5\}$

   [14.1C]

15. Graph: $\{x | x > -1\} \cap \{x | x \leq 2\}$

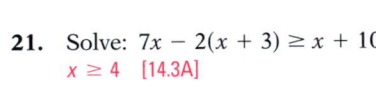

     -5 -4 -3 -2 -1   0   1   2   3   4   5   [14.1C]

16. Solve: $-15x \leq 45$

     $x \geq -3$ [14.2B]

17. Solve: $6x - 9 < 4x + 3(x + 3)$

     $x > -18$ [14.3A]

18. Solve: $5 - 4(x + 9) > 11(12x - 9)$

     $x < \dfrac{1}{2}$ [14.3A]

19. Solve: $-\dfrac{3}{4}x > \dfrac{2}{3}$

     $x < -\dfrac{8}{9}$ [14.2B]

20. Graph: $2x - 3y < 9$

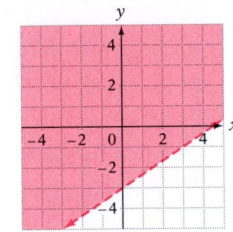

     [14.4A]

21. Solve: $7x - 2(x + 3) \geq x + 10$
     $x \geq 4$ [14.3A]

22. **Floral Delivery**   Florist A charges a \$3 delivery fee plus \$21 per bouquet delivered. Florist B charges a \$15 delivery fee plus \$18 per bouquet delivered. A church wants to supply each resident of a small nursing home with a bouquet for Grandparents Day. Find the number of residents of the nursing home if using florist B is more economical than using florist A.
5 or more   [14.3B]

23. **Landscaping**   The width of a rectangular garden is 12 ft. The length of the garden is $(3x + 5)$ ft. Express as an integer the minimum length of the garden when the area is greater than 276 ft². (The area of a rectangle is equal to its length times its width.)
24 ft   [14.3B]

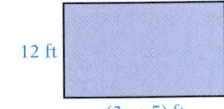

12 ft

$(3x + 5)$ ft

24. **Number Sense**   Six less than a number is greater than twenty-five. Find the smallest integer that will satisfy the inequality.
32   [14.2C]

25. **Grading**   A student's grades on five sociology tests were 68, 82, 90, 73, and 95. What is the lowest score the student can receive on the next test and still be able to attain a minimum of 480 points?
72   [14.2C]

Chapter 14 Test

1. Graph: $\{x | x < 5\} \cap \{x | x > 0\}$

   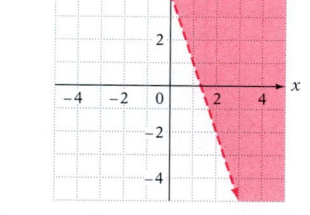

   [14.1C]

2. Use set-builder notation to write the set of positive integers less than 50.
   $\{x | x < 50, x \in \text{positive integers}\}$ [14.1B]

3. Use the roster method to write the set of even positive integers between 3 and 9.
   $A = \{4, 6, 8\}$ [14.1A]

4. Solve: $3(2x - 5) \geq 8x - 9$

   $x \leq -3$ [14.3A]

5. Solve: $x + \dfrac{1}{2} > \dfrac{5}{8}$

   $x > \dfrac{1}{8}$ [14.2A]

6. Graph: $\{x | x > -2\}$

   ![number line graph] [14.1C]

7. Solve: $5 - 3x > 8$

   $x < -1$ [14.3A]

8. Use set-builder notation to write the set of real numbers greater than -23.
   $\{x | x > -23, x \in \text{real numbers}\}$ [14.1B]

9. Graph the solution set of $3x + y > 4$.

   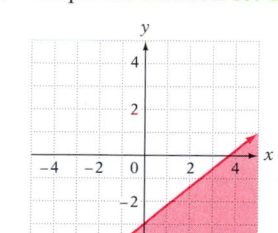

   [14.4A]

10. Graph the solution set of $4x - 5y \geq 15$.

    [14.4A]

11. Find $A \cap B$, given $A = \{6, 8, 10, 12\}$ and $B = \{12, 14, 16\}$.
    $A \cap B = \{12\}$ [14.1A]

12. Solve and graph the solution set of $4 + x < 1$.

    $x < -3$ [14.2A]

13. Solve: $-\dfrac{3}{8}x \leq 5$

    $x \geq -\dfrac{40}{3}$ [14.2B]

14. Solve: $6x - 3(2 - 3x) < 4(2x - 7)$

    $x < -\dfrac{22}{7}$ [14.3A]

15. Solve and graph the solution set of $\frac{2}{3}x \geq 2$.

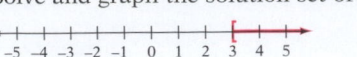

$x \geq 3$ [14.2B]

16. Solve: $2x - 7 \leq 6x + 9$

$x \geq -4$ [14.3A]

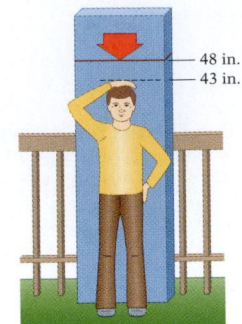

48 in.
43 in.

17. Safety   To ride a certain roller coaster at an amusement park, a person must be at least 48 in. tall. How many inches must a child who is 43 in. tall grow to be eligible to ride the roller coaster?

≥ 5 in.   [14.2C]

18. Geometry   A rectangle is 15 ft long and $(2x - 4)$ ft wide. Express as an integer the maximum width of the rectangle if the area is less than 180 ft². (The area of a rectangle is equal to its length times its width.)

11 ft   [14.3B]

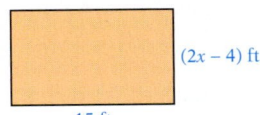

$(2x - 4)$ ft

15 ft

19. Machining   A ball bearing for a rotary engine must have a circumference between 0.1220 in. and 0.1240 in. What are the allowable diameters for the bearing? Round to the nearest ten-thousandth. Recall that $C = \pi d$.

Between 0.0388 in. and 0.0395 in.   [14.2C]

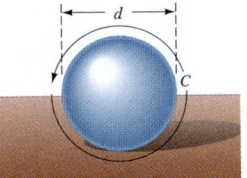

20. Compensation   A stockbroker receives a monthly salary that is the greater of $2500 or $1000 plus 2% of the total value of all stock transactions the broker processes during the month. What dollar amounts of transactions did the broker process in a month for which the broker's salary was $2500?

$\leq$ $75,000   [14.3B]

Cumulative Review Exercises

1. Simplify: $2[5a - 3(2 - 5a) - 8]$

$40a - 28$ [4.2D]

2. Solve: $\dfrac{5}{8} - 4x = \dfrac{1}{8}$

$\dfrac{1}{8}$ [5.2A]

3. Solve: $2x - 3[x - 2(x - 3)] = 2$

4 [5.3B]

4. Simplify: $(-3a)(-2a^3b^2)^2$

$-12a^7b^4$ [9.2B]

5. Simplify: $\dfrac{27a^3b^2}{(-3ab^2)^3}$

$-\dfrac{1}{b^4}$ [9.4A]

6. Divide: $(16x^2 - 12x - 2) \div (4x - 1)$

$4x - 2 - \dfrac{4}{4x - 1}$ [9.5B]

7. Given $f(x) = x^2 - 4x - 5$, find $f(-1)$.

0 [12.1D]

8. Factor: $27a^2x^2 - 3a^2$

$3a^2(3x - 1)(3x + 1)$ [10.4B]

9. Divide: $\dfrac{x^2 - 2x}{x^2 - 2x - 8} \div \dfrac{x^3 - 5x^2 + 6x}{x^2 - 7x + 12}$

$\dfrac{1}{x + 2}$ [11.1C]

10. Subtract: $\dfrac{4a}{2a - 3} - \dfrac{2a}{a + 3}$

$\dfrac{18a}{(2a - 3)(a + 3)}$ [11.3B]

11. Solve: $\dfrac{5y}{6} - \dfrac{5}{9} = \dfrac{y}{3} - \dfrac{5}{6}$

$-\dfrac{5}{9}$ [11.5A]

12. Solve $R = \dfrac{C - S}{t}$ for C.

$C = S + Rt$ [11.6A]

13. Find the slope of the line that passes through the points $(2, -3)$ and $(-1, 4)$.

$-\dfrac{7}{3}$ [12.3B]

14. Find the equation of the line that passes through the point $(1, -3)$ and has slope $-\dfrac{3}{2}$.

$y = -\dfrac{3}{2}x - \dfrac{3}{2}$ [12.4A]

15. Solve by substitution.
$$x = 3y + 1$$
$$2x + 5y = 13$$
$(4, 1)$ [13.2A]

16. Solve by the addition method.
$$9x - 2y = 17$$
$$5x + 3y = -7$$
$(1, -4)$ [13.3A]

17. Find $A \cup B$, given $A = \{0, 1, 2\}$ and $B = \{-10, -2\}$.

$A \cup B = \{-10, -2, 0, 1, 2\}$ [14.1A]

18. Use set-builder notation to write the set of real numbers less than 48.

$\{x \mid x < 48, x \in \text{real numbers}\}$ [14.1B]

19. Graph: $\{x \mid x > 1\} \cup \{x \mid x < -1\}$

$$\xleftarrow{\;\;\; -5 \;\; -4 \;\; -3 \;\; -2 \;\; -1 \;\;\; 0 \;\;\; 1 \;\;\; 2 \;\;\; 3 \;\;\; 4 \;\;\; 5 \;\;\;}$$ [14.1C]

20. Graph the solution set of $\frac{3}{8}x > -\frac{3}{4}$.

$$\xleftarrow{\;\;\; -5 \;\; -4 \;\; -3 \;\; -2 \;\; -1 \;\;\; 0 \;\;\; 1 \;\;\; 2 \;\;\; 3 \;\;\; 4 \;\;\; 5 \;\;\;}$$ [14.2B]

21. Solve: $-\frac{4}{5}x > 12$

$x < -15$ [14.2B]

22. Solve: $15 - 3(5x - 7) < 2(7 - 2x)$

$x > 2$ [14.3A]

23. **Number Sense**   Three-fifths of a number is less than negative fifteen. What integers satisfy this inequality? Write the answer in set-builder notation.

$\{x \mid x \leq -26, x \in \text{integers}\}$ [14.2C]

24. **Rental Agencies**   Company A rents cars for $6 a day and $.25 for every mile driven. Company B rents cars for $15 a day and $.10 per mile. You want to rent a car for six days. What is the maximum number of miles you can drive a Company A car if it is to cost you less than a Company B car?

359 mi   [14.3B]

25. **Conservation**   In a lake, 100 fish are caught, tagged, and then released. Later, 150 fish are caught. Three of these 150 fish are found to have tags. Estimate the number of fish in the lake.

5000 fish   [11.6B]

26. **Geometry**   The measure of the first angle of a triangle is 30° more than the measure of the second angle. The measure of the third angle is 10° more than twice the measure of the second angle. Find the measure of each angle.

65°, 35°, and 80°   [7.1C]

27. Graph: $y = 2x - 1$

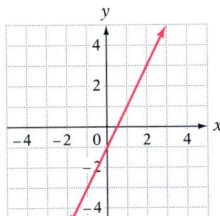

[12.2A]

28. Graph the solution set of $6x - 3y \geq 6$.

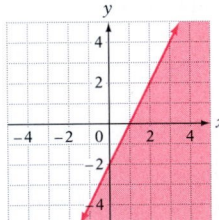

[14.4A]

chapter 15

Radical Expressions

Think back to some of the carnival rides you enjoyed as a child. Was your favorite ride the Ferris wheel, or the tilt-a-whirl? Was it the scrambler, or the carousel? All of these rides involve mathematics and physics. **Exercise 29 on page 774** uses a formula to describe the speed of a child riding a merry-go-round. The formula, which uses a radical expression, is $v = \sqrt{12r}$, where v is the speed in feet per second and r is the distance in feet from the center of the merry-go-round to the rider. You will learn more about radical expressions in this chapter.

OBJECTIVES

Section 15.1
A To simplify numerical radical expressions
B To simplify variable radical expressions

Section 15.2
A To add and subtract radical expressions

Section 15.3
A To multiply radical expressions
B To divide radical expressions

Section 15.4
A To solve an equation containing a radical expression
B To solve application problems

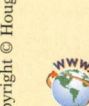

Need help? For online student resources, such as section quizzes, visit this textbook's website at **math.college.hmco.com/students**.

PREP TEST • • •

Do these exercises to prepare for Chapter 15.

1. Evaluate: $-|-14|$
   -14 [3.1C]

2. Simplify: $3x^2y - 4xy^2 - 5x^2y$
   $-2x^2y - 4xy^2$ [4.2A]

3. Solve: $1.5h = 21$
   14 [5.1C]

4. Solve: $3x - 2 = 5 - 2x$
   $\dfrac{7}{5}$ [5.3A]

5. Simplify: $x^3 \cdot x^3$
   x^6 [9.2A]

6. Expand: $(x + y)^2$
   $x^2 + 2xy + y^2$ [9.3D]

7. Expand: $(2x - 3)^2$
   $4x^2 - 12x + 9$ [9.3D]

8. Multiply: $(2 - 3v)(2 + 3v)$
   $4 - 9v^2$ [9.3D]

9. Multiply: $(a - 5)(a + 5)$
   $a^2 - 25$ [9.3D]

10. Simplify: $\dfrac{2x^4y^3}{18x^2y}$
    $\dfrac{x^2y^2}{9}$ [9.4A]

GO FIGURE • • •

Luis, Kim, Reggie, and Dave are standing in line. Dave is not first. Kim is between Luis and Reggie. Luis is between Dave and Kim. Give the order in which the men are standing. Reggie, Kim, Luis, Dave

15.1 Introduction to Radical Expressions

Objective A

To simplify numerical radical expressions

A **square root** of a positive number a is a number whose square is a.

A square root of 16 is 4 because $4^2 = 16$.
A square root of 16 is -4 because $(-4)^2 = 16$.

Every positive number has two square roots, one a positive and one a negative number. The symbol $\sqrt{}$, called a **radical sign,** is used to indicate the positive or **principal square root** of a number. For example, $\sqrt{16} = 4$ and $\sqrt{25} = 5$. The number under the radical sign is called the **radicand.**

When the negative square root of a number is to be found, a negative sign is placed in front of the radical. For example, $-\sqrt{16} = -4$ and $-\sqrt{25} = -5$.

The square of an integer is a **perfect square.** For instance, 49, 81, and 144 are perfect squares. The principal square root of a perfect-square integer is a positive integer.

$$7^2 = 49 \qquad \sqrt{49} = 7$$
$$9^2 = 81 \qquad \sqrt{81} = 9$$
$$12^2 = 144 \qquad \sqrt{144} = 12$$

If a number is not a perfect square, its square root can only be approximated. For example, 2 and 7 are not perfect squares. The square roots of these numbers are *irrational numbers*. Their decimal approximations never terminate or repeat.

$$\sqrt{2} \approx 1.4142135\ldots \qquad \sqrt{7} \approx 2.6457513\ldots$$

Radical expressions that contain radicands that are not perfect squares are frequently written in simplest form. A radical expression is in *simplest form* when the radicand contains no factor greater than 1 that is a perfect square. For instance, $\sqrt{50}$ is not in simplest form because 25 is a perfect-square factor of 50. The radical expression $\sqrt{15}$ is in simplest form because there are no perfect-square factors of 15 that are greater than 1.

The Product Property of Square Roots and a knowledge of perfect squares is used to simplify radicands that are not perfect squares.

> **The Product Property of Square Roots**
>
> If a and b are positive real numbers, then $\sqrt{ab} = \sqrt{a} \cdot \sqrt{b}$.

The chart below shows the square roots of some perfect squares.

> **Square Roots of Perfect Squares**
>
> | | | | |
> |---|---|---|---|
> | $\sqrt{1} = 1$ | $\sqrt{16} = 4$ | $\sqrt{49} = 7$ | $\sqrt{100} = 10$ |
> | $\sqrt{4} = 2$ | $\sqrt{25} = 5$ | $\sqrt{64} = 8$ | $\sqrt{121} = 11$ |
> | $\sqrt{9} = 3$ | $\sqrt{36} = 6$ | $\sqrt{81} = 9$ | $\sqrt{144} = 12$ |

HOW TO Simplify: $\sqrt{72}$

$\sqrt{72} = \sqrt{36 \cdot 2}$ • Write the radicand as the product of **a perfect square and a factor that does not contain a perfect square.**

$\phantom{\sqrt{72}} = \sqrt{36}\sqrt{2}$ • Use the Product Property of Square Roots to write the expression as a product.

$\phantom{\sqrt{72}} = 6\sqrt{2}$ • Simplify.

Objective 15.1A

Vocabulary to Review
irrational numbers   [3.4A]

New Vocabulary
square root
radical sign
principal square root
radicand
perfect square

Discuss the Concepts
Explain the difference between *square root* and *principal square root*. Give an example of a square root that is not a principal square root.
Answers will vary. A square root of a positive number can be positive or negative. A principal square root of a positive number is only positive. For example, the two square roots of 25 are 5 and -5. The principal square root of 25, signified by $\sqrt{25}$, is 5. Thus -5 is a square root that is not a principal square root.

Concept Check
1. Name the two square roots of 64. $-8, 8$
2. Simplify $\sqrt{64}$.   8
3. Explain why the answer to part 1 is -8 and 8 but the answer to part 2 is just 8.
   $\sqrt{}$ means the principal square root.

Discuss the Concepts
Is $\sqrt{12}$ in simplest form? Explain your answer.
No. One of the factors of 12 is 4, which is a perfect square. A radical expression is in simplest form when the radicand contains no factor greater than 1 that is a perfect square.

In-Class Examples (Objective 15.1A)
Simplify.
1. $7\sqrt{48}$ $28\sqrt{3}$
2. $\sqrt{675}$ $15\sqrt{3}$
3. $3\sqrt{12}$ $6\sqrt{3}$

Discuss the Concepts

Explain the difference between *simplifying* and *approximating* principal square roots. For instance:

a. Simplify $\sqrt{20}$. $2\sqrt{5}$

b. Approximate $\sqrt{20}$.
   ≈ 4.472135955

Discuss the Concepts

1. Is $-\sqrt{49}$ a real number? Explain.   Yes; $-\sqrt{49} = -7$

2. Is $\sqrt{-49}$ a real number? Explain.
   No. There is no real number whose square is -49.

Instructor Note

Have students who are having difficulty finding a perfect-square factor write the prime factorization of a number. For instance,

$$\sqrt{288} = \sqrt{2^5 3^2} = \sqrt{(2^4 3^2)2}$$
$$= \sqrt{2^4 3^2}\sqrt{2} = 2^2 \cdot 3\sqrt{2}$$
$$= 12\sqrt{2}$$

Optional Student Activity

The distance from home plate to second base on a major-league baseball field is $\sqrt{16{,}200}$ ft.

1. Simplify $\sqrt{16{,}200}$ ft. $90\sqrt{2}$ ft

2. Determine the distance from home plate to second base to the nearest inch.   1527 in.

Note that 72 must be written as the product of a perfect square and a *factor that does not contain a perfect square*. Therefore, it would not be correct to simplify $\sqrt{72}$ as $\sqrt{9 \cdot 8}$. Although 9 is a perfect-square factor of 72, 8 also contains a perfect-square factor ($8 = 4 \cdot 2$). Therefore, $\sqrt{8}$ is not in simplest form. Remember to find the largest perfect-square factor of the radicand.

$$\sqrt{72} = \sqrt{9 \cdot 8}$$
$$= \sqrt{9} \cdot \sqrt{8}$$
$$= 3\sqrt{8}$$

Not in simplest form

HOW TO   Simplify: $\sqrt{147}$

$$\sqrt{147} = \sqrt{49 \cdot 3}$$
- Write the radicand as the product of **a perfect square** and a factor that does not contain a perfect square.

$$= \sqrt{49}\sqrt{3}$$
- Use the Product Property of Square Roots to write the expression as a product.

$$= 7\sqrt{3}$$
- Simplify.

HOW TO   Simplify: $\sqrt{360}$

$$\sqrt{360} = \sqrt{36 \cdot 10}$$
- Write the radicand as the product of **a perfect square** and a factor that does not contain a perfect square.

$$= \sqrt{36}\sqrt{10}$$
- Use the Product Property of Square Roots to write the expression as a product.

$$= 6\sqrt{10}$$
- Simplify.

From the last example, note that $\sqrt{360} = 6\sqrt{10}$. The two expressions are different representations of the same number. Using a calculator, we find that $\sqrt{360} \approx 18.973666$ and $6\sqrt{10} \approx 6(3.1622777) \approx 18.973666$.

HOW TO   Simplify: $\sqrt{-16}$

Because the square of any real number is positive, there is no real number whose square is -16. $\sqrt{-16}$ is not a real number.

> **Study Tip**
>
> Be sure you understand how to simplify expressions such as those in Example 1 and Example 2, as this skill is a prerequisite for solving quadratic equations in Chapter 16.

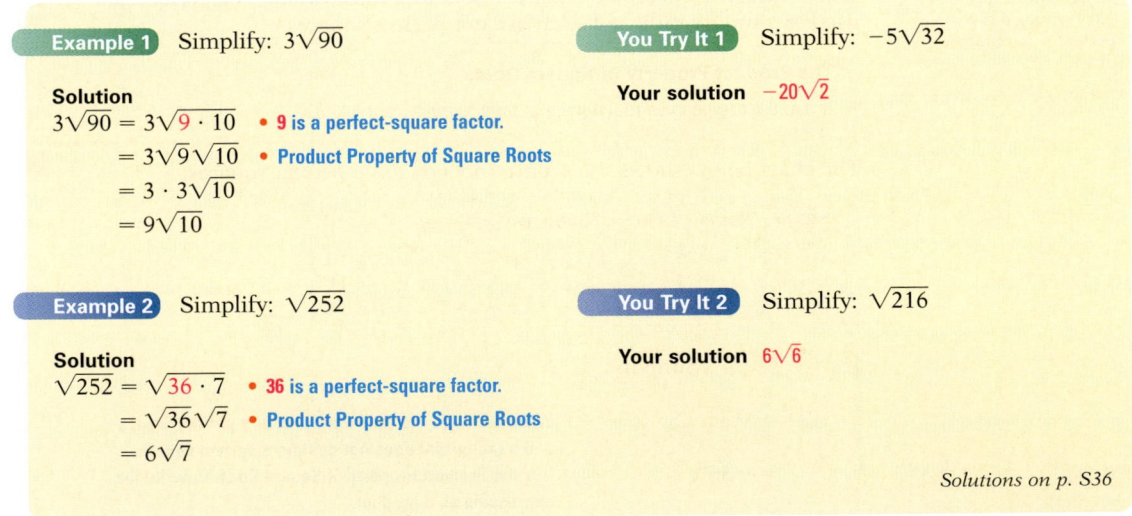

Example 1   Simplify: $3\sqrt{90}$

Solution
$$3\sqrt{90} = 3\sqrt{9 \cdot 10}$$ • **9** is a perfect-square factor.
$$= 3\sqrt{9}\sqrt{10}$$ • **Product Property of Square Roots**
$$= 3 \cdot 3\sqrt{10}$$
$$= 9\sqrt{10}$$

You Try It 1   Simplify: $-5\sqrt{32}$

Your solution $-20\sqrt{2}$

Example 2   Simplify: $\sqrt{252}$

Solution
$$\sqrt{252} = \sqrt{36 \cdot 7}$$ • **36** is a perfect-square factor.
$$= \sqrt{36}\sqrt{7}$$ • **Product Property of Square Roots**
$$= 6\sqrt{7}$$

You Try It 2   Simplify: $\sqrt{216}$

Your solution $6\sqrt{6}$

Solutions on p. S36

Objective B   **To simplify variable radical expressions**

Variable expressions that contain radicals do not always represent real numbers. For example, if $a = -4$, then

$$\sqrt{a^3} = \sqrt{(-4)^3} = \sqrt{-64}$$

and $\sqrt{-64}$ is not a real number.

Now consider the expression $\sqrt{x^2}$. Evaluate this expression for $x = -2$ and $x = 2$.

$$\sqrt{x^2}$$
$$\sqrt{(-2)^2} = \sqrt{4} = 2 = |-2|$$

$$\sqrt{x^2}$$
$$\sqrt{2^2} = \sqrt{4} = 2 = |2|$$

This suggests the following:

For any real number a, $\sqrt{a^2} = |a|$. If $a \geq 0$, then $\sqrt{a^2} = a$.

In order to avoid variable expressions that do not represent real numbers, and so that absolute value signs are not needed for certain expressions, the variables in this chapter will represent *positive* numbers unless otherwise stated.

A variable or a product of variables written in exponential form is a perfect square when each exponent is an even number.

To find the square root of a perfect square, remove the radical sign and multiply each exponent by $\frac{1}{2}$.

HOW TO   Simplify: $\sqrt{a^6}$

$\sqrt{a^6} = a^3$ • Remove the radical sign and multiply the exponent by $\frac{1}{2}$.

A variable radical expression is in simplest form when the radicand contains no factor greater than 1 that is a perfect square.

HOW TO   Simplify: $\sqrt{x^7}$

$$\sqrt{x^7} = \sqrt{x^6 \cdot x}$$ • Write x^7 as the product of **a perfect square** and x.
$$= \sqrt{x^6}\sqrt{x}$$ • Use the Product Property of Square Roots.
$$= x^3\sqrt{x}$$ • Simplify the perfect square.

HOW TO   Simplify: $3x\sqrt{8x^3y^{13}}$

$$3x\sqrt{8x^3y^{13}} = 3x\sqrt{4x^2y^{12}(2xy)}$$ • Write the radicand as the product of **perfect squares** and factors that do not contain a perfect square.

$$= 3x\sqrt{4x^2y^{12}}\sqrt{2xy}$$ • Use the Product Property of Square Roots.
$$= 3x \cdot 2xy^6\sqrt{2xy}$$ • Simplify.
$$= 6x^2y^6\sqrt{2xy}$$

Concept Check

Each of the following is incorrectly simplified. Correct the error in the simplification.
1. $\sqrt{a^{16}} = a^4$ a^8
2. $\sqrt{48x^2} = \sqrt{4x^2}\sqrt{12}$
          $= 2x\sqrt{12}$ $4x\sqrt{3}$

Concept Check

The approximate speed, in feet per second, at which a pole vaulter must run to jump a height of h feet is given by $\sqrt{64h}$. Write this expression in simplest form. What is the approximate speed of a pole vaulter who jumps to a height of 20 ft? Round to the nearest tenth.
$8\sqrt{h}$, 35.8 ft/s

Optional Student Activity

1. Explain in words the procedure for simplifying $\sqrt{27}$.
2. Explain in words the procedure for simplifying $\sqrt{x^{27}}$.

In-Class Examples (Objective 15.1B)
Simplify.
1. $\sqrt{b^{11}}$ $b^5\sqrt{b}$
2. $\sqrt{80a^{15}}$ $4a^7\sqrt{5a}$
3. $8x\sqrt{72x^7y^{19}}$ $48x^4y^9\sqrt{2xy}$
4. $\sqrt{81(x+4)^2}$ $9x + 36$
5. $\sqrt{a^2 + 12a + 36}$ $a + 6$

Instructor Note

As a class exercise, have students verify that $\sqrt{a^2 + b^2} \neq a + b$ by using $a = 3$ and $b = 4$.

Optional Student Activity

The theoretical maximum speed v (in meters per second) of a roller coaster at the bottom of a hill is $v = \sqrt{19.6h}$, where h is the height of the hill in meters. By what factor must the height increase in order to double the speed at the bottom of the hill?

4

HOW TO Simplify: $\sqrt{25(x + 2)^2}$

$$\sqrt{25(x + 2)^2} = 5(x + 2)$$
$$= 5x + 10$$

Example 3

Simplify: $\sqrt{b^{15}}$

Solution
$\sqrt{b^{15}} = \sqrt{b^{14} \cdot b}$ • b^{14} **is a perfect square.**
$= \sqrt{b^{14}} \cdot \sqrt{b}$
$= b^7\sqrt{b}$

You Try It 3

Simplify: $\sqrt{y^{19}}$

Your solution
$y^9\sqrt{y}$

Example 4

Simplify: $\sqrt{24x^5}$

Solution
$\sqrt{24x^5} = \sqrt{4x^4(6x)}$ • **4 and** x^4 **are perfect squares.**
$= \sqrt{4x^4}\sqrt{6x}$
$= 2x^2\sqrt{6x}$

You Try It 4

Simplify: $\sqrt{45b^7}$

Your solution
$3b^3\sqrt{5b}$

Example 5

Simplify: $2a\sqrt{18a^3b^{10}}$

Solution
$2a\sqrt{18a^3b^{10}}$
$= 2a\sqrt{9a^2b^{10}(2a)}$ • **9,** a^2**, and** b^{10} **are perfect squares.**
$= 2a\sqrt{9a^2b^{10}}\sqrt{2a}$
$= 2a \cdot 3ab^5\sqrt{2a}$
$= 6a^2b^5\sqrt{2a}$

You Try It 5

Simplify: $3a\sqrt{28a^9b^{18}}$

Your solution
$6a^5b^9\sqrt{7a}$

Example 6

Simplify: $\sqrt{16(x + 5)^2}$

Solution
$\sqrt{16(x + 5)^2} = 4(x + 5) = 4x + 20$

You Try It 6

Simplify: $\sqrt{25(a + 3)^2}$

Your solution
$5a + 15$

Example 7

Simplify: $\sqrt{x^2 + 10x + 25}$

Solution
$\sqrt{x^2 + 10x + 25} = \sqrt{(x + 5)^2} = x + 5$

You Try It 7

Simplify: $\sqrt{x^2 + 14x + 49}$

Your solution
$x + 7$

Solutions on p. S36

15.1 Exercises

Objective A   To simplify numerical radical expressions

1. Describe in your own words how to simplify a radical expression.

2. Explain why $2\sqrt{2}$ is in simplest form and $\sqrt{8}$ is not in simplest form.

For Exercises 3 to 26, simplify.

3. $\sqrt{16}$
   4

4. $\sqrt{64}$
   8

5. $\sqrt{49}$
   7

6. $\sqrt{144}$
   12

7. $\sqrt{32}$
   $4\sqrt{2}$

8. $\sqrt{50}$
   $5\sqrt{2}$

9. $\sqrt{8}$
   $2\sqrt{2}$

10. $\sqrt{12}$
    $2\sqrt{3}$

11. $6\sqrt{18}$
    $18\sqrt{2}$

12. $-3\sqrt{48}$
    $-12\sqrt{3}$

13. $5\sqrt{40}$
    $10\sqrt{10}$

14. $2\sqrt{28}$
    $4\sqrt{7}$

15. $\sqrt{15}$
    $\sqrt{15}$

16. $\sqrt{21}$
    $\sqrt{21}$

17. $\sqrt{29}$
    $\sqrt{29}$

18. $\sqrt{13}$
    $\sqrt{13}$

19. $-9\sqrt{72}$
    $-54\sqrt{2}$

20. $11\sqrt{80}$
    $44\sqrt{5}$

21. $\sqrt{45}$
    $3\sqrt{5}$

22. $\sqrt{225}$
    15

23. $\sqrt{0}$
    0

24. $\sqrt{210}$
    $\sqrt{210}$

25. $6\sqrt{128}$
    $48\sqrt{2}$

26. $9\sqrt{288}$
    $108\sqrt{2}$

For Exercises 27 to 32, find the decimal approximation rounded to the nearest thousandth.

27. $\sqrt{240}$
    15.492

28. $\sqrt{300}$
    17.321

29. $\sqrt{288}$
    16.971

30. $\sqrt{600}$
    24.495

31. $\sqrt{256}$
    16

32. $\sqrt{324}$
    18

Objective B   To simplify variable radical expressions

For Exercises 33 to 72, simplify.

33. $\sqrt{x^6}$
    x^3

34. $\sqrt{x^{12}}$
    x^6

35. $\sqrt{y^{15}}$
    $y^7\sqrt{y}$

36. $\sqrt{y^{11}}$
    $y^5\sqrt{y}$

37. $\sqrt{a^{20}}$
    a^{10}

38. $\sqrt{a^{16}}$
    a^8

39. $\sqrt{x^4y^4}$
    x^2y^2

40. $\sqrt{x^{12}y^8}$
    x^6y^4

41. $\sqrt{4x^4}$
    $2x^2$

42. $\sqrt{25y^8}$
    $5y^4$

43. $\sqrt{24x^2}$
    $2x\sqrt{6}$

44. $\sqrt{x^3y^{15}}$
    $xy^7\sqrt{xy}$

45. $\sqrt{60x^5}$
    $2x^2\sqrt{15x}$

46. $\sqrt{72y^7}$
    $6y^3\sqrt{2y}$

47. $\sqrt{49a^4b^8}$
    $7a^2b^4$

48. $\sqrt{144x^2y^8}$
    $12xy^4$

49. $\sqrt{18x^5y^7}$
    $3x^2y^3\sqrt{2xy}$

50. $\sqrt{32a^5b^{15}}$
    $4a^2b^7\sqrt{2ab}$

51. $\sqrt{40x^{11}y^7}$
    $2x^5y^3\sqrt{10xy}$

52. $\sqrt{72x^9y^3}$
    $6x^4y\sqrt{2xy}$

53. $\sqrt{80a^9b^{10}}$
    $4a^4b^5\sqrt{5a}$

54. $\sqrt{96a^5b^7}$
    $4a^2b^3\sqrt{6ab}$

55. $2\sqrt{16a^2b^3}$
    $8ab\sqrt{b}$

56. $5\sqrt{25a^4b^7}$
    $25a^2b^3\sqrt{b}$

Quick Quiz (Objective 15.1A)

Simplify.
1. $\sqrt{36}$ 6
2. $\sqrt{18}$ $3\sqrt{2}$
3. $6\sqrt{75}$ $30\sqrt{3}$
4. $\sqrt{11}$ $\sqrt{11}$

Quick Quiz (Objective 15.1B)

Simplify.
1. $\sqrt{x^{10}}$ x^5
2. $\sqrt{x^{15}}$ $x^7\sqrt{x}$
3. $8\sqrt{45x^5y^9}$ $24x^2y^4\sqrt{5xy}$
4. $\sqrt{16(a+5)^2}$ $4a+20$

Suggested Assignment

Exercises 1–71, odds
More challenging problems:
Exercises 73, 74, 77–79

Answers to Writing Exercises

1. Students should explain that to simplify a radical expression, we first write the radicand as the product of a perfect-square factor and a factor that does not contain a perfect square. Take the square root of the perfect-square factor; write the square root in front of the radical sign. The other factor remains under the radical sign.

2. $2\sqrt{2}$ is in simplest form because the radicand does not have a perfect-square factor greater than 1. $\sqrt{8}$ is not in simplest form because the radicand does have a perfect-square factor greater than 1; 4, which is a perfect square, is a factor of 8.

57. $x\sqrt{x^4y^2}$
x^3y

58. $y\sqrt{x^3y^6}$
$xy^4\sqrt{x}$

59. $4\sqrt{20a^4b^7}$
$8a^2b^3\sqrt{5b}$

60. $5\sqrt{12a^3b^4}$
$10ab^2\sqrt{3a}$

61. $3x\sqrt{12x^2y^7}$
$6x^2y^3\sqrt{3y}$

62. $4y\sqrt{18x^5y^4}$
$12x^2y^3\sqrt{2x}$

63. $2x^2\sqrt{8x^2y^3}$
$4x^3y\sqrt{2y}$

64. $3y^2\sqrt{27x^4y^3}$
$9x^2y^3\sqrt{3y}$

65. $\sqrt{25(a+4)^2}$
$5a+20$

66. $\sqrt{81(x+y)^4}$
$9x^2+18xy+9y^2$

67. $\sqrt{4(x+2)^4}$
$2x^2+8x+8$

68. $\sqrt{9(x+2)^8}$
$3(x+2)^4$

69. $\sqrt{x^2+4x+4}$
$x+2$

70. $\sqrt{b^2+8b+16}$
$b+4$

71. $\sqrt{y^2+2y+1}$
$y+1$

72. $\sqrt{a^2+6a+9}$
$a+3$

APPLYING THE CONCEPTS

73. **Automotive Safety** Traffic accident investigators can estimate the speed S, in miles per hour, of a car from the length of its skid mark by using the formula $S = \sqrt{30fl}$, where f is the coefficient of friction (which depends on the type of road surface) and l is the length, in feet, of the skid mark. Say the coefficient of friction is 1.2 and the length of a skid mark is 60 ft.
   a. Determine the speed of the car as a radical expression in simplest form. $12\sqrt{15}$ mph
   b. Write the answer to part **a** as a decimal rounded to the nearest integer.
   46 mph

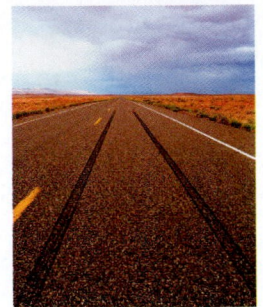

74. **Travel** The distance a passenger in an airplane can see to the horizon can be approximated by $d = 1.2\sqrt{h}$, where d is the distance to the horizon in miles and h is the height of the plane in feet. To the nearest tenth of a mile, what is the distance to the horizon a passenger can see when flying at an altitude of 5000 ft?   84.9 mi

75. If a and b are positive real numbers, does $\sqrt{a+b} = \sqrt{a} + \sqrt{b}$? If not, give an example in which the expressions are not equal.
No. For example, let $a = 16$ and $b = 9$. $\sqrt{a+b} = \sqrt{16+9} = \sqrt{25} = 5$.
$\sqrt{a} + \sqrt{b} = \sqrt{16} + \sqrt{9} = 4 + 3 = 7$.

76. **a.** Find the two-digit perfect square that has exactly nine factors.   36
   b. Find two whole numbers such that their difference is 10, the smaller number is a perfect square, and the larger number is 2 less than a perfect square.   4, 14

77. You are to grade this solution to the problem "Write $\sqrt{72}$ in simplest form." Is the solution correct? If not, what error was made? What is the correct solution? $\sqrt{72} = \sqrt{4}\,\sqrt{18}$
   $= 2\sqrt{18}$
No. 18 contains a perfect-square factor. $6\sqrt{2}$

78. Simplify: **a.** $\sqrt{\sqrt{16}}$ **b.** $\sqrt{\sqrt{81}}$
                    2   3

79. Given $f(x) = \sqrt{2x-1}$, find each of the following. Write your answer in simplest form.
   a. $f(1)$ **b.** $f(5)$ **c.** $f(14)$
       1   3 $3\sqrt{3}$

15.2 Addition and Subtraction of Radical Expressions

Objective A   To add and subtract radical expressions

The Distributive Property is used to simplify the sum or difference of radical expressions with like radicands.

$$5\sqrt{2} + 3\sqrt{2} = (5 + 3)\sqrt{2} = 8\sqrt{2}$$

$$6\sqrt{2x} - 4\sqrt{2x} = (6 - 4)\sqrt{2x} = 2\sqrt{2x}$$

Radical expressions that are in simplest form and have unlike radicands cannot be simplified by the Distributive Property.

$2\sqrt{3} + 4\sqrt{2}$ cannot be simplified by the Distributive Property.

> **HOW TO**   Simplify: $4\sqrt{8} - 10\sqrt{2}$
>
> $$\begin{aligned} 4\sqrt{8} - 10\sqrt{2} &= 4\sqrt{4 \cdot 2} - 10\sqrt{2} \\ &= 4\sqrt{4}\sqrt{2} - 10\sqrt{2} \\ &= 4 \cdot 2\sqrt{2} - 10\sqrt{2} \\ &= 8\sqrt{2} - 10\sqrt{2} \\ &= (8 - 10)\sqrt{2} \\ &= -2\sqrt{2} \end{aligned}$$
>
> • Use the Product Property of Square Roots.
>
> • Simplify the expression by using the Distributive Property.

> **HOW TO**   Simplify: $8\sqrt{18x} - 2\sqrt{32x}$
>
> $$\begin{aligned} 8\sqrt{18x} - 2\sqrt{32x} &= 8\sqrt{9 \cdot 2x} - 2\sqrt{16 \cdot 2x} \\ &= 8\sqrt{9}\sqrt{2x} - 2\sqrt{16}\sqrt{2x} \\ &= 8 \cdot 3\sqrt{2x} - 2 \cdot 4\sqrt{2x} \\ &= 24\sqrt{2x} - 8\sqrt{2x} \\ &= (24 - 8)\sqrt{2x} \\ &= 16\sqrt{2x} \end{aligned}$$
>
> • Use the Product Property of Square Roots.
>
> • Simplify the expression by using the Distributive Property.

Example 1

Simplify: $5\sqrt{2} - 3\sqrt{2} + 12\sqrt{2}$

Solution

$$\begin{aligned} 5\sqrt{2} - 3\sqrt{2} &+ 12\sqrt{2} \\ &= (5 - 3 + 12)\sqrt{2} \quad \text{• Distributive Property} \\ &= 14\sqrt{2} \end{aligned}$$

You Try It 1

Simplify: $9\sqrt{3} + 3\sqrt{3} - 18\sqrt{3}$

Your solution

$-6\sqrt{3}$

Solution on p. S37

Objective 15.2A

Concept Check

Mark each statement as true (T) or false (F).

1. $\sqrt{32} = \sqrt{16}\sqrt{2}$ T
2. $\sqrt{5} = \sqrt{2} + \sqrt{3}$ F
3. $\sqrt{9 + 4} = \sqrt{9} + \sqrt{4}$ F
4. $\sqrt{13} + \sqrt{13} = \sqrt{26}$ F

Instructor Note

Operations with radical expressions are similar to operations with variable expressions. Making this connection for students may help them with simplifying radical expressions.

Discuss the Concepts

1. Can $2\sqrt{3x} + 5\sqrt{6x}$ be simplified? Explain.
No. The terms of the radical expression are in simplest form but have unlike radicands. Therefore, they cannot be simplified.

2. Can $4\sqrt{18x} + 5\sqrt{32x}$ be simplified? Explain.
Yes. First simplify the terms of the radical expression. $4\sqrt{18x} + 5\sqrt{32x} = 12\sqrt{2x} + 20\sqrt{2x} = 32\sqrt{2x}$

In-Class Examples (Objective 15.2A)

Simplify.

1. $7\sqrt{11} + 8\sqrt{11} - 2\sqrt{11}$ $13\sqrt{11}$
2. $9\sqrt{18} + 5\sqrt{50}$ $52\sqrt{2}$
3. $5\sqrt{24x^5} + 3x\sqrt{54x^3}$ $19x^2\sqrt{6x}$
4. $3ab\sqrt{45b} + 2a\sqrt{80b^3} - 7\sqrt{20a^2b^3}$ $3ab\sqrt{5b}$

Example 2

Simplify: $3\sqrt{12} - 5\sqrt{27}$

Solution

$3\sqrt{12} - 5\sqrt{27}$

$= 3\sqrt{4 \cdot 3} - 5\sqrt{9 \cdot 3}$ • **Simplify $\sqrt{12}$ and $\sqrt{27}$.**

$= 3\sqrt{4}\sqrt{3} - 5\sqrt{9}\sqrt{3}$

$= 3 \cdot 2\sqrt{3} - 5 \cdot 3\sqrt{3}$

$= 6\sqrt{3} - 15\sqrt{3}$

$= (6 - 15)\sqrt{3}$ • **Distributive Property**

$= -9\sqrt{3}$

You Try It 2

Simplify: $2\sqrt{50} - 5\sqrt{32}$

Your solution

$-10\sqrt{2}$

Example 3

Simplify: $3\sqrt{12x^3} - 2x\sqrt{3x}$

Solution

$3\sqrt{12x^3} - 2x\sqrt{3x}$

$= 3\sqrt{4x^2 \cdot 3x} - 2x\sqrt{3x}$ • **Simplify $\sqrt{12x^3}$.**

$= 3\sqrt{4x^2}\sqrt{3x} - 2x\sqrt{3x}$

$= 3 \cdot 2x\sqrt{3x} - 2x\sqrt{3x}$

$= 6x\sqrt{3x} - 2x\sqrt{3x}$

$= (6x - 2x)\sqrt{3x}$ • **Distributive Property**

$= 4x\sqrt{3x}$

You Try It 3

Simplify: $y\sqrt{28y} + 7\sqrt{63y^3}$

Your solution

$23y\sqrt{7y}$

Example 4

Simplify: $2x\sqrt{8y} - 3\sqrt{2x^2y} + 2\sqrt{32x^2y}$

Solution

$2x\sqrt{8y} - 3\sqrt{2x^2y} + 2\sqrt{32x^2y}$

$= 2x\sqrt{4 \cdot 2y} - 3\sqrt{x^2 \cdot 2y} + 2\sqrt{16x^2 \cdot 2y}$

$= 2x\sqrt{4}\sqrt{2y} - 3\sqrt{x^2}\sqrt{2y} + 2\sqrt{16x^2}\sqrt{2y}$

$= 2x \cdot 2\sqrt{2y} - 3 \cdot x\sqrt{2y} + 2 \cdot 4x\sqrt{2y}$

$= 4x\sqrt{2y} - 3x\sqrt{2y} + 8x\sqrt{2y}$

$= 9x\sqrt{2y}$

You Try It 4

Simplify: $2\sqrt{27a^5} - 4a\sqrt{12a^3} + a^2\sqrt{75a}$

Your solution

$3a^2\sqrt{3a}$

Solutions on p. S37

15.2 Exercises

Objective A   To add and subtract radical expressions

1. Which of the numbers 2, 9, 20, 25, 50, 81, and 100 are *not* perfect squares?
   2, 20, 50

2. Write down a number that has a perfect-square factor that is greater than 1.   Answers will vary; for instance, 50.

3. ✏ Write a sentence or two that you could e-mail to a friend to explain the concept of a perfect-square factor.

4. Name the perfect-square factors of 540. What number is the largest perfect-square factor of 540?   4, 9, 36; 36

For Exercises 5 to 60, simplify.

5. $2\sqrt{2} + \sqrt{2}$
   $3\sqrt{2}$

6. $3\sqrt{5} + 8\sqrt{5}$
   $11\sqrt{5}$

7. $-3\sqrt{7} + 2\sqrt{7}$
   $-\sqrt{7}$

8. $4\sqrt{5} - 10\sqrt{5}$
   $-6\sqrt{5}$

9. $-3\sqrt{11} - 8\sqrt{11}$
   $-11\sqrt{11}$

10. $-3\sqrt{3} - 5\sqrt{3}$
    $-8\sqrt{3}$

11. $2\sqrt{x} + 8\sqrt{x}$
    $10\sqrt{x}$

12. $3\sqrt{y} + 2\sqrt{y}$
    $5\sqrt{y}$

13. $8\sqrt{y} - 10\sqrt{y}$
    $-2\sqrt{y}$

14. $-5\sqrt{2a} + 2\sqrt{2a}$
    $-3\sqrt{2a}$

15. $-2\sqrt{3b} - 9\sqrt{3b}$
    $-11\sqrt{3b}$

16. $-7\sqrt{5a} - 5\sqrt{5a}$
    $-12\sqrt{5a}$

17. $3x\sqrt{2} - x\sqrt{2}$
    $2x\sqrt{2}$

18. $2y\sqrt{3} - 9y\sqrt{3}$
    $-7y\sqrt{3}$

19. $2a\sqrt{3a} - 5a\sqrt{3a}$
    $-3a\sqrt{3a}$

20. $-5b\sqrt{3x} - 2b\sqrt{3x}$
    $-7b\sqrt{3x}$

21. $3\sqrt{xy} - 8\sqrt{xy}$
    $-5\sqrt{xy}$

22. $-4\sqrt{xy} + 6\sqrt{xy}$
    $2\sqrt{xy}$

23. $\sqrt{45} + \sqrt{125}$
    $8\sqrt{5}$

24. $\sqrt{32} - \sqrt{98}$
    $-3\sqrt{2}$

25. $2\sqrt{2} + 3\sqrt{8}$
    $8\sqrt{2}$

26. $4\sqrt{128} - 3\sqrt{32}$
    $20\sqrt{2}$

27. $5\sqrt{18} - 2\sqrt{75}$
    $15\sqrt{2} - 10\sqrt{3}$

28. $5\sqrt{75} - 2\sqrt{18}$
    $25\sqrt{3} - 6\sqrt{2}$

29. $5\sqrt{4x} - 3\sqrt{9x}$
    $\sqrt{x}$

30. $-3\sqrt{25y} + 8\sqrt{49y}$
    $41\sqrt{y}$

31. $3\sqrt{3x^2} - 5\sqrt{27x^2}$
    $-12x\sqrt{3}$

32. $-2\sqrt{8y^2} + 5\sqrt{32y^2}$
    $16y\sqrt{2}$

33. $2x\sqrt{xy^2} - 3y\sqrt{x^2y}$
    $2xy\sqrt{x} - 3xy\sqrt{y}$

34. $4a\sqrt{b^2a} - 3b\sqrt{a^2b}$
    $4ab\sqrt{a} - 3ab\sqrt{b}$

35. $3x\sqrt{12x} - 5\sqrt{27x^3}$
    $-9x\sqrt{3x}$

36. $2a\sqrt{50a} + 7\sqrt{32a^3}$
    $38a\sqrt{2a}$

37. $4y\sqrt{8y^3} - 7\sqrt{18y^5}$
    $-13y^2\sqrt{2y}$

38. $2a\sqrt{8ab^2} - 2b\sqrt{2a^3}$
    $2ab\sqrt{2a}$

39. $b^2\sqrt{a^5b} + 3a^2\sqrt{ab^5}$
    $4a^2b^2\sqrt{ab}$

40. $y^2\sqrt{x^5y} + x\sqrt{x^3y^5}$
    $2x^2y^2\sqrt{xy}$

Section 15.2

Suggested Assignment
Exercises 1–57, every other odd
More challenging problems:
   Exercises 61, 64

Answers to Writing Exercises

3. A perfect square is the square of an integer. A perfect-square factor of a number is a perfect square that divides the number evenly.

Quick Quiz (Objective 15.2A)
Simplify.
1. $6\sqrt{15} + 7\sqrt{15} - 10\sqrt{15}$ $3\sqrt{15}$
2. $8\sqrt{12} + 2\sqrt{18}$ $16\sqrt{3} + 6\sqrt{2}$
3. $3\sqrt{72x^{15}} + 8x^4\sqrt{8x^7}$ $34x^7\sqrt{2x}$

41. $4\sqrt{2} - 5\sqrt{2} + 8\sqrt{2}$
$7\sqrt{2}$

42. $3\sqrt{3} + 8\sqrt{3} - 16\sqrt{3}$
$-5\sqrt{3}$

43. $5\sqrt{x} - 8\sqrt{x} + 9\sqrt{x}$
$6\sqrt{x}$

44. $\sqrt{x} - 7\sqrt{x} + 6\sqrt{x}$
0

45. $8\sqrt{2} - 3\sqrt{y} - 8\sqrt{2}$
$-3\sqrt{y}$

46. $8\sqrt{3} - 5\sqrt{2} - 5\sqrt{3}$
$3\sqrt{3} - 5\sqrt{2}$

47. $8\sqrt{8} - 4\sqrt{32} - 9\sqrt{50}$
$-45\sqrt{2}$

48. $2\sqrt{12} - 4\sqrt{27} + \sqrt{75}$
$-3\sqrt{3}$

49. $-2\sqrt{3} + 5\sqrt{27} - 4\sqrt{45}$
$13\sqrt{3} - 12\sqrt{5}$

50. $-2\sqrt{8} - 3\sqrt{27} + 3\sqrt{50}$
$11\sqrt{2} - 9\sqrt{3}$

51. $4\sqrt{75} + 3\sqrt{48} - \sqrt{99}$
$32\sqrt{3} - 3\sqrt{11}$

52. $2\sqrt{75} - 5\sqrt{20} + 2\sqrt{45}$
$10\sqrt{3} - 4\sqrt{5}$

53. $\sqrt{25x} - \sqrt{9x} + \sqrt{16x}$
$6\sqrt{x}$

54. $\sqrt{4x} - \sqrt{100x} - \sqrt{49x}$
$-15\sqrt{x}$

55. $3\sqrt{3x} + \sqrt{27x} - 8\sqrt{75x}$
$-34\sqrt{3x}$

56. $5\sqrt{5x} + 2\sqrt{45x} - 3\sqrt{80x}$
$-\sqrt{5x}$

57. $2a\sqrt{75b} - a\sqrt{20b} + 4a\sqrt{45b}$
$10a\sqrt{3b} + 10a\sqrt{5b}$

58. $2b\sqrt{75a} - 5b\sqrt{27a} + 2b\sqrt{20a}$
$-5b\sqrt{3a} + 4b\sqrt{5a}$

59. $x\sqrt{3y^2} - 2y\sqrt{12x^2} + xy\sqrt{3}$
$-2xy\sqrt{3}$

60. $a\sqrt{27b^2} + 3b\sqrt{147a^2} - ab\sqrt{3}$
$23ab\sqrt{3}$

Answers to Writing Exercises

63. $4\sqrt{2a^3b} + 5\sqrt{8a^3b}$

Write each radicand as the product of a perfect square and factors that do not contain a perfect square.
$= 4\sqrt{a^2 \cdot 2ab} + 5\sqrt{4a^2 \cdot 2ab}$

Use the Product Property of Square Roots. Write the perfect square under the first radical sign and all the remaining factors under the second radical sign.
$= 4\sqrt{a^2}\sqrt{2ab} + 5\sqrt{4a^2}\sqrt{2ab}$

Take the square roots of the perfect squares.
$= 4a\sqrt{2ab} + 5 \cdot 2a\sqrt{2ab}$

Simplify.
$= 4a\sqrt{2ab} + 10a\sqrt{2ab}$

Combine like terms.
$= 14a\sqrt{2ab}$

APPLYING THE CONCEPTS

61. Given $G(x) = \sqrt{x + 5} + \sqrt{5x + 3}$, write $G(3)$ in simplest form.
$5\sqrt{2}$

62. Is the equation $\sqrt{a^2 + b^2} = \sqrt{a} + \sqrt{b}$ true for all real numbers a and b?
No

63. ✏️ Explain the steps in simplifying $4\sqrt{2a^3b} + 5\sqrt{8a^3b}$.

64. For each equation, write "ok" if the equation is correct. If the equation is incorrect, correct the right-hand side.
a. $3\sqrt{ab} + 5\sqrt{ab} = 8\sqrt{2ab}$ $8\sqrt{ab}$
b. $7\sqrt{x^3} - 3x\sqrt{x} - x\sqrt{16x} = 0$ ok
c. $5 - 2\sqrt{y} = 3\sqrt{y}$ $5 - 2\sqrt{y}$

15.3 Multiplication and Division of Radical Expressions

Objective A   To multiply radical expressions

The Product Property of Square Roots is used to multiply radical expressions.

$$\sqrt{2x}\,\sqrt{3y} = \sqrt{2x \cdot 3y} = \sqrt{6xy}$$

HOW TO   Simplify: $\sqrt{2x^2}\sqrt{32x^5}$

$$
\begin{aligned}
\sqrt{2x^2}\,\sqrt{32x^5} &= \sqrt{2x^2 \cdot 32x^5} && \bullet \text{ Use the Product Property of Square Roots.}\\
&= \sqrt{64x^7} && \bullet \text{ Multiply the radicands.}\\
&= \sqrt{64x^6 \cdot x} && \bullet \text{ Simplify.}\\
&= \sqrt{64x^6}\,\sqrt{x} = 8x^3\sqrt{x}
\end{aligned}
$$

HOW TO   Simplify: $\sqrt{2x}(x + \sqrt{2x})$

$$
\begin{aligned}
\sqrt{2x}(x + \sqrt{2x}) &= \sqrt{2x}(x) + \sqrt{2x}\sqrt{2x} && \bullet \text{ Use the Distributive Property to}\\
&= x\sqrt{2x} + \sqrt{4x^2} && \quad\text{ remove parentheses.}\\
&= x\sqrt{2x} + 2x && \bullet \text{ Simplify.}
\end{aligned}
$$

Use FOIL to multiply radical expressions with two terms.

HOW TO   Simplify: $(\sqrt{2} - 3x)(\sqrt{2} + x)$

$$
\begin{aligned}
(\sqrt{2} - 3x)(\sqrt{2} + x) &= \sqrt{2 \cdot 2} + x\sqrt{2} - 3x\sqrt{2} - 3x^2 && \bullet \text{ Use the FOIL method to}\\
&= \sqrt{4} + (x - 3x)\sqrt{2} - 3x^2 && \quad\text{ remove parentheses.}\\
&= 2 - 2x\sqrt{2} - 3x^2
\end{aligned}
$$

The expressions $a + b$ and $a - b$, which differ only in the sign of one term, are called **conjugates**. Recall that $(a + b)(a - b) = a^2 - b^2$.

HOW TO   Simplify: $(2 + \sqrt{7})(2 - \sqrt{7})$

$$
\begin{aligned}
(2 + \sqrt{7})(2 - \sqrt{7}) &= 2^2 - (\sqrt{7})^2 && \bullet \; (2 + \sqrt{7})(2 - \sqrt{7}) \text{ is the product of conjugates.}\\
&= 4 - 7 = -3
\end{aligned}
$$

HOW TO   Simplify: $(3 + \sqrt{y})(3 - \sqrt{y})$

$$
\begin{aligned}
(3 + \sqrt{y})(3 - \sqrt{y}) &= 3^2 - (\sqrt{y})^2 && \bullet \; (3 + \sqrt{y})(3 - \sqrt{y}) \text{ is the product of conjugates.}\\
&= 9 - y
\end{aligned}
$$

> **TAKE NOTE**
> For $x > 0$,
> $(\sqrt{x})^2 = x$ because
> $(\sqrt{x})^2 = \sqrt{x} \cdot \sqrt{x} = \sqrt{x^2} = x.$

Example 1   Simplify: $\sqrt{3x^4}\sqrt{2x^2y}\sqrt{6xy^2}$

Solution

$$
\begin{aligned}
&\sqrt{3x^4}\sqrt{2x^2y}\sqrt{6xy^2}\\
&= \sqrt{36x^7y^3} && \bullet \text{ Product Property of Square Roots}\\
&= \sqrt{36x^6y^2 \cdot xy} && \bullet \text{ Simplify.}\\
&= \sqrt{36x^6y^2}\sqrt{xy}\\
&= 6x^3y\sqrt{xy}
\end{aligned}
$$

You Try It 1   Simplify: $\sqrt{5a}\sqrt{15a^3b^4}\sqrt{20b^5}$

Your solution
$10a^2b^4\sqrt{15b}$

Solution on p. S37

Objective 15.3A

New Vocabulary
conjugates

Discuss the Concepts
Describe the procedure for simplifying $\sqrt{3x^3}\sqrt{27x^7}$.
Answers will vary. The simplest form is $9x^5$.

Concept Check
Determine the conjugate of each of the following.
1. $3 + \sqrt{5}$
  $3 - \sqrt{5}$
2. $6 - \sqrt{x}$
  $6 + \sqrt{x}$

Concept Check
Simplify.
1. $(3 + \sqrt{5})(3 - \sqrt{5})$
  4
2. $(6 - \sqrt{x})(6 + \sqrt{x})$
  $36 - x$

In-Class Examples (Objective 15.3A)
Simplify.
1. $\sqrt{30y^5}\sqrt{6y^8}\sqrt{10y^6}$ $30y^9\sqrt{2y}$
2. $\sqrt{8mn}(\sqrt{2m} - \sqrt{3n})$ $4m\sqrt{n} - 2n\sqrt{6m}$
3. $(3\sqrt{5ab} - 7c)(3\sqrt{5ab} + 7c)$ $45ab - 49c^2$
4. $(5\sqrt{a} + 3\sqrt{b})(2\sqrt{a} - \sqrt{b})$
  $10a + \sqrt{ab} - 3b$

Concept Check

Simplify.

1. $\sqrt{5x^3}\sqrt{2x^5}\sqrt{10x^2}$
$10x^5$

2. $\sqrt{7x}(\sqrt{7x} + \sqrt{2})$
$7x + \sqrt{14x}$

3. $(\sqrt{x} - y)(\sqrt{x} + y)$
$x - y^2$

4. $(3\sqrt{x} + 2\sqrt{y})^2$
$9x + 12\sqrt{xy} + 4y$

Example 2

Simplify: $\sqrt{3ab}(\sqrt{3a} + \sqrt{9b})$

Solution
$\sqrt{3ab}(\sqrt{3a} + \sqrt{9b})$
$= \sqrt{9a^2b} + \sqrt{27ab^2}$ • **Distributive Property**
$= \sqrt{9a^2 \cdot b} + \sqrt{9b^2 \cdot 3a}$ • **Simplify.**
$= \sqrt{9a^2}\sqrt{b} + \sqrt{9b^2}\sqrt{3a}$
$= 3a\sqrt{b} + 3b\sqrt{3a}$

You Try It 2

Simplify: $\sqrt{5x}(\sqrt{5x} - \sqrt{25y})$

Your solution
$5x - 5\sqrt{5xy}$

Example 3

Simplify: $(\sqrt{a} - \sqrt{b})(\sqrt{a} + \sqrt{b})$

Solution
$(\sqrt{a} - \sqrt{b})(\sqrt{a} + \sqrt{b})$
$= (\sqrt{a})^2 - (\sqrt{b})^2$ • **Product of conjugates**
$= a - b$

You Try It 3

Simplify: $(2\sqrt{x} + 7)(2\sqrt{x} - 7)$

Your solution
$4x - 49$

Example 4

Simplify: $(2\sqrt{x} - \sqrt{y})(5\sqrt{x} - 2\sqrt{y})$

Solution
$(2\sqrt{x} - \sqrt{y})(5\sqrt{x} - 2\sqrt{y})$
$= 10(\sqrt{x})^2 - 4\sqrt{xy} - 5\sqrt{xy} + 2(\sqrt{y})^2$ • **FOIL**
$= 10x - 9\sqrt{xy} + 2y$

You Try It 4

Simplify: $(3\sqrt{x} - \sqrt{y})(5\sqrt{x} - 2\sqrt{y})$

Your solution
$15x - 11\sqrt{xy} + 2y$

Solutions on p. S37

Objective 15.3B

New Vocabulary
rationalizing the denominator

Objective B **To divide radical expressions**

> **The Quotient Property of Square Roots**
>
> If a and b are positive real numbers, then $\sqrt{\dfrac{a}{b}} = \dfrac{\sqrt{a}}{\sqrt{b}}$ and $\dfrac{\sqrt{a}}{\sqrt{b}} = \sqrt{\dfrac{a}{b}}$.

This property states that the square root of a quotient is equal to the quotient of the square roots.

HOW TO Simplify: $\sqrt{\dfrac{4x^2}{z^6}}$

$\sqrt{\dfrac{4x^2}{z^6}} = \dfrac{\sqrt{4x^2}}{\sqrt{z^6}}$ • **Rewrite the radical expression as the quotient of the square roots.**

$= \dfrac{2x}{z^3}$ • **Simplify.**

Point of Interest

A radical expression that occurs in Einstein's Theory of Relativity is

$$\frac{1}{\sqrt{1 - \frac{v^2}{c^2}}}$$

where v is the velocity of an object and c is the speed of light.

HOW TO   Simplify: $\sqrt{\dfrac{24x^3y^7}{3x^7y^2}}$

$\sqrt{\dfrac{24x^3y^7}{3x^7y^2}} = \sqrt{\dfrac{8y^5}{x^4}}$ • **Simplify the radicand.**

$= \dfrac{\sqrt{8y^5}}{\sqrt{x^4}}$ • **Rewrite the radical expression as the quotient of the square roots.**

$= \dfrac{\sqrt{4y^4 \cdot 2y}}{\sqrt{x^4}}$ • **Simplify.**

$= \dfrac{\sqrt{4y^4}\sqrt{2y}}{\sqrt{x^4}}$

$= \dfrac{2y^2\sqrt{2y}}{x^2}$

The Quotient Property of Square Roots is used to divide radical expressions.

HOW TO   Simplify: $\dfrac{\sqrt{4x^2y}}{\sqrt{xy}}$

$\dfrac{\sqrt{4x^2y}}{\sqrt{xy}} = \sqrt{\dfrac{4x^2y}{xy}}$ • **Use the Quotient Property of Square Roots.**

$= \sqrt{4x}$ • **Simplify the radicand.**

$= \sqrt{4}\sqrt{x}$ • **Simplify the radical expression.**

$= 2\sqrt{x}$

The previous examples all result in radical expressions written in simplest form.

Simplest Form of a Radical Expression

For a radical expression to be in simplest form, three conditions must be met:

1. The radicand contains no factor greater than 1 that is a perfect square.
2. There is no fraction under the radical sign.
3. There is no radical in the denominator of a fraction.

The procedure used to remove a radical from the denominator is called **rationalizing the denominator.**

HOW TO   Simplify: $\dfrac{2}{\sqrt{3}}$

$\dfrac{2}{\sqrt{3}} = \dfrac{2}{\sqrt{3}} \cdot \boxed{\dfrac{\sqrt{3}}{\sqrt{3}}}$ • **To rationalize the denominator, multiply the expression by $\dfrac{\sqrt{3}}{\sqrt{3}}$, which equals 1.**

$= \dfrac{2\sqrt{3}}{(\sqrt{3})^2}$ • **The radicand in the denominator is a perfect square.**

$= \dfrac{2\sqrt{3}}{3}$ • **Simplify.**

Concept Check

Simplify.

1. $\sqrt{\dfrac{9}{x^2}}$ $\dfrac{3}{x}$

2. $\dfrac{\sqrt{32x^{11}}}{\sqrt{2x^5}}$ $4x^3$

Discuss the Concepts

What form of 1 should $\dfrac{5}{\sqrt{6}}$ be multiplied by to rationalize the denominator?

$\dfrac{\sqrt{6}}{\sqrt{6}}$

Instructor Note

Rationalizing the denominator is another case in which the Multiplication Property of One is used. Students do not always believe that $\dfrac{2}{\sqrt{3}}$ and $\dfrac{2\sqrt{3}}{3}$ are equal. Have these students use a calculator to verify, to the limits of the calculator, that the two expressions are equal.

In-Class Examples (Objective 15.3B)

Simplify.

1. $\dfrac{\sqrt{12x^3y^3}}{\sqrt{24xy^6}}$ $\dfrac{x\sqrt{2y}}{2y^2}$

2. $\dfrac{\sqrt{5}}{\sqrt{5} - \sqrt{10}}$ $-1 - \sqrt{2}$

3. $\dfrac{2 + \sqrt{7x}}{6 + \sqrt{7x}}$ $\dfrac{12 + 4\sqrt{7x} - 7x}{36 - 7x}$

Concept Check

1. Identify the error in the following simplification.

$$\frac{\sqrt{3}}{2 + \sqrt{5}} = \frac{\sqrt{3}}{2 + \sqrt{5}} \cdot \frac{\sqrt{5}}{\sqrt{5}}$$

$$= \frac{\sqrt{15}}{2 + 5} = \frac{\sqrt{15}}{7}$$

Multiply numerator and denominator by $2 - \sqrt{5}$. Also, the Distributive Property was not used correctly.

2. Place the correct symbol, $<$, $=$, or $>$, between the two expressions to make a true statement. Do not use a calculator.

a. $\dfrac{1}{3 - \sqrt{8}} \ ? \ 3 + \sqrt{8}$

=

b. $\left(\sqrt{11} + 3\right)^2 \ ? \ 20$

>

c. $\dfrac{9}{3 + \sqrt{5}} \ ? \ 6 - 2\sqrt{5}$

>

Discuss the Concepts

Explain how to determine whether to multiply the numerator and denominator by an expression with one term or by an expression with two terms when rationalizing the denominator of a radical expression.

If the denominator has one term, use one term; if the denominator has two terms, use two terms.

When the denominator contains a radical expression with two terms, rationalize the denominator by multiplying the numerator and denominator by the conjugate of the denominator.

HOW TO Simplify: $\dfrac{\sqrt{2y}}{\sqrt{y} + 3}$

$$\frac{\sqrt{2y}}{\sqrt{y} + 3} = \frac{\sqrt{2y}}{\sqrt{y} + 3} \cdot \frac{\sqrt{y} - 3}{\sqrt{y} - 3}$$

• Multiply the numerator and denominator by $\sqrt{y} - 3$, the conjugate of $\sqrt{y} + 3$.

$$= \frac{\sqrt{2y^2} - 3\sqrt{2y}}{(\sqrt{y})^2 - 3^2} = \frac{y\sqrt{2} - 3\sqrt{2y}}{y - 9}$$

Example 5 Simplify: $\dfrac{\sqrt{4x^2y^5}}{\sqrt{3x^4y}}$

Solution

$$\frac{\sqrt{4x^2y^5}}{\sqrt{3x^4y}} = \sqrt{\frac{4x^2y^5}{3x^4y}} = \sqrt{\frac{4y^4}{3x^2}} = \frac{\sqrt{4y^4}}{\sqrt{3x^2}}$$

$$= \frac{2y^2}{x\sqrt{3}} = \frac{2y^2}{x\sqrt{3}} \cdot \frac{\sqrt{3}}{\sqrt{3}}$$ • Rationalize the denominator.

$$= \frac{2y^2\sqrt{3}}{3x}$$

You Try It 5 Simplify: $\dfrac{\sqrt{15x^6y^7}}{\sqrt{3x^7y^9}}$

Your solution

$$\frac{\sqrt{5x}}{xy}$$

Example 6 Simplify: $\dfrac{\sqrt{2}}{\sqrt{2} + \sqrt{6}}$

Solution

$$\frac{\sqrt{2}}{\sqrt{2} + \sqrt{6}}$$ • Multiply the numerator and denominator by the conjugate of the denominator.

$$= \frac{\sqrt{2}}{\sqrt{2} + \sqrt{6}} \cdot \frac{\sqrt{2} - \sqrt{6}}{\sqrt{2} - \sqrt{6}}$$

$$= \frac{(\sqrt{2})^2 - \sqrt{12}}{2 - 6} = \frac{2 - 2\sqrt{3}}{-4}$$

$$= \frac{2(1 - \sqrt{3})}{-4} = \frac{1 - \sqrt{3}}{-2} = -\frac{1 - \sqrt{3}}{2}$$

You Try It 6 Simplify: $\dfrac{\sqrt{3}}{\sqrt{3} - \sqrt{6}}$

Your solution

$$-1 - \sqrt{2}$$

Example 7 Simplify: $\dfrac{3 - \sqrt{5}}{2 + 3\sqrt{5}}$

Solution

$$\frac{3 - \sqrt{5}}{2 + 3\sqrt{5}} = \frac{3 - \sqrt{5}}{2 + 3\sqrt{5}} \cdot \frac{2 - 3\sqrt{5}}{2 - 3\sqrt{5}}$$ • Rationalize the denominator.

$$= \frac{6 - 9\sqrt{5} - 2\sqrt{5} + 3(\sqrt{5})^2}{4 - 9 \cdot 5}$$

$$= \frac{6 - 11\sqrt{5} + 15}{4 - 45}$$

$$= \frac{21 - 11\sqrt{5}}{-41} = -\frac{21 - 11\sqrt{5}}{41}$$

You Try It 7 Simplify: $\dfrac{5 + \sqrt{y}}{1 - 2\sqrt{y}}$

Your solution

$$\frac{5 + 11\sqrt{y} + 2y}{1 - 4y}$$

Solutions on p. S37

15.3 Exercises

Objective A   To multiply radical expressions

1. ✎ Explain in words and then write in symbols the Product Property of Square Roots.

2. ✎ Give an example to show that $\sqrt{a^2} \neq a$.

For Exercises 3 to 37, simplify.

3. $\sqrt{5} \cdot \sqrt{5}$
   5

4. $\sqrt{11} \cdot \sqrt{11}$
   11

5. $\sqrt{3} \cdot \sqrt{12}$
   6

6. $\sqrt{2} \cdot \sqrt{8}$
   4

7. $\sqrt{x} \cdot \sqrt{x}$
   x

8. $\sqrt{y} \cdot \sqrt{y}$
   y

9. $\sqrt{xy^3} \cdot \sqrt{x^5y}$
   x^3y^2

10. $\sqrt{a^3b^5} \cdot \sqrt{ab^5}$
    a^2b^5

11. $\sqrt{3a^2b^5} \cdot \sqrt{6ab^7}$
    $3ab^6\sqrt{2a}$

12. $\sqrt{5x^3y} \cdot \sqrt{10x^2y}$
    $5x^2y\sqrt{2x}$

13. $\sqrt{6a^3b^2} \cdot \sqrt{24a^5b}$
    $12a^4b\sqrt{b}$

14. $\sqrt{8ab^5} \cdot \sqrt{12a^7b}$
    $4a^4b^3\sqrt{6}$

15. $\sqrt{2}(\sqrt{2} - \sqrt{3})$
    $2 - \sqrt{6}$

16. $3(\sqrt{12} - \sqrt{3})$
    $3\sqrt{3}$

17. $\sqrt{x}(\sqrt{x} - \sqrt{y})$
    $x - \sqrt{xy}$

18. $\sqrt{b}(\sqrt{a} - \sqrt{b})$
    $\sqrt{ab} - b$

19. $\sqrt{5}(\sqrt{10} - \sqrt{x})$
    $5\sqrt{2} - \sqrt{5x}$

20. $\sqrt{6}(\sqrt{y} - \sqrt{18})$
    $\sqrt{6y} - 6\sqrt{3}$

21. $\sqrt{8}(\sqrt{2} - \sqrt{5})$
    $4 - 2\sqrt{10}$

22. $\sqrt{10}(\sqrt{20} - \sqrt{a})$
    $10\sqrt{2} - \sqrt{10a}$

23. $(\sqrt{x} - 3)^2$
    $x - 6\sqrt{x} + 9$

24. $(2\sqrt{a} - y)^2$
    $4a - 4y\sqrt{a} + y^2$

25. $\sqrt{3a}(\sqrt{3a} - \sqrt{3b})$
    $3a - 3\sqrt{ab}$

26. $\sqrt{5x}(\sqrt{10x} - \sqrt{x})$
    $5x\sqrt{2} - x\sqrt{5}$

27. $\sqrt{2ac} \cdot \sqrt{5ab} \cdot \sqrt{10cb}$
    $10abc$

28. $\sqrt{3xy} \cdot \sqrt{6x^3y} \cdot \sqrt{2y^2}$
    $6x^2y^2$

29. $(\sqrt{5} + 3)(2\sqrt{5} - 4)$
    $-2 + 2\sqrt{5}$

30. $(2 - 3\sqrt{7})(5 + 2\sqrt{7})$
    $-32 - 11\sqrt{7}$

31. $(4 + \sqrt{8})(3 + \sqrt{2})$
    $16 + 10\sqrt{2}$

32. $(6 - \sqrt{27})(2 + \sqrt{3})$
    3

33. $(2\sqrt{x} + 4)(3\sqrt{x} - 1)$
    $6x + 10\sqrt{x} - 4$

34. $(5 + \sqrt{y})(6 - 3\sqrt{y})$
    $30 - 9\sqrt{y} - 3y$

35. $(3\sqrt{x} - 2y)(5\sqrt{x} - 4y)$
    $15x - 22y\sqrt{x} + 8y^2$

36. $(5\sqrt{x} + 2\sqrt{y})(3\sqrt{x} - \sqrt{y})$
    $15x + \sqrt{xy} - 2y$

37. $(\sqrt{x} - \sqrt{y})(\sqrt{x} + \sqrt{y})$
    $x - y$

Section 15.3

Suggested Assignment

Exercises 1–69, odds
More challenging problems:
  Exercises 71, 72

Answers to Writing Exercises

1. The Product Property of Square Roots states that the square root of the product of two positive numbers equals the product of the square roots of the two numbers. $\sqrt{ab} = \sqrt{a}\sqrt{b}$, $a > 0$, $b > 0$

2. Any value of a less than 0 will work. For instance, if $a = -6$, then $\sqrt{a^2} = \sqrt{(-6)^2} = \sqrt{36} = 6 \neq a$.

Quick Quiz (Objective 15.3A)

Simplify.

1. $\sqrt{3a^5}\sqrt{6a^3}\sqrt{18a^4}$ $18a^6$
2. $\sqrt{2x}(\sqrt{2x} - \sqrt{8})$ $2x - 4\sqrt{x}$
3. $(4\sqrt{b} - 5)(4\sqrt{b} + 5)$ $16b - 25$
4. $(2\sqrt{x} + 3\sqrt{y})(3\sqrt{x} - \sqrt{y})$ $6x + 7\sqrt{xy} - 3y$

Answers to Writing Exercises

38. $\frac{\sqrt{3}}{3}$ is in simplest form

because the radicand does not contain a perfect-square factor greater than 1, there is no fraction under the radical, and there is no radical expression in the denominator.

$\frac{1}{\sqrt{3}}$ is not in simplest form

because there is a radical expression in the denominator.

39. Because $\frac{\sqrt{5}}{\sqrt{5}} = 1$, multiplying

$\frac{2}{\sqrt{5}}$ by $\frac{\sqrt{5}}{\sqrt{5}}$ is the same as

multiplying by 1, and any number multiplied by 1 is the number itself.

70. Student descriptions should include both cases studied in this section:
(1) To rationalize a denominator of one term, multiply the numerator and denominator of the fraction by the denominator.
(2) To rationalize a denominator of two terms, multiply the numerator and denominator of the fraction by the conjugate of the denominator.
In each case the resulting expression must be simplified.

Objective B   To divide radical expressions

38. ✏ Why is $\frac{\sqrt{3}}{3}$ in simplest form but $\frac{1}{\sqrt{3}}$ not in simplest form?

39. ✏ Why can we multiply $\frac{2}{\sqrt{5}}$ by $\frac{\sqrt{5}}{\sqrt{5}}$ without changing the value of $\frac{2}{\sqrt{5}}$?

For Exercises 40 to 69, simplify.

40. $\dfrac{\sqrt{32}}{\sqrt{2}}$
4

41. $\dfrac{\sqrt{45}}{\sqrt{5}}$
3

42. $\dfrac{\sqrt{98}}{\sqrt{2}}$
7

43. $\dfrac{\sqrt{48}}{\sqrt{3}}$
4

44. $\dfrac{\sqrt{27a}}{\sqrt{3a}}$
3

45. $\dfrac{\sqrt{72x^5}}{\sqrt{2x}}$
$6x^2$

46. $\dfrac{\sqrt{15x^3y}}{\sqrt{3xy}}$
$x\sqrt{5}$

47. $\dfrac{\sqrt{40x^5y^2}}{\sqrt{5xy}}$
$2x^2\sqrt{2y}$

48. $\dfrac{\sqrt{2a^5b^4}}{\sqrt{98ab^4}}$
$\dfrac{a^2}{7}$

49. $\dfrac{\sqrt{48x^5y^2}}{\sqrt{3x^3y}}$
$4x\sqrt{y}$

50. $\dfrac{\sqrt{9xy^2}}{\sqrt{27x}}$
$\dfrac{y\sqrt{3}}{3}$

51. $\dfrac{\sqrt{4x^2y}}{\sqrt{3xy^3}}$
$\dfrac{2\sqrt{3x}}{3y}$

52. $\dfrac{\sqrt{16x^3y^2}}{\sqrt{8x^3y}}$
$\sqrt{2y}$

53. $\dfrac{\sqrt{2}}{\sqrt{8}+4}$
$\dfrac{-1+\sqrt{2}}{2}$

54. $\dfrac{1}{\sqrt{2}-3}$
$-\dfrac{\sqrt{2}+3}{7}$

55. $\dfrac{5}{\sqrt{7}-3}$
$-\dfrac{5\sqrt{7}+15}{2}$

56. $\dfrac{3}{5+\sqrt{5}}$
$\dfrac{15-3\sqrt{5}}{20}$

57. $\dfrac{\sqrt{3}}{5-\sqrt{27}}$
$-\dfrac{5\sqrt{3}+9}{2}$

58. $\dfrac{7}{\sqrt{2}-7}$
$-\dfrac{7\sqrt{2}+49}{47}$

59. $\dfrac{3-\sqrt{6}}{5-2\sqrt{6}}$
$3+\sqrt{6}$

60. $\dfrac{6-2\sqrt{3}}{4+3\sqrt{3}}$
$-\dfrac{42-26\sqrt{3}}{11}$

61. $\dfrac{-6}{4+\sqrt{2}}$
$\dfrac{-12+3\sqrt{2}}{7}$

62. $\dfrac{\sqrt{2}+2\sqrt{6}}{2\sqrt{2}-3\sqrt{6}}$
$-\dfrac{20+7\sqrt{3}}{23}$

63. $\dfrac{2\sqrt{3}-\sqrt{6}}{5\sqrt{3}+2\sqrt{6}}$
$\dfrac{14-9\sqrt{2}}{17}$

64. $\dfrac{3+\sqrt{x}}{2-\sqrt{x}}$
$\dfrac{6+5\sqrt{x}+x}{4-x}$

65. $\dfrac{-\sqrt{15}}{3-\sqrt{12}}$
$\sqrt{15}+2\sqrt{5}$

66. $\dfrac{\sqrt{a}-4}{2\sqrt{a}+2}$
$\dfrac{a-5\sqrt{a}+4}{2a-2}$

67. $\dfrac{\sqrt{xy}}{\sqrt{x}-\sqrt{y}}$
$\dfrac{x\sqrt{y}+y\sqrt{x}}{x-y}$

68. $\dfrac{\sqrt{x}}{\sqrt{x}-\sqrt{y}}$
$\dfrac{x+\sqrt{xy}}{x-y}$

69. $\dfrac{-12}{\sqrt{6}-3}$
$4\sqrt{6}+12$

APPLYING THE CONCEPTS

70. ✏ In your own words, describe the process of rationalizing the denominator.

71. Show that $1+\sqrt{6}$ and $1-\sqrt{6}$ are solutions of the equation $x^2 - 2x - 5 = 0$.   The complete solution is available in the *Instructor's Solutions Manual*.

72. Answer true or false. If the equation is false, correct it.

a. $(\sqrt{y})^4 = y^2$ **b.** $(2\sqrt{x})^3 = 8x\sqrt{x}$ **c.** $(\sqrt{x}+1)^2 = x+1$ **d.** $\dfrac{1}{2-\sqrt{3}} = 2+\sqrt{3}$

True   True   False, $x + 2\sqrt{x} + 1$ True

Quick Quiz (Objective 15.3B)

Simplify.

1. $\dfrac{\sqrt{72x^7}}{\sqrt{2x}}$ $6x^3$

2. $\dfrac{2}{\sqrt{5}}$ $\dfrac{2\sqrt{5}}{5}$

3. $\dfrac{1+\sqrt{6a}}{3+\sqrt{6a}}$ $\dfrac{3+2\sqrt{6a}-6a}{9-6a}$

15.4 Solving Equations Containing Radical Expressions

Objective A To solve an equation containing a radical expression

An equation that contains a variable expression in a radicand is a **radical equation**.

$$\left.\begin{array}{l} \sqrt{x} = 4 \\ \sqrt{x+2} = \sqrt{x-7} \end{array}\right\} \begin{array}{l}\text{Radical} \\ \text{equations}\end{array}$$

The following property of equality, which states that if two numbers are equal then the squares of the numbers are equal, is used to solve radical equations.

> **Property of Squaring Both Sides of an Equation**
>
> If a and b are real numbers and $a = b$, then $a^2 = b^2$.

To solve a radical equation with one radical, use the following procedure.

> **Solving a Radical Equation**
>
> **1.** Write the equation with the radical alone on one side.
> **2.** Square both sides of the equation.
> **3.** Solve for the variable.
> **4.** Check the solution(s) in the original equation.

HOW TO   Solve: $\sqrt{x-2} - 7 = 0$

$$\begin{aligned} \sqrt{x-2} - 7 &= 0 \\ \sqrt{x-2} &= 7 \\ (\sqrt{x-2})^2 &= 7^2 \\ x - 2 &= 49 \\ x &= 51 \end{aligned}$$

- Isolate the radical by adding 7 to both sides of the equation.
- Square both sides of the equation.
- Solve the resulting equation.

Check:
$$\begin{array}{c|c} \sqrt{x-2} - 7 = 0 & \\ \hline \sqrt{51-2} - 7 & 0 \\ \sqrt{49} - 7 & 0 \\ 7 - 7 & 0 \\ 0 &= 0 \quad \text{A true equation} \end{array}$$

The solution is 51.

When both sides of an equation are squared, the resulting equation may have a solution that is not a solution of the original equation. Checking a proposed solution of a radical equation, as we did above, is a necessary step.

HOW TO   Solve: $\sqrt{2x-5} + 3 = 0$

$$\begin{aligned} \sqrt{2x-5} + 3 &= 0 \\ \sqrt{2x-5} &= -3 \\ (\sqrt{2x-5})^2 &= (-3)^2 \\ 2x - 5 &= 9 \\ 2x &= 14 \\ x &= 7 \end{aligned}$$

- Isolate the radical by subtracting 3 from both sides of the equation.
- Square both sides of the equation.
- Solve for *x*.

In-Class Examples (Objective 15.4A)

Solve.

1. $\sqrt{2x} + 5 = 9$ 8

2. $\sqrt{5-4x} - 2 = 3$ −5

3. $5 = 8 - \sqrt{6x}$ $\dfrac{3}{2}$

4. $\sqrt{x+11} + \sqrt{x} = 11$ 25

5. $1 = 9 + \sqrt{2x}$ No solution

Concept Check

Simplify: $(\sqrt{x} + 3)^2$

$x + 6\sqrt{x} + 9$

Optional Student Activity

In your own words, describe the process of solving $\sqrt{x + 7} + \sqrt{x} = 7$.

Answers will vary. Students should indicate that one radical should be isolated as the first step. Both sides of the equation should be squared as the second step, simplifying as you proceed. Next, isolate the remaining radical. Square both sides of the equation again. Solve the resulting equation. Check your answer. The solution is 9.

TAKE NOTE

Any time each side of an equation is squared, you must check the proposed solution of the equation.

Here is the check for the equation on the preceding page.

Check: $\sqrt{2x - 5} + 3 = 0$

$$
\begin{array}{c|c}
\sqrt{2 \cdot 7 - 5} + 3 & 0 \\
\sqrt{14 - 5} + 3 & 0 \\
\sqrt{9} + 3 & 0 \\
3 + 3 & 0 \\
6 \neq 0 &
\end{array}
$$

7 does not check as a solution. The equation has no solution.

Example 1

Solve: $\sqrt{3x} + 2 = 5$

Solution

$\sqrt{3x} + 2 = 5$

$\sqrt{3x} = 3$ • Isolate $\sqrt{3x}$.

$(\sqrt{3x})^2 = 3^2$ • Square both

$3x = 9$ sides.

$x = 3$ • Solve for x.

Check:

$$
\begin{array}{c|c}
\sqrt{3x} + 2 = 5 \\
\sqrt{3 \cdot 3} + 2 & 5 \\
\sqrt{9} + 2 & 5 \\
3 + 2 & 5 \\
5 = 5
\end{array}
$$

The solution is 3.

You Try It 1

Solve: $\sqrt{4x} + 3 = 7$

Your solution

4

Example 2

Solve: $1 = \sqrt{x} - \sqrt{x - 5}$

Solution

When an equation contains two radicals, isolate the radicals one at a time.

$1 = \sqrt{x} - \sqrt{x - 5}$

$1 + \sqrt{x - 5} = \sqrt{x}$ • Isolate $\sqrt{x}$.

$(1 + \sqrt{x - 5})^2 = (\sqrt{x})^2$ • Square both sides.

$1 + 2\sqrt{x - 5} + (x - 5) = x$ • Square the binomial.

$2\sqrt{x - 5} = 4$ • Simplify.

$\sqrt{x - 5} = 2$ • Isolate $\sqrt{x - 5}$.

$(\sqrt{x - 5})^2 = 2^2$ • Square both sides.

$x - 5 = 4$

$x = 9$ • Solve for x.

Check:

$1 = \sqrt{x} - \sqrt{x - 5}$

$$
\begin{array}{c|c}
1 & \sqrt{9} - \sqrt{9 - 5} \\
1 & \sqrt{9} - \sqrt{4} \\
1 & 3 - 2 \\
1 = 1
\end{array}
$$

The solution is 9.

You Try It 2

Solve: $\sqrt{x} + \sqrt{x + 9} = 9$

Your solution

16

Solutions on pp. S37–S38

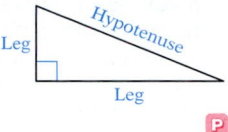

Objective B   **To solve application problems**

A **right triangle** is a triangle that contains a 90° angle. The side opposite the 90° angle is called the **hypotenuse.** The other two sides are called **legs.**

Pythagoras, a Greek mathematician who lived around 550 B.C., is given credit for the Pythagorean Theorem. It states that the square of the hypotenuse of a right triangle is equal to the sum of the squares of the two legs. Actually, this theorem was known to the Babylonians around 1200 B.C.

Pythagoras (c. 580 B.C.–520 B.C.)

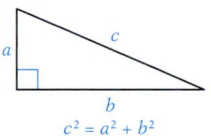
$c^2 = a^2 + b^2$

Vocabulary to Review

right triangle   [7.2A]
hypotenuse   [7.3A]
leg   [7.3A]

Theorems to Review

Pythagorean Theorem   [7.3A]

Optional Student Activity

1. Is the sum of the lengths of the legs of a right triangle always greater than the length of the hypotenuse?
   Yes

2. Use your answer to part 1 to give a reason why, if a and b are positive numbers, $a + b$ is always greater than $\sqrt{a^2 + b^2}$.
   Let a and b represent the lengths of the legs of a right triangle. Then, from part 1, $a + b > c = \sqrt{c^2} = \sqrt{a^2 + b^2}$.

3. From the result to part 2, conclude that for positive numbers, $\sqrt{a^2 + b^2} \neq a + b$.
   From part 2, $a + b > \sqrt{a^2 + b^2}$; therefore, $\sqrt{a^2 + b^2} \neq a + b$.

Point of Interest

The first known proof of this theorem occurs in a Chinese text, *Arithmetic Classic*, which was first written around 600 B.C. (but there are no existing copies) and revised over a period of 500 years. The earliest known copy of this text dates from approximately 100 B.C.

Pythagorean Theorem

If a and b are the lengths of the legs of a right triangle and c is the length of the hypotenuse, then $c^2 = a^2 + b^2$.

Using this theorem, we can find the hypotenuse of a right triangle when we know the two legs. For example, for the triangle at the right, use the formula

$$\text{Hypotenuse} = \sqrt{(\text{leg})^2 + (\text{leg})^2}$$
$$c = \sqrt{a^2 + b^2}$$
$$= \sqrt{(5)^2 + (12)^2}$$
$$= \sqrt{25 + 144}$$
$$= \sqrt{169}$$
$$= 13$$

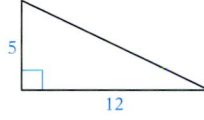

The leg of a right triangle can be found when one leg and the hypotenuse are known. For example, for the triangle at the right, use the formula

$$\text{Leg} = \sqrt{(\text{hypotenuse})^2 - (\text{leg})^2}$$
$$a = \sqrt{c^2 - b^2}$$
$$= \sqrt{(25)^2 - (20)^2}$$
$$= \sqrt{625 - 400}$$
$$= \sqrt{225}$$
$$= 15$$

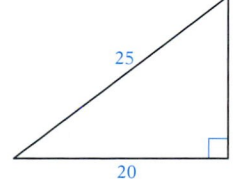

Example 3 and You Try It 3 on the following page illustrate the use of the Pythagorean Theorem. Example 4 and You Try It 4 illustrate other applications of radical equations.

In-Class Examples (Objective 15.4B)

1. Can a straight line 7 in. long be drawn on the bottom of a box that is 5 in. wide, 5 in. long, and 1 in. high?
   Yes. The diagonal of the 5-inch-by-5-inch box is approximately 7.07 in.

2. The maximum distance d in kilometers that you can see from a tall building is given by the formula $d = 1.117\sqrt{10h}$, where h is the height of the building in meters. If a person at the top of the Canadian National Tower in Toronto can see 65.7 km, determine the height of the National Tower to the nearest meter.   346 m

Optional Student Activity

1. The length of a side of the outer square in the diagram below is $2x$ in. The corners of the inner square are connected to the midpoints of the sides of the outer square.

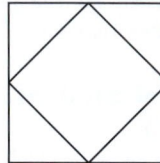

a. What is the length of a side of the inner square?
$x\sqrt{2}$ in.

b. What is the area of the inner square?
$2x^2$ in^2

2. The radius of the circle in the diagram below is 3 in.

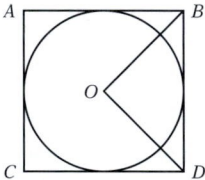

a. What is the ratio of the area of triangle *OBD* to the area of square *ABDC*?
$\dfrac{1}{4}$

b. What is the ratio of the area of triangle OBD to the area of the circle? Leave the answer in terms of π.
$\dfrac{1}{\pi}$

Example 3

A guy wire is attached to a point 20 m above the ground on a telephone pole. The wire is anchored to the ground at a point 8 m from the base of the pole. Find the length of the guy wire. Round to the nearest tenth.

Strategy

To find the length of the guy wire, use the Pythagorean Theorem. One leg is 20 m. The other leg is 8 m. The guy wire is the hypotenuse. Solve the Pythagorean Theorem for the hypotenuse.

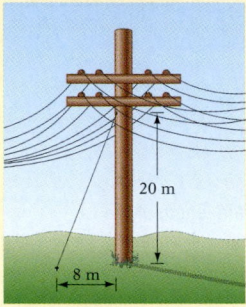

20 m

8 m

Solution
$$c = \sqrt{a^2 + b^2}$$
$$= \sqrt{(20)^2 + (8)^2} \qquad \bullet \; a = 20, \, b = 8$$
$$= \sqrt{400 + 64} = \sqrt{464} \approx 21.5$$

The guy wire has a length of approximately 21.5 m.

Example 4

How far would a submarine periscope have to be above the water to locate a ship 4 mi away? The equation for the distance in miles that the lookout can see is $d = \sqrt{1.5h}$, where h is the height in feet above the surface of the water. Round to the nearest hundredth.

Strategy

To find the height above the water, replace d in the equation with the given value and solve for h.

Solution
$$d = \sqrt{1.5h}$$
$$4 = \sqrt{1.5h} \qquad \bullet \; d = 4$$
$$4^2 = (\sqrt{1.5h})^2$$
$$16 = 1.5h$$
$$10.67 \approx h$$

The periscope must be approximately 10.67 ft above the water.

You Try It 3

A ladder 8 ft long is resting against a building. How high on the building will the ladder reach when the bottom of the ladder is 3 ft from the building? Round to the nearest hundredth.

Your strategy

Your solution
7.42 ft

You Try It 4

Find the length of a pendulum that makes one swing in 2.5 s. The equation for the time for one swing is $T = 2\pi\sqrt{\dfrac{L}{32}}$, where T is the time in seconds and L is the length in feet. Use 3.14 for π. Round to the nearest hundredth.

Your strategy

Your solution
5.07 ft

Solutions on p. S38

15.4 Exercises

Objective A   To solve an equation containing a radical expression

Section 15.4

Suggested Assignment
Exercises 1–29, odds
More challenging problems:
    Exercises 31–35

For Exercises 1 to 22, solve and check.

1. $\sqrt{x} = 5$
25

2. $\sqrt{y} = 7$
49

3. $\sqrt{a} = 12$
144

4. $\sqrt{a} = 9$
81

5. $\sqrt{5x} = 5$
5

6. $\sqrt{4x} + 5 = 2$
No solution

7. $\sqrt{3x} + 9 = 4$
No solution

8. $\sqrt{3x - 2} = 4$
6

9. $\sqrt{5x + 6} = 1$
−1

10. $\sqrt{2x + 1} = 7$
24

11. $\sqrt{5x + 4} = 3$
1

12. $0 = 2 - \sqrt{3 - x}$
−1

13. $0 = 5 - \sqrt{10 + x}$
15

14. $\sqrt{5x + 2} = 0$
$-\dfrac{2}{5}$

15. $\sqrt{3x - 7} = 0$
$\dfrac{7}{3}$

16. $\sqrt{3x} - 6 = -4$
$\dfrac{4}{3}$

17. $\sqrt{x^2 + 5} = x + 1$
2

18. $\sqrt{x^2 - 5} = 5 - x$
3

19. $\sqrt{x + 4} - \sqrt{x - 1} = 1$
5

20. $\sqrt{x} + \sqrt{x - 12} = 2$
No solution

21. $\sqrt{2x + 1} - \sqrt{2x - 4} = 1$
4

22. $\sqrt{3x + 1} - \sqrt{3x - 2} = 1$
1

Objective B   To solve application problems

23. **Baseball**   The infield of a baseball diamond is a square. The distance between successive bases is 90 ft. The pitcher's mound is on the diagonal between home plate and second base at a distance of 60.5 ft from home plate. (See the figure at the right.) Is the pitcher's mound more or less than halfway between home plate and second base?   Less

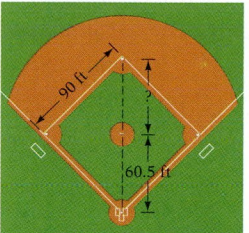

24. **Softball**   The infield of a softball diamond is a square. The distance between successive bases is 60 ft. The pitcher's mound is on the diagonal between home plate and second base at a distance of 46 ft from home plate. Is the pitcher's mound more or less than halfway between home plate and second base?   More

25. **Periscopes**   How far would a submarine periscope have to be above the water to locate a ship 5 mi away? The equation for the distance in miles that the lookout can see is $d = \sqrt{1.5h}$, where h is the height in feet above the surface of the water. Round to the nearest hundredth.   16.67 ft

26. **Building Maintenance**   A 16-foot ladder is leaning against a building. How high on the building will the ladder reach when the bottom of the ladder is 5 ft from the building? (See the figure at the right.) Round to the nearest tenth.   15.2 ft

Quick Quiz (Objective 15.4A)

Solve.

1. $\sqrt{x} = 6$
36

2. $\sqrt{3x + 1} = 4$
5

3. $\sqrt{2x} + 6 = 4$
No solution

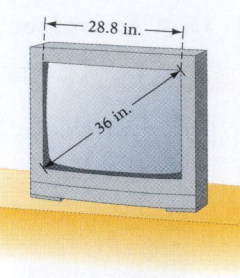

27. **Home Entertainment** The measure of a big-screen television is given by the length of a diagonal across the screen. A 36-inch television has a width of 28.8 in. Find the height of the screen to the nearest tenth of an inch.
    21.6 in.

28. **Home Entertainment** The measure of a television screen is given by the length of a diagonal across the screen. A 33-inch television has a width of 26.4 in. Find the height of the screen to the nearest tenth of an inch.
    19.8 in.

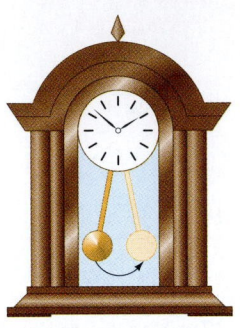

29. **Merry-Go-Rounds** The speed of a child riding a merry-go-round at a carnival is given by the equation $v = \sqrt{12r}$, where v is the speed in feet per second and r is the distance in feet from the center of the merry-go-round to the rider. If a child is moving at 15 ft/s, how far is the child from the center of the merry-go-round?
    18.75 ft

30. **Pendulums** Find the length of a pendulum that makes one swing in 1.5 s. The equation for the time of one swing of a pendulum is $T = 2\pi\sqrt{\dfrac{L}{32}}$, where T is the time in seconds and L is the length in feet. Use 3.14 for π. Round to the nearest hundredth.
    1.83 ft

APPLYING THE CONCEPTS

31. **Geometry** In the coordinate plane, a triangle is formed by drawing lines between the points (0, 0) and (5, 0), (5, 0) and (5, 12), and (5, 12) and (0, 0). Find the perimeter of the triangle.   30 units

32. **Geometry** The hypotenuse of a right triangle is $5\sqrt{2}$ cm, and one leg is $4\sqrt{2}$ cm.
    a. Find the perimeter of the triangle. $12\sqrt{2}$ cm
    b. Find the area of the triangle.   12 cm²

Answers to Writing Exercises

33. If a and b are real numbers and $a^2 = b^2$, then it does not necessarily follow that $a = b$. For example, if $a = 4$ and $b = -4$, then $a^2 = b^2$ but $a \neq b$.

33. ✏ If a and b are real numbers and $a^2 = b^2$, does $a = b$? Explain your answer.

34. **Geometry** Can the Pythagorean Theorem be used to find the length of side c of the triangle at the right? If so, determine c. If not, explain why the theorem cannot be used.
    No. The Pythagorean Theorem is true only for right triangles.

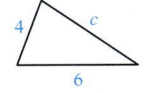

35. **Fountain Design** A circular fountain is being designed for a triangular plaza in a cultural center. The fountain is placed so that each side of the triangle touches the fountain, as shown in the diagram at the right. Find the area of the fountain. The formula for the radius of the circle is given by

$$r = \sqrt{\dfrac{(s - a)(s - b)(s - c)}{s}}$$

where $s = \dfrac{1}{2}(a + b + c)$ and a, b, and c are the lengths of the sides of the triangle. Round to the nearest hundredth.   244.78 ft²

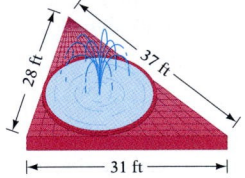

36. Complete each statement using <, =, or >.
    a. For an acute triangle with longest side c, $a^2 + b^2$ __>__ c^2.
    b. For a right triangle with longest side c, $a^2 + b^2$ __=__ c^2.
    c. For an obtuse triangle with longest side c, $a^2 + b^2$ __<__ c^2.

Quick Quiz (Objective 15.4B)

1. An 8-foot ladder is leaning against a building. How high on the building will the ladder reach when the bottom of the ladder is 2.5 ft from the building? Round to the nearest tenth.   7.6 ft

Instructor Note
You might contrast deductive
reasoning with inductive
reasoning, which is presented in
the Focus on Problem Solving in
Chapter 8.

Focus on Problem Solving

Deductive Reasoning   Deductive reasoning uses a rule or statement of fact to reach a conclusion. For instance, if two angles of one triangle are equal to two angles of another triangle, then the two triangles are similar. Thus any time we establish this fact about two triangles, we know that the triangles are similar. Below are two examples of deductive reasoning.

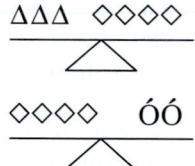

Given that $\triangle\triangle\triangle = \diamond\diamond\diamond\diamond$ and $\diamond\diamond\diamond\diamond = \acute{O}\acute{O}$, then $\triangle\triangle\triangle\triangle\triangle\triangle$ is equivalent to how many Ós?

Because three $\triangle$s = four $\diamond$s and four $\diamond$s = two Ós, three $\triangle$s = two Ós.

Six $\triangle$s is twice three $\triangle$s. We need to find twice two Ós, which is four Ós.

Therefore, $\triangle\triangle\triangle\triangle\triangle\triangle = \acute{O}\acute{O}\acute{O}\acute{O}$.

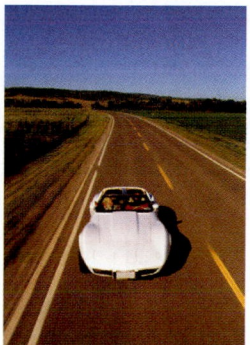

Lomax, Parish, Thorpe, and Wong are neighbors. Each drives a different type of vehicle: a compact car, a sedan, a sports car, or a station wagon. From the following statements, determine which type of vehicle each of the neighbors drives.

1. Although the vehicle owned by Lomax has more mileage on it than does either the sedan or the sports car, it does not have the highest mileage of all four cars. (Use X1 in the chart below to eliminate the possibilities that this statement rules out.)

2. Wong and the owner of the sports car live on one side of the street, and Thorpe and the owner of the compact car live on the other side of the street. (Use X2 to eliminate the possibilities that this statement rules out.)

3. Thorpe owns the vehicle with the most mileage on it. (Use X3 to eliminate the possibilities that this statement rules out.)

TAKE NOTE

To use the chart to solve this problem, write an X in a box to indicate that a possibility has been eliminated. Write a $\checkmark$ to show that a match has been found. When a row or column has 3 X's, a $\checkmark$ is written in the remaining open box in that row or column of the chart.

	Compact	Sedan	Sports Car	Wagon
Lomax	$\checkmark$	X1	X1	X2
Parish	X2	X2	$\checkmark$	X2
Thorpe	X2	X3	X2	$\checkmark$
Wong	X2		X2	

Lomax drives the compact car, Parish drives the sports car, Thorpe drives the station wagon, and Wong drives the sedan.

1. Given that $\ddagger\ddagger = \bullet\bullet\bullet\bullet\bullet$ and $\bullet\bullet\bullet\bullet\bullet = \Lambda\Lambda$, then $\ddagger\ddagger\ddagger\ddagger = $ how many Λs?

2. Given that $\square\square\square\square\square\square = \acute{O}\acute{O}\acute{O}\acute{O}$ and $\acute{O}\acute{O}\acute{O}\acute{O} = \hat{I}\hat{I}$, then $\square\square\square = $ how many $\hat{I}$s?

3. Given that $\square\square\square\square = \Omega\Omega\Omega$ and $\Omega\Omega\Omega = \Delta\Delta$, then $\Delta\Delta\Delta\Delta = $ how many $\square$s?

4. Given that $¥¥¥¥¥ = §§$ and $§§ = \hat{A}\hat{A}\hat{A}$, then $\hat{A}\hat{A}\hat{A}\hat{A}\hat{A}\hat{A} = $ how many ¥s?

**Answers to Focus on
Problem Solving:
Deductive Reasoning**

1. Five
2. One
3. Eight
4. 10
5. Anna: golf
   Kay: sailing
   Megan: tennis
   Nicole: horseback riding
6. Chang: subway
   Nick: taxi
   Pablo: car
   Saul: bus

5. Anna, Kay, Megan, and Nicole decide to travel together during spring break, but they need to find a destination where each of them will be able to participate in her favorite sport (golf, horseback riding, sailing, or tennis). From the following statements, determine the favorite sport of each student.

   a. Anna and the student whose favorite sport is sailing both like to swim, whereas Nicole and the student whose favorite sport is tennis would prefer to scuba-dive.

   b. Megan and the student whose favorite sport is sailing are roommates. Nicole and the student whose favorite sport is golf live by themselves in singles.

6. Chang, Nick, Pablo, and Saul each take a different form of transportation (bus, car, subway, or taxi) from the office to the airport. From the following statements, determine which form of transportation each takes.

   a. Chang spent more on transportation than the fellow who took the bus but less than the fellow who took the taxi.

   b. Pablo, who did not travel by bus and who spent the least on transportation, arrived at the airport after Nick but before the fellow who took the subway.

   c. Saul spent less on transportation than either Chang or Nick.

Projects and Group Activities

Distance to the Horizon

The formula $d = \sqrt{1.5h}$ can be used to calculate the approximate distance d (in miles) that a person can see who uses a periscope h feet above the water. That formula is derived by using the Pythagorean Theorem.

Consider the diagram (not to scale) at the right, which shows Earth as a sphere and the periscope as extending h feet above its surface. From geometry, because AB is tangent to the circle and OA is a radius, triangle AOB is a right triangle. Therefore,

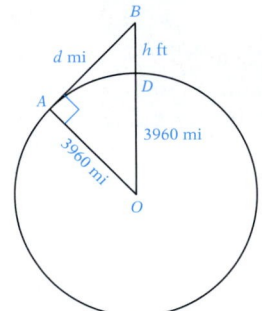

$$(OA)^2 + (AB)^2 = (OB)^2$$

Substituting into this formula, we have

$$3960^2 + d^2 = \left(3960 + \frac{h}{5280}\right)^2$$

$$3960^2 + d^2 = 3960^2 + \frac{2 \cdot 3960}{5280}h + \left(\frac{h}{5280}\right)^2$$

$$d^2 = \frac{3}{2}h + \left(\frac{h}{5280}\right)^2$$

$$d = \sqrt{\frac{3}{2}h + \left(\frac{h}{5280}\right)^2}$$

• Because h is in feet, $\frac{h}{5280}$ is in miles.

At this point, an assumption is made that $\sqrt{\frac{3}{2}h + \left(\frac{h}{5280}\right)^2} \approx \sqrt{1.5h}$, where we have written $\frac{3}{2}$ as 1.5. Thus $d \approx \sqrt{1.5h}$ is used to approximate the distance a person can see using a periscope h feet above the water.

1. Write a paragraph that justifies the assumption that

$$\sqrt{\frac{3}{2}h + \left(\frac{h}{5280}\right)^2} \approx \sqrt{1.5h}$$

(*Suggestion:* Evaluate each expression for various values of h. Because h is the height of a periscope above water, it is unlikely that $h > 25$ ft.)

2. The distance d is the distance from the top of the periscope to A. The distance along the surface of the water is given by arc AD. This distance can be approximated by the equation

$$L \approx \sqrt{1.5h} + 0.306186\left(\sqrt{\frac{h}{5280}}\right)^3$$

Using this formula, calculate L when $h = 10$.

Answers to Projects and Group Activities: Distance to the Horizon

1. Answers will vary. Students should conclude that $(h/5280)^2$ is a very small number and has very little influence on the answer.

2. The distance is approximately 3.9 mi.

Chapter 15 Summary

Key Words

	Examples
A *square root* of a positive number a is a number whose square is a. Every positive number has two square roots, one a positive and one a negative number. The square root of a negative number is not a real number. [15.1A, pp. 753–754]	A square root of 49 is 7 because $7^2 = 49$. A square root of 49 is -7 because $(-7)^2 = 49$. $\sqrt{-9}$ is not a real number.
The symbol $\sqrt{}$ is called a *radical sign* and is used to indicate the positive or *principal square root* of a number. The negative square root of a number is indicated by placing a negative sign in front of the radical. The *radicand* is the expression under the radical sign. [15.1A, p. 753]	$\sqrt{49} = 7$ $-\sqrt{49} = -7$ In the expression $\sqrt{49xy}$, $49xy$ is the radicand.
The square of an integer is a *perfect square*. If a number is not a perfect square, its square root can only be approximated. Such square roots are *irrational numbers*. Their decimal representations never terminate or repeat. [15.1A, p. 753]	1, 4, 9, 16, 25, 36, 49, 64, . . . are examples of perfect squares. 7 is not a perfect square. $\sqrt{7}$ is an irrational number.
Conjugates are expressions with two terms that differ only in the sign of one term. The expressions $a + b$ and $a - b$ are conjugates. [15.3A, p. 763]	$-5 + \sqrt{11}$ and $-5 - \sqrt{11}$ are conjugates. $\sqrt{x} - 3$ and $\sqrt{x} + 3$ are conjugates.
A *radical equation* is an equation that contains a variable expression in a radicand. [15.4A, p. 769]	$\sqrt{2x} + 5 = 9$ is a radical equation. $2x + \sqrt{5} = 9$ is not a radical equation.
A *right triangle* is a triangle that contains a 90° angle. The side opposite the 90° angle is the *hypotenuse*. The other two sides are called *legs*. [15.4B, p. 771]	

Essential Rules and Procedures

	Examples
The Product Property of Square Roots [15.1A, p. 753] If a and b are positive real numbers, then $\sqrt{ab} = \sqrt{a} \cdot \sqrt{b}$. Use the Product Property of Square Roots and a knowledge of perfect squares to simplify radicands that are not perfect squares.	$\sqrt{28} = \sqrt{4 \cdot 7} = \sqrt{4} \cdot \sqrt{7} = 2\sqrt{7}$ $\sqrt{9x^7} = \sqrt{9x^6 \cdot x} = \sqrt{9x^6}\sqrt{x} = 3x^3\sqrt{x}$

Adding or Subtracting Radical Expressions [15.2A, p. 759]
The Distributive Property is used to simplify the sum or difference of radical expressions with like radicands.

$$8\sqrt{2x} - 3\sqrt{2x} = (8 - 3)\sqrt{2x} = 5\sqrt{2x}$$

Multiplying Radical Expressions [15.3A, p. 763]
The Product Property of Square Roots is used to multiply radical expressions.

$$\sqrt{2y}(\sqrt{3} - \sqrt{x}) = \sqrt{6y} - \sqrt{2xy}$$

Use the FOIL method to multiply radical expressions with two terms.

$$(3 - \sqrt{x})(5 + \sqrt{x})$$
$$= 15 + 3\sqrt{x} - 5\sqrt{x} - (\sqrt{x})^2$$
$$= 15 - 2\sqrt{x} - x$$

The Quotient Property of Square Roots [15.3B, p. 764]
If a and b are positive real numbers, then $\sqrt{\dfrac{a}{b}} = \dfrac{\sqrt{a}}{\sqrt{b}}$ and $\dfrac{\sqrt{a}}{\sqrt{b}} = \sqrt{\dfrac{a}{b}}$.

$$\frac{\sqrt{27}}{\sqrt{3}} = \sqrt{\frac{27}{3}} = \sqrt{9} = 3$$

The Quotient Property of Square Roots is used to divide radical expressions.

$$\frac{\sqrt{3x^5y}}{\sqrt{75xy^3}} = \sqrt{\frac{3x^5y}{75xy^3}} = \sqrt{\frac{x^4}{25y^2}} = \frac{x^2}{5y}$$

Simplest Form of a Radical Expression [15.3B, p. 765]
For a radical expression to be in simplest form, three conditions must be met:

1. The radicand contains no factor greater than 1 that is a perfect square.

2. There is no fraction under the radical sign.

3. There is no radical in the denominator of a fraction.

$\sqrt{12}$, $\sqrt{\dfrac{3}{4}}$, and $\dfrac{1}{\sqrt{3}}$ are not in simplest form.

$5\sqrt{3}$ and $\dfrac{\sqrt{3}}{3}$ are in simplest form.

Rationalizing the Denominator [15.3B, p. 765]
The procedure used to remove a radical from the denominator is called **rationalizing the denominator.**

$$\frac{5}{\sqrt{7}} = \frac{5}{\sqrt{7}} \cdot \frac{\sqrt{7}}{\sqrt{7}} = \frac{5\sqrt{7}}{7}$$

Property of Squaring Both Sides of an Equation [15.4A, p. 769]
If a and b are real numbers and $a = b$, then $a^2 = b^2$.

$$\sqrt{x} = 5$$
$$(\sqrt{x})^2 = 5^2$$
$$x = 25$$

Solving a Radical Equation Containing One Radical [15.4A, p. 769]

1. Write the equation with the radical alone on one side.

2. Square both sides of the equation.

3. Solve for the variable.

4. Check the solution(s) in the original equation.

$$\sqrt{2x} - 1 = 5$$
$$\sqrt{2x} = 6 \qquad \bullet \text{ Isolate the radical.}$$
$$(\sqrt{2x})^2 = 6^2 \qquad \bullet \text{ Square both sides.}$$
$$2x = 36$$
$$x = 18 \qquad \bullet \text{ Solve for } x.$$

The solution checks.

Pythagorean Theorem [15.4B, p. 771]
If a and b are lengths of the legs of a right triangle and c is the length of the hypotenuse, then $c^2 = a^2 + b^2$.

Two legs of a right triangle measure 4 cm and 7 cm. Find the length of the hypotenuse.

$$c = \sqrt{a^2 + b^2}$$
$$c = \sqrt{4^2 + 7^2} \qquad \bullet \ a = 4, b = 7$$
$$c = \sqrt{16 + 49}$$
$$c = \sqrt{65}$$

The length of the hypotenuse is $\sqrt{65}$ cm.

Chapter 15 Review Exercises

1. Simplify: $\sqrt{3}(\sqrt{12} - \sqrt{3})$
  3   [15.3A]

2. Simplify: $3\sqrt{18a^5b}$
  $9a^2\sqrt{2ab}$ [15.1B]

3. Simplify: $2\sqrt{36}$
  12   [15.1A]

4. Simplify: $\sqrt{6a}(\sqrt{3a} + \sqrt{2a})$
  $3a\sqrt{2} + 2a\sqrt{3}$ [15.3A]

5. Simplify: $\dfrac{12}{\sqrt{6}}$
  $2\sqrt{6}$ [15.3B]

6. Simplify: $2\sqrt{8} - 3\sqrt{32}$
  $-8\sqrt{2}$ [15.2A]

7. Simplify: $(3 - \sqrt{7})(3 + \sqrt{7})$
  2   [15.3A]

8. Solve: $\sqrt{x + 3} - \sqrt{x} = 1$
  1   [15.4A]

9. Simplify: $\dfrac{2x}{\sqrt{3} - \sqrt{5}}$
  $-x\sqrt{3} - x\sqrt{5}$ [15.3B]

10. Simplify: $-3\sqrt{120}$
  $-6\sqrt{30}$ [15.1A]

11. Solve: $\sqrt{5x} = 10$
  20   [15.4A]

12. Simplify: $5\sqrt{48}$
  $20\sqrt{3}$ [15.1A]

13. Simplify: $\dfrac{\sqrt{98x^7y^9}}{\sqrt{2x^3y}}$
  $7x^2y^4$ [15.3B]

14. Solve: $3 - \sqrt{7x} = 5$
  No solution   [15.4A]

15. Simplify: $6a\sqrt{80b} - \sqrt{180a^2b} + 5a\sqrt{b}$
  $18a\sqrt{5b} + 5a\sqrt{b}$ [15.2A]

16. Simplify: $4\sqrt{250}$
  $20\sqrt{10}$ [15.1A]

17. Simplify: $2x\sqrt{60x^3y^3} + 3x^2y\sqrt{15xy}$
  $7x^2y\sqrt{15xy}$ [15.2A]

18. Simplify: $(4\sqrt{y} - \sqrt{5})(2\sqrt{y} + 3\sqrt{5})$
  $8y + 10\sqrt{5y} - 15$ [15.3A]

19. Simplify: $3\sqrt{12x} + 5\sqrt{48x}$
$26\sqrt{3x}$ [15.2A]

20. Solve: $\sqrt{2x - 3} + 4 = 0$
No solution [15.4A]

21. Simplify: $\dfrac{8}{\sqrt{x} - 3}$
$\dfrac{8\sqrt{x} + 24}{x - 9}$ [15.3B]

22. Simplify: $4y\sqrt{243x^{17}y^9}$

$36x^8y^5\sqrt{3xy}$ [15.1B]

23. Simplify: $y\sqrt{24y^6}$
$2y^4\sqrt{6}$ [15.1B]

24. Solve: $2x + 4 = \sqrt{x^2 + 3}$
-1 [15.4A]

25. Simplify:
$2x^2\sqrt{18x^2y^5} + 6y\sqrt{2x^6y^3} - 9xy^2\sqrt{8x^4y}$
$-6x^3y^2\sqrt{2y}$ [15.2A]

26. Simplify: $\dfrac{16}{\sqrt{a}}$
$\dfrac{16\sqrt{a}}{a}$ [15.3B]

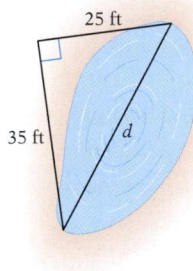

25 ft

35 ft   d

27. **Surveying**   To find the distance across a pond, a surveyor constructs a right triangle as shown at the right. Find the distance d across the pond. Round to the nearest foot.
43 ft [15.4B]

28. **Astronautics**   The weight of an object is related to the distance the object is above the surface of Earth. An equation for this relationship is $d = 4000\sqrt{\dfrac{W_0}{W_d}} - 4000$, where W_0 is the object's weight on the surface of Earth and W_d is the object's weight at a distance of d miles above Earth's surface. If a space explorer weighs 36 lb at a distance of 4000 mi above the surface of Earth, how much does the explorer weigh on the surface of Earth?
144 lb [15.4B]

29. **Tsunamis**   A tsunami is a great sea wave produced by underwater earthquakes or volcanic eruption. The velocity of a tsunami as it approaches land depends on the depth of the water and can be approximated by the equation $v = 3\sqrt{d}$, where d is the depth of the water in feet and v is the velocity of the tsunami in feet per second. Find the depth of the water if the velocity of a tsunami is 30 ft/s.
100 ft [15.4B]

30. **Bicycle Safety**   A bicycle will overturn if it rounds a corner too sharply or too fast. An equation for the maximum velocity at which a cyclist can turn a corner without tipping over is $v = 4\sqrt{r}$, where v is the velocity of the bicycle in miles per hour and r is the radius of the corner in feet. What is the radius of the sharpest corner that a cyclist can safely turn while riding at 20 mph?
25 ft [15.4B]

Chapter 15 Test

1. Simplify: $\sqrt{121x^8y^2}$
   $11x^4y$ [15.1B]

2. Simplify: $\sqrt{3x^2y}\sqrt{6xy^2}\sqrt{2x}$
   $6x^2y\sqrt{y}$ [15.3A]

3. Simplify: $5\sqrt{8} - 3\sqrt{50}$
   $-5\sqrt{2}$ [15.2A]

4. Simplify: $\sqrt{45}$
   $3\sqrt{5}$ [15.1A]

5. Simplify: $\dfrac{\sqrt{162}}{\sqrt{2}}$
   9 [15.3B]

6. Solve: $\sqrt{9x} + 3 = 18$
   25 [15.4A]

7. Simplify: $\sqrt{32a^5b^{11}}$
   $4a^2b^5\sqrt{2ab}$ [15.1B]

8. Simplify: $\dfrac{\sqrt{98a^6b^4}}{\sqrt{2a^3b^2}}$
   $7ab\sqrt{a}$ [15.3B]

9. Simplify: $\dfrac{2}{\sqrt{3} - 1}$
   $\sqrt{3} + 1$ [15.3B]

10. Simplify: $\sqrt{8x^3y}\sqrt{10xy^4}$
    $4x^2y^2\sqrt{5y}$ [15.3A]

11. Solve: $\sqrt{x-5} + \sqrt{x} = 5$

9 [15.4A]

12. Simplify: $3\sqrt{8y} - 2\sqrt{72x} + 5\sqrt{18y}$

$21\sqrt{2y} - 12\sqrt{2x}$ [15.2A]

13. Simplify: $\sqrt{72x^7y^2}$

$6x^3y\sqrt{2x}$ [15.1B]

14. Simplify: $(\sqrt{y} - 3)(\sqrt{y} + 5)$

$y + 2\sqrt{y} - 15$ [15.3A]

15. Simplify: $2x\sqrt{3xy^3} - 2y\sqrt{12x^3y} - 3xy\sqrt{xy}$

$-2xy\sqrt{3xy} - 3xy\sqrt{xy}$ [15.2A]

16. Simplify: $\dfrac{2 - \sqrt{5}}{6 + \sqrt{5}}$

$\dfrac{17 - 8\sqrt{5}}{31}$ [15.3B]

17. Simplify: $\sqrt{a}(\sqrt{a} - \sqrt{b})$

$a - \sqrt{ab}$ [15.3A]

18. Simplify: $\sqrt{75}$

$5\sqrt{3}$ [15.1A]

19. **Pendulums** Find the length of a pendulum that makes one swing in 3 s. The equation for the time of one swing of a pendulum is $T = 2\pi\sqrt{\dfrac{L}{32}}$, where T is the time in seconds and L is the length in feet. Use 3.14 for π. Round to the nearest hundredth.

7.30 ft [15.4B]

20. **Camping** A support rope for a tent is attached to the top of a pole and then secured to the ground as shown in the figure at the right. If the rope is 8 ft long and the pole is 4 ft high, how far, x, from the base of the pole should the rope be secured? Round to the nearest foot.

7 ft [15.4B]

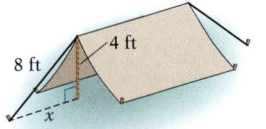

Cumulative Review Exercises

1. Simplify:

$$\left(\frac{2}{3}\right)^2 \cdot \left(\frac{3}{4} - \frac{3}{2}\right) + \left(\frac{1}{2}\right)^2$$

$-\dfrac{1}{12}$ [3.5A]

2. Simplify:

$$-3[x - 2(3 - 2x) - 5x] + 2x$$

$2x + 18$ [4.2D]

3. Solve:
$$2x - 4[3x - 2(1 - 3x)] = 2(3 - 4x)$$
$\dfrac{1}{13}$ [5.3B]

4. Simplify: $(-3x^2y)(-2x^3y^4)$

$6x^5y^5$ [9.2A]

5. Simplify: $\dfrac{12b^4 - 6b^2 + 2}{-6b^2}$

$-2b^2 + 1 - \dfrac{1}{3b^2}$ [9.5A]

6. Given $f(x) = \dfrac{2x}{x - 3}$, find $f(-3)$.

1 [12.1D]

7. Factor: $2a^3 - 16a^2 + 30a$

$2a(a - 5)(a - 3)$ [10.2B]

8. Multiply: $\dfrac{3x^3 - 6x^2}{4x^2 + 4x} \cdot \dfrac{3x - 9}{9x^3 - 45x^2 + 54x}$

$\dfrac{1}{4(x + 1)}$ [11.1B]

9. Subtract: $\dfrac{x + 2}{x - 4} - \dfrac{6}{(x - 4)(x - 3)}$

$\dfrac{x + 3}{x - 3}$ [11.3B]

10. Solve: $\dfrac{x}{2x - 5} - 2 = \dfrac{3x}{2x - 5}$

$\dfrac{5}{3}$ [11.5A]

11. Find the equation of the line that contains the point $(-2, -3)$ and has slope $\dfrac{1}{2}$.

$y = \dfrac{1}{2}x - 2$ [12.4A]

12. Solve by substitution:
$$4x - 3y = 1$$
$$2x + y = 3$$

$(1, 1)$ [13.2A]

13. Solve by the addition method:
$$5x + 4y = 7$$
$$3x - 2y = 13$$
$(3, -2)$ [13.3A]

14. Solve: $3(x - 7) \geq 5x - 12$

$x \leq -\dfrac{9}{2}$ [14.3A]

15. Simplify: $\sqrt{108}$
$6\sqrt{3}$ [15.1A]

16. Simplify: $3\sqrt{32} - 2\sqrt{128}$
$-4\sqrt{2}$ [15.2A]

17. Simplify: $2a\sqrt{2ab^3} + b\sqrt{8a^3b} - 5ab\sqrt{ab}$
$4ab\sqrt{2ab} - 5ab\sqrt{ab}$ [15.2A]

18. Simplify: $\sqrt{2a^9b}\sqrt{98ab^3}\sqrt{2a}$
$14a^5b^2\sqrt{2a}$ [15.3A]

19. Simplify: $\sqrt{3}(\sqrt{6} - \sqrt{x^2})$

$3\sqrt{2} - x\sqrt{3}$ [15.3A]

20. Simplify: $\dfrac{\sqrt{320}}{\sqrt{5}}$

8 [15.3B]

21. Simplify: $\dfrac{3}{2 - \sqrt{5}}$

$-6 - 3\sqrt{5}$ [15.3B]

22. Solve: $\sqrt{3x - 2} - 4 = 0$

6 [15.4A]

23. **Business** The selling price of a book is $29.40. The markup rate used by the book-store is 20%. Find the cost of the book. Use the formula $S = C + rC$, where S is the selling price, C is the cost, and r is the markup rate.

$24.50 [5.2B]

24. **Travel** Two cyclists start from the same point and ride in opposite directions. One cyclist rides 4 mph faster than the other. In 2 h, they are 52 mi apart. Find the rate of the faster cyclist.

15 mph [5.5B]

25. **Number Sense** The sum of two numbers is twenty-one. The product of the two numbers is one hundred four. Find the two numbers.

8, 13 [10.5B]

26. **Work** A small water pipe takes twice as long to fill a tank as does a larger water pipe. With both pipes open, it takes 16 h to fill the tank. Find the time it would take the small pipe working alone to fill the tank.

48 h [11.7A]

27. Solve by graphing: $3x - 2y = 8$
$\qquad\qquad\qquad\quad 4x + 5y = 3$

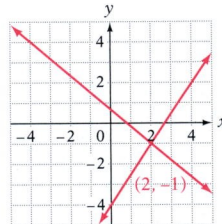

(2, −1) [13.1A]

28. Graph the solution set of $3x + y \le 2$.

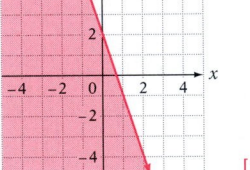

[14.4A]

29. **Number Sense** The square root of the sum of two consecutive integers is equal to 9. Find the smaller integer.

40 [15.4B]

30. **Physics** A stone is dropped from a building and hits the ground 5 s later. How high is the building? The equation for the distance an object falls in T seconds is $T = \sqrt{\dfrac{d}{16}}$, where d is the distance in feet.

400 ft [15.4B]

16

Quadratic Equations

This photo shows employees at the Fender factory in Corona, California, in the process of making guitars. During this stage, the Fender Stratocaster bodies are hand-sanded. The objective of the Fender Musical Instruments Corporation, as with any business, is to earn a profit. Profit is the difference between a company's revenue (the total amount of money the company earns by selling its products or services) and its costs (the total amount of money the company spends in doing business). The **Focus on Problem Solving on page 812** uses quadratic equations to determine the maximum profit a company can earn.

OBJECTIVES

Section 16.1

A To solve a quadratic equation by factoring

B To solve a quadratic equation by taking square roots

Section 16.2

A To solve a quadratic equation by completing the square

Section 16.3

A To solve a quadratic equation by using the quadratic formula

Section 16.4

A To graph a quadratic equation of the form $y = ax^2 + bx + c$

Section 16.5

A To solve application problems

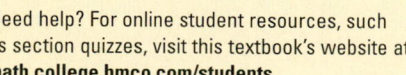

Need help? For online student resources, such as section quizzes, visit this textbook's website at **math.college.hmco.com/students**.

Do these exercises to prepare for Chapter 16.

1. Evaluate $b^2 - 4ac$ when $a = 2$, $b = -3$, and $c = -4$.
   41 [4.1A]

2. Solve: $5x + 4 = 3$
   $-\dfrac{1}{5}$ [5.2A]

3. Factor: $x^2 + x - 12$
   $(x + 4)(x - 3)$ [10.2A]

4. Factor: $4x^2 - 12x + 9$
   $(2x - 3)^2$ [10.4A]

5. Is $x^2 - 10x + 25$ a perfect-square trinomial?
   Yes [10.4A]

6. Solve: $\dfrac{5}{x - 2} = \dfrac{15}{x}$
   3 [11.5A]

7. Graph: $y = -2x + 3$
   [12.2A]

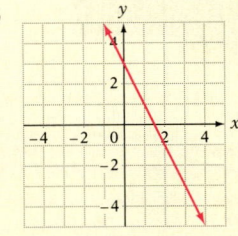

8. Simplify: $\sqrt{28}$
   $2\sqrt{7}$ [15.1A]

9. If a is *any* real number, simplify $\sqrt{a^2}$.
   $|a|$ [15.1B]

10. **Exercising** Walking at a constant speed of 4.5 mph, Lucy and Sam walked from the beginning to the end of a hiking trail. When they reached the end, they immediately started back along the same path at a constant speed of 3 mph. If the round-trip took 2 h, what is the length of the hiking trail?
    3.6 mi [5.5B]

GO FIGURE • • •

Find the value of $\dfrac{1}{\sqrt{1} + \sqrt{2}} + \dfrac{1}{\sqrt{2} + \sqrt{3}} + \dfrac{1}{\sqrt{3} + \sqrt{4}} + \cdots + \dfrac{1}{\sqrt{8} + \sqrt{9}}$.

2

16.1 Solving Quadratic Equations by Factoring or by Taking Square Roots

Objective A   To solve a quadratic equation by factoring

An equation of the form $ax^2 + bx + c = 0$, where $a, b,$ and c are real numbers and $a \neq 0$, is a **quadratic equation.**

$4x^2 - 3x + 1 = 0, a = 4, b = -3, c = 1$
$3x^2 - 4 = 0, a = 3, b = 0, c = -4$
$\dfrac{x^2}{2} - 2x + 4 = 0, a = \dfrac{1}{2}, b = -2, c = 4$

A quadratic equation is also called a **second-degree equation.**

A quadratic equation is in **standard form** when the polynomial is written in descending order and set equal to zero.

Recall that the Principle of Zero Products states that if the product of two factors is zero, then at least one of the factors must be zero.

If $a \cdot b = 0$,
then $a = 0$ or $b = 0$.

The Principle of Zero Products can be used to solve quadratic equations by factoring. Write the equation in standard form, factor the polynomial, apply the Principle of Zero Products, and solve for the variable.

> **HOW TO**   Solve by factoring: $2x^2 - x = 1$
>
> $2x^2 - x = 1$
> $2x^2 - x - 1 = 0$ • Write the equation in standard form.
> $(2x + 1)(x - 1) = 0$ • Factor.
> $2x + 1 = 0 \qquad x - 1 = 0$ • Use the Principle of Zero Products to set each factor equal to zero.
> $2x = -1 \qquad\quad x = 1$ • Rewrite each equation in the form *variable = constant.*
> $x = -\dfrac{1}{2}$
>
> Check:
>
> $$\begin{array}{c|c}
> 2x^2 - x = 1 & 2x^2 - x = 1 \\ \hline
> 2\left(-\dfrac{1}{2}\right)^2 - \left(-\dfrac{1}{2}\right) \;\Big|\; 1 & 2(1)^2 - 1 \;\Big|\; 1 \\
> 2 \cdot \dfrac{1}{4} + \dfrac{1}{2} \;\Big|\; 1 & 2 \cdot 1 - 1 \;\Big|\; 1 \\
> \dfrac{1}{2} + \dfrac{1}{2} \;\Big|\; 1 & 2 - 1 \;\Big|\; 1 \\
> 1 = 1 & 1 = 1
> \end{array}$$
>
> The solutions are $-\dfrac{1}{2}$ and 1.

TAKE NOTE

You should always check your solutions by substituting the proposed solutions back into the *original* equation.

Objective 16.1A

Vocabulary to Review
quadratic equation   [10.5A]
standard form   [10.5A]

New Vocabulary
second-degree equation
double root

Rules to Review
Principle of Zero Products
[10.5A]

Instructor Note
As a class discussion, ask students why the condition $a \neq 0$ is placed on the quadratic equation.

Instructor Note
Quadratic equations were solved by factoring earlier in the text. This objective is completely self-contained so that if this is the first introduction for students, all the necessary information is included here.

Concept Check
Write the following quadratic equation in standard form.
$11x + 10 = 6x^2$
$6x^2 - 11x - 10 = 0$ or
$-6x^2 + 11x + 10 = 0$

Discuss the Concepts
What does the Principle of Zero Products imply about the equation $(x - 4)(3x + 1) = 0$?
Either $x - 4 = 0$ or $3x + 1 = 0$.

Concept Check
Solve.
1. $(x + 1)(x - 2) = 0$
$-1, 2$
2. $(2x - 3)(3x + 5) = 0$
$\dfrac{3}{2}, -\dfrac{5}{3}$
3. $4x(x - 5) = 0$
$0, 5$

In-Class Examples (Objective 16.1A)

Solve by factoring.

1. $x(x - 1) = 12$ $-3, 4$

2. $5x^2 + 9x - 2 = 0$ $\dfrac{1}{5}, -2$

3. $\dfrac{x^2}{2} + \dfrac{5x}{6} + \dfrac{1}{3} = 0$ $-\dfrac{2}{3}, -1$

Discuss the Concepts

Under what conditions will a second-degree equation have a double root?

If both factors of $ax^2 + bx + c$ are identical, the equation will have a double root.

Instructor Note

In an equation in one variable, the highest exponent on the variable is the same as the number of solutions of the equation. For example, for quadratic equations, the highest exponent is 2 and the number of solutions is two. It is for this reason that double roots are emphasized.

HOW TO Solve by factoring: $3x^2 - 4x + 8 = (4x + 1)(x - 2)$

$3x^2 - 4x + 8 = (4x + 1)(x - 2)$

$3x^2 - 4x + 8 = 4x^2 - 7x - 2$
- Multiply the factors on the right side of the equation.

$0 = x^2 - 3x - 10$
- Write the equation in standard form.

$0 = (x - 5)(x + 2)$
- Factor.

$x - 5 = 0 \qquad x + 2 = 0$
- Use the Principle of Zero Products to set each factor equal to zero.

$x = 5 \qquad\qquad x = -2$
- Rewrite each equation in the form *variable* = *constant*.

Check:

$3x^2 - 4x + 8 = (4x + 1)(x - 2)$	
$3(5)^2 - 4(5) + 8$	$(4[5] + 1)(5 - 2)$
$3(25) - 4(5) + 8$	$(20 + 1)(3)$
$75 - 20 + 8$	$(21)(3)$
$63 = 63$	

$3x^2 - 4x + 8 = (4x + 1)(x - 2)$	
$3(-2)^2 - 4(-2) + 8$	$(4[-2] + 1)(-2 - 2)$
$3(4) - 4(-2) + 8$	$(-8 + 1)(-4)$
$12 + 8 + 8$	$(-7)(-4)$
$28 = 28$	

The solutions are 5 and -2.

HOW TO Solve by factoring: $x^2 - 10x + 25 = 0$

$x^2 - 10x + 25 = 0$

$(x - 5)(x - 5) = 0$
- Factor.

$x - 5 = 0 \qquad x - 5 = 0$
- Use the Principle of Zero Products.

$x = 5 \qquad\qquad x = 5$
- Solve each equation for *x*.

The solution is 5.

In this last example, 5 is called a **double root** of the quadratic equation.

Example 1

Solve by factoring: $\dfrac{z^2}{2} - \dfrac{z}{4} - \dfrac{1}{4} = 0$

Solution

$\dfrac{z^2}{2} - \dfrac{z}{4} - \dfrac{1}{4} = 0$

$4\left(\dfrac{z^2}{2} - \dfrac{z}{4} - \dfrac{1}{4}\right) = 4(0)$
- Multiply each side by 4.

$2z^2 - z - 1 = 0$

$(2z + 1)(z - 1) = 0$
- Factor.

$2z + 1 = 0 \qquad z - 1 = 0$
- Principle of Zero Products

$2z = -1 \qquad\qquad z = 1$

$z = -\dfrac{1}{2}$

The solutions are $-\dfrac{1}{2}$ and 1.

You Try It 1

Solve by factoring: $\dfrac{3y^2}{2} + y - \dfrac{1}{2} = 0$

Your solution

$\dfrac{1}{3}, -1$

Solution on p. S39

Objective B

To solve a quadratic equation by taking square roots

Consider a quadratic equation of the form $x^2 = a$. This equation can be solved by factoring.

$$x^2 = 25$$
$$x^2 - 25 = 0$$
$$(x - 5)(x + 5) = 0$$
$$x - 5 = 0 \qquad x + 5 = 0$$
$$x = 5 \qquad x = -5$$

The solutions are 5 and -5. The solutions are plus or minus the same number, which is frequently written by using $\pm$; for example, "the solutions are ± 5." An alternative method of solving this equation is suggested by the fact that ± 5 can be written as $\pm\sqrt{25}$.

> **Principle of Taking the Square Root of Each Side of an Equation**
>
> If $x^2 = a$, then $x = \pm\sqrt{a}$.

HOW TO Solve by taking square roots: $x^2 = 25$

$$x^2 = 25$$
$$\sqrt{x^2} = \sqrt{25}$$
$$x = \pm\sqrt{25} = \pm 5$$

- Take the square root of each side of the equation. Then simplify.

The solutions are 5 and -5.

HOW TO Solve by taking square roots: $3x^2 = 36$

$$3x^2 = 36$$
$$x^2 = 12$$
$$\sqrt{x^2} = \sqrt{12}$$
$$x = \pm\sqrt{12}$$
$$x = \pm 2\sqrt{3}$$

- Solve for x^2.
- Take the square root of each side.
- Simplify.

The solutions are $2\sqrt{3}$ and $-2\sqrt{3}$.

Study Tip

Always check the solution of an equation. Here is a check of the solution $-\dfrac{5}{7}$ for the equation at the right:

$$\begin{array}{c|c} 49y^2 - 25 = 0 \\ \hline 49\left(-\dfrac{5}{7}\right)^2 - 25 & 0 \\ 49\left(\dfrac{25}{49}\right) - 25 & 0 \\ 25 - 25 & 0 \\ 0 & = 0 \end{array}$$

HOW TO Solve by taking square roots: $49y^2 - 25 = 0$

$$49y^2 - 25 = 0$$
$$49y^2 = 25$$
$$y^2 = \frac{25}{49}$$
$$\sqrt{y^2} = \sqrt{\frac{25}{49}}$$
$$y = \pm\frac{5}{7}$$

- Solve for y^2.
- Take the square root of each side.
- Simplify.

The solutions are $\dfrac{5}{7}$ and $-\dfrac{5}{7}$.

Objective 16.1B

Discuss the Concepts
Explain what $x = \pm 6$ means.
x equals either 6 or -6.

Concept Check
What are the two solutions of
$x - 3 = \pm 4$?
7, -1

Optional Student Activity
What are the two solutions of
$x + 3 = \pm\dfrac{2}{3}$?
$-\dfrac{7}{3}, -\dfrac{11}{3}$

Optional Student Activity
Create three equations such that the first has no real number solution, the second has one real number solution, and the third has two real number solutions.

In-Class Examples (Objective 16.1B)
Solve by taking square roots.
1. $x^2 - 8 = 0$ $-2\sqrt{2}, 2\sqrt{2}$
2. $y^2 + 64 = 0$ No real number solution
3. $11(x + 3)^2 = 66$ $-3 + \sqrt{6}, -3 - \sqrt{6}$

An equation that contains the square of a binomial can be solved by taking square roots.

HOW TO   Solve by taking square roots: $2(x - 1)^2 - 36 = 0$

$$2(x - 1)^2 - 36 = 0$$
$$2(x - 1)^2 = 36 \qquad \bullet \text{ Solve for } (x-1)^2.$$
$$(x - 1)^2 = 18$$
$$\sqrt{(x - 1)^2} = \sqrt{18} \qquad \bullet \text{ Take the square root of each side of the equation.}$$
$$x - 1 = \pm\sqrt{18}$$
$$x - 1 = \pm 3\sqrt{2} \qquad \bullet \text{ Simplify.}$$
$$x = 1 \pm 3\sqrt{2} \qquad \bullet \text{ Solve for } x.$$

The solutions are $1 + 3\sqrt{2}$ and $1 - 3\sqrt{2}$.

Example 2

Solve by taking square roots:
$x^2 + 16 = 0$

Solution
$$x^2 + 16 = 0$$
$$x^2 = -16 \qquad \bullet \text{ Solve for } x^2.$$
$$\sqrt{x^2} = \sqrt{-16} \qquad \bullet \text{ Take square roots.}$$

$\sqrt{-16}$ is not a real number.

The equation has no real number solution.

You Try It 2

Solve by taking square roots:
$x^2 + 81 = 0$

Your solution
No real number solution

Example 3

Solve by taking square roots:
$5(y - 4)^2 = 25$

Solution
$$5(y - 4)^2 = 25$$
$$(y - 4)^2 = 5 \qquad \bullet \text{ Solve for } (y-4)^2.$$
$$\sqrt{(y - 4)^2} = \sqrt{5} \qquad \bullet \text{ Take square roots.}$$
$$y - 4 = \pm\sqrt{5} \qquad \bullet \text{ Simplify.}$$
$$y = 4 \pm\sqrt{5} \qquad \bullet \text{ Solve for } y.$$

The solutions are $4 + \sqrt{5}$ and $4 - \sqrt{5}$.

You Try It 3

Solve by taking square roots:
$7(z + 2)^2 = 21$

Your solution
$-2 + \sqrt{3}, -2 - \sqrt{3}$

Solutions on p. S39

16.1 Exercises

Objective A   To solve a quadratic equation by factoring

Section 16.1

Suggested Assignment
Exercises 1–57, every other odd
More challenging problems:
  Exercises 61–67, odds

For Exercises 1 to 4, solve for x.

1. $(x + 3)(x - 5) = 0$
$-3, 5$

2. $x(x - 7) = 0$
$0, 7$

3. $(2x + 5)(3x - 1) = 0$
$-\dfrac{5}{2}, \dfrac{1}{3}$

4. $(x - 4)(2x - 7) = 0$
$\dfrac{7}{2}, 4$

For Exercises 5 to 34, solve by factoring.

5. $x^2 + 2x - 15 = 0$
$-5, 3$

6. $t^2 + 3t - 10 = 0$
$-5, 2$

7. $z^2 - 4z + 3 = 0$
$1, 3$

8. $s^2 - 5s + 4 = 0$
$1, 4$

9. $p^2 + 3p + 2 = 0$
$-1, -2$

10. $v^2 + 6v + 5 = 0$
$-1, -5$

11. $x^2 - 6x + 9 = 0$
3

12. $y^2 - 8y + 16 = 0$
4

13. $12y^2 + 8y = 0$
$0, -\dfrac{2}{3}$

14. $6x^2 - 9x = 0$
$0, \dfrac{3}{2}$

15. $r^2 - 10 = 3r$
$-2, 5$

16. $t^2 - 12 = 4t$
$-2, 6$

17. $3v^2 - 5v + 2 = 0$
$\dfrac{2}{3}, 1$

18. $2p^2 - 3p - 2 = 0$
$-\dfrac{1}{2}, 2$

19. $3s^2 + 8s = 3$
$\dfrac{1}{3}, -3$

20. $3x^2 + 5x = 12$
$\dfrac{4}{3}, -3$

21. $\dfrac{3}{4}z^2 - z = -\dfrac{1}{3}$
$\dfrac{2}{3}$

22. $\dfrac{r^2}{2} = 1 - \dfrac{r}{12}$
$-\dfrac{3}{2}, \dfrac{4}{3}$

23. $4t^2 = 4t + 3$
$-\dfrac{1}{2}, \dfrac{3}{2}$

24. $5y^2 + 11y = 12$
$-3, \dfrac{4}{5}$

25. $4v^2 - 4v + 1 = 0$
$\dfrac{1}{2}$

26. $9s^2 - 6s + 1 = 0$
$\dfrac{1}{3}$

27. $x^2 - 9 = 0$
$-3, 3$

28. $t^2 - 16 = 0$
$-4, 4$

29. $4y^2 - 1 = 0$
$-\dfrac{1}{2}, \dfrac{1}{2}$

30. $9z^2 - 4 = 0$
$-\dfrac{2}{3}, \dfrac{2}{3}$

31. $x + 15 = x(x - 1)$
$-3, 5$

32. $p + 18 = p(p - 2)$
$-3, 6$

33. $r^2 - r - 2 = (2r - 1)(r - 3)$
$1, 5$

34. $s^2 + 5s - 4 = (2s + 1)(s - 4)$
$0, 12$

Objective B   To solve a quadratic equation by taking square roots

For Exercises 35 to 61, solve by taking square roots.

35. $x^2 = 36$
$-6, 6$

36. $y^2 = 49$
$-7, 7$

37. $v^2 - 1 = 0$
$-1, 1$

38. $z^2 - 64 = 0$
$-8, 8$

39. $4x^2 - 49 = 0$
$-\dfrac{7}{2}, \dfrac{7}{2}$

40. $9w^2 - 64 = 0$
$-\dfrac{8}{3}, \dfrac{8}{3}$

Quick Quiz (Objective 16.1A)

Solve by factoring.

1. $x^2 + x - 6 = 0$ $-3, 2$

2. $6x^2 = x + 2$ $-\dfrac{1}{2}, \dfrac{2}{3}$

3. $y^2 = -64$ No real number solution

4. $x(x + 8) = 2x - 9$ -3

41. $9y^2 = 4$
$-\dfrac{2}{3}, \dfrac{2}{3}$

42. $4z^2 = 25$
$-\dfrac{5}{2}, \dfrac{5}{2}$

43. $16v^2 - 9 = 0$
$-\dfrac{3}{4}, \dfrac{3}{4}$

44. $25x^2 - 64 = 0$
$-\dfrac{8}{5}, \dfrac{8}{5}$

45. $y^2 + 81 = 0$
No real number solution

46. $z^2 + 49 = 0$
No real number solution

47. $w^2 - 24 = 0$
$-2\sqrt{6}, 2\sqrt{6}$

48. $v^2 - 48 = 0$
$-4\sqrt{3}, 4\sqrt{3}$

49. $(x - 1)^2 = 36$
$-5, 7$

50. $(y + 2)^2 = 49$
$-9, 5$

51. $2(x + 5)^2 = 8$
$-3, -7$

52. $4(z - 3)^2 = 100$
$-2, 8$

53. $9(x - 1)^2 - 16 = 0$
$-\dfrac{1}{3}, \dfrac{7}{3}$

54. $4(y + 3)^2 - 81 = 0$
$-\dfrac{15}{2}, \dfrac{3}{2}$

55. $49(v + 1)^2 - 25 = 0$
$-\dfrac{2}{7}, -\dfrac{12}{7}$

56. $81(y - 2)^2 - 64 = 0$
$\dfrac{26}{9}, \dfrac{10}{9}$

57. $(x - 4)^2 - 20 = 0$
$4 + 2\sqrt{5}, 4 - 2\sqrt{5}$

58. $(y + 5)^2 - 50 = 0$
$-5 + 5\sqrt{2}, -5 - 5\sqrt{2}$

59. $(x + 1)^2 + 36 = 0$

No real number solution

60. $2\left(z - \dfrac{1}{2}\right)^2 = 12$
$\dfrac{1}{2} + \sqrt{6}, \dfrac{1}{2} - \sqrt{6}$

61. $3\left(v + \dfrac{3}{4}\right)^2 = 36$
$-\dfrac{3}{4} + 2\sqrt{3}, -\dfrac{3}{4} - 2\sqrt{3}$

APPLYING THE CONCEPTS

For Exercises 62 and 63, solve for x.

62. $(3x^2 - 13)^2 = 4$
$-\sqrt{5}, \sqrt{5}, -\dfrac{\sqrt{33}}{3}, \dfrac{\sqrt{33}}{3}$

63. $(6x^2 - 5)^2 = 1$
$-1, 1, -\dfrac{\sqrt{6}}{3}, \dfrac{\sqrt{6}}{3}$

64. Investments   The value A of an initial investment of P dollars after 2 years is given by $A = P(1 + r)^2$, where r is the annual percentage rate earned by the investment. If an initial investment of \$1500 grew to a value of \$1782.15 in 2 years, what was the annual percentage rate?   9%

65. Investments   An initial investment of \$5000 grew to a value of \$5832 in 2 years. Use the formula in Exercise 64 to find the annual percentage rate.
8%

66. Physics   The kinetic energy of a moving body is given by $E = \dfrac{1}{2}mv^2$, where E is the kinetic energy, m is the mass, and v is the velocity in meters per second. What is the velocity of a moving body whose mass is 5 kg and whose kinetic energy is 250 newton-meters?   10 m/s

67. Automotive Safety   On a certain type of street surface, the equation $d = 0.0074v^2$ can be used to approximate the distance d, in feet, a car traveling v miles per hour will slide when its brakes are applied. After applying the brakes, the owner of a car involved in an accident skidded 40 ft. Did the traffic officer investigating the accident issue the car owner a ticket for speeding if the speed limit is 65 mph?   Yes ($v \approx 73.5$ mph)

Quick Quiz (Objective 16.1B)
Solve by taking square roots.

1. $x^2 = 1$ 1, −1

2. $9x^2 - 4 = 0$ $\pm\dfrac{2}{3}$

3. $5x^2 = 5x$ 0, 1

4. $4(y - 1)^2 - 1 = 0$ $\dfrac{1}{2}, \dfrac{3}{2}$

16.2 Solving Quadratic Equations by Completing the Square

Objective A   To solve a quadratic equation by completing the square

 VIDEO & DVD   CD TUTOR   WWW WEB   SSM

Recall that a perfect-square trinomial is the square of a binomial.

Perfect-Square Trinomial		Square of a Binomial
$x^2 + 6x + 9$	$=$	$(x + 3)^2$
$x^2 - 10x + 25$	$=$	$(x - 5)^2$
$x^2 + 8x + 16$	$=$	$(x + 4)^2$

For each perfect-square trinomial, the square of $\frac{1}{2}$ of the coefficient of x equals the constant term.

$$x^2 + 6x + 9, \qquad \left(\frac{1}{2} \cdot 6\right)^2 = 9$$

$$x^2 - 10x + 25, \qquad \left[\frac{1}{2}(-10)\right]^2 = 25$$

$$x^2 + 8x + 16, \qquad \left(\frac{1}{2} \cdot 8\right)^2 = 16$$

Adding to a binomial the constant term that makes it a perfect-square trinomial is called **completing the square.**

HOW TO   Complete the square of $x^2 - 8x$. Write the resulting perfect-square trinomial as the square of a binomial.

$$\left[\frac{1}{2}(-8)\right]^2 = 16$$
• Find the constant term.

$$x^2 - 8x + 16$$
• Complete the square of $x^2 - 8x$ by adding the constant term.

$$x^2 - 8x + 16 = (x - 4)^2$$
• Write the resulting perfect-square trinomial as the square of a binomial.

HOW TO   Complete the square of $y^2 + 5y$. Write the resulting perfect-square trinomial as the square of a binomial.

$$\left(\frac{1}{2} \cdot 5\right)^2 = \left(\frac{5}{2}\right)^2 = \frac{25}{4}$$
• Find the constant term.

$$y^2 + 5y + \frac{25}{4}$$
• Complete the square of $y^2 + 5y$ by adding the constant term.

$$y^2 + 5y + \frac{25}{4} = \left(y + \frac{5}{2}\right)^2$$
• Write the resulting perfect-square trinomial as the square of a binomial.

A quadratic equation that cannot be solved by factoring can be solved by completing the square. When the quadratic equation is in the form $x^2 + bx = c$, add to each side of the equation the term that completes the square on $x^2 + bx$. Factor the perfect-square trinomial, and write it as the square of a binomial. Take the square root of each side of the equation and then solve for x.

Point of Interest

Early mathematicians solved quadratic equations by literally *completing the square.* For these mathematicians, all equations had geometric interpretations. They found that a quadratic equation could be solved by making certain figures into squares. See the second of the Projects and Group Activities at the end of this chapter for an idea of how this was done.

New Vocabulary
completing the square

Instructor Note
This section is very difficult for students. Writing an outline of the procedure on the board and allowing students to use it to solve quadratic equations will help many students. Then erase the procedure and ask students to re-create it on paper for themselves. Go over their procedures to be sure they have a workable plan.

Concept Check
1. What constant term should be added to $x^2 + 12x$ to complete the square?   36
2. What binomial does the answer to part 1 complete the square of? $x + 6$

Discuss the Concepts
What advantage does completing the square have over solving by factoring?
Not all equations can be solved by factoring. All quadratic equations, whether or not they factor, can be solved by completing the square.

In-Class Examples (Objective 16.2A)
Solve by completing the square.
1. $x^2 + 14x = -24$ $-12, -2$
2. $2x^2 + 6x - 1 = 0$ $\dfrac{-3 + \sqrt{11}}{2}, \dfrac{-3 - \sqrt{11}}{2}$
3. $x^2 + 6x + 12 = 0$ No real number solution

Concept Check

If $(x + 6)^2 = 9$, then $x + 6 = ?$

3 or −3

Optional Student Activity

Prepare an e-mail that would help a friend who is having difficulty solving $x^2 - 4x - 5 = 0$ by completing the square.

Answers will vary.

S t u d y   T i p

This is a new skill and one that is difficult for many students. Be sure to do all you need to do in order to be successful at solving quadratic equations by completing the square: Read through the introductory material, work through the How To examples, study the paired examples, and do the You Try Its and check your solutions against the ones in the back of the book. See *AIM for Success*, pages xxxi−xxxii.

HOW TO   Solve by completing the square: $x^2 + 8x - 2 = 0$

$$x^2 + 8x - 2 = 0$$
$$x^2 + 8x = 2$$

$$x^2 + 8x + \left(\frac{1}{2} \cdot 8\right)^2 = 2 + \left(\frac{1}{2} \cdot 8\right)^2$$

$$x^2 + 8x + 16 = 2 + 16$$
$$(x + 4)^2 = 18$$
$$\sqrt{(x + 4)^2} = \sqrt{18}$$

$$x + 4 = \pm\sqrt{18}$$
$$x + 4 = \pm 3\sqrt{2}$$
$$x = -4 \pm 3\sqrt{2}$$

- Add 2 to each side of the equation.
- Complete the square of $x^2 + 8x$. Add $\left(\frac{1}{2} \cdot 8\right)^2$ to each side of the equation.
- Simplify.
- Factor the perfect-square trinomial.
- Take the square root of each side of the equation.
- Solve for x.

Check:

$$\begin{array}{c|c}
x^2 + 8x - 2 = 0 & \\
\hline
(-4 + 3\sqrt{2})^2 + 8(-4 + 3\sqrt{2}) - 2 & 0 \\
16 - 24\sqrt{2} + 18 - 32 + 24\sqrt{2} - 2 & 0 \\
0 = 0 &
\end{array}$$

$$\begin{array}{c|c}
x^2 + 8x - 2 = 0 & \\
\hline
(-4 - 3\sqrt{2})^2 + 8(-4 - 3\sqrt{2}) - 2 & 0 \\
16 + 24\sqrt{2} + 18 - 32 - 24\sqrt{2} - 2 & 0 \\
0 = 0 &
\end{array}$$

The solutions are $-4 + 3\sqrt{2}$ and $-4 - 3\sqrt{2}$.

If the coefficient of the second-degree term is not 1, a step in completing the square is to multiply each side of the equation by the reciprocal of that coefficient.

HOW TO   Solve by completing the square: $2x^2 - 3x + 1 = 0$

$$2x^2 - 3x + 1 = 0$$
$$2x^2 - 3x = -1$$

$$\frac{1}{2}(2x^2 - 3x) = \frac{1}{2} \cdot (-1)$$

$$x^2 - \frac{3}{2}x = -\frac{1}{2}$$

$$x^2 - \frac{3}{2}x + \left[\frac{1}{2}\left(-\frac{3}{2}\right)\right]^2 = -\frac{1}{2} + \left[\frac{1}{2}\left(-\frac{3}{2}\right)\right]^2$$

$$x^2 - \frac{3}{2}x + \frac{9}{16} = -\frac{1}{2} + \frac{9}{16}$$

$$\left(x - \frac{3}{4}\right)^2 = \frac{1}{16}$$

$$\sqrt{\left(x - \frac{3}{4}\right)^2} = \sqrt{\frac{1}{16}}$$

$$x - \frac{3}{4} = \pm\frac{1}{4}$$

$$x = \frac{3}{4} \pm \frac{1}{4}$$

$$x = \frac{3}{4} + \frac{1}{4} = 1 \qquad x = \frac{3}{4} - \frac{1}{4} = \frac{1}{2}$$

- Subtract 1 from each side of the equation.
- In order to complete the square, the coefficient of x^2 must be 1. Multiply each side of the equation by $\frac{1}{2}$.
- Complete the square. Add $\left[\frac{1}{2}\left(-\frac{3}{2}\right)\right]^2$ to each side of the equation.
- Simplify.
- Factor the perfect-square trinomial.
- Take the square root of each side of the equation.
- Solve for x.

The solutions are $\frac{1}{2}$ and 1.

Example 1

Solve by completing the square:
$2x^2 - 4x - 1 = 0$

Solution

$2x^2 - 4x - 1 = 0$

$\qquad 2x^2 - 4x = 1 \qquad$ • Add 1.

$\dfrac{1}{2}(2x^2 - 4x) = \dfrac{1}{2} \cdot 1 \qquad$ • Multiply by $\dfrac{1}{2}$.

$\qquad x^2 - 2x = \dfrac{1}{2} \qquad$ • The coefficient of x^2 is 1.

Complete the square.

$x^2 - 2x + 1 = \dfrac{1}{2} + 1 \qquad$ • $\left[\dfrac{1}{2} \cdot (-2)\right]^2 = [-1]^2 = 1$

$\qquad (x - 1)^2 = \dfrac{3}{2} \qquad$ • Factor.

$\sqrt{(x-1)^2} = \sqrt{\dfrac{3}{2}} \qquad$ • Take square roots.

$\qquad x - 1 = \pm\dfrac{\sqrt{6}}{2} \qquad$ • Simplify.

$\qquad x = 1 \pm \dfrac{\sqrt{6}}{2}$

$x = 1 + \dfrac{\sqrt{6}}{2} \qquad x = 1 - \dfrac{\sqrt{6}}{2}$

$\ = \dfrac{2 + \sqrt{6}}{2} \qquad \ = \dfrac{2 - \sqrt{6}}{2}$

Check:

$$2x^2 - 4x - 1 = 0$$

$$2\left(\dfrac{2 + \sqrt{6}}{2}\right)^2 - 4\left(\dfrac{2 + \sqrt{6}}{2}\right) - 1 \ \Big|\ 0$$

$$2\left(\dfrac{4 + 4\sqrt{6} + 6}{4}\right) - 2(2 + \sqrt{6}) - 1 \ \Big|\ 0$$

$$2 + 2\sqrt{6} + 3 - 4 - 2\sqrt{6} - 1 \ \Big|\ 0$$

$$0 = 0$$

$$2x^2 - 4x - 1 = 0$$

$$2\left(\dfrac{2 - \sqrt{6}}{2}\right)^2 - 4\left(\dfrac{2 - \sqrt{6}}{2}\right) - 1 \ \Big|\ 0$$

$$2\left(\dfrac{4 - 4\sqrt{6} + 6}{4}\right) - 2(2 - \sqrt{6}) - 1 \ \Big|\ 0$$

$$2 - 2\sqrt{6} + 3 - 4 + 2\sqrt{6} - 1 \ \Big|\ 0$$

$$0 = 0$$

The solutions are $\dfrac{2 + \sqrt{6}}{2}$ and $\dfrac{2 - \sqrt{6}}{2}$.

You Try It 1

Solve by completing the square:
$3x^2 - 6x - 2 = 0$

Your solution

$\dfrac{3 + \sqrt{15}}{3}, \dfrac{3 - \sqrt{15}}{3}$

Solution on p. S39

Concept Check

Approximate to the nearest thousandth.

1. $4 + \sqrt{5}$ 6.236

2. $4 - \sqrt{5}$ 1.764

Discuss the Concepts

1. To complete the square on $x^2 + 8x$, what number must be added to the expression? Explain how you arrived at the answer.

2. If you attempt to solve $x^2 - 8x + 16 = 0$ by completing the square, the result after simplifying is $(x - 4)^2 = 0$. Does this mean that the original equation has no solution? If not, what are the solutions of the equation?
No. The equation has a double root, 4.

3. What is the next step when completing the square to solve $x^2 + 9x = 22$?

Example 2

Solve by completing the square:
$x^2 + 4x + 5 = 0$

Solution
$x^2 + 4x + 5 = 0$

$\quad x^2 + 4x = -5$ • Subtract 5.

Complete the square.

$x^2 + 4x + 4 = -5 + 4$ • $\left(\frac{1}{2} \cdot 4\right)^2 = 2^2 = 4$

$\quad\quad (x + 2)^2 = -1$ • Factor.

$\quad \sqrt{(x + 2)^2} = \sqrt{-1}$ • Take square roots.

$\sqrt{-1}$ is not a real number.

The quadratic equation has no real number solution.

You Try It 2

Solve by completing the square:
$x^2 + 6x + 12 = 0$

Your solution
No real number solution

Example 3

Solve $\frac{x^2}{4} + \frac{3x}{2} + 1 = 0$ by completing the square. Approximate the solutions to the nearest thousandth.

Solution
$$\frac{x^2}{4} + \frac{3x}{2} + 1 = 0$$

$$4\left(\frac{x^2}{4} + \frac{3x}{2} + 1\right) = 4(0) \quad \text{• Multiply by 4.}$$

$$x^2 + 6x + 4 = 0$$

$$x^2 + 6x = -4 \quad \text{• Subtract 4.}$$

Complete the square.

$x^2 + 6x + 9 = -4 + 9$ • $\left(\frac{1}{2} \cdot 6\right)^2 = 3^2 = 9$

$\quad\quad (x + 3)^2 = 5$ • Factor.

$\quad \sqrt{(x + 3)^2} = \sqrt{5}$ • Take square roots.

$\quad\quad x + 3 = \pm\sqrt{5}$

$x + 3 = \sqrt{5}$ $x + 3 = -\sqrt{5}$

$\quad x = -3 + \sqrt{5}$ $x = -3 - \sqrt{5}$

$\quad\quad \approx -3 + 2.236$ $\approx -3 - 2.236$

$\quad\quad \approx -0.764$ ≈ -5.236

The solutions are approximately -0.764 and -5.236.

You Try It 3

Solve $\frac{x^2}{8} + x + 1 = 0$ by completing the square. Approximate the solutions to the nearest thousandth.

Your solution
$-1.172, -6.828$

Solutions on p. S39

16.2 Exercises

Objective A   **To solve a quadratic equation by completing the square**

For Exercises 1 to 4, complete the square of each binomial. Write the resulting trinomial as the square of a binomial.

1. $x^2 - 8x$

$x^2 - 8x + 16, (x - 4)^2$

2. $x^2 + 6x$

$x^2 + 6x + 9, (x + 3)^2$

3. $x^2 + 5x$

$x^2 + 5x + \dfrac{25}{4}, \left(x + \dfrac{5}{2}\right)^2$

4. $x^2 - 3x$

$x^2 - 3x + \dfrac{9}{4}, \left(x - \dfrac{3}{2}\right)^2$

For Exercises 5 to 53, solve by completing the square.

5. $x^2 + 2x - 3 = 0$
$-3, 1$

6. $y^2 + 4y - 5 = 0$
$-5, 1$

7. $z^2 - 6z - 16 = 0$
$-2, 8$

8. $w^2 + 8w - 9 = 0$
$-9, 1$

9. $x^2 = 4x - 4$
2

10. $z^2 = 8z - 16$
4

11. $v^2 - 6v + 13 = 0$
No real number
solution

12. $x^2 + 4x + 13 = 0$
No real number
solution

13. $y^2 + 5y + 4 = 0$
$-1, -4$

14. $v^2 - 5v - 6 = 0$
$-1, 6$

15. $w^2 + 7w = 8$
$-8, 1$

16. $y^2 + 5y = -4$
$-1, -4$

17. $v^2 + 4v + 1 = 0$
$-2 + \sqrt{3}, -2 - \sqrt{3}$

18. $y^2 - 2y - 5 = 0$
$1 + \sqrt{6}, 1 - \sqrt{6}$

19. $x^2 + 6x = 5$
$-3 + \sqrt{14}, -3 - \sqrt{14}$

20. $w^2 - 8w = 3$
$4 + \sqrt{19}, 4 - \sqrt{19}$

21. $\dfrac{z^2}{2} = z + \dfrac{1}{2}$
$1 + \sqrt{2}, 1 - \sqrt{2}$

22. $\dfrac{y^2}{10} = y - 2$
$5 + \sqrt{5}, 5 - \sqrt{5}$

23. $p^2 + 3p = 1$
$\dfrac{-3 + \sqrt{13}}{2}, \dfrac{-3 - \sqrt{13}}{2}$

24. $r^2 + 5r = 2$
$\dfrac{-5 + \sqrt{33}}{2}, \dfrac{-5 - \sqrt{33}}{2}$

25. $t^2 - 3t = -2$
$1, 2$

26. $z^2 - 5z = -3$
$\dfrac{5 + \sqrt{13}}{2}, \dfrac{5 - \sqrt{13}}{2}$

27. $v^2 + v - 3 = 0$
$\dfrac{-1 + \sqrt{13}}{2}, \dfrac{-1 - \sqrt{13}}{2}$

28. $x^2 - x = 1$
$\dfrac{1 + \sqrt{5}}{2}, \dfrac{1 - \sqrt{5}}{2}$

29. $y^2 = 7 - 10y$
$-5 + 4\sqrt{2}, -5 - 4\sqrt{2}$

30. $v^2 = 14 + 16v$
$8 + \sqrt{78}, 8 - \sqrt{78}$

31. $r^2 - 3r = 5$
$\dfrac{3 + \sqrt{29}}{2}, \dfrac{3 - \sqrt{29}}{2}$

32. $s^2 + 3s = -1$
$\dfrac{-3 + \sqrt{5}}{2}, \dfrac{-3 - \sqrt{5}}{2}$

33. $t^2 - t = 4$
$\dfrac{1 + \sqrt{17}}{2}, \dfrac{1 - \sqrt{17}}{2}$

34. $y^2 + y - 4 = 0$
$\dfrac{-1 + \sqrt{17}}{2}, \dfrac{-1 - \sqrt{17}}{2}$

35. $x^2 - 3x + 5 = 0$

No real number solution

36. $z^2 + 5z + 7 = 0$

No real number solution

37. $2t^2 - 3t + 1 = 0$

$1, \dfrac{1}{2}$

Quick Quiz (Objective 16.2A)

Solve by completing the square.

1. $x^2 - 6x - 7 = 0$ $-1, 7$

2. $2x^2 - 2x = 1$ $\dfrac{1 - \sqrt{3}}{2}, \dfrac{1 + \sqrt{3}}{2}$

3. $x^2 - 6x + 16 = 6$ No real number solution

38. $2x^2 - 7x + 3 = 0$
$3, \dfrac{1}{2}$

39. $2r^2 + 5r = 3$
$-3, \dfrac{1}{2}$

40. $2y^2 - 3y = 9$
$-\dfrac{3}{2}, 3$

41. $2s^2 = 7s - 6$
$2, \dfrac{3}{2}$

42. $2x^2 = 3x + 20$
$-\dfrac{5}{2}, 4$

43. $2v^2 = v + 1$
$1, -\dfrac{1}{2}$

44. $2z^2 = z + 3$
$-1, \dfrac{3}{2}$

45. $3r^2 + 5r = 2$
$-2, \dfrac{1}{3}$

46. $3t^2 - 8t = 3$
$-\dfrac{1}{3}, 3$

47. $3y^2 + 8y + 4 = 0$
$-2, -\dfrac{2}{3}$

48. $3z^2 - 10z - 8 = 0$
$4, -\dfrac{2}{3}$

49. $4x^2 + 4x - 3 = 0$
$\dfrac{1}{2}, -\dfrac{3}{2}$

50. $4v^2 + 4v - 15 = 0$
$\dfrac{3}{2}, -\dfrac{5}{2}$

51. $6s^2 + 7s = 3$
$\dfrac{1}{3}, -\dfrac{3}{2}$

52. $6z^2 = z + 2$
$\dfrac{2}{3}, -\dfrac{1}{2}$

53. $6p^2 = 5p + 4$
$-\dfrac{1}{2}, \dfrac{4}{3}$

 For Exercises 54 to 59, solve by completing the square. Approximate the solutions to the nearest thousandth.

54. $y^2 + 3y = 5$
$-4.193, 1.193$

55. $w^2 + 5w = 2$
$-5.372, 0.372$

56. $2z^2 - 3z = 7$
$2.766, -1.266$

57. $2x^2 + 3x = 11$
$1.712, -3.212$

58. $4x^2 + 6x - 1 = 0$
$-1.651, 0.151$

59. $4x^2 + 2x - 3 = 0$
$-1.151, 0.651$

APPLYING THE CONCEPTS

60. Explain why the equation $(x - 2)^2 = -4$ does not have a real number solution.

For Exercises 61 to 69, solve.

61. $\dfrac{x^2}{6} - \dfrac{x}{3} = 1$
$1 + \sqrt{7}, 1 - \sqrt{7}$

62. $\sqrt{x + 2} = x - 4$
7

63. $\sqrt{3x + 4} - x = 2$
$0, -1$

64. $\dfrac{x}{3} + \dfrac{3}{x} = \dfrac{8}{3}$
$4 + \sqrt{7}, 4 - \sqrt{7}$

65. $\dfrac{x + 1}{2} + \dfrac{3}{x - 1} = 4$
$4 + \sqrt{3}, 4 - \sqrt{3}$

66. $\dfrac{x - 2}{3} + \dfrac{2}{x + 2} = 4$
$6 + \sqrt{58}, 6 - \sqrt{58}$

67. $4\sqrt{x + 1} - x = 4$
$0, 8$

68. $\sqrt{2x^2 + 7} = x + 2$
$1, 3$

69. $3\sqrt{x - 1} + 3 = x$
$\dfrac{15 + 3\sqrt{17}}{2}$

70. **Basketball** A basketball player shoots at a basket 25 ft away. The height of the ball above the ground at time t is given by $h = -16t^2 + 32t + 6.5$. How many seconds after the ball is released does it hit the basket? *Hint:* When it hits the basket, $h = 10$ ft.
1.88 s

71. **Baseball** After a ball player hits a ball, the height of the ball above the ground can be approximated by the equation $h = -16t^2 + 76t + 5$. When will the ball hit the ground? *Hint:* The ball strikes the ground when $h = 0$ ft.
4.81 s

5 ft

16.3 Solving Quadratic Equations by Using the Quadratic Formula

Objective A   **To solve a quadratic equation by using the quadratic formula**

Any quadratic equation can be solved by completing the square. Applying this method to the standard form of a quadratic equation produces a formula that can be used to solve any quadratic equation.

Solve $ax^2 + bx + c = 0$ by completing the square.

$$ax^2 + bx + c = 0$$

Add the opposite of the constant term to each side of the equation.

$$ax^2 + bx + c + (-c) = 0 + (-c)$$
$$ax^2 + bx = -c$$

Multiply each side of the equation by the reciprocal of a, the coefficient of x^2.

$$\frac{1}{a}(ax^2 + bx) = \frac{1}{a}(-c)$$
$$x^2 + \frac{b}{a}x = -\frac{c}{a}$$

Complete the square by adding $\left(\frac{1}{2}\cdot\frac{b}{a}\right)^2$ to each side of the equation.

$$x^2 + \frac{b}{a}x + \left(\frac{1}{2}\cdot\frac{b}{a}\right)^2 = \left(\frac{1}{2}\cdot\frac{b}{a}\right)^2 - \frac{c}{a}$$
$$x^2 + \frac{b}{a}x + \frac{b^2}{4a^2} = \frac{b^2}{4a^2} - \frac{c}{a}$$

Simplify the right side of the equation.

$$x^2 + \frac{b}{a}x + \frac{b^2}{4a^2} = \frac{b^2}{4a^2} - \left(\frac{c}{a}\cdot\frac{4a}{4a}\right)$$
$$x^2 + \frac{b}{a}x + \frac{b^2}{4a^2} = \frac{b^2}{4a^2} - \frac{4ac}{4a^2}$$
$$x^2 + \frac{b}{a}x + \frac{b^2}{4a^2} = \frac{b^2 - 4ac}{4a^2}$$

Factor the perfect-square trinomial on the left side of the equation.

$$\left(x + \frac{b}{2a}\right)^2 = \frac{b^2 - 4ac}{4a^2}$$

Take the square root of each side of the equation.

$$\sqrt{\left(x + \frac{b}{2a}\right)^2} = \sqrt{\frac{b^2 - 4ac}{4a^2}}$$
$$x + \frac{b}{2a} = \pm\frac{\sqrt{b^2 - 4ac}}{2a}$$

Solve for x.

$$x + \frac{b}{2a} = \frac{\sqrt{b^2 - 4ac}}{2a} \qquad\qquad x + \frac{b}{2a} = -\frac{\sqrt{b^2 - 4ac}}{2a}$$
$$x = -\frac{b}{2a} + \frac{\sqrt{b^2 - 4ac}}{2a} \qquad\qquad x = -\frac{b}{2a} - \frac{\sqrt{b^2 - 4ac}}{2a}$$
$$= \frac{-b + \sqrt{b^2 - 4ac}}{2a} \qquad\qquad\qquad = \frac{-b - \sqrt{b^2 - 4ac}}{2a}$$

The Quadratic Formula

The solutions of $ax^2 + bx + c = 0$, $a \neq 0$, are

$$x = \frac{-b \pm \sqrt{b^2 - 4ac}}{2a}$$

In-Class Examples (Objective 16.3A)

Solve by using the quadratic formula.

1. $4x^2 - 9x - 9 = 0$ $3, -\dfrac{3}{4}$

2. $x^2 - 4x = 1$ $2 - \sqrt{5}, 2 + \sqrt{5}$

3. $2x^2 + x + 5 = 0$ No real number solution

4. $(2x - 1)(x + 2) = 25$ $-\dfrac{9}{2}, 3$

Objective 16.3A

New Formula
Quadratic Formula

Instructor Note
Students wonder why we discuss the different ways in which a quadratic equation can be solved. Explain to students that factoring, when it is easy to determine the factors, is the quickest way. Completing the square has other uses besides being used to derive the quadratic formula. It is rarely used to solve an equation. The quadratic formula is used for those equations that do not factor.

Instructor Note

It may help to give examples of just determining a, b, and c for different equations. Here are some suggestions:

$2x = x^2 - 3$

$4x - 3x^2 + 3 = 0$

$2x^2 - x = 0$

Students will ask whether it matters on which side of the equation a term is moved. To convince these students that it does not matter, have them solve $3x = x^2 - 4$ by rewriting it as $x^2 - 3x - 4 = 0$ and as $-x^2 + 3x + 4 = 0$. An alternative to solving the equations would be to use factoring to show that they are equivalent.

$-x^2 + 3x + 4 = 0$

$\rightarrow -(x^2 - 3x - 4) = 0$

$\rightarrow x^2 - 3x - 4 = 0$

Optional Student Activity

Solve $x^2 + 4x - 12 = 0$ by each of the following methods.

1. Factoring $-6, 2$

2. Completing the square $-6, 2$

3. Quadratic formula $-6, 2$

HOW TO   Solve by using the quadratic formula: $2x^2 = 4x - 1$

$$2x^2 = 4x - 1$$

$$2x^2 - 4x + 1 = 0$$ • Write the equation in standard form.

$$x = \frac{-b \pm \sqrt{b^2 - 4ac}}{2a}$$ • The quadratic formula

$$= \frac{-(-4) \pm \sqrt{(-4)^2 - (4 \cdot 2 \cdot 1)}}{2 \cdot 2}$$ • $a = 2$, $b = -4$, $c = 1$. Replace a, b, and c by their values.

$$= \frac{4 \pm \sqrt{16 - 8}}{4} = \frac{4 \pm \sqrt{8}}{4}$$ • Simplify.

$$= \frac{4 \pm 2\sqrt{2}}{4} = \frac{2 \pm \sqrt{2}}{2}$$

The solutions are $\frac{2 + \sqrt{2}}{2}$ and $\frac{2 - \sqrt{2}}{2}$.

TAKE NOTE

$$\frac{4 \pm 2\sqrt{2}}{4} = \frac{2(2 \pm \sqrt{2})}{2 \cdot 2}$$

$$= \frac{2 \pm \sqrt{2}}{2}$$

Example 1

Solve by using the quadratic formula:
$2x^2 - 3x + 1 = 0$

Solution

$2x^2 - 3x + 1 = 0$ • Standard form

$x = \dfrac{-(-3) \pm \sqrt{(-3)^2 - 4(2)(1)}}{2 \cdot 2}$ • $a = 2$, $b = -3$, $c = 1$

$= \dfrac{3 \pm \sqrt{9 - 8}}{4} = \dfrac{3 \pm \sqrt{1}}{4} = \dfrac{3 \pm 1}{4}$

$x = \dfrac{3 + 1}{4}$ $x = \dfrac{3 - 1}{4}$

$= \dfrac{4}{4} = 1$ $= \dfrac{2}{4} = \dfrac{1}{2}$

The solutions are 1 and $\frac{1}{2}$.

You Try It 1

Solve by using the quadratic formula:
$3x^2 + 4x - 4 = 0$

Your solution

$\dfrac{2}{3}, -2$

Example 2

Solve by using the quadratic formula:
$\dfrac{x^2}{2} = 2x - \dfrac{5}{4}$

Solution $\dfrac{x^2}{2} = 2x - \dfrac{5}{4}$

$4\left(\dfrac{x^2}{2}\right) = 4\left(2x - \dfrac{5}{4}\right)$ • Multiply by 4.

$2x^2 = 8x - 5$

$2x^2 - 8x + 5 = 0$ • Standard form

$x = \dfrac{-(-8) \pm \sqrt{(-8)^2 - 4(2)(5)}}{2 \cdot 2}$ • $a = 2$, $b = -8$, $c = 5$

$= \dfrac{8 \pm \sqrt{64 - 40}}{4} = \dfrac{8 \pm \sqrt{24}}{4}$

$= \dfrac{8 \pm 2\sqrt{6}}{4} = \dfrac{4 \pm \sqrt{6}}{2}$

The solutions are $\dfrac{4 + \sqrt{6}}{2}$ and $\dfrac{4 - \sqrt{6}}{2}$.

You Try It 2

Solve by using the quadratic formula:
$\dfrac{x^2}{4} + \dfrac{x}{2} = \dfrac{1}{4}$

Your solution

$-1 + \sqrt{2}, -1 - \sqrt{2}$

Solutions on p. S40

16.3 Exercises

Objective A   **To solve a quadratic equation by using the quadratic formula**

For Exercises 1 to 38, solve by using the quadratic formula.

1. $x^2 - 4x - 5 = 0$
$-1, 5$

2. $y^2 + 3y + 2 = 0$
$-1, -2$

3. $z^2 - 2z - 15 = 0$
$-3, 5$

4. $v^2 + 5v + 4 = 0$
$-1, -4$

5. $y^2 = 2y + 3$
$-1, 3$

6. $w^2 = 3w + 18$
$-3, 6$

7. $r^2 = 5 - 4r$
$-5, 1$

8. $z^2 = 3 - 2z$
$-3, 1$

9. $2y^2 - y - 1 = 0$
$-\dfrac{1}{2}, 1$

10. $2t^2 - 5t + 3 = 0$
$\dfrac{3}{2}, 1$

11. $w^2 + 3w + 5 = 0$
No real number solution

12. $x^2 - 2x + 6 = 0$
No real number solution

13. $p^2 - p = 0$
$0, 1$

14. $2v^2 + v = 0$
$-\dfrac{1}{2}, 0$

15. $4t^2 - 9 = 0$
$-\dfrac{3}{2}, \dfrac{3}{2}$

16. $4s^2 - 25 = 0$
$-\dfrac{5}{2}, \dfrac{5}{2}$

17. $4y^2 + 4y = 15$
$-\dfrac{5}{2}, \dfrac{3}{2}$

18. $6y^2 + 5y - 4 = 0$
$-\dfrac{4}{3}, \dfrac{1}{2}$

19. $2x^2 + x + 1 = 0$
No real number solution

20. $3r^2 - r + 2 = 0$
No real number solution

21. $\dfrac{1}{2}t^2 - t = \dfrac{5}{2}$
$1 + \sqrt{6}, 1 - \sqrt{6}$

22. $y^2 - 4y = 6$
$2 + \sqrt{10}, 2 - \sqrt{10}$

23. $\dfrac{1}{3}t^2 + 2t - \dfrac{1}{3} = 0$
$-3 + \sqrt{10}, -3 - \sqrt{10}$

24. $z^2 + 4z + 1 = 0$
$-2 + \sqrt{3}, -2 - \sqrt{3}$

25. $w^2 = 4w + 9$
$2 + \sqrt{13}, 2 - \sqrt{13}$

26. $y^2 = 8y + 3$
$4 + \sqrt{19}, 4 - \sqrt{19}$

27. $9y^2 + 6y - 1 = 0$
$\dfrac{-1 + \sqrt{2}}{3}, \dfrac{-1 - \sqrt{2}}{3}$

28. $9s^2 - 6s - 2 = 0$
$\dfrac{1 + \sqrt{3}}{3}, \dfrac{1 - \sqrt{3}}{3}$

29. $4p^2 + 4p + 1 = 0$
$-\dfrac{1}{2}$

30. $9z^2 + 12z + 4 = 0$
$-\dfrac{2}{3}$

31. $\dfrac{x^2}{2} = x - \dfrac{5}{4}$
No real number solution

32. $r^2 = \dfrac{5}{3}r - 2$
No real number solution

33. $4p^2 + 16p = -11$
$\dfrac{-4 + \sqrt{5}}{2}, \dfrac{-4 - \sqrt{5}}{2}$

34. $4y^2 - 12y = -1$
$\dfrac{3 + 2\sqrt{2}}{2}, \dfrac{3 - 2\sqrt{2}}{2}$

35. $4x^2 = 4x + 11$
$\dfrac{1 + 2\sqrt{3}}{2}, \dfrac{1 - 2\sqrt{3}}{2}$

Section 16.3

Suggested Assignment
Exercises 1–47, every other odd
More challenging problems:
   Exercises 49–59, odds

Quick Quiz (Objective 16.3A)

Solve by using the quadratic formula.

1. $x^2 - 3x - 18 = 0$ $-3, 6$

2. $x^2 - 2x = 4$ $1 - \sqrt{5}, 1 + \sqrt{5}$

3. $2x^2 + 3x - 1 = 0$ $\dfrac{-3 - \sqrt{17}}{4}, \dfrac{-3 + \sqrt{17}}{4}$

802

36. $4s^2 + 12s = 3$
$\dfrac{-3 + 2\sqrt{3}}{2}, \dfrac{-3 - 2\sqrt{3}}{2}$

37. $9v^2 = -30v - 23$
$\dfrac{-5 + \sqrt{2}}{3}, \dfrac{-5 - \sqrt{2}}{3}$

38. $9t^2 = 30t + 17$
$\dfrac{5 + \sqrt{42}}{3}, \dfrac{5 - \sqrt{42}}{3}$

For Exercises 39 to 47, solve by using the quadratic formula. Approximate the solutions to the nearest thousandth.

39. $x^2 - 2x - 21 = 0$
5.690, −3.690

40. $y^2 + 4y - 11 = 0$
1.873, −5.873

41. $s^2 - 6s - 13 = 0$
7.690, −1.690

42. $w^2 + 8w - 15 = 0$
1.568, −9.568

43. $2p^2 - 7p - 10 = 0$
4.589, −1.089

44. $3t^2 - 8t - 1 = 0$
2.786, −0.120

45. $4z^2 + 8z - 1 = 0$
0.118, −2.118

46. $4x^2 + 7x + 1 = 0$
−0.157, −1.593

47. $5v^2 - v - 5 = 0$
1.105, −0.905

APPLYING THE CONCEPTS

48. Factoring, completing the square, and using the quadratic formula are three methods of solving quadratic equations. Describe each method, and cite the advantages and disadvantages of using each.

49. Explain why the equation $0x^2 + 3x + 4 = 0$ cannot be solved by the quadratic formula.

50. Solve $x^2 + bx + c = 0$ for x. $\dfrac{-b + \sqrt{b^2 - 4c}}{2}, \dfrac{-b - \sqrt{b^2 - 4c}}{2}$

51. True or false?
a. The equations $x = \sqrt{12 - x}$ and $x^2 = 12 - x$ have the same solutions. False
b. If $\sqrt{a} + \sqrt{b} = c$, then $a + b = c^2$. False
c. $\sqrt{9} = \pm 3$ False
d. $\sqrt{x^2} = |x|$ True

For Exercises 52 to 57, solve.

52. $\sqrt{x + 3} = x - 3$
6

53. $\sqrt{x + 4} = x + 4$
−3, −4

54. $\sqrt{x + 1} = x - 1$
3

55. $\sqrt{x^2 + 2x + 1} = x - 1$
No solution

56. $\dfrac{x}{4} + \dfrac{3}{x} = \dfrac{5}{2}$
$5 + \sqrt{13}, 5 - \sqrt{13}$

57. $\dfrac{x + 1}{5} - \dfrac{4}{x - 1} = 2$
−1, 11

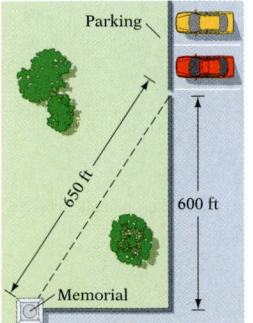

58. **Distance**   An L-shaped sidewalk from the parking lot to a memorial is shown in the figure at the right. The distance directly across the grass to the memorial is 650 ft. The distance to the corner is 600 ft. Find the distance from the corner to the memorial.
250 ft

59. **Travel**   A commuter plane leaves an airport traveling due south at 400 mph. Another plane leaving at the same time travels due east at 300 mph. Find the distance between the two planes after 2 h.
1000 mi

16.4 Graphing Quadratic Equations in Two Variables

Objective A

TAKE NOTE
For the equation
$y = 3x^2 - x + 1$, $a = 3$,
$b = -1$, and $c = 1$.

To graph a quadratic equation of the form $y = ax^2 + bx + c$

An equation of the form $y = ax^2 + bx + c$, $a \neq 0$, is a **quadratic equation in two variables.** Examples of quadratic equations in two variables are shown at the right.

$y = 3x^2 - x + 1$
$y = -x^2 - 3$
$y = 2x^2 - 5x$

For these equations, y is a function of x, and we can write $f(x) = ax^2 + bx + c$. This equation represents a **quadratic function.**

Point of Interest

Mirrors in some telescopes are ground into the shape of a parabola. The mirror at the Palomar Mountain Observatory is 2 ft thick at the ends and weighs 14.75 tons. The mirror has been ground to a true paraboloid (the three-dimensional version of a parabola) to within 0.0000015 in. A possible equation of the mirror is $y = 2640x^2$.

HOW TO Evaluate $f(x) = 2x^2 - 3x + 4$ when $x = -2$.

$f(x) = 2x^2 - 3x + 4$
$f(-2) = 2(-2)^2 - 3(-2) + 4$ • Replace x by -2.
$= 2(4) + 6 + 4 = 18$ • Simplify.

The value of the function when $x = -2$ is 18.

The graph of $y = ax^2 + bx + c$ or $f(x) = ax^2 + bx + c$ is a **parabola.** The graph is U-shaped and opens up when a is positive, and down when a is negative. The graphs of two parabolas are shown below.

TAKE NOTE
One of the equations at the right was written as $y = 2x^2 + 3x - 2$ and the other was written using functional notation as $f(x) = -x^2 + 3x + 2$. Remember that y and $f(x)$ are different symbols for the same quantity.

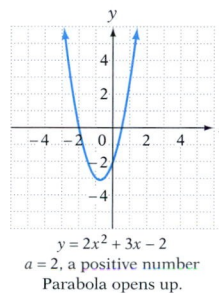
$y = 2x^2 + 3x - 2$
$a = 2$, a positive number
Parabola opens up.

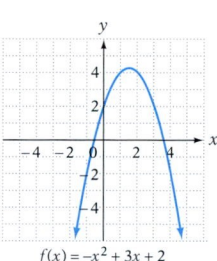
$f(x) = -x^2 + 3x + 2$
$a = -1$, a negative number
Parabola opens down.

HOW TO Graph $y = x^2 - 2x - 3$.

x	y
-2	5
-1	0
0	-3
1	-4
2	-3
3	0
4	5

• Find several solutions of the equation. Because the graph is not a straight line, several solutions must be found in order to determine the U-shape. Record the ordered pairs in a table.

Integrating Technology
One of the Projects and Group Activities at the end of this chapter shows how to graph a quadratic equation by using a graphing calculator. You may want to verify the graphs you draw in this section by drawing them on a graphing calculator.

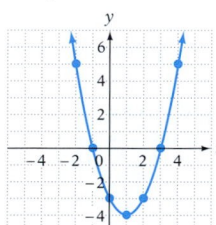

• Graph the ordered-pair solutions on a rectangular coordinate system. Draw a parabola through the points.

Objective 16.4A

New Vocabulary
quadratic equation in two variables
quadratic function
parabola

Concept Check
Complete the following table showing ordered-pair solutions of $y = x^2 - x - 6$.

x	y
-3	6
-2	0
-1	
0	
1	
2	
3	

$(-1, -4)$, $(0, -6)$, $(1, -6)$, $(2, -4)$, $(3, 0)$

Optional Student Activity
1. Create a quadratic function.
2. Make an ordered-pair table using at least seven values for x.
3. Graph the ordered pairs from part 2 on a rectangular coordinate system.
4. Draw a parabola through the points.

Discuss the Concepts
What is the clue in the equation $y = x^2 - 4x + 3$ that the graph will be a parabola and not a straight line?
The exponent 2

In-Class Examples (Objective 16.4A)
Graph.
1. $y = x^2 - 4x + 3$

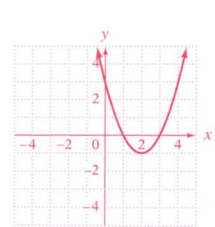

2. $y = -x^2 + 4$

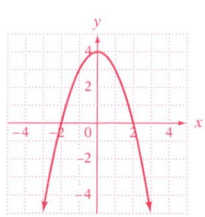

Concept Check

For each quadratic function, determine whether the parabola opens up or opens down.

1. $y = 2x^2 + 3x + 1$
Opens up

2. $y = -2x^2 + 5x + 3$
Opens down

3. $y = x^2 - 4x - 5$
Opens up

4. $y = 4x^2 - 3$
Opens up

5. $y = -x^2 + 2x$
Opens down

6. $y = x^2$
Opens up

Discuss the Concepts

What is the clue in the equation $y = x^2 - 5x + 4$ that the graph of the parabola will open up?
The coefficient of x^2 is positive 1.

Integrating Technology

The first of the Projects and Group Activities at the end of this chapter shows how to use a graphing calculator to draw the graph of a parabola and how to find the x-intercepts.

For the graph of $y = x^2 - 2x - 3$, shown here again below, note that the graph crosses the x-axis at $(-1, 0)$ and $(3, 0)$. This is also confirmed from the table for the graph (see the preceding page). From the table, note that $y = 0$ when $x = -1$ and when $x = 3$. The x-intercepts of the graph are $(-1, 0)$ and $(3, 0)$.

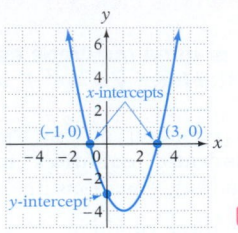

The y-intercept is the point at which the graph crosses the y-axis. At this point, $x = 0$. From the graph, we can see that the y-intercept is $(0, -3)$. This is also confirmed from the table for the graph (see the preceding page). From the table, note that $x = 0$ when $y = -3$.

We can find the x-intercepts algebraically by letting $y = 0$ and solving for x.

$y = x^2 - 2x - 3$
$0 = x^2 - 2x - 3$
$0 = (x + 1)(x - 3)$
$x + 1 = 0 \qquad x - 3 = 0$
$\qquad x = -1 \qquad\qquad x = 3$

• Replace *y* with **0** and solve for *x*.
• This equation can be solved by factoring. However, it will be necessary to use the quadratic formula to solve some quadratic equations.

The x-intercepts are $(-1, 0)$ and $(3, 0)$.

We can find the y-intercept algebraically by letting $x = 0$ and solving for y.

$y = x^2 - 2x - 3$
$y = 0^2 - 2(0) - 3$
$\quad = -3$

• Replace *x* with **0** and simplify.

The y-intercept is $(0, -3)$.

Below is a summary of instructions for graphing a quadratic equation in two variables and for finding the x- and y-intercepts of the graph of a quadratic equation in two variables.

Graph of a Quadratic Equation in Two Variables

To graph a quadratic equation in two variables, find several solutions of the equation. Graph the ordered-pair solutions on a rectangular coordinate system. Draw a parabola through the points.

To find the x-intercepts of the graph of a quadratic equation in two variables, let $y = 0$ and solve for x.

To find the y-intercept, let $x = 0$ and solve for y.

Example 1 Graph $y = x^2 - 2x$.

Solution

x	y
-1	3
0	0
1	-1
2	0
3	3

• Find several solutions of the equation.

• Graph the ordered-pair solutions. Draw a parabola through the points.

You Try It 1 Graph $y = x^2 + 2$.

Your solution

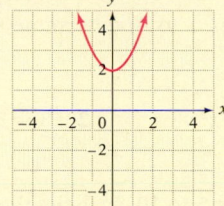

Example 2 Find the x- and y-intercepts of the graph of $y = x^2 - 2x - 5$.

Solution
To find the x-intercepts, let $y = 0$ and solve for x. This gives the equation $0 = x^2 - 2x - 5$, which is not factorable over the integers. Use the quadratic formula.

$x = \dfrac{-b \pm \sqrt{b^2 - 4ac}}{2a}$

$= \dfrac{-(-2) \pm \sqrt{(-2)^2 - 4(1)(-5)}}{2(1)}$ • $a = 1, b = -2,$ $c = -5$

$= \dfrac{2 \pm \sqrt{24}}{2}$

$= \dfrac{2 \pm 2\sqrt{6}}{2}$

$= 1 \pm \sqrt{6}$

The x-intercepts are $(1 - \sqrt{6}, 0)$ and $(1 + \sqrt{6}, 0)$.

To find the y-intercept, let $x = 0$ and solve for y.

$y = x^2 - 2x - 5$
$= 0^2 - 2(0) - 5$ • Replace x with 0.
$= -5$

The y-intercept is $(0, -5)$.

You Try It 2 Find the x- and y-intercepts of the graph of $f(x) = x^2 - 6x + 9$.

Your solution x-intercept: $(3, 0)$
y-intercept: $(0, 9)$

Solutions on p. S40

Optional Student Activity

Exercises such as 26 to 28 in this section can be used to connect technology with algebraic solutions.

1. Graph each of the equations in Exercises 26, 27, and 28. Use the ZERO feature of a graphing calculator to find the x-intercepts of each graph, and verify algebraically that the x-coordinate of the x-intercept is the solution of the corresponding quadratic equation.

2. Graph $y = x^2 - 2x + 3$.
   a. Does the graph have any x-intercepts?   No
   b. Try to solve $0 = x^2 - 2x + 3$ by using the quadratic formula. Does the equation have any real number solutions?   No
   c. Explain how you can tell whether $ax^2 + bx + c = 0$ has any real number solutions by looking at the graph of $y = ax^2 + bx + c$.
   If the graph crosses the x-axis, the equation will have real number solutions.

Section 16.4

Suggested Assignment

Exercises 1–25, odds
More challenging problems:
Exercises 27–37, odds

16.4 Exercises

Objective A To graph a quadratic equation of the form $y = ax^2 + bx + c$

For Exercises 1 to 4, determine whether the graph of the equation opens up or down.

1. $y = -\dfrac{1}{3}x^2$ **2.** $y = x^2 - 2x - 3$ **3.** $y = 2x^2 - 4$ **4.** $f(x) = 3 - 2x - x^2$
Down   Up   Up   Down

For Exercises 5 to 10, evaluate the function for the given value of x.

5. $f(x) = x^2 - 2x + 1; x = 3$ **6.** $f(x) = 2x^2 + x - 1; x = -2$
4   5

7. $f(x) = 4 - x^2; x = -3$ **8.** $f(x) = x^2 + 6x + 9; x = -3$
-5 0

9. $f(x) = -x^2 + 5x - 6; x = -4$ **10.** $f(x) = -2x^2 + 2x - 1; x = -3$
-42 -25

For Exercises 11 to 25, graph.

11. $y = x^2$ **12.** $y = -x^2$ **13.** $y = -x^2 + 1$

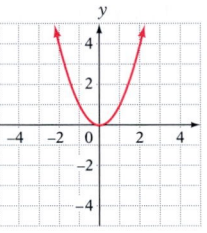 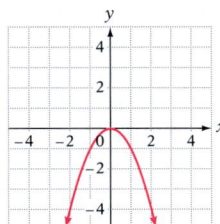 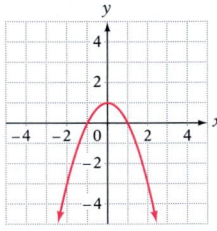

14. $y = x^2 - 1$ **15.** $y = 2x^2$ **16.** $y = \dfrac{1}{2}x^2$

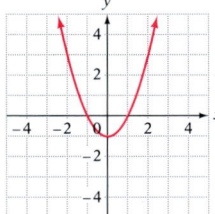 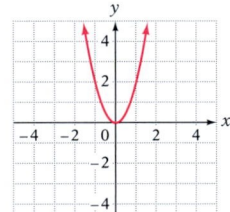 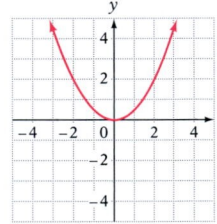

17. $y = -\dfrac{1}{2}x^2 + 1$ **18.** $y = 2x^2 - 1$ **19.** $y = x^2 - 4x$

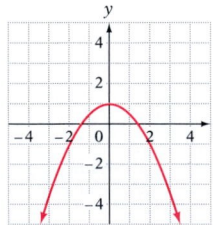 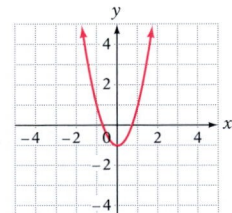 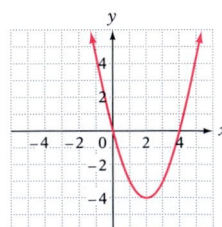

Quick Quiz (Objective 16.4A)

1. Graph: $y = x^2 + 2x - 3$ **2.** Graph: $y = -x^2 - 1$

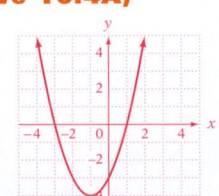 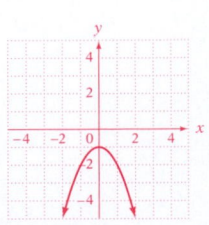

20. $y = x^2 + 4x$

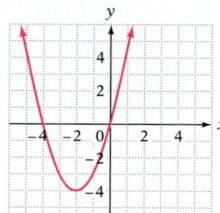

21. $y = x^2 - 2x + 3$

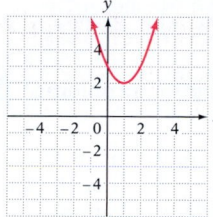

22. $y = x^2 - 4x + 2$

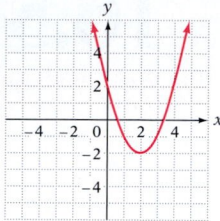

23. $y = -x^2 + 2x + 3$

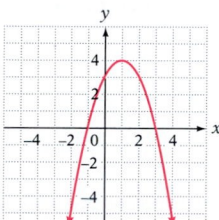

24. $y = -x^2 - 2x + 3$

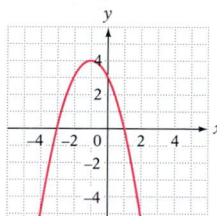

25. $y = -x^2 + 4x - 4$

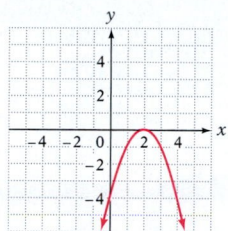

For Exercises 26 to 37, determine the x- and y-intercepts.

26. $y = x^2 - 5x + 6$
$(2, 0), (3, 0); (0, 6)$

27. $y = x^2 + 5x - 6$
$(-6, 0), (1, 0); (0, -6)$

28. $f(x) = 9 - x^2$
$(-3, 0), (3, 0); (0, 9)$

29. $f(x) = x^2 + 12x + 36$
$(-6, 0); (0, 36)$

30. $y = x^2 + 2x - 6$
$(-1 - \sqrt{7}, 0), (-1 + \sqrt{7}, 0); (0, -6)$

31. $f(x) = x^2 + 4x - 2$
$(-2 - \sqrt{6}, 0), (-2 + \sqrt{6}, 0); (0, -2)$

32. $y = x^2 + 2x + 3$

No x-intercepts; $(0, 3)$

33. $y = x^2 - x + 1$

No x-intercepts; $(0, 1)$

34. $f(x) = 2x^2 - x - 3$

$\left(\dfrac{3}{2}, 0\right), (-1, 0); (0, -3)$

35. $f(x) = 2x^2 - 13x + 15$

$\left(\dfrac{3}{2}, 0\right), (5, 0); (0, 15)$

36. $y = 4 - x - x^2$

$\left(\dfrac{-1 - \sqrt{17}}{2}, 0\right), \left(\dfrac{-1 + \sqrt{17}}{2}, 0\right); (0, 4)$

37. $y = 2 - 3x - 3x^2$

$\left(\dfrac{-3 - \sqrt{33}}{6}, 0\right), \left(\dfrac{-3 + \sqrt{33}}{6}, 0\right); (0, 2)$

APPLYING THE CONCEPTS

For Exercises 38 to 41, show that the equation is a quadratic equation in two variables by writing it in the form $y = ax^2 + bx + c$.

38. $y + 1 = (x - 4)^2$
$y = x^2 - 8x + 15$

39. $y - 2 = 3(x + 1)^2$
$y = 3x^2 + 6x + 5$

40. $y - 4 = 2(x - 3)^2$
$y = 2x^2 - 12x + 22$

41. $y + 3 = 3(x - 1)^2$
$y = 3x^2 - 6x$

For Exercises 42 to 45, find the x-intercepts.

42. $y = x^3 - x^2 - 6x$
$(-2, 0), (0, 0), (3, 0)$

43. $y = x^3 - 4x^2 - 5x$
$(-1, 0), (0, 0), (5, 0)$

44. $y = x^3 + x^2 - 4x - 4$
Hint: Factor by grouping.
$(-2, 0), (-1, 0), (2, 0)$

45. $y = x^3 + 3x^2 - x - 3$
Hint: Factor by grouping.
$(-3, 0), (-1, 0), (1, 0)$

Objective 16.5A

Optional Student Activity

Look at Exercises 1 to 6 and Exercises 13 to 18 of this section and determine which type of application each problem is.
Geometry: 1–6   Work: 13–16
Distance-rate: 17–18

Concept Check

The length of a bowling alley is 25 ft more than 10 times the width. The area of the alley is 210 ft^2.

1. Using w to represent the width of the bowling alley, write a variable expression for the length of the bowling alley in terms of w.
   Length = $10w + 25$

2. Write an equation that can be used to find the dimensions of the bowling alley.
   $w(10w + 25) = 210$

3. Solve the equation you wrote in part 2. $w = -6, 3.5$

4. Write the answer to the problem.
   Length: 60 ft; width: 3.5 ft

16.5 Application Problems

Objective A   To solve application problems

The application problems in this section are varieties of those problems solved earlier in the text. Each of the strategies for the problems in this section will result in a quadratic equation.

HOW TO   In 5 h, two campers rowed 12 mi down a stream and then rowed back to their campsite. The rate of the stream's current was 1 mph. Find the rate at which the campers rowed.

> **Strategy for Solving an Application Problem**
>
> 1. Determine the type of problem. For example, is it a distance-rate problem, a geometry problem, or a work problem?

The problem is a distance-rate problem.

> 2. Choose a variable to represent the unknown quantity. Write numerical or variable expressions for all the remaining quantities. These results can be recorded in a table.

The unknown rate of the campers: r

	Distance	÷	Rate	=	Time
Downstream	12	÷	$r + 1$	=	$\dfrac{12}{r + 1}$
Upstream	12	÷	$r - 1$	=	$\dfrac{12}{r - 1}$

> 3. Determine how the quantities are related.

TAKE NOTE

The time going downstream plus the time going upstream is equal to the time of the entire trip.

The total time of the trip was 5 h.

$$\frac{12}{r + 1} + \frac{12}{r - 1} = 5$$

$$(r + 1)(r - 1)\left(\frac{12}{r + 1} + \frac{12}{r - 1}\right) = (r + 1)(r - 1)5$$

$$(r - 1)12 + (r + 1)12 = (r^2 - 1)5$$

$$12r - 12 + 12r + 12 = 5r^2 - 5$$

$$24r = 5r^2 - 5$$

$$0 = 5r^2 - 24r - 5$$

$$0 = (5r + 1)(r - 5)$$

$$5r + 1 = 0 \qquad r - 5 = 0$$

$$5r = -1 \qquad\qquad r = 5$$

$$r = -\frac{1}{5}$$

TAKE NOTE

The solution $r = -\dfrac{1}{5}$ is not possible, because the rate cannot be a negative number.

The rowing rate was 5 mph.

In-Class Examples (Objective 16.5A)

1. It took a boat 2 h more to travel 90 mi against the current than to go 90 mi with the current. The rate of the current was 6 mph. Find the rate of the boat in calm water.   24 mph

2. A tank is emptied by two drains. The smaller drain can empty the tank in twice the amount of time it takes the larger drain to empty the tank. Using both drains, the tank can be emptied in 2 h. How long would it take the larger drain, working alone, to empty the tank?   3 h

Example 1

A painter and the painter's apprentice, working together, can paint a room in 2 h. Working alone, the apprentice requires 3 hours more to paint the room than does the painter working alone. How long does it take the painter, working alone, to paint the room?

Strategy

- This is a work problem.
- Time for the painter to paint the room: t
  Time for the apprentice to paint the room: $t + 3$

	Rate	Time	Part
Painter	$\frac{1}{t}$	2	$\frac{2}{t}$
Apprentice	$\frac{1}{t+3}$	2	$\frac{2}{t+3}$

- The sum of the parts of the task completed must equal 1.

Solution

$$\frac{2}{t} + \frac{2}{t+3} = 1$$

$$t(t+3)\left(\frac{2}{t} + \frac{2}{t+3}\right) = t(t+3) \cdot 1$$

$$(t+3)2 + t(2) = t(t+3)$$
$$2t + 6 + 2t = t^2 + 3t$$
$$4t + 6 = t^2 + 3t$$
$$0 = t^2 - t - 6$$
$$0 = (t-3)(t+2)$$

$t - 3 = 0 \quad t + 2 = 0$
$\quad t = 3 \quad\quad t = -2$

The solution $t = -2$ is not possible.

The time is 3 h.

You Try It 1

The length of a rectangle is 2 m more than the width. The area is 15 m². Find the width.

Your strategy

Your solution
3 m

Solution on p. S40

Concept Check

A pool is filled by two pipes. The smaller pipe takes 8 h longer to fill the pool than does a second, larger pipe. Using both pipes, the pool is filled in 3 h. How long would it take each pipe, working alone, to fill the pool?

1. Using t to represent the time it would take the larger pipe, working alone, to fill the pool, write a variable expression in terms of t for the time it would take the smaller pipe, working alone, to fill the pool.
   Time for smaller pipe = $t + 8$

2. Write an equation that can be used to solve this problem.
   $$\frac{3}{t} + \frac{3}{t+8} = 1$$

3. Solve the equation you wrote in part 2. $t = -6, 4$

4. Write the answer to the problem.
   Working alone, the larger pipe would take 4 h; the smaller pipe would take 12 h.

Section 16.5

Suggested Assignment

Exercises 1–12, odds
More challenging problems:
    Exercises 13–19, odds

16.5 Exercises

> *Objective A*   **To solve application problems**

1. **Geometry**   The height of a triangle is 2 m more than twice the length of the base. The area of the triangle is 20 m². Find the height of the triangle and the length of the base.
   Height: 10 m; length: 4 m

2. **Geometry**   The length of a rectangle is 4 ft more than twice the width. The area of the rectangle is 160 ft². Find the length and width of the rectangle.
   Length: 20 ft; width: 8 ft

3. **Baseball**   The area of the batter's box on a major-league baseball field is 24 ft². The length of the batter's box is 2 ft more than the width. Find the length and width of the batter's box.
   Length: 6 ft; width: 4 ft

   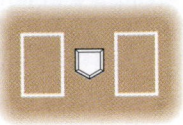

4. **Softball**   The length of the batter's box on a softball field is 1 ft less than twice the width. The area of the batter's box is 15 ft². Find the length and width of the batter's box.
   Length: 5 ft; width: 3 ft

5. **Swimming Pools**   The length of a swimming pool is twice the width. The area of the pool is 5000 ft². Find the length and width of the pool.
   Length: 100 ft; width: 50 ft

   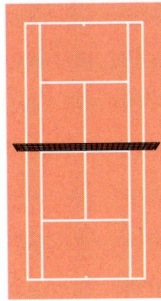

6. **Tennis**   The length of a singles tennis court is 24 ft more than twice the width. The area of the tennis court is 2106 ft². Find the length and width of the court.
   Length: 78 ft; width: 27 ft

7. **Football**   The hang time of a football that is kicked on the opening kick-off is given by $s = -16t^2 + 88t + 4$, where s is the height of the football t seconds after leaving the kicker's foot. What is the hang time of a kickoff that hits the ground without being caught? Round to the nearest tenth.
   5.5 s

8. **Manufacturing**   A square piece of cardboard is to be formed into a box to transport pizzas. The box is formed by cutting 2-inch square corners from the cardboard and folding up the sides, as shown in the figure at the right. If the volume of the box is 512 in³, what are the dimensions of the cardboard?
   20 in. by 20 in.

   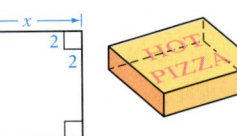

9. **Automotive Safety**   The distance, s, a car needs to come to a stop on a certain surface depends on the velocity, v, in feet per second, of the car when the brakes are applied. The equation is given by $s = 0.0344v^2 - 0.758v$. What is the maximum velocity a car can have when the brakes are applied and stop within 150 ft?
   77 ft/s

10. **Landscaping**   The perimeter of a rectangular garden is 54 ft. The area of the garden is 180 ft². Find the length and width of the garden.
    Width: 12 ft; length: 15 ft

11. **Food Preparation**   The radius of a large pizza is 1 in. less than twice the radius of a small pizza. The difference between the areas of the two pizzas is 33π in². Find the radius of the large pizza.
    7 in.

Quick Quiz (Objective 16.5A)

1. The length of a rectangle is 4 ft more than twice the width. The area of the rectangle is 48 ft². Find the length and width of the rectangle.
Length: 12 ft; width: 4 ft

12. Geometry   The hypotenuse of a right triangle is $\sqrt{13}$ cm. One leg is 1 cm shorter than twice the length of the other leg. Find the lengths of the legs of the right triangle.
2 cm, 3 cm

13. Computer Computations   One computer takes 21 min longer than a second computer to calculate the value of a complex equation. Working together, these computers complete the calculation in 10 min. How long would it take each computer, working separately, to calculate the value?
First computer: 35 min; second computer: 14 min

14. Plumbing   A tank has two drains. One drain takes 16 min longer to empty the tank than does a second drain. With both drains open, the tank is emptied in 6 min. How long would it take each drain, working alone, to empty the tank?
First drain: 24 min; second drain: 8 min

15. Transportation   Using one engine of a ferryboat, it takes 6 h longer to cross a channel than it does using a second engine alone. With both engines operating, the ferryboat can make the crossing in 4 h. How long would it take each engine, working alone, to power the ferryboat across the channel?
First engine: 12 h; second engine: 6 h

16. Masonry   An apprentice mason takes 8 h longer to build a small fireplace than an experienced mason. Working together, they can build the fireplace in 3 h. How long would it take each mason, working alone, to complete the fireplace?
Apprentice mason: 12 h; experienced mason: 4 h

17. Travel   It took a small plane 2 h longer to fly 375 mi against the wind than to fly the same distance with the wind. The rate of the wind was 25 mph. Find the rate of the plane in calm air.
100 mph

18. Travel   It took a motorboat 1 h longer to travel 36 mi against the current than to go 36 mi with the current. The rate of the current was 3 mph. Find the rate of the boat in calm water.
15 mph

APPLYING THE CONCEPTS

19. Food Preparation   If a pizza with a diameter of 8 in. costs $6, what should be the cost of a pizza with a diameter of 16 in. if both pizzas cost the same amount per square inch?
$24

20. Geometry   A wire 8 ft long is cut into two pieces. A circle is formed from one piece, and a square is formed from the other. The total area of both figures is given by $A = \frac{1}{16}(8 - x)^2 + \frac{x^2}{4\pi}$. What is the length of each piece of wire if the total area is 4.5 ft²?
7.507 ft, 0.493 ft

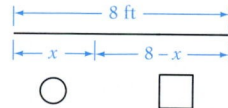

Focus on Problem Solving

Algebraic Manipulation and Graphing Techniques

Problem solving is often easier when we have both algebraic manipulation and graphing techniques at our disposal. Solving quadratic equations and graphing quadratic equations in two variables are used here to solve problems involving profit.

A company's revenue, R, is the total amount of money the company earned by selling its products. The cost, C, is the total amount of money the company spent to manufacture and sell its products. A company's profit, P, is the difference between the revenue and the cost: $P = R - C$. A company's revenue and cost may be represented by equations.

A company manufactures and sells woodstoves. The total monthly cost, in dollars, to produce n woodstoves is $C = 30n + 2000$. Write a variable expression for the company's monthly profit if the revenue, in dollars, obtained from selling all n woodstoves is $R = 150n - 0.4n^2$.

$$P = R - C$$
$$P = 150n - 0.4n^2 - (30n + 2000)$$
$$P = -0.4n^2 + 120n - 2000$$

• Replace R with $150n - 0.4n^2$ and C with $30n + 2000$. Then simplify.

How many woodstoves must the company manufacture and sell in order to make a profit of $6000 a month?

$$P = -0.4n^2 + 120n - 2000$$
$$6000 = -0.4n^2 + 120n - 2000$$

• Substitute 6000 for P.

$$0 = -0.4n^2 + 120n - 8000$$

• Write the equation in standard form.

$$0 = n^2 - 300n + 20{,}000$$

• Divide each side of the equation by -0.4.

$$0 = (n - 100)(n - 200)$$

• Factor.

$$n - 100 = 0 \qquad n - 200 = 0$$

• Solve for n.

$$n = 100 \qquad\quad n = 200$$

The company will make a monthly profit of $6000 if either 100 or 200 woodstoves are manufactured and sold.

The graph of $P = -0.4n^2 + 120n - 2000$ is shown at the right. Note that when $P = 6000$, the values of n are 100 and 200.

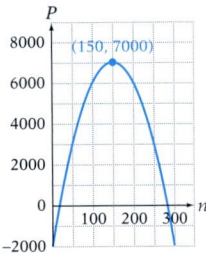

Also note that the coordinates of the highest point on the graph are (150, 7000). This means that the company makes a *maximum* profit of $7000 per month when 150 woodstoves are manufactured and sold.

1. The total cost, in dollars, for a company to produce and sell n guitars per month is $C = 240n + 1200$. The company's revenue, in dollars, from selling all n guitars is $R = 400n - 2n^2$.

   a. How many guitars must the company produce and sell each month in order to make a monthly profit of $1200?

   b. Graph the profit equation. What is the maximum monthly profit the company can make?

Answers to Focus on Problem Solving: Algebraic Manipulation and Graphing Techniques

1a. 20 or 60 guitars

b.

The maximum monthly profit the company can make is $2000.

Projects and Group Activities

Graphical Solutions of Quadratic Equations

A real number x is called a **zero of a function** if the value of the function when evaluated at x is 0. That is, if $f(x) = 0$, then x is called a zero of the function. For instance, evaluating $f(x) = x^2 + x - 6$ when $x = -3$, we have

$$f(x) = x^2 + x - 6$$
$$f(-3) = (-3)^2 + (-3) - 6 \qquad \bullet \text{ Replace } x \text{ with } -3.$$
$$f(-3) = 9 - 3 - 6 = 0$$

For this function, $f(-3) = 0$, so -3 is a zero of the function.

Verify that 2 is a zero of $f(x) = x^2 + x - 6$ by showing that $f(2) = 0$.

The graph of $f(x) = x^2 + x - 6$ is shown at the left. Note that the graph crosses the x-axis at -3 and 2, the two zeros of the function. The points $(-3, 0)$ and $(2, 0)$ are x-intercepts of the graph.

Consider the equation $0 = x^2 + x - 6$, which is $f(x) = x^2 + x - 6$ with $f(x)$ replaced with 0. Solving $0 = x^2 + x - 6$, we have

$$0 = x^2 + x - 6$$
$$0 = (x + 3)(x - 2) \qquad \bullet \text{ Solve by factoring and using the}$$
$$x + 3 = 0 \qquad x - 2 = 0 \qquad \text{Principle of Zero Products.}$$
$$x = -3 \qquad x = 2$$

Observe that the solutions of the equation are the zeros of the function. These important connections among the real zeros of a function, the x-intercepts of its graph, and the solutions of the equation are the basis for using a graphing calculator to solve an equation.

The following method of solving a quadratic equation by using a graphing calculator is based on a TI-83, TI-83 Plus, or TI-84 Plus calculator. Other calculators will require a slightly different approach.

HOW TO Approximate the solutions of $x^2 + 4x = 6$ by using a graphing calculator.

$$x^2 + 4x = 6$$

Write the equation in standard form. $\qquad x^2 + 4x - 6 = 0$

1. Press **Y =** and enter $x^2 + 4x - 6$ for Y1.

2. Press **GRAPH**. If the graph does not appear on the screen, press **ZOOM** 6.

3. Press **2nd** CALC 2. Note that the selection for 2 says **zero**. This will begin the calculation of the zeros of the function, which are the solutions of the equation.

Step 1

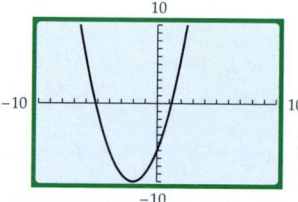

Step 2

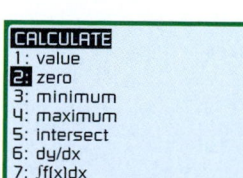

Step 3

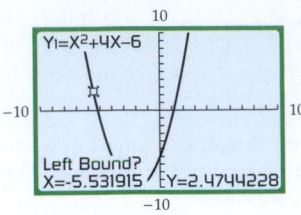

Step 4

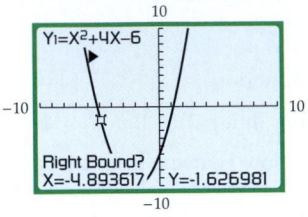

Step 5

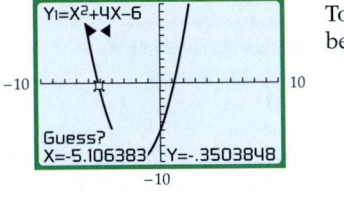

Step 6

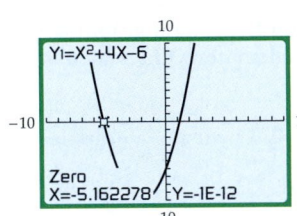

Step 7

4. At the bottom of the screen, you will see **LeftBound?**. This is asking you to move the blinking cursor so that it is to the *left* of the first x-intercept. Use the left arrow key to move the cursor to the left of the first x-intercept. The values of x and y that appear on your calculator may be different from the ones shown here. Just be sure that you are to the left of the x-intercept. When you are done, press ENTER.

5. At the bottom of the screen, you will see **RightBound?**. This is asking you to move the blinking cursor so that it is to the *right* of the x-intercept. Use the right arrow key to move the cursor to the right of the x-intercept. The values of x and y that appear on your calculator may be different from the ones shown here. Just be sure that you are to the right of the x-intercept. When you are done, press ENTER.

6. At the bottom of the screen, you will see **Guess?**. Press ENTER.

7. The zero of the function is approximately -5.162278. Thus one solution of $x^2 + 4x = 6$ is approximately -5.162278. Also note that the value of y is given as $Y1 = ^-1E^-12$. This is the way the calculator writes a number in scientific notation. We would normally write $Y1 = -1 \times 10^{-12}$. This number is very close to zero.

To find the other solution, repeat Steps 3 through 7. The screens are shown below.

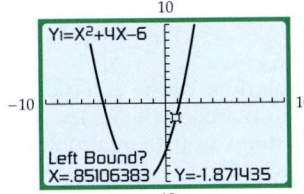

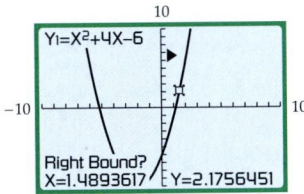

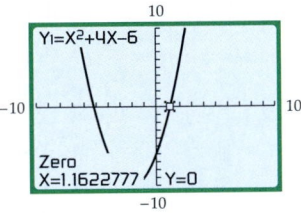

A second zero of the function is approximately 1.1622777. Thus the two solutions of $x^2 + 4x = 6$ are approximately -5.162278 and 1.1622777.

Use a graphing calculator to approximate the solutions of the following equations.

1. $x^2 + 3x - 4 = 0$

2. $x^2 - 4x - 5 = 0$

3. $x^2 + 3.4x = 4.15$

4. $2x^2 - \dfrac{5}{9}x = \dfrac{3}{8}$

5. $\pi x^2 - \sqrt{17}x - 2 = 0$

6. $\sqrt{2}x^2 + x - \sqrt{7} = 0$

Geometric Construction of Completing the Square

Completing the square as a method for solving a quadratic equation has been known for centuries. The Persian mathematician Al-Khwarismi used this method in a textbook written around 825 A.D. The method was very geometric. That is, Al-Khwarismi literally completed a square. To understand how this method works, consider the following geometric shapes: a square whose area is x^2, a rectangle whose area is x, and another square whose area is 1.

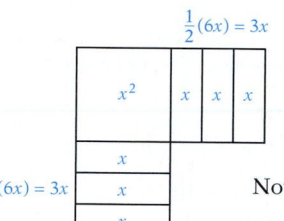

$\frac{1}{2}(6x) = 3x$

$\frac{1}{2}(6x) = 3x$

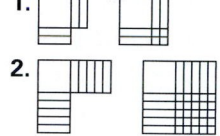

x^2

x

1

Now consider the expression $x^2 + 6x$. From our discussion in this chapter, to complete the square, we added $\left(\frac{1}{2} \cdot 6\right)^2 = 3^2 = 9$ to the expression. The geometric construction that Al-Khwarismi used is shown at the left.

Note that it is necessary to add nine squares to the figure to "complete the square." One of the difficulties of using a geometric method such as this is that it cannot easily be extended to $x^2 - 6x$. There is no way to draw an area of $-6x$! That really did not bother Al-Khwarismi much. Negative numbers were not a significant part of mathematics until well into the 13th century.

9 squares were added

1. Show how Al-Khwarismi would have completed the square for $x^2 + 4x$.

2. Show how Al-Khwarismi would have completed the square for $x^2 + 10x$.

Chapter 16 Summary

Key Words	**Examples**
A *quadratic equation* is an equation that can be written in the form $ax^2 + bx + c = 0$, where a, b, and c are real numbers and $a \neq 0$. [16.1A, p. 787]	$3x^2 - 5x - 3 = 0$ is a quadratic equation. For this equation, $a = 3$, $b = -5$, and $c = -3$.
A quadratic equation is in *standard form* when the polynomial is written in descending order and set equal to zero. [16.1A, p. 787]	$2x - 4 + 5x^2 = 0$ is not in standard form. The same equation in standard form is $5x^2 + 2x - 4 = 0$.
Adding to a binomial the constant term that makes it a perfect-square trinomial is called *completing the square*. [16.2A, p. 793]	Adding to $x^2 - 8x$ the constant term 16 results in a perfect-square trinomial: $x^2 - 8x + 16 = (x - 4)^2$.
An equation of the form $y = ax^2 + bx + c$, $a \neq 0$, is a *quadratic equation in two variables*. [16.4A, p. 803]	$y = 2x^2 + 3x - 4$ is a quadratic equation in two variables.

The graph of an equation of the form $y = ax^2 + bx + c$, $a \neq 0$, is a *parabola*. The graph is U-shaped and opens up when $a > 0$ and opens down when $a < 0$. [16.4A, p. 803]

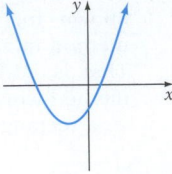

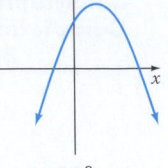

$a > 0$
Parabola opens up

$a < 0$
Parabola opens down

Essential Rules and Procedures

Examples

Solving a Quadratic Equation by Factoring [16.1A, p. 787]
Write the equation in standard form, factor the polynomial, apply the Principle of Zero Products, and solve for the variable.

$$x^2 - 3x = 10$$
$$x^2 - 3x - 10 = 0$$
$$(x + 2)(x - 5) = 0$$
$$x + 2 = 0 \qquad x - 5 = 0$$
$$x = -2 \qquad\quad x = 5$$

Principle of Taking the Square Root of Each Side of an Equation [16.1B, p. 789]
If $x^2 = a$, then $x = \pm\sqrt{a}$.

This principle is used to solve quadratic equations by taking square roots.

$$2x^2 - 36 = 0$$
$$2x^2 = 36$$
$$x^2 = 18$$
$$\sqrt{x^2} = \sqrt{18}$$
$$x = \pm\sqrt{18} = \pm 3\sqrt{2}$$

Solving a Quadratic Equation by Completing the Square [16.2A, p. 793]
When the quadratic equation is in the form $x^2 + bx = c$, add to each side of the equation the term that completes the square on $x^2 + bx$. Factor the perfect-square trinomial, and write it as the square of a binomial. Take the square root of each side of the equation and solve for x.

$$x^2 + 6x = 5$$
$$x^2 + 6x + 9 = 5 + 9$$
$$(x + 3)^2 = 14$$
$$\sqrt{(x + 3)^2} = \sqrt{14}$$
$$x + 3 = \pm\sqrt{14}$$
$$x = -3 \pm \sqrt{14}$$

The Quadratic Formula [16.3A, p. 799]

The solutions of $ax^2 + bx + c = 0$, $a \neq 0$, are

$$x = \frac{-b \pm \sqrt{b^2 - 4ac}}{2a}.$$

$$2x^2 + 3x - 6 = 0$$
$$x = \frac{-b \pm \sqrt{b^2 - 4ac}}{2a}$$
$$= \frac{-3 \pm \sqrt{(3)^2 - 4(2)(-6)}}{2(2)}$$
$$= \frac{-3 \pm \sqrt{9 + 48}}{4} = \frac{-3 \pm \sqrt{57}}{4}$$

Graph of a Quadratic Equation in Two Variables [16.4A, pp. 803–804]
To graph a quadratic equation in two variables, find several solutions of the equation. Graph the ordered-pair solutions on a rectangular coordinate system. Draw a parabola through the points.

To find the x-intercepts of the graph of a quadratic equation in two variables, let $y = 0$ and solve for x.

To find the y-intercept, let $x = 0$ and solve for y.

$y = x^2 - x - 2$

x	y
-2	4
-1	0
0	-2
1	-2
2	0
3	4

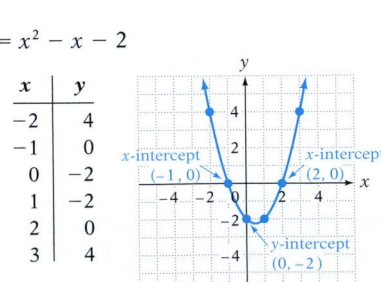

Chapter 16 Review Exercises

1. Solve by factoring: $6x^2 + 13x - 28 = 0$

   $\dfrac{4}{3}, -\dfrac{7}{2}$ [16.1A]

2. Solve by taking square roots:
   $49x^2 = 25$

   $-\dfrac{5}{7}, \dfrac{5}{7}$ [16.1B]

3. Solve by completing the square:
   $x^2 + 2x - 24 = 0$

   $-6, 4$ [16.2A]

4. Solve by using the quadratic formula:
   $x^2 + 5x - 6 = 0$

   $-6, 1$ [16.3A]

5. Solve by completing the square:
   $2x^2 + 5x = 12$

   $-4, \dfrac{3}{2}$ [16.2A]

6. Solve by factoring: $12x^2 + 10 = 29x$

   $2, \dfrac{5}{12}$ [16.1A]

7. Solve by taking square roots:
   $(x + 2)^2 - 24 = 0$

   $-2 - 2\sqrt{6}, -2 + 2\sqrt{6}$ [16.1B]

8. Solve by using the quadratic formula:
   $2x^2 + 3 = 5x$

   $1, \dfrac{3}{2}$ [16.3A]

9. Solve by factoring: $6x(x + 1) = x - 1$

   $-\dfrac{1}{3}, -\dfrac{1}{2}$ [16.1A]

10. Solve by taking square roots:
    $4y^2 + 9 = 0$

    No real number solution [16.1B]

11. Solve by completing the square:
    $x^2 - 4x + 1 = 0$

    $2 - \sqrt{3}, 2 + \sqrt{3}$ [16.2A]

12. Solve by using the quadratic formula:
    $x^2 - 3x - 5 = 0$

    $\dfrac{3 - \sqrt{29}}{2}, \dfrac{3 + \sqrt{29}}{2}$ [16.3A]

13. Solve by completing the square:
    $x^2 + 6x + 12 = 0$

    No real number solution [16.2A]

14. Solve by factoring: $(x + 9)^2 = x + 11$

    $-7, -10$ [16.1A]

15. Solve by taking square roots:

    $\left(x - \dfrac{1}{2}\right)^2 = \dfrac{9}{4}$

    $-1, 2$ [16.1B]

16. Solve by completing the square:
    $4x^2 + 16x = 7$

    $\dfrac{-4 - \sqrt{23}}{2}, \dfrac{-4 + \sqrt{23}}{2}$ [16.2A]

17. Solve by using the quadratic formula:
$x^2 - 4x + 8 = 0$
No real number solution   [16.3A]

18. Solve by using the quadratic formula:
$2x^2 + 5x + 2 = 0$
$-2, -\dfrac{1}{2}$ [16.3A]

19. Graph $y = -3x^2$.

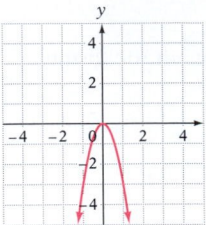

[16.4A]

20. Graph $y = -\dfrac{1}{4}x^2$.

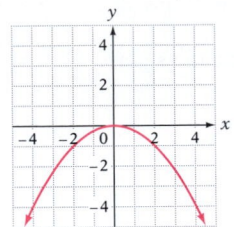

[16.4A]

21. Graph $y = 2x^2 + 1$.

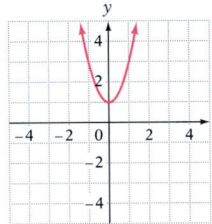

[16.4A]

22. Graph $y = x^2 - 4x + 3$.

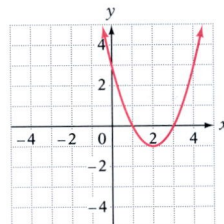

[16.4A]

23. Graph $y = -x^2 + 4x - 5$.

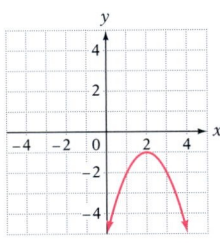

[16.4A]

24. Find the x- and y-intercepts of the graph of
$y = x^2 - 2x - 15$.
x-intercepts: $(-3, 0), (5, 0)$
y-intercept: $(0, -15)$ [16.4A]

25. **Flight**   It took a hawk half an hour longer to fly 70 mi against the wind than to fly 40 mi with the wind. The rate of the wind was 5 mph. Find the rate of the hawk in calm air.
75 mph   [16.5A]

Chapter 16 Test

1. Solve by factoring: $x^2 - 5x - 6 = 0$
   $6, -1$ [16.1A]

2. Solve by factoring: $3x^2 + 7x = 20$
   $-4, \dfrac{5}{3}$ [16.1A]

3. Solve by taking square roots:
   $2(x - 5)^2 - 50 = 0$
   $0, 10$ [16.1B]

4. Solve by taking square roots:
   $3(x + 4)^2 - 60 = 0$
   $-4 + 2\sqrt{5}, -4 - 2\sqrt{5}$ [16.1B]

5. Solve by completing the square:
   $x^2 + 4x - 16 = 0$
   $-2 + 2\sqrt{5}, -2 - 2\sqrt{5}$ [16.2A]

6. Solve by completing the square:
   $x^2 + 3x = 8$
   $\dfrac{-3 + \sqrt{41}}{2}, \dfrac{-3 - \sqrt{41}}{2}$ [16.2A]

7. Solve by completing the square:
   $2x^2 - 6x + 1 = 0$
   $\dfrac{3 + \sqrt{7}}{2}, \dfrac{3 - \sqrt{7}}{2}$ [16.2A]

8. Solve by completing the square:
   $2x^2 + 8x = 3$
   $\dfrac{-4 + \sqrt{22}}{2}, \dfrac{-4 - \sqrt{22}}{2}$ [16.2A]

9. Solve by using the quadratic formula:
   $x^2 + 4x + 2 = 0$
   $-2 + \sqrt{2}, -2 - \sqrt{2}$ [16.3A]

10. Solve by using the quadratic formula:
    $x^2 - 3x = 6$
    $\dfrac{3 + \sqrt{33}}{2}, \dfrac{3 - \sqrt{33}}{2}$ [16.3A]

11. Solve by using the quadratic formula:
$2x^2 - 5x - 3 = 0$

$-\dfrac{1}{2}, 3$ [16.3A]

12. Solve by using the quadratic formula:
$3x^2 - x = 1$

$\dfrac{1 + \sqrt{13}}{6}, \dfrac{1 - \sqrt{13}}{6}$ [16.3A]

13. Graph $y = x^2 + 2x - 4$.

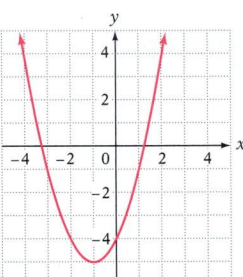

[16.4A]

14. Find the x- and y-intercepts of the graph of $f(x) = x^2 + x - 12$.

x-intercepts: $(-4, 0), (3, 0)$
y-intercept: $(0, -12)$ [16.4A]

15. **Geometry**   The length of a rectangle is 2 ft less than twice the width. The area of the rectangle is 40 ft². Find the length and width of the rectangle.
Length: 8 ft; width: 5 ft   [16.5A]

16. **Travel**   It took a motorboat 1 h longer to travel 60 mi against a current than it took the boat to travel 60 mi with the current. The rate of the current was 1 mph. Find the rate of the boat in calm water.
11 mph   [16.5A]

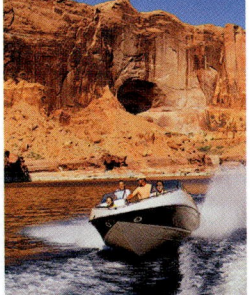

Cumulative Review Exercises

1. Simplify: $2x - 3[2x - 4(3 - 2x) + 2] - 3$
$-28x + 27$ [4.2D]

2. Solve: $-\dfrac{3}{5}x = -\dfrac{9}{10}$
$\dfrac{3}{2}$ [5.1C]

3. Solve: $2x - 3(4x - 5) = -3x - 6$
3 [5.3B]

4. Simplify: $(2a^2b)^2(-3a^4b^2)$
$-12a^8b^4$ [9.2B]

5. Divide: $(x^2 - 8) \div (x - 2)$
$x + 2 - \dfrac{4}{x - 2}$ [9.5B]

6. Factor: $3x^3 + 2x^2 - 8x$
$x(3x - 4)(x + 2)$ [10.3A/10.3B]

7. Divide: $\dfrac{3x^2 - 6x}{4x - 6} \div \dfrac{2x^2 + x - 6}{6x^3 - 24x}$
$\dfrac{9x^2(x - 2)^2}{(2x - 3)^2}$ [11.1C]

8. Subtract: $\dfrac{x}{2(x - 1)} - \dfrac{1}{(x - 1)(x + 1)}$
$\dfrac{x + 2}{2(x + 1)}$ [11.3B]

9. Simplify: $\dfrac{1 - \dfrac{7}{x} + \dfrac{12}{x^2}}{2 - \dfrac{1}{x} - \dfrac{15}{x^2}}$
$\dfrac{x - 4}{2x + 5}$ [11.4A]

10. Find the x- and y-intercepts of the graph of the line $4x - 3y = 12$.
x-intercept: $(3, 0)$; y-intercept: $(0, -4)$ [12.3A]

11. Find the equation of the line that contains the point $(-3, 2)$ and has slope $-\dfrac{4}{3}$.
$y = -\dfrac{4}{3}x - 2$ [12.4A]

12. Solve by substitution:
$3x - y = 5$
$y = 2x - 3$
$(2, 1)$ [13.2A]

13. Solve by the addition method:
$3x + 2y = 2$
$5x - 2y = 14$
$(2, -2)$ [13.3A]

14. Solve: $2x - 3(2 - 3x) > 2x - 5$
$x > \dfrac{1}{9}$ [14.3A]

15. Simplify: $(\sqrt{a} - \sqrt{2})(\sqrt{a} + \sqrt{2})$
$a - 2$ [15.3A]

16. Simplify: $\dfrac{\sqrt{108a^7b^3}}{\sqrt{3a^4b}}$
$6ab\sqrt{a}$ [15.3B]

17. Simplify: $\dfrac{\sqrt{3}}{5 + 2\sqrt{3}}$

$\dfrac{-6 + 5\sqrt{3}}{13}$ [15.3B]

18. Solve: $3 = 8 - \sqrt{5x}$

5   [15.4A]

19. Solve by factoring: $6x^2 - 17x = -5$

$\dfrac{5}{2}, \dfrac{1}{3}$ [16.1A]

20. Solve by taking square roots:
$2(x - 5)^2 = 36$

$5 + 3\sqrt{2}, 5 - 3\sqrt{2}$ [16.1B]

21. Solve by completing the square:
$3x^2 + 7x = -3$

$\dfrac{-7 + \sqrt{13}}{6}, \dfrac{-7 - \sqrt{13}}{6}$ [16.2A]

22. Solve by using the quadratic formula:
$2x^2 - 3x - 2 = 0$

$2, -\dfrac{1}{2}$ [16.3A]

23. **Food Mixtures**   Find the cost per pound of a mixture made from 20 lb of cashews that cost $3.50 per pound and 50 lb of peanuts that cost $1.75 per pound.

$2.25 per pound   [5.5A]

24. **Investments**   A stock investment of 100 shares paid a dividend of $215. At this rate, how many additional shares must the investor own to earn a dividend of $752.50?

250 additional shares   [6.2B]

25. **Travel**   A 720-mile trip from one city to another takes 3 h when a plane is flying with the wind. The return trip, against the wind, takes 4.5 h. Find the rate of the plane in still air and the rate of the wind.

Plane in still air: 200 mph; wind: 40 mph   [13.4A]

26. **Grading**   A student received a 70, a 91, an 85, and a 77 on four tests in a mathematics class. What scores on the fifth test will enable the student to receive a minimum of 400 points?

≥ 77 [14.2C]

27. **Number Sense**   The sum of the squares of three consecutive odd integers is 83. Find the middle odd integer.

-5 or 5   [16.5A]

28. **Exercising**   A jogger ran 7 mi at a constant rate and then reduced the rate by 3 mph and ran an additional 8 mi at the reduced rate. The total time spent jogging the 15 mi was 3 h. Find the jogger's rate for the last 8 mi.

4 mph   [16.5A]

29. Graph the solution set of $2x - 3y > 6$.

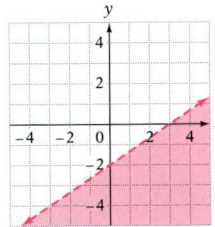

[14.4A]

30. Graph $y = x^2 - 2x - 3$.

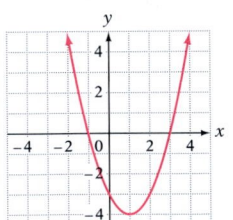

[16.4A]

Final Exam

1. Evaluate $-|-3|$.

   -3 [3.1C]

2. Subtract: $-15 - (-12) - 3$

   -6 [3.2B]

3. Simplify: $-\dfrac{4}{5} - \left(-\dfrac{3}{10}\right)$

   $-\dfrac{1}{2}$ [3.4A]

4. Simplify: $-7 - \dfrac{12 - 15}{2 - (-1)} \cdot (-4)$

   -11 [3.5A]

5. Evaluate $\dfrac{a^2 - 3b}{2a - 2b^2}$ when $a = 3$ and $b = -2$.

   $-\dfrac{15}{2}$ [4.1A]

6. Simplify: $6x - (-4y) - (-3x) + 2y$

   $9x + 6y$ [4.2A]

7. Simplify: $(-15z)\left(-\dfrac{2}{5}\right)$

   $6z$ [4.2B]

8. Simplify: $-2[5 - 3(2x - 7) - 2x]$

   $16x - 52$ [4.2D]

9. Solve: $20 = -\dfrac{2}{5}x$

   -50 [5.1C]

10. Solve: $4 - 2(3x + 1) = 3(2 - x) + 5$

    -3 [5.3B]

11. Write $\dfrac{1}{8}$ as a percent.

    12.5% [6.3B]

12. Find 19% of 80.

    15.2 [6.4A/6.4B]

13. Subtract: $(2x^2 - 5x + 1) - (5x^2 - 2x - 7)$

    $-3x^2 - 3x + 8$ [9.1B]

14. Simplify: $(-3xy^3)^4$

    $81x^4y^{12}$ [9.2B]

15. Multiply: $(3x^2 - x - 2)(2x + 3)$

    $6x^3 + 7x^2 - 7x - 6$ [9.3B]

16. Simplify: $\dfrac{(-2x^2y^3)^3}{(-4xy^4)^2}$

    $-\dfrac{x^4y}{2}$ [9.4A]

17. Divide: $\dfrac{12x^2y - 16x^3y^2 - 20y^2}{4xy^2}$

    $\dfrac{3x}{y} - 4x^2 - \dfrac{5}{x}$ [9.5A]

18. Divide: $(5x^2 - 2x - 1) \div (x + 2)$

    $5x - 12 + \dfrac{23}{x + 2}$ [9.5B]

19. Simplify: $(4x^{-2}y)^2(2xy^{-2})^{-2}$

    $\dfrac{4y^6}{x^6}$ [9.4A]

20. Given $f(t) = \dfrac{t}{t + 1}$, find $f(3)$.

    $\dfrac{3}{4}$ [12.1D]

21. Factor: $x^2 - 5x - 6$

    $(x - 6)(x + 1)$ [10.2A]

22. Factor: $6x^2 - 5x - 6$

    $(3x + 2)(2x - 3)$ [10.3A/10.3B]

23. Factor: $8x^3 - 28x^2 + 12x$
$4x(2x - 1)(x - 3)$ [10.4B]

24. Factor: $25x^2 - 16$
$(5x - 4)(5x + 4)$ [10.4A]

25. Factor: $2a(4 - x) - 6(x - 4)$
$2(a + 3)(4 - x)$ [10.1B]

26. Factor: $75y - 12x^2y$
$3y(5 - 2x)(5 + 2x)$ [10.4B]

27. Solve: $2x^2 = 7x - 3$
$\dfrac{1}{2}, 3$ [10.5A]

28. Multiply: $\dfrac{2x^2 - 3x + 1}{4x^2 - 2x} \cdot \dfrac{4x^2 + 4x}{x^2 - 2x + 1}$
$\dfrac{2(x + 1)}{x - 1}$ [11.1B]

29. Subtract: $\dfrac{5}{x + 3} - \dfrac{3x}{2x - 5}$
$\dfrac{-3x^2 + x - 25}{(2x - 5)(x + 3)}$ [11.3B]

30. Simplify: $x - \dfrac{1}{1 - \dfrac{1}{x}}$
$\dfrac{x^2 - 2x}{x - 1}$ [11.4A]

31. Solve: $\dfrac{5x}{3x - 5} - 3 = \dfrac{7}{3x - 5}$
2 [11.5A]

32. Solve $a = 3a - 2b$ for a.
$a = b$ [11.6A]

33. Find the slope of the line that contains the points $(-1, -3)$ and $(2, -1)$.
$\dfrac{2}{3}$ [12.3B]

34. Find the equation of the line that contains the point $(3, -4)$ and has slope $-\dfrac{2}{3}$.
$y = -\dfrac{2}{3}x - 2$ [12.4A]

35. Solve by substitution:
$y = 4x - 7$
$y = 2x + 5$
$(6, 17)$ [13.2A]

36. Solve by the addition method:
$4x - 3y = 11$
$2x + 5y = -1$
$(2, -1)$ [13.3A]

37. Solve: $4 - x \geq 7$
$x \leq -3$ [14.2B]

38. Solve: $2 - 2(y - 1) \leq 2y - 6$
$y \geq \dfrac{5}{2}$ [14.3A]

39. Simplify: $\sqrt{49x^6}$
$7x^3$ [15.1B]

40. Simplify: $2\sqrt{27a} + 8\sqrt{48a}$
$38\sqrt{3a}$ [15.2A]

41. Simplify: $\dfrac{\sqrt{3}}{\sqrt{5} - 2}$
$\sqrt{15} + 2\sqrt{3}$ [15.3B]

42. Solve: $\sqrt{2x - 3} + 4 = 5$
2 [15.4A]

43. Solve by factoring:
$3x^2 - x = 4$
$-1, \dfrac{4}{3}$ [16.1A]

44. Solve by using the quadratic formula:
$4x^2 - 2x - 1 = 0$
$\dfrac{1 + \sqrt{5}}{4}, \dfrac{1 - \sqrt{5}}{4}$ [16.3A]

45. **Number Sense**   Translate and simplify "the sum of twice a number and three times the difference between the number and two."
    $2x + 3(x - 2); 5x - 6$ [4.3B]

46. **Depreciation**   Because of depreciation, the value of an office machine is now $2400. This is 80% of its original value. Find the original value.
    $3000   [6.4C]

47.  **Meteorology**   The average monthly snowfall amounts, in inches, for January through December in Reno, Nevada, are

    5.8, 5.2, 4.3, 1.2, 0.8, 0, 0, 0, 0, 0.3, 2.4, 4.3

    (*Source:* National Climatic Data Center) Find the mean, median, and mode of the data.
    Mean: 2.025 in., median: 1 in., mode: 0 in.   [8.2A]

48. **Investments**   A sports announcer invested some money at an annual simple interest rate of 4.8%. A second investment, $1000 more than the first, was invested at an annual simple interest rate of 5.6%. The total annual interest earned on the two investments was $472. How much was invested in the 4.8% account?
    $4000   [13.2B]

49. **Food Mixtures**   A grocer mixes 4 lb of peanuts that cost $2 per pound with 2 lb of walnuts that cost $5 per pound. What is the cost per pound of the resulting mixture?
    $3 per pound   [5.5A]

50. **Probability**   Four aces, two kings, and three queens are removed from a deck of cards and placed in a pile. One card is chosen at random from the pile. What is the probability that the card chosen is not a queen?
    $\frac{2}{3}$ [8.3A]

51. **Travel**   At 2 P.M., a small plane had been flying for 1 h when a change of wind direction doubled its average ground speed. The complete 860-kilometer trip took 2.5 h. How far did the plane travel in the first hour?
    215 km   [5.5B]

52. **Geometry**   The angles of a triangle are such that the measure of the second angle is 10° more than the measure of the first angle, and the measure of the third angle is 10° more than the measure of the second angle. Find the measure of each of the three angles.
    50°, 60°, 70°   [7.1C]

53. **Number Sense**   The sum of the squares of three consecutive integers is 50. Find the middle integer.
4 or −4   [16.5A]

54. **Geometry**   The length of a rectangle is 5 m more than the width. The area of the rectangle is 50 m². Find the dimensions of the rectangle.
Width: 5 m; length: 10 m   [10.5B]

55. **Paint Mixtures**   A paint formula requires 2 oz of dye for every 15 oz of base paint. How many ounces of dye are required for 120 oz of base paint?
16 oz   [6.2B]

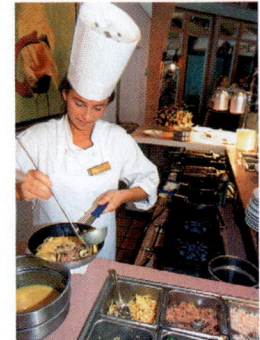

56. **Food Preparation**   It takes a chef 1 h to prepare a dinner. The chef's apprentice can prepare the dinner in 1.5 h. How long would it take the chef and the apprentice, working together, to prepare the dinner?
0.6 h   [11.7A]

57. **Travel**   With the current, a motorboat travels 50 mi in 2.5 h. Against the current, it takes the motorboat twice as long to travel 50 mi. Find the rate of the boat in calm water and the rate of the current.
Boat in calm water: 15 mph; current: 5 mph   [13.4A]

58. **Travel**   Flying against the wind, it took a pilot $\frac{1}{2}$ h longer to travel 500 mi than it took flying with the wind. The rate of the plane in calm air is 225 mph. Find the rate of the wind.
25 mph   [16.5A]

59. Graph the line that has slope $-\frac{1}{2}$ and y-intercept $(0, -3)$.

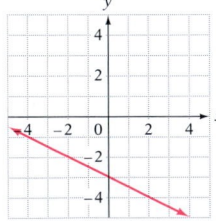

[12.3C]

60. Graph $y = x^2 - 4x + 3$.

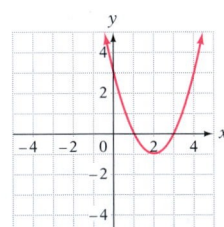

[16.4A]

OBJECTIVES

Section R.1

A To add and subtract fractions
B To multiply and divide fractions

Section R.2

A To add and subtract integers
B To multiply and divide integers

Section R.3

A To add and subtract rational numbers
B To multiply and divide rational numbers
C To evaluate exponential expressions
D To use the Order of Operations
Agreement to simplify expressions

Section R.4

A To solve a first-degree equation in one
variable

Need help? For online student resources, such
as section quizzes, visit this textbook's website at
math.college.hmco.com/students.

R.1 Fractions

Objective A   **To add and subtract fractions**

2.3A Add fractions

When we add two fractions with the same denominator, we add the numerators. The denominator remains the same. Then we write the answer in simplest form.

$$\frac{2}{9} + \frac{4}{9} = \frac{2+4}{9} = \frac{6}{9} = \frac{2}{3}$$

> **Addition of Fractions**
>
> To add fractions with the same denominator, add the numerators and place the sum over the common denominator.
>
> $$\frac{a}{b} + \frac{c}{b} = \frac{a+c}{b}, \qquad \text{where} \quad b \neq 0$$

2.1A Find the least common multiple (LCM)

Before two fractions can be added, the fractions must have the same denominator. To add fractions with different denominators, first rewrite the fractions as equivalent fractions with a common denominator. The common denominator is the least common multiple (LCM) of the denominators of the fractions. The LCM of denominators is sometimes called the least common denominator (LCD).

HOW TO   Add: $\frac{2}{3} + \frac{1}{4}$

$$\frac{2}{3} + \frac{1}{4} = \frac{8}{12} + \frac{3}{12}$$

- **The common denominator is the LCM of 3 and 4, which is 12. Write the fractions as equivalent fractions with the common denominator.**

$$= \frac{8+3}{12} = \frac{11}{12}$$

- **Add the fractions. $\frac{11}{12}$ is in the simplest form.**

2.3B Subtract fractions

When we subtract two fractions with the same denominator, we subtract the numerators. The denominator remains the same. Then we write the answer in simplest form.

$$\frac{7}{9} - \frac{4}{9} = \frac{7-4}{9} = \frac{3}{9} = \frac{1}{3}$$

> **Subtraction of Fractions**
>
> To subtract fractions with the same denominator, subtract the numerators and place the difference over the common denominator.
>
> $$\frac{a}{b} - \frac{c}{b} = \frac{a-c}{b}, \qquad \text{where} \quad b \neq 0$$

To subtract fractions with different denominators, first rewrite the fractions as equivalent fractions with a common denominator. As with addition of fractions, the common denominator is the least common multiple (LCM) of the denominators of the fractions.

HOW TO   Subtract: $\dfrac{2}{3} - \dfrac{5}{8}$

$$\dfrac{2}{3} - \dfrac{5}{8} = \dfrac{16}{24} - \dfrac{15}{24}$$

- The common denominator is the LCM of 3 and 8, which is 24. Write the fractions as equivalent fractions with the common denominator.

$$= \dfrac{16 - 15}{24} = \dfrac{1}{24}$$

- Subtract the fractions. $\dfrac{1}{24}$ is in simplest form.

Example 1   Add: $\dfrac{3}{8} + \dfrac{7}{12}$

Solution

$$\dfrac{3}{8} + \dfrac{7}{12} = \dfrac{9}{24} + \dfrac{14}{24}$$

$$= \dfrac{23}{24}$$

- The LCM of 8 and 12 is 24.
- Add the numerators. Place the sum over the common denominator.

You Try It 1   Add: $\dfrac{5}{12} + \dfrac{9}{16}$

Your solution $\dfrac{47}{48}$

Example 2   Add: $\dfrac{2}{3} + \dfrac{3}{5} + \dfrac{5}{6}$

Solution

$$\dfrac{2}{3} + \dfrac{3}{5} + \dfrac{5}{6}$$

$$= \dfrac{20}{30} + \dfrac{18}{30} + \dfrac{25}{30}$$

$$= \dfrac{63}{30}$$

$$= \dfrac{21}{10} = 2\dfrac{1}{10}$$

- The LCM of 3, 5, and 6 is 30.
- Write equivalent fractions using the LCM.
- Add the numerators. Place the sum over the common denominator.
- Simplify.

You Try It 2   Add: $\dfrac{3}{4} + \dfrac{4}{5} + \dfrac{5}{8}$

Your solution $2\dfrac{9}{10}$

Example 3   Subtract: $\dfrac{11}{16} - \dfrac{5}{12}$

Solution

$$\dfrac{11}{16} - \dfrac{5}{12} = \dfrac{33}{48} - \dfrac{20}{48}$$

$$= \dfrac{13}{48}$$

- The LCM of 16 and 12 is 48.
- Subtract the numerators. Place the difference over the common denominator.

You Try It 3   Subtract: $\dfrac{5}{6} - \dfrac{1}{4}$

Your solution $\dfrac{7}{12}$

Solutions on p. S41

Objective B **To multiply and divide fractions**

To multiply two fractions, multiply the numerators and multiply the denominators.

2.4A Multiply fractions

> **Multiplication of Fractions**
>
> The product of two fractions is the product of the numerators over the product of the denominators.
>
> $$\frac{a}{b} \cdot \frac{c}{d} = \frac{ac}{bd},$$ where $b \neq 0$ and $d \neq 0$

Note that fractions do not need to have the same denominator in order to be multiplied.

HOW TO Multiply: $\dfrac{2}{3} \cdot \dfrac{4}{5}$

$$\frac{2}{3} \cdot \frac{4}{5} = \frac{2 \cdot 4}{3 \cdot 5}$$ • **Multiply the numerators.**
Multiply the denominators.

$$= \frac{8}{15}$$

After multiplying two fractions, write the product in simplest form, as illustrated in the example below.

HOW TO Multiply: $\dfrac{3}{4} \cdot \dfrac{14}{15}$

$$\frac{3}{4} \cdot \frac{14}{15} = \frac{3 \cdot 14}{4 \cdot 15}$$ • **Multiply the numerators.**
Multiply the denominators.

$$= \frac{3 \cdot 2 \cdot 7}{2 \cdot 2 \cdot 3 \cdot 5}$$ • **Express the fraction in simplest form by first writing the prime factorization of each number.**

$$= \frac{\overset{1}{\cancel{3}} \cdot \overset{1}{\cancel{2}} \cdot 7}{\underset{1}{\cancel{2}} \cdot 2 \cdot \underset{1}{\cancel{3}} \cdot 5}$$ • **Divide by the common factors.**

$$= \frac{7}{10}$$ • **Write the product in simplest form.**

Division is defined as multiplication by the reciprocal. Fractions are divided by applying this definition.

2.4B Divide fractions

> **Division of Fractions**
>
> To divide two fractions, multiply the first fraction by the reciprocal of the second fraction.
>
> $$\frac{a}{b} \div \frac{c}{d} = \frac{a}{b} \cdot \frac{d}{c},$$ where $b \neq 0,$ $c \neq 0,$ and $d \neq 0$

HOW TO Divide: $\dfrac{2}{3} \div \dfrac{3}{4}$

$\dfrac{2}{3} \div \dfrac{3}{4} = \dfrac{2}{3} \cdot \dfrac{4}{3}$

- Multiply the first fraction by the reciprocal of the second fraction.

$= \dfrac{2 \cdot 4}{3 \cdot 3}$

- Multiply the numerators. Multiply the denominators.

$= \dfrac{2 \cdot 2 \cdot 2}{3 \cdot 3} = \dfrac{8}{9}$

- There are no common factors in the numerator and denominator.

Example 4 Multiply: $\dfrac{4}{15} \cdot \dfrac{5}{28}$

Solution

$\dfrac{4}{15} \cdot \dfrac{5}{28} = \dfrac{4 \cdot 5}{15 \cdot 28}$

- Multiply the numerators. Multiply the denominators.

$= \dfrac{2 \cdot 2 \cdot 5}{3 \cdot 5 \cdot 2 \cdot 2 \cdot 7}$

- Write the prime factorization of each number.

$= \dfrac{\overset{1}{\cancel{2}} \cdot \overset{1}{\cancel{2}} \cdot \overset{1}{\cancel{5}}}{3 \cdot \underset{1}{\cancel{5}} \cdot \underset{1}{\cancel{2}} \cdot \underset{1}{\cancel{2}} \cdot 7}$

- Divide by the common factors.

$= \dfrac{1}{21}$

- Write the product in simplest form.

You Try It 4 Multiply: $\dfrac{4}{21} \cdot \dfrac{7}{44}$

Your solution $\dfrac{1}{33}$

Example 5 Multiply: $\dfrac{3}{16} \cdot 4$

Solution

$\dfrac{3}{16} \cdot 4 = \dfrac{3}{16} \cdot \dfrac{4}{1}$

- Write 4 as $\dfrac{4}{1}$.

$= \dfrac{3 \cdot 4}{16 \cdot 1}$

- Multiply the fractions.

$= \dfrac{3 \cdot \overset{1}{\cancel{2}} \cdot \overset{1}{\cancel{2}}}{\underset{1}{\cancel{2}} \cdot \underset{1}{\cancel{2}} \cdot 2 \cdot 2 \cdot 1}$

- Divide by the common factors.

$= \dfrac{3}{4}$

- Write the product in simplest form.

You Try It 5 Multiply: $\dfrac{2}{15} \cdot 5$

Your solution $\dfrac{2}{3}$

Example 6 Divide: $\dfrac{3}{4} \div \dfrac{9}{10}$

Solution

$\dfrac{3}{4} \div \dfrac{9}{10} = \dfrac{3}{4} \cdot \dfrac{10}{9}$

- Multiply the first fraction by the reciprocal of the second fraction.

$= \dfrac{3 \cdot 10}{4 \cdot 9}$

$= \dfrac{\overset{1}{\cancel{3}} \cdot \overset{1}{\cancel{2}} \cdot 5}{2 \cdot 2 \cdot \underset{1}{\cancel{3}} \cdot 3}$

$= \dfrac{5}{6}$

You Try It 6 Divide: $\dfrac{1}{6} \div \dfrac{4}{9}$

Your solution $\dfrac{3}{8}$

Solutions on p. S41

R.1 Exercises

To add and subtract fractions

For Exercises 1–36, add or subtract.

1. $\dfrac{2}{7} + \dfrac{1}{7}$

$\dfrac{3}{7}$

2. $\dfrac{3}{11} + \dfrac{5}{11}$

$\dfrac{8}{11}$

3. $\dfrac{1}{2} + \dfrac{1}{2}$

1

4. $\dfrac{1}{3} + \dfrac{2}{3}$

1

5. $\dfrac{3}{8} + \dfrac{7}{8} + \dfrac{1}{8}$

$1\dfrac{3}{8}$

6. $\dfrac{5}{12} + \dfrac{7}{12} + \dfrac{1}{12}$

$1\dfrac{1}{12}$

7. $\dfrac{1}{2} + \dfrac{2}{3}$

$1\dfrac{1}{6}$

8. $\dfrac{2}{3} + \dfrac{1}{4}$

$\dfrac{11}{12}$

9. $\dfrac{3}{14} + \dfrac{5}{7}$

$\dfrac{13}{14}$

10. $\dfrac{7}{10} + \dfrac{3}{5}$

$1\dfrac{3}{10}$

11. $\dfrac{8}{15} + \dfrac{7}{20}$

$\dfrac{53}{60}$

12. $\dfrac{1}{6} + \dfrac{7}{9}$

$\dfrac{17}{18}$

13. $\dfrac{3}{20} + \dfrac{7}{30}$

$\dfrac{23}{60}$

14. $\dfrac{5}{12} + \dfrac{7}{30}$

$\dfrac{13}{20}$

15. $\dfrac{1}{3} + \dfrac{5}{6} + \dfrac{7}{9}$

$1\dfrac{17}{18}$

16. $\dfrac{2}{3} + \dfrac{5}{6} + \dfrac{7}{12}$

$2\dfrac{1}{12}$

17. $\dfrac{2}{3} + \dfrac{1}{5} + \dfrac{7}{12}$

$1\dfrac{9}{20}$

18. $\dfrac{3}{4} + \dfrac{4}{5} + \dfrac{7}{12}$

$2\dfrac{2}{15}$

19. $\dfrac{11}{12} - \dfrac{7}{12}$

$\dfrac{1}{3}$

20. $\dfrac{13}{15} - \dfrac{4}{15}$

$\dfrac{3}{5}$

21. $\dfrac{3}{4} - \dfrac{1}{8}$

$\dfrac{5}{8}$

22. $\dfrac{2}{3} - \dfrac{1}{6}$

$\dfrac{1}{2}$

23. $\dfrac{5}{9} - \dfrac{4}{15}$

$\dfrac{13}{45}$

24. $\dfrac{11}{12} - \dfrac{2}{3}$

$\dfrac{1}{4}$

25. $\dfrac{9}{20} - \dfrac{7}{20}$

$\dfrac{1}{10}$

26. $\dfrac{48}{55} - \dfrac{13}{55}$

$\dfrac{7}{11}$

27. $\dfrac{11}{24} - \dfrac{5}{24}$

$\dfrac{1}{4}$

28. $\dfrac{23}{30} - \dfrac{13}{30}$

$\dfrac{1}{3}$

29. $\dfrac{5}{7} - \dfrac{3}{14}$

$\dfrac{1}{2}$

30. $\dfrac{5}{9} - \dfrac{7}{15}$

$\dfrac{4}{45}$

31. $\dfrac{8}{15} - \dfrac{7}{20}$

$\dfrac{11}{60}$

32. $\dfrac{7}{9} - \dfrac{1}{6}$

$\dfrac{11}{18}$

33. $\dfrac{5}{6} - \dfrac{3}{4}$

$\dfrac{1}{12}$

34. $\dfrac{4}{5} - \dfrac{2}{3}$

$\dfrac{2}{15}$

35. $\dfrac{7}{8} - \dfrac{2}{3}$

$\dfrac{5}{24}$

36. $\dfrac{5}{12} - \dfrac{1}{3}$

$\dfrac{1}{12}$

Objective B   **To multiply and divide fractions**

For Exercises 37–72, multiply or divide.

37. $\dfrac{2}{3} \cdot \dfrac{7}{8}$

$\dfrac{7}{12}$

38. $\dfrac{1}{2} \cdot \dfrac{2}{3}$

$\dfrac{1}{3}$

39. $\dfrac{15}{16} \cdot \dfrac{7}{15}$

$\dfrac{7}{16}$

40. $\dfrac{3}{8} \cdot \dfrac{6}{7}$

$\dfrac{9}{28}$

41. $\dfrac{2}{5} \cdot \dfrac{5}{6}$

$\dfrac{1}{3}$

42. $\dfrac{11}{12} \cdot \dfrac{3}{5}$

$\dfrac{11}{20}$

43. $\dfrac{3}{5} \cdot \dfrac{10}{11}$

$\dfrac{6}{11}$

44. $\dfrac{6}{7} \cdot \dfrac{14}{15}$

$\dfrac{4}{5}$

45. $\dfrac{8}{9} \cdot \dfrac{27}{4}$

6

46. $\dfrac{3}{5} \cdot \dfrac{3}{10}$

$\dfrac{9}{50}$

47. $\dfrac{3}{8} \cdot \dfrac{5}{12}$

$\dfrac{5}{32}$

48. $\dfrac{3}{2} \cdot \dfrac{4}{9}$

$\dfrac{2}{3}$

49. $\dfrac{7}{8} \cdot \dfrac{3}{14}$

$\dfrac{3}{16}$

50. $\dfrac{5}{12} \cdot \dfrac{6}{7}$

$\dfrac{5}{14}$

51. $\dfrac{5}{6} \cdot \dfrac{4}{15}$

$\dfrac{2}{9}$

52. $\dfrac{5}{7} \cdot \dfrac{14}{15}$

$\dfrac{2}{3}$

53. $\dfrac{2}{3} \cdot \dfrac{5}{4} \cdot \dfrac{1}{9}$

$\dfrac{5}{54}$

54. $\dfrac{3}{4} \cdot \dfrac{5}{6} \cdot \dfrac{8}{9}$

$\dfrac{5}{9}$

55. $\dfrac{2}{3} \cdot 6$

4

56. $14 \cdot \dfrac{5}{7}$

10

57. $\dfrac{3}{7} \div \dfrac{3}{2}$

$\dfrac{2}{7}$

58. $\dfrac{3}{7} \div \dfrac{3}{7}$

1

59. $0 \div \dfrac{1}{2}$

0

60. $\dfrac{5}{24} \div \dfrac{15}{36}$

$\dfrac{1}{2}$

61. $\dfrac{2}{15} \div \dfrac{3}{5}$

$\dfrac{2}{9}$

62. $\dfrac{1}{9} \div \dfrac{2}{3}$

$\dfrac{1}{6}$

63. $\dfrac{2}{5} \div \dfrac{4}{7}$

$\dfrac{7}{10}$

64. $\dfrac{3}{8} \div \dfrac{5}{12}$

$\dfrac{9}{10}$

65. $\dfrac{1}{2} \div \dfrac{1}{4}$

2

66. $\dfrac{1}{3} \div \dfrac{1}{9}$

3

67. $\dfrac{4}{15} \div \dfrac{2}{5}$

$\dfrac{2}{3}$

68. $\dfrac{7}{15} \div \dfrac{14}{5}$

$\dfrac{1}{6}$

69. $4 \div \dfrac{2}{3}$

6

70. $\dfrac{2}{3} \div 4$

$\dfrac{1}{6}$

71. $\dfrac{3}{2} \div 3$

$\dfrac{1}{2}$

72. $3 \div \dfrac{3}{2}$

2

 **Integers**

To add and subtract integers

The rule for adding two integers depends on whether the signs of the integers are the same or different.

> **3.2A** Add integers

> **Rule for Adding Two Integers**
>
> **To add two integers with the same sign,** add the absolute values of the numbers. Then attach the sign of the addends.
>
> **To add two integers with different signs,** find the absolute values of the numbers. Subtract the smaller absolute value from the larger absolute value. Then attach the sign of the addend with the larger absolute value.

HOW TO Add: $(-5) + (-11)$

$(-5) + (-11) = -16$　•　The signs of the addends are the same.
Add the absolute values of the numbers.
$|-5| = 5, |-11| = 11, 5 + 11 = 16$
Attach the sign of the addends. (Both addends are negative. The sum is negative.)

HOW TO Add: $-16 + (-32)$

$-16 + (-32) = -48$　•　The signs of the addends are the same.
Add the absolute values of the numbers.
Attach the sign of the addends.

HOW TO Add: $7 + (-20)$

$7 + (-20) = -13$　•　The signs of the addends are different.
Find the absolute values of the numbers.
$|7| = 7, |-20| = 20$
Subtract the smaller absolute value from the larger absolute value.
$20 - 7 = 13$
Attach the sign of the number with the larger absolute value.
$(|-20| > |7|.$ Attach the negative sign.)

HOW TO Add: $82 + (-136)$

$82 + (-136) = -54$　•　The signs are different. Find the difference between the absolute values of the numbers.
$136 - 82 = 54$
Attach the sign of the number with the larger absolute value.

Opposites are used to rewrite subtraction problems as related addition problems. Notice below that the subtraction of two whole numbers is the same as addition of the opposite number.

Subtraction		*Addition of the Opposite*	
$9 - 5$	$=$	$9 + (-5)$	$= 4$
$7 - 4$	$=$	$7 + (-4)$	$= 3$
$8 - 3$	$=$	$8 + (-3)$	$= 5$

Subtraction of integers can be written as the addition of the opposite number. To subtract two integers, rewrite the subtraction expression as the first number plus the opposite of the second number.

3.2B Subtract integers

> **Rule for Subtracting Two Integers**
>
> To subtract two integers, add the opposite of the second integer to the first integer.

HOW TO Subtract: $(-14) - 62$

$(-14) - 62$
$= (-14) + (-62)$ • Rewrite the subtraction operation as the first number plus the opposite of the second number. The opposite of 62 is -62.

$= -76$ • Add.

HOW TO Subtract: $7 - (-5)$

$7 - (-5)$
$= 7 + 5$ • Rewrite the subtraction operation as the first number plus the opposite of the second number. The opposite of -5 is 5.

$= 12$ • Add.

HOW TO Subtract: $9 - 18$

$9 - 18$
$= 9 + (-18)$ • Rewrite the subtraction operation as the first number plus the opposite of the second number. The opposite of 18 is -18.

$= -9$ • Add.

When subtraction occurs several times in an expression, rewrite each subtraction as addition of the opposite, and then add.

HOW TO Subtract: $-23 - 7 - (-5)$

$-23 - 7 - (-5)$
$= -23 + (-7) + 5$ • Rewrite each subtraction as addition of the opposite.
$= -30 + 5$ • Add.
$= -25$

Example 1

Add.

a. $-6 + 17$ **b.** $-15 + (-8)$ **c.** $19 + (-26)$

Solution

a. $-6 + 17 = 11$ • The signs are different. Subtract the absolute values. The sum has the same sign as the number with the larger absolute value.

b. $-15 + (-8) = -23$ • The signs are the same. Add the absolute values. The sum has the same sign as the addends.

c. $19 + (-26) = -7$ • The signs are different. Subtract the absolute values. The sum has the same sign as the number with the larger absolute value.

You Try It 1

Add.

a. $-25 + 13$ **b.** $-41 + (-60)$ **c.** $37 + (-9)$

Your solution

a. -12 **b.** -101 **c.** 28

Example 2

Subtract.

a. $-9 - 18$ **b.** $-25 - (-13)$
c. $17 - (-40)$ **d.** $10 - (-3) - 7 + 6$

Solution

a. $-9 - 18 = -9 + (-18)$ • -9 minus $18 = -9$
  $= -27$ plus the opposite of 18. The opposite of 18 is -18.

b. $-25 - (-13) = -25 + 13$ • -25 minus $-13 =$
  $= -12$ -25 plus the opposite of -13. The opposite of -13 is 13.

c. $17 - (-40) = 17 + 40$ • 17 minus $-40 = 17$
  $= 57$ plus the opposite of -40. The opposite of -40 is 40.

d. $10 - (-3) - 7 + 6$ • Rewrite each
  $= 10 + 3 + (-7) + 6$ subtraction as addition of the opposite.

  $= 13 + (-7) + 6$ • Add the numbers.
  $= 6 + 6$
  $= 12$

You Try It 2

Subtract.

a. $-27 - 18$ **b.** $-34 - (-90)$
c. $8 - 42$ **d.** $-12 + 9 - (-5) - 4$

Your solution

a. -45 **b.** 56 **c.** -34 **d.** -2

Solutions on p. S41

Objective B   **To multiply and divide integers**

The rule for multiplying two integers depends on whether the signs of the integers are the same or different.

3.3A Multiply integers

> **Rule for Multiplying Two Integers**
>
> **To multiply two integers with the same sign,** multiply the absolute values of the numbers. The product is positive.
>
> **To multiply two integers with different signs,** multiply the absolute values of the numbers. The product is negative.

HOW TO   Multiply: $-3(-12)$

$-3(-12) = 36$ • **The signs of the factors are the same (they are both negative). Multiply the absolute values of the factors.**
$|-3| = 3, |-12| = 12, 3(12) = 36$
The product is positive.

HOW TO   Multiply: $-6(30)$

$-6(30) = -180$ • **The signs of the factors are different. Multiply the absolute values of the factors. The product is negative.**

The rule for dividing two integers depends on whether the signs of the integers are the same or different.

3.3B Divide integers

> **Rule for Dividing Two Integers**
>
> **To divide two integers with the same sign,** divide the absolute values of the numbers. The quotient is positive.
>
> **To divide two integers with different signs,** divide the absolute values of the numbers. The quotient is negative.

HOW TO   Divide: $(-24) \div (-8)$

$(-24) \div (-8) = 3$ • **The signs of the numbers are the same. Divide the absolute values of the numbers.**
$|-24| = 24, |-8| = 8, 24 \div 8 = 3$
The quotient is positive.

HOW TO   Divide: $(-44) \div 11$

$(-44) \div 11 = -4$ • **The signs of the numbers are different. Divide the absolute values of the numbers. The quotient is negative.**

Example 3

Multiply.

a. $(-5)(-12)$ **b.** $4(-9)$ **c.** $(-2)(3)(8)(-10)$

Solution

a. $(-5)(-12) = 60$ • The signs are the same. The product is positive.

b. $4(-9) = -36$ • The signs are different. The product is negative.

c. $(-2)(3)(8)(-10)$
$= -6(8)(-10)$
$= -48(-10)$
$= 480$
• Multiply the first two numbers. Then multiply the product by the third number. Continue until all the numbers have been multiplied.

You Try It 3

Multiply.

a. $-25(4)$ **b.** $-4(-61)$ **c.** $(-4)(5)(-3)(-1)$

Your solution

a. -100 **b.** 244 **c.** -60

Example 4

Divide.

a. $-18 \div (-18)$ **b.** $16 \div (-4)$ **c.** $\dfrac{-20}{-5}$

Solution

a. $-18 \div (-18) = 1$ • The signs are the same. The quotient is positive.

b. $16 \div (-4) = -4$ • The signs are different. The quotient is negative.

c. $\dfrac{-20}{-5} = 4$ • The fraction bar can be read "divided by."

$\dfrac{-20}{-5} = (-20) \div (-5)$

The signs are the same. The quotient is positive.

You Try It 4

Divide.

a. $(-30) \div 6$ **b.** $(-50) \div (-25)$ **c.** $\dfrac{32}{-8}$

Your solution

a. -5 **b.** 2 **c.** -4

Solutions on pp. S41–S42

R.2 Exercises

Objective A   **To add and subtract integers**

For Exercises 1–46, add or subtract.

1. $4 + (-9)$
-5

2. $6 + (-7)$
-1

3. $(-5) + (-12)$
-17

4. $(-8) + (-11)$
-19

5. $-5 + 8$
3

6. $-8 + 5$
-3

7. $-14 + (-6)$
-20

8. $-17 + (-3)$
-20

9. $-6 + 6$
0

10. $-19 + 19$
0

11. $64 + (-43)$
21

12. $-78 + 51$
-27

13. $8 - 15$
-7

14. $7 - 10$
-3

15. $-8 - 3$
-11

16. $-10 - 5$
-15

17. $7 - (-1)$
8

18. $4 - (-5)$
9

19. $-9 - (-9)$
0

20. $-13 - (-13)$
0

21. $-10 - 15$
-25

22. $-8 - 7$
-15

23. $(-11) - (-2)$
-9

24. $(-8) - (-5)$
-3

25. $6 - (-16)$
22

26. $4 - (-26)$
30

27. $(-12) - (-6)$
-6

28. $-3 - (-17)$
14

29. $8 - (-8)$
16

30. $(-32) - 46$
-78

31. $45 - 77$
-32

32. $-82 - (-16)$
-66

33. $0 + (-15)$
-15

34. $-18 + 0$
-18

35. $(-21) - (-7)$
-14

36. $-13 - (-4)$
-9

37. $5 - (-6)$
11

38. $12 - (-2)$
14

39. $6 - (-10)$
16

40. $13 - (-5)$
18

41. $7 + 3 - (-3)$
13

42. $(-9) - 8 + (-6)$
23

43. $-2 + (-5) - (-12)$
5

44. $-3 - 5 + 8 - 1$
-1

45. $-4 + 6 - 9 - 2$
-9

46. $7 - (-3) - 6 + 5$
9

Objective B **To multiply and divide integers**

For Exercises 47–94, multiply or divide.

47. $-3 \cdot 7$
−21

48. $-6 \cdot 8$
−48

49. $-5(-7)$
35

50. $-9(-2)$
18

51. $4(-9)$
−36

52. $3(-11)$
−33

53. $-10(5)$
−50

54. $-8(4)$
−32

55. $(-7)(-7)$
49

56. $(-4)(-7)$
28

57. $(-9)(0)$
0

58. $-16(1)$
−16

59. $15(4)$
60

60. $42(3)$
126

61. $-21(6)$
−126

62. $-14(2)$
−28

63. $(-3)(-27)$
81

64. $(-6)(-32)$
192

65. $8(-24)$
−192

66. $7(-30)$
−210

67. $-5 \cdot (17)$
−85

68. $-6 \cdot (22)$
−132

69. $-7(-14)$
98

70. $-4(-62)$
248

71. $2 \cdot (-8) \cdot 5$
−80

72. $5 \cdot 6 \cdot (-1)$
−30

73. $-2(-6)(-3)(4)$
−144

74. $-1(4)(-9)(-2)$
−72

75. $12 \div (-4)$
−3

76. $18 \div (-6)$
−3

77. $(-81) \div (-9)$
9

78. $(-48) \div (-6)$
8

79. $0 \div (-4)$
0

80. $-36 \div 1$
−36

81. $77 \div (-7)$
−11

82. $-50 \div (-10)$
5

83. $\dfrac{36}{-4}$
−9

84. $\dfrac{40}{-8}$
−5

85. $\dfrac{-66}{-3}$
22

86. $\dfrac{-100}{-20}$
5

87. $-84 \div (-6)$
14

88. $-112 \div (-7)$
16

89. $-48 \div 0$
Undefined

90. $(-210) \div (-210)$
1

91. $-126 \div 6$
−21

92. $-160 \div (-5)$
32

93. $(-240) \div 6$
−40

94. $(-96) \div (-8)$
12

R.3 Rational Numbers

Objective A   To add and subtract rational numbers

In this section, operations with rational numbers are discussed. A **rational number** is the quotient of two integers.

Rational Numbers

A rational number is a number that can be written in the form $\frac{a}{b}$, where a and b are integers and $b \neq 0$.

3.4A Add or subtract rational numbers

We begin by reviewing addition of rational numbers in fractional form. If an addend is a fraction containing a negative sign, rewrite the fraction with the negative sign in the numerator. Then add the numerators and place the sum over the common denominator.

HOW TO   Add: $-\dfrac{5}{6} + \dfrac{3}{8}$

$$-\frac{5}{6} + \frac{3}{8} = -\frac{20}{24} + \frac{9}{24}$$
• The LCM of the denominators 6 and 8 is 24. Write each fraction with a denominator of 24.

$$= \frac{-20}{24} + \frac{9}{24}$$
• Write the negative sign in the numerator.

$$= \frac{-20 + 9}{24}$$
• Add the fractions.

$$= \frac{-11}{24}$$
• Simplify the numerator.

$$= -\frac{11}{24}$$
• Write the negative sign in front of the fraction.

HOW TO   Add: $-\dfrac{3}{5} + \left(-\dfrac{1}{3}\right)$

$$-\frac{3}{5} + \left(-\frac{1}{3}\right) = -\frac{9}{15} + \left(-\frac{5}{15}\right)$$
• The LCM of the denominators 5 and 3 is 15. Write each fraction with a denominator of 15.

$$= \frac{-9}{15} + \frac{-5}{15}$$
• Write the negative signs in the numerators.

$$= \frac{-9 + (-5)}{15}$$
• Add the fractions.

$$= \frac{-14}{15}$$
• Simplify the numerator.

$$= -\frac{14}{15}$$
• Write the negative sign in front of the fraction.

To subtract fractions with negative signs, rewrite the fractions with the negative signs in the numerators.

HOW TO Subtract: $-\dfrac{2}{3} - \dfrac{5}{8}$

$$-\dfrac{2}{3} - \dfrac{5}{8} = \dfrac{16}{24} - \dfrac{15}{24}$$ • The LCM of the denominators 3 and 8 is 24. Write each fraction with a denominator of 24.

$$= \dfrac{-16}{24} + \dfrac{-15}{24}$$ • Rewrite subtraction as addition of the opposite. Write the negative signs in the numerators.

$$= \dfrac{-16 + (-15)}{24}$$ • Add the fractions.

$$= \dfrac{-31}{24}$$ • Simplify the numerator.

$$= -\dfrac{31}{24} = -1\dfrac{7}{24}$$ • Write the negative sign in front of the fraction.

HOW TO Subtract: $-\dfrac{1}{6} - \left(-\dfrac{2}{9}\right)$

$$-\dfrac{1}{6} - \left(-\dfrac{2}{9}\right) = -\dfrac{1}{6} + \dfrac{2}{9}$$ • Rewrite subtraction as addition of the opposite.

$$= -\dfrac{3}{18} + \dfrac{4}{18}$$ • Write the fractions as equivalent fractions with a common denominator.

$$= \dfrac{-3}{18} + \dfrac{4}{18}$$ • Write the negative sign in the numerator.

$$= \dfrac{-3 + 4}{18}$$ • Add the fractions.

$$= \dfrac{1}{18}$$ • Simplify the numerator.

The sign rules for adding and subtracting decimals are the same rules used to add and subtract integers.

HOW TO Simplify: $-29.871 + 34.06$

$34.06 - 29.871 = 4.189$ • The signs of the addends are different. Subtract the smaller absolute value from the larger absolute value.

$|34.06| > |-29.871|$
$-29.871 + 34.06 = 4.189$ • Attach the sign of the number with the larger absolute value. The sum is positive.

Recall that the opposite of n is $-n$ and the opposite of $-n$ is n. To find the opposite of a number, change the sign of the number.

HOW TO Simplify: $-3.92 - 21.7$

$$-3.92 - 21.7$$
$$= -3.92 + (-21.7)$$
$$= -25.62$$

- Rewrite subtraction as addition of the opposite. The opposite of 21.7 is -21.7.
- The signs of the addends are the same. Add the absolute values of the numbers. Attach the sign of the addends.

Example 1

Add: $-\dfrac{17}{20} + \dfrac{4}{5}$

Solution

$$-\dfrac{17}{20} + \dfrac{4}{5} = -\dfrac{17}{20} + \dfrac{16}{20}$$

$$= \dfrac{-17}{20} + \dfrac{16}{20}$$

$$= \dfrac{-17 + 16}{20}$$

$$= \dfrac{-1}{20}$$

$$= -\dfrac{1}{20}$$

- Write the fractions with a common denominator.
- Write the negative sign in the numerator.
- Add the fractions.
- Simplify the numerator.
- Write the negative sign in front of the fraction.

You Try It 1

Add: $-\dfrac{1}{4} + \left(-\dfrac{3}{8}\right)$

Your solution

$$-\dfrac{5}{8}$$

Example 2

Subtract: $-\dfrac{7}{20} - \dfrac{1}{5}$

Solution

$$-\dfrac{7}{20} - \dfrac{1}{5} = -\dfrac{7}{20} - \dfrac{4}{20}$$

$$= \dfrac{-7}{20} + \dfrac{-4}{20}$$

$$= \dfrac{-7 + (-4)}{20}$$

$$= \dfrac{-11}{20}$$

$$= -\dfrac{11}{20}$$

- Write the fractions with a common denominator.
- Rewrite subtraction as addition of the opposite. Write the negative signs in the numerators.
- Add the fractions.
- Simplify the numerator.
- Write the negative sign in front of the fraction.

You Try It 2

Subtract: $-\dfrac{3}{8} - \left(-\dfrac{1}{6}\right)$

Your solution

$$-\dfrac{5}{24}$$

Solutions on p. S42

Example 3

Simplify: $-3.97 - (-10.8)$

Solution

$-3.97 - (-10.8)$

$= -3.97 + 10.8$ • Rewrite subtraction as addition of the opposite.

$= 6.83$ • Subtract the absolute values of the numbers. The sum has the same sign as the number with the larger absolute value.

You Try It 3

Simplify: $4.69 - 12.5$

Your solution

-7.81

Solution on p. S42

Objective B To multiply and divide rational numbers

VIDEO & DVD TUTOR WEB SSM

3.4B Multiply or divide rational numbers

The product of two rational numbers written in fractional form is the product of the numerators over the product of the denominators. The sign rules are the same rules used to multiply integers.

> **The product of two numbers with the same sign is positive.**
> **The product of two numbers with different signs is negative.**

HOW TO Multiply: $-\dfrac{3}{8} \cdot \dfrac{4}{15}$

$-\dfrac{3}{8} \cdot \dfrac{4}{15} = -\left(\dfrac{3}{8} \cdot \dfrac{4}{15}\right)$ • The signs are different. The product is negative.

$= -\dfrac{3 \cdot 4}{8 \cdot 15}$ • Multiply the numerators. Multiply the denominators.

$= -\dfrac{3 \cdot 2 \cdot 2}{2 \cdot 2 \cdot 2 \cdot 3 \cdot 5}$ • Write the product is simplest form.

$= -\dfrac{1}{10}$

The sign rules for dividing rational numbers are the same rules used to divide integers.

> **The quotient of two numbers with the same sign is positive.**
> **The quotient of two numbers with different signs is negative.**

HOW TO Divide: $-\dfrac{3}{8} \div \left(-\dfrac{4}{5}\right)$

$-\dfrac{3}{8} \div \left(-\dfrac{4}{5}\right) = \dfrac{3}{8} \div \dfrac{4}{5}$

- **The signs are the same.**
  The quotient is positive.

$= \dfrac{3}{8} \cdot \dfrac{5}{4}$

- **Rewrite division as multiplication by the reciprocal.**

$= \dfrac{3 \cdot 5}{8 \cdot 4}$

- **Multiply the fractions.**

$= \dfrac{3 \cdot 5}{2 \cdot 2 \cdot 2 \cdot 2 \cdot 2}$

$= \dfrac{15}{32}$

The sign rules for multiplying and dividing decimals are the same rules used to multiply and divide integers.

HOW TO Multiply: $(-3.25)(-10.1)$

$(-3.25)(-10.1) = 32.825$

- **The signs are the same.**
  The product is positive.
  Multiply the absolute values of the numbers.

HOW TO Divide: $-29.4 \div 3.5$

$-29.4 \div 3.5 = -8.4$

- **The signs are different.**
  The quotient is negative.
  Divide the absolute values of the numbers.

Example 4

Multiply: $\left(-\dfrac{5}{8}\right)\left(-\dfrac{3}{5}\right)$

Solution

$\left(-\dfrac{5}{8}\right)\left(-\dfrac{3}{5}\right) = \left(\dfrac{5}{8}\right)\left(\dfrac{3}{5}\right)$

- **The signs are the same.**
  The product is positive.

$= \dfrac{5 \cdot 3}{8 \cdot 5}$

- **Multiply the numerators.**
  Multiply the denominators.

$= \dfrac{5 \cdot 3}{2 \cdot 2 \cdot 2 \cdot 5}$

- **Write the product in simplest form.**

$= \dfrac{3}{8}$

You Try It 4

Multiply: $\dfrac{10}{11}\left(-\dfrac{2}{5}\right)$

Your solution

$-\dfrac{4}{11}$

Solution on p. S42

Example 5

Divide: $-\dfrac{9}{16} \div \dfrac{3}{4}$

Solution

$-\dfrac{9}{16} \div \dfrac{3}{4} = -\left(\dfrac{9}{16} \div \dfrac{3}{4}\right)$

$= -\left(\dfrac{9}{16} \cdot \dfrac{4}{3}\right)$

$= -\dfrac{9 \cdot 4}{16 \cdot 3}$

$= -\dfrac{3 \cdot 3 \cdot 2 \cdot 2}{2 \cdot 2 \cdot 2 \cdot 2 \cdot 3}$

$= -\dfrac{3}{4}$

- The signs are different. The quotient is negative.
- Rewrite division as multiplication by the reciprocal.
- Multiply the fractions.

You Try It 5

Divide: $-\dfrac{3}{8} \div \left(-\dfrac{1}{2}\right)$

Your solution

$\dfrac{3}{4}$

Example 6

Multiply: $(-8.9)(0.25)$

Solution

$(-8.9)(0.25) = -2.225$

- The signs are different. The product is negative. Multiply the absolute values of the numbers.

You Try It 6

Multiply: $(-3.6)(-1.45)$

Your solution

5.22

Example 7

Divide: $(-16.2) \div (-3.6)$

Solution

$(-16.2) \div (-3.6) = 4.5$

- The signs are the same. The quotient is positive. Divide the absolute values of the numbers.

You Try It 7

Divide: $5.04 \div (-8.4)$

Your solution

-0.6

Solutions on p. S42

Objective C   **To evaluate exponential expressions**

Recall that an exponent indicates repeated multiplication of the same factor. For example,

$$3^5 = 3 \cdot 3 \cdot 3 \cdot 3 \cdot 3$$

The **exponent,** 5, indicates how many times the **base,** 3, occurs as a factor in the multiplication.

The base of an exponential expression can be any rational number, for example, 0.5^4. To evaluate this expression, write the factor as many times as indicated by the exponent and then multiply.

$$0.5^4 = 0.5(0.5)(0.5)(0.5) = 0.25(0.5)(0.5) = 0.125(0.5) = 0.0625$$

Example 8

Simplify: $\left(-\dfrac{3}{4}\right)^3 \cdot 8^2$

Solution

$\left(-\dfrac{3}{4}\right)^3 \cdot 8^2$

$= \left(-\dfrac{3}{4}\right)\left(-\dfrac{3}{4}\right)\left(-\dfrac{3}{4}\right) \cdot 8 \cdot 8$ • Write each factor the number of times indicated by the exponent.

$= -\left(\dfrac{3}{4} \cdot \dfrac{3}{4} \cdot \dfrac{3}{4} \cdot \dfrac{8}{1} \cdot \dfrac{8}{1}\right)$ • The product is negative.

$= -\dfrac{3 \cdot 3 \cdot 3 \cdot 8 \cdot 8}{4 \cdot 4 \cdot 4 \cdot 1 \cdot 1}$ • Multiply the fractions.

$= -27$ • Simplify.

You Try It 8

Simplify: $\left(\dfrac{2}{9}\right)^2 \cdot (-3)^4$

Your solution

4

Solution on p. S42

Objective D   **To use the Order of Operations Agreement to simplify expressions**

Whenever an expression contains more than one operation, the operations must be performed in a specified order, as listed on the next page in the Order of Operations Agreement.

The Order of Operations Agreement

Step 1 Do all operations inside group symbols. Grouping symbols include parentheses ( ), brackets [ ], and absolute value symbols | |.

Step 2 Simplify any numerical expressions containing exponents.

Step 3 Do multiplication and division as they occur from left to right.

Step 4 Do addition and subtraction as they occur from left to right.

3.5A Use the Order of Operations Agreement

The Order of Operations Agreement is used to simplify the expression in the following example.

HOW TO Simplify: $0.2(2.5 - 5.6) + (1.4)^2$

$$0.2(2.5 - 5.6) + (1.4)^2$$
$$= 0.2(-3.1) + (1.4)^2 \qquad \bullet \text{ Perform operations inside parentheses.}$$
$$= 0.2(-3.1) + 1.96 \qquad \bullet \text{ Simplify the exponential expression.}$$
$$= -0.62 + 1.96 \qquad \bullet \text{ Do the multiplication.}$$
$$= 1.34 \qquad \bullet \text{ Do the addition.}$$

Example 9

Simplify: $3 \div \left(\dfrac{1}{4} - \dfrac{1}{2} \right)^2 - 5$

Solution

$$3 \div \left(\frac{1}{4} - \frac{1}{2} \right)^2 - 5 \qquad \bullet \text{ Use the Order of Operations Agreement.}$$

$$= 3 \div \left(-\frac{1}{4} \right)^2 - 5 \qquad \bullet \text{ Perform the operation inside the parentheses.}$$

$$= 3 \div \frac{1}{16} - 5 \qquad \bullet \text{ Simplify the exponential expression.}$$

$$= 3(16) - 5 \qquad \bullet \text{ Rewrite division as multiplication by the reciprocal.}$$

$$= 48 - 5 \qquad \bullet \text{ Do the multiplication.}$$
$$= 43 \qquad \bullet \text{ Do the subtraction.}$$

You Try It 9

Simplify: $7 \div \left(\dfrac{1}{7} - \dfrac{3}{14} \right) - 9$

Your solution
-107

Solution on p. S42

R.3 Exercises

Objective A   **To add and subtract rational numbers**

For Exercises 1–27, simplify.

1. $-\dfrac{3}{4} + \dfrac{2}{3}$

$-\dfrac{1}{12}$

2. $-\dfrac{5}{12} + \dfrac{3}{8}$

$-\dfrac{1}{24}$

3. $\dfrac{2}{5} + \left(-\dfrac{11}{15}\right)$

$-\dfrac{1}{3}$

4. $\dfrac{1}{4} + \left(-\dfrac{1}{7}\right)$

$\dfrac{3}{28}$

5. $-\dfrac{1}{2} - \dfrac{3}{8}$

$-\dfrac{7}{8}$

6. $-\dfrac{5}{6} - \dfrac{1}{9}$

$-\dfrac{17}{18}$

7. $-\dfrac{3}{10} - \dfrac{4}{5}$

$-1\dfrac{1}{10}$

8. $-\dfrac{5}{12} - \left(-\dfrac{2}{3}\right)$

$\dfrac{1}{4}$

9. $-\dfrac{5}{8} - \left(-\dfrac{7}{12}\right)$

$-\dfrac{1}{24}$

10. $-\dfrac{3}{4} - \left(-\dfrac{5}{16}\right)$

$-\dfrac{7}{16}$

11. $-\dfrac{2}{3} + \left(-\dfrac{1}{12}\right)$

$-\dfrac{3}{4}$

12. $-\dfrac{2}{5} + \left(-\dfrac{4}{15}\right)$

$-\dfrac{2}{3}$

13. $\dfrac{3}{8} + \left(-\dfrac{1}{2}\right) + \dfrac{7}{12}$

$\dfrac{11}{24}$

14. $-\dfrac{7}{12} + \dfrac{2}{3} + \left(-\dfrac{4}{5}\right)$

$-\dfrac{43}{60}$

15. $\dfrac{2}{3} + \left(-\dfrac{5}{6}\right) + \dfrac{1}{4}$

$\dfrac{1}{12}$

16. $-\dfrac{5}{8} + \dfrac{3}{4} + \dfrac{1}{2}$

$\dfrac{5}{8}$

17. $-42.1 - 8.6$

-50.7

18. $-6.57 - 8.933$

-15.503

19. $5.73 - 9.042$
-3.312

20. $-31.894 + 7.5$
-24.394

21. $1.09 - (-8.3)$
9.39

22. $-8 - (-10.37)$
2.37

23. $-19 - (-2.65)$
-16.35

24. $3.18 - 5.72 - 6.4$
-8.94

25. $-12.3 - 4.07 + 6.82$
-9.55

26. $-8.9 + 7.36 - 14.2$
-15.74

27. $-5.6 - (-3.82) - 17.409$
-19.189

28. Without simplifying, which is greater, $\dfrac{5}{8} - \left(-\dfrac{5}{6}\right)$ or $-\dfrac{5}{6} - \dfrac{5}{9}$?

$\dfrac{5}{8} - \left(-\dfrac{5}{6}\right)$, because $\dfrac{5}{8} - \left(-\dfrac{5}{6}\right) = \dfrac{5}{8} + \dfrac{5}{6}$, so the difference is positive, whereas the

difference $-\dfrac{5}{6} - \dfrac{5}{9}$ is negative.

29. Without simplifying, which is greater, $-\dfrac{1}{8} - \dfrac{3}{4}$ or $\dfrac{11}{12} - \left(-\dfrac{1}{4}\right)$?

$\dfrac{11}{12} - \left(-\dfrac{1}{4}\right)$, because $\dfrac{11}{12} - \left(-\dfrac{1}{4}\right) = \dfrac{11}{12} + \dfrac{1}{4}$, so the difference is positive, whereas

the difference $-\dfrac{1}{8} - \dfrac{3}{4}$ is negative.

Objective B   **To multiply and divide rational numbers**

For Exercises 30–68, simplify.

30. $-\dfrac{6}{7} \cdot \dfrac{11}{12}$

$-\dfrac{11}{14}$

31. $\dfrac{3}{8} \cdot \left(-\dfrac{2}{3}\right)$

$-\dfrac{1}{4}$

32. $\dfrac{5}{6} \cdot \left(-\dfrac{2}{5}\right)$

$-\dfrac{1}{3}$

33. $\left(-\dfrac{4}{15}\right)\left(-\dfrac{3}{8}\right)$

$\dfrac{1}{10}$

34. $\left(-\dfrac{3}{4}\right)\left(-\dfrac{2}{9}\right)$

$\dfrac{1}{6}$

35. $-\dfrac{3}{4} \cdot \dfrac{1}{2}$

$-\dfrac{3}{8}$

36. $-\dfrac{8}{15} \cdot \dfrac{5}{12}$

$-\dfrac{2}{9}$

37. $-\dfrac{7}{12} \cdot \dfrac{5}{8} \cdot \dfrac{16}{25}$

$-\dfrac{7}{30}$

38. $\dfrac{5}{12} \cdot \left(-\dfrac{1}{3}\right) \cdot \left(-\dfrac{8}{15}\right)$

$\dfrac{2}{27}$

39. $\left(-\dfrac{3}{5}\right) \cdot \dfrac{1}{2} \cdot \left(-\dfrac{5}{8}\right)$

$\dfrac{3}{16}$

40. $\dfrac{5}{6} \cdot \left(-\dfrac{2}{3}\right) \cdot \dfrac{3}{25}$

$-\dfrac{1}{15}$

41. $12 \cdot \left(-\dfrac{5}{8}\right)$

$-7\dfrac{1}{2}$

42. $24\left(-\dfrac{3}{8}\right)$

-9

43. $-9 \cdot \dfrac{7}{15}$

$-4\dfrac{1}{5}$

44. $\dfrac{1}{3} \cdot (-9)$

-3

45. $-\dfrac{5}{2} \cdot 4$

-10

46. $\dfrac{4}{7} \div \left(-\dfrac{4}{7}\right)$

-1

47. $\left(-\dfrac{3}{8}\right) \div \dfrac{7}{8}$

$-\dfrac{3}{7}$

48. $-\dfrac{5}{16} \div \left(-\dfrac{3}{8}\right)$

$\dfrac{5}{6}$

49. $\left(-\dfrac{3}{4}\right) \div \left(-\dfrac{5}{6}\right)$

$\dfrac{9}{10}$

50. $\dfrac{3}{4} \div (-6)$

$-\dfrac{1}{8}$

51. $-\dfrac{2}{3} \div 8$

$-\dfrac{1}{12}$

52. $\dfrac{5}{12} \div \left(-\dfrac{15}{32}\right)$

$-\dfrac{8}{9}$

53. $\dfrac{3}{8} \div \left(-\dfrac{5}{12}\right)$

$-\dfrac{9}{10}$

54. $-5.2(0.8)$

-4.16

55. $(-2.1)(-0.7)$

1.47

56. $(-6.3)(-2.4)$

15.12

57. $(1.9)(-3.7)$

-7.03

58. $-1.3(4.2)$

-5.46

59. $-8.1(-7.5)$

60.75

60. $1.31(-0.006)$

-0.00786

61. $-10(0.59)$

-5.9

62. $(-100)(4.73)$

-473

63. $27.08 \div (-0.4)$

-67.7

64. $-8.919 \div 0.9$

-9.91

65. $(-3.312) \div (-0.8)$

4.14

66. $84.66 \div (-1.7)$

-49.8

67. $-2.501 \div 0.41$

-6.1

68. $1.003 \div (-0.59)$

-1.7

For Exercises 69–71, divide. Round to the nearest tenth.

69. $-6.824 \div 0.053$

-128.8

70. $0.0416 \div (-0.53)$

-0.1

71. $(-31.792) \div (-0.86)$

37.0

72. Without simplifying, which is greater, $\left(-\dfrac{8}{9}\right)\left(-\dfrac{3}{4}\right)$ or $-\dfrac{5}{16} \div \dfrac{3}{8}$?

$\left(-\dfrac{8}{9}\right)\left(-\dfrac{3}{4}\right)$, because the product is positive, whereas the quotient $-\dfrac{5}{16} \div \dfrac{3}{8}$ is negative.

73. Without simplifying, which is greater, $-\dfrac{5}{6} \div (-5)$ or $-\dfrac{3}{4}\left(\dfrac{2}{9}\right)$?

$-\dfrac{5}{6} \div (-5)$, because the quotient is positive, whereas the product $-\dfrac{3}{4}\left(\dfrac{2}{9}\right)$ is negative.

Objective C   **To evaluate exponential expressions**

For Exercises 74–81, simplify.

74. $\left(-\dfrac{1}{6}\right)^3$
     75. $\left(-\dfrac{2}{7}\right)^3$
     76. $(2.25)^2$
     77. $(3.5)^2$

$-\dfrac{1}{216}$ $-\dfrac{8}{343}$ 5.0625   12.25

78. $\left(\dfrac{4}{5}\right)^4 \cdot \left(-\dfrac{5}{8}\right)^3$
   79. $\left(-\dfrac{9}{11}\right)^2 \cdot \left(\dfrac{1}{3}\right)^4$
   80. $-4 \cdot \left(\dfrac{4}{7}\right)^2 \cdot \left(-\dfrac{3}{4}\right)^3$
   81. $-3 \cdot \left(\dfrac{2}{5}\right)^2 \cdot \left(-\dfrac{1}{6}\right)^2$

$-\dfrac{1}{10}$ $\dfrac{1}{121}$ $\dfrac{27}{49}$ $-\dfrac{1}{75}$

Objective D   **To use the Order of Operations Agreement to simplify expressions**

For Exercises 82–93, simplify.

82. $(0.2)^2 \cdot (-0.5) + 1.72$
     83. $0.3(1.7 - 4.8) + (1.2)^2$
     84. $(1.8)^2 - 2.52 \div (1.8)$

1.7   0.51   1.84

85. $(1.65 - 1.05)^2 \div 0.4 + 0.8$
   86. $\dfrac{7}{12} + \dfrac{5}{6}\left(\dfrac{1}{6} - \dfrac{2}{3}\right)$
   87. $-\dfrac{3}{4}\left(\dfrac{11}{12} - \dfrac{7}{8}\right) + \dfrac{5}{16}$

1.7 $\dfrac{1}{6}$ $\dfrac{9}{32}$

88. $\dfrac{11}{16} - \left(-\dfrac{3}{4}\right)^2 + \dfrac{7}{8}$
   89. $\left(-\dfrac{2}{3}\right)^2 - \dfrac{7}{18} + \dfrac{5}{6}$
   90. $\left(-\dfrac{1}{3}\right)^2 \cdot \left(-\dfrac{9}{4}\right) + \dfrac{3}{4}$

1       $\dfrac{8}{9}$ $\dfrac{1}{2}$

91. $\left(-\dfrac{2}{3}\right)^2 + \left(-\dfrac{1}{6}\right) \div \dfrac{3}{8}$
   92. $\left(\dfrac{1}{3} - \dfrac{5}{6}\right) + \dfrac{7}{8} \div \left(-\dfrac{1}{2}\right)^3$
   93. $\left(-\dfrac{1}{4}\right)^2 \div \left(\dfrac{1}{2} - \dfrac{3}{4}\right) + \dfrac{3}{8}$

0       $-7\dfrac{1}{2}$ $\dfrac{1}{8}$

94. Arrange the expressions in order from greatest value to least value.

$$16 - 3(3 - 8) \div 5$$
$$4(-3) \div [2(6 - 7)^2]$$
$$18 \div (-2) + (-3)^2 - (-15)$$

$16 - 3(3 - 8) \div 5 > 18 \div (-2) + (-3)^2 - (-15) > 4(-3) \div [2(6 - 7)^2] \quad [19 > 15 > -6]$

95. Arrange the expressions in order from greatest value to least value.

$$20 \div (6 - 2^4) + (-5)$$
$$18 \div |2^3 - 9| + (-3)$$
$$16 + 15 \div (-5) - (-4)$$

$16 + 15 \div (-5) - (-4) > 18 \div |2^3 - 9| + (-3) > 20 \div (6 - 2^4) + (-5) \quad [17 > 15 > -7]$

R.4 Equations

Objective A

To solve a first-degree equation in one variable

An **equation** expresses the equality of two mathematical expressions. Each of the equations below is a **first-degree equation in one variable.** *First degree* means that the variable has an exponent of 1.

$$x + 11 = 14$$
$$3a + 5 = 8a$$
$$2(6y - 1) = 3$$

A **solution** of an equation is a number that, when substituted for the variable, results in a true equation.

3 is a solution of the equation $x + 4 = 7$ because $3 + 4 = 7$.
9 is not a solution of the equation $x + 4 = 7$ because $9 + 4 \neq 7$.

> Solve first-degree equations in one variable
> **5.1B, 5.1C, 5.2A, 5.3A, 5.3B**

To **solve an equation** means to find a solution of the equation. In solving an equation, the goal is to rewrite the given equation with the variable alone on one side of the equation and a constant term on the other side of the equation; the constant term is the solution of the equation. The following properties of equations are used to rewrite equations in this form.

Properties of Equations

Addition Property of Equations
The same number can be added to each side of an equation without changing the solution of the equation. In symbols, the equation $a = b$ has the same solution as the equation $a + c = b + c$.

Multiplication Property of Equations
Each side of an equation can be multiplied by the same nonzero number without changing the solution of the equation. In symbols, if $c \neq 0$, then the equation $a = b$ has the same solution as the equation $ac = bc$.

TAKE NOTE

Subtraction is defined as addition of the opposite.

$$a - b = a + (-b)$$

The Addition Property of Equations is used to remove a term from one side of an equation by adding the opposite of that term to each side of the equation. Because subtraction is defined in terms of addition, the Addition Property of Equations also makes it possible to subtract the same number from each side of an equation without changing the solution of the equation.

For example, to solve the equation $t + 9 = -4$, subtract the constant term (9) from each side of the equation.

$$t + 9 = -4$$
$$t + 9 - 9 = -4 - 9$$
$$t = -13$$

Now the variable is alone on one side of the equation and a constant term (-13) is on the other side. The solution is the constant. The solution is -13.

The Multiplication Property of Equations is used to remove a coefficient by multiplying each side of the equation by the reciprocal of the coefficient. Because division is defined in terms of multiplication, each side of an equation can be divided by the same nonzero number without changing the solution of the equation.

For example, to solve the equation $-5q = 120$, divide each side of the equation by the coefficient -5.

$$-5q = 120$$
$$\frac{-5q}{-5} = \frac{120}{-5}$$
$$q = -24$$

Now the variable is alone on one side of the equation and a constant (-24) is on the other side. The solution is the constant. The solution is -24.

In solving more complicated first-degree equations in one variable, use the following sequence of steps.

Steps for Solving a First-Degree Equation in One Variable

Step 1 Use the Distributive Property to remove parentheses.
Step 2 Combine any like terms on the right side of the equation and any like terms on the left side of the equation.
Step 3 Use the Addition Property to rewrite the equation with only one variable term.
Step 4 Use the Addition Property to rewrite the equation with only one constant term.
Step 5 Use the Multiplication Property to rewrite the equation with the variable alone on one side of the equation and a constant on the other side of the equation.

If one of the above steps is not needed to solve a given equation, proceed to the next step.

Example 1

Solve: $3x + 5 - 4x = 6$

Solution

$3x + 5 - 4x = 6$

$5 - x = 6$ • **Step 2**

$5 - 5 - x = 6 - 5$ • **Step 4**

$-x = 1$

$\dfrac{-x}{-1} = \dfrac{1}{-1}$ • **Step 5**

$x = -1$

The solution is -1.

You Try It 1

Solve: $5x + 3 - 7x = 9$

Your solution

-3

Example 2

Solve: $5x + 9 = 23 - 2x$

Solution

$5x + 9 = 23 - 2x$

$5x + 2x + 9 = 23 - 2x + 2x$ • **Step 3**

$7x + 9 = 23$

$7x + 9 - 9 = 23 - 9$ • **Step 4**

$7x = 14$

$\dfrac{7x}{7} = \dfrac{14}{7}$ • **Step 5**

$x = 2$

The solution is 2.

You Try It 2

Solve: $4x + 3 = 7x + 9$

Your solution

-2

Example 3

Solve: $8x - 3(4x - 5) = -2x + 6$

Solution

$8x - 3(4x - 5) = -2x + 6$

$8x - 12x + 15 = -2x + 6$ • **Step 1**

$-4x + 15 = -2x + 6$ • **Step 2**

$-4x + 2x + 15 = -2x + 2x + 6$ • **Step 3**

$-2x + 15 = 6$

$-2x + 15 - 15 = 6 - 15$ • **Step 4**

$-2x = -9$

$\dfrac{-2x}{-2} = \dfrac{-9}{-2}$ • **Step 5**

$x = \dfrac{9}{2}$

The solution is $\dfrac{9}{2}$.

You Try It 3

Solve: $4 - (5x - 8) = 4x + 3$

Your solution

1

Solutions on p. S43

R.4 Exercises

Objective A   **To solve a first-degree equation in one variable**

For Exercises 1–36, solve.

1. $x + 7 = -5$
  −12

2. $9 + b = 21$
  12

3. $-9 = z - 8$
  −1

4. $b - 11 = 11$
  22

5. $-48 = 6z$
  −8

6. $-9a = -108$
  12

7. $-\dfrac{3}{4}x = 15$
  −20

8. $\dfrac{5}{2}x = -10$
  −4

9. $-\dfrac{x}{4} = -2$
  8

10. $\dfrac{2x}{5} = -8$
  −20

11. $3x + 8 = 17$
  3

12. $2 + 5a = 12$
  2

13. $5 = 3x - 10$
  5

14. $4 = 3 - 5x$
  $-\dfrac{1}{5}$

15. $\dfrac{2}{3}x + 5 = 3$
  −3

16. $-\dfrac{1}{2}x + 4 = 1$
  6

17. $2b + 6 - 3b = 4$
  2

18. $3x + 4 - 5x = 8$
  −2

19. $4 - 2b = 2 - 4b$
  −1

20. $4y - 10 = 6 + 2y$
  8

21. $5x - 3 = 9x - 7$
  1

22. $5x + 7 = 8x + 5$
  $\dfrac{2}{3}$

23. $2 - 6y = 5 - 7y$
  3

24. $4b + 15 = 3 - 2b$
  −2

25. $2(x + 1) + 5x = 23$
  3

26. $9n - 15 = 3(2n - 1)$
  4

27. $7a - (3a - 4) = 12$
  2

28. $5(3 - 2y) = 3 - 4y$
  2

29. $9 - 7x = 4(1 - 3x)$
  −1

30. $2(3b + 5) - 1 = 10b + 1$
  2

31. $2z - 2 = 5 - (9 - 6z)$
  $\dfrac{1}{2}$

32. $4a + 3 = 7 - (5 - 8a)$
  $\dfrac{1}{4}$

33. $5(6 - 2x) = 2(5 - 3x)$
  5

34. $4(3y + 1) = 2(y - 8)$
  −2

35. $2(3b - 5) = 4(6b - 2)$
  $-\dfrac{1}{9}$

36. $3(x - 4) = 1 - (2x - 7)$
  4

Appendix

The Metric System of Measurement

International trade, or trade among nations, is a vital and growing segment of business in the world today. The opening of McDonald's restaurants around the globe is testimony to the expansion of international business.

The United States, as a nation, is dependent on world trade. And world trade is dependent on internationally standardized units of measurement: the metric system. The Third International Mathematics and Science Study (TIMSS) compared the performance of half a million students from 41 countries at five different grade levels on tests of their mathematics and science knowledge. One area of mathematics in which the U.S. average was below the international average was measurement, due in large part to the fact that the units cited in the questions were metric units. Because the United States has not yet converted to the metric system, its citizens are less familiar with it.

In this Appendix, we present the metric system of measurement and explain how to convert between different units.

The basic unit of *length,* or distance, in the metric system is the **meter** (m). One meter is approximately the distance from a doorknob to the floor. All units of length in the metric system are derived from the meter. Prefixes to the basic unit denote the length of each unit. For example, the prefix *centi-* means "one-hundredth"; therefore, 1 centimeter is 1 one-hundredth of a meter (0.01 m).

Point of Interest

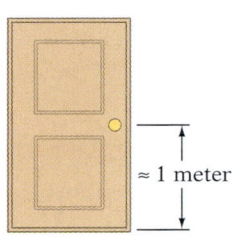

To learn more about these tests, go to **www.ed.gov.** Use the search feature and enter TIMSS.

Point of Interest

Originally, the meter (spelled *metre* in some countries) was defined as $\dfrac{1}{10,000,000}$ of the distance from the equator to the North Pole. Modern scientists have redefined the meter as 1,650,753.73 wavelengths of the orange-red light given off by the element krypton.

kilo-	= 1 000	1 kilometer (km) = 1 000 meters (m)
hecto-	= 100	1 hectometer (hm) = 100 m
deca-	= 10	1 decameter (dam) = 10 m
		1 meter (m) = 1 m
deci-	= 0.1	1 decimeter (dm) = 0.1 m
centi-	= 0.01	1 centimeter (cm) = 0.01 m
milli-	= 0.001	1 millimeter (mm) = 0.001 m

Note in this list that 1000 is written as 1 000, with a space between the 1 and the zeros. **When writing numbers using metric units, each group of three numbers is separated by a space instead of a comma.** A space is also used after each group of three numbers to the right of a decimal point. For example, 31,245.2976 is written 31 245.297 6 in metric notation.

Mass and weight are closely related. *Weight* is a measure of how strongly gravity is pulling on an object. Therefore, an object's weight is less in space than on Earth's surface. However, the amount of material in the object, its *mass,* remains the same. On the surface of Earth, the terms *mass* and *weight* can be used interchangeably.

The basic unit of mass in the metric system is the **gram** (g). If a box that is 1 centimeter long on each side is filled with water, the mass of that water is 1 gram.

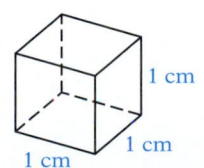

1 gram = the mass of water in a box that is 1 centimeter long on each side

857

The units of mass in the metric system have the same prefixes as the units of length.

$$
\begin{aligned}
1 \text{ kilogram (kg)} &= 1\,000 \text{ grams (g)} \\
1 \text{ hectogram (hg)} &= 100 \text{ g} \\
1 \text{ decagram (dag)} &= 10 \text{ g} \\
1 \text{ gram (g)} &= 1 \text{ g} \\
1 \text{ decigram (dg)} &= 0.1 \text{ g} \\
1 \text{ centigram (cg)} &= 0.01 \text{ g} \\
1 \text{ milligram (mg)} &= 0.001 \text{ g}
\end{aligned}
$$

Weight ≈ 1 gram

The gram is a very small unit of mass. A paperclip weighs about 1 gram. In applications, the kilogram (1 000 grams) is a more useful unit of mass. This textbook weighs about 1 kilogram.

Liquid substances are measured in units of *capacity*.

The basic unit of capacity in the metric system is the **liter** (L). One liter is defined as the capacity of a box that is 10 centimeters long on each side.

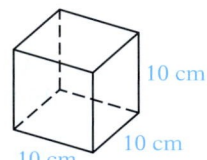

10 cm
10 cm
10 cm

1 liter = the capacity of a box that is 10 centimeters long on each side

The units of capacity in the metric system have the same prefixes as the units of length.

$$
\begin{aligned}
1 \text{ kiloliter (kl)} &= 1\,000 \text{ liters (L)} \\
1 \text{ hectoliter (hl)} &= 100 \text{ L} \\
1 \text{ decaliter (dal)} &= 10 \text{ L} \\
1 \text{ liter (L)} &= 1 \text{ L} \\
1 \text{ deciliter (dl)} &= 0.1 \text{ L} \\
1 \text{ centiliter (cl)} &= 0.01 \text{ L} \\
1 \text{ milliliter (ml)} &= 0.001 \text{ L}
\end{aligned}
$$

Point of Interest

The definition of 1 inch has been changed as a consequence of the wide acceptance of the metric system. One inch is now exactly 25.4 mm.

Converting between units in the metric system involves moving the decimal point to the right or to the left. Listing the units in order from largest to smallest will indicate how many places to move the decimal point in which direction.

To convert 3 800 cm to meters, write the units of length in order from largest to smallest.

km   hm   dam   m   dm   cm   mm

2 positions

• Converting from centimeters to meters requires moving two places to the left.

3 800 cm = 38.00 m

2 places

• Move the decimal point the same number of places in the same direction.

HOW TO Convert 27 kg to grams.

kg hg dag g dg cg mg

3 positions

27 kg = 27 000 g

3 places

- Write the units of mass in order from largest to smallest.
- Converting kilograms to grams requires moving three positions to the right.
- Move the decimal point the same number of places in the same direction.

Example 1

Convert 4.08 m to centimeters.

Solution
Write the units of length from largest to smallest.

km hm dam (m) dm (cm) mm

Converting meters to centimeters requires moving two positions to the right.

4.08 m = 408 cm

You Try It 1

Convert 1 295 m to kilometers.

Your solution
1.295 km

Example 2

Convert 5.93 g to milligrams.

Solution
Write the units of mass from largest to smallest.

kg hg dag (g) dg cg (mg)

Converting grams to milligrams requires moving three positions to the right.

5.93 g = 5 930 mg

You Try It 2

Convert 7 543 g to kilograms.

Your solution
7.543 kg

Example 3

Convert 82 ml to liters.

Solution
Write the units of capacity from largest to smallest.

kl hl dal (L) dl cl (ml)

Converting milliliters to liters requires moving three positions to the left.

82 ml = 0.082 L

You Try It 3

Convert 6.3 L to milliliters.

Your solution
6 300 ml

Solutions on p. S43

Example 4

Convert 9 kl to liters.

Solution

Write the units of capacity from largest to smallest.

(kl) hl dal (L) dl cl ml

Converting kiloliters to liters requires moving three positions to the right.

9 kl = 9 000 L

You Try It 4

Convert 2 kl to liters.

Your solution

2 000 L

Solution on p. S43

Other prefixes in the metric system are becoming more common as a result of technological advances in the computer industry. For example:

$$
\begin{aligned}
\text{tera-} &= 1\ 000\ 000\ 000\ 000 \\
\text{giga-} &= 1\ 000\ 000\ 000 \\
\text{mega-} &= 1\ 000\ 000
\end{aligned}
$$

$$
\begin{aligned}
\text{micro-} &= 0.000\ 001 \\
\text{nano-} &= 0.000\ 000\ 001 \\
\text{pico-} &= 0.000\ 000\ 000\ 001
\end{aligned}
$$

A **bit** is the smallest unit of code that computers can read; it is a <u>bi</u>nary di<u>g</u>it, either a 0 or a 1. Usually bits are grouped into **bytes** of 8 bits. Each byte stands for a letter, a number, or any other symbol we might use in communicating information. For example, the letter W can be represented by 01010111.

The amount of memory in a computer hard drive is measured in terabytes, gigabytes, and megabytes. The speed of a computer used to be measured in microseconds and then nanoseconds, but now speeds are measured in picoseconds.

Here are a few more examples of how these prefixes are used.

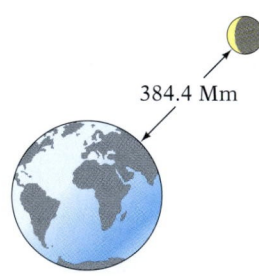

384.4 Mm

The mass of Earth gains 40 Gg (gigagrams) each year from captured meteorites and cosmic dust.

The average distance from Earth to the moon is 384.4 Mm (megameters), and the average distance from Earth to the sun is 149.5 Gm (gigameters).

The wavelength of yellow light is 590 nm (nanometers).

The diameter of a hydrogen atom is about 70 pm (picometers).

There are additional prefixes in the metric system, indicating both larger and smaller numbers. We may hear them more and more often as computer chips hold more and more information, as computers get faster and faster, and as we learn more and more about objects in our universe and beyond that are great distances away.

The U.S. Metric Association (USMA) advocates U.S. conversion to the metric system, which is also referred to as the International System of Units, abbreviated SI. The process of changing measurement units to the metric system is called **metric transition** or **metrication.**

Exercises

1. In the metric system, what is the basic unit of length? of liquid measure? of weight?   Meter, liter, gram

2. ✏️   **a.** Explain how to convert meters to centimeters.
       b. Explain how to convert milliliters to liters.

For Exercises 3–26, name the unit in the metric system that would be used to measure each.

3. The distance from New York to London
   Kilometer
4. The weight of a truck
   Kilogram
5. A person's waist
   Centimeter
6. The amount of coffee in a mug
   Milliliter
7. The weight of a thumbtack
   Gram
8. The amount of water in a swimming pool
   Kiloliter
9. The distance a baseball player hits a baseball
   Meter
10. A person's hat size
    Centimeter
11. The amount of fat in a slice of cheddar cheese
    Gram
12. A person's weight
    Kilogram
13. The maple syrup served with pancakes
    Milliliter
14. The amount of water in a water cooler
    Liter
15. The amount of vitamin C in a vitamin tablet
    Milligram
16. A serving of cereal
    Gram
17. The width of a hair
    Millimeter
18. A person's height
    Centimeter
19. The amount of medication in an aspirin
    Milligram
20. The weight of a lawnmower
    Kilogram
21. The weight of a slice of bread
    Gram
22. The contents of a bottle of salad dressing
    Milliliter
23. The amount of water a family uses monthly
    Kiloliter
24. The newspapers collected at a recycling center
    Kilogram
25. The amount of liquid in a bowl of soup
    Milliliter
26. The distance to the bank
    Kilometer

For Exercises 27–56, convert the given measure.

27. 42 cm = _____ mm
    420
28. 91 cm = _____ mm
    910
29. 360 g = _____ kg
    0.360

30. 1 856 g = _____ kg
    1.856
31. 5 194 ml = _____ L
    5.194
32. 7 285 ml = _____ L
    7.285

33. 2 m = _____ mm
    2 000
34. 8 m = _____ mm
    8 000
35. 217 mg = _____ g
    0.217

36. 34 mg = _____ g
    0.034
37. 4.52 L = _____ ml
    4 520
38. 0.029 7 L = _____ ml
    29.7

39. 8 406 m = _____ km
    8.406
40. 7 530 m = _____ km
    7.530
41. 2.4 kg = _____ g
    2 400

42. 9.2 kg = _____ g
    9 200
43. 6.18 kl = _____ L
    6 180
44. 0.036 kl = _____ L
    36

45. 9.612 km = _____ m
    9 612
46. 2.35 km = _____ m
    2 350
47. 0.24 g = _____ mg
    240

48. 0.083 g = _____ mg
    83
49. 298 cm = _____ m
    2.98
50. 71.6 cm = _____ m
    0.716

51. 2 431 L = _____ kl
    2.431
52. 6 302 L = _____ kl
    6.302
53. 0.66 m = _____ cm
    66

54. 4.58 m = _____ cm
    458
55. 243 mm = _____ cm
    24.3
56. 92 mm = _____ cm
    9.2

57. **a.** Complete the table.

Metric System Prefix	Symbol	Magnitude	Means Multiply the Basic Unit By:
tera-	T	10^{12}	1 000 000 000 000
giga-	G	? 10^9	1 000 000 000
mega-	M	10^6	? 1 000 000
kilo-	? k	? 10^3	1 000
hecto-	h	? 10^2	100
deca-	da	10^1	? 10
deci-	d	$\dfrac{1}{10}$	? 0.1
centi-	? c	$\dfrac{1}{10^2}$	? 0.01
milli-	? m	? $\dfrac{1}{10^3}$	0.001
micro-	μ (mu)	$\dfrac{1}{10^6}$	? 0.000 001
nano-	n	$\dfrac{1}{10^9}$	? 0.000 000 001
pico-	p	? $\dfrac{1}{10^{12}}$	0.000 000 000 001

b. How can the magnitude column in the table above be used to determine how many places to move the decimal point when converting to the basic unit in the metric system?

58. **The Olympics**
  a. One of the events in the summer Olympics is the 50 000-meter walk. How many kilometers do the entrants in this event walk?
  b. One of the events in the winter Olympic Games is the 10 000-meter speed skating event. How many kilometers do the entrants in this event skate?
  a. 50 km   **b.** 10 km

59. **Gemstones**   A carat is a unit of weight equal to 200 mg. Find the weight in grams of a 10-carat precious stone.
2 g

60. **Fabric**   How many pieces of material, each 75 cm long, can be cut from a bolt of fabric that is 6 m long?
8 pieces

61. **Swimming Pools**   An athletic club uses 800 ml of chlorine each day for its swimming pool. How many liters of chlorine are used in a month of 30 days?
24 L

62. Carpentry Each of the four shelves in a bookcase measures 175 cm. Find the cost of the shelves when the price of lumber is $15.75 per meter.
$110.25

63. Containers of Milk The printed label from a container of milk is shown at the right. To the nearest whole number, how many 230-milliliter servings are in the container?
16 servings

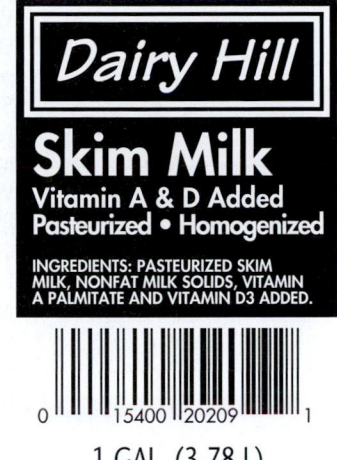

1 GAL. (3.78 L)

64. Cereal A 1.19-kilogram container of Quaker Oats contains 30 servings. Find the number of grams in one serving of the oatmeal. Round to the nearest gram.
40 g

65. Nutritional Supplements A patient is advised to supplement her diet with 2 g of calcium per day. The calcium tablets she purchases contain 500 mg of calcium per tablet. How many tablets per day should the patient take?
4 tablets

66. Education A laboratory assistant is in charge of ordering acid for three chemistry classes of 30 students each. Each student requires 80 ml of acid. How many liters of acid should be ordered? The assistant must order by the whole liter.
8 L

67. Unit Cost A case of 12 one-liter bottles of apple juice costs $19.80. A case of 24 cans, each can containing 340 ml of apple juice, costs $14.50. Which case of apple juice costs less per milliliter?
The case containing 12 one-liter bottles

68. Construction A column assembly is being constructed in a building. The components are shown in the diagram at the right. What length column must be cut?
215.5 cm

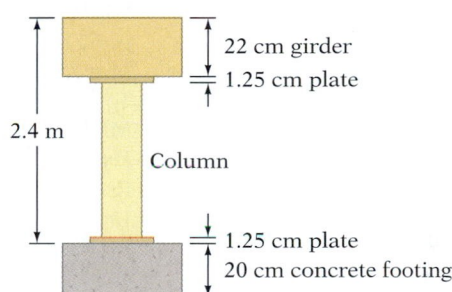

69. Speed of Light The distance between Earth and the sun is 150 000 000 km. Light travels 300 000 000 m in 1 s. How long does it take for light to reach Earth from the sun?
500 s

70. ✏ Why is it necessary to have internationally standardized units of measurement?

APPLYING THE CONCEPTS

71. Business A service station operator bought 85 kl of gasoline for $38,500. The gasoline was sold for $.658 per liter. Find the profit on the 85 kl of gasoline.
$17,430

72. Business For $149.50, a cosmetician buys 5 L of moisturizer and repackages it in 125-milliliter jars. Each jar costs the cosmetician $.55. Each jar of moisturizer is sold for $8.95. Find the profit on the 5 L of moisturizer.
$186.50

73. Business A health food store buys nuts in 10-kilogram containers and repackages the nuts for resale. The store packages the nuts in 200-gram bags, costing $.06 each, and sells them for $2.89 per bag. Find the profit on a 10-kilogram container of nuts costing $75.
$66.50

74. Form two debating teams. One team should argue in favor of changing to the metric system in the United States, and the other should argue against it.

Photo Credits

p. 1, PhotoDisc Blue/Getty Images; **p. 8,** AP/Wide World Photos; **p. 16,** CORBIS; **p. 17,** AP/Wide World Photos; **p. 38,** Dave Bartruff/CORBIS; **p. 64,** Wally McNamee/CORBIS; **p. 66,** Henry Ray Abrams/AFP/Getty Images; **p. 71,** Royalty-Free/CORBIS; **p. 78,** AP/Wide World Photos; **p. 81,** AP/Wide World Photos; **p. 82,** Duomo/CORBIS; **p. 86,** Johnny Buzzerio/CORBIS; **p. 108,** AP/Wide World Photos; **p. 110,** Wally McNamee/CORBIS; **p. 126,** Dennis MacDonald/PhotoEdit, Inc.; **p. 127,** AP/Wide World Photos; **p. 128,** Stephen Frink/CORBIS; **p. 134,** Bettmann/CORBIS; **p. 137,** Frank Siteman/PhotoEdit, Inc.; **p. 155,** AP/Wide World Photos; **p. 170,** Sandor Szabo/EPA/Landov; **p. 175,** Jean Miele/CORBIS; **p. 186,** © Tannen Maury/The Image Works; **p. 201,** © DPA/The Image Works; **p. 213,** AP/Wide World Photos; **p. 244,** Bettmann/CORBIS; **p. 245,** Bob Daemmrich/PhotoEdit, Inc.; **p. 268,** Lester V. Bergman/CORBIS; **p. 274,** Guy Motil/CORBIS; **p. 276,** Shaun Best/Reuters/CORBIS; **p. 279,** A&L Sinibaldi/STONE/Getty Images; **p. 280,** Tony Freeman/PhotoEdit, Inc.; **p. 291,** Lawrence Manning/CORBIS; **p. 291,** Davis Barber/PhotoEdit, Inc.; **p. 313,** Steve Prezant/CORBIS; **p. 328,** Pete Seaward/Getty Images; **p. 332,** © Topham/The Image Works, Inc.; **p. 333,** Ezra Shaw/Getty Images; **p. 361,** PhotoDisc/Getty Images; **p. 375,** Yann Arthus-Bertrand/CORBIS; **p. 379,** Hideo Kurihara/Getty Images; **p. 392,** PhotoDisc/Getty Images; **p. 393,** AP/Wide World Photos; **p. 408,** Royalty-Free/CORBIS; **p. 413,** Patrick Ward/CORBIS; **p. 442,** AP/Wide World Photos; **p. 443,** © Topham/The Image Works; **p. 444,** Roy Morsch/CORBIS; **p. 452,** Royalty-Free/CORBIS; **p. 452,** Royalty-Free/CORBIS; **p. 455,** Jason Reed/Reuters/CORBIS; **p. 456,** Bettmann/CORBIS; **p. 463,** Tony Freeman/PhotoEdit, Inc.; **p. 475,** Jim Richardson/CORBIS; **p. 508,** Duomo/CORBIS; **p. 514,** Duomo/CORBIS; **p. 518,** Royalty-Free/CORBIS; **p. 519,** Pierre Ducharme/Reuters/CORBIS; **p. 557,** Reuters/CORBIS; **p. 559,** Bill Aron/PhotoEdit, Inc.; **p. 564,** Royalty-Free/CORBIS; **p. 568,** Chris Hondros/Newsmakers/Getty Images; **p. 569,** Jonathan Nourok/PhotoEdit, Inc.; **p. 591,** Stephen Frink/CORBIS; **p. 609,** Tom Carter/PhotoEdit, Inc.; **p. 609,** David Young-Wolff/PhotoEdit, Inc.; **p. 610,** Sheldan Collins/CORBIS; **p. 610,** Galen Rowell/CORBIS; **p. 611,** Billy E. Barnes/PhotoEdit, Inc.; **p. 612,** Lee Cohen/CORBIS; **p. 613,** Francis G. Mayer/CORBIS; **p. 623,** AP/Wide World Photos; **p. 629,** Craig Tuttle/CORBIS; **p. 656,** Tony Freeman/PhotoEdit, Inc.; **p. 675,** Jeff Hunter/The Image Bank/Getty Images; **p. 693,** Michael Newman/PhotoEdit, Inc.; **p. 703,** Spencer Grant/PhotoEdit, Inc.; **p. 705,** Eric & David Hosking/CORBIS; **p. 706,** Jeff Greenberg/PhotoEdit, Inc.; **p. 712,** Joel W. Rogers/CORBIS; **p. 717,** Michael Newman/PhotoEdit, Inc.; **p. 731,** Vic Bider/PhotoEdit, Inc.; **p. 732,** Spencer Grant/PhotoEdit, Inc.; **p. 735,** Spencer Grant/PhotoEdit, Inc.; **p. 736,** Free Agents Limited/CORBIS; **p. 746,** PictureArts/CORBIS; **p. 751,** AP/Wide World Photos; **p. 758,** Shaun Egan/Getty Images; **p. 771,** The Granger Collection; **p. 775,** PictureArts/CORBIS; **p. 776,** Robert Brenner/PhotoEdit, Inc.; **p. 785,** AP/Wide World Photos; **p. 786,** Lori Adamski Peek/STONE/Getty Images; **p. 808,** Rudi Von Briel/PhotoEdit, Inc.; **p. 811,** Bonnie Kamin/PhotoEdit, Inc.; **p. 818,** David Ponton/Getty Images; **p. 820,** Grafton Marshall Smith/CORBIS; **p. 825,** Myrleen Ferguson Cate/PhotoEdit, Inc.; **p. 826,** Jeff Greenberg/PhotoEdit, Inc.

Solutions to Chapter 1 "You Try It"

SECTION 1.1

You Try It 1

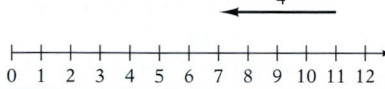

You Try It 2

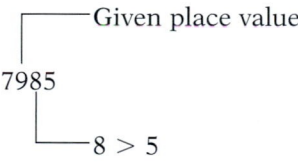

7 is 4 units to the left of 11.

You Try It 3 **a.** $47 > 19$ **b.** $26 > 0$

You Try It 4 0, 3, 17, 52, 68, 94

You Try It 5 Forty-six million thirty-two thousand seven hundred fifteen

You Try It 6 920,008

You Try It 7 $70,000 + 6000 + 200 + 40 + 5$

You Try It 8

```
        ┌──── Given place value
        │
    529,374
        │
        └──── 9 > 5
```

529,374 rounded to the nearest ten-thousand is 530,000.

You Try It 9

```
        ┌──── Given place value
        │
    7985
        │
        └──── 8 > 5
```

7985 rounded to the nearest hundred is 8000.

You Try It 10

Strategy To find the sport named by the greatest number of people, find the largest number given in the circle graph.

Solution The largest number given in the graph is 80.

The sport named by the greatest number of people was football.

You Try It 11

Strategy To find the shorter distance, compare the numbers 347 and 387.

Solution $347 < 387$

The shorter distance is between Los Angeles and San Jose.

You Try It 12

Strategy To determine which state has fewer sanctioned league bowlers, compare the numbers 239,951 and 239,010.

Solution $239,010 < 239,951$

Ohio has fewer sanctioned league bowlers.

You Try It 13

Strategy To find the land area to the nearest thousand square miles, round 3,851,809 to the nearest thousand.

Solution 3,851,809 rounded to the nearest thousand is 3,852,000.

To the nearest thousand, the land area of Canada is 3,852,000 mi².

SECTION 1.2

You Try It 1
$$\begin{aligned} 6285 &\longrightarrow 6000 \\ 3972 &\longrightarrow 4000 \\ 5140 &\longrightarrow +\ 5000 \\ \hline & 15,000 \end{aligned}$$

You Try It 2 The Addition Property of Zero

You Try It 3
$$\begin{aligned} & 111,100,000 \\ & 61,600,000 \\ & 24,100,000 \\ + & 1,600,000 \\ \hline & 198,400,000 \end{aligned}$$

A total of 198,400,000 cases of eggs were produced during the year.

You Try It 4 $x + y + z$
$1692 + 4783 + 5046$

$$\begin{aligned} & {}^{1\,2\,1} \\ & 1692 \\ & 4783 \\ + & 5046 \\ \hline & 11,521 \end{aligned}$$

You Try It 5
$$\begin{array}{c} 13 = b + 6 \\ \hline 13 \ | \ 7 + 6 \\ 13 = 13 \end{array}$$

Yes, 7 is a solution of the equation.

You Try It 6
$$\begin{aligned} & {}^{8\ \ 9\ 9\,12} \\ & 4\cancel{9},\cancel{0}\cancel{0}2 \\ - & 31,865 \\ \hline & 17,137 \end{aligned}$$

Check:
$$\begin{aligned} & 31,865 \\ + & 17,137 \\ \hline & 49,002 \end{aligned}$$

You Try It 7

$$8544 \longrightarrow \quad 9000 \qquad 8544$$
$$3621 \longrightarrow \underline{-4000} \qquad \underline{-3621}$$
$$\qquad \qquad \quad 5000 \qquad \quad 4923$$

You Try It 8   2020: 612 quadrillion Btu
1990: 346 quadrillion Btu

$$\begin{array}{r} 612 \\ -346 \\ \hline 266 \end{array}$$

The difference is 266 quadrillion Btu.

You Try It 9 $x - y$

$7061 - 3229$

$$\begin{array}{r} {\scriptstyle 6\,10\,5\,11} \\ 7\,0\,6\,1\!\!\!/ \\ -3\,2\,2\,9 \\ \hline 3\,8\,3\,2 \end{array}$$

You Try It 10

$$46 = 58 - p$$
$$\begin{array}{c|c} 46 & 58 - 11 \end{array}$$
$$46 \neq 47$$

No, 11 is not a solution of the equation.

You Try It 11

Strategy   To find the total number of fatal accidents during 1991 through 1999:
  • Find the number of fatal accidents each year.
  • Add the nine numbers.

Solution

1991: 3	1994: 2	1997: 4
1992: 2	1995: 3	1998: 5
1993: 4	1996: 3	1999: 6

$3 + 2 + 4 + 2 + 3 + 3 + 4 + 5 + 6 = 32$

During 1991 through 1999, there were 32 fatal accidents on amusement rides.

You Try It 12

Strategy   To find the price, replace C by 148 and M by 74 in the given formula and solve for P.

Solution $P = C + M$
$P = 148 + 74$
$P = 222$
The price of the leather jacket is $222.

You Try It 13

Strategy   Draw a diagram.

60 ft

60 ft

To find the length of fencing needed, use the formula for the perimeter of a rectangle, $P = L + W + L + W$.
$L = 60$ and $W = 60$.

Solution $P = L + W + L + W$
$P = 60 + 60 + 60 + 60$
$P = 240$
240 ft of fencing is needed.

SECTION 1.3

You Try It 1   The average monthly savings in France is $175.

$$\begin{array}{r} 175 \\ \times \ 12 \\ \hline 350 \\ 175 \\ \hline 2100 \end{array}$$

The average annual savings of individuals in France is $2100.

You Try It 2

$$8704 \longrightarrow 9000$$
$$93 \ \longrightarrow 90$$

$9000 \cdot 90 = 810,000$

You Try It 3 $5xy$
$5(20)(60) = 100(60)$
$\qquad \qquad \ = 6000$

You Try It 4 $90(7000) = 630,000$

You Try It 5 $0 \cdot 10 = 0$

You Try It 6

$$7a = 77$$
$$\begin{array}{c|c} 7 \cdot 11 & 77 \end{array}$$
$$77 = 77$$

Yes, 11 is a solution of the equation.

You Try It 7 $2 \cdot 2 \cdot 2 \cdot 3 \cdot 3 \cdot 3 \cdot 3 = 2^3 \cdot 3^4$

You Try It 8 $6^4 = 6 \cdot 6 \cdot 6 \cdot 6 = 36 \cdot 6 \cdot 6$
$\qquad \ = 216 \cdot 6 = 1296$

You Try It 9 $10^8 = 100,000,000$

You Try It 10 $2^4 \cdot 3^2 = (2 \cdot 2 \cdot 2 \cdot 2) \cdot (3 \cdot 3)$
$\qquad \qquad \ = 16 \cdot 9 = 144$

You Try It 11 $x^4 y^2$
$1^4 \cdot 3^2 = (1 \cdot 1 \cdot 1 \cdot 1) \cdot (3 \cdot 3)$
$\qquad \quad \ = 1 \cdot 9$
$\qquad \quad \ = 9$

You Try It 12

$$
\begin{array}{r}
320\ r14 \\
24\overline{)7694} \\
-72 \\
\hline
49 \\
-48 \\
\hline
14 \\
-\ 0 \\
\hline
14
\end{array}
$$

Check: $(320 \cdot 24) + 14 = 7680 + 14$
$$= 7694$$

You Try It 13   The annual expense for food is $7200.

$7200 \div 12 = 600$

The monthly expense for food is $600.

You Try It 14 $216{,}936 \longrightarrow 200{,}000$
$207 \quad\ \longrightarrow 200$

$200{,}000 \div 200 = 1000$

You Try It 15 $\dfrac{x}{y}$

$\dfrac{672}{8} = 84$

You Try It 16 $\dfrac{60}{y} = 2$

$$
\begin{array}{c|c}
\dfrac{60}{12} & 2 \\
\hline
5 & \neq 2
\end{array}
$$

No, 12 is not a solution of the equation.

You Try It 17
$30 \div 1 = 30$
$30 \div 2 = 15$
$30 \div 3 = 10$
$30 \div 4 \qquad$ Does not divide evenly.
$30 \div 5 = 6$
$30 \div 6 = 5 \qquad$ The factors are
$\qquad\qquad\qquad$ repeating.

The factors of 30 are 1, 2, 3, 5, 6, 10, 15, and 30.

You Try It 18

$$
\begin{array}{r}
11 \\
2\overline{)22} \\
2\overline{)44} \\
2\overline{)88}
\end{array}
$$

$88 = 2 \cdot 2 \cdot 2 \cdot 11 = 2^3 \cdot 11$

You Try It 19

$$
\begin{array}{r}
59 \\
5\overline{)295}
\end{array}
$$

$295 = 5 \cdot 59$

You Try It 20

Strategy   To find how many times more expensive a stamp was, divide the cost in 1997 (32) by the cost in 1960 (4).

Solution $32 \div 4 = 8$

A stamp was 8 times more expensive in 1997.

You Try It 21

Strategy   Draw a diagram.

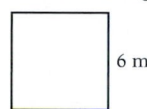
6 m

To find the amount of carpet that should be purchased, use the formula for the area of a square, $A = s^2$, with $s = 6$.

Solution $A = s^2$
$A = 6^2$
$A = 36$

36 m^2 of carpet should be purchased.

You Try It 22

Strategy   To find the speed, replace d by 486 and t by 9 in the given formula and solve for r.

Solution $r = \dfrac{d}{t}$

$r = \dfrac{486}{9}$

$r = 54$

You would need to travel at a speed of 54 mph.

SECTION 1.4

You Try It 1 $4 \cdot (8 - 3) \div 5 - 2 = 4 \cdot 5 \div 5 - 2$
$$= 20 \div 5 - 2$$
$$= 4 - 2$$
$$= 2$$

You Try It 2 $16 + 3(6 - 1)^2 \div 5 = 16 + 3(5)^2 \div 5$
$$= 16 + 3(25) \div 5$$
$$= 16 + 75 \div 5$$
$$= 16 + 15$$
$$= 31$$

You Try It 3 $(a - b)^2 + 5c$
$(7 - 2)^2 + 5(4) = 5^2 + 5(4)$
$$= 25 + 5(4)$$
$$= 25 + 20$$
$$= 45$$

Solutions to Chapter 2 "You Try It"

SECTION 2.1

You Try It 1

	2	3	5
12 =	(2 · 2)	3	
27 =		(3 · 3 · 3)	
50 =	2		(5 · 5)

The LCM = 2 · 2 · 3 · 3 · 3 · 5 · 5
$\qquad$ = 2700.

You Try It 2

	2	3	5
36 =	(2 · 2)	3 · 3	
60 =	2 · 2	(3)	5
72 =	2 · 2 · 2	3 · 3	

The GCF = 2 · 2 · 3 = 12.

You Try It 3

	2	3	5	11
11 =				11
24 =	2 · 2 · 2	3		
30 =	2	3	5	

Because no numbers are circled, the GCF = 1.

SECTION 2.2

You Try It 1 $\dfrac{19}{6}; 3\dfrac{1}{6}$

You Try It 2

$$3\overline{)26} \quad \begin{array}{r} 8 \\ \hline 26 \\ -24 \\ \hline 2 \end{array} \qquad \dfrac{26}{3} = 8\dfrac{2}{3}$$

You Try It 3

$$4\overline{)36} \quad \begin{array}{r} 9 \\ \hline 36 \\ -36 \\ \hline 0 \end{array} \qquad \dfrac{36}{4} = 9$$

You Try It 4 $\quad 9\dfrac{4}{7} = \dfrac{(7 \cdot 9) + 4}{7} = \dfrac{63 + 4}{7} = \dfrac{67}{7}$

You Try It 5 $\quad 3 = \dfrac{3}{1}$

You Try It 6 $\quad 48 \div 8 = 6$

$$\dfrac{5}{8} = \dfrac{5 \cdot 6}{8 \cdot 6} = \dfrac{30}{48}$$

$\dfrac{30}{48}$ is equivalent to $\dfrac{5}{8}$.

You Try It 7 $\quad 8 = \dfrac{8}{1} \qquad 12 \div 1 = 12$

$$8 = \dfrac{8}{1} = \dfrac{8 \cdot 12}{1 \cdot 12} = \dfrac{96}{12}$$

$\dfrac{96}{12}$ is equivalent to 8.

You Try It 8 $\quad \dfrac{21}{84} = \dfrac{3 \cdot 7}{2 \cdot 2 \cdot 3 \cdot 7} = \dfrac{1}{4}$

You Try It 9 $\quad \dfrac{32}{12} = \dfrac{2 \cdot 2 \cdot 2 \cdot 2 \cdot 2}{2 \cdot 2 \cdot 3} = \dfrac{8}{3}$

You Try It 10 $\quad \dfrac{11t}{11} = \dfrac{11 \cdot t}{11} = t$

You Try It 11 The LCM of 9 and 21 is 63.

$$\dfrac{4}{9} = \dfrac{28}{63} \qquad \dfrac{8}{21} = \dfrac{24}{63}$$

$$\dfrac{28}{63} > \dfrac{24}{63}$$

$$\dfrac{4}{9} > \dfrac{8}{21}$$

You Try It 12 The LCM of 24 and 9 is 72.

$$\dfrac{17}{24} = \dfrac{51}{72} \qquad \dfrac{7}{9} = \dfrac{56}{72}$$

$$\dfrac{51}{72} < \dfrac{56}{72}$$

$$\dfrac{17}{24} < \dfrac{7}{9}$$

SECTION 2.3

You Try It 1 $\quad \dfrac{7}{12} + \dfrac{3}{8} = \dfrac{14}{24} + \dfrac{9}{24} = \dfrac{23}{24}$

You Try It 2 $\quad \dfrac{3}{5} + \dfrac{2}{3} + \dfrac{5}{6} = \dfrac{18}{30} + \dfrac{20}{30} + \dfrac{25}{30} = \dfrac{63}{30}$

$$= 2\dfrac{3}{30} = 2\dfrac{1}{10}$$

You Try It 3 $\quad 16 + 8\dfrac{5}{9} = 24\dfrac{5}{9}$

You Try It 4
$$\frac{2}{3} + z = \frac{23}{24}$$

$$\frac{\dfrac{2}{3} + \dfrac{3}{8}}{\dfrac{16}{24} + \dfrac{9}{24}} \left|\ \dfrac{23}{24}\right.$$

$$\frac{25}{24} \neq \frac{23}{25}$$

No, $\dfrac{3}{8}$ is not a solution of $\dfrac{2}{3} + z = \dfrac{23}{24}$.

You Try It 5 $x + y + z$

$$3\frac{5}{6} + 2\frac{1}{9} + 5\frac{5}{12} = 3\frac{30}{36} + 2\frac{4}{36} + 5\frac{15}{36}$$

$$= 10\frac{49}{36}$$

$$= 11\frac{13}{36}$$

You Try It 6 $\dfrac{5}{6} - \dfrac{7}{9} = \dfrac{15}{18} - \dfrac{14}{18} = \dfrac{1}{18}$

You Try It 7 $9\dfrac{7}{8} - 5\dfrac{2}{3} = 9\dfrac{21}{24} - 5\dfrac{16}{24} = 4\dfrac{5}{24}$

You Try It 8 $6 - 4\dfrac{2}{11} = 5\dfrac{11}{11} - 4\dfrac{2}{11} = 1\dfrac{9}{11}$

You Try It 9

Strategy   To find the fraction of the respondents who did not name glazed, filled, or frosted:

- Add the three fractions to find the fraction who named glazed, filled, or frosted.
- Subtract the fraction who named glazed, filled, or frosted from 1, the entire group surveyed.

Solution $\dfrac{2}{5} + \dfrac{8}{25} + \dfrac{3}{20} = \dfrac{40}{100} + \dfrac{32}{100} + \dfrac{15}{100}$

$$= \frac{87}{100}$$

$$1 - \frac{87}{100} = \frac{100}{100} - \frac{87}{100} = \frac{13}{100}$$

$\dfrac{13}{100}$ of the respondents did not name glazed, filled, or frosted as their favorite type of doughnut.

SECTION 2.4

You Try It 1 $\dfrac{y}{10} \cdot \dfrac{z}{7} = \dfrac{y \cdot z}{10 \cdot 7} = \dfrac{yz}{70}$

You Try It 2 $\dfrac{5}{12} \cdot \dfrac{9}{35} \cdot \dfrac{7}{8} = \dfrac{5 \cdot 9 \cdot 7}{12 \cdot 35 \cdot 8}$

$$= \frac{5 \cdot 3 \cdot 3 \cdot 7}{2 \cdot 2 \cdot 3 \cdot 5 \cdot 7 \cdot 2 \cdot 2 \cdot 2}$$

$$= \frac{3}{32}$$

You Try It 3 $\dfrac{8}{9} \cdot 6 = \dfrac{8}{9} \cdot \dfrac{6}{1} = \dfrac{8 \cdot 6}{9 \cdot 1}$

$$= \frac{2 \cdot 2 \cdot 2 \cdot 2 \cdot 3}{3 \cdot 3 \cdot 1} = \frac{16}{3} = 5\frac{1}{3}$$

You Try It 4 $3\dfrac{6}{7} \cdot 2\dfrac{4}{9} = \dfrac{27}{7} \cdot \dfrac{22}{9} = \dfrac{27 \cdot 22}{7 \cdot 9}$

$$= \frac{3 \cdot 3 \cdot 3 \cdot 2 \cdot 11}{7 \cdot 3 \cdot 3} = \frac{66}{7} = 9\frac{3}{7}$$

You Try It 5 $x^4 y^3$

$$\left(2\frac{1}{3}\right)^4 \cdot \left(\frac{3}{7}\right)^3 = \left(\frac{7}{3}\right)^4 \cdot \left(\frac{3}{7}\right)^3$$

$$= \frac{7}{3} \cdot \frac{7}{3} \cdot \frac{7}{3} \cdot \frac{7}{3} \cdot \frac{3}{7} \cdot \frac{3}{7} \cdot \frac{3}{7}$$

$$= \frac{7 \cdot 7 \cdot 7 \cdot 7 \cdot 3 \cdot 3 \cdot 3}{3 \cdot 3 \cdot 3 \cdot 3 \cdot 7 \cdot 7 \cdot 7} = \frac{7}{3} = 2\frac{1}{3}$$

You Try It 6 $\dfrac{5}{6} \div \dfrac{10}{27} = \dfrac{5}{6} \cdot \dfrac{27}{10} = \dfrac{5 \cdot 27}{6 \cdot 10}$

$$= \frac{5 \cdot 3 \cdot 3 \cdot 3}{2 \cdot 3 \cdot 2 \cdot 5} = \frac{9}{4} = 2\frac{1}{4}$$

You Try It 7 $\dfrac{x}{8} \div \dfrac{y}{6} = \dfrac{x}{8} \cdot \dfrac{6}{y}$

$$= \frac{x \cdot 6}{8 \cdot y} = \frac{x \cdot 2 \cdot 3}{2 \cdot 2 \cdot 2 \cdot y} = \frac{3x}{4y}$$

You Try It 8 $4\dfrac{3}{8} \div 3\dfrac{1}{2} = \dfrac{35}{8} \div \dfrac{7}{2} = \dfrac{35}{8} \cdot \dfrac{2}{7} = \dfrac{35 \cdot 2}{8 \cdot 7}$

$$= \frac{5 \cdot 7 \cdot 2}{2 \cdot 2 \cdot 2 \cdot 7} = \frac{5}{4} = 1\frac{1}{4}$$

You Try It 9 $x \div y$

$$2\frac{1}{4} \div 9 = \frac{9}{4} \div \frac{9}{1} = \frac{9}{4} \cdot \frac{1}{9} = \frac{9 \cdot 1}{4 \cdot 9}$$

$$= \frac{3 \cdot 3 \cdot 1}{2 \cdot 2 \cdot 3 \cdot 3} = \frac{1}{4}$$

You Try It 10

$$\frac{2y + 3}{y} = 2$$

$$\frac{2\left(\frac{1}{2}\right) + 3}{\frac{1}{2}} \;\Big|\; 2$$

$$\frac{1 + 3}{\frac{1}{2}} \;\Big|\; 2$$

$$\frac{4}{\frac{1}{2}} \;\Big|\; 2$$

$$4(2) \;\Big|\; 2$$

$$8 \neq 2$$

No, $\frac{1}{2}$ is not a solution of the equation.

You Try It 11

$$\frac{x}{y - z}$$

$$\frac{2\frac{4}{9}}{3 - 1\frac{1}{3}} = \frac{\frac{22}{9}}{\frac{5}{3}} = \frac{22}{9} \div \frac{5}{3} = \frac{22}{9} \cdot \frac{3}{5}$$

$$= \frac{22}{15} = 1\frac{7}{15}$$

You Try It 12

Strategy To find the area, use the formula for the area of a triangle, $A = \frac{1}{2}bh$.

$b = 18$ and $h = 9$.

Solution $A = \frac{1}{2}bh$

$A = \frac{1}{2}(18)(9)$

$A = 81$

81 in² of felt are needed.

You Try It 13

Strategy To find the total cost:
• Multiply the amount of material per sash $\left(1\frac{3}{8}\right)$ by the number of sashes (22) to find the total number of yards of material needed.
• Multiply the total number of yards of material needed by the cost per yard (12).

Solution

$$1\frac{3}{8} \cdot 22 = \frac{11}{8} \cdot \frac{22}{1} = \frac{11 \cdot 22}{8 \cdot 1} = \frac{11 \cdot 2 \cdot 11}{2 \cdot 2 \cdot 2 \cdot 1}$$

$$= \frac{121}{4} = 30\frac{1}{4}$$

$$30\frac{1}{4} \cdot 12 = \frac{121}{4} \cdot \frac{12}{1} = \frac{121 \cdot 12}{4 \cdot 1}$$

$$= \frac{11 \cdot 11 \cdot 2 \cdot 2 \cdot 3}{2 \cdot 2 \cdot 1} = 363$$

The total cost of the material is $363.

SECTION 2.5

You Try It 1 The digit 4 is in the thousandths place.

You Try It 2 $\frac{501}{1000} = 0.501$
(five hundred one thousandths)

You Try It 3 $0.67 = \frac{67}{100}$ (sixty-seven hundredths)

You Try It 4 Fifty-five and six thousand eighty-three ten-thousandths

You Try It 5 806.00491

You Try It 6 $0.065 = 0.0650$
$0.0650 < 0.0802$
$0.065 < 0.0802$

You Try It 7 3.03, 0.33, 0.30, 3.30, 0.03
0.03, 0.30, 0.33, 3.03, 3.30
0.03, 0.3, 0.33, 3.03, 3.3

You Try It 8 Given place value
3.675849
$4 < 5$
3.675849 rounded to the nearest ten-thousandth is 3.6758.

You Try It 9 Given place value
48.907
$0 < 5$
48.907 rounded to the nearest tenth is 48.9.

You Try It 10 Given place value
31.8652
$8 > 5$
31.8652 rounded to the nearest whole number is 32.

You Try It 11

Strategy To determine who had more home runs for every 100 times at bat, compare the numbers 7.03 and 7.09.

Solution $7.09 > 7.03$

Ralph Kiner had more home runs for every 100 times at bat.

You Try It 12

Strategy To determine the average annual precipitation to the nearest inch, round the number 2.65 to the nearest whole number.

Solution 2.65 rounded to the nearest whole number is 3.

To the nearest inch, the average annual precipitation in Yuma is 3 in.

SECTION 2.6

You Try It 1

$$
\begin{array}{r}
{\scriptstyle 1\ 1}\\
8.64\\
52.7\\
+\ 0.39105\\
\hline
61.73105
\end{array}
$$

You Try It 2

$$
\begin{array}{r}
{\scriptstyle 4\ 9\ 10}\\
2\,\cancel{5}.\cancel{0}\,\cancel{0}\\
-\ 4.9\,1\\
\hline
2\,0.0\,9
\end{array}
\qquad
\begin{array}{r}
Check: \quad 4.91\\
+20.09\\
\hline
25.00
\end{array}
$$

You Try It 3

$$
\begin{array}{r}
6.514 \longrightarrow\ 7\\
8.903 \longrightarrow\ 9\\
2.275 \longrightarrow\ +\ 2\\
\hline
18
\end{array}
$$

You Try It 4 $x + y + z$

$7.84 + 3.05 + 2.19$
$= 10.89 + 2.19$
$= 13.08$

You Try It 5

$$
\begin{array}{r}
0.000081\\
\times\ \ \ \ 0.025\\
\hline
405\\
162\\
\hline
0.000002025
\end{array}
$$

You Try It 6

$$
\begin{array}{r}
6.407 \longrightarrow\ \ 6\\
0.959 \longrightarrow\ \times 1\\
\hline
6
\end{array}
$$

You Try It 7 $1.756 \cdot 10^4 = 17{,}560$

You Try It 8 $25xy$

$25(0.8)(0.6) = 20(0.6) = 12$

You Try It 9

$$
\begin{array}{r}
48.2\\
6.53.\overline{)314.74.6}\\
-\ 261\ 2\\
\hline
53\ 54\\
-\ 52\ 24\\
\hline
1\ 30\ 6\\
-\ 1\ 30\ 6\\
\hline
0
\end{array}
$$

You Try It 10

$62.7 \longrightarrow 60$
$3.45 \longrightarrow 3$
$60 \div 3 = 20$

You Try It 11

$$
\begin{array}{r}
6.0391 \approx 6.039\\
86\overline{)519.3700}\\
-516\\
\hline
3\ 3\\
-\ \ \ 0\\
\hline
3\ 37\\
-\ 2\ 58\\
\hline
790\\
-\ 774\\
\hline
160\\
-\ \ 86\\
\hline
74
\end{array}
$$

You Try It 12 $63.7 \div 100 = 0.637$

You Try It 13 $\dfrac{x}{y}$

$$\dfrac{40.6}{0.7} = 58$$

You Try It 14

$$
\begin{array}{r}
0.8\\
5\overline{)4.0}
\end{array}
\qquad
\dfrac{4}{5} = 0.8
$$

You Try It 15

$$
\begin{array}{r}
0.8333\\
6\overline{)5.0000}
\end{array}
\qquad
1\dfrac{5}{6} = 1.8\overline{3}
$$

You Try It 16 $6.2 = 6\dfrac{2}{10} = 6\dfrac{1}{5}$

You Try It 17 $\dfrac{7}{12} \approx 0.5833$

$0.5880 > 0.5833$

$0.588 > \dfrac{7}{12}$

You Try It 18

Strategy   To find the change you receive:
- Multiply the number of stamps (12) by the cost of each stamp (37¢) to find the total cost of the stamps.
- Convert the total cost of the stamps to dollars and cents.
- Subtract the total cost of the stamps from $10.

Solution $12(37) = 444$ The stamps cost 444¢.

444¢ = $4.44   The stamps cost $4.44.

$10.00 - 4.44 = 5.56

You receive $5.56 in change.

You Try It 19

Strategy   To find the profit:
- Divide the number of pounds per 100-pound container (100) by the number of pounds packaged in each bag (2) to find the number of bags sold.
- Multiply the number of bags sold by the selling price per bag (12.50) to find the income from selling the nuts.
- Multiply the number of bags sold by the cost for each bag (.06) to find the total cost of the bags.
- Subtract the cost of the bags and the cost of the nuts (475) from the income.

Solution $100 \div 2 = 50$ Each container makes 50 bags of nuts.

$50(12.50) = 625$ The income from the 50 bags is $625.

$50(.06) = 3$ The total cost of the bags is $3.

$625 - 3 - 475 = 147$

The profit is $147.

You Try It 20

Strategy   To find the insurance premium due, replace B by 276.25 and F by 1.8 in the given formula and solve for P.

Solution $P = BF$

$P = 276.25(1.8)$

$P = 497.25$

The insurance premium due is $497.25.

SECTION 2.7

You Try It 1

$(1.2 - 0.8)^2 + (1.5)(6)$
$= (0.4)^2 + (1.5)(6)$
$= 0.16 + (1.5)(6)$
$= 0.16 + 9$
$= 9.16$

You Try It 2

$\left(\dfrac{1}{2}\right)^3 \cdot \dfrac{7-3}{9-4} + \dfrac{4}{5}$

$= \left(\dfrac{1}{2}\right)^3 \cdot \dfrac{4}{5} + \dfrac{4}{5}$

$= \dfrac{1}{8} \cdot \dfrac{4}{5} + \dfrac{4}{5}$

$= \dfrac{1}{10} + \dfrac{4}{5} = \dfrac{1}{10} + \dfrac{8}{10} = \dfrac{9}{10}$

Solutions to Chapter 3 "You Try It"

SECTION 3.1

You Try It 1

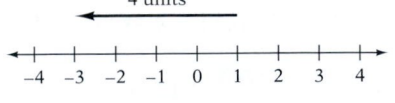

-3 is 4 units to the left of 1.

You Try It 2

A is -5, and C is -3.

You Try It 3

a. 2 is to the right of -5 on the number line.

$2 > -5$

b. -4 is to the left of 3 on the number line.

$-4 < 3$

You Try It 4 $-7, -1, 0, 4, 8$

You Try It 5   **a.** -24 **b.** 13   **c.** b

You Try It 6   **a.** Negative three minus twelve
b. Eight plus negative five

You Try It 7   **a.** $-(-59) = 59$ **b.** $-(y) = -y$

You Try It 8   **a.** $|-8| = 8$ **b.** $|12| = 12$

You Try It 9   **a.** $|0| = 0$ **b.** $-|35| = -35$

You Try It 10 $|-y| = |-2| = 2$

You Try It 11 $|6| = 6, |-2| = 2, -(-1) = 1, -|-8| = -8$
$-8, -4, 1, 2, 6$
$-|-8|, -4, -(-1), |-2|, |6|$

You Try It 12

Strategy   To determine which is closer to blastoff, find the absolute value of each number. The number with the smaller absolute value is closer to zero and, therefore, closer to blastoff.

Solution $|-9| = 9, |-7| = 7$
$7 < 9$
-7 s and counting is closer to blastoff than -9 s and counting.

SECTION 3.2

You Try It 1 $-38 + (-62) = -100$ • The signs of the addends are the same.

You Try It 2 $47 + (-53) = -6$ • The signs of the addends are different.

You Try It 3 $-36 + 17 + (-21) = -19 + (-21)$
$= -40$

You Try It 4 $-154 + (-37) = -191$

You Try It 5 $-x + y$
$-(-3) + (-10) = 3 + (-10) = -7$

You Try It 6 $\dfrac{2 = 11 + a}{2 \mid 11 + (-9)}$ • Replace a by -9.
$2 = 2$
Yes, -9 is a solution of the equation.

You Try It 7 $-35 - (-34)$ • Rewrite "−" as
$= -35 + 34$ "+". The opposite
$= -1$ of −34 is 34.

You Try It 8 $83 - (-29)$ • Rewrite "−" as
$= 83 + 29$ "+". The opposite
$= 112$ of −29 is 29.

You Try It 9   The boiling point of xenon is -108. The melting point of xenon is -112.
$-108 - (-112) = -108 + 112$
$= 4$
The difference is 4°C.

You Try It 10 $-8 - 14$ • Rewrite "−" as
$= -8 + (-14)$ "+". The opposite
$= -22$ of 14 is −14.

You Try It 11 $25 - 68$ • Rewrite "−" as
$= 25 + (-68)$ "+". The opposite
$= -43$ of 68 is −68.

You Try It 12 $-4 - (-3) + 12 - (-7) - 20$
$= -4 + 3 + 12 + 7 + (-20)$
$= -1 + 12 + 7 + (-20)$
$= 11 + 7 + (-20)$
$= 18 + (-20)$
$= -2$

You Try It 13 $x - y$
$-9 - 7 = -9 + (-7)$
$= -16$

You Try It 14 $\dfrac{a - 5 = -8}{-3 - 5 \mid -8}$ • Replace a by -3.
$-3 + (-5) \mid -8$
$-8 = -8$
Yes, -3 is a solution of the equation.

You Try It 15

Strategy   To find the difference, subtract the lowest melting point shown (-259) from the highest melting point shown (181).

Solution $181 - (-259) = 181 + 259$
$= 440$
The difference is 440°C.

You Try It 16

Strategy   To find the temperature, add the increase (10) to the previous temperature (-3).

Solution $-3 + 10 = 7$
The temperature is 7°C.

You Try It 17

Strategy   To find the difference, subtract the lower temperature (-70) from the higher temperature (57).

Solution $57 - (-70) = 57 + 70$
$= 127$

The difference between the average temperatures is 127°F.

You Try It 18

Strategy   To find d, replace a by -6 and b by 5 in the given formula and solve for d.

Solution $d = |a - b|$
$d = |-6 - 5|$
$d = |-11|$
$d = 11$

The distance between the two points is 11 units.

SECTION 3.3

You Try It 1 $-38(51) = -1938$ • **The signs are different. The product is negative.**

You Try It 2 $-7(-8)(9)(-2) = 56(9)(-2)$
$= 504(-2)$
$= -1008$

You Try It 3 $-9y$
$-9(20) = -180$

You Try It 4 $\dfrac{12 = -4a}{12 \mid -4(-3)}$
$12 = 12$

Yes, -3 is a solution of the equation.

You Try It 5 $0 \div (-17) = 0$ • **If $a \neq 0, \dfrac{0}{a} = 0$.**

You Try It 6 $\dfrac{84}{-6} = -14$ • **The signs are different. The quotient is negative.**

You Try It 7   Any number divided by 1 is the number.
$x \div 1 = x$

You Try It 8 $\dfrac{a}{-b}$
$\dfrac{-14}{-(-7)} = \dfrac{-14}{7} = -2$

You Try It 9 $\dfrac{-6}{y} = -2$
$\dfrac{-6}{-3} \mid -2$ • **Replace y by -3.**
$2 \neq -2$

No, -3 is not a solution of the equation.

You Try It 10

Strategy   To find the average daily high temperature:
• Add the seven temperature readings.
• Divide by 7.

Solution
$-7 + (-8) + 0 + (-1) + (-6) + (-11) + (-2) = -35$
$-35 \div 7 = -5$

The average daily high temperature was $-5°$.

SECTION 3.4

You Try It 1
$-\dfrac{5}{12} + \dfrac{5}{8} + \left(-\dfrac{1}{6}\right)$
$= \dfrac{-5}{12} + \dfrac{5}{8} + \dfrac{-1}{6}$ • **Rewrite with negative signs in the numerators.**
$= \dfrac{-10}{24} + \dfrac{15}{24} + \dfrac{-4}{24}$ • **The LCD is 24.**
$= \dfrac{-10 + 15 + (-4)}{24}$ • **Add the numerators.**
$= \dfrac{1}{24}$

You Try It 2 $-\dfrac{5}{6} - \dfrac{7}{9} = \dfrac{-5}{6} - \dfrac{7}{9}$
$= \dfrac{-15}{18} - \dfrac{14}{18}$
$= \dfrac{-15 - 14}{18}$
$= \dfrac{-29}{18}$
$= -\dfrac{29}{18} = -1\dfrac{11}{18}$

You Try It 3 $4.002 - 9.378 = 4.002 + (-9.378)$
$= -5.376$

You Try It 4 $x + y + z$
$-7.84 + (-3.05) + 2.19$
$= -10.89 + 2.19$
$= -8.7$

You Try It 5 $\dfrac{2}{3} - v = \dfrac{11}{12}$
$\dfrac{2}{3} - \left(-\dfrac{1}{4}\right) \mid \dfrac{11}{12}$ • **Replace v by $-\dfrac{1}{4}$.**
$\dfrac{2}{3} + \dfrac{1}{4} \mid \dfrac{11}{12}$
$\dfrac{8}{12} + \dfrac{3}{12} \mid \dfrac{11}{12}$
$\dfrac{11}{12} = \dfrac{11}{12}$

Yes, $-\dfrac{1}{4}$ is a solution of the equation.

You Try It 6

$$-\frac{1}{3}\left(-\frac{5}{12}\right)\left(\frac{8}{15}\right) = \frac{1}{3} \cdot \frac{5}{12} \cdot \frac{8}{15}$$

• The product of two negative fractions is positive.

$$= \frac{1 \cdot 5 \cdot 8}{3 \cdot 12 \cdot 15}$$

$$= \frac{1 \cdot 5 \cdot 2 \cdot 2 \cdot 2}{3 \cdot 2 \cdot 2 \cdot 3 \cdot 3 \cdot 5}$$

$$= \frac{2}{27}$$

You Try It 7

$$\left(3\frac{6}{7}\right)\left(-\frac{4}{9}\right) = -\left(3\frac{6}{7} \cdot \frac{4}{9}\right)$$

• The signs are different. The product is negative.

$$= -\left(\frac{27}{7} \cdot \frac{4}{9}\right)$$

$$= -\left(\frac{27 \cdot 4}{7 \cdot 9}\right)$$

$$= -\left(\frac{3 \cdot 3 \cdot 3 \cdot 4}{7 \cdot 3 \cdot 3}\right)$$

$$= -\frac{12}{7} = -1\frac{5}{7}$$

You Try It 8

$$4 \div \left(-\frac{6}{7}\right) = -\left(\frac{4}{1} \div \frac{6}{7}\right)$$

• The signs are different. The quotient is negative.

$$= -\left(\frac{4}{1} \cdot \frac{7}{6}\right)$$

$$= -\frac{4 \cdot 7}{1 \cdot 6}$$

$$= -\frac{2 \cdot 2 \cdot 7}{1 \cdot 2 \cdot 3} = -\frac{14}{3} = -4\frac{2}{3}$$

You Try It 9

$(-0.7)(-5.8) = 4.06$

• The signs are the same. The product is positive.

You Try It 10

$-25.7 \div 0.31 \approx -82.9$

• The signs are different. The quotient is negative.

You Try It 11 xy

$$\left(-5\frac{1}{8}\right)\left(-\frac{2}{3}\right) = \frac{41}{8} \cdot \frac{2}{3}$$

• The signs are the same. The product is positive.

$$= \frac{41 \cdot 2}{8 \cdot 3}$$

$$= \frac{41 \cdot 2}{2 \cdot 2 \cdot 2 \cdot 3}$$

$$= \frac{41}{12} = 3\frac{5}{12}$$

You Try It 12 $\dfrac{x}{y}$

$$\frac{-40.6}{-0.7} = 58$$

You Try It 13 $25xy$

$$25(-0.8)(0.6) = -20(0.6) = -12$$

You Try It 14

$$-2 = \frac{d}{-0.6}$$

$$-2 \,\bigg|\, \frac{-1.2}{-0.6}$$

• Replace d by -1.2.

$$-2 \neq 2$$

No, -1.2 is not a solution of the equation.

You Try It 15

Strategy To find how many degrees the temperature fell, subtract the lower temperature (-13.33) from the higher temperature (12.78).

Solution

$$12.78 - (-13.33) = 12.78 + 13.33$$
$$= 26.11$$

The temperature fell 26.11°C in the 15-minute period.

SECTION 3.5

You Try It 1

$$8 \div 4 \cdot 4 - (-2)^2 = 8 \div 4 \cdot 4 - 4$$ • Exponents
$$= 2 \cdot 4 - 4$$ • Division
$$= 8 - 4$$ • Multiplication
$$= 4$$ • Subtraction

You Try It 2

$$\left(-\frac{1}{2}\right)^3 \cdot \frac{7-3}{4-9} + \frac{4}{5}$$

$$= \left(-\frac{1}{2}\right)^3 \cdot \frac{4}{-5} + \frac{4}{5}$$ • Simplify above and below the fraction bar.

$$= -\frac{1}{8} \cdot \frac{4}{-5} + \frac{4}{5}$$ • Exponents

$$= \frac{1}{10} + \frac{4}{5}$$ • Multiplication

$$= \frac{1}{10} + \frac{8}{10} = \frac{9}{10}$$ • Addition

You Try It 3 $3a - 4b$

$$3(-2) - 4(5) = -6 - 4(5)$$
$$= -6 - 20$$
$$= -6 + (-20)$$
$$= -26$$

Solutions to Chapter 4 "You Try It"

SECTION 4.1

You Try It 1 -4 is the constant term.

You Try It 2 $2xy + y^2$
$$2(-4)(2) + (2)^2$$
$$= 2(-4)(2) + 4$$
$$= (-8)(2) + 4$$
$$= (-16) + 4$$
$$= -12$$

You Try It 3 $\dfrac{a^2 + b^2}{a + b}$

$$\dfrac{5^2 + (-3)^2}{5 + (-3)} = \dfrac{25 + 9}{5 + (-3)}$$
$$= \dfrac{34}{2}$$
$$= 17$$

You Try It 4 $x^3 - 2(x + y) + z^2$
$$(2)^3 - 2[2 + (-4)] + (-3)^2$$
$$= 8 - 2(-2) + 9$$
$$= 8 + 4 + 9$$
$$= 12 + 9$$
$$= 21$$

SECTION 4.2

You Try It 1 $3a - 2b - 5a + 6b = -2a + 4b$

You Try It 2 $-3y^2 + 7 + 8y^2 - 14 = 5y^2 - 7$

You Try It 3 $-5(4y^2) = -20y^2$

You Try It 4 $-7(-2a) = 14a$

You Try It 5 $-\dfrac{3}{5}\left(-\dfrac{7}{9}a\right) = \dfrac{7}{15}a$

You Try It 6 $5(3 + 7b) = 15 + 35b$

You Try It 7 $(3a - 1)5 = 15a - 5$

You Try It 8 $-8(-2a + 7b) = 16a - 56b$

You Try It 9 $3(12x^2 - x + 8) = 36x^2 - 3x + 24$

You Try It 10 $3(-a^2 - 6a + 7) = -3a^2 - 18a + 21$

You Try It 11 $3y - 2(y - 7x) = 3y - 2y + 14x$
$$= y + 14x$$

You Try It 12
$$-2(x - 2y) - (-x + 3y) = -2x + 4y + x - 3y$$
$$= -x + y$$

You Try It 13
$$3y - 2[x - 4(2 - 3y)] = 3y - 2[x - 8 + 12y]$$
$$= 3y - 2x + 16 - 24y$$
$$= -2x - 21y + 16$$

SECTION 4.3

You Try It 1   the <u>difference between</u> <u>twice</u> n and <u>one-third of</u> n

$$2n - \dfrac{1}{3}n$$

You Try It 2   the <u>quotient of</u> 7 <u>less than</u> b and 15

$$\dfrac{b - 7}{15}$$

You Try It 3   the unknown number: n
the cube of the number: n^3
the total of ten and the cube of the number: $10 + n^3$

$$-4(10 + n^3)$$

You Try It 4   the unknown number: x
the difference between the number and sixty: $x - 60$

$$5(x - 60)$$
$$= 5x - 300$$

You Try It 5   the speed of the older model: s
the speed of the new jet plane is twice the speed of the older model: $2s$

You Try It 6   the length of the longer piece: y
the length of the shorter piece: $6 - y$

Solutions to Chapter 5 "You Try It"

SECTION 5.1

You Try It 1

$$5 - 4x = 8x + 2$$

$$\frac{5 - 4\left(\dfrac{1}{4}\right)}{\;} \left|\; 8\left(\dfrac{1}{4}\right) + 2\right.$$

• Replace x with $\dfrac{1}{4}$.

$$5 - 1 \;\big|\; 2 + 2$$
$$4 = 4$$

Yes, $\dfrac{1}{4}$ is a solution.

You Try It 2

$$10x - x^2 = 3x - 10$$

$$\frac{10(5) - (5)^2 \;\big|\; 3(5) - 10}{50 - 25 \;\big|\; 15 - 10}$$
$$25 \neq 5$$

No, 5 is not a solution.

You Try It 3

$$\frac{5}{6} = y - \frac{3}{8}$$

$$\frac{5}{6} + \frac{3}{8} = y - \frac{3}{8} + \frac{3}{8}$$

$$\frac{29}{24} = y$$

The solution is $\dfrac{29}{24}$.

You Try It 4

$$-\frac{2}{5}x = 6$$

$$\left(-\frac{5}{2}\right)\left(-\frac{2}{5}x\right) = \left(-\frac{5}{2}\right)(6)$$

$$x = -15$$

The solution is -15.

You Try It 5

$$4x - 8x = 16$$
$$-4x = 16$$
$$\frac{-4x}{-4} = \frac{16}{-4}$$
$$x = -4$$

The solution is -4.

You Try It 6

Strategy To find the distance, solve the equation $d = rt$ for d. The time is 3 h. Therefore, $t = 3$. The plane is moving against the wind, which means the headwind is slowing the actual speed of the plane. 250 mph − 25 mph = 225 mph. Thus $r = 225$.

Solution
$$d = rt$$
$$d = 225(3) \qquad • \; r = 225,\ t = 3$$
$$= 675$$

The plane travels 675 mi in 3 h.

SECTION 5.2

You Try It 1

$$5x + 7 = 10$$
$$5x + 7 - 7 = 10 - 7 \qquad • \text{ Subtract 7.}$$
$$5x = 3$$
$$\frac{5x}{5} = \frac{3}{5} \qquad • \text{ Divide by 5.}$$
$$x = \frac{3}{5}$$

The solution is $\dfrac{3}{5}$.

You Try It 2

$$2 = 11 + 3x$$
$$2 - 11 = 11 - 11 + 3x \qquad • \text{ Subtract 11.}$$
$$-9 = 3x$$
$$\frac{-9}{3} = \frac{3x}{3} \qquad • \text{ Divide by 3.}$$
$$-3 = x$$

The solution is -3.

You Try It 3

$$\frac{5}{8} - \frac{2x}{3} = \frac{5}{4}$$

$$\frac{5}{8} - \frac{5}{8} - \frac{2}{3}x = \frac{5}{4} - \frac{5}{8} \qquad • \text{ Recall that } \frac{2x}{3} = \frac{2}{3}x.$$

$$-\frac{2}{3}x = \frac{5}{8}$$

$$-\frac{3}{2}\left(-\frac{2}{3}x\right) = -\frac{3}{2}\left(\frac{5}{8}\right) \qquad • \text{ Multiply by } -\frac{3}{2}.$$

$$x = -\frac{15}{16}$$

The solution is $-\dfrac{15}{16}$.

You Try It 4

$$\frac{2}{3}x + 3 = \frac{7}{2}$$

$$6\left(\frac{2}{3}x + 3\right) = 6\left(\frac{7}{2}\right)$$

$$6\left(\frac{2}{3}x\right) + 6(3) = 6\left(\frac{7}{2}\right) \qquad • \text{ Distributive Property}$$

$$4x + 18 = 21$$

$$4x + 18 - 18 = 21 - 18 \qquad • \text{ Subtract 18.}$$

$$4x = 3$$

$$\frac{4x}{4} = \frac{3}{4} \qquad • \text{ Divide by 4.}$$

$$x = \frac{3}{4}$$

The solution is $\dfrac{3}{4}$.

You Try It 5

$$x - 5 + 4x = 25$$
$$5x - 5 = 25$$
$$5x - 5 + 5 = 25 + 5$$
$$5x = 30$$
$$\frac{5x}{5} = \frac{30}{5}$$
$$x = 6$$

The solution is 6.

You Try It 6

Strategy   Given: $P = 45$
Unknown: D

Solution

$$P = 15 + \frac{1}{2}D$$
$$45 = 15 + \frac{1}{2}D$$
$$45 - 15 = 15 - 15 + \frac{1}{2}D$$
$$30 = \frac{1}{2}D$$
$$2(30) = 2 \cdot \frac{1}{2}D$$
$$60 = D$$

The depth is 60 ft.

SECTION 5.3

You Try It 1

$$5x + 4 = 6 + 10x$$
$$5x - 10x + 4 = 6 + 10x - 10x \quad \bullet \text{ Subtract } 10x.$$
$$-5x + 4 = 6$$
$$-5x + 4 - 4 = 6 - 4 \quad \bullet \text{ Subtract } 4.$$
$$-5x = 2$$
$$\frac{-5x}{-5} = \frac{2}{-5} \quad \bullet \text{ Divide by } -5.$$
$$x = -\frac{2}{5}$$

The solution is $-\frac{2}{5}$.

You Try It 2

$$5x - 10 - 3x = 6 - 4x$$
$$2x - 10 = 6 - 4x \quad \bullet \text{ Combine like terms.}$$
$$2x + 4x - 10 = 6 - 4x + 4x \quad \bullet \text{ Add } 4x.$$
$$6x - 10 = 6$$
$$6x - 10 + 10 = 6 + 10 \quad \bullet \text{ Add } 10.$$
$$6x = 16$$
$$\frac{6x}{6} = \frac{16}{6} \quad \bullet \text{ Divide by } 6.$$
$$x = \frac{8}{3}$$

The solution is $\frac{8}{3}$.

You Try It 3

$$5x - 4(3 - 2x) = 2(3x - 2) + 6$$
$$5x - 12 + 8x = 6x - 4 + 6 \quad \bullet \text{ Distributive Property}$$
$$13x - 12 = 6x + 2$$
$$13x - 6x - 12 = 6x - 6x + 2 \quad \bullet \text{ Subtract } 6x.$$
$$7x - 12 = 2$$
$$7x - 12 + 12 = 2 + 12 \quad \bullet \text{ Add } 12.$$
$$7x = 14$$
$$\frac{7x}{7} = \frac{14}{7} \quad \bullet \text{ Divide by } 7.$$
$$x = 2$$

The solution is 2.

You Try It 4

$$-2[3x - 5(2x - 3)] = 3x - 8$$
$$-2[3x - 10x + 15] = 3x - 8 \quad \bullet \text{ Distributive Property}$$
$$-2[-7x + 15] = 3x - 8$$
$$14x - 30 = 3x - 8 \quad \bullet \text{ Distributive Property}$$
$$14x - 3x - 30 = 3x - 3x - 8 \quad \bullet \text{ Subtract } 3x.$$
$$11x - 30 = -8$$
$$11x - 30 + 30 = -8 + 30 \quad \bullet \text{ Add } 30.$$
$$11x = 22$$
$$\frac{11x}{11} = \frac{22}{11} \quad \bullet \text{ Divide by } 11.$$
$$x = 2$$

The solution is 2.

You Try It 5

Strategy   Given: $F_1 = 45$
$F_2 = 80$
$d = 25$
Unknown: x

Solution

$$F_1 x = F_2(d - x)$$
$$45x = 80(25 - x)$$
$$45x = 2000 - 80x$$
$$45x + 80x = 2000 - 80x + 80x$$
$$125x = 2000$$
$$\frac{125x}{125} = \frac{2000}{125}$$
$$x = 16$$

The fulcrum is 16 ft from the 45-pound force.

SECTION 5.4

You Try It 1

The smaller number: n
The larger number: $12 - n$

The total of three times the smaller number and six	amounts to	seven less than the product of four and the larger number

$$3n + 6 = 4(12 - n) - 7$$
$$3n + 6 = 48 - 4n - 7$$
$$3n + 6 = 41 - 4n$$
$$3n + 4n + 6 = 41 - 4n + 4n$$
$$7n + 6 = 41$$
$$7n + 6 - 6 = 41 - 6$$
$$7n = 35$$
$$\frac{7n}{7} = \frac{35}{7}$$
$$n = 5$$

$$12 - n = 12 - 5 = 7$$

The smaller number is 5.
The larger number is 7.

You Try It 2

Strategy • First integer: n
  Second integer: $n + 1$
  Third integer: $n + 2$
  • The sum of the three integers is -6.

Solution $n + (n + 1) + (n + 2) = -6$
$$3n + 3 = -6$$
$$3n = -9$$
$$n = -3$$
$$n + 1 = -3 + 1 = -2$$
$$n + 2 = -3 + 2 = -1$$

The three consecutive integers are -3, -2, and -1.

You Try It 3

Strategy
To find the number of tickets that you are purchasing, write and solve an equation using x to represent the number of tickets you are purchasing.

Solution

$3.50 plus $17.50 for each ticket	is	$161

$$3.50 + 17.50x = 161$$
$$3.50 - 3.50 + 17.50x = 161 - 3.50$$
$$17.50x = 157.50$$
$$\frac{17.50x}{17.50} = \frac{157.50}{17.50}$$
$$x = 9$$

You are purchasing 9 tickets.

You Try It 4

Strategy
To find the length, write and solve an equation using x to represent the length of the shorter piece and $22 - x$ to represent the length of the longer piece.

Solution

The length of the longer piece	is	4 in. more than twice the length of the shorter piece

$$22 - x = 2x + 4$$
$$22 - x - 2x = 2x - 2x + 4$$
$$22 - 3x = 4$$
$$22 - 22 - 3x = 4 - 22$$
$$-3x = -18$$
$$\frac{-3x}{-3} = \frac{-18}{-3}$$
$$x = 6$$

$$22 - x = 22 - 6 = 16$$

The length of the shorter piece is 6 in.
The length of the longer piece is 16 in.

SECTION 5.5

You Try It 1

Strategy • Pounds of $.55 fertilizer: x

	Amount	Cost	Value
$.80 fertilizer	20	.80	0.80(20)
$.55 fertilizer	x	.55	0.55x
$.75 fertilizer	20 + x	.75	0.75(20 + x)

• The sum of the values before mixing equals the value after mixing.

Solution $0.80(20) + 0.55x = 0.75(20 + x)$
$$16 + 0.55x = 15 + 0.75x$$
$$16 - 0.20x = 15$$
$$-0.20x = -1$$
$$x = 5$$

5 lb of the $.55 fertilizer must be added.

You Try It 2

Strategy • Rate of the first train: r
  Rate of the second train: $2r$

	Rate	Time	Distance
1st train	r	3	3r
2nd train	2r	3	3(2r)

• The sum of the distances traveled by the two trains equals 288 mi.

	Rate	Time	Distance
Out	150	t	$150t$
Back	100	$5 - t$	$100(5 - t)$

Solution

$$3r + 3(2r) = 288$$
$$3r + 6r = 288$$
$$9r = 288$$
$$r = 32$$

$$2r = 2(32) = 64$$

The first train is traveling at 32 mph. The second train is traveling at 64 mph.

You Try It 3

Strategy

• Time spent flying out: t
  Time spent flying back: $5 - t$

• The distance out equals the distance back.

Solution

$$150t = 100(5 - t)$$
$$150t = 500 - 100t$$
$$250t = 500$$
$$t = 2 \quad \text{(The time out was 2 h.)}$$

$$\text{The distance out} = 150t = 150(2)$$
$$= 300 \text{ mi}$$

The parcel of land was 300 mi away.

Solutions to Chapter 6 "You Try It"

SECTION 6.1

You Try It 1

$$\frac{12}{20} = \frac{3}{5}$$

$$12 : 20 = 3 : 5$$
$$12 \text{ to } 20 = 3 \text{ to } 5$$

You Try It 2

$$\frac{20 \text{ bags}}{8 \text{ acres}} = \frac{5 \text{ bags}}{2 \text{ acres}}$$

You Try It 3

$$\frac{\$8.96}{3.5 \text{ lb}}$$

$$8.96 \div 3.5 = 2.56$$

The unit rate is $2.56/lb.

SECTION 6.2

You Try It 1

$$\frac{50}{3} \diagdown \frac{250}{12} \longrightarrow 3 \cdot 250 = 750$$
$$\longrightarrow 50 \cdot 12 = 600$$

$$750 \neq 600$$

The proportion is not true.

You Try It 2

$$\frac{7}{12} = \frac{42}{x}$$

$$12 \cdot 42 = 7 \cdot x \quad \bullet \text{ The cross products}$$
$$504 = 7x \quad \text{ are equal.}$$
$$72 = x$$

You Try It 3

$$\frac{5}{n} = \frac{3}{322}$$

$$n \cdot 3 = 5 \cdot 322 \quad \bullet \text{ The cross products}$$
$$3n = 1610 \quad \text{ are equal.}$$
$$\frac{3n}{3} = \frac{1610}{3}$$
$$n \approx 536.67$$

You Try It 4

$$\frac{4}{5} = \frac{3}{x - 3}$$

$$5 \cdot 3 = 4(x - 3) \quad \bullet \text{ The cross products}$$
$$15 = 4x - 12 \quad \text{ are equal.}$$
$$27 = 4x$$
$$6.75 = x$$

You Try It 5

Strategy

To find the number of gallons, write and solve a proportion using n to represent the number of gallons needed to travel 832 mi.

Solution

$$\frac{396 \text{ mi}}{11 \text{ gal}} = \frac{832 \text{ mi}}{n \text{ gal}}$$
$$11 \cdot 832 = 396 \cdot n$$
$$9152 = 396n$$
$$23.1 \approx n$$

To travel 832 mi, approximately 23.1 gal of gas are needed.

You Try It 6

Strategy

To find the number of defective transmissions, write and solve a proportion using n to represent the number of defective transmissions in 120,000 cars.

Solution

$$\frac{15 \text{ defective transmissions}}{1200 \text{ cars}} = \frac{n \text{ defective transmissions}}{120,000 \text{ cars}}$$
$$1200 \cdot n = 15 \cdot 120,000$$
$$1200n = 1,800,000$$
$$n = 1500$$

1500 defective transmissions would be found in 120,000 cars.

SECTION 6.3

You Try It 1 $110\% = 110\left(\dfrac{1}{100}\right) = \left(\dfrac{110}{100}\right) = 1\dfrac{1}{10}$

$110\% = 110(0.01) = 1.10$

You Try It 2 $16\dfrac{3}{8}\% = 16\dfrac{3}{8}\left(\dfrac{1}{100}\right) = \dfrac{131}{8}\left(\dfrac{1}{100}\right)$

$= \dfrac{131}{800}$

You Try It 3 $0.8\% = 0.8(0.01) = 0.008$

You Try It 4 $0.038 = 0.038(100\%) = 3.8\%$

You Try It 5 $\dfrac{9}{7} = \dfrac{9}{7}(100\%) = \dfrac{900}{7}\% = 128\dfrac{4}{7}\%$

You Try It 6 $1\dfrac{5}{9} = \dfrac{14}{9} = \dfrac{14}{9}(100\%) = \dfrac{1400}{9}\%$

$\approx 155.6\%$

SECTION 6.4

You Try It 1

Strategy   To find the amount, solve the basic percent equation.

Percent $= 33\dfrac{1}{3}\% = \dfrac{1}{3}$, base $= 45$,

amount $= n$

Solution   Percent $\cdot$ base $=$ amount

$\dfrac{1}{3}(45) = n$

$15 = n$

15 is $33\dfrac{1}{3}\%$ of 45.

You Try It 2

Strategy   To find the percent, solve the basic percent equation. Percent $= n$, base $= 40$, amount $= 25$

Solution   Percent $\cdot$ base $=$ amount

$n \cdot 40 = 25$

$\dfrac{40n}{40} = \dfrac{25}{40}$

$n = 0.625 = 62.5\%$

25 is 62.5% of 40.

You Try It 3

Strategy   To find the base, solve the basic percent equation.

Percent $= 16\dfrac{2}{3}\% = \dfrac{1}{6}$, base $= n$,

amount $= 15$

Solution   Percent $\cdot$ base $=$ amount

$\dfrac{1}{6} \cdot n = 15$

$6 \cdot \dfrac{1}{6}n = 15 \cdot 6$

$n = 90$

$16\dfrac{2}{3}\%$ of 90 is 15.

You Try It 4   Percent $= 25$, base $= n$, amount $= 8$

$\dfrac{25}{100} = \dfrac{8}{n}$

$25 \cdot n = 100 \cdot 8$

$25n = 800$

$\dfrac{25n}{25} = \dfrac{800}{25}$

$n = 32$

8 is 25% of 32.

You Try It 5   Percent $= 0.74$, base $= 1200$,
amount $= n$

$\dfrac{0.74}{100} = \dfrac{n}{1200}$

$100 \cdot n = 0.74 \cdot 1200$

$100n = 888$

$\dfrac{100n}{100} = \dfrac{888}{100}$

$n = 8.88$

0.74% of 1200 is 8.88.

You Try It 6   Percent $= n$, base $= 180$, amount $= 54$

$\dfrac{n}{100} = \dfrac{54}{180}$

$n \cdot 180 = 100 \cdot 54$

$180n = 5400$

$\dfrac{180n}{180} = \dfrac{5400}{180}$

$n = 30$

30% of 180 is 54.

You Try It 7

Strategy   To find the percent, use the basic percent equation. Percent $= n$, base $= 4330$, amount $= 649.50$

Solution   Percent $\cdot$ base $=$ amount

$n \cdot 4330 = 649.50$

$\dfrac{4330n}{4330} = \dfrac{649.50}{4330}$

$n = 0.15$

15% of the instructor's salary is deducted for income tax.

You Try It 8

Strategy   To find the number, solve the basic percent equation. Percent = 19% = 0.19, base = 2.4 million, amount = n

Solution   Percent · base = amount
$$0.19 \cdot 2.4 = n$$
$$0.456 = n$$
0.456 million = 456,000

There are approximately 456,000 female surfers in this country.

You Try It 9

Strategy   To find the increase in the hourly wage:
- Find last year's wage. Solve the basic percent equation. Percent = 115% = 1.15, base = n, amount = 30.13
- Subtract last year's wage from this year's wage.

Solution   Percent · base = amount
$$1.15 \cdot n = 30.13$$
$$\frac{1.15n}{1.15} = \frac{30.13}{1.15}$$
$$n = 26.20$$
$$30.13 - 26.20 = 3.93$$

The increase in the hourly wage was $3.93.

SECTION 6.5

You Try It 1

Strategy   To calculate the maturity value:
- Find the simple interest due on the loan by solving the simple interest formula for I.
$$t = \frac{8}{12}, P = 12{,}500, r = 9.5\% = 0.095$$
- Use the formula for the maturity value of a simple interest loan, $M = P + I$.

Solution $I = Prt$
$$I = 12{,}500(0.095)\left(\frac{8}{12}\right)$$
$$I \approx 791.67$$

$M = P + I$
$M = 12{,}500 + 791.67$
$M = 13{,}291.67$

The total amount due on the loan is $13,291.67.

Solutions to Chapter 7 "You Try It"

SECTION 7.1

You Try It 1 $QR + RS + ST = QT$
$$24 + RS + 17 = 62$$
$$41 + RS = 62$$
$$RS = 21$$
$RS = 21$ cm

You Try It 2 $AC = AB + BC$
$$AC = \frac{1}{4}(BC) + BC$$
$$AC = \frac{1}{4}(16) + 16$$
$$AC = 4 + 16$$
$$AC = 20$$
$$AC = 20 \text{ ft}$$

You Try It 3

Strategy   Supplementary angles are two angles whose sum is 180°. To find the supplement, let x represent the supplement of a 129° angle. Write an equation and solve for x.

Solution $x + 129° = 180°$
$$x = 51°$$
The supplement of a 129° angle is a 51° angle.

You Try It 4

Strategy   To find the measure of $\angle a$, write an equation using the fact that the sum of the measures of $\angle a$ and 68° is 118°. Solve for $\angle a$.

Solution $\angle a + 68° = 118°$
              $\angle a = 50°$

The measure of $\angle a$ is 50°.

You Try It 5

Strategy   The angles labeled are adjacent angles of intersecting lines and are therefore supplementary angles. To find x, write an equation and solve for x.

Solution $(x + 16°) + 3x = 180°$
              $4x + 16° = 180°$
              $4x = 164°$
              $x = 41°$

You Try It 6

Strategy $3x = y$ because corresponding angles have the same measure.
              $y + (x + 40°) = 180°$ because adjacent angles of intersecting lines are supplementary angles. Substitute $3x$ for y and solve for x.

Solution $3x + (x + 40°) = 180°$
              $4x + 40° = 180°$
              $4x = 140°$
              $x = 35°$

You Try It 7

Strategy   • To find the measure of angle b, use the fact that $\angle b$ and $\angle x$ are supplementary angles.
              • To find the measure of angle c, use the fact that the sum of the interior angles of a triangle is 180°.
              • To find the measure of angle y, use the fact that $\angle c$ and $\angle y$ are vertical angles.

Solution $\angle b + \angle x = 180°$
              $\angle b + 100° = 180°$
              $\angle b = 80°$

              $\angle a + \angle b + \angle c = 180°$
              $45° + 80° + \angle c = 180°$
              $125° + \angle c = 180°$
              $\angle c = 55°$

              $\angle y = \angle c = 55°$

You Try It 8

Strategy   To find the measure of the third angle, use the fact that the measure of a right angle is 90° and the fact

that the sum of the measures of the interior angles of a triangle is 180°. Write an equation using x to represent the measure of the third angle. Solve the equation for x.

Solution $x + 90° + 34° = 180°$
              $x + 124° = 180°$
              $x = 56°$

The measure of the third angle is 56°.

SECTION 7.2

You Try It 1

Strategy   To find the perimeter, use the formula for the perimeter of a square. Substitute 60 for s and solve for P.

Solution $P = 4s$
              $P = 4(60)$
              $P = 240$

The perimeter of the infield is 240 ft.

You Try It 2

Strategy   To find the perimeter, use the formula for the perimeter of a rectangle. Substitute 11 for L and $8\frac{1}{2}$ for W and solve for P.

Solution $P = 2L + 2W$
              $P = 2(11) + 2\left(8\frac{1}{2}\right)$
              $P = 2(11) + 2\left(\frac{17}{2}\right)$
              $P = 22 + 17$
              $P = 39$

The perimeter of a standard piece of typing paper is 39 in.

You Try It 3

Strategy   To find the circumference, use the circumference formula that involves the diameter. Leave the answer in terms of π.

Solution $C = \pi d$
              $C = \pi(9)$
              $C = 9\pi$

The circumference is 9π in.

You Try It 4

Strategy

To find the number of rolls of wallpaper to be purchased:
• Use the formula for the area of a rectangle to find the area of one wall.
• Multiply the area of one wall by the number of walls to be covered (2).
• Divide the area of wall to be covered by the area one roll of wallpaper will cover (30).

Solution

$A = LW$
$A = 12 \cdot 8 = 96$ • **The area of one wall is 96 ft².**
$2(96) = 192$ • **The area of the two walls is 192 ft².**
$192 \div 30 = 6.4$

Because a portion of a seventh roll is needed, 7 rolls of wallpaper should be purchased.

You Try It 5

Strategy To find the area, use the formula for the area of a circle. An approximation is asked for; use the π key on a calculator. $r = 11$

Solution $A = \pi r^2$
$A = \pi (11)^2$
$A = 121\pi$
$A \approx 380.13$

The area is approximately 380.13 cm².

SECTION 7.3

You Try It 1

Strategy To find the measure of the other leg, use the Pythagorean Theorem.
$a = 2, c = 6$

Solution $a^2 + b^2 = c^2$
$2^2 + b^2 = 6^2$
$4 + b^2 = 36$
$b^2 = 32$
$b = \sqrt{32}$
$b \approx 5.66$

The measure of the other leg is approximately 5.66 m.

You Try It 2

Strategy To find FG, write a proportion using the fact that, in similar triangles, the ratio of corresponding sides equals the ratio of corresponding heights. Solve the proportion for FG.

Solution $\dfrac{AC}{DF} = \dfrac{CH}{FG}$

$\dfrac{10}{15} = \dfrac{7}{FG}$

$10(FG) = 15(7)$
$10(FG) = 105$
$FG = 10.5$

The height FG of triangle DEF is 10.5 m.

You Try It 3

Strategy To determine whether the triangles are congruent, determine whether one of the rules for congruence is satisfied.

Solution $PR = MN$, $QR = MO$, and $\angle QRP = \angle OMN$. Two sides and the included angle of one triangle are equal to two sides and the included angle of the other triangle.

The triangles are congruent by the SAS Rule.

SECTION 7.4

You Try It 1

Strategy To find the volume, use the formula for the volume of a cube. $s = 2.5$

Solution $V = s^3$
$V = (2.5)^3 = 15.625$

The volume of the cube is 15.625 m².

You Try It 2

Strategy To find the volume:
• Find the radius of the base of the cylinder. $d = 8$
• Use the formula for the volume of a cylinder. Leave the answer in terms of π.

Solution $r = \dfrac{1}{2}d = \dfrac{1}{2}(8) = 4$

$V = \pi r^2 h = \pi (4)^2 (22) = \pi (16)(22)$
$= 352\pi$

The volume of the cylinder is 352π ft³.

You Try It 3

Strategy To find the surface area of the cylinder.
• Find the radius of the base of the cylinder. $d = 6$
• Use the formula for the surface area of a cylinder. An approximation is asked for; use the π key on a calculator.

Solution $r = \dfrac{1}{2}d = \dfrac{1}{2}(6) = 3$

$SA = 2\pi r^2 + 2\pi rh$
$SA = 2\pi(3)^2 + 2\pi(3)(8)$
$ = 2\pi(9) + 2\pi(3)(8)$
$ = 18\pi + 48\pi$
$ = 66\pi$
$ \approx 207.35$

The surface area of the cylinder is approximately 207.35 ft².

You Try It 4

Strategy   To find which solid has the larger surface area:
- Use the formula for the surface area of a cube to find the surface area of the cube. $s = 10$
- Find the radius of the sphere. $d = 8$
- Use the formula for the surface area of a sphere to find the surface area of the sphere. Because this number is to be compared to another number, use the π key on a calculator to approximate the surface area.
- Compare the two numbers.

Solution $SA = 6s^2$
$SA = 6(10)^2 = 6(100) = 600$

The surface area of the cube is 600 cm².

$r = \dfrac{1}{2}d = \dfrac{1}{2}(8) = 4$

$SA = 4\pi r^2$

$SA = 4\pi(4)^2 = 4\pi(16) = 64\pi \approx 201.06$

The surface area of the sphere is 201.06 cm².

$600 > 201.06$

The cube has a larger surface area than the sphere.

Solutions to Chapter 8 "You Try It"

SECTION 8.1

You Try It 1

Strategy   **a.** To find the ratio:
- From the graph, find the percent of lane-change accidents and the percent of road-departure accidents.
- Write the ratio in fractional form. Simplify.

b. To find the number of accidents that occurred at intersections:
- From the graph, find the percent of accidents that occurred at intersections.
- Solve the basic percent equation for amount. The base is 4300.

Solution   **a.** Lane-change accidents: 9%
Road-departure accidents: 21%
$\dfrac{9\%}{21\%} = \dfrac{3}{7}$

The ratio is $\dfrac{3}{7}$.

b. Accidents that occurred at intersections: 26% = 0.26

Percent · base = amount
$0.26 \quad \cdot 4300 = n$
$ 1118 = n$

1118 accidents occurred at intersections in Twin Falls in 2005.

SECTION 8.2

You Try It 1

Strategy   To find the mean amount spent by the 12 customers:
- Find the sum of the amounts.
- Divide the sum by the number of customers (12).

Solution $6.26 + 8.23 + 5.09 + 8.11 + 7.50 + 6.69 + 5.66 + 4.89 + 5.25 + 9.36 + 6.75 + 7.05 = 80.84$

$\bar{x} = \dfrac{80.84}{12} \approx 6.74$

The mean amount spent by the 12 customers was $6.74.

You Try It 2

Strategy   To find the median weight loss:
- Arrange the weight losses from smallest to largest.
- Because there is an even number of values, the median is the mean of the middle two numbers.

Solution 10, 14, 16, 16, 22, 27, 29, 31, 31, 40

$$\text{Median} = \frac{22 + 27}{2} = \frac{49}{2} = 24.5$$

The median weight loss was 24.5 lb.

You Try It 3

Strategy To draw the box-and-whiskers plot:
- Find the median, Q_1, and Q_3.
- Use the smallest value, Q_1, the median, Q_3, and the largest value to draw the box-and-whiskers plot.

Solution

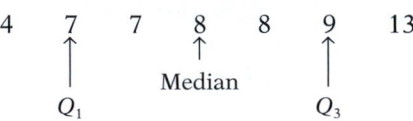

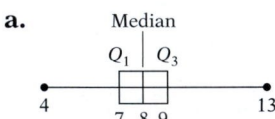

a.

b. Answers about the spread of the data will vary. For example, in You Try It 3, the values in the interquartile range are all very close to the median. They are not so close to the median in Example 3. The whiskers are long with respect to the box in You Try It 3, whereas they are short with respect to the box in Example 3. This shows that the data values outside the interquartile range are closer together in Example 3 than in You Try It 3.

SECTION 8.3

You Try It 1

Strategy To find the probability:
- List the outcomes of the experiment in a systematic way. We will use a table.
- Use the table to count the number of possible outcomes of the experiment.
- Count the number of outcomes of the experiment that are favorable to the event of two "true" answers and one "false" answer.
- Use the probability formula.

Solution

Question 1	Question 2	Question 3
T	T	T
T	T	F
T	F	T
T	F	F
F	T	T
F	T	F
F	F	T
F	F	F

There are 8 possible outcomes:

S = {TTT, TTF, TFT, TFF, FTT, FTF, FFT, FFF}

There are 3 outcomes favorable to the event:

{TTF, TFT, FTT}

Probability of an event
$$= \frac{\text{number of favorable outcomes}}{\text{number of possible outcomes}} = \frac{3}{8}$$

The probability of two "true" answers and one "false" answer is $\frac{3}{8}$.

Solutions to Chapter 9 "You Try It"

SECTION 9.1

You Try It 1
$(-4x^3 + 2x^2 - 8) + (4x^3 + 6x^2 - 7x + 5)$
$= (-4x^3 + 4x^3) + (2x^2 + 6x^2) + (-7x) + (-8 + 5)$
$= 8x^2 - 7x - 3$

You Try It 2
$$\begin{array}{r} 6x^3 \quad\quad + 2x + 8 \\ -9x^3 + 2x^2 - 12x - 8 \\ \hline -3x^3 + 2x^2 - 10x \end{array}$$

You Try It 3
$(-4w^3 + 8w - 8) - (3w^3 - 4w^2 - 2w - 1)$
$= (-4w^3 + 8w - 8)$
$\quad + (-3w^3 + 4w^2 + 2w + 1)$
$= -7w^3 + 4w^2 + 10w - 7$

You Try It 4
$$\begin{array}{r} 13y^3 \quad\quad - 6y - 7 \\ - 4y^2 + 6y + 9 \\ \hline 13y^3 - 4y^2 \quad\quad + 2 \end{array}$$

SECTION 9.2

You Try It 1

$(8m^3n)(-3n^5)$
$= [8(-3)](m^3)(n \cdot n^5)$ • Multiply coefficients. Add
$= -24m^3n^6$ exponents with same base.

You Try It 2

$(12p^4q^3)(-3p^5q^2)$
$= [12(-3)](p^4 \cdot p^5)(q^3 \cdot q^2)$ • Multiply coefficients. Add
$= -36p^9q^5$ exponents with same base.

You Try It 3

$(-3a^4bc^2)^3 = (-3)^{1 \cdot 3}a^{4 \cdot 3}b^{1 \cdot 3}c^{2 \cdot 3}$ • Rule for Simplifying
$= (-3)^3a^{12}b^3c^6$ the Power of a Product
$= -27a^{12}b^3c^6$

You Try It 4

$(-xy^4)(-2x^3y^2)^2 = (-xy^4)[(-2)^{1 \cdot 2}x^{3 \cdot 2}y^{2 \cdot 2}]$ • Rule for
$= (-xy^4)[(-2)^2x^6y^4]$ Simplifying
$= (-xy^4)(4x^6y^4)$ the Power of
$= -4x^7y^8$ a Product

SECTION 9.3

You Try It 1 $(-2y + 3)(-4y) = 8y^2 - 12y$

You Try It 2 $-a^2(3a^2 + 2a - 7) = -3a^4 - 2a^3 + 7a^2$

You Try It 3

$$
\begin{array}{r}
2y^3 + 2y^2 \qquad\quad - 3 \\
3y - 1 \\
\hline
- 2y^3 - 2y^2 \qquad + 3 = -1(2y^3 + 2y^2 - 3) \\
6y^4 + 6y^3 \qquad\quad - 9y \quad = 3y(2y^3 + 2y^2 - 3) \\
\hline
6y^4 + 4y^3 - 2y^2 - 9y + 3
\end{array}
$$

You Try It 4

$(4y - 5)(2y - 3) = 8y^2 - 12y - 10y + 15$
$= 8y^2 - 22y + 15$

You Try It 5

$(3b + 2)(3b - 5) = 9b^2 - 15b + 6b - 10$
$= 9b^2 - 9b - 10$

You Try It 6 $(2a + 5c)(2a - 5c) = 4a^2 - 25c^2$

You Try It 7 $(3x + 2y)^2 = 9x^2 + 12xy + 4y^2$

You Try It 8

Strategy   To find the area, replace the variable r in the equation $A = \pi r^2$ with $(x - 4)$ and solve for A.

Solution $A = \pi r^2$
$A = \pi(x - 4)^2$
$A = \pi(x^2 - 8x + 16)$
$A = \pi x^2 - 8\pi x + 16\pi$

The area of the circle is $(\pi x^2 - 8\pi x + 16\pi)$ ft².

SECTION 9.4

You Try It 1 $(-2x^2)(x^{-3}y^{-4})^{-2}$
$= (-2x^2)(x^6y^8)$ • Rule for Simplifying
$= -2x^8y^8$ the Power of a Product

You Try It 2 $\dfrac{(6a^{-2}b^3)^{-1}}{(4a^3b^{-2})^{-2}}$

$= \dfrac{6^{-1}a^2b^{-3}}{4^{-2}a^{-6}b^4}$ • Rule for Simplifying
                                               the Power of a Product
$= 4^2(6^{-1}a^8b^{-7})$ • Rule for Dividing
$= \dfrac{16a^8}{6b^7} = \dfrac{8a^8}{3b^7}$ Exponential Expressions

You Try It 3 $\left[\dfrac{6r^3s^{-3}}{9r^3s^{-1}}\right]^{-2} = \left[\dfrac{2r^0s^{-2}}{3}\right]^{-2}$

$= \dfrac{2^{-2}s^4}{3^{-2}} = \dfrac{9s^4}{4}$

You Try It 4 $0.000000961 = 9.61 \times 10^{-7}$

You Try It 5 $7.329 \times 10^6 = 7,329,000$

SECTION 9.5

You Try It 1

$\dfrac{24x^2y^2 - 18xy + 6y}{6xy} = \dfrac{24x^2y^2}{6xy} - \dfrac{18xy}{6xy} + \dfrac{6y}{6xy}$

$= 4xy - 3 + \dfrac{1}{x}$

You Try It 2

$$
\begin{array}{r}
x^2 + 2x - 1 \\
2x - 3 \overline{)2x^3 + x^2 - 8x - 3} \\
\underline{2x^3 - 3x^2} \\
4x^2 - 8x \\
\underline{4x^2 - 6x} \\
- 2x - 3 \\
\underline{- 2x + 3} \\
-6
\end{array}
$$

$(2x^3 + x^2 - 8x - 3) \div (2x - 3)$
$= x^2 + 2x - 1 - \dfrac{6}{2x - 3}$

You Try It 3

$$
\begin{array}{r}
x^2 + x - 1 \\
x - 1 \overline{)x^3 + 0x^2 - 2x + 1} \\
\underline{x^3 - x^2} \\
x^2 - 2x \\
\underline{x^2 - x} \\
- x + 1 \\
\underline{- x + 1} \\
0
\end{array}
$$

$(x^3 - 2x + 1) \div (x - 1) = x^2 + x - 1$

Solutions to Chapter 10 "You Try It"

SECTION 10.1

You Try It 1   The GCF is $7a^2$.

$$14a^2 - 21a^4b = 7a^2(2) + 7a^2(-3a^2b)$$
$$= 7a^2(2 - 3a^2b)$$

You Try It 2   The GCF is 9.

$$27b^2 + 18b + 9$$
$$= 9(3b^2) + 9(2b) + 9(1)$$
$$= 9(3b^2 + 2b + 1)$$

You Try It 3
The GCF is $3x^2y^2$.

$$6x^4y^2 - 9x^3y^2 + 12x^2y^4$$
$$= 3x^2y^2(2x^2) + 3x^2y^2(-3x) + 3x^2y^2(4y^2)$$
$$= 3x^2y^2(2x^2 - 3x + 4y^2)$$

You Try It 4
$$2y(5x - 2) - 3(2 - 5x)$$
$$= 2y(5x - 2) + 3(5x - 2) \quad \bullet \; \textbf{5x − 2 is the}$$
$$= (5x - 2)(2y + 3) \qquad\qquad \textbf{common factor.}$$

You Try It 5
$$a^2 - 3a + 2ab - 6b$$
$$= (a^2 - 3a) + (2ab - 6b)$$
$$= a(a - 3) + 2b(a - 3) \quad \bullet \; \textbf{a − 3 is the common factor.}$$
$$= (a - 3)(a + 2b)$$

You Try It 6
$$2mn^2 - n + 8mn - 4$$
$$= (2mn^2 - n) + (8mn - 4)$$
$$= n(2mn - 1) + 4(2mn - 1) \quad \bullet \; \textbf{2mn − 1 is the}$$
$$= (2mn - 1)(n + 4) \qquad\qquad\quad \textbf{common factor.}$$

You Try It 7
$$3xy - 9y - 12 + 4x$$
$$= (3xy - 9y) - (12 - 4x) \quad \bullet \; \textbf{−12 + 4x = −(12 − 4x)}$$
$$= 3y(x - 3) - 4(3 - x) \quad\;\; \bullet \; \textbf{−4(3 − x) = 4(x − 3)}$$
$$= 3y(x - 3) + 4(x - 3) \quad\;\; \bullet \; \textbf{x − 3 is the common}$$
$$= (x - 3)(3y + 4) \qquad\qquad\quad \textbf{factor.}$$

SECTION 10.2

You Try It 1
Find the positive factors of 20 whose sum is 9.

Factors	Sum
1, 20	21
2, 10	12
4, 5	9

$$x^2 + 9x + 20 = (x + 4)(x + 5)$$

You Try It 2
Find the factors of −18 whose sum is 7.

Factors	Sum
+1, −18	−17
−1, +18	17
+2, −9	−7
−2, +9	7
+3, −6	−3
−3, +6	3

$$x^2 + 7x - 18 = (x + 9)(x - 2)$$

You Try It 3
The GCF is $-2x$.

$$-2x^3 + 14x^2 - 12x = -2x(x^2 - 7x + 6)$$

Factor the trinomial $x^2 - 7x + 6$. Find two negative factors of 6 whose sum is −7.

Factors	Sum
−1, −6	−7
−2, −3	−5

$$-2x^3 + 14x^2 - 12x = -2x(x - 6)(x - 1)$$

You Try It 4
The GCF is 3.

$$3x^2 - 9xy - 12y^2 = 3(x^2 - 3xy - 4y^2)$$

Factor the trinomial.

Find the factors of −4 whose sum is −3.

Factors	Sum
+1, −4	−3
−1, +4	3
+2, −2	0

$$3x^2 - 9xy - 12y^2 = 3(x + y)(x - 4y)$$

SECTION 10.3

You Try It 1
Factor the trinomial $2x^2 - x - 3$.

Positive factors of 2: 1, 2   Factors of −3: +1, −3
                                              −1, +3

Trial Factors	Middle Term
$(x + 1)(2x - 3)$	$-3x + 2x = -x$
$(x - 3)(2x + 1)$	$x - 6x = -5x$
$(x - 1)(2x + 3)$	$3x - 2x = x$
$(x + 3)(2x - 1)$	$-x + 6x = 5x$

$$2x^2 - x - 3 = (x + 1)(2x - 3)$$

You Try It 2

The GCF is $-3y$.

$$-45y^3 + 12y^2 + 12y = -3y(15y^2 - 4y - 4)$$

Factor the trinomial $15y^2 - 4y - 4$.

Positive factors of 15:	1, 15	Factors of -4:	1, -4
	3, 5		-1, 4
			2, -2

Trial Factors	*Middle Term*
$(y + 1)(15y - 4)$	$-4y + 15y = 11y$
$(y - 4)(15y + 1)$	$y - 60y = -59y$
$(y - 1)(15y + 4)$	$4y - 15y = -11y$
$(y + 4)(15y - 1)$	$-y + 60y = 59y$
$(y + 2)(15y - 2)$	$-2y + 30y = 28y$
$(y - 2)(15y + 2)$	$2y - 30y = -28y$
$(3y + 1)(5y - 4)$	$-12y + 5y = -7y$
$(3y - 4)(5y + 1)$	$3y - 20y = -17y$
$(3y - 1)(5y + 4)$	$12y - 5y = 7y$
$(3y + 4)(5y - 1)$	$-3y + 20y = 17y$
$(3y + 2)(5y - 2)$	$-6y + 10y = 4y$
$(3y - 2)(5y + 2)$	$6y - 10y = -4y$

$$-45y^3 + 12y^2 + 12y = -3y(3y - 2)(5y + 2)$$

You Try It 3

Factors of -14 [2(-7)]	*Sum*
$+1, -14$	-13
$-1, +14$	13
$+2, -7$	-5
$-2, +7$	5

$$2a^2 + 13a - 7 = 2a^2 - a + 14a - 7$$
$$= (2a^2 - a) + (14a - 7)$$
$$= a(2a - 1) + 7(2a - 1)$$
$$= (2a - 1)(a + 7)$$

$$2a^2 + 13a - 7 = (2a - 1)(a + 7)$$

You Try It 4

The GCF is $5x$.

$$15x^3 + 40x^2 - 80x = 5x(3x^2 + 8x - 16)$$

Factors of -48 [3(-16)]	*Sum*
$+1, -48$	-47
$-1, +48$	47
$+2, -24$	-22
$-2, +24$	22
$+3, -16$	-13
$-3, +16$	13
$+4, -12$	-8
$-4, +12$	8

$$3x^2 + 8x - 16 = 3x^2 - 4x + 12x - 16$$
$$= (3x^2 - 4x) + (12x - 16)$$
$$= x(3x - 4) + 4(3x - 4)$$
$$= (3x - 4)(x + 4)$$

$$15x^3 + 40x^2 - 80x = 5x(3x^2 + 8x - 16)$$
$$= 5x(3x - 4)(x + 4)$$

SECTION 10.4

You Try It 1

$$25a^2 - b^2 = (5a)^2 - b^2$$ • Difference of
$$= (5a + b)(5a - b)$$ two squares

You Try It 2

$$n^4 - 81 = (n^2)^2 - 9^2$$ • Difference of
     two squares

$$= (n^2 + 9)(n^2 - 9)$$ • Difference of
$$= (n^2 + 9)(n + 3)(n - 3)$$ two squares

You Try It 3   Because $16y^2 = (4y)^2$, $1 = 1^2$, and $8y = 2(4y)(1)$, the trinomial is a perfect-square trinomial.

$$16y^2 + 8y + 1 = (4y + 1)^2$$

You Try It 4   Because $x^2 = (x)^2$, $36 = 6^2$, and $15x \neq 2(x)(6)$, the trinomial is not a perfect-square trinomial. Try to factor the trinomial by another method.

$$x^2 + 15x + 36 = (x + 3)(x + 12)$$

You Try It 5   The GCF is $3x$.

$$12x^3 - 75x = 3x(4x^2 - 25)$$
$$= 3x(2x + 5)(2x - 5)$$

You Try It 6

Factor by grouping.

$$a^2b - 7a^2 - b + 7$$
$$= (a^2b - 7a^2) - (b - 7)$$
$$= a^2(b - 7) - (b - 7)$$ • $b - 7$ is the common factor.
$$= (b - 7)(a^2 - 1)$$ • $a^2 - 1$ is the difference
$$= (b - 7)(a + 1)(a - 1)$$ of two squares.

You Try It 7

The GCF is $4x$.

$$4x^3 + 28x^2 - 120x$$
$$= 4x(x^2 + 7x - 30)$$ • Factor the GCF, $4x$.
$$= 4x(x + 10)(x - 3)$$ • Factor the trinomial.

SECTION 10.5

You Try It 1

$2x(x + 7) = 0$

$$2x = 0 \qquad x + 7 = 0$$
$$x = 0 \qquad x = -7$$

• **Principle of Zero Products**

The solutions are 0 and -7.

You Try It 2

$$4x^2 - 9 = 0$$ • **Difference of two squares**
$$(2x - 3)(2x + 3) = 0$$

$$2x - 3 = 0 \qquad 2x + 3 = 0$$ • **Principle of Zero Products**
$$2x = 3 \qquad 2x = -3$$
$$x = \frac{3}{2} \qquad x = -\frac{3}{2}$$

The solutions are $\frac{3}{2}$ and $-\frac{3}{2}$.

You Try It 3

$$(x + 2)(x - 7) = 52$$
$$x^2 - 5x - 14 = 52$$
$$x^2 - 5x - 66 = 0$$
$$(x + 6)(x - 11) = 0$$

$$x + 6 = 0 \qquad x - 11 = 0$$ • **Principle of Zero Products**
$$x = -6 \qquad x = 11$$

The solutions are -6 and 11.

You Try It 4

Strategy First consecutive positive integer: n
Second consecutive positive integer:
$n + 1$

The sum of the squares of the two consecutive positive integers is 61.

Solution
$$n^2 + (n + 1)^2 = 61$$
$$n^2 + n^2 + 2n + 1 = 61$$
$$2n^2 + 2n + 1 = 61$$
$$2n^2 + 2n - 60 = 0$$
$$2(n^2 + n - 30) = 0$$
$$2(n - 5)(n + 6) = 0$$

$$n - 5 = 0 \qquad n + 6 = 0$$ • **Principle of Zero Products**
$$n = 5 \qquad n = -6$$

Because -6 is not a positive integer, it is not a solution.

$$n = 5$$
$$n + 1 = 5 + 1 = 6$$

The two integers are 5 and 6.

You Try It 5

Strategy Width $= x$
Length $= 2x + 4$

The area of the rectangle is 96 in². Use the equation $A = L \cdot W$.

Solution
$$A = L \cdot W$$
$$96 = (2x + 4)x$$
$$96 = 2x^2 + 4x$$
$$0 = 2x^2 + 4x - 96$$
$$0 = 2(x^2 + 2x - 48)$$
$$0 = 2(x + 8)(x - 6)$$

$$x + 8 = 0 \qquad x - 6 = 0$$ • **Principle of Zero Products**
$$x = -8 \qquad x = 6$$

Because the width cannot be a negative number, -8 is not a solution.

$$x = 6$$
$$2x + 4 = 2(6) + 4 = 12 + 4 = 16$$

The length is 16 in. The width is 6 in.

Solutions to Chapter 11 "You Try It"

SECTION 11.1

You Try It 1

$$\frac{6x^5y}{12x^2y^3} = \frac{\overset{1}{\cancel{2}} \cdot \overset{1}{\cancel{3}} \cdot x^5y}{\cancel{2} \cdot 2 \cdot \cancel{3} \cdot x^2y^3} = \frac{x^3}{2y^2}$$

You Try It 2

$$\frac{x^2 + 2x - 24}{16 - x^2} = \frac{\overset{-1}{\cancel{(x - 4)}}(x + 6)}{\underset{1}{\cancel{(4 - x)}}(4 + x)}$$ • $(4 - x) = -1(x - 4)$

$$= -\frac{x + 6}{x + 4}$$

You Try It 3

$$\frac{x^2 + 4x - 12}{x^2 - 3x + 2} = \frac{\overset{1}{\cancel{(x - 2)}}(x + 6)}{(x - 1)\underset{1}{\cancel{(x - 2)}}} = \frac{x + 6}{x - 1}$$

You Try It 4

$$\frac{12x^2 + 3x}{10x - 15} \cdot \frac{8x - 12}{9x + 18} = \frac{3x(4x + 1)}{5(2x - 3)} \cdot \frac{4(2x - 3)}{9(x + 2)}$$ • **Factor.**

$$= \frac{\overset{1}{\cancel{3}}x(4x + 1) \cdot 2 \cdot 2\overset{1}{\cancel{(2x - 3)}}}{5\underset{1}{\cancel{(2x - 3)}} \cdot \underset{1}{\cancel{3}} \cdot 3(x + 2)}$$

$$= \frac{4x(4x + 1)}{15(x + 2)}$$

You Try It 5

$$\frac{x^2 + 2x - 15}{9 - x^2} \cdot \frac{x^2 - 3x - 18}{x^2 - 7x + 6}$$

$$= \frac{(x - 3)(x + 5)}{(3 - x)(3 + x)} \cdot \frac{(x + 3)(x - 6)}{(x - 1)(x - 6)} \qquad \bullet \text{ Factor.}$$

$$= \frac{\overset{-1}{\cancel{(x - 3)}}(x + 5) \cdot \overset{1}{\cancel{(x + 3)}}\overset{1}{\cancel{(x - 6)}}}{\underset{1}{\cancel{(3 - x)}}\underset{1}{\cancel{(3 + x)}} \cdot (x - 1)\underset{1}{\cancel{(x - 6)}}} = -\frac{x + 5}{x - 1}$$

You Try It 6

$$\frac{a^2}{4bc^2 - 2b^2c} \div \frac{a}{6bc - 3b^2}$$

$$= \frac{a^2}{4bc^2 - 2b^2c} \cdot \frac{6bc - 3b^2}{a} \qquad \bullet \text{ Multiply by the reciprocal.}$$

$$= \frac{a^2 \cdot 3b\overset{1}{\cancel{(2c - b)}}}{2bc\underset{1}{\cancel{(2c - b)}} \cdot a} = \frac{3a}{2c}$$

You Try It 7

$$\frac{3x^2 + 26x + 16}{3x^2 - 7x - 6} \div \frac{2x^2 + 9x - 5}{x^2 + 2x - 15}$$

$$= \frac{3x^2 + 26x + 16}{3x^2 - 7x - 6} \cdot \frac{x^2 + 2x - 15}{2x^2 + 9x - 5} \qquad \bullet \text{ Multiply by the reciprocal.}$$

$$= \frac{\overset{1}{\cancel{(3x + 2)}}(x + 8) \cdot \overset{1}{\cancel{(x + 5)}}\overset{1}{\cancel{(x - 3)}}}{\underset{1}{\cancel{(3x + 2)}}\underset{1}{\cancel{(x - 3)}} \cdot (2x - 1)\underset{1}{\cancel{(x + 5)}}} = \frac{x + 8}{2x - 1}$$

SECTION 11.2

You Try It 1

$8uv^2 = 2 \cdot 2 \cdot 2 \cdot u \cdot v \cdot v$
$12uw = 2 \cdot 2 \cdot 3 \cdot u \cdot w$
LCM $= 2 \cdot 2 \cdot 2 \cdot 3 \cdot u \cdot v \cdot v \cdot w = 24uv^2w$

You Try It 2 $m^2 - 6m + 9 = (m - 3)(m - 3)$
$m^2 - 2m - 3 = (m + 1)(m - 3)$
LCM $= (m - 3)(m - 3)(m + 1)$

You Try It 3   The LCM is $36xy^2z$.

$$\frac{x - 3}{4xy^2} = \frac{x - 3}{4xy^2} \cdot \frac{9z}{9z} = \frac{9xz - 27z}{36xy^2z}$$

$$\frac{2x + 1}{9y^2z} = \frac{2x + 1}{9y^2z} \cdot \frac{4x}{4x} = \frac{8x^2 + 4x}{36xy^2z}$$

You Try It 4
The LCM is $(x + 2)(x - 5)(x + 5)$.

$$\frac{x + 4}{x^2 - 3x - 10} = \frac{x + 4}{(x + 2)(x - 5)} \cdot \frac{x + 5}{x + 5}$$

$$= \frac{x^2 + 9x + 20}{(x + 2)(x - 5)(x + 5)}$$

$$\frac{2x}{25 - x^2} = \frac{2x}{-(x^2 - 25)} = -\frac{2x}{(x - 5)(x + 5)} \cdot \frac{x + 2}{x + 2}$$

$$= -\frac{2x^2 + 4x}{(x + 2)(x - 5)(x + 5)}$$

SECTION 11.3

You Try It 1

$$\frac{2x^2}{x^2 - x - 12} - \frac{7x + 4}{x^2 - x - 12}$$

$$= \frac{2x^2 - (7x + 4)}{x^2 - x - 12} = \frac{2x^2 - 7x - 4}{x^2 - x - 12}$$

$$= \frac{(2x + 1)\overset{1}{\cancel{(x - 4)}}}{(x + 3)\underset{1}{\cancel{(x - 4)}}} = \frac{2x + 1}{x + 3}$$

You Try It 2

$$\frac{x^2 - 1}{x^2 - 8x + 12} - \frac{2x + 1}{x^2 - 8x + 12} + \frac{x}{x^2 - 8x + 12}$$

$$= \frac{(x^2 - 1) - (2x + 1) + x}{x^2 - 8x + 12} = \frac{x^2 - 1 - 2x - 1 + x}{x^2 - 8x + 12}$$

$$= \frac{x^2 - x - 2}{x^2 - 8x + 12} = \frac{(x + 1)\overset{1}{\cancel{(x - 2)}}}{\underset{1}{\cancel{(x - 2)}}(x - 6)} = \frac{x + 1}{x - 6}$$

You Try It 3
The LCM of the denominators is $24y$.

$$\frac{z}{8y} - \frac{4z}{3y} + \frac{5z}{4y}$$

$$= \frac{z}{8y} \cdot \frac{3}{3} - \frac{4z}{3y} \cdot \frac{8}{8} + \frac{5z}{4y} \cdot \frac{6}{6} \qquad \bullet \text{ Write each fraction using the LCM as the denominator.}$$

$$= \frac{3z}{24y} - \frac{32z}{24y} + \frac{30z}{24y}$$

$$= \frac{3z - 32z + 30z}{24y} = \frac{z}{24y} \qquad \bullet \text{ Combine the numerators.}$$

You Try It 4

$$2 - x = -(x - 2) \qquad \text{Therefore, } \frac{3}{2 - x} = \frac{-3}{x - 2}.$$

$$\frac{5x}{x - 2} + \frac{3}{2 - x} = \frac{5x}{x - 2} + \frac{-3}{x - 2} \qquad \bullet \text{ The LCM is } x - 2.$$

$$= \frac{5x + (-3)}{x - 2} = \frac{5x - 3}{x - 2} \qquad \bullet \text{ Combine the numerators.}$$

You Try It 5
The LCM is $(3x - 1)(x + 4)$.

$$\frac{4x}{3x - 1} + \frac{9}{x + 4} = \frac{4x}{3x - 1} \cdot \frac{x + 4}{x + 4} + \frac{9}{x + 4} \cdot \frac{3x - 1}{3x - 1}$$

$$= \frac{4x^2 + 16x}{(3x - 1)(x + 4)} + \frac{27x - 9}{(3x - 1)(x + 4)}$$

$$= \frac{(4x^2 + 16x) + (27x - 9)}{(3x - 1)(x + 4)}$$

$$= \frac{4x^2 + 16x + 27x - 9}{(3x - 1)(x + 4)}$$

$$= \frac{4x^2 + 43x - 9}{(3x - 1)(x + 4)}$$

You Try It 6 The LCM is $x - 3$.

$$2 - \frac{1}{x - 3} = 2 \cdot \frac{x - 3}{x - 3} - \frac{1}{x - 3}$$

$$= \frac{2x - 6}{x - 3} - \frac{1}{x - 3}$$

$$= \frac{2x - 6 - 1}{x - 3}$$

$$= \frac{2x - 7}{x - 3}$$

You Try It 7

$$\frac{2}{5 - x} = \frac{-2}{x - 5}$$

The LCM is $(x + 5)(x - 5)$.

$$\frac{2x - 1}{x^2 - 25} + \frac{2}{5 - x} = \frac{2x - 1}{(x + 5)(x - 5)} + \frac{-2}{x - 5}$$

$$= \frac{2x - 1}{(x + 5)(x - 5)} + \frac{-2}{x - 5} \cdot \frac{x + 5}{x + 5}$$

$$= \frac{2x - 1}{(x + 5)(x - 5)} + \frac{-2(x + 5)}{(x + 5)(x - 5)}$$

$$= \frac{2x - 1 + (-2)(x + 5)}{(x + 5)(x - 5)}$$

$$= \frac{2x - 1 - 2x - 10}{(x + 5)(x - 5)}$$

$$= \frac{-11}{(x + 5)(x - 5)}$$

$$= -\frac{11}{(x + 5)(x - 5)}$$

You Try It 8
The LCM is $(3x + 2)(x - 1)$.

$$\frac{2x - 3}{3x^2 - x - 2} + \frac{5}{3x + 2} - \frac{1}{x - 1}$$

$$= \frac{2x - 3}{(3x + 2)(x - 1)} + \frac{5}{3x + 2} \cdot \frac{x - 1}{x - 1}$$

$$\quad - \frac{1}{x - 1} \cdot \frac{3x + 2}{3x + 2}$$

$$= \frac{2x - 3}{(3x + 2)(x - 1)} + \frac{5x - 5}{(3x + 2)(x - 1)}$$

$$\quad - \frac{3x + 2}{(3x + 2)(x - 1)}$$

$$= \frac{(2x - 3) + (5x - 5) - (3x + 2)}{(3x + 2)(x - 1)}$$

$$= \frac{2x - 3 + 5x - 5 - 3x - 2}{(3x + 2)(x - 1)}$$

$$= \frac{4x - 10}{(3x + 2)(x - 1)} = \frac{2(2x - 5)}{(3x + 2)(x - 1)}$$

SECTION 11.4

You Try It 1
The LCM of 3, x, 9, and x^2 is $9x^2$.

$$\frac{\dfrac{1}{3} - \dfrac{1}{x}}{\dfrac{1}{9} - \dfrac{1}{x^2}} = \frac{\dfrac{1}{3} - \dfrac{1}{x}}{\dfrac{1}{9} - \dfrac{1}{x^2}} \cdot \frac{9x^2}{9x^2} = \frac{\dfrac{1}{3} \cdot 9x^2 - \dfrac{1}{x} \cdot 9x^2}{\dfrac{1}{9} \cdot 9x^2 - \dfrac{1}{x^2} \cdot 9x^2}$$

• Multiply by the LCM.

$$= \frac{3x^2 - 9x}{x^2 - 9} = \frac{3x(x - 3)}{(x - 3)(x + 3)} = \frac{3x}{x + 3}$$

You Try It 2
The LCM of x and x^2 is x^2.

$$\frac{1 + \dfrac{4}{x} + \dfrac{3}{x^2}}{1 + \dfrac{10}{x} + \dfrac{21}{x^2}} = \frac{1 + \dfrac{4}{x} + \dfrac{3}{x^2}}{1 + \dfrac{10}{x} + \dfrac{21}{x^2}} \cdot \frac{x^2}{x^2}$$

• Multiply by the LCM.

$$= \frac{1 \cdot x^2 + \dfrac{4}{x} \cdot x^2 + \dfrac{3}{x^2} \cdot x^2}{1 \cdot x^2 + \dfrac{10}{x} \cdot x^2 + \dfrac{21}{x^2} \cdot x^2}$$

• Distributive Property

$$= \frac{x^2 + 4x + 3}{x^2 + 10x + 21} = \frac{(x + 1)(x + 3)}{(x + 3)(x + 7)}$$

$$= \frac{x + 1}{x + 7}$$

You Try It 3
The LCM is $x - 5$.

$$\frac{x + 3 - \dfrac{20}{x - 5}}{x + 8 + \dfrac{30}{x - 5}} = \frac{x + 3 - \dfrac{20}{x - 5}}{x + 8 + \dfrac{30}{x - 5}} \cdot \frac{x - 5}{x - 5}$$

• Multiply by the LCM.

$$= \frac{(x + 3)(x - 5) - \dfrac{20}{x - 5} \cdot (x - 5)}{(x + 8)(x - 5) + \dfrac{30}{x - 5} \cdot (x - 5)}$$

$$= \frac{x^2 - 2x - 15 - 20}{x^2 + 3x - 40 + 30} = \frac{x^2 - 2x - 35}{x^2 + 3x - 10}$$

$$= \frac{(x + 5)(x - 7)}{(x - 2)(x + 5)} = \frac{x - 7}{x - 2}$$

SECTION 11.5

You Try It 1

$$\frac{x}{x + 6} = \frac{3}{x}$$ • The LCM is $x(x + 6)$.

$$\frac{\overset{1}{x(x + 6)}}{1} \cdot \frac{x}{x + 6} = \frac{x(x + 6)}{1} \cdot \frac{3}{x}$$ • Multiply by the LCM.

$$x^2 = (x + 6)3$$ • Simplify.
$$x^2 = 3x + 18$$
$$x^2 - 3x - 18 = 0$$
$$(x + 3)(x - 6) = 0$$ • Factor.

$$x + 3 = 0 \qquad x - 6 = 0$$ • Principle of
$$x = -3 \qquad x = 6$$ Zero Products

Both -3 and 6 check as solutions.
The solutions are -3 and 6.

You Try It 2

$$\frac{5x}{x + 2} = 3 - \frac{10}{x + 2}$$ • The LCM is $x + 2$.

$$\frac{(x + 2)}{1} \cdot \frac{5x}{x + 2} = \frac{(x + 2)}{1}\left(3 - \frac{10}{x + 2}\right)$$ • Clear denominators.

$$\frac{\overset{1}{x + 2}}{1} \cdot \frac{5x}{x + 2} = \frac{x + 2}{1} \cdot 3 - \frac{\overset{1}{x + 2}}{1} \cdot \frac{10}{x + 2}$$

$$5x = (x + 2)3 - 10$$ • Solve for x.
$$5x = 3x + 6 - 10$$
$$5x = 3x - 4$$
$$2x = -4$$
$$x = -2$$

-2 does not check as a solution.
The equation has no solution.

SECTION 11.6

You Try It 1

$$5x - 2y = 10$$
$$5x - 5x - 2y = -5x + 10$$ • Subtract $5x$.
$$-2y = -5x + 10$$
$$\frac{-2y}{-2} = \frac{-5x + 10}{-2}$$ • Divide by -2.
$$y = \frac{5}{2}x - 5$$

You Try It 2

$$s = \frac{A + L}{2}$$

$$2 \cdot s = 2\left(\frac{A + L}{2}\right)$$ • Multiply by 2.

$$2s = A + L$$
$$2s - A = A - A + L$$ • Subtract A.
$$2s - A = L$$

You Try It 3

$$S = a + (n - 1)d$$
$$S = a + nd - d$$
$$S - a = a - a + nd - d$$ • Subtract a.
$$S - a = nd - d$$
$$S - a + d = nd - d + d$$ • Add d.
$$S - a + d = nd$$
$$\frac{S - a + d}{d} = \frac{nd}{d}$$ • Divide by d.
$$\frac{S - a + d}{d} = n$$

You Try It 4

$$S = rS + C$$
$$S - rS = rS - rS + C$$ • Subtract rS.
$$S - rS = C$$
$$(1 - r)S = C$$ • Factor.
$$\frac{(1 - r)S}{1 - r} = \frac{C}{1 - r}$$ • Divide by $1 - r$.
$$S = \frac{C}{1 - r}$$

SECTION 11.7

You Try It 1

Strategy • Time for one printer to complete the job: t

	Rate	*Time*	*Part*
1st printer	$\dfrac{1}{t}$	2	$\dfrac{2}{t}$
2nd printer	$\dfrac{1}{t}$	5	$\dfrac{5}{t}$

• The sum of the parts of the task completed must equal 1.

Solution

$$\frac{2}{t} + \frac{5}{t} = 1$$

$$t\left(\frac{2}{t} + \frac{5}{t}\right) = t \cdot 1$$

$$2 + 5 = t$$
$$7 = t$$

Working alone, one printer takes 7 h to print the payroll.

You Try It 2

Strategy
- Rate sailing across the lake: r
  Rate sailing back: $3r$

	Distance	Rate	Time
Across	6	r	$\dfrac{6}{r}$
Back	6	$3r$	$\dfrac{6}{3r}$

- The total time for the trip was 2 h.

Solution

$$\frac{6}{r} + \frac{6}{3r} = 2$$

$$3r\left(\frac{6}{r} + \frac{6}{3r}\right) = 3r(2) \quad \bullet \text{ Multiply by the LCM, } 3r.$$

$$3r \cdot \frac{6}{r} + 3r \cdot \frac{6}{3r} = 6r$$

$$18 + 6 = 6r \quad \bullet \text{ Solve for } r.$$

$$24 = 6r$$

$$4 = r$$

The rate across the lake was 4 km/h.

Solutions to Chapter 12 "You Try It"

SECTION 12.1

You Try It 1

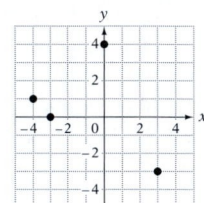

You Try It 2 $A(4, -2)$, $B(-2, 4)$.
The abscissa of D is 0.
The ordinate of C is 0.

You Try It 3

$$\frac{x - 3y = -14}{\begin{array}{c|c} -2 - 3(4) & -14 \\ -2 - 12 & -14 \\ -14 = -14 \end{array}}$$

Yes, $(-2, 4)$ is a solution of
$x - 3y = -14$.

You Try It 4 $x + 2y = 4$
$2y = -x + 4$
$y = -\dfrac{1}{2}x + 2$

You Try It 5

$\{(145, 140), (140, 125), (150, 130), (165, 150), (140, 130), (165, 160)\}$

No, the relation is not a function. The two ordered pairs (140, 125) and (140, 130) have the same first coordinate but different second coordinates.

You Try It 6 Determine the ordered pairs defined by the equation. Replace x in $y = \dfrac{1}{2}x + 1$ with the given values and solve for y. $\{(-4, -1), (0, 1), (2, 2)\}$ Yes, y is a function of x.

You Try It 7 $H(x) = \dfrac{x}{x - 4}$

$$H(8) = \frac{8}{8 - 4} \quad \bullet \text{ Replace } x \text{ with 8.}$$

$$H(8) = \frac{8}{4} = 2$$

SECTION 12.2

You Try It 1

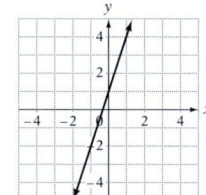

You Try It 2

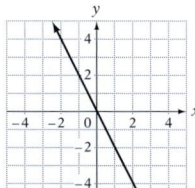

You Try It 3

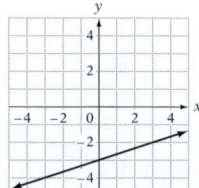

You Try It 4 $5x - 2y = 10$ • Solve for y.

$$-2y = -5x + 10$$

$$y = \frac{5}{2}x - 5$$

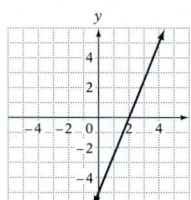

You Try It 5 $x - 3y = 9$ • Solve for y.

$$-3y = -x + 9$$

$$y = \frac{1}{3}x - 3$$

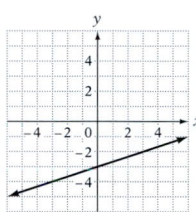

You Try It 6

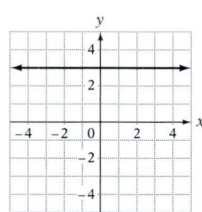

You Try It 7

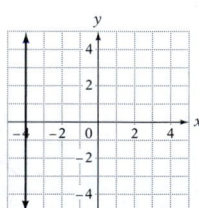

You Try It 8

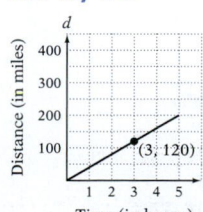

The ordered pair (3, 120) means that in 3 h the car will have traveled 120 mi.

SECTION 12.3

You Try It 1   x-intercept: y-intercept:

$$y = 2x - 4 \qquad (0, b)$$
$$0 = 2x - 4 \qquad b = -4$$
$$-2x = -4 \qquad (0, -4)$$
$$x = 2$$
$$(2, 0)$$

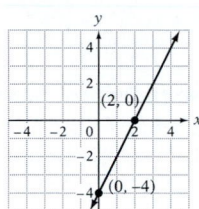

You Try It 2   Let $P_1 = (1, 4)$ and $P_2 = (-3, 8)$.

$$m = \frac{y_2 - y_1}{x_2 - x_1} = \frac{8 - 4}{-3 - 1} = \frac{4}{-4} = -1$$

The slope is -1.

You Try It 3   Let $P_1 = (-1, 2)$ and $P_2 = (4, 2)$.

$$m = \frac{y_2 - y_1}{x_2 - x_1} = \frac{2 - 2}{4 - (-1)} = \frac{0}{5} = 0$$

The slope is 0.

You Try It 4 $$m = \frac{8650 - 6100}{1 - 4} = \frac{2550}{-3}$$

$$m = -850$$

A slope of -850 means that the value of the car is decreasing at a rate of $850 per year.

You Try It 5   y-intercept $= (0, b) = (0, -1)$

$$m = -\frac{1}{4}$$

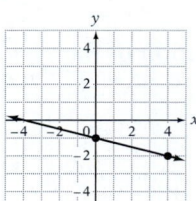

You Try It 6   Solve the equation for y.

$$x - 2y = 4$$

$$-2y = -x + 4$$

$$y = \frac{1}{2}x - 2$$

y-intercept $= (0, b) = (0, -2)$

$m = \dfrac{1}{2}$

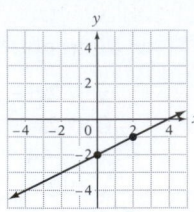

SECTION 12.4

You Try It 1 Because the slope and y-intercept are known, use the slope-intercept formula, $y = mx + b$.

$y = mx + b$

$y = \dfrac{5}{3}x + 2$ • $m = \dfrac{5}{3}; b = 2$

You Try It 2 $m = \dfrac{3}{4}$ $(x_1, y_1) = (4, -2)$

$y - y_1 = m(x - x_1)$

$y - (-2) = \dfrac{3}{4}(x - 4)$

$y + 2 = \dfrac{3}{4}x - 3$

$y = \dfrac{3}{4}x - 5$

The equation of the line is $y = \dfrac{3}{4}x - 5$.

You Try It 3 Find the slope of the line between the two points.

$m = \dfrac{y_2 - y_1}{x_2 - x_1} = \dfrac{1 - (-2)}{3 - (-6)} = \dfrac{3}{9} = \dfrac{1}{3}$

Use the point-slope formula.

$y - y_1 = m(x - x_1)$

$y - (-2) = \dfrac{1}{3}[x - (-6)]$ • $y_1 = -2;$ $x_1 = -6$

$y + 2 = \dfrac{1}{3}x + 2$

$y = \dfrac{1}{3}x$

You Try It 4

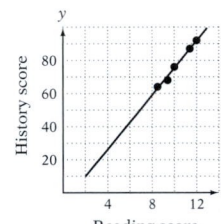

The slope of the line means that the grade on the history test increases 8.3 points for each 1-point increase in the grade on the reading test.

Solutions to Chapter 13 "You Try It"

SECTION 13.1

You Try It 1

$2x - 5y = 8$		$-x + 3y = -5$	
$2(-1) - 5(-2)$	8	$-(-1) + 3(-2)$	-5
$-2 + 10$	8	$1 + (-6)$	-5
$8 = 8$		$-5 = -5$	

Yes, $(-1, -2)$ is a solution of the system of equations.

You Try It 2

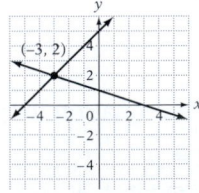

The solution is $(-3, 2)$.

You Try It 3

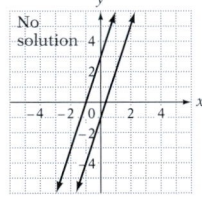

The lines are parallel. The system of equations is inconsistent and does not have a solution.

SECTION 13.2

You Try It 1 (1) $7x - y = 4$
(2) $3x + 2y = 9$

Solve Equation (1) for y.

$7x - y = 4$

$-y = -7x + 4$

$y = 7x - 4$

Substitute in Equation (2).

$$3x + 2y = 9$$
$$3x + 2(7x - 4) = 9 \qquad \bullet \ y = 7x - 4$$
$$3x + 14x - 8 = 9$$
$$17x - 8 = 9$$
$$17x = 17$$
$$x = 1$$

Substitute in Equation (1).

$$7x - y = 4$$
$$7(1) - y = 4 \qquad \bullet \ x = 1$$
$$7 - y = 4$$
$$-y = -3$$
$$y = 3$$

The solution is (1, 3).

You Try It 2   (1) $3x - y = 4$
(2) $y = 3x + 2$

$$3x - y = 4$$
$$3x - (3x + 2) = 4 \qquad \bullet \ y = 3x + 2$$
$$3x - 3x - 2 = 4$$
$$-2 = 4$$

This is not a true equation. The system of equations is inconsistent and therefore has no solution.

You Try It 3   (1) $y = -2x + 1$
(2) $6x + 3y = 3$

$$6x + 3y = 3$$
$$6x + 3(-2x + 1) = 3 \qquad \bullet \ y = -2x + 1$$
$$6x - 6x + 3 = 3$$
$$3 = 3$$

The system of equations is dependent. The solutions are the ordered pairs that satisfy the equation $y = -2x + 1$.

You Try It 4

Strategy   • Amount invested at 6.5%: x
Amount invested at 4.5%: y

	Principal	Rate	Interest
Amount at 6.5%	x	0.065	0.065x
Amount at 4.5%	y	0.045	0.045y

• The sum of the two investments is $330,000: $x + y = 330{,}000$. The interest earned at 6.5% equals the interest earned at 4.5%: $0.065x = 0.045y$

Solution
(1) $x + y = 330{,}000$
(2) $0.065x = 0.045y$

Solve Equation (2) for y.

$$(3) \qquad y = \frac{13}{9} x$$

Replace y with $\frac{13}{9} x$ in Equation (1) and solve for x.

$$x + y = 330{,}000$$
$$x + \frac{13}{9} x = 330{,}000 \qquad \bullet \ y = \frac{13}{9} x$$
$$\frac{22}{9} x = 330{,}000$$
$$x = 135{,}000$$

Replace x with 135,000 in Equation (3) and solve for y.

$$y = \frac{13}{9} x$$
$$= \frac{13}{9} (135{,}000) = 195{,}000 \qquad \bullet \ x = 135{,}000$$

$135,000 should be invested at 6.5% and $195,000 should be invested at 4.5%.

SECTION 13.3

You Try It 1   (1) $x - 2y = 1$
(2) $2x + 4y = 0$

Eliminate y.

$$2(x - 2y) = 2 \cdot 1 \qquad \bullet \ \textbf{Multiply by 2.}$$
$$2x + 4y = 0$$

$$2x - 4y = 2$$
$$2x + 4y = 0$$

Add the equations.

$$4x = 2$$
$$x = \frac{2}{4} = \frac{1}{2}$$

Replace x in Equation (2).

$$2\left(\frac{1}{2}\right) + 4y = 0 \qquad \bullet \ x = \frac{1}{2}$$
$$1 + 4y = 0$$
$$4y = -1$$
$$y = -\frac{1}{4}$$

The solution is $\left(\frac{1}{2}, -\frac{1}{4}\right)$.

You Try It 2   (1) $2x - 3y = 4$
(2) $-4x + 6y = -8$

Eliminate y.

$$2(2x - 3y) = 2 \cdot 4 \quad \bullet \text{ Multiply by 2.}$$
$$-4x + 6y = -8$$

$$4x - 6y = 8$$
$$-4x + 6y = -8$$

Add the equations.

$$0x + 0y = 0$$
$$0 = 0$$

The system of equations is dependent. The solutions are the ordered pairs that satisfy the equation $2x - 3y = 4$.

You Try It 3   (1) $4x + 5y = 11$
                  (2) $3y = x + 10$

Write equation (2) in the form $Ax + By = C$.

$$3y = x + 10$$
$$-x + 3y = 10$$

Eliminate x.

$$4x + 5y = 11$$
$$4(-x + 3y) = 4 \cdot 10 \quad \bullet \text{ Multiply by 4.}$$

$$4x + 5y = 11$$
$$-4x + 12y = 40$$

Add the equations.

$$17y = 51$$
$$y = 3$$

Replace y in Equation (1).

$$4x + 5y = 11$$
$$4x + 5 \cdot 3 = 11 \quad \bullet \; y = 3$$
$$4x + 15 = 11$$
$$4x = -4$$
$$x = -1$$

The solution is $(-1, 3)$.

SECTION 13.4

You Try It 1

    Strategy   • Rate of the current: c
                      Rate of the canoeist in calm water: r

	Rate	Time	Distance
With current	$r + c$	3	$3(r + c)$
Against current	$r - c$	5	$5(r - c)$

• The distance traveled with the current is 15 mi. The distance

traveled against the current is 15 mi.

    Solution

$$3(r + c) = 15 \qquad \frac{1}{3} \cdot 3(r + c) = \frac{1}{3} \cdot 15 \quad \bullet \text{ Multiply by } \frac{1}{3}.$$

$$5(r - c) = 15 \qquad \frac{1}{5} \cdot 5(r - c) = \frac{1}{5} \cdot 15 \quad \bullet \text{ Multiply by } \frac{1}{5}.$$

$$r + c = 5$$
$$r - c = 3$$
$$\overline{2r = 8}$$
$$r = 4$$

$$r + c = 5$$
$$4 + c = 5 \quad \bullet \; r = 4$$
$$c = 1$$

The rate of the current is 1 mph.
The rate of the canoeist in calm water is 4 mph.

You Try It 2

    Strategy   • Cost of an orange tree: x
                      Cost of a grapefruit tree: y

First purchase:

	Amount	Unit Cost	Value
Orange trees	25	x	$25x$
Grapefruit trees	20	y	$20y$

Second purchase:

	Amount	Unit Cost	Value
Orange trees	20	x	$20x$
Grapefruit trees	30	y	$30y$

• The total cost of the first purchase was $290. The total cost of the second purchase was $330.

    Solution

$$25x + 20y = 290 \qquad 4(25x + 20y) = 4 \cdot 290$$
$$\bullet \text{ Multiply by 4.}$$

$$20x + 30y = 330 \qquad -5(20x + 30y) = -5 \cdot 330$$
$$\bullet \text{ Multiply by } -5.$$

$$100x + 80y = 1160$$
$$-100x - 150y = -1650$$
$$\overline{-70y = -490}$$
$$y = 7$$

$$25x + 20y = 290$$
$$25x + 20(7) = 290 \quad \bullet \; y = 7$$
$$25x + 140 = 290$$
$$25x = 150$$
$$x = 6$$

The cost of an orange tree is $6.
The cost of a grapefruit tree is $7.

Solutions to Chapter 14 "You Try It"

SECTION 14.1

You Try It 1 $A = \{-9, -7, -5, -3, -1\}$

You Try It 2 $A = \{1, 3, 5, \ldots\}$

You Try It 3 $A \cup B = \{-2, -1, 0, 1, 2, 3, 4\}$

You Try It 4 $C \cap D = \{10, 16\}$

You Try It 5 $A \cap B = \varnothing$

You Try It 6 $\{x \mid x < 59, x \in \text{positive even integers}\}$

You Try It 7 $\{x \mid x > -3, x \in \text{real numbers}\}$

You Try It 8   The solution set is the numbers greater than -2.

You Try It 9   The solution set is the numbers greater than -1 and the numbers less than -3.

You Try It 10   The solution set is the numbers less than or equal to 4 and greater than or equal to -4.

You Try It 11   The solution set is the real numbers.

SECTION 14.2

You Try It 1
$$x + 2 < -2$$
$$x + 2 - 2 < -2 - 2 \quad \bullet \text{ Subtract 2.}$$
$$x < -4$$

You Try It 2
$$5x + 3 > 4x + 5$$
$$5x - 4x + 3 > 4x - 4x + 5 \quad \bullet \text{ Subtract } 4x.$$
$$x + 3 > 5$$
$$x + 3 - 3 > 5 - 3 \quad \bullet \text{ Subtract 3.}$$
$$x > 2$$

You Try It 3
$$-3x > -9$$
$$\frac{-3x}{-3} < \frac{-9}{-3} \quad \bullet \text{ Divide by } -3.$$
$$x < 3$$

You Try It 4
$$-\frac{3}{4}x \geq 18$$
$$-\frac{4}{3}\left(-\frac{3}{4}x\right) \leq -\frac{4}{3}(18) \quad \bullet \text{ Multiply by } -\frac{4}{3}.$$
$$x \leq -24$$

You Try It 5

Strategy   To find the selling prices, write and solve an inequality using p to represent the possible selling prices.

Solution
$$0.70p > 314$$
$$p > 448.571 \quad \bullet \text{ Divide by 0.70.}$$

The dealer will make a profit with any selling price greater than or equal to $448.58.

SECTION 14.3

You Try It 1
$$5 - 4x > 9 - 8x$$
$$5 - 4x + 8x > 9 - 8x + 8x \quad \bullet \text{ Add } 8x.$$
$$5 + 4x > 9$$
$$5 - 5 + 4x > 9 - 5 \quad \bullet \text{ Subtract 5.}$$
$$4x > 4$$
$$\frac{4x}{4} > \frac{4}{4} \quad \bullet \text{ Divide by 4.}$$
$$x > 1$$

You Try It 2
$$8 - 4(3x + 5) \leq 6(x - 8)$$
$$8 - 12x - 20 \leq 6x - 48 \quad \bullet \text{ Distributive Property}$$
$$-12 - 12x \leq 6x - 48$$
$$-12 - 12x - 6x \leq 6x - 6x - 48 \quad \bullet \text{ Subtract } 6x.$$
$$-12 - 18x \leq -48$$
$$-12 + 12 - 18x \leq -48 + 12 \quad \bullet \text{ Add 12.}$$
$$-18x \leq -36$$
$$\frac{-18x}{-18} \geq \frac{-36}{-18} \quad \bullet \text{ Divide by } -18.$$
$$x \geq 2$$

You Try It 3

Strategy To find the maximum number of miles:
- Write an expression for the cost of each car, using x to represent the number of miles driven during the week.
- Write and solve an inequality.

Solution

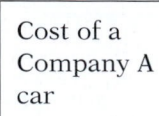

Cost of a Company A car	is less than	cost of a Company B car

$$8(7) + 0.10x < 10(7) + 0.08x$$
$$56 + 0.10x < 70 + 0.08x$$
$$56 + 0.10x - 0.08x < 70 + 0.08x - 0.08x \quad \bullet \text{ Subtract}$$
$$56 + 0.02x < 70 \qquad\qquad\qquad\quad 0.08x.$$
$$56 - 56 + 0.02x < 70 - 56 \qquad \bullet \text{ Subtract 56.}$$
$$0.02x < 14$$
$$\frac{0.02x}{0.02} < \frac{14}{0.02} \qquad\qquad \bullet \text{ Divide by 0.02.}$$
$$x < 700$$

The maximum number of miles is 699 mi.

SECTION 14.4

You Try It 1
$$x - 3y < 2$$
$$x - x - 3y < -x + 2 \qquad \bullet \text{ Subtract } x.$$
$$-3y < -x + 2$$
$$\frac{-3y}{-3} > \frac{-x + 2}{-3} \qquad \bullet \text{ Divide by } -3.$$
$$y > \frac{1}{3}x - \frac{2}{3}$$

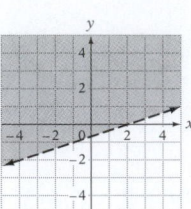

You Try It 2
$$2x - 4y \le 8$$
$$2x - 2x - 4y \le -2x + 8 \qquad \bullet \text{ Subtract } 2x.$$
$$-4y \le -2x + 8$$
$$\frac{-4y}{-4} \ge \frac{-2x + 8}{-4} \qquad \bullet \text{ Divide by } -4.$$
$$y \ge \frac{1}{2}x - 2$$

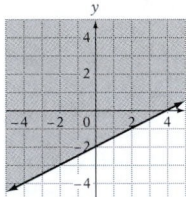

You Try It 3 $x < 3$

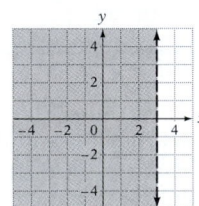

Solutions to Chapter 15 "You Try It"

SECTION 15.1

You Try It 1
$$-5\sqrt{32} = -5\sqrt{16 \cdot 2} \qquad \bullet \text{ 16 is a perfect square.}$$
$$= -5\sqrt{16}\sqrt{2}$$
$$= -5 \cdot 4\sqrt{2} = -20\sqrt{2}$$

You Try It 2
$$\sqrt{216} = \sqrt{36 \cdot 6} \qquad \bullet \text{ 36 is a perfect square.}$$
$$= \sqrt{36}\sqrt{6} = 6\sqrt{6}$$

You Try It 3
$$\sqrt{y^{19}} = \sqrt{y^{18} \cdot y} \qquad \bullet \; y^{18} \text{ is a perfect square.}$$
$$= \sqrt{y^{18}}\sqrt{y} = y^9\sqrt{y}$$

You Try It 4
$$\sqrt{45b^7} = \sqrt{9b^6 \cdot 5b} \qquad \bullet \; 9b^6 \text{ is a perfect square.}$$
$$= \sqrt{9b^6}\sqrt{5b} = 3b^3\sqrt{5b}$$

You Try It 5
$$3a\sqrt{28a^9b^{18}} = 3a\sqrt{4a^8b^{18}(7a)} \qquad \bullet \; 4a^8b^{18} \text{ is a perfect square.}$$
$$= 3a\sqrt{4a^8b^{18}}\sqrt{7a}$$
$$= 3a \cdot 2a^4b^9\sqrt{7a} = 6a^5b^9\sqrt{7a}$$

You Try It 6 $\sqrt{25(a + 3)^2} = 5(a + 3) = 5a + 15$

You Try It 7 $\sqrt{x^2 + 14x + 49} = \sqrt{(x + 7)^2} = x + 7$

SECTION 15.2

You Try It 1
$9\sqrt{3} + 3\sqrt{3} - 18\sqrt{3} = (9 + 3 - 18)\sqrt{3} = -6\sqrt{3}$

You Try It 2
$2\sqrt{50} - 5\sqrt{32}$

$\quad = 2\sqrt{25 \cdot 2} - 5\sqrt{16 \cdot 2}$ • **Simplify the radicands.**

$\quad = 2\sqrt{25}\sqrt{2} - 5\sqrt{16}\sqrt{2}$

$\quad = 2 \cdot 5\sqrt{2} - 5 \cdot 4\sqrt{2}$

$\quad = 10\sqrt{2} - 20\sqrt{2}$

$\quad = (10 - 20)\sqrt{2}$ • **Distributive Property**

$\quad = -10\sqrt{2}$

You Try It 3
$y\sqrt{28y} + 7\sqrt{63y^3}$

$\quad = y\sqrt{4 \cdot 7y} + 7\sqrt{9y^2 \cdot 7y}$ • **Simplify the radicands.**

$\quad = y\sqrt{4}\sqrt{7y} + 7\sqrt{9y^2}\sqrt{7y}$

$\quad = y \cdot 2\sqrt{7y} + 7 \cdot 3y\sqrt{7y}$

$\quad = 2y\sqrt{7y} + 21y\sqrt{7y}$

$\quad = (2y + 21y)\sqrt{7y}$ • **Distributive Property**

$\quad = 23y\sqrt{7y}$

You Try It 4
$2\sqrt{27a^5} - 4a\sqrt{12a^3} + a^2\sqrt{75a}$

$\quad = 2\sqrt{9a^4 \cdot 3a} - 4a\sqrt{4a^2 \cdot 3a} + a^2\sqrt{25 \cdot 3a}$

$\quad = 2\sqrt{9a^4}\sqrt{3a} - 4a\sqrt{4a^2}\sqrt{3a}$
$\qquad + a^2\sqrt{25}\sqrt{3a}$

$\quad = 2 \cdot 3a^2\sqrt{3a} - 4a \cdot 2a\sqrt{3a} + a^2 \cdot 5\sqrt{3a}$

$\quad = 6a^2\sqrt{3a} - 8a^2\sqrt{3a} + 5a^2\sqrt{3a} = 3a^2\sqrt{3a}$

SECTION 15.3

You Try It 1
$\sqrt{5a}\sqrt{15a^3b^4}\sqrt{20b^5}$

$\quad = \sqrt{1500a^4b^9} = \sqrt{100a^4b^8 \cdot 15b}$

$\quad = \sqrt{100a^4b^8} \cdot \sqrt{15b}$

$\quad = 10a^2b^4\sqrt{15b}$

You Try It 2
$\sqrt{5x}(\sqrt{5x} - \sqrt{25y})$

$\quad = \sqrt{25x^2} - \sqrt{125xy}$ • **Distributive Property**

$\quad = \sqrt{25x^2} - \sqrt{25 \cdot 5xy}$

$\quad = \sqrt{25x^2} - \sqrt{25}\sqrt{5xy}$

$\quad = 5x - 5\sqrt{5xy}$

You Try It 3
$(2\sqrt{x} + 7)(2\sqrt{x} - 7)$ • **Product of conjugates**

$\quad = 4(\sqrt{x})^2 - 7^2$

$\quad = 4x - 49$

You Try It 4
$(3\sqrt{x} - \sqrt{y})(5\sqrt{x} - 2\sqrt{y})$

$\quad = 15(\sqrt{x})^2 - 6\sqrt{xy} - 5\sqrt{xy} + 2(\sqrt{y})^2$ • **FOIL**

$\quad = 15(\sqrt{x})^2 - 11\sqrt{xy} + 2(\sqrt{y})^2$

$\quad = 15x - 11\sqrt{xy} + 2y$

You Try It 5
$\dfrac{\sqrt{15x^6y^7}}{\sqrt{3x^7y^9}} = \sqrt{\dfrac{15x^6y^7}{3x^7y^9}} = \sqrt{\dfrac{5}{xy^2}} = \dfrac{\sqrt{5}}{\sqrt{xy^2}}$

$\quad = \dfrac{\sqrt{5}}{y\sqrt{x}} = \dfrac{\sqrt{5}}{y\sqrt{x}} \cdot \dfrac{\sqrt{x}}{\sqrt{x}}$ • **Rationalize the denominator.**

$\quad = \dfrac{\sqrt{5x}}{xy}$

You Try It 6
$\dfrac{\sqrt{3}}{\sqrt{3} - \sqrt{6}} = \dfrac{\sqrt{3}}{\sqrt{3} - \sqrt{6}} \cdot \dfrac{\sqrt{3} + \sqrt{6}}{\sqrt{3} + \sqrt{6}}$ • **Rationalize the denominator.**

$\quad = \dfrac{3 + \sqrt{18}}{3 - 6} = \dfrac{3 + 3\sqrt{2}}{-3}$

$\quad = \dfrac{3(1 + \sqrt{2})}{-3} = -1(1 + \sqrt{2})$

$\quad = -1 - \sqrt{2}$

You Try It 7
$\dfrac{5 + \sqrt{y}}{1 - 2\sqrt{y}} = \dfrac{5 + \sqrt{y}}{1 - 2\sqrt{y}} \cdot \dfrac{1 + 2\sqrt{y}}{1 + 2\sqrt{y}}$ • **Rationalize the denominator.**

$\quad = \dfrac{5 + 10\sqrt{y} + \sqrt{y} + 2(\sqrt{y})^2}{1 - 4y}$

$\quad = \dfrac{5 + 11\sqrt{y} + 2y}{1 - 4y}$

SECTION 15.4

You Try It 1
$\sqrt{4x} + 3 = 7$

$\qquad \sqrt{4x} = 4$ • **Isolate $\sqrt{4x}$.**

$\qquad (\sqrt{4x})^2 = 4^2$ • **Square both sides.**

$\qquad 4x = 16$

$\qquad x = 4$ • **Solve for x.**

Check:

$$\begin{array}{c|c} \sqrt{4x} + 3 = 7 & \\ \hline \sqrt{4 \cdot 4} + 3 & 7 \\ \sqrt{16} + 3 & 7 \\ 4 + 3 & 7 \\ 7 = 7 & \end{array}$$

The solution is 4.

You Try It 2

$$\sqrt{x} + \sqrt{x + 9} = 9$$

$$\sqrt{x} = 9 - \sqrt{x + 9} \qquad \bullet \text{ Isolate } \sqrt{x}.$$

$$(\sqrt{x})^2 = (9 - \sqrt{x + 9})^2 \qquad \bullet \text{ Square both sides.}$$

$$x = 81 - 18\sqrt{x + 9} + (x + 9)$$

$$-90 = -18\sqrt{x + 9}$$

$$5 = \sqrt{x + 9} \qquad \bullet \text{ Isolate } \sqrt{x + 9}.$$

$$5^2 = (\sqrt{x + 9})^2 \qquad \bullet \text{ Square both sides.}$$

$$25 = x + 9$$

$$16 = x \qquad \bullet \text{ Solve for } x.$$

Check:

$$\sqrt{x} + \sqrt{x + 9} = 9$$

$$\begin{array}{c|c} \sqrt{16} + \sqrt{16 + 9} & 9 \\ \sqrt{16} + \sqrt{25} & 9 \\ 4 + 5 & 9 \\ 9 = 9 \end{array}$$

The solution is 16.

You Try It 3

Strategy　To find the distance, use the Pythagorean Theorem. The hypotenuse is the length of the ladder. One leg is the distance from the bottom of the ladder to the base of the building. The vertical distance along the building from the ground to the top of the ladder is the unknown leg.

Solution
$$a = \sqrt{c^2 - b^2} \qquad \bullet \ c = 8, b = 3$$
$$= \sqrt{(8)^2 - (3)^2}$$
$$= \sqrt{64 - 9}$$
$$= \sqrt{55}$$
$$\approx 7.42$$

The distance is approximately 7.42 ft.

You Try It 4

Strategy　To find the length of the pendulum, replace T in the equation with the given value and solve for L.

Solution
$$T = 2\pi\sqrt{\frac{L}{32}}$$

$$2.5 = 2(3.14)\sqrt{\frac{L}{32}} \qquad \bullet \ T = 2.5$$

$$2.5 = 6.28\sqrt{\frac{L}{32}}$$

$$\frac{2.5}{6.28} = \sqrt{\frac{L}{32}}$$

$$\left(\frac{2.5}{6.28}\right)^2 = \left(\sqrt{\frac{L}{32}}\right)^2$$

$$\frac{6.25}{39.4384} = \frac{L}{32}$$

$$(32)\left(\frac{6.25}{39.4384}\right) = (32)\left(\frac{L}{32}\right)$$

$$\frac{200}{39.4384} = L$$

$$5.07 \approx L$$

The length of the pendulum is approximately 5.07 ft.

Solutions to Chapter 16 "You Try It"

SECTION 16.1

You Try It 1

$$\frac{3y^2}{2} + y - \frac{1}{2} = 0$$

$$2\left(\frac{3y^2}{2} + y - \frac{1}{2}\right) = 2(0)$$ • **Multiply each side by 2.**

$$3y^2 + 2y - 1 = 0$$

$$(3y - 1)(y + 1) = 0$$ • **Factor.**

$3y - 1 = 0$	$y + 1 = 0$	• **Principle of**
$3y = 1$	$y = -1$	**Zero Products**
$y = \dfrac{1}{3}$		

The solutions are $\frac{1}{3}$ and -1.

You Try It 2 　$x^2 + 81 = 0$

$$x^2 = -81$$ • **Solve for x^2.**

$$\sqrt{x^2} = \sqrt{-81}$$ • **Take square roots.**

$\sqrt{-81}$ is not a real number.

The equation has no real number solution.

You Try It 3 　$7(z + 2)^2 = 21$

$$(z + 2)^2 = 3$$ • **Solve for $(z + 2)^2$.**

$$\sqrt{(z + 2)^2} = \sqrt{3}$$ • **Take square roots.**

$$z + 2 = \pm\sqrt{3}$$

$$z = -2 \pm \sqrt{3}$$ • **Solve for z.**

The solutions are $-2 + \sqrt{3}$ and $-2 - \sqrt{3}$.

SECTION 16.2

You Try It 1
$$3x^2 - 6x - 2 = 0$$

$$3x^2 - 6x = 2$$ • **Add 2.**

$$\frac{1}{3}(3x^2 - 6x) = \frac{1}{3} \cdot 2$$ • **Multiply by $\frac{1}{3}$.**

$$x^2 - 2x = \frac{2}{3}$$

Complete the square.

$$x^2 - 2x + 1 = \frac{2}{3} + 1$$ • $\left[\frac{1}{2}(-2)\right]^2 = [-1]^2 = 1$

$$(x - 1)^2 = \frac{5}{3}$$ • **Factor.**

$$\sqrt{(x - 1)^2} = \sqrt{\frac{5}{3}}$$ • **Take square roots.**

$$x - 1 = \pm\sqrt{\frac{5}{3}}$$ • **Simplify.**

$$x = 1 \pm \sqrt{\frac{5}{3}}$$

$$x = 1 \pm \frac{\sqrt{15}}{3}$$

$$x = \frac{3 \pm \sqrt{15}}{3}$$

The solutions are $\frac{3 + \sqrt{15}}{3}$ and $\frac{3 - \sqrt{15}}{3}$.

You Try It 2
$$x^2 + 6x + 12 = 0$$

$$x^2 + 6x = -12$$ • **Subtract 12.**

$$x^2 + 6x + 9 = -12 + 9$$ • $\left(\frac{1}{2} \cdot 6\right)^2 = 3^2 = 9$

$$(x + 3)^2 = -3$$ • **Factor.**

$$\sqrt{(x + 3)^2} = \sqrt{-3}$$ • **Take square roots.**

$\sqrt{-3}$ is not a real number.

The quadratic equation has no real number solution.

You Try It 3
$$\frac{x^2}{8} + x + 1 = 0$$

$$8\left(\frac{x^2}{8} + x + 1\right) = 8(0)$$ • **Multiply by 8.**

$$x^2 + 8x + 8 = 0$$

$$x^2 + 8x = -8$$ • **Subtract 8.**

$$x^2 + 8x + 16 = -8 + 16$$ • $\left(\frac{1}{2} \cdot 8\right)^2 = 4^2 = 16$

$$(x + 4)^2 = 8$$ • **Factor.**

$$\sqrt{(x + 4)^2} = \sqrt{8}$$ • **Take square roots.**

$$x + 4 = \pm\sqrt{8}$$

$$x + 4 = \pm 2\sqrt{2}$$

$$x = -4 \pm 2\sqrt{2}$$

$x = -4 + 2\sqrt{2}$	$x = -4 - 2\sqrt{2}$
$\approx -4 + 2(1.414)$	$\approx -4 - 2(1.414)$
$\approx -4 + 2.828$	$\approx -4 - 2.828$
≈ -1.172	≈ -6.828

The solutions are approximately -1.172 and -6.828.

SECTION 16.3

You Try It 1
$3x^2 + 4x - 4 = 0$
$a = 3, b = 4, c = -4$
$$x = \frac{-(4) \pm \sqrt{(4)^2 - 4(3)(-4)}}{2 \cdot 3}$$
$$= \frac{-4 \pm \sqrt{16 + 48}}{6}$$
$$= \frac{-4 \pm \sqrt{64}}{6} = \frac{-4 \pm 8}{6}$$
$$x = \frac{-4 + 8}{6} \qquad x = \frac{-4 - 8}{6}$$
$$= \frac{4}{6} = \frac{2}{3} \qquad = \frac{-12}{6} = -2$$

The solutions are $\frac{2}{3}$ and -2.

You Try It 2
$$\frac{x^2}{4} + \frac{x}{2} = \frac{1}{4}$$
$$4\left(\frac{x^2}{4} + \frac{x}{2}\right) = 4\left(\frac{1}{4}\right) \qquad \text{• Multiply by 4.}$$
$$x^2 + 2x = 1$$
$$x^2 + 2x - 1 = 0 \qquad \text{• Standard form}$$
$$a = 1, b = 2, c = -1$$
$$x = \frac{-(2) \pm \sqrt{(2)^2 - 4(1)(-1)}}{2 \cdot 1}$$
$$= \frac{-2 \pm \sqrt{4 + 4}}{2} = \frac{-2 \pm \sqrt{8}}{2}$$
$$= \frac{-2 \pm 2\sqrt{2}}{2} = -1 \pm \sqrt{2}$$

The solutions are $-1 + \sqrt{2}$ and $-1 - \sqrt{2}$.

SECTION 16.4

You Try It 1 $y = x^2 + 2$

x	y
−2	6
−1	3
0	2
1	3
2	6

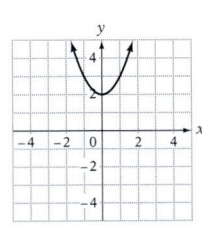

You Try It 2 To find the x-intercept, let $f(x) = 0$ and solve for x.
$$f(x) = x^2 - 6x + 9$$
$$0 = x^2 - 6x + 9$$
$$0 = (x - 3)(x - 3) \qquad \text{• Factor.}$$
$$x - 3 = 0 \qquad x - 3 = 0 \qquad \text{• Principle of}$$
$$x = 3 \qquad x = 3 \qquad \text{Zero Products}$$

The equation has a double root.

There is only one x-intercept.

The x-intercept is $(3, 0)$.

To find the y-intercept, evaluate the function at $x = 0$.
$$f(x) = x^2 - 6x + 9$$
$$f(0) = 0^2 - 6(0) + 9 = 9$$

The y-intercept is $(0, 9)$.

SECTION 16.5

You Try It 1

Strategy
- This is a geometry problem.
- Width of the rectangle: W
  Length of the rectangle: $W + 2$
- Use the equation $A = L \cdot W$.

Solution
$$A = L \cdot W$$
$$15 = (W + 2)W \qquad \text{• } A = 15, L = W + 2$$
$$15 = W^2 + 2W$$
$$0 = W^2 + 2W - 15$$
$$0 = (W + 5)(W - 3) \qquad \text{• Factor.}$$
$$W + 5 = 0 \qquad W - 3 = 0 \qquad \text{• Principle of Zero}$$
$$W = -5 \qquad W = 3 \qquad \text{Products}$$

The solution -5 is not possible.
The width is 3 m.

Solutions to Chapter R "You Try It"

SECTION R.1

You Try It 1

$\dfrac{5}{12} + \dfrac{9}{16} = \dfrac{20}{48} + \dfrac{27}{48}$
- The LCM of 12 and 16 is 48.

$= \dfrac{47}{48}$
- Add the numerators. Place the sum over the common denominator.

You Try It 2

$\dfrac{3}{4} + \dfrac{4}{5} + \dfrac{5}{8}$
- The LCM of 4, 5, and 8 is 40. Write equivalent fractions using the LCM.

$= \dfrac{30}{40} + \dfrac{32}{40} + \dfrac{25}{40}$

$= \dfrac{87}{30}$
- Add the numerators. Place the sum over the common denominator.

$= \dfrac{29}{10} = 2\dfrac{9}{10}$
- Simplify.

You Try It 3

$\dfrac{5}{6} - \dfrac{1}{4} = \dfrac{10}{12} - \dfrac{3}{12}$
- The LCM of 6 and 4 is 12.

$= \dfrac{7}{12}$
- Subtract the numerators. Place the difference over the common denominator.

You Try It 4

$\dfrac{4}{21} \cdot \dfrac{7}{44} = \dfrac{4 \cdot 7}{21 \cdot 44}$
- Multiply the numerators. Multiply the denominators.

$= \dfrac{2 \cdot 2 \cdot 7}{3 \cdot 7 \cdot 2 \cdot 2 \cdot 11}$
- Write the prime factorization of each number.

$= \dfrac{\overset{1}{\cancel{2}} \cdot \overset{1}{\cancel{2}} \cdot \overset{1}{\cancel{7}}}{3 \cdot \cancel{7} \cdot \cancel{2} \cdot \cancel{2} \cdot 11}$
- Divide by the common factors.

$= \dfrac{1}{33}$
- Write the fraction in simplest form.

You Try It 5

$\dfrac{2}{15} \cdot 5 = \dfrac{2}{15} \cdot \dfrac{5}{1}$
- Write 5 as $\dfrac{5}{1}$.

$= \dfrac{2 \cdot 5}{15 \cdot 1}$
- Multiply the fractions.

$= \dfrac{2 \cdot \overset{1}{\cancel{5}}}{3 \cdot \cancel{5} \cdot 1}$
- Divide by the common factors.

$= \dfrac{2}{3}$
- Write the fraction in simplest form.

You Try It 6

$\dfrac{1}{6} \div \dfrac{4}{9} = \dfrac{1}{6} \cdot \dfrac{9}{4}$
- Multiply the first fraction by the reciprocal of the second fraction.

$= \dfrac{1 \cdot 9}{6 \cdot 4}$

$= \dfrac{1 \cdot \overset{1}{\cancel{3}} \cdot 3}{2 \cdot \cancel{3} \cdot 2 \cdot 2}$

$= \dfrac{3}{8}$

SECTION R.2

You Try It 1

a. $-25 + 13 = -12$
- The signs are different. Subtract the absolute values. The sum has the same sign as the number with the larger absolute value.

b. $-41 + (-60) = -101$
- The signs are the same. Add the absolute values. The sum has the same sign as the addends.

c. $37 + (-9) = 28$
- The signs are different. Subtract the absolute values. The sum has the same sign as the number with the larger absolute value.

You Try It 2

a. $-27 - 18 = -27 + (-18)$
$\qquad = -45$
- -27 minus 18 $= -27$ plus the opposite of 18. The opposite of 18 is -18.

b. $-34 - (-90) = -34 + 90$
$\qquad = 56$
- -34 minus $-90 = -34$ plus the opposite of -90. The opposite of -90 is 90.

c. $8 - 42 = 8 + (-42)$
$\qquad = -34$
- 8 minus $42 = 8$ plus the opposite of 42. The opposite of 42 is -42.

d. $-12 + 9 - (-5) - 4$
$\quad = -12 + 9 + 5 + (-4)$
- Rewrite each subtraction as addition of the opposite.

$\quad = -3 + 5 + (-4)$
- Add the numbers.
$\quad = 2 + (-4)$
$\quad = -2$

You Try It 3

a. $-25(4) = -100$
- The signs are different. The product is negative.

b. $-4(-61) = 244$
- The signs are the same. The product is positive.

c. $(-4)(5)(-3)(-1)$
$= (-20)(-3)(-1)$
$= 60(-1)$
$= -60$

- Multiply the first two numbers. Then multiply the product by the third number. Continue until all the numbers have been multiplied.

You Try It 4

a. $(-30) \div 6 = -5$

- The signs are different. The quotient is negative.

b. $(-50) \div (-25) = 2$

- The signs are the same. The quotient is positive.

c. $\dfrac{32}{-8} = -4$

- The fraction bar can be read "divided by."

$$\dfrac{32}{-8} = (32) \div (-8)$$

The signs are different. The quotient is negative.

SECTION R.3

You Try It 1

$-\dfrac{1}{4} + \left(-\dfrac{3}{8}\right) = -\dfrac{2}{8} + \left(-\dfrac{3}{8}\right)$

- Write the fractions with a common denominator.

$= \dfrac{-2}{8} + \dfrac{-3}{8}$

- Write the negative signs in the numerators.

$= \dfrac{-2 + (-3)}{8}$

- Add the fractions.

$= \dfrac{-5}{8}$

- Simplify the numerator.

$= -\dfrac{5}{8}$

- Write the negative sign in front of the fraction.

You Try It 2

$-\dfrac{3}{8} - \left(-\dfrac{1}{6}\right) = -\dfrac{3}{8} + \dfrac{1}{6}$

- Rewrite subtraction as addition of the opposite.

$= -\dfrac{9}{24} + \dfrac{4}{24}$

- Write the fractions with a common denominator.

$= \dfrac{-9}{24} + \dfrac{4}{24}$

- Write the negative sign in the numerator.

$= \dfrac{-9 + 4}{24}$

- Add the fractions.

$= \dfrac{-5}{24}$

- Simplify the numerator.

$= -\dfrac{5}{24}$

- Write the negative sign in front of the fraction.

You Try It 3

$4.69 - 12.5 = 4.69 + (-12.5)$

- Rewrite subtraction as addition of the opposite.

$= -7.81$

- Subtract the absolute values of the numbers. The sum has the same sign as the number with the larger absolute value.

You Try It 4

$\dfrac{10}{11}\left(-\dfrac{2}{5}\right) = -\left(\dfrac{10}{11} \cdot \dfrac{2}{5}\right)$

- The signs are different. The product is negative.

$= -\dfrac{10 \cdot 2}{11 \cdot 5}$

- Multiply the numerators. Multiply the denominators.

$= -\dfrac{2 \cdot 5 \cdot 2}{11 \cdot 5}$

- Write the product in simplest form.

$= -\dfrac{4}{11}$

You Try It 5

$-\dfrac{3}{8} \div \left(-\dfrac{1}{2}\right) = \dfrac{3}{8} \div \dfrac{1}{2}$

- The signs are the same. The quotient is positive.

$= \dfrac{3}{8} \cdot \dfrac{2}{1}$

- Rewrite division as multiplication by the reciprocal.

$= \dfrac{3 \cdot 2}{8 \cdot 1}$

- Multiply the fractions.

$= \dfrac{3 \cdot 2}{2 \cdot 2 \cdot 2 \cdot 1}$

$= \dfrac{3}{4}$

You Try It 6

$(-3.6)(-1.45) = 5.22$

- The signs are the same. The product is positive. Multiply the absolute values of the numbers.

You Try It 7

$5.04 \div (-8.4) = -0.6$

- The signs are different. The quotient is negative. Divide the absolute values of the numbers.

You Try It 8

$\left(\dfrac{2}{9}\right)^2 \cdot (-3)^4$

$= \left(\dfrac{2}{9}\right)\left(\dfrac{2}{9}\right) \cdot (-3)(-3)(-3)(-3)$

- Write each factor the number of times indicated by the exponent.

$= \left(\dfrac{2}{9}\right)\left(\dfrac{2}{9}\right) \cdot (3)(3)(3)(3)$

- The product is positive.

$= \dfrac{2 \cdot 2 \cdot 3 \cdot 3 \cdot 3 \cdot 3}{3 \cdot 3 \cdot 3 \cdot 3}$

- Multiply.

$= 4$

You Try It 9

$7 \div \left(\dfrac{1}{7} - \dfrac{3}{14}\right) - 9$

- Use the Order of Operations Agreement.

$= 7 \div \left(-\dfrac{1}{14}\right) - 9$

- Perform the operation inside the parentheses.

$= 7(-14) - 9$

- Rewrite division as multiplication by the reciprocal.

$= -98 - 9$

- Do the multiplication.

$= -98 + (-9)$

- Do the subtraction.

$= -107$

SECTION R.4

You Try It 1

$$5x + 3 - 7x = 9$$
$$3 - 2x = 9 \qquad \bullet \text{ Step 2}$$
$$3 - 3 - 2x = 9 - 3 \qquad \bullet \text{ Step 4}$$
$$-2x = 6$$
$$\frac{-2x}{-2} = \frac{6}{-2} \qquad \bullet \text{ Step 5}$$
$$x = -3$$

The solution is -3.

You Try It 2

$$4x + 3 = 7x + 9$$
$$4x - 7x + 3 = 7x - 7x + 9 \qquad \bullet \text{ Step 3}$$
$$-3x + 3 = 9$$
$$-3x + 3 - 3 = 9 - 3 \qquad \bullet \text{ Step 4}$$
$$-3x = 6$$
$$\frac{-3x}{-3} = \frac{6}{-3} \qquad \bullet \text{ Step 5}$$
$$x = -2$$

The solution is -2.

You Try It 3

$$4 - (5x - 8) = 4x + 3$$
$$4 - 5x + 8 = 4x + 3 \qquad \bullet \text{ Step 1}$$
$$-5x + 12 = 4x + 3 \qquad \bullet \text{ Step 2}$$
$$-5x - 4x + 12 = 4x - 4x + 3 \qquad \bullet \text{ Step 3}$$
$$-9x + 12 = 3$$
$$-9x + 12 - 12 = 3 - 12 \qquad \bullet \text{ Step 4}$$
$$-9x = -9$$
$$\frac{-9x}{-9} = \frac{-9}{-9} \qquad \bullet \text{ Step 5}$$
$$x = 1$$

The solution is 1.

Solutions to Appendix "You Try It"

You Try It 1
1 295 m = 1.295 km

You Try It 2
7 543 g = 7.543 kg

You Try It 3
6.3 L = 6 300 ml

You Try It 4
2 kl = 2 000 L

Answers to Chapter 1 Selected Exercises

PREP TEST

1. 8   **2.** 1 2 3 4 5 6 7 8 9 10   **3.** a and D; b and E; c and A; d and B; e and F; f and C   **4.** 0   **5.** Fifty

SECTION 1.1

3.
```
+--+--●--+--+--+--+--+--+--+--+--+--+-->
0  1  2  3  4  5  6  7  8  9 10 11 12
```
5.
```
+--+--+--+--+--+--+--+--+--+--●--+--+-->
0  1  2  3  4  5  6  7  8  9 10 11 12
```
7.
```
+--+--+--+--●--+--+--+--+--+--+--+--+-->
0  1  2  3  4  5  6  7  8  9 10 11 12
```
9. 5   **11.** 5   **13.** 0   **15.** $27 < 39$ **17.** $0 < 52$ **19.** $273 > 194$
21. $2761 < 3857$ **23.** $4610 > 4061$ **25.** $8005 < 8050$ **27.** 11, 14, 16, 21, 32   **29.** 13, 48, 72, 84, 93
31. 26, 49, 77, 90, 106   **33.** 204, 399, 662, 736, 981   **35.** 307, 370, 377, 3077, 3700   **37.** Five hundred eight
39. Six hundred thirty-five   **41.** Four thousand seven hundred ninety   **43.** Fifty-three thousand six hundred fourteen
45. Two hundred forty-six thousand fifty-three   **47.** Three million eight hundred forty-two thousand nine hundred five
49. 496   **51.** 53,340   **53.** 502,140   **55.** 9706   **57.** 5,012,907   **59.** 8,005,010   **61.** $7000 + 200 + 40 + 5$
63. $500,000 + 30,000 + 2000 + 700 + 90 + 1$ **65.** $5000 + 60 + 4$ **67.** $20,000 + 300 + 90 + 7$
69. $400,000 + 2000 + 700 + 8$ **71.** $8,000,000 + 300 + 10 + 6$ **73.** 7110   **75.** 5000   **77.** 28,600
79. 7000   **81.** 94,000   **83.** 630,000   **85.** 350,000   **87.** 72,000,000   **89.** Billy Hamilton   **91.** *Fiddler on the Roof*   **93.** Two tablespoons of peanut butter   **95.** St. Louis to San Diego   **97.** Neptune   **99. a.** 67 min
b. 2000   **101.** 160,000 acres   **103. a.** 1985   **b.** Decrease   **105.** 300,000 km/s   **107.** 999; 10,000
109. a. True   **b.** False

SECTION 1.2

3. 1,383,659   **5.** 6043   **7.** 112,152   **9.** 12,548   **11.** 199,556   **13.** 327,473   **15.** 168,574   **17.** 7947
19. 99,637   **21.** 1872 students   **23.** 15,000; 15,040   **25.** 1,400,000; 1,388,917   **27.** 2000; 1998
29. 307,000; 329,801   **31.** 1272   **33.** 12,150   **35.** 89,900   **37.** 1572   **39.** 14,591   **41.** 56,010
43. The Commutative Property of Addition   **45.** The Associative Property of Addition   **47.** The Addition Property of Zero   **49.** 28   **51.** 4   **53.** 15   **55.** Yes   **57.** No   **59.** Yes   **63.** 416   **65.** 188   **67.** 464
69. 208   **71.** 3557   **73.** 2836   **75.** 1437   **77.** 20,148   **79.** 1618   **81.** 7378   **83.** 17,548   **85.** 15 ft
87. 2000; 2136   **89.** 40,000; 38,283   **91.** 35,000; 31,195   **93.** 100,000; 125,665   **95.** 13   **97.** 643
99. 355   **101.** 5211   **103.** 766   **105.** 18,231   **107.** Yes   **109.** No   **111.** Yes   **113.** 210   **115.** 901
117. 370 calories   **119.** 78 m   **121.** 43 in.   **123.** 560 ft   **125.** 43 orbits   **127.** 10,818 seats   **129.** $1645
131. 20,000 mi   **133.** January to February; 24 cars   **135.** $13,275   **137.** $261,000   **139.** 350 mph
141. a. 9571 drivers   **b.** 4211 drivers   **143.** No   **145.** 11   **147. a.** Always true   **b.** Always true

SECTION 1.3

3. 1143   **5.** 46,963   **7.** 470,152   **9.** 48,493   **11.** 324,438   **13.** 3,206,160   **15.** 1500   **17.** 2000
19. 0   **21.** qrs **23.** 1,200,000; 1,244,653   **25.** 1,200,000; 1,138,134   **27.** 42,000; 46,935   **29.** 6,300,000;
6,491,166   **31.** 14,880   **33.** 3255   **35.** 1800   **37.** 3082   **39.** The Multiplication Property of One
41. The Commutative Property of Multiplication   **43.** 30   **45.** 0   **47.** Yes   **49.** No   **51.** Yes   **53.** $2^3 \cdot 7^5$
55. $2^2 \cdot 3^3 \cdot 5^4$ **57.** c^2 **59.** $x^3 y^3$ **61.** 32   **63.** 1,000,000   **65.** 200   **67.** 9000   **69.** 0   **71.** 540
73. 144   **75.** 512   **77.** a^4 **79.** 24   **81.** 320   **83.** 225   **87.** 307   **89.** 309 r4   **91.** 2550   **93.** 21 r9
95. 147 r38   **97.** 200 r8   **99.** 404 r34   **101.** 16 r97   **103.** 907   **105.** 881 r1   **107.** $\dfrac{c}{d}$ **109.** 800; 776
111. 5000; 5129   **113.** 500; 493 r37   **115.** 1500; 1516   **117.** 48   **119.** Undefined   **121.** 9800   **123.** Yes
125. No   **127.** 1, 2, 5, 10   **129.** 1, 2, 3, 4, 6, 12   **131.** 1, 2, 4, 8   **133.** 1, 13   **135.** 1, 2, 3, 6, 9, 18
137. 1, 5, 25   **139.** 1, 2, 4, 7, 8, 14, 28, 56   **141.** 1, 2, 4, 7, 14, 28   **143.** 1, 2, 3, 4, 6, 8, 12, 16, 24, 48
145. 1, 2, 3, 6, 9, 18, 27, 54   **147.** 2^4 **149.** $2^2 \cdot 3$ **151.** $3 \cdot 5$ **153.** $2^3 \cdot 5$ **155.** Prime   **157.** $5 \cdot 13$
159. $2^2 \cdot 7$ **161.** $2 \cdot 3 \cdot 7$ **163.** $3 \cdot 17$ **165.** $2 \cdot 23$ **167.** 460 calories   **169.** 4325 gal   **171. a.** 78 m
b. 360 m²   **173.** 96 ft   **175.** 576 ft²   **177.** 59,136 cm²   **179.** $16,000   **181.** $6840   **183.** 9 h
185. $21   **187.** Approximation   **189.** 222

SECTION 1.4

3. 4   **5.** 29   **7.** 13   **9.** 19   **11.** 11   **13.** 6   **15.** 61   **17.** 54   **19.** 19   **21.** 24   **23.** 186
25. 39   **27.** 18   **29.** 14   **31.** 14   **33.** 2   **35.** 57   **37.** 8   **39.** 68   **41.** 16
43. $12 + (9 - 5) \cdot 3 > 11 + (8 + 4) \div 6$
45. $4 + 3 \cdot 12 > 81 - 8^2 > 27 \div 9 + 8 > 5(10 - 2) \div 4 > 2(1 + 4)^2 \div 10 > 50 - 6(8)$

CHAPTER 1 REVIEW EXERCISES*

1. [1.1A]   **2.** 10,000 [1.3B]   **3.** 2583 [1.2B]   **4.** $3^2 \cdot 5^4$ [1.3B]

5. 1389 [1.2A]   **6.** 38,700 [1.1C]   **7.** $247 > 163$ [1.1A]   **8.** 32,509 [1.1B]   **9.** 700 [1.3A]

10. 2607 [1.3C]   **11.** 4048 [1.2B]   **12.** 1500 [1.2A]   **13.** 1, 2, 5, 10, 25, 50 [1.3D]   **14.** Yes [1.2B]

15. 18 [1.4A]   **16.** The Commutative Property of Addition [1.2A]   **17.** Four million nine hundred twenty-seven

thousand thirty-six [1.1B]   **18.** 675 [1.3B]   **19. a.** 16 times more   **b.** 61 times more [1.3E]   **20.** 67 r70

[1.3C]   **21.** 2636 [1.3A]   **22.** 137 [1.2B]   **23.** $2 \cdot 3^2 \cdot 5$ [1.3D]   **24.** 80 [1.3C]   **25.** 1 [1.3A]   **26.** 10

[1.4A]   **27.** 932 [1.2A]   **28.** 432 [1.3A]   **29.** 56 [1.4A]   **30.** Kareem Abdul-Jabbar [1.2C]   **31.** $182,000

[1.3E]   **32. a.** 74 m   **b.** 300 m² [1.2C, 1.3E]   **33. a.** 1960s   **b.** 4,792,000 students [1.2C]   **34.** 42 [1.3E]

35. $449 [1.2C]

CHAPTER 1 TEST

1. 329,700 [1.3A]   **2.** 16,000 [1.3B]   **3.** 4029 [1.2B]   **4.** $x^4 y^3$ [1.3B]   **5.** Yes [1.2A]   **6.** 3000 [1.1C]
7. $7177 < 7717$ [1.1A]   **8.** 8490 [1.1B]   **9.** Three hundred eighty-two thousand nine hundred four [1.1B]
10. 2000 [1.2A]   **11.** 11,008 [1.3A]   **12.** 2,400,000 [1.3A]   **13.** 1, 2, 4, 23, 46, 92 [1.3D]   **14.** $2^4 \cdot 3 \cdot 5$
[1.3D]   **15.** 30,866 [1.2B]   **16.** The Commutative Property of Addition [1.2A]   **17.** 897 [1.3C]   **18.** 26 [1.4A]
19. $13,900 [1.2C]   **20.** 71 [1.4A]   **21.** $3000 + 900 + 70 + 2$ [1.1B]   **22.** 56 [1.4A]   **23.** 7 [1.2A]
24. 720 [1.3E]   **25.** $556 [1.2C]   **26. a.** 96 cm   **b.** 576 cm² [1.3E]   **27.** $4456 [1.2C]   **28. a.** 2001–2002
b. 125,000 vehicles [1.2C]   **29.** $960 [1.3E]   **30.** $11 [1.3E]

Answers to Chapter 2 Selected Exercises

PREP TEST

1. 20 [1.3A]   **2.** 120 [1.3A]   **3.** 9 [1.3A]   **4.** 10 [1.2A]   **5.** 7 [1.2B]   **6.** 2 r3 [1.3C]   **7.** 36,900
[1.1C]   **8.** Four thousand seven hundred ninety-one [1.1B]   **9.** 6842 [1.1B]   **10.** 1, 2, 3, 4, 6, 12 [1.3C]
11. 59 [1.4A]   **12.** 7 [1.2A]   **13.** $44 < 48$ [1.1A]

SECTION 2.1

1. 40   **3.** 24   **5.** 30   **7.** 12   **9.** 24   **11.** 60   **13.** 56   **15.** 9   **17.** 32   **19.** 36   **21.** 660
23. 9384   **25.** 24   **27.** 30   **29.** 24   **31.** 576   **33.** 1680   **35.** 1   **37.** 3   **39.** 5   **41.** 25   **43.** 1
45. 4   **47.** 4   **49.** 6   **51.** 4   **53.** 1   **55.** 7   **57.** 5   **59.** 8   **61.** 1   **63.** 25   **65.** 7   **67.** 8
69. Two composite numbers are relatively prime if they do not have any common factor except the factor 1. Examples: 4
and 5, 8 and 9, and 16 and 21.

SECTION 2.2

1. $\dfrac{4}{5}$ **3.** $\dfrac{1}{4}$ **5.** $\dfrac{4}{3}$; $1\dfrac{1}{3}$ **7.** $\dfrac{13}{5}$; $2\dfrac{3}{5}$ **9.** $3\dfrac{1}{4}$ **11.** 4   **13.** $2\dfrac{7}{10}$ **15.** 7   **17.** $1\dfrac{8}{9}$ **19.** $2\dfrac{2}{5}$ **21.** 18

*Note: The numbers in brackets following the answers in the Chapter Review Exercises are a reference to the objective that
corresponds to that problem. For example, the reference [1.2A] stands for Section 1.2, Objective A. This notation will be
used for all Prep Tests, Chapter Review Exercises, Chapter Tests, and Cumulative Review Exercises throughout the text.

23. $2\frac{2}{15}$ **25.** 1   **27.** $9\frac{1}{3}$ **29.** $\frac{9}{4}$ **31.** $\frac{11}{2}$ **33.** $\frac{14}{5}$ **35.** $\frac{47}{6}$ **37.** $\frac{7}{1}$ **39.** $\frac{33}{4}$ **41.** $\frac{31}{3}$

43. $\frac{55}{12}$ **45.** $\frac{8}{1}$ **47.** $\frac{64}{5}$ **49.** $\frac{6}{12}$ **51.** $\frac{9}{24}$ **53.** $\frac{6}{51}$ **55.** $\frac{24}{32}$ **57.** $\frac{108}{18}$ **59.** $\frac{30}{90}$ **61.** $\frac{14}{21}$

63. $\frac{42}{49}$ **65.** $\frac{8}{18}$ **67.** $\frac{28}{4}$ **69.** $\frac{1}{4}$ **71.** $\frac{3}{4}$ **73.** $\frac{1}{6}$ **75.** $\frac{8}{33}$ **77.** 0   **79.** $\frac{7}{6}$ **81.** 1   **83.** $\frac{3}{5}$

85. $\frac{4}{15}$ **87.** $\frac{3}{5}$ **89.** $\frac{2m}{3}$ **91.** $\frac{y}{2}$ **93.** $\frac{2a}{3}$ **95.** c **97.** $6k$ **99.** $\frac{3}{8} < \frac{2}{5}$ **101.** $\frac{3}{4} < \frac{7}{9}$

103. $\frac{2}{3} > \frac{7}{11}$ **105.** $\frac{17}{24} > \frac{11}{16}$ **107.** $\frac{7}{15} > \frac{5}{12}$ **109.** $\frac{5}{9} > \frac{11}{21}$ **111.** $\frac{7}{12} < \frac{13}{18}$ **113.** $\frac{4}{5} > \frac{7}{9}$ **115.** $\frac{9}{16} > \frac{5}{9}$

117. $\frac{5}{8} < \frac{13}{20}$ **119.** $\frac{1}{8}$ **121.** $\frac{5}{6}$ **123.** $\frac{3}{4}$ **125.** Location   **127.** $\frac{2}{25}$

SECTION 2.3

1. $\frac{9}{11}$ **3.** 1   **5.** $1\frac{2}{3}$ **7.** $1\frac{1}{6}$ **9.** $\frac{16}{b}$ **11.** $\frac{9}{c}$ **13.** $\frac{11}{x}$ **15.** $\frac{11}{12}$ **17.** $\frac{11}{12}$ **19.** $1\frac{7}{12}$ **21.** $2\frac{2}{15}$

23. $15\frac{2}{3}$ **25.** $5\frac{2}{3}$ **27.** $15\frac{1}{20}$ **29.** $10\frac{7}{36}$ **31.** $7\frac{5}{12}$ **33.** $\frac{3}{4}$ **35.** $6\frac{5}{24}$ **37.** $2\frac{5}{24}$ **39.** $1\frac{2}{5}$

41. $1\frac{13}{18}$ **43.** $1\frac{5}{24}$ **45.** $11\frac{2}{3}$ **47.** $14\frac{3}{4}$ **49.** Yes   **51.** $\frac{1}{6}$ **53.** $\frac{1}{6}$ **55.** $\frac{5}{d}$ **57.** $\frac{5}{n}$ **59.** $\frac{1}{14}$

61. $\frac{1}{2}$ **63.** $\frac{1}{4}$ **65.** $2\frac{1}{3}$ **67.** $6\frac{3}{4}$ **69.** $1\frac{1}{12}$ **71.** $3\frac{3}{8}$ **73.** $5\frac{1}{9}$ **75.** $2\frac{3}{4}$ **77.** $1\frac{17}{24}$ **79.** $4\frac{19}{24}$

81. $1\frac{7}{10}$ **83.** $\frac{5}{24}$ **85.** $6\frac{5}{12}$ **87.** $\frac{1}{3}$ **89.** $\frac{1}{6}$ **91.** $1\frac{1}{9}$ **93.** $2\frac{2}{9}$ **95.** Yes   **97.** You now own $1\frac{3}{4}$

acres of the property.   **99.** $7\frac{3}{4}$ more hours of community service are still required of you.   **101.** The boxer must gain

$6\frac{3}{4}$ lb during the third and fourth weeks.   **103. a.** The response "5 meals" was given most frequently.

b. $\frac{23}{100}$ of the adult population cooks two or fewer dinners at home per week.   **c.** $\frac{49}{100}$ of the adult population cooks five

or more dinners at home per week. This is less than $\frac{1}{2}$ of the people.   **105.** The difference is $\frac{3}{32}$ in.   **107.** You need

$29\frac{1}{2}$ ft of fencing.   **109.** $55\frac{3}{4}$ ft of wood beams must be purchased.   **111.** Austin's jump was $1\frac{1}{3}$ ft higher than

Osborn's jump.   **113.** No, because the parts are not equal in size.

SECTION 2.4

3. $\frac{3}{5}$ **5.** $\frac{4}{5}$ **7.** 0   **9.** $\frac{63}{xy}$ **11.** $\frac{1}{9}$ **13.** 1   **15.** 6   **17.** 0   **19.** $\frac{1}{2}$ **21.** 19   **23.** 42   **25.** $5\frac{1}{2}$ **27.** $\frac{7}{10}$

29. $1\frac{4}{5}$ **31.** A typical household spends \$13,000 on housing per year.   **33.** $\frac{7}{48}$ **35.** $3\frac{1}{2}$ **37.** $\frac{1}{5}$ **39.** $\frac{1}{6}$

41. $3\frac{2}{3}$ **43.** No   **45.** $\frac{9}{16}$ **47.** $\frac{5}{128}$ **49.** $\frac{4}{45}$ **51.** $1\frac{1}{7}$ **53.** $\frac{16}{81}$ **55.** $\frac{2}{3}$ **57.** $1\frac{11}{14}$ **59.** 0   **61.** 8

63. $\frac{1}{8}$ **65.** Undefined   **67.** $\frac{bd}{30}$ **69.** $5\frac{1}{3}$ **71.** $\frac{1}{2}$ **73.** $5\frac{2}{7}$ **75.** $1\frac{29}{31}$ **77.** $1\frac{1}{5}$ **79.** $\frac{7}{26}$ **81.** $\frac{1}{12}$

83. 48   **85.** There are 32 servings in the box.   **87.** $\frac{3}{4}$ **89.** $\frac{1}{6}$ **91.** 6   **93.** $\frac{18}{35}$ **95.** $1\frac{7}{25}$ **97.** $3\frac{3}{11}$

99. 17   **101.** 1   **103.** No   **105.** There are 30 min in four chukkers.   **107.** There are $16\frac{1}{2}$ ft in one rod. There are

198 in. in one rod.   **109.** The average couple spends 234 h cleaning house each year.   **111.** The developer plans to

build 30 houses on the property.   **113.** When it is closed, the dimensions of the board are 14 in. by 7 in. by $1\frac{3}{4}$ in.

115. a. The average teenage boy drinks $23\frac{1}{3}$ cans of soda per week.   **b.** The average teenage boy consumes 3500 calories each week in soda.   **c.** The average teenage boy drinks 7 more cans of soda per week than the average teenage girl.

117. 96 m² of canvas was needed.   **119.** The area is $42\frac{1}{2}$ yd².   **121.** Two bags of seed should be purchased.

123. The rate of the hiker is $3\frac{1}{2}$ mph.   **125.** The distance between the two cities is 1250 mi.

SECTION 2.5

1. Thousandths   **3.** Ten-thousandths   **5.** Hundredths   **7.** 0.3   **9.** 0.21   **11.** 0.461   **13.** 0.093
15. $\frac{1}{10}$ **17.** $\frac{47}{100}$ **19.** $\frac{289}{1000}$ **21.** $\frac{9}{100}$ **23.** Thirty-seven hundredths   **25.** Nine and four tenths
27. Fifty-three ten-thousandths   **29.** Forty-five thousandths   **31.** Twenty-six and four hundredths   **33.** 3.0806
35. 407.03   **37.** 246.024   **39.** 73.02684   **41.** $0.7 > 0.56$ **43.** $3.605 > 3.065$ **45.** $9.004 < 9.04$
47. $9.31 > 9.031$ **49.** $4.6 < 40.6$ **51.** $0.07046 > 0.07036$ **53.** 0.609, 0.66, 0.696, 0.699   **55.** 1.237, 1.327,
1.372, 1.732   **57.** 21.78, 21.805, 21.87, 21.875   **59.** 5.4   **61.** 30.0   **63.** 413.60   **65.** 6.062   **67.** 97
69. 5440   **71.** 0.0236   **73.** The weight is 0.18 oz.   **75.** The distance is 26.2 mi.   **77.** Barry Sanders has the
greatest average number of yards per carry.   **79.** The length of the race is 42.2 km.   **81. a.** $2.40   **b.** $3.60
c. $6.00   **d.** $7.00   **e.** $4.70   **f.** $2.40   **g.** $2.40   **83.** For example: **a.** 0.15   **b.** 1.05   **c.** 0.001

SECTION 2.6

1. 65.9421   **3.** 190.857   **5.** 21.26   **7.** 21.26   **9.** 2.768   **11.** 56.361   **13.** 53.67   **15.** 12; 12.325
17. 40; 33.63   **19.** 0.3; 0.303   **21.** 40; 38.618   **23.** 137.505   **25.** 24.53   **27.** 11.789   **29.** 1.70   **31.** 0.03316
33. 4250   **35.** 67,100   **37.** 8.0; 7.5537   **39.** 70; 68.5936   **41.** 30; 32.1485   **43.** 50.16   **45.** 48
47. 32.3   **49.** 67.7   **51.** 6.3   **53.** 5.8   **55.** 0.81   **57.** 0.08   **59.** 5.278   **61.** 0.4805   **63.** 10; 11.17
65. 1; 1.16   **67.** 50; 58.90   **69.** 6; 7.20   **71.** 132   **73.** 4.06   **75.** 3.8   **77.** 0.375   **79.** $0.\overline{72}$ **81.** $0.58\overline{3}$
83. 1.75   **85.** 1.5   **87.** $4.1\overline{6}$ **89.** 2.25   **91.** $3.\overline{8}$ **93.** $\frac{1}{5}$ **95.** $\frac{3}{4}$ **97.** $\frac{1}{8}$ **99.** $2\frac{1}{2}$ **101.** $4\frac{11}{20}$ **103.** $1\frac{18}{25}$

105. $\frac{9}{200}$ **107.** $\frac{9}{10} > 0.89$ **109.** $\frac{4}{5} < 0.803$ **111.** $0.444 < \frac{4}{9}$ **113.** $0.13 > \frac{3}{25}$ **115.** $\frac{5}{16} > 0.312$

117. $\frac{10}{11} > 0.909$ **119.** Your monthly salary is $3968.25.   **121.** The cost is $.37 per can.   **123. a.** There are 53.446
million children in grades K–12.   **b.** There are 40.49 million more children being educated in public school.
125. It costs $3.42 to operate the motor for 90 h.   **127. a.** The increase is 3.7 trillion cigarettes.   **b.** The cigarette
consumption was 2.5 times greater in 2000 than in 1960.   **129.** You have $505.17 left in the budget for the remainder of
the month.   **131.** The bookkeeper's total income for the week is $808.50.   **133.** The bill is approximately $69.
135. The profit is $91.80.   **137.** The cost is $13.30.   **139.** The perimeter is 15.5 in.   **141.** The area is 14.625 in².
143. The perimeter is 13.95 m.   **145.** The markup is $578.62.   **147.** The cost is $.48 per mile.   **149.** The cost is
$.54.

SECTION 2.7

1. $1\frac{1}{5}$ **3.** $\frac{5}{36}$ **5.** $\frac{11}{32}$ **7.** 1   **9.** 4   **11.** $\frac{8}{9}$ **13.** 1.72   **15.** 1.84   **17.** 2.04   **19.** $1\frac{3}{10}$ **21.** $1\frac{1}{9}$

23. $1\frac{15}{16}$ **25.** $\frac{1}{2}$ **27.** 1   **29.** 18.09   **31.** 30.5   **33.** $0; \frac{9}{16}$

CHAPTER 2 REVIEW EXERCISES

1. $9\frac{1}{2}$ [2.2A] **2.** $2\frac{5}{6}$ [2.3B] **3.** $1\frac{1}{2}$ [2.4B] **4.** 5.034 [2.5A] **5.** $\frac{7}{25}$ [2.6D] **6.** $2\frac{2}{3}$ [2.4A]

7. $8.039 < 8.31$ [2.5B] **8.** $\frac{3}{5} > \frac{7}{15}$ [2.2C] **9.** 150 [2.1A] **10.** 91,800 [2.6B] **11.** $3\frac{1}{3}$ [2.4A]

12. $\frac{10}{7}; 1\frac{3}{7}$ [2.2A] **13.** $\frac{3}{7} < 0.429$ [2.6D] **14.** $\frac{3}{5}$ [2.4C] **15.** $\frac{32}{72}$ [2.2B] **16.** $\frac{1}{3}$ [2.4A] **17.** $\frac{2}{7}$ [2.7A]

18. 21 [2.1B] **19.** 0.0142 [2.6C] **20.** 0.1 [2.6C] **21.** $\frac{5}{6}$ [2.4B] **22.** 0.11 [2.6C] **23.** 440 [2.6A]

24. 2.4622 [2.6B] **25.** 50.743 [2.6A] **26.** $2\frac{1}{4}$ [2.4A] **27.** $9\frac{1}{12}$ [2.3A] **28.** $\frac{2}{7}$ [2.2B] **29.** $4\frac{7}{10}$ [2.3B]

30. The projected increase is 1.2 million workers. [2.6E] **31.** The wrestler must gain $6\frac{1}{4}$ lb during the third and fourth weeks. [2.3C] **32.** The employee can assemble 192 units during an 8-hour day. [2.4D] **33.** The employee is due $150 in overtime pay. [2.4D] **34.** The final velocity is 496 ft/s. [2.4D]

CHAPTER 2 TEST

1. $2\frac{4}{7}$ [2.2A] **2.** $3\frac{11}{12}$ [2.3B] **3.** $22\frac{1}{2}$ [2.4A] **4.** $\frac{7}{12}$ [2.4A] **5.** 90 [2.1A] **6.** 9.033 [2.5A]

7. $2\frac{11}{32}$ [2.4A] **8.** $\frac{19}{5}$ [2.2A] **9.** $\frac{7}{9}$ [2.4B] **10.** $4.003 < 4.009$ [2.5B] **11.** 7 [2.4C] **12.** 18 [2.1B]

13. $\frac{1}{6}$ [2.3B] **14.** $\frac{4}{5}$ [2.2B] **15.** $2\frac{17}{24}$ [2.3A] **16.** $\frac{5}{6} > \frac{11}{15}$ [2.2C] **17.** $3\frac{16}{25}$ [2.7A] **18.** $0.22 < \frac{2}{9}$ [2.6D]

19. 6.051 [2.5C] **20.** $1\frac{1}{2}$ [2.4B] **21.** 22.753 [2.6A] **22.** 70 [2.6A] **23.** 14.497 [2.6A] **24.** 64 [2.6B]

25. 0.8496 [2.6C] **26.** $\frac{12}{28}$ [2.2B] **27.** The gross was $40.8 million greater. [2.6E] **28.** Ten more hours of community serivce are still required of you. [2.3C] **29.** The employee can assemble 80 units in 6 h. [2.4C] **30.** The stockholders' equity is $20.6 million. [2.6E] **31.** The perimeter is 18.5 m. [2.6E]

CUMULATIVE REVIEW EXERCISES

1. 0.03879 [2.6C] **2.** 15 [1.4A] **3.** $2^5 \cdot 5 \cdot 7$ [1.3D] **4.** 8,072,092 [1.1B] **5.** $\frac{7}{11} < \frac{4}{5}$ [2.2C]

6. 36 [2.1B] **7.** $\frac{1}{7}$ [2.3B] **8.** 1900 [1.2A] **9.** 11,272 [1.2A] **10.** $1\frac{1}{2}$ [2.4B]

11. 55.42 [2.6A] **12.** $\frac{1}{7}$ [2.4A] **13.** 1600 [1.3B] **14.** $2^2 \cdot 5 \cdot 13$ [1.3D] **15.** 0.76 [2.6D]

16. 20,000 [1.2B] **17. a.** Sweden mandates more vacation days. **b.** Austria mandates 1.5 times more vacation days than Switzerland. [2.6E] **18.** Undefined [1.3C] **19.** $\frac{19}{21}$ [2.3A] **20.** $7\frac{1}{28}$ [2.3B] **21.** 39 [1.4A]

22. 17 [1.4A] **23.** $\frac{3}{10}$ [2.4C] **24.** $\frac{3}{28}$ [2.4A] **25.** 2.8 [2.6C] **26.** You would burn 40 more calories. [1.3E]

27. The projected increase in the population is 1,740,000 people. [1.2C] **28. a.** On average, a salesperson works 46.5 h per week. **b.** The average salesperson spends more time face-to-face selling. [2.6E] **29.** The distance traveled by the bicyclist is $4\frac{1}{8}$ mi. [2.4D] **30.** The cost is $3.12 per visit. [2.6E]

Answers to Chapter 3 Selected Exercises

PREP TEST

1. $54 > 45$ [1.1A] **2.** 4 units [1.1A] **3.** 15,847 [1.2A] **4.** 3779 [1.2B] **5.** 26,432 [1.3A] **6.** 6 [1.3C]

7. $1\frac{4}{15}$ [2.3A] **8.** $\frac{7}{16}$ [2.3B] **9.** 11.058 [2.6A] **10.** 3.781 [2.6A] **11.** $\frac{2}{5}$ [2.4A] **12.** $\frac{5}{9}$ [2.4B]

13. 9.4 [2.6B] **14.** 0.4 [2.6C] **15.** 31 [1.4A]

SECTION 3.1

1. **3.**

5. **7.** **9.** 1 **11.** -1 **13.** 3

15. A is -4. C is -2. **17.** A is -7. D is -4. **19.** $-2 > -5$ **21.** $3 > -7$ **23.** $-42 < 27$ **25.** $53 > -46$

27. $-51 < -20$ **29.** $-131 < 101$ **31.** $-7, -2, 0, 3$ **33.** $-5, -3, 1, 4$ **35.** $-4, 0, 5, 9$ **37.** $-10, -7, -5, 4, 12$

39. $-11, -7, -2, 5, 10$ **41.** -45 **43.** 88 **45.** $-n$ **47.** d **49.** The opposite of negative thirteen

51. The opposite of negative p **53.** Five plus negative ten **55.** Negative fourteen minus negative three

57. Negative thirteen minus eight **59.** m plus negative n **61.** 7 **63.** 61 **65.** -46 **67.** 73 **69.** z **71.** $-p$

73. 4 **75.** 9 **77.** 11 **79.** 12 **81.** 23 **83.** -27 **85.** 25 **87.** -41 **89.** -93 **91.** 10 **93.** 8

95. 6 **97.** $|-12| > |8|$ **99.** $|6| < |13|$ **101.** $|-1| < |-17|$ **103.** $|x| = |-x|$ **105.** $-|6|, -(4), |-7|, -(-9)$

107. $-9, -|-7|, -(5), |4|$ **109.** $-|10|, -|-8|, -(-2), -(-3), |5|$ **111.** The wind chill factor is $-9°$F.

113. The cooling power is $-35°$F. **115.** A temperature of $-30°$F with a 5-mph wind would feel colder.

117. Stock B showed the least net change. **119.** The loss was greater during the third quarter. **121.** $11, -11$

123. $-6, -5, -4, -3, -2, -1, 0, 1, 2, 3, 4, 5, 6$ **125. a.** -2 and 6 **b.** -2 and 8

SECTION 3.2

3. -11 **5.** -5 **7.** 8 **9.** -4 **11.** -2 **13.** -9 **15.** 1 **17.** -15 **19.** 0 **21.** -21 **23.** -14

25. 19 **27.** -5 **29.** -30 **31.** 9 **33.** -12 **35.** -28 **37.** -13 **39.** -18 **41.** 11 **43.** 1

45. $x + (-7)$ **47. a.** $-\$85,509,000,000$ **b.** $-\$76,987,000,000$ **c.** $-\$68,071,000,000$ **49.** 5 **51.** -2

53. -11 **55.** -17 **57.** The Addition Property of Zero **59.** The Associative Property of Addition **61.** 0

63. 18 **65.** No **67.** Yes **69.** No **73.** -3 **75.** -13 **77.** 7 **79.** 0 **81.** -17 **83.** -3

85. 12 **87.** 27 **89.** -106 **91.** -67 **93.** -6 **95.** -15 **97.** $-t - r$ **99.** $82°$C **101.** -9

103. 11 **105.** 0 **107.** -138 **109.** 26 **111.** 13 **113.** -8 **115.** 5 **117.** 2 **119.** -6 **121.** 12

123. -3 **125.** 18 **127.** Yes **129.** No **131.** Yes **133. a.** The difference in elevation is 7046 m.

b. The difference in elevation is 6051 m. **135.** The difference between the highest and lowest elevations is smallest in

Europe. **137.** The difference is $86°$. **139.** The average temperature is $36°$ colder at 30,000 ft.

141. The golfer's score was -3. **143.** 19 **145. a.** Sometimes true **b.** Always true

SECTION 3.3

3. -24 **5.** 6 **7.** 18 **9.** -20 **11.** -16 **13.** 25 **15.** 0 **17.** 42 **19.** -128 **21.** 208

23. -243 **25.** -115 **27.** 238 **29.** -96 **31.** -210 **33.** -224 **35.** -40 **37.** 180 **39.** $-qr$

41. a. The annual net income would be $-\$188,400,000$. **b.** The annual net income would be $-\$330,800,000$.

c. The annual net income would be $-\$8,400,000$. **43.** The Multiplication Property of One **45.** The Associative

Property of Multiplication **47.** -6 **49.** 1 **51.** -24 **53.** -60 **55.** 357 **57.** -56

59. -1600 **61.** No **63.** No **65.** Yes **67.** -6 **69.** 8 **71.** -49 **73.** 8 **75.** -11 **77.** 14

79. 13 **81.** 1 **83.** 26 **85.** 23 **87.** -110 **89.** 111 **91.** $\dfrac{-9}{x}$ **93.** The average monthly net income was

$-\$236,000$. **95.** -9 **97.** 9 **99.** -6 **101.** 6 **103.** Yes **105.** No **107.** Yes **109.** The average score

was -3.　**111.** The average record low temperature is $-62°F$.　**113.** The average daily low temperature was $-4°$.
115. The wind chill factor is $-45°F$.　**117.** $-16, 32, -64$　**119.** $-125, -625, -3125$　**121. a.** 81　**b.** -17
123. $-3, -2, -1$

SECTION 3.4

1. $-\dfrac{5}{24}$　**3.** $-\dfrac{19}{24}$　**5.** $\dfrac{5}{26}$　**7.** $\dfrac{7}{24}$　**9.** $-\dfrac{19}{60}$　**11.** $-1\dfrac{3}{8}$　**13.** $\dfrac{3}{4}$　**15.** $-\dfrac{47}{48}$　**17.** $\dfrac{3}{8}$　**19.** $-\dfrac{7}{60}$　**21.** $\dfrac{13}{24}$

23. -3.4　**25.** -8.89　**27.** -8.0　**29.** -0.68　**31.** -181.51　**33.** 2.7　**35.** -20.7　**37.** -37.19

39. -34.99　**41.** $-\dfrac{5}{48}$　**43.** $-1\dfrac{5}{36}$　**45.** $1\dfrac{3}{10}$　**47.** -649.36　**49.** 31.09　**51.** $-\dfrac{1}{6}$　**53.** $-1\dfrac{5}{24}$

55. -1.159　**57.** -25.665　**59.** $-1\dfrac{1}{4}$　**61.** $\dfrac{5}{36}$　**63.** -2.163　**65.** Yes　**67.** No　**69.** No　**71.** $-\dfrac{3}{8}$

73. $\dfrac{1}{10}$　**75.** $-\dfrac{4}{9}$　**77.** $-\dfrac{7}{26}$　**79.** 10　**81.** $-\dfrac{7}{30}$　**83.** $-\dfrac{2}{3}$　**85.** $4\dfrac{2}{7}$　**87.** $-\dfrac{8}{9}$　**89.** $\dfrac{9}{10}$　**91.** $-\dfrac{1}{6}$

93. 28.14　**95.** -7.84　**97.** 0.117　**99.** 9.91　**101.** -84.3　**103.** -49.8　**105.** -1.7　**107.** $-\dfrac{1}{12}$

109. $-\dfrac{1}{21}$　**111.** $-1\dfrac{1}{24}$　**113.** -131.328　**115.** -25.4　**117.** -0.5　**119.** $-\dfrac{7}{48}$　**121.** $-17\dfrac{1}{2}$　**123.** $\dfrac{1}{6}$

125. $-3\dfrac{2}{3}$　**127.** -388.72　**129.** $\dfrac{1}{12}$　**131.** -48　**133.** 10.5　**135.** -1.7　**137.** Yes　**139.** No　**141.** No

143. Yes　**145.** The temperature fell $32.22°C$.　**147.** The difference is $35.438°C$.　**149. a.** True　**b.** True
c. False　**d.** False

SECTION 3.5

1. -3　**3.** -6　**5.** -5　**7.** -12　**9.** -3　**11.** 19　**13.** 2　**15.** 1　**17.** 14　**19.** 42　**21.** -13　**23.** -12

25. -6　**27.** 0.21　**29.** -0.96　**31.** -0.29　**33.** -1　**35.** 0　**37.** $-\dfrac{5}{8}$　**39.** 2　**41.** 1　**43.** 15　**45.** 32

47. 1　**49.** 1　**51.** 5　**53.** 28　**55.** $\dfrac{13}{18}$　**57.** -4　**59.** No

CHAPTER 3 REVIEW EXERCISES

1. Eight minus negative one [3.1B]　**2.** -36 [3.1C]　**3.** 200 [3.3A]　**4.** -9 [3.3B]　**5.** -14 [3.2A]　**6.** 13 [3.1B]
7. [3.1A]　**8.** -98.38 [3.4A]　**9.** 17 [3.3B]　**10.** -210 [3.3B]

11. -2 [3.2B]　**12.** -18 [3.3A]　**13.** -1 [3.2A]　**14.** -72 [3.3A]　**15.** -4 [3.5A]　**16.** -2 [3.2B]

17. 13 [3.2B]　**18.** $\dfrac{2}{7}$ [3.4B]　**19.** Yes [3.2B]　**20.** 14 [3.2B]　**21.** 0 [3.3B]　**22.** -60 [3.3A]　**23.** -12 [3.2A]

24. 5 [3.5A]　**25.** $-8 > -10$ [3.1A]　**26.** 9 [3.5A]　**27.** 27 [3.1C]　**28.** -2.8 [3.4B]　**29.** $-\dfrac{5}{48}$ [3.4A]

30. A temperature of $-12°C$ is colder. [3.1D]　**31.** The boiling point of neon is $-238°C$. [3.3C]　**32.** The temperature
is $-3°C$ after the increase. [3.4C]　**33.** 12 [3.2C]

CHAPTER 3 TEST

1. Negative three plus negative five [3.1B]　**2.** -34 [3.1C]　**3.** 18 [3.2B]　**4.** -20 [3.2A]　**5.** 24 [3.3A]

6. $-\dfrac{7}{18}$ [3.4A]　**7.** 12 [3.3B]　**8.** 2 [3.2A]　**9.** $16 > -19$ [3.1A]　**10.** -2 [3.2B]　**11.** -3 [3.2B]

12. 49 [3.1B]　**13.** -250 [3.3A]　**14.** $-|5|, -(3), |-9|, -(-11)$ [3.1C]　**15.** No [3.2B]　**16.** -3 [3.1A]

17. 0 [3.3B]　**18.** 19 [3.5A]　**19.** -25 [3.1B]　**20.** -11.613 [3.4A]　**21.** -11 [3.2B]　**22.** 24 [3.3B]

23. 10 [3.5A]　**24.** -7 [3.3B]　**25.** 60 [3.3A]　**26.** -107 [3.5A]　**27.** -27 [3.4B]　**28.** -1.53 [3.4B]

29. -10 [3.2B]　**30.** The temperature is $4.5°C$ after the increase. [3.4C]　**31.** The wind chill factor is $-64°F$. [3.3C]
32. The high temperature was $-5°C$. [3.2C]　**33.** 16 units [3.2C]

CUMULATIVE REVIEW EXERCISES

1. 5 [3.2B]   **2.** 12,000 [1.3A]   **3.** 32.3 [2.6C]   **4.** 2 [1.5A]   **5.** −82 [3.1C]   **6.** 309,480 [1.1B]
7. 2400 [1.3A]   **8.** 21 [3.3B]   **9.** −11 [3.2B]   **10.** −40 [3.2A]   **11.** 1, 2, 4, 11, 22, 44 [1.3D]   **12.** 1 [2.4A]
13. 630,000 [1.1C]   **14.** 1300 [1.2A]   **15.** 9 [3.2B]   **16.** −2500 [3.3A]   **17.** 8.77 [2.6A]   **18.** $5\frac{2}{3}$ [2.3A]
19. −32 [3.5A]   **20.** −4 [3.3B]   **21.** $1\frac{1}{5}$ [2.4B]   **22.** $-62 < 26$ [3.1A]   **23.** 126 [3.3A]   **24.** 4.14 [3.4B]
25. $2^5 \cdot 7^2$ [1.3B]   **26.** 47 [1.5A]   **27.** 10,062 [1.2A]   **28.** −26 [3.2B]   **29.** 5000 [1.2B]   **30.** 2025 [1.3B]
31. The land area was 1,722,685 mi² after the purchase. [1.2C]   **32.** Albert Einstein was 76 years old when he died.
[1.2C]   **33.** The amount that remains to be paid is $14,200. [1.2C]   **34.** The total cost of the land is $92,250. [1.3E]
35. The temperature is −5°C after the increase. [3.2C]   **36. a.** The difference is 168°F.   **b.** The difference is greatest
for Alaska. [3.2C]   **37.** Your sales for the fourth quarter must be $24,900. [1.2C]   **38.** The golfer's score is −8. [3.2C]

Answers to Chapter 4 Selected Exercises

PREP TEST

1. 3 [3.2B]   **2.** 4 [3.3B]   **3.** $\frac{1}{12}$ [3.4A]   **4.** $-\frac{4}{9}$ [2.4A]   **5.** $\frac{3}{10}$ [3.4B]   **6.** −16 [3.5A]   **7.** $\frac{8}{27}$ [2.4A]
8. 48 [1.4A]   **9.** 1 [1.4A]   **10.** 12 [1.4A]

SECTION 4.1

1. $2x^2, 5x, \underline{-8}$ **3.** $-a^4, \underline{6}$ **5.** $7x^2y, 6xy^2$ **7.** $1, -9$ **9.** $1, -4, -1$ **13.** 10   **15.** 32   **17.** 21   **19.** 16
21. −9   **23.** 41   **25.** −7   **27.** 13   **29.** −15   **31.** 41   **33.** 1   **35.** 5   **37.** 1   **39.** 57   **41.** 5
43. 8   **45.** −3   **47.** −2   **49.** −4   **51.** 225   **53.** 60   **55.** 4   **57.** 81   **59.** $n^x > x^n$ if $x \geq n + 1$

SECTION 4.2

3. $14x$ **5.** $5a$ **7.** $-6y$ **9.** $7 - 3b$ **11.** $5a$ **13.** $-2ab$ **15.** $5xy$ **17.** 0   **19.** $-\frac{5}{6}x$ **21.** $6.5x$

23. $0.45x$ **25.** $7a$ **27.** $-14x^2$ **29.** $-\frac{11}{24}x$ **31.** $17x - 3y$ **33.** $-2a - 6b$ **35.** $-3x - 8y$ **37.** $-4x^2 - 2x$

39. $12x$ **41.** $-21a$ **43.** $6y$ **45.** $8x$ **47.** $-6a$ **49.** $12b$ **51.** $-15x^2$ **53.** x^2 **55.** a **57.** x **59.** n
61. x **63.** y **65.** $3x$ **67.** $-2x$ **69.** $-8a^2$ **71.** $8y$ **73.** $4y$ **75.** $-2x$ **77.** $6a$ **79.** $-x - 2$
81. $8x - 6$ **83.** $-2a - 14$ **85.** $-6y + 24$ **87.** $35 - 21b$ **89.** $2 - 5y$ **91.** $15x^2 + 6x$ **93.** $2y - 18$
95. $-15x - 30$ **97.** $-6x^2 - 28$ **99.** $-6y^2 + 21$ **101.** $3x^2 - 3y^2$ **103.** $-4x + 12y$ **105.** $-6a^2 + 7b^2$

107. $4x^2 - 12x + 20$ **109.** $\frac{3}{2}x - \frac{9}{2}y + 6$ **111.** $-12a^2 - 20a + 28$ **113.** $12x^2 - 9x + 12$ **115.** $10x^2 - 20xy - 5y^2$

117. $-8b^2 + 6b - 9$ **119.** $a - 7$ **121.** $-11x + 13$ **123.** $-4y - 4$ **125.** $-2x - 16$ **127.** $14y - 45$
129. $a + 7b$ **131.** $6x + 28$ **133.** $5x - 75$ **135.** $4x - 4$ **137.** $2x - 9$ **139. a.** False. For example,
$8 \div 2 \neq 2 \div 8$ **b.** False. For example, $(12 \div 4) \div 2 \neq 12 \div (4 \div 2)$ **c.** False. For example, $(9 - 2) - 3 \neq 9 - (2 - 3)$
d. False. For example, $10 - 4 \neq 4 - 10$ **141.** No. 0 does not have a multiplicative inverse.

SECTION 4.3

1. $8 + y$ **3.** $t + 10$ **5.** $z + 14$ **7.** $x^2 - 20$ **9.** $\frac{3}{4}n + 12$ **11.** $8 + \frac{n}{4}$ **13.** $3(y + 7)$ **15.** $t(t + 16)$

17. $\frac{1}{2}x^2 + 15$ **19.** $5n^3 + n^2$ **21.** $r - \frac{r}{3}$ **23.** $x^2 - (x + 17)$ **25.** $9(z + 4)$ **27.** $12 - x$ **29.** $\frac{2}{3}x$ **31.** $\frac{2x}{9}$

33. $11x - 8$ **35.** $(x + 2) - 9; x - 7$ **37.** $\dfrac{7}{5 + x}$ **39.** $5 + \dfrac{1}{2}(x + 3); \dfrac{1}{2}x + \dfrac{13}{2}$ **41.** $(2x - 4) + x; 3x - 4$

43. $(x - 5)7; 7x - 35$ **45.** $\dfrac{2x + 5}{x}$ **47.** $x - (3x - 8); -2x + 8$ **49.** $3x + x; 4x$ **51.** $(x + 6) + 5; x + 11$

53. $x - (x + 10); -10$ **55.** $\dfrac{1}{6}x + \dfrac{4}{9}x; \dfrac{11}{18}x$ **57.** $\dfrac{x}{3} + x; \dfrac{4}{3}x$ **59.** Number of e-mails: A; number of spam e-mails: $\dfrac{1}{2}A$

61. Length of one piece: S; length of second piece: $12 - S$ **63.** Distance traveled by the faster car: x; distance traveled

by the slower car: $200 - x$ **65.** Number of bones in your body: N; number of bones in your foot: $\dfrac{1}{4}N$

67. Salary needed in San Francisco: S; salary needed in Daytona Beach: $\dfrac{1}{2}S$ **69.** $2x$

CHAPTER 4 REVIEW EXERCISES

1. $3x^2 - 24x - 21$ [4.2C] **2.** $11x$ [4.2A] **3.** $8a - 4b$ [4.2A] **4.** $-5n$ [4.2B] **5.** 79 [4.1A]
6. $10x - 35$ [4.2C] **7.** $12y^2 + 8y - 10$ [4.2C] **8.** $-6a$ [4.2B] **9.** $-42x^2$ [4.2B] **10.** $-63 - 36x$ [4.2C]
11. $-5y$ [4.2A] **12.** -4 [4.1A] **13.** $-6x - 1$ [4.2D] **14.** $-40a + 40$ [4.2D] **15.** $24y + 30$ [4.2D]
16. $9c - 5d$ [4.2A] **17.** $20x$ [4.2B] **18.** $7x + 46$ [4.2D] **19.** 29 [4.1A] **20.** $-9r + 8s$ [4.2A]
21. 50 [4.1A] **22.** 28 [4.1A] **23.** $-4x^2 + 6x$ [4.2A] **24.** $-90x + 25$ [4.2D] **25.** $28a^2 - 8a + 12$ [4.2C]
26. $-4x + 20$ [4.2D] **27.** $\dfrac{2}{3}(x + 10)$ [4.3A] **28.** $x - 6$ [4.3A] **29.** $x + 2x; 3x$ [4.3B]
30. $3x + 5(x - 1); 8x - 5$ [4.3B] **31.** Number of American League cards: A; number of National League cards:
$5A$ [4.3C] **32.** Number of calories in an apple: a; number of calories in a candy bar: $2a + 8$ [4.3C]

CHAPTER 4 TEST

1. $5x$ [4.2A] **2.** $-6x^2 + 21y^2$ [4.2C] **3.** $-x + 6$ [4.2D] **4.** $-7x + 33$ [4.2D] **5.** $-9x - 7y$ [4.2A]
6. 22 [4.1A] **7.** $2x$ [4.2B] **8.** $7x + 38$ [4.2D] **9.** $-10x^2 + 15x - 30$ [4.2C] **10.** $-2x - 5y$ [4.2A]
11. 3 [4.1A] **12.** $3x$ [4.2B] **13.** y^2 [4.2A] **14.** $-4x + 8$ [4.2C] **15.** $-10a$ [4.2B]
16. $2x + y$ [4.2D] **17.** $36y$ [4.2B] **18.** $15 - 35b$ [4.2C] **19.** $a^2 - b^2$ [4.3A]
20. $10(x - 3) = 10x - 30$ [4.3B] **21.** $x + 2x^2$ [4.3B] **22.** $\dfrac{6}{x} - 3$ [4.3B] **23.** $b - 7b$ [4.3A] **24.** Speed of
return throw: s; speed of fastball: $2s$ [4.3C] **25.** Shorter piece: x; longer piece: $4x - 3$ [4.3C]

CUMULATIVE REVIEW EXERCISES

1. -7 [3.2A] **2.** 5 [3.2B] **3.** 24 [3.3A] **4.** -5 [3.3B] **5.** $2 \cdot 5 \cdot 11$ [1.3D] **6.** $\dfrac{11}{48}$ [3.4A]

7. $-\dfrac{1}{6}$ [3.4B] **8.** $\dfrac{1}{4}$ [3.4B] **9.** 1300 [1.2A] **10.** -5 [3.5A] **11.** $-\dfrac{27}{26}$ [3.5A] **12.** 16 [4.1A]

13. $5x^2$ [4.2A] **14.** $-7a - 10b$ [4.2A] **15.** 8.357 [2.5A] **16.** $5.101 > 5.013$ [2.5B] **17.** $24 - 6x$ [4.2C]
18. $6y - 18$ [4.2C] **19.** 10 [2.6A] **20.** 8.7 [2.5C] **21.** $-8x^2 + 12y^2$ [4.2C] **22.** $-9y^2 + 9y + 21$ [4.2C]
23. $-7x + 14$ [4.2D] **24.** $5x - 43$ [4.2D] **25.** $17x - 24$ [4.2D] **26.** $-3x + 21y$ [4.2D]

27. $\dfrac{1}{2}b + b$ [4.3A] **28.** $\dfrac{10}{y - 2}$ [4.3A] **29.** $8 - \dfrac{x}{12}$ [4.3B] **30.** $x + (x + 2); 2x + 2$ [4.3B]

31. 9 in. [2.6E] **32.** Speed of dial-up connection: s; speed of DSL connection: $10s$ [4.3C]

Answers to Chapter 5 Selected Exercises

PREP TEST

1. -4 [3.2B] **2.** 1 [3.4B] **3.** -10 [3.4B] **4.** 1 [3.4B] **5.** $7y$ [4.2A] **6.** -9 [4.2A] **7.** -5 [4.1A]

SECTION 5.1

3. Yes **5.** No **7.** No **9.** Yes **11.** No **13.** Yes **15.** No **17.** Yes **19.** No **23.** 2 **25.** 15
27. 6 **29.** 3 **31.** 0 **33.** −7 **35.** −7 **37.** −12 **39.** −5 **41.** 15 **43.** 9 **45.** 14 **47.** −1
49. 1 **51.** $-\frac{1}{2}$ **53.** $-\frac{3}{4}$ **55.** $\frac{1}{12}$ **57.** $-\frac{7}{12}$ **61.** −3 **63.** 0 **65.** −2 **67.** 9 **69.** 80 **71.** 0
73. −7 **75.** 12 **77.** −18 **79.** 15 **81.** −20 **83.** 0 **85.** $\frac{8}{3}$ **87.** $\frac{1}{3}$ **89.** $-\frac{1}{2}$ **91.** $-\frac{3}{2}$ **93.** $\frac{15}{7}$
95. 4 **97.** 3 **99.** The runner will travel 3 mi. **101.** Marcella's average rate of speed is 36 mph. **103.** It would take Palmer 2.5 h to walk the course. **105.** The two joggers will meet 40 min after they start. **107.** It will take them 0.5 h. **109. a.** Answers will vary. **b.** Answers will vary.

SECTION 5.2

1. 3 **3.** 6 **5.** −1 **7.** −3 **9.** 2 **11.** 2 **13.** 5 **15.** −3 **17.** 6 **19.** 3 **21.** 1 **23.** 6 **25.** −7
27. 0 **29.** $\frac{3}{4}$ **31.** $\frac{4}{9}$ **33.** $\frac{1}{3}$ **35.** $-\frac{1}{2}$ **37.** $-\frac{3}{4}$ **39.** $\frac{1}{3}$ **41.** $-\frac{1}{6}$ **43.** 1 **45.** 1 **47.** 0 **49.** $\frac{13}{10}$
51. $\frac{2}{5}$ **53.** $-\frac{4}{3}$ **55.** $-\frac{3}{2}$ **57.** 18 **59.** 8 **61.** −16 **63.** 25 **65.** $\frac{3}{4}$ **67.** $\frac{3}{8}$ **69.** $\frac{16}{9}$ **71.** $\frac{1}{18}$
73. $\frac{15}{2}$ **75.** $-\frac{18}{5}$ **77.** 2 **79.** 3 **81.** $x = 7$ **83.** $y = 3$ **85.** 19 **87.** −1 **89.** −11 **91.** The initial velocity is 8 ft/s. **93.** The depreciated value will be $38,000 after 2 years. **95.** The approximate length is 31.8 in.
97. The distance the car will skid is 168 ft. **99.** The estimated population is 51,000 people. **101.** $a = 7$ **103.** 385

SECTION 5.3

1. 2 **3.** 3 **5.** −1 **7.** 2 **9.** −2 **11.** −3 **13.** 0 **15.** −1 **17.** −3 **19.** −1 **21.** 4 **23.** $\frac{2}{3}$
25. $\frac{5}{6}$ **27.** $\frac{3}{4}$ **29.** −17 **31.** 41 **33.** 8 **35.** 1 **37.** 4 **39.** −1 **41.** −1 **43.** 24 **45.** 495
47. $\frac{1}{2}$ **49.** $-\frac{1}{3}$ **51.** $\frac{10}{3}$ **53.** $-\frac{1}{4}$ **55.** 0 **57.** −1 **59.** A force of 25 lb must be applied to the other end.
61. The fulcrum is 6 ft from the 180-pound person. **63.** The fulcrum is 10 ft from the 128-pound acrobat.
65. The minimum force to move the rock is 34.6 lb. **67.** The break-even point is 260 barbecues. **69.** The break-even point is 520 desk lamps. **71.** The break-even point is 3000 softball bats. **73.** No solution **75.** 0

SECTION 5.4

1. $x - 15 = 7$; 22 **3.** $7x = -21$; −3 **5.** $9 - x = 7$; 2 **7.** $5 - 2x = 1$; 2 **9.** $2x + 5 = 15$; 5 **11.** $4x - 6 = 22$; 7
13. $3(4x - 7) = 15$; 3 **15.** $3x = 2(20 - x)$; 8, 12 **17.** $2x - (14 - x) = 1$; 5, 9 **19.** 15, 17, 19 **21.** −1, 1, 3
23. 4, 6 **25.** 5, 7 **27.** The processor speed of the newer personal computer is $4.2\overline{6}$ GHz. **29.** The lengths of the sides are 6 ft, 6 ft, and 11 ft. **31.** The union member worked 168 h during March. **33.** 37 h of labor were required to paint the house. **35.** There are 1024 vertical pixels. **37.** 3 h of labor were required for the job. **39.** The shorter piece is 3 ft; the longer piece is 9 ft. **41.** The larger scholarship is $5000.

SECTION 5.5

1. The amount of $1 herbs is 20 oz. **3.** The mixture will cost $1.84 per pound. **5.** The amount of caramel is 3 lb.
7. The amount of olive oil is 2 c; the amount of vinegar is 8 c. **9.** The cost of the mixture is $3.00 per ounce.
11. To make the mixture, 16 oz of the alloy are needed. **13.** The amount of almonds is 37 lb; the amount of walnuts is 63 lb. **15.** There were 228 adult tickets sold. **17.** The cost per pound of the sugar-coated cereal is $.70. **19.** The first plane is traveling at a rate of 105 mph; the second plane is traveling at a rate of 130 mph. **21.** The planes will be 3000 km apart at 11 A.M. **23.** In 2 h, the cabin cruiser will be alongside the motorboat. **25.** The corporate offices are 120 mi from the airport. **27.** The rate of the car is 68 mph. **29.** The distance between the airports was 300 mi.

31. The planes will pass each other 2.5 h after the plane leaves Seattle.   **33.** The cyclists will meet after 1.5 h.
35. The bus overtakes the car 180 mi from the starting point.   **37.** The campers turned around at 10:15 A.M.

CHAPTER 5 REVIEW EXERCISES

1. 21 [5.1B]   **2.** 10 [5.3B]   **3.** 7 [5.2A]   **4.** No [5.1A]   **5.** 20 [5.1C]   **6.** −2 [5.3B]

7. −4 [5.1B]   **8.** 4 [5.3A]   **9.** −1 [5.3B]   **10.** 4 [5.3A]   **11.** $-\dfrac{5}{2}$ [5.3A]   **12.** $\dfrac{5}{4}$ [5.1C]

13. 10 [5.2A]   **14.** $\dfrac{4}{3}$ [5.3B]   **15.** The force is 24 lb. [5.3C]   **16.** The average speed on the winding road was 32 mph. [5.5B]   **17.** The number is 2. [5.4B]   **18.** The number is 10. [5.4B]   **19.** The amount of cranberry juice is 7 qt; the amount of apple juice is 3 qt. [5.5A]   **20.** The three integers are −1, 0, 1. [5.4A]   **21.** $5n − 4 = 16$; 4 [5.4A]   **22.** The height of the Eiffel Tower is 1063 ft. [5.4B]   **23.** The temperature is 37.8°C. [5.2B]   **24.** The jet overtakes the propeller-driven plane 600 mi from the starting point. [5.5B]   **25.** The numbers are 8 and 13. [5.4A]
26. The farmer harvested 25,300 bushels of corn last year. [5.4B]

CHAPTER 5 TEST

1. −5 [5.3A]   **2.** −5 [5.1B]   **3.** −3 [5.2A]   **4.** 2 [5.3B]   **5.** No [5.1A]   **6.** 5 [5.2A]   **7.** $-\dfrac{1}{2}$ [5.2A]

8. $-\dfrac{1}{3}$ [5.3B]   **9.** 2 [5.3A]   **10.** −12 [5.1C]   **11.** 95 [5.2A]   **12.** $\dfrac{16}{5}$ [5.3A]   **13.** −3 [5.2A]

14. 11 [5.3B]   **15.** The amount of rye flour is 10 lb; the amount of wheat flour is 5 lb. [5.5A]   **16.** 200 calculators were produced. [5.2B]   **17.** The numbers are 10, 12, and 14. [5.4A]   **18.** 4000 clocks were made during the month. [5.2B]   **19.** $3x − 15 = 27$; 14 [5.4A]   **20.** The rate of the snowmobile was 6 mph. [5.5B]   **21.** The company makes 110 25-inch TVs each day. [5.4B]   **22.** The smaller number is 8; the larger number is 10. [5.4A]   **23.** The distance between the airports is 360 mi. [5.5B]   **24.** The time required is 11.5 s. [5.2B]   **25.** The final temperature is 60°C. [5.3C]

CUMULATIVE REVIEW EXERCISES

1. 6 [3.2B]   **2.** −48 [3.3A]   **3.** $-\dfrac{19}{48}$ [3.4A]   **4.** −2 [3.4B]   **5.** 54 [3.5A]   **6.** 24 [3.5A]

7. 6 [4.1A]   **8.** $−17x$ [4.2A]   **9.** $−5a − 2b$ [4.2A]   **10.** $2x$ [4.2B]   **11.** $36y$ [4.2B]
12. $2x^2 + 6x − 4$ [4.2C]   **13.** $−4x + 14$ [4.2D]   **14.** $6x − 34$ [4.2D]   **15.** Yes [5.1A]   **16.** No [5.1A]

17. $\dfrac{11}{18}$ [3.5A]   **18.** −25 [5.1C]   **19.** −3 [5.2A]   **20.** 3 [5.2A]   **21.** 0.047383 [2.6B]   **22.** 7 [1.2B]

23. 13 [5.3B]   **24.** 2 [5.3B]   **25.** −3 [5.3A]   **26.** $\dfrac{1}{2}$ [5.3A]   **27.** The final temperature is 60°C. [5.3C]
28. $12 − 5x = −18$; 6 [5.4A]   **29.** The area of the garage is 600 ft². [5.4B]   **30.** 20 lb of oat flour are needed for the mixture. [5.5A]   **31.** $3n + 4$ [4.3B]   **32.** The number is 2. [5.4B]   **33.** The length of the track is 120 m. [5.5B]

Answers to Chapter 6 Selected Exercises

PREP TEST

1. $\dfrac{4}{5}$ [2.2B]   **2.** 24.8 [2.6C]   **3.** 4×33 [1.1A/1.3A]   **4.** $\dfrac{19}{100}$ [2.4A]   **5.** 0.23 [2.6B]   **6.** 47 [2.6B]

7. 2850 [2.6B]   **8.** 4000 [2.6C]   **9.** 8000 [2.6C]   **10.** 62.5 [3.4B]   **11.** $66\dfrac{2}{3}$ [2.2A]   **12.** 1.75 [2.6C]

SECTION 6.1

1. $\frac{2}{3}$, 2:3, 2 to 3 **3.** $\frac{3}{8}$, 3:8, 3 to 8 **5.** $\frac{9}{2}$, 9:2, 9 to 2 **7.** $\frac{1}{2}$, 1:2, 1 to 2 **9.** The ratio is $\frac{13}{14}$. **11.** The ratio is $\frac{2}{3}$.

13. $\frac{25 \text{ mi}}{1 \text{ h}}$ **15.** $\frac{\$3.28}{3 \text{ bars}}$ **17.** $\frac{9 \text{ children}}{4 \text{ families}}$ **19.** \$3225/month **21.** 38.4 mi/gal **23.** \$1.344/oz

25. The ratio is $\frac{1}{12}$. **27.** The population density of Australia is 6.6 people/mi². The population density of India is 824.1 people/mi². The population density of the U.S. is 80.7 people/mi². **29.** \$18/h

SECTION 6.2

1. Not true **3.** True **5.** True **7.** True **9.** True **11.** True **13.** 10 **15.** 2.4 **17.** 4.5 **19.** 17.14 **21.** 25.6 **23.** 20.83 **25.** 4.35 **27.** 10.97 **29.** 1.15 **31.** 38.73 **33.** 0.5 **35.** 10 **37.** 0.43 **39.** −1.6 **41.** 6.25 **43.** 32 **45.** 6.2 **47.** 5.8 **49.** The actual length is 0.0052 in. **51.** 24 robes can be made from 26 yd of material. **53.** At this rate, the property tax on a home appraised at \$280,000 is \$6720. **55.** At the same rate, the car would travel 406 mi on 14.5 gal of gasoline. **57.** Fourteen grapefruit cost \$4.48. **59.** Using this estimate, 438 light fixtures are necessary. **61.** It will take the dieter 60 weeks to lose 36 lb. **63.** At this rate, 2750 defects would be found in 125,000 cars. **65.** At the same rate, the executive will drive 198,000 mi in 3 years. **67.** An additional \$1500 would have to be invested. **69.** In 4 h, 9 in. of the candle will burn. **71.** Using this recommendation, the yearly midmanagement salary would be \$126,000. **73.** Yes **75.** No

SECTION 6.3

3. $\frac{1}{20}$, 0.05 **5.** $\frac{3}{10}$, 0.30 **7.** $\frac{5}{2}$, 2.50 **9.** $\frac{7}{25}$, 0.28 **11.** $\frac{7}{20}$, 0.35 **13.** $\frac{29}{100}$, 0.29 **15.** $\frac{1}{9}$ **17.** $\frac{3}{8}$

19. $\frac{2}{3}$ **21.** $\frac{1}{15}$ **23.** $\frac{1}{200}$ **25.** $\frac{1}{16}$ **27.** 0.073 **29.** 0.158 **31.** 0.003 **33.** 1.212 **35.** 0.6214

37. 0.0825 **39.** $\frac{6}{25}$ **41.** 37% **43.** 2% **45.** 12.5% **47.** 136% **49.** 96% **51.** 7% **53.** 83%

55. 33.3% **57.** 44.4% **59.** 45% **61.** 250% **63.** 16.7% **65.** 68% **67.** $56\frac{1}{4}$% **69.** $262\frac{1}{2}$%

71. $283\frac{1}{3}$% **73.** $23\frac{1}{3}$% **75.** $22\frac{2}{9}$%

SECTION 6.4

1. 8 **3.** 0.075 **5.** $16\frac{2}{3}$% **7.** 37.5% **9.** 100 **11.** 1200 **13.** 51.895 **15.** 13 **17.** 2.7%

19. 400% **21.** 7.5 **23.** 65 **25.** 25% **27.** 75 **29.** 12.5% **31.** 400 **33.** 19.5 **35.** 14.8% **37.** 62.62 **39.** 5 **41.** 45 **43.** 300% **45.** The estimated safe-life use of the brakes is 50,000 mi. **47.** There are 1,242,424 workers in the Arkansas labor force. **49.** 12% of the alarms received were false alarms. **51.** The increase in population is 145.5% of the 2000 population of Kern County. **53.** The U.S. total turkey production for that year was 7 million pounds. **55.** The farmer would receive a tax credit of \$12,750. **57.** There are 0.015 g of clobetasol propionate in a 30-gram tube of the cream. **59.** The sun's diameter is 10,875% of Earth's diameter. **61.** 7944 of the tested computer boards were not defective. **63.** 12,151 more faculty members described their political views as liberal. **65.** No **67.** Your salary after the third year is less than it would have been if you had received a 6% raise each year.

SECTION 6.5

3. a. \$25, \$50, \$75, \$100, \$125 **b.** \$150 **c.** \$175 **d.** \$200 **e.** \$225 **5.** The simple interest due on the loan is \$273.70. **7.** The simple interest due on the loan is \$6750. **9.** The interest owed is \$20. **11.** The simple interest owed is \$1440. **13.** The maturity value of the loan is \$164,250. **15.** The maturity value of the loan is \$15,061.51. **17.** The annual simple interest rate is 7.7%. **19.** The annual simple interest rate earned on the investment is 9.125%.

CHAPTER 6 REVIEW EXERCISES

1. $\frac{1}{1}$, 1:1, 1 to 1 [6.1A]   **2.** $\frac{2 \text{ roof supports}}{1 \text{ ft}}$ [6.1A]   **3.** $15.70/h [6.1A]   **4.** $\frac{8}{15}$ [6.1A]   **5.** 1.6 [6.2A]

6. $\frac{5 \text{ lb}}{4 \text{ trees}}$ [6.1A]   **7.** 57 mph [6.1A]   **8.** 6.86 [6.2A]   **9.** $\frac{8}{25}$ [6.3A]   **10.** 0.22 [6.3A]   **11.** $\frac{1}{4}$, 0.25 [6.3A]

12. $\frac{17}{500}$ [6.3A]   **13.** 17.5% [6.3B]   **14.** 128.6% [6.3B]   **15.** 280% [6.3B]   **16.** 21 [6.4A/6.4B]

17. 500% [6.4A/6.4B]   **18.** 66.7% [6.4A/6.4B]   **19.** 30 [6.4A/6.4B]   **20.** 15.3 [6.4A/6.4B]

21. 562.5 [6.4A/6.4B]   **22.** 5.625% [6.4A/6.4B]   **23.** The ratio is $\frac{2}{5}$. [6.1A]   **24.** $12,000 must be invested to earn $780 in dividends. [6.2B]   **25.** 2.75 lb of plant food should be used. [6.2B]   **26.** The other attorney receives $64,000. [6.2B]   **27.** 34.0% of the tourists will be visiting China. [6.4C]   **28.** The company spent $8400 for advertising. [6.4C]   **29.** 3952 of the phones were not defective. [6.4C]   **30.** Cable households spend 36.5% of the week watching TV. [6.4C]   **31.** The simple interest is $31.81. [6.5A]   **32.** The maturity value of the loan is $10,630. [6.5A]   **33.** The airline would sell 196 tickets for an airplane that has 175 seats. [6.4C]

CHAPTER 6 TEST

1. $\frac{1}{8}$, 1:8, 1 to 8 [6.1A]   **2.** $\frac{1 \text{ oz}}{4 \text{ cookies}}$ [6.1A]   **3.** 0.6 mi/min [6.1A]   **4.** $\frac{2}{1}$ [6.1A]   **5.** 0.75 [6.2A]

6. 2 ft/s [6.1A]   **7.** 476.67 ft²/h [6.1A]   **8.** 3.56 [6.2A]   **9.** 0.864 [6.3A]   **10.** 40% [6.3B]

11. 125% [6.3B]   **12.** $\frac{5}{6}$ [6.3A]   **13.** $\frac{8}{25}$ [6.3A]   **14.** 118% [6.3B]   **15.** 90 [6.4A/6.4B]

16. 49.64 [6.4A/6.4B]   **17.** 56.25% [6.4A/6.4B]   **18.** 200 [6.4A/6.4B]   **19.** The ratio is $\frac{33}{38}$. [6.1A]

20. The sales tax is $3136. [6.2B]   **21.** 243,750 voters would vote. [6.2B]   **22.** The room is 50 ft long. [6.2B]   **23.** 33 accidents are expected. [6.4C]   **24.** The student answered 82.2% of the questions correctly. [6.4C]   **25.** The dollar increase in the weekly wage over last year is $80. [6.4C]   **26.** The number of fat grams in the beef burger is 600% of the number in the soy burger. [6.4C]   **27.** The maturity value of the loan is $41,520.55. [6.5A]

CUMULATIVE REVIEW EXERCISES

1. 57 [3.5A]   **2.** 625 [1.3B]   **3.** $3\frac{31}{36}$ [2.3B]   **4.** $1\frac{2}{3}$ [2.7A]   **5.** -114 [3.3B]   **6.** 22 [3.5A]

7. 4 [5.2A]   **8.** 3 [5.3B]   **9.** $\frac{23}{24}$ [3.4A]   **10.** 100,500 [2.6B]   **11.** 21 [3.5A]   **12.** $-\frac{3}{8}$ [3.5A]

13. 4 [3.2B]   **14.** $8a - 3$ [4.2D]   **15.** -12 [5.1C]   **16.** $-4y^2 - 3y$ [4.2A]   **17.** $1\frac{25}{62}$ [2.4B]

18. $\frac{3}{10}$ [6.1A]   **19.** $3885/month [6.1A]   **20.** 24.954 [2.6A]   **21.** 32 [6.2A]   **22.** $\frac{10}{11}$ [2.4C]

23. 8.3% [6.4A/6.4B]   **24.** 25.2 [6.4A/6.4B]   **25.** The difference is $28.67. [2.6E]   **26.** The number is 12. [5.4A]   **27.** $4x - 3(x + 2)$; $x - 6$ [4.3B]   **28.** There are 64 mi left to drive. [1.2C]   **29.** The new balance is $265.48. [2.6E]   **30.** $\frac{4}{15}$ of the job remains to be finished on the third day. [2.3C]   **31.** $\frac{1}{10}$ of the population aged 75–84 is affected by Alzheimer's disease. [6.3A]   **32.** The car traveled 35 mi per gallon of gas. [6.1A]   **33.** The rpm of the engine in third gear is 3750. [5.4B]

Answers to Chapter 7 Selected Exercises

PREP TEST

1. 56 [1.4A]   **2.** 56.52 [4.1A]   **3.** 113.04 [4.1A]   **4.** 43 [5.1B]   **5.** 51 [5.1B]   **6.** 14.4 [6.2A]

SECTION 7.1

1. 40°; acute **3.** 115°; obtuse **5.** 90°; right **7.** 28° **9.** 18° **11.** 14 cm **13.** 28 ft **15.** 30 m
17. 86° **19.** 71° **21.** 30° **23.** 36° **25.** 127° **27.** 116° **29.** 20° **31.** 20° **33.** 20° **35.** 141°
37. 106° **39.** 11° **41.** $\angle a = 38°$, $\angle b = 142°$ **43.** $\angle a = 47°$, $\angle b = 133°$ **45.** 20° **47.** 47° **49.** $\angle x = 155°$,
$\angle y = 70°$ **51.** $\angle a = 45°$, $\angle b = 135°$ **53.** $90° - x$ **55.** 60° **57.** 35° **59. a.** 1° **b.** 179° **63.** 360°

SECTION 7.2

1. Hexagon **3.** Pentagon **5.** Scalene **7.** Equilateral **9.** Obtuse **11.** Acute **13.** 56 in. **15.** 14 ft
17. 47 mi **19.** 8π cm; 25.13 cm **21.** 11π mi; 34.56 mi **23.** 17π ft; 53.41 ft **25.** 17.4 cm **27.** 8 cm
29. 24 m **31.** 48.8 cm **33.** 17.5 in. **35.** 1.5π in. **37.** 226.19 cm **39.** 60 ft **41.** 44 ft **43.** 120 ft
45. 10 in. **47.** 12 in. **49.** 2.55 cm **51.** 13.19 ft **53.** 50.27 ft **55.** 39,935.93 km **57.** 60 ft^2
59. 20.25 in^2 **61.** 546 ft^2 **63.** 16π cm^2; 50.27 cm^2 **65.** 30.25π mi^2; 95.03 mi^2 **67.** 72.25π ft^2; 226.98 ft^2
69. 156.25 cm^2 **71.** 570 in^2 **73.** 192 in^2 **75.** 13.5 ft^2 **77.** 330 cm^2 **79.** 25π in^2 **81.** 10,000π in^2
83. 126 ft^2 **85.** 7500 yd^2 **87.** 10 in. **89.** 20 m **91.** 2 qt **93.** \$74 **95.** 113.10 in^2 **97.** \$638
99. 216 m^2 **101.** 4:9 **105. a.** Sometimes true **b.** Sometimes true **c.** Always true **d.** Always true
e. Always true **f.** Always true

SECTION 7.3

1. 5 in. **3.** 8.6 cm **5.** 11.2 ft **7.** 4.5 cm **9.** 12.7 yd **11.** 8.5 cm **13.** 24.3 cm **15.** $\frac{1}{2}$ **17.** $\frac{3}{4}$
19. 7.2 cm **21.** 3.3 m **23.** 12 m **25.** 12 in. **27.** 56.3 cm^2 **29.** 18 ft **31.** 16 m **33.** Yes, SAS Rule
35. Yes, SSS Rule **37.** Yes, ASA Rule **39.** Yes, SAS Rule **41.** No **43.** No

SECTION 7.4

1. 840 in^3 **3.** 15 ft^3 **5.** 4.5π cm^3; 14.14 cm^3 **7.** 34 m^3 **9.** 15.625 in^3 **11.** 36π ft^3 **13.** 8143.01 cm^3
15. 392.70 cm^3 **17.** 216 m^3 **19.** 2.5 ft **21.** 4.00 in. **23.** Length: 5 in.; width: 5 in. **25.** 75.40 m^3
27. 94 m^2 **29.** 56 m^2 **31.** 96π in^2; 301.59 in^2 **33.** 184 ft^2 **35.** 69.36 m^2 **37.** 225π cm^2 **39.** 402.12 in^2
41. 6π ft^2 **43.** 297 in^2 **45.** 3 cm **47.** 11 cans **49.** 456 in^2 **51.** 22.53 cm^2 **53. a.** Always true
b. Never true **c.** Sometimes true

CHAPTER 7 REVIEW EXERCISES

1. $\angle x = 22°$, $\angle y = 158°$ [7.1C] **2.** 24 in. [7.3B] **3.** No [7.3C] **4.** 68° [7.1B] **5.** Yes, by the SAS Rule
[7.3C] **6.** 7.21 in. [7.3A] **7.** 44 cm [7.1A] **8.** 19° [7.1A] **9.** 32 in^2 [7.2B] **10.** 96 cm^3 [7.4A]
11. 42 in. [7.2A] **12.** $\angle a = 138°$, $\angle b = 42°$ [7.1B] **13.** 220 ft^2 [7.4B] **14.** 9.75 ft [7.3A] **15.** 42.875 in^3
[7.4A] **16.** 148° [7.1A] **17.** 39 ft^3 [7.4A] **18.** 95° [7.1C] **19.** 8 cm [7.2B] **20.** 288π mm^3 [7.4A]
21. 21.5 cm [7.2A] **22.** 4 cans [7.4B] **23.** 208 yd [7.2A] **24.** 90.25 m^2 [7.2B] **25.** 276 m^2 [7.2B]

CHAPTER 7 TEST

1. 7.55 cm [7.3A] **2.** Congruent, SAS [7.3C] **3.** 111 m^2 [7.2B] **4.** 42 ft^2 [7.2B] **5.** $\frac{784\pi}{3}$ cm^3 [7.4A]
6. 75 m^2 [7.4B] **7.** 4618.14 cm^3 [7.4A] **8.** 159 in^2 [7.2B] **9.** 34 ft [7.2A] **10.** 100° [7.1A]
11. 34° [7.1B] **12.** Octagon [7.2A] **13.** Not necessarily congruent [7.3C] **14.** 168 ft^3 [7.4A]
15. 8.06 m [7.3A] **16.** 143° [7.1B] **17.** 500π cm^2 [7.4B] **18.** 61° [7.1C] **19.** 6.67 ft [7.3B]
20. 4.27 ft [7.3B] **21.** 20 m [7.2A] **22.** 26 cm [7.2A] **23.** 51.6 ft [7.3A] **24.** 102° [7.1C]
25. 113° [7.1A]

CUMULATIVE REVIEW EXERCISES

1. 204 [6.4A/6.4B] **2.** 1, 2, 3, 6, 13, 26, 39, 78 [1.3D] **3.** $\frac{5}{6}$ [2.4B] **4.** 18 [3.1C] **5.** 12.8 [2.6C]
6. -2 [3.2B] **7.** 131° [7.1B] **8.** 26 cm [7.3A] **9.** $1\frac{11}{30}$ [2.3A] **10.** $-7b$ [4.2B]

11. -28 [3.5A]   **12.** $\dfrac{1}{2}$ [5.3B]   **13.** $2a$ [4.2A]   **14.** 75 [3.3A]   **15.** $7x + 18$ [4.2D]

16. -2 [4.1A]   **17.** $\dfrac{4}{5}$ [4.1A]   **18.** $2 \cdot 3 \cdot 13$ [1.3D]   **19.** 5 [5.3A]   **20.** 37.5% [6.3B]

21. $8(2n); 16n$ [4.3B]   **22.** The two cars are traveling 50 mph and 55 mph. [5.5B]   **23.** The simple interest is
$1313.01. [6.5A]   **24.** 24 oz of the silver alloy that costs $3.50 per ounce must be used. [5.5A]   **25.** The executive
used the cell phone for 87 min. [5.4B]   **26.** The sales tax on a $75 purchase is $4.50. [6.2B]   **27.** The increase in the
value of the imports is $.08 trillion. [2.6E]   **28.** The height of the box is 3 ft. [7.4A]   **29.** The depth when the
pressure is 35 lb/in² is 40 ft. [5.2B]   **30.** It takes these elevators 8.8 s to travel from the fifth to the 25th floor. [6.2B]

Answers to Chapter 8 Selected Exercises

PREP TEST

1. 48.0% was bill-related mail. [6.4A/6.4B]   **2. a.** The greatest cost increase is between 2009 and 2010.   **b.** Between
these years, there was an increase of $5318. [1.2C]   **3. a.** The ratio is $\dfrac{5}{3}$.   **b.** The ratio is $1:1$. [6.1A]

4. a. 3.9, 3.9, 4.2, 4.5, 5.2, 5.5, 7.1 [2.5B]   **b.** The average is 4.9 million viewers per night. [2.6E]

5. a. 4500 women are in the Marine Corps. [6.4C]   **b.** $\dfrac{1}{20}$ of the women in the military are in the Marine Corps. [6.3A]

SECTION 8.1

1. 128 units are required to graduate with a degree in accounting.   **3.** 35.2% of the units required to graduate are taken
in accounting.   **5.** $1,085,000,000 is spent on TV game machines.   **7.** $\dfrac{2}{25}$ of the total money spent was spent on
accessories.   **9.** 39 million passenger cars were produced worldwide.   **11.** 28% of the passenger cars were produced in
Asia.   **13.** Women ages 75 and up have the lowest recommended number of Calories.

SECTION 8.2

1. a. Median   **b.** Mean   **c.** Mode   **d.** Median   **e.** Mode   **f.** Mean   **3.** The mean number of seats filled is
381.5625 seats. The median number of seats filled is 394.5 seats. Since each number occurs only once, there is no mode.
5. The mean cost is $45.615. The median cost is $45.855.   **7.** The mean monthly rate is $403.625. The median monthly
rate is $404.50.   **9. a.** The mean life expectancy is 70.8 years.   **b.** The median life expectancy is 72 years.
13. a. 25%   **b.** 75%   **c.** 75%   **d.** 25%   **15.** Lowest = $46,596; highest = $82,879; Q_1 = $56,067; Q_3 = $66,507;
median = $61,036; range = $36,283; interquartile range = $10,440   **17. a.** There were 40 adults who had cholesterol
levels above 217.   **b.** There were 60 adults who had cholesterol levels below 254.   **c.** There are 20 cholesterol levels in
each quartile.   **d.** 25% of the adults had cholesterol levels not more than 198.   **19. a.** Range = 4.39 million metric
tons; Q_1 = 0.56 million metric tons; Q_3 = 2.10 million metric tons; interquartile range = 1.54 million metric tons
b.

0.41   2.10   4.80
0.56 0.815

c. 4.80   **21. a.** No, the difference in the means is not greater than 1 in.

b. The difference in the medians is 0.3 in.   **c.**

1.8   3.0   5.95   7.8
2.15

0.5   1.55 2.7   4.55   6.4

23. Answers will vary. For example, 55, 55, 55, 55, 55, or 50, 55, 55, 55, 60

SECTION 8.3

1. {(HHHH), (HHHT), (HHTT), (HHTH), (HTTT), (HTHH), (HTTH), (HTHT), (TTTT), (TTTH), (TTHH), (THHH), (TTHT), (THHT), (THTT), (THTH)} **3.** {(1, 1), (1, 2), (1, 3), (1, 4), (2, 1), (2, 2), (2, 3), (2, 4), (3, 1), (3, 2), (3, 3), (3, 4), (4, 1), (4, 2), (4, 3), (4, 4)} **5. a.** {1, 2, 3, 4, 5, 6, 7, 8} **b.** {1, 2, 3} **7. a.** The probability that the sum is 5 is $\frac{1}{9}$.
b. The probability that the sum is 15 is 0. **c.** The probablity that the sum is less than 15 is 1. **d.** The probability that the sum is 2 is $\frac{1}{36}$. **9. a.** The probability that the number is divisible by 4 is $\frac{1}{4}$. **b.** The probability that the number is a multiple of 3 is $\frac{1}{3}$. **11.** The probability of throwing a sum of 5 is greater. **13. a.** The probability is $\frac{4}{11}$ that the letter *I* is drawn. **b.** The probability of choosing an *S* is greater. **15. a.** The probability is $\frac{1}{3}$ that the marble chosen is green. **b.** The probability of choosing a red marble is greater. **17.** The probability is $\frac{8}{47}$ that the paper earned a B grade. **19.** The probability is 0.81 that an employee has a group health insurance plan.

CHAPTER 8 REVIEW EXERCISES

1. The agencies spent $349 million on maintaining websites. [8.1A] **2.** The ratio is $\frac{9}{8}$. [8.1A] **3.** 8.9% of the total amount of money was spent by NASA. [8.1A] **4.** The difference was 50 days. [8.1A] **5.** The percent is 50%. [8.1A] **6. a.** The Southeast had the lowest number of days of full operation. **b.** This region had 30 days of full operation. [8.1A] **7.** The mean is 214.$\overline{54}$. The median is 210. [8.2A] **8.** The mean is 7.17 lb. The median is 7.05 lb. [8.2A] **9.** The modal response was "good." [8.2A] **10.** [8.2B]

11. There are 12 elements in the sample space. [8.3A] **12.** The probability is $\frac{1}{500}$. [8.3A] **13.** $Q_1 = 1$; $Q_3 = 3$ [8.2B] **14.** The probability is $\frac{1}{6}$. [8.3A] **15.** The probability is $\frac{5}{14}$. [8.3A] **16. a.** The mean is 91.6 heartbeats per minute. The median is 93.5 heartbeats per minute. The mode is 96 heartbeats per minute. [8.2A] **b.** The range is 36 heartbeats per minute. The interquartile range is 15 heartbeats per minute. [8.2B]

CHAPTER 8 TEST

1. There were 355 more films rated R. [8.1A] **2.** The number of PG-13 films was 16 times the number of NC-17 films. [8.1A] **3.** The percent of films rated G was 5.6%. [8.1A] **4.** The student enrollment increased the least during the 1990s. [8.1A] **5.** The increase in the enrollment was 11 million students. [8.1A] **6.** The mean score was 152.875. [8.2A] **7.** The median response time was 14 min. [8.2A] **8.** The modal response was "very good." [8.2A] **9.** The first quartile is 9.8. [8.2B] **10. a.** The range is 22 days. **b.** The median is 14 vacation days. [8.2B] **11.** [8.2B] **12.** The sample space is [(N, D, Q), (N, Q, D), (D, N, Q), (D, Q, N),

(Q, N, D), (Q, D, N)]. [8.3A] **13.** The probability is $\frac{4}{31}$. [8.3A] **14.** The probability is $\frac{1}{3}$. [8.3A]

15. The probability is $\frac{1}{8}$. [8.3A] **16.** The probability is $\frac{2}{3}$. [8.3A] **17.** The probability is $\frac{5}{12}$. [8.3A]

18. a. The mean is 2.53 days. **b.** The median is 2.55 days. [8.2A] **c.** [8.2B]

CUMULATIVE REVIEW EXERCISES

1. 540 [1.3B]   **2.** 14 [1.4A]   **3.** 120 [2.1A]   **4.** $\frac{5}{12}$ [2.2B]   **5.** $12\frac{3}{40}$ [2.3A]   **6.** $4\frac{17}{24}$ [2.3B]

7. 2 [2.4A]   **8.** $\frac{64}{85}$ [2.4B]   **9.** $8\frac{1}{4}$ [2.7A]   **10.** 209.305 [2.5A]   **11.** 2.82348 [2.6B]   **12.** 16.67 [2.6D]

13. 26.4 mpg [6.1A]   **14.** 3.2 [6.2A]   **15.** 80% [6.3B]   **16.** 80 [6.4A/6.4B]   **17.** 16.34 [6.4A/6.4B]

18. 40% [6.4A/6.4B]   **19.** The salesperson's income for the week was $650. [6.4C]   **20.** The cost is $407.50. [6.2B]

21. The interest due is $3750. [6.5A]   **22.** The price is $279. [6.4C]   **23.** The amount budgeted for food is $570.

[8.1A]   **24.** The difference is 12 points. [8.1A]   **25.** The mean high temperature is 69.6°F. [8.2A]   **26.** The

probability is $\frac{5}{36}$. [8.3A]

Answers to Chapter 9 Selected Exercises

PREP TEST

1. 1 [3.2B]   **2.** −18 [3.3A]   **3.** $\frac{2}{3}$ [3.3B]   **4.** 48 [4.1A]   **5.** 0 [1.3C]   **6.** No [4.1A]

7. $5x^2 - 9x - 6$ [4.2A]   **8.** 0 [4.2A]   **9.** $-6x + 24$ [4.2C]   **10.** $-7xy + 10y$ [4.2D]

SECTION 9.1

1. Yes   **3.** No   **5.** Yes   **7.** Yes   **9.** Binomial   **11.** Trinomial   **13.** None of these   **15.** Binomial

17. $-2x^2 + 3x$ **19.** $y^2 - 8$ **21.** $5x^2 + 7x + 20$ **23.** $x^3 + 2x^2 - 6x - 6$ **25.** $2a^3 - 3a^2 - 11a + 2$ **27.** $5x^2 + 8x$

29. $7x^2 + xy - 4y^2$ **31.** $3a^2 - 3a + 17$ **33.** $5x^3 + 10x^2 - x - 4$ **35.** $3r^3 + 2r^2 - 11r + 7$ **37.** $4x$

39. $3y^2 - 4y - 2$ **41.** $-7x - 7$ **43.** $4x^3 + 3x^2 + 3x + 1$ **45.** $y^3 + 5y^2 - 2y - 4$ **47.** $-y^2 - 13xy$

49. $2x^2 - 3x - 1$ **51.** $-2x^3 + x^2 + 2$ **53.** $3a^3 - 2$ **55.** $4y^3 + 2y^2 + 2y - 4$ **57.** $x^2 + 9x - 11$

SECTION 9.2

3. $30x^3$ **5.** $-42c^6$ **7.** $9a^7$ **9.** x^3y^4 **11.** $-10x^9y$ **13.** $12x^7y^8$ **15.** $-6x^3y^5$ **17.** x^4y^5z **19.** $a^3b^5c^4$

21. $-30a^5b^8$ **23.** $6a^5b$ **25.** $40y^{10}z^6$ **27.** $x^3y^3z^2$ **29.** $-24a^3b^3c^3$ **31.** $8x^7yz^6$ **33.** $30x^6y^8$ **35.** $-36a^3b^2c^3$

37. x^{15} **39.** x^{14} **41.** x^8 **43.** y^{12} **45.** $-8x^6$ **47.** x^4y^6 **49.** $9x^4y^2$ **51.** $-243x^{15}y^{10}$ **53.** $-8x^7$

55. $24x^8y^7$ **57.** a^4b^6 **59.** $64x^{12}y^3$ **61.** $-18x^3y^4$ **63.** $-8a^7b^5$ **65.** $-54a^9b^3$ **67.** $12x^2$ **69.** $2x^6y^2 + 9x^4y^2$

71. 0   **73.** $17x^4y^8$ **75.** True   **77.** False. $(x^2)^5 = x^{2\cdot5} = x^{10}$ **79.** No. $2^{(3^2)}$ is larger.

SECTION 9.3

1. 3, 3, $12x - 15$ **3.** $x^2 - 2x$ **5.** $-x^2 - 7x$ **7.** $3a^3 - 6a^2$ **9.** $-5x^4 + 5x^3$ **11.** $-3x^5 + 7x^3$ **13.** $12x^3 - 6x^2$

15. $6x^2 - 12x$ **17.** $3x^2 + 4x$ **19.** $-x^3y + xy^3$ **21.** $2x^4 - 3x^2 + 2x$ **23.** $2a^3 + 3a^2 + 2a$ **25.** $3x^6 - 3x^4 - 2x^2$

27. $-6y^4 - 12y^3 + 14y^2$ **29.** $-2a^3 - 6a^2 + 8a$ **31.** $6y^4 - 3y^3 + 6y^2$ **33.** $x^3y - 3x^2y^2 + xy^3$ **35.** $x^3 + 4x^2 + 5x + 2$

37. $a^3 - 6a^2 + 13a - 12$ **39.** $-2b^3 + 7b^2 + 19b - 20$ **41.** $-6x^3 + 31x^2 - 41x + 10$ **43.** $x^3 - 3x^2 + 5x - 15$

45. $x^4 - 4x^3 - 3x^2 + 14x - 8$ **47.** $15y^3 - 16y^2 - 70y + 16$ **49.** $5a^4 - 20a^3 - 5a^2 + 22a - 8$

51. $y^4 + 4y^3 + y^2 - 5y + 2$ **53.** $x^2 + 4x + 3$ **55.** $a^2 + a - 12$ **57.** $y^2 - 5y - 24$ **59.** $y^2 - 10y + 21$

61. $2x^2 + 15x + 7$ **63.** $3x^2 + 11x - 4$ **65.** $4x^2 - 31x + 21$ **67.** $3y^2 - 2y - 16$ **69.** $9x^2 + 54x + 77$

71. $21a^2 - 83a + 80$ **73.** $6a^2 - 25ab + 14b^2$ **75.** $2a^2 - 11ab - 63b^2$ **77.** $100a^2 - 100ab + 21b^2$

79. $15x^2 + 56xy + 48y^2$ **81.** $14x^2 - 97xy - 60y^2$ **83.** $56x^2 - 61xy + 15y^2$ **85.** $y^2 - 25$ **87.** $4x^2 - 9$

89. $9x^2 - 49$ **91.** $16 - 9y^2$ **93.** $x^2 + 2x + 1$ **95.** $9a^2 - 30a + 25$ **97.** $4a^2 + 4ab + b^2$ **99.** $x^2 - 4xy + 4y^2$

101. $25x^2 + 20xy + 4y^2$ **103.** The area of the rectangle is $(10x^2 - 35x)$ ft².   **105.** The area of the square is

$(4x^2 + 4x + 1)$ km².   **107.** The area of the triangle is $(4x^2 + 10x)$ m².   **109.** The area is $(60w + 3000)$ yd².

111. $x^4 + 2x^3 - 5x^2 - 6x + 9$ **113.** $12x^2 - x - 20$ **115.** $x^3 - 7x^2 - 7$

SECTION 9.4

1. 7, 5, 2 **3.** $\dfrac{1}{25}$ **5.** 64 **7.** $\dfrac{1}{27}$ **9.** 2 **11.** $\dfrac{1}{x^2}$ **13.** a^6 **15.** $\dfrac{4}{x^7}$ **17.** $\dfrac{2}{3z^2}$ **19.** $5b^8$ **21.** $\dfrac{x^2}{3}$ **23.** 1

25. -1 **27.** y^4 **29.** a^3 **31.** p^4 **33.** $2x^3$ **35.** $2k$ **37.** m^5n^2 **39.** $\dfrac{3r^2}{2}$ **41.** $-\dfrac{2a}{3}$ **43.** $\dfrac{1}{y^5}$ **45.** $\dfrac{1}{a^6}$

47. $\dfrac{1}{3x^3}$ **49.** $\dfrac{2}{3x^5}$ **51.** $\dfrac{y^4}{x^2}$ **53.** $\dfrac{2}{5m^3n^8}$ **55.** $\dfrac{1}{p^3q}$ **57.** $\dfrac{1}{2y^3}$ **59.** $\dfrac{7xz}{8y^3}$ **61.** $\dfrac{p^2}{2m^3}$ **63.** $-\dfrac{8x^3}{y^6}$ **65.** $\dfrac{9}{x^2y^4}$

67. $\dfrac{2}{x^4}$ **69.** $-\dfrac{5}{a^8}$ **71.** $-\dfrac{a^5}{8b^4}$ **73.** $\dfrac{10y^3}{x^4}$ **75.** $\dfrac{1}{2x^3}$ **77.** $\dfrac{3}{x^3}$ **79.** $\dfrac{1}{2x^2y^6}$ **81.** $\dfrac{1}{x^6y}$ **83.** $\dfrac{a^4}{y^{10}}$ **85.** $-\dfrac{1}{6x^3}$

87. $-\dfrac{a^2b}{6c^2}$ **89.** $-\dfrac{7b^6}{a^2}$ **91.** $\dfrac{s^8t^4}{4r^{12}}$ **93.** $\dfrac{125p^3}{27m^{15}n^6}$ **97.** 3.24×10^{-9} **99.** 3×10^{-18} **101.** 3.2×10^{16}

103. 1.22×10^{-19} **105.** 5.47×10^8 **107.** 0.000167 **109.** 68,000,000 **111.** 0.0000305 **113.** 0.00000000102

115. 6.023×10^{23} **117.** 3.7×10^{-6} **119.** 1×10^{-9} **121.** 1.6×10^{-19} **123.** $\dfrac{1}{4}, \dfrac{1}{2}, 1, 2, 4$ **125.** $4, 2, 1, \dfrac{1}{2}, \dfrac{1}{4}$

127. False. $(2a)^{-3} = \dfrac{1}{8a^3}$ **129.** False. $(2 + 3)^{-1} = (5)^{-1} = \dfrac{1}{5}$

SECTION 9.5

1. $2a - 5$ **3.** $6y + 4$ **5.** $x - 2$ **7.** $-x + 2$ **9.** $x^2 + 3x - 5$ **11.** $x^4 - 3x^2 - 1$ **13.** $xy + 2$ **15.** $-3y^3 + 5$

17. $3x - 2 + \dfrac{1}{x}$ **19.** $-3x + 7 - \dfrac{6}{x}$ **21.** $4a - 5 + 6b$ **23.** $9x + 6 - 3y$ **25.** $(x + 2), (x - 3)$ **27.** $b - 7$

29. $y - 5$ **31.** $2y - 7$ **33.** $2y + 6 + \dfrac{25}{y - 3}$ **35.** $x - 2 + \dfrac{8}{x + 2}$ **37.** $3y - 5 + \dfrac{20}{2y + 4}$ **39.** $6x - 12 + \dfrac{19}{x + 2}$

41. $b - 5 - \dfrac{24}{b - 3}$ **43.** $3x + 17 + \dfrac{64}{x - 4}$ **45.** $5y + 3 + \dfrac{1}{2y + 3}$ **47.** $4a + 1$ **49.** $2a + 9 + \dfrac{33}{3a - 1}$

51. $x^2 - 5x + 2$ **53.** $x^2 + 5$ **55.** $3ab$

CHAPTER 9 REVIEW EXERCISES

1. $8b^2 - 2b - 15$ [9.3C] **2.** $21y^2 + 4y - 1$ [9.1A] **3.** $x^4y^8z^4$ [9.2A] **4.** $\dfrac{2x^3}{3}$ [9.4A]

5. $-8x^3 - 14x^2 + 18x$ [9.3A] **6.** $-\dfrac{1}{2a}$ [9.4A] **7.** $16u^{12}v^{16}$ [9.2B] **8.** 64 [9.2B] **9.** $2x^2 + 3x - 8$ [9.1B]

10. $\dfrac{b^6}{a^4}$ [9.4A] **11.** $-108x^{18}$ [9.2B] **12.** $25y^2 - 70y + 49$ [9.3D] **13.** $100a^{15}b^{13}$ [9.2B]

14. $4b^4 + 12b^2 - 1$ [9.5A] **15.** $-\dfrac{1}{16}$ [9.4A] **16.** $13y^3 - 12y^2 - 5y - 1$ [9.1B] **17.** $-x + 2 + \dfrac{1}{x + 3}$ [9.5B]

18. $2ax - 4ay - bx + 2by$ [9.3C] **19.** $6y^3 + 17y^2 - 2y - 21$ [9.3B] **20.** $b^2 + 5b + 2 + \dfrac{7}{b - 7}$ [9.5B]

21. $8a^3b^3 - 4a^2b^4 + 6ab^5$ [9.3A] **22.** $4a^2 - 25b^2$ [9.3D] **23.** $12b^5 - 4b^4 - 6b^3 - 8b^2 + 5$ [9.3B]

24. $2x^3 + 9x^2 - 3x - 12$ [9.1A] **25.** $-4y + 8$ [9.5A] **26.** $a^2 - 49$ [9.3D] **27.** 3.756×10^{10} [9.4B]

28. 14,600,000 [9.4B] **29.** $-54a^{13}b^5c^7$ [9.2A] **30.** $2y - 9$ [9.5B] **31.** $\dfrac{x^4y^6}{9}$ [9.4A]

32. $10a^2 + 31a - 63$ [9.3C] **33.** 1.27×10^{-7} [9.4B] **34.** 0.0000000000032 [9.4B]

35. The area is $(2w^2 - w)$ ft². [9.3E] **36.** The area is $(9x^2 - 12x + 4)$ in². [9.3E]

CHAPTER 9 TEST

1. $4x^3 - 6x^2$ [9.3A] **2.** $4x - 1 + \dfrac{3}{x^2}$ [9.5A] **3.** $-\dfrac{4}{x^6}$ [9.4A] **4.** $-6x^3y^6$ [9.2A] **5.** $x - 1 + \dfrac{2}{x + 1}$ [9.5B]

6. $x^3 - 7x^2 + 17x - 15$ [9.3B] **7.** $-8a^6b^3$ [9.2B] **8.** $\dfrac{9y^{10}}{x^{10}}$ [9.4A] **9.** $a^2 + 3ab - 10b^2$ [9.3C]

10. $4x^4 - 2x^2 + 5$ [9.5A] **11.** $x + 7$ [9.5B] **12.** $6y^4 - 9y^3 + 18y^2$ [9.3A] **13.** $-4x^4 + 8x^3 - 3x^2 - 14x + 21$

[9.3B]   **14.** $16y^2 - 9$ [9.3D]   **15.** a^4b^7 [9.2A]   **16.** $8ab^4$ [9.4A]   **17.** $4a - 7$ [9.5A]

18. $-5a^3 + 3a^2 - 4a + 3$ [9.1B]   **19.** $4x^2 - 20x + 25$ [9.3D]   **20.** $2x + 3 + \dfrac{2}{2x - 3}$ [9.5B]

21. $-2x^3$ [9.4A]   **22.** $10x^2 - 43xy + 28y^2$ [9.3C]   **23.** $3x^3 + 6x^2 - 8x + 3$ [9.1A]   **24.** 3.02×10^{-9} [9.4B]

25. The area of the circle is $(\pi x^2 - 10\pi x + 25\pi)$ m². [9.3E]

CUMULATIVE REVIEW EXERCISES

1. $\dfrac{5}{144}$ [3.4A]   **2.** $\dfrac{5}{3}$ [3.5A]   **3.** $\dfrac{25}{11}$ [3.5A]   **4.** $-\dfrac{22}{9}$ [4.1A]   **5.** $5x - 3xy$ [4.2A]   **6.** $-9x$ [4.2B]

7. $-18x + 12$ [4.2D]   **8.** -16 [5.1C]   **9.** -16 [5.3A]   **10.** 15 [5.3B]   **11.** 22% [6.4A/6.4B]

12. $4b^3 - 4b^2 - 8b - 4$ [9.1A]   **13.** $3y^3 + 2y^2 - 10y$ [9.1B]   **14.** a^9b^{15} [9.2B]   **15.** $-8x^3y^6$ [9.2A]

16. $6y^4 + 8y^3 - 16y^2$ [9.3A]   **17.** $10a^3 - 39a^2 + 20a - 21$ [9.3B]   **18.** $15b^2 - 31b + 14$ [9.3C]   **19.** $\dfrac{1}{2b^2}$ [9.4A]

20. $a - 7$ [9.5B]   **21.** 0.0000609 [9.4B]   **22.** $8x - 2x = 18; 3$ [5.4B]   **23.** The cost is $.13 per ounce. [5.5A]

24. The car overtakes the cyclist 25 mi from the starting point. [5.5B]   **25.** The length is 15 m and the width is 6 m. [7.2A]

Answers to Chapter 10 Selected Exercises

PREP TEST

1. $2 \cdot 3 \cdot 5$ [1.3D]   **2.** $-12y + 15$ [4.2C]   **3.** $-a + b$ [4.2C]   **4.** $-3a + 3b$ [4.2D]   **5.** 0 [5.1C]

6. $-\dfrac{1}{2}$ [5.2A]   **7.** $x^2 - 2x - 24$ [9.3C]   **8.** $6x^2 - 11x - 10$ [9.3C]   **9.** x^3 [9.4A]   **10.** $3x^3y$ [9.4A]

SECTION 10.1

3. $5(a + 1)$ **5.** $8(2 - a^2)$ **7.** $4(2x + 3)$ **9.** $6(5a - 1)$ **11.** $x(7x - 3)$ **13.** $a^2(3 + 5a^3)$ **15.** $y(14y + 11)$

17. $2x(x^3 - 2)$ **19.** $2x^2(5x^2 - 6)$ **21.** $4a^5(2a^3 - 1)$ **23.** $xy(xy - 1)$ **25.** $3xy(xy^3 - 2)$ **27.** $xy(x - y^2)$

29. $5y(y^2 - 4y + 1)$ **31.** $3y^2(y^2 - 3y - 2)$ **33.** $3y(y^2 - 3y + 8)$ **35.** $a^2(6a^3 - 3a - 2)$ **37.** $ab(2a - 5ab + 7b)$

39. $2b(2b^4 + 3b^2 - 6)$ **41.** $x^2(8y^2 - 4y + 1)$ **43.** $(a + z)(y + 7)$ **45.** $(a - b)(3r + s)$ **47.** $(m - 7)(t - 7)$

49. $(4a - b)(2y + 1)$ **51.** $(x + 2)(x + 2)$ **53.** $(p - 2)(p - 3r)$ **55.** $(a + 6)(b - 4)$ **57.** $(2z - 1)(z + y)$

59. $(2v - 3y)(4v + 7)$ **61.** $(2x - 5)(x - 3)$ **63.** $(y - 2)(3y - a)$ **65.** $(3x - y)(y + 1)$ **67.** $(3s + t)(t - 2)$

69. a. 28   **b.** 496   **71. a.** $r^2(\pi - 2)$ **b.** $2r^2(4 - \pi)$ **c.** $r^2(4 - \pi)$

SECTION 10.2

1. the same   **3.** $(x + 1)(x + 2)$ **5.** $(x + 1)(x - 2)$ **7.** $(a + 4)(a - 3)$ **9.** $(a - 1)(a - 2)$ **11.** $(a + 2)(a - 1)$

13. $(b - 3)(b - 3)$ **15.** $(b + 8)(b - 1)$ **17.** $(y + 11)(y - 5)$ **19.** $(y - 2)(y - 3)$ **21.** $(z - 5)(z - 9)$

23. $(z + 8)(z - 20)$ **25.** $(p + 3)(p + 9)$ **27.** $(x + 10)(x + 10)$ **29.** $(b + 4)(b + 5)$ **31.** $(x + 3)(x - 14)$

33. $(b + 4)(b - 5)$ **35.** $(y + 3)(y - 17)$ **37.** $(p + 3)(p - 7)$ **39.** Nonfactorable over the integers

41. $(x - 5)(x - 15)$ **43.** $(p + 3)(p + 21)$ **45.** $(x + 2)(x + 19)$ **47.** $(x + 9)(x - 4)$ **49.** $(a + 4)(a - 11)$

51. $(a - 3)(a - 18)$ **53.** $(z + 21)(z - 7)$ **55.** $(c + 12)(c - 15)$ **57.** $(p + 9)(p + 15)$ **59.** $(c + 2)(c + 9)$

61. $(x + 15)(x - 5)$ **63.** $(x + 25)(x - 4)$ **65.** $(b - 4)(b - 18)$ **67.** $(a + 45)(a - 3)$ **69.** $(b - 7)(b - 18)$

71. $(z + 12)(z + 12)$ **73.** $(x - 4)(x - 25)$ **75.** $(x + 16)(x - 7)$ **77.** $3(x + 2)(x + 3)$ **79.** $-(x - 2)(x + 6)$

81. $a(b + 8)(b - 1)$ **83.** $x(y + 3)(y + 5)$ **85.** $-2a(a + 1)(a + 2)$ **87.** $4y(y + 6)(y - 3)$ **89.** $2x(x^2 - x + 2)$

91. $6(z + 5)(z - 3)$ **93.** $3a(a + 3)(a - 6)$ **95.** $(x + 7y)(x - 3y)$ **97.** $(a - 5b)(a - 10b)$ **99.** $(s + 8t)(s - 6t)$

101. Nonfactorable over the integers   **103.** $z^2(z + 10)(z - 8)$ **105.** $b^2(b + 2)(b - 5)$ **107.** $3y^2(y + 3)(y + 15)$

109. $-x^2(x + 1)(x - 12)$ **111.** $3y(x + 3)(x - 5)$ **113.** $-3x(x - 3)(x - 9)$ **115.** $(x - 3y)(x - 5y)$

117. $(a - 6b)(a - 7b)$ **119.** $(y + z)(y + 7z)$ **121.** $3y(x + 21)(x - 1)$ **123.** $3x(x + 4)(x - 3)$

125. $4z(z + 11)(z - 3)$ **127.** $4x(x + 3)(x - 1)$ **129.** $5(p + 12)(p - 7)$ **131.** $p^2(p + 12)(p - 3)$
133. $(t - 5s)(t - 7s)$ **135.** $(a + 3b)(a - 11b)$ **137.** $y(x + 6)(x - 9)$ **139.** $-36, 36, -12, 12$
141. $22, -22, 10, -10$ **143.** $6, 10, 12$ **145.** $6, 10, 12$ **147.** $4, 6$

SECTION 10.3

1. $(x + 1)(2x + 1)$ **3.** $(y + 3)(2y + 1)$ **5.** $(a - 1)(2a - 1)$ **7.** $(b - 5)(2b - 1)$ **9.** $(x + 1)(2x - 1)$
11. $(x - 3)(2x + 1)$ **13.** $(t + 2)(2t - 5)$ **15.** $(p - 5)(3p - 1)$ **17.** $(3y - 1)(4y - 1)$ **19.** Nonfactorable over
the integers **21.** $(2t - 1)(3t - 4)$ **23.** $(x + 4)(8x + 1)$ **25.** Nonfactorable over the integers
27. $(3y + 1)(4y + 5)$ **29.** $(a + 7)(7a - 2)$ **31.** $(b - 4)(3b - 4)$ **33.** $(z - 14)(2z + 1)$ **35.** $(p + 8)(3p - 2)$
37. $2(x + 1)(2x + 1)$ **39.** $5(y - 1)(3y - 7)$ **41.** $x(x - 5)(2x - 1)$ **43.** $b(a - 4)(3a - 4)$ **45.** Nonfactorable
over the integers **47.** $-3x(x + 4)(x - 3)$ **49.** $4(4y - 1)(5y - 1)$ **51.** $z(2z + 3)(4z + 1)$ **53.** $y(2x - 5)(3x + 2)$
55. $5(t + 2)(2t - 5)$ **57.** $p(p - 5)(3p - 1)$ **59.** $2(z + 4)(13z - 3)$ **61.** $2y(y - 4)(5y - 2)$
63. $yz(z + 2)(4z - 3)$ **65.** $3a(2a + 3)(7a - 3)$ **67.** $y(3x - 5y)(3x - 5y)$ **69.** $xy(3x - 4y)(3x - 4y)$
71. $(2x - 3)(3x - 4)$ **73.** $(b + 7)(5b - 2)$ **75.** $(3a + 8)(2a - 3)$ **77.** $(z + 2)(4z + 3)$ **79.** $(2p + 5)(11p - 2)$
81. $(y + 1)(8y + 9)$ **83.** $(6t - 5)(3t + 1)$ **85.** $(b + 12)(6b - 1)$ **87.** $(3x + 2)(3x + 2)$ **89.** $(2b - 3)(3b - 2)$
91. $(3b + 5)(11b - 7)$ **93.** $(3y - 4)(6y - 5)$ **95.** $(3a + 7)(5a - 3)$ **97.** $(2y - 5)(4y - 3)$ **99.** $(2z + 3)(4z - 5)$
101. Nonfactorable over the integers **103.** $(2z - 5)(5z - 2)$ **105.** $(6z + 5)(6z + 7)$ **107.** $(x + y)(3x - 2y)$
109. $(a + 2b)(3a - b)$ **111.** $(y - 2z)(4y - 3z)$ **113.** $-(z - 7)(z + 4)$ **115.** $-(x - 1)(x + 8)$
117. $3(x + 5)(3x - 4)$ **119.** $4(2x - 3)(3x - 2)$ **121.** $a^2(5a + 2)(7a - 1)$ **123.** $5(b - 7)(3b - 2)$
125. $(x - 7y)(3x - 5y)$ **127.** $3(8y - 1)(9y + 1)$ **129.** $-(x - 1)(x + 21)$ **133.** $x(x - 1)$ **135.** $(2y + 1)(y + 3)$
137. $(4y - 3)(y - 3)$ **139.** $-5, 5, -1, 1$ **141.** $-5, 5, -1, 1$ **143.** $-9, 9, -3, 3$

SECTION 10.4

1. a. Answers will vary. For instance, $x^2 - 25$. **b.** Answers will vary. For instance, $x^2 + 6x + 9$. **3.** $(x + 2)(x - 2)$
5. $(a + 9)(a - 9)$ **7.** $(y + 1)^2$ **9.** $(a - 1)^2$ **11.** $(2x + 1)(2x - 1)$ **13.** $(x^3 + 3)(x^3 - 3)$ **15.** Nonfactorable over
the integers **17.** $(x + y)^2$ **19.** $(2a + 1)^2$ **21.** $(3x + 1)(3x - 1)$ **23.** $(1 + 8x)(1 - 8x)$ **25.** Nonfactorable over
the integers **27.** $(3a + 1)^2$ **29.** $(b^2 + 4a)(b^2 - 4a)$ **31.** $(2a - 5)^2$ **33.** $(3a - 7)^2$ **35.** $(5z + y)(5z - y)$
37. $(ab + 5)(ab - 5)$ **39.** $(5x + 1)(5x - 1)$ **41.** $(2a - 3b)^2$ **43.** $(2y - 9z)^2$ **45.** $\left(\dfrac{1}{x} + 2\right)\left(\dfrac{1}{x} - 2\right)$
47. $(3ab - 1)^2$ **49.** $2(2y + 1)(2y - 1)$ **51.** $3a(a + 1)^2$ **53.** $(m^2 + 16)(m + 4)(m - 4)$ **55.** $(x + 1)(9x + 4)$
57. $4y^2(2y + 3)^2$ **59.** $(y^4 + 9)(y^2 + 3)(y^2 - 3)$ **61.** $(5 - 2p)^2$ **63.** $(4x - 3 + y)(4x - 3 - y)$
65. $(x - 2 + y)(x - 2 - y)$ **67.** $5(x + 1)(x - 1)$ **69.** $x(x + 2)^2$ **71.** $x^2(x + 7)(x - 5)$ **73.** $5(b + 3)(b + 12)$
75. Nonfactorable over the integers **77.** $2y(x + 11)(x - 3)$ **79.** $x(x^2 - 6x - 5)$ **81.** $3(y^2 - 12)$
83. $(2a + 1)(10a + 1)$ **85.** $y^2(x + 1)(x - 8)$ **87.** $5(a + b)(2a - 3b)$ **89.** $-2(x + 5)(x - 5)$ **91.** $b^2(a - 5)^2$
93. $ab(3a - b)(4a + b)$ **95.** $3a(2a - 1)^2$ **97.** $3(81 + a^2)$ **99.** $2a(2a - 5)(3a - 4)$ **101.** $a(2a + 5)^2$
103. $3b(3a - 1)^2$ **105.** $-6(x - 2)(x + 4)$ **107.** $x^2(x + y)(x - y)$ **109.** $2a(3a + 2)^2$ **111.** $-b(3a - 2)(2a + 1)$
113. $2x^2(x - 8)(2x - 3)$ **115.** $x^2(x + 5)(x - 5)$ **117.** $(a^2 + 4)(a + 2)(a - 2)$ **119.** $-3y^2(2y + 5)(4y - 3)$
121. $2(x - 3)(2a - b)$ **123.** $(a - b)(y + 1)(y - 1)$ **125.** $(a + b)(a - b)(x - y)$ **127.** $12, -12$ **129.** $16, -16$
131. $10, -10$ **133.** 12 **135.** 18

SECTION 10.5

3. $-3, -2$ **5.** $7, 3$ **7.** $0, 5$ **9.** $0, 9$ **11.** $0, -\dfrac{3}{2}$ **13.** $0, \dfrac{2}{3}$ **15.** $-2, 5$ **17.** $-9, 9$ **19.** $-\dfrac{7}{2}, \dfrac{7}{2}$
21. $-\dfrac{1}{3}, \dfrac{1}{3}$ **23.** $-2, -4$ **25.** $-7, 2$ **27.** $-\dfrac{1}{2}, 5$ **29.** $-\dfrac{1}{3}, -\dfrac{1}{2}$ **31.** $0, 3$ **33.** $0, 7$ **35.** $-1, -4$
37. $2, 3$ **39.** $\dfrac{1}{2}, -4$ **41.** $\dfrac{1}{3}, 4$ **43.** $3, 9$ **45.** $-2, 9$ **47.** $-1, -2$ **49.** $-9, 5$ **51.** $-7, 4$ **53.** $-2, -3$
55. $-8, 9$ **57.** $1, 4$ **59.** $-5, 2$ **61.** The number is 6. **63.** The numbers are 2 and 4. **65.** The numbers are 4
and 5. **67.** The numbers are 3 and 7. **69.** There will be 12 consecutive numbers. **71.** There are 6 teams in the
league. **73.** The object will hit the ground 3 s later. **75.** The golf ball will return to the ground 3.75 s later.

77. The length is 15 in. The width is 5 in.   **79.** The height of the triangle is 14 m.   **81.** The dimensions of the type area are 4 in. by 7 in.   **83.** The radius of the original circular lawn was approximately 3.81 ft.   **85.** 3, 48   **87.** $-\frac{3}{2}$, -5 **89.** 0, 9

CHAPTER 10 REVIEW EXERCISES

1. $(b - 10)(b - 3)$ [10.2A]   **2.** $(x - 3)(4x + 5)$ [10.1B]   **3.** Nonfactorable over the integers [10.3A]   **4.** $5x(x^2 + 2x + 7)$ [10.1A]   **5.** $7y^3(2y^6 - 7y^3 + 1)$ [10.1A]   **6.** $(y + 9)(y - 4)$ [10.2A]   **7.** $(2x - 7)(3x - 4)$ [10.3A]   **8.** $3ab(4a + b)$ [10.1A]   **9.** $(a^3 + 10)(a^3 - 10)$ [10.4A]   **10.** $n^2(n - 3)(n + 1)$ [10.2B]   **11.** $(6y - 1)(2y + 3)$ [10.3A]   **12.** $2b(2b - 7)(3b - 4)$ [10.4B]   **13.** $(3y^2 + 5z)(3y^2 - 5z)$ [10.4A]   **14.** $(c + 6)(c + 2)$ [10.2A]   **15.** $(6a - 5)(3a + 2)$ [10.3B]   **16.** $\frac{1}{4}$, -7 [10.5A]   **17.** $4x(x - 6)(x + 1)$ [10.2B]   **18.** $3(a - 7)(a + 2)$ [10.2B]   **19.** $(a - 12)(2a + 5)$ [10.3B]   **20.** 7, -3 [10.5A]   **21.** $(3a - 5b)(7x + 2y)$ [10.1B]   **22.** $(ab + 1)(ab - 1)$ [10.4A]   **23.** $(2x + 5)(5x + 2y)$ [10.1B]   **24.** $5(x - 3)(x + 2)$ [10.2B]   **25.** $3(x + 6)^2$ [10.4B]   **26.** $(x - 5)(3x - 2)$ [10.3B]   **27.** The length is 100 yd. The width is 60 yd. [10.5B]   **28.** The distance is 20 ft. [10.5B]   **29.** The width of the frame is 1.5 in. [10.5B]   **30.** The length of a side of the original garden plot was 20 ft. [10.5B]

CHAPTER 10 TEST

1. $(b + 6)(a - 3)$ [10.1B]   **2.** $2y^2(y - 8)(y + 1)$ [10.2B]   **3.** $4(x + 4)(2x - 3)$ [10.3B]   **4.** $(2x + 1)(3x + 8)$ [10.3A]   **5.** $(a - 16)(a - 3)$ [10.2A]   **6.** $2x(3x^2 - 4x + 5)$ [10.1A]   **7.** $(x + 5)(x - 3)$ [10.2A]   **8.** $-\frac{1}{2}$, $\frac{1}{2}$ [10.5A]   **9.** $5(x^2 - 9x - 3)$ [10.1A]   **10.** $(p + 6)^2$ [10.4A]   **11.** 3, 5 [10.5A]   **12.** $3(x + 2y)^2$ [10.4B]   **13.** $(b + 4)(b - 4)$ [10.4A]   **14.** $3y^2(2x + 1)(x + 1)$ [10.3B]   **15.** $(p + 3)(p + 2)$ [10.2A]   **16.** $(x - 2)(a + b)$ [10.1B]   **17.** $(p + 1)(x - 1)$ [10.1B]   **18.** $3(a + 5)(a - 5)$ [10.4B]   **19.** Nonfactorable over the integers [10.3A]   **20.** $(x - 12)(x + 3)$ [10.2A]   **21.** $(2a - 3b)^2$ [10.4A]   **22.** $(2x + 7y)(2x - 7y)$ [10.4A]   **23.** $\frac{3}{2}$, -7 [10.5A]   **24.** The two numbers are 7 and 3. [10.5B]   **25.** The length is 15 cm. The width is 6 cm. [10.5B]

CUMULATIVE REVIEW EXERCISES

1. 7 [3.2B]   **2.** 4 [3.5A]   **3.** -7 [4.1A]   **4.** $15x^2$ [4.2B]   **5.** 12 [4.2D]   **6.** $\frac{2}{3}$ [5.1C]   **7.** $\frac{7}{4}$ [5.3A]   **8.** 3 [5.3B]   **9.** 45 [6.4A/6.4B]   **10.** $9a^6b^4$ [9.2B]   **11.** $x^3 - 3x^2 - 6x + 8$ [9.3B]   **12.** $4x + 8 + \frac{21}{2x - 3}$ [9.5B]   **13.** $\frac{y^6}{x^8}$ [9.4A]   **14.** $(a - b)(3 - x)$ [10.1B]   **15.** $5xy^2(3 - 4y^2)$ [10.1A]   **16.** $(x - 7y)(x + 2y)$ [10.2A]   **17.** $(p - 10)(p + 1)$ [10.2A]   **18.** $3a(3a + 2)(2a + 5)$ [10.4B]   **19.** $(6a + 7b)(6a - 7b)$ [10.4A]   **20.** $(2x + 7y)^2$ [10.4A]   **21.** $(3x + 7)(3x - 2)$ [10.3A]   **22.** $2(3x - 4y)^2$ [10.4B]   **23.** $(x - 3)(3y - 2)$ [10.1B]   **24.** $\frac{2}{3}$, -7 [10.5A]   **25.** The shorter piece is 4 ft long. The longer piece is 6 ft long. [5.4B]   **26.** The discount rate is 40%. [5.2B/6.3B]   **27.** $\angle a = 72°$; $\angle b = 108°$ [7.1B]   **28.** The distance to the resort is 168 mi. [5.5B]   **29.** The integers are 10, 12, and 14. [5.4A]   **30.** The length of the base of the triangle is 12 in. [10.5B]

Answers to Chapter 11 Selected Exercises

PREP TEST

1. 36 [2.1A]   **2.** $\frac{3x}{y^3}$ [9.4A]   **3.** $-\frac{5}{36}$ [3.4A]   **4.** $-\frac{10}{11}$ [3.4B]   **5.** No [1.3C]   **6.** $\frac{19}{8}$ [5.2A]   **7.** 130° [7.1B]   **8.** $(x - 6)(x + 2)$ [10.2A]   **9.** $(2x - 3)(x + 1)$ [10.3A]   **10.** 9:40 A.M. [5.5B]

SECTION 11.1

3. $\dfrac{3}{4x}$ **5.** $\dfrac{1}{x+3}$ **7.** -1 **9.** $\dfrac{2}{3y}$ **11.** $-\dfrac{3}{4x}$ **13.** $\dfrac{a}{b}$ **15.** $-\dfrac{2}{x}$ **17.** $\dfrac{y-2}{y-3}$ **19.** $\dfrac{x+5}{x+4}$ **21.** $\dfrac{x+4}{x-3}$

23. $-\dfrac{x+2}{x+5}$ **25.** $\dfrac{2(x+2)}{x+3}$ **27.** $\dfrac{2x-1}{2x+3}$ **29.** $-\dfrac{x+7}{x+6}$ **31.** $\dfrac{35ab^2}{24x^2y}$ **33.** $\dfrac{4x^3y^3}{3a^2}$ **35.** $\dfrac{3}{4}$ **37.** ab^2

39. $\dfrac{x^2(x-1)}{y(x+3)}$ **41.** $\dfrac{y(x-1)}{x^2(x+10)}$ **43.** $-ab^2$ **45.** $\dfrac{x+5}{x+4}$ **47.** 1 **49.** $-\dfrac{n-10}{n-7}$ **51.** $\dfrac{x(x+2)}{2(x-1)}$ **53.** $-\dfrac{x+2}{x-6}$

55. $\dfrac{x+5}{x-12}$ **59.** $\dfrac{7a^3y^2}{40bx}$ **61.** $\dfrac{4}{3}$ **63.** $\dfrac{3a}{2}$ **65.** $\dfrac{x^2(x+4)}{y^2(x+2)}$ **67.** $\dfrac{x(x-2)}{y(x-6)}$ **69.** $-\dfrac{3by}{ax}$ **71.** $\dfrac{(x+6)(x-3)}{(x+7)(x-6)}$

73. 1 **75.** $-\dfrac{x+8}{x-4}$ **77.** $\dfrac{2n+1}{2n-3}$ **81.** $\dfrac{x}{x+8}$ **83.** $\dfrac{n-2}{n+3}$

SECTION 11.2

1. $24x^3y^2$ **3.** $30x^4y^2$ **5.** $8x^2(x+2)$ **7.** $6x^2y(x+4)$ **9.** $36x(x+2)^2$ **11.** $6(x+1)^2$

13. $(x-1)(x+2)(x+3)$ **15.** $(2x+3)^2(x-5)$ **17.** $(x-1)(x-2)$ **19.** $(x+2)(x-3)(x+4)$

21. $(x+1)(x+4)(x-7)$ **23.** $(x+4)(x-6)(x+6)$ **25.** $(x+3)(x-10)(x-8)$ **27.** $(x-3)(3x-2)(x+2)$

29. $(x+2)(x-3)$ **31.** $(x+1)(x-5)$ **33.** $(x-1)(x-2)(x-3)(x-6)$ **35.** $\dfrac{5}{ab^2},\dfrac{6b}{ab^2}$ **37.** $\dfrac{15y^2}{18x^2y},\dfrac{14x}{18x^2y}$

39. $\dfrac{ay+5a}{y^2(y+5)},\dfrac{6y}{y^2(y+5)}$ **41.** $\dfrac{a^2y+7a^2}{y(y+7)^2},\dfrac{ay}{y(y+7)^2}$ **43.** $\dfrac{b}{y(y-4)},-\dfrac{b^2y}{y(y-4)}$ **45.** $-\dfrac{3y-21}{(y-7)^2},\dfrac{2}{(y-7)^2}$ **47.** $\dfrac{2y^2}{y^2(y-3)},$

$\dfrac{3}{y^2(y-3)}$ **49.** $\dfrac{x^3+4x^2}{(2x-1)(x+4)},\dfrac{2x^2+x-1}{(2x-1)(x+4)}$ **51.** $\dfrac{3x^2+15x}{(x+5)(x-5)},\dfrac{4}{(x+5)(x-5)}$ **53.** $\dfrac{x^2-1}{(x-3)(x+5)(x+1)},$

$\dfrac{x^2-3x}{(x-3)(x+5)(x+1)}$ **55.** $\dfrac{800}{10^5},\dfrac{9}{10^5}$ **57.** $\dfrac{x^3-x}{x^2-1},\dfrac{x}{x^2-1}$ **59.** $\dfrac{3c^2-3cd}{3(6c+d)(c+d)(c-d)},$ $\dfrac{6cd+d^2}{3(6c+d)(c+d)(c-d)}$

SECTION 11.3

1. $\dfrac{11}{y^2}$ **3.** $-\dfrac{7}{x+4}$ **5.** $\dfrac{8x}{2x+3}$ **7.** $\dfrac{5x+7}{x-3}$ **9.** $\dfrac{2x-5}{x+9}$ **11.** $\dfrac{-3x-4}{2x+7}$ **13.** $\dfrac{1}{x+5}$ **15.** $\dfrac{1}{x-6}$ **17.** $\dfrac{3}{2y-1}$

19. $\dfrac{1}{x-5}$ **23.** $\dfrac{4y+5x}{xy}$ **25.** $\dfrac{19}{2x}$ **27.** $\dfrac{5}{12x}$ **29.** $\dfrac{19x-12}{6x^2}$ **31.** $\dfrac{52y-35x}{20xy}$ **33.** $\dfrac{13x+2}{15x}$ **35.** $\dfrac{7}{24}$

37. $\dfrac{x+90}{45x}$ **39.** $\dfrac{x^2+2x+2}{2x^2}$ **41.** $\dfrac{2x^2+3x-10}{4x^2}$ **43.** $\dfrac{-x^2-4x+4}{x+4}$ **45.** $\dfrac{4x+7}{x+1}$ **47.** $\dfrac{4x^2+9x+9}{24x^2}$

49. $\dfrac{3x-1-2xy-3y}{xy^2}$ **51.** $\dfrac{20x^2+28x-12xy+9y}{24x^2y^2}$ **53.** $\dfrac{9x^2-3x-2xy-10y}{18xy^2}$ **55.** $\dfrac{7x-23}{(x-3)(x-4)}$

57. $\dfrac{-y-33}{(y+6)(y-3)}$ **59.** $\dfrac{3x^2+20x-8}{(x-4)(x+6)}$ **61.** $\dfrac{3(4x^2+5x-5)}{(x+5)(2x+3)}$ **63.** $\dfrac{-4x+5}{x-6}$ **65.** $\dfrac{2(y+2)}{(y+4)(y-4)}$ **67.** $-\dfrac{4x}{(x+1)^2}$

69. $\dfrac{2x-1}{(1+x)(1-x)}$ **71.** $\dfrac{14}{(x-5)^2}$ **73.** $\dfrac{-2(x+7)}{(x+6)(x-7)}$ **75.** $\dfrac{x-4}{x-6}$ **77.** $\dfrac{2x+1}{x-1}$ **79.** $\dfrac{-3(x^2+8x+25)}{(x-3)(x+7)}$

81. a. $\dfrac{20{,}400}{x}$ dollars **b.** $\dfrac{102{,}000}{x(x+5)}$ dollars **c.** \$136

SECTION 11.4

1. $\dfrac{x}{x-3}$ **3.** $\dfrac{2}{3}$ **5.** $\dfrac{y+3}{y-4}$ **7.** $\dfrac{2(2x+13)}{5x+36}$ **9.** $\dfrac{x+2}{x+3}$ **11.** $\dfrac{x-6}{x+5}$ **13.** $\dfrac{-x+2}{x+1}$ **15.** $x-1$ **17.** $\dfrac{1}{2x-1}$

19. $\dfrac{x-3}{x+5}$ **21.** $\dfrac{x-7}{x-8}$ **23.** $\dfrac{2y-1}{2y+1}$ **25.** $\dfrac{x-2}{2x-5}$ **27.** $\dfrac{-x-1}{4x-3}$ **29.** $\dfrac{x+1}{2(5x-2)}$ **31.** $\dfrac{5}{3}$ **33.** $-\dfrac{1}{x-1}$

35. $\dfrac{y+4}{2(y-2)}$ **37.** $\dfrac{x+1}{x-1}$ **39.** $\dfrac{y^2+x^2}{xy}$

SECTION 11.5

3. 3   **5.** 1   **7.** 9   **9.** 1   **11.** $\dfrac{1}{4}$ **13.** 1   **15.** -3 **17.** $\dfrac{1}{2}$ **19.** 8   **21.** 5   **23.** -1 **25.** 5

27. No solution   **29.** 4, 2   **31.** $-\dfrac{3}{2}$, 4   **33.** 3   **35.** 4   **37.** 0   **39.** $-\dfrac{2}{5}$ **41.** 0, $-\dfrac{2}{3}$

SECTION 11.6

1. $y = -3x + 10$ **3.** $y = 4x - 3$ **5.** $y = -\dfrac{3}{2}x + 3$ **7.** $y = \dfrac{2}{5}x - 2$ **9.** $y = -\dfrac{2}{7}x + 2$ **11.** $y = -\dfrac{1}{3}x + 2$

13. $y = 3x + 8$ **15.** $y = -\dfrac{2}{3}x - 3$ **17.** $x = -6y + 10$ **19.** $x = \dfrac{1}{2}y + 3$ **21.** $x = -\dfrac{3}{4}y + 3$ **23.** $x = 4y + 3$

25. $t = \dfrac{d}{r}$ **27.** $T = \dfrac{PV}{nR}$ **29.** $l = \dfrac{P - 2w}{2}$ **31.** $b_1 = \dfrac{2A - hb_2}{h}$ **33.** $h = \dfrac{3V}{A}$ **35.** $S = C - Rt$ **37.** $P = \dfrac{A}{1 + rt}$

39. $w = \dfrac{A}{S + 1}$ **41. a.** $S = \dfrac{F + BV}{B}$ **b.** The required selling price is \$180.   **c.** The required selling price is \$75.

SECTION 11.7

3. It will take 2 h to fill the fountain with both sprinklers working.   **5.** With both skiploaders working together, it would take 3 h to remove the earth.   **7.** With both computers working, it would take 30 h to solve the problem.   **9.** It would take 30 min to cool the room with both air conditioners working.   **11.** It would take the second pipeline 90 min to fill the tank.   **13.** It would take the apprentice 15 h to construct the wall.   **15.** It will take the second technician 3 h to complete the wiring.   **17.** It would have taken one of the welders 40 h to complete the welds.   **19.** It would have taken one machine 28 h to fill the boxes.   **21.** The jogger ran 16 mi in 2 h.   **23.** The rate of travel in the congested area was 20 mph.   **25.** The rate of the jogger was 8 mph. The rate of the cyclist was 20 mph.   **27.** The rate of the jet is 360 mph.   **29.** Camille's walking rate is 4 mph.   **31.** The rate of the car is 48 mph.   **33.** The rate of the wind is 20 mph.   **35.** The rate of the gulf current is 6 mph.   **37.** The rate of the trucker for the first 330 mi was 55 mph.   **39.** The bus usually travels 60 mph.

CHAPTER 11 REVIEW EXERCISES

1. $\dfrac{b^3 y}{10ax}$ [11.1C]   **2.** $\dfrac{7x + 22}{60x}$ [11.3B]   **3.** $\dfrac{2xy}{5}$ [11.1B]   **4.** $\dfrac{2xy}{3(x + y)}$ [11.1C]   **5.** $\dfrac{x - 2}{3x - 10}$ [11.4A]

6. $-\dfrac{x + 6}{x + 3}$ [11.1A]   **7.** $\dfrac{2x^4}{3y^7}$ [11.1A]   **8.** 62 [11.5A]   **9.** $\dfrac{(3y - 2)^2}{(y - 1)(y - 2)}$ [11.1C]   **10.** $x = \dfrac{5}{3a - 1}$ [11.6A]

11. 8 [11.5A]   **12.** $\dfrac{x^2 + 3y}{xy}$ [11.3B]   **13.** $y = -\dfrac{5}{4}x + 5$ [11.6A]   **14.** $\dfrac{by^3}{6ax^2}$ [11.1B]   **15.** $\dfrac{x}{x - 7}$ [11.4A]

16. $\dfrac{3x^2 - x}{(6x - 1)(2x + 3)(3x - 1)}, \dfrac{24x^3 - 4x^2}{(6x - 1)(2x + 3)(3x - 1)}$ [11.2B]   **17.** $a = \dfrac{T - 2bc}{2b + 2c}$ [11.6A]   **18.** 2 [11.5A]

19. $\dfrac{2x + 1}{3x - 2}$ [11.4A]   **20.** $\dfrac{x^2 + 5}{(x - 5)(x - 2)}$ [11.3B]   **21.** $c = \dfrac{100m}{i}$ [11.6A]   **22.** No solution [11.5A]

23. $\dfrac{1}{x^2}$ [11.1C]   **24.** $\dfrac{2y - 3}{5y - 7}$ [11.3B]   **25.** $\dfrac{1}{x + 3}$ [11.3A]   **26.** $(5x - 3)(2x - 1)(4x - 1)$ [11.2A]

27. $y = -\dfrac{4}{9}x + 2$ [11.6A]   **28.** $\dfrac{2x + 1}{x + 2}$ [11.1B]   **29.** 5 [11.5A]   **30.** $\dfrac{3x - 1}{x - 5}$ [11.3B]   **31.** 10 [11.5A]

32. 12 [11.5A]   **33.** It would take 6 h to fill the pool. [11.7A]   **34.** The rate of the car is 45 mph. [11.7B]

35. The rate of the wind is 20 mph. [11.7B]

CHAPTER 11 TEST

1. $\dfrac{x^2 - 4x + 5}{(x + 3)(x - 2)}$ [11.3B]   **2.** -1 [11.5A]   **3.** $\dfrac{(2x - 1)(x - 5)}{(x + 3)(2x + 5)}$ [11.1B]   **4.** $\dfrac{2x^3}{3y^3}$ [11.1A]   **5.** $t = \dfrac{d - s}{r}$ [11.6A]

6. 2 [11.5A]   **7.** $-\dfrac{x + 5}{x + 1}$ [11.1A]   **8.** $3(2x - 1)(x + 1)$ [11.2A]   **9.** $\dfrac{5}{(2x - 1)(3x + 1)}$ [11.3B]   **10.** $\dfrac{x + 5}{x + 4}$ [11.1C]

11. $\dfrac{x-3}{x-2}$ [11.4A] **12.** $\dfrac{3x+6}{x(x-2)(x+2)}, \dfrac{x^2}{x(x-2)(x+2)}$ [11.2B] **13.** $\dfrac{2}{x+5}$ [11.3A] **14.** $y=\dfrac{3}{8}x-2$ [11.6A]

15. No solution [11.5A] **16.** $\dfrac{x+1}{x^3(x-2)}$ [11.1B] **17.** $\dfrac{6b^4x}{ay^3}$ [11.1C] **18.** $\dfrac{1}{x^2y}$ [11.3A] **19.** It will take 3 h until

the run can be opened. [11.7A] **20.** It would take 4 h to fill the pool. [11.7A] **21.** The rate of the wind is 20 mph.

[11.7B] **22.** The rate of the current is 5 mph. [11.7B]

CUMULATIVE REVIEW EXERCISES

1. $\dfrac{31}{30}$ [2.7A] **2.** 21 [4.1A] **3.** $5x-2y$ [4.2A] **4.** $-8x+26$ [4.2D] **5.** $-\dfrac{9}{2}$ [5.2A] **6.** -12 [5.3B]

7. 10 [6.4A/6.4B] **8.** a^3b^7 [9.2A] **9.** $a^2+ab-12b^2$ [9.3C] **10.** $3b^3-b+2$ [9.5A] **11.** x^2+2x+4 [9.5B]

12. $(3x-1)(4x+1)$ [10.3A] **13.** $(y-6)(y-1)$ [10.2A] **14.** $a(a+5)(2a-3)$ [10.3A]

15. $4(b+5)(b-5)$ [10.4B] **16.** $-3, \dfrac{5}{2}$ [10.5A] **17.** $\dfrac{2x^3}{3y^5}$ [11.1A] **18.** $-\dfrac{x-2}{x+5}$ [11.1A] **19.** 1 [11.1C]

20. $\dfrac{3}{(2x-1)(x+1)}$ [11.3B] **21.** $\dfrac{x+3}{x+5}$ [11.4A] **22.** 4 [11.5A] **23.** 3 [11.5A] **24.** $t=\dfrac{f-v}{a}$ [11.6A]

25. $5x-13=-8; x=1$ [5.4A] **26.** The school-age population is 50 million people. [6.4C] **27.** The base is 10 in.

The height is 6 in. [10.5B] **28.** The cost of a \$5000 policy is \$80. [6.2B] **29.** It would take both pipes 6 min to fill

the tank. [11.7A] **30.** The rate of the current is 2 mph. [11.7B]

Answers to Chapter 12 Selected Exercises

PREP TEST

1. 3 [3.5A] **2.** -1 [4.1A] **3.** $-3x+12$ [4.2C] **4.** -2 [5.2A] **5.** $x=5$ [5.2A] **6.** $y=-2$ [5.2A]

7. $-4x+5$ [9.5A] **8.** 4 [6.2A] **9.** $y=\dfrac{3}{5}x-3$ [11.6A] **10.** $y=-\dfrac{1}{2}x-5$ [11.6A]

SECTION 12.1

1. 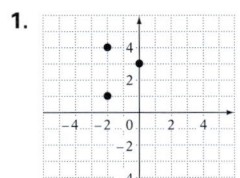 **3.** 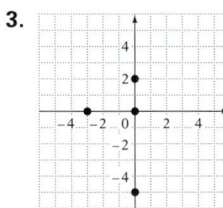 **5.** 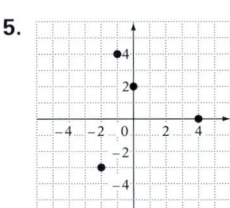 **7.** $A(2, 3), B(4, 0), C(-4, 1), D(-2, -2)$

9. $A(-2, 5), B(3, 4), C(0, 0), D(-3, -2)$ **11. a.** $2, -4$ **b.** $1, -3$ **15.** Yes **17.** No **19.** No **21.** No

23.  **25.** 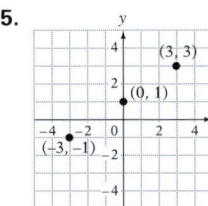 **27.**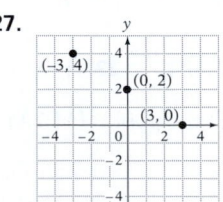

29. $\{(24, 600), (32, 750), (22, 430), (15, 300), (4.4, 68), (17, 370), (15, 310), (4.4, 55)\}$; No

31. $\{(390, 0.115), (591, 0.073), (517, 0.077), (576, 0.068), (605, 0.064)\}$; Yes **33.** Yes **35.** No **37.** Yes **39.** 8

41. 9 **43.** 2 **45.** -1 **47.** 22 **49.** $-\dfrac{3}{2}$ **51.** -7

SECTION 12.2

1.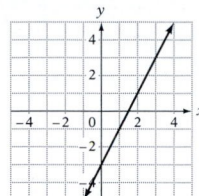

3.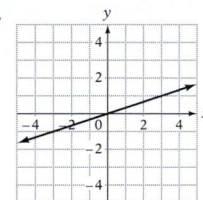

5.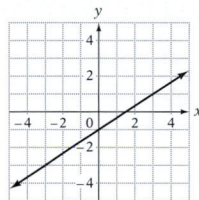

7.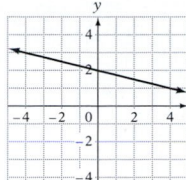

9.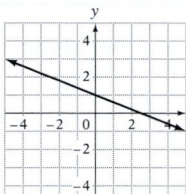

11.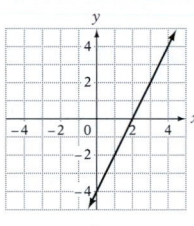

13.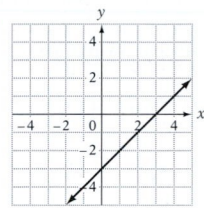

15.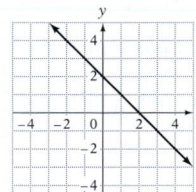

17.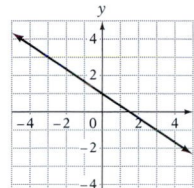

19.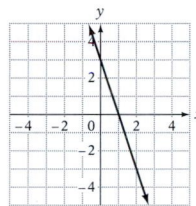

21.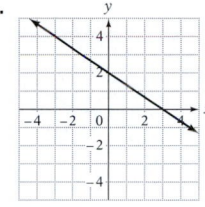

23.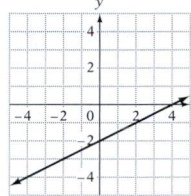

25.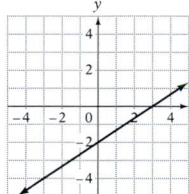

27.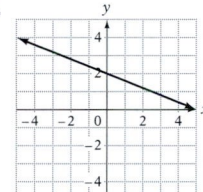

29.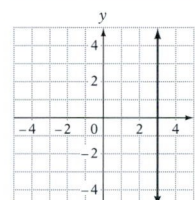

31.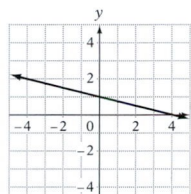

33.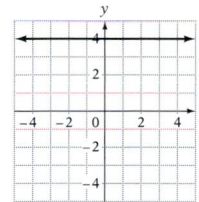

35.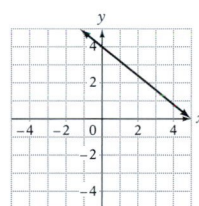

37. After flying for 3 min, the helicopter is 3.5 mi away from the victims.

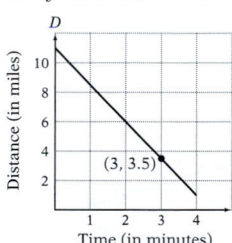
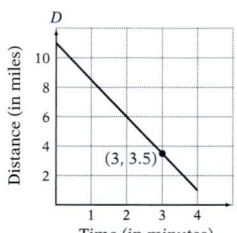

39. A dog 6 years old is equivalent in age to a human 40 years old.

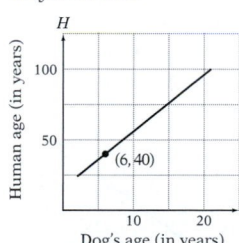

43. Increases; 3; 3   **45. a.** The cost is $.99.   **b.** The cost is $1.74.

SECTION 12.3

1. $(3, 0), (0, -3)$ **3.** $(2, 0), (0, -6)$ **5.** $(10, 0), (0, -2)$ **7.** $(-4, 0), (0, 12)$ **9.** $(0, 0), (0, 0)$ **11.** $(6, 0), (0, 3)$

13. 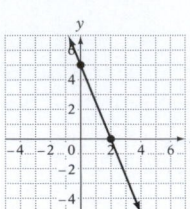 **15.** 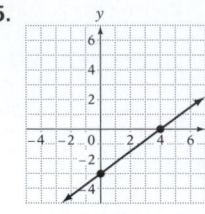 **17.** 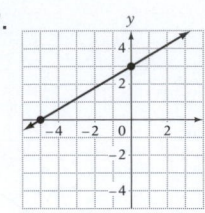 **21.** -2 **23.** $\dfrac{1}{3}$ **25.** $-\dfrac{5}{2}$

27. Undefined **29.** 0 **31.** $-\dfrac{1}{3}$ **33.** Neither **35.** Neither **37.** Parallel **39.** Neither **41.** $m = 33$. The worldwide sales of camera-phones are increasing by 33 million units per year. **43.** $m = -180$. The value of the car is decreasing \$180 for each additional 1000 mi the car is driven. **45.** $m = \dfrac{2}{3}$; $(0, -2)$ **47.** $m = -\dfrac{2}{5}$; $(0, 2)$

49. $m = \dfrac{1}{4}$; $(0, 0)$ **51.** 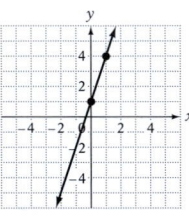 **53.** 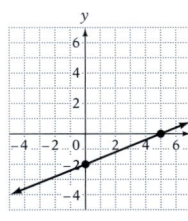 **55.**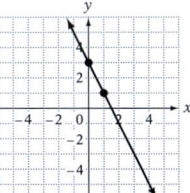

57. 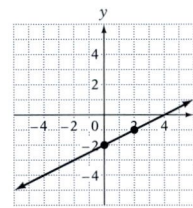 **59.** 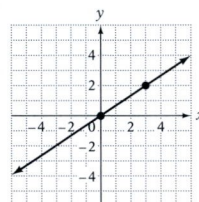 **61.** 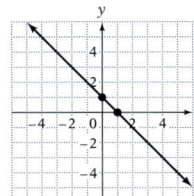 **63.**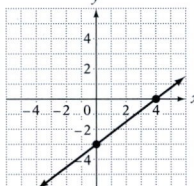

65. 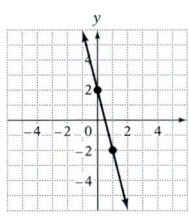 **67.** Yes

SECTION 12.4

3. $y = 2x + 2$ **5.** $y = -3x - 1$ **7.** $y = \dfrac{1}{3}x$ **9.** $y = \dfrac{3}{4}x - 5$ **11.** $y = -\dfrac{3}{5}x$ **13.** $y = \dfrac{1}{4}x + \dfrac{5}{2}$

15. $y = 2x - 3$ **17.** $y = -2x - 3$ **19.** $y = \dfrac{2}{3}x$ **21.** $y = \dfrac{1}{2}x + 2$ **23.** $y = -\dfrac{3}{4}x - 2$ **25.** $y = \dfrac{3}{4}x + \dfrac{5}{2}$

27. The tennis player is using 1.55 g of carbohydrates/min.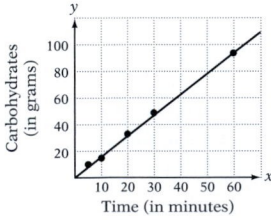

29. The percent of music purchased in stores is decreasing 3% per year.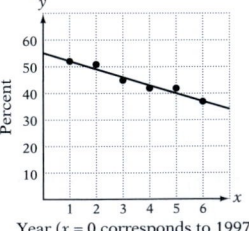

31. No **33.** Yes **35.** $-\dfrac{3}{2}$ **37.** -5 **39.** $y = -\dfrac{2}{3}x + \dfrac{5}{3}$

CHAPTER 12 REVIEW EXERCISES

1. a.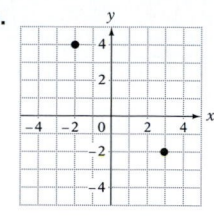

2.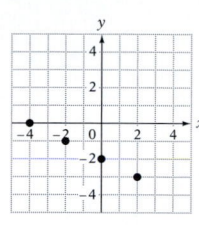

[12.1B]

3. $y = -\dfrac{8}{3}x + \dfrac{1}{3}$ [12.4B]   **4.** $y = -\dfrac{5}{2}x + 16$ [12.4A]

b. -2

c. -4 [12.1A]

5.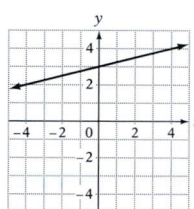

[12.2A]

6.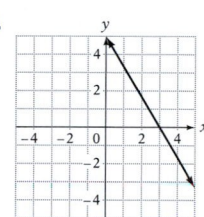

[12.2B]

7. Neither [12.3B]   **8.** -1 [12.1D]

9. $y = -\dfrac{2}{3}x + \dfrac{11}{3}$ [12.4B]   **10.** Yes [12.1C]   **11.** $\dfrac{7}{11}$ [12.3B]   **12.** $(8, 0), (0, -12)$ [12.3A]   **13.** 0 [12.3B]

14.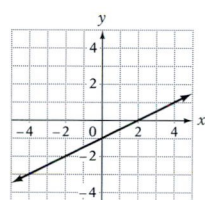

[12.3C]

15.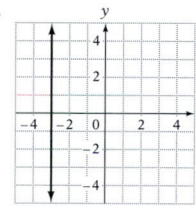

[12.2B]

16.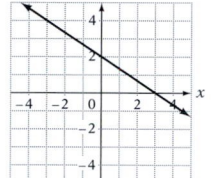

[12.3C]

17.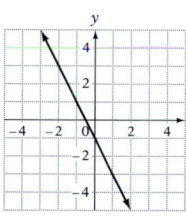

[12.2A]

18.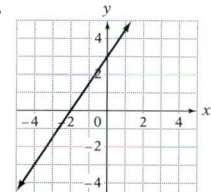

[12.3C]

19.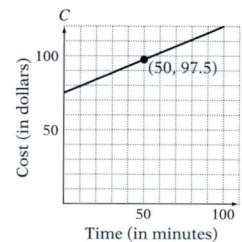

[12.2B]

20. $\{(55, 95), (57, 101), (53, 94), (57, 98), (60, 100), (61, 105), (58, 97), (54, 95)\}$; No [12.1C]

21. The cost of 50 min of access time for 1 month is $97.50.

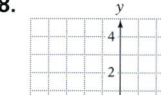

[12.2C]

22. The average annual telephone bill for a family is increasing by $34 per year.

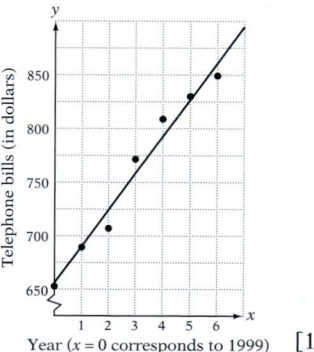

[12.4C]

CHAPTER 12 TEST

1. $(3, -3)$ [12.1B]   **2.**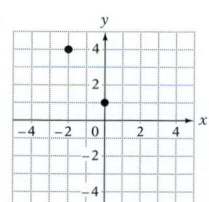

[12.1B]

3. Yes [12.1C]   **4.** 6 [12.1D]   **5.** 3 [12.1D]

6. {(3.5, 25), (4.0, 30), (5.2, 45), (5.0, 38), (4.0, 42), (6.3, 12), (5.4, 34)}; No [12.1C]   **7.**

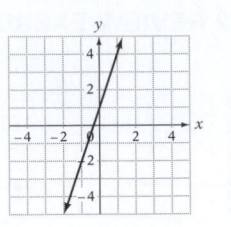

[12.2A]

8. 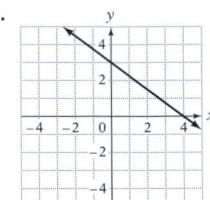 [12.2A]   **9.** 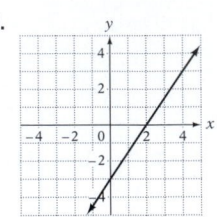 [12.2B]   **10.** 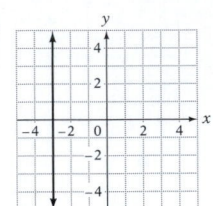 [12.2B]

11. 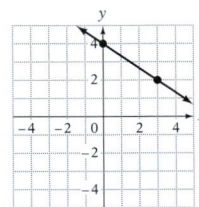 [12.3C]   **12.** 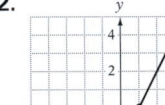 [12.3C]   **13.** After 1 s, the speed of the ball is 96 ft/s. [12.2C]

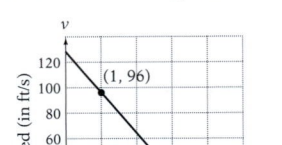

14. $m = 0.46$. The average hourly wage is increasing by \$.46 per year. [12.3B]

15. 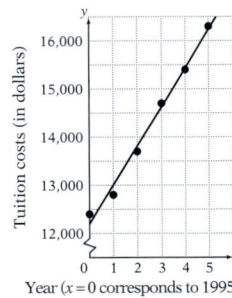   The average annual tuition for a private 4-year college is increasing \$809 per year. [12.4C]

16. $(2, 0)$, $(0, -3)$ [12.3A]   **17.** $(-2, 0)$, $(0, 1)$ [12.3A]   **18.** 2 [12.3B]   **19.** Parallel [12.3B]

20. Undefined [12.3B]   **21.** $-\dfrac{2}{3}$ [12.3B]   **22.** $y = 3x - 1$ [12.4A]   **23.** $y = \dfrac{2}{3}x + 3$ [12.4A]

24. $y = -\dfrac{5}{8}x - \dfrac{7}{8}$ [12.4B]   **25.** $y = -\dfrac{2}{7}x - \dfrac{4}{7}$ [12.4B]

CUMULATIVE REVIEW EXERCISES

1. -12 [3.5A]   **2.** $-\dfrac{5}{8}$ [4.1A]   **3.** $f(-2) = -\dfrac{2}{3}$ [12.1D]   **4.** $\dfrac{3}{2}$ [5.2A]   **5.** $\dfrac{19}{18}$ [5.3B]   **6.** $\dfrac{1}{15}$ [6.3A]

7. $-32x^8y^7$ [9.2B]   **8.** $-3x^2$ [9.4A]   **9.** $x + 3$ [9.5B]   **10.** $5(x + 2)(x + 1)$ [10.2B]   **11.** $(a + 2)(x + y)$ [10.1A]

12. 4 and -2 [10.5A]   **13.** $\dfrac{x^3(x + 3)}{y(x + 2)}$ [11.1B]   **14.** $\dfrac{3}{x + 8}$ [11.3A]   **15.** 2 [11.5A]   **16.** $y = \dfrac{4}{5}x - 3$ [11.6A]

17. $(-2, -5)$ [12.1B]   **18.** 0 [12.3B]   **19.** $y = \dfrac{1}{2}x - 2$ [12.4A]   **20.** $y = -3x + 2$ [12.4A]

21. $y = 2x + 2$ [12.4A]   **22.** $y = \dfrac{2}{3}x - 3$ [12.4A]   **23.** The probability is $\dfrac{2}{3}$. [8.3A]   **24.** The angles measure 46°,

43°, and 91° [7.1C]   **25.** The value of the home is \$110,000. [6.2B]   **26.** It would take $3\dfrac{3}{4}$ h for both, working

together, to wire the garage. [11.7A]

27. 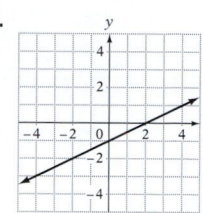 [12.2A]

28. 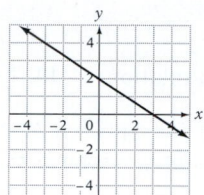 [12.3C]

Answers to Chapter 13 Selected Exercises

PREP TEST

1. $y = \frac{3}{4}x - 6$ [11.6A]   **2.** 1000 [5.3B]   **3.** $33y$ [4.2D]   **4.** $10x - 10$ [4.2D]   **5.** Yes [12.1B]

6. $(4, 0), (0, -3)$ [12.3A]   **7.** Yes [12.3B]   **8.** [12.3C]

9. The hikers will be side-by-side 1.5 h after the second hiker starts. [5.5B]

SECTION 13.1

1. I: c; II: a; III: b   **3.** $(2, -1)$ **5.** The ordered-pair solutions of $y = -\frac{3}{2}x + 1$ **7.** No solution   **9.** $(-2, 4)$

11. Yes   **13.** No   **15.** Yes   **17.** Yes

19. 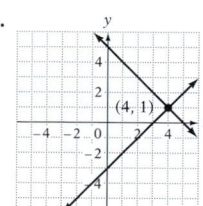   **21.** 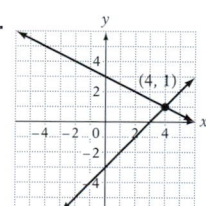   **23.** 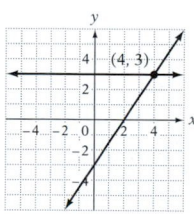   **25.**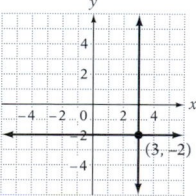

27. 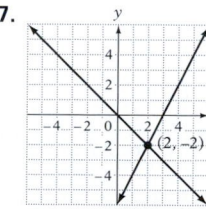   **29.** 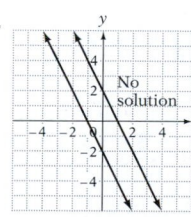   **31.** 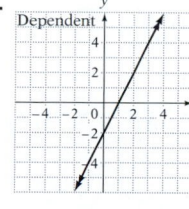 The ordered-pair solutions of $y = 2x - 2$

33. 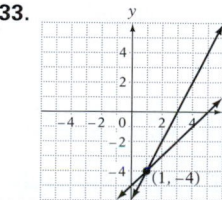   **35.** 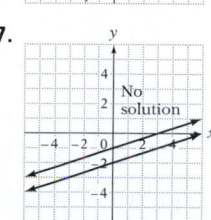   **37.**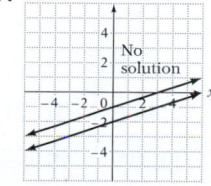

39. a. Sometimes true   **b.** Always true   **c.** Never true   **d.** Always true

43. Answers will vary.

SECTION 13.2

3. $(2, 1)$ **5.** $(4, 1)$ **7.** $(-1, 1)$ **9.** No solution   **11.** No solution   **13.** $\left(-\dfrac{3}{4}, -\dfrac{3}{4}\right)$ **15.** $(1, 1)$ **17.** $(2, 0)$

19. $(1, -2)$ **21.** $(0, 0)$ **23.** Dependent. The solutions satisfy the equation $2x - y = 2$.   **25.** $(-4, -2)$ **27.** $(10, 31)$

29. $(3, -10)$ **31.** $(-22, -5)$ **33.** The amounts invested should be $1900 at 5% and $1600 at 7.5%.

35. The amounts invested were $2400 at 9% and $3600 at 6%.   **37.** The amounts invested should be $4400 at 8% and

$1600 at 11%.   **39.** The amount invested at 6.5% was $21,000.   **41.** The amounts invested were $12,000 at 8% and

$8000 at 7%.   **43.** The amount invested in the trust deed was $3750.   **45.** The gas dryer becomes more economical

after 185 loads of clothes.   **47.** 1   **49.** The assertion is not correct. The system of equations is independent. The

solution is $(0, 2)$.   **51.** The research consultant's investment is $45,000.   **53.** Simple interest: $400; compounded

monthly: $415.00; compounded daily: $416.39

SECTION 13.3

1. $(5, -1)$ **3.** $(1, 3)$ **5.** $(1, 1)$ **7.** $(3, -2)$ **9.** Dependent. The solutions satisfy the equation $2x - y = 1$.

11. $(3, 1)$ **13.** Dependent. The solutions satisfy the equation $2x - 3y = 1$.   **15.** $\left(\dfrac{2}{3}, \dfrac{1}{2}\right)$ **17.** $(2, 0)$ **19.** $(0, 0)$

21. $(5, -2)$ **23.** $\left(\dfrac{32}{19}, -\dfrac{9}{19}\right)$ **25.** $\left(\dfrac{7}{4}, -\dfrac{5}{16}\right)$ **27.** $(1, -1)$ **29.** No solution   **31.** $(3, 1)$ **33.** $(-1, 2)$

35. $(1, 1)$ **39.** $A = 3; B = -1$ **41. a.** 1   **b.** $\dfrac{3}{2}$ **c.** 4

SECTION 13.4

1. The rate of the whale in calm water was 35 mph. The rate of the current was 5 mph.   **3.** The rowing rate in calm

water was 14 km/h. The rate of the current was 6 km/h.   **5.** The rate of the Learjet was 525 mph. The rate of the wind

was 35 mph.   **7.** The rate of the helicopter in calm air was 225 mph. The rate of the wind was 45 mph.   **9.** The rate

of the canoeist in calm water was 6 mph. The rate of the current was 1 mph.   **11.** The cost per pound of the wheat flour

was $.65. The cost per pound of the rye flour was $.70.   **13.** The cost per foot of redwood was $3.50. The cost per foot of

pine was $1.25.   **15.** The cost per pound of cinnamon tea was $3. The cost per pound of spice tea was $3.50.

17. 1 nickel and 2 dimes or 3 nickels and 1 dime are in the bank.   **19.** 12.5 acres of good land and 87.5 acres of bad land

were bought.

CHAPTER 13 REVIEW EXERCISES

1. Yes   [13.1A]   **2.** No   [13.1A]   **3.**

[13.1A]

4.

The solutions are the ordered-pair solutions of $y = 2x - 4$.

[13.1A]

5. 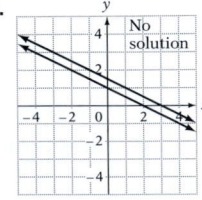   [13.1A]   **6.** $(-1, 1)$ [13.2A]   **7.** $(1, 6)$ [13.2A]   **8.** $(-3, 1)$ [13.3A]   **9.** $\left(-\dfrac{5}{6}, \dfrac{1}{2}\right)$ [13.3A]

10. No solution   [13.2A]   **11.** $(1, 6)$ [13.2A]   **12.** $(1, -5)$ [13.3A]   **13.** No solution   [13.3A]   **14.** Dependent.

The solutions satisfy the equation $y = -\dfrac{4}{3}x + 4$.   [13.2A]   **15.** $(-1, -3)$ [13.2A]   **16.** Dependent. The solutions satisfy

the equation $3x + y = -2$.   [13.3A]   **17.** $\left(\dfrac{2}{3}, -\dfrac{1}{6}\right)$ [13.3A]   **18.** The rate of the sculling team in calm water was

9 mph. The rate of the current was 3 mph. [13.4A]   **19.** 1300 $6 shares were purchased, and 200 $25 shares were purchased. [13.4B]   **20.** The rate of the flight crew in calm air was 125 km/h. The rate of the wind was 15 km/h. [13.4A]   **21.** The rate of the plane in calm air was 105 mph. The rate of the wind was 15 mph. [13.4A]   **22.** The number of ads requiring $.25 postage was 130. The number of ads requiring $.45 postage was 60. [13.4B]   **23.** The amounts invested are $7000 at 7% and $5000 at 8.5%. [13.2B]   **24.** There were originally 350 bushels of lentils and 200 bushels of corn in the silo. [13.4B]   **25.** The amounts invested were $165,000 at 5.4% and $135,000 at 6.6%. [13.2B]

CHAPTER 13 TEST

1. Yes [13.1A]   **2.** Yes [13.1A]   **3.**

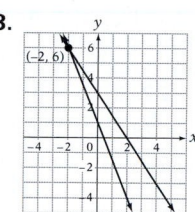

[13.1A]   **4.** (3, 1) [13.2A]   **5.** (1, −1) [13.2A]

6. (2, −1) [13.2A]   **7.** $\left(\dfrac{22}{7}, -\dfrac{5}{7}\right)$ [13.2A]   **8.** No solution [13.2A]   **9.** (2, 1) [13.3A]   **10.** $\left(\dfrac{1}{2}, -1\right)$ [13.3A]   **11.** Dependent. The solutions satisfy the equation $x + 2y = 8$. [13.3A]   **12.** (2, −1) [13.3A]   **13.** (1, −2) [13.3A]   **14.** The rate of the plane in calm air was 100 mph. The rate of the wind was 20 mph. [13.4A]   **15.** The price of a reserved-seat ticket was $10. The price of a general-admission ticket was $6. [13.4B]   **16.** The amounts invested were $15,200 at 6.4% and $12,800 at 7.6%. [13.2B]

CUMULATIVE REVIEW EXERCISES

1. $\dfrac{3}{2}$ [4.1A]   **2.** $-\dfrac{3}{2}$ [5.1C]   **3.** 7 [2.1D]   **4.** $-6a^3 + 13a^2 - 9a + 2$ [9.3B]   **5.** $-2x^5y^2$ [9.4A]

6. $2b - 1 + \dfrac{1}{2b - 3}$ [9.5B]   **7.** $-\dfrac{4y}{x^3}$ [9.4A]   **8.** $4y^2(xy - 4)(xy + 4)$ [10.4B]   **9.** 4, −1 [10.5A]

10. $x - 2$ [11.1C]   **11.** $\dfrac{x^2 + 2}{(x + 2)(x - 1)}$ [11.3B]   **12.** $\dfrac{x - 3}{x + 1}$ [11.4A]   **13.** $-\dfrac{1}{5}$ [11.5A]   **14.** $r = \dfrac{A - P}{Pt}$ [11.6A]

15. x-intercept: (6, 0); y-intercept: (0, −4) [12.3A]   **16.** $-\dfrac{7}{5}$ [12.3B]   **17.** $y = -\dfrac{3}{2}x$ [12.4A]

18. Yes [13.1A]   **19.** (−6, 1) [13.2A]   **20.** (4, −3) [13.3A]   **21.** The amounts invested should be $3750 at 9.6% and $5000 at 7.2%. [13.2B]   **22.** The rate of the passenger train is 56 mph. The rate of the freight train is 48 mph. [5.5B]   **23.** The side of the original square is 8 in. [10.5B]   **24.** The rate of the wind is 30 mph. [13.4A]

25.

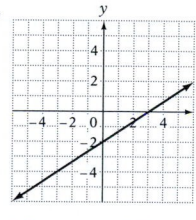

[12.2B]

26.

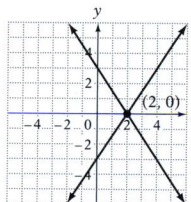

[13.1A]

27. The rate of the motorboat in calm water is 14 mph. [13.4A]   **28.** Registered voters comprise 76% of the voting-age population. [6.4C]

Answers to Chapter 14 Selected Exercises

PREP TEST

1. < [3.1A]   **2.** $-7x + 15$ [4.2D]   **3.** The same number can be added to each side of an equation without changing the solution of the equation. [5.1B]   **4.** Each side of an equation can be multiplied by the same nonzero number without changing the solution of the equation. [5.1C]   **5.** There are 0.45 lb of fat in 3 lb of this grade of hamburger. [6.4C]

6. $-\dfrac{1}{2}$ [5.2A] **7.** $-\dfrac{8}{3}$ [5.2A] **8.** 2 [5.3B] **9.**

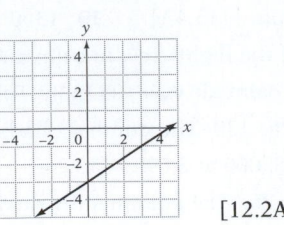

[12.2A]

10.

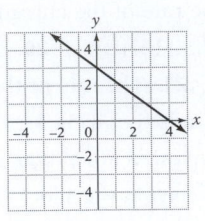

[12.2B]

SECTION 14.1

3. $A = \{16, 17, 18, 19, 20, 21\}$ **5.** $A = \{9, 11, 13, 15, 17\}$ **7.** $A = \{b, c\}$ **9.** $A \cup B = \{3, 4, 5, 6\}$
11. $A \cup B = \{-10, -9, -8, 8, 9, 10\}$ **13.** $A \cup B = \{a, b, c, d, e, f\}$ **15.** $A \cup B = \{1, 3, 7, 9, 11, 13\}$
17. $A \cap B = \{4, 5\}$ **19.** $A \cap B = \varnothing$ **21.** $A \cap B = \{c, d, e\}$ **23.** $\{x \mid x > -5, x \in$ negative integers$\}$
25. $\{x \mid x > 30, x \in$ integers$\}$ **27.** $\{x \mid x > 5, x \in$ even integers$\}$ **29.** $\{x \mid x > 8, x \in$ real numbers$\}$
31. ⟨+++++++(+++→ $-5\,-4\,-3\,-2\,-1\ 0\ 1\ 2\ 3\ 4\ 5$ **33.** ⟨+++++++]+++→ $-5\,-4\,-3\,-2\,-1\ 0\ 1\ 2\ 3\ 4\ 5$ **35.** ⟨+++++)+++→ $-5\,-4\,-3\,-2\,-1\ 0\ 1\ 2\ 3\ 4\ 5$
37. ⟨+++(++++++→ $-5\,-4\,-3\,-2\,-1\ 0\ 1\ 2\ 3\ 4\ 5$ **39.** ⟨+++]+++++→ $-5\,-4\,-3\,-2\,-1\ 0\ 1\ 2\ 3\ 4\ 5$ **41. a.** Never true **b.** Always true
c. Always true **43. a.** Yes **b.** Yes

SECTION 14.2

1. $x < 2$ ⟨++++++)+++→ $-5\,-4\,-3\,-2\,-1\ 0\ 1\ 2\ 3\ 4\ 5$ **3.** $x > 3$ ⟨++++++++(+→ $-5\,-4\,-3\,-2\,-1\ 0\ 1\ 2\ 3\ 4\ 5$
5. $n \geq 3$ ⟨++++++++[++→ $-5\,-4\,-3\,-2\,-1\ 0\ 1\ 2\ 3\ 4\ 5$ **7.** $x \leq -4$ ⟨+]+++++++→ $-5\,-4\,-3\,-2\,-1\ 0\ 1\ 2\ 3\ 4\ 5$ **9.** $x < 1$ **11.** $x \leq -3$
13. $y \geq -9$ **15.** $x < 12$ **17.** $x \geq 5$ **19.** $x < -11$ **21.** $x \leq 10$ **23.** $x \geq -6$ **25.** $x > 2$ **27.** $d < -\dfrac{1}{6}$
29. $x \geq -\dfrac{31}{24}$ **31.** $x < \dfrac{5}{8}$ **33.** $x < \dfrac{5}{4}$ **35.** $x > \dfrac{5}{24}$ **37.** $x < -3.8$ **39.** $x \leq -1.2$ **41.** $x < 5.6$
43. ⟨+++++++)++→ $-5\,-4\,-3\,-2\,-1\ 0\ 1\ 2\ 3\ 4\ 5$ $x < 4$ **45.** ⟨++++++[+++→ $-5\,-4\,-3\,-2\,-1\ 0\ 1\ 2\ 3\ 4\ 5$ $y \geq 3$
47. ⟨++++++]++++→ $-5\,-4\,-3\,-2\,-1\ 0\ 1\ 2\ 3\ 4\ 5$ $x \leq 1$ **49.** ⟨++++)++++++→ $-5\,-4\,-3\,-2\,-1\ 0\ 1\ 2\ 3\ 4\ 5$ $x < -1$
51. ⟨+)++++++++→ $-5\,-4\,-3\,-2\,-1\ 0\ 1\ 2\ 3\ 4\ 5$ $b < -4$ **53.** $y \leq 0$ **55.** $x > \dfrac{2}{7}$ **57.** $x \leq -\dfrac{5}{2}$ **59.** $x < 16$ **61.** $x \geq 16$
63. $x \geq -14$ **65.** $x \leq 21$ **67.** $x > 0$ **69.** $x \leq -\dfrac{12}{7}$ **71.** $x > \dfrac{2}{3}$ **73.** $x \leq \dfrac{2}{3}$ **75.** $x \leq 2.3$ **77.** $x < -3.2$
79. $x \leq 5$ **81.** The team must win 11 or more games to be eligible for the tournament. **83.** The service organization
must collect more than 440 lb of aluminum cans on the fourth drive to collect the bonus. **85.** The person needs 50 or
more milligrams of additional vitamin C to satisfy the recommended daily allowance. **87.** No, the student cannot earn
an A grade. **89.** $\{c \mid c > 0\}$ **91.** $\{c \mid c \in$ real numbers$\}$ **93.** $\{c \mid c > 0\}$

SECTION 14.3

1. $x < 4$ **3.** $x < -4$ **5.** $x \geq 1$ **7.** $x < 5$ **9.** $x < 0$ **11.** $x < 20$ **13.** $y \leq \dfrac{5}{2}$ **15.** $x < \dfrac{25}{11}$ **17.** $n \leq \dfrac{11}{18}$
19. $x \geq 6$ **21.** In 1 month, the agent expects to make sales totaling $20,000 or less. **23.** A person must use more
than 60 min to exceed $10. **25.** The amount of artificial flavors that can be added is less than or equal to 8 oz.
27. The distance to the ski resort must be greater than 38 mi. **29.** $\{1, 2\}$ **31.** $\{3, 4, 5\}$ **33.** $\{x \mid x \in$ real numbers$\}$

SECTION 14.4

1.

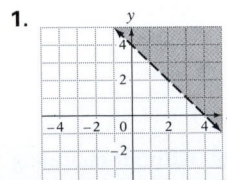

3.

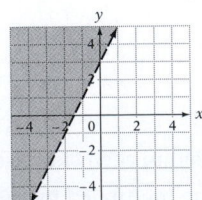

5.

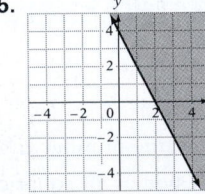

7.

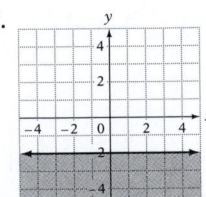

9. 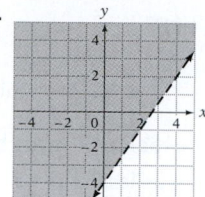 **11.** 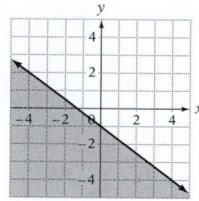 **13.** 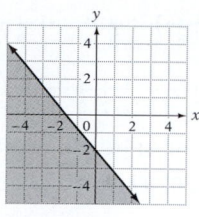 **15.**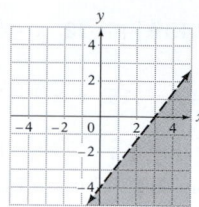

17. 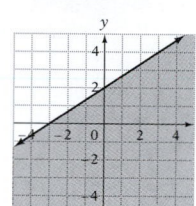 **19.** 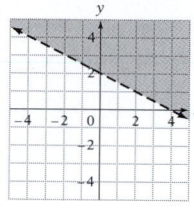 **21.** 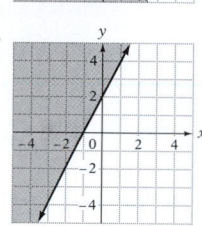 **23.** $x \leq 3$

CHAPTER 14 REVIEW EXERCISES

1. $x > 18$ [14.2A]   **2.** $A \cap B = \varnothing$ [14.1A]   **3.** $\{x \mid x > -8, x \in$ odd integers$\}$ [14.1B]

4. $A \cup B = \{2, 4, 6, 8, 10\}$ [14.1A]   **5.** $A = \{1, 3, 5, 7\}$ [14.1A]   **6.** $x \geq 4$ [14.3A]

7.   [14.1C]   **8.** $x \geq -4$ [14.3A]   **9.**

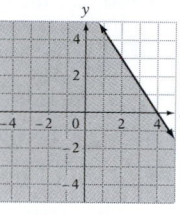
[14.4A]

10.

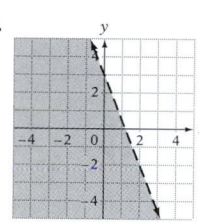
[14.4A]

11. $\{x \mid x > 3, x \in$ real numbers$\}$ [14.1B]   **12.** $x > 2$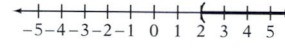

[14.2A]   **13.** $A \cap B = \{1, 5, 9\}$ [14.1A]   **14.** ⟵┼┼┼┼┼┼┼┼┼┼⟶ -5-4-3-2-1 0 1 2 3 4 5 [14.1C]   **15.** ⟵┼┼┼┼┼┼┼┼┼┼⟶ -5-4-3-2-1 0 1 2 3 4 5

[14.1C]   **16.** $x \geq -3$ [14.2B]   **17.** $x > -18$ [14.3A]   **18.** $x < \dfrac{1}{2}$ [14.3A]   **19.** $x < -\dfrac{8}{9}$ [14.2B]

20.

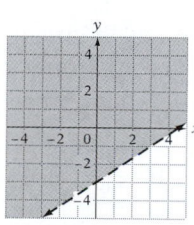
[14.4A]

21. $x \geq 4$ [14.3A]   **22.** For florist B to be more economical, there must be five or more residents in the nursing home. [14.3B]

23. The minimum length is 24 ft. [14.3B]   **24.** 32 is the smallest integer that satisfies the inequality. [14.2C]

25. 72 is the lowest score the student can receive and still attain a minimum of 480 points. [14.2C]

CHAPTER 14 TEST

1. ⟵┼┼┼┼┼┼┼┼┼┼⟶ -5-4-3-2-1 0 1 2 3 4 5 [14.1C]   **2.** $\{x \mid x < 50, x \in$ positive integers$\}$ [14.1B]   **3.** $A = \{4, 6, 8\}$ [14.1A]

4. $x \leq -3$ [14.3A]   **5.** $x > \dfrac{1}{8}$ [14.2A]   **6.** ⟵┼┼┼┼┼┼┼┼┼┼⟶ -5-4-3-2-1 0 1 2 3 4 5 [14.1C]   **7.** $x < -1$ [14.3A]

8. $\{x \mid x > -23, x \in$ real numbers$\}$ [14.1B]   **9.** 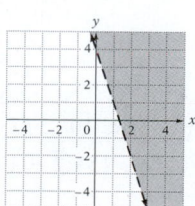 [14.4A]   **10.** 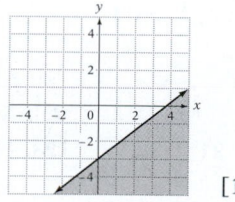 [14.4A]

11. $A \cap B = \{12\}$ [14.1A]   **12.** $x < -3$ 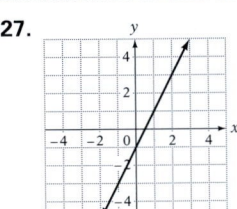   [14.2A]   **13.** $x \geq -\dfrac{40}{3}$ [14.2B]

14. $x < -\dfrac{22}{7}$ [14.3A]   **15.** $x \geq 3$ 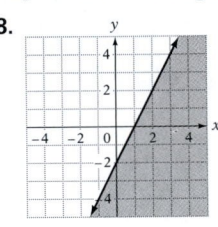   [14.2B]   **16.** $x \geq -4$ [14.3A]   **17.** The child must grow 5 in. or more. [14.2C]   **18.** The width must be less than or equal to 11 ft. [14.3B]   **19.** The diameter must be between 0.0388 in. and 0.0395 in. [14.2C]   **20.** The total value of the stock processed by the broker was less than or equal to $75,000. [14.3B]

CUMULATIVE REVIEW EXERCISES

1. $40a - 28$ [4.2D]   **2.** $\dfrac{1}{8}$ [5.2A]   **3.** 4   [5.3B]   **4.** $-12a^7b^4$ [9.2B]   **5.** $-\dfrac{1}{b^4}$ [9.4A]

6. $4x - 2 - \dfrac{4}{4x - 1}$ [9.5B]   **7.** 0   [12.1D]   **8.** $3a^2(3x + 1)(3x - 1)$ [10.4B]   **9.** $\dfrac{1}{x + 2}$ [11.1C]

10. $\dfrac{18a}{(2a - 3)(a + 3)}$ [11.3B]   **11.** $-\dfrac{5}{9}$ [11.5A]   **12.** $C = S + Rt$ [11.6A]   **13.** $-\dfrac{7}{3}$ [12.3B]

14. $y = -\dfrac{3}{2}x - \dfrac{3}{2}$ [12.4A]   **15.** $(4, 1)$ [13.2A]   **16.** $(1, -4)$ [13.3A]   **17.** $A \cup B = \{-10, -2, 0, 1, 2\}$ [14.1A]

18. $\{x \,|\, x < 48, x \in \text{real numbers}\}$ [14.1B]   **19.** [14.1C]

20. [14.2B]   **21.** $x < -15$ [14.2B]   **22.** $x > 2$ [14.3A]

23. $\{x \,|\, x \leq -26, x \in \text{integers}\}$ [14.2C]   **24.** The maximum number of miles is 359 mi. [14.3B]   **25.** There are an estimated 5000 fish in the lake. [11.6B]   **26.** The angle measures are 65°, 35°, and 80°. [7.1C]

27.   [12.2A]   **28.**   [14.4A]

Answers to Chapter 15 Selected Exercises

PREP TEST

1. -14 [3.1C]   **2.** $-2x^2y - 4xy^2$ [4.2A]   **3.** 14   [5.1C]   **4.** $\dfrac{7}{5}$ [5.3A]   **5.** x^6 [9.2A]

6. $x^2 + 2xy + y^2$ [9.3D]   **7.** $4x^2 - 12x + 9$ [9.3D]   **8.** $4 - 9v^2$ [9.3D]   **9.** $a^2 - 25$ [9.3D]   **10.** $\dfrac{x^2y^2}{9}$ [9.4A]

SECTION 15.1

3. 4   **5.** 7   **7.** $4\sqrt{2}$ **9.** $2\sqrt{2}$ **11.** $18\sqrt{2}$ **13.** $10\sqrt{10}$ **15.** $\sqrt{15}$ **17.** $\sqrt{29}$ **19.** $-54\sqrt{2}$ **21.** $3\sqrt{5}$
23. 0   **25.** $48\sqrt{2}$ **27.** 15.492   **29.** 16.971   **31.** 16   **33.** x^3 **35.** $y^7\sqrt{y}$ **37.** a^{10} **39.** x^2y^2 **41.** $2x^2$
43. $2x\sqrt{6}$ **45.** $2x^2\sqrt{15x}$ **47.** $7a^2b^4$ **49.** $3x^2y^3\sqrt{2xy}$ **51.** $2x^5y^3\sqrt{10xy}$ **53.** $4a^4b^5\sqrt{5a}$ **55.** $8ab\sqrt{b}$
57. x^3y **59.** $8a^2b^3\sqrt{5b}$ **61.** $6x^2y^3\sqrt{3y}$ **63.** $4x^3y\sqrt{2y}$ **65.** $5a + 20$ **67.** $2x^2 + 8x + 8$ **69.** $x + 2$
71. $y + 1$ **73. a.** The speed of the car was $12\sqrt{15}$ mph.   **b.** 46 mph   **75.** No. For example, let $a = 16$ and $b = 9$.
$\sqrt{a + b} = \sqrt{16 + 9} = \sqrt{25} = 5$. $\sqrt{a} + \sqrt{b} = \sqrt{16} + \sqrt{9} = 4 + 3 = 7$.   **77.** No. 18 contains a perfect-square factor. $6\sqrt{2}$
79. a. 1   **b.** 3   **c.** $3\sqrt{3}$

SECTION 15.2

1. 2, 20, and 50   **5.** $3\sqrt{2}$ **7.** $-\sqrt{7}$ **9.** $-11\sqrt{11}$ **11.** $10\sqrt{x}$ **13.** $-2\sqrt{y}$ **15.** $-11\sqrt{3b}$ **17.** $2x\sqrt{2}$
19. $-3a\sqrt{3a}$ **21.** $-5\sqrt{xy}$ **23.** $8\sqrt{5}$ **25.** $8\sqrt{2}$ **27.** $15\sqrt{2} - 10\sqrt{3}$ **29.** $\sqrt{x}$ **31.** $-12x\sqrt{3}$

33. $2xy\sqrt{x} - 3xy\sqrt{y}$ **35.** $-9x\sqrt{3x}$ **37.** $-13y^2\sqrt{2y}$ **39.** $4a^2b^2\sqrt{ab}$ **41.** $7\sqrt{2}$ **43.** $6\sqrt{x}$ **45.** $-3\sqrt{y}$
47. $-45\sqrt{2}$ **49.** $13\sqrt{3} - 12\sqrt{5}$ **51.** $32\sqrt{3} - 3\sqrt{11}$ **53.** $6\sqrt{x}$ **55.** $-34\sqrt{3x}$ **57.** $10a\sqrt{3b} + 10a\sqrt{5b}$
59. $-2xy\sqrt{3}$ **61.** $5\sqrt{2}$

SECTION 15.3

3. 5 **5.** 6 **7.** x **9.** x^3y^2 **11.** $3ab^6\sqrt{2a}$ **13.** $12a^4b\sqrt{b}$ **15.** $2 - \sqrt{6}$ **17.** $x - \sqrt{xy}$ **19.** $5\sqrt{2} - \sqrt{5x}$
21. $4 - 2\sqrt{10}$ **23.** $x - 6\sqrt{x} + 9$ **25.** $3a - 3\sqrt{ab}$ **27.** $10abc$ **29.** $-2 + 2\sqrt{5}$ **31.** $16 + 10\sqrt{2}$
33. $6x + 10\sqrt{x} - 4$ **35.** $15x - 22y\sqrt{x} + 8y^2$ **37.** $x - y$ **41.** 3 **43.** 4 **45.** $6x^2$ **47.** $2x^2\sqrt{2y}$
49. $4x\sqrt{y}$ **51.** $\dfrac{2\sqrt{3x}}{3y}$ **53.** $\dfrac{-1 + \sqrt{2}}{2}$ **55.** $-\dfrac{5\sqrt{7} + 15}{2}$ **57.** $-\dfrac{5\sqrt{3} + 9}{2}$ **59.** $3 + \sqrt{6}$ **61.** $\dfrac{-12 + 3\sqrt{2}}{7}$
63. $\dfrac{14 - 9\sqrt{2}}{17}$ **65.** $\sqrt{15} + 2\sqrt{5}$ **67.** $\dfrac{x\sqrt{y} + y\sqrt{x}}{x - y}$ **69.** $4\sqrt{6} + 12$

SECTION 15.4

1. 25 **3.** 144 **5.** 5 **7.** No solution   **9.** -1 **11.** 1 **13.** 15 **15.** $\dfrac{7}{3}$ **17.** 2 **19.** 5 **21.** 4

23. The pitcher's mound is less than halfway between home plate and second base.   **25.** The periscope must be 16.67 ft above the water.   **27.** The height of the screen is 21.6 in.   **29.** The distance of the child from the center is 18.75 ft. **31.** The perimeter of the triangle is 30 units.   **35.** The area of the fountain is approximately 244.78 ft^2.

CHAPTER 15 REVIEW EXERCISES

1. 3 [15.3A]   **2.** $9a^2\sqrt{2ab}$ [15.1B]   **3.** 12 [15.1A]   **4.** $3a\sqrt{2} + 2a\sqrt{3}$ [15.3A]   **5.** $2\sqrt{6}$ [15.3B]
6. $-8\sqrt{2}$ [15.2A]   **7.** 2 [15.3A]   **8.** 1 [15.4A]   **9.** $-x\sqrt{3} - x\sqrt{5}$ [15.3B]   **10.** $-6\sqrt{30}$ [15.1A]
11. 20 [15.4A]   **12.** $20\sqrt{3}$ [15.1A]   **13.** $7x^2y^4$ [15.3B]   **14.** No solution [15.4A]
15. $18a\sqrt{5b} + 5a\sqrt{b}$ [15.2A]   **16.** $20\sqrt{10}$ [15.1A]   **17.** $7x^2y\sqrt{15xy}$ [15.2A]   **18.** $8y + 10\sqrt{5y} - 15$ [15.3A]
19. $26\sqrt{3x}$ [15.2A]   **20.** No solution [15.4A]   **21.** $\dfrac{8\sqrt{x} + 24}{x - 9}$ [15.3B]   **22.** $36x^8y^5\sqrt{3xy}$ [15.1B]
23. $2y^4\sqrt{6}$ [15.1B]   **24.** -1 [15.4A]   **25.** $-6x^3y^2\sqrt{2y}$ [15.2A]   **26.** $\dfrac{16\sqrt{a}}{a}$ [15.3B]   **27.** The distance across the pond is approximately 43 ft. [15.4B]   **28.** The explorer weighs 144 lb on the surface of Earth. [15.4B]
29. The depth of the water is 100 ft. [15.4B]   **30.** The radius of the corner is 25 ft. [15.4B]

CHAPTER 15 TEST

1. $11x^4y$ [15.1B]   **2.** $6x^2y\sqrt{y}$ [15.3A]   **3.** $-5\sqrt{2}$ [15.2A]   **4.** $3\sqrt{5}$ [15.1A]   **5.** 9 [15.3B]
6. 25 [15.4A]   **7.** $4a^2b^5\sqrt{2ab}$ [15.1B]   **8.** $7ab\sqrt{a}$ [15.3B]   **9.** $\sqrt{3} + 1$ [15.3B]   **10.** $4x^2y^2\sqrt{5y}$ [15.3A]
11. 9 [15.4A]   **12.** $21\sqrt{2y} - 12\sqrt{2x}$ [15.2A]   **13.** $6x^3y\sqrt{2x}$ [15.1B]   **14.** $y + 2\sqrt{y} - 15$ [15.3A]
15. $-2xy\sqrt{3xy} - 3xy\sqrt{xy}$ [15.2A]   **16.** $\dfrac{17 - 8\sqrt{5}}{31}$ [15.3B]   **17.** $a - \sqrt{ab}$ [15.3A]   **18.** $5\sqrt{3}$ [15.1A]
19. The length of the pendulum is 7.30 ft. [15.4B]   **20.** The rope should be secured about 7 ft from the base of the pole. [15.4B]

CUMULATIVE REVIEW EXERCISES

1. $-\dfrac{1}{12}$ [3.5A]   **2.** $2x + 18$ [4.2D]   **3.** $\dfrac{1}{13}$ [5.3B]   **4.** $6x^5y^5$ [9.2A]   **5.** $-2b^2 + 1 - \dfrac{1}{3b^2}$ [9.5A]
6. 1 [12.1D]   **7.** $2a(a - 5)(a - 3)$ [10.2B]   **8.** $\dfrac{1}{4(x + 1)}$ [11.1B]   **9.** $\dfrac{x + 3}{x - 3}$ [11.3B]   **10.** $\dfrac{5}{3}$ [11.5A]
11. $y = \dfrac{1}{2}x - 2$ [12.4A]   **12.** $(1, 1)$ [13.2A]   **13.** $(3, -2)$ [13.3A]   **14.** $x \le -\dfrac{9}{2}$ [14.3A]   **15.** $6\sqrt{3}$ [15.1A]
16. $-4\sqrt{2}$ [15.2A]   **17.** $4ab\sqrt{2ab} - 5ab\sqrt{ab}$ [15.2A]   **18.** $14a^5b^2\sqrt{2a}$ [15.3A]   **19.** $3\sqrt{2} - x\sqrt{3}$ [15.3A]
20. 8 [15.3B]   **21.** $-6 - 3\sqrt{5}$ [15.3B]   **22.** 6 [15.4A]   **23.** The cost of the book is $24.50. [5.2B]
24. The rate of the faster cyclist is 15 mph. [5.5B]   **25.** The numbers are 8 and 13. [10.5B]   **26.** It would take the

small pipe, working alone, 48 h.   [11.7A]   **27.** 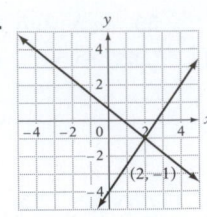   [13.1A]   **28.** 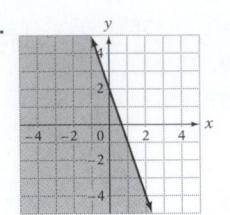   [14.4A]

29. The smaller integer is 40.   [15.4B]   **30.** The height of the building is 400 ft.   [15.4B]

Answers to Chapter 16 Selected Exercises

PREP TEST

1. 41   [4.1A]   **2.** $-\dfrac{1}{5}$ [5.2A]   **3.** $(x + 4)(x - 3)$ [10.2A]   **4.** $(2x - 3)^2$ [10.4A]   **5.** Yes   [10.4A]   **6.** 3

[11.5A]   **7.**   **8.** $2\sqrt{7}$ [15.1A]   **9.** $|a|$ [15.1B]   **10.** 3.6 mi   [5.5A]

[12.2A]

SECTION 16.1

1. $-3, 5$ **3.** $-\dfrac{5}{2}, \dfrac{1}{3}$ **5.** $-5, 3$ **7.** $1, 3$ **9.** $-2, -1$ **11.** 3 **13.** $-\dfrac{2}{3}, 0$ **15.** $-2, 5$ **17.** $\dfrac{2}{3}, 1$

19. $-3, \dfrac{1}{3}$ **21.** $\dfrac{2}{3}$ **23.** $-\dfrac{1}{2}, \dfrac{3}{2}$ **25.** $\dfrac{1}{2}$ **27.** $-3, 3$ **29.** $-\dfrac{1}{2}, \dfrac{1}{2}$ **31.** $-3, 5$ **33.** $1, 5$ **35.** $-6, 6$

37. $-1, 1$ **39.** $-\dfrac{7}{2}, \dfrac{7}{2}$ **41.** $-\dfrac{2}{3}, \dfrac{2}{3}$ **43.** $-\dfrac{3}{4}, \dfrac{3}{4}$ **45.** No real number solution   **47.** $-2\sqrt{6}, 2\sqrt{6}$ **49.** $-5, 7$

51. $-7, -3$ **53.** $-\dfrac{1}{3}, \dfrac{7}{3}$ **55.** $-\dfrac{12}{7}, -\dfrac{2}{7}$ **57.** $4 - 2\sqrt{5}, 4 + 2\sqrt{5}$ **59.** No real number solution

61. $-\dfrac{3}{4} - 2\sqrt{3}, -\dfrac{3}{4} + 2\sqrt{3}$ **63.** $-1, -\dfrac{\sqrt{6}}{3}, \dfrac{\sqrt{6}}{3}, 1$ **65.** The annual percentage rate is 8%.   **67.** Yes, because

$v \approx 73.5$ mph.

SECTION 16.2

1. $x^2 - 8x + 16, (x - 4)^2$ **3.** $x^2 + 5x + \dfrac{25}{4}, \left(x + \dfrac{5}{2}\right)^2$ **5.** $-3, 1$ **7.** $-2, 8$ **9.** 2 **11.** No real number

solution   **13.** $-4, -1$ **15.** $-8, 1$ **17.** $-2 - \sqrt{3}, -2 + \sqrt{3}$ **19.** $-3 - \sqrt{14}, -3 + \sqrt{14}$ **21.** $1 - \sqrt{2}, 1 + \sqrt{2}$

23. $\dfrac{-3 - \sqrt{13}}{2}, \dfrac{-3 + \sqrt{13}}{2}$ **25.** $1, 2$ **27.** $\dfrac{-1 - \sqrt{13}}{2}, \dfrac{-1 + \sqrt{13}}{2}$ **29.** $-5 - 4\sqrt{2}, -5 + 4\sqrt{2}$

31. $\dfrac{3 - \sqrt{29}}{2}, \dfrac{3 + \sqrt{29}}{2}$ **33.** $\dfrac{1 - \sqrt{17}}{2}, \dfrac{1 + \sqrt{17}}{2}$ **35.** No real number solution   **37.** $\dfrac{1}{2}, 1$ **39.** $-3, \dfrac{1}{2}$

41. $\dfrac{3}{2}, 2$ **43.** $-\dfrac{1}{2}, 1$ **45.** $-2, \dfrac{1}{3}$ **47.** $-2, -\dfrac{2}{3}$ **49.** $-\dfrac{3}{2}, \dfrac{1}{2}$ **51.** $-\dfrac{3}{2}, \dfrac{1}{3}$ **53.** $-\dfrac{1}{2}, \dfrac{4}{3}$ **55.** $-5.372, 0.372$

57. $-3.212, 1.712$ **59.** $-1.151, 0.651$ **61.** $1 - \sqrt{7}, 1 + \sqrt{7}$ **63.** $-1, 0$ **65.** $4 - \sqrt{3}, 4 + \sqrt{3}$ **67.** $0, 8$

69. $\dfrac{15 + 3\sqrt{17}}{2}$ **71.** The ball hits the ground approximately 4.81 s after it is hit.

SECTION 16.3

1. $-1, 5$ **3.** $-3, 5$ **5.** $-1, 3$ **7.** $-5, 1$ **9.** $-\dfrac{1}{2}, 1$ **11.** No real number solution   **13.** $0, 1$ **15.** $-\dfrac{3}{2}, \dfrac{3}{2}$

17. $-\dfrac{5}{2}, \dfrac{3}{2}$ **19.** No real number solution   **21.** $1 - \sqrt{6}, 1 + \sqrt{6}$ **23.** $-3 - \sqrt{10}, -3 + \sqrt{10}$

25. $2 - \sqrt{13}, 2 + \sqrt{13}$ **27.** $\dfrac{-1 - \sqrt{2}}{3}, \dfrac{-1 + \sqrt{2}}{3}$ **29.** $-\dfrac{1}{2}$ **31.** No real number solution   **33.** $\dfrac{-4 - \sqrt{5}}{2}, \dfrac{-4 + \sqrt{5}}{2}$

35. $\dfrac{1 - 2\sqrt{3}}{2}, \dfrac{1 + 2\sqrt{3}}{2}$ **37.** $\dfrac{-5 - \sqrt{2}}{3}, \dfrac{-5 + \sqrt{2}}{3}$ **39.** $-3.690, 5.690$ **41.** $-1.690, 7.690$ **43.** $-1.089, 4.589$

45. $-2.118, 0.118$ **47.** $-0.905, 1.105$ **51. a.** False   **b.** False   **c.** False   **d.** True   **53.** $-4, -3$

55. No solution   **57.** $-1, 11$ **59.** The planes are 1000 mi apart after 2 h.

SECTION 16.4

1. Down   **3.** Up   **5.** 4   **7.** -5 **9.** -42 **11.** 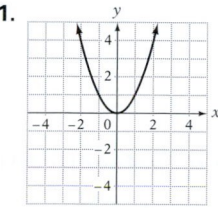   **13.**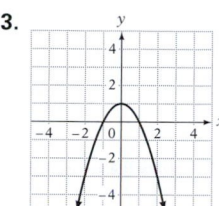

15. 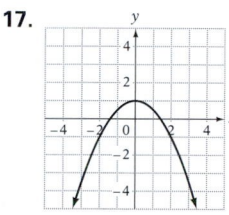   **17.** 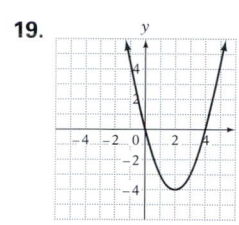   **19.** 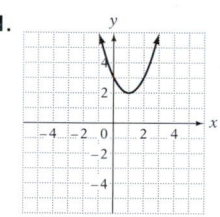   **21.**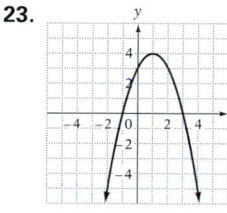

23. 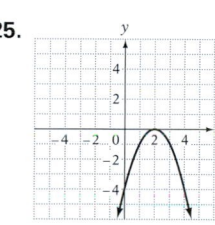   **25.** 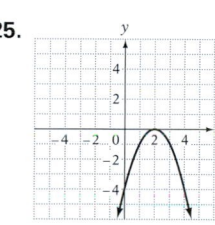   **27.** $(-6, 0), (1, 0); (0, -6)$ **29.** $(-6, 0); (0, 36)$

31. $(-2 - \sqrt{6}, 0), (-2 + \sqrt{6}, 0); (0, -2)$ **33.** No x-intercepts; $(0, 1)$ **35.** $\left(\dfrac{3}{2}, 0\right), (5, 0); (0, 15)$

37. $\left(\dfrac{-3 - \sqrt{33}}{6}, 0\right), \left(\dfrac{-3 + \sqrt{33}}{6}, 0\right); (0, 2)$ **39.** $y = 3x^2 + 6x + 5$ **41.** $y = 3x^2 - 6x$ **43.** $(-1, 0), (0, 0), (5, 0)$

45. $(-3, 0), (-1, 0), (1, 0)$

SECTION 16.5

1. The length is 4 m. The height is 10 m.   **3.** The width is 4 ft. The length is 6 ft.   **5.** The width is 50 ft. The length is 100 ft.   **7.** The hang time of the football is approximately 5.5 s.   **9.** The maximum velocity is 77 ft/s.   **11.** The large pizza has a radius of 7 in.   **13.** The first computer can solve the equation in 35 min. The second computer can solve the equation in 14 min.   **15.** The first engine would take 12 h. The second engine would take 6 h.   **17.** The rate of the plane in calm air is 100 mph.   **19.** The 16-inch pizza should cost $24.

CHAPTER 16 REVIEW EXERCISES

1. $-\dfrac{7}{2}, \dfrac{4}{3}$ [16.1A]   **2.** $-\dfrac{5}{7}, \dfrac{5}{7}$ [16.1B]   **3.** $-6, 4$ [16.2A]   **4.** $-6, 1$ [16.3A]   **5.** $-4, \dfrac{3}{2}$ [16.2A]

6. $\dfrac{5}{12}, 2$ [16.1A]   **7.** $-2 - 2\sqrt{6}, -2 + 2\sqrt{6}$ [16.1B]   **8.** $1, \dfrac{3}{2}$ [16.3A]   **9.** $-\dfrac{1}{2}, -\dfrac{1}{3}$ [16.1A]

10. No real number solution [16.1B]   **11.** $2 - \sqrt{3}, 2 + \sqrt{3}$ [16.2A]   **12.** $\dfrac{3 - \sqrt{29}}{2}, \dfrac{3 + \sqrt{29}}{2}$ [16.3A]

13. No real number solution [16.2A] **14.** −10, −7 [16.1A] **15.** −1, 2 [16.1B]

16. $\dfrac{-4 - \sqrt{23}}{2}, \dfrac{-4 + \sqrt{23}}{2}$ [16.2A] **17.** No real number solution [16.3A] **18.** $-2, -\dfrac{1}{2}$ [16.3A]

19.

[16.4A]

20.

[16.4A]

21.

[16.4A]

22.

[16.4A]

23.

[16.4A]

24. $(-3, 0), (5, 0); (0, -15)$ [16.4A] **25.** The rate of the hawk in calm air is 75 mph. [16.5A]

CHAPTER 16 TEST

1. −1, 6 [16.1A] **2.** $-4, \dfrac{5}{3}$ [16.1A] **3.** 0, 10 [16.1B] **4.** $-4 - 2\sqrt{5}, -4 + 2\sqrt{5}$ [16.1B]

5. $-2 - 2\sqrt{5}, -2 + 2\sqrt{5}$ [16.2A] **6.** $\dfrac{-3 - \sqrt{41}}{2}, \dfrac{-3 + \sqrt{41}}{2}$ [16.2A] **7.** $\dfrac{3 - \sqrt{7}}{2}, \dfrac{3 + \sqrt{7}}{2}$ [16.2A]

8. $\dfrac{-4 - \sqrt{22}}{2}, \dfrac{-4 + \sqrt{22}}{2}$ [16.2A] **9.** $-2 - \sqrt{2}, -2 + \sqrt{2}$ [16.3A] **10.** $\dfrac{3 - \sqrt{33}}{2}, \dfrac{3 + \sqrt{33}}{2}$ [16.3A]

11. $-\dfrac{1}{2}, 3$ [16.3A] **12.** $\dfrac{1 - \sqrt{13}}{6}, \dfrac{1 + \sqrt{13}}{6}$ [16.3A] **13.**

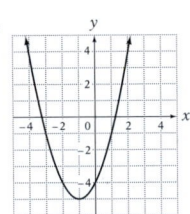

[16.4A]

14. $(-4, 0), (3, 0); (0, -12)$ [16.4A] **15.** The width is 5 ft. The length is 8 ft. [16.5A]

16. The rate of the boat in calm water is 11 mph. [16.5A]

CUMULATIVE REVIEW EXERCISES

1. $-28x + 27$ [4.2D] **2.** $\dfrac{3}{2}$ [5.1C] **3.** 3 [5.3B] **4.** $-12a^8b^4$ [9.2B] **5.** $x + 2 - \dfrac{4}{x - 2}$ [9.5B]

6. $x(3x - 4)(x + 2)$ [10.3A/10.3B] **7.** $\dfrac{9x^2(x - 2)^2}{(2x - 3)^2}$ [11.1C] **8.** $\dfrac{x + 2}{2(x + 1)}$ [11.3B] **9.** $\dfrac{x - 4}{2x + 5}$ [11.4A]

10. $(3, 0); (0, -4)$ [12.3A] **11.** $y = -\dfrac{4}{3}x - 2$ [12.4A] **12.** $(2, 1)$ [13.2A] **13.** $(2, -2)$ [13.3A]

14. $x > \dfrac{1}{9}$ [14.3A] **15.** $a - 2$ [15.3A] **16.** $6ab\sqrt{a}$ [15.3B] **17.** $\dfrac{-6 + 5\sqrt{3}}{13}$ [15.3B] **18.** 5 [15.4A]

19. $\dfrac{1}{3}, \dfrac{5}{2}$ [16.1A] **20.** $5 - 3\sqrt{2}, 5 + 3\sqrt{2}$ [16.1B] **21.** $\dfrac{-7 - \sqrt{13}}{6}, \dfrac{-7 + \sqrt{13}}{6}$ [16.2A] **22.** $-\dfrac{1}{2}, 2$ [16.3A]

23. The cost of the mixture is $2.25 per pound. [5.5A] **24.** 250 additional shares are required. [6.2B] **25.** The rate of the plane in still air is 200 mph. The rate of the wind is 40 mph. [13.4A] **26.** The score on the last test must be 77 or better. [14.2C] **27.** The middle integer can be −5 or 5. [16.5A] **28.** The rate for the last 8 mi was 4 mph. [16.5A]

29.

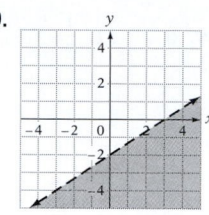

[14.4A]

30.

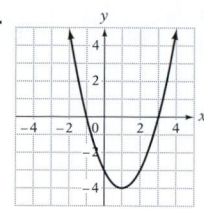

[16.4A]

FINAL EXAM

1. -3 [3.1C]   **2.** -6 [3.2B]   **3.** $-\dfrac{1}{2}$ [3.4A]   **4.** -11 [3.5A]   **5.** $-\dfrac{15}{2}$ [4.1A]   **6.** $9x + 6y$ [4.2A]

7. $6z$ [4.2B]   **8.** $16x - 52$ [4.2D]   **9.** -50 [5.1C]   **10.** -3 [5.3B]   **11.** 12.5% [6.3B]

12. 15.2 [6.4A/6.4B]   **13.** $-3x^2 - 3x + 8$ [9.1B]   **14.** $81x^4y^{12}$ [9.2B]   **15.** $6x^3 + 7x^2 - 7x - 6$ [9.3B]

16. $-\dfrac{x^4y}{2}$ [9.4A]   **17.** $\dfrac{3x}{y} - 4x^2 - \dfrac{5}{x}$ [9.5A]   **18.** $5x - 12 + \dfrac{23}{x + 2}$ [9.5B]   **19.** $\dfrac{4y^6}{x^6}$ [9.4A]   **20.** $\dfrac{3}{4}$ [12.1D]

21. $(x - 6)(x + 1)$ [10.2A]   **22.** $(3x + 2)(2x - 3)$ [10.3A/10.3B]   **23.** $4x(2x - 1)(x - 3)$ [10.4B]

24. $(5x + 4)(5x - 4)$ [10.4A]   **25.** $2(a + 3)(4 - x)$ [10.1B]   **26.** $3y(5 + 2x)(5 - 2x)$ [10.4B]   **27.** $\dfrac{1}{2}, 3$ [10.5A]

28. $\dfrac{2(x + 1)}{x - 1}$ [11.1B]   **29.** $\dfrac{-3x^2 + x - 25}{(x + 3)(2x - 5)}$ [11.3B]   **30.** $\dfrac{x^2 - 2x}{x - 1}$ [11.4A]   **31.** 2 [11.5A]   **32.** $a = b$ [11.6A]

33. $\dfrac{2}{3}$ [12.3B]   **34.** $y = -\dfrac{2}{3}x - 2$ [12.4A]   **35.** $(6, 17)$ [13.2A]   **36.** $(2, -1)$ [13.3A]   **37.** $x \le -3$ [14.2B]

38. $y \ge \dfrac{5}{2}$ [14.3A]   **39.** $7x^3$ [15.1B]   **40.** $38\sqrt{3a}$ [15.2A]   **41.** $\sqrt{15} + 2\sqrt{3}$ [15.3B]   **42.** 2 [15.4A]

43. $-1, \dfrac{4}{3}$ [16.1A]   **44.** $\dfrac{1 - \sqrt{5}}{4}, \dfrac{1 + \sqrt{5}}{4}$ [16.3A]   **45.** $2x + 3(x - 2), 5x - 6$ [4.3B]   **46.** The original value

was $3000. [6.4C]   **47.** The mean is 2.025 in., the median is 1 in., and the mode is 0 in. [8.2A]   **48.** The amount

invested in the 4.8% account is $4000. [13.2B]   **49.** The cost for the mixture is $3 per pound. [5.5A]

50. The probability is $\dfrac{2}{3}$. [8.3A]   **51.** The distance traveled in the first hour was 215 km. [5.5B]   **52.** The angles

are 50°, 60°, and 70°. [7.1C]   **53.** The middle integer can be -4 or 4. [16.5A]   **54.** The width is 5 m. The length is

10 m. [10.5B]   **55.** 16 oz of dye are required. [6.2B]   **56.** Working together, it would take them 36 min or 0.6 h.

[11.7A]   **57.** The rate of the boat in calm water is 15 mph. The rate of the current is 5 mph. [13.4A]   **58.** The rate of

the wind is 25 mph. [16.5A]   **59.**

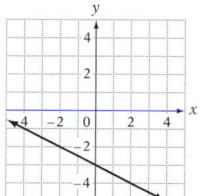

[12.3C]

60.

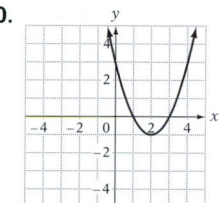

[16.4A]

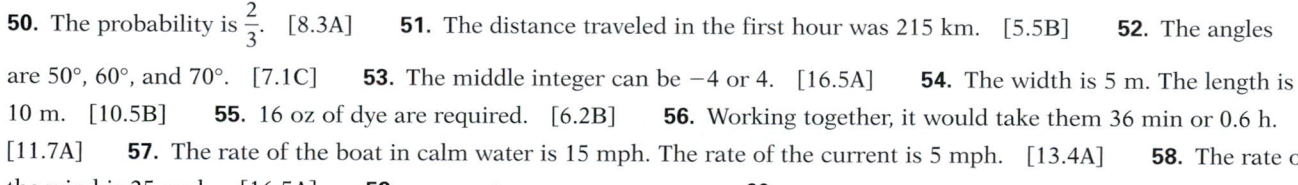

Answers to Chapter R Selected Exercises

SECTION R.1

1. $\dfrac{3}{7}$ **3.** 1 **5.** $1\dfrac{3}{8}$ **7.** $1\dfrac{1}{6}$ **9.** $\dfrac{13}{14}$ **11.** $\dfrac{53}{60}$ **13.** $\dfrac{23}{60}$ **15.** $1\dfrac{17}{18}$ **17.** $1\dfrac{9}{20}$ **19.** $\dfrac{1}{3}$ **21.** $\dfrac{5}{8}$ **23.** $\dfrac{13}{45}$

25. $\dfrac{1}{10}$ **27.** $\dfrac{1}{4}$ **29.** $\dfrac{1}{2}$ **31.** $\dfrac{11}{60}$ **33.** $\dfrac{1}{12}$ **35.** $\dfrac{5}{24}$ **37.** $\dfrac{7}{12}$ **39.** $\dfrac{7}{16}$ **41.** $\dfrac{1}{3}$ **43.** $\dfrac{6}{11}$ **45.** 6 **47.** $\dfrac{5}{32}$

49. $\dfrac{3}{16}$ **51.** $\dfrac{2}{9}$ **53.** $\dfrac{5}{54}$ **55.** 4 **57.** $\dfrac{2}{7}$ **59.** 0 **61.** $\dfrac{2}{9}$ **63.** $\dfrac{7}{10}$ **65.** 2 **67.** $\dfrac{2}{3}$ **69.** 6 **71.** $\dfrac{1}{2}$

SECTION R.2

1. -5 **3.** -17 **5.** 3 **7.** -20 **9.** 0 **11.** 21 **13.** -7 **15.** -11 **17.** 8 **19.** 0 **21.** -25
23. -9 **25.** 22 **27.** -6 **29.** 16 **31.** -32 **33.** -15 **35.** -14 **37.** 11 **39.** 16 **41.** 13 **43.** 5
45. -9 **47.** -21 **49.** 35 **51.** -36 **53.** -50 **55.** 49 **57.** 0 **59.** 60 **61.** -126 **63.** 81
65. -192 **67.** -85 **69.** 98 **71.** -80 **73.** -144 **75.** -3 **77.** 9 **79.** 0 **81.** -11 **83.** -9
85. 22 **87.** 14 **89.** Undefined **91.** -21 **93.** -40

SECTION R.3

1. $-\dfrac{1}{12}$ **3.** $-\dfrac{1}{3}$ **5.** $-\dfrac{7}{8}$ **7.** $-1\dfrac{1}{10}$ **9.** $-\dfrac{1}{24}$ **11.** $-\dfrac{3}{4}$ **13.** $\dfrac{11}{24}$ **15.** $\dfrac{1}{12}$ **17.** -50.7 **19.** -3.312

21. 9.39 **23.** -16.35 **25.** -9.55 **27.** -19.189 **29.** $\dfrac{11}{12} - \left(-\dfrac{1}{4}\right)$, because $\dfrac{11}{12} - \left(-\dfrac{1}{4}\right) = \dfrac{11}{12} + \dfrac{1}{4}$, so the

difference is positive, whereas the difference $-\dfrac{1}{8} - \dfrac{3}{4}$ is negative. **31.** $-\dfrac{1}{4}$ **33.** $\dfrac{1}{10}$ **35.** $-\dfrac{3}{8}$ **37.** $-\dfrac{7}{30}$

39. $\dfrac{3}{16}$ **41.** $-7\dfrac{1}{2}$ **43.** $-4\dfrac{1}{5}$ **45.** -10 **47.** $-\dfrac{3}{7}$ **49.** $\dfrac{9}{10}$ **51.** $-\dfrac{1}{12}$ **53.** $-\dfrac{9}{10}$ **55.** 14.7

57. -7.03 **59.** 60.75 **61.** -5.9 **63.** -67.7 **65.** 4.14 **67.** -6.1 **69.** -128.8 **71.** 37.0

73. $-\dfrac{5}{6} \div (-5)$, because the quotient is positive, whereas the product $-\dfrac{3}{4}\left(\dfrac{2}{9}\right)$ is negative. **75.** $-\dfrac{8}{343}$ **77.** 12.25

79. $\dfrac{1}{121}$ **81.** $-\dfrac{1}{75}$ **83.** 0.51 **85.** 1.7 **87.** $\dfrac{9}{32}$ **89.** $\dfrac{8}{9}$ **91.** 0 **93.** $\dfrac{1}{8}$

95. $16 + 15 \div (-5) - (-4) > 18 \div |2^3 - 9| + (-3) > 20 \div (6 - 2^4) + (-5)$ $[17 > 15 > -7]$

SECTION R.4

1. -12 **3.** -1 **5.** -8 **7.** -20 **9.** 8 **11.** 3 **13.** 5 **15.** -3 **17.** 2 **19.** -1 **21.** 1 **23.** 3

25. 3 **27.** 2 **29.** -1 **31.** $\dfrac{1}{2}$ **33.** 5 **35.** $-\dfrac{1}{9}$

Answers to Appendix Selected Exercises

1. Meter, liter, gram **3.** Kilometer **5.** Centimeter **7.** Gram **9.** Meter **11.** Gram **13.** Milliliter
15. Milligram **17.** Millimeter **19.** Milligram **21.** Gram **23.** Kiloliter **25.** Milliliter **27.** 420 mm
29. 0.360 kg **31.** 5.194 L **33.** 2 000 mm **35.** 0.217 g **37.** 4 520 ml **39.** 8.406 km **41.** 2 400 g
43. 6 180 L **45.** 9 612 m **47.** 240 mg **49.** 2.98 m **51.** 2.431 kl **53.** 66 cm **55.** 24.3 cm

57. a. Column 2: k, c, m; column 3: 10^9, 10^3, 10^2, $\dfrac{1}{10^3}$, $\dfrac{1}{10^{12}}$; column 4: 1 000 000, 10, 0.1, 0.01, 0.000 001, 0.000 000 001

59. The weight is 2 g. **61.** 24 L of chlorine are used in a month of 30 days. **63.** There are 16 servings in the
container. **65.** The patient should take 4 tablets per day. **67.** The case containing 12 one-liter bottles costs less per
milliliter. **69.** It takes light 500 s to reach Earth from the sun. **71.** The profit is $17,430. **73.** The profit is $66.50.

Glossary

abscissa The first number in an ordered pair. It measures a horizontal distance and is also called the first coordinate. [12.1]

absolute value of a number The distance of a number from zero on the number line. [3.1]

acute angle An angle whose measure is between 0° and 90°. [7.1]

acute triangle A triangle that has three acute angles. [7.2]

addend In addition, one of the numbers added. [1.2]

addition The process of finding the total of two numbers. [1.2]

addition method An algebraic method of finding an exact solution of a system of linear equations, in which the equations in the system are added. [13.3]

additive inverses Numbers that are the same distance from zero on the number line, but on opposite sides; also called opposites. [3.2]

adjacent angles Two angles that share a common side. [7.1]

alternate exterior angles Two nonadjacent angles that are on opposite sides of the transversal and outside the parallel lines. [7.1]

alternate interior angles Two nonadjacent angles that are on opposite sides of the transversal and between the parallel lines. [7.1]

analytic geometry Geometry in which a coordinate system is used to study the relationships between variables. [12.1]

angle The figure formed when two rays start at the same point; it is measured in degrees. [1.2]

area A measure of the amount of surface in a region. [7.2]

axes The two number lines that form a rectangular coordinate system; also called coordinate axes. [12.1]

bar graph A graph that represents data by the height of the bars. [1.1]

base In exponential notation, the factor that is multiplied the number of times shown by the exponent. [1.3]

base of a triangle The side of a triangle that the triangle rests on. [2.4]

basic percent equation The equation that states that percent times base equals amount. [6.4]

binomial A polynomial of two terms. [9.1]

binomial factor A factor that has two terms. [10.1]

borrowing In subtraction, taking a unit from the next larger place value in the minuend and adding it to the number in the given place value in order to make that number larger than the number to be subtracted from it. [1.2]

box-and-whiskers plot A graph that shows the smallest value in a set of numbers, the first quartile, the median, the third quartile, and the greatest value. [8.2]

broken-line graph A graph that represents data by the positions of the lines and shows trends and comparisons. [1.1]

center of a circle The point from which all points on the circle are equidistant. [7.2]

center of a sphere The point from which all points on the surface of the sphere are equidistant. [7.4]

circle A plane figure in which all points are the same distance from point O, which is the figure's center. [7.2]

circle graph A graph that represents data by the size of the sectors. [1.1]

circumference The distance around a circle. [7.2]

clearing denominators Removing denominators from an equation that contains fractions by multiplying each side of the equation by the LCM of the denominators. [5.2]

coefficient The number part of a variable term; for example, the 2 in the variable expression $2x$. [4.1]

combining like terms Adding like terms of a variable expression. [4.2]

common factor A number that is a factor of two or more numbers. [2.1]

common multiple A number that is a multiple of two or more numbers. [2.1]

complementary angles Two angles whose sum is 90°. [7.1]

completing the square Adding to a binomial the constant term that makes it a perfect-square trinomial. [16.2]

complex fraction A fraction whose numerator or denominator contains one or more fractions. [2.4]

composite number A number that has whole number factors besides 1 and itself. For instance, 10 has whole number factors of 2 and 5. [1.3]

congruent objects Objects that have the same shape and the same size. [7.3]

congruent triangles Triangles that have the same shape and the same size. [7.3]

conjugates Binomial expressions that differ only in the sign of a term; for example, the expressions $a + b$ and $a - b$. [15.3]

consecutive even integers Even integers that follow one another in order. [5.4]

consecutive integers Integers that follow one another in order. [5.4]

consecutive odd integers Odd integers that follow one another in order. [5.4]

constant term A term that includes no variable part. [4.1]

coordinate axes The two number lines that form a rectangular coordinate system. [12.1]

coordinates of a point The numbers in an ordered pair that is associated with a point. [12.1]

corresponding angles Two angles that are on the same side of the transversal and are both acute angles or are both obtuse angles. [7.1]

cross product In a proportion, either the product of the numerator on the left side of the proportion times the denominator on the right, or the product of the denominator on the left side of the proportion times the numerator on the right. [6.2]

cube A rectangular solid in which all six faces are squares. [7.4]

data Numerical information. [8.1]

decimal A number written in decimal notation. [2.5]

decimal notation Notation in which a number is written with a whole number part, a decimal point, and a decimal part. [2.5]

decimal part In decimal notation, that part of the number that appears to the right of the decimal point. [2.5]

decimal point In decimal notation, the point that separates the whole number part of a number from the decimal part. [2.5]

degree A unit used to measure angles. [1.2]

degree of a polynomial in one variable For a polynomial in one variable, the largest exponent that appears on a variable in the expression. [9.1]

denominator The part of a fraction that appears below the fraction bar. [2.2]

dependent system A system of equations that has an infinite number of solutions. [13.1]

dependent variable In a function, the variable whose value depends on the value of another variable, known as the independent variable. [12.1]

descending order A polynomial in one variable arranged so that the exponents on the variable terms decrease from left to right. For example, the polynomial $9x^5 - 2x^4 + 7x^3 + x^2 - 8x + 1$. [9.1]

diameter of a circle A line segment with endpoints on the circle and passing through the center. [7.2]

diameter of a sphere A line segment with endpoints on the sphere and passing through the center. [7.4]

difference In subtraction, the result of subtracting two numbers. [1.2]

difference of two squares A polynomial of the form $a^2 - b^2$. [10.4]

dividend In division, the number into which the divisor is divided to yield the quotient. [1.3]

division The process of finding the quotient of two numbers. [1.3]

divisor In division, the number that is divided into the dividend to yield the quotient. [1.3]

domain The set of first coordinates of the ordered pairs in a relation. [12.1]

double-bar graph A graph used to display data for purposes of comparison. [1.1]

double root Two equal roots of a quadratic equation. [16.1]

element of a set One of the objects in a set. [14.1]

empirical probability Probability expressed as the ratio of the number of observations of an event to the total number of observations. [8.3]

empty set The set that contains no elements; also called the null set. [14.1]

equation A statement of the equality of two mathematical expressions. [1.2]

equilateral triangle A triangle that has three sides of equal length; the three angles are also of equal measure. [7.2]

equivalent equations Equations that have the same solution. [5.1]

equivalent fractions Equal fractions with different denominators; for example, 2/3 and 4/6. [2.2]

evaluating a function Replacing x in $f(x)$ with some value and then simplifying the numerical expression that results. [12.1]

evaluating a variable expression Replacing the variable or variables in an expression with numbers and then simplifying the resulting numerical expression. [1.2]

even integer An integer that is divisible by 2. [5.4]

event One or more outcomes of an experiment. [8.3]

expanded form The form of the number 46,208 when written as $40{,}000 + 6000 + 200 + 0 + 8$. [1.1]

experiment Any activity that has an observable outcome. [8.3]

exponent In exponential notation, the raised number that indicates how many times the base is taken as a factor. [1.3]

exponential form The expression of a number to a power, indicated by an exponent. [1.3]

exterior angle of a triangle The angle adjacent to an interior angle of a triangle. [7.1]

factor by grouping The process of grouping and factoring terms of a polynomial in such a way that a common binomial factor is found. [10.1]

factor completely The process of writing a polynomial as a product of factors that are nonfactorable over the integers. [10.2]

factor of a number In multiplication, a number being multiplied. [1.3]

factor a polynomial The process of writing the polynomial as a product of other polynomials. [10.1]

factor a trinomial of the form $x^2 + bx + c$ To express the trinomial as the product of two binomials. [10.2]

factored form The expression of a number as a product of its factors; for example, $2 \cdot 2 \cdot 2 \cdot 2 \cdot 2$. [1.3]

favorable outcomes The outcomes of an experiment that satisfy the requirements of a particular event. [8.3]

first coordinate The first number in an ordered pair. It measures a horizontal distance and is also called the abscissa. [12.1]

first-degree equation in two variables An equation of the form $y = mx + b$, where m is the coefficient and b is a constant; also called a linear equation in two variables or a linear function. [12.2]

first quartile In a set of numbers, the number below which one-quarter of the data lie. [8.2]

FOIL A method of finding the product of two binomials; it ensures that each term of one binomial is multiplied by each term of the other binomial. [9.3]

formula An equation that expresses a relationship among variables; for example, $A = LW$. [11.6]

fraction The notation used to represent the number of equal parts of a whole. [2.2]

fraction bar The horizontal line that separates the numerator of a fraction from the denominator. [2.2]

function A relation in which no two ordered pairs that have the same first coordinate have different second coordinates. [12.1]

functional notation The notation $f(x)$, which is used to designate a function and represents the value of the function at x. [12.1]

geometric solid A figure in space. [7.4]

graph of an equation in two variables A graph of the ordered-pair solutions of an equation in two variables. [12.2]

graph of an ordered pair The dot drawn at the coordinates of the point in the plane. [12.1]

graph a point in the plane To place a dot at the location given by the ordered pair; also called plotting a point. [12.1]

graph of a relation The graph of the ordered pairs that belong to the relation. [12.1]

graph of a whole number A heavy dot placed directly above a number on the number line. [1.1]

greater than The meaning of the symbol $>$. [1.1]

greatest common factor (GCF) The largest common factor of two or more numbers. [2.1]

greatest common factor (GCF) of two or more monomials The product of the GCF of the coefficients and the common variable factors of the monomials. [10.1]

half-plane The solution set of an inequality in two variables. [14.4]

height of a parallelogram The distance between parallel sides of a parallelogram. [7.2]

height of a triangle In a triangle, a line segment perpendicular to the base from the opposite vertex. [2.4]

hypotenuse The side opposite the right angle in a right triangle. [7.3]

improper fraction A fraction in which the numerator is greater than or equal to the denominator. [2.2]

inconsistent system A system of equations that has no solution. [13.1]

independent system A system of equations that has one solution. [13.1]

independent variable In a function, the variable that varies independently and whose value determines the value of the dependent variable. [12.1]

inequality An expression that contains the symbol $>$, $<$, $\geq$ (is greater than or equal to), or $\leq$ (is less than or equal to). [1.1]

integers The numbers . . . , -3, -2, -1, 0, 1, 2, 3, ... [3.1]

interest Money paid for the privilege of using someone else's money. [6.5]

interest rate The percent used to determine the amount of interest to be paid. [6.5]

interior angle of a triangle An angle within the region enclosed by a triangle. [7.1]

interquartile range The difference between the third quartile and the first quartile. [8.2]

intersecting lines Lines that cross at a point in the plane. [1.2]

intersection of sets A and B The set that contains the elements that are common to both A and B. [14.1]

inverting a fraction The process of interchanging the numerator and denominator of a fraction. [2.4]

irrational number A number whose decimal representation never repeats or terminates. [3.4]

isosceles triangle A triangle that has two sides of equal length; the angles opposite the equal sides are of equal measure. [7.2]

least common denominator (LCD) The least common multiple of the denominators of two or more fractions. [2.2]

least common multiple (LCM) The smallest common multiple of two or more numbers. [2.1]

least common multiple (LCM) of two or more polynomials The polynomial of least degree that contains all the factors of each polynomial. [11.2]

legs of a right triangle The two shortest sides of a right triangle. [7.3]

less than The meaning of the symbol $<$. [1.1]

like terms Terms of a variable expression that have the same variable part. [4.2]

line A geometric figure that extends indefinitely in two directions in a plane; it has no width. [1.2]

line of best fit A line drawn to approximate data that are graphed as points in a coordinate system. [12.4]

line segment Part of a line; it has two endpoints. [1.2]

linear equation in two variables An equation of the form $y = mx + b$, where m and b are constants; also called a linear function or a first-degree equation in two variables. [12.2]

linear function A function that can be expressed in the form $f(x) = mx + b$. [12.2]

linear model A first-degree equation that is used to describe a relationship between quantities. [12.4]

literal equation An equation that contains more than one variable. [11.6]

maturity value of a loan The principal of a loan plus the interest owed on it. [6.5]

mean The sum of a set of values divided by the number of those values; also known as the average value. [8.2]

median The average that separates a list of values in such a way that the number of values below it is the same as the number of values above it. [8.2]

minuend In subtraction, the number from which another number (the subtrahend) is subtracted. [1.2]

mixed number A number greater than 1 that has a whole number part and a fractional part. [2.2]

mode In a set of numbers, the value that occurs most frequently. [8.2]

monomial A number, a variable, or a product of numbers and variables; a polynomial of one term. [9.1]

multiples of a number The products of a number and the numbers 1, 2, 3, 4, ... [2.1]

multiplication The process of finding the product of two numbers. [1.3]

multiplicative inverse The reciprocal of a nonzero number. [2.4]

natural numbers The numbers 1, 2, 3, . . . ; also called the positive integers. [1.1]

negative integers The numbers . . . , −4, −3, −2, −1. [3.1]

negative numbers The numbers less than zero. [3.1]

negative slope A property of a line that slants downward to the right. [12.3]

nonfactorable over the integers A polynomial that does not factor using only integers. [10.2]

null set The set that contains no elements; also called the empty set. [14.1]

number line A line on which points are marked off at regular, evenly spaced intervals and are labeled with ordered numbers. [1.1]

numerator The part of a fraction that appears above the fraction bar. [2.2]

numerical coefficient The number part of a variable term; for example, the 2 in the variable expression $2x$. [4.1]

obtuse angle An angle whose measure is between 90° and 180°. [7.1]

obtuse triangle A triangle in which one angle measures more than 90°. [7.2]

odd integer An integer that is not divisible by 2. [5.4]

opposite numbers Two numbers that are the same distance from zero on the number line, but on opposite sides. [3.1]

opposite of a polynomial The polynomial created when the sign of each term of the original polynomial is changed. [9.1]

Order of Operations Agreement A set of rules that tells us in what order to perform the operations that occur in a numerical expression. [1.4]

ordered pair A pair of numbers, such as (a, b), that can be used to identify a point in the plane determined by the axes of a rectangular coordinate system. [12.1]

ordinate The second number in an ordered pair. It measures a vertical distance and is also called the second coordinate. [12.1]

origin The point of intersection of the two coordinate axes that form a rectangular coordinate system. [12.1]

parabola The graph of a quadratic equation in two variables. [16.4]

parallel lines Lines that never meet; the distance between them is always the same. [1.2/7.3]

parallelogram A quadrilateral that has equal and parallel opposite sides. [7.2]

percent The word used to mean "parts per hundred." [6.3]

perfect square The square of an integer. [15.1]

perfect-square trinomial A trinomial that is a product of a binomial and itself. [10.4]

perimeter The distance around a plane figure. [1.2]

period In a number written in standard form, each group of digits separated from other digits by a comma or commas. [1.1]

perpendicular lines Intersecting lines that form right angles. [7.1]

pictograph A graph in which the data are displayed using pictures or symbols. [1.1]

place value The value associated with the position of a digit in a number; it indicates the value of the digit. [1.1]

place-value chart A chart that indicates the place value of every digit in a number. [1.1]

plane A flat surface that extends forever in all directions. [1.2]

plane figure A figure that lies entirely in a plane. [1.2]

plot a point in the plane To place a dot at the location given by an ordered pair; also called graphing a point. [12.1]

point-slope formula The formula that states that if (x_1, y_1) is a point on a line with slope m, then $y - y_1 = m(x - x_1)$. [12.4]

polygon A closed figure determined by three or more line segments that lie in a plane. [1.2]

polynomial A variable expression in which the terms are monomials. [9.1]

positive integers The numbers 1, 2, 3, 4, . . . ; also called the natural numbers. [3.1]

positive numbers The numbers greater than zero. [3.1]

positive slope A property of a line that slants upward to the right. [12.3]

prime factorization The expression of a number as the product of numbers whose only whole number factors are 1 and themselves. [1.3]

prime number A positive number other than 1, such as 5 or 13, whose only whole number factors are 1 and itself. [1.3]

prime polynomial A polynomial that is nonfactorable over the integers. [10.2]

principal The amount of money originally deposited or borrowed. [6.5]

principal square root The positive square root of a number. [15.1]

probability A number from 0 to 1 that tells us how likely it is that a certain outcome of an experiment will happen. [8.3]

product In multiplication, the result of multiplying two numbers. [1.3]

proper fraction A fraction in which the numerator is less than the denominator. [2.2]

proportion An equation that states the equality of two ratios or rates. [6.2]

Pythagorean Theorem The theorem that states that the square of the hypotenuse of a right triangle is equal to the sum of the squares of the two legs. [15.4]

quadrant One of the four regions into which the two axes of a rectangular coordinate system divide a plane. [12.1]

quadratic equation An equation of the form $ax^2 + bx + c = 0$, where a, b, and c are constants and a is not equal to zero; also called a second-degree equation. [10.5/11.1]

quadratic equation in two variables An equation of the form $y = ax^2 + bx + c$, where a is not equal to zero. [16.4]

quadratic function A function of the form $f(x) = ax^2 + bx + c$, where a is not equal to zero. [16.4]

quadrilateral A four-sided closed figure. [1.2]

quotient In division, the result of dividing the divisor into the dividend. [1.3]

radical equation An equation that contains a variable expression in a radicand. [15.4]

radical sign The symbol $\sqrt{}$, which is used to indicate the positive, or principal, square root of a number. [15.1]

radicand In a radical expression, the expression under the radical sign. [15.1]

radius of a circle A line segment going from the center of a circle to a point on the circle. [7.2]

radius of a sphere A line segment going from the center of a sphere to a point on the sphere. [7.4]

range of a relation The set of second coordinates of the ordered pairs in a relation. [12.1]

range of a set of data In a set of numbers, the difference between the largest and smallest values. [8.2]

rate The quotient of two quantities that have different units. [6.1]

rate of work That part of a task that is completed in one unit of time. [11.7]

ratio The quotient of two quantities that have the same unit. [6.1]

rational expression A fraction in which the numerator and denominator are polynomials. [11.1]

rational number A number that can be written in the form $\dfrac{a}{b}$, where a and b are integers and b is not equal to zero. [3.4]

rationalizing the denominator The procedure used to remove a radical from the denominator of a fraction. [15.3]

ray A geometric figure that starts at a point and extends indefinitely in one direction. [1.2]

real numbers The rational numbers and the irrational numbers taken together. [3.4]

reciprocal of a fraction The word used to describe a fraction with the numerator and denominator interchanged. [2.4]

reciprocal of a rational expression A rational expression in which the numerator and denominator have been interchanged. [11.1]

rectangle A parallelogram that has four right angles. [1.2]

rectangular coordinate system A system formed by two number lines, one horizontal and one vertical, that intersect at the zero point of each line. [12.1]

rectangular solid A solid in which all six faces are rectangles. [7.4]

regular polygon A polygon in which each side has the same length and each angle has the same measure. [7.2]

relation Any set of ordered pairs. [12.1]

remainder In division, the quantity left over when it is not possible to separate objects or numbers into a whole number of equal groups. [1.3]

repeating decimal A decimal in which a block of one or more digits repeats forever. [2.6]

right angle A 90° angle. [1.2]

right triangle A triangle that contains one right angle. [7.2]

roster method The method of writing a set by enclosing a list of the elements of the set in braces. [14.1]

rounding Giving an approximate value of an exact number. [1.1]

sample space All possible outcomes of an experiment. [8.3]

scalene triangle A triangle that has no sides of equal length; no two of its angles are of equal measure. [7.2]

scatter diagram A graph of collected data as points in a coordinate system. [12.4]

scientific notation A notation in which a number is expressed as the product of two factors, one a number between 1 and 10 and the other a power of 10. [9.4]

second coordinate The second number in an ordered pair. It measures a vertical distance and is also called the ordinate. [12.1]

second-degree equation An equation of the form $ax^2 + bx + c = 0$, where a, b, and c are constants and a is not equal to zero; also called a quadratic equation. [16.1]

set A collection of objects. [14.1]

set-builder notation A method of designating a set that makes use of a variable and a certain property that only elements of that set possess. [14.1]

sides of a polygon The line segments that form the polygon. [1.2]

similar objects Objects that have the same shape but not necessarily the same size. [7.3]

similar triangles Triangles that have the same shape but not necessarily the same size. [7.3]

simple interest Interest computed on the original principal. [6.5]

simplest form of a fraction The form of a fraction in which the numerator and denominator contain no common factors other than 1. [2.2]

simplest form of a rate A rate is in simplest form when the numbers that make up the rate have no common factor. [6.1]

simplest form of a ratio A ratio is in simplest form when the two numbers do not have a common factor. [6.1]

simplest form of a rational expression The form of a rational expression in which the numerator and denominator have no common factors other than 1. [11.1]

simplifying a variable expression Combining like terms of an expression by adding their numerical coefficients. [4.2]

slope The measure of the slant of a line, symbolized by m. [12.3]

slope-intercept form The form of an equation of a straight line written as $y = mx + b$. [12.3]

solid An object that exists in space. [7.4]

solution of an equation A number that, when substituted for the variable in an equation, results in a true equation. [1.2]

solution of an equation in two variables An ordered pair whose coordinates make an equation in two variables a true statement. [12.1]

solution set of an inequality A set of numbers each element of which, when substituted for the variable in a variable inequality, results in a true inequality. [14.1]

solution of a system of equations in two variables An ordered pair that is a solution of each equation in a system of equations. [13.1]

solving an equation Finding a solution of the equation. [5.1]

sphere A solid in which all points are the same distance from point O, which is the sphere's center. [7.4]

square A rectangle that has four equal sides. [7.2]

square root One of two identical factors of a number; for example, 3 is one of two identical factors of 9. [15.1]

standard form of a linear equation in two variables The form of an equation in two variables when it is written as $Ax + By = C$, where A and B are coefficients and C is a constant. [12.2]

standard form of a number The form of a number when it is written using the digits 0, 1, 2, . . . , 9. An example is 46,208. [1.1]

standard form of a quadratic equation The form of a quadratic equation when it is written with the polynomial in descending order and equal to zero. [10.5/11.1]

statistics The branch of mathematics concerned with data, or numerical information. [8.1]

straight angle An angle whose measure is 180°. [7.1]

substitution method An algebraic method of finding an exact solution of a system of equations, in which one variable is expressed in terms of another variable. [13.2]

subtraction The process of finding the difference between two numbers. [1.2]

subtrahend In subtraction, the number that is subtracted from another number (the minuend). [1.2]

sum In addition, the total of the numbers being added. [1.2]

supplementary angles Two angles whose sum is 180°. [7.1]

system of equations Two or more equations considered together. [13.1]

terminating decimal A decimal that has a finite number of digits after the decimal point, which means that it comes to an end and does not go on forever. [2.6]

terms of a variable expression The addends of a variable expression. [4.1]

theoretical probability A fraction consisting of the number of favorable outcomes of an experiment in the numerator and the total number of possible outcomes of the experiment in the denominator. [8.3]

third quartile In a set of numbers, the number above which one-quarter of the data lie. [8.2]

transversal A line intersecting two other lines at two different points. [7.1]

triangle A three-sided closed figure. [1.2]

trinomial A polynomial of three terms. [9.1]

undefined slope The slope of a vertical line. [12.3]

uniform motion The motion of a moving object whose speed and direction do not change. [5.1]

union of sets A and B The set that contains all the elements of A and all the elements of B. [14.1]

unit rate A rate in which the number in the denominator is 1. [6.1]

value of a function at x The result of evaluating a variable expression, represented by the symbol $f(x)$. [12.1]

value of a variable The number assigned to a variable. [4.1]

variable A letter of the alphabet used to represent a number that is unknown or that can change. [1.2]

variable expression An expression that contains one or more variables. [1.2]

variable part In a variable term, the variable or variables and their exponents. [4.1]

variable term A term composed of a numerical coefficient and a variable part. [4.1]

vertex The point at which the rays of an angle meet. [7.1]

vertical angles Two angles that are on opposite sides of the intersection of two lines. [7.1]

volume A measure of the amount of space inside a closed surface. [7.4]

whole numbers The numbers 0, 1, 2, 3, 4, ... [1.1]

whole number part In decimal notation, that part of the number that appears to the left of the decimal point. [2.5]

x-coordinate The abscissa, or first coordinate, of an ordered pair in an xy-coordinate system. [12.1]

x-intercept The point at which a graph crosses the x-axis. [12.3]

xy-coordinate system A rectangular coordinate system in which the horizontal axis is labeled x and the vertical axis is labeled y. [12.1]

y-coordinate The ordinate, or second coordinate, of an ordered pair in an xy-coordinate system. [12.1]

y-intercept The point at which a graph crosses the y-axis. [12.3]

zero slope The slope of a horizontal line. [12.3]

Index

A

Abscissa, 625
Absolute value, 180–181
  Order of Operations Agreement and, 848
Abundant number, 53, 707
Acute angle, 380
Acute triangle, 394
Addends, 19
Addition, 19
  applications of, 30–32
  Associative Property of, 22, 189, 251
  carrying in, 20
  Commutative Property of, 22, 189, 252
  of decimals, 139, 140, 141, 218, 219, 842
  Distributive Property and, 251
  of fractions, 98–101, 215–216, 218–219, 828, 829, 841, 843
  of integers, 187–191, 834–835, 836
  Inverse Property of, 189, 252
  of mixed numbers, 99–100, 101
  negative numbers in, 841, 842, 843
  Order of Operations Agreement and, 67–68, 159, 231, 848
  of polynomials, 477–478
  properties of, 22, 189, 251–252, 324–325
  of radical expressions, 759–760
  of rational expressions, 583–587
  of rational numbers, 215–216, 218–219, 841, 842–843
  sign rules for, 188, 834
  verbal phrases for, 19, 31, 261
  of whole numbers, 19–24
  zero in, 22, 189, 252
Addition method, for solving systems of equations, 695–698
Addition Property of Equations, 282–283, 853–854
Addition Property of Inequalities, 725–726, 733
Addition Property of Zero, 22, 189, 252
Additive identity, 325
Additive inverse, 189, 252
Adjacent angles, 380, 382
Algebraic fraction(s), *see* Rational expression(s)
Al-Khwarismi, 815
Alternate exterior angles, 383
Alternate interior angles, 383
Amount
  in basic percent equation, 351–352, 353
  in value mixture equation, 315
Amount-to-base ratio, 353
Analytic geometry, 625
Angle(s), 29, 378
  acute, 380
  adjacent, 380, 382
  complementary, 379
  corresponding, 383

  exterior, 383, 385
  interior, 383, 385
  of intersecting lines, 382–384
  measure of, 29, 378–380
  naming of, 378
  obtuse, 380
  right, 29, 379
  sides of, 378
  straight, 380
  supplementary, 380, 382
  of triangles, 385–386
  vertex of, 378
  vertical, 382
Angle-Side-Angle Rule (ASA), 416
Apothem, 410
Application problems
  area, 402, 809
  current or wind, 286–287, 607, 701–702, 808
  decimals in, 134, 149–150, 224
  factoring in, 553–554
  formulas in, 32, 55–58, 120–121, 150, 295, 303
  fractions in, 104, 120–121
  inequalities in, 728, 734
  integers in, 181, 195–196, 208, 308–310, 553
  investments, 689–690
  levers, 303
  linear functions in, 642
  line of best fit, 660
  motion, 285–287, 317–318, 607–608, 665–666, 701–702, 808
  percent, 355–356
  perimeter, 397
  polynomials in, 488
  profit, 812
  proportions in, 341–342, 366
  quadratic equations in, 808–809, 812
  radical equations in, 771–772
  rational numbers in, 224
  right triangles in, 772, 776–777
  simple interest, 361–362, 689–690
  slope in, 650, 651, 665
  statistical graphs in, 11–12, 31, 445–446
  strategy for, 11–12, 71, 808
  subtraction in, 31
  systems of equations in, 701–704
  translating verbal problems, 264
  in two variables, 642, 701–704
  units in, 507–508
  value mixtures, 315–316, 703–704
  variable in, 264
  whole numbers in, 11–12, 31–32, 55–58
  work, 605–606, 809
Approximately equal to (≈), 144
Approximation
  by rounding decimals, 132–133, 140, 141, 144
  by rounding whole numbers, 6–7, 21

  of solution of system of equations, 708–709
  of square root, 412
  symbol for, 144
  see also Estimate
Area, 56, 398
  of circle, 401
  of parallelogram, 399
  of rectangle, 56, 398–399
  of square, 56–57, 398, 399
  surface area, 424–426
  of trapezoid, 400–401
  of triangle, 120, 400
  units of, 56, 398, 507
Arrow, on number line, 187
ASA (Angle-Side-Angle) Rule, 416
Associative Property
  of Addition, 22, 189, 251
  of Multiplication, 44, 204, 253
Average, 449–451, 741–743
Average high temperature, 324
Average speed, 508
Axes, coordinate, 625
Axis of symmetry, 432

B

Bar graph, 9, 445
Bars, musical, 164
Base
  of cone, 421
  of cylinder, 421
  of exponential expression, 46
    fractional, 113
    rational, 847
  of parallelogram, 399
  in percent equation, 351, 352
  of pyramid, 421
  of trapezoid, 400
  of triangle, 120, 400
Basic percent equation, 351–353
  applications of, 355–356
Binary digit, 860
Binomial(s), 477
  expanding powers of, 487, 509
  multiplication of, 486–488
  nonfactorable over the integers, 543, 545
  prime, 528
  square of, 487, 509, 543–544, 790, 793
Binomial factor(s)
  in factoring by grouping, 523–524
  in factoring completely, 545–546
  of trinomials, 527–530, 535–538, 543–545
Bit, 860
Body mass index (BMI), 614
Borrowing, 25–26
Box-and-whiskers plot, 452–454
Braces, in roster method, 719
Bracket(s)
  Distributive Property and, 256

I1

TI-30X IIS

6 [Ab/c] 2 [Ab/c] 3 [+] 3 [Ab/c] 4 [ENTER]

Operations on fractions
$6\frac{2}{3} + \frac{3}{4} = 7\frac{5}{12}$

```
6⌴2⌴3+3⌴4
                7⌴5/12
```

The value of π

```
π
3.141592654
```

13 [∧] 4 [ENTER]

Power of a number
(See Note 1 below.)

```
13^4
                28561
```

[2nd] [√] 36 [)] [ENTER]

Square root of a number

```
√(36)
                      6
```

7 [x^2] [ENTER]

Square a number

```
7²
                     49
```

Access operations in blue

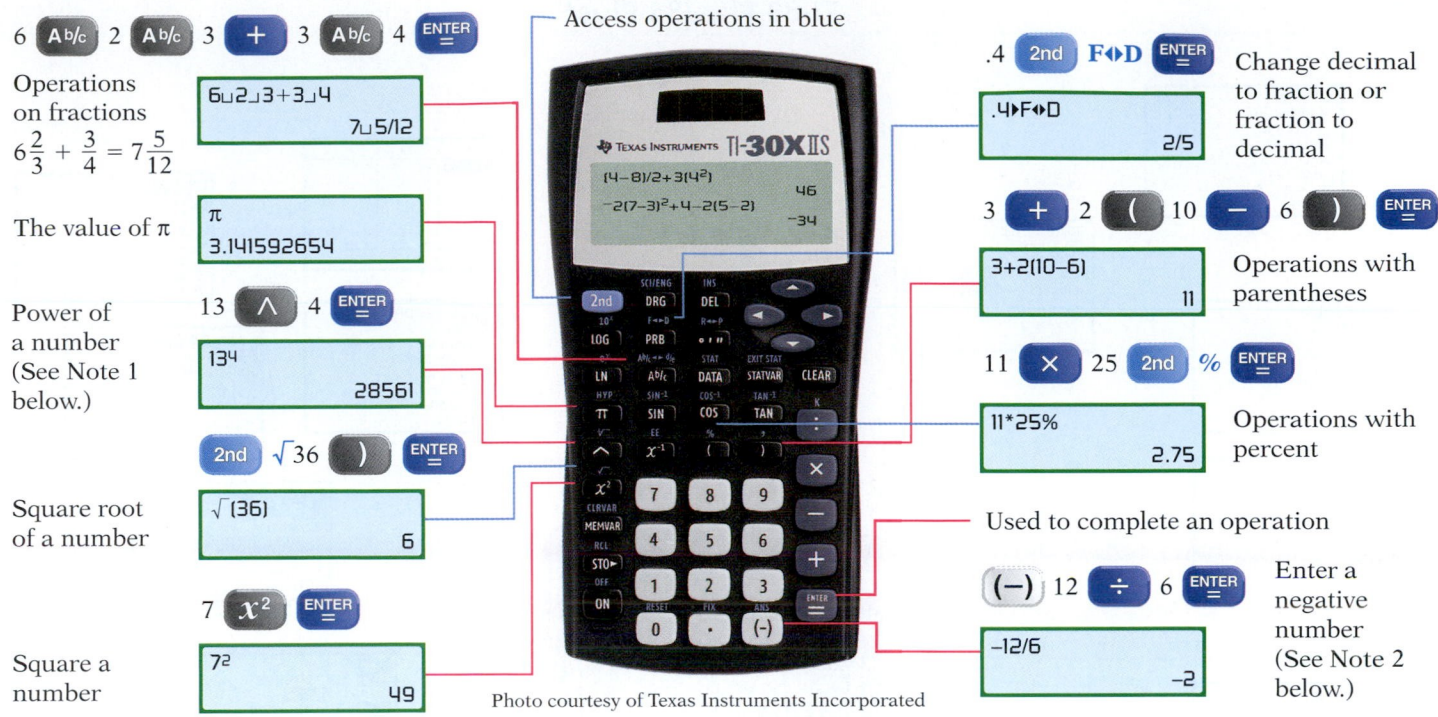

Photo courtesy of Texas Instruments Incorporated

.4 [2nd] [F◆D] [ENTER]

Change decimal to fraction or fraction to decimal

```
.4▶F◆D
                    2/5
```

3 [+] 2 [(] 10 [−] 6 [)] [ENTER]

Operations with parentheses

```
3+2(10−6)
                     11
```

11 [×] 25 [2nd] [%] [ENTER]

Operations with percent

```
11*25%
                   2.75
```

Used to complete an operation

[(−)] 12 [÷] 6 [ENTER]

Enter a negative number
(See Note 2 below.)

```
−12/6
                     −2
```

fx-300MS

[√] 36 [=]

Square root of a number

```
√36
                      6
```

6 [a b/c] 2 [a b/c] 3 [+] 3 [a b/c] 4 [=]

Operations on fractions
$6\frac{2}{3} + \frac{3}{4} = 7\frac{5}{12}$

```
6⌴2⌴3+3⌴4
                7⌴5⌴12
```

7 [x^2] [=]

Square a number

```
7²
                     49
```

[(−)] 12 [÷] 6 [=]

Enter a negative number
(See Note 2 below.)

```
−12÷6
                     −2
```

Access operations in gold

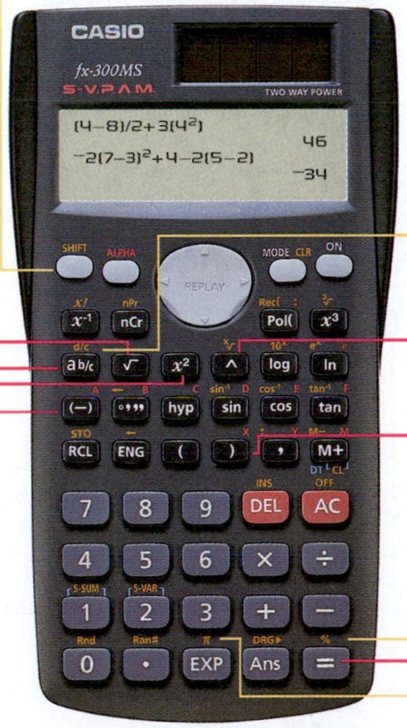

Photo courtesy of Casio, Inc.

.4 [=] [d/c]

Change decimal to fraction

```
.4
                    2⌴5
```

13 [∧] 4 [=]

Power of a number
(See Note 1 below.)

```
13^4
                28561
```

3 [+] 2 [(] 10 [−] 6 [)] [=]

Operations with parentheses

```
3+2(10−6)
                     11
```

11 [×] 25 [%] [=]

Operations with percent

```
11x25%
                   2.75
```

Used to complete an operation

[SHIFT] [π] [=]

The value of π

```
π
3.141592654
```

NOTE 1: Some calculators use the [y^x] key to calculate a power. For those calculators, enter 13 [y^x] 4 [=] to evaluate 13^4.

NOTE 2: Some calculators use the [+/−] key to enter a negative number. For those calculators, enter 12 [+/−] [÷] 6 [=] to calculate $-12 \div 6$.

TI-83 Plus/84 Plus*

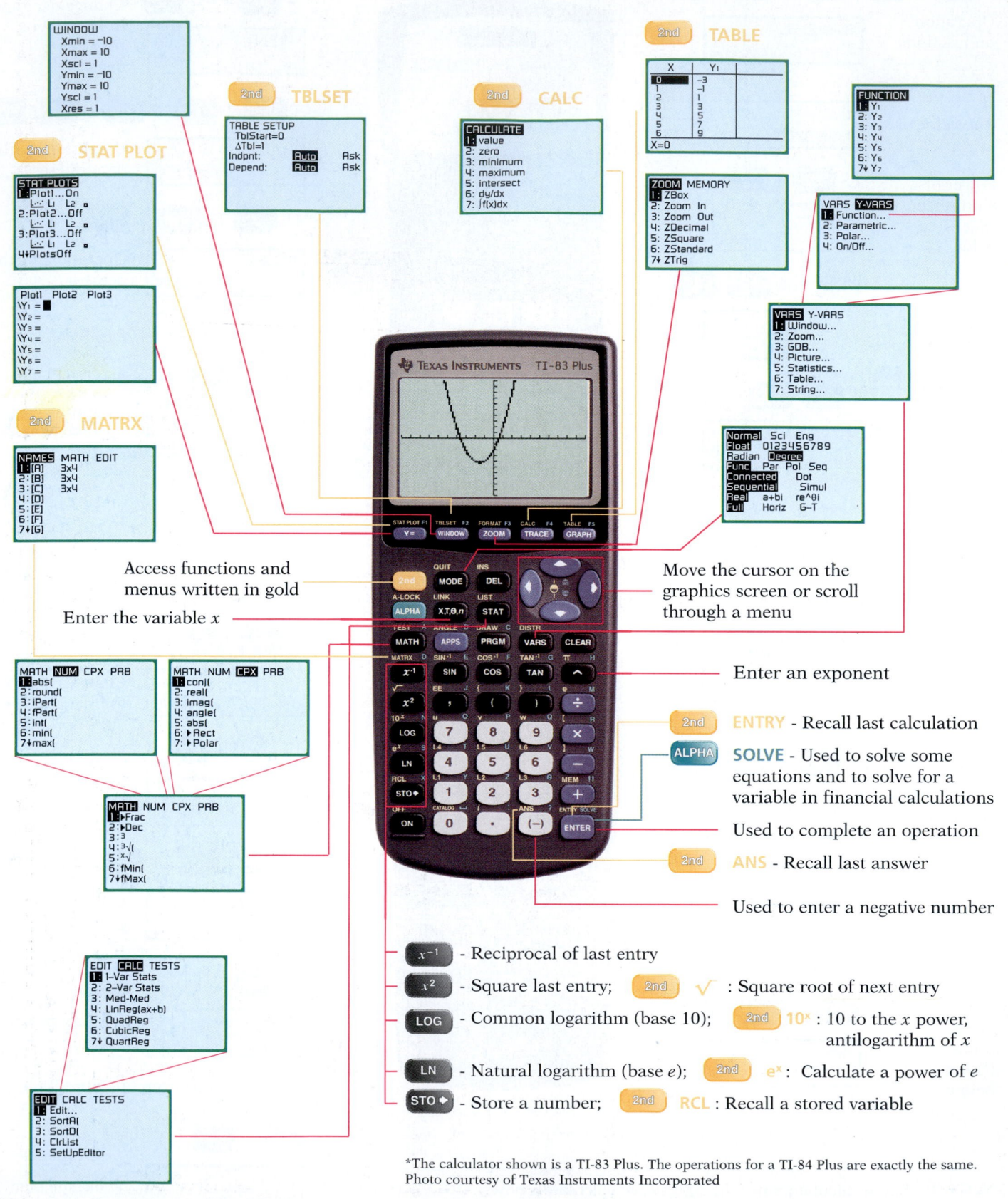

WINDOW
Xmin = -10
Xmax = 10
Xscl = 1
Ymin = -10
Ymax = 10
Yscl = 1
Xres = 1

2nd TABLE

X	Y₁
0	-3
1	-1
2	1
3	3
4	5
5	7
6	9
X=0

FUNCTION
1: Y₁
2: Y₂
3: Y₃
4: Y₄
5: Y₅
6: Y₆
7↓ Y₇

2nd TBLSET
TABLE SETUP
TblStart=0
ΔTbl=1
Indpnt: **Auto** Ask
Depend: **Auto** Ask

2nd CALC
CALCULATE
1: value
2: zero
3: minimum
4: maximum
5: intersect
6: dy/dx
7: ∫f(x)dx

2nd STAT PLOT
STAT PLOTS
1: Plot1...On
   L₁ L₂
2: Plot2...Off
   L₁ L₂
3: Plot3...Off
   L₁ L₂
4↓ PlotsOff

Plot1 Plot2 Plot3
\Y₁ = ■
\Y₂ =
\Y₃ =
\Y₄ =
\Y₅ =
\Y₆ =
\Y₇ =

ZOOM MEMORY
1: ZBox
2: Zoom In
3: Zoom Out
4: ZDecimal
5: ZSquare
6: ZStandard
7↓ ZTrig

VARS Y-VARS
1: Function...
2: Parametric...
3: Polar...
4: On/Off...

VARS Y-VARS
1: Window...
2: Zoom...
3: GDB...
4: Picture...
5: Statistics...
6: Table...
7: String...

2nd MATRX
NAMES MATH EDIT
1: [A]   3x4
2: [B]   3x4
3: [C]   3x4
4: [D]
5: [E]
6: [F]
7↓ [G]

Normal Sci Eng
Float 0123456789
Radian **Degree**
Func Par Pol Seq
Connected   Dot
Sequential   Simul
Real a+bi re^θi
Full Horiz G–T

MATH NUM CPX PRB
1: abs(
2: round(
3: iPart(
4: fPart(
5: int(
6: min(
7↓ max(

MATH NUM CPX PRB
1: conj(
2: real(
3: imag(
4: angle(
5: abs(
6: ▶Rect
7: ▶Polar

MATH NUM CPX PRB
1: ▶Frac
2: ▶Dec
3: ³
4: ³√(
5: ˣ√
6: fMin(
7↓ fMax(

Access functions and menus written in gold

Enter the variable *x*

Move the cursor on the graphics screen or scroll through a menu

Enter an exponent

2nd ENTRY - Recall last calculation

ALPHA SOLVE - Used to solve some equations and to solve for a variable in financial calculations

Used to complete an operation

2nd ANS - Recall last answer

Used to enter a negative number

EDIT CALC TESTS
1: 1-Var Stats
2: 2-Var Stats
3: Med-Med
4: LinReg(ax+b)
5: QuadReg
6: CubicReg
7↓ QuartReg

EDIT CALC TESTS
1: Edit...
2: SortA(
3: SortD(
4: ClrList
5: SetUpEditor

x⁻¹ - Reciprocal of last entry

x² - Square last entry; **2nd** √ : Square root of next entry

LOG - Common logarithm (base 10); **2nd 10ˣ** : 10 to the *x* power, antilogarithm of *x*

LN - Natural logarithm (base *e*); **2nd eˣ** : Calculate a power of *e*

STO▶ - Store a number; **2nd RCL** : Recall a stored variable

*The calculator shown is a TI-83 Plus. The operations for a TI-84 Plus are exactly the same.
Photo courtesy of Texas Instruments Incorporated